A NOTE ABOUT THE CONTENTS OF THIS CUSTOM BOOK

Welcome to **MECH2100 Machine Design**.

The chapters in this custom book have been chosen specifically to meet your course requirements from *Machine Elements in Mechanical Design 5th Edition* by Mott.

All chapters in the source book are included in this compilation except for **Chapter 13 Tolerances and Fits** and **Chapter 18 Springs** and this is reflected in the **Contents**.

The **chapter numbers, page numbers and index page references** in this custom book are the same as the original source book.

We hope that you will find this book easy to follow and that your reading experience is a positive one.

MECH2100 MACHINE DESIGN

This custom book is compiled for University of Queensland from:

Machine Elements in Mechanical Design
5th Edition
Mott

Pearson Australia
707 Collins St
Melbourne VIC 3008
Ph: 03 9811 2400
www.pearson.com

Project Management Team Leader:	Jill Gillies
Key Account Manager:	Danielle Woods
Project Manager:	Linda Chryssavgis
Production Controller:	Brad Smith

ISBN: 978 1 4886 1603 7

CONTENTS

Chapter 13 Tolerances and Fits is not included in this custom edition

Chapter 18 Springs is not included in this custom edition

PREFACE

The objective of this book is to provide the concepts, procedures, data, and decision analysis techniques necessary to design machine elements commonly found in mechanical devices and systems. Students completing a course of study using this book should be able to execute original designs for machine elements and integrate the elements into a system composed of several elements.

This process requires a consideration of the performance requirements of an individual element and of the interfaces between elements as they work together to form a system. For example, a gear must be designed to transmit power at a given speed. The design must specify the number of teeth, pitch, tooth form, face width, pitch diameter, material, and method of heat treatment. But the gear design also affects, and is affected by, the mating gear, the shaft carrying the gear, and the environment in which it is to operate. Furthermore, the shaft must be supported by bearings, which must be contained in a housing. Thus, the designer should keep the complete system in mind while designing each individual element. This book will help the student approach design problems in this way.

This text is designed for those interested in practical mechanical design. The emphasis is on the use of readily available materials and processes and appropriate design approaches to achieve a safe, efficient design. It is assumed that the person using the book will be the designer, that is, the person responsible for determining the configuration of a machine or a part of a machine. Where practical, all design equations, data, and procedures needed to make design decisions are specified.

It is expected that students using this book will have a good background in statics, strength of materials, college algebra, and trigonometry. Helpful, but not required, would be knowledge of kinematics, industrial mechanisms, dynamics, materials, and manufacturing processes.

Among the important features of this book are the following:

1. It is designed to be used at the undergraduate level in a first course in machine design.
2. The large list of topics allows the instructor some choice in the design of the course. The format is also appropriate for a two-course sequence and as a reference for mechanical design project courses.
3. Students should be able to extend their efforts into topics not covered in classroom instruction because explanations of principles are straightforward and include many example problems.
4. The practical presentation of the material leads to feasible design decisions and is useful to practicing designers.
5. The text advocates and demonstrates use of computer spreadsheets in cases requiring long, laborious solution procedures. Using spreadsheets allows the designer to make decisions and to modify data at several points within the problem while the computer performs all computations. See Chapter 6 on columns, Chapter 9 on spur gears, Chapter 12 on shafts, Chapter 13 on shrink fits, and Chapter 18 on spring design. Other computer-aided calculation software can also be used.
6. References to other books, standards, and technical papers assist the instructor in presenting alternate approaches or extending the depth or breadth of treatment.
7. Lists of Internet sites pertinent to topics in this book are included at the end of most chapters to assist readers in accessing additional information or data about commercial products.
8. In addition to the emphasis on original design of machine elements, much of the discussion covers commercially available machine elements and devices, since many design projects require an optimum combination of new, uniquely designed parts and purchased components.
9. For some topics the focus is on aiding the designer in selecting commercially available components, such as rolling contact bearings, flexible couplings, ball screws, electric motors, belt drives, chain drives, clutches, and brakes.
10. Computations and problem solutions use both the International System of Units (SI) and the U.S. Customary System (inch-pound-second) approximately equally. The basic reference for the usage of SI units is IEEE/ASTM-SI-10 *Standard for Use of the International System of Units (SI): The Modern Metric System*, which has replaced ASTM E380 and ANSI/IEEE Standard 268-1992.
11. Extensive appendices are included along with detailed tables in many chapters to help the reader to make real design decisions, using only this text. Several new appendix tables featuring commercially available structural shapes in smaller sizes and many in purely metric dimensions are included in this edition to give instructors and students more options for completing design problems.

MECHANICAL DESIGN SOFTWARE PROVIDED WITH THE BOOK

The design of machine elements inherently involves extensive procedures, complex calculations, and many design decisions. Data must be found from numerous charts and tables. Furthermore, design is typically iterative, requiring the designer to try several options for any given element, leading to the repetition of design calculations with new data or new design decisions. This is especially true for complete mechanical devices containing several components as the interfaces between components are considered. Changes to one component often require changes to mating elements. Use of computer-aided mechanical design software can facilitate the design process by performing many of the tasks while leaving the major design decisions to the creativity and judgment of the designer or engineer.

We emphasize that users of computer software must have a solid understanding of the principles of design and stress analysis to ensure that design decisions are based on reliable foundations. We recommend that the software be used only after mastering a given design methodology by careful study and using manual techniques.

Included on a CD with this book is the MDESIGN-MOTT mechanical design software created by the TEDATA Company. Derived from the successful MDESIGN mec software produced for the European market, the U.S. version of MDESIGN-MOTT employs standards and design methods that are typically in use in North America. Many of the textual aids and design procedures come directly from *Machine Elements in Mechanical Design*.

Topics for which the MDESIGN-MOTT software can be used as a supplement to this book include:

Beam stress analysis	Beam deflections	Mohr's circle	Columns
Belt drives	Chain drives	Spur gears	Helical gears
Bevel gears	Wormgearing	Shafts	Keys
Power screws	Springs	Rolling contact bearings	Plain surface bearings
Bolted connections	Fasteners	Clutches	Brakes

A special icon as shown above is placed in the margins at places in this book where use of the software is pertinent. Also, the Solutions Manual, available only to instructors using this book in scheduled classes, includes guidance for use of the software.

The MDESIGN-MOTT software contains a total of 66 different modules organized into 15 classifications. Of those, 33 modules are directly relevant to topics addressed in this book. The remaining 33 modules are valuable additions for the designer of a diverse array of machinery and products, particularly when designing for global markets or when working with design staffs in countries outside the United States. These 33 modules typically apply the ISO system of units and are based on industry standards and general practices commonly used in Europe and Canada. The strong movement in the United States toward global sourcing of materials and products and the use of multinational design teams makes these modules highly valuable during the lifelong career of designers and engineers.

FEATURES OF THE FIFTH EDITION

The practical approach to designing machine elements in the context of complete mechanical designs is retained and refined in this edition. An extensive amount of updating has been accomplished through the inclusion of new photographs of commercially available machine components, new design data for some elements, new or revised standards, new end-of-chapter references, listings of Internet sites, and some completely new elements. A second color has been added to enhance the visual attractiveness of the book and to highlight prominent features of charts, graphs, and technical illustrations.

The following list summarizes the primary features and the updates.

1. The three-part structure that was introduced in the third edition has been maintained.
    - Part I (Chapters 1–6) focuses on reviewing and upgrading readers' understanding of design philosophies, the principles of strength of materials, the design properties of materials, combined stresses, design for different types of loading, and the analysis and design of columns.
    - Part II (Chapters 7–15) is organized around the concept of the design of a complete power-transmission system, covering some of the primary machine elements such as belt drives, chain drives, gears, shafts, keys, couplings, seals, and rolling contact bearings. These topics are tied together to emphasize both their interrelationships and their unique characteristics. Chapter 15, **Completion of the Design of a Power Transmission**, is a guide through detailed design decisions such as the overall layout, detail drawings, tolerances, and fits.
    - Part III (Chapters 16–22) presents methods of analysis and design of several important machine elements that were not pertinent to the design of a power transmission. These chapters can be covered in any order or can be used as reference material for general design projects. Covered here are plain surface bearings, linear motion elements, fasteners, springs, machine frames, bolted connections, welded joints, electric motors, controls, clutches, and brakes.
2. **The Big Picture, You Are the Designer**, and **Objectives** features introduced in earlier editions are maintained and refined. Feedback about these features from users, both students and instructors, has been enthusiastically

favorable. They help readers to draw on their own experiences and to appreciate what competencies they will acquire from the study of each chapter. Constructivist theories of learning espouse this approach.

3. Lists of Internet sites and printed references have been updated and edited in every chapter. Many new entries have been added. A large number of illustrations have been revised and upgraded in their appearance.
4. Some of the new or updated topics from individual chapters are summarized here.
   - In Chapter 1, **The Nature of Mechanical Design,** the discussion of the topic of *robust design* has been expanded.
   - Chapter 2, **Materials in Mechanical Design**, is refined, notably through updates to discussion of flexural strength, flexural modulus, and wear resistance. An expanded discussion of materials property data available in the book should aid readers in problem solving and design projects. New tables for listing materials commonly used in commercially available shapes are included. To serve the global nature of machine design, an extensive table of designations for steel and aluminum alloys from several countries has been added. A major change in the designations for steel alloys is implemented to align with the SAE numbering designations in place of the formerly used AISI system. In recent years, the AISI has discontinued its involvement in steel designation systems. For most commonly used alloys, the actual number designations are the same in both systems. Therefore, the primary change is simply to use 'SAE' in place of 'AISI.' The discussion of heat treating has been enhanced with emphasis on case hardening. Additional descriptions of white iron, powder metals, aluminum casting and forging alloys, magnesium, nickel-based alloys, titanium alloys, and brasses and bronzes are included. Enhanced discussion of advanced engineering composites includes more SI data, nanocomposites, and design approaches. Materials selection using decision analysis techniques has been expanded.
   - Chapter 3, **Stress and Deformation Analysis**, has added material on lug joints and curved beams along with an Internet-based approach to finding stress concentration factors.
   - Chapter 4, **Combined Stresses and Mohr's Circle,** has expanded discussion of the von Mises stress for biaxial stress systems and three-dimensional stress systems. Problem sets have been refined to gain better distribution of different types of stress conditions throughout the range of possible combinations on Mohr's circle.
   - Chapter 5, **Design for Different Types of Loading**, is upgraded and refined in the topics of endurance strength under low-cycle fatigue and high-cycle fatigue and mechanisms of fatigue failure. A new graph illustrating representative endurance strengths has been added along with the exponential form of the stress-life equation. Separate graphs for SI units and U.S. Customary units are now shown for the variation of endurance strength modified for surface condition. The discussion of failure modes has been reorganized to focus initially on the most frequently used methods, with other techniques shown later. A set of recommendations for design and materials processing under fatigue loading has been added. A new section is added on the Smith Diagram approach for showing the effect of mean stress on fatigue, a method often used in Europe. The *damage accumulation method* for varying stress amplitudes is refined.
   - In Chapter 7, **Belt Drives and Chain Drives**, significant new material on synchronous belt drive designs in both SI and U.S. units has been added along with more related end-of-chapter practice problems. Common metric sizes for V-belts, chains, and sprockets are included.
   - In Chapter 8, **Kinematics of Gears,** the geometry of metric module-type gearing has been greatly expanded and integrated with the discussion of spur, helical, bevel, and wormgearing. A new table has been added to assist in calculating key geometric features of gears and gear teeth. Discussions of velocity ratios, train values, and devising gear trains have been combined and additional guidelines and design methods are shown.
   - Chapter 9, **Spur Gear Design**, is refined in accordance with revised AGMA standards, especially in relation to new quality numbers in the A_v system from AGMA Standard 2015 that is more in line with the ISO system that is used widely internationally. Numerous elements of the primary spur gear and helical gear design standard, AGMA Standard 2001-D04 have been implemented. The sections on gear material selection have been refined. A glossary of terms has been added that helps to facilitate problem solving. Discussions of gear lubricants and typical viscosity grades have been updated.
   - Chapter 10, **Helical Gears, Bevel Gears, and Wormgearing,** has been updated along similar lines as discussed for Chapter 9 on Spur Gear Design to reflect revised AGMA standards.
   - In Chapter 11, **Keys, Couplings, and Seals,** new information is provided for metric sizes of keys, common materials specified for keys, and example rating data for Ringfeder® locking assemblies. The design procedure for parallel keys has been refined.
   - In Chapter 12, **Shaft Design**, the procedure for the design of a shaft has been refined and streamlined. A table of example data for the torque capacity of selected flexible shaft sizes has been added.
   - In Chapter 14, **Rolling Contact Bearings,** the bearing selection procedure has been closely tied to the

use of the extensive tables of data in the MDESIGN-MOTT software included with this book. Only example data are shown in the book to illustrate the primary factors involved such as basic dynamic load ratings and common designation systems. The software includes data for numerous styles and sizes of bearings from prominent international bearing manufacturers. More discussion of bearing materials is included for steels, ceramics, Monel, titanium/nickel alloys, and plastics.

- Chapter 16, **Plain Surface Bearings**, includes refined data on *pV factors* for boundary-lubricated bearings and common lubricants. Analysis of plain bearing performance under oscillating motion has been added.
- In Chapter 17, **Linear Motion Elements,** data have been added for metric sizes of trapezoidal power screws.
- The former Chapter 19 on **Springs** has been relocated to become Chapter 18. The former Chapter 18, **Fasteners**, is now Chapter 19 to align it more logically with Chapter 20, **Frames, Bolted Connections, and Welded Joints.**
- In Chapter 21, **Electric Motors,** several new photos of commercially available motor styles have been added, including one showing an exploded view of a typical AC induction motor.
- **The Appendix** has many updated tables for material properties of steels, cast irons, aluminum alloys, zinc and magnesium alloys, plastics, nickel-based alloys, titanium alloys, bronzes, brasses, and other copper alloys. Several new tables of data for section properties of commercially available shapes in smaller sizes and in pure metric dimensions have been added to provide a wider array of choices for problem-solving and design.

ACKNOWLEDGMENTS

My appreciation is extended to all who provided helpful suggestions for improvements to this book. I thank the editorial staff of Pearson, those who provided illustrations, and the many users of the book, both instructors and students, with whom I have had discussions. Special appreciation goes to my colleagues at the University of Dayton, Professors David Myszka, Joseph Untener, Michael Kozak, Philip Doepker, and Robert Wolff. I would like to thank Dexter C. Hulse, University of Cincinnati; Wilson Liang, Purdue University Fort Wayne; Jennifer Love, Northeastern University; and Dipo Onipede Jr., Ph.D., Penn State University Behrend, for their helpful reviews of this revision. I also thank those who provided thoughtful reviews of the prior edition. I especially thank my students—past and present—for their encouragement and their positive feedback about this book.

Robert L. Mott

PRINCIPLES OF DESIGN AND STRESS ANALYSIS

As you complete the first six chapters of this book, you will gain an understanding of design philosophies, and you will build on earlier-learned principles of strength of materials, materials science, and manufacturing processes. The competencies gained from these chapters are useful throughout the book and in general machine design or product design projects.

Chapter 1: The Nature of Mechanical Design helps you see the big picture of the process of mechanical design. Several examples are shown from different industry sectors: consumer products, manufacturing systems, construction equipment, agricultural equipment, transportation equipment, ships, and space systems. The responsibilities of designers are discussed, along with an illustration of the iterative nature of the design process. Units and conversions complete the chapter.

Chapter 2: Materials in Mechanical Design emphasizes the design properties of materials. Much of this chapter is probably review for you, but it is presented here to emphasize the importance of material selection to the design process and to explain the data for materials presented in the Appendices.

Chapter 3: Stress and Deformation Analysis is a review of the basic principles of stress and deflection analysis. It is essential that you understand the basic concepts summarized here before proceeding with later material. Reviewed are direct tensile, compressive, and shearing stresses; bending stresses; and torsional shear stresses.

Chapter 4: Combined Stresses and Mohr's Circle is important because many general design problems and the design of machine elements covered in later chapters of the book involve combined stresses. You may have covered these topics in a course in strength of materials.

Chapter 5: Design for Different Types of Loading is an in-depth discussion of design factors, fatigue, and many of the details of stress analysis as used in this book.

Chapter 6: Columns discusses the long, slender, axially loaded members that tend to fail by buckling rather than by exceeding the yield, ultimate, or shear stress of the material. Special design and analysis methods are reviewed here.

CHAPTER ONE

THE NATURE OF MECHANICAL DESIGN

THE BIG PICTURE — The Nature of Mechanical Design

Discussion Map

- To design mechanical components and devices, you must be competent in the design of individual elements that comprise the system.
- But you must also be able to integrate several components and devices into a coordinated, robust system that meets your customer's needs.

Discover

Think, now, about the many fields in which you can use mechanical design:

What are some of the products of those fields?

What kinds of materials are used in the products?

What are some of the unique features of the products?

How were the components made?

How were the parts of the products assembled?

Consider consumer products, construction equipment, agricultural machinery, manufacturing systems, and transportation systems on the land, in the air, in space, and on and under water.

In this book, you will find the tools to learn the principles of ***Machine Elements in Mechanical Design***.

Design of machine elements is an integral part of the larger and more general field of mechanical design. Designers and design engineers create devices or systems to satisfy specific needs. Mechanical devices typically involve moving parts that transmit power and accomplish specific patterns of motion. Mechanical systems are composed of several mechanical devices.

Therefore, to design mechanical devices and systems, you must be competent in the design of individual machine elements that comprise the system. But you must also be able to integrate several components and devices into a coordinated, robust system that meets your customer's needs. From this logic comes the name of this book, *Machine Elements in Mechanical Design*.

Think about the many fields in which you can use mechanical design. Discuss these fields with your instructor and with your colleagues who are studying with you. Talk with people who are doing mechanical design in local industries. Try to visit their companies if possible, or meet designers and design engineers at meetings of professional societies. Consider the following fields where mechanical products are designed and produced.

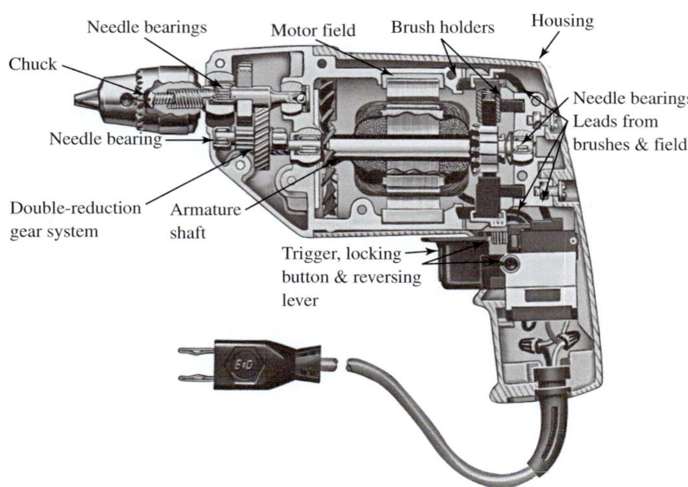

FIGURE 1–1 Cutaway view of a hand drill
[Courtesy of Standard Black & Decker (U.S.) Inc. New Britain, CT]

- ***Consumer products:*** Household appliances (can openers, food processors, mixers, toasters, vacuum cleaners, clothes washers), lawn mowers, chain saws, power tools, garage door openers, air-conditioning systems, and many others. See Figures 1–1 and 1–2 for a few examples of commercially available products.
- ***Manufacturing systems:*** Material handling devices, conveyors, cranes, transfer devices, industrial robots, machine tools, automated assembly systems, special-purpose processing systems, forklift trucks, and packaging equipment. See Figures 1–3 and 1–4.
- ***Construction equipment:*** Tractors with front-end loaders or backhoes, mobile cranes, power shovels, earthmovers, graders, dump trucks, road pavers, concrete mixers, powered nailers and staplers, compressors, and many others. See Figures 1–5 and 1–6.
- ***Agricultural equipment:*** Tractors, harvesters (for corn, wheat, tomatoes, cotton, fruit, and many other crops), rakes, hay balers, plows, disc harrows, cultivators, and conveyors. See Figures 1–6, 1–7, and 1–8.
- ***Transportation equipment:*** (a) Automobiles, trucks, and buses, which include hundreds of mechanical devices such as suspension components (springs, shock absorbers, and struts); door and window operators; windshield wiper mechanisms; steering systems; hood and trunk latches and hinges; clutch and braking systems; transmissions; driveshafts; seat adjusters; and numerous parts of the engine systems. (b) Aircraft, which include retractable landing gear, flap and rudder actuators, cargo-handling devices, seat reclining mechanisms, dozens of latches, structural components, and door operators. See Figures 1–9 and 1–10.
- ***Ships:*** Winches to haul up the anchor, cargo-handling cranes, rotating radar antennas, rudder steering gear, drive gearing and driveshafts, and the numerous sensors and controls for operating on-board systems.
- ***Space systems:*** Satellite systems, the space shuttle, the space station, and launch systems, which contain numerous mechanical systems such as devices to

FIGURE 1–2 Chain saw
(Anatoliy Kosolapov/Shutterstock)

(a) Conveyor system drive mechanism

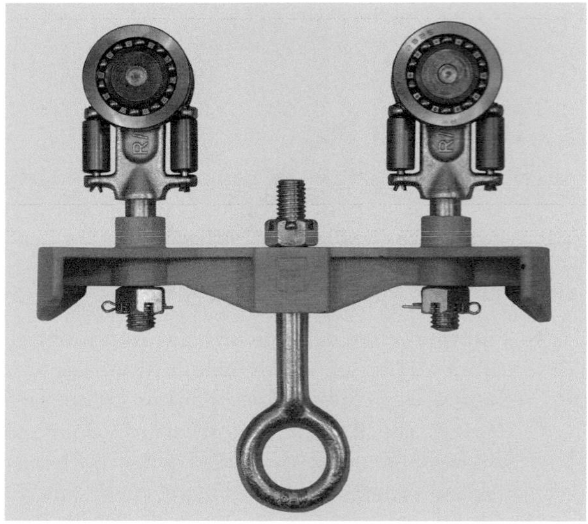

(b) Carrier assembly with rollers that ride inside the rectangular tubular rail section

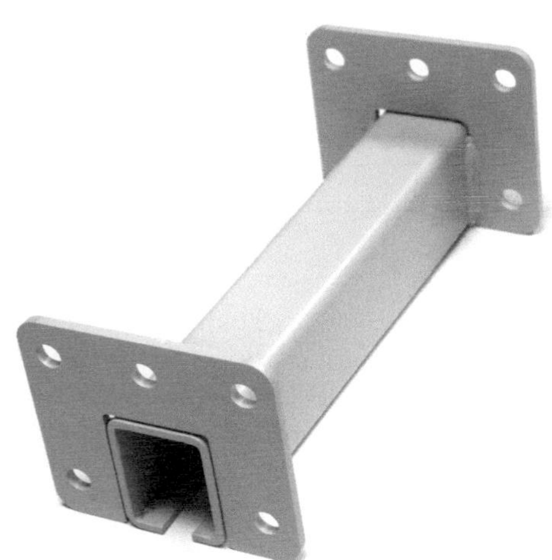

(c) Conveyor rail section

FIGURE 1–3 Chain conveyor system components
(Richards-Wilcox, Aurora, IL)

FIGURE 1–4 Robotic work cell in an automotive assembly plant
(© Jim West/Alamy)

FIGURE 1–5 Construction crane on a building site
(Joseph S.L. Tan Matt/Shutterstock)

FIGURE 1–6 Tractor with a backhoe and a front-end-loader for construction applications
(Alekcey/Shutterstock)

FIGURE 1–7 Tractor with tilling equipment for farm applications
(Chas/Shutterstock)

FIGURE 1–8 Heavy-duty tractor for farm, highway, and commercial application
(John Gordon/Shutterstock)

FIGURE 1–9 Aircraft wing showing aileron actuation mechanism
(© RTimages/Fotolia)

FIGURE 1–10 Aircraft landing gear
(Mikhail Starodubov/Shutterstock)

deploy antennas, hatches, docking systems, robotic arms, vibration control devices, devices to secure cargo, positioning devices for instruments, actuators for thrusters, and propulsion systems.

How many examples of mechanical devices and systems can you add to these lists?

What are some of the unique features of the products in these fields?

What kinds of mechanisms are included?

What kinds of materials are used in the products?

How were the components made?

How were the parts assembled into the complete products?

In this book, you will find the tools to learn the principles of *Machine Elements in Mechanical Design*. In the introduction to each chapter, we include a brief scenario called *You Are the Designer*. The purpose of these scenarios is to stimulate your thinking about the material presented in the chapter and to show examples of realistic situations in which you may apply it.

YOU ARE THE DESIGNER

Consider, now, that you are the designer responsible for the design of a new consumer product, such as the hand drill for a home workshop shown in Figure 1–1. What kind of technical preparation would you need to complete the design? What steps would you follow? What information would you need? How would you show, by calculation, that the design is safe and that the product will perform its desired function?

The general answers to these questions are presented in this chapter. As you complete the study of this book, you will learn about many design techniques that will aid in your design of a wide variety of machine elements. You will also learn how to integrate several machine elements into a mechanical system by considering the relationships between and among elements. ■

1–1 OBJECTIVES OF THIS CHAPTER

After completing this chapter, you will be able to:

1. Recognize examples of mechanical systems in which the application of the principles discussed in this book is necessary to complete their design.
2. List what design skills are required to perform competent mechanical design.
3. Describe the importance of integrating individual machine elements into a more comprehensive mechanical system.
4. Describe the main elements of the *product realization process*.
5. Write statements of *functions* and *design requirements* for mechanical devices.
6. Establish a set of criteria for evaluating proposed designs.
7. Work with appropriate units in mechanical design calculations both in the U.S. Customary Unit System and in SI metric units.
8. Distinguish between *force* and *mass*, and express them properly in both unit systems.
9. Present design calculations in a professional, neat, and orderly manner that can be understood and evaluated by others knowledgeable in the field of machine design.

1–2 THE DESIGN PROCESS

The ultimate objective of mechanical design is to produce a useful product that satisfies the needs of a customer and that is safe, efficient, reliable, economical, and practical to manufacture. Think broadly when answering the question, “Who is the customer for the product or system I am about to design?” Consider the following scenarios:

- ***You are designing a can opener for the home market.*** The ultimate customer is the person who will purchase the can opener and use it in the kitchen of a home. Other customers may include the designer of the packaging for the opener, the manufacturing staff who must produce the opener economically, and service personnel who repair the unit.
- ***You are designing a piece of production machinery for a manufacturing operation.*** The customers include the manufacturing engineer who is responsible for the production operation, the operator of the machine, the staff who install the machine, and the maintenance personnel who must service the machine to keep it in good running order.
- ***You are designing a powered system to open a large door on a passenger aircraft.*** The customers include the person who must operate the door in normal service or in emergencies, the people who must pass through the door during use, the personnel who manufacture

the opener, the installers, the aircraft structure designers who must accommodate the loads produced by the opener during flight and during operation, the service technicians who maintain the system, and the interior designers who must shield the opener during use while allowing access for installation and maintenance.

It is essential that you know the desires and expectations of all customers before beginning product design. Marketing professionals are often employed to manage the definition of customer expectations, but designers will likely work with them as a part of a product development team.

Numerous approaches are available that guide designers through the complete process of product design and methods for creating new, innovative products. Some are oriented toward large complex products such as aircraft, automobiles, and multifunction machine tools. It is advisable for a company to select one method that is suitable to their particular style of products or to create one that meets their special needs. The following discussion identifies the salient features of some of the approaches and the listed references and Internet sites provide more details. Some of the listed methods are applied in combination.

- ***Axiomatic design.*** See References 14, 15, and 18 and Internet site 8. Axiomatic design methods implement a process where developers think functionally first, followed by the innovative creation of the physical embodiment of a product to meet customer requirements along with the process needed to produce the product.
- ***Quality function deployment (QFD).*** See Reference 8 and Internet sites 9 and 10. QFD is a quality system that espouses understanding customer requirements and uses quality systems thinking to maximize positive quality that adds value. The process also includes use of the "House of Quality" matrix described in Reference 8.
- ***Design for six sigma (DFSS).*** See References 18–20 and Internet sites 11 and 16. The objective of Six Sigma Quality is to reduce output variation that will result in no more than 3.4 defective parts per million (PPM). The term *six sigma or 6Σ* refers to a distribution of performance measures, in which products produced are within upper and lower specification limits that are six process standard deviations from the mean.
- ***TRIZ.*** See References 21–23 and Internet sites 12–15. TRIZ is an acronym for a Russian phrase that translates into English as "Theory of Inventive Problem Solving." Developed in 1946 in Russia by Genrich Altshuller and colleagues, the process is applied throughout the world to create and to improve products, services, and systems. TRIZ is a problem-solving method based on logic and data, not intuition, which accelerates the project team's ability to solve problems creatively.
- ***Total design.*** See Reference 13. An integrated approach to product engineering using a systematic and disciplined process to create innovative products that satisfy customer needs.
- ***The engineering design process—embodiment design.*** See Reference 26. A comprehensive process involving need identification, concept selection, decision making, detail design, modeling and simulation, design for manufacturing, robust design, and several other elements.
- ***Failure modes and effects analysis (FMEA).*** See Reference 24 and Internet site 17. An analysis technique which facilitates the identification of potential problems in the design of a product by examining the effects of lower level failures. Recommended actions or compensating provisions are made to reduce the likelihood of the problem occurring or mitigating the risk if problems do occur. The process evolved from military and NASA procedures designed to enhance the reliability of products and systems. MIL-STD-1629A defines accepted methods used in military and commercial industry. The prevalent standard in the automotive industry is SAE J1739.
- ***Product design for manufacture and assembly.*** See Reference 27. A product design methodology with a heavy emphasis on how the components and the assembled product are to be manufactured to achieve a low-cost, high-quality design. Included are design for die casting, forging, powder metal processing, sheet metalworking, machining, injection molding, and many other processes.

It is also important to consider how the design process fits with all functions that must happen to deliver a satisfactory product to the customer and to service the product throughout its life cycle. In fact, it is important to consider how the product will be disposed of after it has served its useful life. The total of all such functions that affect the product is sometimes called the *product realization process* or *PRP*. (See References 3 and 10.) Some of the factors included in PRP are as follows:

- Marketing functions to assess customer requirements
- Research to determine the available technology that can reasonably be used in the product
- Availability of materials and components that can be incorporated into the product
- Product design and development
- Performance testing
- Documentation of the design
- Vendor relationships and purchasing functions
- Consideration of global sourcing of materials and global marketing
- Work-force skills
- Physical plant and facilities available
- Capability of manufacturing systems
- Production planning and control of production systems
- Production support systems and personnel
- Quality systems requirements
- Operation and maintenance of the physical plant

- Distribution systems to get products to the customer
- Sales operations and time schedules
- Cost targets and other competitive issues
- Customer service requirements
- Environmental concerns during manufacture, operation, and disposal of the product
- Legal requirements
- Availability of financial capital

Can you add to this list?

You should be able to see that the design of a product is but one part of a comprehensive process. In this book, we will focus more carefully on the design process itself, but the producibility of your designs must always be considered. This simultaneous consideration of product design and manufacturing process design is often called *concurrent engineering.* Note that this process is a subset of the larger list given previously for the product realization process. Other major books discussing general approaches to mechanical design are listed as References 6, 7, and 12–26.

1–3 SKILLS NEEDED IN MECHANICAL DESIGN

Product engineers and mechanical designers use a wide range of skills and knowledge in their daily work, including the following:

1. Sketching, technical drawing, and 2D and 3D computer-aided design
2. Properties of materials, materials processing, and manufacturing processes
3. Applications of chemistry such as corrosion protection, plating, and painting
4. Statics, dynamics, strength of materials, kinematics, and mechanisms
5. Oral communication, listening, technical writing, and teamwork skills
6. Fluid mechanics, thermodynamics, and heat transfer
7. Fluid power, the fundamentals of electrical phenomena, and industrial controls
8. Experimental design, performance testing of materials and mechanical systems, and use of computer-aided engineering software
9. Creativity, problem solving, and project management
10. Stress analysis
11. Specialized knowledge of the behavior of machine elements such as gears, belt drives, chain drives, shafts, bearings, keys, splines, couplings, seals, springs, connections (bolted, riveted, welded, adhesive), electric motors, linear motion devices, clutches, and brakes

It is expected that you will have acquired a high level of competence in items 1–5 in this list prior to beginning the study of this text. The competencies in items 6–8 are typically acquired in other courses of study either before, concurrently, or after the study of design of machine elements. Item 9 represents skills that are developed continuously throughout your academic study and through experience. Studying this book will help you acquire significant knowledge and skills for the topics listed in items 10 and 11.

1–4 FUNCTIONS, DESIGN REQUIREMENTS, AND EVALUATION CRITERIA

Section 1–2 emphasized the importance of carefully identifying the needs and expectations of the customer prior to beginning the design of a mechanical device. You can formulate these by producing clear, complete statements of *functions*, *design requirements*, and *evaluation criteria*:

- ***Functions*** tell what the device must do, using general, nonquantitative statements that employ action phrases such as *to support a load*, *to lift a crate*, *to transmit power*, or *to hold two structural members together*.
- ***Design requirements*** are detailed, usually quantitative statements of *expected performance levels*, *environmental conditions in which the device must operate*, *limitations on space or weight*, or *available materials and components that may be used*.
- ***Evaluation criteria*** are statements of *desirable qualitative characteristics* of a design that assist the designer in deciding which alternative design is optimum—that is, the design that maximizes benefits while minimizing disadvantages.

Together these elements can be called the *specifications* for the design.

Most designs progress through a cycle of activities as outlined in Figure 1–11. You should typically propose more than one possible alternative design concept. This is where creativity is exercised to produce truly novel designs. Each design concept must satisfy the functions and design requirements. A critical evaluation of the desirable features, advantages, and disadvantages of each design concept should be completed. Then a rational decision analysis technique should use the evaluation criteria to decide which design concept is the optimum and, therefore, should be produced. See Reference 25 and Internet site 18.

The final block in the design flowchart is the detailed design, and the primary focus of this book is on that part of the overall design process. It is important to recognize that a significant amount of activity precedes the detailed design.

Example of Functions, Design Requirements, and Evaluation Criteria

Consider that you are the designer of a speed reducer that is part of the power transmission for a small tractor. The tractor's engine operates at a fairly high speed, while the drive for the wheels must rotate more slowly and transmit a higher torque than is available at the output of the engine.

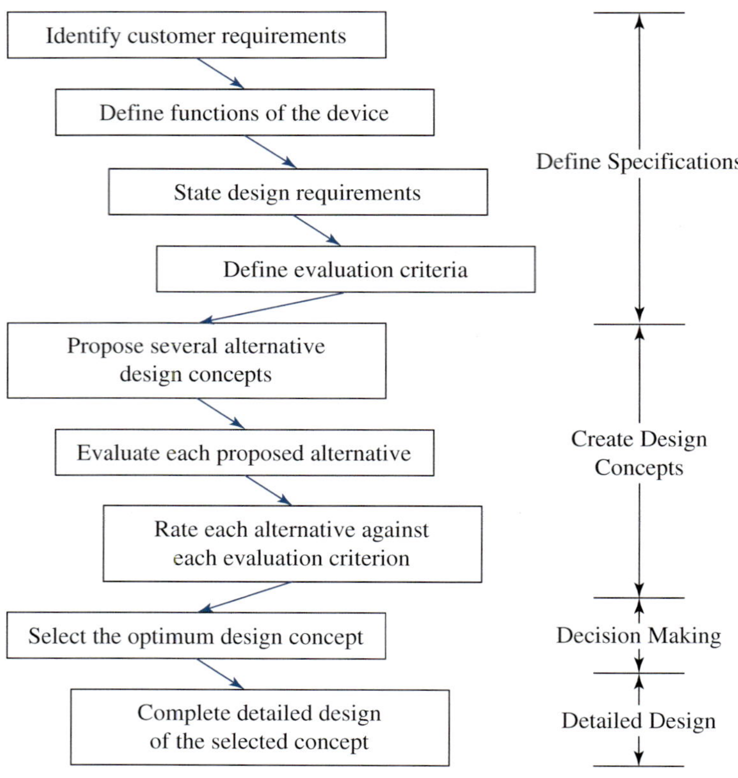

FIGURE 1–11 Steps in the design process

To begin the design process, let us list the *functions* of the speed reducer. What is it supposed to do? Some answers to this question are as follows:

Functions

1. To receive power from the tractor's engine through a rotating shaft.
2. To transmit the power through machine elements that reduce the rotational speed to a desired value.
3. To deliver the power at the lower speed to an output shaft that ultimately drives the wheels of the tractor.

Now the *design requirements* should be stated. The following list is hypothetical, but if you were on the design team for the tractor, you would be able to identify such requirements from your own experience and ingenuity and/or by consultation with fellow designers, marketing staff, manufacturing engineers, service personnel, suppliers, and customers.

The product realization process calls for personnel from all of these functions to be involved from the earliest stages of design.

Design Requirements

1. The reducer must transmit 15.0 hp.
2. The input is from a two-cylinder gasoline engine with a rotational speed of 2000 rpm.
3. The output delivers the power at a rotational speed in the range of 290 to 295 rpm.
4. A mechanical efficiency of greater than 95% is desirable.
5. The minimum output torque capacity of the reducer should be 3050 pound-inches (lb·in).
6. The reducer output is connected to the driveshaft for the wheels of a farm tractor. Moderate shock will be encountered.
7. The input and output shafts must be in-line.
8. The reducer is to be fastened to a rigid steel frame of the tractor.
9. Small size is desirable. The reducer must fit in a space no larger than 20 in × 20 in, with a maximum height of 24 in.
10. The tractor is expected to operate 8 hours (h). per day, 5 days per week, with a design life of 10 years.
11. The reducer must be protected from the weather and must be capable of operating anywhere in the United States at temperatures ranging from 0 to 130°F.
12. Flexible couplings will be used on the input and output shafts to prohibit axial and bending loads from being transmitted to the reducer.
13. The production quantity is 10 000 units per year.
14. A moderate cost is critical to successful marketing.
15. All government and industry safety standards must be met.

Careful preparation of function statements and design requirements will ensure that the design effort is focused on the desired results. Much time and money can be wasted on designs that, although technically sound, do not meet design requirements. Design requirements should include everything that is needed, but at the same time they should offer ample opportunity for innovation.

Evaluation criteria should be developed by all members of a product development team to ensure that the interests of all concerned parties are considered. Often weights are assigned to the criteria to reflect their relative importance.

Safety must always be the paramount criterion. Different design concepts may have varying levels of inherent safety in addition to meeting stated safety requirements as noted in the design requirements list. Designers and engineers are legally liable if a person is injured because of a design error. You must consider any reasonably foreseeable uses of the device and ensure safety of those operating it or those who may be close by.

Achieving a high overall performance should also be a high priority. Certain design concepts may have desirable features not present on others.

The remaining criteria should reflect the special needs of a particular project. The following list gives examples of possible evaluation criteria for the small tractor.

Evaluation Criteria

1. Safety (the relative inherent safety over and above stated requirements)
2. Performance (the degree to which the design concept exceeds requirements)
3. Ease of manufacture
4. Ease of service or replacement of components
5. Ease of operation
6. Low initial cost
7. Low operating and maintenance costs
8. Small size and low weight
9. Low noise and vibration; smooth operation
10. Use of readily available materials and purchased components
11. Prudent use of both uniquely designed parts and commercially available components
12. Appearance that is attractive and appropriate to the application

1–5 EXAMPLE OF THE INTEGRATION OF MACHINE ELEMENTS INTO A MECHANICAL DESIGN

Mechanical design is the process of designing and/or selecting mechanical components and putting them together to accomplish a desired function. Of course, machine elements must be compatible, must fit well together, and must perform safely and efficiently. The designer must consider not only the performance of the element being designed at a given time but also the elements with which it must interface.

To illustrate how the design of machine elements must be integrated with a larger mechanical design, let us consider the design of a speed reducer for the small tractor discussed in Section 1–4. Suppose that, to accomplish the speed reduction, you decide to design a double-reduction, spur gear reducer. You specify four gears, three shafts, six bearings, and a housing to hold the individual elements in proper relation to each other, as shown in Figure 1–12.

The primary elements of the speed reducer in Figure 1–12 are:

1. The input shaft (shaft 1) is to be connected to the power source, a gasoline engine whose output shaft rotates at 2000 rpm. A flexible coupling is to be employed to minimize difficulties with alignment.

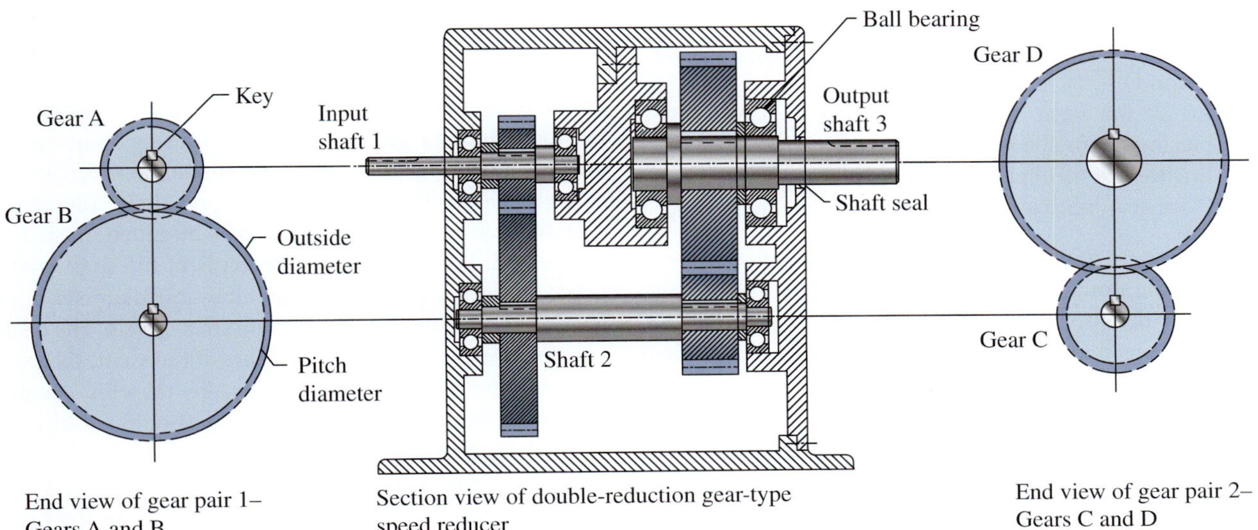

FIGURE 1–12 Conceptual design for a speed reducer

2. The first pair of gears, A and B, causes a reduction in the speed of the intermediate shaft (shaft 2) proportional to the ratio of the numbers of teeth in the gears. Gears B and C are both mounted to shaft 2 and rotate at the same speed.
3. A key is used at the interface between the hub of each gear and the shaft on which it is mounted to transmit torque between the gear and the shaft.
4. The second pair of gears, C and D, further reduces the speed of gear D and the output shaft (shaft 3) to the range of 290 to 295 rpm.
5. The output shaft is to carry a chain sprocket (not shown). The chain drive ultimately is to be connected to the drive wheels of the tractor.
6. Each of the three shafts is supported by two ball bearings, making them statically determinate and allowing the analysis of forces and stresses using standard principles of mechanics.
7. The bearings are held in a housing that is to be attached to the frame of the tractor. Note the manner of holding each bearing so that the inner race rotates with the shaft, while the outer race is held stationary.
8. Seals are shown on the input and output shafts to prohibit contaminants from entering the housing.
9. Other parts of the housing are shown schematically. Details of how the active elements are to be installed, lubricated, and aligned are only suggested at this stage of the design process to demonstrate feasibility. One possible assembly process could be as follows:
   - Start by placing the gears, keys, spacers, and bearings on their respective shafts.
   - Then insert shaft 1 into its bearing seat on the left side of the housing.
   - Insert the left end of shaft 2 into its bearing seat while engaging the teeth of gears A and B.
   - Install the center bearing support to provide support for the bearing at the right side of shaft 1.
   - Install shaft 3 by placing its left bearing into the seat on the center bearing support while engaging gears C and D.
   - Install the right side cover for the housing while placing the final two bearings in their seats.
   - Ensure careful alignment of the shafts.
   - Place gear lubricant in the lower part of the housing.

Figures 9–31 to 9–33 show three examples of commercially available double-reduction gear reducers where you can see these details.

The arrangement of the gears, the placement of the bearings so that they straddle the gears, and the general configuration of the housing are also design decisions. The design process cannot rationally proceed until these kinds of decisions are made. Notice that the sketch of Figure 1–12 is where *integration* of the elements into a whole design begins. When the overall design is conceptualized, the design of the individual machine elements in the speed reducer can proceed. As each element is discussed, scan the relevant chapters of the book. Part II of this book, Chapters 7–15, provides details for the elements of the reducer. You should recognize that you have already made many design decisions by rendering such a sketch. First, you chose *spur gears* rather than helical gears, a worm and wormgear, or bevel gears. In fact, other types of speed-reduction devices—belt drives, chain drives, or many others—could be appropriate.

Gears

For the gear pairs, you must specify the number of teeth in each gear, the pitch (size) of the teeth, the pitch diameters, the face width, and the material and its heat treatment. These specifications depend on considerations of strength and wear of the gear teeth and the motion requirements (Chapters 8 and 9). You must also recognize that the gears must be mounted on shafts in a manner that ensures proper location of the gears, adequate torque transmitting capability from the gears to the shafts (as through keys), and safe shaft design.

Shafts

Having designed the gear pairs, you next consider the shaft design (Chapter 12). The shaft is loaded in bending and torsion because of the forces acting at the gear teeth. Thus, its design must consider strength and rigidity, and it must permit the mounting of the gears and bearings. Shafts of varying diameters may be used to provide shoulders against which to seat the gears and bearings. There may be keyseats cut into the shaft (Chapter 11). The input and output shafts will extend beyond the housing to permit coupling with the engine and the drive axle. The type of coupling must be considered, as it can have a dramatic effect on the shaft stress analysis (Chapter 11). Seals on the input and output shafts protect internal components (Chapter 11).

Bearings

Design of the bearings (Chapter 14) is next. If rolling contact bearings are to be used, you will probably select commercially available bearings from a manufacturer's catalog, rather than design a unique one. You must first determine the magnitude of the loads on each bearing from the shaft analysis and the gear designs. The rotational speed and reasonable design life of the bearings and their compatibility with the shaft on which they are to be mounted must also be considered. For example, on the basis of the shaft analysis, you could specify the minimum allowable diameter at each bearing seat location to ensure safe stress levels. The bearing selected to support a particular part of the shaft, then, must have a bore (inside diameter) no smaller than the safe diameter of the shaft. Of course, the bearing should not be grossly larger than necessary. When a specific bearing is selected, the diameter of the shaft at the bearing seat location and allowable tolerances must be specified, according to the bearing manufacturer's recommendations, to achieve proper operation and life expectancy of the bearing.

Keys

Now the keys (Chapter 11) and the keyseats can be designed. The diameter of the shaft at the key determines the key's basic size (width and height). The torque that must be transmitted is used in strength calculations to specify key length and material. Once the working components are designed, the housing design can begin.

Housing

The housing design process must be both creative and practical. What provisions should be made to mount the bearings accurately and to transmit the bearing loads safely through the case to the structure on which the speed reducer is mounted? How will the various elements be assembled into the housing? How will the gears and bearings be lubricated? What housing material should be used? Should the housing be a casting, a weldment, or an assembly of machined parts?

The design process as outlined here implies that the design can progress in sequence: from the gears to the shafts, to the bearings, to the keys and couplings, and finally to the housing. It would be rare, however, to follow this logical path only once for a given design. Usually the designer must go back many times to adjust the design of certain components affected by changes in other components. This process, called *iteration,* continues until an acceptable overall design is achieved. Frequently prototypes are developed and tested during iteration.

Chapter 15 shows how all of the machine elements are finally integrated into a unit.

1–6 COMPUTATIONAL AIDS IN THIS BOOK

Because of the usual need for several iterations and because many of the design procedures require long, complex calculations, spreadsheets, mathematical analysis software, computer programs, or programmable calculators are often useful in performing the design analysis. Interactive spreadsheets or programs allow you, the designer, to make design decisions during the design process. In this way, many trials can be made in a short time, and the effects of changing various parameters can be investigated. Spreadsheets using Microsoft Excel are used most frequently as examples in this book for computer-aided design and analysis calculations.

In addition to using personally prepared computational aids like spreadsheets or mathematical analysis software, it is important to access commercially available calculation software and use it productively. One example that is tailored to this book is included for each user on a CD packaged at the back of the book. Called MDESIGN, the software includes a total of 66 modules divided among 15 categories that have been updated since the previous edition of this book. Approximately half of the modules are closely related to the study of topics in this book and icons are shown in the margins for those sections where a particular module is applicable. Some modules include very large databases of commercially available products such as belt drive and chain drive components and rolling contact bearings. The MDESIGN software includes an overview and introduction that should be read before using it. Part of that reads:

- ***Users of computer software and calculation aids must have a solid understanding of the relevant principles of design and stress analysis to ensure that design decisions are based on reliable foundations.***
- ***Software should be used only after mastering a given design methodology by careful study and practicing manual techniques.***

1–7 DESIGN CALCULATIONS

As you study this book and as you progress in your career as a designer, you will make many design calculations. It is important to record the calculations neatly, completely, and in an orderly fashion. You may have to explain to others how you approached the design, which data you used, and which assumptions and judgments you made. In some cases, someone else will actually check your work when you are not there to comment on it or to answer questions. Also, an accurate record of your design calculations is often useful if changes in design are likely. In all of these situations, you are going to be asked to communicate your design to someone else in written and graphic form.

To prepare a careful design record, you will usually take the following steps:

1. Identify the machine element being designed and the nature of the design calculation.
2. Draw a sketch of the element, showing all features that affect performance or stress analysis.
3. Show in a sketch the forces acting on the element (the free-body diagram), and provide other drawings to clarify the actual physical situation.
4. Identify the kind of analysis to be performed, such as stress due to bending, deflection of a beam, buckling of a column, and so on.
5. List all given data and assumptions.
6. Write the formulas to be used in symbol form, and clearly indicate the values and units of the variables involved. If a formula is not well known to a potential reader of your work, give the source. The reader may want to refer to it to evaluate the appropriateness of the formula.
7. Solve each formula for the desired variable.
8. Insert data, check units, and perform computations.
9. Judge the reasonableness of the result.
10. If the result is not reasonable, change the design decisions and recompute. Perhaps a different geometry or material would be more appropriate.
11. When a reasonable, satisfactory result has been achieved, specify the final values for all important design parameters, using standard sizes, convenient dimensions, readily available materials, and so on.

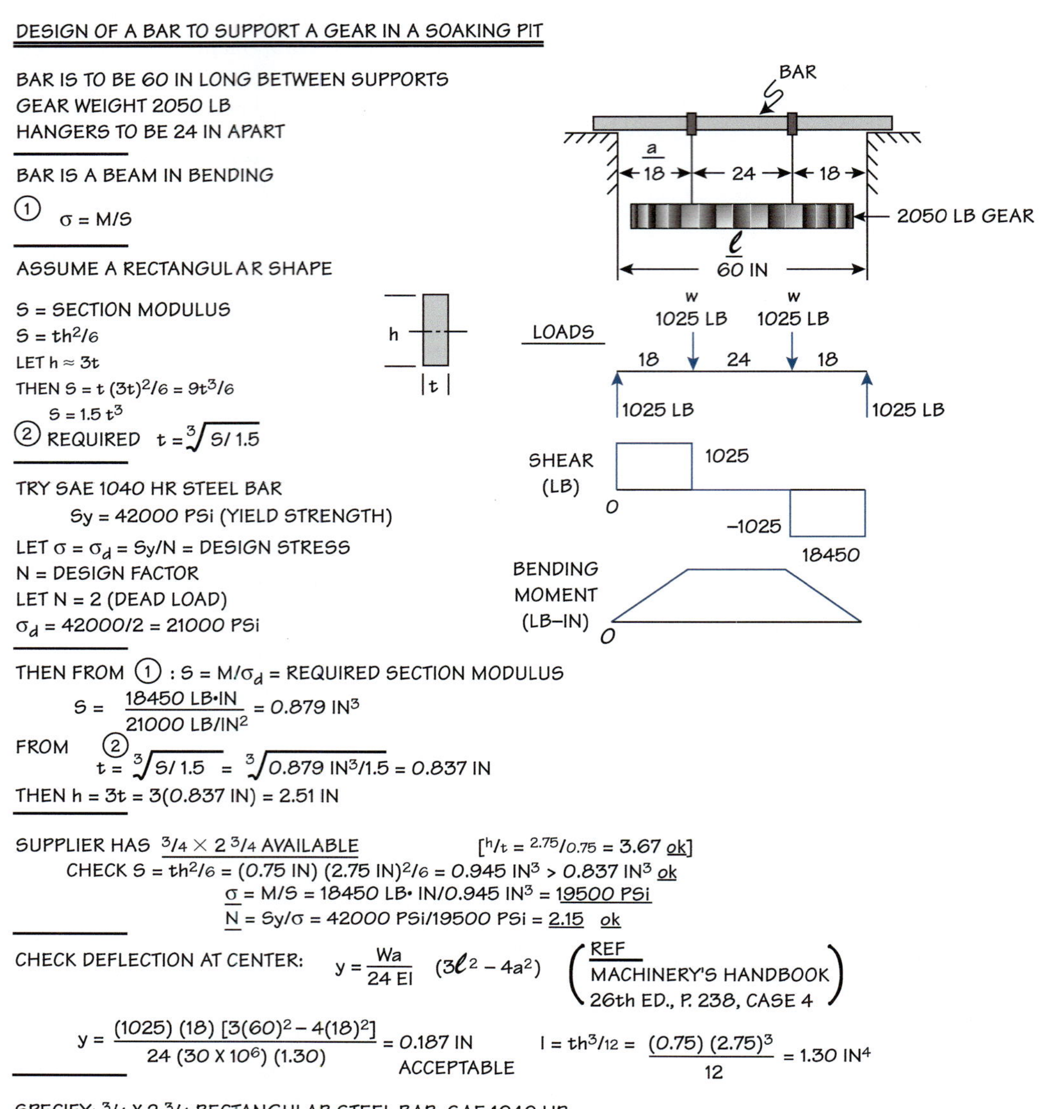

FIGURE 1–13 Sample design calculation

Figure 1–13 shows a sample design calculation. A beam is to be designed to span a 60-in pit to support a large gear weighing 2050 pounds (lb). The design assumes that a rectangular shape is to be used for the cross section of the beam. Other practical shapes could have been used. The objective is to compute the required dimensions of the cross section, considering both stress and deflection. A material for the beam is also chosen. Refer to Chapter 3 for a review of stress due to bending.

1–8 PREFERRED BASIC SIZES, SCREW THREADS, AND STANDARD SHAPES

One responsibility of a designer is to specify the final dimensions for load-carrying members. After completing the analyses for stress and deformation (strain), the designer will know the minimum acceptable values for dimensions that will ensure that the member will meet performance requirements. The designer then typically specifies the final dimensions to be standard or convenient values that will facilitate the purchase of materials and the manufacture of the parts. This section presents some guides to aid in these decisions and specifications.

Preferred Basic Sizes

Table A2–1 lists preferred basic sizes for fractional-inch, decimal-inch, and metric sizes.[1] You should choose one of these preferred sizes as the final part of your design. An

[1]Throughout this book, some references to tables and figures have the letter A included in their numbers; these tables and figures are in the appendices in the back of the book. For example, Table A2–1 is the first table in Appendix 2; Figure A15–4 is the fourth figure in Appendix 15. These tables and figures are clearly identified in their captions in the appendices.

example is at the end of the sample design calculation shown in Figure 1–13. You may, of course, specify another size if there is a sound functional reason.

American Standard Screw Threads

Threaded fasteners and machine elements having threaded connections are manufactured according to standard dimensions to ensure interchangeability of parts and to permit convenient manufacture with standard machines and tooling. Table A2–2 gives the dimensions of American Standard Unified threads. Sizes smaller than 1/4 in are given numbers from 0 to 12, while fractional-inch sizes are specified for 1/4 in and larger sizes. Two series are listed: UNC is the designation for coarse threads, and UNF designates fine threads. Standard designations are as follows:

6–32 UNC	(number size 6, 32 threads per inch, coarse thread)
12–28 UNF	(number size 12, 28 threads per inch, fine thread)
$\frac{1}{2}$–13 UNC	(fractional size 1/2 in, 13 threads per inch, coarse thread)
$1\frac{1}{2}$–12 UNF	(fractional size $1\frac{1}{2}$ in, 12 threads per inch, fine thread)

The *pitch* of a screw thread, *P*, is the distance between corresponding points on two adjacent threads and it is the distance the screw would move axially when the screw is turned one complete revolution. In the American Standard Unified thread system,

$$P = 1/n = 1/\text{number of threads per inch}$$

For example, for the 1/2–13 UNC thread, $P = 1/n = 1/13 = 0.0769\,\text{in}$. For metric screw threads, described next, the pitch is included as part of the thread designation system. It is given as an axial distance between adjacent threads in mm.

Given in the tables are the basic major diameter (*D*), the number of threads per inch (*n*), and the tensile stress area (A_t), found from

➪ Tensile Stress Area for Threads

$$A_t = 0.7854\left(D - \frac{0.9743}{n}\right)^2 \qquad \textbf{(1–1)}$$

When a threaded member is subjected to direct tension, the tensile stress area is used to compute the average tensile stress. It is based on a circular area computed from the mean of the pitch diameter and the minor diameter of the threaded member.

Metric Screw Threads

Table A2–3 gives similar dimensions for metric threads. Standard metric thread designations are of the form

$$\text{M10} \times 1.5$$

where M stands for metric

The following number is the basic major diameter, *D*, in mm

The last number is the pitch, *P*, between adjacent threads in mm

The tensile stress area for metric threads is computed from the following equation and is based on a slightly different diameter. (See Reference 11, page 1483.)

$$A_t = 0.7854(D - 0.9382P)^2 \qquad \textbf{(1–2)}$$

Thus, the designation above would denote a metric thread with a basic major diameter of $D = 10.0$ mm and a pitch of $P = 1.5$ mm. Note that pitch $= 1/n$. The tensile stress area for this thread is 58.0 mm^2.

Commercially Available Shapes for Load-Carrying Members

An extensive array of shapes for the cross sections of load-carrying members is included in the 19 tables of Appendix 15. Many are available in either steel or aluminum.

Both U.S. Customary sizes and metric sizes are included with sizes ranging from quite small [about 10 mm or 3/8 in (0.375 in)] to large (up to 300 mm or 24 in). Note that the metric sizes listed are designed specifically in metric dimensions, rather than being *soft converted* from inch-sizes.

You are advised to refer first to the introduction that precedes the tables at the start of this appendix where the units, basic shape descriptions, available materials, and available sizes are given. This will guide you to appropriate tables for your application. In general, the following types of shapes are included:

- Tables A15–1 to A15–3: Angles or L-shapes
- Tables A15–4 to A15–8: Channels or C-shapes
- Tables A15–9 to A15–13: I-beam shapes, such as wide-flange (W) shapes, American Standard S-shapes, Aluminum Association I-beam shapes, European standard I-beam shapes
- Tables A15–14 to A15–16: Hollow square and rectangular tubing
- Tables A15–17 to A15–19: Pipe and hollow mechanical tubing

In many tables, links to Internet sites are given to allow searching for a much larger set of sizes for the types of shapes in the given table. You are encouraged to use this feature for problems and design exercises in this book.

Footnotes for many tables list the density, taken to be *mass density,* for steel and/or aluminum as an aid in computing the mass or weight of a given design. Note that:

$$\text{Mass} = \text{Density} \times \text{Volume, where}$$

$$\text{Density} = \text{mass/unit volume}(\text{kg/m}^3 \text{ or } \text{lb}_\text{m}/\text{in}^3)$$

On earth, it is typically sufficiently accurate to assume that the U.S. Customary force unit, lb_f, is numerically equal to the mass unit, lb_m. For example, if density is given to be

0.283 lb_m/ft^3, it is reasonable to say that the weight density (also called specific weight) is 0.283 lb_f/ft^3. To obtain weight in the metric system, it is necessary to multiply the mass by g, the acceleration due to gravity. The nominal value of $g = 32.2\ ft/s^2$ or $9.81\ m/s^2$. See Section 1–10 for additional discussion of the relationship between mass and weight.

Typical section properties listed for the shapes are:

- A—Cross-sectional area
- I—Moment of inertia, sometimes called the second moment of the area; used for beam analysis and design
- S—Section modulus. Note that $S = I/c$, where c is the distance from the neutral axis to the outermost part of the section; used for beam analysis and design
- r—Radius of gyration. Note that $r = \sqrt{I/A}$; used for column analysis
- J—Polar moment of inertia; used for torsional analysis and design
- Z_p—Polar section modulus. Note that $Z_p = J/c$, where c is the outside radius of the pipe or tube; used for torsional analysis and design

Structural Shapes—Designations

Steel manufacturers provide a large array of standard structural shapes that are efficient in the use of material and that are convenient for specification and installation into building structures or machine frames. Included, as shown in Table 1–1, are standard angles (L-shapes), channels (C-shapes), wide-flange beams (W-shapes), American Standard beams (S-shapes), structural tubing, and pipe. Note that the W-shapes and the S-shapes are often referred to in general conversation as "I-beams" because the shape of the cross section looks like the capital letter I. See Reference 2.

Materials used for structural shapes are typically called *structural steels,* and their characteristics and properties are described more fully in Chapter 2. Refer to Appendix 7 for typical strength data. Rolled W-shapes are most readily available in ASTM A992, A572 Grade 50, or A36. S-shapes and C-shapes are typically made from ASTM A572 Grade 50 or A36. ASTM A36 should be specified for steel angles and plates. Hollow structural shapes (HSS) are most readily available in ASTM A500.

Aluminum structural shapes are most often made from extruded 6061-T6 alloy.

Angles (L-Shapes)

Table A15–1 shows sketches of the typical shapes of steel angles having equal or unequal leg lengths. Called *L-shapes* because of the appearance of the cross section, angles are often used as tension members of trusses and towers, framing members for machine structures, lintels over windows and doors in construction, stiffeners for large plates used in housings and beams, brackets, and ledge-type supports for equipment. Some refer to these shapes as "angle iron." The U.S. standard designation takes the following form, using one example size:

$$L4 \times 3 \times \tfrac{1}{2}$$

where L refers to the L-shape

4 is the length of the longer leg

3 is the length of the shorter leg

$\frac{1}{2}$ is the thickness of the legs

Dimensions are in inches

Smaller angles in U.S. units are listed in Table A15–2. Angles made to purely metric dimensions are listed in Table A15–3 over a range of sizes from 10 mm to 100 mm. Many of the listed shapes are available in either aluminum or steel.

Channels (C-Shapes)

See Table A15–4 for the appearance of American Standard channels and their geometric properties. Channels are used in applications similar to those described for angles. The flat web and the two flanges provide a generally stiffer shape than angles.

The sketch at the top of the table shows that channels have tapered flanges and webs with constant thickness. The slope of the flange taper is approximately 2 inches in 12 inches, and this makes it difficult to attach other members to the flanges. Special tapered washers are available to facilitate fastening. Note the designation of the x- and y-axes in the sketch, defined with the web of the channel vertical which gives it the characteristic C-shape. This is most important when using channels as beams or columns. The x-axis is located on the horizontal axis of symmetry, while the dimension x, given in the table, locates the y-axis relative to the back of the web. The centroid is at the intersection of the x- and y-axes.

The form of the U.S. standard designation for channels is

$$C15 \times 50$$

where C indicates that it is a standard C-shape

15 is the nominal (and actual) depth in inches with the web vertical

50 is the weight per unit length in lb/ft

Most other channels have flanges with uniform thicknesses that are often made by rolling flat sheets into the C-shape or by extrusion. Smaller channels in U.S. units are listed in Table A15–5. Aluminum angles made to European standards in purely metric dimensions are listed in Table A15–7, and steel metric channels are listed in Table A15–8.

Channels made to Aluminum Association standard shapes in U.S. units are listed in Table A15–6. Several designation systems are in common use. Some give the web height, leg length, and thickness. In this book, we use a form similar to that for standard steel shapes. For example,

$$C4 \times 1.738$$

Where C indicates the basic shape

4 indicates the depth of the shape (web height) in inches

1.738 indicates the weight per unit length in lb/ft

TABLE 1–1 Designations for Steel and Aluminum Shapes

Name of shape	Shape	Symbol	Example designation and Appendix table
Angle		L	L4 × 3 × $\frac{1}{2}$ Table A15–1
Channel		C	C15 × 50 Table A15–4
Wide-flange beam		W	W14 × 43 Table A15–9
American Standard beam		S	S10 × 35 Table A15–10
Structural tubing—square		HSS	4 × 4 × $\frac{1}{4}$ Table A15–14
Structural tubing—rectangular		HSS	6 × 4 × $\frac{1}{4}$ Table A15–14
Pipe		PIPE	4-inch standard weight 4-inch Schedule 40 Table A15–17
Aluminum Association channel		C	C4 × 1.738 Table A15–6
Aluminum Association I-beam		I	18 × 6.181 Table A15–11

I-Beam Shapes

Wide-Flange Shapes (W-Shapes). Refer to Table A15–9, which illustrates the most common shape used for beams. W-shapes have relatively thin webs and somewhat thicker, flat flanges with constant thickness. Most of the area of the cross section is in the flanges, farthest away from the horizontal centroidal axis (x-axis), thus making the moment of inertia very high for a given amount of material. Note that the properties of moment of inertia and section modulus are very much higher with respect to the x-axis than they are for the y-axis. Therefore, W-shapes are typically used in the orientation shown in the sketch in Table A15–9. Also, these shapes are best when used in pure bending without twisting because they are quite flexible in torsion.

The standard designation for steel W-shapes carries much information. Consider the following example:

$$\text{W14} \times 43$$

where W indicates that it is a W-shape

14 is the nominal depth in inches

43 is the weight per unit length in lb/ft

The term *depth* is the standard designation for the vertical height of the cross section when placed in the orientation shown in Table A15–9. Note from the data in the table that the actual depth is often different from the nominal depth. For the W14 × 43, the actual depth is 13.66 in.

All sizes listed in Table A15–9, along with many more (see Reference 2), are available in steel. A few of the listed sizes are also available in aluminum.

American Standard Beams (S-Shapes). Table A15–10 shows the properties for S-shapes. Much of the discussion given for W-shapes applies to S-shapes as well. Note that, again, the weight per foot of length is included in the designation such as the S10 × 35, which weighs 35 lb/ft. For most, but not all, of the S-shapes, the actual depth is the same as the nominal depth. The flanges of the S-shapes are tapered at a slope of approximately 2 inches in 12 inches, similar to the flanges of the C-shapes. The x- and y-axes are defined as shown with the web vertical.

Often wide-flange shapes (W-shapes) are preferred over S-shapes because of their relatively wide flanges, the constant thickness of the flanges, and the generally higher section properties for a given weight and depth.

A few of the sizes for S-shapes are available in aluminum.

Aluminum Association I-Beam Shapes. Aluminum Association standard I-beam shapes in U.S. units are listed in Table A15–11. Several designation systems are in common use. Some give the web height, flange width, and the thickness for either the web or the flange. In this book, we use a form similar to that for standard steel I-shapes. For example,

$$\text{I8} \times 6.181$$

Where I indicates the basic I-shape

8 indicates the depth of the shape (web height) in inches

6.181 indicates the weight per unit length in lb/ft

Other I-Beam Shapes. Table A15–12 lists small extruded aluminum shapes. European standard steel shapes in purely metric units are listed in Table A15–13.

Hollow Tubing (Square and Rectangular)

Square tubing and rectangular tubing are very useful in machine structures because they provide good section properties for members loaded as beams in bending and for torsional loading (twisting) because of the closed cross section. The flat sides often facilitate fastening of members together or the attachment of equipment to the structural members. Some frames are welded into an integral unit that functions as a stiff space-frame. Square tubing makes an efficient section for columns.

See Table A15–14 for the appearance and properties for hollow steel structural shapes. These shapes, often called HSS, are usually formed from flat sheet and welded along the length. The section properties account for the corner radii. Note the sketches showing the x- and y-axes. The standard designation takes the form

$$\text{HSS } 6 \times 4 \times \tfrac{1}{4}$$

where 6 is the depth of the longer side in inches

4 is the width of the shorter side in inches

$\frac{1}{4}$ is the wall thickness in inches

Note that for standard HSS shapes as listed in Reference 2, the *design wall thickness*, t_w, from Table A15–14 should be used. That value, *smaller* than the nominal size used in the designation, is used to compute the listed section properties.

Table A15–15 lists smaller steel and aluminum square and rectangular tubing in inch-sizes from 0.375 in (3/8 in) to 3.00 in depth. Metric sizes for square and rectangular tubing are listed in Table A15–16 from 20 mm to 300 mm depth.

Pipe and Hollow Circular Tubing

Hollow circular sections, commonly called *pipe,* are very efficient for use as beams, torsion members, and columns. The placement of the material uniformly away from the center of the pipe enhances the moment of inertia for a given amount of material and gives the pipe uniform properties with respect to all axes through the center of the cross section. The closed cross-sectional shape gives it high strength and stiffness in torsion as well as in bending.

Table A15–17 gives the properties for American National Standard Schedule 40 welded and seamless wrought steel pipe. This type of pipe is often used to transport water and other fluids, but it also performs well in structural applications. Note that the actual inside and outside diameters are somewhat different from the nominal size, except for the very large sizes. Construction pipe is often called *Standard Weight Pipe,* and it has the same dimensions as the Schedule 40 pipe for sizes from 1/2 in to 10 in. Other "schedules" and "weights" of pipe are available with larger and smaller wall thicknesses.

Other hollow circular sections are commonly available that are referred to as *tubing*. These sections are available in carbon steel, alloy steel, stainless steel, aluminum, copper, brass, titanium, and other materials. See References 1, 2, 5, and 9 for a variety of types and sizes of pipe and tubing.

Tubing is typically specified by its outside diameter and wall thickness, with the wall thickness sometimes given as a standard *gauge*. Table A15–18 lists tubing in U.S. sizes in steel and aluminum with outside diameters from 0.50 in to 5.0 in and various wall thicknesses. Table A15–19 lists tubing in metric sizes in steel and aluminum with outside diameters from 10 mm to 150 mm in various wall thicknesses.

1–9 UNIT SYSTEMS

We will perform computations in this book by using either the U.S. Customary Unit System (inch-pound-second) or the International System (SI). Table 1–2 lists the typical units used in the study of machine design. *SI*, the abbreviation for "Le Système International d'Unités," is the standard for metric units throughout the world. (See Reference 4.) For convenience, the term *SI units* will be used instead of *metric units*.

Prefixes applied to the basic units indicate order of magnitude. Only those prefixes listed in Table 1–3, which differ by a factor of 1000, should be used in technical calculations. The final result for a quantity should be reported as a number between 0.1 and 10 000, times some multiple of 1000. Then the unit with the appropriate prefix should be specified. Table 1–4 lists examples of proper SI notation.

Sometimes you have to convert a unit from one system to another. Appendix 16 provides tables of conversion factors. Also, you should be familiar with the typical order of magnitude of the quantities encountered in machine design so that you can judge the reasonableness of design calculations (see Table 1–5 for several examples).

TABLE 1–2 Typical Units used in Machine Design

Quantity	U.S. Customary unit	SI unit
Length or distance	inch (in) foot (ft)	meter (m) millimeter (mm)
Area	square inch (in^2)	square meter (m^2) or square millimeter (mm^2)
Force	pound (lb) kip (K) (1000 lb)	newton (N) ($1\ N = 1\ kg \cdot m/s^2$)
Mass Time Angle Temperature	slug (lb · s^2/ft) second (s) degree (°) degrees Fahrenheit (°F)	kilogram (kg) second (s) radian (rad) or degree (°) degrees Celsius (°C)
Torque or moment	pound-inch (lb · in) or pound-foot (lb · ft)	newton-meter (N · m)
Energy or work	pound-inch (lb · in)	joule (J) $1\ J = 1\ N \cdot m$
Power	horsepower (hp) ($1\ hp = 550\ lb \cdot ft/s$)	watt (W) or kilowatts (kW) ($1\ W = 1\ J/s = 1\ N \cdot m/s$)
Stress, pressure, or modulus of elasticity	pounds per square inch (lb/in^2, or psi) kips per square inch (K/in^2, or ksi)	pascal (Pa) ($1\ Pa = 1\ N/m^2$) kilopascal (kPa) ($1\ kPa = 10^3\ Pa$) megapascal (MPa) ($1\ MPa = 10^6\ Pa$) gigapascal (GPa) ($1\ GPa = 10^9\ Pa$)
Section modulus Moment of inertia	inches cubed (in^3) inches to the fourth power (in^4)	meters cubed (m^3) or millimeters cubed (mm^3) meters to the fourth power (m^4) or millimeters to the fourth power (mm^4)
Rotational speed	revolutions per min (rpm)	radians per second (rad/s)

TABLE 1–3 Prefixes used with SI Units

Prefix	SI symbol	Factor
micro-	μ	$10^{-6} = 0.000001$
milli-	m	$10^{-3} = 0.001$
kilo-	k	$10^3 = 1000$
mega-	M	$10^6 = 1000000$
giga-	G	$10^9 = 1000000000$

TABLE 1–4 Quantities Expressed in SI Units

Computed result	Reported result
0.001 65 m	1.65×10^{-3} m, or 1.65 mm
32 540 N	32.54×10^3 N, or 32.54 kN
1.583×10^5 W	158.3×10^3 W, or 158.3 kW; or 0.1583×10^6 W; or 0.1583 MW
2.07×10^{11} Pa	207×10^9 Pa, or 207 GPa

TABLE 1–5 Typical Order of Magnitude for Commonly Encountered Quantities

Quantity	U.S. Customary unit	SI unit
Dimensions of a wood standard 2 × 4	1.50 in × 3.50 in	38 mm × 89 mm
Moment of inertia of a 2 × 4 (3.50-in side vertical)	5.36 in^4	2.23×10^6 mm^4, or 2.23×10^{-6} m^4
Section modulus of a 2 × 4 (3.50-in side vertical)	3.06 in^3	5.02×10^4 mm^3, or 5.02×10^{-5} m^3
Force required to lift 1.0 gal of gasoline	6.01 lb	26.7 N
Density of water	1.94 slugs/ft^3	1000 kg/m^3, or 1.0 Mg/m^3
Compressed air pressure in a factory	100 psi	690 kPa
Yield point of SAE 1040 hot-rolled steel	42 000 psi, or 42 ksi	290 MPa
Modulus of elasticity of steel	30 000 000 psi, or 30×10^6 psi	207 GPa

Example Problem 1–1 Express the diameter of a shaft in millimeters if it is measured to be 2.755 in.

Solution Table A16 gives the conversion factor for length to be 1.00 in = 25.4 mm. Then

$$\text{Diameter} = 2.755\,\text{in}\,\frac{25.4\ \text{mm}}{1.00\,\text{in}} = 69.98\ \text{mm}$$

Example Problem 1–2 An electric motor is rotating at 1750 revolutions per minute (rpm). Express the speed in radians per second (rad/s).

Solution A series of conversions is required.

$$\text{Rotational speed} = \frac{1750\ \text{rev}}{\text{min}}\,\frac{2\pi\ \text{rad}}{\text{rev}}\,\frac{1\,\text{min}}{60\,\text{s}} = 183.3\,\text{rad/s}$$

1–10 DISTINCTION AMONG WEIGHT, FORCE, AND MASS

Distinction must be made among the terms *force*, *mass*, and *weight*. *Mass* is the quantity of matter in a body. A *force* is a push or pull applied to a body that results in a change in the body's motion or in some deformation of the body. Clearly these are two different physical phenomena, but the distinction is not always understood. The units for force and mass used in this text are listed in Table 1–2.

The term *weight*, as used in this book, refers to the amount of *force* required to support a body against the influence of gravity. Thus, in response to "What is the weight of 75 kg of steel?" we would use the relationship between force and mass from physics:

➩ Weight/Mass Relationship

$$F = ma \quad \text{or} \quad w = mg$$

where F = force
m = mass
a = acceleration
w = weight
g = acceleration due to gravity

We will use

$$g = 32.2\,\text{ft/s}^2 \quad \text{or} \quad g = 9.81\,\text{m/s}^2$$

Then, to compute the weight,

$$w = mg = 75\,\text{kg}(9.81\,\text{m/s}^2)$$
$$w = 736\,\text{kg}\cdot\text{m/s}^2 = 736\,\text{N}$$

Remember that, as shown in Table 1–2, the newton (N) is equivalent to 1.0 kg · m/s^2. In fact, the newton is defined as the force required to give a mass of 1.0 kg an acceleration of 1.0 m/s^2. In our example, then, we would say that the 75-kg mass of steel has a weight of 736 N.

REFERENCES

1. Aluminum Association. *Aluminum Design Manual.* Washington, DC: Aluminum Association, 2005.
2. American Institute of Steel Construction. *Steel Construction Manual.* 13th ed. Chicago: American Institute of Steel Construction, 2005.
3. American Society of Mechanical Engineers. *Integrating the Product Realization Process (PRP) into the Undergraduate Curriculum.* New York: American Society of Mechanical Engineers, 1995.
4. American Society for Testing and Materials. *IEEE/ASTM SI-10 Standard for Use of the International System of Units (SI): The Modern Metric System.* West Conshohocken, PA: American Society for Testing and Materials, 2002.
5. Avallone, Eugene, Theodore Baumeister, and Ali Sadegh, eds. *Marks' Standard Handbook for Mechanical Engineers.* 11th ed. New York: McGraw-Hill, 2007.
6. Dym, Clive L., and Patrick Little. *Engineering Design: A Project-Based Introduction.* 3rd ed. New York: John Wiley & Sons, 2009.
7. Ertas, Atila, and Jesse C. Jones. *The Engineering Design Process.* 2nd ed. New York: John Wiley & Sons, 1997. Discussion of the design process from definition of design objectives through product certification and manufacture.
8. Hauser, John R., and Don Clausing. "The House of Quality." *Harvard Business Review* (May–June 1988): 63–73. Discusses Quality Function Deployment.
9. Mott, Robert L. *Applied Fluid Mechanics.* 6th ed. Upper Saddle River, NJ: Pearson/Prentice Hall, 2006.
10. National Research Council. *Improving Engineering Design: Designing for Competitive Advantage.* Washington, DC: National Academy Press, 1991. Describes the Product Realization Process (PRP).
11. Oberg, Erik, Franklin D. Jones, Holbrook L. Horton, and Henry H. Ryffell. *Machinery's Handbook.* 28th ed. New York: Industrial Press, 2008.
12. Pahl, Gerhard, Wolfgang Beitz, Jörg Feldhusen, and K. H. Grote. *Engineering Design: A Systematic Approach.* 3rd ed. London: Springer-Verlag, 2007.
13. Pugh, Stuart. *Total Design: Integrated Methods for Successful Product Engineering.* Upper Saddle River, NJ: Pearson/Prentice Hall, 1991.
14. Suh, Nam Pyo. *Axiomatic Design: Advances and Applications.* New York: Oxford University Press, 2001.
15. Suh, Nam Pyo. *The Principles of Design.* New York: Oxford University Press, 1990.
16. Ullman, David G. *The Mechanical Design Process.* 4th ed. New York: McGraw-Hill, 2010.
17. Taguchi, Genichi, Rajesh Jugulum, and Shin Taguchi. *Computer Based Robust Engineering: Essentials for DFSS.* Milwaukee, WI: ASQ Quality Press, 2005.
18. Yang, Kai, and Basem S. El-Haik. *Design for Six Sigma: A Roadmap for Product Development.* Brighton, MA: Axiomatic Design Solutions, 2009.
19. Jugulum, Rajesh, and Philip Samuel. *Design for Lean Six Sigma.* Brighton, MA: Axiomatic Design Solutions, 2009.
20. Wilson, Graham. *Six Sigma and the Product Development Cycle.* New York: Elsevier, Inc., 2006.
21. Altshuller, Genrich. *40 Principles: TRIZ Keys to Technical Innovation.* Worcester, MA: Technical Innovation Center, Inc., 1998.
22. Fey, Victor. *Innovation on Demand: New Product Development Using TRIZ.* Cambridge, England: Cambridge University Press, 2006.
23. Royzen, Zinovy. *Designing and Manufacturing Better Products Faster Using TRIZ.* 7th ed. Seattle, WA: TRIZ Consulting, 2010.
24. Mikulak, Raymond J., Robin McDermott, and Michael Beauregard. *The Basics of FMEA.* 2nd ed. Boca Raton, FL: CRC Press, 2009.
25. Kepner, Charles H., and Benjamin B. Tregoe. *The New Rational Manager.* Princeton, NJ: Kepner-Tregoe, Inc., 1998.
26. Dieter, George. *Engineering Design.* 4th ed. New York: McGraw-Hill, 2009.
27. Boothroyd, Geoffrey, Peter Dewhust, and Winston A. Knight *Product Design for Manufacture and Assembly.* 3rd ed. Boca Raton, FL: CRC Press, 2011.

INTERNET SITES FOR GENERAL MECHANICAL DESIGN

Included here are Internet sites that can be used in many of the chapters of this book and in general design practice to identify commercial suppliers of machine elements and standards for design or to perform stress analyses. Later chapters include sites specific to the topics covered there.

1. **American National Standards Institute (ANSI)** www.ansi.org A private, nonprofit organization that administers and coordinates the U.S. voluntary standardization and conformity assessment system.
2. **Global Engineering Documents** http://global.ihs.com A database of standards and publications offered by many standards-developing organizations such as ASME, ASTM, and ISO.
3. **GlobalSpec** www.globalspec.com A database of a wide variety of technical products and services that provides for searching by technical specifications, access to supplier information, and comparison of suppliers for a given product. The Mechanical Components category includes many of the topics addressed in this book.
4. **MDSOLIDS** www.mdsolids.com Educational software for strength of materials topics, including beams, flexure, torsion members, columns, axial structures, statically indeterminate structures, trusses, section properties, and Mohr's circle analysis. This software may serve as a review tool for the prerequisite knowledge needed in this book.
5. **Engineering.com** www.engineering.com/software/Downloads/tabic/74/Default.aspx Software for various engineering applications including beam analysis programs *Beam Analyzer* and *Beam Boy Analysis Tool.*
6. **Softpedia.com** www.softpedia.com/get/Science-CAD/Beam-Analysis-Tool.shtml Software for various technical fields including *Beam Analysis Tool.*
7. **Power Transmission Home Page** www.powertransmission.com Clearinghouse on the Internet for buyers, users, and sellers of power transmission products and services. Included are gears, gear drives, belt drives, chain drives, bearings, clutches, brakes, and many other machine elements covered in this book.

INTERNET SITES FOR INNOVATION AND MANAGING COMPLEX DESIGN PROJECTS

8. **Axiomatic Design Solutions, Inc.** www.axiomaticdesign.com An integrated method of combining customer needs and functional requirements to produce detailed design parameters, the final embodiment of the design solution, and the process needed to produce the design efficiently.
9. **QFD Institute** www.qfdi.org An organization promoting the Quality Function Deployment (QFD) method for product design and quality systems management.
10. **Applied Marketing Science (AMS), Inc.** www.ams-inc.com Consulting firm providing voice of the customer (VOC) insights and using Quality Function Deployment (QFD) based on the "House of Quality" matrix described in Reference 8.
11. **iSixSigma.com** www.isixsigma.com The mission of iSixSigma.com is to provide a free information resource to help business professionals successfully implement Lean Six Sigma and business process improvement tools and methodologies within their organizations.
12. **The TRIZ Journal.** www.triz-journal.com A site offering case studies, theoretical articles, TRIZ history, symposia, classes, and books to help designers learn TRIZ.
13. **Technical Innovation Center, Inc.** www.triz.org Provider of TRIZ consulting, training, and publishing services.
14. **The Altshuller Institute for TRIZ Studies** www.aitriz.org Provider of services to the world TRIZ community, authorized by Genrikh Altshuller, the founder of TRIZ.
15. **Triz Consulting, Inc.** www.trizconsulting.com Provider of TRIZ training for systematic innovation, consulting, and publications.
16. **Systems2win.com** www.systems2win.com Provider of Excel templates for Lean Six Sigma to Define, Measure, Analyze, Improve, and Control quality (DMAIC 6Σ tools).
17. **FMEA-FMECA.com** www.fmea-fmeca.com An Internet source of information about the Failure Modes and Effects Analysis (FMEA) and Failure Modes and Effects with Criticality Analysis (FMECA). Source for books, software, forms, examples, and other resources.
18. **Kepner-Tregoe, Inc.** www.kepner-tregoe.com Kepner-Tregoe provides consulting and training services to help clients implement their strategies by embedding problem-solving, decision-making, and project execution methods. Publisher of the book cited in Reference 25.

PROBLEMS

Functions and Design Requirements

For the devices described in Problems 1–14, write a set of functions and design requirements in a similar manner to those in Section 1–4. You or your instructor may add more specific information to the general descriptions given.

1. The hood latch for an automobile
2. A hydraulic jack used for car repair
3. A portable crane to be used in small garages and homes
4. A machine to crush soft-drink or beer cans
5. An automatic transfer device for a production line
6. A device to raise a 55-gallon (gal) drum of bulk materials and dump the contents into a hopper
7. A paper feed device for a copier
8. A conveyor to elevate and load gravel into a truck
9. A crane to lift building materials from the ground to the top of a building during construction
10. A machine to insert toothpaste tubes into cartons
11. A machine to insert 24 cartons of toothpaste into a shipping container
12. A gripper for a robot to grasp a spare tire assembly and insert it into the trunk of an automobile on an assembly line
13. A table for positioning a weldment in relation to a robotic welder
14. A garage door opener

Units and Conversions

For Problems 15–28, perform the indicated conversion of units. (Refer to Appendix 16 for conversion factors.) Express the results with the appropriate prefix as illustrated in Tables 1–3 and 1–4.

15. Convert a shaft diameter of 1.75 in to mm.
16. Convert the length of a conveyor from 46 ft to meters.
17. Convert the torque developed by a motor of 12 550 lb · in to N · m.
18. A wide-flange steel-beam shape, W12 × 14, has a cross-sectional area of 4.12 in^2. Convert the area to mm^2.
19. The W12 × 14 beam shape has a section modulus of 14.8 in^3. Convert it to mm^3.
20. The W12 × 14 beam shape has a moment of inertia of 88.0 in^4. Convert it to mm^4.
21. What standard steel equal leg angle would have a cross-sectional area closest to (but greater than) 750 mm^2? See Appendix 15.
22. An electric motor is rated at 7.5 hp. What is its rating in watts (W)?
23. A vendor lists the ultimate tensile strength of a steel to be 127 000 psi. Compute the strength in MPa.
24. Compute the weight of a steel shaft, 35.0 mm in diameter and 675 mm long. (See Appendix 3 for the density of steel.)
25. A torsional spring requires a torque of 180 lb · in to rotate it 35°. Convert the torque to N · m and the rotation to radians. If the *scale of the spring* is defined as the applied torque per unit of angular rotation, compute the spring scale in both unit systems.
26. To compute the energy used by a motor, multiply the power that it draws by the time of operation. Consider a motor that draws 12.5 hp for 16 h/day, five days per week. Compute the energy used by the motor for one year. Express the result in ft · lb and W · h.
27. One unit used for fluid viscosity in Chapter 16 of this book is the *reyn,* defined as 1.0 lb · s/in^2. If a lubricating oil has a viscosity of 3.75 reyn, convert the viscosity to the standard units in the U.S. Customary System lb · s/ft^2 and in the SI (N · s/m^2).
28. The life of a bearing supporting a rotating shaft is expressed in number of revolutions. Compute the life of a bearing that rotates 1750 rpm continuously for 24 h/day for five years.

CHAPTER
TWO

MATERIALS IN MECHANICAL DESIGN

THE BIG PICTURE Materials in Mechanical Design

Discussion Map

- You must understand the behavior of materials to make good design decisions and to communicate with suppliers and manufacturing staff.

Discover

Examine consumer products, industrial machinery, automobiles, and construction machinery.

What materials are used for the various parts?

Why do you think those materials were specified?

How were they processed?

What material properties were important to the decisions to use particular materials?

Examine the appendices tables, and refer to them later as you read about specific materials.

This chapter summarizes the design properties of a variety of materials. The appendices include data for many examples of these materials in many conditions.

It is the designer's responsibility to specify suitable materials for each component of a mechanical device. Your initial efforts in specifying a material for a particular component of a mechanical design should be directed to the basic kind of material to be used. Keep an open mind until you have specified the functions of the component, the kinds and magnitudes of loads it will carry, and the environment in which it must operate. Your selection of a material must consider its physical and mechanical properties and match them to the expectations placed on it. First consider the following classes of materials:

Metals and their alloys	Plastics	Composites
Elastomers	Woods	Ceramics and glasses

Each of these classes contains a large number of specific materials covering a wide range of actual properties. However, you probably know from your experience the general behavior of each kind and have some feel for the applications in which each is typically used. Most of the applications considered in the study of design of machine elements in this book use metal alloys, plastics, and composites.

Satisfactory performance of machine components and systems depends greatly on the materials that the designer specifies. As a designer, you must understand how materials behave, what properties of the material affect the performance of the parts, and how you should interpret the large amounts of data available on material properties. Your ability to effectively communicate your specifications for materials with suppliers, purchasing agents, metallurgists, manufacturing process personnel, heat treatment personnel, plastics molders, machinists, and quality assurance specialists often has a strong influence on the success of a design.

Explore what kinds of materials are used in consumer products, industrial machinery, automobiles, construction machinery, and other devices and systems that you come into contact with each day. Make judgments about why each material was specified for a particular application. Where do you see steel being used? Contrast that usage with where aluminum or other nonferrous materials are used. How are the products produced? Can you find different parts that are machined, cast, forged, roll-formed, and welded? Why do you think those processes were specified for those particular products?

Document several applications for plastics and describe the different forms that are available and that have been made by different manufacturing processes. Which are made by plastic molding processes, vacuum forming, blow molding, and others? Can you identify parts made from composite materials that have a significant amount of high-strength fibers embedded in a plastic matrix? Check out sporting goods and parts of cars, trucks, and airplanes.

From the products that you found from the exploration outlined previously, identify the basic properties of the materials that were important to the designers: strength, rigidity (stiffness), weight (density), corrosion resistance, appearance, machinability, weldability, ease of forming, cost, and others.

This chapter focuses on material selection and the use of material property data in design decisions, rather than on the metallurgy or chemistry of the materials. One of the uses of the information in this chapter is as a glossary of terms that you can use throughout the book; important terms are given in *italic* type. Also, there are numerous references to Appendices 3 through 13, where tables of data for material properties are given. Go there now and see what kinds of data are provided. Then you can study the tables in more depth as you read the text. Note that many of the problems that you will solve in this book and the design projects that you complete will use data from these tables.

The subject of this chapter is very broad and it is not possible to include in this book all the detail you may need for each design situation that arises as you study this book or in your career. Huge amounts of additional information are available on the Internet and the numerous sites listed at the end of this chapter give suggestions that may lead to what you need. Many of the listed sites are referred to within this chapter, but some are not and they provide alternate ways for you to search for what you need.

Now apply some of what you have gained from **The Big Picture** exploration to a specific design situation as outlined in **You Are the Designer**, which follows.

YOU ARE THE DESIGNER

You are part of a team responsible for the design of an electric lawn mower for the household market. One of your tasks is to specify suitable materials for the various components. Consider your own experience with such lawn mowers and think what materials would be used for these key components: *wheels, axles, housing,* and *blade.* What are their functions? What conditions of service will each encounter? What is one reasonable type of material for each component and what general properties should it have? How could they be manufactured? Possible answers to these questions follow.

Wheels

Function: Support the weight of the mower. Permit easy, rolling movement. Provide for mounting on an axle. Ensure safe operation on flat or sloped lawn surfaces.

Conditions of service: Must operate on grass, hard surfaces, and soft earth. Exposed to water, lawn fertilizers, and general outdoor conditions. Will carry moderate loads. Requires an attractive appearance.

One reasonable material: One-piece plastic wheel incorporating the tire, rim, and hub. Must have good strength, stiffness, toughness, and wear resistance.

Manufacturing method: Plastic injection molding

Axles

Function: Transfer the weight of mower from the housing to the wheels. Allow rotation of the wheels. Maintain location of the wheels relative to the housing.

Conditions of service: Exposure to general outdoor conditions. Moderate loads.

One possible material: Steel rod with provisions for mounting wheels and attaching to housing. Requires moderate strength, stiffness, and corrosion resistance.

Manufacturing method: Commercially available cylindrical rod. Possibly machining.

Housing

Function: Support, safely enclose, and protect operating components, including the blade and motor. Accommodate the attachment of two axles and a handle. Permit cut grass to exit the cutting area.

Conditions of service: Moderate loads and vibration due to motor. Possible shock loads from wheels. Multiple attachment points for axles, handle, and motor. Exposed to wet grass and general outdoor conditions. Requires attractive appearance.

One possible material: Heavy-duty plastic with good strength, stiffness, impact resistance, toughness, and weather resistance.

Manufacturing method: Plastic injection molding. May require machining for holes and mounting points for the motor.

Blade

Function: Cut blades of grass and weeds while rotating at high speed. Facilitate connection to motor shaft. Operate safely when foreign objects are encountered, such as stones, sticks, or metal pieces.

Conditions of service: Normally moderate loads. Occasional shock and impact loads. Must be capable of sharpening a portion of the blade to ensure clean cutting of grass. Maintain sharpness for reasonable time during use.

One possible material: Steel with high strength, stiffness, impact resistance, toughness, and corrosion resistance.

Manufacturing method: Stamping from flat steel strip. Machining and/or grinding for cutting edge. ■

This simplified example of the material selection process should help you to understand the importance of the information provided in this chapter about the behavior of materials commonly used in the design of machine elements. A more comprehensive discussion of material selection occurs at the end of the chapter.

2–1 OBJECTIVES OF THIS CHAPTER

After completing this chapter, you will be able to:

1. State the types of material properties that are important to the design of mechanical devices and systems.
2. Define the following terms: *tensile strength, yield strength, proportional limit, elastic limit, modulus of elasticity in tension, ductility and percent elongation, shear strength, Poisson's ratio, modulus of elasticity in shear, hardness, machinability, impact strength, creep density, coefficient of thermal expansion, thermal conductivity,* and *electrical resistivity.*
3. Describe the nature of *carbon and alloy steels,* the number-designation system for steels, and the effect of several kinds of alloying elements on the properties of steels.
4. Describe the manner of designating the condition and heat treatment of steels, including *hot rolling, cold drawing, annealing, normalizing, through-hardening, tempering,* and *case hardening by flame hardening, induction hardening,* and *carburizing.*
5. Describe *stainless steels* and recognize many of the types that are commercially available.
6. Describe *structural steels* and recognize many of their designations and uses.
7. Describe *cast irons* and several kinds of *gray iron, ductile iron,* and *malleable iron.*
8. Describe *powdered metals* and their properties and uses.
9. Describe several types of *tool steels* and *carbides* and their typical uses.
10. Describe *aluminum alloys* and their conditions, such as *strain hardening* and *heat treatment.*
11. Describe the nature and typical properties of *zinc, titanium, copper, brass, bronze,* and *nickel-based alloys.*
12. Describe several types of *plastics,* both *thermosetting* and *thermoplastic,* and their typical properties and uses.
13. Describe several kinds of *composite materials* and their typical properties and uses.
14. Implement a rational material selection process.

2–2 PROPERTIES OF MATERIALS

Machine elements are very often made from one of the metals or metal alloys such as steel, aluminum, cast iron, zinc, titanium, or bronze. This section describes the important properties of materials as they affect mechanical design.

Strength, elastic, and ductility properties for metals, plastics, and other types of materials are usually determined from a *tensile test* in which a sample of the material, typically in the form of a round or flat bar, is clamped between jaws and pulled slowly until it breaks in tension. The magnitude of the force on the bar and the corresponding change in length (strain) are monitored and recorded continuously during the test. Because the stress in the bar is equal to the applied force divided by the area, stress is proportional to the applied force. The data from such tensile tests are often shown on *stress–strain diagrams,* such as those shown in Figures 2–1 and 2–2. In the following paragraphs, several strength, elastic, and ductility properties of metals are defined.

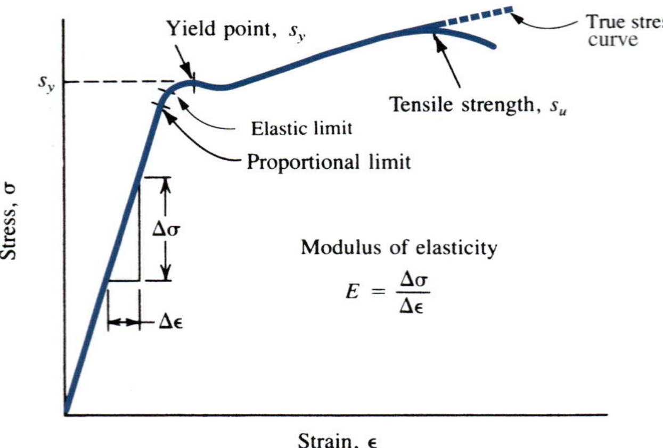

FIGURE 2–1 Typical stress–strain diagram for steel

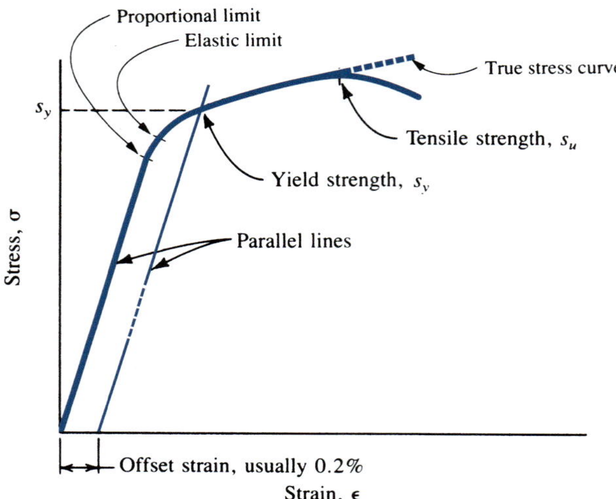

FIGURE 2–2 Typical stress–strain diagram for aluminum and other metals having no yield point

Tensile Strength, s_u

The peak of the stress–strain curve is considered the *ultimate tensile strength* (s_u), sometimes called the *ultimate strength* or simply the *tensile strength*. At this point during the test, the highest *apparent stress* on a test bar of the material is measured. As shown in Figures 2–1 and 2–2, the curve appears to drop off after the peak. However, notice that the instrumentation used to create the diagrams is actually plotting *load versus deflection* rather than *true stress versus strain*. The apparent stress is computed by dividing the load by the original cross-sectional area of the test bar. After the peak of the curve is reached, there is a pronounced decrease in the bar's diameter, referred to as *necking down*. Thus, the load acts over a smaller area, and the *actual stress* continues to increase until failure. It is very difficult to follow the reduction in diameter during the necking-down process, so it has become customary to use the peak of the curve as the tensile strength, although it is a more conservative value.

Yield Strength, s_y

That portion of the stress–strain diagram where there is a large increase in strain with little or no increase in stress is called the *yield strength* (s_y). This property indicates that the material has, in fact, yielded or elongated plastically, permanently, and to a large degree. If the point of yielding is quite noticeable, as it is in Figure 2–1, the property is called the *yield point* rather than the yield strength. This is typical of a plain carbon hot-rolled steel.

Figure 2–2 shows the stress–strain diagram that is typical of a nonferrous metal such as aluminum or titanium or of certain high-strength steels. Notice that there is no pronounced yield point, but the material has actually yielded at or near the stress level indicated as s_y. That point is determined by the *offset method*, in which a line is drawn parallel to the straight-line portion of the curve and is offset to the right by a set amount, usually 0.20% strain (0.002 in/in). The intersection of this line and the stress–strain curve defines the material's yield strength. In this book, the term *yield strength* will be used for s_y, regardless of whether the material exhibits a true yield point or whether the offset method is used.

Proportional Limit

That point on the stress–strain curve where it deviates from a straight line is called the *proportional limit*. That is, at or above that stress value, stress is no longer proportional to strain. Below the proportional limit, Hooke's law applies: Stress is proportional to strain. In mechanical design, materials are rarely used at stresses above the proportional limit.

Elastic Limit

At some point, called the *elastic limit*, a material experiences some amount of plastic strain and thus will not return to its original shape after release of the load. Below that level, the material behaves completely elastically. The proportional limit and the elastic limit lie quite close to the yield strength. Because they are difficult to determine, they are rarely reported.

Modulus of Elasticity in Tension, E

For the part of the stress–strain diagram that is straight, stress is proportional to strain, and the value of E, *the modulus of elasticity*, is the constant of proportionality. That is,

➪ Modulus of Elasticity in Tension

$$E = \frac{\text{stress}}{\text{strain}} = \frac{\sigma}{\epsilon} \quad \textbf{(2–1)}$$

This is the slope of the straight-line portion of the diagram. The modulus of elasticity indicates the stiffness of the material, or its resistance to deformation.

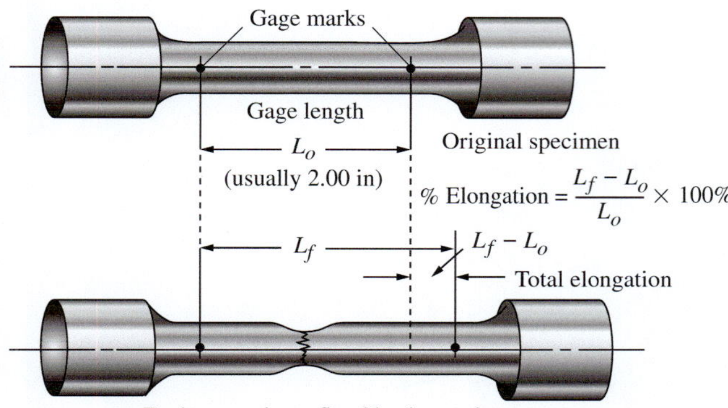

FIGURE 2–3 Measurement of percent elongation

Ductility and Percent Elongation

Ductility is the degree to which a material will deform before ultimate fracture. The opposite of ductility is *brittleness*. When ductile materials are used in machine members, impending failure is detected easily, and sudden failure is unlikely. Also, ductile materials normally resist the repeated loads on machine elements better than brittle materials.

The usual measure of ductility is the *percent elongation* of the material after fracture in a standard tensile test. Figure 2–3 shows a typical standard tensile specimen before and after the test. Before the test, gage marks are placed on the bar, usually 2.00 in apart. Then, after the bar is broken, the two parts are fitted back together, and the final length between the gage marks is measured. The percent elongation is the difference between the final length and the original length divided by the original length, converted to a percentage. That is,

➪ Percent Elongation

$$\text{percent elongation} = \frac{L_f - L_o}{L_o} \times 100\% \qquad (2\text{–}2)$$

The percent elongation is assumed to be based on a gage length of 2.00 in unless some other gage length is specifically indicated. Tests of structural steels often use a gage length of 8.00 in.

Theoretically, a material is considered ductile if its percent elongation is greater than 5% (lower values indicate brittleness). For practical reasons, it is advisable to use a material with a value of 12% or higher for machine members subject to repeated loads or shock or impact.

Percent reduction in area is another indication of ductility. To find this value, compare the original cross-sectional area with the final area at the break for the tensile test specimen.

Shear Strength, s_{ys} and s_{us}

Both the yield strength and the ultimate strength in shear (s_{ys} and s_{us}, respectively) are important properties of materials. Unfortunately, these values are seldom reported. We will use the following estimates:

➪ Estimates for S_{ys} and S_{us}

$$s_{ys} = s_y/2 = 0.50\, s_y = \text{yield strength in shear} \qquad (2\text{–}3)$$

$$s_{us} = 0.75 s_u = \text{ultimate strength in shear} \qquad (2\text{–}4)$$

Poisson's Ratio, ν

When a material is subjected to a tensile strain, there is a simultaneous shortening of the cross-sectional dimensions perpendicular to the direction of the tensile strain. The ratio of the shortening strain to the tensile strain is called *Poisson's ratio*, usually denoted by ν, the Greek letter nu. (The Greek letter mu, μ, is sometimes used for this ratio.) Poisson's ratio is illustrated in Figure 2–4. Typical ranges of values for Poisson's ratio are shown in Table 2–1.

Modulus of Elasticity in Shear, *G*

The *modulus of elasticity in shear* (G) is the ratio of shearing stress to shearing strain. This property indicates a material's stiffness under shear loading—that is, the resistance to shear deformation. There is a simple relationship between E, G, and Poisson's ratio:

➪ Modulus of Elasticity in Shear

$$G = \frac{E}{2(1 + \nu)} \qquad (2\text{–}5)$$

This equation is valid within the elastic range of the material.

Non-Destructive Measurement of Elastic Modulus and Poisson's Ratio

While most determinations for the elastic modulus in tension and in shear and the related property of Poisson's ratio are done using mechanical destructive testing as previously described, a nondestructive method exists that can be used for most metals, ceramics, and glasses. Internet site 29 describes a method that uses a measurement of the velocity of sound in the material using an ultrasonic pulse-echo technique along with a thickness measuring device. A sample of material approximately 12 mm (0.50 in) thick with smooth parallel end surfaces is used. After measuring the sound velocity, simple calculations enable determination of Poisson's ratio, Young's modulus (elastic modulus in tension), and shear modulus.

Flexural Strength and Flexural Modulus

For some materials, particularly plastics, additional properties in bending are obtained, called *flexural strength* and *flexural modulus*. Typical testing fixtures are shown in Figure 2–5 and testing should be done according to the recognized standards from either ASTM or ISO. ASTM

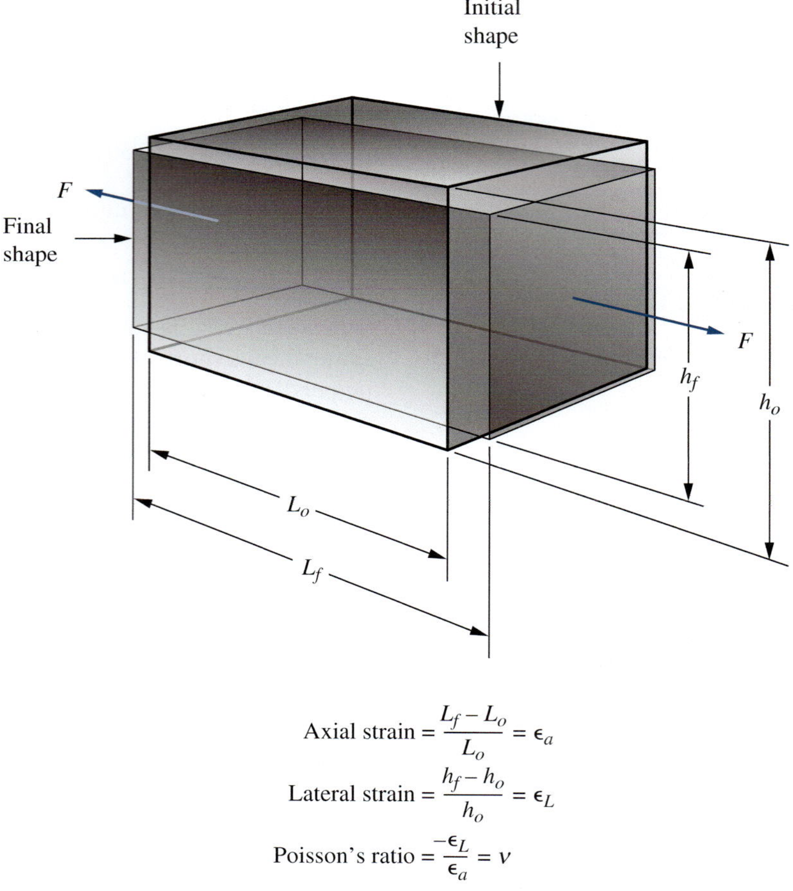

$$\text{Axial strain} = \frac{L_f - L_o}{L_o} = \epsilon_a$$

$$\text{Lateral strain} = \frac{h_f - h_o}{h_o} = \epsilon_L$$

$$\text{Poisson's ratio} = \frac{-\epsilon_L}{\epsilon_a} = \nu$$

FIGURE 2–4 Illustration of Poisson's ratio for an element in tension

TABLE 2–1 Approximate Values of Poisson's Ratio, ν

Concrete	0.10–0.25	Aluminum (most alloys)	0.33
Glass	0.24	Brass	0.33
Ductile iron	0.27	Copper	0.33
Gray cast iron	0.21	Zinc	0.33
Plastics	0.20–0.40	Phosphor bronze	0.35
Carbon and alloy steel	0.29	Magnesium	0.35
Stainless steel (18-8)	0.30	Lead	0.43
Titanium	0.30	Rubber, elastomers	~0.50

Notes: Values are approximate and vary somewhat with specific composition. Rubber and elastomers approach a limiting value of 0.50.

D790[1] and ISO 178[2] use the 3-point bending principle [Part (a) of Figure 2–5], wherein a flat specimen is placed on two cylindrical or knife-edge supports with a known span, L, between them. The bending force is applied at mid-span ($L/2$). The specimen for plastic materials is typically a flat strip with the dimensions shown. Its section modulus, S, is needed for bending stress calculations. From Appendix 1, $S = bh^2/6$. The maximum bending moment occurs at the point of load application and has the value, $M_{max} = PL/4$. Then the flexural (bending) stress is:

$$\sigma = \frac{M_{max}}{S} = \frac{PL/4}{bh^2/6} = \frac{1.5PL}{bh^2} \tag{2–6}$$

The stress at break or after a certain specified percentage of deflection is reported as the flexural strength. One

[1]ASTM International. *Standard Test Method for Flexural Properties of Unreinforced and Reinforced Plastics and Electrical Insulating Materials, Standard D790.*West Conshohocken, PA: ASTM International, DOI: 10.1520/D790-10, www.astm.org, 2010.

[2]International Standards Organization (ISO). *Plastics—Determination of Flexural Properties.* ISO 178:2001.

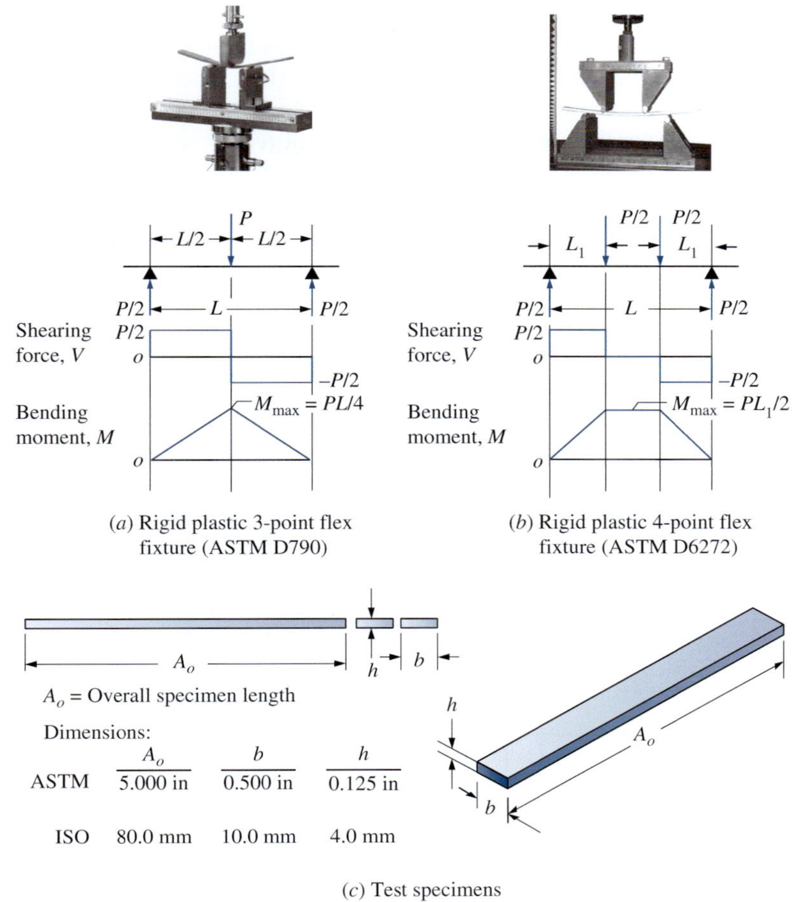

(*a*) Rigid plastic 3-point flex fixture (ASTM D790)

(*b*) Rigid plastic 4-point flex fixture (ASTM D6272)

(*c*) Test specimens

FIGURE 2–5 Test Fixtures for Flexural Testing (Instron, Norwood, MA)

disadvantage of this method is that shearing stress exists along with bending stress and that may affect the results for some specimens.

Flexural modulus is measured as the slope of the load-deflection curve taken during the test at the straight-line portion, if any. Some plastic materials exhibit a nonlinear slope for the curve and other techniques should be used. One uses a *secant* line drawn between two agreed-upon points on the curve.

The 4-point bending principle, as used in ASTM D6272[3] and ISO 14125[4], overcomes the shearing stress issue described above because the highest stressed part of the beam experiences no shearing force. See Part (b) of Figure 2–5. The distances, L_1, from the supports to each load are the same. Then, $M_{max} = PL_1/2$ and the flexural stress is:

$$\sigma = \frac{M_{max}}{S} = \frac{PL_1/2}{bh^2/6} = \frac{3PL_1}{bh^2} \qquad (2\text{–}7)$$

[3]ASTM International. *Standard Test Method for Flexural Properties of Unreinforced and Reinforced Plastics and Electrical Insulating Materials by Four-Point Bending, Standard D6272.* West Conshohocken, PA: ASTM International, DOI: 10.1520/D6272-10, www.astm.org, 2010.

[4]International Standards Organization (ISO). *Fibre-Reinforced Plastic Composites—Determination of Flexural Properties.* ISO 14125:1998/Cor.1:2001(E).

Hardness

The resistance of a material to indentation by a penetrator is an indication of its *hardness.* Hardness in a steel indicates wear resistance as well as strength. Wear resistance will be discussed in later chapters, particularly with regard to gear teeth. Several types of devices, procedures, and penetrators measure hardness; the Brinell hardness tester and the Rockwell hardness tester are most frequently used for machine elements. For steels, the Brinell hardness tester employs a hardened steel ball 10 mm in diameter as the penetrator under a load of 3000-kg force. The load causes a permanent indentation in the test material, and the diameter of the indentation is related to the Brinell hardness number, which is abbreviated BHN or HB. The actual quantity being measured is the load divided by the contact area of the indentation. For steels, the value of HB ranges from approximately 100 for an annealed, low-carbon steel to more than 700 for high-strength, high-alloy steels in the as-quenched condition. In the high ranges, above HB 500, the penetrator is sometimes made of tungsten carbide rather than steel. For softer metals, a 500-kg load is used.

The Rockwell hardness tester uses a hardened steel ball with a 1/16-in diameter under a load of 100-kg force for softer metals, and the resulting hardness is listed as Rockwell

B, R_B, or HRB. For harder metals, such as heat-treated alloy steels, the Rockwell C scale is used. A load of 150-kg force is placed on a diamond penetrator (a *brale* penetrator) made in a sphero-conical shape. Rockwell C hardness is sometimes referred to as R_C or HRC. Many other Rockwell scales are used.

The Brinell and Rockwell methods are based on different parameters and lead to quite different numbers. However, since they both measure hardness, there is a correlation between them, as noted in Appendix 17 and Figure 2–6. It is also important to note that, especially for highly hardenable alloy steels, there is a nearly linear relationship between the Brinell hardness number and the tensile strength of the steel, according to the equation

➪ Approximate Relationship between Hardness and Strength for Steel

$$0.50(\text{HB}) = \text{approximate tensile strength (ksi)} \quad \textbf{(2–8)}$$

This relationship is shown in Figure 2–6.

To compare the hardness scales with the tensile strength, consider Table 2–2. Note that there is some overlap between the HRB and HRC scales. Normally, HRB is used for the softer metals and ranges from approximately 60 to 100, whereas HRC is used for harder metals and ranges from 20 to 65. Using HRB numbers above 100 or HRC numbers below 20 is not recommended. Those shown in Table 2–2 are for comparison purposes only.

The Vickers hardness test, preferred in some European countries and elsewhere, is similar to the Brinell test except for the nature of the penetrator (square-based diamond pyramid) and the applied load, typically 50 kg. Appendix 17 includes the Vickers hardness for comparison with Brinell and Rockwell hardness values. See also Reference 38.

Whereas Rockwell B and C scales are used for substantial components made from steels, other Rockwell hardness scales can be used for different materials as listed in Table 2–3.

For plastics, rubbers, and elastomers, it is typical to use the Shore or IRHD (International rubber hardness degree) methods. Table 2–4 lists some of the more popular scales. Also used for some plastics are the Rockwell R, L, M, E, K, and α scales. These vary by the size and geometry of the indenter and the applied force.

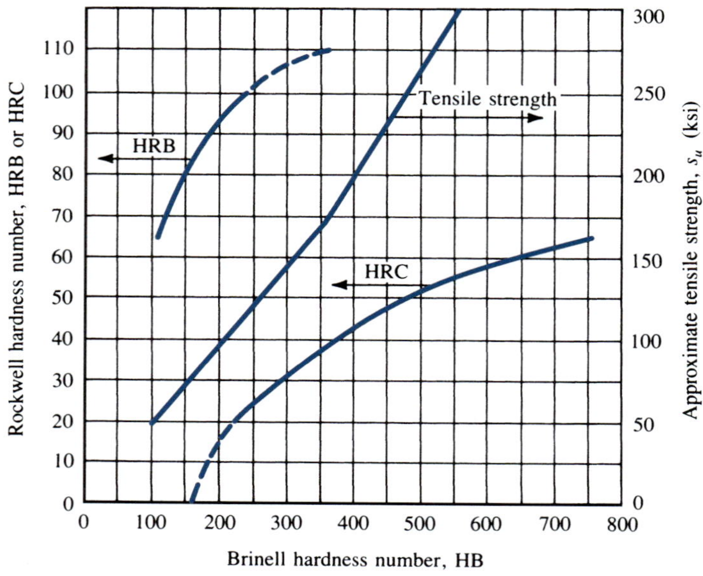

FIGURE 2–6 Hardness conversions

TABLE 2–2 Comparison of Hardness Scales with Tensile Strength

Material and condition	Hardness			Tensile strength	
	HB	HRB	HRC	ksi	MPa
1020 annealed	121	70		60	414
1040 hot-rolled	144	79		72	496
4140 annealed	197	93	13	95	655
4140 OQT 1000	341	109	37	168	1160
4140 OQT 700	461		49	231	1590

TABLE 2–3 Rockwell Hardness Scales and their Uses

Scale	Symbol	Typical uses
A	HRA	Thin steel, hardened steel with shallow case
B	HRB	Softer steels such as low carbon, annealed or hot-rolled; softer aluminum, copper, cast irons
C	HRC	Harder steels such as heat-treated alloy steels, tool steels, titanium
D	HRD	Medium-depth case-hardened steels, harder cast irons
E	HRE	Harder bearing metals, aluminum, magnesium, cast irons
F	HRF	Soft thin sheet metals, annealed copper and zinc alloys
G	HRG	Beryllium copper, phosphor bronze, softer cast irons
H	HRH	Aluminum, zinc, lead
K	HRK	Softer bearing metals, plastics, rubbers, and other soft materials. Scales L, M, P, R, S, V, and α also available
15 N	HR15N	Similar to A scale but for thinner materials or thin case hardening
30 N	HR30N	Similar to C scale but for thinner materials or thin case hardening
45 N	HR45N	Similar to D scale but for thinner materials or thin case hardening
15 T	HR15T	Similar to B scale but for thinner materials
30 T	HR30T	Similar to F scale but for thinner materials
45 T	HR45T	Similar to G scale but for thinner materials

Notes: Measurement devices use different indenter sizes and shapes and applied forces. For details, consult standards ASTM E18, ISO 6508.

TABLE 2–4 Hardness Measurement for Plastics, Rubbers, and Elastomers

Shore method

Scale of durometer	Symbol	Typical uses
A	Shore A	Soft vulcanized natural rubbers, elastomers (e.g., neoprene), leather, wax, felt
B	Shore B	Moderately hard rubber as used for printer rolls and platens
C	Shore C	Medium hard rubbers and plastics
D	Shore D	Hard rubbers and plastics such as vinyl sheets, Plexiglas, laminate countertops
DO	Shore DO	Very dense textile windings
O	Shore O	Soft rubber such as artgum; textile windings
OO	Shore OO	Low density textile windings; sponge rubber
OOO	Shore OOO	Soft plastic foams
T	Shore T	Medium density textiles on spools
M	Shore M	Rubber O-rings and thin sheet rubber

IRHD method (International rubber hardness degree)

Name	Typical uses
IRHD Micro	Small: O-rings, small components, thin materials
IRHD Macro L	Soft: Readings up to 35 IRHD L
IRHD Macro N	Normal: Readings from 30 IRHD N to 100 IRHD N

Notes: Measurement devices use different indenter sizes and shapes and applied forces. For details, consult standards ASTM D2240, ISO 868, or DIN 53505.

Other hardness measurement methods include Knoop, Universal, and rebound hardness.

Wear in Mechanical Devices

Wear is an issue whenever two components operate with relative motion between them or when liquids or solids impinge on a surface at high velocity. The action of one surface on another tends to develop material loss, scoring, or roughening of the mating surfaces. Wear is often cumulative and can eventually render the components incapable of delivering their expected performance. See References 7, 26, and 28. Failure modes include developing:

- Excessive clearances between the mating objects
- Noise
- Unacceptable appearance of either mating surface
- Increased friction at the mating surfaces
- Progressive rise in temperature at the mating surfaces
- Seizure resulting in complete stoppage of the device

Designers must take steps to minimize wear by proper material selection, surface preparation during manufacture, lubrication of the mating surfaces, periodic inspection, and remedial steps to overcome the effects of wear in early stages.

While not really a material property itself, it is pertinent to mention wear in this chapter because the characteristics of the materials at the mating surfaces determine the potential for wear to occur and to what degree. Conditions that the designer must control include:

- Chemical compatibility of the materials of the two contacting surfaces, of any lubricant present, and the environment in which the equipment is to operate to inhibit corrosion that would build up on mating surfaces, cause roughening, increase friction, and initiate material removal.
- The hardness of the materials at the contacting surfaces; typically harder materials have greater ability to resist wear, although exceptions occur.
- The smoothness of the surface finish at areas of contact to minimize the tendency for microscopic hills and valleys to come into contact.
- The physical size of the mating area over which applied forces act to control the effective pressure between mating bodies to a manageable level.
- The inherent coefficient of friction between the mating surfaces at operating conditions.
- The type of lubricant used, if any, and its ability to effectively separate the sliding bodies while under load.
- Control of the temperature of the mating components and of any lubricant used because elevated temperatures can reduce the effectiveness of the lubricant and lower the strength of the materials in contact.
- The cleanliness of the lubricant and of the surfaces themselves because even small solid particles between the mating surfaces can initiate scoring and roughening of the surfaces.

Several kinds of wear can exist.

- ***Erosive wear***—The displacement of particles from a surface due to impact of solids or liquids on a surface as might occur on equipment subjected to wind and rain or the insides of pipes, elbows, and other fittings carrying liquids or gases containing solids, called slurries.
- ***Abrasive wear***—The mechanical tearing of particles from one material by the action of the mating material resulting in the loss of mass from one or both materials.
- ***Adhesive wear***—The tendency of one material to adhere to the mating material and subsequently remove particles by breaking the adhesion and making the surfaces rougher.
- ***Fretting wear***—Cyclical relative motion of two tightly joined parts under high surface pressure as might occur with connectors, fasteners, clamps, and similar situations where even vibration can initiate fretting wear.
- ***Surface fatigue***—Progressive damage caused by creating high contact stresses between mating components that eventually lead to fatigue failure of one of the mating components. Examples include the action of rolling balls or rollers on the inner or outer races of bearings, treads of wheels rolling on rails or flat surfaces, and supports for beams where high stresses and relative motion can occur.

It is virtually impossible to predict the amount of wear that may occur in a given situation and testing of the proposed design is recommended. At times, designers can benefit from test data from samples of materials under controlled loads and mating conditions. At least a measure of relative merit or ranking of certain combinations of materials can be determined. Standardized testing can be done using either of two methods described by ASTM International (See Internet site 5):

1. ASTM Standard G65-04 Standard Test Method for Measuring Abrasion Using the Dry Sand/Rubber Wheel Apparatus, DOI: 10.1520/G0065-04.
    a. Conditions are standardized for abrasive particle size, shape, and hardness; the magnitude of the stress imposed by the particle; and the frequency of contact of the abrasive particles to create scratching abrasion. Volume loss from the test materials is measured.
2. ASTM Standard G132-96(2007) Standard Test Method for Pin Abrasion Testing, DOI: 10.1520/G0132-96R07.
    a. Relative motion is established between flat surfaces of two materials while maintaining a measurable force between them. Mass loss from the test materials is measured.

Machinability

Machinability is related to the ease with which a material can be machined to a good surface finish with reasonable tool life. Production rates are directly affected by machinability. It is difficult to define measurable properties related to machinability, so machinability is usually reported in comparative terms, relating the performance of a given material with some standard.

Toughness, Impact Energy

Toughness is the ability of a material to absorb applied energy without failure. Parts subjected to suddenly applied loads, shock, or impact need a high level of toughness. Several methods are used to measure the amount of energy required to break a particular specimen made from a material of interest. The energy absorption value from such tests is often called *impact energy* or *impact resistance*. However, it is important to note that the actual value is highly dependent on the nature of the test sample, particularly its geometry. It is not possible to use the test results in a quantitative way when making design calculations. Rather, the impact energy for several candidate materials for a particular application can be compared with each other as a qualitative indication of their toughness. The final design should be tested under real service conditions to verify its ability to survive safely during expected use.

For metals and plastics, two methods of determining impact energy, *Izod* and *Charpy*, are popular, with data often reported in the literature from vendors of the material. Figure 2–7 shows sketches of the dimensions of standard specimens and the manner of loading. In each method, a pendulum with a heavy mass carrying a specially designed striker is allowed to fall from a known height. The striker contacts the specimen with a high velocity at the bottom of the pendulum's arc; therefore, the pendulum possesses a known amount of kinetic energy. The specimen is typically broken during the test, taking some of the energy from the pendulum but allowing it to pass through the test area. The testing machine is configured to measure the final height to which the pendulum swings and to indicate the amount of energy removed. That value is reported in energy units of J (Joules or N · m) or ft · lb. Some highly ductile metals and many plastics do not break during the test, and the result is then reported as *No Break*.

The standard *Izod* test employs a square specimen with a V-shaped notch carefully machined 2.0 mm (0.079 in) deep according to specifications in ASTM standard D 256.[5] The specimen is clamped in a special vise with the notch aligned with the top edge of the vise. The striker contacts the specimen at a height of 22 mm above the notch, loading it as a cantilever in bending. When used for plastics, the width dimension can be different from that shown in Figure 2–7. This obviously changes the total amount of energy that the specimen will absorb during fracture. Therefore, the data for impact energy are divided by the actual width of the specimen, and the results are reported in units of N · m/m or ft · lb/in. Also, some vendors and customers may agree to test the material with the notch facing away from the striker rather than toward it as shown in Figure 2–7. This gives a measure of the material's impact energy with less influence from the notch.

The *Charpy* test also uses a square specimen with a 2.0 mm (0.079 in) deep notch, but it is centered along the length. The specimen is placed against a rigid anvil without

[5]ASTM International. *Standard Test Methods for Determining the Izod Pendulum Impact Resistance of Plastics, Standard D256*. West Conshohocken, PA: ASTM International, DOI: 10.1520/D0256-10, 2010.

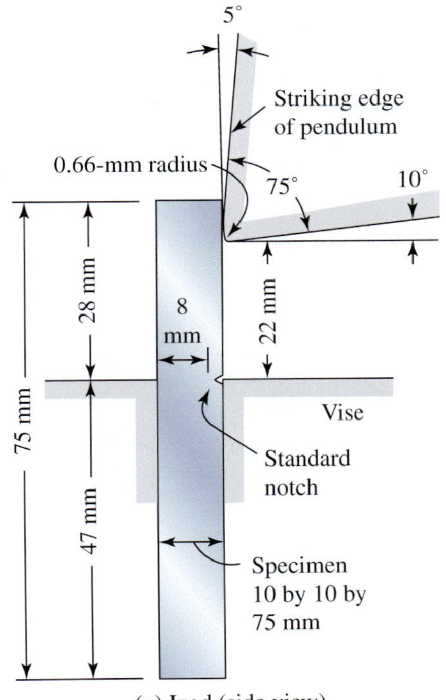

(*a*) Izod (side view)

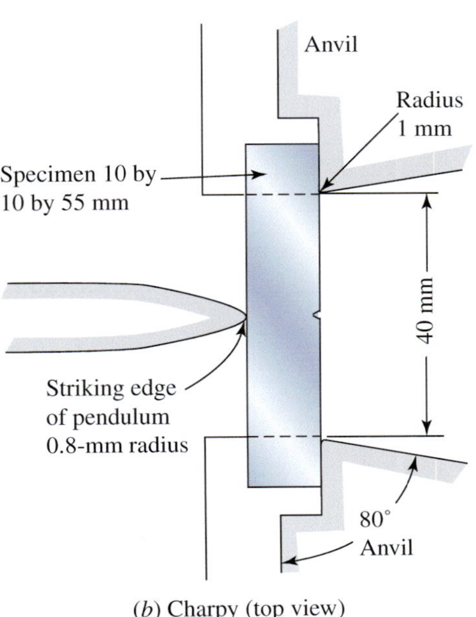

(*b*) Charpy (top view)

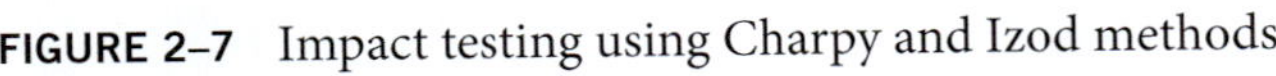

FIGURE 2–7 Impact testing using Charpy and Izod methods

being clamped. See ASTM standard A 370[6] for the specific geometry and testing procedure. The notch faces away from the place where the striker contacts the specimen. The loading can be described as the bending of a simply supported beam. The Charpy test is most often used for testing metals.

Another impact testing method used for some plastics, composites, and completed products is the *drop-weight* tester. Here a known mass is elevated vertically above the test specimen to a specified height. Thus, it has a known amount of potential energy. Allowing the mass to fall freely imparts a predictable amount of kinetic energy to the specimen clamped to a rigid base. The initial energy, the manner of support, the specimen geometry, and the shape of the striker (called a *tup*) are critical to the results found. One standard method, described in ASTM D 3763,[7] employs a spherical tup with a diameter of 12.7 mm (0.50 in). The tup usually pierces the specimen. The apparatus is typically equipped with sensors that measure and plot the load versus deflection characteristics dynamically, giving the designer much information about how the material behaves during an impact event. Summary data reported typically include maximum load, deflection of the specimen at the point of maximum load, and the energy dissipated up to the maximum load point. The energy is calculated by determining the area under the load-deflection diagram. The appearance of the test specimen is also described, indicating whether fracture occurred and whether it was a ductile or brittle fracture.

Fatigue Strength or Endurance Strength

Parts subjected to repeated applications of loads or to stress conditions that vary with time over several thousands or millions of cycles fail because of the phenomenon of *fatigue.* Materials are tested under controlled cyclic loading to determine their ability to resist such repeated loads. The resulting data are reported as the *fatigue strength,* also called the *endurance strength* of the material. (See Chapter 5.)

Creep

When materials are subjected to high loads continuously, they may experience progressive elongation over time. This phenomenon, called *creep,* should be considered for metals operating at high temperatures. You should check for creep when the operating temperature of a loaded metal member exceeds approximately 0.3 (T_m) where T_m is the melting temperature expressed as an absolute temperature. (See Reference 25.) Creep can be important for critical members in internal combustion engines, furnaces, steam turbines, gas turbines, nuclear reactors, or rocket engines. The stress can be tension, compression, flexure, or shear. (See Reference 12.)

Figure 2–8 shows the typical behavior of metals that creep. The vertical axis is the creep strain, in units such as in/in or mm/mm, over that which occurs initially as the load is applied. The horizontal axis is time, typically measured in hours because creep develops slowly over a long term. During the primary portion of the creep strain versus time curve, the rate of increase in strain initially rises with a rather steep slope that then decreases. The slope is constant (straight line) during the secondary portion of the curve. Then the slope increases in the tertiary portion that precedes the ultimate fracture of the material.

Creep is measured by subjecting a specimen to a known steady load, possibly through application of a dead weight, while the specimen is heated and maintained at a uniform temperature. Data for strain versus time are taken at least into the secondary creep stage and possibly all the way to fracture to determine the creep rupture strain. Testing over a range of temperatures gives a family of curves that are useful for design.

Creep can occur for many plastics even at or near room temperature. Figure 2–9 shows one way that creep data are

[6]ASTM International. *Standard Test Methods and Definitions for Mechanical Testing of Steel Products, Standard A370.* West Conshohocken, PA: ASTM International, DOI: 10.1520/A0370-09AE01, 2009.

[7]ASTM International. *Standard Test Methods for High Speed Puncture of Plastics Using Load and Displacement Sensors, Standard D3763.* West Conshohocken, PA: ASTM International, DOI: 10.1520/D3763-08, 2008.

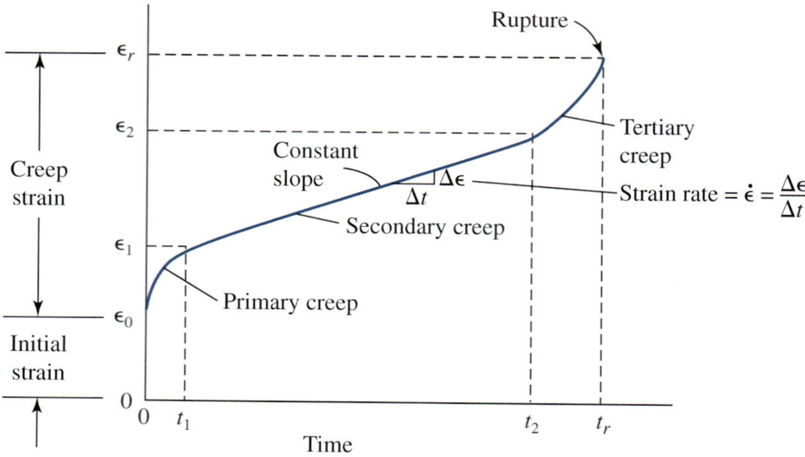

FIGURE 2–8 Typical creep behavior

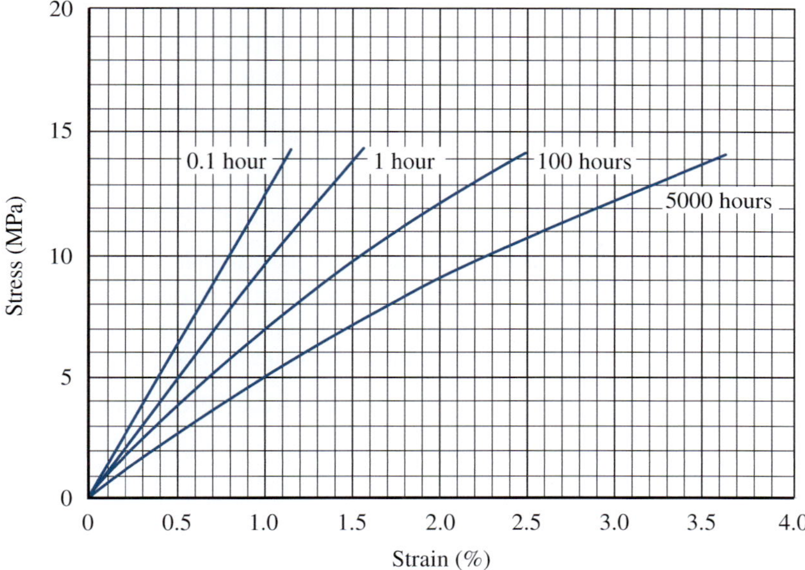

FIGURE 2–9 Example of stress versus strain as a function of time for nylon 66 plastic at 23°C (73°F) (DuPont Polymers, Wilmington, DE)

displayed for plastic materials. (See Reference 14.) It is a graph of applied stress versus strain in the member with data shown for a specific temperature of the specimen. The curves show the amount of strain that would be developed within the specified times at increasing stress levels. For example, if this material were subjected to a constant stress of 5.0 MPa for 5000 hours, the total strain would be 1.0%. That is, the specimen would elongate by an amount 0.01 times the original length. If the stress were 10.0 MPa for 5000 hours, the total strain would be approximately 2.25%. The designer must take this creep strain into account to ensure that the product performs satisfactorily over time.

Example Problem 2–1 A solid circular bar has a diameter of 5.0 mm and a length of 250 mm. It is made from nylon 66 plastic (30% Glass, 50% R.H.) and subjected to a steady tensile load of 240 N. Compute the elongation of the bar immediately after the load is applied and after 5000 hours (approximately seven months). See Appendix 13 and Figure 2–9 for properties of the nylon.

Solution The stress and deflection immediately after loading will first be computed using fundamental equations of strength of materials:

$$\sigma = F/A \text{ and } \delta = FL/EA$$

See Chapter 3 for a review of strength of materials.

Then creep data from Figure 2–9 will be applied to determine the elongation after 5000 hours.

Results *Stress:*

The cross-sectional area of the bar is

$$A = \pi D^2/4 = \pi(5.0\text{ mm})^2/4 = 19.63\text{ mm}^2$$

$$\sigma = \frac{F}{A} = \frac{240\text{ N}}{19.63\text{ mm}^2} = 12.2\text{ N/mm}^2 = 12.2\text{ MPa}$$

Appendix 13 lists the tensile strength for nylon 66 to be 102 MPa. Therefore, the rod is safe from fracture.

Elongation:

The tensile modulus of elasticity for nylon 66 is found from Appendix 13 to be $E = 5500$ MPa. Then the initial elongation is,

$$\delta = \frac{FL}{EA} = \frac{(240\text{ N})(250\text{ mm})}{(5500\text{ N/mm}^2)(19.63\text{ mm}^2)} = 0.556\text{ mm}$$

Creep:

Referring to Figure 2–9 we find that when a tensile stress of 12.2 MPa is applied to the nylon 66 plastic for 5000 hours, a total strain of approximately 2.95% occurs. This can be expressed as

$$\epsilon = 2.95\% = 0.0295\ \text{mm/mm} = \delta/L$$

Then,

$$\delta = \epsilon L = (0.0295\ \text{mm/mm})\ (250\text{mm}) = 7.375\ \text{mm}$$

Comment This is approximately seven times as much deformation as originally experienced when the load was applied. So designing with the reported value of modulus of elasticity is not appropriate when stresses are applied continuously for a long time. We can now compute an apparent modulus of elasticity, E_{app}, for this material at the 5000 hours service life.

$$E_{app} = \sigma/\epsilon = (12.2\ \text{MPa})/(0.0295\ \text{mm/mm}) = 414\ \text{MPa}$$

Relaxation

A phenomenon related to creep occurs when a member under stress is captured under load, giving it a certain fixed length and a fixed strain. Over time, the stress in the member would decrease, exhibiting a behavior called *relaxation*. This is important in such applications as clamped joints, press-fit parts, and springs installed with a fixed deflection. Figure 2–10 shows the comparison between creep and relaxation. For stresses below approximately 1/3 of the ultimate tensile strength of the material at any temperature, the apparent modulus in either creep or relaxation at any time of loading may be considered similar for engineering purposes. Furthermore, values for apparent modulus are the same for tension, compression, or flexure. (See Reference 14.) Analysis of relaxation is complicated by the fact that as the stress decreases, the rate of creep also decreases. Additional material data beyond that typically reported would be required to accurately predict the amount of relaxation at any given time. Testing under realistic conditions is recommended.

Physical Properties

Here we will discuss density, coefficient of thermal expansion, thermal conductivity, and electrical resistivity.

Density. *Density* is defined as the mass per unit volume of a material. Its usual units are kg/m^3 in the SI and lb/in^3 in the U.S. Customary Unit System, where the pound unit is taken to be pounds-mass. The Greek letter rho (ρ) is the symbol for density.

In some applications, the term *specific weight* or *weight density* is used to indicate the weight per unit volume of a

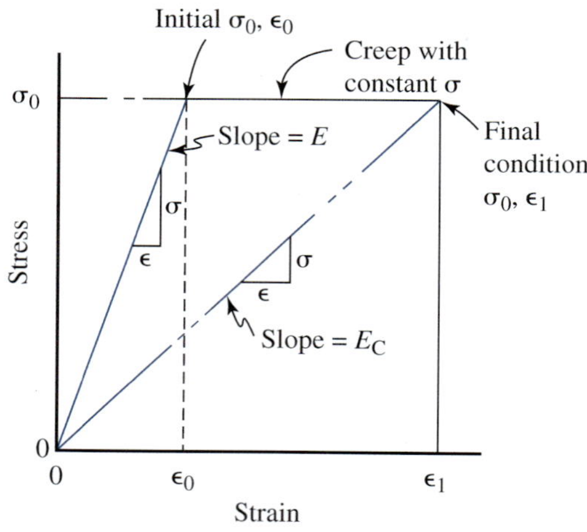

$E = \dfrac{\sigma_0}{\epsilon_0}$ = Tensile modulus

$E_C = \dfrac{\sigma_0}{\epsilon_1}$ = Creep modulus

(*a*) Creep behavior

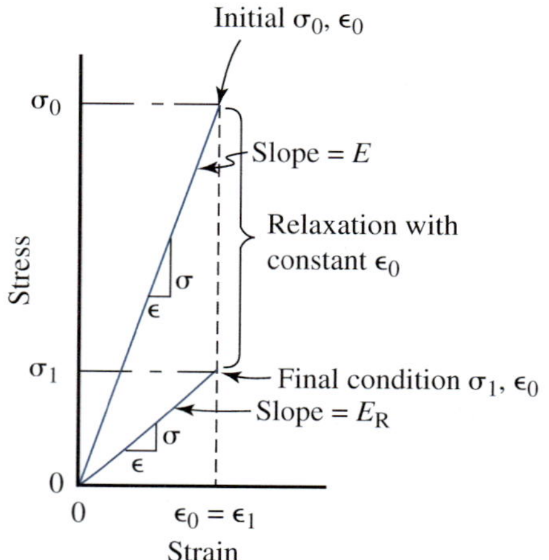

$E = \dfrac{\sigma_0}{\epsilon_0}$ = Tensile modulus

$E_R = \dfrac{\sigma_1}{\epsilon_0}$ = Relaxation modulus

(*b*) Relaxation behavior

FIGURE 2–10 Comparison of creep and relaxation (DuPont Polymers, Wilmington, DE)

material. Typical units are N/m^3 in the SI and lb/in^3 in the U.S. Customary Unit System, where the pound is taken to be pounds-force. The Greek letter gamma (γ) is the symbol for specific weight.

Coefficient of Thermal Expansion. The *coefficient of thermal expansion* is a measure of the change in length of a material subjected to a change in temperature. It is defined by the relation

Coefficient of Thermal Expansion

$$\alpha = \frac{\text{change in length}}{L_o(\Delta T)} = \frac{\text{strain}}{(\Delta T)} = \frac{\epsilon}{(\Delta T)} \qquad \textbf{(2–9)}$$

where L_o = original length
ΔT = change in temperature

Virtually all metals and plastics expand with increasing temperature, but different materials expand at different rates. For machines and structures containing parts of more than one material, the different rates can have a significant effect on the performance of the assembly and on the stresses produced.

Thermal Conductivity. *Thermal conductivity* is the property of a material that indicates its ability to transfer heat. Where machine elements operate in hot environments or where significant internal heat is generated, the ability of the elements or of the machine's housing to transfer heat away can affect machine performance. For example, wormgear speed reducers typically generate frictional heat due to the rubbing contact between the worm and the wormgear teeth. If not adequately transferred, heat causes the lubricant to lose its effectiveness, allowing rapid gear-tooth wear.

Electrical Resistivity. For machine elements that conduct electricity while carrying loads, the electrical resistivity of the material is as important as its strength. *Electrical resistivity* is a measure of the resistance offered by a given thickness of a material; it is measured in ohm-centimeters ($\Omega \cdot$cm). *Electrical conductivity*, a measure of the capacity of a material to conduct electric current, is sometimes used instead of resistivity. It is often reported as a percentage of the conductivity of a reference material, usually the International Annealed Copper Standard.

2–3 CLASSIFICATION OF METALS AND ALLOYS

Various industry associations take responsibility for setting standards for the classification of metals and alloys. Each has its own numbering system, convenient to the particular metal covered by the standard. But this leads to confusion at times when there is overlap between two or more standards and when widely different schemes are used to denote the metals. Order has been brought to the classification of metals by the use of the Unified Numbering Systems (UNS) as defined in the Standard E 527-07, *Standard Practice for Numbering Metals and Alloys (UNS)*, by the American Society for Testing and Materials, or ASTM. (See Reference 23.) Besides listing materials under the control of ASTM itself, the UNS coordinates designations of the following:

The Aluminum Association (AA)

The American Iron and Steel Institute (AISI)

The Copper Development Association (CDA)

The Society of Automotive Engineers (SAE)

The primary series of numbers within UNS are listed in Table 2–5, along with the organization having responsibility for assigning numbers within each series.

Recent Developments in Steel Designations

The most-used type of material for machine elements in this book is steel. Section 2–5 presents significant information on the various types of steel and their conditions as it relates to design decisions. For many years, two organizations, AISI and SAE, have played major roles in cataloging steels and in specifying the designation systems used in the United States. This work was coordinated with steel producers, professional societies, and design engineers in several steel-using sectors of industry. To a large extent, the designation systems of AISI and SAE were identical. The automotive industry tended to use the SAE system while a broad array of other industries used the AISI system. In recent years, the AISI has discontinued the practice of assigning steel designation numbers, and the SAE now plays the lead role.

In parallel with these developments, standards-writing committees of SAE, in cooperation with ASTM, began a process of broadening the designation systems for wrought and rolled steels beyond the series listed in Table 2–5. New systems for series G, H, and K have been developed and are described in SAE Standard J402 (R) *New Steel Designation System for Wrought or Rolled Steel.* These new steel designations are the same in both the SAE and UNS systems as described in SAE Standard J402 and the joint SAE J1086/ASTM E527 publication. Many of the alloys listed within series G, H, and K retain the familiar four-digit AISI and SAE designations as *a part* of the UNS number. Details of the new steel designations will be discussed in Section 2–5.

Aerospace Materials System (AMS). The development of the AMS system for designating materials resulted from the needs for special materials to meet the unique conditions encountered in aerospace applications. Aerospace Materials Specifications (AMS), published by SAE, are complete specifications that are generally adequate for procurement purposes. Most of the AMS designations pertain to materials intended for aerospace applications. The specifications may include mechanical property requirements significantly more severe than those for grades of steel having similar compositions but intended for other applications. Processing requirements, such as for consumable electrode remelting, are common in AMS steels.

TABLE 2–5 Unified Numbering System (UNS)

Number series	Types of metals and alloys	Responsible organization
Nonferrous metals and alloys		
A00001–A99999	Aluminum and aluminum alloys	AA
C00001–C99999	Copper and copper alloys	CDA
E00001–E99999	Rare earth metals and alloys	ASTM
L00001–L99999	Low-melting metals and alloys	ASTM
M00001–M99999	Miscellaneous nonferrous metals and alloys	ASTM
N00001–N99999	Nickel and nickel alloys	SAE
P00001–P99999	Precious metals and alloys	ASTM
R00001–R99999	Reactive and refractory metals and alloys	SAE
Z00001–Z99999	Zinc and zinc alloys	ASTM
Ferrous metals and alloys		
D00001–D99999	Steels; mechanical properties specified	SAE
F00001–F99999	Cast irons and cast steels	ASTM
[1]G00001–G99999	Carbon and alloy steels	SAE
[1]H00001–H99999	H-steels; specified hardenability	SAE
J00001–J99999	Cast steels (except tool steels)	ASTM
[1]K00001–K99999	Miscellaneous steels and ferrous alloys	SAE
S00001–S99999	Heat- and corrosion-resistant (stainless) steels	ASTM
T00001–T99999	Tool steels	SAE

Note: [1]The series G, H, and K systems have been changed as described in Section 2–5. For a time, both the form listed in this table and the new system may be seen in practice.

AMS standards can be accessed through Internet site 4 for SAE International by selecting the *Standards* tab, then *Standards by Industry*, then *Aerospace*. At that point, if you are looking for a standard for a particular steel alloy, use the keyword search and enter the SAE alloy designation. For example, searching on "SAE 4340" will bring up several standards for the application of this alloy with SAE 4340 in red. Standards deal with this material in bars, forgings, tubing, sheet, strip, plate, welding wire, castings, and several more.

Note that the lists of SAE and AMS standards on the SAE website include many other types of materials: *Adhesives, Sealants, Fabrics, Ceramics, Composites, Foams, Polymers,* and many more.

Material Property Data Used in This Book

The following sections of this chapter provide much additional detail about the types of metals, plastics, and composites for which data are given in the appendix and which are used in example problems, problems for your solution, and design projects. Here, we give an overview of those data and you are advised to peruse the appendix tables and those included in this section to become acquainted with the type of data available. This extensive set of data provides you with many options when selecting a material for a given design and gives you experience comparing their properties.

Appendices

A–3 Properties of carbon and alloy steels
A–4 Properties of heat-treated steels
A–5 Properties of carburized steels
A–6 Properties of stainless steels
A–7 Properties of structural steels
A–8 Properties of cast iron
A–9 Properties of aluminum
A–10 Properties of die cast zinc and magnesium alloys
A–11 Properties of nickel-based alloys and titanium alloys
A–12 Properties of bronze, brass, and other copper alloys
A–13 Properties of selected plastics

Typical information and property data given for the materials are:

- Material name, type, and condition
  - For metals, conditions are typically hot-rolled, cold-drawn, annealed, normalized, or heat-treated. It is essential to specify the condition in order to fully predict the properties
- Tensile strength; sometimes called ultimate strength
  - For plastics, flexural strength and flexural modulus are also given
- Yield strength

- Percent elongation, used as a measure of the ductility of metals
- Impact strength for some metals and plastics
- Physical property data: modulus of elasticity, density; sometimes given in footnotes

Other pertinent data are given for particular types of materials.

Metals Used for Commercially Available Shapes

Many designs for components of machines are fabricated from standard forms produced by metal suppliers, including round bars, hexagonal bars, square bars, rectangular bars, thin flat strip, and thicker sheets or plates. Other designs call for use of commercially available shapes having good strength and stiffness properties in bending or torsion. Appendix 15 gives 19 tables of dimensional data for several kinds of commercially available shapes:

1. Angles or L-shapes
2. Channels or C-shapes
3. I-beam shapes
4. Hollow tubing: round, pipe, square, and rectangular

Suppliers of such shapes typically offer only selected types of materials in a few conditions as stock products. Table 2–6

TABLE 2–6 Common Metals and Alloys Used for Commercially Available Rolled, Welded, and Extruded Shapes

Ref.	Material & alloy	Shapes	Tensile strength (ksi)	Tensile strength (MPa)	Yield strength (ksi)	Yield strength (MPa)
		Carbon and alloy steels				
a	ASTM A36	Rolled shapes; S-beams, angles, channels, plates, bars	58	400	36	248
b	ASTM A500 GR-B	Tubing; round, square, rectangular	58	400	42	290
c	ASTM A992	Rolled W-shapes	65	448	50	345
d	1018 HR	Bars, hexagons, squares, rectangles	55	379	40	276
e	1018 CD	Bars, hexagons, squares, rectangles	70	483	60	414
f	1045 HR	Bars, hexagons, squares, rectangles	90	621	55	379
g	1045 CD	Bars, hexagons, squares, rectangles	90	621	75	517
h	1117 HR	Bars, hexagons, squares, rectangles	65	448	40	276
i	1117 CD	Bars, hexagons, squares, rectangles	80	552	65	448
j	1141 HR	Bars, hexagons, squares, rectangles	95	655	55	379
k	1141 CD	Bars, hexagons, squares, rectangles	100	690	85	586
l	4140 Annealed	Bars, hexagons, squares, rectangles	95	655	54	372
m	8620 HR	Bars, hexagons, squares, rectangles	80	552	60	414
n	8620 CD	Bars, hexagons, squares, rectangles	95	655	80	552
		Stainless steels				
o	304 Annealed	Bars, hexagons, squares, rectangles, tubing (round, square, rectangular)	85	586	35	241
p	316 Annealed	Bars, hexagons, squares, rectangles, tubing (round, square, rectangular)	80	552	30	207
		Aluminum				
q	2011-T3	Bars, hexagons, squares, rectangles	55	379	43	296
r	2017-T4	Bars, hexagons, squares, rectangles	62	427	40	276
s	2024-T4	Bars, hexagons, squares, rectangles	68	469	47	324
t	3003-H14	Sheet, plate, tubing (round, square, rectangular)	22	152	21	145
u	6061-T6	Bars, hexagons, squares, rectangles, tubing (round, square, rectangular), extruded structural shapes	45	310	40	276
v	6063-T6	Bars, hexagons, squares, rectangles, tubing (round, square, rectangular), extruded structural shapes	35	241	31	214

Notes: Properties are typical but not guaranteed. Other materials are available.
Obtain detailed information from vendors.
All listed alloys are ductile with percent elongation >10%.

lists some of the more commonly used materials with their properties, for carbons and alloy steels, stainless steels, and aluminum alloys. These data will be useful to you for design projects where you are asked to specify shape, size, and material for a component. While it is possible to obtain such products in other materials, it is usually less costly to specify standard alloys. Footnotes to the tables suggest suppliers offering the given shapes and gives their website URLs. You should go to those sites to determine actual materials offered.

Structural shapes (L-shapes, C-shapes, and I-beam shapes) used for building construction and some large-scale manufacturing equipment are typically made from structural steel classified by ASTM designations, described in more detail in Section 2–8.

Steel and Aluminum Designations from Different Parts of the World. While this book uses material designations from the United States and other parts of North America, users of this book may work in any part of the world. Producers of commercially available shapes in other countries will use material designations developed for their own industries and are typically different from U.S. designations. However, many will make alloys that are compositionally similar to U.S. alloys. Table 2–7 lists some common

TABLE 2–7 Comparisons of Designations for Steels and Aluminums from Different Parts of the World

Ref.	USA	European Union (Euronorm, EN)	Germany (W-Number)[1]	Germany (DIN)	UK (BS)	Japan (JIS)	China (GB)
	(AISI, SAE, ASTM)			Carbon Steels			
a	A36	S235JRG2	1.0122	S235JRG2	S235JRG2	—	Q235B
b	1015	C15	1.0401	C15	080A15	S15Ck	699-15
c	1018	C18D	1.1141	CK15	040A15	S15	ML3-715
d	1045	C45	1.0503	C45	060A47	S45C	699-45
	(AISI, SAE)			Alloy Steels			
e	1213	11SMn30	1.0715	11SMn30	230M07	SUM22	—
f	4130	25CrMo4	1.7218	25CrM04	708A30	SCM 420	ML30CrMoA
g	4140	42CrMo4	1.7225	42CrMo4	708A42	SCM 440H	ML42CrMo
h	4340	34CrNiM06	1.6582	34CrNiM06	817M40	SNCM 447	ML40CrNiMoA
i	6150	50CrV4	1.7222	50CrV4	735A50	SUP10	50CrVA
j	8620	20NiCrMo2-2	1.6523	20NiCrMo2	805M20	SNCM 220 (H)	20CrNiMo
	(AISI, SAE)			Stainless Steels			
k	304	X2CrNi18-10	1.4301	X5CrNi18-10	304S 18	SUS 304	0Cr18Ni9
l	316	X5CrNiMo17-12-2	1.4401	X5CrNiMo17-12-2	316S 29	SUS 316	0Cr17Ni12Mo2
m	321	X6CrNiTi18-10	1.4541	X6CrNiTi18-10	321S 31	SUS321	0Cr18Ni10Ti
n	430	X6Cr17	1.4016	X6Cr17	430S 17	SUS430	ML1Cr17
o	17-4PH	X5CrNiCuNb17-4-4	1.4542	X5CrNiCuNb17-4-4	17Cr4Ni	SUS630	0Cr17Ni4Cu4Nb
	Aluminum Assoc.			Aluminum Alloys			
p	1100	Al99.0Cu	—	—	—	—	—
q	2014	AlCu4SiMg	3.1255	AlCuSiMn	L.93, L.94	—	—
r	2024	AlCu4Mg1	3.1355	AlCuMg2	L.97, L.98	—	—
s	6061	AlMg1SiCu	—	—	H20	—	—
t	6063	AlMgSi	3.3206	AlMgSi0.5	H19	—	—
u	7075	AlZn6MgCu	3.4365	AlZnMgCu1.5	L.95, L.96	—	—

Notes: [1]Werkstoff Number

The given examples are a small sample of the thousands of available alloys, heat treatments, and formulations. Exact comparisons are not practical. Obtain data from vendors for designations and properties before use.

Sources: Parker Steel Company, Toledo, Ohio, www.MetricMetai.com

Steel Strip Company, United Kingdom, www.steelstrip.co.uk/international_equivalents.htm

Key to Metals North America, St. Louis, Missouri, www.keytometals.com

All Metals & Forge, Parsippany, NJ, www.steelforge.com

Aluminum Association, Arlington, VA, www.aluminum.org

materials and their designations in Germany, the United Kingdom (UK–England), other parts of the European Union (sometimes called Euronorm or EN), Japan, and China.

2–4 VARIABILTY OF MATERIAL PROPERTIES DATA

Tables of data such as those shown in Appendices 3 through 13 normally report single values for the strength, the modulus of elasticity (stiffness), or the percent elongation (ductility) of a particular material at a particular condition created by heat treatment or by the manner in which it was formed. It is important for you to understand the limitations of such data in making design decisions. You should seek information about the bases for the reported data.

Some tables of data report *guaranteed minimum values* for tensile strength, yield strength, and other values. This might be the case when you are using data obtained from a particular supplier. With such data, you should feel confident that the material that actually goes into your product has at least the reported strength. The supplier should be able to provide actual test data and statistical analyses used to determine the reported minimum strengths. Alternatively, you could arrange to have the actual materials to be used in a project tested to determine their minimum strength values. Such tests are costly, but they may be justified in critical designs.

Other tables of data report *typical values* for material properties. Thus, most batches of material (greater than 50%) delivered will have the stated values or greater. However, about 50% will have lower values, and this fact will affect your confidence in specifying a particular material and heat treatment if strength is critical. In such cases, you are advised to use higher than average design factors in your calculations of allowable (design) strength. (See Chapter 5.)

Using the guaranteed minimum values for strength in design decisions would be the safest approach. However, it is very conservative because most of the material actually delivered would have strengths significantly greater than the listed values.

One way to make the design more favorable is to acquire data for the statistical distribution of strength values taken for many samples. Then applications of probability theories can be used to specify suitable conditions for the material with a reasonable degree of confidence that the parts will perform according to specifications. Figure 2–11 illustrates some of the basic concepts of statistical distribution. The variation of strength over the entire population of samples is often assumed to have a normal distribution around some mean or average value. If you used a strength value that is one standard deviation (1σ) below the mean, 84% of the products would survive. At two standard deviations, greater than 97% would survive; at three standard deviations, more than 99.8%; and at four standard deviations, more than 99.99%.

As the designer, you must carefully judge the reliability of the data that you use. Ultimately, you should evaluate the reliability of the final product by considering the actual

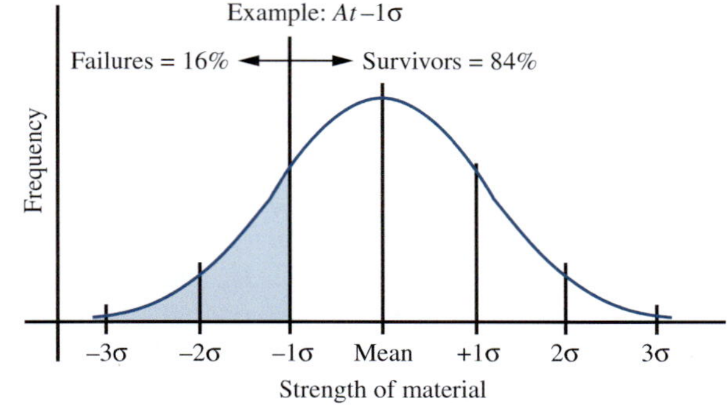

Assuming normal distribution of strength:

Stress level	% Surviving
Mean	50
-1σ	84
-2σ	97
-3σ	99.8
-4σ	99.99

FIGURE 2–11 Normal statistical distribution of material strength

variations in material properties, the manufacturing considerations that may affect performance, and the interactions of various components with each other. There is more discussion on this point in Chapter 5.

2–5 CARBON AND ALLOY STEEL

Steel is possibly the most widely used material for machine elements because of its properties of high strength, high stiffness, durability, and relative ease of fabrication. Many types of steels are available. This section will discuss the methods used for designating steels and will describe the most frequently used types.

The term *steel* refers to an alloy of iron, carbon, manganese, and one or more other significant elements. Carbon has a very strong effect on the strength, hardness, and ductility of any steel alloy. The other elements affect hardenability, toughness, corrosion resistance, machinability, and strength retention at high temperatures. The primary alloying elements present in the various alloy steels are sulfur, phosphorus, silicon, nickel, chromium, molybdenum, and vanadium.

Designation Systems

As mentioned in Section 2–3, the designation systems for steels are managed by either SAE International or ASTM International. The wrought or rolled steels typically used for machine elements are most likely to be selected from SAE grades. Structural steels and many cast metals carry ASTM grade designations. It was pointed out earlier that, until the recent past, both SAE and AISI issued wrought or rolled steel designations that were substantially identical, but AISI no longer performs this function. However, it is likely that much published literature and commercial product information will continue to use the term *AISI* until the new systems are phased in.

Furthermore, SAE, in cooperation with ASTM, is promoting an expanded system for designating wrought steels and the new system is placed within the Unified Numbering System (UNS). Refer back to Table 2–5. See SAE Standard J402 (R) *New Steel Designation System for Wrought or Rolled Steel.* Three UNS series for steels, G, H, and K, are being changed to conform to the new system that provides for specified modifications of the chemistry of the alloys and may also define special processing methods or limits on material properties.

The system starts with the familiar four-digit designation system that has been used for several decades under both the SAE and AISI systems, as shown in Figure 2–12. The first two digits indicate the specific alloy group that identifies the primary alloying elements other than carbon in the steel as shown in Table 2–8. The last two digits

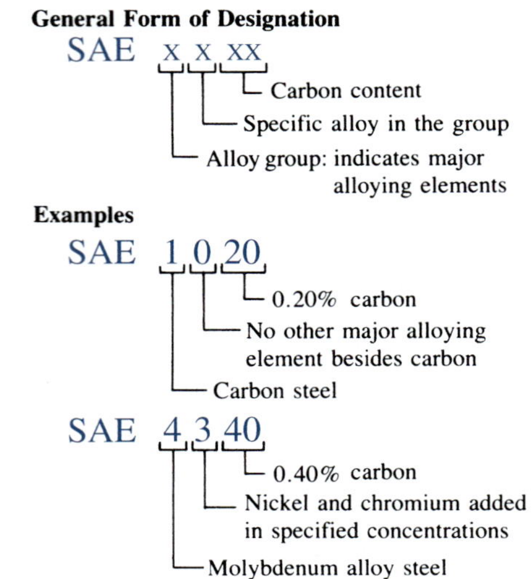

FIGURE 2–12 Steel designation system

TABLE 2–8 Alloy Groups in the SAE Numbering System

10xx	Plain carbon steel: No significant alloying element except carbon and manganese; less than 1.0% manganese. Also called *nonresulfurized.*
11xx	Free-cutting steel: Resulfurized. Sulfur content (typically 0.10%) improves machinability.
12xx	Free-cutting steel: Resulfurized and rephosphorized. Presence of increased sulfur and phosphorus improves machinability and surface finish.
12Lxx	Free-cutting steel: Lead added to 12xx steel further improves machinability.
13xx	Manganese steel: Nonresulfurized. Presence of approximately 1.75% manganese increases hardenability.
15xx	Carbon steel: Nonresulfurized; greater than 1.0% manganese.
23xx	Nickel steel: Nominally 3.5% nickel.
25xx	Nickel steel: Nominally 5.0% nickel.
31xx	Nickel-chromium steel: Nominally 1.25% Ni; 0.65% Cr.
33xx	Nickel-chromium steel: Nominally 3.5% Ni; 1.5% Cr.
40xx	Molybdenum steel: 0.25% Mo.
41xx	Chromium-molybdenum steel: 0.95% Cr; 0.2% Mo.
43xx	Nickel-chromium-molybdenum steel: 1.8% Ni; 0.5% or 0.8% Cr; 0.25% Mo.
44xx	Molybdenum steel: 0.5% Mo.
46xx	Nickel-molybdenum steel: 1.8% Ni; 0.25% Mo.
48xx	Nickel-molybdenum steel: 3.5% Ni; 0.25% Mo.
5xxx	Chromium steel: 0.4% Cr.
51xx	Chromium steel: Nominally 0.8% Cr.
51100	Chromium steel: Nominally 1.0% Cr; bearing steel, 1.0% C.
52100	Chromium steel: Nominally 1.45% Cr; bearing steel, 1.0% C.
61xx	Chromium-vanadium steel: 0.50–1.10% Cr; 0.15% V.
86xx	Nickel-chromium-molybdenum steel: 0.55% Ni; 0.5% Cr; 0.20% Mo.
87xx	Nickel-chromium-molybdenum steel: 0.55% Ni; 0.5% Cr; 0.25% Mo.
92xx	Silicon steel: 2.0% silicon.
93xx	Nickel-chromium-molybdenum steel: 3.25% Ni; 1.2% Cr; 0.12% Mo.

indicate the amount of carbon in the steel in hundredths of a percent. For example, when the last two digits are 20, the alloy includes approximately 0.20% carbon, and it is often said that the steel has *20 points of carbon.* Small variations are allowed, say from 0.18% to 0.23%.

Carbon content is part of the designation system because the primary mechanical properties of strength, hardness, and ductility are strongly dependent on it. As carbon content increases, strength and hardness also increase during processing and heat treatment. Ductility typically decreases so designers must balance strength and ductility when specifying a steel and its condition. It is useful to categorize steels into four broad classes, *low carbon, medium carbon, high carbon,* and *bearing steels.*

- *Low carbon:* Less than 30 points of carbon. Relatively low strength but good formability. Used where high strength is not required.
- *Medium carbon:* From 30 to 50 points of carbon. Higher strength and hardness with moderate ductility. A large percentage of machine elements are made from steels in this class.
- *High carbon:* 50 to 95 points of carbon. Even higher strength and hardness but often with lower ductility. The high hardness provides good wear properties leading to uses such as knives requiring a durable cutting edge, parts subject to abrasion, tools, chisels, and agricultural implements.
- *Bearing steels:* typically 100 points of carbon or higher ($\geq$1.0%). Often specified for balls, rollers, and races of rolling contact bearings because of the very high contact stresses developed in operation.

In this book, when standard specifications for a steel alloy are intended, without modification, we will use the traditional designation as shown in Figure 2–12: for example, SAE 1020 or SAE 4340 steel. See also Appendix 3. Designers, manufacturing professionals, and engineers sometimes refer to only the number in conversation, such as 1020 steel and 4340 steel, or describe them colloquially as *plain carbon steel* and *chrome-moly* steel. Of course, when specifying a steel on design drawings or purchase orders, the complete, formal designations should be used.

The New UNS Designations for G, H, and K Steels. SAE Standard J402 defines a 12-character designation using the format:

GXXXX-XXX-XXXX

where,

- The lead letter may be G for carbon and alloy steels, H for steels with specific hardenability, or K for miscellaneous steels and ferrous alloys as previously listed in the UNS
- The first set of four Xs are the traditional four-digit SAE designation described earlier
- The second set of three Xs denote modifications to the standard chemistry
- The final set of four Xs refer to special production methods, quality standards, or processing

Details of the chemistry and other requirements of various alloys are contained in other SAE standards, including:

- J403 *Chemical Compositions of SAE Carbon Steels;* for carbon steels in the 10XX, 11XX, 12XX, and 15XX series (see Table 2–8)
- J404 *Chemical Compositions of SAE Alloy Steels;* for alloy steels in the remaining series
- J1086 *Numbering Metals and Alloys*
- J1249 *Former SAE Standard and Former SAE-Ex Steels*
- J1268 *Hardenability Bands for Carbon and Alloy H Steels*
- J1868 *Restricted Hardenability Bands for Selected Alloy Steels*

The place of steels in the overall UNS is documented in the standard jointly published by SAE and ASTM: SAE HS-1086/ASTM DS 56H *Metals and Alloys in the Unified Numbering System.* See Internet sites 4 and 5 for additional information and to acquire these standards.

The new designation system permits steel manufacturers to modify the standard chemistry for commonly used steels to meet specific customer needs without creating an entirely new four-digit designation number. Performance requirements for press-forming, roll-forming, abrasion resistance, environmental corrosion resistance, cleanliness of the steel, and similar special needs can be assured more easily and reliably with such a system. These situations often arise in production operations in the automotive, aerospace, appliance, construction equipment, agricultural equipment, manufacturing equipment, energy production equipment, and similar industries.

A *bearing steel* nominally contains 1.0% carbon. Common grades are 50100, 51100, and 52100; the usual four-digit designation is replaced by five digits, indicating 100 points of carbon.

Alloy Groups

As indicated in Table 2–8, sulfur, phosphorus, and lead improve the machinability of steels and are added in significant amounts to the 11xx, 12xx, and 12Lxx grades. These grades are used for screw machine parts requiring high production rates where the resulting parts are not subjected to high stresses or wear conditions. In the other alloys, these elements are controlled to a very low level because of their adverse effects, such as increased brittleness.

Nickel improves the toughness, hardenability, and corrosion resistance of steel and is included in most of the alloy steels. Chromium improves hardenability, wear and abrasion resistance, and strength at elevated temperatures. In high concentrations, chromium provides significant corrosion resistance, as discussed in the section on stainless steels.

TABLE 2–9 Uses of Some Steels

SAE number	Applications
1015	Formed sheet-metal parts; machined parts (may be carburized)
1030	General-purpose, bar-shaped parts, levers, links, keys
1040	Shafts, gears
1080	Springs; agricultural equipment parts subjected to abrasion (rake teeth, disks, plowshares, mower teeth)
1112	Screw machine parts
4140	Gears, shafts, forgings
4340	Gears, shafts, parts requiring good through-hardening
4640	Gears, shafts, cams
5150	Heavy-duty shafts, springs, gears
6150	Gears, forgings, shafts, springs
8650	Gears, shafts
9260	Springs

Molybdenum also improves hardenability and high-temperature strength.

The steel selected for a particular application must be economical and must provide optimum properties of strength, ductility, toughness, machinability, and formability. Frequently, metallurgists, manufacturing engineers, and heat-treatment specialists are consulted. (See also References 3, 8, 24, and 25.) See also Section 2–18.

Table 2–9 lists some common steels used for machine parts, with typical applications listed for the alloys. You should benefit from the decisions of experienced designers when specifying materials.

AHSS—Advanced High-Strength Steels

Significant developments are being made in the field of steels used in automotive and other transportation-related fields that depend on high strength with good ductility. Many emerging products are called *advanced high-strength steels* (AHSS). Goals for their development include safety enhancement, cost reduction, durability, reduced environmental impact, and increased fuel economy for vehicles. AHSS steels are characterized by unique microstructural phases other than ferrite and pearlite, including martensite, austenite, and/or retained austenite in quantities sufficient to produce unique mechanical properties with high strength, enhanced formability, and additional stretchability as compared with more traditional high strength-low alloy (HSLA) steels. See Reference 37 and Internet site 40. Types of AHSS steels include:

- Dual phase (DP)
- Transformation induced plasticity (TRIP)
- Complex phase (CP)
- Martensitic (MS)

Examples of grades are (numbers give yield strength/ultimate tensile strength in MPa):

DP 300/500 (30–34% el.)
TRIP 450/800 (26–32% el.)
CP 700/800 (10–15% el.)
MS 950/1200 (5–7% el.)
HSLA 350/450 (23–27% el.)[1]
DP 350/600 (24–30% el.)
DP 500/800 (14–20% el.)
DP 700/1000 (12–17% el.)
MS 1250/1520 (4–6% el.)

[1]Listed for comparison

2–6 CONDITIONS FOR STEELS AND HEAT TREATMENT

The final properties of steels are dramatically affected by the way the steels are produced. Some processes involve mechanical working, such as rolling to a particular shape or drawing through dies. In machine design, many bar-shaped parts, shafts, wire, and structural members are produced in these ways. But most machine parts, particularly those carrying heavy loads, are heat-treated to produce high strength with acceptable toughness and ductility.

Carbon steel bar and sheet forms are usually delivered in the *as-rolled condition*; that is, they are rolled at an elevated temperature that eases the rolling process. The rolling can also be done cold to improve strength and surface finish. Cold-drawn bar and wire have the highest strength of the worked forms, along with a very good surface finish. However, when a material is designated to be *as-rolled*, it should be assumed that it was hot-rolled.

Heat Treating

Heat treating is any process in which steel is subjected to elevated temperatures to modify its properties. Of the several processes available, those most used for machine steels are annealing, normalizing, through-hardening (quench

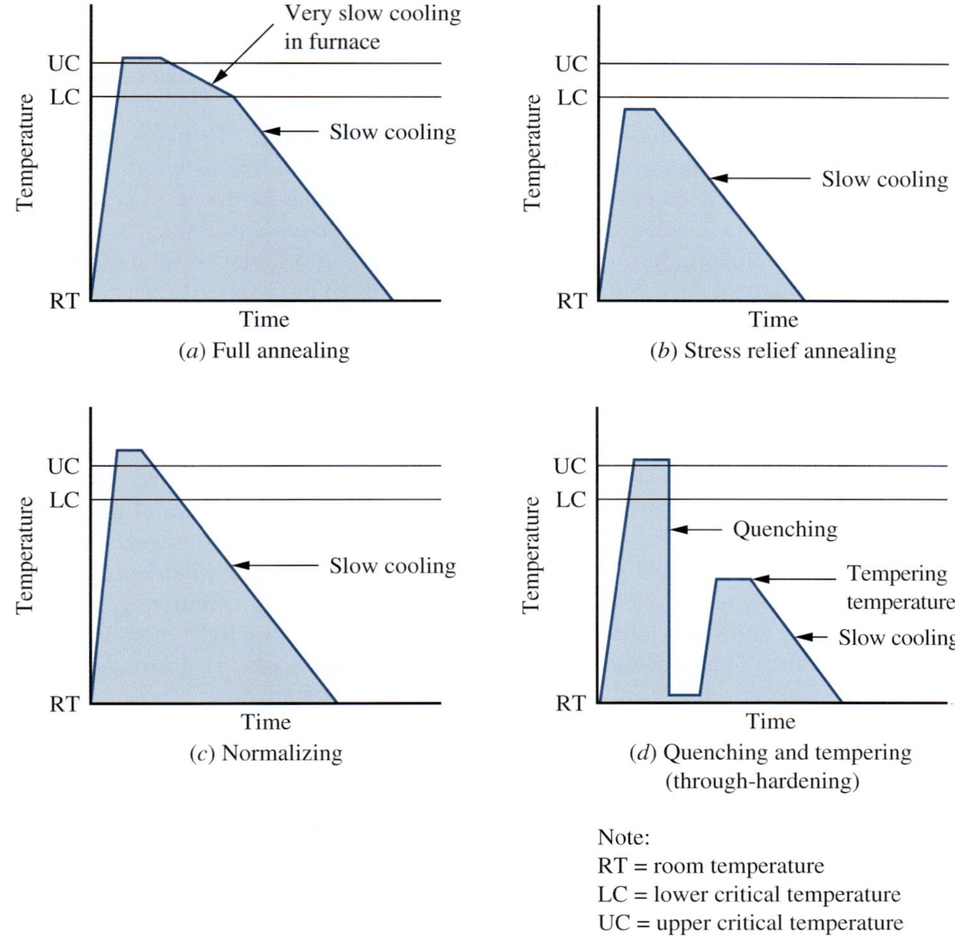

FIGURE 2–13 Heat treatments for steel (*Note*: RT = room temperature, LC = lower critical temperature, UC = upper critical temperature)

and temper), and case hardening. (See References 6 and 16–18.)

Figure 2–13 shows the temperature–time cycles for these heat-treatment processes. The symbol RT indicates normal room temperature, and LC refers to the lower critical temperature at which the transformation of ferrite to austenite begins during the heating of the steel. At the upper critical temperature (UC), the transformation is complete. These temperatures vary with the composition of the steel. For most medium-carbon (0.30–0.50% carbon) steels, UC is approximately 1500°F (822°C). References giving detailed heat-treatment process data should be consulted.

Annealing. *Full annealing* [Figure 2–13(a)] is performed by heating the steel above the upper critical temperature and holding it until the composition is uniform. Then the steel is cooled very slowly in the furnace to below the lower critical temperature. Slow cooling to room temperature outside the furnace completes the process. This treatment produces a soft, low-strength form of the material, free of significant internal stresses. Parts are frequently cold-formed or machined in the annealed condition.

Stress relief annealing [Figure 2–13(b)] is often used following welding, machining, or cold-forming to relieve residual stresses and thereby minimize subsequent distortion. The steel is heated to approximately 1000°F to 1200°F (540°C–650°C), held to achieve uniformity, and then slowly cooled in still air to room temperature.

Normalizing. *Normalizing* [Figure 2–13(c)] is performed in a similar manner to annealing, but at a higher temperature above the transformation range where austenite is formed, approximately 1600°F (870°C). The result is a uniform internal structure in the steel and somewhat higher strength than annealing produces. Machinability and toughness are usually improved over the as-rolled condition.

Through-Hardening and Quenching and Tempering. *Through-hardening* [Figure 2–13(d)] is accomplished by heating the steel to above the transformation range where austenite forms and then rapidly cooling it in a *quenching* medium. The rapid cooling causes the formation of martensite, the hard, strong form of steel. The degree to which martensite forms depends on the alloy's composition. An alloy containing a minimum of 80% of its structure in the martensite form over the entire cross section has *high hardenability*. This is an important property to look for when selecting a steel requiring high strength and hardness. The common quenching liquid media are water, brine, and special mineral oils. Formulations of water-soluble polyalkylene glycol liquid organic polymer (PAG) with corrosion-resisting additives often replace other quenchants to provide tailored control of the rate of quenching to reduce distortion and residual stresses. They can be used either in immersion systems or applied as a spray for either steels or aluminum alloys. One brand name is UCON™ (a trademark of The Dow Chemical Company) that offers ten different formulations.

Air or other gases may also be used for quenching of some materials. Inert gases such as helium, nitrogen, hydrogen, and argon can be used in an enclosed tank, typically under low pressure (vacuum), to inhibit oxidation. Careful specification of the quenchants is critical to ensure compatibility with the particular material being heat-treated. The selection of a quenching medium depends on the rate at which cooling should proceed. Most machine steels use either oil or water quenching.

Tempering is usually performed immediately after quenching and involves reheating the steel to a temperature of 400°F to 1300°F (200°C–700°C) and then slowly cooling it in air back to room temperature. This process modifies the steel's properties: Tensile strength and yield strength decrease with increasing tempering temperature, whereas ductility improves, as indicated by an increase in the percent elongation. Thus, the designer can tailor the properties of the steel to meet specific requirements. Furthermore, the steel in its as-quenched condition has high internal stresses and is usually quite brittle. Machine parts should normally be tempered at 700°F (370°C) or higher after quenching.

To illustrate the effects of tempering on the properties of steels, several charts in Appendix 4 show graphs of strength versus tempering temperature. Included in these charts are tensile strength, yield point, percent elongation, percent reduction of area, and hardness number HB, all plotted in relation to tempering temperature. Note the difference in the shape of the curves and the absolute values of the strength and hardness when comparing the plain carbon SAE 1040 steel with the alloy steel SAE 4340. Although both have the same nominal carbon content, the alloy steel reaches a much higher strength and hardness. Note also the as-quenched hardness in the upper right part of the heading of the charts; it indicates the degree to which a given alloy can be hardened. When the case-hardening processes (described next) are used, the as-quenched hardness becomes very important.

Appendix 3 lists the range of properties that can be expected for several grades of carbon and alloy steels. The alloys are listed with their SAE numbers and conditions. For the heat-treated conditions, the designation reads, for example, SAE 4340 OQT 1000, which indicates that the alloy was oil-quenched and tempered at 1000°F. Expressing the properties at the 400°F and 1300°F tempering temperatures indicates the end-points of the possible range of properties that can be expected for that alloy. To specify a strength between these limits, you could refer to graphs such as those shown in Appendix 4, or you could determine the required heat-treatment process from a specialist. For the purposes of material specification in this book, a rough interpolation between given values will be satisfactory. As noted before, you should seek more specific data for critical designs.

Case Hardening. In many cases, the bulk of the part requires only moderate strength although the surface must have a very high hardness. In gear teeth, for example, high surface hardness is necessary to resist wear as the mating teeth come into contact several million times during the expected life of the gears. At each contact, a high stress develops at the surface of the teeth. For applications such as this, *case hardening* is used; the surface (or *case*) of the part is given a high hardness to a depth of perhaps 0.010 to 0.040 in (0.25–1.00 mm), although the interior of the part (the *core*) is affected only slightly, if at all. The advantage of surface hardening is that as the surface receives the required wear-resisting hardness, the core of the part remains in a more ductile form, resistant to impact and fatigue. The processes used most often for case hardening are flame hardening, induction hardening, carburizing, nitriding, cyaniding, and carbo-nitriding. (See References 16–18.)

Figure 2–14 shows a drawing of a typical case-hardened gear-tooth section, clearly showing the hard case surrounding the softer, more ductile core. Case hardening is used in applications requiring high wear and abrasion resistance in normal service (gear teeth, crane wheels, wire-rope sheaves, and heavy-duty shafts).

The most commonly used processes for case hardening are described in the following list.

1. ***Flame hardening and induction hardening:*** The processes of flame hardening and induction hardening involve the rapid heating of the surface of the part for a limited time so that a small, controlled depth of the material reaches the transformation range. Upon immediate quenching, only that part above the transformation range produces the high level of martensite required for high hardness.

   Flame hardening uses a concentrated flame impinging on a localized area for a controlled amount of time to heat the part, followed by quenching in a bath or by a stream of water or oil. *Induction hardening* is a process in which the part is surrounded by a coil through which high-frequency electric current is passed. Because of the electrical conductivity of the steel, current is *induced* primarily near the surface of the part. The resistance of the material to the flow of

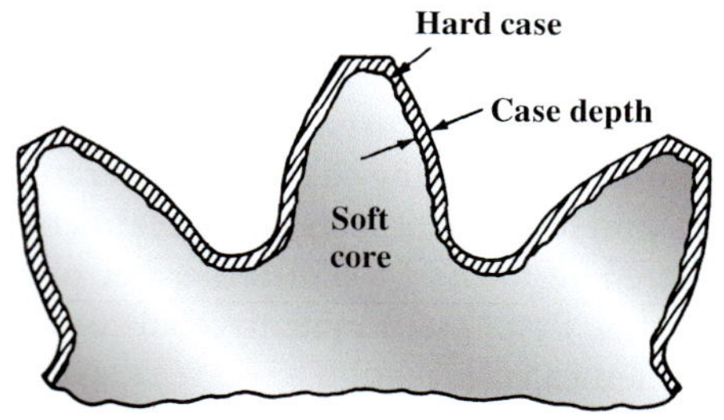

FIGURE 2–14 Typical case-hardened gear-tooth section

current results in a heating effect. Controlling the electrical power and the frequency of the induction system, and the time of exposure, determines the depth to which the material reaches the transformation temperature. Rapid quenching after heating hardens the surface. (See Reference 17.)

The design of the devices for heating the part is critical, especially when localized hardening is desired, as for gear teeth and selective hardening of an area subjected to abrasive wear. For gears of relatively small overall size and small teeth, the impinging flame or the induction heating coil may encircle the entire gear, allowing all teeth to be hardened at the same time. For larger gears with larger teeth, this may be impractical and heating may be tooth-by-tooth or over a small segment of the gear. A system for immediate quenching after heating must be provided. The most critical area for hardening of gear teeth is typically the flanks where the two meshing teeth make contact and transmit the driving forces. High contact stresses where the two curved surfaces meet must be resisted. Pitting resistance of the teeth, as discussed in Chapter 9, is directly related to the hardness of the flanks and that is the most prominent failure mode for gears that are to be used for many thousands of hours of expected life. The area where the root of the tooth blends with the involute tooth form of the flank experiences the highest bending stress. For gears that are used intermittently, bending fatigue is often the predominant mode of failure. Ensuring that the fillet area is adequately hardened will optimize the fatigue life.

Note that for flame or induction hardening to be effective, the material must have a good hardenability. Usually the goal of case hardening is to produce a case hardness in the range of Rockwell C hardness HRC 55 to 60 (Brinell hardness approximately HB 550 to 650). Therefore, the material must be capable of being hardened to the desired level. Carbon and alloy steels with fewer than 30 points of carbon typically cannot meet this requirement. Thus, the alloy steels with 40 points or more of carbon are the usual types given flame- or induction-hardening treatments.

Typical steel materials specified are SAE 1045, 1552, 4140, 4150, 4340, and 5150.

2. ***Carburizing, nitriding, cyaniding, and carbo-nitriding:*** The remaining case-hardening processes—carburizing, nitriding, cyaniding, and carbo-nitriding—actually alter the composition of the surface of the material by exposing it to carbon-bearing gases, liquids, or solids at high temperatures that produce carbon and diffuse it into the surface of the part. The concentration and the depth of penetration of carbon depend on the nature of the carbon-bearing material and the time of exposure. Nitriding and cyaniding typically result in very hard, thin cases that are good for general wear resistance. Where high load-carrying capability in addition to wear resistance is required, as with gear teeth, carburizing is preferred because of the thicker case.

   Several steels are produced as carburizing grades. Among these are 1015, 1020, 1022, 1117, 1118, 4118, 4320, 4620, 4820, and 8620. Appendix 5 lists the expected properties of these carburized steels. Note when evaluating a material for use that the core properties determine its ability to withstand prevailing stresses, and the case hardness indicates its wear resistance. Carburizing, properly done, will virtually always produce a case hardness from HRC 55 to 64 (Rockwell C hardness) or from HB 550 to 700 (Brinell hardness).

   Carburizing has several variations that allow the designer to tailor the properties to meet specific requirements. The exposure to the carbon atmosphere takes place at a temperature of approximately 1700°F (920°C) and usually takes eight hours. Immediate quenching achieves the highest strength, although the case is somewhat brittle. Normally, a part is allowed to cool slowly after carburizing. It is then reheated to approximately 1500°F (815°C) and then quenched. A tempering at the relatively low temperature of either 300°F or 450°F (150°C or 230°C) follows, to relieve stresses induced by quenching. As shown in Appendix 5, the higher tempering temperature lowers the core strength and the case hardness by a small amount, but in general it improves the part's toughness. The process just described is *single quenching and tempering.*

   When a part is quenched in oil and tempered at 450°F, for example, the condition is *case hardening by carburizing, SOQT 450.* Reheating after the first quench and quenching again further refines the case and core properties; this process is *case hardening by carburizing, DOQT 450.* These conditions are listed in Appendix 5.

Cautions for Heat-Treating Operations. Several methods of performing heat-treating operations exist and the choice can affect the final quality and performance of the finished part. Consultation with heat-treating specialists is recommended and careful specifications should be agreed upon. Issues that arise include:

- Final range of acceptable hardness
- Distribution of hardness over critical surfaces of the parts
- Case depth from case-hardening processes
- Microstructure of heat-treated zones in either the case or the core of the part
- Distortion of the part as a result of heat-treating
- Residual stresses developed during heat-treating, particularly residual tensile stresses that reduce fatigue life of the part
- Cracking in critical areas that may affect fatigue life
- Appearance of the part after heat-treating

A wide range of equipment is used for heat-treating; a few are described briefly here. All include heating in some kind of furnace or by flame or induction heating at temperatures up to 1250°C (2300°F), controlled for the individual types of metals being treated. Heating by immersion in molten salts is also sometimes used. Then, parts are quenched by any of several means:

1. Transfer to liquid quenching tanks manually, by gravity dropping, or by conveyor or robotic handling. The quenching liquid may be at room temperature or elevated and it may be still or agitated.
2. Air quenching either in ambient air or with high-velocity blowers or enclosed chambers
3. Impinging of the quenchant on the part by flow from a nozzle
4. High-temperature vacuum carburizing in enclosed tanks with high-pressure gas quenching using inert gases such as helium, argon, hydrogen, or neon.

Careful support of the part during heating and quenching is often needed to reduce distortion. Special racks, shelving, and material-handling equipment may be needed.

Shot Peening for Favorable Residual Stresses. As mentioned earlier in the discussion of machining, grinding, forming, and heat-treating steels, unfavorable tensile residual stresses are often created that exacerbate problems of crack formation where high applied tensile stress is also present in operation. The combination hastens the onset of fatigue failure. *Shot peening* is a secondary process that can mitigate this problem by producing favorable residual *compressive stresses* near the surface of a part.

Shot peening is a process in which fine steel or cast iron shot is projected at high velocity on critical surfaces of a part. The multiple impacts plastically deform the surface and after completing the process, residual compressive stresses remain. Thus, when operating under conditions where applied tensile stresses are produced, the net result is a lower tensile stress which would reduce the likelihood of initiating fatigue cracks. Cases of increasing fatigue life by 100% by shot peening have been reported. Many variables are involved and testing is recommended to quantify the benefits in any given situation. Products for which shot peening have been used include gears, shafts, helical compression and tension springs, leaf springs, and turbine blades.

Steel Alloys for Castings. Often, the composition of steel alloys for casting is similar to some already discussed here for wrought forms of steel produced primarily by rolling processes. However, because casting results in different internal structures from those of rolled forms, special designations are made by ASTM International. General classes, in increasing level of performance, are *carbon, low alloy, high alloy,* and *special alloy*. Material selection must consider operational conditions for a given application, particularly:

- Structural strength in tension, compression, and shear
- Toughness required to resist impact loading
- High temperature operation
- Need for wear resistance
- Need for pressure containment

Table 2–10 gives some of the pertinent standards for cast steel alloys. You should refer to the indicated ASTM

TABLE 2–10 Cast Carbon and Alloy Steels

ASTM designation	Description
General applications	
A27/A27M-08	Carbon steel; Heat-treated to a range of tensile strengths Grades: 60–30 (415–205), 65–35 (450–240), 70–36 (485–250), and 70–40 (485–275) Each set of numbers is tensile strength–yield strength in ksi (MPa)
A915/A915M-08	Carbon and alloy steels; similar composition to standard wrought steels Grades: SC1020, SC1025, SC1040, SC1045, SC4130, SC4140, SC4330, SC4340, SC8620, SC8625, SC8630 Grade numbers match similarly named steel in wrought form
A128/A128M-93(2007)	Hadfield manganese steel castings
A148/A148M-08	High strength carbon, alloy, and martensitic stainless steels for castings
Pressure-containing parts	
A757/A757M-00(2004) and A352/352M-06	Carbon and alloy, ferritic and martensitic, for low temperature
A351/A351M-10	General pressure-containing service Some grades are suitable for high temperatures or corrosive environments
A216/A216M-08	Carbon steel for high temperature; weldable
A389/A389M-08	Alloy steel for high temperatures

specification for details of strength limits and other performance factors for the indicated steel alloys. Note that the compositions of steels in ASTM A915/915M-08 are designed to be similar to commonly used wrought steels designated in the SAE system. However, properties may be different because of the production processes.

2–7 STAINLESS STEELS

The term *stainless steel* characterizes the high level of corrosion resistance offered by alloys in this group. To be classified as a stainless steel, the alloy must have a chromium content of at least 10%. Most have 12% to 18% chromium. (See Reference 9.)

The SAE designates most stainless steels by its 200, 300, and 400 series.

The three main groups of stainless steels are austenitic, ferritic, and martensitic. *Austenitic* stainless steels fall into the SAE 200 and 300 series. They are general-purpose grades with moderate strength. Most are not heat-treatable, and their final properties are determined by the amount of working, with the resulting temper referred to as 1/4 hard, 1/2 hard, 3/4 hard, and full hard. These alloys are nonmagnetic and are typically used in food processing equipment.

SAE 304 and 316 are most often used in commercially available bars, hexagons, squares rectangles, and tubing as listed in Table 2–6.

Ferritic stainless steels belong to the SAE 400 series, designated as 405, 409, 430, 446, and so on. They are magnetic and perform well at elevated temperatures, from 1300°F to 1900°F (700°C–1040°C), depending on the alloy. They are not heat-treatable, but they can be cold-worked to improve properties. Typical applications include heat exchanger tubing, petroleum refining equipment, automotive trim, furnace parts, and chemical equipment.

Martensitic stainless steels are also members of the SAE 400 series, including 403, 410, 414, 416, 420, 431, and 440 types. They are magnetic, can be heat-treated, and have higher strength than the 200 and 300 series, while retaining good toughness. Typical uses include turbine engine parts, cutlery, scissors, pump parts, valve parts, surgical instruments, aircraft fittings, and marine hardware.

There are many other grades of stainless steels, many of which are proprietary to particular manufacturers. A group used for high-strength applications in aerospace, marine, and vehicular applications is of the precipitation-hardening type. They develop very high strengths with heat treatments at relatively low temperatures, from 900°F to 1150°F (480°C–620°C). This characteristic helps to minimize distortion during treatment. Some examples are 17-4PH, 15-5PH, 17-7PH, PH15-7Mo, and AMS362 stainless steels.

2–8 STRUCTURAL STEEL

Most structural steels are designated by ASTM numbers established by the American Society for Testing and Materials. One common grade is ASTM A36, which has a minimum yield point of 36 000 psi (248 MPa) and is very ductile. It is basically a low-carbon, hot-rolled steel available in sheet, plate, bar, and structural shapes such as some wide-flange beams, American Standard beams, channels, and angles. The geometric properties of some of each of these sections are listed in Appendix 15.

Most wide-flange beams (W-shapes) are currently made using ASTM A992 structural steel, which has a yield point of 50 to 65 ksi (345 to 448 MPa) and a minimum tensile strength of 65 ksi (448 MPa). An additional requirement is that the maximum ratio of the yield point to the tensile strength is 0.85. This is a highly ductile steel, having a minimum of 21% elongation in a 2.00-inch gage length. Using this steel instead of the lower strength ASTM A36 steel typically allows smaller, lighter structural members at little or no additional cost.

Hollow structural sections (HSS) are typically made from ASTM A500 steel that is cold-formed and either welded or made seamless. Included are round tubes and square and rectangular shapes. Note in Appendix 7 that there are different strength values for round tubes as compared with the shaped forms. Also, several strength grades can be specified. Some of these HSS products are made from ASTM A501 hot-formed steel having properties similar to the ASTM A36 hot-rolled steel shapes.

Many higher-strength grades of structural steel are available for use in construction, vehicular, and machine applications. They provide yield points in the range from 42 000 to 100 000 psi (290–700 MPa). Some of these grades, referred to as *high-strength, low-alloy (HSLA) steels*, are ASTM A242, A440, A514, A572, and A913.

Appendix 7 lists the properties of several structural steels.

2–9 TOOL STEELS

The term *tool steels* refers to a group of steels typically used for cutting tools, punches, dies, shearing blades, chisels, and similar uses. The numerous varieties of tool steel materials have been classified into seven general types as shown in Table 2–11. Whereas most uses of tool steels are related to the field of manufacturing engineering, they are also pertinent to machine design where the ability to maintain a keen edge under abrasive conditions is required (Type H and F). Also, some tool steels have rather high shock resistance which may be desirable in machine components such as parts for mechanical clutches, pawls, blades, guides for moving materials, and clamps (Types S, L, F, and W). (See Reference 10 for a more extensive discussion of tool steels.)

2–10 CAST IRON

Large gears, machine structures, brackets, linkage parts, and other important machine parts are made from cast iron. The several types of grades available span wide ranges of strength, ductility, machinability, wear resistance, and cost. These features are attractive in many applications. The three most commonly used types of cast iron are gray iron, ductile iron, and malleable iron. Appendix 8 (U.S. units) and 8A

TABLE 2–11 Examples of Tool Steel Types

General type	Type symbol	Specific types: Major alloying elements	Examples: AISI No.	Examples: UNS No.	Typical uses (and other common alloys)
High-speed	M	Molybdenum	M2 M10 M42	T11302 T11310 T11342	General-purpose tool steels for cutting tools and dies for forging, extrusion, bending, drawing, and piercing (M1, M3, M4–M7, M30, M34, M36, M41–M47)
	T	Tungsten	T1 T15	T12001 T12015	Similar to uses for M-types (T2, T4, T5, T6, T8)
Hot-worked	H	Chromium	H10	T20810	Cold-heading dies, shearing knives, aircraft parts, low-temperature extrusion and die-casting dies (H1–H19)
		Tungsten	H21	T20821	Higher-temperature dies, hot shearing knives (H20–H39)
		Molybdenum	H42	T20842	Applications that tend to produce high wear (H40–H59)
Cold-worked	D	High-carbon, high-chromium	D2	T30402	Stamping dies, punches, gages (D3–D5, D7)
	A	Medium-alloy, air-hardening	A2	T30102	Punches, thread-rolling dies, die-casting dies (A3–A10)
	O	Oil-hardening	O1	T31501	Taps, reamers, broaches, gages, jigs and fixtures, bushings, machine tool arbors, tool shanks (O2, O6, O7)
Shock-resisting	S		S1	T41901	Chisels, pneumatic tools, heavy-duty punches, machine parts subject to shocks (S2, S4–S7)
Molded steels	P		P2	T51602	Plastic molding dies, zinc die-casting dies (P3–P6, P20, P21)
Special-purpose	L	Low-alloy types	L2	T61202	Tooling and machine parts requiring high toughness (L3, L6)
	F	Carbon-tungsten types	F1	T60601	Similar to L-types but with higher abrasion resistance (F2)
Water-hardened	W		W1	T72301	General-purpose tool and die uses, vise and chuck jaws, hand tools, jigs and fixtures, punches (W2, W5)

(SI Units) list the properties of several types and grades of cast iron. See also Reference 13.

Gray iron is available in grades having tensile strengths ranging from 20 000 to 60 000 psi (138–414 MPa). Its ultimate compressive strength is much higher, three to five times as high as the tensile strength. One disadvantage of gray iron is that it is brittle and therefore should not be used in applications where impact loading is likely. But it has excellent wear resistance, is relatively easy to machine, has good vibration damping ability, and can be surface-hardened. Applications include engine blocks, gears, brake parts, and machine bases. The gray irons are rated by the ASTM specification A48-03 (2008) and A48M in classes 20, 25, 30, 40, 50, and 60, where the number refers to the minimum tensile strength in kips/in^2(ksi). For example, class 40 gray iron has a minimum tensile strength of 40 ksi or 40 000 psi (276 MPa). Because it is brittle, gray iron does not exhibit the property of yield strength.

Malleable iron is a group of heat-treatable cast irons with moderate to high strength, high modulus of elasticity (stiffness), good machinability, and good wear resistance. The five-digit designation roughly indicates the yield strength and the expected percent elongation of the iron. For example, Grade 40010 has a yield strength of 40 ksi (276 MPa) and a 10% elongation. The strength properties listed in Appendix 8 are for the non-heat-treated condition. Higher strengths would result from heat treating. See ASTM specifications A 47-99 (2009) and A 220-99 (2009).

Ductile irons have higher strengths than the gray irons and, as the name implies, are more ductile. However, their ductility is still much lower than that of typical steels. A three-part grade designation is used for ductile iron in the ASTM A536-84 (2009) specification. The first number refers to the tensile strength in ksi, the second is the yield strength in ksi, and the third is the approximate percent elongation. For example, the grade 80-55-06 has a tensile strength of 80 ksi (552 MPa), a yield strength of 55 ksi (379 MPa), and a 6% elongation in 2.00 in. Higher-strength cast parts, such as crankshafts and gears, are made from ductile iron.

Austempered ductile iron (ADI) is an alloyed and heat-treated ductile iron. (See Reference 13.) It has attractive properties that lead to its use in transportation equipment, industrial machinery, and other applications where the low cost, good machinability, high damping characteristics, good wear resistance, and near-net-shape advantages of casting offer special benefits. Examples are drive train gears, parts for constant velocity joints, and suspension components. ASTM Standard A897/A897M-06 lists

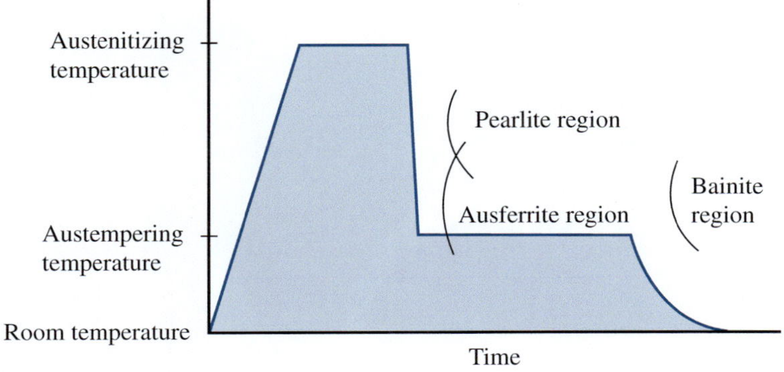

FIGURE 2–15 Heat treatment cycle for austempered ductile iron (ADI)

five grades of ADI ranging in tensile strength from 125 ksi (850 MPa) to 230 ksi (1600 MPa). Yield strengths range from 80 ksi (550 MPa) to 185 ksi (1300 MPa). Ductility decreases with increasing strength and hardness with percent elongation values in the range from approximately 10% to less than 1%. ADI begins as a conventional ductile iron with careful control of composition and the casting process to produce a sound, void-free casting. Small amounts of copper, nickel, and molybdenum are added to enhance the metal's response to the special heat treatment cycle shown in Figure 2–15. It is heated to the austenitizing temperature (1550°F–1750°F, or 843°C–954°C) depending on the composition. It is held at that temperature for one to three hours as the material becomes fully austenitic. A rapid quench follows in a medium at 460°F to 750°F (238°C–400°C), and the casting is held at this temperature for one-half to four hours. This is the *austempering* part of the cycle during which all of the material is converted to a mixture of mostly austenite and ferrite, sometimes called *ausferrite*. It is important that neither pearlite nor bainite form during this cycle. The casting is then allowed to cool to room temperature.

An emerging form of ADI, called *carbidicaustempered ductile iron (CADI),* is produced by alloying conventional ADI with carbide stabilizers such as chromium, molybdenum, or titanium. CADI exhibits significantly higher wear and abrasion resistance while retaining a good level of toughness. Applications are found in railroad rolling stock, earthmoving equipment, agricultural machinery, materials processing equipment, crushers, pump components, and bulk material conveyors. In these applications, it competes well with *white iron,* described next. Another good application is piping that transports aerated dry bulk materials or slurries containing suspended abrasive constituents. Wear is particularly a problem in elbows for such systems.

White iron is produced by rapidly chilling a casting of either gray iron or ductile iron during the solidification process. Chilling is typically applied to selected areas that will experience heavy abrasion because chilled areas become very hard and have high wear resistance. Chilling does not allow the carbon in the iron to precipitate out during solidification, resulting in the white appearance. Areas away from the chilling medium solidify more slowly and acquire the normal properties of the base iron. One disadvantage of the chilling process is that the white iron is somewhat brittle. White iron can be alloyed with chromium and molybdenum to achieve high hardness ranging from 65 to 70 HRC. ASTM Standard A532/A532M-10 covers the composition and heat-treatment for white iron. Typical applications include liners of ball or roll mills used to crush aggregates or other materials into powder form, the balls and rolls themselves, pulveriser components, clay-mixing equipment, brick-molding processes, and wear parts of crushers and material-handling devices.

2–11 POWDERED METALS

Making parts with intricate shapes by powder metallurgy can sometimes eliminate the need for extensive machining. Metal powders are available in many formulations whose properties approach those of the wrought form of the metal. The processing involves preparing a preform by compacting the powder in a die under high pressure. Sintering at a high temperature to fuse the powder into a uniform mass is the next step. Re-pressing is sometimes done to improve properties or dimensional accuracy of the part. Typical parts made by the powder metallurgy (PM) process are gears, gear segments, cams, eccentrics, and various machine parts having oddly shaped holes or projections. Dimensional tolerances of 0.001 in to 0.005 in (0.025–0.125 mm) are typical.

One disadvantage of PM parts is that they are sometimes brittle and should not be used in applications where high-impact loading is expected. Another important application is in sintered bearings, which are made to a relatively low density with consequent high porosity. The bearing is impregnated with a lubricant that may be sufficient for the life of the part. This type of material is discussed further in Chapter 16.

Manufacturers of metal powders have many proprietary formulations and grades. However, the Metal Powder Industries Federation (MPIF) is promoting standardization of materials. Figure 2–16 shows photographs of some powder metal parts. (See Reference 6 and Internet site 9.)

FIGURE 2–16 Example of powder metal components. (Courtesy of the Metal Powder Industries Federation, Princeton, NJ)

A variety of metals can be used for PM parts, samples of which are listed in Table 2–12. Several processes are applied to achieve particular properties and to meet production goals. Brief descriptions are given here.

- Press and Sinter
  - Constituent powders in the proper percentage for the desired material are blended uniformly.
  - Blended metal powders are fed into a die.
  - A press compacts the powders into the desired shape, producing a *green part* having relatively low strength; the part is ejected from the die.
  - A heating process, called *sintering,* metallurgically bonds the grains of powder into a solid, strong form.
  - Optional manufacturing steps may be performed as needed, such as:
    - Repressing to achieve refined properties and greater dimensional accuracy
    - Machining of features that could not be made in the die
    - Heat-treating to achieve higher mechanical strength
    - Plating or other surface treatments
- Powder Forging
  - The steps described for the press and sinter process are completed as before. However, the geometry is not yet in final form or size.
  - As the part is withdrawn from the sintering furnace, it is coated with a lubricant, transferred to a forging press in a close-form die where it is hot worked, causing plastic flow of the material to its final shape.
  - The forging process refines the strength of the part, nearly equaling that of wrought forms of similar metals, while tending to orient the internal structure in favorable directions according to the anticipated loads.
- Isostatic Pressing
  - The blended metal powders are placed into a hermetic chamber and restrained by a flexible membrane.
  - A high pressure is exerted on the membrane compressing the powder uniformly and producing a high density.

TABLE 2–12 Materials Used for Powder Metallurgy (PM) Parts

Material type	Designation and composition	Tensile strength (ksi)	Tensile strength (MPa)	Density (lb_m/in^3)	Density (kg/m^3)
Ferrous metals					
Carbon steel	F-0008 0.8% C	54	370	0.249	6900
	F-0008-HT 0.8% C	85	590	0.249	6900
Copper steel	FC-0208 2% Cu, 0.8% C	60	410	0.242	6700
	FC-0208-HT 2% Cu, 0.8% C	85	590	0.242	6700
Nickel steel	FN-0205 2% Ni, 0.5% C	57	390	0.257	7100
	FN-0205-HT 2% Ni, 0.5% C	145	1000	0.257	7100
Low-alloy steel	FL-4405 0.85% Mo, 0.5% C	66	460	0.257	7100
	FL-4405-HT 0.85% Mo, 0.5% C	160	1100	0.257	7100
Low-alloy steel hybrid	FLN-4405 2% Ni, 0.85% Mo, 0.5% C	80	550	0.255	7050
	FLN-4405-HT 2% Ni, 0.85% Mo, 0.5% C	170	1170	0.255	7050
Diffusion-alloyed steel	FD-0205 1.75% Ni, 1.5% Cu, 0.5% Mo, 0.5% C	78	540	0.251	6950
	FD-0205-HT 1.75% Ni, 1.5% Cu, 0.5% Mo, 0.5% C	130	900	0.251	6950
Sinter-hardened steel	FLC-4608-HT 1.75% Ni, 1.5% Cu, 0.5% Mo, 0.5% C	100	690	0.253	7000
Stainless steel	SS-316N2 17% Cr, 12% Ni, 2.5% M	60	410	0.235	6500
	SS-430N2 17% Cr	60	410	0.257	7100
Copper	C-0000 0.2% maximum other	Electrical applications		0.289	8000
Bronze	CTG-1001 10% Sn, 1% Graphite	Oil-impregnated bearings		0.224	6200
Red brass	CZ-1000 10% Zn	11	76	0.285	7900
Yellow brass	CZ-3000 30% Zn	16	110	0.275	7600
Nickel silver	CNZ-1818 18% Ni, 18% Zn	20	138	0.285	7900
Aluminum	Aluminum 4% Cu, 0.5% Mg, 1% Si	32	221	0.094	2600

Note: C = Carbon, Cu = Copper, Ni = Nickel, Mo = Molybdenum, Cr = Chromium, Sn = Tin, Zn = Zinc, Mg = Magnesium, Si = Silicon

Source: MPIF Standard 35, Materials Standards for P/M Structural Parts, Metal Powder Industries Federation.

    - Compressing at room temperature is called *cold isostatic pressing.*
    - Compressing at elevated temperatures is called *hot isostatic pressing,* which accomplishes the forming to the desired shape and simultaneously sintering the powders, thus eliminating one step in the process.
  - The ability to make larger, more complex parts and the use of a wider range of powder types are advantages of using the isostatic pressing processes.
  - The density of the completed part is typically more uniform as compared with the press and sinter process and the resulting density and mechanical properties are very close to those obtained in wrought forms of metals of similar composition.
- Metal Injection Molding (MIM)
  - Blended metal powders are enhanced with binders and additives such as waxes and thermoplastic polymers.
  - The mixture is fed into an injection molding machine, similar to plastics injection molding systems to create the *green part.*
  - The green part is extracted from the mold and sintered as with other powder metal processes.
  - More complex shapes can be made with MIM than with other processes and dimensional accuracy is improved.
  - Mechanical properties of the finished metal part are nearly equal to those of wrought forms of similar composition.

Production of powder metal products often requires expensive tooling, presses, and special equipment, lending its use for high volume or high-value products. Typical industries using powder metal products are:

Aircraft	Space vehicles	Satellites	Military equipment
Medical devices	Automotive	Appliances	Power tools
Hydraulic systems	Agricultural machinery	Weapons	Business machines
Motor components	Recreational equipment	Consumer products	
Computer peripheral components	Chemical processing equipment		

Guidelines for Specifying Powder Metal Parts. While it is not practical to define all possible conditions, the following discussion presents the framework in which powder metal products fall. Consultation with an experienced supplier is recommended.

1. Part size is typically limited to those with a projected surface area of compressed powder of 50 in^2 (32 000 mm^2) based on press capacity.
2. Part length should be less than 6 in (300 mm) but greater than 0.06 in (1.5 mm).
3. The length/diameter ratio should be less than 5.
4. The length/wall thickness ratio should be less than 8.
5. The part shape must be such that it can be formed by linear pressing by a ram.
6. No undercut, threaded, or re-entrant forms can be made.
7. Density variations should be expected from normal press and sinter processing for relatively long parts.
8. Relatively high production volumes are normally required to amortize tooling and equipment costs.
9. Consultation with the supplier is recommended on powder composition, processing method, and performance testing of the finished product.
10. Some PM parts may retain some porosity so the product may not be pressure tight without subsequent impregnation of sealing materials.

2–12 ALUMINUM

Aluminum is widely used for structural and mechanical applications. Chief among its attractive properties are light weight, good corrosion resistance, relative ease of forming and machining, and pleasing appearance. Its density is approximately one-third that of steel. However, its strength is somewhat lower, also. (See References 1, 2, 12, and 23.) Table 2–13 lists the commonly used alloy groups for wrought aluminums.

The standard designations for aluminum alloys listed by the Aluminum Association use a four-digit system. The first digit indicates the alloy type according to the major alloying element. The second digit, if it is other than zero, indicates modifications of another alloy or limits placed on impurities in the alloy. The presence of impurities is particularly important for electrical conductors. Within each group are several specific alloys, indicated by the last two digits in the designation.

Table 2–14 lists 11 of the 50 or more available aluminum alloys along with the forms in which they are typically produced and some of their major applications. See also Table 2–6 that lists six aluminum alloys used for commercially available shapes. These tables should aid you in selecting suitable alloys for particular applications.

The mechanical properties of the aluminum alloys are highly dependent on their condition. For this reason, the specification of an alloy is incomplete without a reference to its *temper*. The following list describes the usual tempers given to aluminum alloys. Note that some alloys respond to

TABLE 2–13 Wrought Aluminum Alloy Groups

Alloy designations (by major alloying element)
1xxx 99.00% or greater aluminum content
2xxx Copper
3xxx Manganese
4xxx Silicon
5xxx Magnesium
6xxx Magnesium and silicon
7xxx Zinc

TABLE 2–14 Common Aluminum Alloys and their Uses

Alloy	Applications	Forms
1060	Chemical equipment and tanks	Sheet, plate, tube
1350	Electrical conductors	Sheet, plate, tube, rod, bar, wire, pipe, shapes
2014	Aircraft structures and vehicle frames	Sheet, plate, tube, rod, bar, wire, shapes, forgings
2024	Aircraft structures, wheels, machine parts	Sheet, plate, tube, rod, bar, wire, shapes, rivets
2219	Parts subjected to high temperatures (to 600°F)	Sheet, plate, tube, rod, bar, shapes, forgings
3003	Chemical equipment, tanks, cooking utensils, architectural parts	Sheet, plate, tube, rod, bar, wire, shapes, pipe, rivets, forgings
5052	Hydraulic tubes, appliances, sheet-metal fabrications	Sheet, plate, tube, rod, bar, wire, rivets
6061	Structures, vehicle frames and parts, marine uses	All forms
6063	Furniture, architectural hardware	Tube, pipe, extruded shapes
7001	High-strength structures	Tube, extruded shapes
7075	Aircraft and heavy-duty structures	All forms except pipe

heat treating, and others are processed by strain hardening. *Strain hardening* is controlled cold working of the alloy, in which increased working increases hardness and strength while reducing ductility. Commonly available tempers are described next.

F (as-fabricated): No special control of properties is provided. Actual limits are unknown. This temper should be accepted only when the part can be thoroughly tested prior to service.

O (annealed): A thermal treatment that results in the softest and lowest strength condition. Sometimes specified to obtain the most workable form of the alloy. The resulting part can be heat-treated for improved properties if it is made from alloys in the 2xxx, 4xxx, 6xxx, or 7xxx series. Also, the working itself may provide some improvement in properties similar to that produced by strain hardening for alloys in the 1xxx, 3xxx, and 5xxx series.

H (strain-hardened): A process of cold working under controlled conditions that produces improved, predictable properties for alloys in the 1xxx, 3xxx, and 5xxx groups. The greater the amount of cold work, the higher the strength and hardness, although the ductility is decreased. The *H* designation is followed by two or more digits (usually 12, 14, 16, or 18) that indicate progressively higher strength. Temper H18 is called *full hard*, produced by approximately 75% reduction in area during the strain hardening process. H16 is 3/4 hard; H14 is 1/2 hard; and H12 is 1/4 hard. H19 extra-hard temper is available and results in harder and stronger properties than H18. Other designations are used.

T (heat-treated): A series of controlled heating and cooling processes applied to alloys in the 2xxx, 4xxx, 6xxx, and 7xxx groups. The letter *T* is followed by one or more numbers to indicate specific processes. The more common designations for mechanical and structural products are T3, T4, and T6.

Property data for aluminum alloys are included in Appendix 9. Because these data are typical values, not guaranteed values, the supplier should be consulted for data at the time of purchase.

For mechanical design applications, alloy 6061 is one of the most versatile types. Note that it is available in virtually all forms, has good strength and corrosion resistance, and is heat-treatable to obtain a wide variety of properties. It also has good weldability. In its softer forms, it is easily formed and worked. Then, if higher strength is required, it can be heat-treated after forming. However, it has low machinability.

Aluminum Casting Alloys

Aluminum casting alloys are designated by a system of three digits followed by a decimal point and one digit to the right: xxx.x. The first digit represents the alloying element in the base aluminum having the highest concentration as follows:

1xx.x—99.00% minimum aluminum
2xx.x—Copper
3xx.x—Silicon with added copper and/or magnesium
4xx.x—Silicon
5xx.x—Magnesium
7xx.x—Zinc
8xx.x—Tin

The second two digits represent a particular composition of alloying elements, typically some combination of those listed above. The last digit after the decimal point indicates the form of the product being described, as follows,

0—Casting, with specification of sand, permanent mold, or die cast

1—Ingot form, as supplied to the caster with typical control of nonspecified elements

2—Ingot form, with tighter limits of specified elements

Temper designations must also be applied to the alloy, typically *F (as-cast), O (annealed),* or *Tx (heat-treated),* where *x* is one or more numbers indicating the nature of the heat-treatment similar to those used for wrought alloys. Tempers T4 and T6 are most popular.

Mechanical properties data for three popular alloys, 204.0-T4, 206.0-T6, and 356.0-T6, for permanent mold castings are listed in Appendix 9. At least 75 other alloys are available, such as 201, 222, 242, 319, 360, 380, 413, 443, 444, 512, 518, 535, 712, 713, 771, 850, 851, and 852. The 850 series alloys are used primarily for bearings in applications similar to those for which bronzes are used. For special purposes, standard alloys are modified by slightly different compositions, specifications of tighter limits on constituents, or specifying minimum or maximum amounts of trace elements. Designations for such alloys are preceded by a letter (A, B, C, D, and so forth).

Aluminum Forging Alloys

Many of the same alloys used in wrought form are also used for forging stock. Examples are general-purpose alloys 2014, 2024, 5052, and 6061. Alloys 7075, 7175, and 7145 offer much higher strengths and are often used in aerospace applications. Alloys 2218 and 2025 have high fatigue strengths and are applied in rotating engine parts, aircraft propellers, and machine parts that are cyclically loaded. Alloys 2219, 2618, and 4032 are designed for high-temperature operation as experienced in engines. Class I forging alloys are typically supplied in extruded form with wide tolerances in *F* or *O* tempers as rounds, squares, or rectangles. Class II forging stock is wrought and cold finished with much tighter tolerances, sometimes as high as ten times tighter. See Internet site 7.

2–13 ZINC ALLOYS AND MAGNESIUM

Zinc Alloys

Zinc is the fourth most commonly used metal in the world. Much of it is in the form of zinc galvanizing used as a corrosion inhibitor for steels, but very large quantities of zinc alloys are used in castings and for bearing materials. See Reference 30 and Internet site 10. Figure 2–17 shows two examples of zinc die-cast parts having complex shapes and many cast-in holes, bosses, and other features, many of which can be used without further machining.

High production quantities are made using zinc pressure die casting, which results in very smooth surfaces and excellent dimensional accuracy. A variety of coating processes can be used to produce desirable finish appearance and to inhibit corrosion. Although the as-cast parts have inherently good corrosion resistance, the performance in some environments can be enhanced with chromate or phosphate treatments or anodizing. Painting and chrome plating are also used to produce a wide variety of attractive surface finishes.

In addition to die casting, zinc products are often made by permanent mold casting, graphite permanent mold casting, sand casting, and shell-mold casting. Other, less frequently used processes are investment casting, low-pressure permanent mold casting, centrifugal casting, continuous casting, and rubber-mold casting. Plaster-mold casting is often used for prototyping. Continuous casting is used to produce standard shapes (rod, bar, tube, and slabs). Prototypes or finished products can then be machined from these shapes.

Zinc alloys typically contain aluminum and a small amount of magnesium. Some alloys include copper or nickel. The performance of the final products can be very sensitive to small amounts of other elements, and maximum

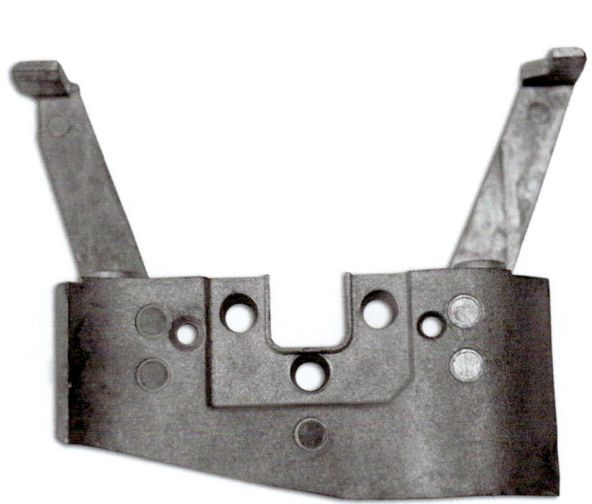

(*a*) Camera support bracket for an interactive whiteboard

(*b*) Steering wheel lock housing

FIGURE 2–17 Examples of zinc die-cast parts (Photo courtesy of North American Die Casting Association, Wheeling, IL)

limits are placed on the content of iron, lead, cadmium, and tin in some alloys.

The most widely used zinc casting alloy is called *alloy No. 3*, sometimes referred to as *Zamak 3*. It has 4% aluminum and 0.035% magnesium. Another form, called *alloy No. 5* (*Zamak 5*), is often used in Europe and it also has 4% aluminum with 0.055% magnesium and 1.0% copper.

A group of alloys having higher aluminum content are the *ZA-alloys*, with ZA-8, ZA-12, and ZA-27 in wide use. The numbers 8, 12, and 27 indicate the nominal percentage of aluminum in the alloy, far higher than the 4% in the Zamak alloys. This results in higher strength, particularly in die-cast form. Each of the ZA alloys can be sand cast, permanent mold cast, or die cast. Percent elongation of cast zinc materials are relatively low, ranging from 2% to 13% depending on the alloy and casting method. Brittle behavior is to be expected. Appendix 10–1 gives a summary of the typical properties of these alloys.

Two additional alloys, No. 2 and No 7, are available. No. 2 is sometimes called KIRKSITE and is good for gravity casting. No. 7 is modified from No. 3 with lower magnesium content which gives it improved fluidity to fill more complex castings or those requiring modest forming during assembly.

Magnesium Alloys

One of the lightest metals used in common practice, magnesium is applied in many applications in automotive, military vehicles and equipment, motorcycles, construction tools, chainsaws, luggage, business equipment, ladders, consumer products, computers, cell phones, other electronic devices, and appliances. Its density of approximately 1800 kg/m^3 (0.065 lb$_m$/in^3) is two-thirds that of aluminum and about one-fifth that of steel. Magnesium has high inherent damping capacity to absorb vibration energy providing quieter operation of equipment when used for housings and enclosures. It is readily cast using die casting (the most prevalent choice), permanent mold casting, sand casting, and squeeze casting. However, strengths are relatively low and its modulus of elasticity, typically 45 GPa (6.5310^6 psi), is the lowest of common metals.

The most widely used alloy is AZ91, where A refers to aluminum and Z refers to zinc, the two major alloying elements. A series of three AM alloys contain aluminum and magnesium. The AS alloys contain aluminum and silicon that give them better high temperature performance. Appendix 10–2 gives values for mechanical properties of AZ91, AM-60, AM-50, and AM-20 alloys.

2–14 NICKEL-BASED ALLOYS AND TITANIUM

Nickel-Based Alloys

Nickel-based alloys are often specified for applications where high corrosion resistance and/or high strength at elevated temperatures are required. Examples are turbine engine components, rocket motors, furnace parts, chemical processing systems, marine-based systems, aerospace systems, valves, pumps, heat exchangers, food processing equipment, pharmaceutical processing systems, nuclear power systems, and air pollution equipment.

Appendix 11–1 gives typical properties of three nickel-based alloys listed according to their UNS designations in the NXXXXX series. Most nickel-based alloys are not heat-treated; however, cold working is employed to increase strength as is evident for the N04400 alloy in Appendix 11–1.

Alloys N06600 and N06110 retain approximately 70% to 80% of room-temperature strength up to 1200°F (649°C). Some alloys are usable up to 2000°F (1093°C). Alloys are often grouped under the categories of *corrosion resistant* or *heat resistant*. While nickel makes up 50% or more of these alloys, the major alloying elements include chromium, molybdenum, cobalt, iron, copper, and niobium. See Internet sites 28–30 for some producers of these alloys. Note the brand names offered by these companies and the extensive set of property data available on their websites.

Creep performance is an important property for the nickel-based alloys because they often operate under significant loads and stresses at high temperatures where creep rupture is a possible failure mode. Also, the modulus of elasticity decreases with increasing temperature, which can result in unacceptable deformations that must be considered in the design phase. Producers publish creep and elastic modulus versus temperature data for their materials in addition to the basic strength data.

Titanium Alloys

The applications of titanium include aerospace structures and components, chemical tanks and processing equipment, fluids-handling devices, and marine hardware. Titanium has very good corrosion resistance and a high strength-to-weight ratio. Its stiffness and density are between those of steel and aluminum; its modulus of elasticity is approximately 16×10^6 psi (110 GPa), and its density 0.160 lb/in^3 (4.429 kg/m^3). Typical yield strengths range from 25 to 175 ksi (172–1210 MPa). Disadvantages of titanium include relatively high cost and difficult machining.

The classification of titanium alloys usually falls into four types: commercially pure alpha titanium, alpha alloys, alpha–beta alloys, and beta alloys. The term *alpha* refers to the hexagonal, close-packed, metallurgical structure that forms at low temperatures, and *beta* refers to the high-temperature, body-centered, cubic structure.

Generally speaking, the alpha–beta alloys and the beta alloys are stronger forms of titanium. They are heat-treatable for close control of their properties. Since several alloys are available, a designer can tailor the properties to meet special needs for formability, machinability, forgeability, corrosion resistance, high-temperature strength, weldability, and creep resistance, as well as basic room-temperature strength and ductility. Alloy Ti-6Al-4V contains 6% aluminum and 4% vanadium and is used in a variety of aerospace applications.

Appendix 11–2 lists the properties of:

- Three commercially pure alpha alloys in wrought form: Ti-35A, Ti-50A, and Ti-65. For each, the number represents the nominal tensile strength in ksi.
- Alpha alloy Ti-0.2Pd in wrought form
- Beta alloy Ti-3A1-13V-11Cr in two forms of heat-treatment
- Alpha-beta alloy Ti-6Al-4V in one annealed and one heat-treated form.

2–15 COPPER, BRASS, AND BRONZE

Copper is widely used in its nearly pure form for electrical and plumbing applications because of its high electrical conductivity and good corrosion resistance. It is rarely used for machine parts because of its relatively low strength compared with that of its alloys, *brass* and *bronze*. (See Reference 6 and Internet site 8.)

Brass is a family of alloys of copper and zinc, with the content of zinc ranging from about 5% to 40%. Brass is often used in marine applications because of its resistance to corrosion in salt water. Many brass alloys also have excellent machinability and are used as connectors, fittings, and other parts made on screw machines. *Yellow brass* contains about 30% or more of zinc and often contains a significant amount of lead to improve machinability. *Red brass* contains 5% to 15% zinc. Some alloys also contain tin, lead, nickel, or aluminum.

Bronze is a class of alloys of copper with several different elements, one of which is usually tin. They are useful in gears, bearings, and other applications where good strength and high wear resistance are desirable.

Wrought bronze alloys are available in four types:

Phosphor bronze: Copper-tin-phosphorus alloy

Leaded phosphor bronze: Copper-tin-lead-phosphorus alloy

Aluminum bronze: Copper-aluminum alloy

Silicon bronze: Copper-silicon alloy

Cast bronze alloys have four main types:

Tin bronze: Copper-tin alloy

Leaded tin bronze: Copper-tin-lead alloy

Nickel tin bronze: Copper-tin-nickel alloy

Aluminum bronze: Copper-aluminum alloy

The cast alloy called *manganese bronze* is actually a high-strength form of brass because it contains zinc, the characteristic alloying element of the brass family. Manganese bronze contains copper, zinc, tin, and manganese.

In the UNS, copper alloys are designated by the letter C, followed by a five-digit number. Numbers from 10000 to 79900 refer to wrought alloys; 80000 to 99900 refer to casting alloys. See Appendix 12 for typical properties.

Wrought forms of brass and bronze include circular rods, flat bars, hexagonal bars, sheet, strip, and wire. Over 120 different wrought alloys are available, listed on Internet site 8. Four common types of wrought brasses are:

***Extra high-leaded brass*—C35600** Copper-lead-tin alloy

Uses include: Gears, gear wheels, and general machined parts

***Free-cutting brass*—C36000** Copper-lead-tin alloy

Uses include: Screw machine parts, gears, and roll-threaded parts

***Free-cutting Muntz metal*—C37000** Copper-tin alloy with a small amount of lead

Uses include: Screw machine parts in high-production operations

***Naval brass*—C46400** Copper-tin-lead-zinc alloy

Uses include: Marine hardware, propeller shafts, valve stems, nuts, bolts, and rivets

***Phosphor bronze*—C54400** Copper-lead-tin-zinc-phosphorous alloy

Uses include: Gears, bushings, stamped parts, drawn parts, machine components

***Silicon bronze*—C65500** Copper (97%)-silicon (3%)

Uses include: Electrical fittings, marine hardware, heavy screws, bolts, propeller shafts

Manganese bronze—C67500 Copper-zinc-iron-tin-manganese alloy

Uses include: Clutch discs, pump parts, shafting, aircraft parts, forgings, valve stems

Casting alloys are made using sand casting, permanent mold casting, and centrifugal casting. See Internet site 8 for properties of over 90 casting alloys. A few common types are:

***Manganese bronze*—C86200** Copper-zinc, aluminum, iron, manganese alloy

Uses include: Marine castings, rudders, gun mounts, large gears, cams, machine parts

***Bearing bronze*—C93200** Copper-tin-lead-zinc alloy

Uses include: Bearings and bushings for heavy equipment, pump impellers, hose fittings

***Aluminum bronze*—C95400** Copper-aluminum-iron alloy

Uses include: Valve seats, gears, worm wheels, bearings, landing gear parts, welding tips

***Copper-nickel-zinc alloy*—C96200**

Uses include: Pump housings, valve bodies, marine hardware, and plumbing components

***Nickel silver*—C97300** Copper-zinc-nickel alloy

Uses include: Camera parts, optical equipment, jewelry, fasteners, drawn shapes

Conditions for Copper Alloys. Copper alloys typically have a wide range of properties, heavily dependent on thermal treatment, strain hardening by cold working, or a combination of both. Thermal treatments include annealing, solution heat-treating, and precipitation hardening.

Strain hardening is categorized by many systems, one of which is called *cold-worked tempers (H)*, in which control of the degree of working produces predictable strength and ductility properties. Working tends to align the grains of the alloy's structure and higher degrees of working create finer grain sizes, a preferred condition. The range of cold-worked tempers is typically given in ten steps with the initial condition being annealed.

H00 1/8 hard	H04 Full hard	H12 Special spring
H01 1/4 hard	H06 Extra hard	H13 Ultra spring
H02 1/2 hard	H08 Spring	H14 Super spring
H03 3/4 hard	H10 Extra spring	

Other conditions or tempers include:

annealed tempers (O)	*cold-worked and stress relieved (HR)*
average grain size control tempers (OS)	*cold-worked and order-strengthened (HT)*
mill-hardened tempers (TM)	*solution heat-treated (TB)*
precipitation hardened tempers (TF)	*solution heat-treated and cold-worked (TD)*
precipitation hardened and cold-worked (TL)	*precipitation hardened and stress relieved (TR)*
Cold-worked and precipitation hardened (TH)	

There is a strong relationship between the condition of copper alloys and the ease of which they can be fabricated, especially the tempers involving cold working. In general, as the degree of cold working increases, the strength is greater, but the ductility is lower indicated by the percent elongation. Figure 2–18 illustrates the general nature of this situation. The designer's responsibility is to specify a condition that has adequate strength while retaining sufficient ductility to permit forming and fabricating using such methods as:

machining	thread rolling	press forming	brake forming
roll forming	drawing	forging	blanking
stamping	coining	punching	upsetting
cold heading	knurling	shearing	swaging
spinning	soldering	welding	brazing

2–16 PLASTICS

Plastics include a wide variety of materials formed of large molecules called *polymers*. The thousands of different plastics are created by combining different chemicals to form long molecular chains.

One method of classifying plastics is by the terms *thermoplastic* and *thermosetting*. In general, the *thermoplastic* materials can be formed repeatedly by heating or molding because their basic chemical structure is unchanged from its initial linear form. *Thermosetting* plastics do undergo some change during forming and result in a structure in which the molecules are cross-linked and form a network of interconnected molecules. Some designers recommend the terms *linear* and *cross-linked* in place of the more familiar *thermoplastic* and *thermosetting*.

Listed next are several thermoplastics and several thermosets that are used for load-carrying parts and that are therefore of interest to the designer of machine elements. These listings show the main advantages and uses of a sample of the many plastics available. Appendix 13 lists typical properties.

Thermoplastics

- *Polyamide (PA or Nylon):* Good strength, wear resistance, and toughness; wide range of possible properties depending on fillers and formulations. Used for structural parts, mechanical devices such as gears and bearings, and parts needing wear resistance.
- *Acrylonitrile-butadiene-styrene (ABS):* Good impact resistance, rigidity, moderate strength. Used for housings, helmets, cases, appliance parts, pipe, and pipe fittings.

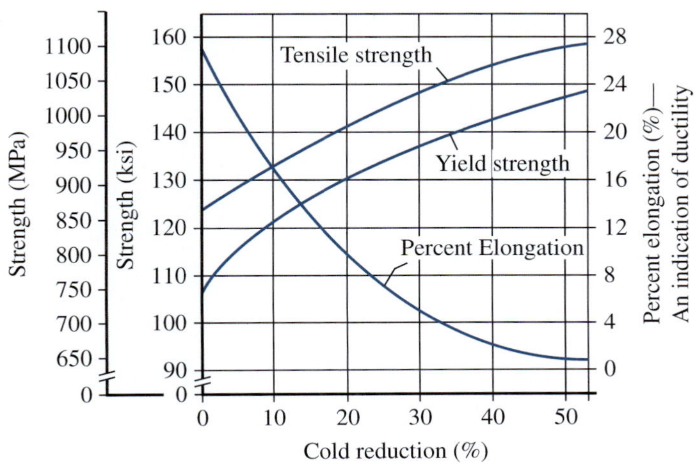

FIGURE 2–18 General characteristics of the variation of strength and ductility of brass alloys as related to the degree of cold working

- *Polycarbonate:* Excellent toughness, impact resistance, and dimensional stability. Used for cams, gears, housings, electrical connectors, food processing products, helmets, and pump and meter parts.
- *Acrylic:* Good weather resistance and impact resistance; can be made with excellent transparency or translucent or opaque with color. Used for glazing, lenses, signs, and housings.
- *Polyvinyl chloride (PVC):* Good strength, weather resistance, and rigidity. Used for pipe, electrical conduit, small housings, ductwork, and moldings.
- *Polyimide (PI):* Good strength and wear resistance; very good retention of properties at elevated temperatures up to 500°F. Used for bearings, seals, rotating vanes, and electrical parts.
- *Acetal:* High strength, stiffness, hardness, and wear resistance; low friction; good weather resistance and chemical resistance. Used for gears, bushings, sprockets, conveyor parts, and plumbing products.
- *Polyurethane elastomer:* A rubberlike material with exceptional toughness and abrasion resistance; good heat resistance and resistance to oils. Used for wheels, rollers, gears, sprockets, conveyor parts, and tubing.
- *Thermoplastic polyester resin (PET):* Polyethylene terephthalate (PET) resin with fibers of glass and/or mineral. Very high strength and stiffness, excellent resistance to chemicals and heat, excellent dimensional stability, and good electrical properties. Used for pump parts, housings, electrical parts, motor parts, auto parts, oven handles, gears, sprockets, and sporting goods.
- *Polyether-ester elastomer:* Flexible plastic with excellent toughness and resilience, high resistance to creep, impact, and fatigue under flexure, good chemical resistance. Remains flexible at low temperatures and retains good properties at moderately elevated temperatures. Used for seals, belts, pump diaphragms, protective boots, tubing, springs, and impact absorbing devices. High modulus grades can be used for gears and sprockets.

Thermosets

- *Phenolic:* High rigidity, good moldability and dimensional stability, very good electrical properties. Used for load-carrying parts in electrical equipment, switchgear, terminal strips, small housings, handles for appliances and cooking utensils, gears, and structural and mechanical parts. Alkyd, allyl, and amino thermosets have properties and uses similar to those of the phenolics.
- *Polyester:* Known as *fiber glass* when reinforced with glass fibers; high strength and stiffness, good weather resistance. Used for housings, structural shapes, and panels.

Special Considerations for Selecting Plastics

A particular plastic is often selected for a combination of properties, such as light weight, flexibility, color, strength, stiffness, chemical resistance, low friction characteristics, or transparency. Table 2–15 lists the primary plastic materials used for six different types of applications. References 14, 19, 29, and 33 provide us extensive comparative studies of the design properties of plastics.

While most of the same definitions of design properties described in Section 2–2 of this chapter can be used for plastics as well as metals, a significant amount of additional information is typically needed to specify a suitable plastic material. Some of the special characteristics of plastics follow. The charts shown in Figures 2–19 to 2–22 are examples only and are not meant to indicate the general nature of the performance of the given type of material. There is a wide range of properties among the many formulations of plastics

TABLE 2–15 Applications of Plastic Materials

Applications	Desired properties	Suitable plastics
Housings, containers, ducts	High impact strength, stiffness, low cost, formability, environmental resistance, dimensional stability	ABS, polystyrene, polypropylene, PET, polyethylene, cellulose acetate, acrylics
Low friction—bearings, slides	Low coefficient of friction; resistance to abrasion, heat, corrosion	TFE fluorocarbons, nylon, acetals
High-strength components, gears, cams, rollers	High tensile and impact strength, stability at high temperatures, machinable	Nylon, phenolics, TFE-filled acetals, PET, polycarbonate
Chemical and thermal equipment	Chemical and thermal resistance, good strength, low moisture absorption	Fluorocarbons, polypropylene, polyethylene, epoxies, polyesters, phenolics
Electrostructural parts	Electrical resistance, heat resistance, high impact strength, dimensional stability, stiffness	Allyls, alkyds, aminos, epoxies, phenolics, polyesters, silicones, PET
Light-transmission components	Good light transmission in transparent and translucent colors, formability, shatter resistance	Acrylics, polystyrene, cellulose acetate, vinyls

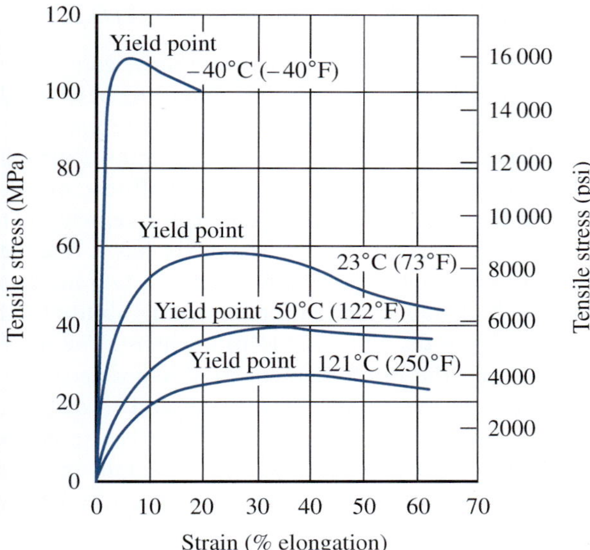

FIGURE 2–19 Stress–strain curves for nylon 66 at four temperatures (DuPont Polymers, Wilmington, DE)

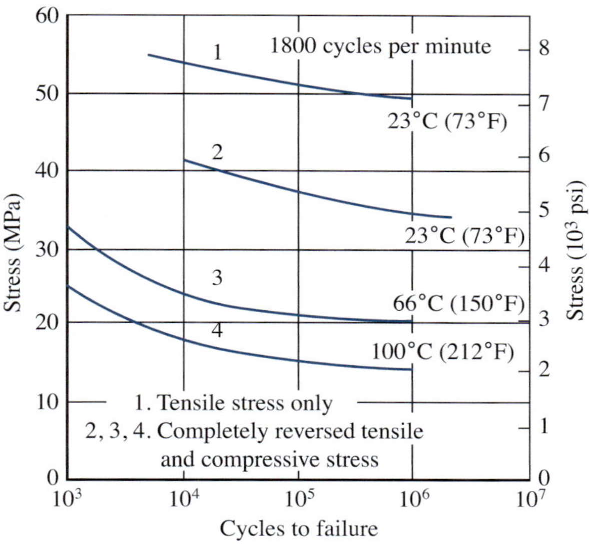

FIGURE 2–21 Fatigue stress versus number of cycles to failure for an acetal resin plastic (DuPont Polymers, Wilmington, DE)

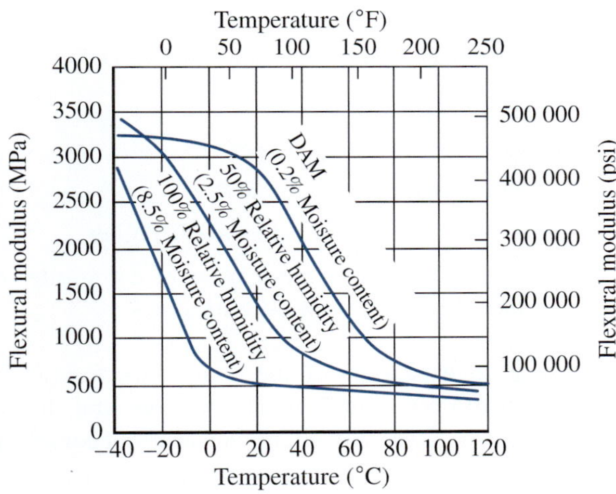

FIGURE 2–20 Effect of temperature and humidity on the flexural modulus of nylon 66 (DuPont Polymers, Wilmington, DE)

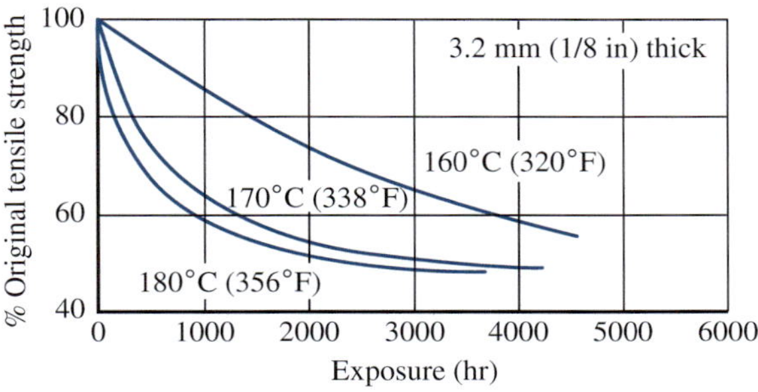

FIGURE 2–22 Effect of exposure to elevated temperature on a thermoplastic polyester resin (PET) (DuPont Polymers, Wilmington, DE)

even within a given class. Consult the extensive amount of design guidance available from vendors of the plastic materials.

1. Most properties of plastics are highly sensitive to temperature. In general, tensile strength, compressive strength, elastic modulus, and impact failure energy decrease significantly as the temperature increases. Figure 2–19 shows the tensile strength of nylon 66 at four temperatures. Note also the rather different shapes of the stress–strain curves. The slope of the curve at any point indicates the elastic modulus, and you can see a large variation for each curve.
2. Many plastics absorb a considerable amount of moisture from the environment and exhibit dimensional changes and degradation of strength and stiffness properties as a result. See Figure 2–20 that shows the flexural modulus versus temperature for a nylon in dry air, 50% relative humidity (RH), and 100% RH. A consumer product may well experience a major part of this range. At a temperature of 20°C (68°F, near room temperature), the flexural modulus would decrease dramatically from approximately 2900 MPa to about 500 MPa as humidity changes from dry air to 100% RH. The product may also see a temperature range from 0°C (32°F, freezing point of water) to 40°C (104°F). Over this range, the flexural modulus for the nylon at 50% RH would decrease from approximately 2300 MPa to 800 MPa.
3. Components that carry loads continuously must be designed to accommodate creep or relaxation. See Figures 2–8 and 2–9 and Example Problem 2–1.

4. Fatigue resistance data of a plastic must be acquired for the specific formulation used and at a representative temperature. Chapter 5 gives more information about fatigue. Figure 2–21 shows the fatigue stress versus number of cycles to failure for an acetal resin plastic. Curve 1 is at 23°C (73°F, near room temperature) with cyclic loading in tension only as when a tensile load is applied and removed many times. Curve 2 is at the same temperature, but the loading is completely reversed tension and compression as would be experienced with a rotating beam or shaft loaded in bending. Curve 3 is the reversed bending load at 66°C (150°F), and Curve 4 is the same loading at 100°C (212°F) to show the effect of temperature on fatigue data.
5. Processing methods can have large effects on the final dimensions and properties of parts made from plastics. Molded plastics shrink significantly during solidification and curing. Parting lines produced where mold halves meet may affect strength. The rate of solidification may be widely different in a given part depending on the section thicknesses, the complexity of the shape and the location of sprues that deliver molten plastic into the mold. The same material can produce different properties depending on whether it is processed by injection molding, extrusion, blow molding, or machining from a solid block or bar. See Reference 29.
6. Resistance to chemicals, weather, and other environmental conditions must be checked.
7. Plastics may exhibit a change in properties as they age, particularly when subjected to elevated temperatures. Figure 2–22 shows the reduction in tensile strength for a thermoplastic polyester resin when subjected to temperatures from 160°C (320°F) to 180°C (356°F) for a given number of hours of exposure. The reduction can be as much as 50% in as little time as 2000 hours (12 weeks).
8. Flammability and electrical characteristics must be considered. Some plastics are specially formulated for high flammability ratings as called for by Underwriters Laboratory and other agencies.
9. Plastics used for food storage or processing must meet U.S. Food and Drug Administration standards.

2–17 COMPOSITE MATERIALS

Composites are materials having two or more constituents blended in such a way that results in bonding between the materials. To form a composite, a discrete *reinforcement* material is distributed in a second continuous material called a *matrix* so that it adds strength, stiffness, or other desirable properties. Typically, the reinforcement material is strong and stiff while the matrix has a relatively low density and specific weight. When the two materials bond together, much of the load-carrying ability of the composite is produced by the reinforcement. The matrix serves to hold the reinforcement in a favorable orientation relative to the manner of loading and to distribute the loads to the reinforcement. The result is a somewhat optimized composite that has high strength and high stiffness with low weight. See References 6, 20, and 34 and Internet sites 19–26 and 31.

Designers can tailor the properties of composite materials to meet the specific needs of a particular application by the selection of each of several variables that determine the performance of the final product. Among the factors under the designer's control are the following:

1. Matrix resin or metal
2. Type of reinforcing fibers
3. Amount of fiber contained in the composite
4. Orientation of the fibers
5. Number of individual layers used
6. Overall thickness of the material
7. Orientation of the layers relative to each other
8. Combination of two or more types of composites or other materials into a composite structure

Table 2–16 lists some of the composites formed by combinations of resins and fibers and their general characteristics and uses. A virtually unlimited variety of composite materials can be produced by combining different matrix materials with different reinforcements in different forms and in different orientations.

TABLE 2–16 Examples of Composite Materials and their Uses

Type of composite	Typical applications
Glass/epoxy	Automotive and aircraft parts, tanks, sporting goods, printed wiring boards
Boron/epoxy	Aircraft structures and stabilizers, sporting goods
Graphite/epoxy	Aircraft and spacecraft structures, sporting goods, agricultural equipment, material handling devices, medical devices
Aramid/epoxy	Filament-wound pressure vessels, aerospace structures and equipment, protective clothing, automotive components
Glass/polyester	Sheet-molding compound (SMC), body panels for trucks and cars, large housings

Examples of Finished Products Made from Composite Materials

The number and variety of applications for composite materials is large and growing. The following items are but a sampling of these applications.

Consumer Products and Recreation: Sporting goods such as tennis rackets, snow skis, snowboards, water skis, surfboards, baseball bats, hockey sticks, vaulting poles, and golf clubs; numerous products having the familiar fiberglass housings and panels; boat hulls and other onboard equipment; medical systems and prosthetic devices.

Ground Transportation Equipment: Bicycle frames, wheels, and seats; automotive and truck body panels and support structures, air ducts, air bags, driveshafts, springs for high-performance sports cars and trucks, floor pans, pickup truck beds, and bumpers.

Aircraft and Aerospace Systems: Fuselage panels and internal structural elements, wings, control surfaces (ailerons, spoilers, tails, rudders), floor systems, engine cowls and nacelles, landing gear doors, cargo compartment structure and fittings, interior sidewalls, trim, partitions, ceiling panels, dividers, environmental control system ducting, stowage bins, lavatory structure systems and fixtures, airfoils (blades) in the compressor section of turbine engines, rocket nozzles, helicopter rotors, propellers, and onboard tanks for storing water and wastewater.

Industrial Facilities: Storage tanks and pressure vessels for chemical, agricultural, and petroleum processing, piping for chemicals and through corrosive environments, septic systems, wastewater treatment facilities, chemical cleaning and plating systems, pulp and paper making equipment, portable tanks for trucks and railroad applications, environmental treatment equipment, protective clothing and helmets, food processing and storage systems, mining, and material handling systems.

Electrical and Electronic Systems: Printed circuit boards, printed wiring boards, surface mount cards, packaging of electronic components, and switching system components.

Building Construction: Structural shapes, exterior panels, roofing and decking systems, doors, window frames, equipment housings, gutters and downspouts, cooling towers, bridges and walkways, piping systems, and ductwork.

Classifications of Composite Materials by Matrix

One method of classifying composite materials is by the type of matrix material. Three general classifications are used as described next, along with typical matrix materials, uses, and matrix-reinforcement combinations.

Polymer Matrix Composites (PMC)

Thermoplastics: Polyethylenes, Polyamides (nylons), polystyrenes, polypropylenes, polycarbonates, polyetheretherketones (PEEK), polyphenylene sulfides (PPS), polyvinyl chloride (PVC)

Thermosets: Polyesters, epoxies, phenolics, polyimides (PI), vinyl esters, silicones

PMCs are used for their high strength and stiffness, low density, and relatively low cost in aerospace, automotive, marine, chemical, electrical, and sporting applications. Common PMC composites include polyester-glass (conventional fiber-glass), epoxy-glass, polyimide-glass, epoxy-aramid, epoxy-carbon, PEEK-carbon, and PPS-carbon.

Metal Matrix Composites (MMC) Aluminum (Al), titanium (Ti), magnesium (Mg), iron (Fe), copper (Cu), nickel (Ni), and alloys of these metals with themselves and with molybdenum (Mo), cesium (Ce), boron (B).

MMCs are preferred for high strength, high stiffness, abrasion resistance, dimensional stability, electrical and thermal conductivity, ability to operate in high temperatures, and toughness and are applied typically in aerospace and engine applications. Examples of MMC composites include Al-SiC (silicon carbide), Ti-SiC, Al-B, Al-C (carbon), Al-graphite, Mg-SiC, and Al-Al_2O_3 (aluminum oxide).

Ceramic Matrix Composites (CMC) Silicon carbide, silicon nitride, alumina, zirconia, glass-ceramic, glass, carbon, graphite.

CMCs are preferred for high strength, high stiffness, high fracture toughness relative to ceramics alone, ability to operate at high temperatures, and low thermal expansion and are attractive for furnaces, engines, and aerospace applications. Common CMC composites include carbon-carbon (C-C), silicon carbide-carbon (SiC-C), silicon carbide-silicon carbide (SiC-SiC), glass ceramic-silicon carbide, silicon carbide-lithium aluminosilicate (SiC-LAS), and silicon carbide-calcium aluminosilicate (SiC-CAS). Where the same basic material is listed as both the matrix and the reinforcement, the reinforcement is of a different form such as whiskers, chopped fibers, or strands to achieve the preferred properties.

Forms of Reinforcement Materials

Many forms of reinforcement materials are used as listed here.

- Continuous fiber strand consisting of many individual filaments bound together
- Chopped strands in short lengths (0.75 to 50 mm or 0.03 to 2.00 in)
- Chopped longer strands randomly spread in the form of a mat
- Chopped longer strands aligned with the principal directions of the load path
- Roving: A group of parallel strands
- Woven fabric made from roving or strands

- Braided sleeves
- Metal filaments or wires
- Metal, glass, or mica flakes
- Single crystal whiskers of materials such as graphite, silicon carbide, and copper.

Types of Reinforcement Materials

Reinforcement fibers come in many types based on both organic and inorganic materials. The following are some of the more popular types:

- Glass fibers in five different types:
  A-glass: Good chemical resistance because it contains alkalis such as sodium oxide
  C-glass: Special formulations for even higher chemical resistance than A-glass
  E-glass: Widely used glass with good electrical insulating ability and good strength
  S-glass: High-strength, high-temperature glass
  D-glass: Better electrical properties than E-glass
- Quartz fibers and high-silica glass: Good properties at high temperatures up to 2000°F (1095°C)
- Carbon fibers made from PAN-base carbon (PAN is polyacrylonitrile): Approximately 95% carbon with very high modulus of elasticity
- Graphite fibers: Greater than 99% carbon and an even higher modulus of elasticity than carbon; the stiffest fibers typically used in composites
- Boron coated onto tungsten fibers: Good strength and a higher modulus of elasticity than glass
- Silicon carbide coated onto tungsten fibers: Strength and stiffness similar to those of boron/tungsten, but with higher temperature capability
- Aramid fibers: A member of the polyamide family of polymers; higher strength and stiffness with lower density as compared with glass; very flexible. (Aramid fibers produced by the DuPont Company carry the name *Kevlar*™.)

Processing of Composites

One method that is frequently used to produce composite products is first to place layers of sheet-formed fabrics on a form having the desired shape and then to impregnate the fabric with wet resin. Each layer of fabric can be adjusted in its orientation to produce special properties of the finished article. After the lay-up and resin impregnation are completed, the entire system is subjected to heat and pressure while a curing agent reacts with the base resin to produce cross-linking that binds all of the elements into a three-dimensional, unified structure. The polymer binds to the fibers and holds them in their preferred position and orientation during use.

An alternative method of fabricating composite products starts with a process of preimpregnating the fibers with the resin material to produce strands, tape, braids, or sheets. The resulting form, called a *prepreg*, can then be stacked into layers or wound onto a form to produce the desired shape and thickness. The final step is the curing cycle as described for the wet process. See Internet sites 23, 24, and 31.

Injection molding and extrusion can be used for composites made from thermoplastics in conventional plastics processing equipment, provided the reinforcement fibers are short so the material can flow properly. Glass fiber reinforcement is most commonly used.

Sheet molding compounds (SMC) are used in compression molding of panels and shells used for automotive body panels, industrial and business machine cabinets, watercraft, and electrical parts. Complex three-dimensional products can be made in final form or near-net-shape using SMC materials. Glass reinforcement fibers [typically in lengths from 1/2 in to 1.0 in (12 mm to 25 mm) and from 10% to 60% volume fraction] are randomly mixed with a polymer resin (often polyester), catalysts, colorants, thickeners, release agents, and inert fillers. Plastic films of nylon or polyethylene cover the mixture top and bottom as a sandwich and the combination is rolled to the desired thickness. At the time of use, the sheets are placed into a mold and shaped and cured simultaneously under heat and pressure. The resulting product has good strength and stiffness properties and excellent surface appearance. See Internet site 32. Where higher mechanical properties are required, SMC sheets can be made from an epoxy matrix and carbon fiber reinforcements. (See Internet sites 31 and 32.)

Pultrusion is a process in which the fiber reinforcement is coated with resin as it is pulled through a heated die to produce a continuous form in the desired shape. This process is used to produce rod, tubing, structural shapes (I-beams, channels, angles, and so on), tees, and hat sections used as stiffeners in aircraft structures, truck and trailer bodies, and similar large structures.

Filament winding is used to make pipe, pressure vessels, rocket motor cases, instrument enclosures, and other cylindrical containers. The continuous filament can be placed in a variety of patterns, including helical, axial, and circumferential, to produce desired strength and stiffness characteristics.

Selection of Composite Materials

Selection of which matrix, reinforcement, and processing method to use for a component made from composites is indeed a complex process. However, some generalizations can be made depending on two major design criteria, *cost* and *performance.*

When *low cost* dominates the design criteria, fiberglass is the predominant choice. Common fiberglass made from thermosetting polyester resin and glass fiber reinforcement is most often used in boats, piping, housings, canisters, and structural components.

For *high-volume applications* that can be made from injection molding or extrusion, low-cost designs can be made using thermoplastics such as polyamides and polycarbonates with short glass reinforcements. Higher strength or stiffness properties may result when using PEEK, PPS, or PVC.

When *high performance* is required, carbon fiber in an epoxy matrix is the predominant choice. Typical applications include aircraft structural components; high-end sporting goods such as tennis rackets, golf clubs, and fishing poles; and race car components.

See also the discussion of *Materials Selection* in this chapter.

Advantages of Composites

Designers typically seek to produce products that are safe, strong, stiff, lightweight, and highly tolerant of the environment in which the product will operate. Composites often excel in meeting these objectives when compared with alternative materials such as metals, wood, and unreinforced plastics. Two parameters that are used to compare materials are *specific strength* and *specific modulus*, defined as follows:

> **Specific strength** ***is the ratio of the tensile strength of a material to its specific weight.***
>
> **Specific modulus** ***is the ratio of the modulus of elasticity of a material to its specific weight.***

Because the modulus of elasticity is a measure of the stiffness of a material, the specific modulus is sometimes called *specific stiffness.*

Although obviously not a length, both of these quantities have the *unit* of length, derived from the ratio of the units for strength or modulus of elasticity and the units for specific weight. In the U.S. Customary System, the units for tensile strength and modulus of elasticity are lb/in^2, whereas specific weight (weight per unit volume) is in lb/in^3. Thus, the unit for specific strength or specific modulus is inches. In SI units, strength and modulus are expressed in N/m^2 (pascals), whereas specific weight is in N/m^3. Then the unit for specific strength or specific modulus is meters.

Table 2–17 gives comparisons of the specific strength and specific stiffness of selected composite materials with certain steel, aluminum, and titanium alloys. Figure 2–23 shows a comparison of these materials using bar charts. Figure 2–24 is a plot of these data with specific strength on the vertical axis and specific modulus on the horizontal axis. When weight is critical, the ideal material will lie in the upper-right part of this chart. Note that data in these charts and figures are for composites having the reinforcement

TABLE 2–17 Comparison of Specific Strength and Specific Modulus for Selected Materials

Material	Modulus of elasticity, E (psi)	(GPa)	Tensile strength, s_u (ksi)	(MPa)	Specific weight, γ (lb/in^3)	(kN/m^3)	Specific strength (in)	(m)	Specific modulus (in)	(m)
Metals										
Steel										
SAE 1020 HR	30.0×10^6	207	55	379	0.283	76.8	0.194×10^6	0.494×10^4	1.06×10^8	2.69×10^6
SAE 5160 OQT 700	30.0×10^6	207	263	1813	0.283	76.8	0.929×10^6	2.36×10^4	1.06×10^8	2.69×10^6
Aluminum										
6061-T6	10.6×10^6	69.0	45	310	0.100	27.1	0.450×10^6	1.14E04	1.00×10^8	2.54E + 06
7075-T6	10.4×10^6	71.7	83	572	0.100	27.1	0.830×10^6	2.11E04	1.04×10^8	2.64E + 06
Titanium										
Ti-6Al-4V	16.5×10^6	114	160	1103	0.160	43.4	1.00×10^6	2.54×10^4	1.03×10^8	2.62×10^6
Composites										
Glass/epoxy	4.00×10^6	28	114	786	0.061	16.6	1.87×10^6	4.75×10^4	0.656×10^8	1.67×10^6
34% fiber content										
Aramid/epoxy	11.0×10^6	76	200	1379	0.050	13.6	4.00×10^6	10.2×10^4	2.20×10^8	5.59×10^6
60% fiber content										
Boron/epoxy	30.0×10^6	207	270	1862	0.075	20.4	3.60×10^6	9.14×10^4	4.00×10^8	10.2×10^6
60% fiber content										
Graphite/epoxy	19.7×10^6	136	278	1917	0.057	15.5	4.88×10^6	12.4×10^4	3.46×10^8	8.78×10^6
62% fiber content										
Graphite/epoxy	48.0×10^6	331	160	1103	0.058	15.7	2.76×10^6	7.01×10^4	8.28×10^8	21.0×10^6
Ultrahigh modulus										

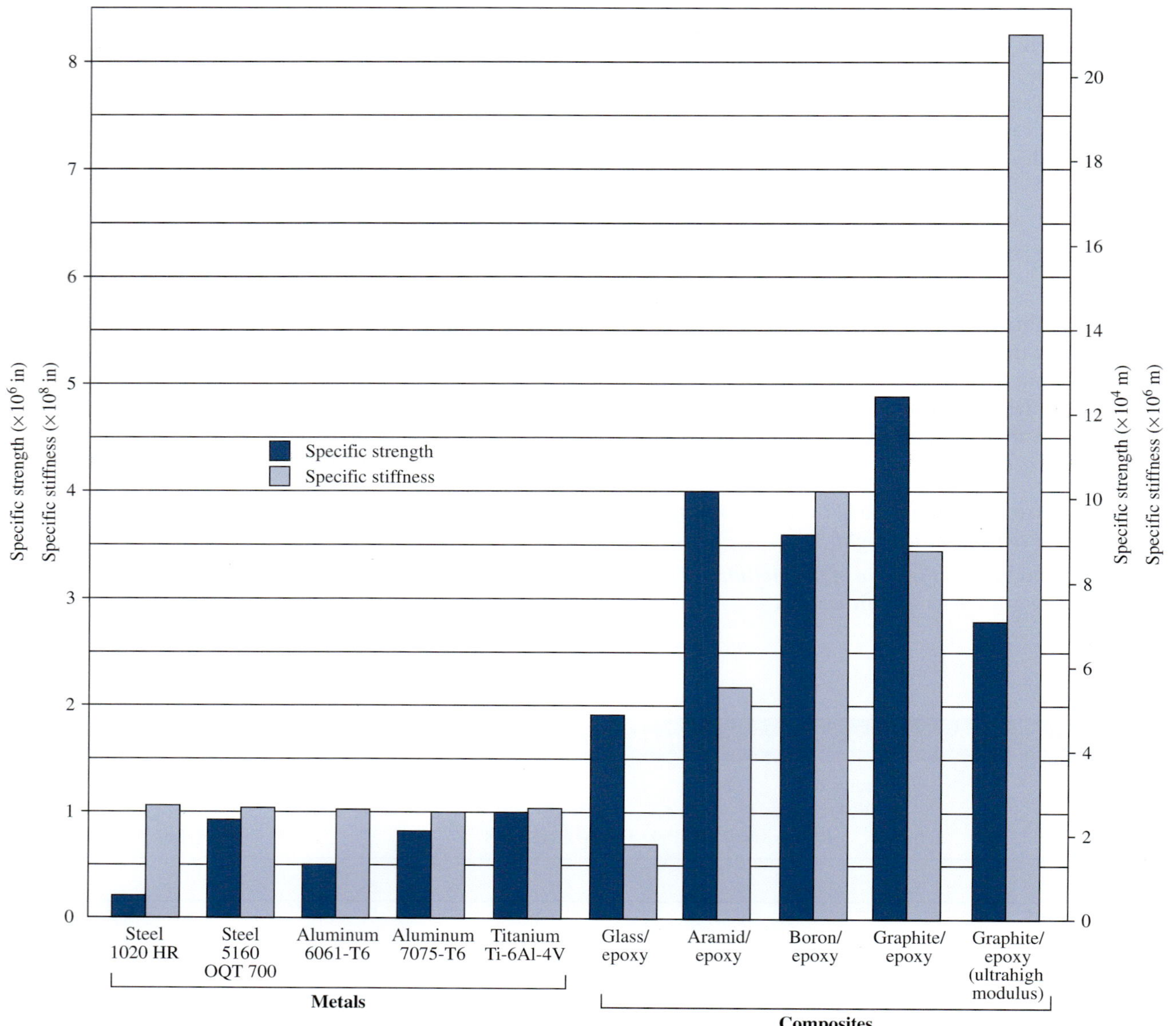

FIGURE 2–23 Comparison of specific strength and specific stiffness for selected metals and composites

materials aligned in the most favorable direction to withstand the applied loads.

Advantages of composites can be summarized as follows:

1. Specific strengths for composite materials can range as high as five times those of high-strength steel alloys. See Table 2–17 and Figures 2–23 and 2–24.
2. Specific modulus values for composite materials can be as high as eight times those for steel, aluminum, or titanium alloys. See Table 2–17 and Figures 2–23 and 2–24.
3. Composite materials typically perform better than steel or aluminum in applications where cyclic loads can lead to the potential for fatigue failure.
4. Where impact loads and vibrations are expected, composites can be specially formulated with materials that provide high toughness and a high level of damping.
5. Some composites have much higher wear resistance than metals.
6. Careful selection of the matrix and reinforcement materials can provide superior corrosion resistance.
7. Dimensional changes due to changes in temperature are typically much less for composites than for metals.
8. Because composite materials have properties that are highly directional, designers can tailor the placement of reinforcing fibers in directions that provide the required strength and stiffness under the particular loading conditions to be encountered.
9. Composite structures can often be made in complex shapes in one piece, thus reducing the number of parts in a product and the number of fastening operations required. The elimination of joints typically improves the reliability of such structures as well.
10. Composite structures are typically made in their final form directly or in a near-net shape, thus reducing the number of secondary operations required.

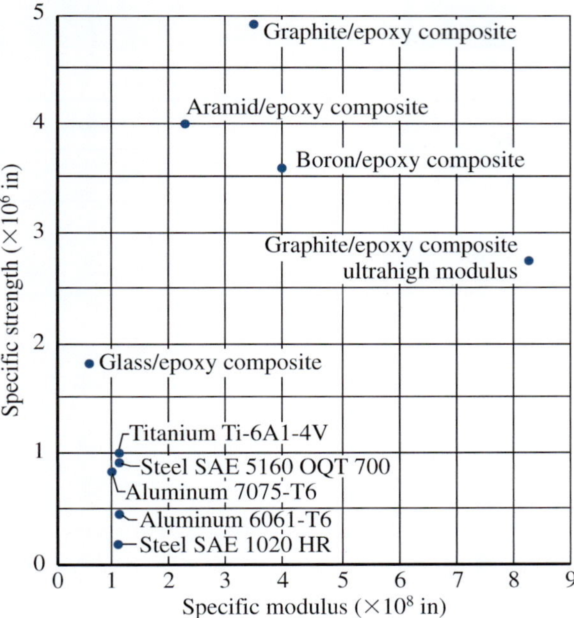

FIGURE 2–24 Specific strength versus specific modulus for selected metals and composites

Limitations of Composites

Designers must balance many properties of materials in their designs while simultaneously considering manufacturing operations, costs, safety, life, and service of the product. Listed next are some of the major concerns when using composites.

1. Material costs for composites are typically higher than for many alternative materials.
2. Fabrication techniques are quite different from those used to shape metals. New manufacturing equipment may be required, along with additional training for production operators.
3. The performance of products made from some composite production techniques is subject to a wider range of variability than the performance of products made from most metal fabrication techniques.
4. The operating temperature limits for composites having a polymeric matrix are approximately 500°F (260°C). (But ceramic or metal matrix composites can be used at higher temperatures such as those found in engines.)
5. The properties of composite materials are not isotropic: Properties vary dramatically with the direction of the applied loads. Designers must account for these variations to ensure safety and satisfactory operation under all expected types of loading.
6. At this time, many designers lack understanding of the behavior of composite materials and the details of predicting failure modes. Whereas major advancements have been made in certain industries such as the aerospace and recreational equipment fields, there is a need for more general understanding about designing with composite materials.
7. The analysis of composite structures requires detailed knowledge of more properties of the materials than would be required for metals.
8. Inspection and testing of composite structures are typically more complicated and less precise than for metal structures. Special nondestructive techniques may be required to ensure that there are no major voids in the final product that could seriously weaken the structure. Testing of the complete structure may be required rather than testing of simply a sample of the material because of the interaction of different parts on each other and because of the directionality of the material properties.
9. Repair and maintenance of composite structures are serious concerns. Some of the initial production techniques require special environments of temperature and pressure that may be difficult to reproduce in the field when damage repair is required. Bonding of a repaired area to the parent structure may also be difficult.

Laminated Composite Construction

Many structures using composite materials are made from several layers of the basic material containing both the matrix and the reinforcing fibers. The manner in which the layers are oriented relative to one another affects the final properties of the completed structure.

As an illustration, consider that each layer is made from a set of parallel strands of the reinforcing filler material, such as E-glass fibers, embedded in the resin matrix, such as polyester. As mentioned previously, in this form, the material is sometimes called a *prepreg*, indicating that the reinforcement has been preimpregnated with the matrix prior to the forming of the structure and the curing of the assembly. To produce the maximum strength and stiffness in a particular direction, several layers or plies of the prepreg could be laid on top of one another with all of the fibers aligned in the direction of the expected tensile load. This is called a *unidirectional laminate*. After curing, the laminate would have a very high strength and stiffness when loaded in the direction of the strands, called the *longitudinal* direction. However, the resulting product would have a very low strength and stiffness in the direction perpendicular to the fiber direction, called the *transverse* direction. If any off-axis loads are encountered, the part may fail or deform significantly. Table 2–18 gives sample data for a unidirectional laminated, carbon/epoxy composite.

To overcome the lack of off-axis strength and stiffness, laminated structures should be made with a variety of orientations of the layers. One popular arrangement is shown in Figure 2–25. Naming the longitudinal direction of the surface layer the *0° ply*, this structure is referred to as

$$0°, 90°, +45°, -45°, -45°, +45°, 90°, 0°$$

The symmetry and the balance of this type of layering technique result in more nearly uniform properties in

TABLE 2–18 Examples of the Effect of Laminate Construction on Strength and Stiffness

	Tensile strength				Modulus of elasticity			
	Longitudinal		Transverse		Longitudinal		Transverse	
Laminate type	ksi	MPa	ksi	MPa	10^6 psi	GPa	10^6 psi	GPa
Unidirectional	200	1380	5	34	21	145	1.6	11
Quasi-isotropic	80	552	80	552	8	55	8	55

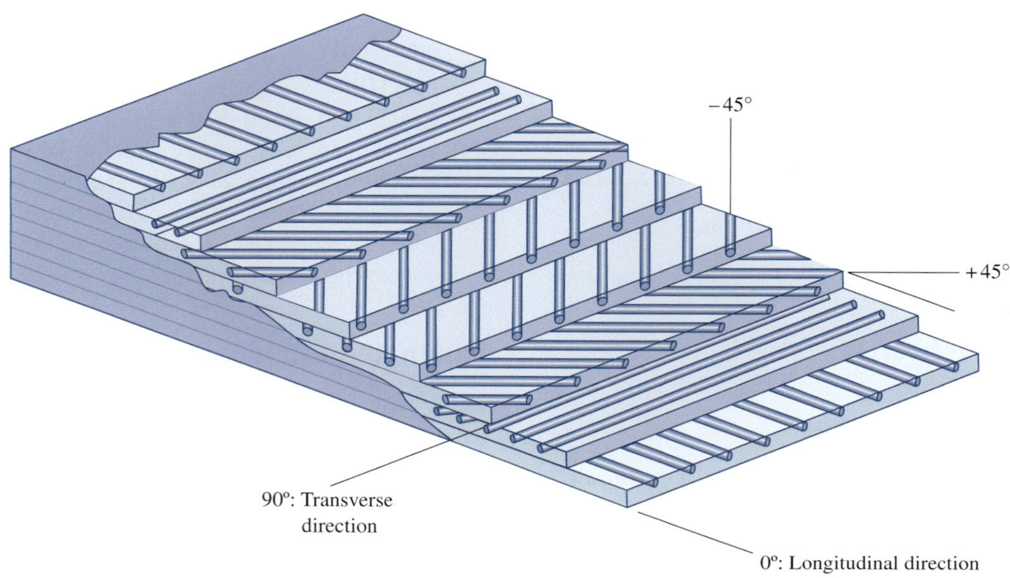

FIGURE 2–25 Multilayer, laminated, composite construction designed to produce quasi-isotropic properties

two directions. The term *quasi-isotropic* is sometimes used to describe such a structure. Note that the properties perpendicular to the faces of the layered structure (through the thickness) are still quite low because fibers do not extend in that direction. Also, the strength and the stiffness in the primary directions are somewhat lower than if the plies were aligned in the same direction. Table 2–18 also shows sample data for a quasi-isotropic laminate compared with one having unidirectional fibers in the same matrix.

Laminate Orientation Code. The example shown earlier of how an eight-ply laminated composite structure, shown in Figure 2–25, is produced is convenient but that method becomes complicated and extensive for structures made from a large number of plies. The *laminate orientation code* seeks to simplify and shorten the designation for a given fabricated composite structure. Note in the previous example that the eight plies are symmetrical about the midplane between plies four and five. This is typical for many practical designs of composite structures and it is unnecessary to repeat the names of all plies. At times, two or more adjacent plies of the same orientation are used and it is simpler to indicate the number of identical plies. Some of the conventions followed in the code are:

1. The code for a given structure is enclosed within square brackets [].
2. The naming of the plies starts at the bottom, the ply that is in contact with the form tool.
3. The degree symbol (°) is not shown in the code.
4. Successive plies are separated by a forward slash (/).
5. No spaces are used.
6. When + and − plies of the same angle are adjacent, the ± symbol is used.
7. The + sign is not applied to a single ply.
8. The 0° direction is defined in the part specifications and is typically either the primary axis of the completed component or the direction of the largest applied load. Often plies with a larger number of reinforcement fibers are aligned with the 0° direction.
9. When two or more plies have the same orientation of reinforcement fibers, a subscript indicates the number of plies.

10. For symmetrical structures, only half of the plies are defined and the subscript *s* is added to the final bracket.
11. For a structure having an odd number of plies, but which is otherwise symmetrical, the orientation of the central ply is written with a horizontal bar over it.

Some examples are:

(a) Symmetrical, as in Figure 2–25:

- $0°, 90°, +45°, -45°, -45°, +45°, 90°, 0° \longrightarrow [0/90/\pm 45]_s$

(b) Multiple plies at same orientation:

- $0°, 0°, 90°, 90°, +45°, -45°, -45°, +45°, 90°, 90°, 0°, 0° \longrightarrow [0_2/90_2/\pm 45]_s$

(c) Odd number of plies:

- $0°, 90°, +45°, -45°, 90°, -45°, +45°, 90°, 0° \longrightarrow [0/90/\pm 45/\overline{90}]_s$

(d) Nonsymmetrical:

- $90°, 90°, +30°, +30°, -30°, -30°, 0°, 0° \longrightarrow [90_2/30_2/-30_2/0_2]$

(e) Repeating sets of plies with the same orientation:

- $0°, 90°, +45°, 0°, 90°, +45°, 0°, 90°, +45° \longrightarrow [0/90/45]_3$

Predicting Composite Properties

The following discussion summarizes some of the important variables needed to define the properties of a composite. The subscript *c* refers to the composite, *m* refers to the matrix, and *f* refers to the fibers. The strength and the stiffness of a composite material depend on the elastic properties of the fiber and matrix components. But another parameter is the relative volume of the composite composed of fibers, V_f, and that composed of the matrix material, V_m. That is,

V_f = volume fraction of fiber in the composite

V_m = volume fraction of matrix in the composite

Note that for a unit volume, $V_f + V_m = 1$; thus, $V_m = 1 - V_f$.

We will use an ideal case to illustrate the way in which the strength and the stiffness of a composite can be predicted. Consider a composite with unidirectional, continuous fibers aligned in the direction of the applied load. The fibers are typically much stronger and stiffer than the matrix material. Furthermore, the matrix will be able to undergo a larger strain before fracture than the fibers can. Figure 2–26 shows these phenomena on a plot of stress versus strain for the fibers and the matrix. We will use the following notation for key parameters from Figure 2–26:

s_{uf} = ultimate strength of fiber

ϵ_{uf} = strain in the fiber corresponding to its ultimate strength

σ'_m = stress in the matrix at the same strain as ϵ_{uf}

The ultimate strength of the composite, s_{uc}, is at some intermediate value between s_{uf} and σ'_m, depending on the volume fraction of fiber and matrix in the composite. That is,

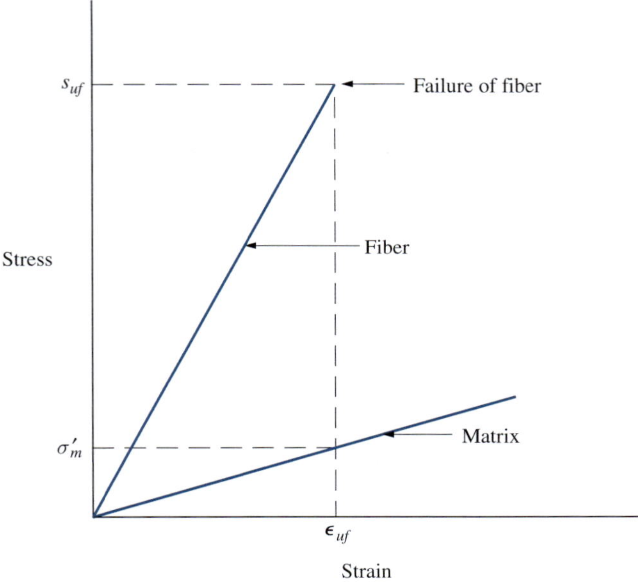

FIGURE 2–26 Stress versus strain for fiber and matrix materials

Rule of Mixtures for Ultimate Strength

$$s_{uc} = s_{uf} V_f + \sigma'_m V_m \qquad (2\text{–}10)$$

At any lower level of stress, the relationship among the overall stress in the composite, the stress in the fibers, and the stress in the matrix follows a similar pattern:

Rule of Mixtures for Stress in a Composite

$$\sigma_c = \sigma_f V_f + \sigma_m V_m \qquad (2\text{–}11)$$

Figure 2–27 illustrates this relationship on a stress–strain diagram.

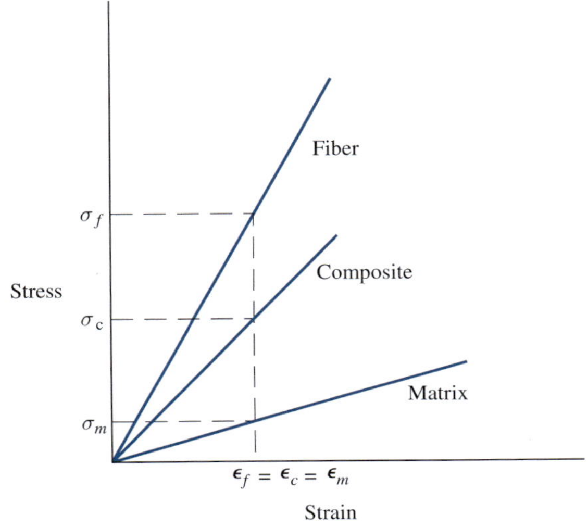

FIGURE 2–27 Relationship among stresses and strains for a composite and its fiber and matrix materials

TABLE 2–19 Example Properties of Matrix and Reinforcement Materials

	Tensile strength		Tensile modulus		Specific weight	
	ksi	MPa	10^6 psi	GPa	lb/in^3	kN/m^3
Matrix materials:						
Polyester	10	69	0.40	2.76	0.047	12.7
Epoxy	18	124	0.56	3.86	0.047	12.7
Aluminum	45	310	10.0	69	0.100	27.1
Titanium	170	1170	16.5	114	0.160	43.4
Reinforcement materials:						
S-glass	600	4140	12.5	86.2	0.09	24.4
Carbon-PAN	470	3240	33.5	231	0.064	17.4
Carbon-PAN (high-strength)	820	5650	40	276	0.065	17.7
Carbon (high-modulus)	325	2200	100	690	0.078	21.2
Aramid	500	3450	19.0	131	0.052	14.1

Both sides of Equation (2–11) can be divided by the strain at which these stresses occur. Since for each material, $\sigma/\epsilon = E$, the modulus of elasticity for the composite can be shown as

➪ Rule of Mixtures for Modulus of Elasticity

$$E_c = E_f V_f + E_m V_m \tag{2–12}$$

The density of a composite can be computed in a similar fashion:

➪ Rule of Mixtures for Density of a Composite

$$\rho_c = \rho_f V_f + \rho_m V_m \tag{2–13}$$

As mentioned previously (Section 2–2), density is defined as mass per unit volume. A related property, *specific weight,* is defined as weight per unit volume and is denoted by the symbol γ (the Greek letter gamma). The relationship between density and specific weight is simply $\gamma = \rho g$, where g is the acceleration due to gravity. Multiplying each term in Equation (2–13) by g gives the formula for the specific weight of a composite:

➪ Rule of Mixtures for Specific Weight of a Composite

$$\gamma_c = \gamma_f V_f + \gamma_m V_m \tag{2–14}$$

The forms of Equations (2–10) through (2–14) are examples of *the rules of mixtures.*

Table 2–19 lists example values for the properties of some matrix and filler materials. Remember that wide variations can occur in such properties, depending on the exact formulation and the condition of the materials.

Example Problem 2–2 Compute the expected properties of ultimate tensile strength, modulus of elasticity, and specific weight of a composite made from unidirectional strands of carbon-PAN fibers in an epoxy matrix. The volume fraction of fibers is 30%. Use data from Table 2–19.

Solution

Objective: Compute the expected values of s_{uc}, E_c, and γ_c for the composite.

Given: Matrix-epoxy: $s_{um} = 18$ ksi; $E_m = 0.56\times10^6$ psi; $\gamma_m = 0.047$ lb/in^3.

Fiber-carbon-PAN: $s_{uf} = 470$ ksi; $E_f = 33.5\times10^6$ psi; $\gamma_f = 0.064$ lb/in^3.

Volume fraction of fiber: $V_f = 0.30$, and $V_m = 1.0 - 0.30 = 0.70$.

Analysis and Results: The ultimate tensile strength, s_{uc}, is computed from Equation (2–10):

$$s_{uc} = s_{uf}V_f + \sigma'_m V_m$$

To find σ'_m, we first find the strain at which the fibers would fail at s_{uf}. Assume that the fibers are linearly elastic to failure. Then

$$\epsilon_f = s_{uf}/E_f = (470\times10^3 \text{ psi})/(33.5\times10^6 \text{ psi}) = 0.014$$

At this same strain, the stress in the matrix is

$$\sigma'_m = E_m\epsilon = (0.56\times10^6\ \text{psi})(0.014) = 7840\ \text{psi}$$

Then, in Equation (2–10),

$$s_{uc} = (470\ 000\ \text{psi})(0.30) + (7840\ \text{psi})(0.70) = 146\ 500\ \text{psi}$$

The modulus of elasticity computed from Equation (2–12):

$$E_c = E_f V_f + E_m V_m = (33.5\times10^6)(0.30) + (0.56\times10^6)(0.70)$$

$$E_c = 10.4\times10^6\ \text{psi}$$

The specific weight is computed from Equation (2–14):

$$\gamma_c = \gamma_f V_f + \gamma_m V_m = (0.064)(0.30) + (0.047)(0.70) = 0.052\ \text{lb/in}^3$$

Summary of Results

$$s_{uc} = 146\ 500\ \text{psi}$$

$$E_c = 10.4\times10^6\ \text{psi}$$

$$\gamma_c = 0.052\ \text{lb/in}^3$$

Comment Note that the resulting properties for the composite are intermediate between those for the fibers and the matrix.

Design Guidelines for Members Made from Composites

The most important difference between designing with metals and designing with composites is that metals are typically taken to be homogeneous with isotropic strength and stiffness properties, whereas composites are decidedly *not* homogeneous or isotropic.

The failure modes of composite materials are complex. Tensile failure when the load is in-line with continuous fibers occurs when the individual fibers break. If the composite is made with shorter, chopped fibers, failure occurs when the fibers are pulled free from the matrix. Tensile failure when the load is perpendicular to continuous fibers occurs when the matrix itself fractures. If the fibers are in a woven form, or if a mat having shorter, randomly oriented fibers is used, other failure modes, such as fiber breakage or pullout, prevail. Such composites would have more nearly equal properties in any direction, or, as shown in Figure 2–25, multilayer laminate construction can be used.

Thus, an important design guideline to produce optimum strength is as follows:

Align the fibers with the direction of the load.

Another important failure mode is *interlaminar shear*, in which the plies of a multilayer composite separate under the action of shearing forces. The following is another design guideline:

Avoid or minimize shear loading, if possible.

Connections to composite materials are sometimes difficult to accomplish and provide places where fractures or fatigue failure could initiate. The manner of forming composites often allows the integration of several components into one part. Brackets, ribs, flanges, and the like, can be molded in along with the basic form of the part. The design guideline, then, is the following:

Combine several components into an integral structure.

When high panel stiffness is desired to resist flexure, as in beams or in broad panels such as floors, the designer can take advantage of the fact that the most effective material is near the outside surfaces of the panel or beam shape. Placing the high-strength fibers on these outer layers while filling the core of the shape with a light, yet rigid, material produces an efficient design in terms of weight for a given strength and stiffness. See Internet sites 25 and 26. Figure 2–28 illustrates some examples of such designs. Another design guideline follows:

Use light core material covered with strong composite layers.

Because most composites use a polymeric material for the matrix, the temperatures that they can withstand are limited. Both strength and stiffness decrease as temperature increases. The polyimides provide better high-temperature properties [up to 600°F (318°C)] than most other polymer matrix materials. Epoxies are typically limited to from 250°F to 350°F (122°C–178°C). Any application above room temperature should be checked with material suppliers. The following is a design guideline:

Avoid high temperatures.

When high temperatures [above 350°F (178°C)] must be encountered, matrix materials other than polymers should be considered, leading to this guideline:

Consider using ceramic matrix composites (CMCs) or metal matrix composites (MMCs) when high temperatures are encountered.

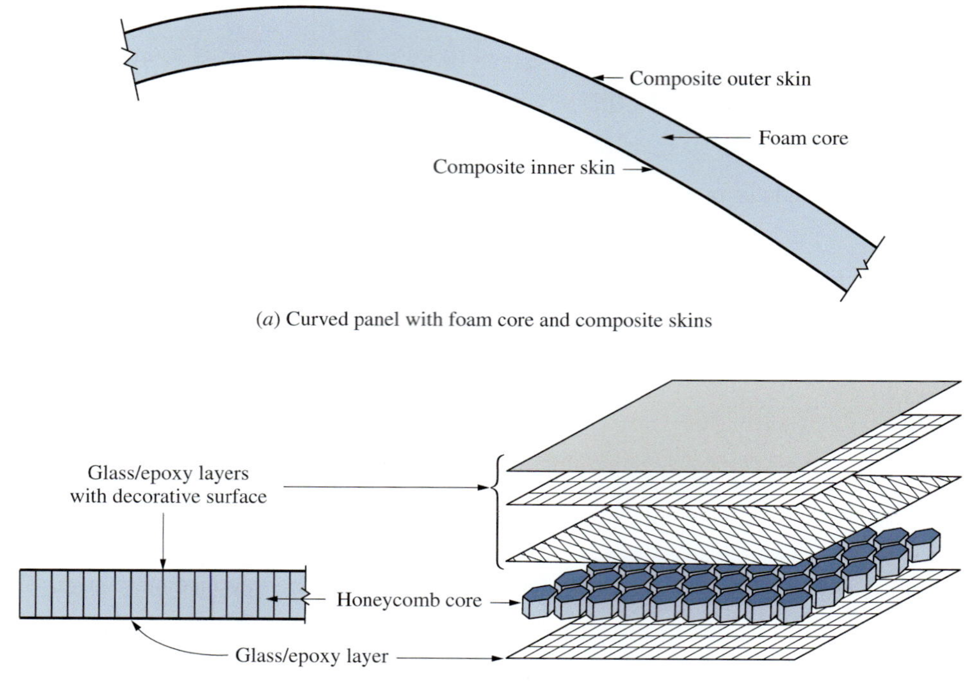

FIGURE 2–28 Laminated panels with lightweight cores

When using sandwich structures such as those illustrated in Figure 2–28, it is essential to have strong adhesive bonding between the various cover layers and with the core material or honeycomb to inhibit interlaminar shear or pealing. Some choices for adhesives for composites are polyester, epoxy, cyanoacrylate, polyurethane, silicone, and anaerobic adhesives (similar to thread-locking compounds). In addition, the surfaces of materials to be joined may be treated to enhance adhesion and the resulting strength of the bond.

Select carefully adhesives used to bond core materials with composite surface layers and prepare properly the contact surfaces.

When a structure requires wide panels, sometimes called plates, the resulting stiffness is a function of the thickness of the wall (i.e., the number of plies in the composite) and the overall geometry. A balance can be sought by adding stiffeners to the panel to break the wide surface into a series of smaller areas. The stiffeners can be molded into the panel or applied separately with adhesives or mechanical fasteners. Three popular shapes for stiffeners are the *hat section, channel,* and *Z-section* that can be made of composite material using the pultrusion process or roll-formed from metal sheet. See Figure 2–29.

Consider using stiffeners on large panels to achieve adequate rigidity with minimum weight.

As described earlier in this section, many different fabrication techniques are used for composite materials. The shape can dictate a part's manufacturing technique. This is a good reason to implement the principles of concurrent engineering and adopt another design guideline:

Involve manufacturing considerations early in the design.

References 34 and 35 present broad and in-depth coverage of the design and manufacture of composite structures with emphasis on the applications of the principles.

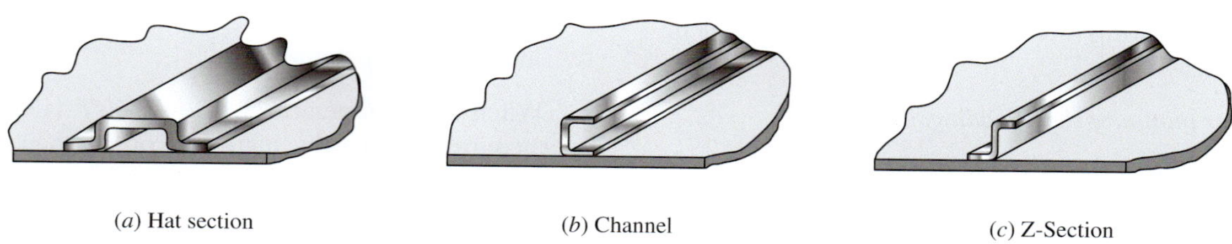

FIGURE 2–29 Panel stiffening techniques

Reference 36 is one of the landmark documents defining analytical methods of design and analysis of composites.

The complete analysis of composite structures requires detailed knowledge of the mechanical and physical properties of the composite materials at the layer stage, at the finished laminate stage, during curing, at joints, at fastening points, and around complex geometric forms. Because of the anisotropic nature of composites, data for strength, modulus of elasticity in tension and shear, and Poisson's ratio in each of the three dimensions, x, y, and z are required for reliable analysis. Furthermore, many of these properties are nonlinear and dependent on detailed knowledge of loading patterns and magnitudes, and environmental conditions (temperature, moisture exposure, chemical exposure, and so forth). Dynamic response and fatigue resistance should also be analyzed.

Some of the material property data can be acquired from suppliers such as those listed in Internet sites 23–26 and 31–32. Databases and research reports from organizations such as those listed in Internet sites 19–22 and 33 are good sources of information on material properties and processing methods.

Analysis of composites structures inherently requires computer-aided design and analysis tools, including finite-element analysis, because of the special nature of the three-dimensional, nonlinear analysis required for reliable results. Several commercially available software products are tailored to composite analysis, such as those mentioned in Internet site 34.

Nanotechnology Applications in Materials. The world of materials science, engineering, and technology is rapidly changing with the commercialization of *nanotechnology*, the practice of applying tiny particles of matter. The term *nanotechnology* is derived from the measure of length as small as 1×10^{-9} meters (1.0 *nanometer* or 1.0 nm). While arbitrary, some consider a nanomaterial particle to be any with a characteristic dimension from 1 to 100 nanometers. Applying this technology is having major influence on developments in physics, biology, and chemistry in fields such as materials science, electronics, biotechnology, food processing, environmental technology, medical technology, sensors, coatings for surface protection, tire reinforcements, solar cells, highly efficient batteries and energy storage devices, adhesives, and so-called *smart materials* that change their characteristics in response to electrical stimuli, temperature, stress, strain, and so forth. See Reference 21.

One type of nanomaterial that is involved in much materials research and development is the carbon nanotube (CNT), very fine layers of graphite oriented in cylindrical form with diameters of only 1 to 3 nm and lengths from fine whiskers, platelets, and powders up to fibers 20 microns (20 μm) long. Reported strengths of CNTs range to 20 times that of high-alloy steels with approximately one-sixth the weight. Thus its strength to weight ratio is approximately 120 times higher. CNTs also have excellent electrical conductivity allowing their use in a variety of electrical and electronic systems. Another form of nanomaterial is the set of organic nanoparticles, called *dendrimers*. Some of the emerging applications for nanomaterials include:

- Creation of a series of *nanocomposites* (NCs), in which CNTs or other nanoparticles are blended with common materials to enhance their strength, stiffness, or toughness. Examples are:
  - Adding CNTs to CMCs (ceramic matrix composites) enhances the toughness of the matrix while retaining the other desirable properties of high hardness, wear resistance, and high temperature resistance. They are called CMNCs.
  - Adding CNTs to PMCs (polymer matrix composites) produce PMNCs that have enhanced matrix strength and stiffness and higher overall capabilities of the composite. An issue is the adherence of the polymers to the surface of the carbon nanotubes and the homogeneous distribution of the nanoparticles in the matrix.
  - Adding CNTs to a MMCs (metal matrix composite) also enhances the strength and stiffness of the composite, which becomes a MMNC.
- Blending CNTs with common metal powders (iron, copper, magnesium, aluminum, and others) prior to sintering enhances the performance of powder metal products.
- Internet site 35 describes high performance materials that use CNTs to enhance polymer matrix composites. PMNCs are produced as prepregs along with woven fabrics and unidirectional fibers of carbon, E-glass, S-glass, aramid, and others. Applications are in aerospace, marine, sports equipment, and motor sports.
- Internet site 36 describes multiple applications in nanotechnology, nanomaterials, and nanoelectronics. A series of nanotools and nanodevices are produced to aid in the manipulation of matter at the nano and atomic regime for fabrication, metrology, lithography, chemical vapor deposition, 3D printing, and nanofluidics—the study of the behavior of fluids in nanoscale applications.
- Internet site 37 describes a process for applying surface coatings of nanoparticles of aluminum oxide to provide enhanced scratch resistance of hard surfaces, plastics, architectural, and automotive applications.
- Internet site 38 describes a process of applying patented steel alloys with microstructural grain sizes in the 10 to 100 nm size. A thermal spray technique is used to apply the coatings using atomized powder, cored wire, or electrode forms to produce coating thicknesses in the 5 to 30 mil (0.13–0.75 mm) range. Also, weld overlays in thicknesses from 1/8 in to 1/2 in (3–12 mm) are produced. Applications include coating shovel buckets, hydraulic cylinder rods, inside surfaces of piping systems that carry abrasive materials, storage bins, conveyor components, and similar products to reduce significantly wear, corrosion, and erosion.

2–18 MATERIALS SELECTION

One of the most important tasks for a designer is the specification of the material from which any individual component of a product is to be made. The decision must consider a huge number of factors, many of which have been discussed in this chapter.

The process of material selection must commence with a clear understanding of the functions and design requirements for the product and the individual component. Refer to Section 1–4 for a discussion of these concepts. Then, the designer should consider the interrelationships among the following:

- The functions of the component
- The component's shape
- The material from which the component is to be made
- The manufacturing process used to produce the component

Overall requirements for the performance of the component must be detailed. This includes, for example:

- The nature of the forces applied to the component
- The types and magnitudes of stresses created by the applied forces
- The allowable deformation of the component at critical points
- Interfaces with other components of the product
- The environment in which the component is to operate
- Physical size and weight of the component
- Aesthetics expected for the component and the overall product
- Cost targets for the product as a whole and this component in particular
- Anticipated manufacturing processes available

A much more detailed list may be made with more knowledge of specific conditions.

From the results of the exercises described previously, you should develop a list of key material properties that are important. Examples often include:

1. Strength as indicated by ultimate tensile strength, yield strength, compressive strength, fatigue strength, shear strength, and others
2. Stiffness as indicated by the tensile modulus of elasticity, shear modulus of elasticity, or flexural modulus
3. Weight and mass as indicated by specific weight or density
4. Ductility as indicated by the percent elongation
5. Toughness as indicated by the impact energy (Izod, Charpy, etc.)
6. Creep performance data
7. Corrosion resistance and compatibility with the environment
8. Cost for the material
9. Cost to process the material

A list of candidate materials should then be created using your knowledge of the behavior of several material types, successful similar applications, and emerging materials technologies. A rational decision analysis should be applied to determine the most suitable types of materials from the list of candidates. This could take the form of a matrix in which data for the properties just listed for each candidate material are entered and ranked. An analysis of the complete set of data will aid in making the final decision.

Decision Analysis Process. An example of how decision analysis may be used to aid in deciding which material to specify in a given application is shown in Figure 2–30. The process described here is relatively simple and is sometimes called the *criteria rating process.* The decision analysis chart, sometimes called a matrix, is at the top, listing five candidate materials being considered for the production of a set of 300 wheels for use on special material handling carts. The left column lists five criteria for evaluation. Observe also the three sets of supporting information below the matrix:

1. **Notes about the proposed materials:** Describes pertinent mechanical properties (from the appendix) and qualitative evaluations of the materials' corrosion resistance, cost of purchasing the materials, and costs to process the materials. With more information available, actual costs could be listed for a more detailed analysis. Note that the ratings for the processing costs for both magnesium and zinc are affected by the proposed use of die casting that entails relatively high cost for die design and fabrication for only 300 parts to be produced. The low rating for processing titanium is based on the relative difficulty of machining titanium compared with steel or aluminum.
2. **Application:** Gives more detail about the use of the wheels and the environment in which the carts will be operated, including the quantity to be produced, weather, and other environmental issues, general nature of the loading in regard to its level and whether shock loading is expected, and the relationship with the mating tire. This is an abbreviated list and additional detail would be expected such as actual loads to be carried, proposed size, mounting details, and so forth.
3. **Desirable characteristics:** Support the criteria for judging the proposed materials.

MATERIAL SELECTION—DECISION A NALYSIS CHART EXAMPLE							
Desired property	*Candidate materials*						
	Steel	Aluminum	Magnesium	Zinc	Titanium		
	Rating [1 to 10 scale]						
Yield strength (High is better)	9	6	3	7	10		
Modulus of elasticity (High is better)	10	4	2	6	8		
Ductility (High is better)	10	8	7	4	5		
Density (Low is better)	3	8	10	5	6		
Corrosion resistance (High is better)	4	10	7	7	10		
Material cost (Low is better)	10	7	7	7	2		
Processing cost (Low is better)	10	10	5	5	5		
Totals:	56	53	41	41	46		
Ranking:	1	2	4	4	3		
Notes about the proposed materials:	s_y (MPa)	*E* (GPa)	*% Elong.*	*Density* (kg/m^3)	*Corr. res.*	*Matl cost*	*Proc. cost*
1. Steel: Wrought SAE 3140 OQT 1000	920	207	17	7680	Low	Low	Low
2. Aluminum wrought 6061-T6	276	69	12	2774	High	Moderate	Low
3. Magnesium: Die-cast AM-50	120	45	10	1770	Moderate	Moderate	High
4. Zinc: Die-cast ZA-12	320	82.7	5.5	6000	Moderate	Moderate	High
5. Titanium: Wrought Ti-6Ai-4V Quenched and aged at 1000°F	1030	114	7	4432	High	Very high	High

Application: Wheel hub for four-wheel cart used in material handling within a factory environment

1. Quantity: 300 pieces
2. Exposed to weather in a variety of climates throughout North America
3. Wheel hub; light shock loading; moderate level of load
4. Must maintain good fit with solid tire made from urethane polymer

Desirable characteristics:

1. High strength to survive factory conditions
2. High ductility to provide high impact strength and toughness
3. High stiffness to exhibit rigidity during operation
4. Low weight and mass to achieve low inertia and effort to accelerate cart
5. Low overall cost of acquisition and service

FIGURE 2–30 Example of a decision analysis technique to aid decision making for materials selection

In the matrix, each material is rated against each desired property on a scale of 1 to 10. Admittedly, these ratings are subjective, but they are based on the data given and the nature of the application. Then the seven ratings for each material are summed and the total is listed. Finally, the sums for the five candidate materials are ranked from highest (rank = 1) to lowest.

> ***Note that the decision analysis matrix is* not a decision-making device; *rather it is an* aid to decision making.**

It is still the responsibility of the designer to make the final decision. However, the results suggest that either steel or aluminum would perform well in this application relative to the other three suggested materials. More careful examination of the comparative ratings of steel and aluminum is recommended and perhaps additional data collection can be done. Questions can be asked such as:

- Are all the criteria of equal weight? If not, weighting factors can be applied to the list of criteria. This is commonly done.
- How can the lower corrosion resistance of the steel be managed: painting, plating, or other secondary processes?
- How do those processes affect the final overall cost for the steel wheels?
- Is the relatively low rating for modulus of elasticity for aluminum an important factor? Does it affect the final design for the wheels, perhaps by requiring thicker sections for critical features? How does this affect the final overall cost for the aluminum wheels?

- Is the lighter projected weight of the aluminum due to lower density compared with steel a strong advantage? Is this judgment firm?
- Should more detailed designs of the wheels be pursued for both steel and aluminum before committing to a final decision?

We leave the process at this point, pending answers to questions such as these, but the value of using the decision-analysis process should be evident to compare the performance of a variety of proposed materials using several parameters.

More comprehensive materials selection processes are described in References 3–6, 25, 27, 31, and 32. Internet sites 1, 2, 14–17, and 33 provide vast amounts of data for material properties that can be used in combination with the appendix to generate suitable candidate materials.

Other Considerations for Materials Selection. Most of the discussion in this book deals with metallic materials that are common choices for machine elements such as gears, bearings, shafts, and springs. For design in general, a wider list of types of materials from which a designer may choose for a particular application should be considered. Refer back to earlier sections of this chapter for additional information. Consider the following list of six primary classes of materials with a few examples for each listed as well:

- ***Metals*** Steel, aluminum, copper alloys, titanium, magnesium, cast iron, zinc
- ***Polymers*** PA, ABS, PC, acrylic, PVC, acetal, polyurethane, PET, polyester
- ***Ceramics*** Silicon nitride, zirconia, silicon carbide, alumina
- ***Glasses*** Silica glass, borosilicate glass, soda glass
- ***Elastomers*** Natural rubber, butyl rubber, neoprene, isoprene, silicones
- ***Hybrids*** Composites, foams, sandwich structures, honeycomb structures

The sheer number of different materials from which to choose makes material selection a daunting task. Specialized approaches, as described in References 3–6, offer significant guidance in the selection process. Furthermore, computer software packages are available to permit rapid searching and sorting based on numerous parameters to produce refined lists of candidate materials with quantitative data about their performance, cost, producibility, or other important criteria. Two examples of selection aids help to illustrate the application of this method. See Internet site 39 for more information about one set of software packages.

When designing a machine element or a structure, both strength and stiffness must be considered and performance requirements in each area must be met. Some designs are *strength limited (e.g., yield strength* or *tensile strength)* while others are *stiffness limited (e.g., tensile modulus of elasticity* or *shear modulus of elasticity)*. In addition, most design projects seek minimal or optimal weight and mass because performance of the overall system tends to be improved and there is often a strong relationship between mass and cost. For these reasons, designers should look for materials that have both high strength and low mass and also high stiffness and low mass. The material's mass is represented by density.

A good approach to meeting this design goal is to consider the *strength to density ratio* and the *stiffness to density ratio* of candidate materials. The software mentioned in Internet site 39 allows the determination of both ratios for virtually any material in its database and the graphic display of *strength versus density* and *stiffness versus density* charts. Figures 2–31 and 2–32 show the basic concept with general orders of magnitude indicated for comparison. The better performing materials would tend to have both high strength and low density but trade-offs are inevitable. The shaded areas show the approximate range of data for materials within a given class or type of material.

This is but a small example of the capabilities of industry-oriented material selection software packages. Additional searches and graphic charts can display almost any combination of properties such as:

- Cost of the material and its processing
- Other mechanical properties: elongation, hardness, fatigue strength, and fracture toughness
- Thermal properties: conductivity, specific heat, thermal expansion, and maximum service temperature
- Optical properties: transparent or opaque?
- Corrosion and related properties: resistance to acids, alkalis, freshwater, seawater, organic solvents, and ultraviolet light (UV); flammability; and oxidation
- Electrical properties: conductor or insulator?

Extensive tree-structures of processes appropriate to each material, within the families of *shaping, joining,* and *surfacing/internal structure modification*, are also part of the databases. Examples are:

- ***Shaping***: Casting, machining, molding, rapid prototyping, and others
- ***Joining***: Welding, fastening, adhesives, soldering, brazing, and others
- ***Surfacing/internal structure modification***: Heat-treating, induction hardening, flame hardening, carburizing, coating, anodizing, polishing, painting, plating, and others

Specialized searches and analyses: In addition to general searches, special modules of some software help to focus

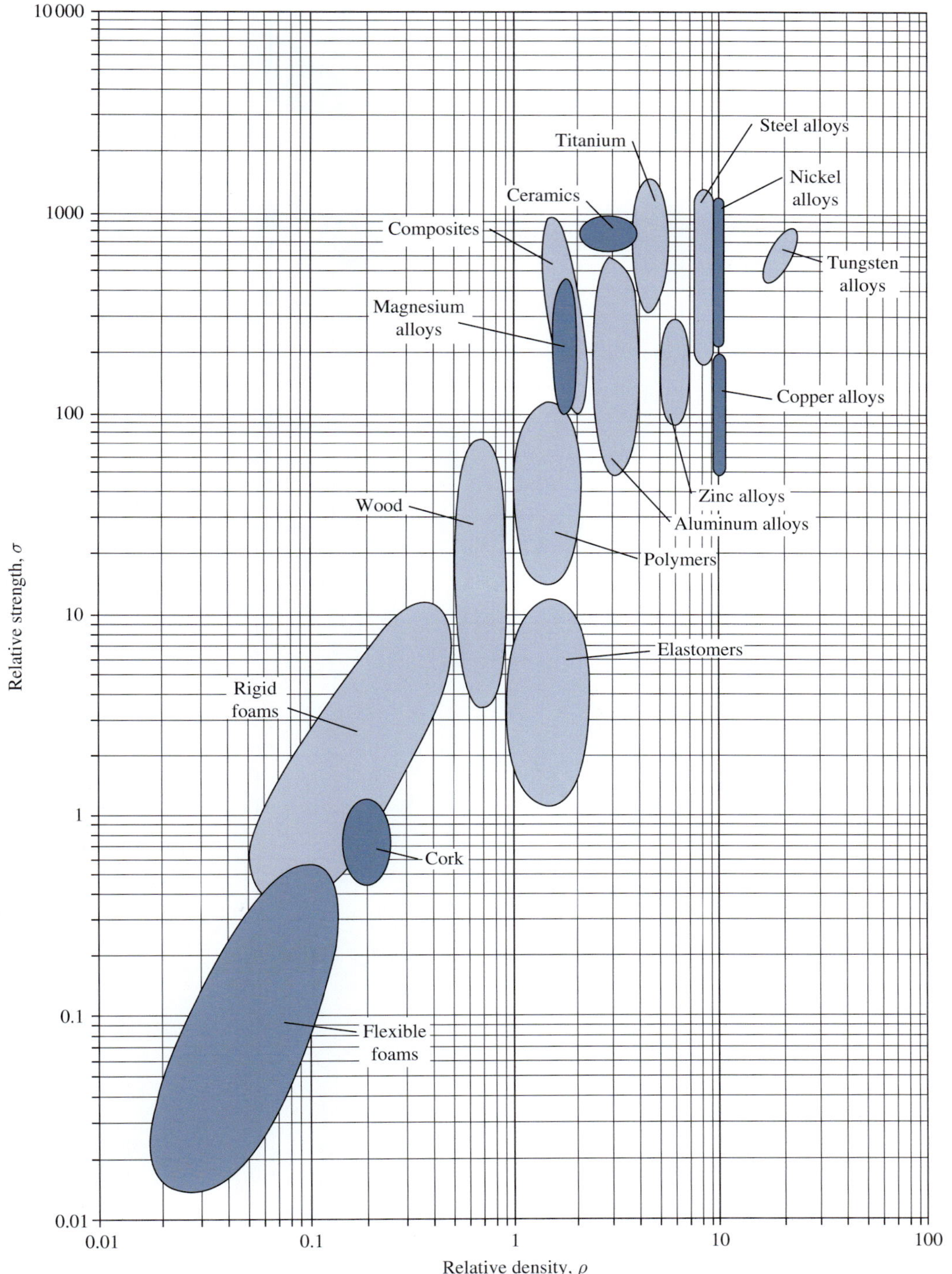

FIGURE 2–31 Relative Strength versus Relative Density

on certain attributes and applications of materials. Among them are:

- Materials and the Environment—Eco Audits (See Reference 5 and Internet site 39.)
  - Energy required to produce a material, energy to transport a product made from a particular material, and energy needed during use
  - Emissions, disposal issues
  - Product life cycle cost
- Aerospace materials
- Materials for medical devices
- Metals and alloys
- Polymeric materials

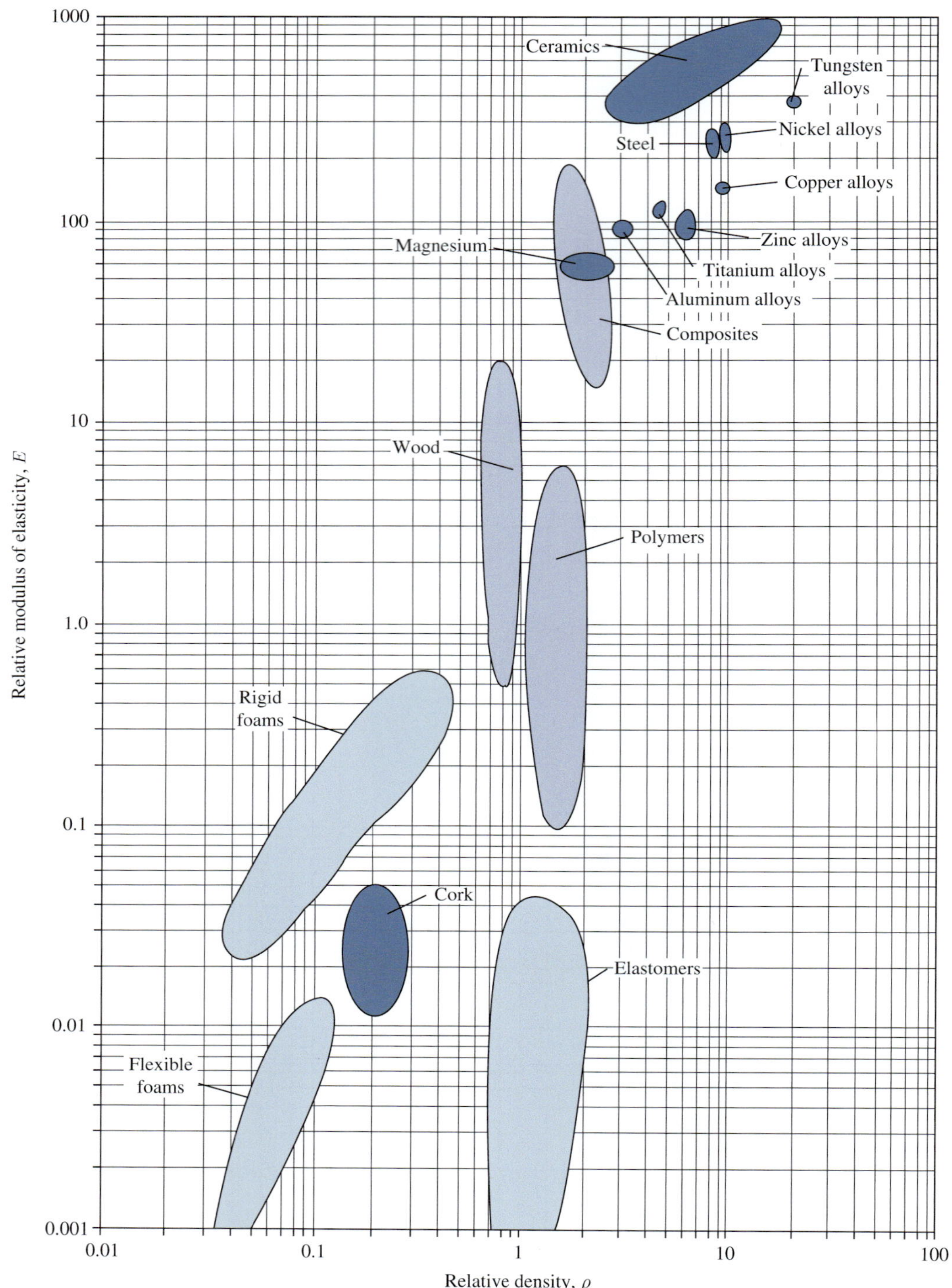

FIGURE 2–32 Relative Modulus of Elasticity versus Relative Density

REFERENCES

1. Aluminum Association. *Aluminum Design Manual, 2010.* Washington, DC: The Aluminum Association, 2010. Includes specifications for aluminum structures, design guide, material properties, section properties, design aids, and illustrative design examples.
2. Aluminum Association. *Aluminum Standards and Data, 2009.* Washington, DC: The Aluminum Association, 2009. Includes data on mechanical and physical properties of commercial aluminum alloys. Available in both U.S. Customary and Metric SI units.
3. Ashby, Michael F. *Materials Selection in Mechanical Design*. 4th ed. Oxford, England: Butterworth-Heinemann, 2010.
4. Ashby, Michael F. *Materials: Engineering Science, Processing, and Design*. 2nd ed. Oxford, England: Butterworth-Heinemann, 2010.

5. Ashby, Michael F. *Materials: Materials and the Environment: Eco-informed Material Choice.* Oxford, England: Butterworth-Heinemann, 2009.
6. ASM International. *ASM Handbook. Vol. 1, Properties and Selection: Iron, Steels, and High-Performance Alloys* (1990). *Vol. 2, Properties and Selection: Nonferrous Alloys and Special-Purpose Materials* (1990). *Vol. 3, Alloy Phase Diagrams* (1992). *Vol. 4, Heat Treating* (1991). *Vol. 7, Powder Metal Technologies and Applications* (1998). *Vol. 11, Failure Analysis and Prevention* (2002). *Vol. 19, Fatigue and Fracture* (2004). *Vol. 20, Materials Selection and Design* (1997). *Vol. 21, Composites* (2001). Materials Park, OH: ASM International.
7. ASM International. *Mechanical Wear Fundamentals and Testing.* 2nd ed. Materials Park, OH: ASM International, 2004.
8. ASM International. *ASM Specialty Handbook: Carbon and Alloy Steels.* Edited by J. R. Davis. Materials Park, OH: ASM International, 1996.
9. ASM International. *ASM Specialty Handbook: Stainless Steels.* Edited by J. R. Davis. Materials Park, OH: ASM International, 1994.
10. ASM International. *ASM Specialty Handbook: Tool Materials.* Edited by J. R. Davis. Materials Park, OH: ASM International, 1995.
11. ASM International. *ASM Specialty Handbook: Heat-Resistant Materials.* Edited by J. R. Davis. Materials Park, OH: ASM International, 1997.
12. ASM International. *ASM Specialty Handbook: Aluminum and Aluminum Alloys.* Edited by J. R. Davis. Materials Park, OH: ASM International, 1993.
13. ASM International. *ASM Specialty Handbook: Cast Irons.* Edited by J. R. Davis. Materials Park, OH: ASM International, 1996.
14. ASM International. *ASM Engineered Materials Handbook: Engineering Plastics.* Materials Park, OH: ASM International, 1988.
15. ASM International. *Atlas of Fatigue Curves.* Edited by H. E. Boyer. Materials Park, OH: ASM International, 1986.
16. ASM International. *Surface Hardening of Steels.* Edited by J. R. Davis. Materials Park, OH: ASM International, 2002.
17. ASM International. *Practical Induction Heat Treating.* Materials Park, OH: ASM International, 2001.
18. ASM International. *Heat Treatment of Gears.* Materials Park, OH: ASM International, 2000.
19. ASM International. *Materials Science of Polymers for Engineers.* 2nd ed. Materials Park, OH: ASM International, 2003.
20. ASM International. *Advanced Structural Materials.* Materials Park, OH: ASM International, 2010. Reviews of recently developed classes of materials and their applications.
21. ASM International. *Nanomaterials Handbook.* Edited by Yury Gogotsi. Materials Park, OH: ASM International, 2010. Overview of nanomaterials field and their applications to ceramics, metals, polymers, and biomaterials.
22. ASM International. *Biomaterials Fabrication and Processing Handbook.* Materials Park, OH: ASM International, 2010.
23. ASTM International. *Standard Practice for Numbering Metals and Alloys in the Unified Numbering System (UNS).* West Conshohocken, PA: ASTM International Standard, DOI: 10.1520/E0527-07, www.astm.org, 2007.
24. Bethlehem Steel Corporation. *Modern Steels and Their Properties.* Bethlehem, PA: Bethlehem Steel Corporation, 1980.
25. Budinski, Kenneth G. *Engineering Materials: Properties and Selection.* 9th ed. Upper Saddle River, NJ: Prentice Hall, 2010.
26. Budinski, Kenneth G. *Surface Engineering for Wear Resistance.* Upper Saddle River, NJ: Prentice Hall, 1988.
27. Callister, William D. *Materials Science and Engineering: An Introduction.* 7th ed. New York: Wiley Higher Education, 2007.
28. Davis, Joseph R. *Surface Engineering for Corrosion and Wear Resistance.* London, England: Maney Publishing, 2001.
29. DuPont Engineering Polymers. *Design Guide—Module I: General Design Principles for DuPont Engineering Polymers.* Wilmington, DE: E. I. du Pont de Nemours and Company, 2000.
30. INTERZINC. *Zinc Casting: A Systems Approach.* Algonac, MI: INTERZINC.
31. Lesko, J. *Industrial Design Materials and Manufacturing.* 2nd ed. New York: John Wiley & Sons, 2008.
32. Shackelford, James F., William Alexander, and Jun S. Park. *CRC Practical Handbook of Materials Selection.* Boca Raton, FL: CRC Press, 1995.
33. Strong, A. Brent. *Plastics: Materials and Processing.* 3rd ed. Upper Saddle River, NJ: Pearson/Prentice Hall, 2006.
34. Strong, A. Brent. *Fundamentals of Composites Manufacturing: Materials, Methods, and Applications.* 2nd ed. Dearborn, MI: Society of Manufacturing Engineers, 2008.
35. Barbero, Ever J. *Introduction to Composite Materials Design.* 2nd ed. Boca Raton, FL: Taylor and Francis, CRC Press, 2011.
36. Tsai, Stephen W. *Composites Design.* Palo Alto, CA: Think Composites, Inc. 1988.
37. Keeler, Stuart (editor). *Advanced High Strength Steel Application Guidelines.* Middletown, OH: WorldAutoSteel, 2011.
38. Oberg, Eric, H. L. Horton, and H. H. Ryffel. *Meachinery's Handbook,* 28th ed. New York: Industrial Press, 2008.

INTERNET SITES RELATED TO DESIGN PROPERTIES OF MATERIALS

The sites listed here provide additional details about the subject of this chapter, beyond what is included here. Most, but not all, sites are mentioned in the text. You should consider perusing this list to find sites that may help you with questions about materials whenever they arise.

1. AZoM.com (The A to Z of Materials) www.azom.com Materials information resource for the design community. No cost, databases for metals, ceramics, polymers, and composites. Can also search by keyword, application, or industry type.
2. Matweb www.matweb.com Database of material properties for many metals, plastics, ceramics, and other engineering materials.
3. ASM International www.asminternational.org The society for materials engineers and scientists, a worldwide network dedicated to advancing industry, technology, and applications of metals and other materials.
4. SAE International www.sae.org The engineering society for advancing mobility for the benefit of mankind. A resource for

technical information and standards used in designing self-propelled vehicles of all kinds. Site includes access to SAE and AMS (Aerospace Materials Standards) standards on metals, polymers, composites, and many other materials, along with components and subsystems of vehicles such as steering, suspension, engines, and many others.

5. ASTM International www.astm.org Formerly known as the American Society for Testing and Materials. Develops and sells standards for material properties, testing procedures, and numerous other technical standards.
6. Aluminum Association www.aluminum.org The association of the aluminum industry. Provides numerous publications that can be purchased.
7. Alcoa, Inc. www.alcoa.com A producer of aluminum and fabricated products. Website can be searched for properties of specific alloys.
8. Copper Development Association www.copper.org Provides a large database of properties of wrought and cast copper, copper alloys, brasses, and bronzes. Allows searching for appropriate alloys for typical industrial uses based on several performance characteristics. Select the *Resources* tab and then *Properties.*
9. Metal Powder Industries Federation www.mpif.org The international trade association representing the powder metal producers. Standards and publications related to the design and production of products using powder metals.
10. INTERZINC www.interzinc.com A market development and technology transfer group dedicated to increasing awareness of zinc casting alloys. The site contains numerous references to aid designers in using zinc effectively, such as alloy selection, zinc casting processes, design guides, and case histories. The organization sponsors the Zinc Challenge competition for college and university students to design novel products using zinc die-casting processes.
11. Association for Iron & Steel Technology (AIST) www.aist.org An organization that advances the technical development, production, processing, and application of iron and steel.
12. American Iron and Steel Institute www.steel.org AISI serves as the voice of the North American steel industry, promotes the use of steel, and has played a role in the development and application of steels and steelmaking technology.
13. Techstreet www.techstreet.com A part of Thompson Reuters Corporation, it provides engineers, librarians, and technical professionals with efficient, convenient, reliable access to mission-critical industry codes and standards from over 350 sources including AGMA, ASM, ANSI, ASME, ASTM, DOD, DIN, ISO, MIL, SAE, and SME.
14. DuPont Engineering Plastics www2.dupont.com/plastics Information on DuPont plastics and their properties and processing. Database by type of plastic or application. The DuPont publication, *General Design Principles for DuPont Engineering Polymers—Module I*, can be downloaded from this site through the *Products* tab and selecting the *Crastin PBT* material. Other modules of the design guide can be downloaded for the *Zytel/ Minlon, Delron, Rynite PET,* and *Hytrel TPE* materials. The design handbook for the use of *Vespel* polymer for bushings, bearings, and seals can also be downloaded.
15. PolymerPlace.com www.polymerplace.com A web-based portal for information on the development of products and applications from plastics materials and processes.
16. Plastics Technology Online www.ptonline.com The online resource for the Plastics Technology magazine. Information about plastics materials and processes.
17. PLASPEC Global www.ptonline.com/plaspec Affiliated with Plastics Technology Online. Provides a materials selection database, articles, and information about plastics, injection molding, extrusion, blow molding, tooling, and auxiliary equipment.
18. Society of Plastics Engineers www.4spe.org SPE promotes scientific and engineering knowledge and education about plastics and polymers worldwide. Site includes an online Plastics Encyclopedia.
19. Composites World www.gardnerweb.com/composites Online site for *High Performance Magazine* and *Composites Technology Magazine.* Articles on current practices in composites technology with emphasis on engineering, design, and manufacturing solutions.
20. National Center for Advanced Materials Performance (NCAMP) www.niar.wichita.edu/coe/ncamp.asp Part of the National Institute for Aviation Research (NIAR) in cooperation with National Aeronautics and Space Agency (NASA) and the Federal Aviation Administration, NCAMP promotes validation and quality assurance of composite materials to be applied in the commercial and military aviation industry.
21. Center of Excellence for Composites & Advanced Materials (CECAM) www.niar.wichita.edu/coe/cecam.asp An academic team of universities led by Wichita State University and the NIAR (see Internet site 20) to perform research, testing, certification, and technology transfer in composites technology; to promote coordination and cooperation among the FAA, aircraft manufacturers, materials suppliers, and airline companies; and the education of the aircraft manufacturing and maintenance workforces.
22. Center for Advanced Materials in Transport Aircraft Structures (AMTAS) http://depts.washington.edu/amtas An FAA supported center of excellence based at the University of Washington that addresses the engineering and science issues associated with safety, regulation, and product certification of advanced materials and structures. It also establishes engineering standards and develops an advanced materials and structures knowledge base.
23. A&P Technology www.braider.com A producer of precision braided textiles for use in the composites industry, including braided carbon fiber and fiberglass biaxial sleevings, biaxial tapes, unidirectional tapes, and bias braided fabrics.
24. Cytec Engineered Materials (CEM) www.cytecengineered-materials.com A provider of technologically advanced materials for composites used in commercial aerospace, military aircraft, automotive, marine, and high-performance industrial markets. Products include prepreg systems, woven fabrics, tape, roving, tooling, carbon fiber reinforcements, and thermoplastics.
25. Triumph Group, Inc. www.triumphgroup.com Producer of composite systems used for floor panels, employing light-weight cores with fiberglass or carbon graphite skins.
26. General Plastics Manufacturing Company www.general-plastics.com Producer of high-performance *Last-A-Form*® cellular solid polyurethane and polyisocyanurate systems for use in light-weight cores for composite structures, other industrial applications, and graphics products.

27. Steel Founders' Society of America www.sfsa.org/sfsa/sfscast.php A trade association representing steel foundries that promotes the use of cast steel products and disseminates information about steel castings, materials, and casting processes. The site provides general overviews of castings and working with foundries.
28. Special Metals Corporation www.specialmetals.com Producer of nickel-based alloys with brand names INCONEL, INCOLOY, NIMONIC, UDIMET, MONEL, and NILO.
29. Allegheny Ludlum Corporation www.alleghenyludlum.com Producer of nickel-based alloys with brand names AL, ALLCORR, and ALTEMP.
30. Haynes International, Inc. www.haynesintl.com Producer of nickel- and cobalt-based alloys with brand names HASTELLOY, HAYNES, and ULTIMET.
31. Hexcel Corporation www.hexcel.com/products Producer of carbon fibers and matrix products for the composites industry, including prepregs, core materials, fabrics, laminates, sheet molding compounds, and systems for film infusion technology, resin transfer molding, and resin film infusion.
32. IDI Composites International www.idicomposites.com/sheet-mold-compounds.php A producer of sheet molding compounds and bulk molding compounds.
33. CINDAS LLC www.cindasdata.com Developer of databases of materials properties: Aerospace Structural Metals Database (ASMD), Damage Tolerant Design Handbook (DTDH), Structural Alloys Handbook (SAH), Microelectronics Packaging Materials Database (MPMD), and Thermophysical Properties of Matter Database (TPMD).
34. Think Composites, Inc. www.thingcomposites.com A consulting firm specializing in design and analysis of structures made from advanced composites, based on the work documented in Reference 36. Develops and markets software and provides education and training programs in this field. Software products include GENLAM (strength and stiffness of laminates), LAMRANK (ranking of the performance of laminates), MIC-MAC (micro- and macromechanics analysis of composites), and 3DBEAM (design and analysis of three-dimensional beams).
35. Zyvex Performance Materials www.zyvexpro.com Producer of advanced high-performance materials using nanotechnology.
36. Sigma-Aldrich www.sigmaaldrch.com/materials Producer of nanomaterials for the materials development market. Included are carbon nanotubes, dendrimers, nanocomposites, and polymer-clay nanocomposites.
37. Nanophase Technologies Corporation www.nanophase.com Producer of a family of nanomaterial technologies for automotive, electronics, plastics, textiles, exterior coatings, and other industries.
38. Nanosteel Company www.nanosteelco.com Producer of NanoSteel alloys that are applied as coatings or overlays to surfaces to enhance their hardness and resistance to wear, erosion, and corrosion.
39. Granta Material Intelligence www.grantadesign.com Producer and marketer of numerous products and services related to the effective management of materials property data and software to support materials selection in industry and academia. Included is the CES EduPack materials selection software using concepts developed by Professor Michael F. Ashby.
40. WorldAutoSteel www.owrldautosteel.org The automotive group of the World Steel Association advances and communicates steel's unique ability to meet the automotive industry's needs and challenges in a sustainable and environmentally responsible way. Publisher of *Advanced High Strength Steel Application Guidelines*, available for download from the site.

PROBLEMS

1. Define *ultimate tensile strength.*
2. Define *yield point.*
3. Define *yield strength* and tell how it is measured.
4. What types of materials would have a yield point?
5. What is the difference between proportional limit and elastic limit?
6. Define *Hooke's law.*
7. What property of a material is a measure of its stiffness?
8. What property of a material is a measure of its ductility?
9. If a material is reported to have a percent elongation in a 2.00-in gage length of 2%, is it ductile?
10. Define *Poisson's ratio.*
11. If a material has a tensile modulus of elasticity of 114 GPa and a Poisson's ratio of 0.33, what is its modulus of elasticity in shear?
12. A material is reported to have a Brinell hardness of 525. What is its approximate hardness on the Rockwell C scale?
13. A steel is reported to have a Brinell hardness of 450. What is its approximate tensile strength?

For Problems 14–17, describe what is wrong with each statement.

14. "After annealing, the steel bracket had a Brinell hardness of 750."
15. "The hardness of that steel shaft is HRB 120."
16. "The hardness of that bronze casting is HRC 12."
17. "Based on the fact that this aluminum plate has a hardness of HB 150, its approximate tensile strength is 75 ksi."
18. Name two tests used to measure impact energy.
19. What are the principal constituents in steels?
20. What are the principal alloying elements in SAE 4340 steel?
21. How much carbon is in SAE 4340 steel?
22. What is the typical carbon content of a low-carbon steel? Of a medium-carbon steel? Of a high-carbon steel?
23. How much carbon does a bearing steel typically contain?
24. What is the main difference between SAE 1213 steel and SAE 12L13 steel?
25. Name four materials that are commonly used for shafts.
26. Name four materials that are typically used for gears.
27. Describe the properties desirable for the auger blades of a post hole digger, and suggest a suitable material.
28. Appendix 3 lists SAE 5160 OQT 1000. Describe the basic composition of this material, how it was processed, and its properties in relation to other steels listed in that table.
29. If a shovel blade is made from SAE 1040 steel, would you recommend flame hardening to give its edge a surface hardness of HRC 40? Explain.
30. Describe the differences between through-hardening and carburizing.
31. Describe the process of induction hardening.
32. Name 10 steels used for carburizing. What is their approximate carbon content prior to carburizing?
33. What types of stainless steels are nonmagnetic?
34. What is the principal alloying element that gives a stainless steel corrosion resistance?

35. Of what material is a typical wide-flange beam made?
36. With regard to structural steels, what does the term *HSLA* mean? What strengths are available in HSLA steel?
37. Name three types of cast iron.
38. Describe the following cast iron materials according to type, tensile strength, yield strength, ductility, and stiffness:
    ASTM A48, Grade 30
    ASTM A536, Grade 100-70-03
    ASTM A47, Grade 32510
    ASTM A220, Grade 70003
39. Describe the process of making parts from powdered metals.
40. What properties are typical for parts made from Zamak 3 zinc casting alloy?
41. What are the typical uses for Group D tool steels?
42. What does the suffix *O* in aluminum 6061-O represent?
43. What does the suffix *H* in aluminum 3003-H14 represent?
44. What does the suffix *T* in aluminum 6061-T6 represent?
45. Name the aluminum alloy and condition that has the highest strength of those listed in Appendix 9.
46. Which is one of the most versatile aluminum alloys for mechanical and structural uses?
47. Name three typical uses for titanium alloys.
48. What is the principal constituent of bronze?
49. Describe the bronze having the UNS designation C86200.
50. Name two typical uses for bronze in machine design.
51. Describe the difference between thermosetting plastics and thermoplastics.
52. Suggest a suitable plastic material for each of the following uses:
    (a) Gears
    (b) Football helmets
    (c) Transparent shield
    (d) Structural housing
    (e) Pipe
    (f) Wheels
    (g) Electrical switch-gear, structural part
53. Name eight factors over which the designer has control when specifying a composite material.
54. Define the term *composite.*
55. Name four base resins often used for composite materials.
56. Name four types of reinforcement fibers used for composite materials.
57. Name three types of composite materials used for sporting equipment, such as tennis rackets, golf clubs, and skis.
58. Name three types of composite materials used for aircraft and aerospace structures.
59. What base resin and reinforcement are typically used for sheet-molding compound (SMC)?
60. For what applications are sheet-molding compounds used?
61. Describe six forms in which reinforcing fibers are produced.
62. Describe *wet processing* of composite materials.
63. Describe *preimpregnated materials.*
64. Describe the production processing of sheet-molding compounds.
65. Describe *pultrusion*, and list four shapes produced by this process.
66. Describe *filament winding* and four types of products made by this process.
67. Define the term *specific strength* as it is applied to structural materials.
68. Define the term *specific stiffness* as it is applied to structural materials.
69. Discuss the advantages of composite materials relative to metals with regard to specific strength and specific stiffness.
70. Compare the specific strength of SAE 1020 hot-rolled steel with that of SAE 5160 OQT 700 steel, the two aluminum alloys 6061-T6 and 7075-T6, and titanium Ti-6Al-4V.
71. Compare the specific stiffness of SAE 1020 hot-rolled steel with that of SAE 5160 OQT 700 steel, the two aluminum alloys 6061-T6 and 7075-T6, and titanium Ti-6Al-4V.
72. Compare the specific strengths of each of the five composite materials shown in Figure 2–23 with that of SAE 1020 hot-rolled steel.
73. Compare the specific stiffness of each of the five composite materials shown in Figure 2–23 with that of SAE 1020 hot-rolled steel.
74. Describe the general construction of a composite material identified as $[0/+30/-30/90]$.
75. List and discuss six design guidelines for the application of composite materials.
76. Why is it desirable to form a composite material in layers or plies with the angle of orientation of the different plies in different directions?
77. Why is it desirable to form a composite structural element with relatively thin skins of the stronger composite material over a core of light foam?
78. Describe why concurrent engineering and early manufacturing involvement are important when you are designing parts made from composite materials.
79. Describe a PMC material.
80. Describe a MMC material.
81. Describe a CMC material.

Problems 82–90. For composites made with the given matrix material and reinforcement fiber, which of the three classifications of composites mentioned in Questions 79, 80, and 81 is it in?

	Matrix material	Reinforcement material
82.	PEEK	Carbon
83.	SiC	LAS
84.	C	C
85.	Mg	SiC
86.	PPS	C
87.	SiC	CAS
88.	Al	B
89.	PI	glass
90.	epoxy	aramid

91. Recommend a type of composite and materials for a high-performance application such as an aircraft structural component.
92. Recommend a type of composite and materials and a processing method for a high-volume application with low per-part cost.
93. Name the predominant choice of composite type and materials for low-cost applications.

Problems 94–96. For the given specification for a laminate-type composite, describe the layers.

94. $[0/\pm 45/\pm 30]_s$
95. $[90_3/\pm 45/0_2]_s$
96. $[0_4/90_2/\pm 30/0_2/90/\pm 60/0_2]$
97. Describe a nanometer.
98. Describe a CNT.
99. Describe how CNTs are used in a MMNC and what benefits can be expected compared with a MMC.
100. Describe how CNTs are used in a CMNC and what benefits can be expected compared with a CMC.

SUPPLEMENTARY PROBLEMS

1. List the approximate value of Poisson's ratio for the following materials: (a) carbon and alloy steel; (b) aluminum; (c) lead; (d) gray cast iron; (e) concrete; (f) elastomers.
2. Describe and compare the methods for measuring flexural strength and flexural modulus for rigid plastics using the 3-point bending method and the 4-point bending method.
3. List five kinds of wear.
4. Name four types of carbon and alloy steel commonly used for commercially available shapes.
5. Name two types of stainless steel commonly used for commercially available shapes.
6. Name four types of aluminum alloys commonly used for commercially available shapes.
7. Name three U.S. organizations whose names are commonly used in designations for steels used in machine design.
8. Name the U.S. organization chiefly responsible for assigning aluminum alloy designations.
9. A U.S. designer specifies SAE 4140 steel for a machine component. Name the alloy designation if the same component were to be produced in: (a) Germany; (b) the United Kingdom; (c) the European Union; (d) China; (e) Japan.
10. Repeat Problem 9 for SAE 1045 steel.
11. Repeat Problem 9 for SAE 430 stainless steel.
12. Repeat Problem 9, parts a–c, for aluminum alloy 7075.
13. List four liquid media used as quenchants during heat-treating operations.
14. Describe the process of shot peening and the benefits of using it.
15. List four carbon and alloy casting steels used in general applications.
16. List four carbon and alloy casting steels used for parts that must sustain pressurized fluids.
17. Describe the cast material called *CADI* and list four typical applications.
18. Describe the cast material called *white iron* and list four typical applications.
19. Describe the *press and sinter* procedure for producing machine components from powder metallurgy technology.
20. Describe the *isostatic pressing* procedure for producing machine components from powder metallurgy technology.
21. Describe the *metal injection molding* procedure for producing machine components from powder metallurgy technology.
22. Describe the *powder forging* procedure for producing machine components from powder metallurgy technology.
23. List four carbon or alloy powdered steel designations and give their tensile strength rating in the heat-treated condition.
24. List powdered metal designations for the following types of metals and give their tensile strength rating: (a) nickel silver; (b) bronze; (c) copper; (d) aluminum.
25. State the typical limit on part size for powder metal parts?
26. Name at least two aluminum casting alloys in each of the series: 200, 300, 400, 500, 700, and 800.
27. Name three general-purpose aluminum forging alloys.
28. Name the most widely used zinc casting alloy and give its typical tensile strength.
29. Name the three widely used *ZA* zinc alloys and describe the significance of the *A* in the designation name. For each named alloy, give its typical tensile strength.
30. Give two major reasons for specifying nickel-based alloys for a machine component.
31. Specify at least one brass or bronze alloy for the following applications: (a) bearings; (b) gears; (c) screw machine parts; (d) marine hardware; (e) pump housings; (f) aircraft parts.
32. Describe the conditions for copper alloys designated by: (a) *H04*; (b) *H02*; (c) *H01*; d) *H08*.
33. Describe what the temper designation *TD* means.
34. Using Figure 2–18, describe the effect of the degree of cold work between 10% and 40% on the properties of brass alloys.
35. List the six general classifications of materials that designers may specify for machine components.
36. Give three examples of hybrid materials.
37. Using Figure 2–32, rank the following general classifications of materials with regard to their strength: (a) natural materials (wood, rubber); (b) ceramics; (c) polymers; (d) metals; (e) cork; (f) foams; (g) composites; (h) elastomers.
38. Using Figure 2–32, rank the following general classifications of materials with regard to their stiffness as indicated by its Young's modulus: (a) natural materials (wood, rubber); (b) ceramics; (c) polymers; (d) metals; (e) cork; (f) foams; (g) composites; (h) elastomers.
39. Using Figure 2–32, rank the following general classifications of materials with regard to their density: (a) natural materials (wood, rubber); (b) ceramics; (c) polymers; (d) metals; (e) cork; (f) foams; (g) composites; (h) elastomers.
40. Compare the three lists generated for Problems 37, 38, and 39.

INTERNET-BASED ASSIGNMENTS

1. Use the Matweb website to determine at least three appropriate materials for a shaft design. An alloy steel is preferred with a minimum yield strength of 150 ksi (1035 MPa) and a good ductility as represented by an elongation of 10% or greater.
2. Use the Matweb website to determine at least three appropriate plastic materials for use as a cam. The materials should have good strength properties and a high toughness.
3. Use the DuPont Plastics website to determine at least three appropriate plastic materials for use as a cam. The materials should have good strength properties and a high toughness.
4. Use the DuPont Plastics website to determine at least three appropriate plastic materials for use as a housing for an industrial product. Moderate strength, high rigidity, and high toughness are required.
5. Use the Alcoa website to determine at least three appropriate aluminum alloys for a mechanical component that requires moderate strength, good machinability, and good corrosion resistance.
6. Use the INTERZINC website to determine at least three appropriate zinc casting alloys for a structural component that requires good strength and that is recommended for die casting.
7. Use the Copper Development Association website to recommend at least three copper alloys for a wormgear. Good strength and ductility are desirable along with good wear properties.
8. Use the Copper Development Association website to recommend at least three copper alloys for a bearing application. Moderate strength and good friction and wear properties are required.
9. Locate the description of the ASTM Standard A992 structural steel that is commonly used for rolled-steel beam shapes. Determine how to acquire a copy of the standard.

CHAPTER THREE

STRESS AND DEFORMATION ANALYSIS

THE BIG PICTURE — Stress and Deformation Analysis

Discussion Map

- As a designer, you are responsible for ensuring the safety of the components and systems you design.
- You must apply your prior knowledge of the principles of strength of materials.

Discover

How could consumer products and machines fail?

Describe some product failures you have seen.

This chapter presents a brief review of the fundamentals of stress analysis. It will help you design products that do not fail, and it will prepare you for other topics later in this book.

A designer is responsible for ensuring the safety of the components and systems that he or she designs. Many factors affect safety, but one of the most critical aspects of design safety is that the level of stress to which a machine component is subjected must be safe under reasonably foreseeable conditions. This principle implies, of course, that nothing actually breaks. Safety may also be compromised if components are permitted to deflect excessively, even though nothing breaks.

You have already studied the principles of strength of materials to learn the fundamentals of stress analysis. Thus, at this point, you should be competent to analyze load-carrying members for stress and deflection due to direct tensile and compressive loads, direct shear, torsional shear, and bending.

Think, now, about consumer products and machines with which you are familiar, and try to explain how they *could fail.* Of course, we do not expect them to fail, because most such products are well designed. But some do fail. Can you recall any? How did they fail? What were the operating conditions when they failed? What was the material of the components that failed? Can you visualize and describe the kinds of loads that were placed on the components that failed? Were they subjected to bending, tension, compression, shear, or torsion? Could there have been more than one type of stress acting at the same time? Are there evidences of accidental overloads? Should such loads have been anticipated by the designer? Could the failure be due to the manufacture of the product rather than its design?

Talk about product and machine failures with your associates and your instructor. Consider parts of your car, home appliances, lawn maintenance equipment, or equipment where you have worked. If possible, bring failed components to the meetings with your associates, and discuss the components and their failure.

Most of this book emphasizes developing special methods to analyze and design machine elements. These methods are all based on the fundamentals of stress analysis, and it is assumed that you have completed a course in strength of materials. This chapter presents a brief review of the fundamentals. (See References 2–4.)

YOU ARE THE DESIGNER

You are the designer of a utility crane that might be used in an automotive repair facility, in a manufacturing plant, or on a mobile unit such as a truck bed. Its function is to raise heavy loads.

A schematic layout of one possible configuration of the crane is shown in Figure 3–1. It is comprised of four primary load-carrying members, labeled 1, 2, 3, and 4. These members are connected to each other with pin-type joints at A, B, C, D, E, and F. The load is applied to the end of the horizontal boom, member 3. Anchor points for the crane are provided at joints A and B that carry the loads from the crane to a rigid structure. Note that this is a simplified view of the crane showing only the primary structural components and the forces in the plane of the applied load. The crane would also need stabilizing members in the plane perpendicular to the drawing.

You will need to analyze the kinds of forces that are exerted on each of the load-carrying members before you can design them. This calls for the use of the principles of statics in which you should have already gained competence. The following discussion provides a review of some of the key principles you will need in this course.

Your work as a designer proceeds as follows:

1. Analyze the forces that are exerted on each load-carrying member using the principles of statics.

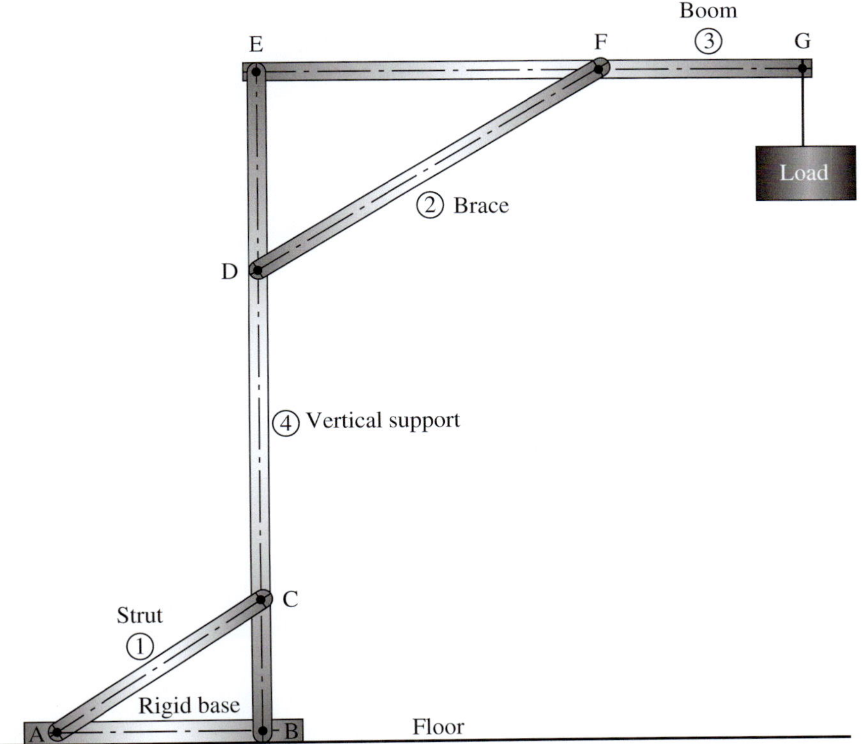

FIGURE 3–1 Schematic layout of a crane

2. Identify the kinds of stresses that each member is subjected to by the applied forces.
3. Propose the general shape of each load-carrying member and the material from which each is to be made.
4. Complete the stress analysis for each member to determine its final dimensions.

Let's work through steps 1 and 2 now as a review of statics. You will improve your ability to do steps 3 and 4 as you perform several practice problems in this chapter and in Chapters 4 and 5 by reviewing strength of materials and adding competencies that build on that foundation.

Force Analysis

One approach to the force analysis is outlined here.

1. Consider the entire crane structure as a free-body with the applied force acting at point G and the reactions acting at support points A and B. See Figure 3–2, which shows these forces and important dimensions of the crane structure.
2. Break the structure apart so that each member is represented as a free-body diagram, showing all forces acting at each joint. See the result in Figure 3–3.
3. Analyze the magnitudes and directions of all forces.

Comments are given here to summarize the methods used in the static analysis and to report results. You should work through the details of the analysis yourself or with colleagues to ensure that you can perform such calculations. All of the forces are directly proportional to the applied force F. We will show the results with an assumed value of $F = 10.0$ kN (approximately 2250 lb).

Step 1: The pin joints at A and B can provide support in any direction. We show the x and y components of the reactions in Figure 3–2. Then, proceed as follows:

1. Sum moments about B to find $R_{Ay} = 2.667\ F = 26.67$ kN
2. Sum forces in the vertical direction to find $R_{By} = 3.667\ F = 36.67$ kN.

At this point we need to recognize that the strut AC is pin-connected at each end and carries loads only at its ends. Therefore, it is a *two-force member,* and the direction of the total force, R_A, acts along the member itself. Then R_{Ay} and R_{Ax} are the rectangular components of R_A as shown in the lower left of Figure 3–2. We can then say that

$$\tan(33.7°) = R_{Ay}/R_{Ax}$$

and then

$$R_{Ax} = R_{Ay}/\tan(33.7°) = 26.67\ \text{kN}/\tan(33.7°) = 40.0\ \text{kN}$$

The total force, R_A, can be computed from the Pythagorean theorem,

$$R_A = \sqrt{R_{Ax}^2 + R_{Ay}^2} = \sqrt{(40.0)^2 + (26.67)^2} = 48.07\ \text{kN}$$

This force acts along the strut AC, at an angle of 33.7° above the horizontal, and it is the force that tends to shear the pin in joint A. The force at C on the strut AC is also 48.07 kN acting upward to the right to balance R_A on the two-force member as shown in Figure 3–3. Member AC is therefore in pure tension.

We can now use the sum of the forces in the horizontal direction on the entire structure to show that $R_{Ax} = R_{Bx} = 40.0$ kN. The resultant of R_{Bx} and R_{By} is 54.3 kN acting at an angle of 42.5°

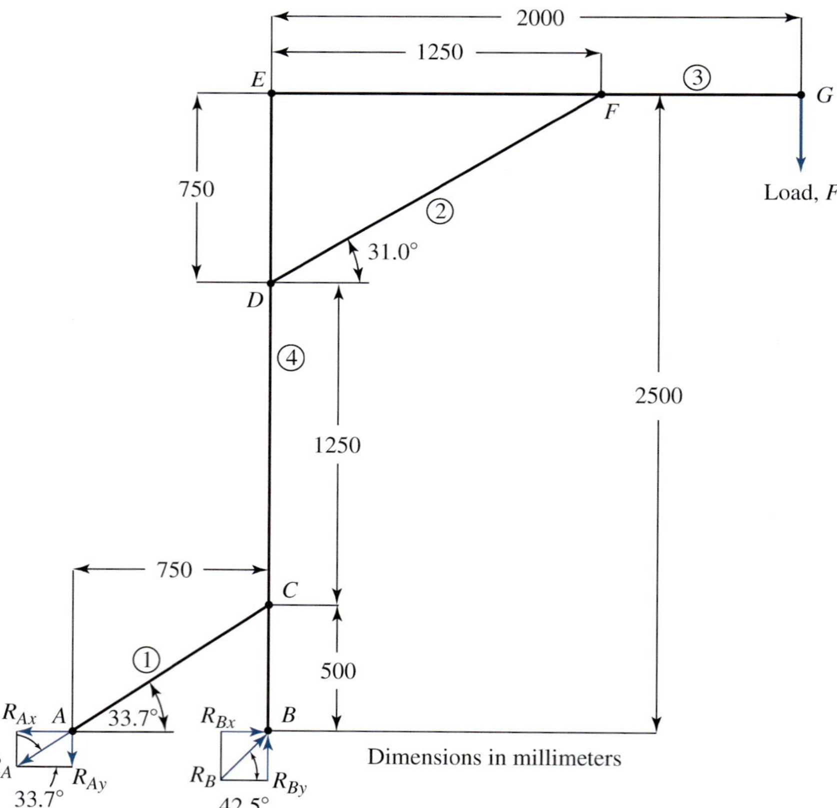

FIGURE 3–2 Free-body diagram of complete crane structure

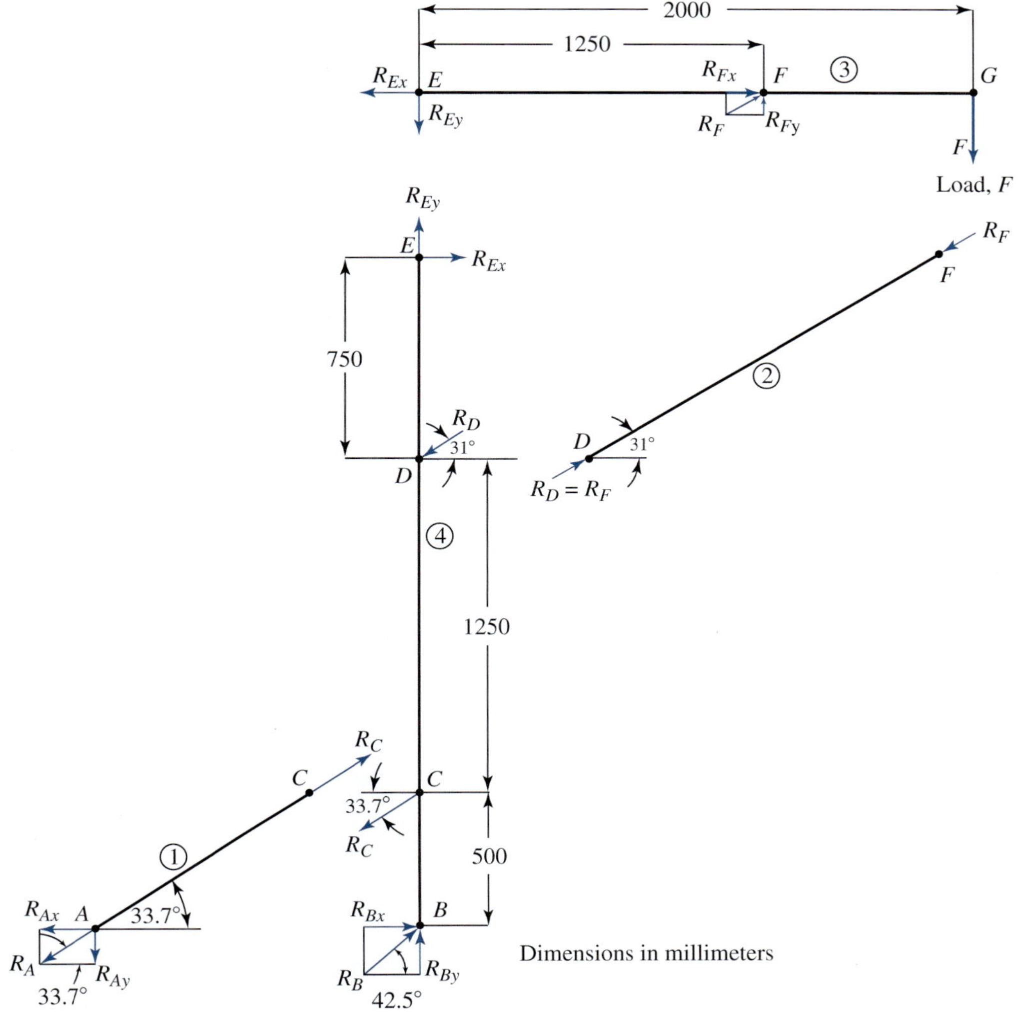

FIGURE 3–3 Free-body diagrams of each component of the crane

above the horizontal, and it is the total shearing force on the pin in joint *B*. See the diagram in the lower right of Figure 3–2.

Step 2: The set of free–body diagrams is shown in Figure 3–3.

Step 3: Now consider the free-body diagrams of all of the members in Figure 3–3. We have already discussed member 1, recognizing it as a two–force member in tension carrying forces R_A and R_C equal to 48.07 kN. The reaction to R_C acts on the vertical member 4.

Now note that member 2 is also a two-force member, but it is in compression rather than tension. Therefore, we know that the forces on points *D* and *F* are equal and that they act in line with member 2, 31.0° with respect to the horizontal. The reactions to these forces, then, act at point *D* on the vertical support, member 4, and at point *F* on the horizontal boom, member 3. We can find the value of R_F by considering the free-body diagram of member 3. You should be able to verify the following results using the methods already demonstrated.

$$R_{Fy} = 1.600\ F = (1.600)(10.0\text{ kN}) = 16.00\text{ kN}$$
$$R_{Fx} = 2.667\ F = (2.667)(10.0\text{ kN}) = 26.67\text{ kN}$$
$$R_F = 3.110\ F = (3.110)(10.0\text{ kN}) = 31.10\text{ kN}$$
$$R_{Ey} = 0.600\ F = (0.600)(10.0\text{ kN}) = 6.00\text{ kN}$$
$$R_{Ex} = 2.667\ F = (2.667)(10.0\text{ kN}) = 26.67\text{ kN}$$
$$R_E = 2.733\ F = (2.733)(10.0\text{ kN}) = 27.33\text{ kN}$$

Now all forces on the vertical member 4 are known from earlier analyses using the principle of action-reaction at each joint.

Types of Stresses on Each Member

Consider again the free-body diagrams in Figure 3–3 to visualize the kinds of stresses that are created in each member. This will lead to the use of particular kinds of stress analysis as the design process is completed. Members 3 and 4 carry forces perpendicular to their long axes and, therefore, they act as beams in bending. Figure 3–4 shows these members with the additional shearing force and bending moment diagrams. You should have learned to prepare such diagrams in the prerequisite study of strength of materials. The following is a summary of the kinds of stresses in each member.

Member 1: The strut is in pure tension.

Member 2: The brace is in pure compression. Column buckling should be checked.

Member 3: The boom acts as a beam in bending. The right end between *F* and *G* is subjected to bending stress and vertical shear stress. Between *E* and *F* there is bending and shear combined with an axial tensile stress.

Member 4: The vertical support experiences a complex set of stresses depending on the segment being considered as described here.

Between *E* and *D*: Combined bending stress, vertical shear stress, and axial tension.

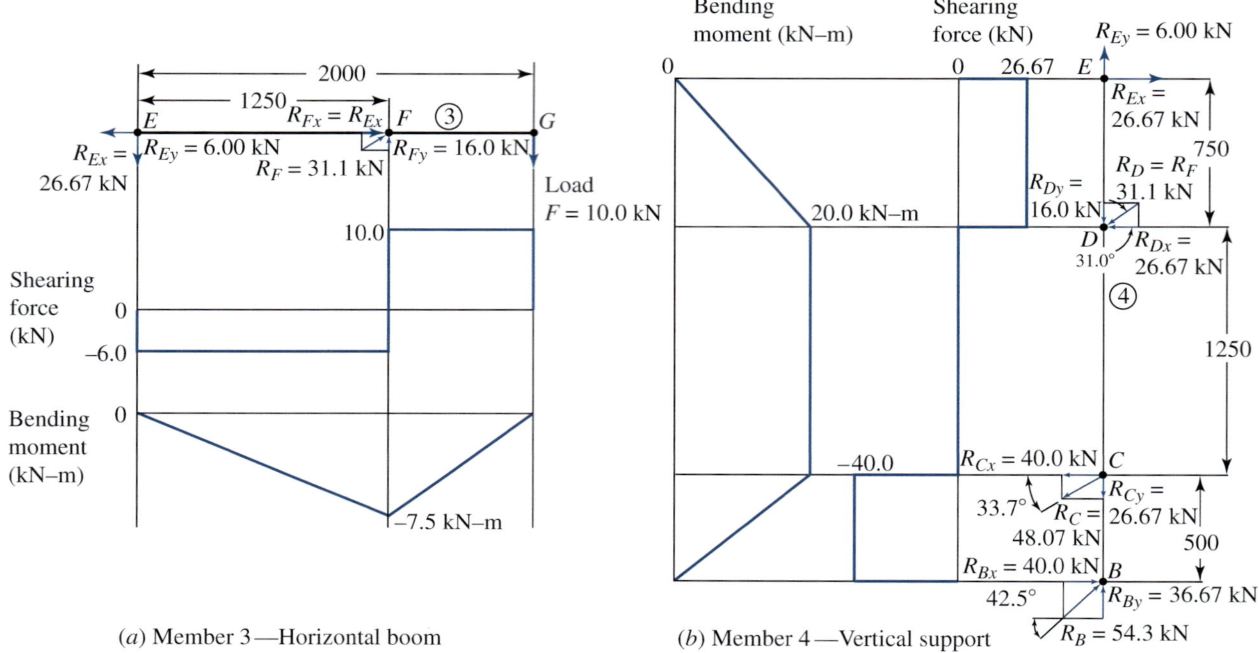

(a) Member 3—Horizontal boom (b) Member 4—Vertical support

FIGURE 3–4 Shearing force and bending moment diagrams for members 3 and 4

Between *D* and *C*: Combined bending stress and axial compression.
Between *C* and *B*: Combined bending stress, vertical shear stress, and axial compression.

Pin Joints: The connections between members at each joint must be designed to resist the total reaction force acting at each, computed in the earlier analysis. In general, each connection will likely include a cylindrical pin connecting two parts. The pin will typically be in direct shear. ■

3–1 OBJECTIVE OF THIS CHAPTER

After completing this chapter, you will:

1. Have reviewed the principles of stress and deformation analysis for several kinds of stresses, including the following:

   Direct tension and compression
   Direct shear
   Torsional shear for both circular and noncircular sections
   Vertical shearing stresses in beams
   Bending

2. Be able to interpret the nature of the stress at a point by drawing the *stress element* at any point in a load-carrying member for a variety of types of loads.
3. Have reviewed the importance of the *flexural center* of a beam cross section with regard to the alignment of loads on beams.
4. Have reviewed beam-deflection formulas.
5. Be able to analyze beam-loading patterns that produce abrupt changes in the magnitude of the bending moment in the beam.
6. Be able to use the principle of superposition to analyze machine elements that are subjected to loading patterns that produce combined stresses.
7. Be able to properly apply stress concentration factors in stress analyses.

3–2 PHILOSOPHY OF A SAFE DESIGN

In this book, every design approach will ensure that the stress level is below yield in ductile materials, automatically ensuring that the part will not break under a static load. For brittle materials, we will ensure that the stress levels are well below the ultimate tensile strength. We will also analyze deflection where it is critical to safety or performance of a part.

Two other failure modes that apply to machine members are fatigue and wear. *Fatigue* is the response of a part subjected to repeated loads (see Chapter 5). *Wear* is discussed within the chapters devoted to the machine elements, such as gears, bearings, and chains, for which it is a major concern.

3–3 REPRESENTING STRESSES ON A STRESS ELEMENT

One major goal of stress analysis is to determine *the point* within a load-carrying member that is subjected to the highest stress level. You should develop the ability to visualize a

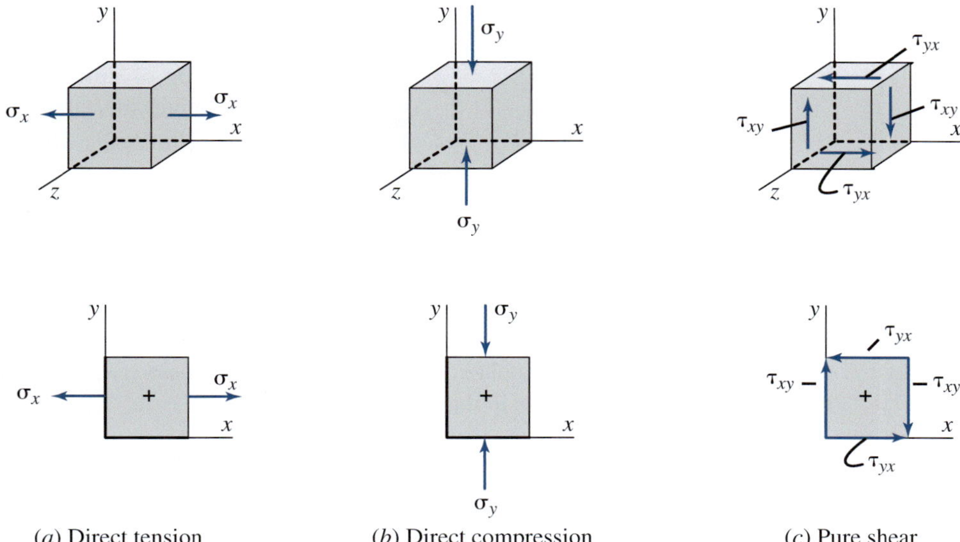

FIGURE 3–5 Stress elements for three types of stresses

stress element, a single, infinitesimally small cube from the member in a highly stressed area, and to show vectors that represent the kind of stresses that exist on that element. The orientation of the stress element is critical, and it must be aligned with specified axes on the member, typically called x, y, and z.

Figure 3–5 shows three examples of stress elements with three basic fundamental kinds of stress: tensile, compressive, and shear. Both the complete three-dimensional cube and the simplified, two-dimensional square forms for the stress elements are shown. The square is one face of the cube in a selected plane. The sides of the square represent the projections of the faces of the cube that are perpendicular to the selected plane. It is recommended that you visualize the cube form first and then represent a square stress element showing stresses on a particular plane of interest in a given problem. In some problems with more general states of stress, two or three square stress elements may be required to depict the complete stress condition.

Tensile and compressive stresses, called *normal stresses*, are shown acting perpendicular to opposite faces of the stress element. Tensile stresses tend to pull on the element, whereas compressive stresses tend to crush it.

Shear stresses are created by direct shear, vertical shear in beams, or torsion. In each case, the action on an element subjected to shear is a tendency to *cut* the element by exerting a stress downward on one face while simultaneously exerting a stress upward on the opposite, parallel face. This action is that of a simple pair of shears or scissors. But note that if only one pair of shear stresses acts on a stress element, it will not be in equilibrium. Rather, it will tend to spin because the pair of shear stresses forms a couple. To produce equilibrium, a second pair of shear stresses on the other two faces of the element must exist, acting in a direction that opposes the first pair.

In summary, shear stresses on an element will always be shown as two pairs of equal stresses acting on (parallel to) the four sides of the element. Figure 3–5(c) shows an example.

Sign Convention for Shear Stresses

This book adopts the following convention:

> ***Positive shear stresses tend to rotate the element in a clockwise direction.***
>
> ***Negative shear stresses tend to rotate the element in a counterclockwise direction.***

A double subscript notation is used to denote shear stresses in a plane. For example, in Figure 3–5(c), drawn for the x–y plane, the pair of shear stresses, τ_{xy}, indicates a shear stress acting on the element face that is perpendicular to the x-axis and parallel to the y-axis. Then τ_{yx} acts on the face that is perpendicular to the y-axis and parallel to the x-axis. In this example, τ_{xy} is positive and τ_{yx} is negative.

3–4 DIRECT STRESSES: TENSION AND COMPRESSION

Stress can be defined as the internal resistance offered by a unit area of a material to an externally applied load. *Normal stresses* (σ) are either *tensile* (positive) or *compressive* (negative).

For a load-carrying member in which the external load is uniformly distributed across the cross-sectional area of the member, the magnitude of the stress can be calculated from the direct stress formula:

⇨ Direct Tensile or Compressive Stress

$$\sigma = \text{force}/\text{area} = F/A \qquad \textbf{(3–1)}$$

The units for stress are always *force per unit area*, as is evident from Equation (3–1). Common units in the U.S. Customary system and the SI metric system follow.

U.S. Customary Units	*SI Metric Units*
$lb/in^2 = psi$	$N/m^2 = pascal = Pa$
$kips/in^2 = ksi$	$N/mm^2 = megapascal$
Note: $1.0 kip = 1000 lb$	$= 10^6 Pa = MPa$
$1.0 ksi = 1000 psi$	

The conditions on the use of Equation (3–1) are as follows:

1. The load-carrying member must be straight.
2. The line of action of the load must pass through the centroid of the cross section of the member.
3. The member must be of uniform cross section near where the stress is being computed.
4. The material must be homogeneous and isotropic.
5. In the case of compression members, the member must be short to prevent buckling. The conditions under which buckling is expected are discussed in Chapter 6.

Example Problem 3–1 A tensile force of 9500 N is applied to a 12-mm-diameter round bar, as shown in Figure 3–6. Compute the direct tensile stress in the bar.

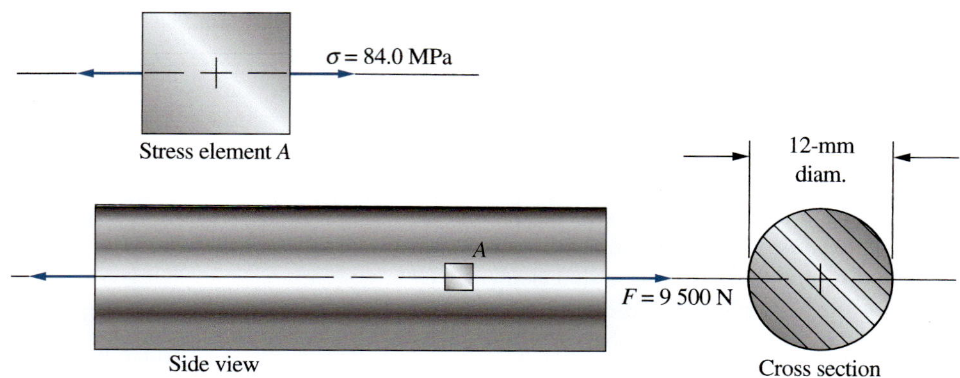

Figure 3–6 Tensile stress in a round bar

Solution

Objective: Compute the tensile stress in the round bar.

Given: Force $= F = 9500$ N; diameter $= D = 12$ mm.

Analysis: Use the direct tensile stress formula, Equation (3–1): $\sigma = F/A$. Compute the cross-sectional area from $A = \pi D^2/4$.

Results:

$$A = \pi D^2/4 = \pi(12\text{ mm})^2/4 = 113\text{ mm}^2$$

$$\sigma = F/A = (9500\text{ N})/(113\text{ mm}^2) = 84.0\text{ N/mm}^2 = 84.0\text{ MPa}$$

Comment: The results are shown on stress element A in Figure 3–6, which can be taken to be anywhere within the bar because, ideally, the stress is uniform on any cross section. The cube form of the element is as shown in Figure 3–5 (a).

Example Problem 3–2 For the round bar subjected to the tensile load shown in Figure 3–6, compute the total deformation if the original length of the bar is 3600 mm. The bar is made from a steel having a modulus of elasticity of 207 GPa.

Solution

Objective: Compute the deformation of the bar.

Given: Force $= F = 9500$ N; diameter $= D = 12$ mm.
Length $= L = 3600$ mm; $E = 207$ GPa

Analysis: From Example Problem 3–1, we found that $\sigma = 84.0$ MPa. Use Equation (3–3).

Results:

$$\delta = \frac{\sigma L}{E} = \frac{(84.0 \times 10^6\text{ N/m}^2)(3600\text{ mm})}{(207 \times 10^9\text{ N/m}^2)} = 1.46\text{ mm}$$

3-5 DEFORMATION UNDER DIRECT AXIAL LOADING

The following formula computes the stretch due to a direct axial tensile load or the shortening due to a direct axial compressive load:

$$\delta = FL/EA \tag{3-2}$$

➪ Deformation Due to Direct Axial Load

where δ = total deformation of the member carrying the axial load

F = direct axial load

L = original total length of the member

E = modulus of elasticity of the material

A = cross-sectional area of the member

Noting that $\sigma = F/A$, we can also compute the deformation from

$$\delta = \sigma L/E \tag{3-3}$$

3-6 DIRECT SHEAR STRESS

Direct shear stress occurs when the applied force tends to cut through the member as scissors or shears do or when a punch and a die are used to punch a slug of material from a sheet. Another important example of direct shear in machine design is the tendency for a key to be sheared off at the section between the shaft and the hub of a machine element when transmitting torque. Figure 3–7 shows the action.

The method of computing direct shear stress is similar to that used for computing direct tensile stress because

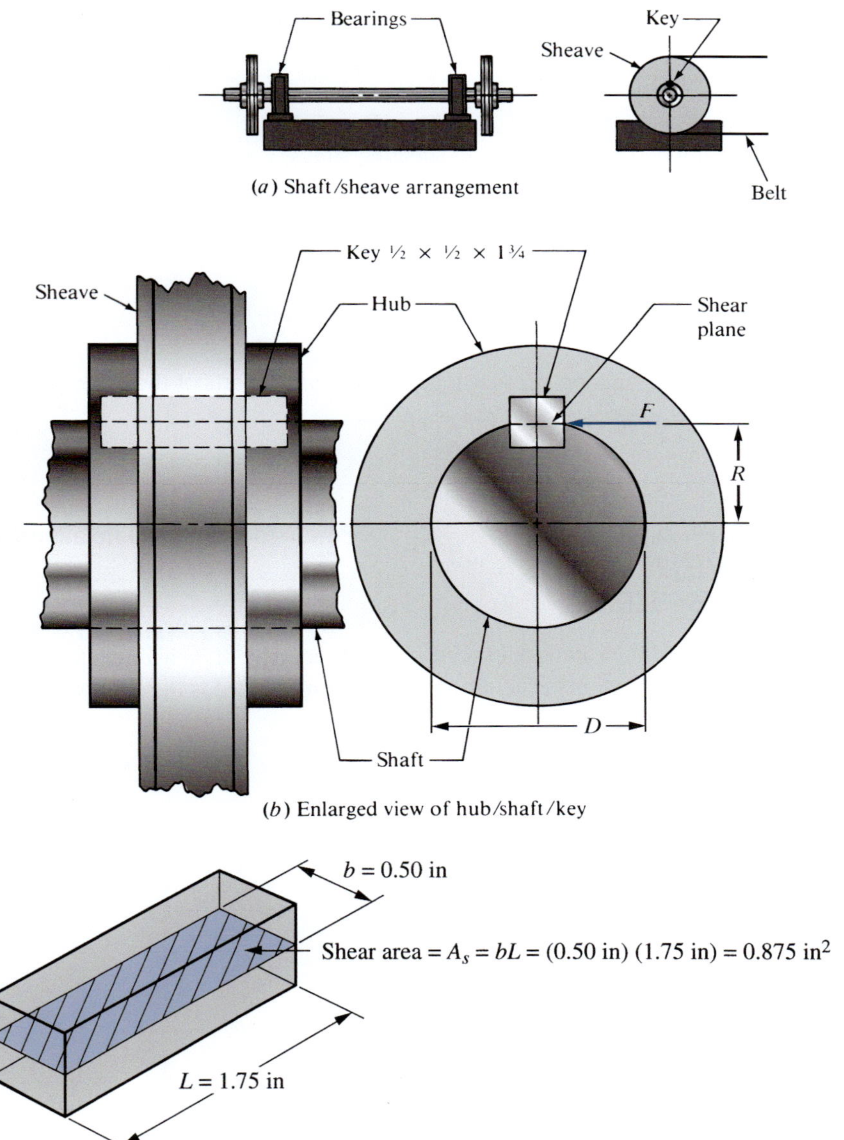

FIGURE 3–7 Direct shear on a key

the applied force is assumed to be uniformly distributed across the cross section of the part that is resisting the force. But the kind of stress is *shear stress* rather than *normal stress*. The symbol used for shear stress is the Greek letter tau (τ). The formula for direct shear stress can thus be written

➩ Direct Shear Stress

$$\tau = \text{shearing force/area in shear} = F/A_s \qquad (3\text{–}4)$$

This stress is more properly called the *average shearing* stress, but we will make the simplifying assumption that the stress is uniformly distributed across the shear area.

Example Problem 3–3 Figure 3–7 shows a shaft carrying two sheaves that are keyed to the shaft. Part (b) shows that a force F is transmitted from the shaft to the hub of the sheave through a square key. The shaft has a diameter of 2.25 in and transmits a torque of 14 063 lb · in. The key has a square cross section, 0.50 in on a side, and a length of 1.75 in. Compute the force on the key and the shear stress caused by this force.

Solution

Objective: Compute the force on the key and the shear stress.

Given: Layout of shaft, key, and hub shown in Figure 3–7.
Torque $= T =$ 14 063 lb · in; key dimensions $= 0.5 \times 0.5 \times 1.75$ in.
Shaft diameter $= D = 2.25$ in; radius $= R = D/2 = 1.125$ in.

Analysis: Torque T = force $F \times$ radius R. Then $F = T/R$.
Use equation (3–4) to compute shearing stress: $\tau = F/A_s$.
Shear area is the cross section of the key at the interface between the shaft and the hub: $A_s = bL$.

Results: $F = T/R = (14\ 063\ \text{lb}\cdot\text{in})/(1.125\ \text{in}) = 12\ 500\ \text{lb}$
$A_s = bL = (0.50\ \text{in})(1.75\ \text{in}) = 0.875\ \text{in}^2$
$\tau = F/A = (12\ 500\ \text{lb})/(0.875\ \text{in}^2) = 14\ 300\ \text{lb/in}^2$

Comment: This level of shearing stress will be uniform on all parts of the cross section of the key.

3–7 RELATIONSHIP AMONG TORQUE, POWER, AND ROTATIONAL SPEED

The relationship among the power *(P)*, the rotational speed *(n)*, and the torque *(T)* in a shaft is described by the equation

➩ Power-Torque-Speed Relationship

$$T = P/n \qquad (3\text{–}5)$$

In SI units, power is expressed in the unit of *watt* (W) or its equivalent, *newton meter per second* (N · m/s), and the rotational speed is in *radians per second* (rad/s).

Example Problem 3–4 Compute the amount of torque in a shaft transmitting 750 W of power while rotating at 183 rad /s. (*Note:* This is equivalent to the output of a 1.0-hp, 4-pole electric motor, operating at its rated speed of 1750 rpm. See Chapter 21.)

Solution

Objective: Compute the torque T in the shaft.

Given: Power $= P = 750\ \text{W} = 750\ \text{N}\cdot\text{m/s}$.
Rotational speed $= n = 183\ \text{rad/s}$.

Analysis: Use Equation (3–5).

Results: $T = P/n = (750\ \text{N}\cdot\text{m/s})/(183\ \text{rad/s})$
$T = 4.10\ \text{N}\cdot\text{m/rad} = 4.10\ \text{N}\cdot\text{m}$

Comments — In such calculations, the unit of N·m/rad is dimensionally correct, and some advocate its use. Most, however, consider the radian to be dimensionless, and thus torque is expressed in N·m or other familiar units of force times distance.

In the U.S. Customary Unit System, power is typically expressed as *horsepower*, equal to 550 ft·lb/s. The typical unit for rotational speed is rpm, or revolutions per minute. But the most convenient unit for torque is the pound-inch (lb·in). Considering all of these quantities and making the necessary conversions of units, we use the following formula to compute the torque (in lb·in) in a shaft carrying a certain power P (in hp) while rotating at a speed of n rpm.

➪ *P-T-n* Relationship for U.S. Customary Units

$$T = 63\,000\,P/n \qquad (3\text{–}6)$$

The resulting torque will be in pound-inches. You should verify the value of the constant, 63 000.

Example Problem 3–5 Compute the torque on a shaft transmitting 1.0 hp while rotating at 1750 rpm. Note that these conditions are approximately the same as those for which the torque was computed in Example Problem 3–4 using SI units.

Solution

Objective — Compute the torque in the shaft.

Given — $P = 1.0$ hp; $n = 1750$ rpm.

Analysis — Use Equation (3–6).

Results — $T = 63\,000\,P/n = [63\,000(1.0)]/1750 = 36.0$ lb·in

3–8 TORSIONAL SHEAR STRESS

When a *torque*, or twisting moment, is applied to a member, it tends to deform by twisting, causing a rotation of one part of the member relative to another. Such twisting causes a shear stress in the member. For a small element of the member, the nature of the stress is the same as that experienced under direct shear stress. However, in *torsional shear*, the distribution of stress is not uniform across the cross section.

The most frequent case of torsional shear in machine design is that of a round circular shaft transmitting power. Chapter 12 covers shaft design.

Torsional Shear Stress Formula

When subjected to a torque, the outer surface of a solid round shaft experiences the greatest shearing strain and therefore the largest torsional shear stress. See Figure 3–8. The value of the maximum torsional shear stress is found from

➪ Maximum Torsional Shear Stress in a Circular Shaft

$$\tau_{max} = Tc/J \qquad (3\text{–}7)$$

where c = radius of the shaft to its outside surface
J = polar moment of inertia

See Appendix 1 for formulas for J.

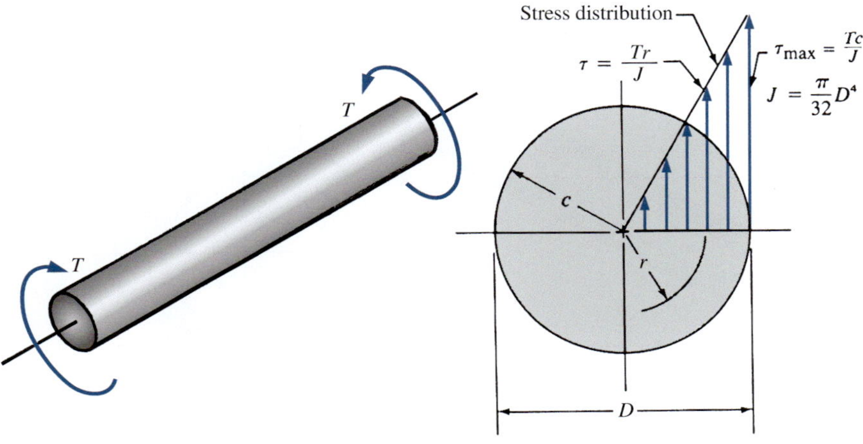

FIGURE 3–8 Stress distribution in a solid shaft

Example Problem 3–6 Compute the maximum torsional shear stress in a shaft having a diameter of 10 mm when it carries a torque of 4.10 N·m.

Solution

Objective: Compute the torsional shear stress in the shaft.

Given: Torque $= T = 4.10$ N·m; shaft diameter $= D = 10$ mm.
$c =$ radius of the shaft $= D/2 = 5.0$ mm.

Analysis: Use Equation (3–7) to compute the torsional shear stress: $\tau_{max} = Tc/J$.
J is the polar moment of inertia for the shaft: $J = \pi D^4/32$ (see Appendix 1).

Results:

$$J = \pi D^4/32 = [(\pi)(10\text{ mm})^4]/32 = 982\text{ mm}^4$$

$$\tau_{max} = \frac{(4.10\text{ N}\cdot\text{m})(5.0\text{ mm})}{982\text{ mm}^4}\frac{10^3\text{ mm}}{\text{m}} = 20.9\text{ N/mm}^2 = 20.9\text{ MPa}$$

Comment: The maximum torsional shear stress occurs at the outside surface of the shaft around its entire circumference.

If it is desired to compute the torsional shear stress at some point inside the shaft, the more general formula is used:

General Formula for Torsional Shear Stress

$$\tau = Tr/J \quad (3\text{–}8)$$

where r = radial distance from the center of the shaft to the point of interest

Figure 3–8 shows graphically that this equation is based on the linear variation of the torsional shear stress from zero at the center of the shaft to the maximum value at the outer surface.

Equations (3–7) and (3–8) apply also to hollow shafts (Figure 3–9 shows the distribution of shear stress). Again note that the maximum shear stress occurs at the outer surface. Also note that the entire cross section carries a relatively high stress level. As a result, the hollow shaft is more efficient. Notice that the material near the center of the solid shaft is not highly stressed.

For design, it is convenient to define the *polar section modulus*, Z_p:

Polar Section Modulus

$$Z_p = J/c \quad (3\text{–}9)$$

Then the equation for the maximum torsional shear stress is

$$\tau_{max} = T/Z_p \quad (3\text{–}10)$$

Formulas for the polar section modulus are also given in Appendix 1. This form of the torsional shear stress equation is useful for design problems because the polar section modulus is the only term related to the geometry of the cross section.

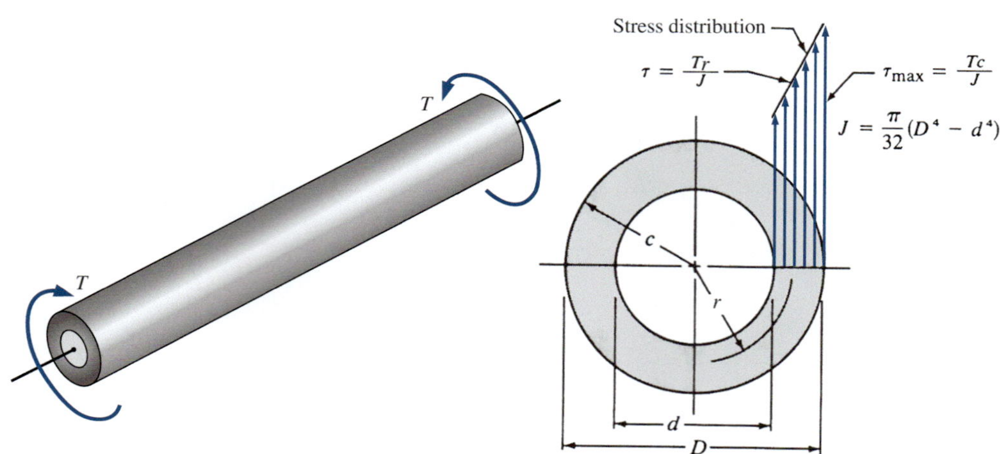

FIGURE 3–9 Stress distribution in a hollow shaft

3–9 TORSIONAL DEFORMATION

When a shaft is subjected to a torque, it undergoes a twisting in which one cross section is rotated relative to other cross sections in the shaft. The angle of twist is computed from

➯ Torsional Deformation

$$\theta = TL/GJ \qquad (3\text{–}11)$$

where θ = angle of twist (radians)
L = length of the shaft over which the angle of twist is being computed
G = modulus of elasticity of the shaft material in *shear*

Example Problem 3–7 Compute the angle of twist of a 10-mm-diameter shaft carrying 4.10 N·m of torque if it is 250 mm long and made of steel with G = 80 GPa. Express the result in both radians and degrees.

Solution

Objective: Compute the angle of twist in the shaft.

Given: Torque = T = 4.10 N·m; length = L = 250 mm.
Shaft diameter = D = 10 mm; G = 80 GPa.

Analysis: Use Equation (3–11). For consistency, let $T = 4.10 \times 10^3$ N·mm and $G = 80 \times 10^3$ N/mm^2. From Example Problem 3–6, J = 982 mm^4.

Results:

$$\theta = \frac{TL}{GJ} = \frac{(4.10 \times 10^3\ \text{N}\cdot\text{mm})(250\ \text{mm})}{(80 \times 10^3\ \text{N/mm}^2)(982\ \text{mm}^4)} = 0.013\ \text{rad}$$

Using π rad = 180°,

$$\theta = (0.013\ \text{rad})(180°/\pi\ \text{rad}) = 0.75°$$

Comment: Over the length of 250 mm, the shaft twists 0.75°.

3–10 TORSION IN MEMBERS HAVING NONCIRCULAR CROSS SECTIONS

The behavior of members having noncircular cross sections when subjected to torsion is radically different from that for members having circular cross sections. However, the factors of most use in machine design are the maximum stress and the total angle of twist for such members. The formulas for these factors can be expressed in similar forms to the formulas used for members of circular cross section (solid and hollow round shafts).

The following two formulas can be used:

➯ Torsional Shear Stress

$$\tau_{max} = T/Q \qquad (3\text{–}12)$$

➯ Deflection for Noncircular Sections

$$\theta = TL/GK \qquad (3\text{–}13)$$

Note that Equations (3–12) and (3–13) are similar to Equations (3–10) and (3–11), with the substitution of Q for Z_p and K for J. Refer to Figure 3–10 for the methods of determining the values for K and Q for several types of cross sections useful in machine design. These values are appropriate only if the ends of the member are free to deform. If either end is fixed, as by welding to a solid structure, the resulting stress and angular twist are quite different. (See References 1–3, 6, and 7.)

Example Problem 3–8 A 2.50-in-diameter shaft carrying a chain sprocket has one end milled in the form of a square to permit the use of a hand crank. The square is 1.75 in on a side. Compute the maximum shear stress on the square part of the shaft when a torque of 15 000 lb·in is applied.

Also, if the length of the square part is 8.00 in, compute the angle of twist over this part. The shaft material is steel with $G = 11.5 \times 10^6$ psi.

Solution

Objective: Compute the maximum shear stress and the angle of twist in the shaft.

Cross-sectional shape	K = for use in $\theta = TL/GK$ Q = for use in $\tau = T/Q$	Colored dot (•) denotes location of τ_{max}
Square 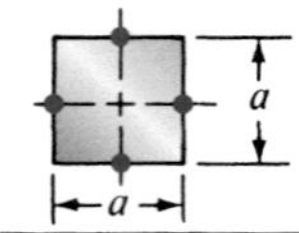	$K = 0.141a^4$ $Q = 0.208a^3$	τ_{max} at midpoint of each side
Rectangle 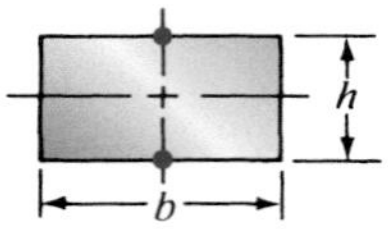	$K = bh^3\left[\frac{1}{3} - 0.21\frac{h}{b}\left(1 - \frac{(h/b)^4}{12}\right)\right]$ $Q = \frac{bh^2}{[3 + 1.8(h/b)]}$	(Approximate; within ≈ 5%) τ_{max} at midpoint of long sides
Triangle (equilateral) 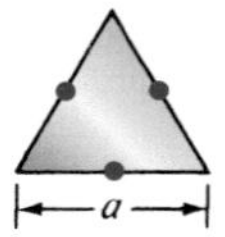	$K = 0.0217a^4$ $Q = 0.050a^3$	
Shaft with One Flat 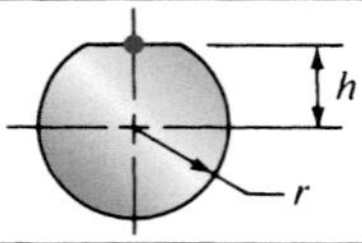	$K = C_1r^4$ $Q = C_2r^3$	
Shaft with Two Flats 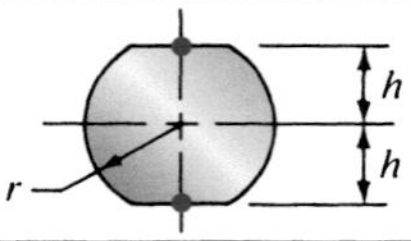	$K = C_3r^4$ $Q = C_4r^3$	
Hollow Rectangle t (uniform)	$K = \frac{2t(a-t)^2(b-t)^2}{(a+b-2t)}$ $Q = 2t(a - t)(b - t)$ Gives average stress; good approximation of maximum stress if t is small—thin-walled tube Inner corners should have generous fillets	
Split Tube 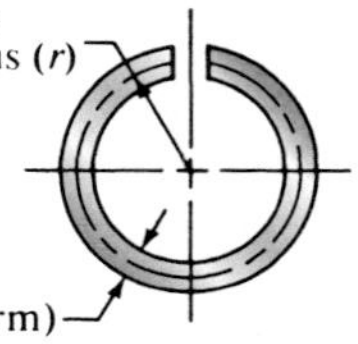 Mean radius (r) t (uniform)	$K = 2\pi rt^3/3$ $Q = \frac{4\pi^2r^2t^2}{(6\pi r + 1.8t)}$ t must be small—thin-walled tube	

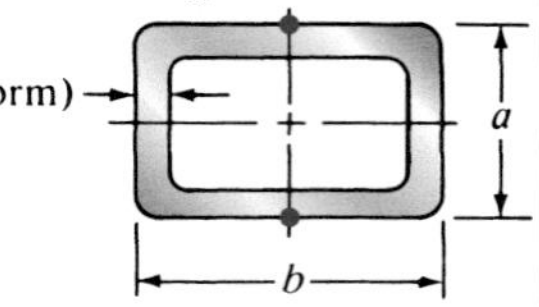

Shaft with One Flat:

h/r	0	0.2	0.4	0.6	0.8	1.0
C_1	0.30	0.51	0.78	1.06	1.37	1.57
C_2	0.35	0.51	0.70	0.92	1.18	1.57

Shaft with Two Flats:

h/r	0.5	0.6	0.7	0.8	0.9	1.0
C_3	0.44	0.67	0.93	1.19	1.39	1.57
C_4	0.47	0.60	0.81	1.02	1.25	1.57

FIGURE 3–10 Methods for determining values for K and Q for several types of cross sections

Given Torque = T = 15000 lb·in; length = L = 8.00 in.
The shaft is square; thus, a = 1.75 in.
$G = 11.5 \times 10^6$ psi.

Analysis Figure 3–10 shows the methods for calculating the values for Q and K for use in Equations (3–12) and (3–13).

Results $Q = 0.208a^3 = (0.208)(1.75\text{ in})^3 = 1.115\text{ in}^3$
$K = 0.141a^4 = (0.141)(1.75\text{ in})^4 = 1.322\text{ in}^4$

Now the stress and the deflection can be computed.

$$\tau_{max} = \frac{T}{Q} = \frac{15\,000\text{ lb}\cdot\text{in}}{(1.115\text{ in}^3)} = 13\,460\text{ psi}$$

$$\theta = \frac{TL}{GK} = \frac{(15\,000\text{ lb}\cdot\text{in})(8.00\text{ in})}{(11.5 \times 10^6\text{ lb/in}^2)(1.322\text{ in}^4)} = 0.0079\text{ rad}$$

Convert the angle of twist to degrees:

$$\theta = (0.0079\text{ rad})(180°/\pi\text{ rad}) = 0.452°$$

Comments Over the length of 8.00 in, the square part of the shaft twists 0.452°. The maximum shear stress is 13 460 psi, and it occurs at the midpoint of each side as shown in Figure 3–10.

3–11 TORSION IN CLOSED, THIN-WALLED TUBES

A general approach for closed, thin-walled tubes of virtually any shape uses Equations (3–12) and (3–13) with special methods of evaluating K and Q. Figure 3–11 shows such a tube having a constant wall thickness. The values of K and Q are

$$K = 4A^2t/U \quad \textbf{(3–14)}$$

$$Q = 2tA \quad \textbf{(3–15)}$$

where A = area enclosed by the median boundary (indicated by the dashed line in Figure 3–11)

t = wall thickness (which must be uniform and thin)

U = length of the median boundary

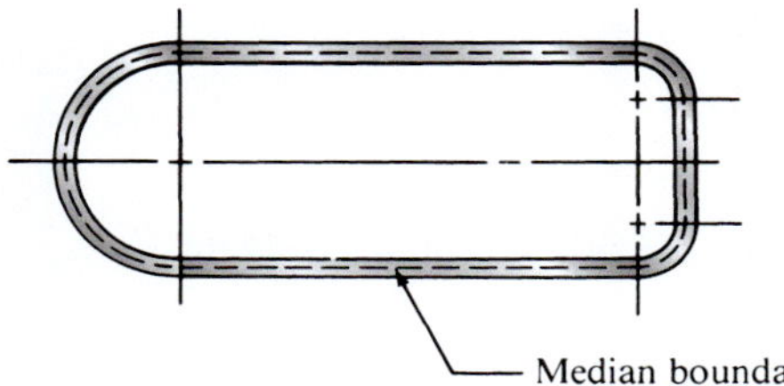

FIGURE 3–11 Closed, thin-walled tube with a constant wall thickness

The shear stress computed by this approach is the *average stress* in the tube wall. However, if the wall thickness t is small (a thin wall), the stress is nearly uniform throughout the wall, and this approach will yield a close approximation of the maximum stress. For the analysis of tubular sections having nonuniform wall thickness, see References 1–3, 6, and 7.

To design a member to resist torsion only, or torsion and bending combined, it is advisable to select hollow tubes, either round or rectangular, or some other closed shape. They possess good efficiency both in bending and in torsion.

3–12 OPEN TUBES AND A COMPARISON WITH CLOSED TUBES

The term *open tube* refers to a shape that appears to be tubular but is not completely closed. For example, some tubing is manufactured by starting with a thin, flat strip of steel that is roll-formed into the desired shape (circular, rectangular, square, and so on). Then the seam is welded along the entire length of the tube. It is interesting to compare the properties of the cross section of such a tube before and after it is welded. The following example problem illustrates the comparison for a particular size of circular tubing.

Example Problem 3–9 Figure 3–12 shows a tube before [Part (b)] and after [Part (a)] the seam is welded. Compare the stiffness and the strength of each shape.

Solution Objective Compare the torsional stiffness and the strength of the closed tube of Figure 3–12(a) with those of the open-seam (split) tube shown in Figure 3–12(b).

Given The tube shapes are shown in Figure 3–12. Both have the same length, diameter, and wall thickness, and both are made from the same material.

Analysis Equation (3–13) gives the angle of twist for a noncircular member and shows that the angle is inversely proportional to the value of K. Similarly, Equation (3–11) shows that the angle of twist for a hollow circular tube is inversely proportional to the polar moment of inertia J. All other terms in

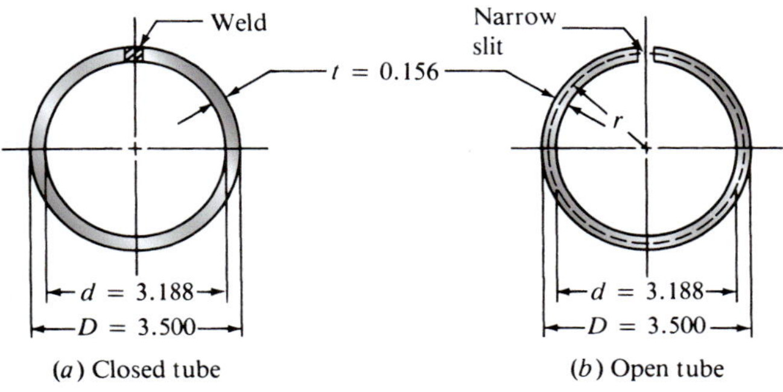

FIGURE 3–12 Comparison of closed and open tubes

the two equations are the same for each design. Therefore, the ratio of θ_{open} to θ_{closed} is equal to the ratio J/K. From Appendix 1, we find

$$J = \pi(D^4 - d^4)/32$$

From Figure 3–10, we find

$$K = 2\pi r t^3/3$$

Using similar logic, Equations (3–12) and (3–10) show that the maximum torsional shear stress is inversely proportional to Q and Z_p for the open and closed tubes, respectively. Then we can compare the strengths of the two forms by computing the ratio Z_p/Q. By Equation (3–9), we find that

$$Z_p = J/c = J/(D/2)$$

The equation for Q for the split tube is listed in Figure 3–10.

Results We make the comparison of torsional stiffness by computing the ratio J/K. For the closed, hollow tube,

$$J = \pi(D^4 - d^4)/32$$
$$J = \pi(3.500^4 - 3.188^4)/32 = 4.592 \text{ in}^4$$

For the open tube before the slit is welded, from Figure 3–10,

$$K = 2\pi r t^3/3$$
$$K = [(2)(\pi)(1.672)(0.156)^3]/3 = 0.0133 \text{ in}^4$$
$$\text{Ratio} = J/K = 4.592/0.0133 = 345$$

Then we make the comparison of the strengths of the two forms by computing the ratio Z_p/Q.

The value of J has already been computed to be 4.592 in^4. Then

$$Z_p = J/c = J/(D/2) = (4.592 \text{ in}^4)/[(3.500 \text{ in})/2] = 2.624 \text{ in}^3$$

For the open tube,

$$Q = \frac{4\pi^2 r^2 t^2}{(6\pi r + 1.8t)} = \frac{4\pi^2(1.672 \text{ in})^2(0.156 \text{ in})^2}{[6\pi(1.672 \text{ in}) + 1.8(0.156 \text{ in})]} = 0.0845 \text{ in}^3$$

Then the strength comparison is

$$\text{Ratio} = Z_p/Q = 2.624/0.0845 = 31.1$$

Comments Thus, for a given applied torque, the slit tube would twist 345 times as much as the closed tube. The stress in the slit tube would be 31.1 times higher than in the closed tube. Also note that if the material for the tube is thin, it will likely buckle at a relatively low stress level, and the tube will collapse suddenly. This comparison shows the dramatic superiority of the closed form of a hollow section to an open form. A similar comparison could be made for shapes other than circular.

3–13 VERTICAL SHEARING STRESS

A beam carrying loads transverse to its axis will experience shearing forces, denoted by V. In the analysis of beams, it is usual to compute the variation in shearing force across the entire length of the beam and to draw the *shearing force diagram*. Then the resulting vertical shearing stress can be computed from

➪ Vertical Shearing Stress in Beams

$$\tau = VQ/It \qquad \textbf{(3–16)}$$

where I = rectangular moment of inertia of the cross section of the beam

t = thickness of the section at the place where the shearing stress is to be computed

Q = *firstmoment*, with respect to the overall centroidal axis, *of the area* of that part of the cross section that lies away from the axis where the shearing stress is to be computed.

To calculate the value of Q, we define it by the following equation,

➪ First Moment of the Area

$$Q = A_p\bar{y} \qquad \textbf{(3–17)}$$

where A_p = that part of the area of the section above the place where the stress is to be computed

$\bar{y}$ = distance from the neutral axis of the section to the centroid of the area A_p

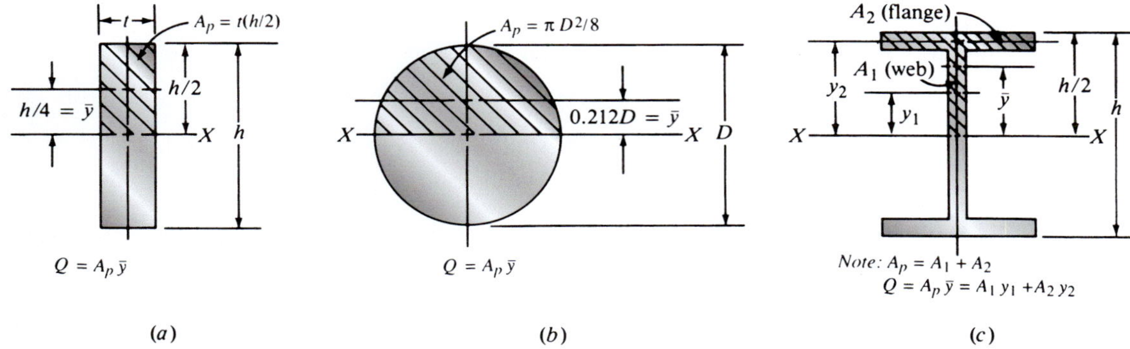

FIGURE 3–13 Illustrations of A_p and $\bar{y}$ used to compute Q for three shapes

In some books or references, and in earlier editions of this book, Q was called the *statical moment*. Here we will use the term *first moment of the area*.

For most section shapes, the maximum vertical shearing stress occurs at the centroidal axis. Specifically, if the thickness is not less at a place away from the centroidal axis, then it is assured that the maximum vertical shearing stress occurs at the centroidal axis.

Figure 3–13 shows three examples of how Q is computed in typical beam cross sections. In each, the maximum vertical shearing stress occurs at the neutral axis.

Example Problem 3–10 Figure 3–14 shows a simply supported beam carrying two concentrated loads. The shearing force diagram is shown, along with the rectangular shape and size of the cross section of the beam. The stress distribution is parabolic, with the maximum stress occurring at the neutral axis. Use Equation (3–16) to compute the maximum shearing stress in the beam.

Solution Objective: Compute the maximum shearing stress τ in the beam in Figure 3–14.

Given: The beam shape is rectangular: $h = 8.00$ in; $t = 2.00$ in.
Maximum shearing force $= V = 1000$ lb at all points between A and B.

Analysis: Use Equation (3–16) to compute τ. V and t are given. From Appendix 1,

$$I = th^3/12$$

The value of the first moment of the area Q can be computed from Equation (3–17). For the rectangular cross section shown in Figure 3–13(a), $A_p = t(h/2)$ and $\bar{y} = h/4$. Then

$$Q = A_p\bar{y} = (th/2)(h/4) = th^2/8$$

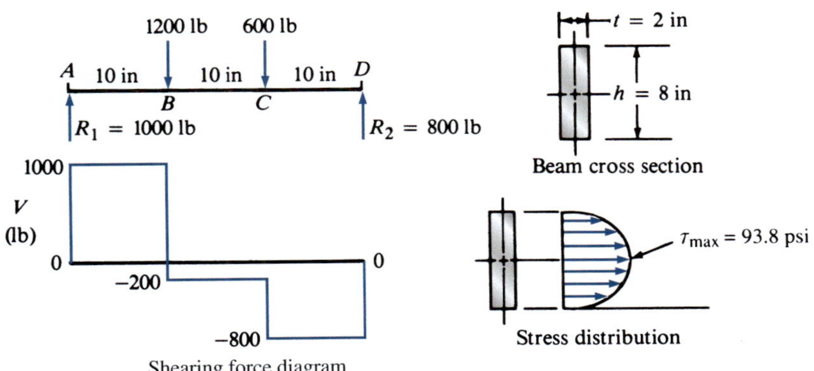

FIGURE 3–14 Shearing force diagram and vertical shearing stress for beam

Results $I = th^3/12 = (2.0\text{ in})(8.0\text{ in})^3/12 = 85.3\text{ in}^4$

$Q = A_p\bar{y} = th^2/8 = (2.0\text{ in})(8.0\text{ in})^2/8 = 16.0\text{ in}^3$

Then the maximum shearing stress is

$$\tau = \frac{VQ}{It} = \frac{(1000\text{ lb})(16.0\text{ in}^3)}{(85.3\text{ in}^4)(2.0\text{ in})} = 93.8\text{ lb/in}^2 = 93.8\text{ psi}$$

Comments The maximum shearing stress of 93.8 psi occurs at the neutral axis of the rectangular section as shown in Figure 3–14. The stress distribution within the cross section is generally parabolic, ending with zero shearing stress at the top and bottom surfaces. This is the nature of the shearing stress everywhere between the left support at *A* and the point of application of the 1200-lb load at *B*. The maximum shearing stress at any other point in the beam is proportional to the magnitude of the vertical shearing force at the point of interest.

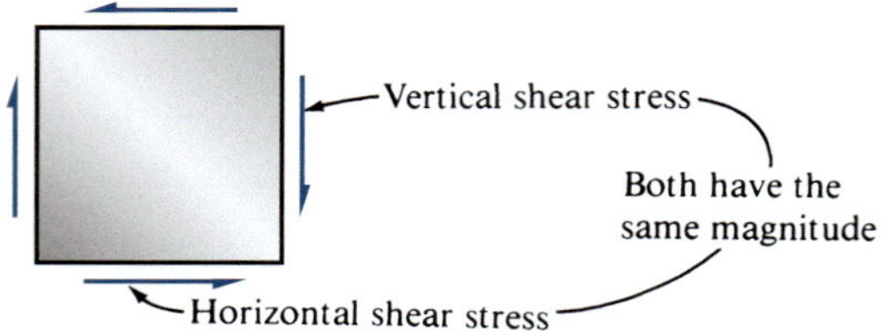

FIGURE 3–15 Shear stresses on an element

Note that the vertical shearing stress is equal to the *horizontal shearing stress* because any element of material subjected to a shear stress on one face must have a shear stress of the same magnitude on the adjacent face for the element to be in equilibrium. Figure 3–15 shows this phenomenon.

In most beams, the magnitude of the vertical shearing stress is quite small compared with the bending stress (see the following section). For this reason, it is frequently not computed at all. Those cases where it is of importance include the following:

1. When the material of the beam has a relatively low shear strength (such as wood).
2. When the bending moment is zero or small (and thus the bending stress is small), for example, at the ends of simply supported beams and for short beams.
3. When the thickness of the section carrying the shearing force is small, as in sections made from rolled sheet, some extruded shapes, and the web of rolled structural shapes such as wide-flange beams.

3–14 SPECIAL SHEARING STRESS FORMULAS

Equation (3–16) can be cumbersome because of the need to evaluate the first moment of the area *Q*. Several commonly used cross sections have special, easy-to-use formulas for the maximum vertical shearing stress:

➪ τ_{max} for Rectangle

$$\tau_{max} = 3V/2A\ \text{(exact)} \qquad \textbf{(3–18)}$$

where A = total cross-sectional area of the beam

➪ τ_{max} for Circle

$$\tau_{max} = 4V/3A\ \text{(exact)} \qquad \textbf{(3–19)}$$

➪ τ_{max} for I-Shape

$$\tau_{max} \simeq V/th\ \text{(approximate: about 15\% low)} \qquad \textbf{(3–20)}$$

where t = web thickness
h = height of the web (e.g., a wide-flange beam)

➪ τ_{max} for Thin-walled Tube

$$\tau_{max} \simeq 2V/A\ \text{(approximate: a little high)} \qquad \textbf{(3–21)}$$

In all of these cases, the maximum shearing stress occurs at the neutral axis.

Example Problem 3–11 Compute the maximum shearing stress in the beam described in Example Problem 3–10 using the special shearing stress formula for a rectangular section.

Solution Objective Compute the maximum shearing stress τ in the beam in Figure 3–14.

Given The data are the same as stated in Example Problem 3–10 and as shown in Figure 3–14.

Analysis Use Equation (3–18) to compute $\tau = 3V/2A$. For the rectangle, $A = th$.

Results $$\tau_{max} = \frac{3V}{2A} = \frac{3(1000\text{ lb})}{2[(2.0\text{ in})(8.0\text{ in})]} = 93.8\text{ psi}$$

Comment This result is the same as that obtained for Example Problem 3–10, as expected.

3–15 STRESS DUE TO BENDING

A *beam* is a member that carries loads transverse to its axis. Such loads produce bending moments in the beam, which result in the development of bending stresses. Bending stresses are *normal stresses*, that is, either tensile or compressive. The maximum bending stress in a beam cross section will occur in the part farthest from the neutral axis of the section. At that point, the *flexure formula* gives the stress:

➩ Flexure Formula for Maximum Bending Stress

$$\sigma = Mc/I \tag{3–22}$$

where M = magnitude of the bending moment at the section
I = moment of inertia of the cross section with respect to its neutral axis
c = distance from the neutral axis to the outermost fiber of the beam cross section

The magnitude of the bending stress varies linearly within the cross section from a value of zero at the neutral axis, to the maximum tensile stress on one side of the neutral axis, and to the maximum compressive stress on the other side. Figure 3–16 shows a typical stress distribution in a beam cross section. Note that the stress distribution is independent of the shape of the cross section.

Note that *positive bending* occurs when the deflected shape of the beam is concave upward, resulting in compression on the upper part of the cross section and tension on the lower part. Conversely, *negative bending* causes the beam to be concave downward.

The flexure formula was developed subject to the following conditions:

1. The beam must be in pure bending. Shearing stresses must be zero or negligible. No axial loads are present.
2. The beam must not twist or be subjected to a torsional load.

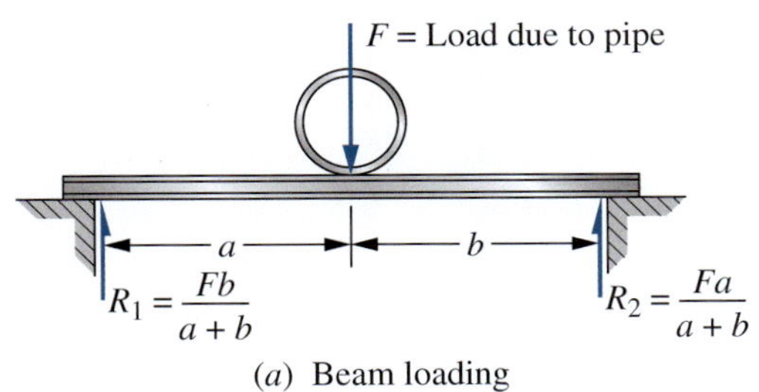

(*a*) Beam loading

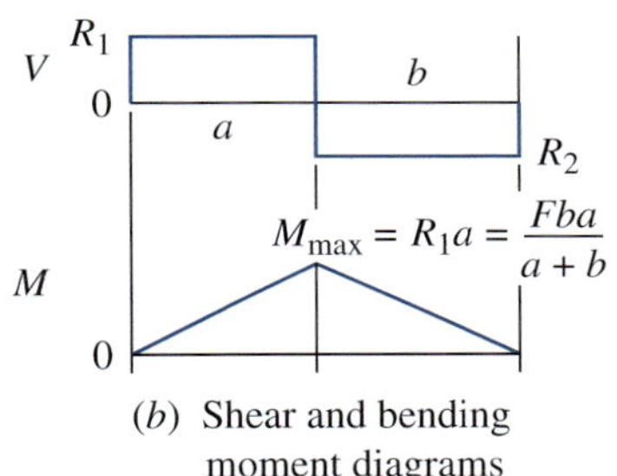

(*b*) Shear and bending moment diagrams

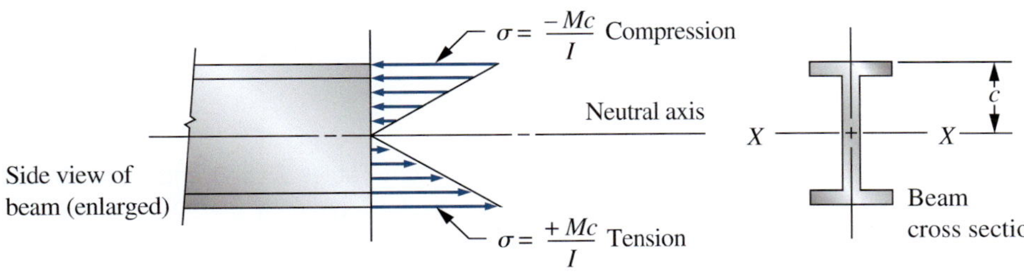

(*c*) Stress distribution on beam section

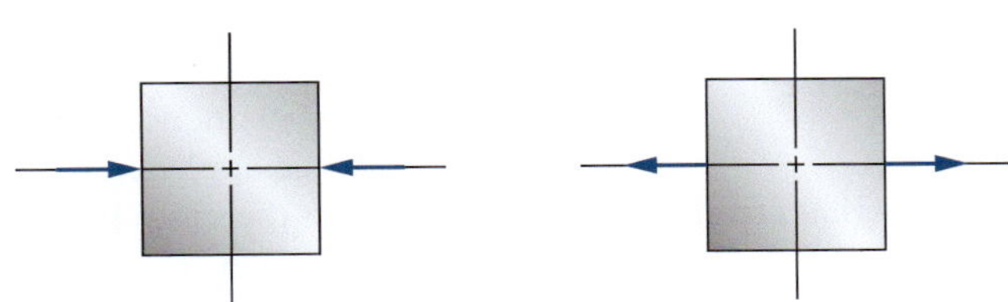

(*d*) Stress element in compression in top part of beam

(*e*) Stress element in tension in bottom part of beam

FIGURE 3–16 Typical bending stress distribution in a beam cross section

3. The material of the beam must obey Hooke's law.
4. The modulus of elasticity of the material must be the same in both tension and compression.
5. The beam is initially straight and has a constant cross section.
6. Any plane cross section of the beam remains plane during bending.
7. No part of the beam shape fails because of local buckling or wrinkling.

If condition 1 is not strictly met, you can continue the analysis by using the method of combined stresses presented in Chapter 4. In most practical beams, which are long relative to their height, shear stresses are sufficiently small as to be negligible. Furthermore, the maximum bending stress occurs at the outermost fibers of the beam section, where the shear stress is in fact zero. A beam with varying cross section, which would violate condition 5, can be analyzed by the use of stress concentration factors discussed later in this chapter.

For design, it is convenient to define the term *section modulus*, S, as

$$S = I/c \qquad (3\text{–}23)$$

The flexure formula then becomes

➯ Flexure Formula

$$\sigma = M/S \qquad (3\text{–}24)$$

Since I and c are geometrical properties of the cross section of the beam, S is also. Then, in design, it is usual to define a design stress, σ_d, and, with the bending moment known, solve for S:

➯ Required Section Modulus

$$S = M/\sigma_d \qquad (3\text{–}25)$$

This results in the required value of the section modulus. From this, the required dimensions of the beam cross section can be determined.

Example Problem 3–12 For the beam shown in Figure 3–16, the load F due to the pipe is 12 000 lb. The distances are $a = 4$ ft and $b = 6$ ft. Determine the required section modulus for the beam to limit the stress due to bending to 30 000 psi, the recommended design stress for a typical structural steel in static bending. Then specify the lightest suitable steel beam.

Solution

Objective: Compute the required section modulus S for the beam in Figure 3–16.

Given: The layout and the loading pattern are shown in Figure 3–16.

Lengths: Overall length $= L = 10$ ft; $a = 4$ ft; $b = 6$ ft.

Load $= F = 12\,000$ lb.

Design stress $= \sigma_d = 30\,000$ psi.

Analysis: Use Equation (3–25) to compute the required section modulus S. Compute the maximum bending moment that occurs at the point of application of the load using the formula shown in Part (b) of Figure 3–16.

Results:

$$M_{max} = R_1\,a = \frac{Fba}{a+b} = \frac{(12\,000\text{ lb})(6\text{ ft})(4\text{ ft})}{(6\text{ ft} + 4\text{ ft})} = 28\,800\text{ lb}\cdot\text{ft}$$

$$S = \frac{M}{\sigma_d} = \frac{28\,800\text{ lb}\cdot\text{ft}}{30\,000\text{ lb/in}^2}\,\frac{12\text{ in}}{\text{ft}} = 11.5\text{ in}^3$$

Comments: A steel beam section can now be selected from Tables A15–9 and A15–10 that has at least this value for the section modulus. The lightest section, typically preferred, is the W8 × 15 wide-flange shape with $S = 11.8\text{ in}^3$.

3–16 FLEXURAL CENTER FOR BEAMS

A beam section must be loaded in a way that ensures symmetrical bending; that is, there must be no tendency for the section to twist under the load. Figure 3–17 shows several shapes that are typically used for beams having a vertical axis of symmetry. If the line of action of the loads on such sections passes through the axis of symmetry, then there is no tendency for the section to twist, and the flexure formula applies.

When there is no vertical axis of symmetry, as with the sections shown in Figure 3–18, care must be exercised in placement of the loads. If the line of action of the loads were shown as F_1 in the figure, the beam would twist and bend, so the flexure formula would not give accurate results for the stress in the section. For such sections, the load must be

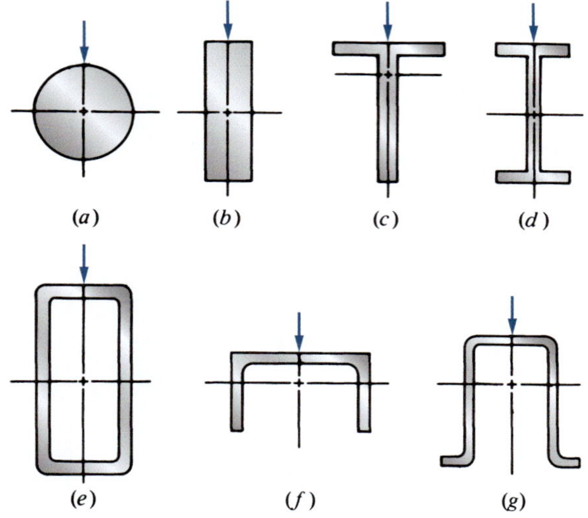

FIGURE 3–17 Symmetrical sections. A load applied through the axis of symmetry results in pure bending in the beam.

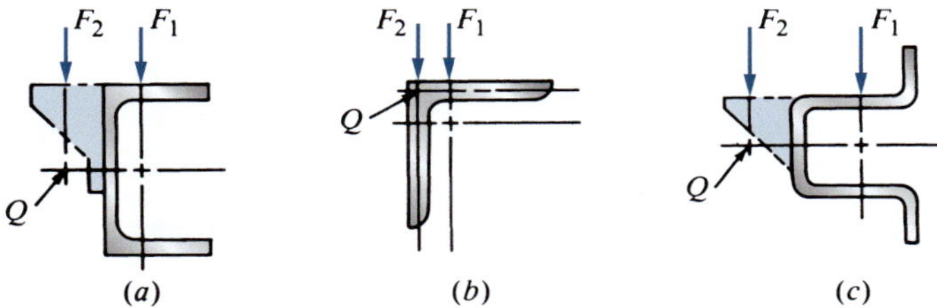

FIGURE 3–18 Nonsymmetrical sections. A load applied as at F_1 would cause twisting; loads applied as at F_2 through the flexural center Q would cause pure bending.

placed in line with the *flexural center,* sometimes called the *shear center.* Figure 3–18 shows the approximate location of the flexural center for these shapes (indicated by the symbol Q). Applying the load in line with Q, as shown with the forces labeled F_2, would result in pure bending. A table of formulas for the location of the flexural center is available (see Reference 7).

3–17 BEAM DEFLECTIONS

The bending loads applied to a beam cause it to deflect in a direction perpendicular to its axis. A beam that was originally straight will deform to a slightly curved shape. In most cases, the critical factor is either the maximum deflection of the beam or its deflection at specific locations.

Consider the double-reduction speed reducer shown in Figure 3–19. The four gears (A, B, C, and D) are mounted on three shafts, each of which is supported by two bearings. The action of the gears in transmitting power creates a set of forces that in turn act on the shafts to cause bending. One component of the total force on the gear teeth acts in a direction that tends to separate the two gears. Thus, gear A is forced upward, while gear B is forced downward. For good gear performance, the net deflection of one gear relative to the other should not exceed 0.005 in (0.13 mm) for medium-sized industrial gearing.

To evaluate the design, there are many methods of computing shaft deflections. We will review briefly those methods using deflection formulas, superposition, and a general analytical approach.

A set of formulas for computing the deflection of beams at any point or at selected points is useful in many practical problems. Appendix 14 includes several cases.

For many additional cases, superposition is useful if the actual loading can be divided into parts that can be computed by available formulas. The deflection for each loading is computed separately, and then the individual deflections are summed at the points of interest.

Many commercially available computer software programs allow the modeling of beams having rather complex loading patterns and varying geometry. The results include reaction forces, shearing force and bending moment diagrams, and deflections at any point. It is important that you understand the principles of beam deflection, studied in strength of materials and reviewed here, so that you can apply such programs accurately and interpret the results carefully.

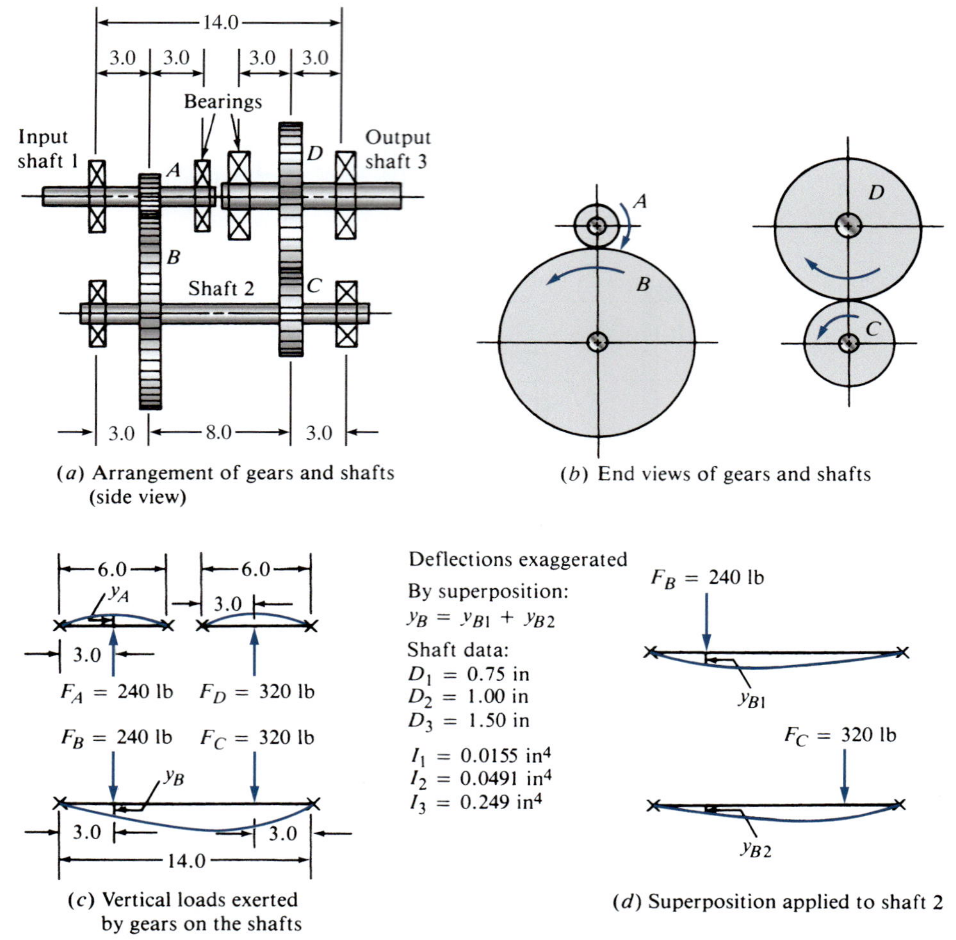

FIGURE 3–19 Shaft deflection analysis for a double-reduction speed reducer

Example Problem 3–13 For the two gears, A and B, in Figure 3–19, compute the relative deflection between them in the plane of the paper that is due to the forces shown in Part (c). These *separating forces*, or *normal forces*, are discussed in Chapters 9 and 10. It is customary to consider the loads at the gears and the reactions at the bearings to be concentrated. The shafts carrying the gears are steel and have uniform diameters as listed in the figure.

Solution

Objective Compute the relative deflection between gears A and B in Figure 3–19.

Given The layout and loading pattern are shown in Figure 3–19. The separating force between gears A and B is 240 lb. Gear A pushes downward on gear B, and the reaction force of gear B pushes upward on gear A. Shaft 1 has a diameter of 0.75 in and a moment of inertia of 0.0155 in^4. Shaft 2 has a diameter of 1.00 in and a moment of inertia of 0.0491 in^4. Both shafts are steel. Use $E = 30 \times 10^6$ psi.

Analysis Use the deflection formulas from Appendix 14 to compute the upward deflection of shaft 1 at gear A and the downward deflection of shaft 2 at gear B. The sum of the two deflections is the total deflection of gear A with respect to gear B.

Case (a) from Table A14–1 applies to shaft 1 because there is a single concentrated force acting at the midpoint of the shaft between the supporting bearings. We will call that deflection y_A.

Shaft 2 is a simply supported beam carrying two nonsymmetrical loads. No single formula from Appendix 14 matches that loading pattern. But we can use superposition to compute the deflection of the shaft at gear B by considering the two forces separately as shown in Part (d) of Figure 3–19. Case (b) from Table A14–1 is used for each load.

We first compute the deflection at B due only to the 240-lb force, calling it y_{B1}. Then we compute the deflection at B due to the 320-lb force, calling it y_{B2}. The total deflection at B is $y_B = y_{B1} + y_{B2}$.

Results The deflection of shaft 1 at gear A is

$$y_A = \frac{F_A L_1^3}{48\,EI} = \frac{(240)(6.0)^3}{48(30 \times 10^6)(0.0155)} = 0.0023\text{ in}$$

The deflection of shaft 2 at B due only to the 240-lb force is

$$y_{B1} = -\frac{F_B a^2 b^2}{3\,EI_2\,L_2} = -\frac{(240)(3.0)^2(11.0)^2}{3(30 \times 10^6)(0.0491)(14)} = -0.0042\text{ in}$$

The deflection of shaft 2 at B due only to the 320-lb force at C is

$$y_{B2} = -\frac{F_c bx}{6\,EI_2\,L_2}(L_2^2 - b^2 - x^2)$$

$$y_{B2} = -\frac{(320)(3.0)(3.0)}{6(30 \times 10^6)(0.0491)(14)}[(14)^2 - (3.0)^2 - (3.0)^2]$$

$$y_{B2} = -0.0041\text{ in}$$

Then the total deflection at gear B is

$$y_B = y_{B1} + y_{B2} = -0.0042 - 0.0041 = -0.0083\text{ in}$$

Because shaft 1 deflects upward and shaft 2 deflects downward, the total relative deflection is the sum of y_A and y_B:

$$y_{total} = y_A + y_B = 0.0023 + 0.0083 = 0.0106\text{ in}$$

Comment This deflection is very large for this application. How could the deflection be reduced?

3–18 EQUATIONS FOR DEFLECTED BEAM SHAPE

The general principles relating the deflection of a beam to the loading on the beam and its manner of support are presented here. The result will be a set of relationships among the load, the vertical shearing force, the bending moment, the slope of the deflected beam shape, and the actual deflection curve for the beam. Figure 3–20 shows diagrams for these five factors, with θ as the slope and y indicating deflection of the beam from its initial straight position. The product of modulus of elasticity and the moment of inertia, EI, for the beam is a measure of its stiffness or resistance to bending deflection. It is convenient to combine EI with the slope and deflection values to maintain a proper relationship, as discussed next.

One fundamental concept for beams in bending is

$$\frac{M}{EI} = \frac{d^2y}{dx^2}$$

where M = bending moment
x = position on the beam measured along its length
y = deflection

Thus, if it is desired to create an equation of the form $y = f(x)$ (i.e., y as a function of x), it would be related to the other factors as follows:

$$y = f(x)$$

$$\theta = \frac{dy}{dx}$$

$$\frac{M}{EI} = \frac{d^2y}{dx^2}$$

$$\frac{V}{EI} = \frac{d^3y}{dx^3}$$

$$\frac{w}{EI} = \frac{d^4y}{dx^4}$$

where w = general term for the load distribution on the beam

The last two equations follow from the observation that there is a derivative (slope) relationship between shear and bending moment and between load and shear.

In practice, the fundamental equations just given are used in reverse. That is, the load distribution as a function of x is known, and the equations for the other factors are derived by successive integrations. The results are

$$w = f(x)$$

$$V = \int w\,dx + V_0$$

$$M = \int V\,dx + M_0$$

where V_0 and M_0 = constants of integration evaluated from the boundary conditions

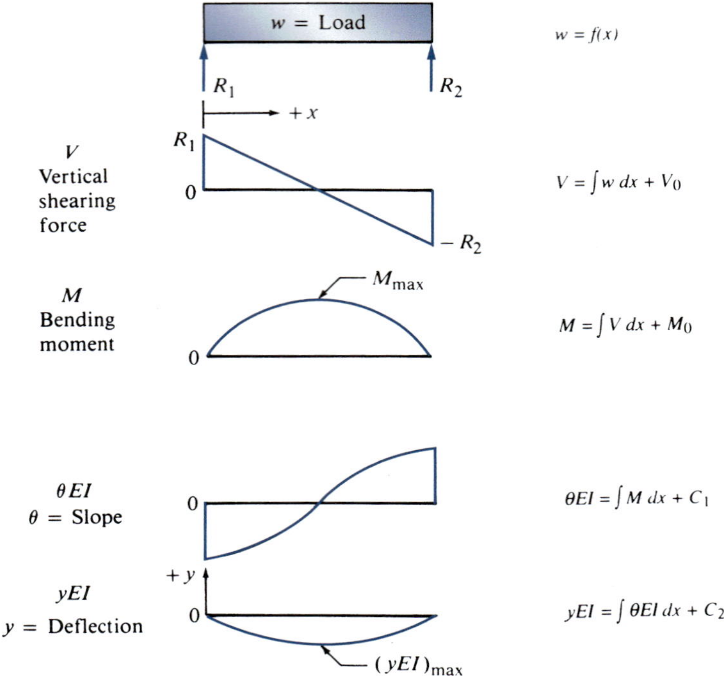

FIGURE 3–20 Relationships of load, vertical shearing force, bending moment, slope of deflected beam shape, and actual deflection curve of a beam

In many cases, the load, shear, and bending moment diagrams can be drawn in the conventional manner, and the equations for shear or bending moment can be created directly by the principles of analytic geometry. With M as a function of x, the slope and deflection relations can be found:

$$\theta EI = \int M\,dx + C_1$$

$$yEI = \int \theta EI\,dx + C_2$$

The constants of integration must be evaluated from boundary conditions. Texts on strength of materials show the details. (See Reference 4.)

3–19 BEAMS WITH CONCENTRATED BENDING MOMENTS

Figures 3–16 and 3–20 show beams loaded only with concentrated forces or distributed loads. For such loading in any combination, the moment diagram is continuous. That is, there are no points of abrupt change in the value of the bending moment. Many machine elements such as cranks, levers, helical gears, and brackets carry loads whose line of action is offset from the centroidal axis of the beam in such a way that a concentrated moment is exerted on the beam.

Figures 3–21, 3–22, and 3–23 show three different examples where concentrated moments are created on machine elements. The bell crank in Figure 3–21 pivots around point O and is used to transfer an applied force to a different line of action. Each arm behaves similar to a cantilever beam, bending with respect to an axis through the pivot. For analysis, we can isolate an arm by making an imaginary cut through the pivot and showing the reaction force at the pivot pin and the internal moment in the arm. The shearing force and

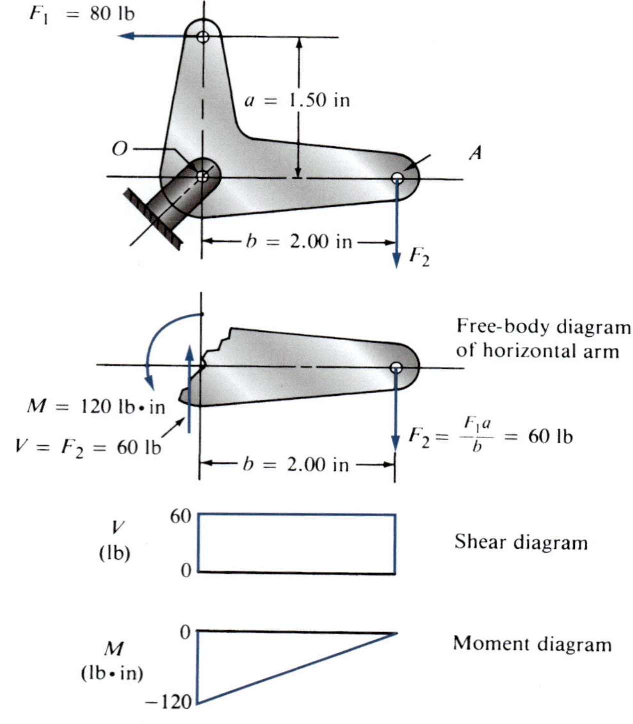

FIGURE 3–21 Bending moment in a bell crank

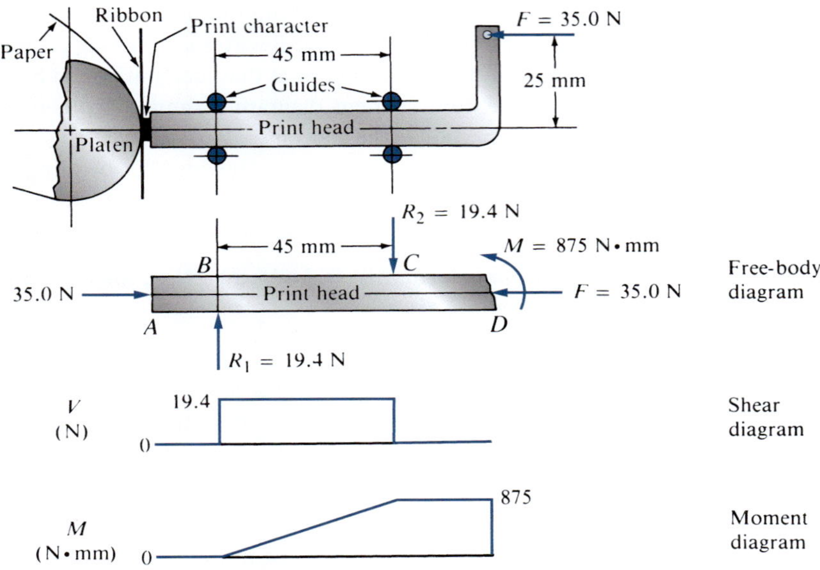

FIGURE 3–22 Bending moment on a print head

bending moment diagrams included in Figure 3–21 show the results, and Example Problem 3–14 gives the details of the analysis. Note the similarity to a cantilever beam with the internal concentrated moment at the pivot reacting to the force, F_2, acting at the end of the arm.

Figure 3–22 shows a print head for an impact-type printer in which the applied force, F, is offset from the neutral axis of the print head itself. Thus the force creates a concentrated bending moment at the right end where the vertical lever arm attaches to the horizontal part. The free-body diagram shows the vertical arm cut off and an internal axial force and moment replacing the effect of the extended arm. The concentrated moment causes the abrupt change in the value of the bending moment at the right end of the arm as shown in the bending moment diagram. Example Problem 3–15 gives the details of the analysis.

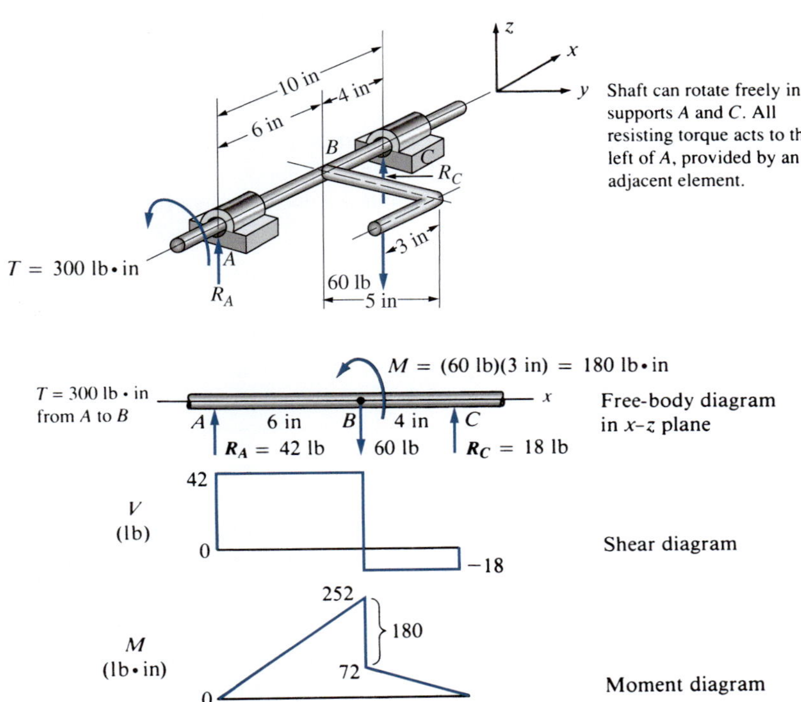

FIGURE 3–23 Bending moment on a shaft carrying a crank

Figure 3–23 shows an isometric view of a crankshaft that is actuated by the vertical force acting at the end of the crank. One result is an applied torque that tends to rotate the shaft *ABC* clockwise about its *x*-axis. The reaction torque is shown acting at the forward end of the crank. A second result is that the vertical force acting at the end of the crank creates a twisting moment in the rod attached at *B* and thus tends to bend the shaft *ABC* in the *x–z* plane. The twisting moment is treated as a concentrated moment acting at *B* with the resulting abrupt change in the bending moment at that location as can be seen in the bending moment diagram. Example Problem 3–16 gives the details of the analysis.

When drawing the bending moment diagram for a member to which a concentrated moment is applied, the following sign convention will be used.

When a concentrated bending moment acts on a beam in a counterclockwise direction, the moment diagram drops; when a clockwise concentrated moment acts, the moment diagram rises.

Example Problem 3–14 The bell crank shown in Figure 3–21 is part of a linkage in which the 80-lb horizontal force is transferred to F_2 acting vertically. The crank can pivot about the pin at *O*. Draw a free-body diagram of the horizontal part of the crank from *O* to *A*. Then draw the shearing force and bending moment diagrams that are necessary to complete the design of the horizontal arm of the crank.

Solution

Objective: Draw the free-body diagram of the horizontal part of the crank in Figure 3–21. Draw the shearing force and bending moment diagrams for that part.

Given: The layout from Figure 3–21.

Analysis: Use the entire crank first as a free body to determine the downward force F_2 that reacts to the applied horizontal force F_1 of 80 lb by summing moments about the pin at *O*.

Then create the free-body diagram for the horizontal part by breaking it through the pin and replacing the removed part with the internal force and moment acting at the break.

Results: We can first find the value of F_2 by summing moments about the pin at *O* using the entire crank:

$$F_1 \cdot a = F_2 \cdot b$$

$$F_2 = F_1(a/b) = 80\text{ lb}(1.50/2.00) = 60\text{ lb}$$

Below the drawing of the complete crank, we have drawn a sketch of the horizontal part, isolating it from the vertical part. The internal force and moment at the cut section are shown. The externally applied downward force F_2 is reacted by the upward reaction at the pin. Also, because F_2 causes a moment with respect to the section at the pin, an internal reaction moment exists, where

$$M = F_2 \cdot b = (60\text{ lb})(2.00\text{ in}) = 120\text{ lb}\cdot\text{in}$$

The shear and moment diagrams can then be shown in the conventional manner. The result looks much like a cantilever that is built into a rigid support. The difference here is that the reaction moment at the section through the pin is developed in the vertical arm of the crank.

Comments: Note that the shape of the moment diagram for the horizontal part shows that the maximum moment occurs at the section through the pin and that the moment decreases linearly as we move out toward point *A*. As a result, the shape of the crank is optimized, having its largest cross section (and section modulus) at the section of highest bending moment. You could complete the design of the crank using the techniques reviewed in Section 3–15.

Example Problem 3–15 Figure 3–22 represents a print head for a computer printer. The force *F* moves the print head toward the left against the ribbon, imprinting the character on the paper that is backed up by the platen. Draw the free-body diagram for the horizontal portion of the print head, along with the shearing force and bending moment diagrams.

Solution

Objective: Draw the free-body diagram of the horizontal part of the print head in Figure 3–22. Draw the shearing force and bending moment diagrams for that part.

Given: The layout from Figure 3–22.

Analysis The horizontal force of 35 N acting to the left is reacted by an equal 35 N horizontal force produced by the platen pushing back to the right on the print head. The guides provide simple supports in the vertical direction. The applied force also produces a moment at the base of the vertical arm where it joins the horizontal part of the print head.

We create the free-body diagram for the horizontal part by breaking it at its right end and replacing the removed part with the internal force and moment acting at the break. The shearing force and bending moment diagrams can then be drawn.

Results The free-body diagram for the horizontal portion is shown below the complete sketch. Note that at the right end (section *D*) of the print head, the vertical arm has been removed and replaced with the internal horizontal force of 35.0 N and a moment of 875 N · mm caused by the 35.0 N force acting 25 mm above it. Also note that the 25 mm-moment arm for the force is taken from the line of action of the force *to the neutral axis of the horizontal part.* The 35.0 N reaction of the platen on the print head tends to place the head in compression over the entire length. The rotational tendency of the moment is reacted by the couple created by R_1 and R_2 acting 45 mm apart at *B* and *C*.

Below the free-body diagram is the vertical shearing force diagram in which a constant shear of 19.4 N occurs only between the two supports.

The bending moment diagram can be derived from either the left end or the right end. If we choose to start at the left end at *A*, there is no shearing force from *A* to *B*, and therefore there is no change in bending moment. From *B* to *C*, the positive shear causes an increase in bending moment from zero to 875 N·mm. Because there is no shear from *C* to *D*, there is no change in bending moment, and the value remains at 875 N · mm. The counterclockwise-directed concentrated moment at *D* causes the moment diagram to drop abruptly, closing the diagram.

Example Problem 3–16 Figure 3–23 shows a crank in which it is necessary to visualize the three-dimensional arrangement. The 60-lb downward force tends to rotate the shaft *ABC* around the *x*-axis. The reaction torque acts only at the end of the shaft outboard of the bearing support at *A*. Bearings *A* and *C* provide simple supports. Draw the complete free-body diagram for the shaft *ABC*, along with the shearing force and bending moment diagrams.

Solution Objective Draw the free-body diagram of the shaft *ABC* in Figure 3–23. Draw the shearing force and bending moment diagrams for that part.

Given The layout from Figure 3–23.

Analysis The analysis will take the following steps:

1. Determine the magnitude of the torque in the shaft between the left end and point *B* where the crank arm is attached.
2. Analyze the connection of the crank at point *B* to determine the force and moment transferred to the shaft *ABC* by the crank.
3. Compute the vertical reactions at supports *A* and *C*.
4. Draw the shearing force and bending moment diagrams considering the concentrated moment applied at point *B*, along with the familiar relationships between shearing force and bending moments.

Results The free-body diagram is shown as viewed looking at the *x–z* plane. Note that the free body must be in equilibrium in all force and moment directions. Considering first the torque (rotating moment) about the *x*-axis, note that the crank force of 60 lb acts 5.0 in from the axis. The torque, then, is

$$T = (60\text{ lb})(5.0\text{ in}) = 300\text{ lb}\cdot\text{in}$$

This level of torque acts from the left end of the shaft to section *B*, where the crank is attached to the shaft.

Now the loading at *B* should be described. One way to do so is to visualize that the crank itself is separated from the shaft and is replaced with a force and moment caused by the crank. First, the downward force of 60 lb pulls down at *B*. Also, because the 60-lb applied force acts 3.0 in to the left of *B*, it causes a concentrated moment in the *x–z plane* of 180 lb · in to be applied at *B*.

Both the downward force and the moment at *B* affect the magnitude and direction of the reaction forces at *A* and *C*. First, summing moments about *A*,

$$(60\ \text{lb})(6.0\ \text{in}) - 180\ \text{lb}\cdot\text{in} - R_C(10.0\ \text{in}) = 0$$

$$R_C = [(360 - 180)\ \text{lb}\cdot\text{in}]/(10.0\ \text{in}) = 18.0\ \text{lb upward}$$

Now, summing moments about C,

$$(60\ \text{lb})(4.0\ \text{in}) + 180\ \text{lb}\cdot\text{in} - R_A(10.0\ \text{in}) = 0$$

$$R_A = [(240 + 180)\ \text{lb}\cdot\text{in}]/(10.0\ \text{in}) = 42.0\ \text{lb upward}$$

Now the shear and bending moment diagrams can be completed. The moment starts at zero at the simple support at *A*, rises to 252 lb · in at *B* under the influence of the 42-lb shear force, then drops by 180 lb · in due to the counterclockwise concentrated moment at *B*, and finally returns to zero at the simple support at *C*.

Comments In summary, shaft *ABC* carries a torque of 300 lb · in from point *B* to its left end. The maximum bending moment of 252 lb · in occurs at point *B* where the crank is attached. The bending moment then suddenly drops to 72 lb · in under the influence of the concentrated moment of 180 lb · in applied by the crank.

3–20 COMBINED NORMAL STRESSES: SUPERPOSITION PRINCIPLE

When the same cross section of a load-carrying member is subjected to both a direct tensile or compressive stress and a stress due to bending, the resulting normal stress can be computed by the method of superposition. The formula is

$$\sigma = \pm Mc/I \pm F/A \qquad (3\text{–}26)$$

where *tensile stresses are positive and compressive stresses are negative.*

An example of a load-carrying member subjected to combined bending and axial tension is shown in Figure 3–24. It shows a beam subjected to a load applied downward and to the right through a bracket below the beam. Resolving the load into horizontal and vertical components shows that its effect can be broken into three parts:

1. The vertical component tends to place the beam in bending with tension on the top and compression on the bottom.
2. The horizontal component, because it acts away from the neutral axis of the beam, causes bending with tension on the bottom and compression on the top.
3. The horizontal component causes direct tensile stress across the entire cross section.

We can proceed with the stress analysis by using the techniques from the previous section to prepare the shearing force and bending moment diagrams and then using Equation (3–26) to combine the effects of the bending stress and the direct tensile stress at any point. The details are shown within Example Problem 3–17.

Example Problem 3–17 The cantilever beam in Figure 3–24 is a steel American Standard beam, S6 × 12.5. The force *F* is 10 000 lb, and it acts at an angle of 30° below the horizontal, as shown. Use $a = 24$ in and $e = 6.0$ in. Draw the free-body diagram and the shearing force and bending moment diagrams for the beam. Then compute the maximum tensile and maximum compressive stresses in the beam and show where they occur.

Solution Objective Determine the maximum tensile and compressive stresses in the beam.

Given The layout from Figure 3–24(a). Force = F = 10000 lb; angle $\theta = 30°$.
The beam shape: S6 × 12.5; length = a = 24 in.
Section modulus = S = 7.37 in^3; area = A = 3.67 in^2 (Table A15–10).
Eccentricity of the load = e = 6.0 in from the neutral axis of the beam to the line of action of the horizontal component of the applied load.

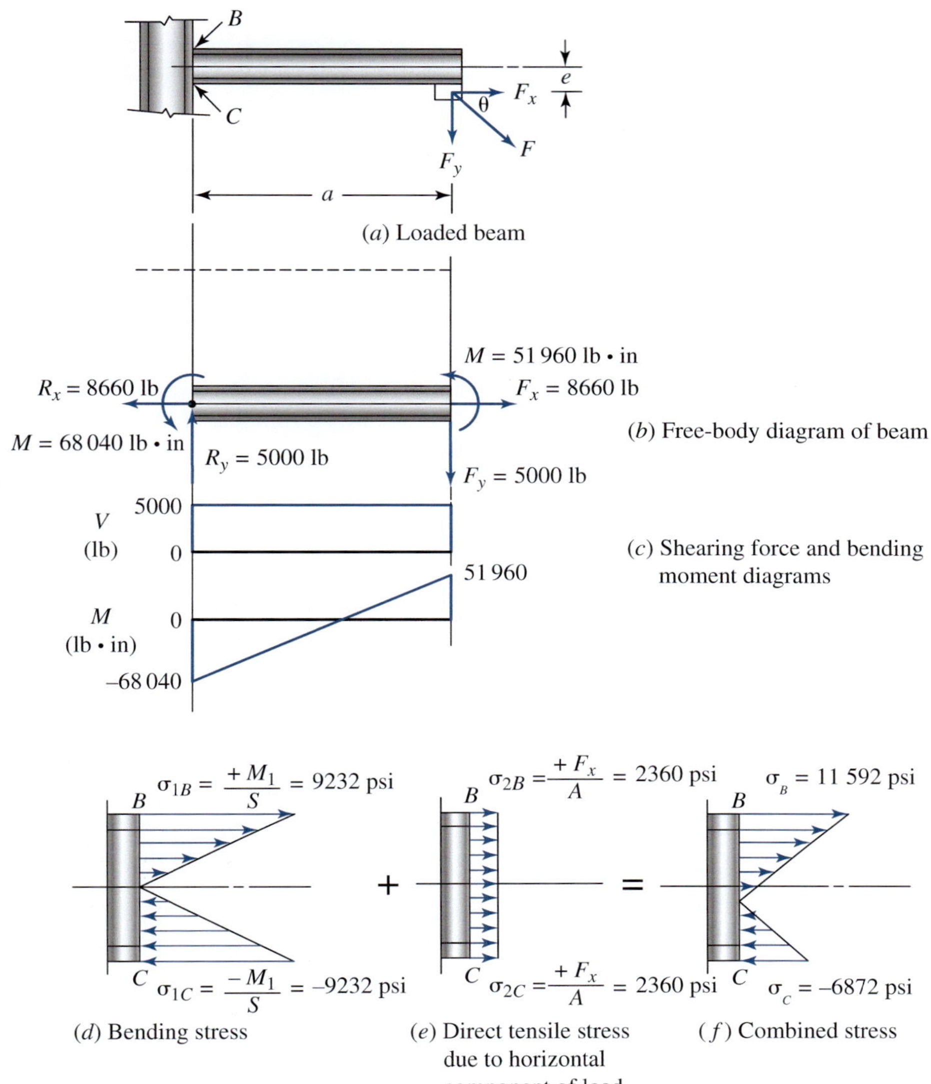

FIGURE 3–24 Beam subjected to combined stresses

Analysis The analysis takes the following steps:

1. Resolve the applied force into its vertical and horizontal components.
2. Transfer the horizontal component to an equivalent loading at the neutral axis having a direct tensile force and a moment due to the eccentric placement of the force.
3. Prepare the free-body diagram using the techniques from Section 3–19.
4. Draw the shearing force and bending moment diagrams and determine where the maximum bending moment occurs.
5. Complete the stress analysis at that section, computing both the maximum tensile and maximum compressive stresses.

Results The components of the applied force are:

$$F_x = F\cos(30^\circ) = (10\,000 \text{ lb})[\cos(30^\circ)] = 8660 \text{ lb acting to the right}$$

$$F_y = F\sin(30^\circ) = (10\,000 \text{ lb})[\sin(30^\circ)] = 5000 \text{ lb acting downward}$$

The horizontal force produces a counterclockwise concentrated moment at the right end of the beam with a magnitude of:

$$M_1 = F_x(6.0 \text{ in}) = (8660 \text{ lb})(6.0 \text{ in}) = 51\,960 \text{ lb}\cdot\text{in}$$

The free-body diagram of the beam is shown in Figure 3–24(*b*).

Figure 3–24(*c*) shows the shearing force and bending moment diagrams.

The maximum bending moment, 68 040 lb in, occurs at the left end of the beam where it is attached firmly to a column.

The bending moment, taken alone, produces a tensile stress (+) on the top surface at point *B* and a compressive stress (−) on the bottom surface at *C*. The magnitudes of these stresses are:

$$\sigma_1 = \pm\ M/S = \pm\ (68\ 040\ \text{lb in})\ /\ (7.37\ \text{in}^3) = \pm\ 9232\ \text{psi}$$

Figure 3–24(*d*) shows the stress distribution due only to the bending stress.

Now we compute the tensile stress due to the axial force of 8660 lb.

$$\sigma_2 = F_x/A = (8660\ \text{lb})/(3.67\ \text{in}^2) = 2360\ \text{psi}$$

Figure 3–24(*e*) shows this stress distribution, uniform across the entire section.

Next, let's compute the combined stress at *B* on the top of the beam.

$$\sigma_B = +\ \sigma_1 + \sigma_2 = 9232\ \text{psi} + 2360\ \text{psi} = 11\ 592\text{—psi Tensile}$$

At *C* on the bottom of the beam, the stress is:

$$\sigma_C = -\sigma_1 + \sigma_2 = -9232\ \text{psi} + 2360\ \text{psi} = -6872\text{—psi Compressive}$$

Figure 3–24(*f*) shows the combined stress condition that exists on the cross section of the beam at its left end at the support. It is a superposition of the component stresses shown in Figure 3–24(*d*) and (*e*).

3–21 STRESS CONCENTRATIONS

The formulas reviewed earlier for computing simple stresses due to direct tensile and compressive forces, bending moments, and torsional moments are applicable under certain conditions. One condition is that the geometry of the member is uniform throughout the section of interest.

In many typical machine design situations, inherent geometric discontinuities are necessary for the parts to perform their desired functions. For example, as shown in Figure 12–2, shafts carrying gears, chain sprockets, or belt sheaves usually have several diameters that create a series of shoulders that seat the power transmission members and support bearings. Grooves in the shaft allow the installation of retaining rings. Keyseats milled into the shaft enable keys to drive the elements. Similarly, tension members in linkages may be designed with retaining ring grooves, radial holes for pins, screw threads, or reduced sections.

Any of these geometric discontinuities will cause the actual maximum stress in the part to be higher than the simple formulas predict. Defining *stress concentration factors* as the factors by which the actual maximum stress exceeds the nominal stress, σ_{nom} or τ_{nom}, predicted from the simple equations allows the designer to analyze these situations. The symbol for these factors is K_t. In general, the K_t factors are used as follows:

$$\sigma_{max} = K_t\sigma_{nom} \quad or \quad \tau_{max} = K_t\tau_{nom} \qquad (3\text{–}27)$$

depending on the kind of stress produced for the particular loading. The value of K_t depends on the shape of the discontinuity, the specific geometry, and the type of stress.

Figure 3–25 shows two examples of how geometric discontinuities affect the stress levels in mechanical components, creating stress concentrations. Part (a) shows a drawing of a model of a beam made from a flat plate using a photoelastic material that reacts to stress levels by producing light and dark areas proportional to variations in stresses when viewed through a polarizing filter. Each dark line, called a *fringe,* indicates a change in stress of a certain amount.

The beam is simply supported at its ends and carries a concentrated load at its middle, where the breadth of the section is made larger. To the right side of the middle, there is a reduction in the breadth with a circular fillet. To the left side of the middle, the breadth of the section is much smaller and the circular fillet is also used. You should visualize that for this simply supported beam, the maximum bending moment occurs at the middle and it decreases linearly to zero at the supports. The predicted maximum bending stress would be in the left portion where the cross section is smaller and toward the right end of that portion where the moment is higher. The fringe lines verify this observation because the number of fringes increases as you look toward the right end of the smaller section. However, note that near where the fillet is located, there are many more, closely spaced fringe lines. This indicates that the local stress around this geometric discontinuity is much higher than would be predicted by simple bending stress calculations. This illustrates the concept of *stress concentrations.* A similar pattern exists in the right portion of the beam, but the overall stress levels are lower because the section size is larger.

Another detail that can be observed from the beam in Figure 3–25(a) is in the area under where the force is applied. The numerous, closely spaced fringes there illustrate the concept of *contact stress* that can be very high locally. Special care should be applied in the design of that part of

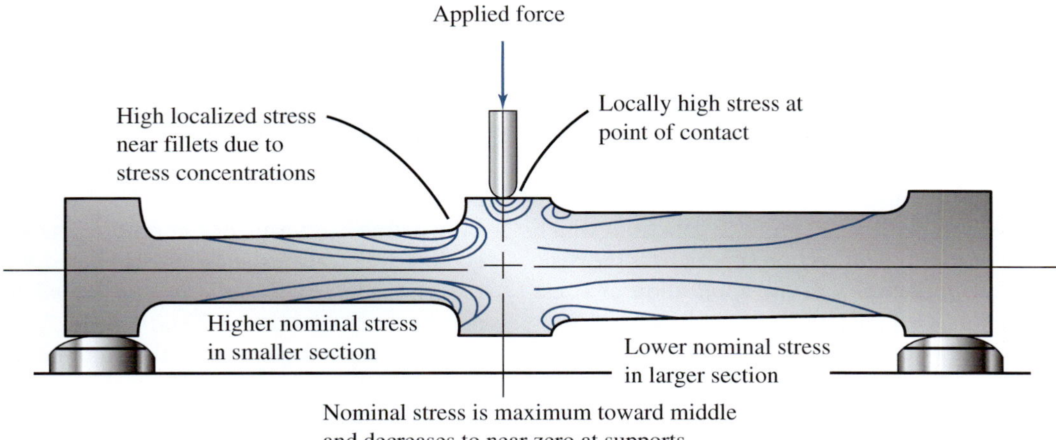

(*a*) Beam in bending with load at middle applied to enlarged section; reduced sections with circular fillets on either side; left side smaller than right side

Plate minimum width at bottom of notches is equal for all three plates so nominal maximum stresses are equal in all three plates

Uniform applied stress field in plates away from notches

Sharper notches create higher localized stresses due to stress concentrations

(*b*) Flat plates with uniform tensile stress field applied; reduced sections at middle all have equal widths; increasingly sharper notch curvature in three plates

FIGURE 3–25 Sketches of stress distributions near stress concentrations at circular fillets in a beam in bending and in notched flat plates in tension

the beam to withstand the high contact stress. This concept is also pertinent to stresses in gear teeth (Chapters 9 and 10) and rolling contact bearings (Chapter 14).

Figure 3–25(b) shows a different loading and stress concentration phenomenon. Three flat plates are each subjected to identical levels of uniformly distributed tensile stresses in the larger part of each plate toward the ends. Each plate has an equal-width, reduced-breadth section at the middle but with quite different geometry for each. The top plate's width is reduced very gradually and the stress in the material increases essentially only due to the reduction in cross-sectional area. In the plate in the middle, the reduction is made by machining a circular notch with a relatively large radius. The stress at the notch again increases because of the reduction in cross-sectional area, but there is also an increased stress level near the two notches because of the sudden change in geometry. The larger number of fringe lines is the indication of a *stress concentration* near the notches. The lower plate also contains a notch, but it is much narrower and the radius at its bottom is much sharper than that for the middle plate. Here, there are even more fringe lines, more closely spaced, indicating a much higher local stress near the small notches.

Stress concentrations are generally highly local phenomena, affecting only the region close to the geometric discontinuity. The entire component must be examined to determine where maximum stresses may occur and often two or more areas must be analyzed to ensure that no failure occurs. More is said about this in Chapter 5.

As a designer, you are responsible for applying stress concentration factors correctly and careful judgment is required. Numerous geometries and loading cases have been evaluated analytically, empirically, and by the use of finite element stress analysis methods to produce data from which values for K_t can be found. This section summarizes the basic principles and shows examples of the application of stress concentration factors. General guidelines are listed next.

General Guidelines for the Use of Stress Concentration Factors

1. Stress concentration factors must always be applied for brittle materials because locally high stresses typically lead to fracture.
2. For ductile materials under static tensile loading, it is typical to ignore stress concentrations in small local areas because the ductility permits yielding that redistributes the stress over a larger area and ultimate failure does not occur. However, some designers prefer to ensure that no yielding occurs anywhere in the component.
3. Areas of ductile members under cyclical compressive stress are unlikely to produce failure unless the local maximum stress exceeds the ultimate compressive stress. At lower stresses, a small amount of material near the highest stress area may yield and permit the load to be redistributed over a larger area.
4. Flexing a member in tension tends to cause excessively high local stresses where there are geometric discontinuities, where small internal voids occur, or where surface imperfections occur such as cracks, tool marks, nicks, roughness, corrosion, plating, or scratches. Small cracks appear that will grow to cover larger areas over time, leading to ultimate failure. This type of failure, called *fatigue*, is discussed further in Chapter 5. ***Therefore, always apply stress concentration factors in areas of cyclical tensile stress where fatigue failures initiate.***
5. Designing components to avoid the occurrence of factors listed in item 4 is recommended. Maintaining smooth surfaces and homogeneous internal structure of the material and avoiding damage in manufacturing and handling are critical.
6. Stress concentration factors are highest in regions of acutely abrupt changes in geometry. Therefore, good design will call for more gradual changes such as by providing generous fillet radii and blending mating surfaces smoothly.

Reference 5 is arguably the most comprehensive source of data for stress concentration factors. Reference 7 also contains much useful data.

Stress concentration factors are often presented in graphical form to aid designers in visualizing the effect of specific geometrical decisions on local stresses. Examples of the general forms of such graphs are shown in Figure 3–26. Note that the curves shown are approximate and should not be relied upon for final designs. The graphs show values for K_t on the vertical axis. The horizontal axis is typically defined in terms of the ratio of the sizes for primary geometrical features such as diameters, widths, and fillet radii. *It is essential that you understand the basis for the nominal stress for use in Equation (3–27).* For each case in Figure 3–26, the equation for the nominal stress is reported. It is typical for the geometry of the smaller section to be used to compute nominal stress because that will give the largest stress in the region around the discontinuity before accounting for the specific geometry. However, exceptions occur. Additional discussion of the four general types of graphs follows.

(a) ***Stepped flat plate in tension:*** The magnitude of the ratio of the larger width, w, to the smaller width, t, is a major factor and it is typically shown as a family of curves like the three in the figure. The horizontal axis is the ratio of r/t, where r is the fillet radius at the step. Note that the larger the ratio, the smaller the value of K_t. Small fillet radii should be avoided if possible because the stress concentration factor typically increases exponentially for very small r/t ratios. *The nominal stress is based on the applied force, F, divided by the minimum cross-sectional area in the smaller section as shown in the figure.*

(b) ***Round bar in bending:*** A family of curves is shown for the ratio of the larger diameter, D, to the smaller diameter, d, similar to the discussion for Part (a). The horizontal axis is the ratio of r/d and, again, it is obvious that small fillet radii should be avoided. *The nominal stress is based on the applied bending moment, M, divided by the section modulus of the smaller cross section as shown in the figure.*

(c) ***Flat plate in tension with a central hole:*** This case assumes that the applied force is uniformly distributed to all parts of the plate in areas away from the hole. The maximum stress rises in the near vicinity of the hole and it occurs at the top and bottom of the hole. The horizontal axis of the chart is the ratio of d/w, where d is the hole diameter and w is the width of the plate. Note that the plate thickness is called t. *The nominal stress is based on the applied force divided by the net area found at the cross section through the hole.* The figure shows that $A_{net} = (w - d)t$.

(d) ***Circular bar in torsion with a central hole:*** *The stress concentration factor in this case is based on the gross polar section modulus for the full cross section of the bar.* This is one of the few cases where the gross section is used instead of the net section. The primary reason is the complexity of the calculation for the polar section

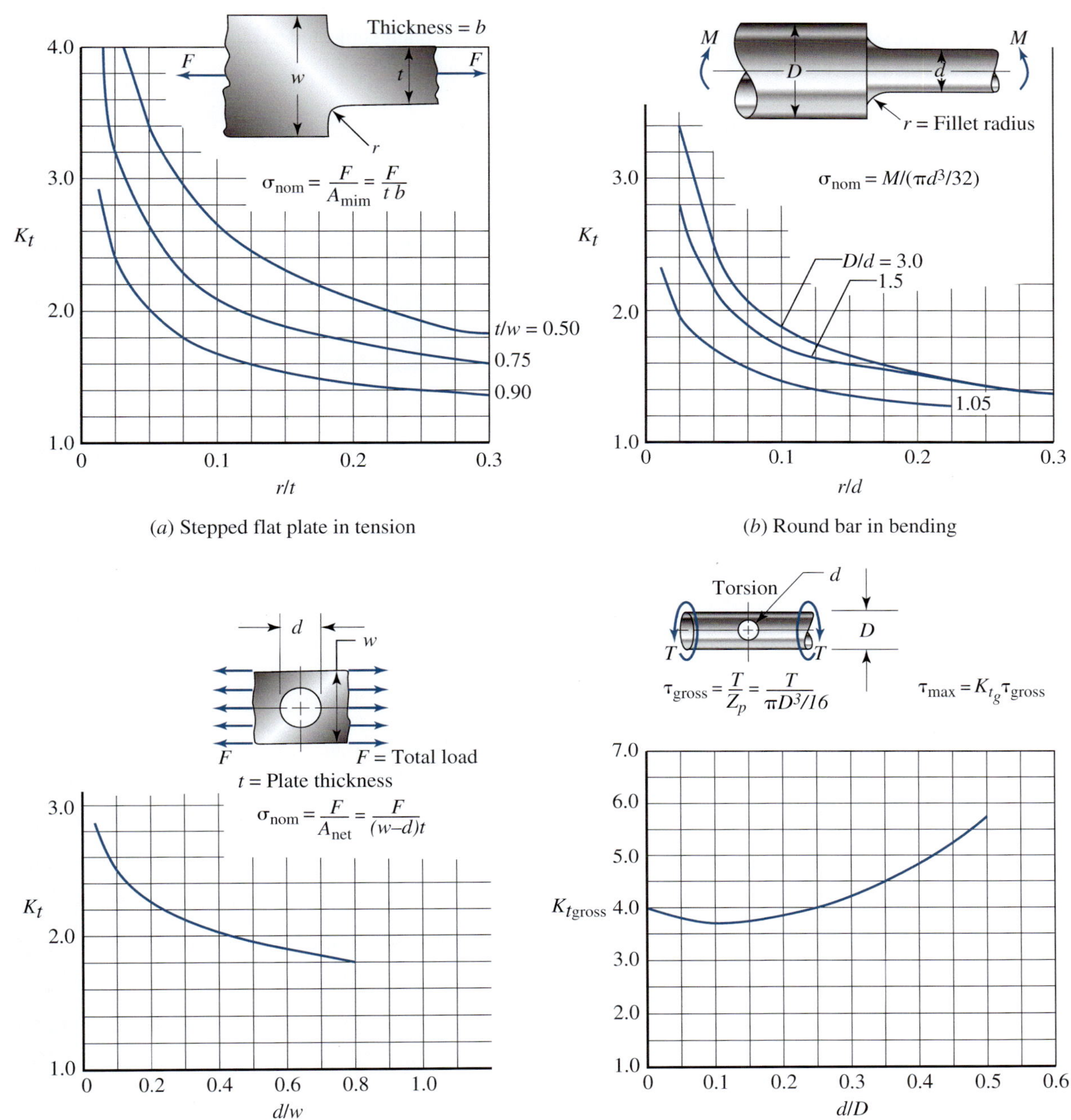

FIGURE 3–26 General Forms of Stress Concentration Factor Curves [Approximate Data-Not for Final Designs]

modulus around the hole. The result is that the value of K_t accounts for both the reduction in polar section modulus and the increase in the stress due to the discontinuity around the hole, resulting in rather large values for K_t. The horizontal axis is the ratio of d/D, where D is the full diameter of the bar and d is the diameter of the hole.

Refer to Internet site 3 for K_t data for problem solving in this book. Four categories of shapes, listed below, are included on the site with multiple kinds of stresses, typically axial tension, bending, and torsion.

1. Rectangular bars with several types of notches, holes, slots, and changes in thickness (17 geometry types; 31 cases)
2. Round bars and shafts with steps in diameter, grooves, keyseats, and holes (8 geometry types; 23 cases)
3. Infinite or semi-infinite plates (3 geometry types; 7 cases)
4. Special shapes (3 geometry types; 6 cases)

Example Problem 3–18 Compute the maximum stress in a stepped flat plate subjected to an axial tensile force of 9800 N. The geometry is shown in Figure 3–27.

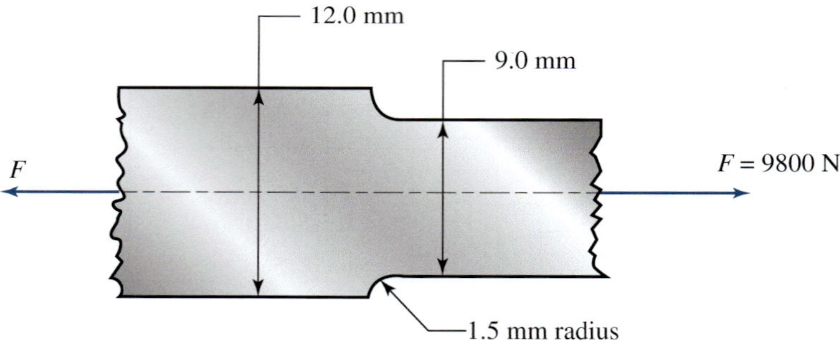

FIGURE 3–27 Stepped flat plate for example problem 3–18

Solution Objective Compute the maximum stress in the stepped flat plate in Figure 3–27.

Given The layout from Figure 3–27. Force $= F = 9800$ N

Using the notation from Figure 3–26(a): widths $w = 12.0$ mm; $t = 9.0$ mm

Plate thickness: $b = 6.0$ mm

Fillet radius at the step: $r = 1.50$ mm

Analysis The presence of the change in width at the step causes a stress concentration to occur. The geometry is of the type shown in Figure 3–26(a) and, for illustration of the method, we will use that chart to determine the value of the stress concentration factor K_t. That value is used in Equation (3–27) to compute the maximum stress.

Results The values of the ratios t/w and r/t are required:

$$t/w = 9.0\ \text{mm}/12.0\ \text{mm} = 0.75$$

$$r/t = 1.5\ \text{mm}/9.0\ \text{mm} = 0.167$$

The value of $K_t = 1.83$ can be read from Figure 3–26(a).

The nominal stress is computed for the small section having a cross section of 6.0 mm by 9.0 mm.

$$A_{net} = t \cdot b = (9.0\ \text{mm})(6.0\ \text{mm}) = 54.0\ \text{mm}^2$$

The nominal stress is:

$$\sigma_{nom} = F/A_{net} = (9800\ \text{N})/(54.0\ \text{mm}^2) = 181.5\,\text{N/mm}^2 = 181.5\ \text{MPa}$$

The maximum stress in the area of the fillet at the step is:

$$\sigma_{max} = K_t\sigma_{nom} = (1.83)(181.5\ \text{MPa}) = 332\ \text{MPa}$$

Comments The maximum stress of 332 MPa occurs in the fillet area at both the top and bottom of the small section. To the right of the fillet a short distance, the local stress reduces to the nominal value of 181.5 MPa. The stress in the larger section is obviously much smaller because of the larger cross-sectional area of the plate. Note that specifying a smaller fillet radius would cause a much larger maximum stress because the curve for K_t increases sharply as the ratio of r/t decreases. Modestly smaller maximum stress would be produced for larger fillet radii.

We used Figure 3–26(a) to illustrate the process for finding the approximate value for the stress concentration factor. It is recommended that Internet site 3 be used for problem solving in this book because of the more accurate data and the much larger array of types of geometry and loading cases.

Stress Concentration Factors for Lug Joints. A common machine element is the familiar *clevis joint* or *lug joint,* sketched in Figure 3–28. For this discussion, the bar-shaped plate is called the *lug* that may have either a flat, square-ended design as shown or a rounded end that is sometimes used to provide clearance for rotation of the lug. The generic name for the lug in stress analysis is *flat plate with a central hole with a load applied through a pin in tension.* Figure 3–29

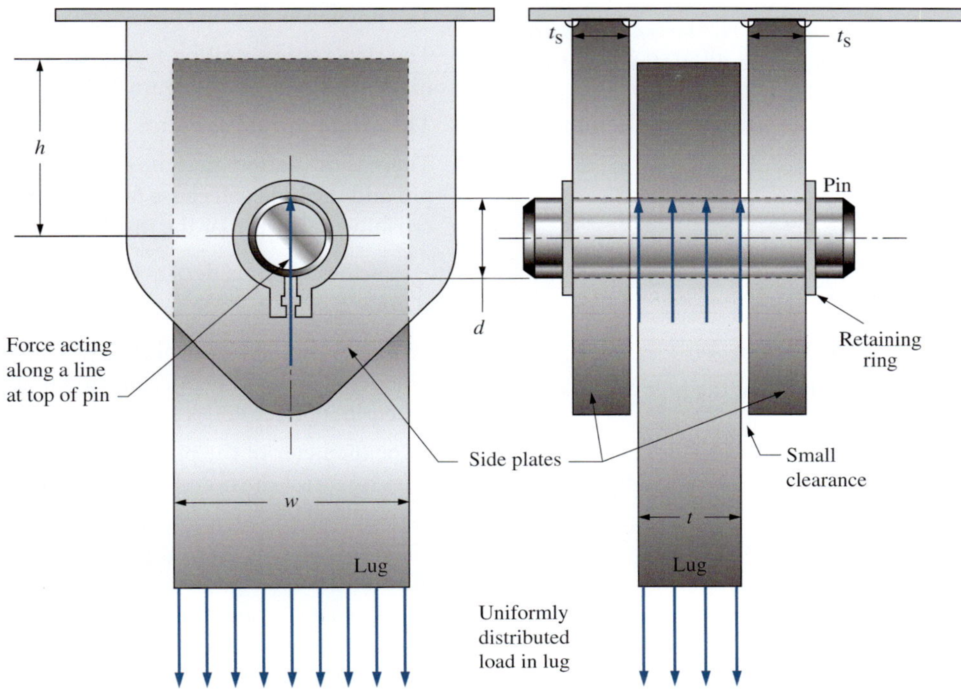

FIGURE 3–28 Clevis joint showing a pin–lug connection including a flat plate with a central hole as used in Figure 3–29.

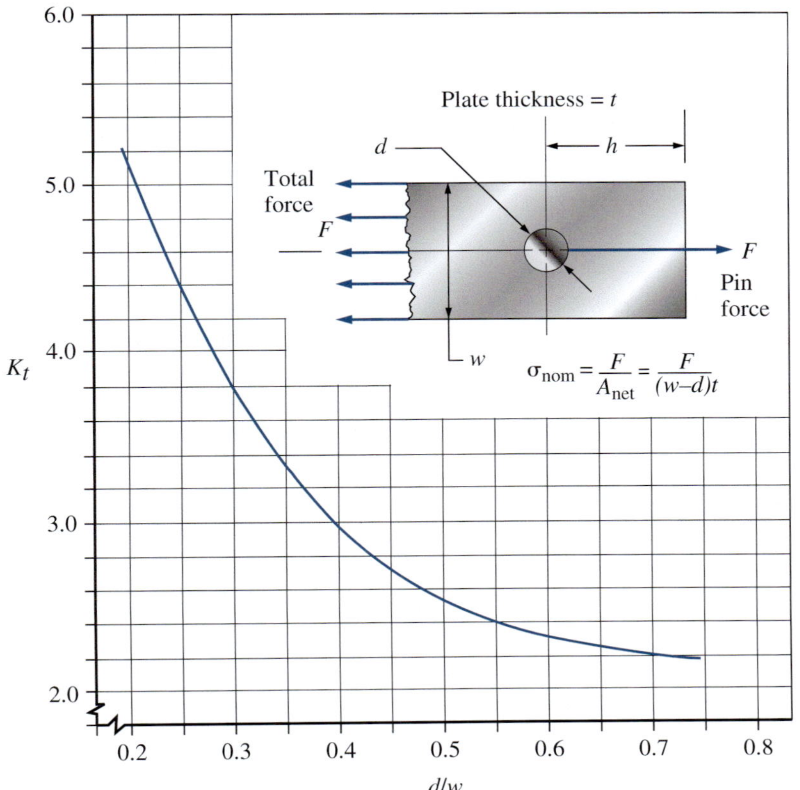

FIGURE 3–29 Stress concentration factor for flat plate with tensile load applied through a pin

shows approximate data for the stress concentration factors that occur around the hole for typical conditions, described below. The plate is assumed to have a uniform distribution of the load across any cross section well away from the hole. The critical section with regard to axial tensile stress in the plate occurs in the section through the hole where the minimum net cross-sectional area occurs. The stress concentration factor accounts for the geometric discontinuity at this section.

In general, the axial tensile stress near the hole in the lug depends on several factors:

1. The width and thickness of the lug and the size and placement of the hole relative to the end of the lug
2. The clearances between the faces of the lug and the side plates of the clevis
3. The diameter and length of the pin
4. The clearance between the pin and the hole
5. The materials of the pin and the lug

A typical lug connection should have the following features:

- a square-ended lug
- close-fitting pin: $d_{hole} \cong d_{pin}(1.002)$
- $t/d \leqq 0.5$ [t = thickness of lug parallel to axis of pin; d = nominal hole diameter]
- Both pin and lug have the same modulus of elasticity; $E_{pin}/E_{lug} = 1.0$
- $h/w = 1.0$
  - h = distance from centerline of hole to the top of the lug
  - w = width of the lug across the location of the hole

Problems in this book will assume the conditions listed above and use the values for K_t from Internet site 3. The curve in Figure 3–29 gives approximate data for K_t versus the ratio d/w.

Reference 5 includes extensive data that allow consideration of factors different from those assumed above. Very loose fitting pins should be avoided because:

- The pin makes a virtual line contact on the top of the hole
- The stress concentration factor for the lug is somewhat higher than shown in Figure 3–29
- The maximum stress occurs at the line of contact
- There is the danger that the contact stress between the pin and the hole will cause failure in bearing, a local crushing of the pin or the inside surface of the hole.

Other factors to consider in the design of a clevis or lug joint are:

1. The pin should be as short as practical to minimize the tendency for it to bend, causing nonuniform contact of the pin in the hole.
2. The diameter of the pin must be adequate to resist failure by direct shear and to keep bending deflection to an acceptable level.
3. The distance, h, from the centerline of the hole to the top of the lug should be nominally equal to the width of the plate, w, and as large as practical to minimize the bending stresses in the upper part of the lug and to prevent shearing tear-out of the lug.
4. The bearing stress between the pin surface and the inside of the hole of the lug must be acceptable, typically less than $0.9s_y$.
5. Similarly, the bearing stress between the pin surface and the holes in the side plates must be acceptable.
6. The clearance between the faces of the lug and the side plates of the clevis should be small to minimize bending of the pin.

Example Problem 3–19 The lug for a clevis joint of the type shown in Figure 3–28 is being designed for an applied force of 8.75 kN. The steel pin nominal diameter has been specified to be $d = 12.0$ mm. Specify the dimensions w, h, and t for the steel lug to be consistent with the parameters listed in this section and a d/w ratio of 0.40. Also specify the nominal clearance between the pin and the hole in the lug. Then determine the expected maximum stress in the lug.

Objective Specify w, h, and t and the nominal clearance.

Given $d = 12.0$ mm; $d/w = 0.40$. $F = 8.75$ kN

Analysis *Lug thickness, t*: Recommended $t/d \leq 0.50$. Using the upper limit gives:
$t = 0.50\,d = (0.50)(12.0\text{ mm}) = 6.0\text{ mm}$

Lug width, w: Using $d/w = 0.40, w = d/0.40 = 12\text{ mm}/0.40 = 30.0\text{ mm}$
End distance, h: Recommended ratio $h/w = 1.0$. Then,

$$h = w = 30.0\text{ mm}$$

Close-fitting pin: $d_{hole} \cong d_{pin}(1.002) = 12.0\text{ mm}(1.002) = 12.024\text{ mm}$

Maximum stress: $\sigma_{max} = K_t \times \sigma_{nom}$

$$\sigma_{nom} = F/(w - d)t = [8.75\ \text{kN}/(30.0 - 12.0)(6.0)\text{mm}^2](1000\ \text{N/kN})$$

$$\sigma_{nom} = 81.8\ \text{N/mm}^2 = 81.8\ \text{MPa}$$

To find K_t: At $d/w = 0.40$ in Figure 3–28, $K_t \cong 3.0$. Then

$$\sigma_{max} = K_t \times \sigma_{nom} = (3.0)(81.8\ \text{MPa}) = 243\ \text{MPa}$$

Results The design details for the lug of the clevis joint in Figure 3–28 are:

$$d = 12.0\ \text{mm};\ w = 30.0\ \text{mm};\ t = 6.0\ \text{mm};\ h = 30.0\ \text{mm}$$

Nominal clearance between pin and hole: $\delta = 0.024$ mm
Maximum stress in lug at hole: $\sigma_{max} = 243$ MPa

3–22 CURVED BEAMS

The analysis of stress due to bending in beams summarized earlier in this chapter is limited to beams that are straight or very nearly so. It was noted that when a beam, that is originally straight, is bent:

1. The radius of curvature of the neutral axis of the beam is very large
2. Plane sections perpendicular to the axis of the beam remain plane as the beam bends and they rotate in opposite directions under the influence of the bending moment placing part of the section in compression and part in tension
3. The location of the neutral axis of the cross section of the beam is:
   a. Where the bending stress is zero
   b. Coincident with the centroidal axis of the cross section of the beam
4. The maximum tensile and compressive stresses occur at the outermost fibers of the beam
5. There is a linear variation of the bending stress from the place of maximum compression to the place of maximum tension
6. Under these conditions, the maximum bending stress is computed from the familiar flexure formula:
   a. $\sigma = Mc/I$
   b. where M is the bending moment at the section; c is the distance from the neutral axis to the outermost fiber; and I is the moment of inertia of the cross section with respect to its centroidal axis

The flexure formula applies reasonably well when the ratio of the radius of curvature, R, to the depth of the cross section is larger than 10. That is $R/h > 10$ (Reference 6).

When a component has a smaller radius of curvature, the maximum stress due to bending is significantly higher than that which would be predicted from the flexure formula and different analysis procedures apply. There are many practical cases where smaller radii of curvature for beams occur. Examples include:

- Crane hooks used for lifting loads
- Clamps such as the C-clamp
- Open, C-section frames for machinery such as punch presses
- Building beams that are curved for architectural and aesthetic reasons
- Parts of hand tools such as saws, pliers, and wire cutters
- Structural or functional members of machinery where the curvature of the member allows clearance of obstructions
- Parts of some types of furniture such as tables and chairs
- Automotive frames, suspension elements, and actuation arms for window lifts and seats

Major features and the general nature of the results of the analysis for curved beams include:

1. When the applied bending moment tends to *straighten the beam* by increasing the radius of curvature, it is considered to be *positive*.
2. When the applied bending moment tends to decrease the radius of curvature, it is considered to be *negative*.
3. The stress distribution within the cross section of the member is *not* linear; rather it is hyperbolic. Figure 3–30 shows the general pattern of the stress distribution for the case of a curved beam with a rectangular cross section when a positive bending moment is applied. Note the following:
   a. The inside surface is placed in tension and the maximum tensile stress occurs on that surface.
   b. The outside surface is placed in compression and the maximum compressive stress occurs at that surface.

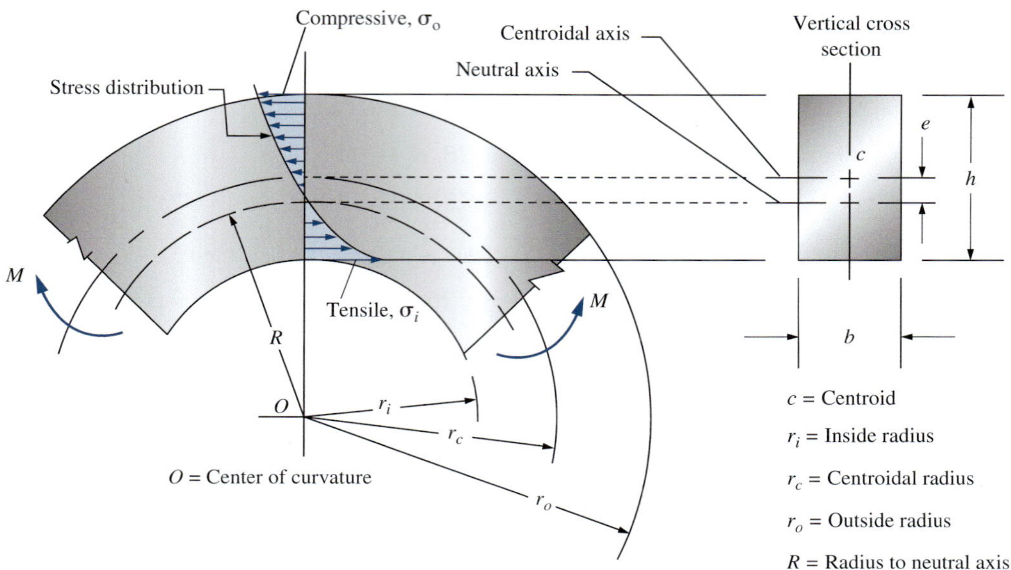

FIGURE 3–30 Segment of a curved beam carrying a positive bending moment

   c. The place within the cross section where the stress becomes zero is called the *neutral axis* as it is for straight beams. However, the neutral axis *is not* coincident with the centroidal axis of the shape of the cross section. For the case shown in Figure 3–30, it is displaced toward the center of curvature by an amount called *e*.

4. The *curved beam formula* given below for the stress distribution has been derived in References 1, 2, and 6.

$$\sigma = \frac{M(R - r)}{Ar(r_c - R)} \tag{3–28}$$

where:

A = cross-sectional area of the shape
M = bending moment applied

The three radii are measured from the center of curvature of the beam:

r to the place where the stress is to be computed

r_c to the centroid of the area

R to the *neutral axis* of the cross section

R depends on the shape of the cross section using the *area shape factor, ASF,* found from the relations shown in Figure 3–31. Then compute:

$$R = A/ASF \tag{3–29}$$

The quantity $(r_c - R)$ in the denominator of Equation (3–28) is the radial distance from the centroidal axis of the cross section to the neutral axis. This value, called *e*, is typically a very small number, requiring that it be computed with high accuracy, at least three significant figures.

General Procedure for Analyzing Curved Beams Carrying a Pure Bending Moment

This procedure is used to compute the maximum tensile and compressive stresses for a curved beam at the inside (at r_i) and outside (at r_o) surfaces. The results are then compared to determine which value is the true maximum stress.

1. Determine the value of the applied bending moment, *M*, including its sign.
2. For the cross-sectional area of the beam:
   a. Compute the total area, *A*.
   b. Compute the location of the centroid of the area.
   c. Determine the value of four radii from the center of curvature of the beam; r_i, r_o, r_c, and R, using Equation (3–29).
   d. Compute the area shape factor, *ASF*. See Figure 3–31.
3. Compute the stress at the outside surface from:

$$\sigma_o = \frac{M(R - r_o)}{Ar_o(r_c - R)} \tag{3–30}$$

4. Compute the stress at the inside surface from:

$$\sigma_1 = \frac{M(R - r_i)}{Ar_i(r_c - R)} \tag{3–31}$$

5. Compare σ_o and σ_i to determine the maximum value.

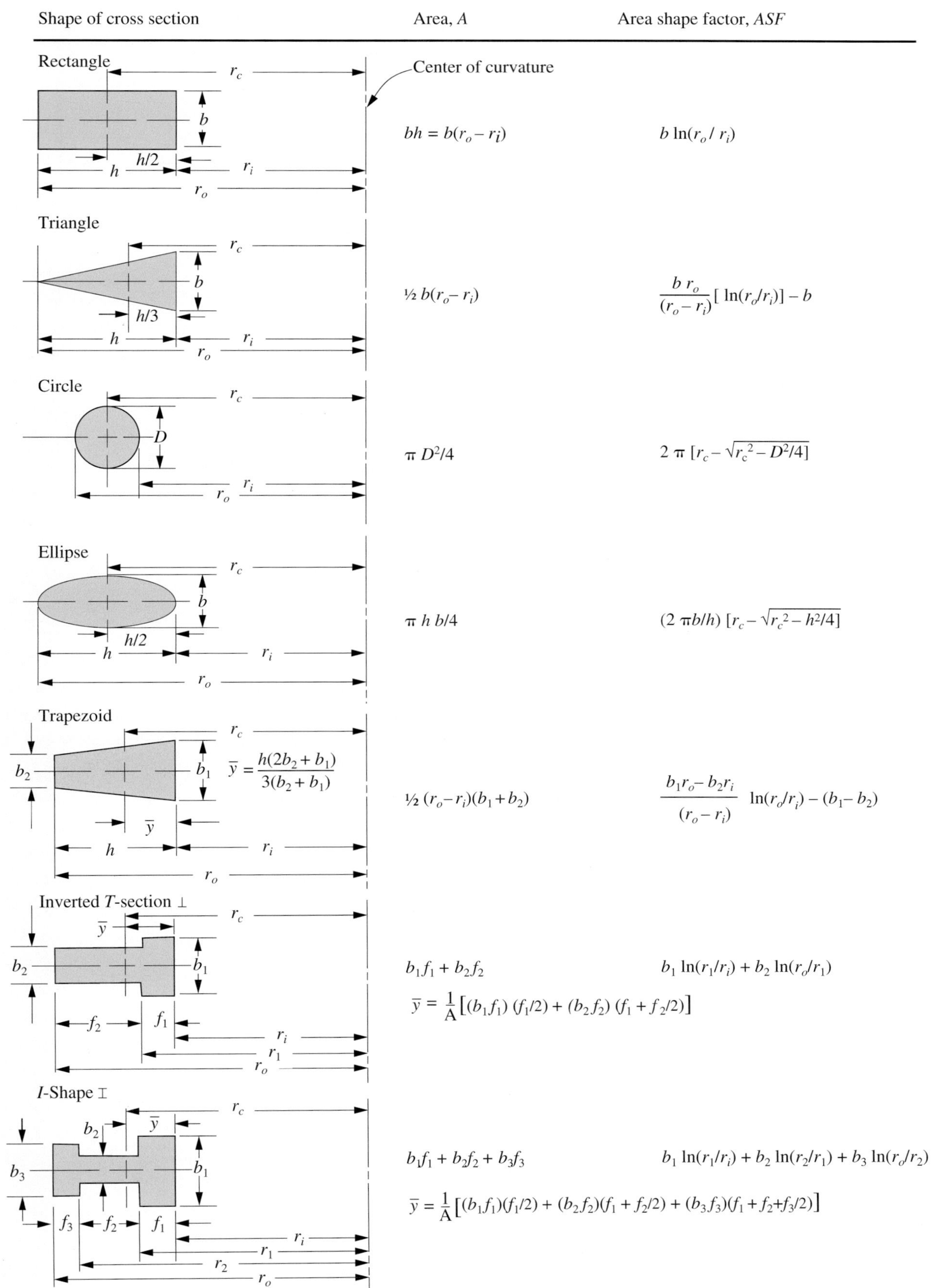

FIGURE 3–31 Area shape factors for selected cross sections of curved bars

Shape of cross section	Area, A	Area shape factor, ASF
Semicircle—On outside of area 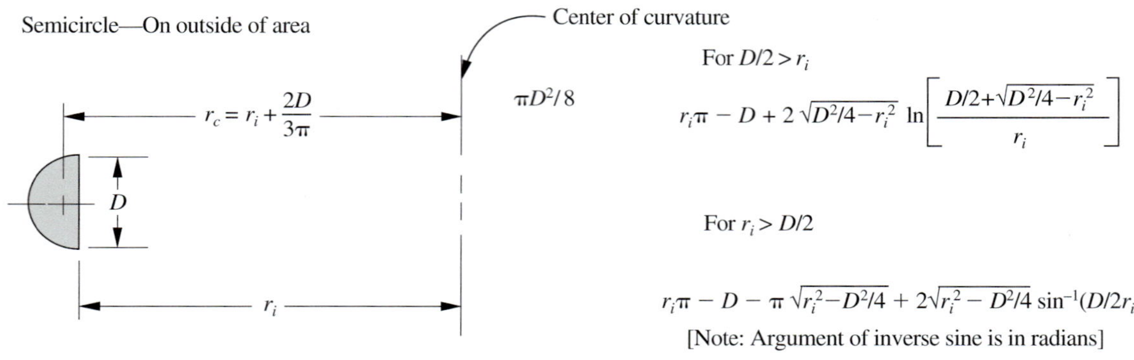	$\pi D^2/8$	For $D/2 > r_i$: $r_i\pi - D + 2\sqrt{D^2/4 - r_i^2}\,\ln\left[\dfrac{D/2 + \sqrt{D^2/4 - r_i^2}}{r_i}\right]$ For $r_i > D/2$: $r_i\pi - D - \pi\sqrt{r_i^2 - D^2/4} + 2\sqrt{r_i^2 - D^2/4}\,\sin^{-1}(D/2r_i)$ [Note: Argument of inverse sine is in radians]
Semicircle—On inside of area 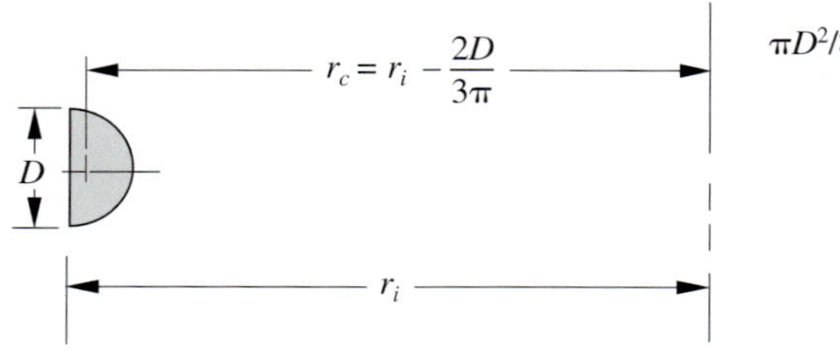	$\pi D^2/8$	$r_i\pi + D - \pi\sqrt{r_i^2 - D^2/4} - 2\sqrt{r_i^2 - D^2/4}\,\sin^{-1}(D/r_i)$ [Note: Argument of inverse sine is in radians]

FIGURE 3–31 *(Continued)*

Example Problem 3–20 A curved bar has a rectangular cross section 15.0 mm thick by 25.0 mm deep as shown in Figure 3–32. The bar is bent into a circular arc producing an inside radius of 25.0 mm. For an applied bending moment of +400 N · m, compute the maximum tensile and compressive stresses in the bar.

Solution Objective Compute the maximum tensile and compressive stresses.

Given $M = +400$ N · m that tends to straighten the bar.
$r_i = 25.0$ mm.

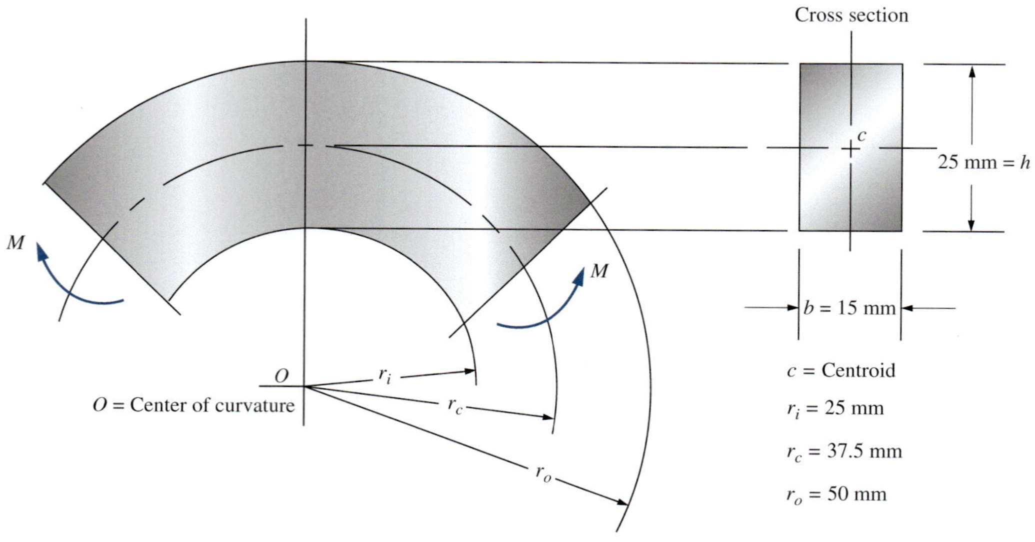

FIGURE 3–32 Curved bar with a rectangular cross section for Example Problem 3–20

Analysis Apply Equations (3–30) and (3–31).

Results First compute the cross-sectional area:

$$A = bh = (15.0 \text{ mm})(25.0 \text{ mm}) = 375.0 \text{ mm}^2$$

Now compute the quantities involving radii. See Figure 3–32 for related dimensions:

$$r_o = r_i + h = 25.0 \text{ mm} + 25.0 \text{ mm} = 50.0 \text{ mm}$$
$$r_c = r_i + h/2 = 25.0 \text{ mm} + (25.0/2) \text{ mm} = 37.5 \text{ mm}$$
$$R = A/ASF$$

From Figure 3–31, for a rectangular cross section:

$$ASF = b \cdot \ln(r_o/r_i) = (15 \text{ mm})[\ln(50.0/25.0)] = 10.3972 \text{ mm}$$

Then, $R = A/ASF = (375 \text{ mm}^2)/10.3972 \text{ mm} = 36.0674 \text{ mm}$

Quantities needed in the stress equations include:

$$r_c - R = 37.5 \text{ mm} - 36.0674 \text{ mm} = 1.4326 \text{ mm}$$

This is the distance e as shown in Figure 3–30.

$$R - r_o = 36.0674 \text{ mm} - 50.0 \text{ mm} = -13.9326 \text{ mm}$$
$$R - r_i = 36.0674 \text{ mm} - 25.0 \text{ mm} = 11.0674 \text{ mm}$$

This is the distance from the inside surface to the neutral axis.

Stress at outer surface, using Equation (3–30):

$$\sigma_o = \frac{M(R - r_o)}{Ar_o(r_c - R)} = \frac{(400 \text{ N}\cdot\text{m})(-13.9326 \text{ mm})[1000 \text{ mm/m}]}{(375 \text{ mm}^2)(50.0 \text{ mm})(1.4326 \text{ mm})}$$

$$\sigma_o = -207.5 \text{ N/mm}^2 = -207.5 \text{ MPa}$$

This is the maximum compressive stress in the bar.

Stress at inner surface, using Equation (3–31):

$$\sigma_i = \frac{M(R - r_i)}{Ar_i(r_c - R)} = \frac{(400 \text{ N}\cdot\text{m})(11.0674 \text{ mm})[1000 \text{ mm/m}]}{(375 \text{ mm}^2)(25.0 \text{ mm})(1.4326 \text{ mm})}$$

$$\sigma_i = 329.6 \text{ N/mm}^2 = 329.6 \text{ MPa}$$

This is the maximum tensile stress in the bar.

Comment The stress distribution between the outside and the inside is similar to that shown in Figure 3–30.

Cross Sections Comprised of Two or More Shapes. Designers often work toward optimizing the shape of the cross section of curved beams to more nearly balance the maximum tensile and compressive stresses when the material is isotropic. Similarly, for materials like cast irons having different strengths in compression and tension, the goal is to achieve a nearly equal design factor for both the tensile and compressive parts of the section. The cross-sectional shape of the curved beam may also be modified to optimize the casting or machining processes or to make the shape more pleasing. Adjusting the shape and dimensions of the cross section of the curved beam can accomplish these goals. Reaching these goals typically results in a cross section that is a composite of two or more of the standard shapes shown in Figure 3–31. Examples of the composite shapes are shown in Figure 3–33.

The analysis process introduced earlier can be modified to consider a composite shape for the cross section of the curved beam.

Procedure for Analyzing Curved Beams with Composite Cross-sectional Shapes Carrying a Pure Bending Moment

This procedure is used to compute the maximum tensile and compressive stresses for a curved beam at the inside (at r_i) and outside (at r_o) surfaces. The results are then compared to determine which value is the true maximum stress.

1. Determine the value of the applied bending moment, M, including its sign.
2. For the composite cross-sectional area of the beam:
   a. Determine the inside radius, r_i, and the outside radius, r_o.
   b. Separate the composite area into two or more parts that are shapes from Figure 3–31.
   c. Compute the area of each component part, A_i, and the total area, A.

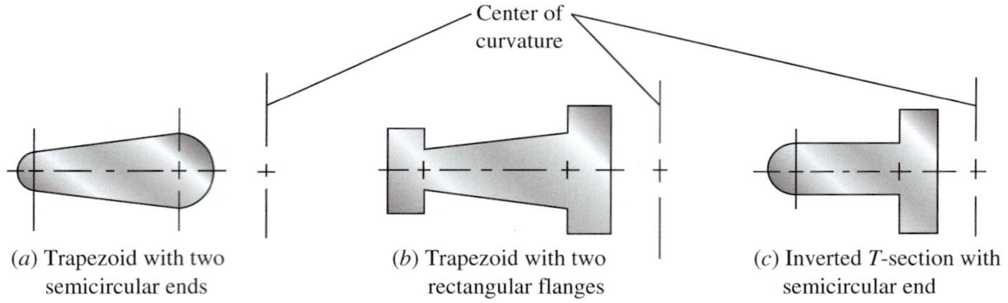

FIGURE 3–33 Examples of composite shapes for cross sections of curved beams

d. Locate of the centroid of each component area.
e. Compute the radius of the centroid of the composite area, r_c.
f. Compute the value of the area shape factor, ASF_i, for each component area using the equations in Figure 3-31.
g. Compute the radius, R, from the center of curvature to the neutral axis from:

$$R = A/\Sigma(ASF_i)$$

3. Compute the stress at the outside surface from:

$$\sigma_o = \frac{M(R - r_o)}{Ar_o(r_c - R)} \quad \textbf{(3–30)}$$

4. Compute the stress at the inside surface from:

$$\sigma_i = \frac{M(R - r_i)}{Ar_i(r_c - R)} \quad \textbf{(3–31)}$$

5. Compare σ_o and σ_i to determine the maximum value.

Example Problem 3–21 A curved beam has the shape shown in Figure 3–34 and is subjected to a pure bending moment of −640 lb·in. The inside radius is 2.90 in. Compute the maximum tensile and compressive stresses in the beam.

Solution

Objective: Compute the maximum tensile and compressive stresses.

Given: $M = -640$ lb·in that tends to decrease the radius of curvature.
$r_i = 2.90$ in (see Figure 3–34 for related dimensions).

Analysis: Apply Equations (3–30) and (3–31).

Results: Separate the shape into a composite of an inverted T-section 1 and a semicircular area 2. Compute cross-sectional area for each part:

$$A_1 = b_1 f_1 + b_2 f_2 = (0.80\text{ in})(0.20\text{ in}) + (1.0\text{ in})(0.40\text{ in})$$

$$A_1 = 0.560\text{ in}^2$$

$$A_2 = \pi D^2/8 = \pi(0.40\text{ in})^2/8 = 0.06283\text{ in}^2$$

$$\text{Total area} = A = A_1 + A_2 = 0.560 + 0.06283 = 0.62283\text{ in}^2$$

Now locate the centroid of each area with respect to the inside surface; from Figure 3–31:

$$\bar{y}_1 = [1/A][(b_1 f_1)(f_1/2) + (b_2 f_2)(f_1 + f_2/2)]$$

$$\bar{y}_1 = [1/0.56\text{ in}^2][(0.80)(0.20)(0.10) + (1.0)(0.40)(0.70)]\text{ in}^3$$

$$\bar{y}_1 = 0.5286\text{ in}$$

$$\bar{y}_2 = 1.20\text{ in} + 2D/3\pi = 1.20\text{ in} + (2)(0.40\text{ in})/(3)(\pi)$$

$$\bar{y}_2 = 1.2849\text{ in}$$

Now locate the centroid of the composite area:

$$\bar{y}_c = [1/A][A_1\bar{y}_1 + A_2\bar{y}_2]$$

$$\bar{y}_c = [1/0.62283\text{ in}^2][(0.560)(0.5286) + (0.06283)(1.2849)]\text{ in}^3$$

$$\bar{y}_c = 0.60489\text{ in}$$

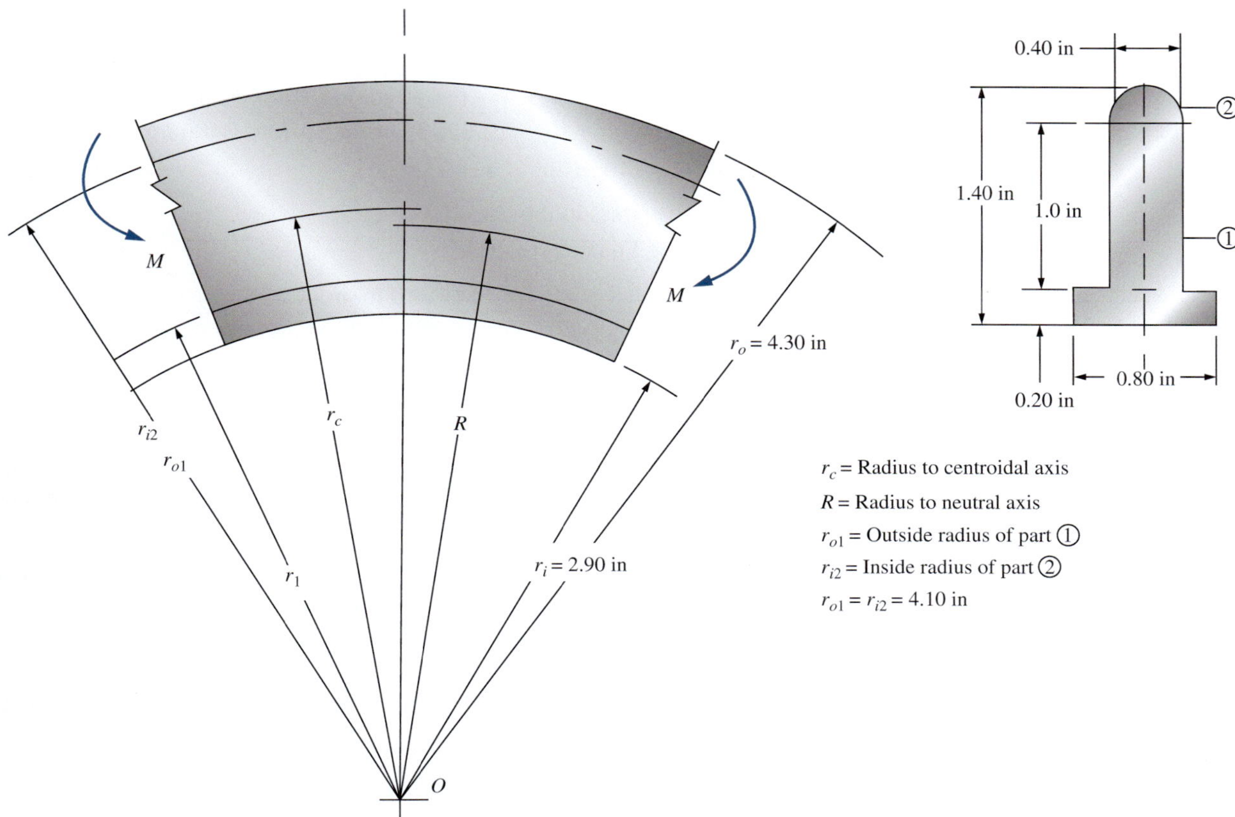

FIGURE 3–34 Curved beam with a composite shape carrying a negative bending moment for example problem 3–21

Now define the pertinent radii from the center of curvature:

$$r_o = r_i + 1.40 \text{ in} = 2.90 \text{ in} + 1.40 \text{ in} = 4.30 \text{ in}$$

$$r_c = r_i + \bar{y}_c = 1.40 \text{ in} + 0.60489 \text{ in} = 3.5049 \text{ in}$$

Compute the value of the area shape factor, ASF_i, for each part:

For the T-shape: $ASF_1 = b_1 \ln(r_1/r_i) + b_2 \ln(r_{o1}/r_1)$

$$ASF_1 = (0.80 \text{in}) \ln(3.1/2.9) + (0.40 \text{in}) \ln(4.1/3.1)$$

$$ASF_1 = 0.16519 \text{ in}$$

For the semicircular area, we must determine the relationship between r_{i2} and $D/2$:

$$r_{i2} = 4.10 \text{ in at the base of the semicircle}$$

$$D/2 = (0.40 \text{ in})/2 = 0.20 \text{ in}$$

Because $r_{i2} > D/2$, we use the equation,

$$ASF_2 = r_i\pi - D - \pi\sqrt{r_{i2}^2 - D^2/4} + 2\sqrt{r_{i2}^2 - D^2/4}\,[\sin^{-1}(D/2r_i)]$$

[Note: Argument of inverse sine is in radians]

$$ASF_2 = (4.10)\pi - 0.40 - \pi\sqrt{(4.1)^2 - (0.40)^2/4}$$
$$+ 2\sqrt{(4.1)^2 - (0.40)^2/4}\,[\sin^{-1}(0.40)/(2)(4.1)] = 0.015016 \text{ in}$$

Compute the radius, R, from the center of curvature to the neutral axis from:

$$R = A/\Sigma(ASF_i) = 0.62283 \text{ in}^2/(0.16519 + 0.015016) \text{ in}$$

$$R = 3.4563 \text{ in}$$

Now compute:

$$r_c - R = 3.5049 \text{ in} - 3.4563 \text{ in} = 0.0486 \text{ in} = e$$

$$R - r_o = 3.4563 \text{ in} - 4.30 \text{ in} = -0.8437 \text{ in}$$

$$R - r_i = 3.4563 \text{ in} - 2.90 \text{ in} = 0.5563 \text{ in}$$

This locates the neutral axis from the inner surface.

Now compute the stress at outer surface, using Equation (3–30):

$$\sigma_o = \frac{M(R - r_o)}{Ar_o(r_c - R)} = \frac{(-640\ \text{lb}\cdot\text{in})(-0.8437\ \text{in})}{(0.62283\ \text{in}^2)(4.30\ \text{in})(0.0486\ \text{in})}$$

$$\sigma_o = 4149\ \text{lb/in}^2 = 4149\ \text{psi}$$

This is the maximum tensile stress in the bar.

Stress at inner surface, using Equation (3–31):

$$\sigma_i = \frac{M(R - r_i)}{Ar_i(r_c - R)} = \frac{(-640\ \text{lb}\cdot\text{in})(0.5563\ \text{in})}{(0.62283\ \text{in}^2)(2.90\ \text{in})(0.0486\ \text{in})}$$

$$\sigma_i = -4056\ \text{lb/in}^2 = -4056\ \text{psi}$$

This is the maximum compressive stress in the bar.

Comment This problem demonstrated the process for analyzing a composite section. Note that the maximum tensile and compressive stresses are very nearly equal for this design, a desirable condition for homogeneous, isotropic materials.

Curved Beams with Combined Bending Moment and Normal Load. Each of the example curved beams considered thus far carried only a moment that tended to either increase or decrease the radius of curvature for the beam. Only the bending stress was computed. However, many curved beams carry a combination of a moment and a normal load acting perpendicular to the cross section of the beam. Examples are a crane hook and a C-clamp.

The complete stress analysis requires that the normal stress, $\sigma = P/A$, be added to the bending stress computed from Equations (3–30) and (3–31). This process is discussed in Section 3–20, "Combined Normal Stresses: Superposition Principle." Apply Equation (3–26) as illustrated in Example Problem 3–17.

Alternate Approaches to Analyzing Stresses in Curved Beams. An alternate approach to analyzing curved beams is reported in Reference 5. Using the concept of stress concentration factors, data are reported for an equivalent K_t value for curved beams having five different shapes: circular, elliptical, hollow circular, rectangular, and I-shapes for one specific set of proportions. The controlling parameter is r_c/c, where r_c is the radius to the centroidal axis of the shape and c is one-half of the total height of each section. Then K_t is used in the same manner that is discussed in Section 3–21. That is,

$$K_t = \frac{\sigma_{\text{max}}}{\sigma_{\text{nom}}} \quad \text{or} \quad \sigma_{\text{max}} = K_t\sigma_{\text{nom}} = K_t(Mc/I)$$

Note that this is *not* strictly a stress concentration in the sense that there is no geometric discontinuity or abrupt change in shape or size of the cross section. However, the approach gives a reasonable approximation of the maximum stress in the curved beam.

Reference 7 presents a similar approach to the analysis of curved beams but data are presented in tabular form. Again, the primary parameter is r_c/c which is used to compute the eccentricity, e, the radial distance from the centroidal axis to the true neutral axis. Then two factors are determined from the table: k_o for the stress at the outside surface and k_i for the stress at the inside surface. Then,

$$\sigma_{o\text{-max}} = k_o\,\sigma_{\text{nom}} = k_o(Mc/I) \quad \text{and}$$

$$\sigma_{i\text{-max}} = k_i\,\sigma_{\text{nom}} = k_i(Mc/I)$$

3–23 NOTCH SENSITIVITY AND STRENGTH REDUCTION FACTOR

The amount by which a load-carrying member is weakened by the presence of a stress concentration (notch), considering both the material and the sharpness of the notch, is defined as

K_f = fatigue strength reduction factor

$$K_f = \frac{\text{endurance limit of a notch-free specimen}}{\text{endurance limit of a notched specimen}}$$

This factor could be determined by actual test. However, it is typically found by combining the stress concentration factor, K_t, defined in Section 3-21, and a material factor called the *notch sensitivity*, q. We define

$$q = (K_f - 1)/(K_t - 1) \qquad \textbf{(3–28)}$$

When q is known, K_f can be computed from

$$K_f = 1 + q\,(K_t - 1) \qquad \textbf{(3-29)}$$

Values of q range from zero to 1.0, and therefore K_f varies from 1.0 to K_t. Under repeated bending loads, very ductile steels typically exhibit values of q from 0.5 to 0.7. High-strength steels with hardness approximately HB 400 ($s_u \cong 200$ ksi or 1400 MPa) have values of q from 0.90 to 0.95. (See Reference 2 for further discussion of values of q.)

Because reliable values of q are difficult to obtain, the problems in this book will typically assume that $q = 1.0$ and $K_f = K_t$, the safest, most conservative value.

REFERENCES

1. Boresi, Arthur P., and Richard J. Schmidt. *Advanced Mechanics of Materials.* 6th ed. New York: John Wiley & Sons, 2003.
2. Hibbler, Russell C. *Mechanics of Materials.* 8th ed. Upper Saddle River, NJ: Pearson/Prentice Hall, 2011.
3. Huston, Ronald, and Harold Josephs. *Practical Stress Analysis in Engineering Design.* 3rd ed. Boca Raton, FL: CRC Press, Taylor & Francis Group, 2009.
4. Mott, Robert L. *Applied Strength of Materials.* 5th ed. Upper Saddle River, NJ: Pearson/Prentice Hall, 2008.
5. Pilkey, Walter D., and Deborah F. Pilkey. *Peterson's Stress Concentration Factors.* 3rd ed. New York: John Wiley & Sons, 2008.
6. Timoshenko, S. *Strength of Materials, Part II—Advanced Theory and Problems.* New York: D. Van Nostrand Co., 1930.
7. Young, Warren C., and Richard G. Budynas. *Roark's Formulas for Stress and Strain.* 7th ed. New York: McGraw-Hill, 2002.

INTERNET SITES RELATED TO STRESS AND DEFORMATION ANALYSIS

1. BEAM 2D-Stress Analysis 3.1 www.orandsystems.com Software for mechanical, structural, civil, and architectural designers providing detailed analysis of statically indeterminate and determinate beams.
2. MDSolids www.mdsolids.com Educational software devoted to introductory mechanics of materials. Includes modules on basic stress and strain; beam and strut axial problems; trusses; statically indeterminate axial structures; torsion; determinate beams; section properties; general analysis of axial, torsion, and beam members; column buckling; pressure vessels; and Mohr's circle transformations.
3. eFatigue.com www.efatigue.com/constantamplitude/stress-concentration/ An Internet source for consultation and information related to design of machine components subjected to fatigue loading. The page cited above permits online calculation of stress concentration factors for 60 cases covering a variety of geometry types and loading conditions.

PROBLEMS

Direct Tension and Compression

1. A tensile member in a machine structure is subjected to a steady load of 4.50 kN. It has a length of 750 mm and is made from a steel tube having an outside diameter of 18 mm and an inside diameter of 12 mm. Compute the tensile stress in the tube and the axial deformation.
2. Compute the stress in a round bar having a diameter of 10.0 mm and subjected to a direct tensile force of 3500 N.
3. Compute the stress in a rectangular bar having cross-sectional dimensions of 10.0 mm by 30.0 mm when a direct tensile force of 20.0 kN is applied.
4. A link in a packaging machine mechanism has a square cross section 0.40 in on a side. It is subjected to a tensile force of 860 lb. Compute the stress in the link.
5. Two circular rods support the 3800 lb weight of a space heater in a warehouse. Each rod has a diameter of 0.375 in and carries 1/2 of the total load. Compute the stress in the rods.
6. A tensile load of 5.00 kN is applied to a square bar, 12 mm on a side and having a length of 1.65 m. Compute the stress and the axial deformation in the bar if it is made from (a) SAE 1020 hot-rolled steel, (b) SAE 8650 OQT 1000 steel, (c) ductile iron A536(60-40-18), (d) aluminum 6061-T6, (e) titanium Ti-6Al-4V, (f) rigid PVC plastic, and (g) phenolic plastic.
7. An aluminum rod is made in the form of a hollow square tube, 2.25 in outside, with a wall thickness of 0.120 in. Its length is 16.0 in. What axial compressive force would cause the tube to shorten by 0.004 in? Compute the resulting compressive stress in the aluminum.
8. Compute the stress in the middle portion of rod *AC* in Figure P3–8 if the vertical force on the boom is 2500 lb. The rod is rectangular, 1.50 in by 3.50 in.

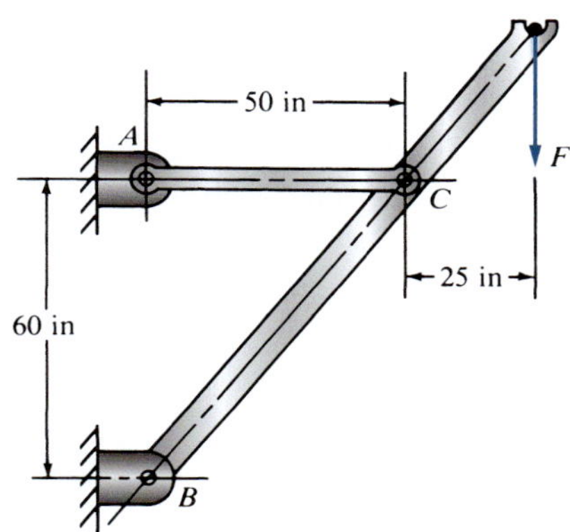

FIGURE P3–8 (Problems 8, 16, 56, and 80)

9. Compute the forces in the two angled rods in Figure P3–9 for an applied force, $F = 1500$ lb, if the angle θ is 45°.
10. If the rods from Problem 9 are circular, determine their required diameter if the load is static and the allowable stress is 18 000 psi.
11. Repeat Problems 9 and 10 if the angle θ is 15°.

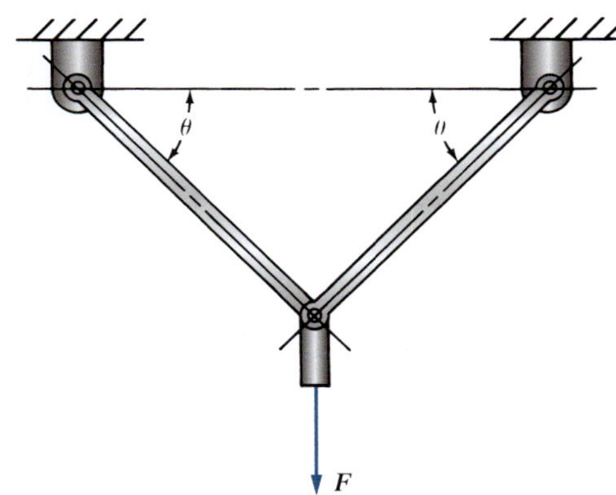

FIGURE P3–9 (Problems 9, 10, 11, 17, and 18)

12. Figure P3–12 shows a small truss spanning between solid supports and suspending a 10.5 kN load. The cross sections for the three main types of truss members are shown. Compute the stresses in all of the members of the truss near their midpoints away from the connections. Consider all joints to be pinned.

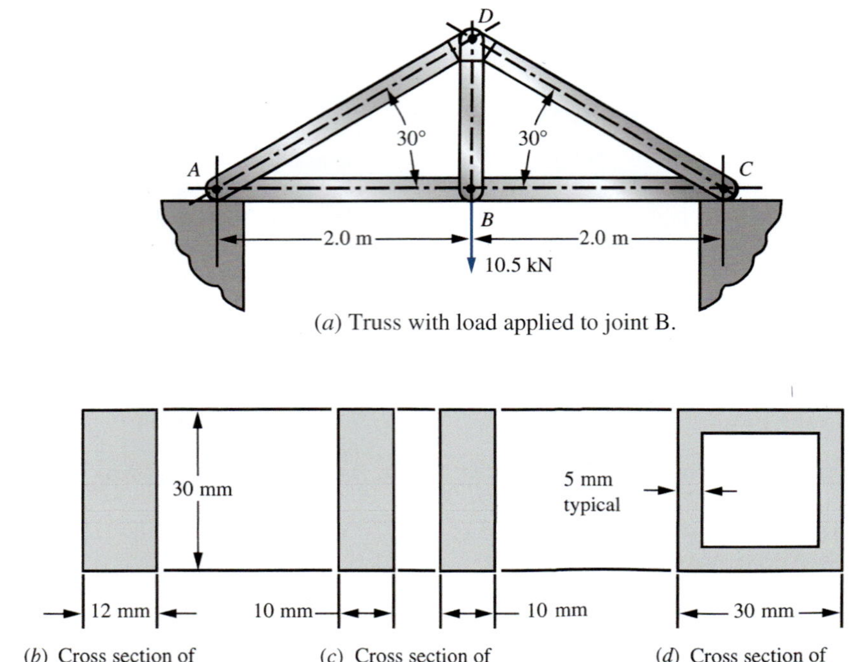

(a) Truss with load applied to joint B.

(b) Cross section of members AB, BC

(c) Cross section of member BD

(d) Cross section of members AD, CD

FIGURE P3–12 (Problem 12)

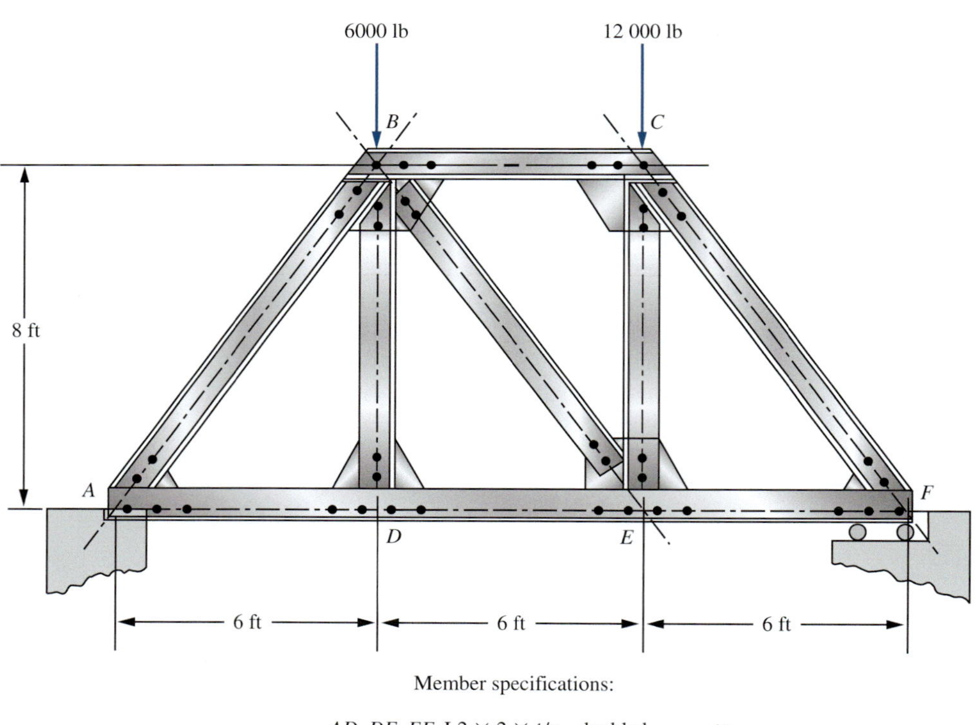

Member specifications:

AD, DE, EF L2 × 2 × 1/8 – doubled

BD, CE, BE L2 × 2 × 1/8 – single

AB, BC, CF C3 × 4.1 – doubled

FIGURE P3–13 (Problem 13)

13. The truss shown in Figure P3–13 spans a total space of 18.0 ft and carries two concentrated loads on its top chord. The members are made from standard steel angle and channel shapes as indicated in the figure. Consider all joints to be pinned. Compute the stresses in all members near their midpoints away from the connections.

14. Figure P3–14 shows a short leg for a machine that carries a direct compression load. Compute the compressive stress if the

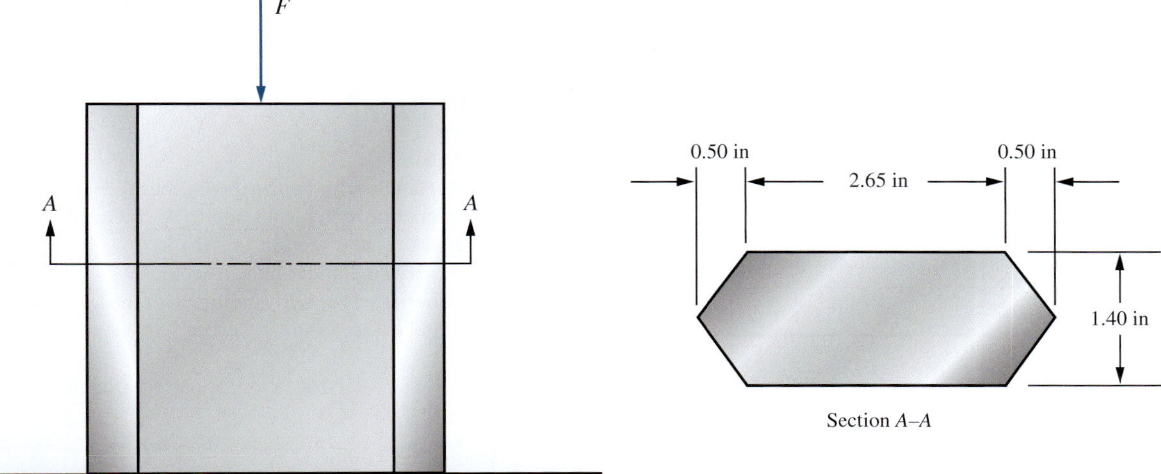

FIGURE P3–14 (Problem 14)

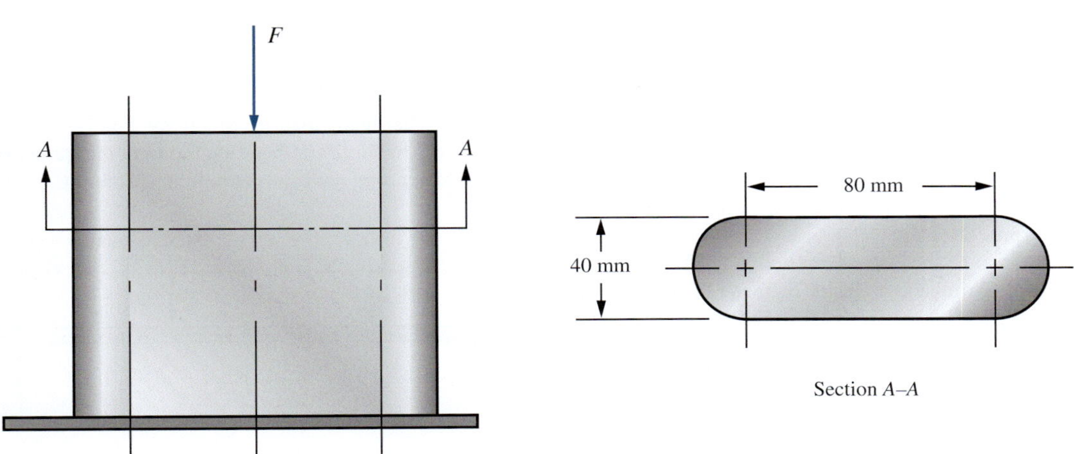

FIGURE P3–15 (Problem 15)

cross section has the shape shown and the applied force is $F = 52\,000\,\text{lb}$.

15. Consider the short compression member shown in Figure P3–15. Compute the compressive stress if the cross section has the shape shown and the applied load is 640 kN.

Direct Shear Stress

16. Refer to Figure P3–8. Each of the pins at A, B, and C has a diameter of 0.50 in and is loaded in double shear. Compute the shear stress in each pin.
17. Compute the shear stress in the pins connecting the rods shown in Figure P3–9 when a load of $F = 1500$ lb is carried. The pins have a diameter of 0.75 in. The angle $\theta = 40°$.
18. Repeat Problem 17, but change the angle to $\theta = 15°$.
19. Refer to Figure 3–7. Compute the shear stress in the key if the shaft transmits a torque of 1600 N·M. The shaft diameter is 60 mm. The key is square with $b = 12\,\text{mm}$, and it has a length of 45 mm.
20. A punch is attempting to cut a slug having the shape shown in Figure P3–20 from a sheet of aluminum having a thickness of 0.060 in. Compute the shearing stress in the aluminum when a force of 52 000 lb is applied by the punch.
21. Figure P3–21 shows the shape of a slug that is to be cut from a sheet of steel having a thickness of 2.0 mm. If the punch exerts a force of 225 kN, compute the shearing stress in the steel.

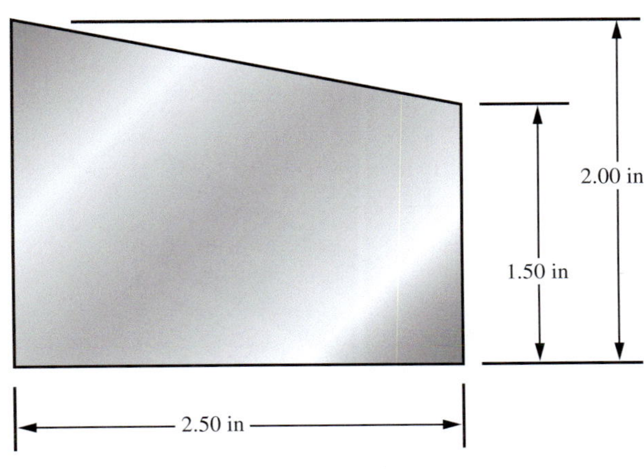

FIGURE P3–20 (Problem 20)

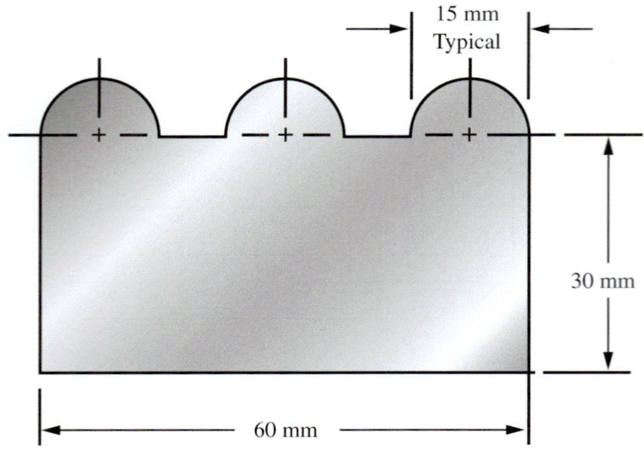

FIGURE P3–21 (Problem 21)

Torsion

22. Compute the torsional shear stress in a circular shaft with a diameter of 50 mm that is subjected to a torque of 800 N · m.
23. If the shaft of Problem 22 is 850 mm long and is made of steel, compute the angle of twist of one end in relation to the other.
24. Compute the torsional shear stress due to a torque of 88.0 lb · in in a circular shaft having a 0.40-in diameter.
25. Compute the torsional shear stress in a solid circular shaft having a diameter of 1.25 in that is transmitting 110 hp at a speed of 560 rpm.
26. Compute the torsional shear stress in a hollow shaft with an outside diameter of 40 mm and an inside diameter of 30 mm when transmitting 28 kilowatts (kW) of power at a speed of 45 rad/s.
27. Compute the angle of twist for the hollow shaft of Problem 26 over a length of 400 mm. The shaft is steel.

Noncircular Members in Torsion

28. A square steel bar, 25 mm on a side and 650 mm long, is subjected to a torque of 230 N · m. Compute the shear stress and the angle of twist for the bar.
29. A 3.00 in-diameter steel bar has a flat milled on one side, as shown in Figure P3–29. If the shaft is 44.0 in long and carries a torque of 10 600 lb · in, compute the stress and the angle of twist.
30. A commercial steel supplier lists rectangular steel tubing having outside dimensions of 4.00 by 2.00 in and a wall thickness of 0.109 in. Compute the maximum torque that can be applied to such a tube if the shear stress is to be limited to 6000 psi. For this torque, compute the angle of twist of the tube over a length of 6.5 ft.

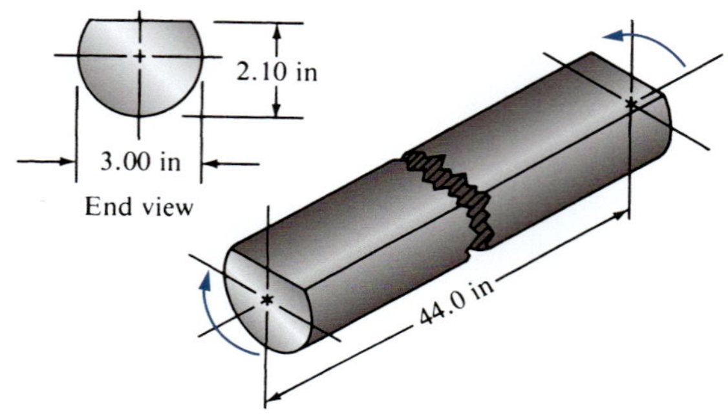

FIGURE P3–29 (Problem 29)

Beams

31. A beam is simply supported and carries the load shown in Figure P3–31. Specify suitable dimensions for the beam if it is steel and the stress is limited to 18 000 psi, for the following shapes:
    (a) Square
    (b) Rectangle with height three times the width
    (c) Rectangle with height one-third the width
    (d) Solid circular section
    (e) American Standard beam section
    (f) American Standard channel with the legs down
    (g) Standard steel pipe

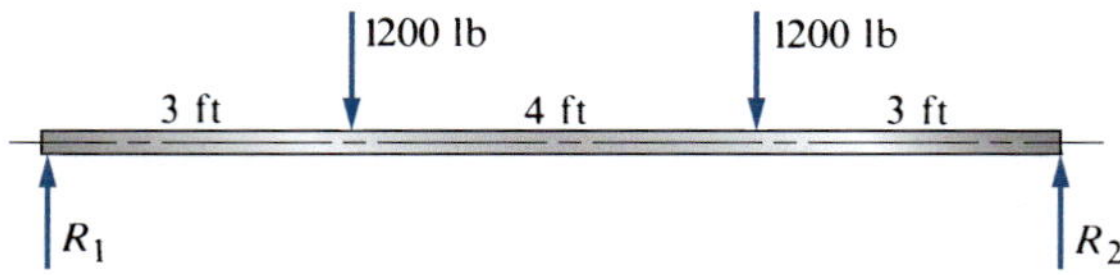

FIGURE P3–31 (Problems 31, 32, and 33)

32. For each beam of Problem 31, compute its weight if the steel weighs 0.283 lb/in^3.
33. For each beam of Problem 31, compute the maximum deflection and the deflection at the loads.
34. For the beam loading of Figure P3–34, draw the complete shearing force and bending moment diagrams, and determine the bending moments at points *A*, *B*, and *C*.

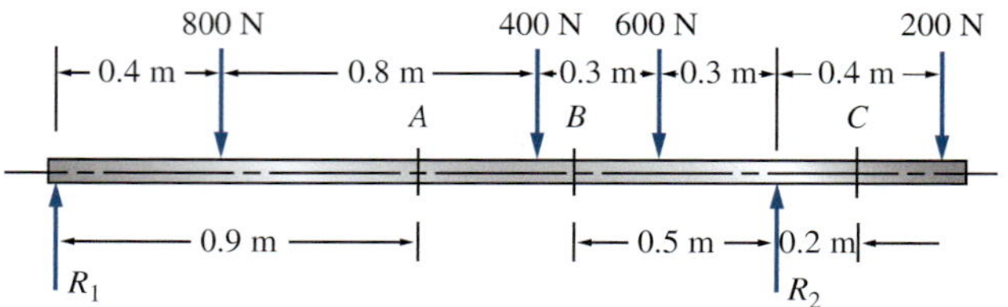

FIGURE P3–34 (Problems 34 and 35)

35. For the beam loading of P3–34, design the beam choosing a commercially available shape in standard SI units from Appendix 15 with the smallest cross-sectional area that will limit the bending stress to 100 MPa.
36. Figure P3–36 shows a beam made from 4 in schedule 40 steel pipe. Compute the deflection at points *A* and *B* for two cases: (a) the simple cantilever and (b) the supported cantilever.

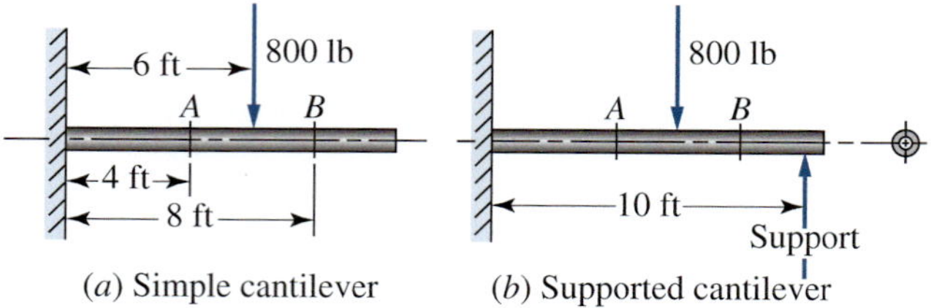

FIGURE P3–36 (Problem 36)

37. Select an aluminum I-beam shape to carry the load shown in Figure P3–37 with a maximum stress of 12 000 psi. Then compute the deflection at each load.
38. Figure P3–38 represents a wood joist for a platform, carrying a uniformly distributed load of 120 lb/ft and two concentrated loads applied by some machinery. Compute the maximum stress due to bending in the joist and the maximum vertical shear stress.

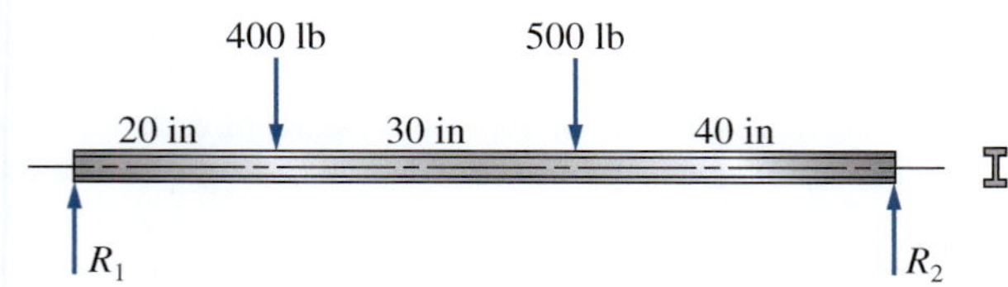

FIGURE P3–37 (Problem 37)

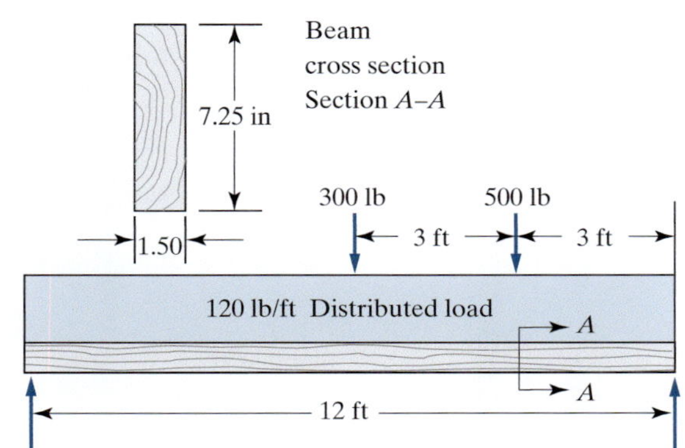

FIGURE P3–38 (Problem 38)

Beams with Concentrated Bending Moments

For Problems 39 through 50, draw the free-body diagram of only the horizontal beam portion of the given figures. Then draw the complete shear and bending moment diagrams. Where used, the symbol X indicates a simple support capable of exerting a reaction force in any direction but having no moment resistance. For beams having unbalanced axial loads, you may specify which support offers the reaction.

39. Use Figure P3–39.
40. Use Figure P3–40.
41. Use Figure P3–41.
42. Use Figure P3–42.
43. Use Figure P3–43.
44. Use Figure P3–44.
45. Use Figure P3–45.
46. Use Figure P3–46.
47. Use Figure P3–47.

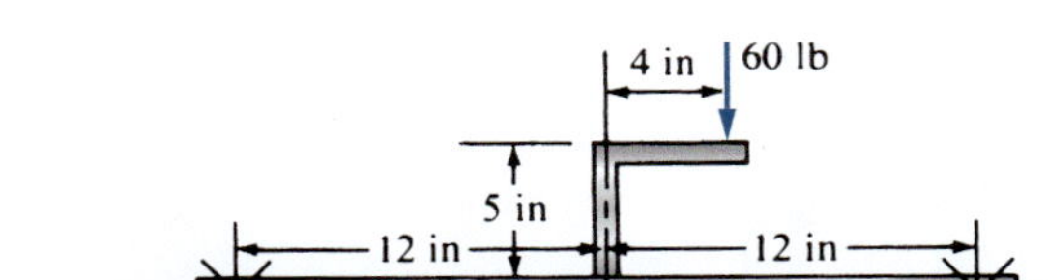

FIGURE P3–39 (Problems 39 and 57)

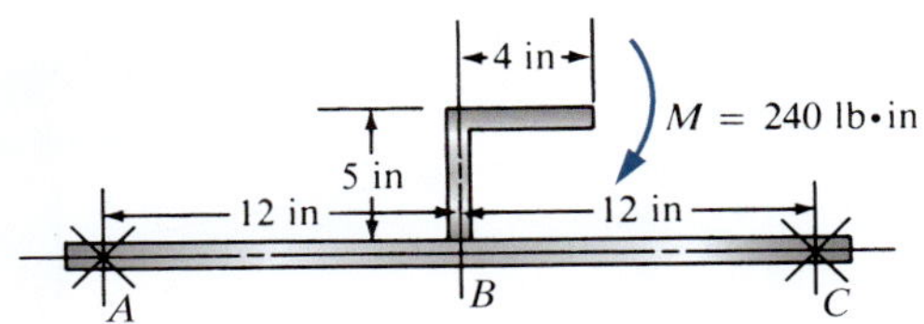

FIGURE P3–40 (Problem 40)

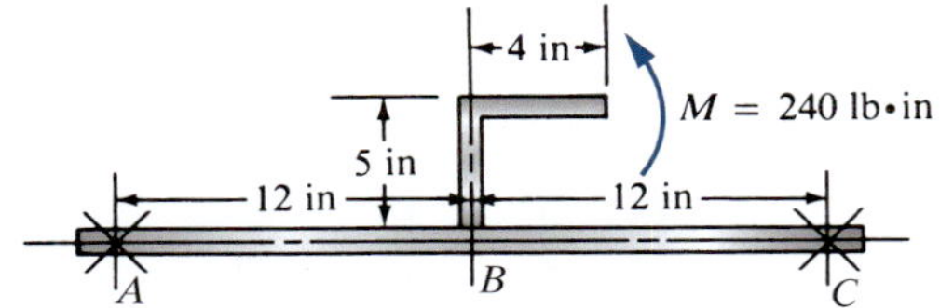

FIGURE P3–41 (Problem 41)

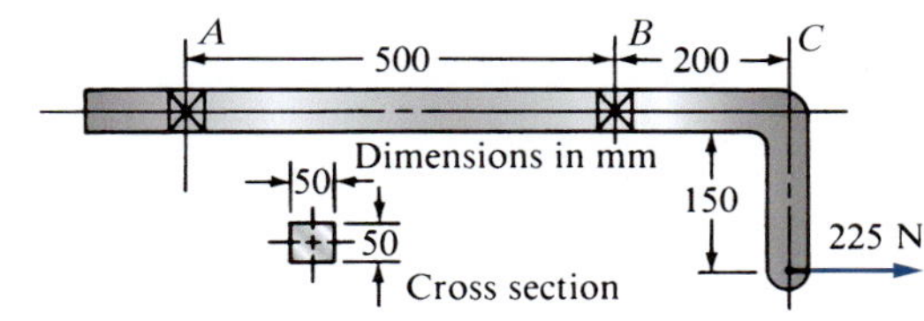

FIGURE P3–42 (Problems 42 and 58)

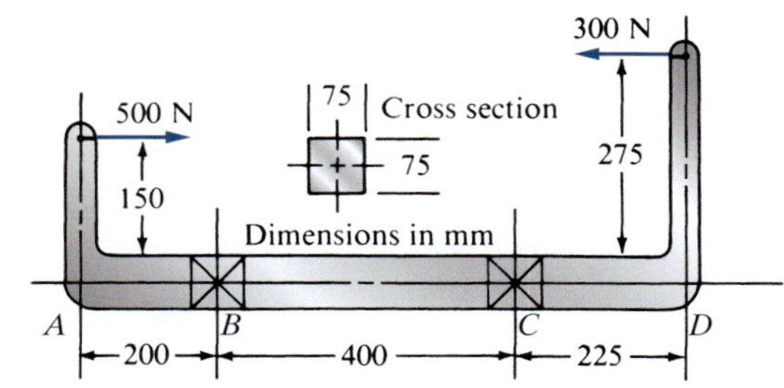

FIGURE P3–43 (Problems 43 and 59)

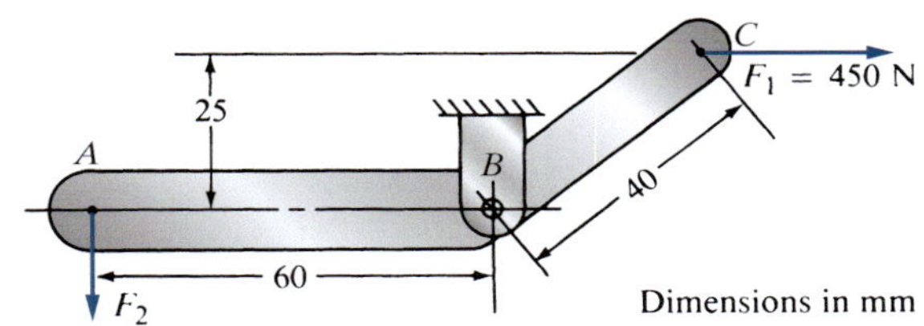

FIGURE P3–44 (Problem 44)

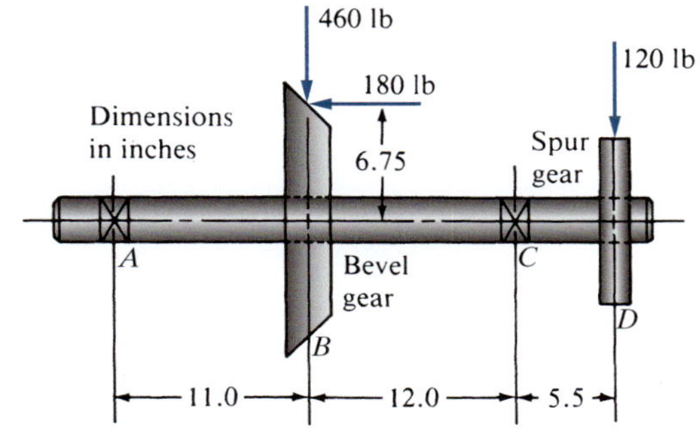

FIGURE P3–45 (Problem 45)

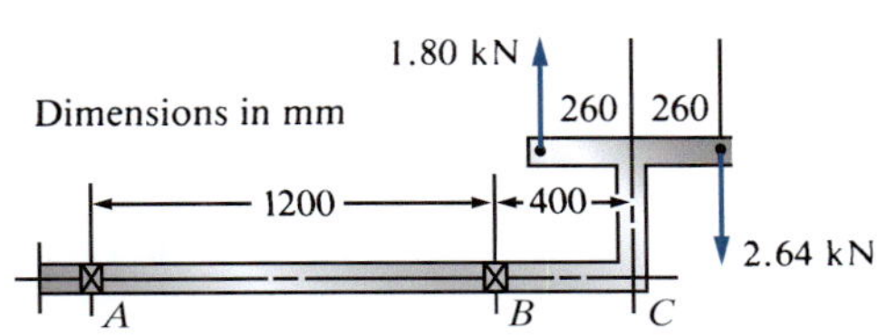

FIGURE P3–46 (Problem 46)

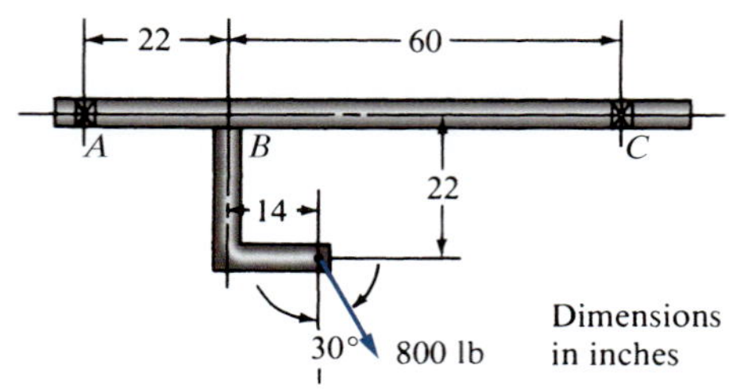

FIGURE P3–47 (Problems 47 and 60)

For Problems 48–50, draw the free-body diagram of the main shaft portion, labeled *A*, *B*, and *C*. Include any unbalanced torque on the shaft that tends to rotate it about the *z*-axis. In each case, the reaction to the unbalanced torque is taken at the right end of the shaft labeled *C*. Then draw the complete shearing force and bending moment diagrams for loading in the *y*–*z* plane. Also prepare a graph of the torque in the shaft as a function of position along the shaft from *A* to *C*.

48. Use Figure P3–48.
49. Use Figure P3–49.
50. Use Figure P3–50.

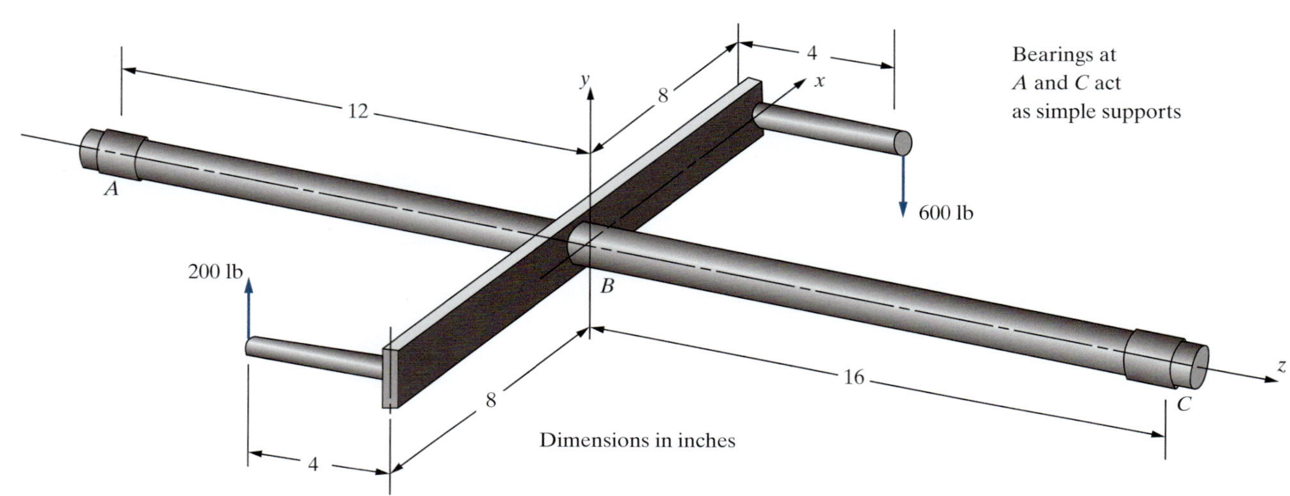

FIGURE P3–48 (Problem 48)

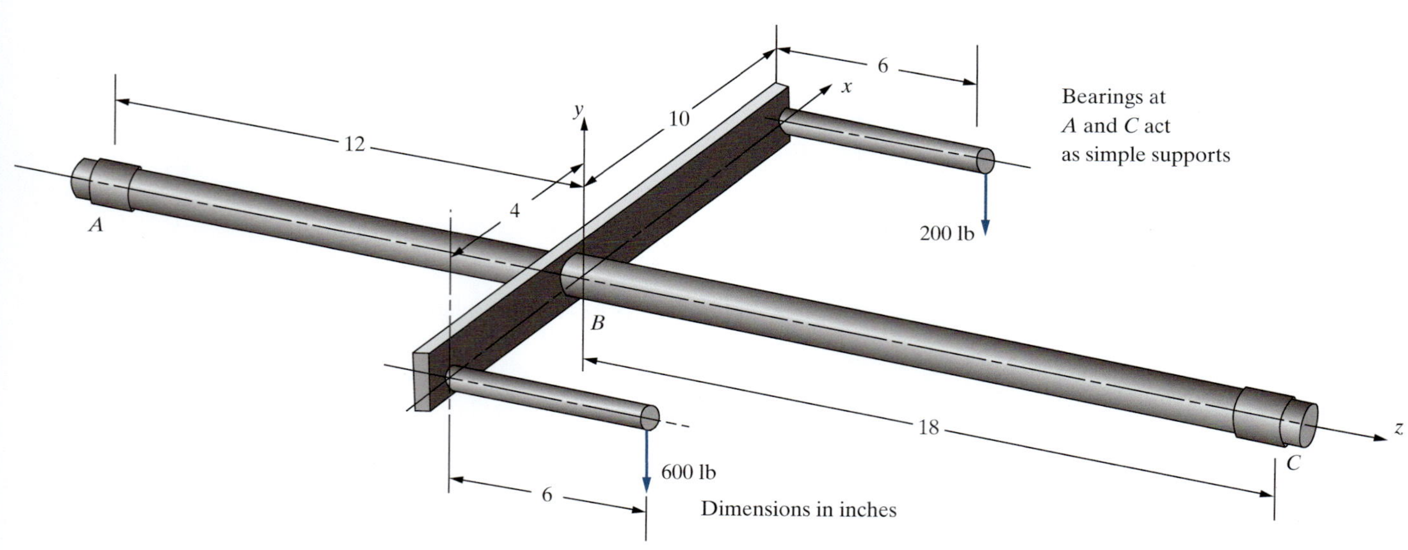

FIGURE P3–49 (Problem 49)

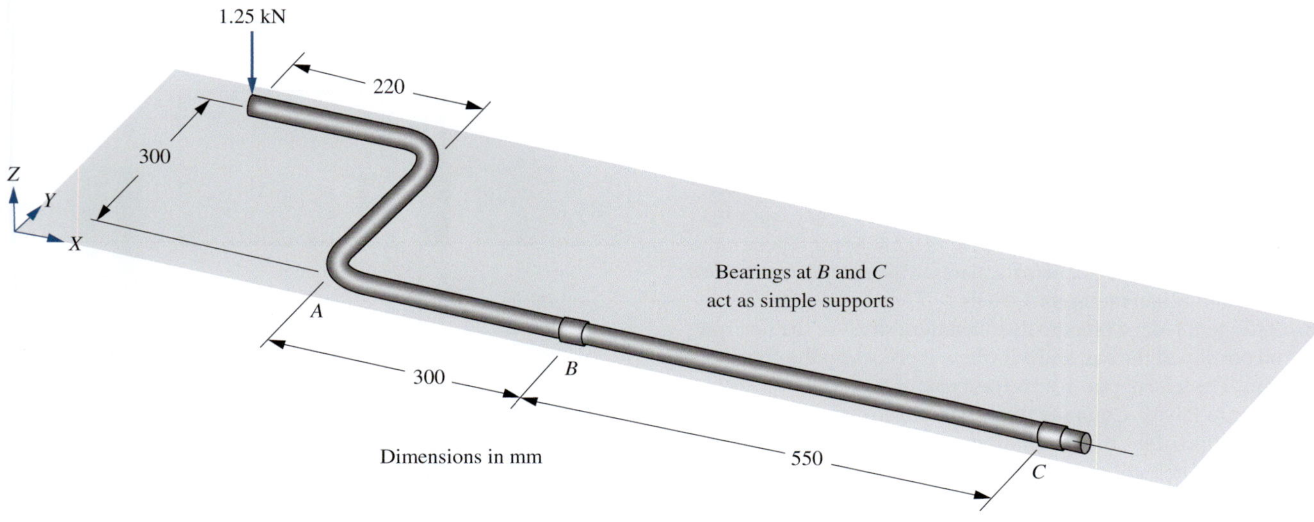

FIGURE P3–50 (Problem 50)

Combined Normal Stresses

51. Compute the maximum tensile stress in the bracket shown in Figure P3–51.

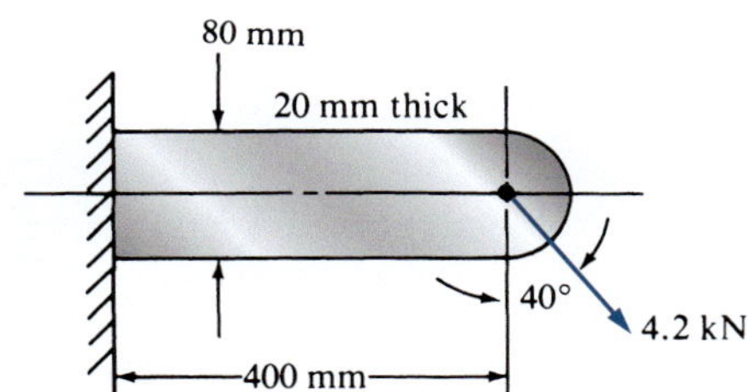

FIGURE P3–51 (Problem 51)

52. Compute the maximum tensile and compressive stresses in the horizontal beam shown in Figure P3–52.

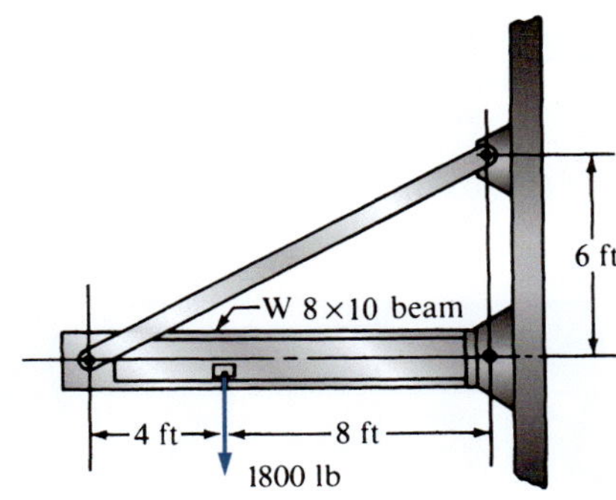

FIGURE P3–52 (Problem 52)

53. For the lever shown in Figure P3–53(a), compute the stress at section A near the fixed end. Then redesign the lever to the tapered form shown in Part (b) of the figure by adjusting only the height of the cross section at sections *B* and *C* so that they have no greater stress than section *A*.

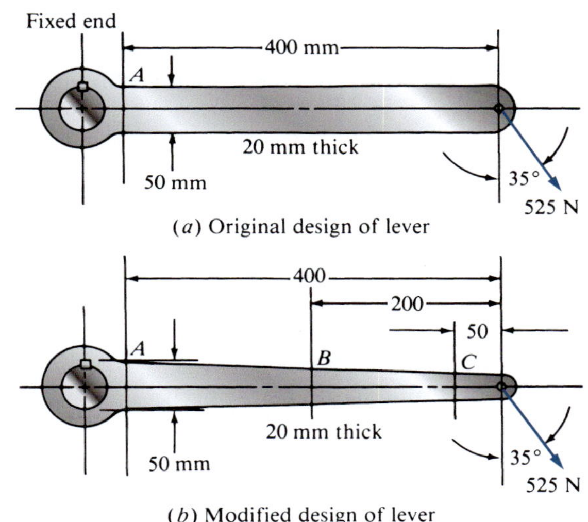

FIGURE P3–53 (Problem 53)

54. Compute the maximum tensile stress at sections *A* and *B* on the crane boom shown in Figure P3–54.

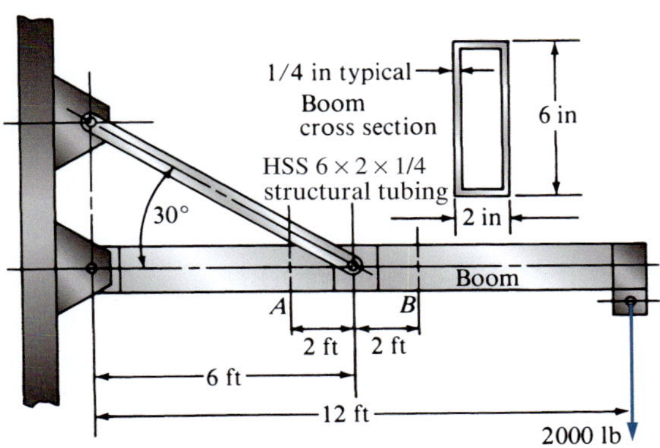

FIGURE P3–54 (Problem 54)

55. Refer to Figure 3–22. Compute the maximum tensile stress in the print head just to the right of the right guide. The head has a rectangular cross section, 5.0 mm high in the plane of the paper and 2.4 mm thick.
56. Refer to Figure P3–8. Compute the maximum tensile and compressive stresses in the member *B-C* if the load *F* is 1800 lb. The cross section of *B-C* is a HSS 6 × 4 × 1/4 rectangular tube.
57. Refer to P3–39. The vertical member is to be made from steel with a maximum allowable stress of 12 000 psi. Specify the required size of a standard square cross section if sizes are available in increments of 1/16 in.
58. Refer to P3–42. Compute the maximum stress in the horizontal portion of the bar, and tell where it occurs on the cross section. The left support resists the axial force.
59. Refer to P3–43. Compute the maximum stress in the horizontal portion of the bar, and indicate where it occurs on the cross section. The right support resists the unbalanced axial force.
60. Refer to P3–47. Specify a suitable diameter for a solid circular bar to be used for the top horizontal member, which is supported in the bearings. The left bearing resists the axial load. The allowable normal stress is 25 000 psi.

Stress Concentrations

61. Figure P3–61 shows a valve stem from an engine subjected to an axial tensile load applied by the valve spring. For a force of 1.25 kN, compute the maximum stress at the fillet under the shoulder.

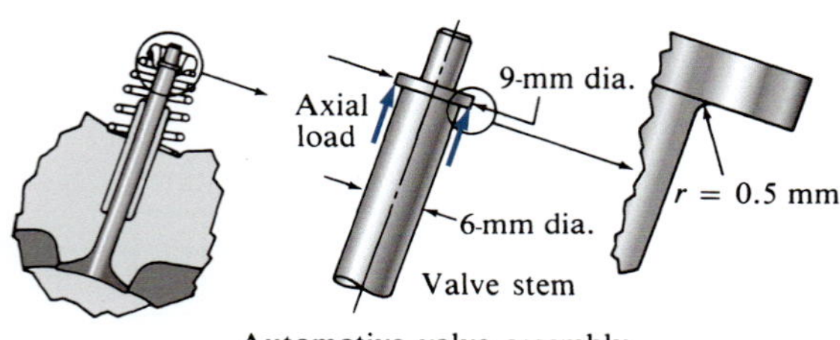

FIGURE P3–61 (Problem 61)

62. The conveyor fixture shown in Figure P3–62 carries three heavy assemblies (1200 lb each). Compute the maximum stress in the fixture, considering stress concentrations at the fillets and assuming that the load acts axially.
63. For the flat plate in tension in Figure P3–63, compute the stress at each hole, assuming that the holes are sufficiently far apart that their effects do not interact.

For Problems 64 through 68, compute the maximum stress in the member, considering stress concentrations.

64. Use Figure P3–64.
65. Use Figure P3–65.
66. Use Figure P3–66.
67. Use Figure P3–67.
68. Use Figure P3–68.

Problems of a General Nature

69. Figure P3–69 shows a horizontal beam supported by a vertical tension link. The cross sections of both the beam and the link are 20 mm square. All connections use 8.00 mm-diameter cylindrical pins in double shear. Compute the tensile stress in member *A-B*, the stress due to bending in *C-D*, and the shearing stress in the pins *A* and *C*.

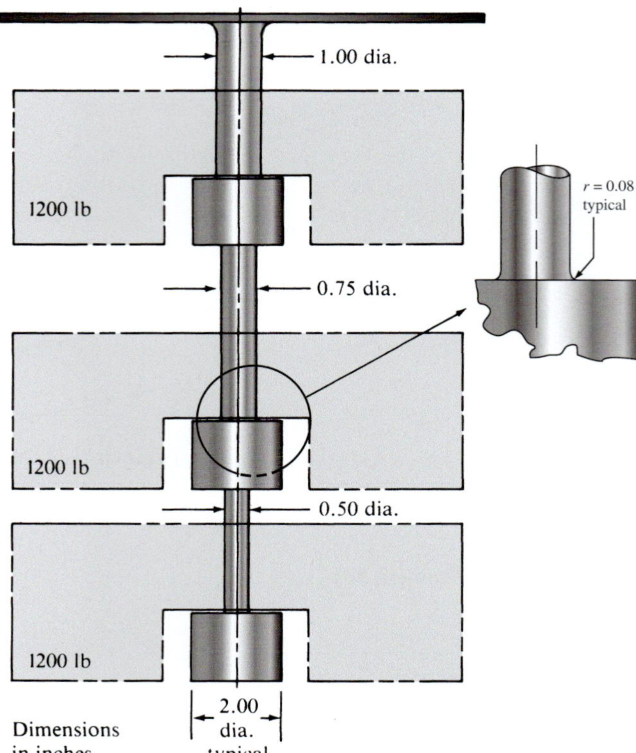

FIGURE P3–62 (Problem 62)

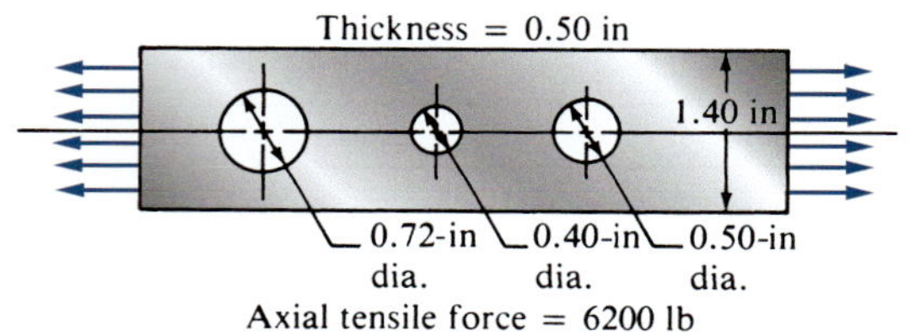

FIGURE P3–63 (Problem 63)

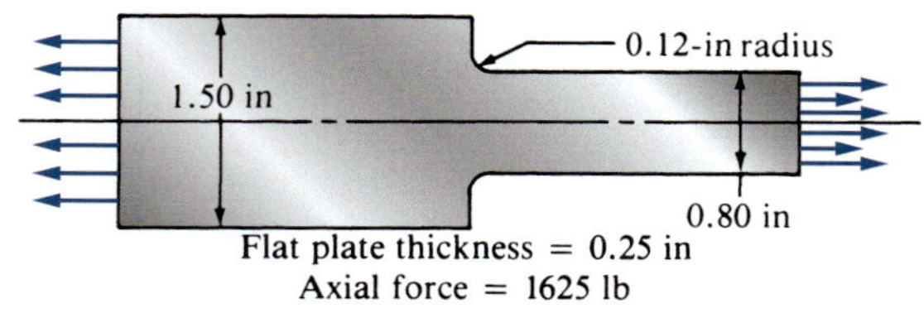

FIGURE P3–64 (Problem 64)

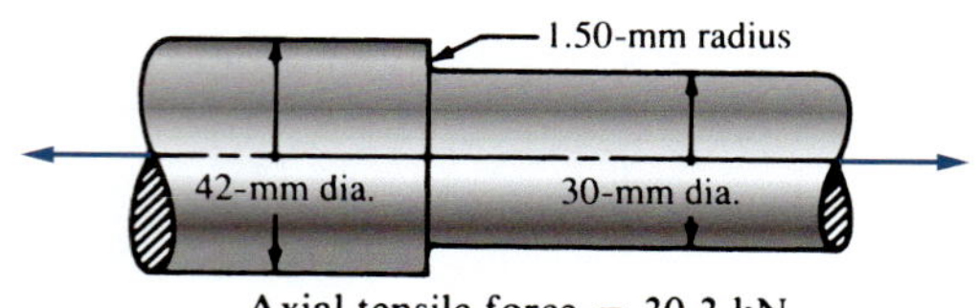

FIGURE P3–65 (Problem 65)

70. Figure P3–70 shows a tapered flat bar that has a uniform thickness of 20 mm. The depth tapers from h_1 = 40 mm near the load to h_2 = 20 mm at each support. Compute the stress due to bending in the bar at points spaced 40 mm apart from the support to the load. Let the load *P* = 5.0 kN.

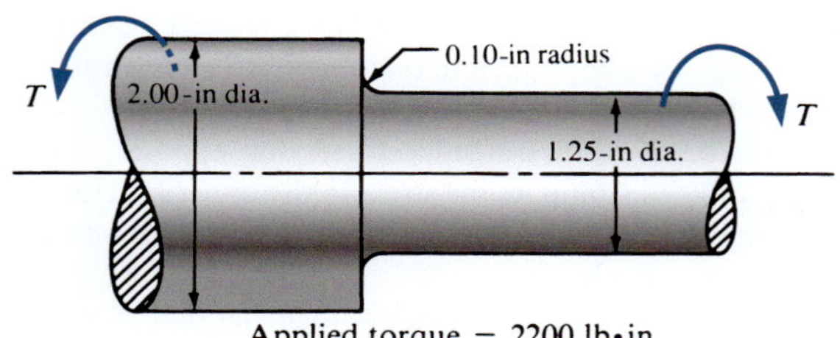

FIGURE P3–66 (Problem 66)

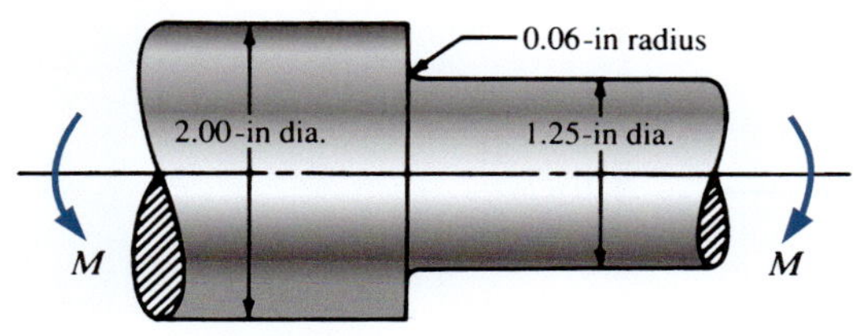

FIGURE P3–67 (Problem 67)

71. For the flat bar shown in Figure P3–70, compute the stress in the middle of the bar if a hole of 25 mm diameter is drilled directly under the load on the horizontal centerline. The load is $P = 5.0$ kN. See data in Problem 70.
72. The beam shown in Figure P3–72 is a stepped, flat bar having a constant thickness of 1.20 in. It carries a single concentrated

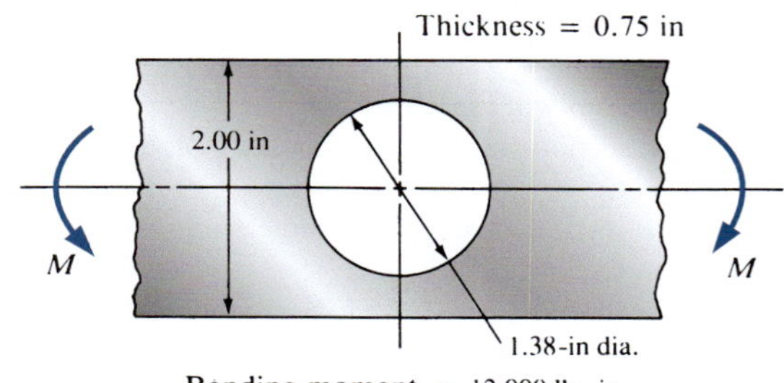

FIGURE P3–68 (Problem 68)

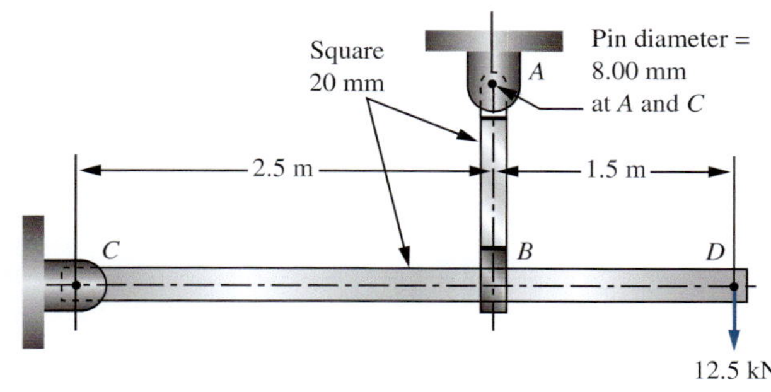

FIGURE P3–69 (Problem 69)

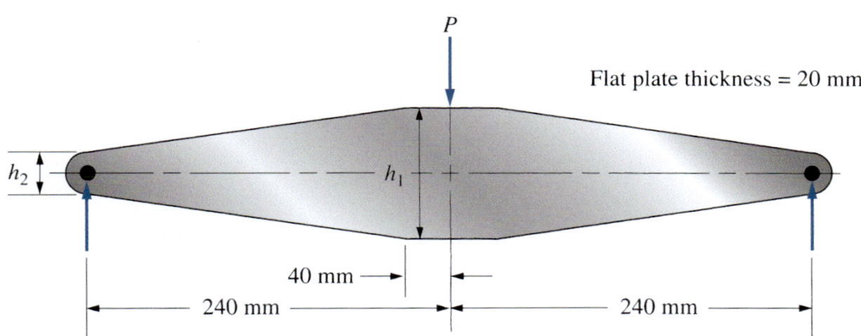

FIGURE P3–70 Tapered flat bar for Problems 70 and 71

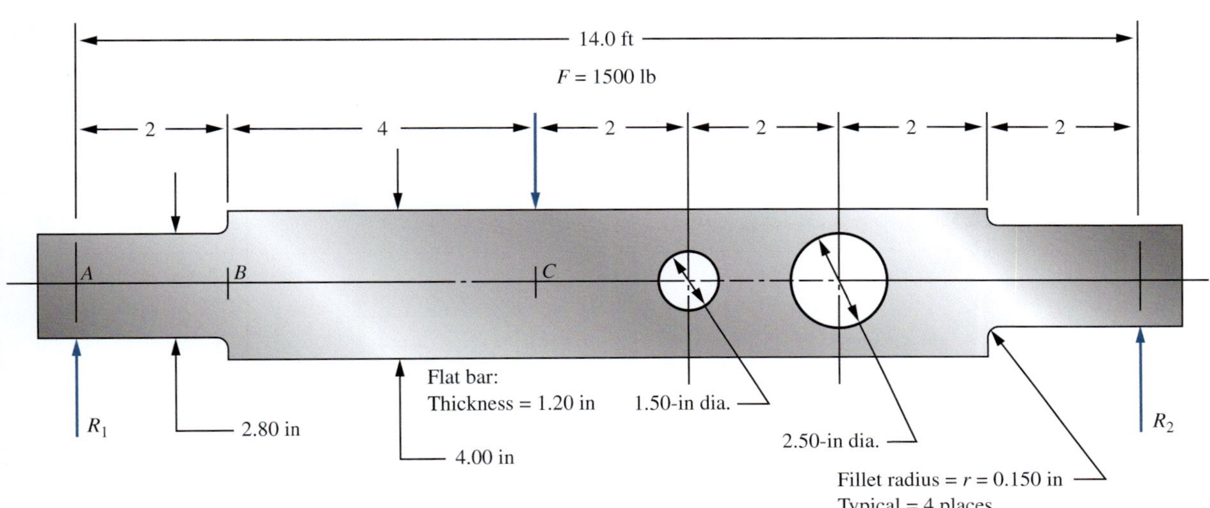

FIGURE P3–72 (Problem 72)

load at C of 1500 lb. Compare the stresses at the following locations:

(a) In the vicinity of the load

(b) At the section through the smaller hole to the right of section C

(c) At the section through the larger hole to the right of section C

(d) Near section B where the bar changes height

73. Figure P3–73 shows a stepped, flat bar having a constant thickness of 8.0 mm. It carries three concentrated loads as shown. Let $P = 200$ N, $L_1 = 180$ mm, $L_2 = 80$ mm, and $L_3 = 40$ mm. Compute the maximum stress due to bending, and state where it occurs. The bar is braced against lateral bending and twisting. Note that the dimensions in the figure are not drawn to scale.

74. Figure P3–74 shows a bracket carrying opposing forces of $F = 2500$ N. Compute the stress in the upper horizontal part through one of the holes as at B. Use $d = 15.0$ mm for the diameter of the holes.

75. Repeat Problem 74, but use a hole diameter of $d = 12.0$ mm.

76. Figure P3–76 shows a lever made from a rectangular bar of steel. Compute the stress due to bending at the fulcrum (20 in from the pivot) and at the section through the bottom hole. The diameter of each hole is 1.25 in.

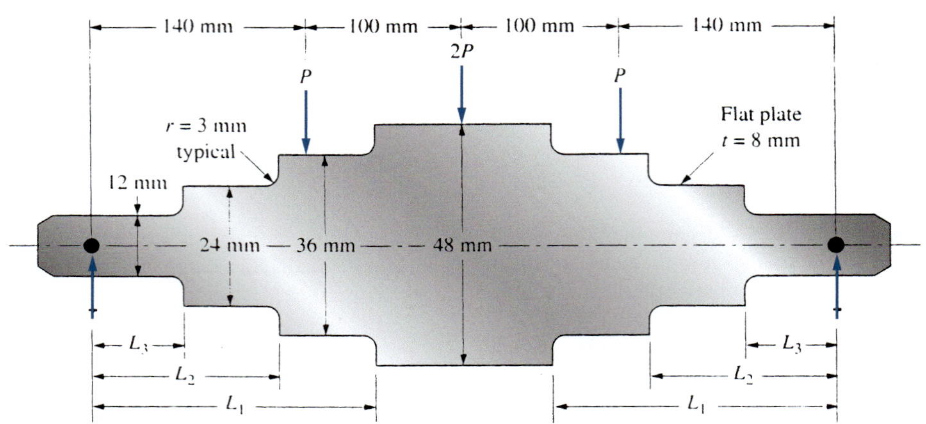

FIGURE P3–73 Stepped flat bar for Problem 73

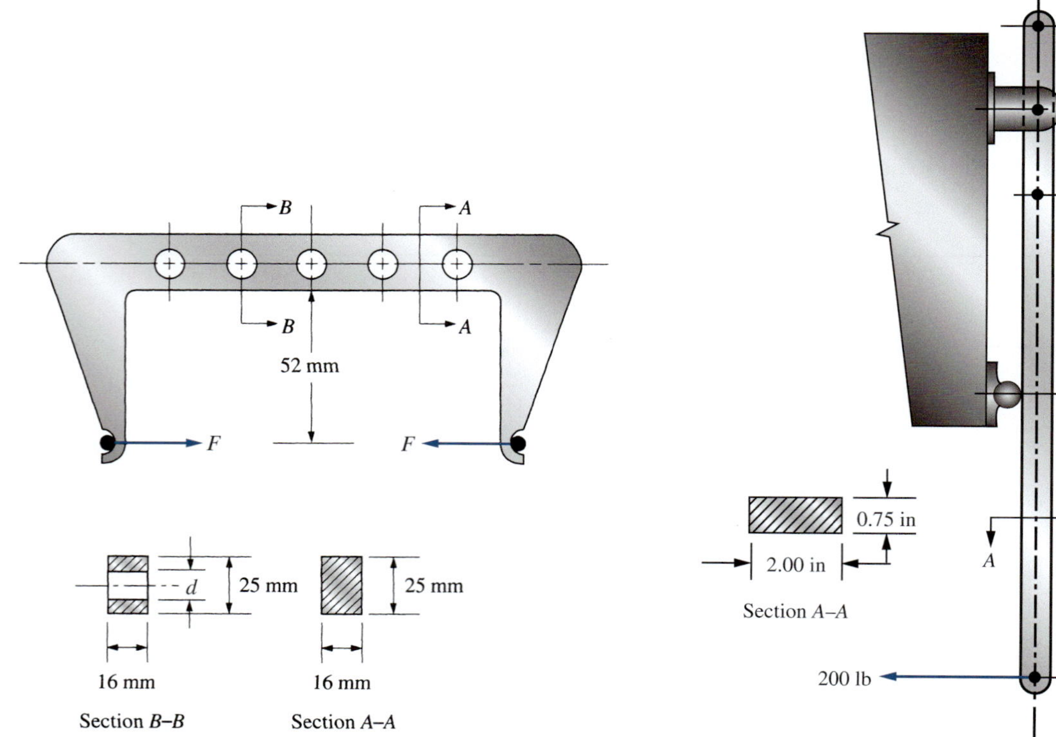

FIGURE P3–74 Bracket for Problems 74 and 75

FIGURE P3–76 Lever for Problems 76 and 77

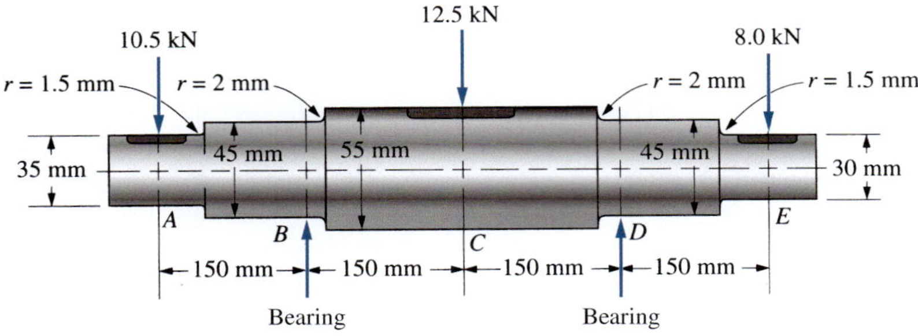

FIGURE P3–78 Data for Problem 78

77. For the lever in P3–76, determine the maximum stress if the attachment point is moved to each of the other two holes.
78. Figure P3–78 shows a shaft that is loaded only in bending. Bearings are located at points B and D to allow the shaft to rotate. Pulleys at A, C, and E carry cables that support loads from below while allowing the shaft to rotate. Compute the maximum stress due to bending in the shaft considering stress concentrations.

Lug Joints

79. For the vertical beam support shown in Figure P3–69, design the clevis joint at point A according to the recommendations described in Section 3–21 on stress concentrations. Use the given pin diameter, $d = 8.0$ mm, and lug width, $w = 20.0$ mm, using the terminology shown in Figure 3–28. The primary design decisions are the thickness of the lug, t, and the materials for the lug and the pin. Work toward a design factor of $N = 5$ based on ultimate strength for both tension in the lug and shearing of the pin.
80. Repeat Problem 3–79 for Joint A in Figure P3–8, using the data from Problem 3–8. All dimensions for the pin and the lug of the clevis are to be specified.

Curved Beams

81. A hanger is made from ASTM A36 structural steel bar with a square cross section, 10 mm on a side, as shown in Figure P3–81. The radius of curvature is 150 mm to the inside surface of the bar. Determine the load F that would cause yielding of the steel.
82. A coping saw frame shown in Figure P3–82 is made from SAE 1020 CD steel. A screw thread in the handle draws the blade of the saw into a tension of 120 N. Determine the resulting

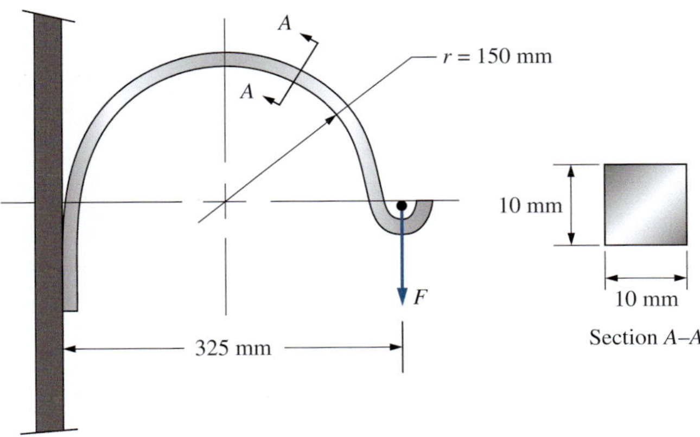

FIGURE P3–81 Hanger for Problem 3–81

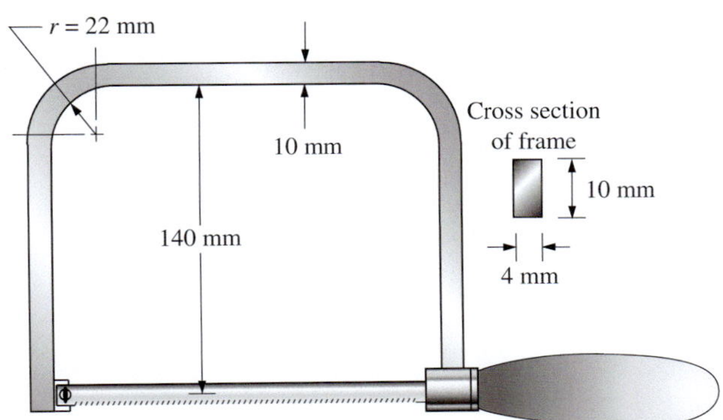

FIGURE P3–82 Coping saw frame for Problem 3–82

design factor based on yield strength in the area of the corner radii of the frame.

83. Repeat Problem 3–82 for the hacksaw frame shown in Figure P3–83 when the tensile force in the blade is 480 N.

84. Figure P3–84 shows a hand garden tool used to break up soil. Compute the force applied to the end of one prong that would cause yielding in the curved area. The tool is made from cast aluminum, alloy 356.0-T6.

85. Figure P3–85 shows a basketball backboard and goal attached to a steel pipe that is firmly cemented into the ground. The force, $F = 230$ lb, represents a husky player hanging from the back of the rim. Compute the design factor based on yield strength for the pipe if it is made from ASTM A53 Grade B structural steel.

86. The C-clamp in Figure P3–86 is made of cast zinc, ZA12. Determine the force that the clamp can exert for a design factor of 3 based on ultimate strength in either tension or compression.

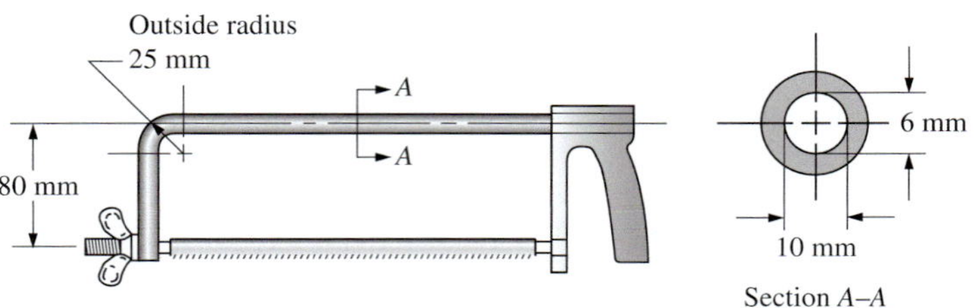

FIGURE P3–83 Hacksaw frame for Problem 3–83

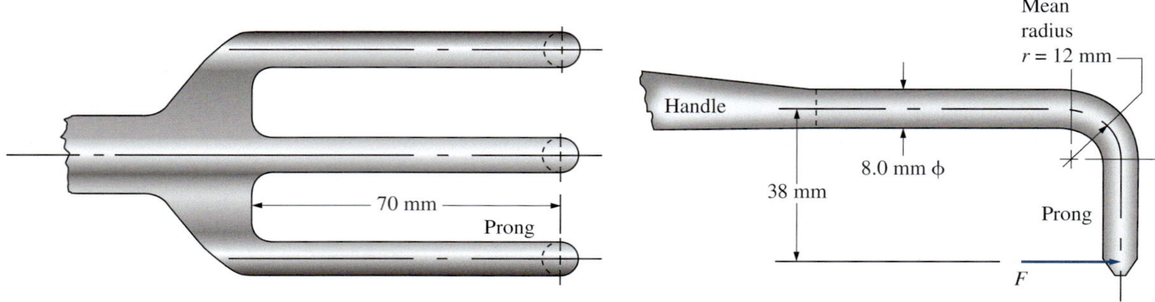

FIGURE P3–84 Garden tool for Problem 3–84

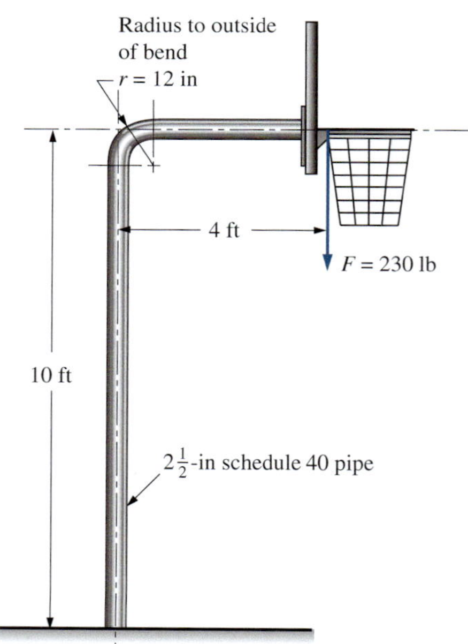

FIGURE P3–85 Basketball backboard for Problem 3–85

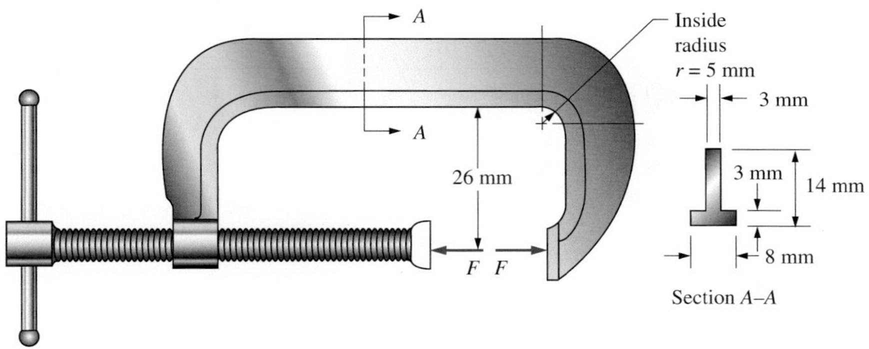

FIGURE P3–86 C-clamp for problem 3–86

CHAPTER FOUR

COMBINED STRESSES AND MOHR'S CIRCLE

THE BIG PICTURE — Combined Stresses and Mohr's Circle

Discussion Map

- You must build your ability to analyze more complex parts and loading patterns.

Discover

Find products around you that have complex geometries or loading patterns.

Discuss these products with your colleagues.

This chapter helps you analyze complex objects to determine maximum stresses. We will use *Mohr's circle*, a graphical tool for stress analysis, as an aid in understanding how stresses vary within a load-carrying member.

In Chapter 3, you reviewed the basic principles of stress and deformation analysis, practiced the application of those principles to machine design problems, and solved some problems by superposition when two or more types of loads caused normal, either tensile or compressive, stresses.

But what happens when the loading pattern is more complex?

Many practical machine components experience combinations of normal and shear stresses. Sometimes the pattern of loading or the geometry of the component causes the analysis to be very difficult to solve directly using the methods of basic stress analysis.

Look around you and identify products, parts of structures, or machine components that have a more complex loading or geometry. Perhaps some of those identified in **The Big Picture** for Chapter 3 have this characteristic.

Discuss how the selected items are loaded, where the maximum stresses are likely to occur, and how the loads and the geometry are related. Did the designer tailor the shape of the object to be able to carry the applied loads in an efficient manner? How are the shape and the size of critical parts of the item related to the expected stresses?

When we move on to **Chapter 5: Design for Different Types of Loading,** we will need tools to determine the magnitude and the direction of maximum shear stresses or maximum principal (normal) stresses.

Completing this chapter will help you develop a clear understanding of the distribution of stress in a load-carrying member, and it will help you determine the maximum stresses, either normal or shear, so that you can complete a reliable design or analysis.

Some of the techniques of combining stresses require the application of fairly involved equations. A graphical tool, *Mohr's circle*, can be used as an aid in completion of the analysis. Applied properly, the method is precise and should aid you in understanding how the stresses vary within a complex load-carrying member. It should also help you correctly use commercially available stress analysis software.

YOU ARE THE DESIGNER

Your company is designing a special machine to test a high-strength fabric under prolonged exposure to a static load to determine whether it continues to deform a greater amount with time. The tests will be run at a variety of temperatures requiring a controlled environment around the test specimen. Figure 4–1 shows the general layout of one proposed design. Two rigid supports are available at the rear of the machine with a 24-in gap between them. The line of action of the load on the test fabric is centered on this gap and 15.0 in out from the middle of the supports. You are asked to design a bracket to hold the upper end of the load frame.

Assume that one of your design concepts uses the arrangement shown in Figure 4–2. Two circular bars are bent 90°. One end of

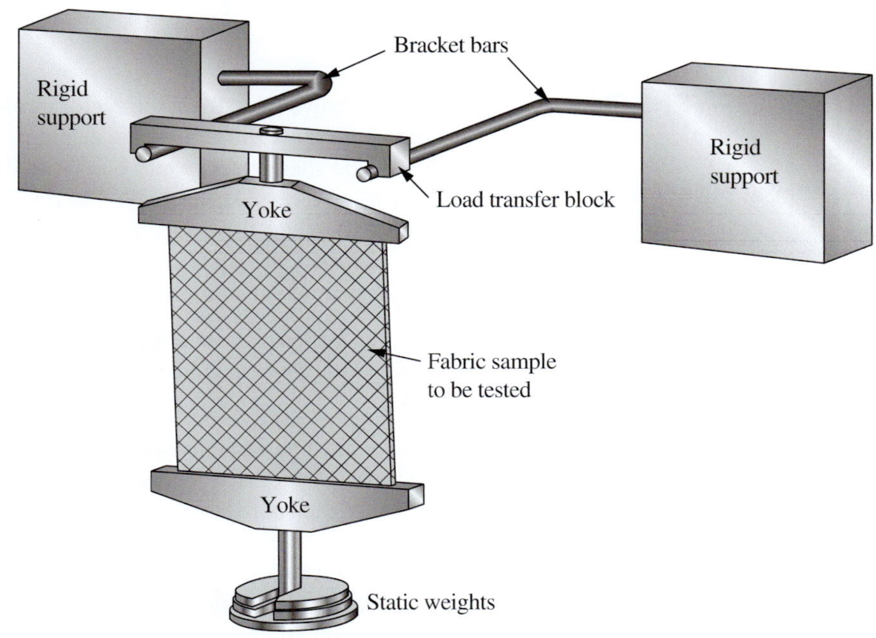

FIGURE 4–1 General concept for a load frame for testing fabric strength

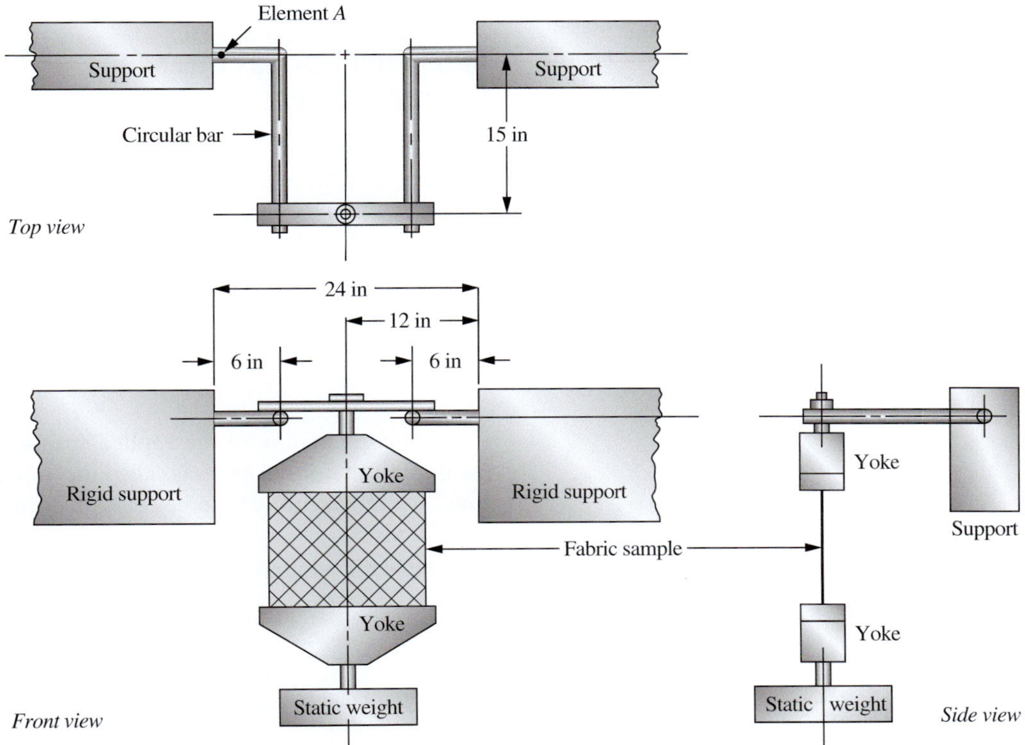

FIGURE 4–2 Proposed bracket design

each bar is securely welded to the vertical support surface. A flat load transfer block is attached across the outboard end of each bar so that the load is shared evenly by the two bars.

One of your design problems is to determine the maximum stress that exists in the bent bars to ensure that they are safe. What kinds of stress are developed in the bars? Where are the stresses likely to be the greatest? How can the magnitude of the stresses be computed? Note that the part of the bar near its attachment to the support has a combination of stresses exerted on it.

Consider the element on the top surface of the bar, labeled element *A* in Figure 4–2. The moment caused by the force acting at an extension of 6.0 in from the support places element *A* in tension due to the bending action. The torque caused by the force acting 15.0 in out from the axis of the bar at its point of support creates a torsional shear stress on element *A*. Both of these stresses act in the *x*–*y* plane, subjecting element *A* to a combined normal and shear stress. How do you analyze such a stress condition? How do the tensile and shear stresses act together? What are the maximum normal stress and the maximum shear stress on element *A*, and where do they occur?

You would need such answers to complete the design of the bars. The material in this chapter will enable you to complete the necessary analyses. ■

4–1 OBJECTIVES OF THIS CHAPTER

After completing this chapter, you will be able to:

1. Illustrate a variety of combined stresses on stress elements.
2. Analyze a load-carrying member subjected to combined stress to determine the maximum normal stress and the maximum shear stress on any given element.
3. Determine the directions in which the maximum stresses are aligned.
4. Determine the state of stress on an element in any specified direction.
5. Draw the complete Mohr's circle as an aid in completing the analyses for the maximum stresses.

4–2 GENERAL CASE OF COMBINED STRESS

To visualize the general case of combined stress, it is helpful to consider a small element of the load-carrying member on which combined normal and shear stresses act. For this discussion we will consider a two-dimensional stress condition, as illustrated in Figure 4–3. The *x*- and *y*-axes are aligned with corresponding axes on the member being analyzed.

The normal stresses, σ_x and σ_y, could be due to a direct tensile force or due to bending. If the normal stresses were compressive (negative), the vectors would be pointing in the opposite sense, into the stress element.

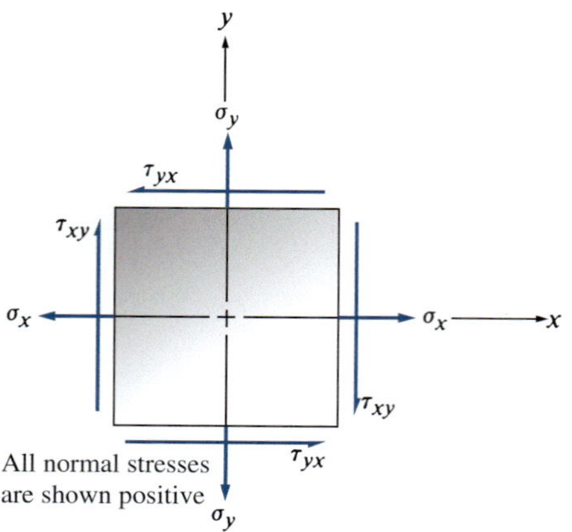

FIGURE 4–3 General two-dimensional stress element

The shear stress could be due to direct shear, torsional shear, or vertical shear stress. The double-subscript notation helps to orient the direction of shear stresses. For example, τ_{xy} indicates the shear stress acting on the element face that is perpendicular to the *x*-axis and parallel to the *y*-axis.

A positive shear stress is one that tends to rotate the stress element clockwise.

In Figure 4–3, τ_{xy} is positive, and τ_{yx} is negative. Their magnitudes must be equal to maintain the element in equilibrium.

It is necessary to determine the magnitudes and the signs of each of these stresses in order to show them properly on the stress element. Example Problem 4–1, which follows the definition of principal stresses, illustrates the process.

With the stress element defined, the objectives of the remaining analysis are to determine the maximum normal stress, the maximum shear stress, and the planes on which these stresses occur. The governing formulas follow. (See Reference 1 for the derivations.)

Maximum Normal Stresses: Principal Stresses

The combination of the applied normal and shear stresses that produces the maximum normal stress is called the *maximum principal stress*, σ_1. The magnitude of σ_1 can be computed from the following equation:

➭ Maximum Principal Stress

$$\sigma_1 = \frac{\sigma_x + \sigma_y}{2} + \sqrt{\left(\frac{\sigma_x - \sigma_y}{2}\right)^2 + \tau_{xy}^2} \tag{4–1}$$

The combination of the applied stresses that produces the minimum normal stress is called the *minimum principal stress*, σ_2. Its magnitude can be computed from

➭ Minimum Principal stress

$$\sigma_2 = \frac{\sigma_x + \sigma_y}{2} - \sqrt{\left(\frac{\sigma_x - \sigma_y}{2}\right)^2 + \tau_{xy}^2} \tag{4–2}$$

Particularly in experimental stress analysis, it is important to know the orientation of the principal stresses. The angle of inclination of the planes on which the principal stresses act, called the *principal planes*, can be found from

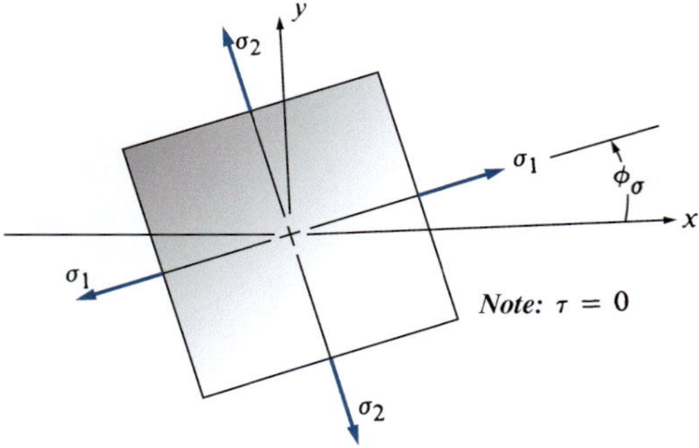

FIGURE 4–4 Principal stress element

➪ **Angle for Principal Stress Element**

$$\phi_\sigma = \frac{1}{2}\arctan\,[2\tau_{xy}/(\sigma_x - \sigma_y)] \qquad \textbf{(4–3)}$$

The angle ϕ_σ is measured from the positive x-axis of the original stress element to the maximum principal stress, σ_1. Then the minimum principal stress, σ_2, is on the plane 90° from σ_1.

When the stress element is oriented as discussed so that the principal stresses are acting on it, the shear stress is zero. The resulting stress element is shown in Figure 4–4.

Maximum Shear Stress

On a different orientation of the stress element, the maximum shear stress will occur. Its magnitude can be computed from

➪ **Maximum shear stress**

$$\tau_{max} = \sqrt{\left(\frac{\sigma_x - \sigma_y}{2}\right)^2 + \tau_{xy}^2} \qquad \textbf{(4–4)}$$

The angle of inclination of the element on which the maximum shear stress occurs is computed as follows:

➪ **Angle for Maximum Shear Stress Element**

$$\phi_\tau = \frac{1}{2}\arctan\,[-(\sigma_x - \sigma_y)/2\tau_{xy}] \qquad \textbf{(4–5)}$$

The angle between the principal stress element and the maximum shear stress element is always 45°.

On the maximum shear stress element, there will be normal stresses of equal magnitude acting perpendicular to the planes on which the maximum shear stresses are acting. These normal stresses have the value

➪ **Average Normal Stress**

$$\sigma_{avg} = (\sigma_x + \sigma_y)/2 \qquad \textbf{(4–6)}$$

Note that this is the *average* of the two applied normal stresses. The resulting maximum shear stress element is shown in Figure 4–5. Note, as stated above, that the angle between the principal stress element and the maximum shear stress element is always 45°.

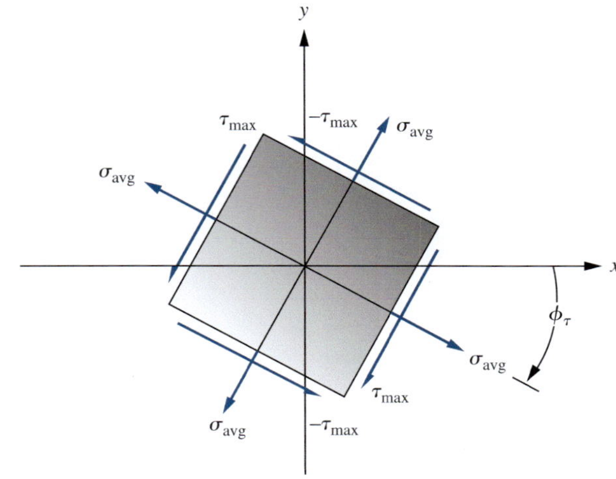

FIGURE 4–5 Maximum shear stress element

Summary and General Procedure for Analyzing Combined Stresses

The following list gives a summary of the techniques presented in this section; it also outlines the general procedure for applying the techniques to a given stress analysis problem.

GENERAL PROCEDURE FOR COMPUTING PRINCIPAL STRESSES AND MAXIMUM SHEAR STRESSES

1. Decide for which point you want to compute the stresses.
2. Clearly specify the coordinate system for the object, the free-body diagram, and the magnitude and direction of forces.
3. Compute the stresses on the selected point due to the applied forces, and show the stresses acting on a stress element at the desired point with careful attention to directions. Figure 4–3 is a model for how to show these stresses.
4. Compute the principal stresses on the point and the directions in which they act. Use Equations (4–1), (4–2), and (4–3).
5. Draw the stress element on which the principal stresses act, and show its orientation relative to the original x-axis. It is recommended that the principal stress element be drawn beside the original stress element to illustrate the relationship between them.
6. Compute the maximum shear stress on the element and the orientation of the plane on which it acts. Also, compute the normal stress that acts on the maximum shear stress element. Use Equations (4–4), (4–5), and (4–6).
7. Draw the stress element on which the maximum shear stress acts, and show its orientation to the original x-axis. It is recommended that the maximum shear stress element be drawn beside the maximum principal stress element to illustrate the relationship between them.
8. The resulting set of three stress elements will appear as shown in Figure 4–6.

The following example problem illustrates the use of this procedure.

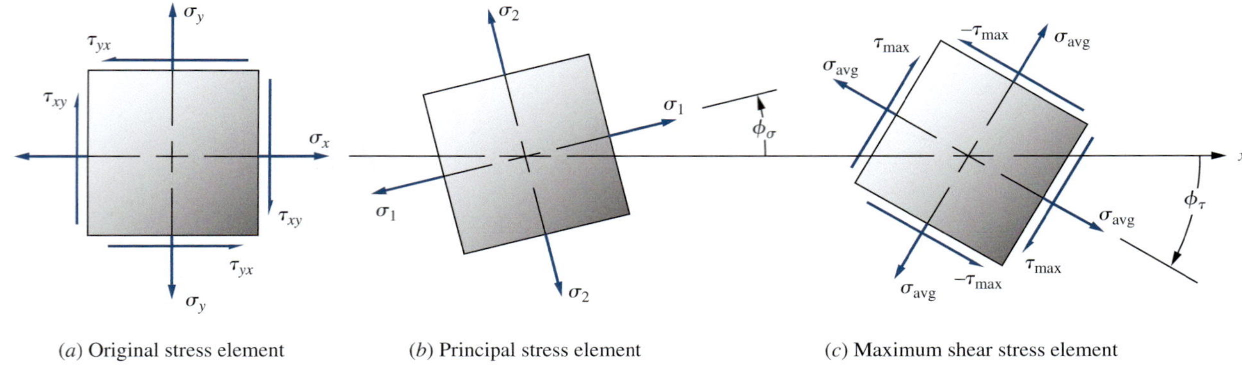

(a) Original stress element (b) Principal stress element (c) Maximum shear stress element

FIGURE 4–6 Relationships among original stress element, principal stress element, and maximum shear stress element for a given loading

---

Example Problem 4–1 The shaft shown in Figure 4–7 is supported by two bearings and carries two V-belt sheaves. The tensions in the belts exert horizontal forces on the shaft, tending to bend it in the *x–z* plane. Sheave *B* exerts a clockwise torque on the shaft when viewed toward the origin of the coordinate system along the *x*-axis. Sheave *C* exerts an equal but opposite torque on the shaft. For the loading condition shown, determine the principal stresses and the maximum shear stress on element *K* on the front surface of the shaft (on the positive *z*-side) just to the right of sheave *B*. Follow the general procedure for analyzing combined stresses given in this section.

Solution Objective Compute the principal stresses and the maximum shear stresses on element *K*.

Given Shaft and loading pattern shown in Figure 4–7.

Analysis Use the general procedure for analyzing combined stresses.

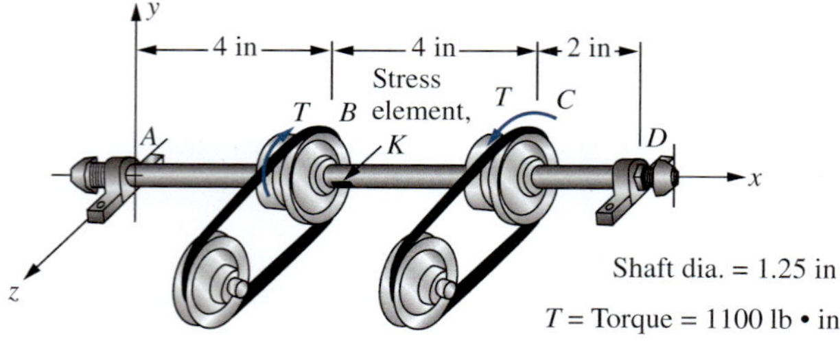

(a) Pictorial view of shaft

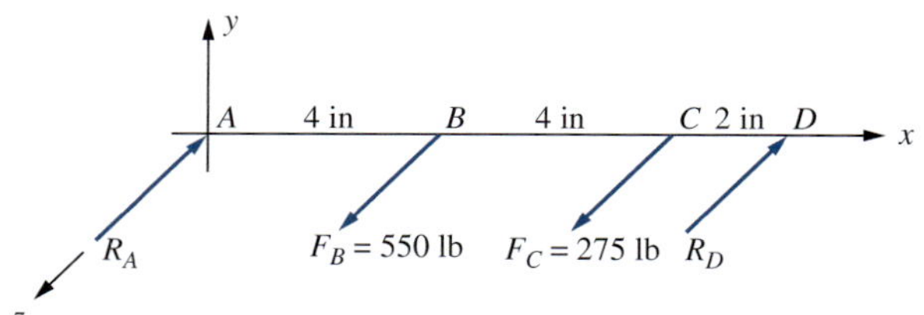

(b) Forces acting on shaft at *B* and *C* caused by belt drives

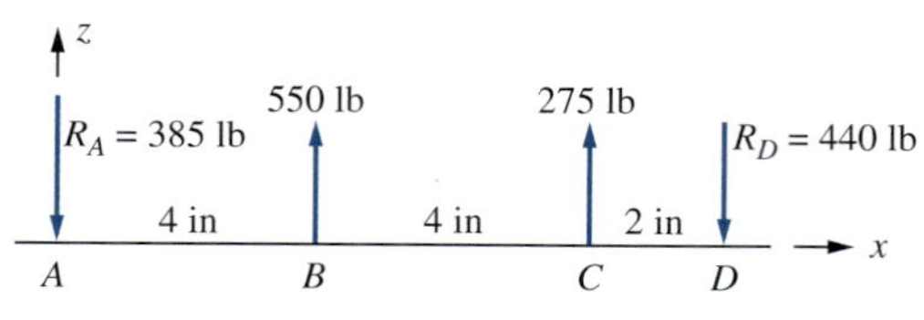

(c) Normal view of forces on shaft in *x–z* plane with reactions at bearings

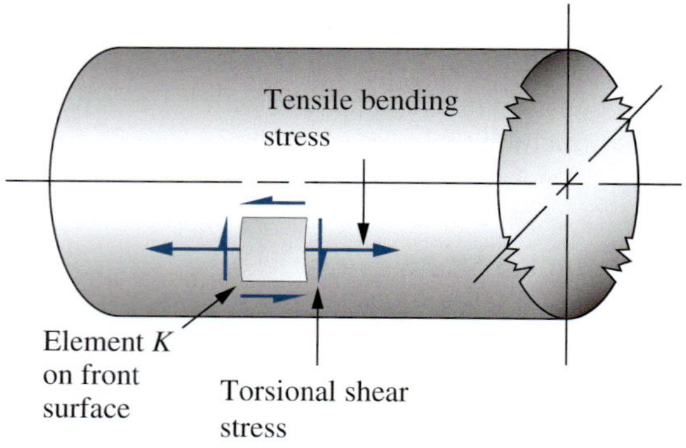

(d) Enlarged view of element *K* on front of shaft

FIGURE 4–7 Shaft supported by two bearings and carrying two V-belt sheaves

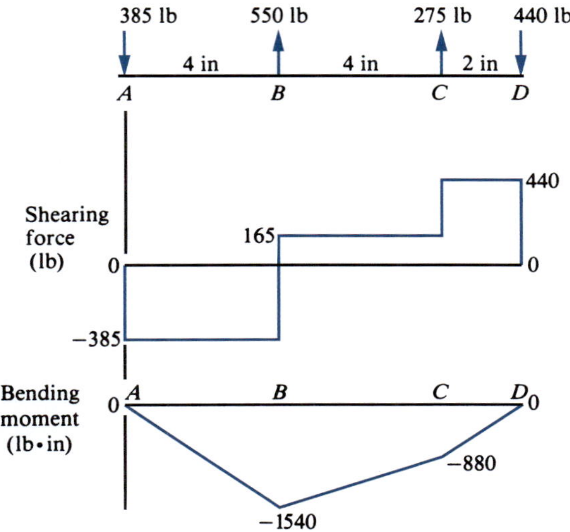

FIGURE 4–8 Shearing force and bending moment diagrams for the shaft

Results Element K is subjected to bending that produces a tensile stress acting in the x-direction. Also, there is a torsional shear stress acting at K. Figure 4–8 shows the shearing force and bending moment diagrams for the shaft and indicates that the bending moment at K is 1540 lb·in. The bending stress is therefore

$$\sigma_x = M/S$$
$$S = \pi D^3/32 = [\pi(1.25\text{ in})^3]/32 = 0.192\text{ in}^3$$
$$\sigma_x = (1540\text{ lb}\cdot\text{in})/(0.192\text{ in}^3) = 8030\text{ psi}$$

The torsional shear stress acts on element K in a way that causes a downward shear stress on the right side of the element and an upward shear stress on the left side. This action results in a tendency to rotate the element in a *clockwise* direction, which is the *positive* direction for shear stresses according to the standard convention. Also, the notation for shear stresses uses double subscripts. For example, τ_{xy} indicates the shear stress acting on the face of an element that is perpendicular to the x-axis and parallel to the y-axis. Thus, for element K,

$$\tau_{xy} = T/Z_p$$
$$Z_p = \pi D^3/16 = \pi(1.25\text{ in})^3/16 = 0.383\text{ in}^3$$
$$\tau_{xy} = (1100\text{ lb}\cdot\text{in})/(0.383\text{ in}^3) = 2870\text{ psi}$$

The values of the normal stress, σ_x, and the shear stress, τ_{xy}, are shown on the stress element K in Figure 4–9. Note that the stress in the y-direction is zero for this loading. Also, the value of the shear stress, τ_{yx}, must be equal to τ_{xy}, and it must act as shown in order for the element to be in equilibrium.

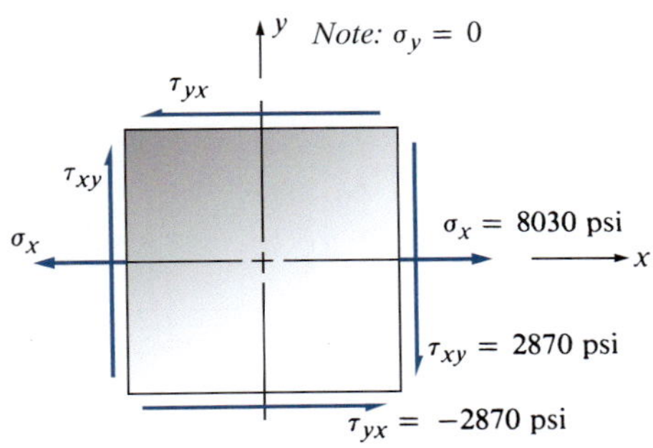

FIGURE 4–9 Stresses on element K

We can now compute the principal stresses on the element, using Equations (4–1) through (4–3). The maximum principal stress is

$$\sigma_1 = \frac{\sigma_x + \sigma_y}{2} + \sqrt{\left(\frac{\sigma_x - \sigma_y}{2}\right)^2 + \tau_{xy}^2} \tag{4–1}$$

$$\sigma_1 = (8030/2) + \sqrt{(8030/2)^2 + (2870)^2}$$

$$\sigma_1 = 4015 + 4935 = 8950 \text{ psi}$$

The minimum principal stress is

$$\sigma_2 = \frac{\sigma_x + \sigma_y}{2} - \sqrt{\left(\frac{\sigma_x - \sigma_y}{2}\right)^2 + \tau_{xy}^2} \tag{4–2}$$

$$\sigma_2 = (8030/2) - \sqrt{(8030/2)^2 + (2870)^2}$$

$$\sigma_2 = 4015 - 4935 = -920 \text{ psi (compression)}$$

The direction in which the maximum principal stress acts is

$$\phi_\sigma = \frac{1}{2}\arctan\,[2\tau_{xy}/(\sigma_x - \sigma_y)] \tag{4–3}$$

$$\phi_\sigma = \frac{1}{2}\arctan\,[(2)(2870)/(8030)] = 17.8^\circ$$

The positive sign calls for a *clockwise* rotation of the element.

The principal stresses can be shown on a stress element as illustrated in Figure 4–10(b). Note that the element is shown in relation to the original element to emphasize the direction of the principal stresses in relation to the original *x*-axis. The positive sign for ϕ_σ indicates that the principal stress element is rotated *clockwise* from its original position.

Now the maximum shear stress element can be defined, using Equations (4–4) through (4–6):

$$\tau_{max} = \sqrt{\left(\frac{\sigma_x - \sigma_y}{2}\right)^2 + \tau_{xy}^2} \tag{4–4}$$

$$\tau_{max} = \sqrt{(8030/2)^2 + (2870)^2}$$

$$\tau_{max} = \pm 4935 \text{ psi}$$

The two pairs of shear stresses, $+\tau_{max}$ and $-\tau_{max}$, are equal in magnitude but opposite in direction.

The orientation of the element on which the maximum shear stress acts is found from Equation (4–5):

$$\phi_\tau = \frac{1}{2}\arctan\,[-(\sigma_x - \sigma_y)/2\tau_{xy}] \tag{4–5}$$

$$\phi_\tau = \frac{1}{2}\arctan\,(-8030/[(2)(2870)]) = -27.2^\circ$$

The negative sign calls for a *counterclockwise* rotation of the element.

There are equal normal stresses acting on the faces of this stress element, which have the value of

$$\sigma_{avg} = (\sigma_x + \sigma_y)/2 \tag{4–6}$$

$$\sigma_{avg} = 8030/2 = 4015 \text{ psi}$$

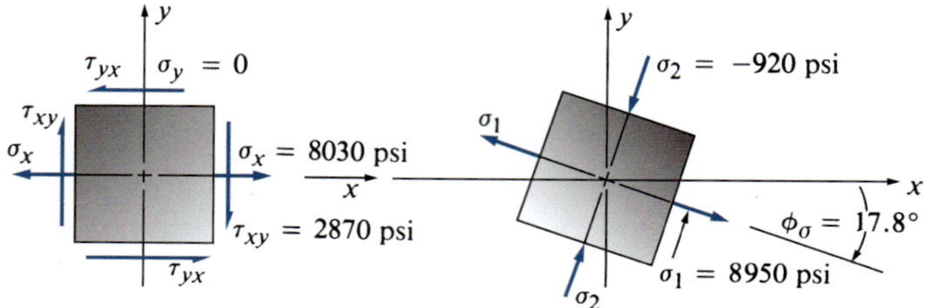

(*a*) Original stress element

(*b*) Principal stress element

FIGURE 4–10 Principal stress element

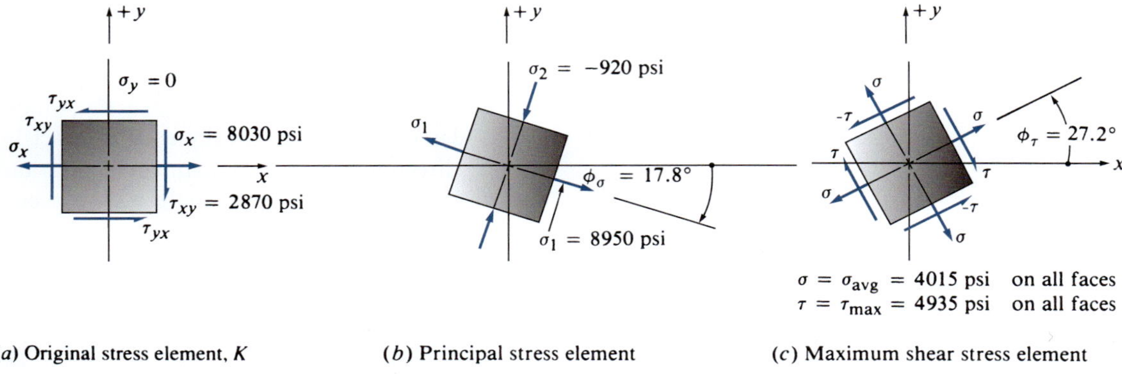

(*a*) Original stress element, *K* (*b*) Principal stress element (*c*) Maximum shear stress element

FIGURE 4–11 Relation of maximum shear stress element to the original stress element and the principal stress element

Comments Figure 4–11(c) shows the stress element on which the maximum shear stress acts in relation to the original stress element. Note that the angle between this element and the principal stress element is 45°. It is typical to show all three stress elements side by side as shown in Figure 4–11.

Examine the results of Example Problem 4–1. The maximum principal stress, $\sigma_1 = 8950$ psi, is 11 percent greater than the value of $\sigma_x = 8030$ psi computed for the bending stress in the shaft acting in the x-direction. The maximum shear stress, $\tau_{max} = 4935$ psi, is 72 percent greater than the computed applied torsional shear stress of $\tau_{xy} = 2870$ psi. You will see in Chapter 5 that either the maximum normal stress or the maximum shear stress is often required for accurate failure prediction and for safe design decisions. The angles of the final stress elements also predict the alignment of the most damaging stresses that can be an aid in experimental stress analysis and the analysis of actual failed components.

4–3 ANALYSIS OF COMPLEX LOADING CONDITIONS AND VON MISES STRESS

The examples shown in this chapter involve relatively simple part geometries and loading conditions for which the necessary stress analysis can be performed using familiar methods of statics and strength of materials. If more complex geometries or loading conditions are involved, you may not be able to complete the required analysis to create the original stress element from which the Mohr's circle is constructed as shown in section 4–4.

Consider, for example, a cast wheel for a high-performance racing car. The geometry would likely involve webs or spokes of a unique design connecting the hub to the rim to create a lightweight wheel. The loading would be a complex combination of torsion, bending, and compression generated by the cornering action of the wheel.

One method of analysis of such a load-carrying member would be accomplished by experimental stress analysis using strain gages or photoelastic techniques. The results would identify the stress levels at selected points in certain specified directions that could be used as the input to the construction of the Mohr's circle for critical points in the structure.

Another method of analysis would involve the modeling of the geometry of the wheel as a *finite-element model*. The three-dimensional model would be divided into several hundred small-volume elements. Points of support and restraint would be defined on the model, and then external loads would be applied at appropriate points. The complete data set would be input to a special type of computer analysis program called *finite-element analysis*. The output from the program lists the stresses and the deflection for each of the elements. These data can be plotted on the computer model so that the designer can visualize the stress distribution within the model. Most such programs list the principal stresses and the maximum shear stress for each element, eliminating the need to actually draw the Mohr's circle. A special form of stress, called the *von Mises stress*, is often computed by combining the principal stresses. (The von Mises stress is introduced next and used later in Section 5–11 as one of the methods for analyzing designs.) Several different finite-element analysis programs are commercially available for use on personal computers, on engineering work stations, or on mainframe computers.

von Mises Stress for Biaxial Stresses

With an understanding of the general case of combined biaxial stress from Section 4–2, we can now present a very useful concept called the *von Mises stress*, named in honor of R. von Mises who developed the concept in 1913. Additional contributions by H. Hencky in 1925 to its use led to another term that is often used for the method, *the von Mises-Hencky method.*

We define the *von Mises stress*, called σ', as a special combination of the maximum and minimum principal stresses, σ_1 and σ_2, using the equation

$$\sigma' = \sqrt{\sigma_1^2 + \sigma_2^2 - \sigma_1\sigma_2} \quad (4\text{–}7)$$

Under static loading conditions, failure is predicted when $\sigma' > s_y$.

As noted in Section 4–2, the von Mises stress is larger than any of its components. Most finite-element analysis software packages produce the *von Mises stress* as part of the typical output calculations because of its importance in predicting failure. More will be discussed in Chapter 5 in the sections **Predictions of Failure** and **Design Analysis Methods**.

Alternate Form for the von Mises Stress

It is sometimes convenient to express the von Mises stress in terms of the stresses on the original stress element rather than completing the entire Mohr's circle process. That form of the equation is

$$\sigma' = \sqrt{\sigma_x^2 + \sigma_y^2 - \sigma_x\sigma_y + 3\tau_{xy}^2} \quad (4\text{–}8)$$

Three-Dimensional Stresses and the von Mises Stress

The most general representation of stress at a point is three-dimensional (triaxial), visualized as a cube-like structure shown in Figure 4–12 on an x-, y-, and z-axis system. The following characteristics describe the system of stresses that can act on the cubic element.

1. Any of the six faces of the cube can be subjected to a normal stress, either tension or compression. The normal stress vectors act in pairs on parallel opposite faces to either pull or push on the sides of the cube.
2. Similarly, two shearing stresses can act on any face, perpendicular to each other.

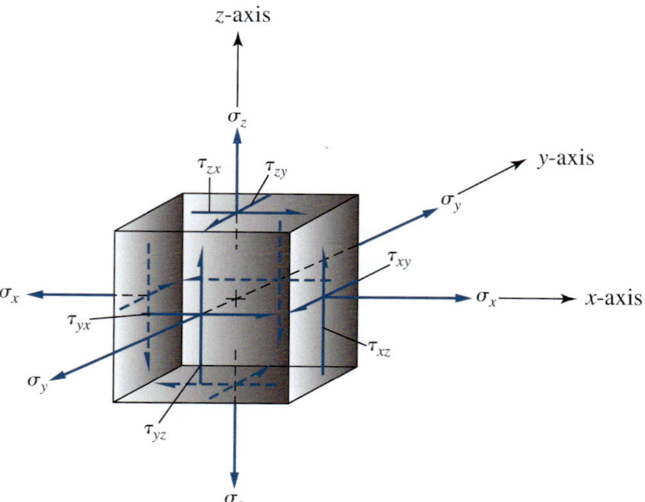

FIGURE 4–12 Three-dimensional stress element with normal and shear stresses on all faces

a. Note that each shearing stress on a given face has an equal counterpart on the parallel opposite face to create the shearing action.
b. The pairs of shearing stresses acting on perpendicular faces are numerically equal but act in opposite directions in order to maintain equilibrium.
c. Only shearing stresses on the visible faces are labeled in the figure.

Therefore, the set of possible stresses acting on the element is

Normal stresses:	σ_x	σ_y	σ_z
Shearing stresses:	τ_{xy} and τ_{yx}	τ_{xz} and τ_{zx}	τ_{yz} and τ_{zy}

The values of these stresses are most commonly determined from finite-element analysis. While the derivation is not presented here, the three principal stresses, σ_1, σ_2, σ_3, can be determined. These are ordered by magnitude such that $\sigma_1 > \sigma_2 > \sigma_3$. Then the von Mises stress for triaxial loading can be shown to be

$$\sigma' = (\sqrt{2}/2)[\sqrt{(\sigma_2-\sigma_1)^2 + (\sigma_3-\sigma_1)^2 + (\sigma_3-\sigma_2)^2}] \quad (4\text{–}9)$$

Again, for static loading, failure is predicted when $\sigma' > s_y$.

4–4 MOHR'S CIRCLE

The process of computing the principal stresses and the maximum shear stress shown in Example Problem 4–1 may seem somewhat abstract. These same results can be obtained using a method called *Mohr's circle*, which is discussed next. This method uses a combination of a graphical aid and simple calculations. With practice, the use of Mohr's circle should provide you with a more intuitive feel for the variations in stress that exist at a point in relation to the angle of orientation of the stress element. In addition, it provides a streamlined approach to determining the stress condition on any plane of interest. Because of the many terms and signs involved, and the many calculations required in the computation of the principal stresses and the maximum shear stress, there is a rather high probability of error. Using the graphic aid Mohr's circle helps to minimize errors and gives a better "feel" for the stress condition at the point of interest.

After Mohr's circle is constructed, it can be used for the following:

1. Finding the maximum and minimum principal stresses and the directions in which they act.
2. Finding the maximum shear stresses and the orientation of the planes on which they act.
3. Finding the value of the normal stresses that act on the planes where the maximum shear stresses act.
4. Finding the values of the normal and shear stresses that act on an element with any orientation.

The data needed to construct Mohr's circle are, of course, the same as those needed to compute the preceding values, because the graphical approach is an exact analogy to the computations.

If the normal and shear stresses that act on any two mutually perpendicular planes of an element are known, the circle can be constructed and any of items 1 through 4 can be found.

Mohr's circle is actually a plot of the combinations of normal and shearing stresses that exist on a stress element for all possible angles of orientation of the element. This method is particularly valuable in experimental stress analysis work because the results obtained from many types of standard strain gage instrumentation techniques give the necessary inputs for the creation of Mohr's circle. (See Reference 1.) When the principal stresses and the maximum shear stress are known, the complete design and analysis can be done, using the various theories of failure discussed in Chapter 5.

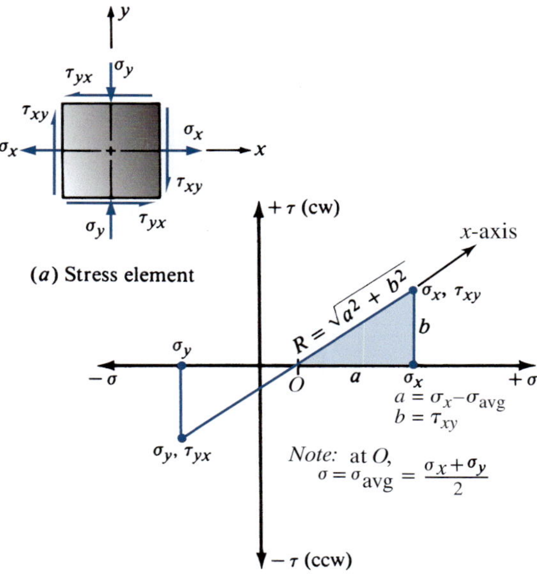

FIGURE 4–13 Partially completed Mohr's circle, Steps 1–7

PROCEDURE FOR CONSTRUCTING MOHR'S CIRCLE

1. Perform the stress analysis to determine the magnitudes and directions of the normal and shear stresses acting at the point of interest.
2. Draw the stress element at the point of interest as shown in Figure 4–13(a). Normal stresses on any two mutually perpendicular planes are drawn with tensile stresses positive—projecting outward from the element. Compressive stresses are negative—directed inward on the face. Note that the *resultants* of all normal stresses acting in the chosen directions are plotted. Shear stresses are considered to be positive if they tend to rotate the element in a *clockwise* (cw) direction, and negative otherwise.

   Note that on the stress element illustrated, σ_x is positive, σ_y is negative, τ_{xy} is positive, and τ_{yx} is negative. This assignment is arbitrary for the purpose of illustration. In general, any combination of positive and negative values could exist.
3. Refer to Figure 4–13(b). Set up a rectangular coordinate system in which the positive horizontal axis represents positive (tensile) normal stresses, and the positive vertical axis represents positive (clockwise) shear stresses. Thus, the plane created will be referred to as the *σ-τ plane.*
4. Plot points on the σ-τ plane corresponding to the stresses acting on the faces of the stress element. If the element is drawn in the *x-y* plane, the two points to be plotted are σ_x, τ_{xy} and σ_y, τ_{yx}. These are two points on Mohr's circle.
5. Draw the line connecting the two points. The resulting line crosses the σ-axis at the center of Mohr's circle at the average of the two applied normal stresses, where

$$\sigma_{avg} = (\sigma_x + \sigma_y)/2$$

The center of Mohr's circle is called *O* in Figure 4–13.

6. Now it is important to note that the line from the center of Mohr's circle, at *O*, and through the first plotted point, (σ_x, τ_{xy}), represents the *x*-axis from the original stress element. ***Draw and label an extension of this line at this time.*** The direction of this line, of course, corresponds to a known direction on the actual component being analyzed. Angles of rotation for the principle stress element and the maximum shear stress element, found from Steps 11–14 later in this process, will be measured from this *x*-axis.
7. Note in Figure 4–13 that a right triangle has been formed, having the sides *a*, *b*, and *R*, where

$$R = \sqrt{a^2 + b^2}$$

   By inspection, we can see that

$$a = \sigma_x - \sigma_{avg}$$

$$b = \tau_{xy}$$

   Also compute the angle α within the triangle between lines *a* and *R*. You can see that

$$\alpha = \text{Tan}^{-1}(b/a) \quad \text{or} \quad \alpha = \text{Sin}^{-1}(b/R)$$

   we can now proceed with the construction of the circle.
8. Draw the complete circle with the center at *O* and a radius of *R*, as shown in Figure 4–14.
9. Find σ_1, the maximum principal stress, at the point where the circle crosses the σ-axis at the right. Note that $\sigma_1 = \sigma_{avg} + R$. Because the Mohr's circle represents a plot of all possible combinations of normal and shear stresses on the element for any angle of orientation, it stands to reason that σ_1 lies at the right end of the horizontal diameter.
10. Find σ_2, the minimum principal stress, at the left end of the horizontal diameter. Note that $\sigma_2 = \sigma_{avg} - R$.
11. Find τ_{max}, the maximum shear stress, at the top end of the vertical diameter of the circle. By observation, $\tau_{max} = R$, the radius of the circle. Note also that the coordinates for the point at the top of the circle are $(\sigma_{avg}, \tau_{max})$.

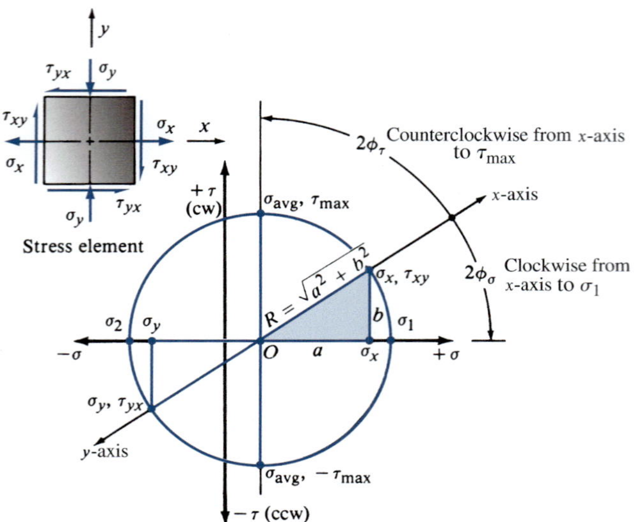

FIGURE 4–14 Completed Mohr's circle, Steps 8–14

At this point in the process, it is important to realize that angles on Mohr's circle are actually double the true angles.

The following steps define the general method for finding the angles of orientation of the principal stress element and the maximum shear stress element in relation to the original x-axis, located in Step 6 and shown in Figure 4–14. Now we can observe that the line from the center of the circle to the second plotted point, (σ_y, τ_{yx}), represents the y-axis from the original stress element. Of course, the x-axis and the y-axis are truly 90° apart, whereas they are 180° apart on Mohr's circle, illustrating the double-angle phenomenon.

12. Find the angle called $2\varphi_\sigma$, always measured ***from the x-axis to the σ-axis***, and ***note the direction—either clockwise or counterclockwise***. In the *current problem*, we can observe that $2\varphi_\sigma = \alpha$, the angle found in Step 7, and the rotation is clockwise. ***However, in other problems, the angle must be found from the geometry of the circle.*** (This is illustrated in example practice problems that follow.) Now compute $\varphi_\sigma = 2\varphi_\sigma/2$.

13. Draw the *principal stress element* as shown in Figure 4–15(b). In Part (a) of the figure, we have reproduced the *original stress element* to indicate the direction of the x-axis and we draw a new element in relation to that axis at an angle of φ_σ. The new element must be rotated in the same direction, clockwise or counterclockwise, as observed in Step 12.
    a. On the face of the element found from the rotation, draw the vector σ_1.
    b. Then show the vector σ_1 acting on the opposite parallel face in the opposite sense to indicate the tension or compression stress.
    c. Draw two σ_2 vectors on the other, perpendicular faces, in the proper sense.
    d. Note that ***on the principal stress element, the shear stress is always zero***. This is evident from the Mohr's circle where the coordinates of the principal stresses are always $(\sigma_1, 0)$ and $(\sigma_2, 0)$, indicating zero shear stress.

14. Find the angle called $2\varphi_\tau$, always measured ***from the x-axis to the τ_{max}-axis***, and ***note the direction—either clockwise or counterclockwise***. In the *current problem*, we can observe that $2\varphi_\tau = 90° - \alpha$, the angle found in Step 7, and the rotation is counterclockwise. ***However, in other problems, the angle must be found from the geometry of the circle.*** Now compute $\varphi_\tau = 2\varphi_\tau/2$.

15. Draw the *maximum shear stress element* as shown in Figure 4–15(c). Here, we draw a new element in relation to the x-axis at an angle of φ_τ. The new element must be rotated in the direction, clockwise or counterclockwise, as observed in Step 14.
    a. On the face of the element found from the rotation, draw the vector τ_{max}.
    b. Then show the vector τ_{max} acting on the opposite parallel face in the opposite sense to indicate the shearing action.
    c. Draw two $-\tau_{max}$ vectors on the other, perpendicular faces. The combination of all four shearing vectors must indicate equilibrium of the element.
    d. Note that ***on the maximum shear stress element, the average normal stress acts on all four faces.*** This is evident from the Mohr's circle where the coordinates of the maximum shear stress is always $(\sigma_{avg}, \tau_{max})$ and for the minimum shear stress it is always $(\sigma_{avg}, -\tau_{max})$.

16. The final step in the process of preparing Mohr's circle is to summarize the primary results, typically: $\sigma_1, \sigma_2, \tau_{max}, \sigma_{avg}, \varphi_\sigma$, and φ_τ.

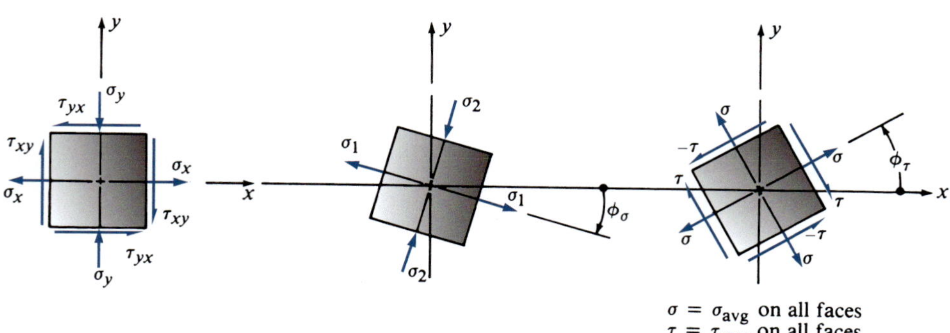

FIGURE 4–15 Display of results from Mohr's circle

We will now illustrate the construction of Mohr's circle by using the same data as in Example Problem 4–1, in which the principal stresses and the maximum shear stress were computed directly from the equations.

Example Problem 4–2 The shaft shown in Figure 4–7 is supported by two bearings and carries two V-belt sheaves. The tensions in the belts exert horizontal forces on the shaft, tending to bend it in the *x–z* plane. Sheave *B* exerts a clockwise torque on the shaft when viewed toward the origin of the coordinate system along the *x*-axis. Sheave *C* exerts an equal but opposite torque on the shaft. For the loading condition shown, determine the principal stresses and the maximum shear stress on element *K* on the front surface of the shaft (on the positive *z*-side) just to the right of sheave *B*. Use the procedure for constructing Mohr's circle in this section.

Solution

Objective: Determine the principal stresses and the maximum shear stresses on element *K*.

Given: Shaft and loading pattern shown in Figure 4–7.

Analysis: Use the *Procedure for Constructing Mohr's Circle*. Some intermediate results will be taken from the solution to Example Problem 4–1 and from Figures 4–7, 4–8, and 4–9.

Results: ***Steps 1 and 2.*** The stress analysis for the given loading was completed in Example Problem 4–1. Figure 4–16 is identical to Figure 4–9 and represents the results of Step 2 of the Mohr's circle procedure, the original stress element.

Steps 3–6. Figure 4–17 shows the results. The first point plotted was

$$\sigma_x = 8030 \text{ psi}, \tau_{xy} = 2870 \text{ psi}$$

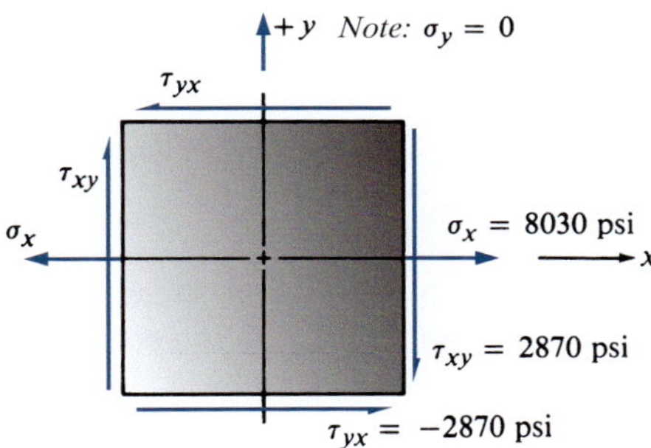

FIGURE 4–16 Stresses on element *K*

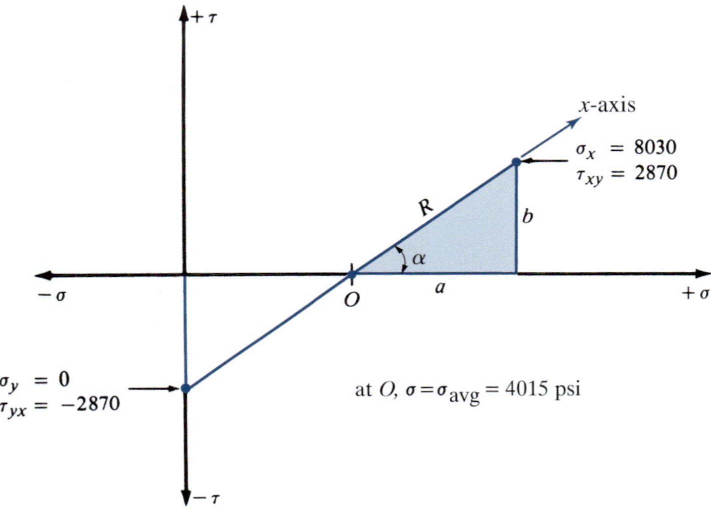

FIGURE 4–17 Partially completed Mohr's circle

The second point was plotted at

$$\sigma_y = 0 \text{ psi}, \ \tau_{yx} = -2870 \text{ psi}$$

Then a line was drawn between them, crossing the σ-axis at O. The value of the stress at O is

$$\sigma_{avg} = (\sigma_x + \sigma_y)/2 = (8030 + 0)/2 = 4015 \text{ psi}$$

Extend the line through the point (σ_x, τ_{xy}) and label it, x-axis.

Step 7. We compute the values for a, b, and R from

$$a = (\sigma_x - \sigma_{avg}) = (8030 - 4018) = 4015 \text{ psi}$$

$$b = \tau_{xy} = 2870 \text{ psi}$$

$$R = \sqrt{a^2 + b^2} = \sqrt{(4015)^2 + (2870)^2} = 4935 \text{ psi}$$

Step 8. Figure 4–18 shows the completed Mohr's circle. The circle has its center at O and the radius R. Note that the circle passes through the two points originally plotted. It must do so because the circle represents all possible states of stress on the element K.

Step 9. The maximum principal stress is at the right side of the circle.

$$\sigma_1 = \sigma_{avg} + R$$

$$\sigma_1 = 4015 + 4935 = 8950 \text{ psi}$$

Step 10. The minimum principal stress is at the left side of the circle.

$$\sigma_2 = \sigma_{avg} - R$$

$$\sigma_2 = 4015 - 4935 = -920 \text{ psi}$$

Step 11. At the top of the circle,

$$\sigma = \sigma_{avg} = 4015 \text{ psi}$$

$$\tau = \tau_{max} = R = 4935 \text{ psi}$$

The value of the normal stress on the element that carries the maximum shear stress is the same as the coordinate of O, the center of the circle.

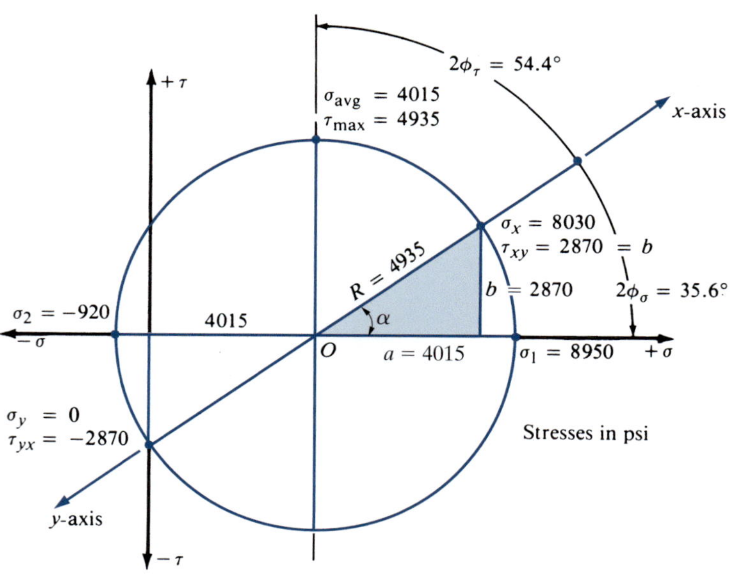

FIGURE 4–18 Completed Mohr's circle

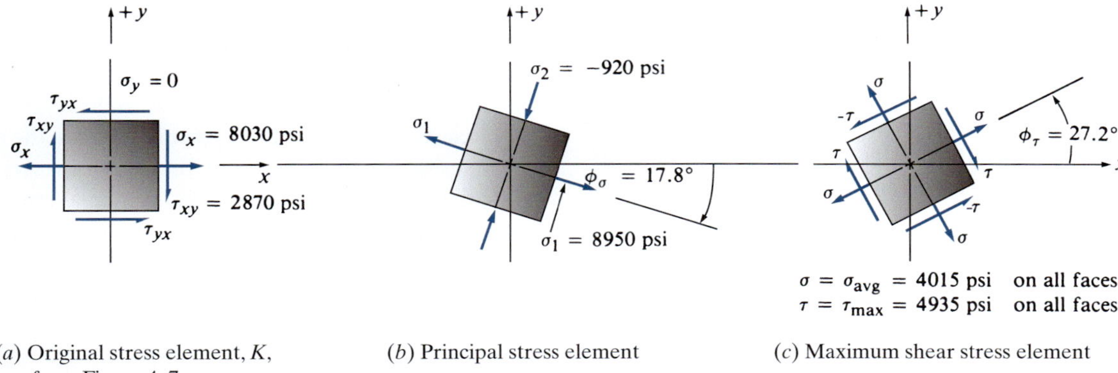

(*a*) Original stress element, *K*, from Figure 4–7

(*b*) Principal stress element

(*c*) Maximum shear stress element

FIGURE 4–19 Results from Mohr's circle analysis

Step 12. Compute the angles α, $2\phi_\sigma$, and then ϕ_σ. Use the circle as a guide.

$$\alpha = 2\phi_\sigma = \arctan(b/a) = \arctan(2870/4015) = 35.6°$$
$$\phi_\sigma = 35.6°/2 = 17.8°$$

Note that ϕ_σ must be measured *clockwise* from the original *x*-axis to the direction of the line of action of σ_1 for this set of data. The principal stress element will be rotated in the same direction as part of step 13.

Step 13. See Figure 4–19 where the original stress element has been reproduced. Now draw the principal stress element to the right (Part (b) of the figure). Rotate the element 17.8° *clockwise* from the *x*-axis as shown. Then draw the vectors for the maximum principal stress, $\sigma_1 = 8950$ psi (tensile), on that face and the one opposite. Complete the element by drawing vectors for $\sigma_2 = -920$ psi (compressive) on the other two faces. Label the element as shown in the figure.

Step 14. Compute the angle $2\phi_\tau$ and then ϕ_τ. From the circle we see that

$$2\phi_\tau = 90° - \alpha = 90° - 35.6° = 54.4°$$
$$\phi_\tau = 54.4°/2 = 27.2°$$

Note that the stress element on which the maximum shear stress acts must be rotated *counterclockwise* from the orientation of the original element for this set of data.

Step 15. In Figure 4–19, we draw the maximum shear stress element to the right of the principal stress element. It is rotated *counterclockwise* 27.2° from the *x*-axis.

a. On that face we show the positive maximum shearing stress, $\tau_{max} = 4935$ psi.
b. Then we complete τ_{max} by drawing an equal shearing stress vector on the opposite face acting in the opposite direction so that the two vectors tend to rotate the element clockwise, indicating a positive shearing stress.
c. Now we can draw the two vectors that make up $-\tau_{max}$ on the other two faces. The set of four vectors now put the element in rotational equilibrium.
d. The element is completed by drawing four equal vectors, σ_{avg}, on all four faces of the element.

Step 16. Summary of results:

- Maximum principal stress $= \sigma_1 = 8950$ psi—Tension
- Minimum principal stress $= \sigma_2 = -920$ psi—Compression
- Maximum shear stress $= \tau_{max} = 4935$ psi
- Angle of rotation of principal stress element $= \varphi_\sigma = 17.8°$ clockwise below the *x*-axis
- Angle of rotation of the maximum shear stress element $= \sigma_\tau = 27.2°$ counterclockwise above the *x*-axis
- Figure 4–19 shows the resulting stress elements. These are identical to those shown in Figure 4–11 from the hand calculations performed in Example Problem 4–1.

4–5 MOHR'S CIRCLE PRACTICE PROBLEMS

To a person seeing the construction of Mohr's circle for the first time, it may seem long and involved. But with practice under a variety of combinations of normal and shear stresses, you should be able to execute the 16 steps quickly and accurately.

Table 4–1 gives four sets of data (Example Problems 4–3 through 4–6) for normal and shear stresses in the x–y plane. You are advised to complete the Mohr's circle for each before looking at the solutions in Figures 4–20 through 4–23. From the circle, determine the two principal stresses, the maximum shear stress, and the planes on which these stresses act. Then draw the given stress element, the principal stress element, and the maximum shear stress element, all oriented properly with respect to the x- and y-directions. Note that each problem results in the x-axis being in a different quadrant.

TABLE 4–1 Practice problems for Mohr's circle

Example Problem	x-axis	σ_x	σ_y	τ_{xy}	Fig. No.
4–3	1st quadrant	+10.0 ksi	+4.0 ksi	+5.0 ksi	4–20
4–4	2nd quadrant	−80 MPa	+20 MPa	+50 MPa	4–21
4–5	3rd quadrant	−80 MPa	+20 MPa	−50 MPa	4–22
4–6	4th quadrant	+10.0 ksi	−2.0 ksi	−4.0 ksi	4–23

Example Problem 4–3

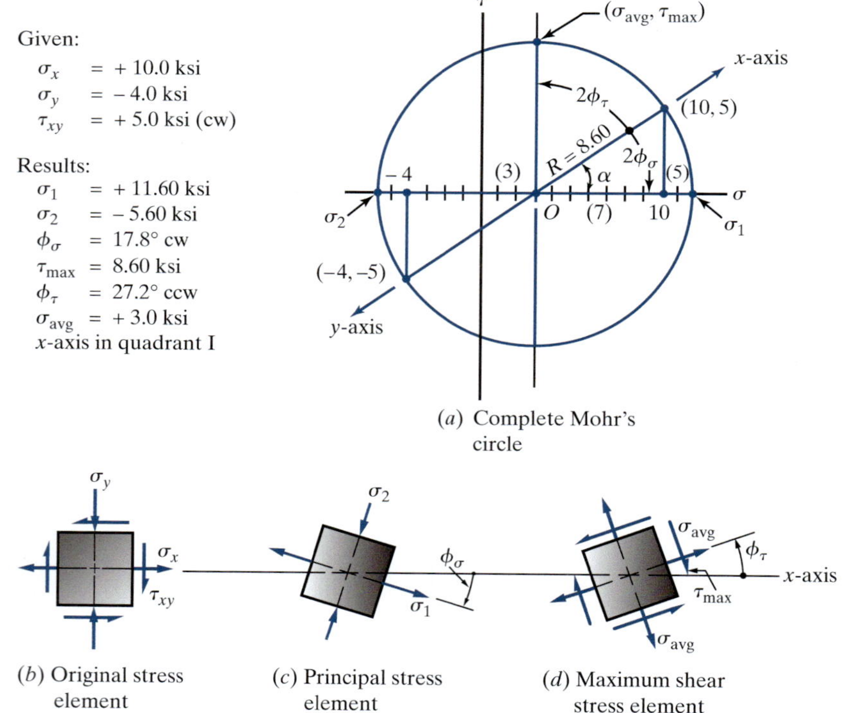

FIGURE 4–20 Solution for Example Problem 4–3. x-axis in 1st quadrant

Example Problem 4–4

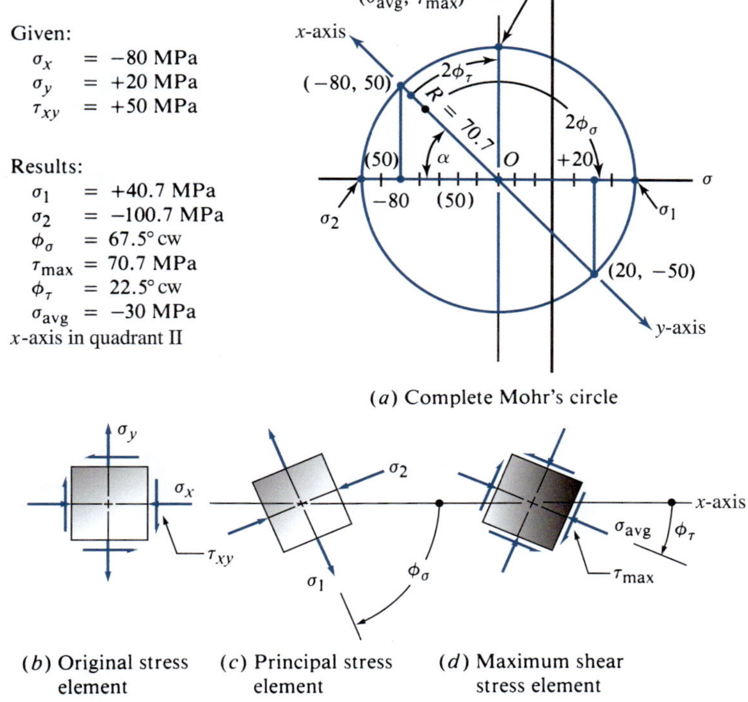

FIGURE 4–21 Solution for Example Problem 4–4. x-axis in 2nd quadrant

Example Problem 4–5

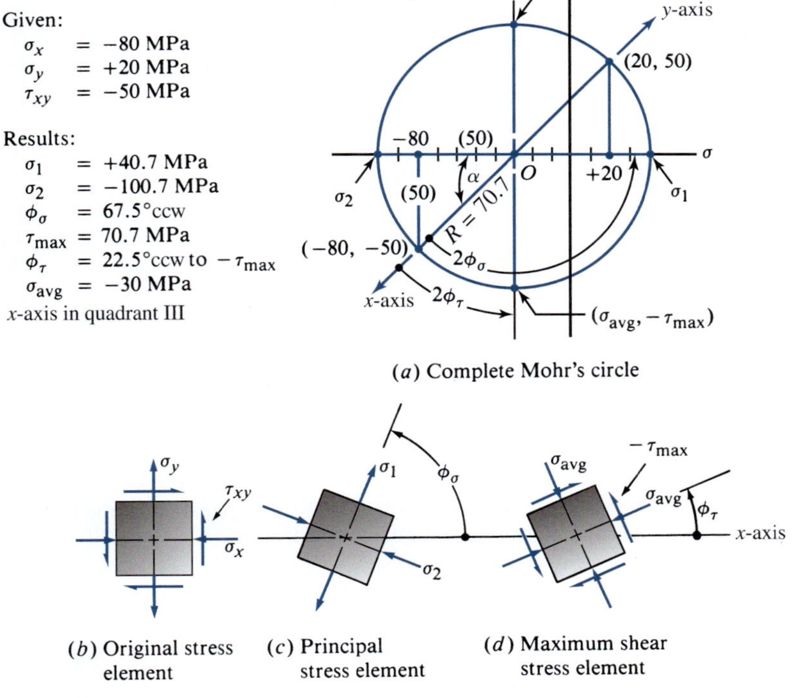

FIGURE 4–22 Solution for Example Problem 4–5. x-axis in 3rd quadrant

Example Problem 4–6

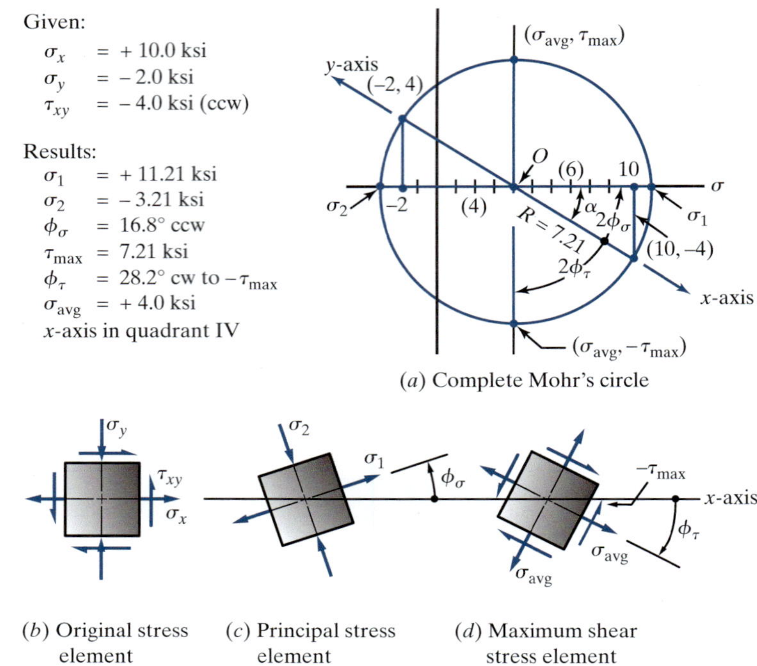

FIGURE 4–23 Solution for Example Problem 4–6. x-axis in 4th quadrant

4–6 CASE WHEN BOTH PRINCIPAL STRESSES HAVE THE SAME SIGN

Remember that all of the problems presented thus far have been plane stress problems, also called *biaxial stress* problems, because stresses are acting in only two directions within one plane. Obviously, real load-carrying members are three-dimensional objects. The assumption here is that if no stress is given for the third direction, it is zero. In most cases, the solutions as given will produce the true maximum shear stress, along with the two principal stresses for the given plane. This will always be true if the two principal stresses have opposite signs—that is, if one is tensile and the other is compressive.

But the true maximum shear stress on the element will not be found if the two principal stresses are of the same sign. In such cases, you must consider the three-dimensional case.

Familiar examples of real products in which two principal stresses have the same sign are various forms of pressure vessels. A hydraulic cylinder with closed ends contains fluids under high pressure that tend to burst the walls of the cylinder. In strength of materials, you learned that the outer surfaces of the walls of such cylinders are subjected to tensile stresses in two directions: (1) tangent to its circumference and (2) axially, parallel to the axis of the cylinder. The stress perpendicular to the wall at the outer surface is zero.

Figure 4–24 shows the stress condition on an element of the surface of the cylinder. The tangential stress, also called *hoop stress*, is aligned with the x-direction and is labeled σ_x.

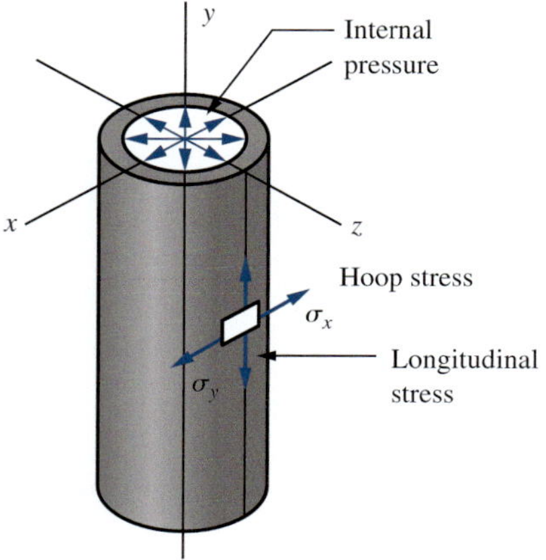

FIGURE 4–24 Thin-walled cylinder subjected to pressure with its ends closed

The axially directed stress, also called *longitudinal stress*, acts in line with the y-direction and is labeled σ_y.

In strength of materials, you learned that if the wall of the cylinder is relatively thin, the maximum hoop stress is

$$\sigma_x = pD/2t$$

where p = internal pressure in the cylinder
D = diameter of the cylinder
t = thickness of the cylinder wall

Also, the longitudinal stress is

$$\sigma_y = pD/4t$$

Both stresses are tensile, and the hoop stress is twice as large as the longitudinal stress.

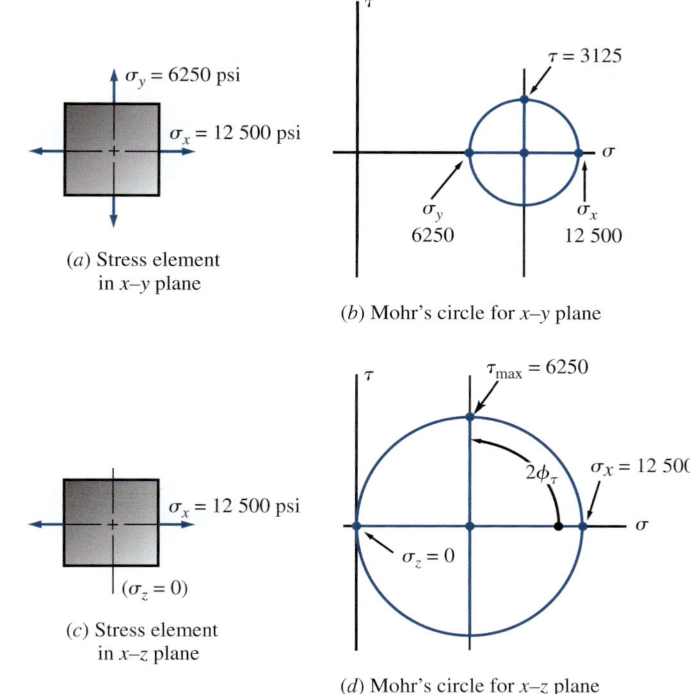

FIGURE 4–25 Stress analysis for a thin-walled cylinder

The analysis would be similar for any kind of thin-walled cylindrical vessel carrying an internal pressure. Examples are storage tanks for compressed gases, pipes carrying moving fluids under pressure, and the familiar beverage can that releases internal pressure when the top is popped open.

Let's use the hydraulic cylinder as an example for illustrating the special use of Mohr's circle when both principal stresses have the same sign. Consider that Figure 4–24 shows a cylinder with closed ends carrying an internal pressure of 500 psi. The wall thickness is $t = 0.080$ in, and the diameter of the cylinder is $D = 4.00$ in. The ratio of $D/t = 50$ indicates that the cylinder can be considered thin-walled. Any ratio over 20 is typically considered to be thin-walled.

The computed hoop and longitudinal stresses in the wall are

➪ Hoop Stress

$$\sigma_x = \frac{pD}{2t} = \frac{(500\text{ psi})(4.0\text{ in})}{(2)(0.080\text{ in})} = 12\,500\text{ psi (tension)}$$

➪ Longitudinal Stress

$$\sigma_y = \frac{pD}{4t} = \frac{(500\text{ psi})(4.0\text{ in})}{(4)(0.080\text{ in})} = 6250\text{ psi (tension)}$$

There are no shear stresses applied in the x- and y-directions.

Figure 4–25(a) shows the stress element for the x–y plane, and Part (b) shows the corresponding Mohr's circle. Because there are no applied shear stresses, σ_x and σ_y are the principal stresses for the plane. The circle would predict the maximum shear stress to be equal to the radius of the circle, 3125 psi.

But notice Part (c) of the figure. We could have chosen the x–z plane for analysis, instead of the x–y plane. The stress in the z-direction is zero because it is perpendicular to the free face of the element. Likewise, there are no shear stresses on this face. The Mohr's circle for this plane is shown in Part (d) of the figure. The maximum shear stress is equal to the radius of the circle, 6250 psi, or *twice* as much as would be predicted from the x–y plane. This approach should be used any time the two principal stresses in a biaxial stress problem have the same sign.

In summary, on a general three-dimensional stress element, there will be one orientation of the element in which there are no shear stresses acting. The normal stresses on the three perpendicular faces are then the three principal stresses. If we call these stresses σ_1, σ_2, and σ_3, taking care to order them such that $\sigma_1 > \sigma_2 > \sigma_3$, then the maximum shear stress on the element will always be

$$\tau_{max} = \frac{\sigma_1 - \sigma_3}{2}$$

Figure 4–26 shows the three-dimensional element.

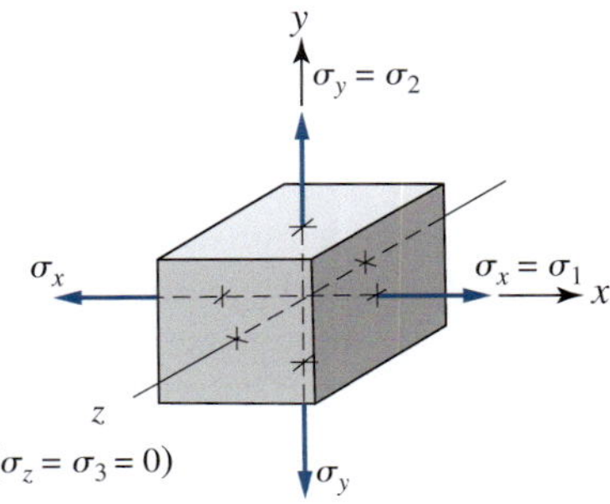

FIGURE 4–26 Three-dimensional stress element

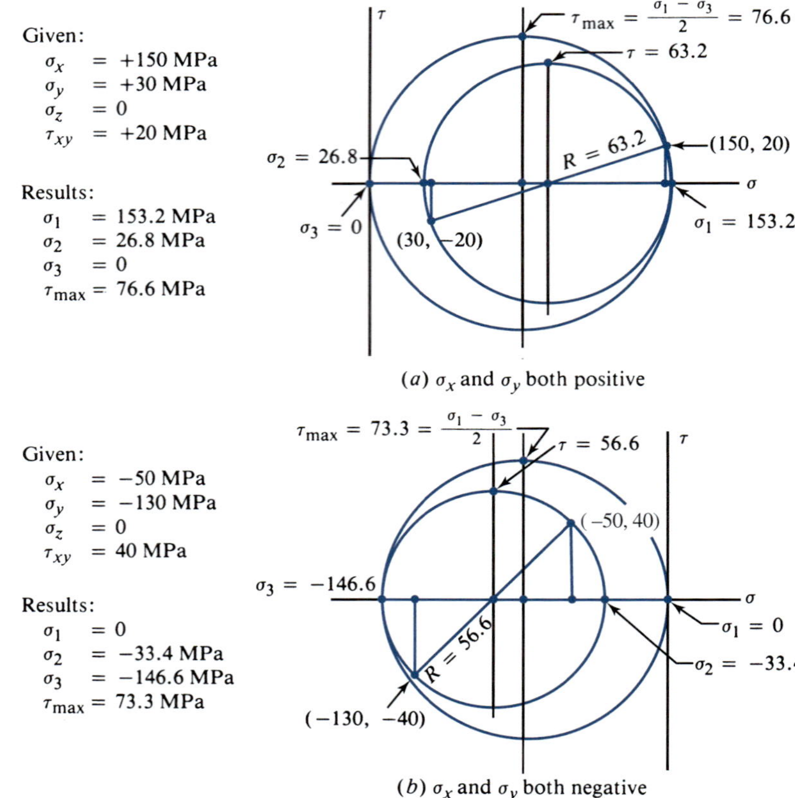

FIGURE 4–27 Mohr's circle for cases in which two principal stresses have the same sign

For the cylinder of Figure 4–24, we can conclude that

$$\sigma_1 = \sigma_x = 12\,500 \text{ psi}$$

$$\sigma_2 = \sigma_y = 6250 \text{ psi}$$

$$\sigma_3 = \sigma_z = 0$$

$$\tau_{max} = (\sigma_1 - \sigma_3)/2 = (12\,500 - 0)/2 = 6250 \text{ psi}$$

Figure 4–27 shows two additional examples in which the two principal stresses in the given plane have the same sign. Then the zero stress in the third direction is added to the diagram, and the new Mohr's circle is superimposed on the original one. This serves to illustrate that the maximum shear stress will occur on the Mohr's circle having the largest radius.

4–7 MOHR'S CIRCLE FOR SPECIAL STRESS CONDITIONS

Mohr's circle is used here to demonstrate the relationship among the applied stresses, the principal stresses, and the maximum shear stress for the following special cases:

Pure uniaxial tension

Pure uniaxial compression

Pure torsional shear

Uniaxial tension combined with torsional shear

These are important, frequently encountered stress conditions, and they will be used in later chapters to illustrate failure theories and design methods. These failure theories are based on the values of the principal stresses and the maximum shear stress.

Pure Uniaxial Tension

The stress condition produced in all parts of a standard tensile test specimen is pure uniaxial tension. Figure 4–28 shows the stress element and the corresponding Mohr's circle. Note that the maximum principal stress, σ_1, is equal to the applied stress, σ_x; the minimum principal stress, σ_2, is zero; and the maximum shear stress, τ_{max}, is equal to $\sigma_x/2$.

Pure Uniaxial Compression

Figure 4–29 shows pure uniaxial compression as it would be produced by a standard compression test. Mohr's circle shows that $\sigma_1 = 0$; $\sigma_2 = \sigma_x$ (a negative value); and the magnitude of the maximum shear stress is $\tau_{max} = \sigma_x/2$.

Pure Torsion

Figure 4–30 shows that the Mohr's circle for this special case has its center at the origin of the $\sigma-\tau$ axes and that the

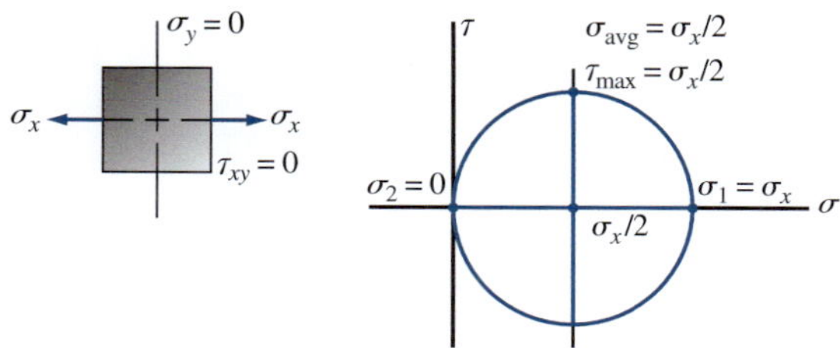

FIGURE 4–28 Mohr's circle for pure uniaxial tension

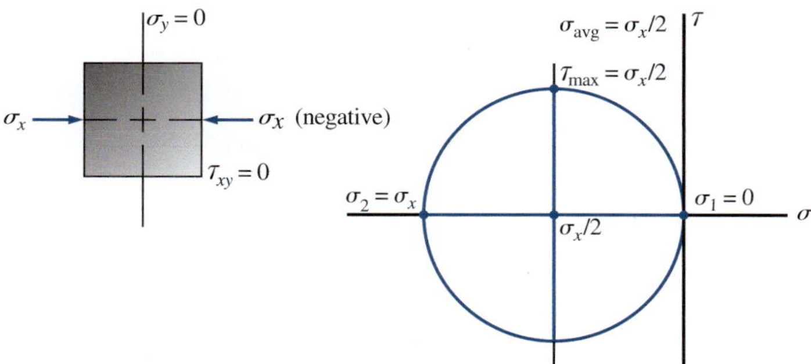

FIGURE 4–29 Mohr's circle for pure uniaxial compression

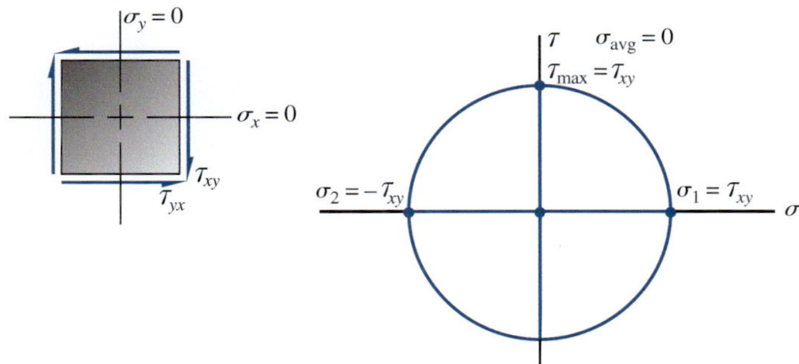

FIGURE 4–30 Mohr's circle for pure torsional shear

radius of the circle is equal to the value of the applied shear stress, τ_{xy}. Therefore, $\tau_{max} = \tau_{xy}$; $\sigma_1 = \tau_{xy}$; and $\sigma_2 = -\tau_{xy}$.

Uniaxial Tension Combined with Torsional Shear

This is an important special case because it describes the stress condition in a rotating shaft carrying bending loads while simultaneously transmitting torque. This is the type of stress condition on which the procedure for designing shafts, presented in Chapter 12, is based. If the applied stresses are called σ_x and τ_{xy}, the Mohr's circle in Figure 4–31 shows that

$$\tau_{max} = R = \text{radius of circle} = \sqrt{(\sigma_x/2)^2 + \tau_{xy}^2} \quad \textbf{(4–10)}$$

$$\sigma_1 = \sigma_x/2 + R = \sigma_x/2 + \sqrt{(\sigma_x/2)^2 + \tau_{xy}^2} \quad \textbf{(4–11)}$$

$$\sigma_2 = \sigma_x/2 - R = \sigma_x/2 - \sqrt{(\sigma_x/2)^2 + \tau_{xy}^2} \quad \textbf{(4–12)}$$

A convenient and useful concept called *equivalent torque* can be developed from Equation (4–10) for the special case of a body subjected to only bending and torsion.

An example is shown in Figure 4–32, where a circular bar is loaded at one end by a downward force and a torsional moment. The force causes bending in the bar with the maximum moment at the point where the bar is attached to the support. The moment causes a tensile stress on the top of the bar in the x-direction at the point called A, where the magnitude of the stress is

$$\sigma_x = M/S \quad \textbf{(4–13)}$$

where S = section modulus of the round bar

Now the torsional moment causes torsional shear stress in the x–y plane at point A having a magnitude of

$$\tau_{xy} = T/Z_p \quad \textbf{(4–14)}$$

where Z_p = polar section modulus of the bar

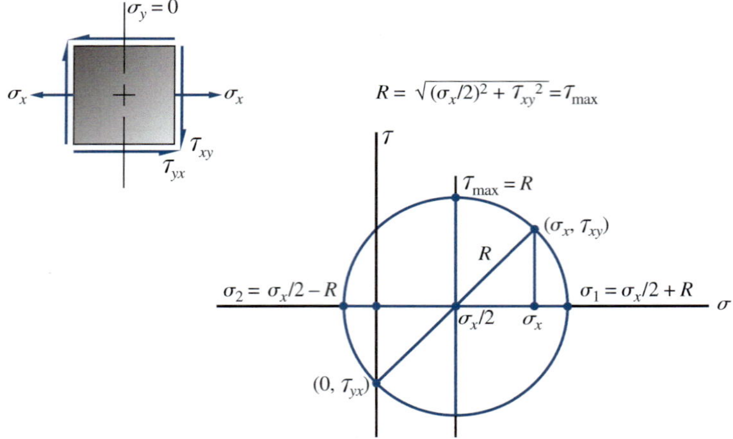

FIGURE 4–31 Mohr's circle for uniaxial tension combined with torsional shear

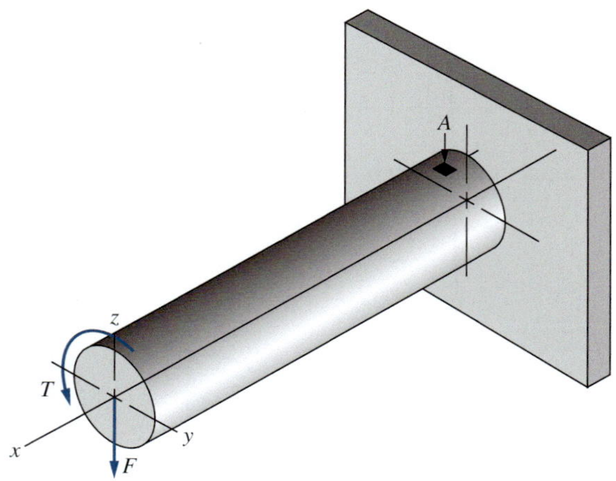

FIGURE 4–32 Circular bar in bending and torsion

Point A then is subjected to a tensile stress combined with shear, the special case shown in the Mohr's circle of Figure 4–31. The maximum shear stress can be computed from Equation (4–10). If we substitute Equations (4–13) and (4–14) in Equation (4–10), we get

$$\tau_{max} = \sqrt{(M/2S)^2 + (T/Z_p)^2} \quad \textbf{(4–15)}$$

Note from Appendix 1 that $Z_p = 2S$. Equation (4–15) can then be written as

$$\tau_{max} = \frac{\sqrt{M^2 + T^2}}{Z_p} \quad \textbf{(4–16)}$$

It is convenient to define the quantity in the numerator of this equation to be the *equivalent torque*, T_e. Then the equation becomes

$$\tau_{max} = T_e/Z_p \quad \textbf{(4–17)}$$

REFERENCE

Mott, Robert L. *Applied Strength of Materials*. 5th ed. Upper Saddle River, NJ: Prentice Hall, 2008.

INTERNET SITES

The following is a short list of the numerous companies that develop and provide finite-element analysis (FEA) software for a wide variety of applications including static and dynamic structural analysis, thermal analysis, dynamic performance of mechanical systems, vibration analysis, computational fluid dynamics analysis, and other computer-aided engineering (CAE) capabilities.

1. ADINA R&D, Inc. www.adina.com
2. Autodesk Algor Simulation www.algor.com
3. ANSYS, Inc. www.ansys.com
4. MSC Software, Inc. www.mscsoftware.com
5. NEi Software, Inc. www.nenastran.com

PROBLEMS

For the sets of given stresses on an element given in Table 4–2, draw a complete Mohr's circle, find the principal stresses and the maximum shear stress, and draw the principal stress element and the

TABLE 4–2 Given Stresses for Problems 1–30

Problem	σ_x	σ_y	τ_{xy}
1	20 ksi	0 ksi	10 ksi
2	−85 ksi	40 ksi	30 ksi
3	40 ksi	−40 ksi	−30 ksi
4	−80 ksi	40 ksi	−30 ksi
5	−120 ksi	40 ksi	−20 ksi
6	−120 ksi	40 ksi	20 ksi
7	60 ksi	−40 ksi	−35 ksi
8	120 ksi	−40 ksi	100 ksi
9	−100 MPa	0 MPa	80 MPa
10	−250 MPa	80 MPa	−110 MPa
11	50 MPa	−80 MPa	40 MPa
12	150 MPa	−80 MPa	−40 MPa
13	−150 MPa	80 MPa	−40 MPa
14	0 MPa	0 MPa	40 MPa
15	250 MPa	−80 MPa	0 MPa
16	50 MPa	−80 MPa	−30 MPa
17	400 MPa	−300 MPa	200 MPa
18	−120 MPa	180 MPa	−80 MPa
19	−30 MPa	20 MPa	40 MPa
20	220 MPa	−120 MPa	0 MPa
21	40 ksi	0 ksi	0 ksi
22	0 ksi	0 ksi	40 ksi
23	38 ksi	−25 ksi	−18 ksi
24	55 ksi	0 ksi	0 ksi
25	22 ksi	0 ksi	6.8 ksi
26	−4250 psi	3250 psi	2800 psi
27	300 MPa	100 MPa	80 MPa
28	250 MPa	150 MPa	40 MPa
29	−840 kPa	−335 kPa	−120 kPa
30	−325 kPa	−50 kPa	−60 kPa

maximum shear stress element. Any stress components not shown are assumed to be zero.

31. Refer to Figure 3–23. For the shaft aligned with the *x*-axis, create a stress element on the bottom of the shaft just to the left of section *B*. Then draw the Mohr's circle for that element. Use $D = 0.50$ in.
32. Refer to Figure P3–48. For the shaft *ABC*, create a stress element on the bottom of the shaft just to the right of section *B*. The torque applied to the shaft at *B* is resisted at support *C* only. Draw the Mohr's circle for the stress element. Use $D = 1.50$ in.
33. Repeat Problem 32 for the shaft in Figure P3–49 $D = 0.50$ in.
34. Refer to Figure P3–50. For the shaft *ABC*, create a stress element on the bottom of the shaft just to the left of section *B*. The torque applied to the shaft by the crank is resisted at support *B* only. Draw the Mohr's circle for the stress element. Use $D = 50$ mm.
35. A short cylindrical bar having a diameter of 4.00 in is subjected to an axial compressive force of 75 000 lb and a torsional moment of 20000 lb · in. Draw a stress element on the surface of the bar. Then draw the Mohr's circle for the element.
36. A torsion bar is used as a suspension element for a vehicle. The bar has a diameter of 20 mm. It is subjected to a torsional moment of 450 N · m and an axial tensile force of 36.0 kN. Draw a stress element on the surface of the bar, and then draw the Mohr's circle for the element.

CHAPTER FIVE

DESIGN FOR DIFFERENT TYPES OF LOADING

THE BIG PICTURE Design for Different Types of Loading

Discussion Map

- This chapter provides additional tools you can use to design load-carrying components that are safe and reasonably efficient in their use of materials.
- You must learn how to classify the kind of loading the component is subjected to: *static, repeated and reversed, fluctuating, shock,* or *impact.*
- You will learn to identify the appropriate analysis techniques based on the type of load and the type of material.

Discover

Identify components of real products or structures that are subjected to static loads.

Identify components that are subjected to equal, repeated loads that reverse directions.

Identify components that experience fluctuating loads that vary with time.

Identify components that are loaded with shock or impact, such as being struck by a hammer or dropped onto a hard surface.

Using the techniques you learn in this chapter will help you to complete a wide variety of design tasks.

For the concepts considered in this chapter, the big picture encompasses a huge array of examples in which you will build on the principles of strength of materials that you reviewed in Chapters 3 and 4 and extend them from the analysis mode to the design mode. Several steps are involved, and you must learn to make rational judgments about the appropriate method to apply to complete the design.

In this chapter, you will learn how to do the following:

1. Recognize the manner of loading for a part: Is it static, repeated and reversed, fluctuating, shock, or impact?
2. Select the appropriate method to analyze the stresses produced.
3. Determine the strength property for the material that is appropriate to the kind of loading and to the

kind of material: Is the material a metal or a non-metal? Is it brittle or ductile? Should the design be based on the yield strength, ultimate tensile strength, compressive strength, endurance strength, or some other material property?

4. Specify a suitable *design factor*, often called a *factor of safety*.
5. Design a wide variety of load-carrying members to be safe under their particular expected loading patterns.

The following paragraphs show by example some of the situations to be studied in this chapter.

An ideal *static load* is one that is applied slowly and is never removed. Some loads that are applied slowly and removed and replaced very infrequently can also be considered to be static. What examples can you think of for products or their components that are subjected to static loads? Consider load-carrying members of structures, parts of furniture pieces, and brackets or support rods holding equipment in your home or in a business or factory. Try to identify specific examples, and describe them to your colleagues. Discuss how the load is applied and which parts of the load-carrying member are subjected to the higher stress levels. Some of the examples that you discovered during **The Big Picture** discussion for Chapter 3 could be used again here.

Fluctuating loads are those that vary during the normal service of the product. They typically are applied for quite a long time so the part experiences many thousands or millions of cycles of stress during its expected life. There are many examples in consumer products around your home, in your car, in commercial buildings, and in manufacturing facilities. Consider virtually anything that has moving parts. Again, try to identify specific examples, and describe them to your colleagues. How does the load fluctuate? Is it applied and then completely removed each cycle? Or is there always some level of mean or average load with an alternating load superimposed on it? Does the load swing from a positive maximum value to a negative minimum value of equal magnitude during each cycle of loading? Consider parts with rotating shafts, such as engines or agricultural, production, and construction machinery.

Consider products that have failed. You may have identified some from **The Big Picture** discussion for Chapter 3. Did they fail the first time they were used? Or did they fail after some fairly long service? Why do you think they were able to operate for some time before failure?

Can you find components that failed suddenly because the material was brittle, such as cast iron, some ceramics, or some plastics? Can you find others that failed only after some considerable deformation? Such failures are called *ductile fractures*.

What were the consequences of the failures that you have found? Was anyone hurt? Was there damage to some other valuable component or property? Or was the failure simply an inconvenience? What was the order of magnitude of cost related to the failure? The answer to some of these questions can help you make rational decisions about design factors to be used in your designs.

It is the designer's responsibility to ensure that a machine part is safe for operation under reasonably foreseeable conditions. This requires that a stress analysis be performed in which the predicted stress levels in the part are compared with the *design stress*, or that level of stress permitted under the operating conditions.

The stress analysis can be performed either analytically or experimentally, depending on the degree of complexity of the part, the knowledge about the loading conditions, and the material properties. The designer must be able to verify that the stress to which a part is subjected is safe.

The manner of computing the design stress depends on the manner of loading and on the type of material. Loading types include the following:

Static
Repeated and reversed
Fluctuating
Shock or impact
Random

Material types are many and varied. Among the metallic materials, the chief classification is between *ductile* and *brittle* materials. Other considerations include the manner of forming the material (casting, forging, rolling, machining, and so on), the type of heat treatment, the surface finish, the physical size, the environment in which it is to operate, and the geometry of the part. Different factors must be considered for plastics, composites, ceramics, wood, and others.

This chapter outlines methods of analyzing load-carrying machine parts to ensure that they are safe. Several different cases are described in which knowledge of the combinations of material types and loading patterns leads to the determination of the appropriate method of analysis. It will then be your job to apply these tools correctly and judiciously as you continue your career.

YOU ARE THE DESIGNER

Recall the task presented at the start of Chapter 4, in which you were the designer of a bracket to hold a fabric sample during a test to determine its long-term stretch characteristics. Figure 4–2 showed a proposed design.

Now you are asked to continue this design exercise by selecting a material from which to make the two bent circular bars that are welded to the rigid support. Also, you must specify a suitable diameter for the bars when a certain load is applied to the test material. ■

5–1 OBJECTIVES OF THIS CHAPTER

After completing this chapter, you will be able to:

1. Identify various kinds of loading commonly encountered by machine parts, including *static, repeated and reversed, fluctuating, shock or impact*, and *random.*
2. Define the term *stress ratio* and compute its value for the various kinds of loading.
3. Define the concept of *fatigue.*
4. Define the material property of *endurance strength* and determine estimates of its magnitude for different materials.
5. Recognize the factors that affect the magnitude of endurance strength.
6. Define the term *design factor.*
7. Specify a suitable value for the design factor.
8. Define the *maximum normal stress theory of failure* and the *modified Mohr method* for design with brittle materials.
9. Define the *maximum shear stress theory of failure.*
10. Define the *distortion energy theory*, also called the *von Mises theory* or the *Mises-Hencky theory.*
11. Describe the *Goodman method* and apply it to the design of parts subjected to fluctuating stresses.
12. Consider *statistical approaches, finite life*, fracture mechanics, and *damage accumulation methods* for design.

5–2 TYPES OF LOADING AND STRESS RATIO

The primary factors to consider when specifying the type of loading to which a machine part is subjected are the manner of variation of the load and the resulting variation of stress with time. Stress variations are characterized by four key values:

1. Maximum stress, σ_{max}
2. Minimum stress, σ_{min}
3. Mean (average) stress, σ_m
4. Alternating stress, σ_a (*stress amplitude*)

The maximum and minimum stresses are usually computed from known information by stress analysis or finite-element methods, or they are measured using experimental stress analysis techniques. Then the mean and alternating stresses can be computed from

$$\sigma_m = (\sigma_{max} + \sigma_{min})/2 \tag{5–1}$$

$$\sigma_a = (\sigma_{max} - \sigma_{min})/2\sigma \tag{5–2a}$$

$$\sigma_a = (\sigma_{max} - \sigma_m) \tag{5–2b}$$

The behavior of a material under varying stresses is dependent on the manner of the variation. One method used to characterize the variation is called *stress ratio.* Two types of stress ratios are commonly used, defined as

$$\text{Stress ratio } R = \frac{\text{minimum stress}}{\text{maximum stress}} = \frac{\sigma_{min}}{\sigma_{max}}$$
$$\text{Stress ratio } A = \frac{\text{alternating stress}}{\text{mean stress}} = \frac{\sigma_a}{\sigma_m} \tag{5–3}$$

Stress ratio R is used in this book.

Static Stress

When a part is subjected to a load that is applied slowly, without shock, and is held at a constant value, the resulting stress in the part is called *static stress.* An example is the load on a structure due to the dead weight of the building materials. Figure 5–1 shows a diagram of stress versus time for static loading. Because $\sigma_{max} = \sigma_{min}$, the stress ratio for static stress is $R = 1.0$.

Static loading can also be assumed when a load is applied and is removed slowly and then reapplied, if the number of load applications is small, that is, under a few thousand cycles of loading.

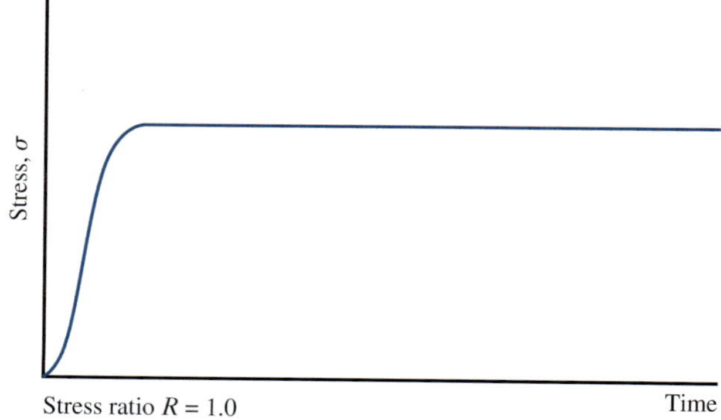

FIGURE 5–1 Static stress

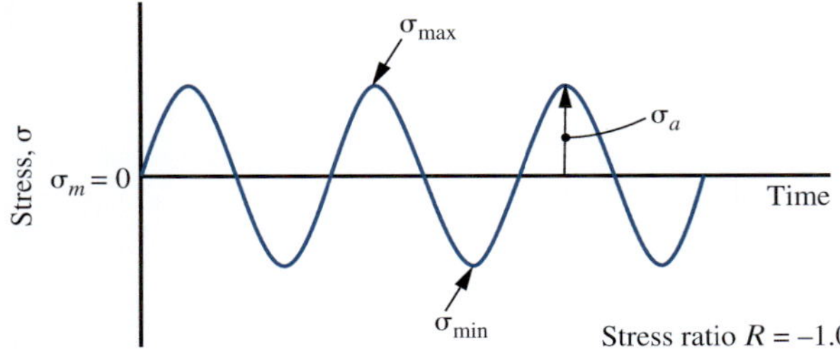

FIGURE 5–2 Repeated, reversed stress or pure oscillation

Repeated and Reversed Stress—Pure Oscillation

A stress reversal occurs when a given element of a load-carrying member is subjected to a certain level of tensile stress followed by the *same level* of compressive stress. If this stress cycle is repeated many thousands of times, the stress is called *repeated and reversed* or *pure oscillation*. Figure 5–2 shows the diagram of stress versus time for repeated and reversed stress. Because $\sigma_{min} = -\sigma_{max}$, the stress ratio is $R = -1.0$, and the mean stress is zero.

An important example in machine design is a rotating circular shaft loaded in bending such as that shown in Figure 5–3. In the position shown, an element on the bottom of the shaft experiences tensile stress while an element on the top of the shaft sees a compressive stress of equal magnitude. As the shaft is rotated 180° from the given position, these two elements experience a complete reversal of stress. Now if the shaft continues to rotate, all parts of the shaft that are in bending see repeated, reversed stress. This is a description of the classic loading case of *reversed bending*.

This type of loading is often called *fatigue loading*, and a machine of the type shown in Figure 5–3 is called a *standard R. R. Moore fatigue test device*. Such machines are used to test materials for their ability to resist repeated loads. The material property called *endurance strength* is measured in this manner. More is said later in this chapter about endurance strength. Actually, reversed bending is only a special case of fatigue loading, since any stress that varies with time can lead to fatigue failure of a part.

Fluctuating Stress—Pulsating Stress

When a load-carrying member is subjected to an alternating stress with a nonzero mean, the loading produces *fluctuating stress*, sometimes called *pulsating stress*. Figure 5–4 shows four diagrams of stress versus time for this type of stress. Differences among the four diagrams occur in whether the various stress levels are positive (tensile) or negative (compressive). ***Any varying stress with a nonzero mean is considered a fluctuating stress***. Figure 5–4 also shows the possible ranges of values for the stress ratio R for the given loading patterns.

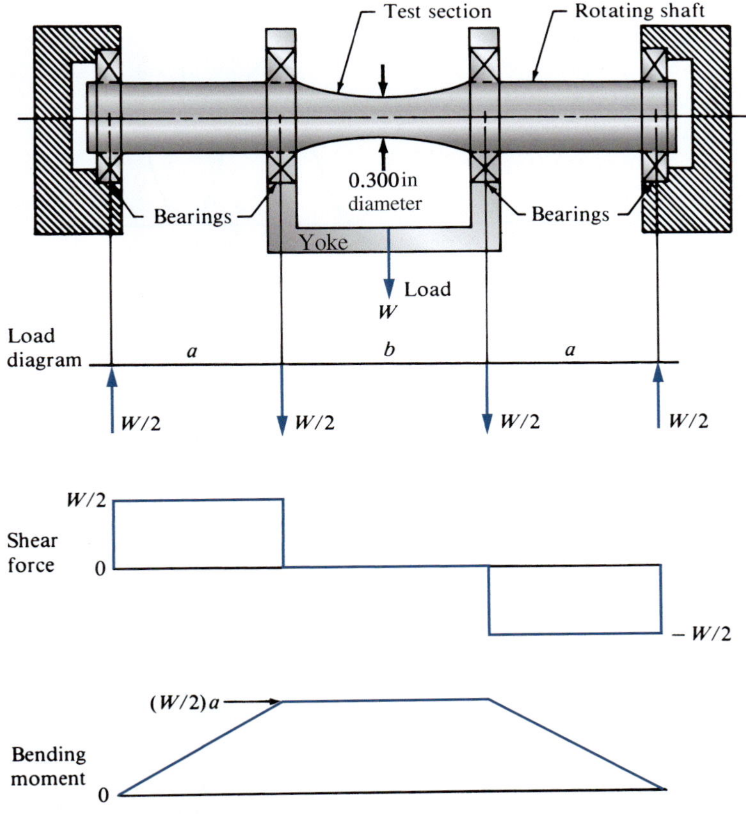

FIGURE 5–3 R. R. Moore fatigue test device; example of reversed bending

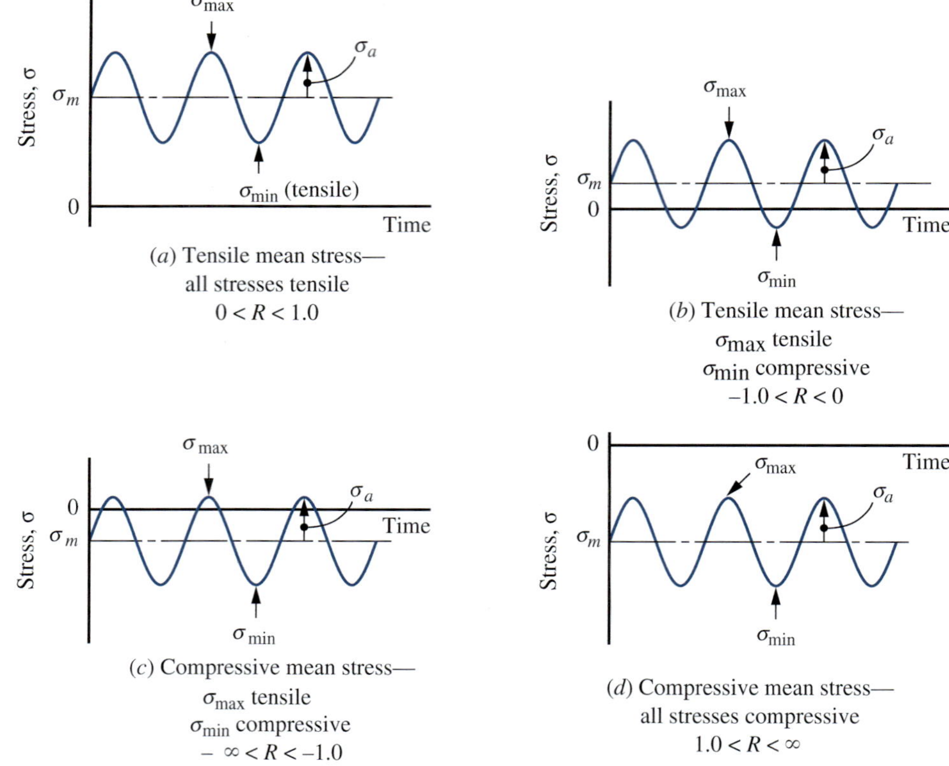

FIGURE 5–4 Fluctuating stresses—pulsating stresses

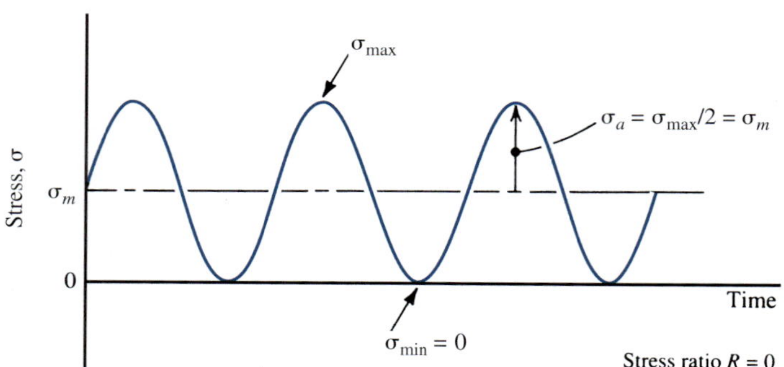

FIGURE 5–5 Repeated, one-direction stress, a special case of fluctuating stress or pure pulsating stress

A special, frequently encountered case of fluctuating stress is *repeated, one-direction stress,* or pure pulsating stress, in which the load is applied and removed many times. As shown in Figure 5–5, the stress varies from zero to a maximum with each cycle. Then, by observation,

$$\sigma_{min} = 0$$

$$\sigma_m = \sigma_a = \sigma_{max}/2$$

$$R = \sigma_{min}/\sigma_{max} = 0$$

An example of a machine part subjected to fluctuating stress of the type shown in Figure 5–4(a) is shown in Figure 5–6, in which a reciprocating cam follower feeds spherical balls one at a time from a chute. The follower is held against the eccentric cam by a flat spring loaded as a cantilever. When the follower is farthest to the left, the spring is deflected from its free (straight) position by an amount $y_{min} = 3.0$ mm. When the follower is farthest to the right, the spring is deflected to $y_{max} = 8.0$ mm. Then, as the cam continues to rotate, the spring sees the cyclic loading between the minimum and maximum values. Point *A* at the base of the spring on the convex side experiences the varying tensile stresses of the type shown in Figure 5–4(a). Example Problem 5–1 completes the analysis of the stress in the spring at point *A*.

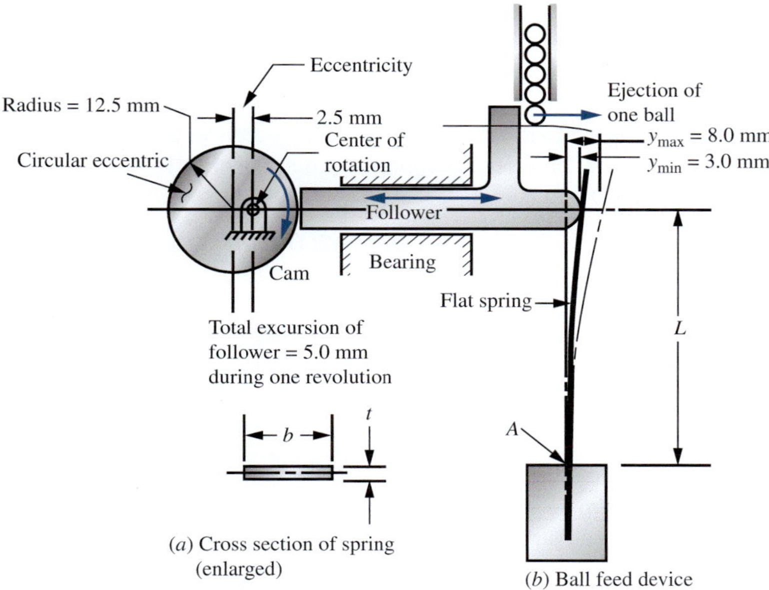

FIGURE 5–6 Example of cyclic loading in which the flat spring is subjected to fluctuating stress

Example Problem 5–1 For the flat steel spring shown in Figure 5–6, compute the maximum stress, the minimum stress, the mean stress, and the alternating stress. Also compute the stress ratio, R. The length L is 65 mm. The dimensions of the spring cross section are $t = 0.80$ mm and $b = 6.0$ mm.

Solution Objective Compute the maximum, minimum, mean, and alternating tensile stresses in the flat spring. Compute the stress ratio, R.

Given Layout shown in Figure 5–6. The spring is steel: $L = 65$ mm.

Spring cross-section dimensions: $t = 0.80$ mm and $b = 6.0$ mm.

Maximum deflection of the spring at the follower = 8.0 mm.

Minimum deflection of the spring at the follower = 3.0 mm.

Analysis Point A at the base of the spring experiences the maximum tensile stress. Determine the force exerted on the spring by the follower for each level of deflection using the formulas from Table A14–2, Case (a). Compute the bending moment at the base of the spring for each deflection. Then compute the stresses at point A using the bending stress formula, $\sigma = Mc/I$. Use Equations (5–1), (5–2), and (5–3) for the mean and alternating stresses and R.

Results Case (a) of Table A14–2 gives the following formula for the amount of deflection of a cantilever for a given applied force:

$$y = PL^3/3EI$$

Solve for the force as a function of deflection:

$$P = 3EIy/L^3$$

Appendix 3 gives the modulus of elasticity for steel to be $E = 207$ GPa. The moment of inertia, I, for the spring cross section is found from

$$I = bt^3/12 = (6.00\text{ mm})(0.80\text{ mm})^3/12 = 0.256\text{ mm}^4$$

Then the force on the spring when the deflection y is 3.0 mm is

$$P = \frac{3(207 \times 10^9\text{ N/m}^2)(0.256\text{ mm}^4)(3.0\text{ mm})}{(65\text{ mm})^3}\frac{(1.0\text{ m}^2)}{(10^6\text{ mm}^2)} = 1.74\text{ N}$$

The bending moment at the support is

$$M = P \cdot L = (1.74\text{ N})(65\text{ mm}) = 113\text{ N}\cdot\text{mm}$$

The bending stress at point A caused by this moment is

$$\sigma = \frac{Mc}{I} = \frac{(113\ \text{N}\cdot\text{mm})(0.40\ \text{mm})}{0.256\ \text{mm}^4} = 176\ \text{N/mm}^2 = 176\ \text{MPa}$$

This is the lowest stress that the spring sees in service, and therefore $\sigma_{\text{min}} = 176$ MPa.

Because the force on the spring is proportional to the deflection, the force exerted when the deflection is 8.0 mm is

$$P = (1.74\ \text{N})(8.0\ \text{mm})/(3.0\ \text{mm}) = 4.63\ \text{N}$$

The bending moment is

$$M = P\cdot L = (4.63\ \text{N})(65\ \text{mm}) = 301\ \text{N}\cdot\text{mm}$$

The bending stress at point A is

$$\sigma = \frac{Mc}{I} = \frac{(301\ \text{N}\cdot\text{mm})(0.40\ \text{mm})}{0.256\ \text{mm}^4} = 470\ \text{N/mm}^2 = 470\ \text{MPa}$$

This is the maximum stress that the spring sees, and therefore $\sigma_{\text{max}} = 470$ MPa.

Now the mean stress can be computed:

$$\sigma_m = (\sigma_{\text{max}} + \sigma_{\text{min}})/2 = (470 + 176)/2 = 323\ \text{MPa}$$

Finally, the alternating stress is

$$\sigma_a = (\sigma_{\text{max}} - \sigma_{\text{min}})/2 = (470 - 176)/2 = 147\ \text{MPa}$$

The stress ratio is found using Equation (5–3):

$$\text{Stress ratio } R = \frac{\text{minimum stress}}{\text{maximum stress}} = \frac{\sigma_{\text{min}}}{\sigma_{\text{max}}} = \frac{176\ \text{MPa}}{470\ \text{MPa}} = 0.37$$

Comments The sketch of stress versus time shown in Figure 5–4(a) illustrates the form of the fluctuating stress on the spring. In Section 5–8, you will see how to design parts subjected to this kind of stress.

Shock or Impact Loading

Loads applied suddenly and rapidly cause shock or impact. Examples include a hammer blow, a weight falling onto a structure, and the action inside a rock crusher. The design of machine members to withstand shock or impact involves an analysis of their energy-absorption capability, a topic not considered in this book. (See References 3, 6, 10, and 12.)

Random Loading

When varying loads are applied that are not regular in their amplitude, the loading is called *random.* Statistical analysis is used to characterize random loading for purposes of design and analysis. This topic is not covered in this book. See Reference 11.

5–3 ENDURANCE STRENGTH AND MECHANISMS OF FATIGUE FAILURE

Whenever a machine element is subjected to cyclical loading characterized by the patterns like those shown in Figures 5-2, 5-4, and 5-5, the loading is generally called *fatigue loading,* common to machinery components. Yield strength and ultimate tensile strength of a material are not adequate to represent the ability of materials to resist fatigue loading. This section presents the concept of *endurance strength,* sometimes called *fatigue strength,* that must be used in such cases. Components may fail at stress levels lower than ultimate or yield strengths after experiencing applied stresses for several cycles. Fatigue failures are often classified as either *low-cycle fatigue* (LCF) or *high-cycle fatigue* (HCF) because the mechanism of failure is different for each. While no specific dividing line can be defined, designers often use up to 1000 cycles (10^3) for LCF and higher numbers of cycles—up to infinite life—as HCF.

Fatigue failures start at small surface cracks, internal imperfections, or even at grain boundaries in the material in areas subjected to tensile stress. With repeated applications of stress, the cracks grow and progress to larger areas of the cross section. Eventually, the component fails, often suddenly and catastrophically. Such failures frequently occur within areas of stress concentration such as keyways or grooves in shafts, steps in the size of a cross section, notches, or other geometric discontinuities as discussed in Section 3–21. Even surface roughness from machining or accidental nicks and scratches can serve as points of crack initiation. Therefore, designers must consider the possibility of fatigue failure when sizing critical sections of components. Manufacturers must also understand this phenomenon and produce parts with good finishes that are free from damage. End users of critical components must also handle them with care.

In low-cycle fatigue, local stresses experience high strain levels, approaching or exceeding the strain at yield of the material. Such events may be due to accidental overloading, or infrequently encountered situations during fabrication of a component, installation into an assembly, shock during transportation or handling, evasive maneuvers, takeoffs or landings of aircraft, launch of a ship or spacecraft, initial testing, seismic shock during an earthquake, or operating for prolonged periods near the limits of the capability of a system. The high strain may cause microscopic cracks that progress to ultimate failure. Prediction of the life of a component under such conditions falls under the analysis procedure called *fracture mechanics* that requires extensive knowledge of the geometry of the crack and the ability to characterize how a specific material behaves in the highly localized region of high strain and stress around the crack. A life prediction method called *strain-life* is used. References 1, 2, 7, and 9 and Internet sites 1–8 contain extensive detail about these topics. This book will not cover the topic of fracture mechanics encountered in low-cycle fatigue and will concentrate on designing to prevent high-cycle fatigue.

The endurance strength of a material under high-cycle fatigue loading is determined from tests that apply cyclic patterns of stress for long periods of time, and data are obtained for the number of cycles to failure for a given stress level. As expected, higher stress levels produce failure at fewer numbers of cycles and lower stresses permit higher numbers of cycles—up to a point. For many common materials used in machinery, a stress level is reached where a virtually unlimited number of cycles of stress can be applied without fatigue failure. This stress level is called the *endurance strength*, *fatigue strength*, or *fatigue limit* of the material. In this book, we use the symbol, s_n, for this property.

Data for endurance strength are reported on charts such as that shown in Figure 5–7, called a *stress-life diagram*. The vertical axis is the *stress amplitude*, σ_a, as defined by equation (5–2) and shown in Figure 5–2, and it is assumed that the same amplitude occurs for each cycle for many thousands of cycles. The horizontal axis is the number of cycles to failure, *N*. Both axes are logarithmic scales, resulting in the data plotting as straight lines. The data are average values of endurance strength through the scattered data points taken from numerous tests at each stress level. The transition from the sloped line to the horizontal line at the fatigue limit for any given material typically occurs at approximately one million cycles (10^6), and the curves shown are idealized, showing the break to be

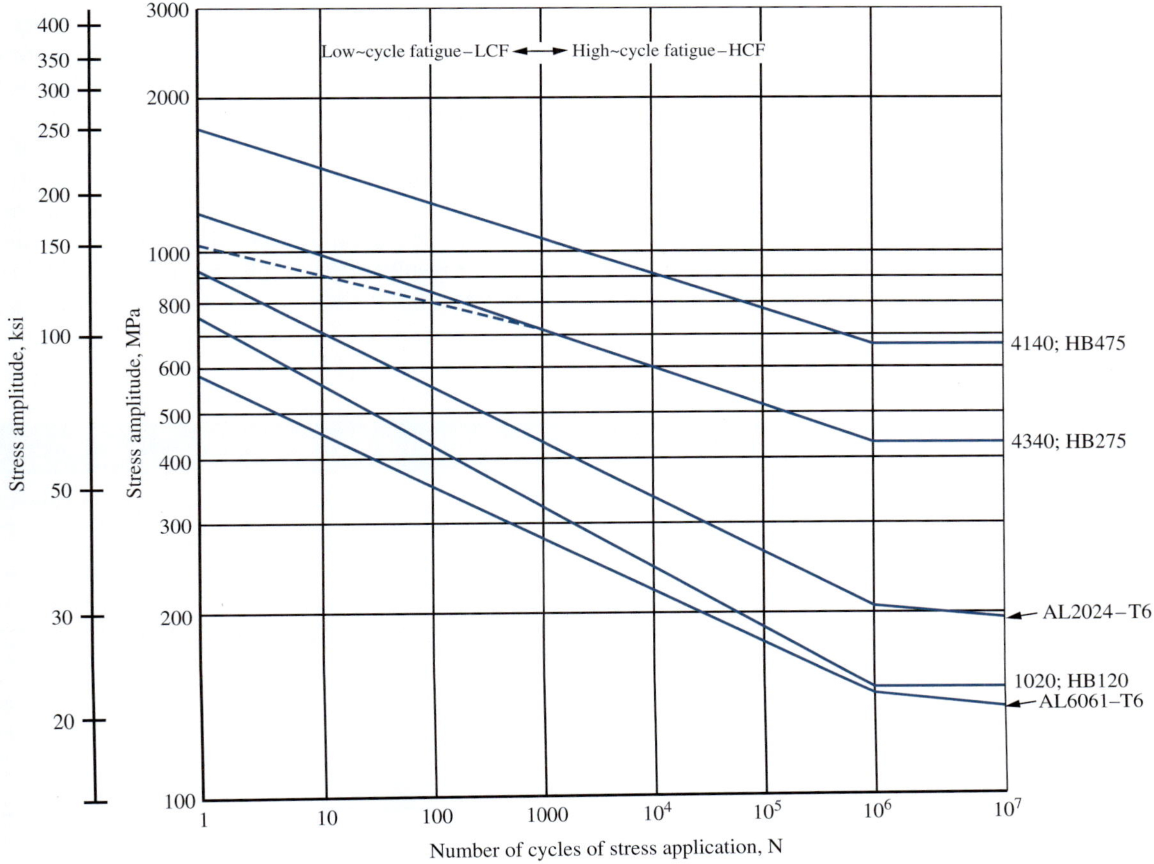

FIGURE 5–7 Representative endurance strengths

sharp. The following equation represents the sloped portion of the curve.

$$s_a = s_n(N)^b \quad (5\text{–}4)$$

where

s_a = stress amplitude level for a given number of cycles to failure
N = number of cycles to failure at a given stress level
s_n = fatigue limit or endurance strength of the material
b = exponent related to the slope of the curve

Data for these properties for many materials are presented in References 1, 2, 7, and 9, and Internet sites 1–8 describe software programs that contain sizable databases of such data. Table 5–1 shows the data for the five selected materials shown in Figure 5–7 for one plain carbon steel, two alloy steels, and two aluminum alloys, taken from Internet site 1.

The last column for *curve intercept*, s'_f, represents the stress value where the curve intersects the vertical axis. This value has no further use as discussed later for low-cycle fatigue.

Note the difference between the curves for the three steels and those for the two aluminum alloys. The steels exhibit a true fatigue limit resulting in the horizontal line to the right of 10^6 cycles, and should never fail by fatigue at higher numbers of cycles of loading. The curves for aluminum continue to drop after 10^6 cycles, although at a much reduced slope. Therefore, data for endurance strengths of aluminum, many other nonferrous metals, and some very high-strength ferrous metals are quoted as a value of s_n at a stated number of cycles, typically 10^6 or 10^7. For higher numbers of cycles, additional data should be sought.

The rotating bending test as shown in Figure 5–3 has been used for many years to acquire endurance strength data and much of the reported data are based on this test. The specimen has a small diameter (typically 0.30 in or 7.62 mm) and is highly polished to eliminate any effect of surface texture. The shape and manner of loading produces pure bending with zero shearing stress and no stress concentrations in the central section. The magnitude of the load can be varied to produce a desired stress level and the shaft is rotated until it breaks. The total number of revolutions to failure is recorded. The test produces the classic repeated and reversed stress shown in Figure 5–2 having zero mean stress, stress amplitude of s_a, and stress ratio $R = -1.0$.

In recent times, other methods have gained favor, particularly programmable servo-controlled axial tension testing devices. Test specimens can be loaded in many different patterns, simulating any of the conditions shown in Figures 5–2, 5–4, 5–5, and others. When the load is reversed and repeated, seemingly similar to the rotating bending test, the stress cycle shown in Figure 5–2 is produced, resulting in a mean stress of zero and a stress ratio, $R = -1.0$. However, an important difference in the behavior of the material occurs because the stress distribution created in the specimens is ideally uniform across the entire section. Note that for rotating bending, only the outermost part of the cylindrical specimen sees the maximum stress and the stress decreases linearly to zero at the center of the bar. Fatigue failures are more likely to initiate in regions of high tensile stress. Because in the axial load test all of the material is subjected to the highest stress, reported endurance strength data are typically lower than those for the rotating bending test by approximately 20%. This situation is discussed more in Section 5–4.

The servo-controlled testing devices are also used to evaluate the effect of loading that produces a fluctuating or pulsating cyclical stress pattern with a nonzero mean stress such as those shown in Figures 5–4 and 5–5. The *mean stress effect* is discussed later.

References 4 and 11 include many tables of data for fatigue strengths of materials along with additional detail on the nature of fatigue failures. Data for the endurance strength should be used wherever it is available, either from test results or from reliable published data. However, such data are not always available. Reference 5 suggests the

TABLE 5–1 Materials and Fatigue Property Data for Curves shown in Figure 5–7

Material	Condition	Ultimate Strength, s_u		Fatigue Limit, s_n		Curve Slope, b	Curve Intercept, s'_f	
		MPa	ksi	MPa	ksi		MPa	ksi
Carbon and alloy steels								
1020	HB120	393	57	142	21	−0.121	754	109
4340	HB275	1048	152	430	62	−0.075	1211	176
4140	HB475	2033	295	663	96	−0.070	1745	253
Aluminum								
6061	T6	310	45	138	20	−0.102	565	82
2024	T6	476	69	205	30	−0.110	938	136

following approximation for the basic rotating bending endurance strength for wrought steel having $s_u \leq 1500\,\text{MPa}$ (220 ksi).

$$\text{Endurance strength} = s_n = 0.50(\text{ultimate tensile strength}) = 0.50\,s_u \quad (5\text{–}5)$$

Low-cycle Fatigue

That part of Figure 5–7 to the left of the line for 1000 cycles is the low-cycle fatigue region and the discussion above does not apply. In fact, the parts of the lines from $N = 1000$ down to $N = 1$ are not used at all, except for providing a convenient way of drawing the sloped part of the curves—a straight line from the curve intercept to the fatigue limit. Failure at a single cycle ($N = 1$), of course, occurs at the ultimate tensile strength for the material, s_u. Some designers add a supplementary line from $N = 1000$ to $N = 1$ as shown in dashed form for 4340 steel. However, it is recommended that the strain-life technique mentioned earlier be used in this region.

5–4 ESTIMATED ACTUAL ENDURANCE STRENGTH, s'_n

If the actual material characteristics or operating conditions for a machine part are different from those for which the basic endurance strength was determined, the fatigue strength must be reduced from the reported value. Some of the factors that decrease the endurance strength are discussed in this section. The discussion relates only to the endurance strength for materials subjected to reversed and repeated bending stress. Cases involving endurance strength in shear are discussed separately in Section 5–8.

We begin by presenting a procedure for estimating the *actual endurance strength*, s'_n, for the material for the part being designed. It involves applying several factors to the basic endurance strength for the material. Additional elaboration on the factors follows.

PROCEDURE FOR ESTIMATING ACTUAL ENDURANCE STRENGTH, s'_n

1. Specify the material for the part and determine its ultimate tensile strength, s_u, considering its condition, as it will be used in service.
2. Specify the manufacturing process used to produce the part with special attention to the condition of the surface in the most highly stressed area.
3. Use Figure 5–8 to estimate the modified endurance strength, s_n.
4. Apply a material factor, C_m, from the following list.

   Wrought steel: $C_m = 1.00$
   Cast steel: $C_m = 0.80$
   Powdered steel: $C_m = 0.76$
   Malleable cast iron: $C_m = 0.80$
   Gray cast iron: $C_m = 0.70$
   Ductile cast iron: $C_m = 0.66$
5. Apply a type-of-stress factor: $C_{st} = 1.0$ for bending stress; $C_{st} = 0.80$ for axial tension.
6. Apply a reliability factor, C_R, from Table 5–2.
7. Apply a size factor, C_s, using Figure 5–9 and Table 5–3 as guides.
8. Compute the estimated actual endurance strength, s'_n, from

$$s'_n = s_n(C_m)(C_{st})(C_R)(C_s) \quad (5\text{–}6)$$

These are the only factors that will be used consistently in this book. If data for other factors can be determined from additional research, they should be multiplied as additional terms in Equation (5–6). In most cases, we suggest accounting for other factors for which reasonable data cannot be found by adjusting the value of the design factor as discussed in Section 5–7.

Stress concentrations caused by sudden changes in geometry are, indeed, likely places for fatigue failures to occur. Care should be taken in the design and manufacture of cyclically loaded parts to keep stress concentration factors to a low value. We will apply stress concentration factors to the computed stress rather than to the endurance strength. See Section 5–8.

While 12 factors affecting endurance strength are discussed in the following section, note that the procedure just given includes only the first five. They are *surface finish, material factor, type-of-stress factor, reliability factor*, and *size factor*. The others are mentioned to alert you to the variety of conditions you should investigate as you complete a design. However, generalized data are difficult to acquire for all factors. Special testing or additional literature searching should be done when conditions exist for which no data are provided in this book. The end-of-chapter references contain a huge amount of such information (see References 2, 4, 7, 9, 11, and 13–16).

Surface Finish

Any deviation from a polished surface reduces endurance strength because the rougher surface provides sites where locally increased stresses or irregularities in the material structure promote the initiation of microscopic cracks that can progress to fatigue failures. Manufacturing processes, corrosion, and careless handling produce detrimental surface roughening.

Figure 5–8, adapted from data in Reference 8, shows estimates for the endurance strength s_n compared with the ultimate tensile strength of wrought steels for several practical surface conditions. The data first estimate the endurance strength for the polished specimen to be 0.50 times the ultimate strength and then apply a factor related to the surface condition. U.S. Customary units are used in Figure 5–8(a)

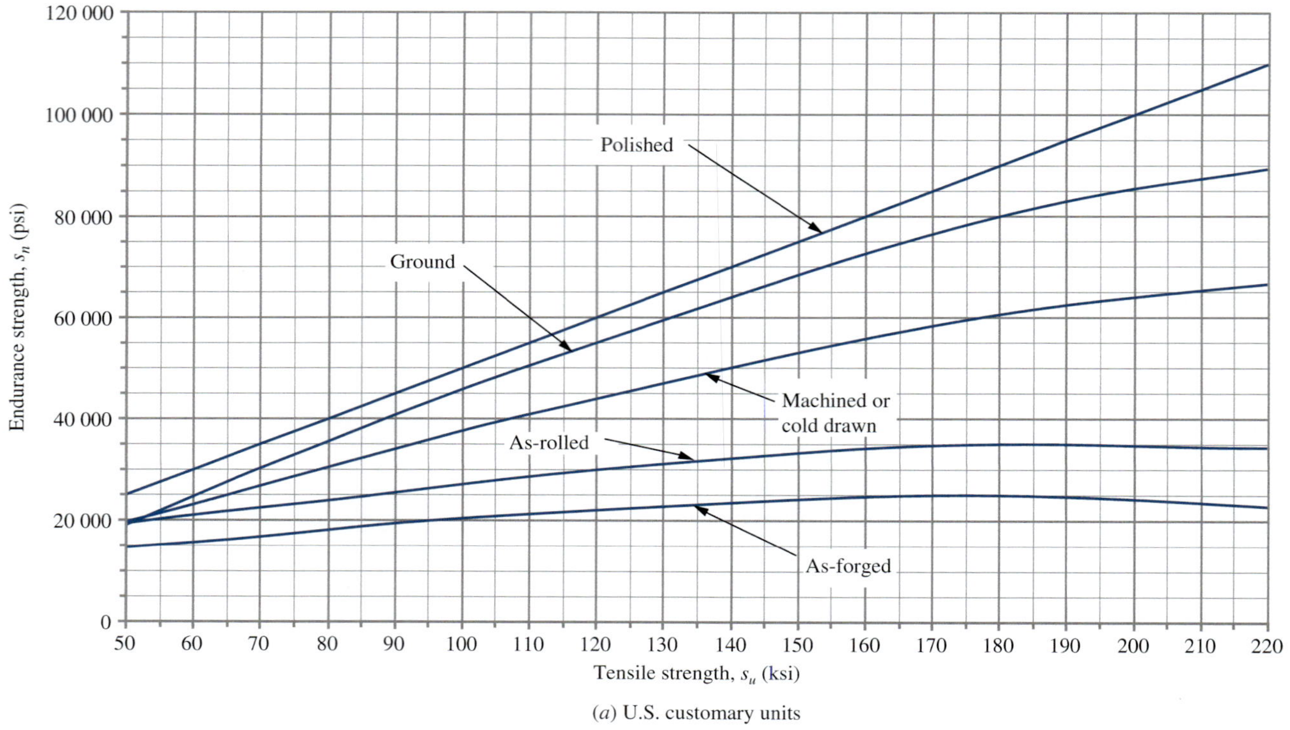

(*a*) U.S. customary units

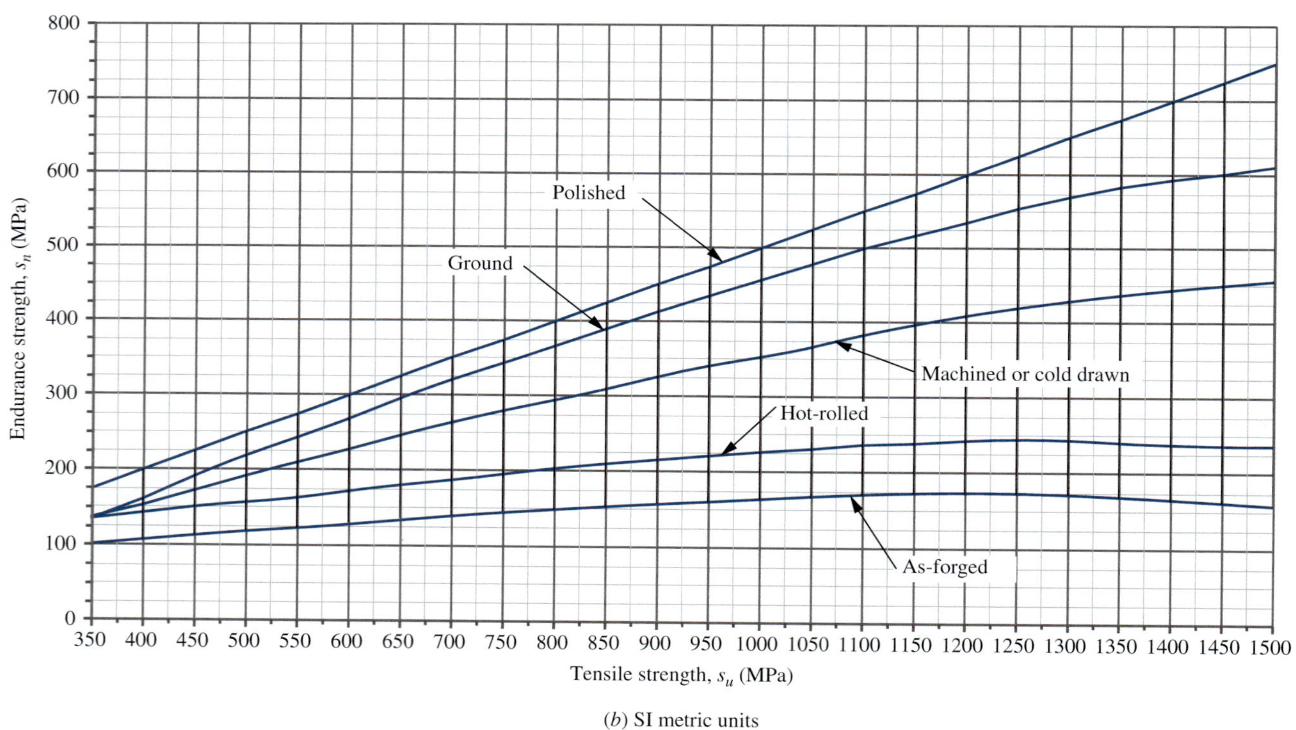

(*b*) SI metric units

FIGURE 5–8 Endurance strength modified for surface condition versus tensile strength for wrought steel

while SI units are shown in Figure 5–8(b). Project vertically from the s_u axis to the appropriate curve and then horizontally to the endurance strength axis.

The data from Figure 5–8 should not be extrapolated for $s_u > 220$ ksi or 1500 MPa without specific testing as empirical data reported in Reference 4 are inconsistent at higher strength levels.

Note that the curve labeled *Polished* is actually the straight line, $s_n = 0.50s_u$, implying a factor of 1.0 because endurance strength test specimens are polished.

Ground surfaces are fairly smooth and reduce the endurance strength by a factor of approximately 0.90 for $s_u < 160$ ksi (1100 MPa), decreasing to about 0.80 for $s_u = 220$ ksi (1500 MPa). Machining or cold drawing produce a somewhat rougher surface because of tooling marks resulting in a reduction factor in the range of 0.80 to 0.60 over the range of strengths shown. The outer part of a hot-rolled steel has a roughened oxidized scale that produces a reduction factor from 0.72 to 0.30 if a part is used in the as-rolled condition. For a forged part, not subsequently machined, the factor ranges from 0.57 to 0.20.

From these data, it should be obvious that you must give special attention to surface finish for critical surfaces exposed to fatigue loading in order to benefit from the steel's basic strength. Also, critical surfaces of fatigue-loaded parts must be protected from nicks, scratches, and corrosion because they drastically reduce fatigue strength.

Material Factors, C_m

Metal alloys having similar chemical composition can be wrought, cast, or made by powder metallurgy to produce the final form. Wrought materials are usually rolled or drawn, and they typically have higher endurance strength than cast materials. The grain structure of many cast materials or powder metals and the likelihood of internal flaws and inclusions tend to reduce their endurance strength. Reference 5 provides data from which the *material factors* listed in step 4 of the procedure outlined previously are taken.

Type-of-Stress Factor, C_{st}

As discussed in Section 5–3, most endurance strength data are obtained from tests using a rotating cylindrical bar subjected to repeated and reversed bending in which the outer part experiences the highest stress. Stress levels decrease linearly to zero at the center of the bar. Because fatigue cracks usually initiate in regions of high tensile stress, a relatively small proportion of the material experiences such stresses. Contrast this with the case of a bar subjected to direct axial tensile stress for which *all* of the material experiences the maximum stress. There is a higher statistical probability that local flaws anywhere in the bar may start fatigue cracks. The result is that the endurance strength of a material subjected to repeated and reversed axial stress is approximately 80% of that from repeated and reversed bending. In this book, we assume that basic endurance strength data are obtained from rotating bending tests and recommend the factors $C_{st} = 1.0$ for bending stress and $C_{st} = 0.80$ for axial loading.

TABLE 5–2 Approximate Reliability Factors, C_R

Desired reliability	C_R
0.50	1.0
0.90	0.90
0.99	0.81
0.999	0.75

Reliability Factor, C_R

The data for endurance strength for steel shown in Figure 5–8 represent average values derived from many tests of specimens having the appropriate ultimate strength and surface conditions. Naturally, there is variation among the data points; that is, half are higher and half are lower than the reported values on the given curve. The curve, then, represents a reliability of 50%, indicating that half of the parts would fail. Obviously, it is advisable to design for a higher reliability, say, 90%, 99%, or 99.9%. A factor can be used to estimate a lower endurance strength that can be used for design to produce the higher reliability values. Ideally, a statistical analysis of actual data for the material to be used in the design should be obtained. By making certain assumptions about the form of the distribution of strength data, Reference 8 reports the values in Table 5–2 as approximate reliability factors, C_R.

Size Factor, C_s—Circular Sections in Rotating Bending

Recall that the basic endurance strength data were taken for a specimen with a circular cross section that has a diameter of 0.30 in (7.62 mm) and that it was subjected to repeated and reversed bending while rotating. Therefore, each part of the surface is subjected to the maximum tensile bending stress with each revolution. Furthermore, the most likely place for fatigue failure to initiate is in the zone of maximum tensile stress within a small distance of the outer surface.

Data from References 5 and 8 show that as the diameter of a rotating circular bending specimen increases, the endurance strength decreases because the stress gradient (change in stress as a function of radius) places a greater proportion of the material in the highly stressed region. Figure 5–9 and Table 5–3 show the size factor to be used in this book, adapted from Reference 8. These data can be used for either solid or hollow circular sections.

Size Factor, C_s—Other Conditions

We need different approaches to determining the size factor when a part with a circular section is subjected to repeated and reversed bending but it is *not rotating*, or if the part

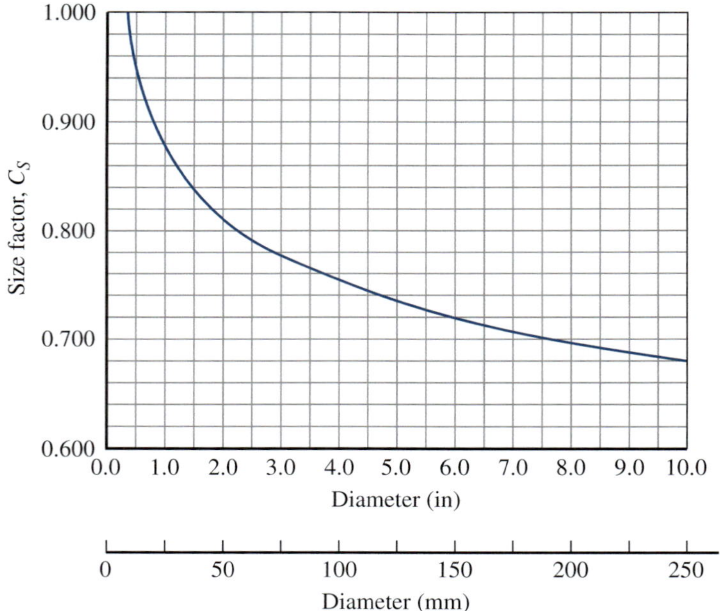

FIGURE 5–9 Size factor, C_s

TABLE 5–3 Size Factors

U.S. customary units	
Size range	**For *D* in inches**
$D \leq 0.30$	$C_s = 1.0$
$0.30 < D \leq 2.0$	$C_s = (D/0.3)^{-0.11}$
$2.0 < D < 10.0$	$C_s = 0.859 - 0.02125D$
SI units	
Size range	**For *D* in mm**
$D \leq 7.62$	$C_s = 1.0$
$7.62 < D \leq 50$	$C_s = (D/7.62)^{-0.11}$
$50 < D < 250$	$C_s = 0.859 - 0.000837D$

has a noncircular cross section. Here we show a procedure adapted from Reference 8 that focuses on the volume of the part that experiences 95% or more of the maximum stress. It is in this volume that fatigue failure is most likely to be initiated. Furthermore, in order to relate the physical size of such sections to the size factor data in Figure 5–9, we develop an equivalent diameter, D_e.

When the parts in question have a uniform geometry over the length of interest, the volume is the product of the length and the cross-sectional area. We can compare different shapes by considering a unit length for each and looking only at the areas. As a base, let's begin by determining an expression for that part of a circular section subjected to 95% or more of the maximum rotating bending stress, calling this area, A_{95}. Because the stress is directly proportional to the radius, we need the area of the thin ring between the outside surface with the full diameter D and a circle whose diameter is $0.95D$, as shown in Figure 5-10(a). Then,

$$A_{95} = (\pi/4)[D^2 - (0.95D)^2] = 0.0766D^2 \quad \textbf{(5–7)}$$

You should demonstrate that this same equation applies to a hollow circular section as shown in Figure 5–10(b). This verifies that the data for size factor shown in Figure 5–9 and Table 5–3 apply directly to either the solid or hollow circular sections when they experience rotating bending.

Nonrotating Circular Section in Repeated and Reversed Flexure. Now consider a solid circular section that does not rotate but that is flexed back and forth in repeated and reversed bending. Only the top and bottom segments beyond a radius of $0.475D$ experience 95% or higher of the maximum bending stress as shown in Figure 5–10(c). By using properties of a segment of a circle, it can be shown that

$$A_{95} = 0.0105D^2 \quad \textbf{(5–8)}$$

Now we can determine the *equivalent diameter*, D_e, for this area by equating equations (5–7) and (5–8) while designating the diameter in equation (5–7) as D_e and then solving for D_e.

$$0.0766D_e^2 = 0.0105D^2$$
$$D_e = 0.370D \quad \textbf{(5–9)}$$

This same equation applies to a hollow circular section. The diameter, D_e, can be used in Figure 5–9 or in Table 5–3 to find the size factor.

Rectangular Section in Repeated and Reversed Flexure. The A_{95} area is shown in Figure 5–10(d) as the

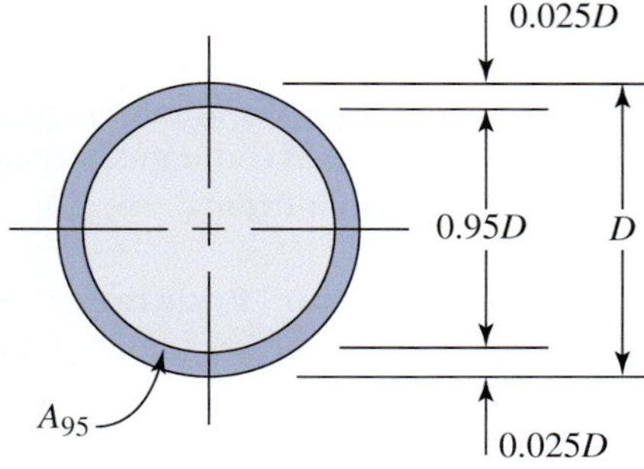

(a) Solid circular section-rotating

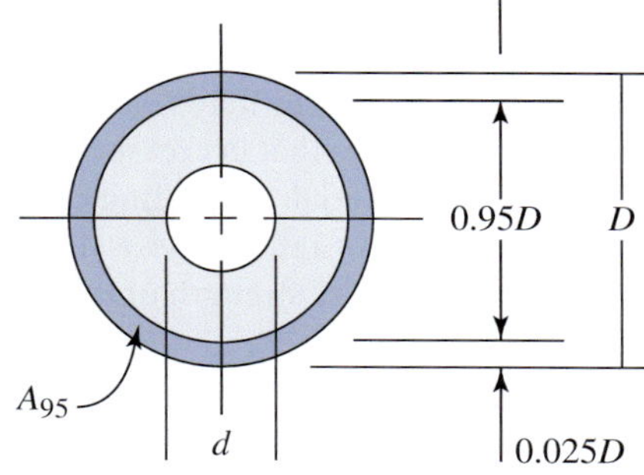

(b) Hollow circular section-rotating

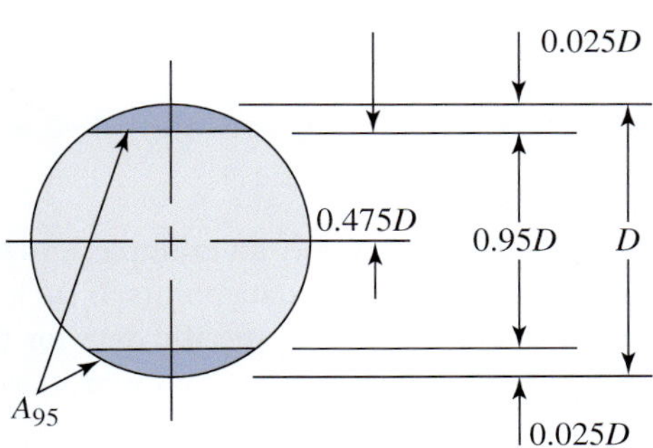

(c) Solid circular section not rotating

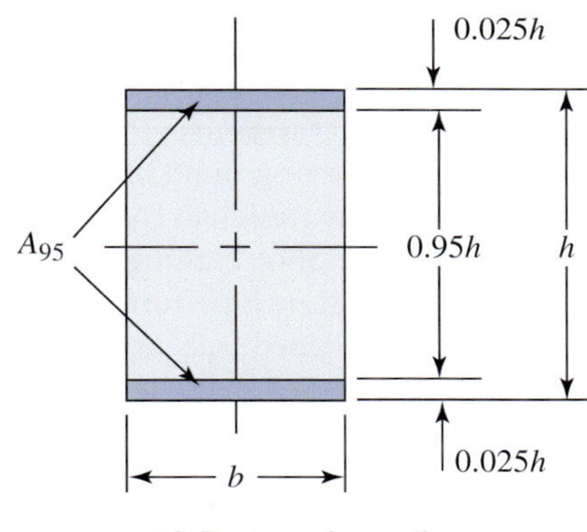

(d) Rectangular section

FIGURE 5–10 Geometry of sections for computing A_{95} area

two strips having a thickness of $0.025h$ at the top and bottom of the section. Therefore,

$$A_{95} = 0.05hb$$

Equating this to A_{95} for a circular section gives,

$$0.0766D_e^2 = 0.05hb \qquad \textbf{(5–10)}$$
$$D_e = 0.808\sqrt{hb}$$

This diameter can be used in Figure 5–9 or in Table 5–3 to find the size factor.

Other shapes can be analyzed in a similar manner.

Any Shape in Repeated Direct Axial Tensile Stress. This special case is based on the concept that the greatest likelihood of initiating a fatigue failure is in zones of highest tensile stress. For bending and torsion, the highest tensile stress occurs in the outermost parts of the cross section and that is the basis for the data in Figure 5–9. However, for axial tensile stress, all parts of the cross section experience the same level of tensile stress and, therefore, all parts are equally susceptible to initiation of a fatigue crack.

> ***For this case, use C_s = 1.0 regardless of the size of the member.***

Other Factors

The following factors are not included quantitatively in problem solutions in this book because of the difficulty of finding generalized data. However, you should consider each one as you engage in future designs and seek additional data as appropriate.

Flaws. Internal flaws of the material, especially likely in cast parts, are places in which fatigue cracks initiate. Critical parts can be inspected by x-ray techniques for internal flaws. If they are not inspected, a higher-than-average design factor should be specified for cast parts, and a lower endurance strength should be used.

Temperature. Most materials have a lower endurance strength at high temperatures. The reported values are typically for room temperatures. Operation above 500°F (260°C) will reduce the endurance strength of most steels. See Reference 8.

Nonuniform Material Properties. Many materials have different strength properties in different directions because of the manner in which the material was processed. Rolled sheet or bar products are typically stronger in the direction of rolling than they are in the transverse direction. Fatigue tests are likely to have been run on test bars oriented in the stronger direction. Stressing of such materials in the transverse direction may result in lower endurance strength.

Nonuniform properties are also likely to exist in the vicinity of welds because of incomplete weld penetration, slag inclusions, and variations in the geometry of the part at the weld. Also, welding of heat-treated materials may alter the strength of the material because of local annealing near the weld. Some welding processes may result in the production of residual tensile stresses that decrease the effective endurance strength of the material. Annealing or normalizing after welding is often used to relieve these stresses, but the effect of such treatments on the strength of the base material must be considered.

Residual Stresses. Fatigue failures typically initiate at locations of relatively high tensile stress. Any manufacturing process that tends to produce residual tensile stress will decrease the endurance strength of the component. Welding has already been mentioned as a process that may produce residual tensile stress. Grinding and machining, especially with high material removal rates, also cause undesirable residual tensile stresses. Critical areas of cyclically loaded components should be machined or ground in a gentle fashion.

Processes that produce residual *compressive* stresses can prove to be beneficial. Shot blasting and peening are two such methods. *Shot blasting* is performed by directing a high-velocity stream of hardened balls or pellets at the surface to be treated. *Peening* uses a series of hammer blows on the surface. Crankshafts, springs, gears and other cyclically loaded machine parts can benefit from these methods.

Corrosion and Environmental Factors. Endurance strength data are typically measured with the specimen in air. Operating conditions that expose a component to water, salt solutions, or other corrosive environments can significantly reduce the effective endurance strength. Corrosion may cause harmful local surface roughness and may also alter the internal grain structure and chemistry of the material. Steels exposed to hydrogen are especially affected adversely.

Nitriding. Nitriding is a surface-hardening process for alloy steels in which the material is heated to 950°F (514°C) in a nitrogen atmosphere, typically ammonia gas, followed by slow cooling. Improvement of endurance strength of 50% or more can be achieved with nitriding.

Effect of Stress Ratio on Endurance Strength. Figure 5–11 shows the general variation of endurance-strength data for a given material when the stress ratio R varies from -1.0 to $+1.0$, covering the range of cases including the following:

- Repeated, reversed stress (Figure 5–3); $R = -1.0$
- Partially reversed fluctuating stress with a tensile mean stress [Figure 5–4(b)]; $-1.0 < R < 0$
- Repeated, one-direction tensile stress (Figure 5–6); $R = 0$
- Fluctuating tensile stress [Figure 5–4(a)]; $0 < R < 1.0$
- Static stress (Figure 5–1); $R = 1$

Note that Figure 5–11 is only an example, and it should not be used to determine actual data points. If such data are desired for a particular material, specific data for that material must be found either experimentally or in published literature.

The most damaging kind of stress among those listed is the repeated, reversed stress with $R = -1$. (See Reference 4.) Recall that the rotating shaft in bending as shown in Figure 5–3 is an example of a load-carrying member subjected to a stress ratio $R = -1$.

Fluctuating stresses with a compressive mean stress as shown in Parts (c) and (d) of Figure 5–4 do not significantly affect the endurance strength of the material because fatigue failures tend to originate in regions of tensile stress.

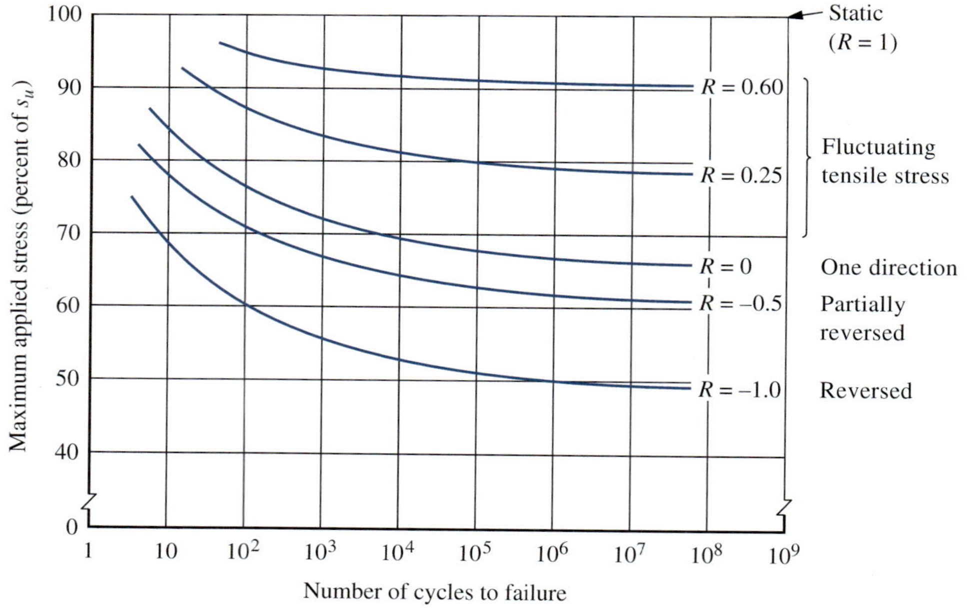

FIGURE 5–11 Effect of stress ratio R on endurance strength of a material

Note that the curves of Figure 5–11 show estimates of the endurance strength, s_n, as a function of the ultimate tensile strength for steel. These data apply to ideal polished specimens and do not include any of the other factors discussed in this section. For example, the curve for $R = -1.0$ (reversed bending) shows that the endurance strength for steel is approximately 0.5 times the ultimate strength ($0.50 \times s_u$) for large numbers of cycles of loading (approximately 10^5 or higher). This is a good general estimate for steels. The chart also shows that types of loads producing R greater than -1.0 but less than 1.0 have less of an effect on the endurance strength. This illustrates that using data from the reversed bending test is the most conservative.

We will not use Figure 5–11 directly for problems in this book because our procedure for estimating the actual endurance strength starts with the use of Figure 5–8, which presents data from reversed bending tests. Therefore, the effect of stress ratio is already included. Section 5–8 includes methods of analysis for loading cases in which the fluctuating stress produces a stress ratio different from $R = -1.0$.

5–5 EXAMPLE PROBLEMS FOR ESTIMATING ACTUAL ENDURANCE STRENGTH

This section shows two examples that demonstrate the application of the *Procedure for Estimating Actual Endurance Strength, s_n'*, that is presented in the previous section.

Example Problem 5–2 Estimate the actual endurance strength of SAE 1050 cold-drawn steel when used in a circular shaft subjected to rotating bending only. The shaft will be machined to a diameter of approximately 1.75 in.

Solution

Objective: Compute the estimated actual endurance strength of the shaft material.

Given: SAE 1050 cold-drawn steel, machined.
Size of section: $D = 1.75\text{ in.}$
Type of stress: Reversed, repeated bending.

Analysis: Use the Procedure for Estimating Actual Endurance Strength, s_n'

Step 1: The ultimate tensile strength: $s_u = 100$ ksi from Appendix 3.

Step 2: Diameter is machined.

Step 3: From Figure 5–8, $s_n = 38\text{ ksi}$

Step 4: Material factor for wrought steel: $C_m = 1.0$

Step 5: Type-of-stress factor for reversed bending: $C_{st} = 1.0$

Step 6: Specify a desired reliability of 0.99. Then $C_R = 0.81$ (Design decision)

Step 7: Size factor for circular section with $D = 1.75\text{ in.}$
From Figure 5–9, $C_s = 0.83$.

Step 8: Use Equation (5–6) to compute the estimated actual endurance strength.

$$s_n' = s_n(C_m)(C_{st})(C_R)(C_s) = 38\text{ ksi}(1.0)(1.0)(0.81)(0.83) = 25.5\text{ ksi}$$

Comments: This is the level of stress that would be expected to produce fatigue failure in a rotating shaft due to the action of reversed bending. It accounts for the basic endurance strength of the wrought SAE 1050 cold-drawn material, the effect of the machined surface, the size of the section, and the desired reliability.

Example Problem 5–3 Estimate the actual endurance strength of cast steel having an ultimate strength of 120 ksi when used in a bar subjected to a reversed, repeated, bending load. The bar will be machined to a rectangular cross section, 1.50 in wide $\times$ 2.00 in high.

Solution

Objective: Compute the estimated actual endurance strength of the bar material.

Given: Cast steel, machined: $s_u = 120$ ksi.
Size of section: $b = 1.50$ in., $h = 2.00$ in rectangular
Type of stress: Repeated, reversed bending.

Analysis Use the Procedure for Estimating Actual Endurance Strength, s'_n.

Step 1: The ultimate tensile strength is given to be $s_u = 120$ ksi.

Step 2: Surfaces are machined.

Step 3: From Figure 5–8, $s_n = 44$ ksi

Step 4: Material factor for cast steel: $C_m = 0.80$

Step 5: Type-of-stress factor for bending: $C_{st} = 1.00$

Step 6: Specify a desired reliability of 0.99. Then $C_R = 0.81$ (Design decision)

Step 7: Size factor for rectangular section: First use Equation (5–10) to determine the equivalent diameter,

$$D_e = 0.808\sqrt{hb} = 0.808\sqrt{(2.00\,\text{in})(1.50\,\text{in})} = 1.40\text{ in}$$

Then from Figure 5–9, $C_s = 0.85$.

Step 8: Use Equation (5–6) to compute the estimated actual endurance strength.

$$s'_n = s_n(C_m)(C_{st})(C_R)(C_s) = 44\text{ ksi}(0.80)(1.00)(0.81)(0.85) = 24.2\text{ ksi}$$

5–6 DESIGN PHILOSOPHY

It is the designer's responsibility to ensure that a machine part is safe for operation under reasonably foreseeable conditions. You should evaluate carefully the application in which the component is to be used, the environment in which it will operate, the nature of applied loads, the types of stresses to which the component will be exposed, the type of material to be used, and the degree of confidence you have in your knowledge about the application. Some general considerations are:

1. ***Application.*** Is the component to be produced in large or small quantities? What manufacturing techniques will be used to make the component? What are the consequences of failure in terms of danger to people and economic cost? How cost-sensitive is the design? Are small physical size or low weight important? With what other parts or devices will the component interface? For what life is the component being designed? Will the component be inspected and serviced periodically? How much time and expense for the design effort can be justified?
2. ***Environment.*** To what temperature range will the component be exposed? Will the component be exposed to electrical voltage or current? What is the potential for corrosion? Will the component be inside a housing? Will guarding protect access to the component? Is low noise important? What is the vibration environment?
3. ***Loads.*** Identify the nature of loads applied to the component being designed in as much detail as practical. Consider all modes of operation, including startup, shut down, normal operation, and foreseeable overloads. The loads should be characterized as *static, repeated and reversed, fluctuating, shock*, or *impact* as discussed in Section 5–2. Key magnitudes of loads are the *maximum, minimum*, and *mean*. Variations of loads over time should be documented as completely as practical. Will high mean loads be applied for extended periods of time, particularly at high temperatures, for which creep must be considered? This information will influence the details of the design process.
4. ***Types of Stresses.*** Considering the nature of the loads and the manner of supporting the component, what kinds of stresses will be created: direct tension, direct compression, direct shear, bending, or torsional shear? Will two or more kinds of stresses be applied simultaneously? Are stresses developed in one direction (*uniaxially*), two directions (*biaxially*), or three directions (*triaxially*)? Is buckling likely to occur?
5. ***Material.*** Consider the required material properties of yield strength, ultimate tensile strength, ultimate compressive strength, endurance strength, stiffness, ductility, toughness, creep resistance, corrosion resistance, and others in relation to the application, loads, stresses, and the environment. Will the component be made from a ferrous metal such as plain carbon, alloy, stainless, or structural steel, or cast iron? Or will a nonferrous metal such as aluminum, brass, bronze, titanium, magnesium, or zinc be used? Is the material brittle (percent elongation $<5\%$) or ductile (percent elongation $>5\%$)? Ductile materials are highly preferred for components subjected to fatigue, shock, or impact loads. Will plastics be used? Is the application suitable for a composite material? Should you consider other nonmetals such as ceramics or wood? Are thermal or electrical properties of the material important?
6. ***Confidence.*** How reliable are the data for loads, material properties, and stress calculations? Are controls for manufacturing processes adequate to ensure that the component will be produced as designed with regard to dimensional accuracy, surface finish, and final as-made

material properties? Will subsequent handling, use, or environmental exposure create damage that can affect the safety or life of the component? These considerations will affect your decision for the design factor, *N*, to be discussed in the next section.

All design approaches must define the relationship between the applied stresses on a component and the strength of the material from which it is to be made, considering the conditions of service. The strength basis for design can be yield strength in tension, compression, or shear; ultimate strength in tension, compression, or shear; endurance strength; or some combination of these. The goal of the design process is to achieve a suitable *design factor*, *N* (sometimes called a *factor of safety*) that ensures the component is safe. That is, the strength of the material must be greater than the applied stresses. Design factors are discussed in the next section.

The sequence of design analysis will be different depending on what has already been specified and what is left to be determined. For example,

1. ***Geometry of the component and the loading are known:*** We apply the desired design factor, *N*, to the actual expected stress to determine the required strength of the material. Then a suitable material can be specified.
2. ***Loading is known and the material for the component has been specified:*** We compute a *design stress* by applying the desired design factor, *N*, to the appropriate strength of the material. This is the maximum allowable stress to which any part of the component can be exposed. We can then complete the stress analysis to determine what shape and size of the component will ensure that stresses are safe.
3. ***Loading is known, and the material and the complete geometry of the component have been specified:*** We compute both the expected maximum applied stress and the design stress. By comparing these stresses, we can determine the resulting design factor, *N*, for the proposed design and judge its acceptability. A redesign may be called for if the design factor is either too low (unsafe) or too high (over designed).

Practical Considerations. While ensuring that a component is safe, the designer is expected to also make the design practical to produce, considering several factors.

- Each design decision should be tested against the cost of achieving it.
- Material availability must be checked.
- Manufacturing considerations may affect final specifications for overall geometry, dimensions, tolerances, or surface finish.
- In general, components should be as small as practical unless operating conditions call for larger size or weight.
- After computing the minimum acceptable dimension for a feature of a component, standard or preferred sizes should be specified using normal company practice or tables of preferred sizes such as those listed in Appendix 2.
- Before a design is committed to production, tolerances on all dimensions and acceptable surface finishes must be specified so the manufacturing engineer and the production technician can specify suitable manufacturing processes.
- Surface finishes should only be as smooth as required for the function of a particular area of a component, considering appearance, effects on fatigue strength, and whether or not the area mates with another component. Producing smoother surfaces increases cost dramatically. See Chapter 13.
- Tolerances should be as large as possible while maintaining acceptable performance of the component. The cost to produce smaller tolerances rises dramatically. See Chapter 13.
- The final dimensions and tolerances for some features may be affected by the need to mate with other components. Proper clearances and fits must be defined, as discussed in Chapter 13. Another example is the mounting of a commercially available bearing on a shaft for which the bearing manufacturer specifies the nominal size and tolerances for the bearing seat on the shaft. Chapter 16 gives guidelines for clearances between the moving and stationary parts where either boundary or hydrodynamic lubrication is used.
- Will any feature of the component be subsequently painted or plated, affecting the final dimensions?

Deformations. Machine elements can also fail because of excessive deformation or vibration. From your study of strength of materials, you should be able to compute deformations due to axial tensile or compressive loads, bending, torsion, or changes in temperature. Some of the basic concepts are reviewed in Chapter 3. For more complex shapes or loading patterns, computer-based analysis techniques such as finite-element analysis or beam analysis software are important aids.

Criteria for failure due to deformation are often highly dependent on the machine's use. Will excessive deformation cause two or more members to touch when they should not? Will the desired precision of the machine be compromised? Will the part look or feel too flexible (flimsy)? Will parts vibrate excessively or resonate at the frequencies experienced during operation? Will rotating shafts exhibit a critical speed during operation, resulting in wild oscillations of parts carried by the shaft?

This chapter will not pursue the quantitative analysis of deformation, leaving that to be your responsibility as the design of a machine evolves. Later chapters do address some critical cases such as the interference fit between two mating parts (Chapter 13), the position of the teeth of one gear relative to its mating gear (Chapter 9), the radial clearance between a journal bearing and the shaft rotating within it (Chapter 16), and the deformation of springs (Chapter 18). Also, Section 5–9, as a part of the general design procedure, suggests some guidelines for limiting deflections.

5–7 DESIGN FACTORS

The term *design factor, N,* is a measure of the relative safety of a load-carrying component. In most cases, the strength of the material from which the component is to be made is divided by the design factor to determine a *design stress*, σ_d, sometimes called the *allowable stress*. Then the actual stress to which the component is subjected should be less than the design stress. For some kinds of loading, it is more convenient to set up a relationship from which the design factor, N, can be computed from the actual applied stresses and the strength of the material. Still in other cases, particularly for the case of the buckling of columns, as discussed in Chapter 6, the design factor is applied to the *load* on the column rather than the strength of the material.

Section 5–8 presents methods for computing the design stress or design factor for several different kinds of loading and materials.

The designer must determine what a reasonable value for the design factor should be in any given situation. Often the value of the design factor or the design stress is governed by codes established by standards-setting organizations such as the American Society of Mechanical Engineers, the American Gear Manufacturers Association, the U.S. Department of Defense, the Aluminum Association, or the American Institute of Steel Construction. For structures, local or state building codes often prescribe design factors or design stresses. Some companies have adopted their own policies specifying design factors based on past experience with similar conditions.

In the absence of codes or standards, the designer must use judgment to specify the desired design factor. Part of the design philosophy, discussed in Section 5–6, identified issues such as the nature of the application, environment, nature of the loads on the component to be designed, stress analysis, material properties, and the degree of confidence in data used in the design processes. All of these considerations affect the decision about what value for the design factor is appropriate. This book will use the following guidelines.

Ductile Materials

1. **N = 1.25 to 2.0** Design of structures under static loads for which there is a high level of confidence in all design data.
2. **N = 2.0 to 2.5.** Design of machine elements under dynamic loading with average confidence in all design data. (Typically used in problem solutions in this book.)
3. **N = 2.5 to 4.0.** Design of static structures or machine elements under dynamic loading with uncertainty about loads, material properties, stress analysis, or the environment.
4. **N = 4.0 or higher.** Design of static structures or machine elements under dynamic loading with uncertainty about some combination of loads, material properties, stress analysis, or the environment. The desire to provide extra safety to critical components may also justify these values.

Brittle Materials

5. **N = 3.0 to 4.0.** Design of structures under static loads for which there is a high level of confidence in all design data.
6. **N = 4.0 to 8.0.** Design of static structures or machine elements under dynamic loading with uncertainty about loads, material properties, stress analysis, or the environment.

The following sections (5–8 and 5–9) provide guidance on the introduction of the design factor into the design process with particular attention to the selection of the strength basis for the design and the computation of the design stress. In general, design for static loading involves applying the design factor to the yield strength or ultimate strength of the material. Dynamic loading requires the application of the design factor to the endurance strength using the methods described in Section 5–4 to estimate the actual endurance strength for the conditions under which the component is operating.

5–8 PREDICTIONS OF FAILURE AND DESIGN ANALYSIS METHODS

Designers should understand the various ways that load-carrying components can fail in order to complete a design that ensures that failure *does not occur*. Several different methods of predicting failure are available, and it is the designer's responsibility to select the one most appropriate to the conditions of the project. In this section, we describe the methods that have found a high level of use in the field and discuss the situations in which each is applicable. The factors involved are the nature of the load (static, repeated and reversed, or fluctuating), the type of material involved (ductile or brittle), and the amount of design effort and analysis that can be justified by the nature of the component or product being designed.

This section describes the most frequently used methods for stress analysis and design used in this book. Following sections present a generalized design analysis method and show several design examples. Later parts of the chapter include other design-related approaches that may be useful in your career.

The following design methods define the relationship between applied stresses on a component and the strength of the material from which it is made based on relevant types of loading. The strength basis for design can be yield strength, ultimate strength, endurance strength, or some combination of these. The goal of the design process is to achieve a suitable design factor, N, that ensures that the component is safe.

Failure Prediction Method	Uses
1. Maximum normal stress	Uniaxial static stress on brittle materials
2. Yield strength	Uniaxial static stress on ductile materials
3. Maximum shear stress	Direct shear, vertical shear, and torsional shear on ductile materials
4. Endurance strength	Reversed/repeated normal or shear stresses on ductile materials
5. Goodman	Fluctuating stress on ductile materials

1. Maximum Normal Stress Method for Uniaxial Static Stress on Brittle Materials

The maximum normal stress theory states that a material will fracture when the maximum normal stress (either tension or compression) exceeds the ultimate strength of the material as obtained from a standard tensile or compressive test. Its use is limited, namely for brittle materials under pure uniaxial static tension or compression. When applying this theory, any stress concentration factor at the region of interest should be applied to the computed stress because brittle materials do not yield and therefore cannot redistribute the increased stress.

The following equations apply the maximum normal stress theory to design.

For tensile stress: $$K_t\sigma < \sigma_d = s_{ut}/N \quad (5\text{–}11)$$

For compressive stress: $$K_t\sigma < \sigma_d = s_{uc}/N \quad (5\text{–}12)$$

Note that many brittle materials such as gray cast iron have a significantly higher compressive strength than tensile strength.

2. Yield Strength Method for Uniaxial Static Normal Stresses on Ductile Materials

This is a simple application of the principle of yielding in which a component is carrying a direct tensile or compressive load in the manner similar to the conditions of the standard tensile or compressive test for the material. Failure is predicted when the actual applied stress exceeds the yield strength. Stress concentrations can normally be neglected for static stresses on ductile materials because the higher stresses near the stress concentrations are highly localized. When the local stress on a small part of the component reaches the yield strength of the material, it does in fact yield. In the process, the stress is redistributed to other areas and the component is still safe.

The following equations apply the yield strength principle to design.

For tensile stress: $$\sigma < \sigma_d = s_{yt}/N \quad (5\text{–}13)$$

For compressive stress: $$\sigma < \sigma_d = s_{yc}/N \quad (5\text{–}14)$$

For most wrought ductile metals, $s_{yt} = s_{yc}$.

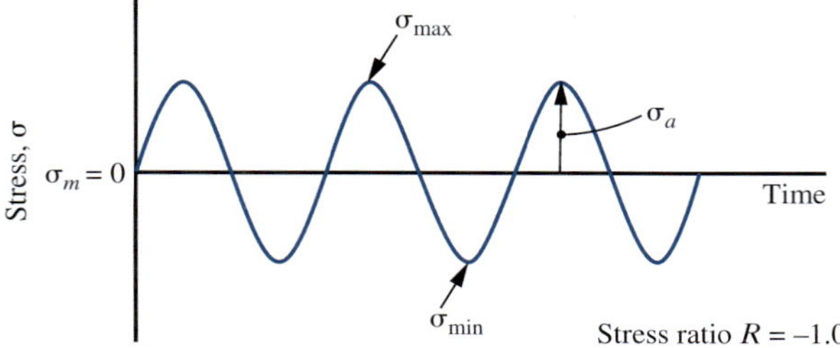

FIGURE 5–12 Repeated, reversed stress or pure oscillation

3. Maximum Shear Stress Method for Static Shear Stress on Ductile Materials

The maximum shear stress method of failure prediction states that a ductile material begins to yield when the maximum shear stress in a load-carrying component exceeds that in a tensile-test specimen when yielding begins. A Mohr's circle analysis for the uniaxial tension test, discussed in Section 4–7, shows that the maximum shear stress is one-half of the applied tensile stress. At yield, then, $s_{sy} = s_y/2$. We use this approach in this book to estimate s_{sy}. Then, for design, use

$$\tau_{max} < \tau_d = s_{sy}/N = 0.5s_y/N = s_y/2N \quad (5\text{–}15)$$

The maximum shear stress method of failure prediction has been shown by experimentation to be somewhat conservative for ductile materials subjected to a combination of normal and shear stresses. It is relatively easy to use and is often chosen by designers. For more precise analysis, the distortion energy method, discussed later, is preferred.

4. Endurance Strength Method for Repeated and Reversed Normal or Shear Stresses on Ductile Material

Figure 5–12 shows the special case of repeated and reversed stresses. Note that the magnitudes of the stresses in the positive and negative zones are equal, resulting in a zero mean stress. That is, for normal stresses,

$$\sigma_{min} = -\sigma_{max} \quad \text{and} \quad \sigma_m = 0$$

This is the same type of stress shown in Figure 5–2 and is the basis for the determination of the endurance strength of the material, s_n. Also, the repeated stress makes the component subject to fatigue failure and stress concentration factors must be considered. Therefore, the design equation for this case is:

$$K_t\sigma_{max} < \sigma_d = s_n'/N \quad (5\text{–}16)$$

Similarly, for reversed and repeated shearing stresses and using the maximum shear stress principle to obtain an estimate for the endurance strength in shear,

$$s_{sn}' = 0.5s_n' \text{(estimate for endurance strength in shear)}$$

$$K_t\tau_{max} < \tau_d = s_{sn}'/N = 0.5s_n'/N \quad (5\text{–}17)$$

5. Goodman Method for Fluctuating Stresses on Ductile Materials

Explanation of the Goodman method is required before showing the final design equations used for this very important case. It is perhaps the most-often used analysis for

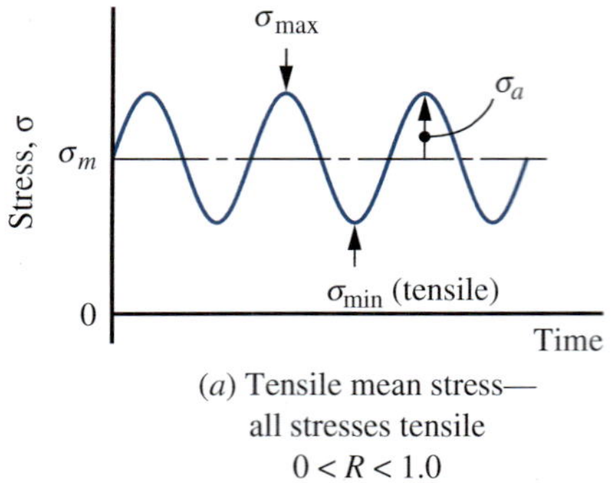

(a) Tensile mean stress—
all stresses tensile
$0 < R < 1.0$

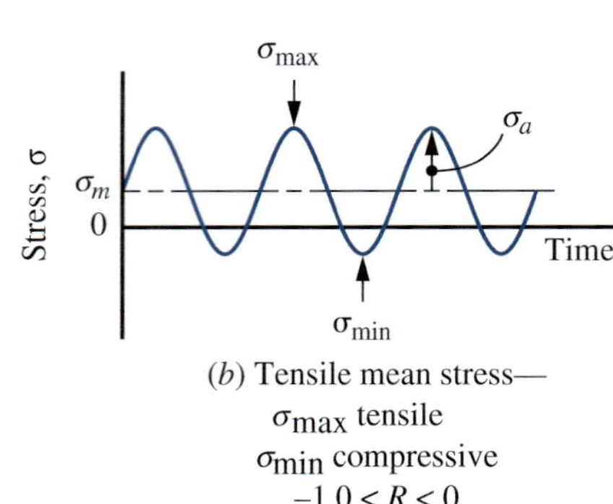

(b) Tensile mean stress—
σ_{max} tensile
σ_{min} compressive
$-1.0 < R < 0$

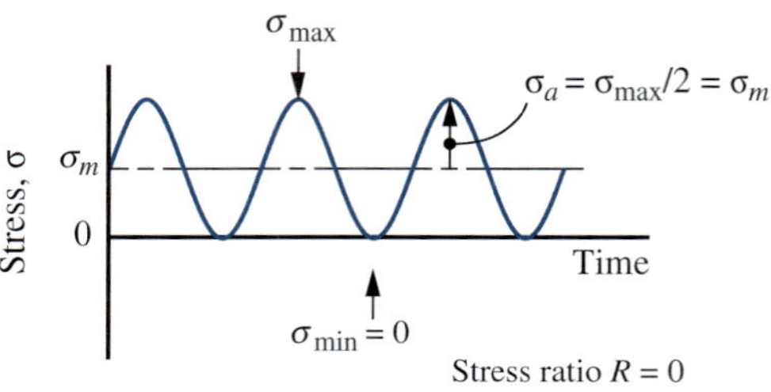

(c) Repeated, one-direction stress, a special case of
fluctuating stress or pure pulsating stress

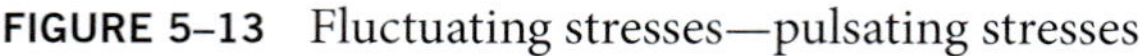

FIGURE 5–13 Fluctuating stresses—pulsating stresses

machine components subjected to cyclical loads. Figure 5–13 shows the general pattern of stress application for fluctuating stresses. The key characteristics are:

- The stress cycles between the maximum, σ_{max}, and minimum, σ_{min}, values
- The mean stress is not zero: $\sigma_m = (\sigma_{max} + \sigma_{min})/2 \neq 0$
- The stress amplitude is: $\sigma_a = \sigma_{max} - \sigma_m$ or $\sigma_a = (\sigma_{max} - \sigma_{min})/2$

It is helpful to depict the Goodman method in the graph shown in Figure 5–14, plotting stress amplitude, σ_a, on the vertical axis and mean stress, σ_m, on the horizontal axis. We first consider only the case when both σ_a and σ_m are positive or tensile. It is recommended that any stress concentration factor be applied to the alternating stress component. However, it is not necessary to apply a stress concentration factor to the mean stress because experimental evidence shows that the presence of a stress concentration does not affect the contribution of mean stress to fatigue failure.

The maximum permissible value of σ_a is the estimated actual endurance strength, s_n', shown on the vertical axis. The maximum permissible mean stress with no cyclical load is taken to be the ultimate strength, s_u. (Yielding is added as a criterion later.) A line, called the Goodman line, is then drawn between these two points. Ideally, combinations of mean and alternating stresses that lie under the Goodman line are considered to be safe. Testing shows that the predominance of failure points lie above this line. The equation of the Goodman line is,

$$\frac{K_t\sigma_a}{s_n'} + \frac{\sigma_m}{s_u} = 1 \tag{5–18}$$

For design, a design factor is applied to both the endurance strength and the ultimate strength and a new line, called the *safe stress line*, is drawn between those points creating a *safe zone* beneath it. The equation for the safe stress line is,

$$\frac{K_t\sigma_a}{s_n'} + \frac{\sigma_m}{s_u} = \frac{1}{N} \tag{5–19}$$

This is the design equation we will use in this book for fluctuating stresses. However, the possibility of yielding must be considered as described next.

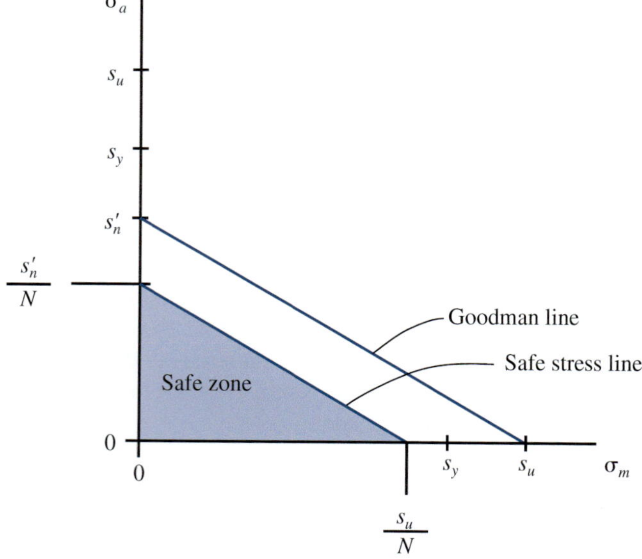

FIGURE 5–14 Illustration of the Goodman method for fluctuating stresses

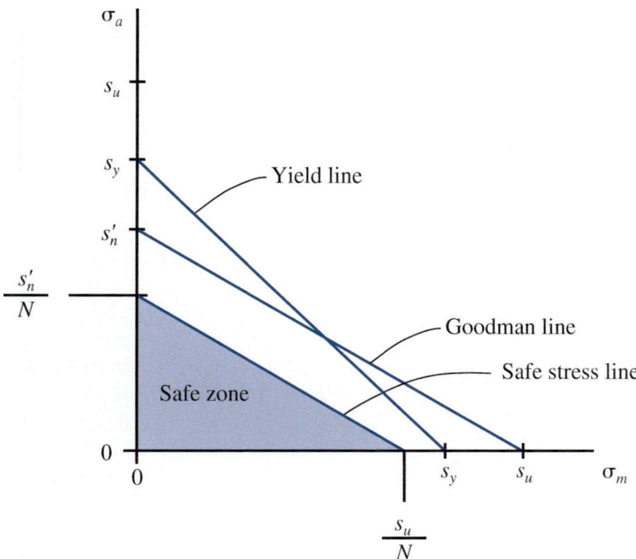

FIGURE 5–15 Illustration of the Goodman method for fluctuating stresses with yield line added

Checking for Early Cycle Yielding. As drawn in Figure 5–14, the Goodman line appears to allow a pure mean stress above the yield strength of the material which is unacceptable. Therefore, an additional criterion based on preventing yielding is used with Equation 5–19. Figure 5–15 shows the result. The yield strength of the material, s_y is added to both the vertical and horizontal axes and a line is drawn between them. Note that the lower right part of this line extends under the Goodman line, indicating that some possible combinations of σ_a and σ_m could result in a total maximum stress greater than yield. It was stated earlier that fatigue cracks are typically initiated in zones of high tensile stress and correspondingly high tensile strains. Therefore, it is desirable to limit any possible stress to below the yield strength. Furthermore, this statement implies that any stress concentration should be considered for both the stress amplitude and the mean stress to ensure that even infrequently applied high stresses will not cause damaging strains that could precipitate fatigue cracks. Applying a design factor to the yield line results in a design equation to protect against yield as,

$$\frac{K_t\sigma_a}{s_y} + \frac{K_t\sigma_m}{s_y} = \frac{1}{N} \quad \textbf{(5–20)}$$

In most cases, Equation 5–18 governs the design. Therefore, we recommend that the design first be based on that criterion, achieving an acceptable design factor. Then, the proposed design should be checked against yielding. It is convenient to solve Equation 5–20 for the design factor and use the following equation,

$$N = \frac{s_y}{K_t(\sigma_a + \sigma_m)} \quad \textbf{(5–21)}$$

If the resulting design factor from yield analysis is greater than that for the Goodman criterion, the design is safe.

The Goodman method can also be applied to cases where shearing stresses occur by simply substituting shearing stresses and corresponding shearing strengths for the normal stresses used in Equations 5–20 and 5–21. Those equations are reported in the **Summary of Design Equations** that follows.

Summary of Design Equations

1. Maximum normal stress method for static normal stress on brittle materials

Tensile stress: $K_t\sigma < \sigma_d = s_{ut}/N$ **(5–11)**

Compressive stress: $K_t\sigma < \sigma_d = s_{uc}/N$ **(5–12)**

2. Yield strength method for static normal stress on ductile materials

Tensile stress: $\sigma < \sigma_d = s_{yt}/N$ **(5–13)**

Compressive stress: $\sigma < \sigma_d = s_{yc}/N$ **(5–14)**

3. Maximum shear stress method for static shear stress on ductile materials

Shear stress: $\tau_{max} < \tau_d = s_{sy}/N = 0.50s_y/N$ **(5–15)**

4. Endurance strength method for reversed and repeated normal or shear stresses on ductile materials

Normal stress: $K_t\sigma_{max} < \sigma_d = s'_n/N$ **(5–16)**

Shear stress:

$s'_{sn} = 0.5s'_n$ (estimate for endurance strength in shear)

$K_t\tau_{max} < \tau_d = s'_{sn}/N = 0.5s'_n/N$ **(5–17)**

5. Goodman method for fluctuating stress on ductile materials

Normal stress design equation

$$\frac{K_t\sigma_a}{s'_n} + \frac{\sigma_m}{s_u} = \frac{1}{N} \quad \textbf{(5–20)}$$

Check yielding in tension

$$N = \frac{s_y}{K_t(\sigma_a + \sigma_m)} \quad \textbf{(5–21)}$$

Shear stress design equation

$$\frac{K_t\tau_a}{s'_{sn}} + \frac{\tau_m}{s_{su}} = \frac{1}{N} \quad \textbf{(5–22)}$$

Check yielding in shear

$$N = \frac{s_{sy}}{K_t(\tau_a + \tau_m)} = \frac{s_{y/2}}{K_t(\tau_a + \tau_m)} \quad \textbf{(5–23)}$$

5–9 GENERAL DESIGN PROCEDURE

The earlier parts of this chapter have provided guidance related to the many factors involved in design of machine elements that must be safe when carrying the applied loads.

This section brings these factors together so that you can complete the design. The general design procedure described here is meant to give you a feel for the process. It is not practical to provide a completely general procedure. You will have to adapt it to the specific situations that you encounter.

The procedure is set up assuming that the following factors are known or can be specified or estimated:

- General design requirements: Objectives and limitations on size, shape, weight, desired precision, and so forth.
- Nature of the loads to be carried.
- Types of stresses produced by the loads.
- Type of material from which the element is to be made.
- General description of the manufacturing process to be used, particularly with regard to the surface finish that will be produced.
- Desired reliability.

GENERAL DESIGN PROCEDURE

1. Specify the objectives and limitations, if any, of the design, including desired life, size, shape, and appearance.
2. Determine the environment in which the element will be placed, considering such factors as corrosion potential and temperature.
3. Determine the nature and characteristics of the loads to be carried by the element, such as
   - Static, dead, slowly applied loads.
   - Dynamic, live, varying, repeated loads that may potentially cause fatigue failure.
   - Shock or impact loads.
4. Determine the magnitudes for the loads and the operating conditions, such as
   - Maximum expected load.
   - Minimum expected load.
   - Mean and alternating levels for fluctuating loads.
   - Frequency of load application and repetition.
   - Expected number of cycles of loading.
5. Analyze how loads are to be applied to determine the type of stresses produced, such as direct normal stress, bending stress, direct shear stress, torsional shear stress, or some combination of stresses.
6. Propose the basic geometry for the element, paying particular attention to:
   - Its ability to carry the applied loads safely.
   - Its ability to transmit loads to appropriate support points. Consider the *load paths*.
   - The use of efficient shapes according to the nature of the loads and the types of stresses encountered. This applies to the general shape of the element and to each of its cross sections. Achieving efficiency involves optimizing the amount and the type of material involved. Section 20–2 gives some suggestions for efficient design of frames and members in bending and torsion.
   - Providing appropriate attachments to supports and to other elements in the machine or structure.
   - Providing for the positive location of other components that may be installed on the element being designed. This may call for shoulders, grooves, holes, retaining rings, keys and keyseats, pins, or other means of fastening or holding parts.
7. Propose the method of manufacturing the element with particular attention to the precision required for various features and the surface finish that is desired. Will it be cast, machined, ground, or polished, or produced by some other process? These design decisions have important impacts on the performance of the element, its ability to withstand fatigue loading, and the cost to produce it.
8. Specify the material from which the element is to be made, along with its condition. For metals the specific alloy should be specified, and the condition could include such processing factors as hot rolling, cold drawing, and a specific heat treatment. For nonmetals, it is often necessary to consult with vendors to specify the composition and the mechanical and physical properties of the desired material. Consult Chapter 2 and Section 20–2 for additional guidance.
9. Determine the expected properties of the selected material, for example
   - Ultimate tensile strength, s_u.
   - Ultimate compressive strength, s_{uc}, if appropriate.
   - Yield strength, s_y.
   - Ductility as represented by percent elongation.
   - Stiffness as represented by modulus of elasticity, E or G.
10. Specify an appropriate design factor, N, for the stress analysis using the guidelines discussed in Section 5–7.
11. Determine which stress analysis method outlined in Section 5–8 applies to the design being completed.
12. Compute the appropriate design stress for use in the stress analysis. If fatigue loading is involved, the actual expected endurance strength of the material should be computed as outlined in Section 5–4. This requires the consideration of the expected size of the section, the type of material to be used, the nature of the stress, and the desired reliability. Because the size of the section is typically unknown at the start of the design process, an estimate must be made to allow the inclusion of a reasonable size factor, C_s. You should check the estimate at the end of the design process to verify that reasonable values were assumed at this stage of the design.
13. Determine the nature of any stress concentrations that may exist in the design at places where geometry changes occur. Stress analysis should be considered at all such places because of the likelihood of localized high tensile stresses that may produce fatigue failure. If the geometry of the element in these areas is known, determine the appropriate stress concentration factor, K_t. If the geometry is not yet known, it is advisable to estimate the expected magnitude of K_t. The estimate must then be checked at the end of the design process.
14. Complete the required stress analyses at all points where the stress may be high and at changes of cross section to determine the minimum acceptable dimensions for critical areas.

15. Specify suitable, convenient dimensions for all features of the element. Many design decisions are required, such as
    - The use of preferred basic sizes as listed in Table A2–1.
    - The size of any part that will be installed on or attached to the element being analyzed. Examples of this are shown in Chapter 12 on shaft design where gears, chain sprockets, bearings, and other elements are to be installed on the shafts. But many machine elements have similar needs to accommodate mating elements.
    - Elements should not be significantly oversized without good reason in order to achieve an efficient overall design.
    - Sometimes the manufacturing process to be used has an effect on the dimensions. For example, a company may have a preferred set of cutting tools for use in producing the elements. Casting, rolling, or molding processes often have limitations on the dimensions of certain features such as the thickness of ribs, radii produced by machining or bending, variation in cross section within different parts of the element, and convenient handling of the element during manufacture.
    - Consideration should be given to the sizes and shapes that are commercially available in the desired material. This could result in significant cost reductions both in material and in processing.
    - Sizes should be compatible with standard company practices if practical.
16. After completing all necessary stress analyses and proposing the basic sizes for all features, check all assumptions made earlier in the design to ensure that the element is still safe and reasonably efficient. See Steps 7, 12, and 13.
17. Specify suitable tolerances for all dimensions, considering the performance of the element, its fit with mating elements, the capability of the manufacturing process, and cost. Chapter 13 may be consulted. The use of computer–based tolerance-analysis techniques may be appropriate.
18. Check to determine whether some part of the component may deflect excessively. If that is an issue, complete an analysis of the deflection of the element as designed to this point. Sometimes there are known limits for deflection based on the operation of the machine of which the element being designed is a part. In the absence of such limits, the following guidelines may be applied based on the degree of precision desired:

    Deflection of a Beam Due to Bending

General machine part:	0.000 5 to 0.003 in/in of beam length
Moderate precision:	0.000 01 to 0.000 5 in/in
High precision:	0.000 001 to 0.000 01 in/in

    Deflection (Rotation) Due to Torsion

General machine part:	0.001° to 0.01°/in of length
Moderate precision:	0.000 02° to 0.000 4°/in
High precision:	0.000 001° to 0.000 02°/in

    See also Section 20–2 for additional suggestions for efficient design. The results of the deflection analysis may cause you to redesign the component. Typically, when high stiffness and precision are required, deflection, rather than strength, will govern the design.
19. Document the final design with drawings and specifications.
20. Maintain a careful record of the design analyses for future reference. Keep in mind that others may have to consult these records whether or not you are still involved in the project.

5–10 DESIGN EXAMPLES

Example design problems are shown here to give you a feel for the application of the process outlined in Section 5–9. It is not practical to illustrate all possible situations, and you must develop the ability to adapt the design procedure to the specific characteristics of each problem. Also note that there are many possible solutions to any given design problem. The selection of a final solution is the responsibility of you, the designer.

In most design situations, a great deal more information will be available than is given in the problem statements in this book. But, often, you will have to seek out that information. We will make certain assumptions in the examples that allow the design to proceed. In your job, you must ensure that such assumptions are appropriate. The design examples focus on only one or a few of the components of the given systems. In real situations, you must ensure that each design decision is compatible with the totality of the design.

Design Example 5–1 A large electrical transformer is to be suspended from a roof truss of a building. The total weight of the transformer is 32 000 lb. Design the means of support.

Solution

Objective: Design the means of supporting the transformer.

Given: The total load is 32 000 lb. The transformer will be suspended below a roof truss inside a building. The load can be considered to be static. It is assumed that it will be protected from the weather and that temperatures are not expected to be severely cold or hot in the vicinity of the transformer.

Basic Design Decisions

Two straight, cylindrical rods will be used to support the transformer, connecting the top of its casing to the bottom chord of the roof truss. The ends of the rod will be threaded to allow them to be secured by nuts or by threading them into tapped holes. This design example will be concerned only with the rods. It is assumed that appropriate attachment points are available to allow the two rods to share the load equally during service. However, it is possible that only one rod will carry the entire load at some point during installation. Therefore, each rod will be designed to carry the full 32 000 lb.

We will use steel for the rods, and because neither weight nor physical size is critical in this application, a plain, medium-carbon steel will be used. We specify SAE 1040 cold-drawn steel. From Appendix 3, we find that it has a yield strength of 71 ksi and moderately high ductility as represented by its 12% elongation. The rods should be protected from corrosion by appropriate coatings.

The objective of the design analysis that follows is to determine the size of the rod.

Analysis

The rods are to be subjected to direct normal tensile stress. Assuming that the threads at the ends of the rods are cut or rolled into the nominal diameter of the rods, the critical place for stress analysis is in the threaded portion.

Use the direct tensile stress formula, Equation (3-1): $\sigma = F/A$. We will first compute the design stress and then compute the required cross-sectional area to maintain the stress in service below that value. Finally, a standard thread will be specified from the data in Chapter 19 on fasteners.

Method 2 from Section 5–8 applies for computing the design stress because the rod is made from a ductile steel and it carries a static load. The design stress is

$$\sigma_d = s_y/N \tag{5–22}$$

We specify a design factor of $N = 3$, because it is typical for general machine design and because there is some uncertainty about the actual installation procedures that may be used (see Section 5–7). Then

$$\sigma_d = s_y/N = (71\,000\text{ psi})/3 = 23\,667\text{ psi}$$

Results

In the basic tensile stress equation $\sigma = F/A$, we know F, and we will let $\sigma = \sigma_d$. Then the required cross-sectional area is

$$A = F/\sigma_d = (32\,000\text{ lb})/(23\,667\text{ lb/in}^2) = 1.35\text{ in}^2$$

A standard size thread will now be specified from the data in Chapter 19 on fasteners. You should be familiar with such data from earlier courses. Table A2–2(b) lists the tensile stress area for American Standard threads. A $1\frac{1}{2}$–6 UNC thread ($1\frac{1}{2}$-in-diameter rod with 6 threads per in) has a tensile stress area of 1.405 in^2 which should be satisfactory for this application.

Comments

The final design specifies a $1\frac{1}{2}$-in-diameter rod made from SAE 1040 cold-drawn steel with $1\frac{1}{2}$–6 UNC threads machined on each end to allow the attachment of the rods to the transformer and to the truss.

Design Example 5–2

A part of a conveyor system for a production operation is shown in Figure 5–16. Design the pin that connects the horizontal bar to the fixture. The empty fixture weighs 85 lb. A cast iron engine block weighing 225 lb is hung on the fixture to carry it from one process to another, where it is then removed. It is expected that the system will experience many thousands of cycles of loading and unloading of the engine blocks.

Solution

Objective

Design the pin for attaching the fixture to the conveyor system.

Given

The general arrangement is shown in Figure 5–16. The fixture places a shearing load that is alternately 85 lb and 310 lb (85+225) on the pin many thousands of times in the expected life of the system.

Basic Design Decisions

It is proposed to make the pin from SAE 1020 cold-drawn steel. Appendix 3 lists $s_y = 51$ ksi and $s_u = 61$ ksi. The steel is ductile with 15% elongation. This material is inexpensive, and it is not necessary to achieve a particularly small size for the pin.

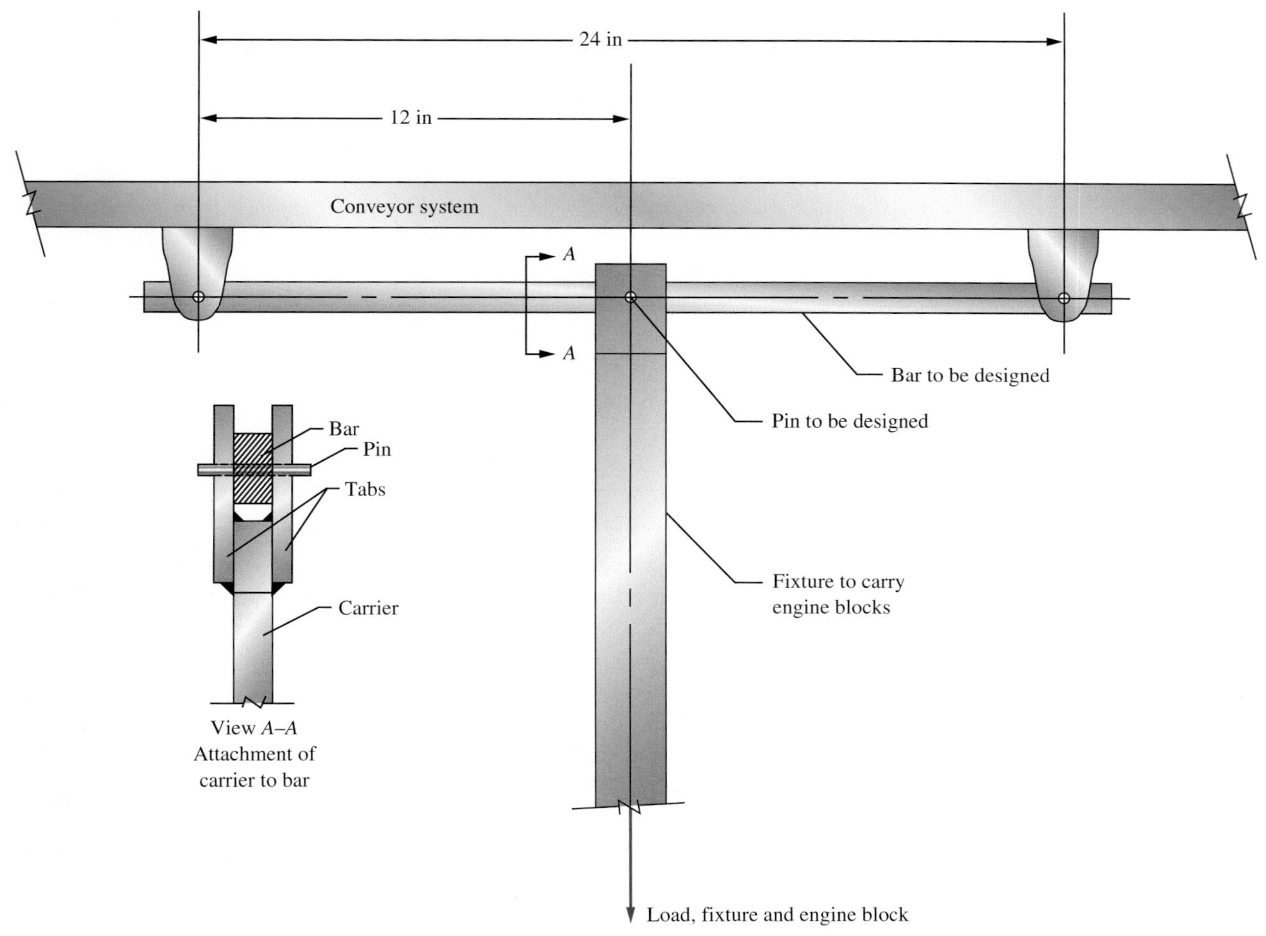

FIGURE 5–16 Conveyor system

The connection of the fixture to the bar is basically a clevis joint with two tabs at the top of the fixture, one on each side of the bar. There will be a close fit between the tabs and the bar to minimize bending action on the pin. Also, the pin will be a fairly close fit in the holes while still allowing rotation of the fixture relative to the bar.

Analysis Method 5 from Section 5–8 applies for completing the design analysis because fluctuating shearing stresses are experienced by the pin. Therefore, we will have to determine relationships for the mean and alternating stresses (τ_m and τ_a) in terms of the applied loads and the cross-sectional area of the bar. Note that the pin is in double shear, so two cross sections resist the applied shearing force. In general, $\tau = F/2A$.

Now we will use the basic forms of Equations (5–1) and (5–2) to compute the values for the mean and alternating forces on the pin:

$$F_m = (F_{\text{max}} + F_{\text{min}})/2 = (310 + 85)/2 = 198 \text{ lb}$$

$$F_a = (F_{\text{max}} - F_{\text{min}})/2 = (310 - 85)/2 = 113 \text{ lb}$$

The stresses will be found from $\tau_m = F_m/2A$ and $\tau_a = F_a/2A$.

The material strength values needed in Equation (5–22) for Method 5 are

$$s_{su} = 0.75 s_u = 0.75(51 \text{ ksi}) = 38.3 \text{ ksi} = 38\,300 \text{ psi}$$

$$s'_{sn} = 0.500 s'_n$$

We must find the value of s'_n using the method from Section 5–4. We find from Figure 5–8 that $s_n = 21$ ksi for the machined pin having a value of $s_u = 61$ ksi. It is expected that the pin will be fairly small, so we will use $C_s = 1.0$. The material is wrought steel rod, so $C_m = 1.0$. Let's use $C_{st} = 1.0$ to be conservative because there is little information about such factors for direct shearing stress. A high reliability is desired for this application, so let's use $C_R = 0.75$ to produce a reliability of 0.999 (see Table 5–2). Then

$$s'_n = C_R(s_n) = (0.75)(21 \text{ ksi}) = 15.75 \text{ ksi} = 15\,750 \text{ psi}$$

Finally,

$$s'_{sn} = 0.500\,s'_n = 0.500(15\,750 \text{ psi}) = 7875 \text{ psi}$$

We can now apply Equation (5–22) from Method 1:

$$\frac{1}{N} = \frac{\tau_m}{s_{su}} + \frac{K_t\tau_a}{s'_{sn}}$$

Because the pins will be of uniform diameter, $K_t = 1.0$.

Substituting $\tau_m = F_m/2A$ and $\tau_a = F_a/2A$ found earlier gives

$$\frac{1}{N} = \frac{F_m}{2As_{su}} + \frac{F_a}{2As'_{sn}}$$

Let's use $N = 4$ because mild shock can be expected.

Note that we now know all factors in this equation except the cross-sectional area of the pin, A. We can solve for the required area:

$$A = \frac{N}{2}\left[\frac{F_m}{s_{su}} + \frac{F_a}{s'_{sn}}\right]$$

Finally, we can compute the minimum allowable pin diameter, D, from $A = \pi D^2/4$ and $D = \sqrt{4A/\pi}$.

Results

The required area is

$$A = \frac{4}{2}\left[\frac{198 \text{ lb}}{38\,300 \text{ lb/in}^2} + \frac{113 \text{ lb}}{7875 \text{ lb/in}^2}\right] = 0.0390 \text{ in}^2$$

Now the required diameter is

$$D = \sqrt{4A/\pi} = \sqrt{4(0.0390 \text{ in}^2/\pi)} = 0.223 \text{ in}$$

Final Design Decisions and Comments

The computed value for the minimum required diameter for the pin, 0.223 in, is quite small. Other considerations such as bearing stress and wear at the surfaces that contact the tabs of the fixture and the bar indicate that a larger diameter would be preferred. Let's specify $D = 0.50$ in for the pin at this location. The pin will be of uniform diameter within the area of the bar and the tabs. It should extend beyond the tabs, and it could be secured with cotter pins or retaining rings.

This completes the design of the pin. But the next design example deals with the horizontal bar for this same system. There are pins at the conveyor hangers to support the bar. They would also have to be designed. However, note that each of these pins carries only half the load of the pin in the fixture connection. These pins would experience less relative motion as well, so wear should not be so severe. Thus, let's use pins with $D = 3/8$ in $= 0.375$ in at the ends of the horizontal bar.

Design Example 5–3

A part of a conveyor system for a production operation is shown in Figure 5–16. The complete system will include several hundred hanger assemblies like this one. Design the horizontal bar that extends between two adjacent conveyor hangers and that supports a fixture at its midpoint. The empty fixture weighs 85 lb. A cast iron engine block weighing 225 lb is hung on the fixture to carry it from one process to another, where it is then removed. It is expected that the bar will experience several thousand cycles of loading and unloading of the engine blocks. Design Example 5–2 considered this same system with the objective of specifying the diameter of the pins. The pin at the middle of the horizontal bar where the fixture is hung has been specified to have a diameter of 0.50 in. Those at each end where the horizontal bar is connected to the conveyor hangers are each 0.375 in.

Solution Objective Design the horizontal bar for the conveyor system.

Given The general arrangement is shown in Figure 5–16. The bar is simply supported at points 24 in apart. A vertical load that is alternately 85 lb and 310 lb (85 + 225) is applied at the middle of the bar through the pin connecting the fixture to the bar. The load will cycle between these two values many thousands of times in the expected life of the bar. The pin at the middle of the bar has a diameter of 0.50 in, while the pins at each end are 0.375 in.

Basic Design Decisions It is proposed to make the bar from steel in the form of a rectangular bar with the long dimension of its cross section vertical. Cylindrical holes will be machined on the neutral axis of the bar at the support points and at its center to receive cylindrical pins that will attach the bar to the conveyor carriers and to the fixture. Figure 5–17 shows the basic design for the bar.

The thickness of the bar, *t*, should be fairly large to provide a good bearing surface for the pins and to ensure lateral stability of the bar when subjected to the bending stress. A relatively thin bar would tend to buckle along its top surface where the stress is compressive. As a design decision, we will use a thickness of $t = 0.50$ in. The design analysis will determine the required height of the

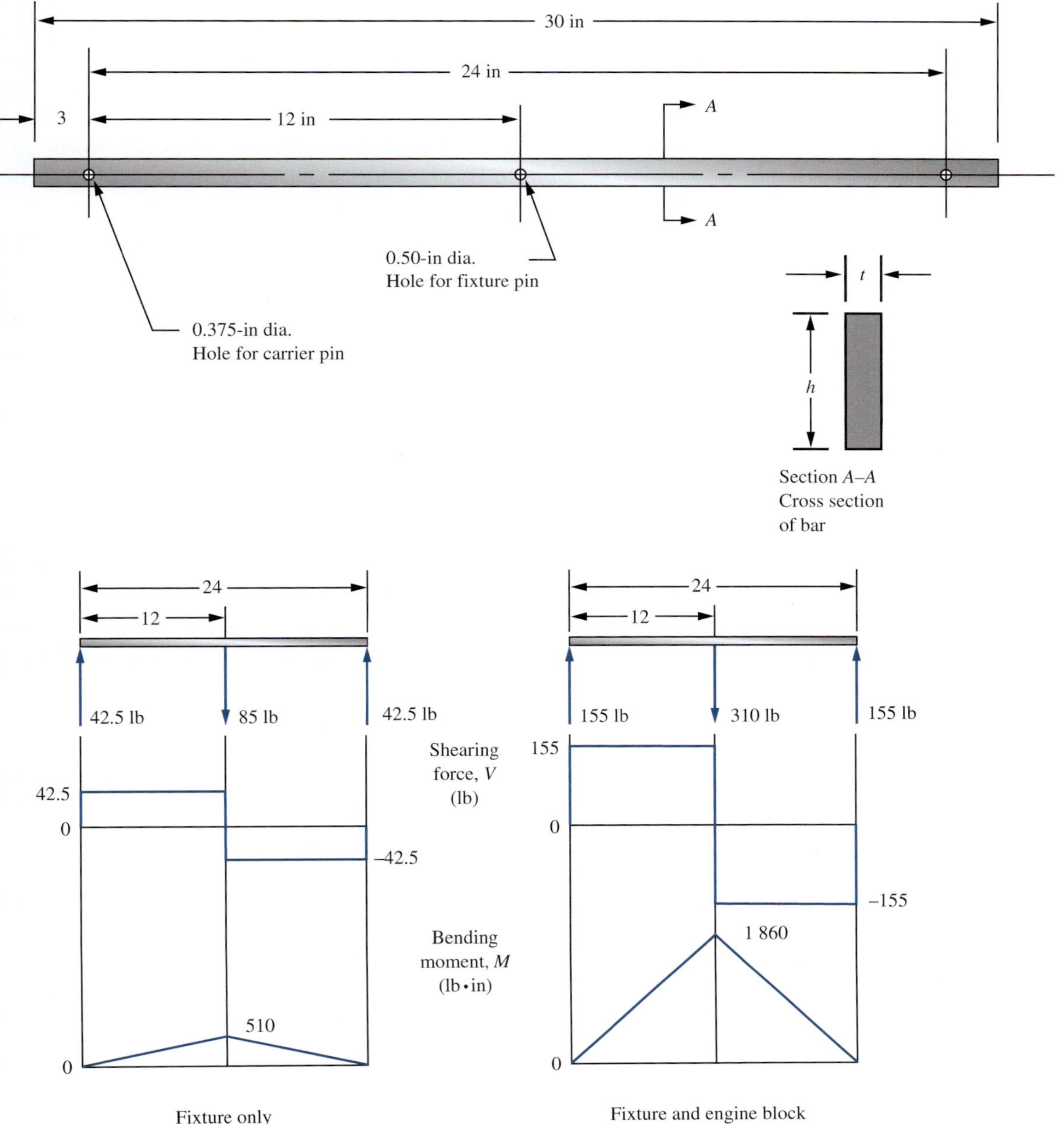

FIGURE 5–17 Basic design of the horizontal bar and the load, shearing force, and bending moment diagrams

bar, h, assuming that the primary mode of failure is stress due to bending. Other possible modes of failure are discussed in the comments at the end of this example.

An inexpensive steel is desirable because several hundred bars will be made. We specify SAE 1020 hot-rolled steel having a yield strength of $s_y = 30$ ksi and an ultimate strength of $s_u = 55$ ksi (Appendix 3).

Analysis Method 5 from Section 5–8 applies for completing the design analysis because fluctuating normal stress due to bending is experienced by the bar. Equation (5–20) will be used:

$$\frac{1}{N} = \frac{\sigma_m}{s_u} + \frac{K_t \sigma_a}{s'_n}$$

In general, the bending stress in the bar will be computed from the flexure formula:

$$\sigma = M/S$$

where M = bending moment

S = section modulus of the cross section of the bar = $th^2/6$ (App. 1)

Our approach will be to first determine the values for both the mean and the alternating bending moments experienced by the bar at its middle. Then the yield and endurance strength values for the steel will be found. Reference 5 in Chapter 3 indicates that a small hole, with diameter d, in a plate-beam does not weaken the beam if the ratio d/h is less than 0.50. That is, if $d/h < 0.50$, $K_t = 1.0$. We will make that assumption and check it later. Finally, Equation (5–20) includes the design factor, N. Based on the application conditions, let's use $N = 4$ as advised in item 4 in Section 5–7 because the actual use pattern for this conveyor system in a factory environment is somewhat uncertain and shock loading is likely.

Bending Moments Figure 5–17 shows the shearing force and bending moment diagrams for the bar when carrying just the fixture and then both the fixture and the engine block. The maximum bending moment occurs at the middle of the bar where the load is applied. The values are $M_{max} = 1860$ lb·in with the engine block on the fixture and $M_{min} = 510$ lb·in for the fixture alone. Now the values for the mean and alternating bending moments are calculated using modified forms of Equations (5–1) and (5–2):

$$M_m = (M_{max} + M_{min})/2 = (1860 + 510)/2 = 1185 \text{ lb·in}$$
$$M_a = (M_{max} - M_{min})/2 = (1860 - 510)/2 = 675 \text{ lb·in}$$

The stresses will be found from $\sigma_m - M_m/S$ and $\sigma_a = M_a/S$.

Material Strength Values The material strength properties required are the ultimate strength s_u and the estimated actual endurance strength s'_n. We know that the ultimate strength $s_u = 55$ ksi. We now find s'_n using the method outlined in Section 5–4.

Size factor, Cs: From Section 5–4, Equation (5–10) defines an equivalent diameter, D_e, for the rectangular section as,

$$D_e = 0.808\sqrt{ht}$$

We have specified the thickness of the bar to be $t = 0.50$ in. The height is unknown at this time. As an estimate, let's assume $h \approx 2.0$ in. Then,

$$D_e = 0.808\sqrt{ht} = 0.808\sqrt{(2.0)(0.50)} = 0.808 \text{ in}$$

We can now use Figure 5–9 or the equations in Table 5–3 to find $C_s = 0.90$. This value should be checked later after a specific height dimension is proposed.

Material factor, C_m: Use $C_m = 1.0$ for the wrought, hot-rolled steel.

Stress-type factor, C_{st}: Use $C_{st} = 1.0$ for repeated bending stress.

Reliability factor, C_R: A high reliability is desired. Let's use $C_R = 0.75$ to achieve a reliability of 0.999 as indicated in Table 5–2.

The value of $s_n = 20$ ksi is found from Figure 5–8 for hot-rolled steel having an ultimate strength of 55 ksi.

Now, applying Equation (5–6) from Section 5–4, we have

$$s'_n = (C_m)(C_{st})(C_R)(C_s)\ s_n = (1.0)(1.0)(0.75)(0.90)(20\text{ ksi}) = 13.5\text{ ksi}$$

Solution for the Required Section Modulus At this point, we have specified all factors in Equation (5–20) except the section modulus of the cross section of the bar that is involved in each expression for stress as shown above. We will now solve the equation for the required value of S.

Recall that we showed earlier that $\sigma_m = M_m/S$ and $\sigma_a = M_a/S$. Then

$$\frac{1}{N} = \frac{\sigma_m}{s_u} + \frac{K_t\sigma_a}{s'_n} = \frac{M_m}{Ss_u} + \frac{K_tM_a}{Ss'_n} = \frac{1}{S}\left[\frac{M_m}{s_u} + \frac{K_tM_a}{s'_n}\right]$$

$$S = N\left[\frac{M_m}{s_u} + \frac{K_tM_a}{s'_n}\right] = 4\left[\frac{1185\text{ lb}\cdot\text{in}}{55\,000\text{ lb/in}^2} + \frac{1.0(675\text{ lb}\cdot\text{in})}{13\,500\text{ lb/in}^2}\right]$$

$$S = 0.286\text{ in}^3$$

Results

The required section modulus has been found to be $S = 0.286\text{ in}^3$. We observed earlier that $S = th^2/6$ for a solid rectangular cross section, and we decided to use this form to find an initial estimate for the required height of the section, h. We have specified $t = 0.50$ in. Then the estimated minimum acceptable value for the height h is

$$h = \sqrt{6S/t} = \sqrt{6(0.286\text{ in}^3)/(0.50\text{ in})} = 1.85\text{ in}$$

The table of preferred basic sizes in the decimal-inch system (Table A2–1) recommends $h = 2.00$ in. We should first check the earlier assumption that the ratio $d/h < 0.50$ at the middle of the bar. The actual ratio is

$$d/h = (0.50\text{ in})/(2.00\text{ in}) = 0.25(\text{okay})$$

This indicates that our earlier assumption that $K_t = 1.0$ is correct. Also, our assumed value of $C_s = 0.90$ is correct because the actual height, $h = 2.0$ in, is identical to our assumed value.

We will now compute the actual value for the section modulus of the cross section with the hole in it. See Figure A15–6 in the appendix.

$$S = \frac{t(h^3 - d^3)}{6h} = \frac{(0.50\text{ in})[(2.00\text{ in})^3 - (0.50\text{ in})^3]}{6(2.00\text{ in})} = 0.328\text{ in}^3$$

This value is larger than the minimum required value of 0.286 in^3. Therefore, the size of the cross section is satisfactory with regard to stress due to bending.

Final Design Decisions and Comments

In summary, the following are the design decisions for the horizontal bar of the conveyor hanger shown in Figure 5–17.

1. ***Material:*** SAE 1020 hot-rolled steel.
2. ***Size:*** Rectangular cross section. Thickness $t = 0.50$ in; height $h = 2.00$ in.
3. ***Overall design:*** Figure 5–17 shows the basic features of the bar.
4. ***Other considerations:*** Remaining to be specified are the tolerances on the dimensions for the bar and the finishing of its surfaces. The potential for corrosion should be considered and may call for paint, plating, or some other corrosion protection. The size of the cross section can likely be used with the as-received tolerances on thickness and height, but this is somewhat dependent on the design of the fixture that holds the engine block and the conveyor hangers. So the final tolerances will be left open pending later design decisions. The holes in the bar for the pins should be designed to produce a close sliding fit with the pins, and the details of specifying the tolerances on the hole diameters for such a fit are discussed in Chapter 13. See also the discussion on lug joints in Section 3–21.
5. ***Other possible modes of failure:*** The analysis used in this problem assumed that failure would occur due to the bending stresses in the rectangular bar. The dimensions were specified to preclude this from happening. Other possible modes are discussed here:

   a. *Deflection of the bar as an indication of stiffness:* The type of conveyor system described in this problem should not be expected to have extreme rigidity because moderate deflection of members should not impair its operation. However, if the horizontal bar deflects so much that it appears to be rather flexible, it would be deemed unsuitable. This is a subjective judgment. We can use Case (a) in Table A14–1 to compute the deflection.

$$y = FL^3/48EI$$

In this design,

$F = 310$ lb = maximum load on the bar

$L = 24.0$ in = distance between supports

$E = 30 \times 10^6$ psi = modulus of elasticity of steel

$I = th^3/12$ = moment of inertia of the cross section

$I = (0.50 \text{ in})(2.00 \text{ in})^3/12 = 0.333 \text{ in}^4$

Then

$$y = \frac{(310 \text{ lb})(24.0 \text{ in})^3}{48(30 \times 10^6 \text{ lb/in}^2)(0.333 \text{ in}^4)} = 0.0089 \text{ in}$$

This value seems satisfactory. In Section 5–9, some guidelines were given for deflection of machine elements. One stated that bending deflections for general machine parts should be limited to the range of 0.0005 to 0.003 in/in of beam length. For the bar in this design, the ratio of y/L can be compared to this range:

$$y/L = (0.0089 \text{ in})/(24.0 \text{ in}) = 0.0004 \text{ in/in of beam length}$$

Therefore, this deflection is well within the recommended range.

b. *Buckling of the bar:* When a beam with a tall, thin, rectangular cross section is subjected to bending, it would be possible for the shape to distort due to buckling before the bending stresses would cause failure of the material. This is called *elastic instability,* and a complete discussion is beyond the scope of this book. However, Reference 14 shows a method of computing the critical buckling load for this kind of loading. The pertinent geometrical feature is the ratio of the thickness t of the bar to its height h. It can be shown that the bar as designed will not buckle.

c. *Bearing stress on the inside surfaces of the holes in the beam:* Pins transfer loads between the bar and the mating elements in the conveyor system. It is possible that the bearing stress at the pin–hole interface could be large, leading to excessive deformation or wear. Reference 4 in Chapter 3 indicates that the allowable bearing stress for a steel pin in a steel hole is $0.90s_y$.

$$\sigma_{bd} = 0.90s_y = 0.90(30\,000 \text{ psi}) = 27\,000 \text{ psi}$$

The actual bearing stress at the center hole is found using the projected area, D_pt.

$$\sigma_b = F/D_pt = (310 \text{ lb})/(0.50 \text{ in})(0.50 \text{ in}) = 1240 \text{ psi}$$

Thus the pin and hole are very safe for bearing.

Design Example 5–4

A bracket is made by welding a rectangular bar to a circular rod as shown in Figure 5–18. Design the bar and the rod to carry a static load of 250 lb.

Solution

Objective: The design process will be divided into two parts:

1. Design the rectangular bar for the bracket.
2. Design the circular rod for the bracket.

Rectangular Bar

Given: The bracket design is shown in Figure 5–18. The rectangular bar carries a load of 250 lb vertically downward at its end. Support is provided by the weld at its left end where the loads are transferred to the circular rod. The bar acts as a cantilever beam, 12 in long. The design task is to specify the material for the bar and the dimensions of its cross section.

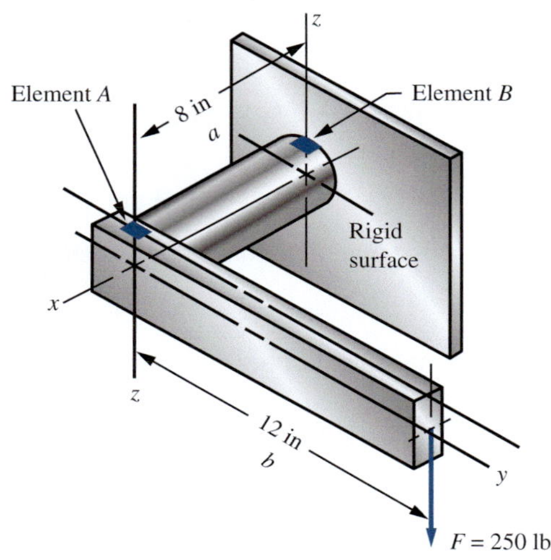

FIGURE 5–18 Bracket design

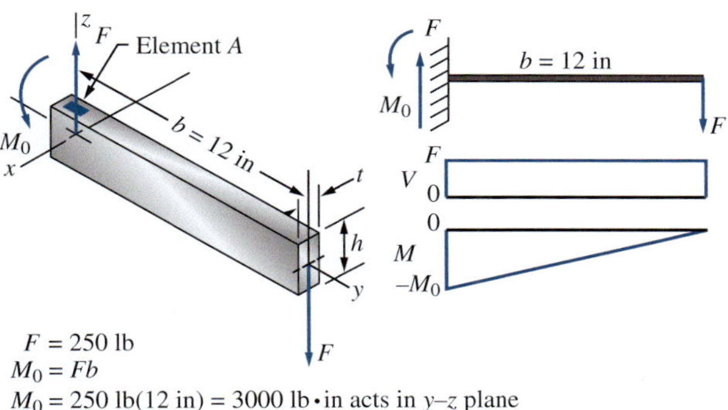

FIGURE 5–19 Free-body diagram of bar

Basic Design Decisions We will use steel for both parts of the bracket because of its relatively high stiffness, the ease of welding, and the wide range of strengths available. Let's specify SAE 1340 annealed steel having $s_y = 63$ ksi and $s_u = 102$ ksi (Appendix 3). The steel is highly ductile, with a 26% elongation.

The objective of the design analysis that follows is to determine the size of the cross section of the rectangular bar. Assuming that the loading and processing conditions are well known, we will use a design factor of $N = 2$ because of the static load.

Analysis and Results The free-body diagram of the cantilever bar is shown in Figure 5–19, along with the shearing force and bending moment diagrams. This should be a familiar case, leading to the judgment that the maximum tensile stress occurs at the top of the bar near to where it is supported by the circular rod. This point is labeled element A in Figure 5–19. The maximum bending moment there is $M = 3000$ lb·in. The stress at A is

$$\sigma_A = M/S$$

where S = section modulus of the cross section of the bar.

We will first compute the minimum allowable value for S and then determine the dimensions for the cross section.

Method 2 from Section 5–8 applies because of the static loading. We will first compute the design stress from

$$\sigma_d = s_y/N$$

$$\sigma_d = s_y/N = (63\ 000 \text{ psi})/2 = 31\ 500 \text{ psi}$$

Now we must ensure that the expected maximum stress $\sigma_A = M/S$ does not exceed the design stress. We can substitute $\sigma_A = \sigma_d$ and solve for S.

$$S = M/\sigma_d = (3000 \text{ lb}\cdot\text{in})/(31\ 500 \text{ lb/in}^2) = 0.095 \text{ in}^3$$

The relationship for S is

$$S = th^2/6$$

As a design decision, let's specify the approximate proportion for the cross-sectional dimensions to be $h = 3t$. Then

$$S = th^2/6 = t(3t)^2/6 = 9t^3/6 = 1.5t^3$$

The required minimum thickness is then

$$t = \sqrt[3]{S/1.5} = \sqrt[3]{(0.095\ \text{in}^3)/1.5} = 0.399\ \text{in}$$

The nominal height of the cross section should be, approximately,

$$h = 3t = 3(0.399\ \text{in}) = 1.20\ \text{in}$$

Final Design Decisions and Comments

In the fractional-inch system, preferred sizes are selected to be $t = 3/8\ \text{in} = 0.375\ \text{in}$ and $h = 1\frac{1}{4}\ \text{in} = 1.25\ \text{in}$ (see Table A2–1). Note that we chose a slightly smaller value for t but a slightly larger value for h. We must check to see that the resulting value for S is satisfactory.

$$S = th^2/6 = (0.375\ \text{in})(1.25\ \text{in})^2/6 = 0.0977\ \text{in}^3$$

This is larger than the required value of $0.095\ \text{in}^3$, so the design is satisfactory.

Circular Rod

Given

The bracket design is shown in Figure 5–18. The design task is to specify the material for the rod and the diameter of its cross section.

Basic Design Decisions

Let's specify SAE 1340 annealed steel, the same as that used for the rectangular bar. Its properties are $s_y = 63$ ksi and $s_u = 102$ ksi.

Analysis and Results

Figure 5–20 is the free-body diagram for the rod. The rod is loaded at its left end by the reactions at the end of the rectangular bar, namely, a downward force of 250 lb and a moment of 3000 lb·in. The figure shows that the moment acts as a torque on the circular rod, and the 250-lb force causes bending with a maximum bending moment of 2000 lb·in at the right end. Reactions are provided by the weld at its right end where the loads are transferred to the support. The rod then is subjected to a combined stress due to torsion and bending. Element B on the top of the rod is subjected to the maximum combined stress.

The manner of loading on the circular rod is identical to that analyzed earlier in Section 4–6. It was shown that when only bending and torsional shear occur, a procedure called the *equivalent torque method* can be used to complete the analysis. First we define the equivalent torque, T_e:

$$T_e = \sqrt{M^2 + T^2} = \sqrt{(2000)^2 + (3000)^2} = 3606\ \text{lb·in}$$

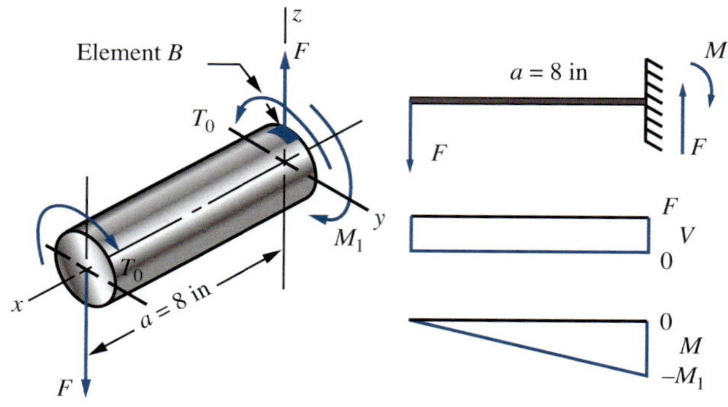

$F = 250$ lb

$M_1 = Fa = 250$ lb(8 in) = 2000 lb·in (acts in y–z plane)

$T_0 = Fb = 250$ lb(12 in) = 3000 lb·in (see Figure 5–19. $T_0 = M_0$)

FIGURE 5–20 Free-body diagram of rod

Then the shear stress in the bar is

$$\tau = T_e/Z_p$$

where Z_p = polar section modulus

For a solid circular rod,

$$Z_p = \pi D^3/16$$

Our approach is to determine the design shear stress and T_e and then solve for Z_p. Method 3 using the maximum shear stress theory of failure can be applied. The design shear stress is

$$\tau_d = 0.50 s_y/N = (0.5)(63\ 000 \text{ psi})/2 = 15\ 750 \text{ psi}$$

We let $\tau = \tau_d$ and solve for Z_p:

$$Z_p = T_e/\tau_d = (3606 \text{ lb}\cdot\text{in})/(15\ 750 \text{ lb/in}^2) = 0.229 \text{ in}^3$$

Now that we know Z_p, we can compute the required diameter from

$$D = \sqrt[3]{16Z_p/\pi} = \sqrt[3]{16(0.229 \text{ in}^3)/\pi} = 1.053 \text{ in}$$

This is the minimum acceptable diameter for the rod.

Final Design Decisions and Comments The circular rod is to be welded to the side of the rectangular bar, and we have specified the height of the bar to be $1\frac{1}{4}$ in. Let's specify the diameter of the circular rod to be machined to 1.10 in. This will allow welding all around its periphery.

5–11 ADDITIONAL DESIGN ANALYSIS METHODS

Presented here are several additional design analysis methods to supplement those presented in Section 5–8 and included within the design procedure shown and demonstrated in Sections 5–9 and 5–10. It is imperative for you as a designer to carefully decide which analysis method is consistent with the nature of the configuration, the material, and the manner of loading of the actual component being designed.

Some of the added methods build from the discussion on Mohr's circle and combined stresses from Chapter 4 with regard to:

- Maximum and minimum principal stresses
- Maximum shear stress
- Special stress conditions
- Analysis of complex loading conditions

The concepts presented here are typically integrated within modern finite-element analysis (FEA) software packages and the discussion in this section should help you understand better the bases for the results generated from such software. See Internet sites 1–8.

Modified Mohr Method for Biaxial Static Stress on Brittle Materials

When stresses are applied in more than one direction or when normal stress and shear stress are applied simultaneously, it is necessary to compute the principal stresses, σ_1 and σ_2, using Mohr's circle or the equations in Chapter 4. *Stress concentrations should be included in the applied stresses before preparing Mohr's circle for brittle materials.*

For safety, the *combination* of the two principal stresses must lie within the area shown in Figure 5–21 that graphically depicts the *modified Mohr theory*. The graph is a plot of the maximum principal stress, σ_1, on the horizontal axis (abscissa) and the minimum principal stress, σ_2, on the vertical axis (ordinate).

Note that the failure criteria depend on the quadrant in which the principal stresses lie. In the first quadrant, both principal stresses are tensile, and failure is predicted when either one exceeds the ultimate tensile strength, s_{ut}, of the material. Similarly, in the third quadrant, both principal stresses are compressive, and failure is predicted when either one exceeds the ultimate compressive strength, s_{uc}, of the material.

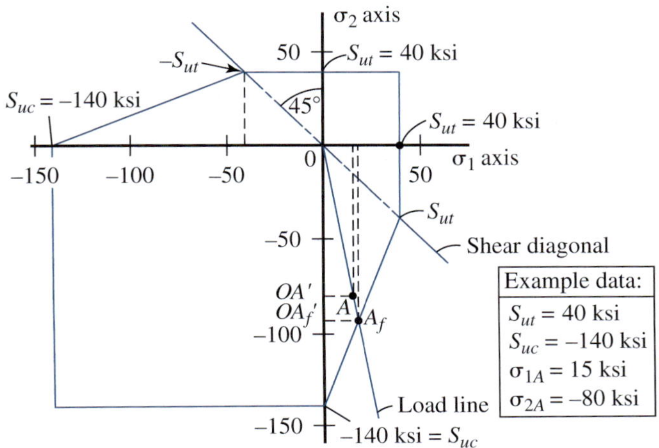

FIGURE 5–21 Modified Mohr diagram with example data and a load line plotted

The failure lines for the second and fourth quadrants are more complex and have been derived semiempirically to correlate with test data. The ultimate tensile strength lines are extended from the first quadrant into the second and fourth quadrants to the point where each intersects the *shear diagonal*, drawn at 45° through the origin. Then the failure line proceeds at an angle to the ultimate compressive strength.

For design, because of the many different shapes and dimensions of safe-stress zones in Figure 5–21, it is suggested that a rough plot be made of the pertinent part of the modified Mohr diagram from actual material strength data. Then the actual values of σ_1 and σ_2 can be plotted to ensure that they lie within the safe zone of the diagram.

A *load line* can be an aid in determining the design factor, N, using the modified Mohr diagram. The assumption is made that stresses increase proportionally as loads increase. Apply the following steps for an example stress state, A, for which $\sigma_{1A} = 15$ ksi and $\sigma_{2A} = -80$ ksi. The material is Grade 40A gray cast iron having $s_{ut} = 40$ ksi and $s_{uc} = 140$ ksi.

1. Draw the modified Mohr diagram as shown in Figure 5–21.
2. Plot point A at $(15, -80)$.
3. Draw the load line from the origin through point A until it intersects the failure line on the diagram at the point labeled A_f.
4. Determine the distances $OA = 81.4$ ksi and $OA_f = 96.0$ ksi by scaling the diagram.
5. Compute the design factor from $N = OA_f/OA = 96.0/81.4 = 1.18$.

Distortion Energy Method for Static Stress on Ductile Materials Using von Mises Stress

The distortion energy method has been shown to be the best predictor of failure for ductile materials under static loads or completely reversed normal, shear, or combined stresses. It is typically applied using the von Mises stress, σ', which was described in Section 4–3. Recall the following relationship for σ', based on the principle stresses (σ_1, σ_2).

von Mises Stress for Biaxial Stress State. Given the principle stresses:

$$\sigma' = \sqrt{\sigma_1^2 + \sigma_2^2 - \sigma_1\sigma_2} \qquad (5\text{–}23)$$

It is helpful to visualize the distortion energy failure prediction method by plotting a failure line on a graph with σ_1 on the horizontal axis and σ_2 on the vertical axis as shown in Figure 5–22. The failure line is an ellipse centered at the origin and passing through the yield strength on each axis, in both the tensile and compressive regions. It is necessary that the material has equal values for yield strength in tension and compression for direct use of this method. The numerical scales on the graph are normalized to the yield strength so the ellipse passes through $s_y/\sigma_1 = 1.0$ on the σ_1 axis and similarly on the other axes. *Combinations of principal stresses that lie within the distortion energy ellipse are predicted to be safe, while those outside would predict failure.*

For design, the design factor, N, can be applied to the yield strength. Then use

$$\sigma' < \sigma_d = s_y/N \qquad (5\text{–}24)$$

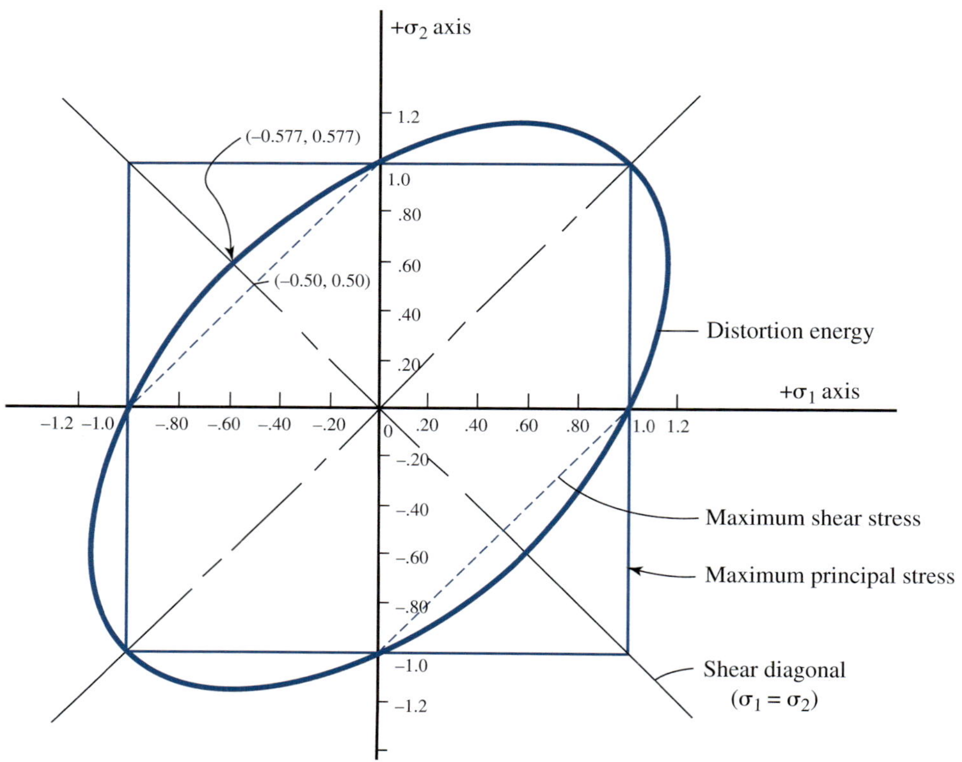

FIGURE 5–22 Distortion energy method compared with maximum shear stress and maximum principal stress methods

For comparison, the failure prediction lines for the maximum shear stress method are also shown in Figure 5–22. With data showing that the distortion energy method is the best predictor, it can be seen that the maximum shear stress method is generally conservative and that it coincides with the distortion energy ellipse at six points. In other regions, it is as much as 16% low. Note the 45° diagonal line through the second and fourth quadrants, called the *shear diagonal.* It is the locus of points where $\sigma_1 = \sigma_2$ and its intersection with the failure ellipse is at the point $(-0.577, 0.577)$ in the second quadrant. This predicts failure when the shear stress is $0.577s_y$. The maximum shear stress method predicts failure at $0.50s_y$, thus quantifying the conservatism of the maximum shear stress method.

Also shown in Figure 5–22 are the failure prediction lines for the maximum principal stress method. It is coincident with the maximum shear stress lines in the first and third quadrants for which the two principal stresses have the same sign, either tensile (+) or compressive (−). Therefore, it, too, is conservative in these regions. *But note that it is dangerously nonconservative in the second and fourth quadrants.*

Fluctuating Combined Stresses—Maximum Shear Stress Method

The approach presented here is similar to the Goodman method described previously, but the effect of the combined stresses is first determined by using Mohr's circle.

For the maximum shear stress theory, draw two Mohr's circles, one for the mean stresses and one for the alternating stresses. From the first circle, determine maximum mean shear stress, $(\tau_m)_{\max}$. From the second circle, determine the maximum alternating shear stress, $(\tau_a)_{\max}$. Then use these values in the design equation

$$\frac{K_t(\tau_a)_{\max}}{s'_{sn}} + \frac{(\tau_m)_{\max}}{s_{su}} = \frac{1}{N} \quad (5\text{–}25)$$

In the absence of shear strength data, use the estimates, $s'_{sn} = 0.577 s'_n$ and $s_{su} = 0.75 s_u$.

Fluctuating Combined Stresses—Distortion Energy Method

For the distortion energy theory, draw two Mohr's circles, one for the mean stresses and one for the alternating stresses. From these circles, determine the maximum and minimum principal stresses. Then compute the von Mises stresses for both the mean and the alternating components from

$$\sigma'_m = \sqrt{\sigma_{1m}^2 + \sigma_{2m}^2 - \sigma_{1m}\sigma_{2m}}$$

$$\sigma'_a = \sqrt{\sigma_{1a}^2 + \sigma_{2a}^2 - \sigma_{1a}\sigma_{2a}}$$

The Goodman equation then becomes

$$\frac{K_t\sigma'_a}{s'_n} + \frac{\sigma'_m}{s_u} = \frac{1}{N} \quad (5\text{–}26)$$

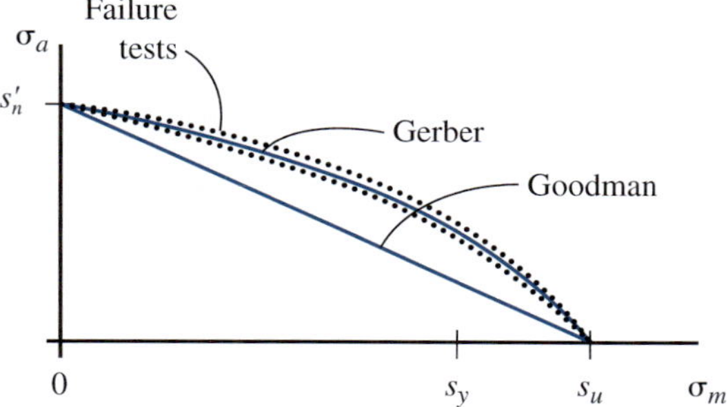

FIGURE 5–23 Comparison of Gerber and Goodman methods for fluctuating stresses on ductile materials

Gerber Method for Fluctuating Stresses on Ductile Materials

Those interested in a more precise predictor of fatigue failure typically use the Gerber method, shown in Figure 5–23. The axes are the same as those used for the Goodman method with alternating stress level, σ_a, on the vertical axis and mean stress, σ_m, on the horizontal axis. The Gerber line is a parabolic curve that passes generally through the array of experimental data points for failures under specific combinations of mean and alternating stresses. It starts at s'_n and ends at s_u as the Goodman line does as shown in the illustration. It can be seen that the Goodman line is slightly conservative but it has the advantage of lying entirely below the array of failure points and analysis is somewhat simpler. See References 5 and 8 for more details on the Gerber method.

Smith Diagram for Showing the Effect of Mean Stress on Fatigue

Some stress analysts, for example many in Germany, use the *Smith Diagram* to map a region of acceptable combinations of mean and alternating stresses. Additional materials property data are required to apply this method as compared with the Goodman approach defined earlier and used in this book. Figure 5–24 shows one approach to drawing the Smith Diagram for bending stress. Data required are:

1. Fatigue limit strength for completely reversed and repeated bending stress ($R = -1$), called s_n in this book. Some call this loading pattern *pure oscillation*, $s_{n(-1)}$, as shown in Figure 5–2. The mean stress is zero $[\sigma_m = 0]$ and the stress oscillates between $+\sigma_{\max}$ and $-\sigma_{\min}$ having the same absolute value. The alternating stress, $\sigma_a = \sigma_{\max}$.
2. Fatigue limit strength for repeated, one-direction bending stress ($R = 0$), sometimes called *pure pulsating stress*, $s_{n(0)}$, and shown in Figure 5–5. The stress oscillates between the value of $\sigma_{\max}$ and $\sigma_{\min} = 0$. The alternating stress and the mean stress are equal: $\sigma_a = \sigma_m = \sigma_{\max}/2$.
3. Ultimate strength, s_u.

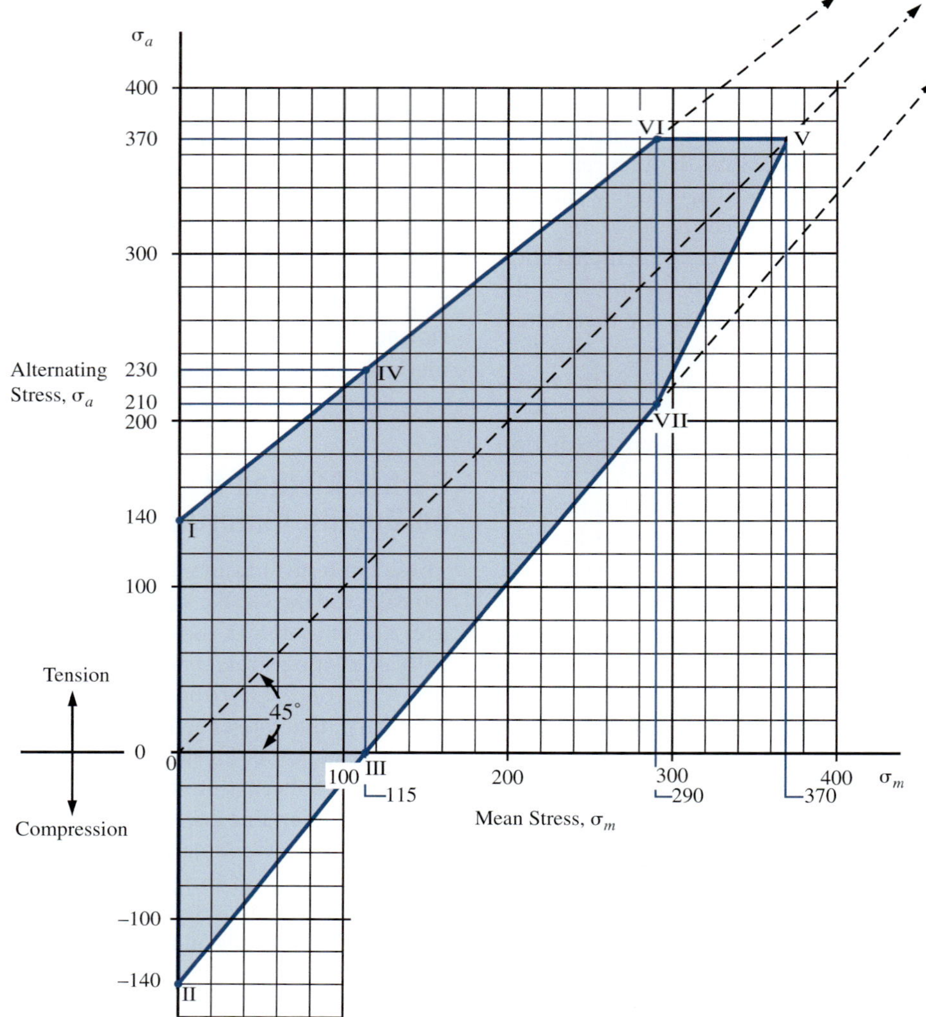

FIGURE 5–24 Example of a Smith diagram for fatigue due to fluctuating bending stress

The steps used to draw the diagram in Figure 5–24 are:

1. Draw the vertical axis for alternating stress, σ_a, and the horizontal axis for mean stress, σ_m.
2. Draw a dashed line at 45° from the origin. This represents the line of increasing mean stress.
3. On the vertical axis, plot two points: Point I at $+\sigma_{max} = s_{n(-1)}$ and Point II at $-\sigma_{min} = -s_{n(-1)}$ for the pure oscillation case.
4. Plot data for the pure pulsating case: Point III at $s_{n(0)}/2$, 0 and Point IV at $s_{n(0)}/2$, $s_{n(0)}$.
5. Draw a line from Point I through Point IV toward its intersection with the 45° line.
6. Draw another line from Point II through Point III toward its intersection with the 45° line.
7. Draw a horizontal line from the value of the ultimate strength on the vertical axis to its intersection with the 45° line, calling that intersection Point V.
8. Call the intersection of the line from Step 7 and the line from Step 5 Point VI.
9. Drop a vertical line from Point VI to its intersection with the line from Step 6, calling that intersection Point VII.
10. Draw a line from Point V to Point VII.
11. The resulting polygon from Points I → VI → V → VII → II → I encloses an area of theoretically safe combinations of mean and alternating stresses.
12. The upper part of the polygon from Points I → VI → V is a plot of the maximum stress that can be applied to a component.
13. The lower part of the polygon from Points V → VII → II is a plot of the minimum stress corresponding to any maximum stress directly vertically above.
14. For any vertical line drawn through the upper and lower parts of the polygon:
    a. The top intersection is $+\sigma_{max}$
    b. The intersection with the dashed line is σ_m
    c. The lower intersection is σ_{min}
15. Actual allowable applied stresses, of course, would be lower than these points according to the desired design factor, *N*.

The example data used for drawing Figure 5–24 are:

Ultimate strength $= s_u = 370\,\text{MPa}$

Fatigue strength for pure oscillation, $s_{n(-1)} = 140\,\text{MPa}$

Fatigue strength for pure pulsation, $s_{n(0)} = 230\,\text{MPa}$ and $s_{n(0)}/2 = 115\,\text{MPa}$

Point coordinates:

	Horizontal coordinate	Vertical coordinate	
I	0	140 MPa	[Step 3]
II	0	−140 MPa	[Step 3]
III	115 MPa	0	[Step 4]
IV	115 MPa	230 MPa	[Step 4]
V	370 MPa	370 MPa	[Step 7]
VI	290 MPa (read from diagram)	370 MPa	[Step 8]
VII	290 MPa	210 MPa (read from diagram)	[Step 9]

This method is an approximation of data from fatigue tests from many combinations of mean and alternating stresses. A modification of this method is to limit any applied stress to the yield strength of the material instead of the ultimate strength. Other charts of similar nature are often drawn for torsional shear stresses and for alternating direct tension–compression.

5–12 RECOMMENDED DESIGN AND PROCESSING FOR FATIGUE LOADING

Throughout this chapter, we have described factors that influence the fatigue life of components subjected to cyclical loads. This section presents some summary recommendations for designers when specifying size, shape, and processing techniques for a given component. See Reference 15 for additional methods and supporting data.

1. Where critical stress levels are encountered, prepare the surface with low roughness and specify a processing method that will produce well rounded valleys between peaks. Examples are to use a turning tool for cylindrical parts with a broad tip radius or a ball end mill with a large diameter for milling operations.
2. Design parts having inherent abrupt changes in geometry with low values of stress concentrations. An important example is to provide large radii at sites of change in diameter or width of a component. Consult Section 3–21 and Figure 3–26.
3. Design the part so the predominant direction of processing is parallel to the major axis of the part.
4. Increase the fatigue strength of critical areas of a component by using processing methods that leave compressive residual stresses. Examples include shot peening, cold working by rolling or burnishing.
5. Avoid processing methods that produce tensile residual stresses that decrease the fatigue strength of the material. Examples are (a) aggressive grinding operations that rapidly remove large amounts of material, (b) some heat-treating operations that use rapid quenching from high temperatures, and (c) some welding processes near the heat-affected zone. If such processes are used, supplementary processing is recommended to relieve the tensile residual stress.
6. If processing leaves surface irregularities such as burning, cracks, decarburized materials, or rough welds, take additional steps to remove those irregularities by light finishing cuts, electropolishing, or lapping.
7. Control the entire path of manufacturing processes to ensure that (a) design specifications are met, (b) material properties are within the values used in design analyses, and (c) parts are handled carefully to prohibit accidental surface damage from nicks, scratches, or corrosion.

5–13 STATISTICAL APPROACHES TO DESIGN

The design approaches presented in this chapter are somewhat deterministic in the sense that data are taken to be discrete values, and analyses use the data to determine specific results. The method of accounting for uncertainty with regard to the data themselves lies with the selection of an acceptable value for the design factor represented by the final design decision. Obviously this is a subjective judgment. Often decisions that are made to ensure the safety of a design cause many designs to be quite conservative.

Competitive pressures call for ever more efficient, less conservative designs. Throughout this book, recommendations are made to seek more reliable data for loads, material properties, and environmental factors, providing more confidence in the results of design analyses and allowing lower values for the design factor as discussed in Section 5–7. A more robust and reliable product results from testing samples of the actual material to be used in the product; performing extensive measurements of loads to be experienced; investing in more detailed performance testing, experimental stress analysis, and finite-element analysis; exerting more careful control of manufacturing processes; and life testing of prototype products in realistic conditions where possible. All of these measures typically come with significant additional costs, and difficult decisions must be made about whether or not to implement them.

In combination with the approaches listed previously, a greater use of statistical methods (also called *stochastic methods*) is emerging to account for the inevitable variability of data by determining the mean values of critical parameters from several sets of data and quantifying the variability using the concepts of distributions and standard deviations. References 5 and 11 provide guidance in these methods. Section 2–4 provided a modest discussion of this approach to account for the variability of materials property data.

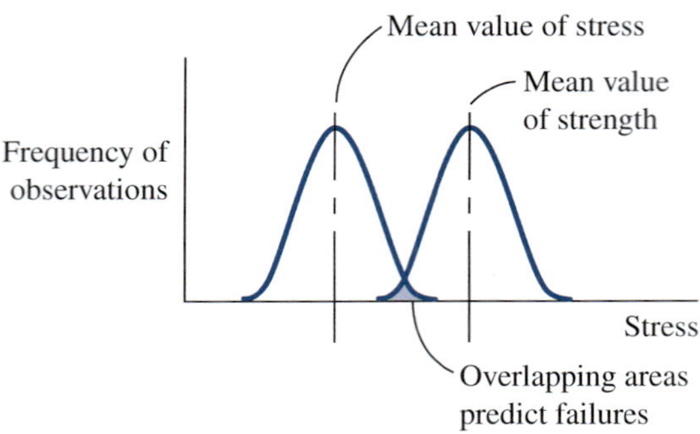

FIGURE 5–25 Illustration of statistical variation in failure potential

Industries such as automotive, aerospace, construction equipment, and machine tools devote considerable resources to acquiring useful data for operating conditions that will aid designers in producing more efficient designs.

Samples of Statistical Terminology and Tools

- Statistical methods analyze data to present useful information about the source of the data.
- Stochastic methods apply probability theories to characterize variability in the data.
- Data sets can be analyzed to determine the mean (average), the range of variation, and the standard deviation.
- Inferences can be made about the nature of the distribution of the data, such as normal or lognormal.
- Linear regression and other means of curve fitting can be employed to represent a set of data by mathematical functions.
- The distributions for applied loads and stresses can be compared with the distribution for the strength of the material to determine to what degree they overlap and the probability that a certain number of failures will occur. See Figure 5–25.
- The reliability of a component or a complete product can be quantified.
- The optimum assignment of tolerances can be made to reasonably ensure satisfactory performance of a product while allowing as broad a range of tolerances as practical.

5–14 FINITE LIFE AND DAMAGE ACCUMULATION METHOD

The fatigue design methods described thus far in this chapter have the goal to design a component for infinite life by using the estimated actual endurance strength as the basis for design and assuming that this value is the *endurance limit.* Applied repeated stress levels below this value will provide infinite life. Furthermore, the analyses were based on the assumption that the loading pattern was uniform over the life of the component. That is, the mean and alternating stresses do not vary over time.

There are, however, many common examples for which a finite life is adequate for the application and where the loading pattern does vary with time. Consider the following.

Finite Life Examples

First, let's discuss the concept of finite life. Refer to the endurance strength curves in Figure 5–7 in Section 5–3. Data are plotted on a stress versus number of cycles to failure graph (σ *vs.* N) with both axes using logarithmic scales. For the materials that exhibit an endurance limit, it can be seen that the limit occurs for a life of approximately 10^6 cycles. How long would it take to accumulate 1 million cycles of load applications? Here are some examples.

Bicycle brake lever: Assume that the brake is applied every 5.0 minutes while riding 4.0 hours per day every day of each year. It would take more than 57 years to apply the brake 1 million times.

Lawn mower height-adjustment mechanism: Consider a lawn mower used by a commercial lawn maintenance company. Assume that the cutting height for the mower is adjusted to accommodate varying terrain three times per mowing and that the mower is used 40 times per week for all 12 months per year. It would take 160 years to accumulate 1 million cycles of load on the height adjustment mechanism.

Automotive lift in a service station: Assume that the service technician lifts four automobiles per hour, 10 hours per day, 6 days per week, each week of the year. It would take more than 80 years to accumulate 1 million cycles of load on the lift mechanism.

Each of these examples indicates that it may be appropriate to design the load-carrying members of the example systems for something less than infinite life. But many industrial examples do require design for infinite life. Here is an example.

Parts feeding device: On an automated assembly system, a feeding device inserts 120 parts per minute. If the system operates 16 hours per day, 6 days per week, each week of the year, it would take only 8.7 days to accumulate 1 million cycles of loading. It would see 35.9 million cycles in one year.

When it can be justified to design for a finite life less than the number of cycles corresponding to the endurance limit, you will need data similar to that in Figure 5–7 for the actual material to be used in the component. Testing the material yourself is preferred, but it would be a time-consuming and costly exercise to acquire sufficient data to make statistically valid σ-N curves. References 2–4, 9, 11 and 13 may provide suitable data, or an additional literature search may be required. Once reliable data are identified, use the endurance strength at the specified number of cycles as the starting point for computing the estimated actual endurance strength as described in Section 5–5. Then use that value in subsequent analyses as described in Section 5–8.

Varying Stress Amplitude Examples

Here we are looking for examples where the component experiences cyclical loading for a large number of cycles but for which the amplitude of the stress varies over time.

Bicycle brake lever: Let's reconsider the braking action on a bicycle. Sometimes you need to bring the bike to a stop very quickly from a high speed, requiring a rather high force on the brake lever. Other times you may apply a lighter force to simply slow down a bit to safely negotiate a curve.

Automotive suspension member: Suspension parts such as a strut, spring, shock absorber, control arm, or fastener pass loads from the wheel to the frame of a car. The magnitude of the load depends on vehicle speed, the condition of the road, and driver action. Roads may be smoothly paved, potholed, or rough surfaced gravel. The vehicle may even be driven off-road where violent peaks of stress will be encountered.

Machine tool drive system: Consider the lifetime of a milling machine. Its primary function is to cut metal, and it takes a certain amount of torque to drive the cutter depending on the machinability of the material, the depth of cut, and the feed rate. Surely the torque will vary significantly from job to job. During part of its operating time, there may be no cutting action at all as a new part is positioned or as the operation completes one cut and adjusts before making another. At times the cutter will encounter locally harder material, requiring higher torque for a short period of time.

Cranes, power shovels, bulldozers, and other construction equipment: Here there are obviously varying loads as the equipment is used for numerous tasks, such as hoisting large steel beams or small bracing members, digging through hard clay or soft sandy soil, rapidly grading a hillside or performing the final smoothing of a driveway, or encountering a tree stump or large rock.

How would you determine the loads that these devices would experience over time? One method involves building a prototype and instrumenting critical elements with strain gages, load cells, or accelerometers. Then the system would be "put through its paces" over a wide range of tasks while loads and stresses are recorded as a function of time. Similar vehicles could be monitored to determine the frequency that different kinds of loading would be encountered over its expected life. Combining such data would produce a record from which the total number of cycles of stress at any given level can be estimated. Statistical techniques such as spectrum analysis, Fast Fourier Transform analysis, and time compression analysis produce charts that summarize stress amplitude and frequency data that are useful for fatigue and vibration analysis. See Reference 11 for extended discussion of these techniques.

Damage Accumulation Method

The principle of *damage accumulation* is based on the assumption that any given level of stress applied for one cycle of loading contributes to a certain amount of damage to a component. Refer again to Figure 5–7 and observe the *S-N* curve for the alloy steel, SAE 4340 at a hardness of HB275 and having an ultimate strength of 1048 MPa (152 ksi). Reading a few points from that curve gives the following set of data:

Stress amplitude	Number of cycles to failure
710 MPa (103 ksi)	1.0×10^3
600 MPa (87 ksi)	1.0×10^4
505 MPa (73 ksi)	1.0×10^5
415 MPa (60 ksi)	1.0×10^6 [Considered the endurance limit, s_n]
<415 MPa	∞

There is a theoretically infinite life for stresses below the endurance limit and no damage occurs.

Now, let's say that the component experiences 480 cycles of stress at the 600 MPa (87 ksi) level. It can be said that it suffers a damage of the ratio of $480/(1.0 \times 10^4) = 0.048$. Being subjected to 250 cycles of stress at 710 MPa (103 ksi) produces a damage of $250/(1.0 \times 10^3) = 0.250$. The cumulative damage of both events is the sum, $0.048 + 0.250 = 0.298$. When the sum of these types of calculations is equal to 1.0, it is predicted that the component would fail. Therefore, this brief example predicts that approximately 30% of the component's life has been accumulated.

This kind of logic can be used to predict the total life of a component subjected to a sequence of loading levels. Let n_i represent the number of cycles of a specific stress level experienced by a component. Let N_i be the number of cycles to failure for this stress level as indicated by an σ-N curve such as those shown in Figure 5–7. Then the damage contribution from this loading is

$$D_i = n_i/N_i$$

When several stress levels are experienced for different numbers of cycles, the cumulative damage, D_c, can be represented as,

$$D_c = \sum_{i=1}^{i=k}(n_i/N_i) \tag{5–27}$$

Failure is predicted when $D_c \geq 1.0$. This process is called the *Miner linear cumulative-damage rule* or simply *Miner's rule* in honor of his work in 1945. An example problem will now demonstrate the application of Miner's rule.

The example data in Figure 5–7 and listed above are taken for completely reversed and repeated bending stress on a small [0.30 in (7.62 mm)] diameter circular bar with a mirror polished surface as used in the standard R. R. Moore test device (Figure 5–3). For other conditions (the typical situation), the *S-N* curve can be adjusted by first computing the actual expected endurance strength of the material, s'_n, using Equation 5–6 from Section 5–4 as has been done throughout this chapter. Then a ratio of s'_n/s_n can be applied to stress data from the *S-N* curves before determining the number of cycles to failure. The process is illustrated in Example Problem 5–5.

Example Problem 5–5 Determine the cumulative damage experienced by a ground circular rod, 38 mm diameter, subjected to the combination of the cycles of loading and varying levels of reversed, repeated bending stress shown in Table 5–4.

The bar is made from SAE 4340, HB275 alloy steel with an ultimate strength of 1048 MPa and a fatigue limit of 430 MPa. Use Figure 5–7 for the *S-N* curve.

TABLE 5–4 Loading Pattern for Example Problem 5–5

Stress level		Number of cycles
MPa	ksi	n_i
650	94.3	2 000
600	87.0	3 000
500	72.5	10 000
350	50.8	25 000
300	43.5	15 000

Solution

Given SAE 4340, HB 275 alloy steel rod. $s_u = 1048$ MPa $D = 38$ mm. Ground surface.
Endurance strength data (*S-N*) shown in Figure 5–7 $s_n = 430$ MPa.
Loading is reversed, repeated bending. Load history shown in Table 5–4.

Analysis First adjust *S-N* data for actual conditions using methods of Section 5–4. Use Miner's rule to estimate portion of life used by loading pattern.

Results For SAE 4340, HB 275, $s_u = 1048$ MPa
From Figure 5–8, basic $s_n = 475$ MPa for ground surface
Material factor, $C_m = 1.00$ for wrought steel
Type-of-stress factor, $C_{st} = 1.0$ for reversed, rotating bending stress
Reliability factor, $C_R = 0.81$ (Table 5–2) for $R = 0.99$ (Design decision)
Size factor, $C_s = 0.84$ (Figure 5–9 and Table 5–3 for $D = 38$ mm)
Estimated actual endurance strength, s'_n – Computed:

$$s'_n = s_n \, C_m \, C_{st} \, C_R \, C_s = (475 \text{ MPa})(1.0)(1.0)(0.81)(0.84) = 323 \text{ MPa}$$

This is the estimate for the actual endurance limit of the steel. In Figure 5–7, the endurance limit for the standard specimen is 430 MPa. The ratio of the actual to the standard data is $323/430 = 0.75$. We can now adjust the entire *S-N* curve by this factor. The result is shown in Figure 5–26.

Now we can read the number of cycles of life, N_i, corresponding to each of the given loading levels from Table 5–4. The combined data for the number of applied load cycles, n_i, and the life cycles, N_i, are now used in Miner's rule, Equation 5–27, to determine the cumulative damage, D_c. Results are shown in Table 5–5.

TABLE 5–5 Cumulative Life Calculations for Example Problem 5–5

Stress level		Number of cycles	Life cycles	
MPa	ksi	n_i	N_i	n_i/N_i
650	94.3	2 000	1.1×10^4	0.182
600	87.0	3 000	1.8×10^4	0.167
500	72.5	10 000	5.8×10^4	0.172
350	50.8	25 000	5.6×10^5	0.045
300	43.5	15 000	∞	0.000
			Total	0.566

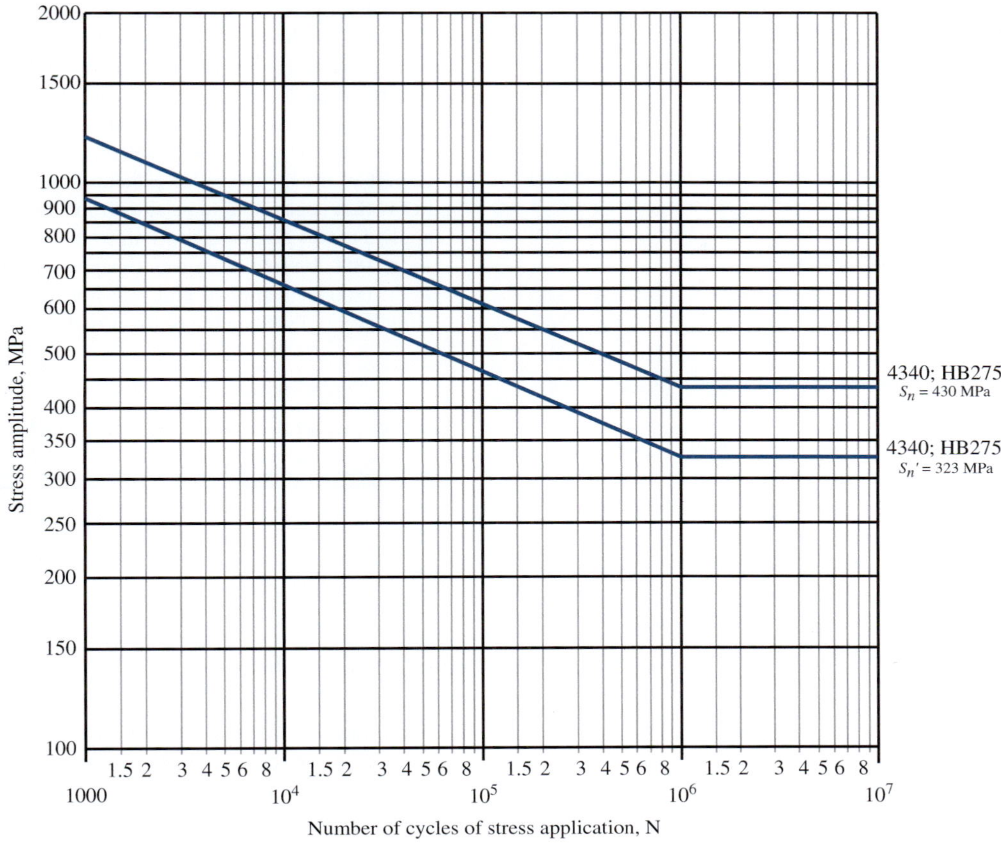

FIGURE 5–26 Endurance strength for Example Problem 5–5

Comment We can conclude from this number that approximately 57% of the life of the component has been accumulated by the given loading. For these data, the greatest damage occurs from the 650 MPa loading for 2 000 cycles. An almost equal amount of damage is caused by the 500 MPa loading for 10 000 cycles. Note that the cycles of loading at 300 MPa contributed nothing to the damage because they are below the endurance limit of the steel.

REFERENCES

1. Anderson, Ted L. *Fracture Mechanics: Fundamentals and Applications*. 3rd ed. Boca Raton, FL: CRC Press, 2005.
2. ASM International. *ASM Handbook Volume 19, Fatigue and Fracture*. Materials Park, OH: ASM International, 1996.
3. Balandin, Dimitri V., Nikolaj N. Bolotnik, and Walter D. Pilkey. *Optimal Protection from Impact, Shock and Vibration*. London, UK: Taylor & Francis, 2001.
4. Boyer, Howard E. *Atlas of Fatigue Curves*. Materials Park, OH: ASM International, 1986.
5. Budynas, Richard G., and Keith J. Nisbett. *Shigley's Mechanical Engineering Design*. 9th ed. New York: McGraw-Hill, 2011.
6. deSilva, Clarence W. *Vibration and Shock Handbook*. Boca Raton, FL: CRC Press, 2005.
7. Dowling, Norman E. *Mechanical Behavior of Materials*. 3rd ed. Upper Saddle River, NJ: Pearson/Prentice-Hall, 2007.
8. Juvinall, Robert C., and Kurt M. Marshek. *Fundamentals of Machine Component Design*. 4th ed. New York: John Wiley & Sons, 2005.
9. Meyers, Marc A., and K. Krishan Chawla. *Mechanical Behavior of Materials*. 2nd ed. Cambridge, UK: Cambridge University Press, 2009.
10. Piersoll, Allan, and Thomas Paez. *Harris' Shock and Vibration Handbook*. 6th ed. New York, NY: McGraw-Hill, 2010.
11. SAE International. *SAE Fatigue Design Handbook*. 3rd ed. Warrendale, PA: SAE International, 1997.
12. Spotts, Merhyle F., Terry E. Shoup, and Lee E. Hornberger. *Design of Machine Elements*. 8th ed. Upper Saddle River, NJ: Pearson/Prentice-Hall, 2004.
13. Stephens, Ralph I., Ali Fatemi, Robert R. Stephens, and Henry O. Fuchs. *Metal Fatigue in Engineering*. 2nd ed. New York: John Wiley & Sons, 2001.
14. Young, Warren C., Richard G. Budynas, and Ali Sadegh. *Roark's Formulas for Stress and Strain*. 8th ed. New York, NY: McGraw-Hill, 2012.
15. Zahavi, Eliaju, and Vladimir Torbilo. *Fatigue Design—Life Expectancy of Machine Parts*. Boca Raton, FL: CRC Press, 1996.
16. Huston, Ronald, and Harold Josephs. *Practical Stress Analysis in Engineering Design*. 3rd ed. Boca Raton, FL: CRC Press, 2009.

INTERNET SITES

1. eFatigue LLC https://efatigue.com An online fatigue calculator.
2. Safe Technology www.safetechnology.com A developer of the fe-safe™ line of software for fatigue analysis that can be linked with finite-element models. Uses several analysis methods such as stress-life, strain-life, von Mises stress, multiaxial fatigue, and damage accumulation.
3. HBM-nCode www.ncode.com Provider of software and measurement systems for acquiring and processing data for fatigue analysis. Systems also work with finite-element analysis data systems. Uses stress-life, strain-life, multiaxial, weld analysis, and other methods.
4. ProsigLtd www.prosig.com/dats/optfat.html A producer of the DATS Fatigue Life and Analysis software along with hardware for measurement and testing of vibration and noise of vehicles and industrial equipment.
5. Comsol Multiphysics http://www.comsol.com/products/multiphysics/ A producer of simulation software covering many types of physical phenomena including structural mechanics, thermal analysis, computational fluid mechanics, and others. An overview of the range of structural mechanics analysis methods is illustrated in the short, free YouTube video at http://www.youtube.com/watch?v=uEjJb-KAEoE.
6. AutoFEM Software, LLP www.autofemsoft.com A producer of finite-element analysis software linked with AutoCAD. Includes five modules: static, fatigue, frequency, buckling, and thermal analyses. Offers free AutoFEM Analysis Lite for users of AutoCAD with somewhat restricted capabilities.
7. ANSYS www.ansys.com Structural analysis finite-element software that includes linear and nonlinear static and dynamic analysis of structures and mechanical devices. Also includes the nCode Design Life fatigue analysis software.
8. MSC Software www.mscsoftware.com Provider of desktop versions of many types of MD NASTRAN software for static and dynamic analysis of structures and moving devices, including fatigue analysis.

PROBLEMS

Stress Ratio

For each of Problems 1–9, draw a sketch of the variation of stress versus time, and compute the maximum stress, minimum stress, mean stress, alternating stress, and stress ratio, R. For Problems 6–9, analyze the beam at the place where the maximum stress would occur at any time in the cycle.

1. A link in a mechanism is made from a round bar having a diameter of 10.0 mm. It is subjected to a tensile force that varies from 3500 to 500 N in a cyclical fashion as the mechanism runs.
2. A strut in a space frame has a rectangular cross section of 10.0 mm by 30.0 mm. It sees a load that varies from a tensile force of 20.0 kN to a compressive force of 8.0 kN.
3. A link in a packaging machine mechanism has a square cross section of 0.40 in on a side. It is subjected to a load that varies from a tensile force of 860 lb to a compressive force of 120 lb.
4. A circular rod with a diameter of 3/8 in supports part of a storage shelf in a warehouse. As products are loaded and unloaded, the rod is subjected to a tensile load that varies from 1800 to 150 lb.
5. A part of a latch for a car door is made from a circular rod having a diameter of 3.0 mm. With each actuation, it sees a tensile force that varies from 780 to 360 N.
6. A part of the structure for a factory automation system is a beam that spans 30.0 in as shown in Figure P5–6. Loads are applied at two points, each 8.0 in from a support. The left load $F_1 = 1800$ lb remains constantly applied, while the right load $F_2 = 1800$ lb is applied and removed frequently as the machine cycles. Evaluate the beam at both B and C.

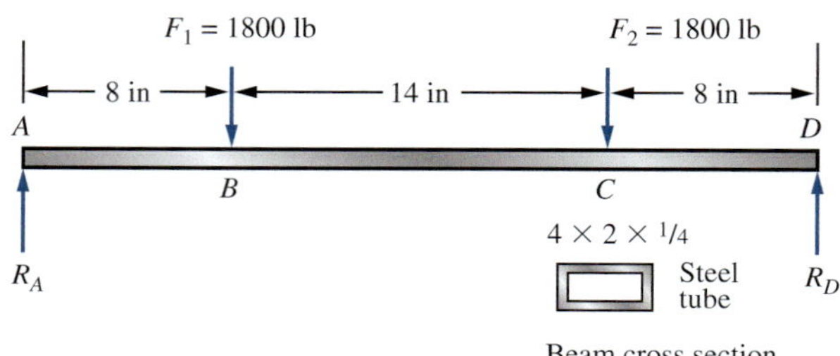

FIGURE P5–6 (Problems 6 and 23)

7. A cantilevered boom is part of an assembly machine and is made from an American Standard steel beam, S4 × 7.7. A tool with a weight of 500 lb moves continuously from the end of the 60-in beam to a point 10 in from the support.
8. A part of a bracket in the seat assembly of a bus is shown in Figure P5–8. The load varies from 1450 to 140 N as passengers enter and exit the bus.

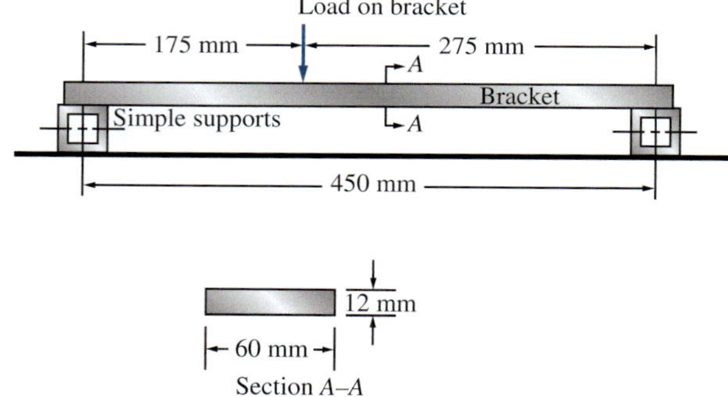

FIGURE P5–8 Seat bracket (Problems 8, 19, and 20)

9. A flat steel strip is used as a spring to maintain a force against part of a cabinet latch in a commercial printer as shown in Figure P5–9. When the cabinet door is open, the spring is deflected by the latch pin by an amount $y_1 = 0.25$ mm. The pin causes the deflection to increase to 0.40 mm when the door is closed.

Endurance Strength

For Problems 10–14, use the method outlined in Section 5–4 to determine the expected actual endurance strength for the material.

10. Compute the estimated endurance strength for a 0.75-in-diameter rod made from SAE 1040 cold-drawn steel. It is to be subjected to repeated and reversed bending stress. A reliability of 99% is desired.

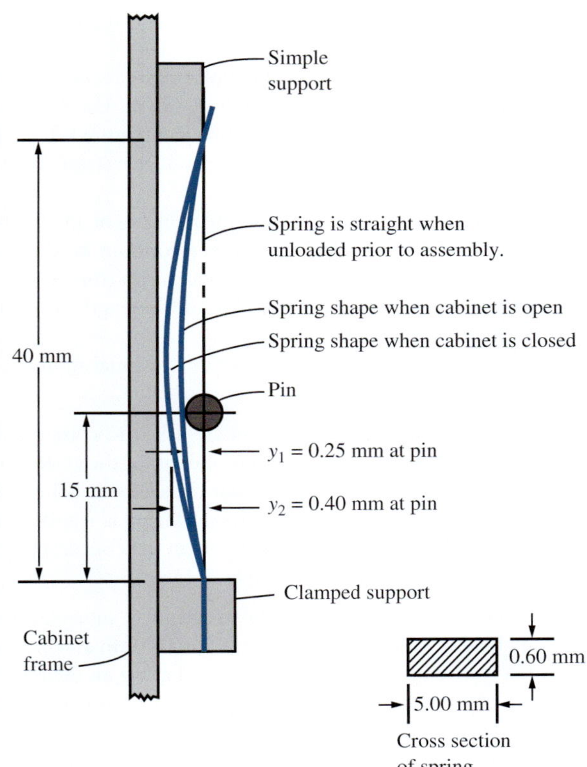

FIGURE P5–9 Cabinet latch spring (Problems 9 and 22)

11. Compute the estimated actual endurance strength for SAE 5160 OQT 1300 steel rod with a diameter of 20.0 mm. It is to be machined and subjected to repeated and reversed bending stress. A reliability of 99.9% is desired.
12. Compute the estimated actual endurance strength for SAE 4130 WQT 1300 steel bar with a rectangular cross section of 20.0 mm by 60 mm. It is to be machined and subjected to repeated and reversed bending stress. A reliability of 99% is desired.
13. Compute the estimated actual endurance strength for SAE 301 stainless steel rod, 1/2 hard, with a diameter of 0.60 in. It is to be machined and subjected to repeated axial tensile stress. A reliability of 99.9% is desired.
14. Compute the estimated actual endurance strength for a machined rectangular steel bar (ASTM A242) 3.5 in high by 0.375 in thick, subjected to repeated and reversed bending stress. A reliability of 99% is desired.

Design and Analysis

15. A link in a mechanism is to be subjected to a tensile force that varies from 3500 to 500 N in a cyclical fashion as the mechanism runs. It has been decided to use SAE 1040 cold-drawn steel. Complete the design of the link, specifying a suitable cross-sectional shape and dimensions.
16. A circular rod is to support part of a storage shelf in a warehouse. As products are loaded and unloaded, the rod is subjected to a tensile load that varies from 1800 to 150 lb. Specify a suitable shape, material, and dimensions for the rod.
17. A strut in a space frame sees a load that varies from a tensile force of 20.0 kN to a compressive force of 8.0 kN. Specify a suitable shape, material, and dimensions for the strut.
18. A part of a latch for a car door is to be made from a straight bar. With each actuation, it sees a tensile force that varies from 780 to 360 N. Small size is important. Complete the design, and specify a suitable shape, material, and dimensions for the rod.
19. A part of a bracket in the seat assembly of a bus is shown in Figure P5–8. The load varies from 1450 to 140 N as passengers enter and exit the bus. The bracket is made from SAE 1020 hot-rolled steel. Determine the resulting design factor.
20. For the bus seat bracket described in Problem 19 and shown in Figure P5–8, propose an alternate design for the bracket, different from that shown in the figure, to achieve a lighter design with a design factor of approximately 4.0.
21. A cantilevered boom is part of an assembly machine. A tool with a weight of 500 lb moves continuously from the end of the 60-in beam to a point 10 in from the support. Specify a suitable design for the boom, giving the material, the cross-sectional shape, and the dimensions.
22. A flat steel strip is used as a spring to maintain a force against part of a cabinet latch in a commercial printer as shown in Figure P5–9. When the cabinet door is open, the spring is deflected by the latch pin by an amount $y_1 = 0.25$ mm. The pin causes the deflection to increase to 0.40 mm when the door is closed. Specify a suitable material for the spring if it is made to the dimensions shown in the figure.
23. A part of the structure for a factory automation system is a beam that spans 30.0 in as shown in Figure P5–6. Loads are applied at two points, each 8.0 in from a support. The left load $F_1 = 1800$ lb remains constantly applied, while the right load $F_2 = 1800$ lb is applied and removed frequently as the machine cycles. If the rectangular tube is made from ASTM A500 Grade B steel, is the proposed design satisfactory? Improve the design to achieve a lighter beam.
24. Figure P5–24 shows a hydraulic cylinder that pushes a heavy tool during the outward stroke, placing a compressive load of 400 lb in the piston rod. During the return stroke, the rod pulls on the tool with a force of 1500 lb. Compute the resulting design factor for the 0.60-in-diameter rod when subjected to this pattern of forces for many cycles. The material is SAE 4130 WQT 1300 steel. If the resulting design factor is much different from 4.0, determine the size of rod that would produce $N = 4.0$.

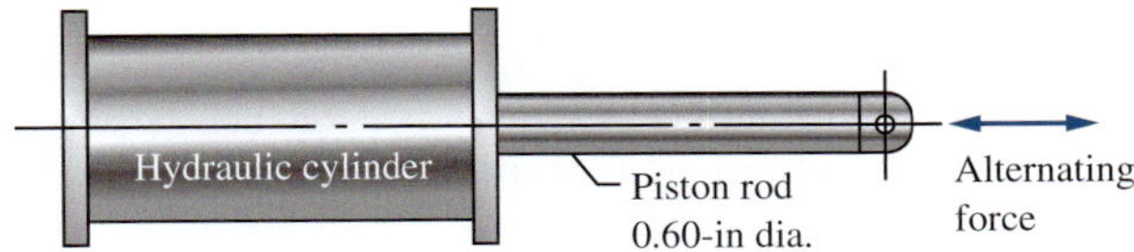

FIGURE P5–24 (Problem 24)

25. The cast iron cylinder shown in Figure P5–25 carries only an axial compressive load of 75 000 lb. (The torque $T = 0$.) Compute the design factor if it is made from gray cast iron, Grade 40A, having a tensile ultimate strength of 40 ksi and a compressive ultimate strength of 140 ksi.
26. Repeat Problem 25, except using a tensile load with a magnitude of 12 000 lb.
27. Repeat Problem 25, except using a load that is a combination of an axial compressive load of 75 000 lb and a torsion of 20 000 lb · in.
28. The shaft shown in Figure P5–28 is supported by bearings at each end, which have bores of 20.0 mm. Design the shaft to carry the given load if it is steady and the shaft is stationary.

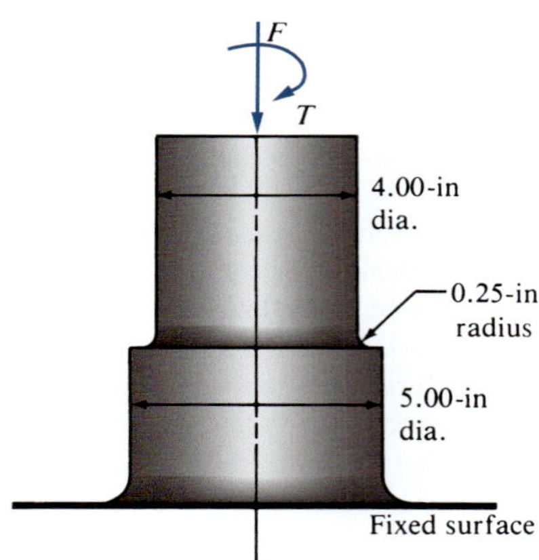

FIGURE P5–25 (Problems 25, 26, and 27)

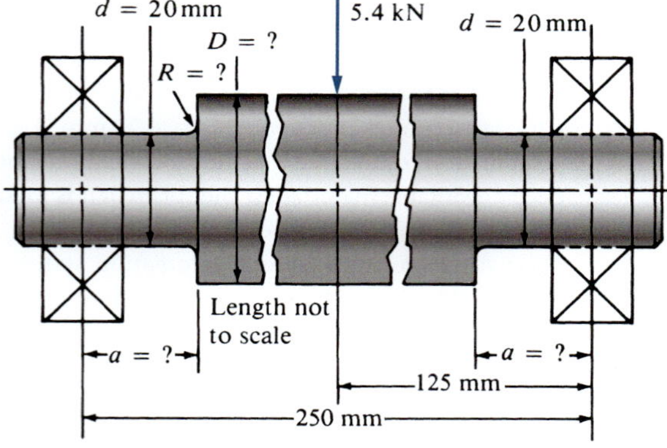

FIGURE P5–28 (Problems 28, 29, and 30)

Make the dimension a as large as possible while keeping the stress safe. Determine the required diameter in the middle portion. The maximum fillet permissible is 2.0 mm. Use SAE 1137 cold-drawn steel. Use a design factor of 3.

29. Repeat Problem 28, except using a rotating shaft.
30. Repeat Problem 28, except using a shaft that is rotating and transmitting a torque of 150 N · m from the left bearing to the middle of the shaft. Also, there is a profile keyseat at the middle under the load.
31. Figure P5–31 shows a proposed design for a seat support. The vertical member is to be a standard hollow circular shape selected from either Appendix 15–17 (steel pipe) or 15–18 (mechanical tubing). The two loads are static and act simultaneously. The material is similar to SAE 1020 hot-rolled steel. Use a design factor of 3.

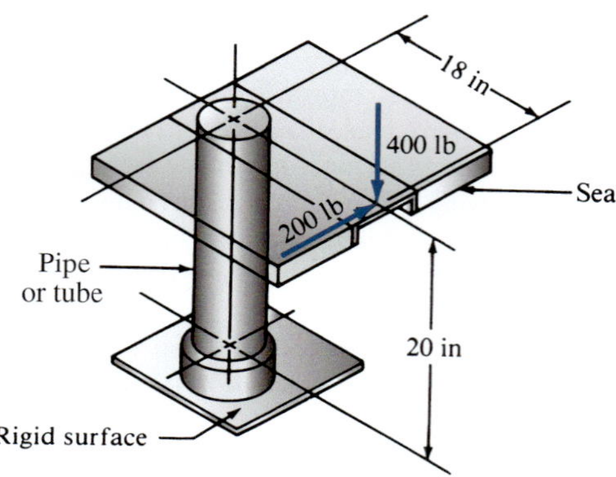

FIGURE P5–31 (Problem 31)

32. A torsion bar is to have a solid circular cross section. It is to carry a fluctuating torque from 30 to 65 N · m. Use SAE 4140 OQT 1000 for the bar, and determine the required diameter for a design factor of 2. Attachments produce a stress concentration of 2.5 near the ends of the bar.
33. Determine the required size for a square bar to be made from SAE 1213 cold-drawn steel. It carries a constant axial tensile load of 1500 lb and a bending load that varies from zero to a maximum of 800 lb at the center of the 48-in length of the bar. Use a design factor of 3.
34. Repeat Problem 33, but add a constant torsional moment of 1200 lb · in to the other loads.

In some of the following problems, you are asked to compute the design factor resulting from the design proposed for the given loading. Unless stated otherwise, assume that the element being analyzed has a machined surface. If the design factor is significantly different from $N = 3$, redesign the component to achieve approximately $N = 3$. (See the figures in Chapter 3.)

35. A tensile member in a machine structure is subjected to a steady load of 4.50 kN. It has a length of 750 mm and is made from a steel tube, SAE 1040 hot-rolled, having an outside diameter of 18 mm and an inside diameter of 12 mm. Compute the resulting design factor.
36. A steady tensile load of 5.00 kN is applied to a square bar, 12 mm on a side and having a length of 1.65 m. Compute the stress in the bar and the resulting design factor if it is made from (a) SAE 1020 hot-rolled steel; (b) SAE 8650 OQT 1000 steel; (c) ductile iron A536 (60-40-18); (d) aluminum alloy 6061-T6; (e) titanium alloy Ti-6Al-4V, annealed; (f) rigid PVC plastic; and (g) phenolic plastic.
37. An aluminum rod, made from alloy 6061-T6, is made in the form of a hollow square tube, 2.25 in outside with a wall thickness of 0.125 in. Its length is 16.0 in. It carries an axial compressive force of 12 600 lb. Compute the resulting design factor. Assume that the tube does not buckle.
38. Compute the design factor in the middle portion only of the rod AC in Figure P3–8 if the steady vertical force on the boom is 2500 lb. The rod is rectangular, 1.50 in by 3.50 in, and is made from SAE 1144 cold-drawn steel.
39. Compute the forces in the two angled rods in Figure P3–9 for a steady applied force, $F = 1500$ lb, if the angle θ is 45°. Then design the middle portion of each rod to be circular and made from SAE 1040 hot-rolled steel. Specify a suitable diameter.
40. Repeat Problem 39 if the angle θ is 15°.
41. Figure 3–27 shows a portion of a circular bar that is subjected to a repeated and reversed force of 7500 N. If the bar is made from SAE OQT 1000, compute the resulting design factor.
42. Compute the torsional shear stress in a circular shaft having a diameter of 50 mm when subjected to a torque of 800 N · m. If the torque is completely reversed and repeated, compute the resulting design factor. The material is SAE 1040 WQT 1000.
43. If the torque in Problem 42 fluctuates from zero to the maximum of 800 N · m, compute the resulting design factor.
44. Compute the torsional shear stress in a circular shaft 0.40 in in diameter that is due to a steady torque of 88.0 lb · in. Specify a suitable aluminum alloy for the rod.
45. Compute the required diameter for a solid circular shaft if it is transmitting a maximum of 110 hp at a speed of 560 rpm. The torque varies from zero to the maximum. There are no other significant loads on the shaft. Use SAE 4130 WQT 700.

46. Specify a suitable material for a hollow shaft with an outside diameter of 40 mm and an inside diameter of 30 mm when transmitting 28 kilowatts (kW) of steady power at a speed of 45 radians per second (rad/s).
47. Repeat Problem 46 if the power fluctuates from 15 to 28 kW.
48. Figure P5–48 shows part of a support bar for a heavy machine, suspended on springs to soften applied loads. The tensile load on the bar varies from 12 500 lb to a minimum of 7500 lb. Rapid cycling for many million cycles is expected. The bar is made from SAE 6150 OQT 1300 steel. Compute the design factor for the bar in the vicinity of the hole.

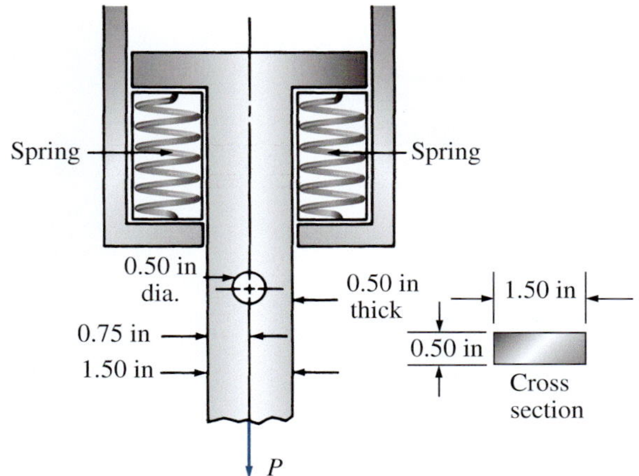

FIGURE P5–48 (Problem 48)

49. Figure P3–61 shows a valve stem from an engine subjected to an axial tensile load applied by the valve spring. The force varies from 0.80 to 1.25 kN. Compute the resulting design factor at the fillet under the shoulder. The valve is made from SAE 8650 OQT 1300 steel.
50. A conveyor fixture shown in Figure P3–62 carries three heavy assemblies (1200 lb each). The fixture is machined from SAE 1144 OQT 900 steel. Compute the resulting design factor in the fixture, considering stress concentrations at the fillets and assuming that the load acts axially. The load will vary from zero to the maximum as the conveyor is loaded and unloaded.
51. For the flat plate in tension in Figure P3–63, compute the minimum resulting design factor, assuming that the holes are sufficiently far apart that their effects do not interact. The plate is machined from stainless steel, UNS S17400 in condition H1150. The load is repeated and varies from 4000 to 6200 lb.

For Problems 52–56, select a suitable material for the member, considering stress concentrations, for the given loading to produce a design factor of $N = 3$.

52. Use Figure P3–64. The load is steady. The material is to be some grade of gray cast iron, ASTM A48.
53. Use Figure P3–65. The load varies from 20.0 to 30.3 kN. The material is to be titanium.
54. Use Figure P3–66. The torque varies from zero to 2200 lb. in. The material is to be steel.
55. Use Figure P3–67. The bending moment is steady. The material is to be ductile iron, ASTM A536.
56. Use Figure P3–68. The bending moment is completely reversed. The material is to be stainless steel.
57. Figure P5–57 shows part of an automatic screwdriver designed to drive several million screws. The maximum torque required to drive a screw is 100 lb · in. Compute the design factor for the proposed design if the part is made from SAE 8740 OQT 1000.
58. The beam in Figure P5–58 carries two steady loads, $P = 750$ lb. Evaluate the design factor that would result if the beam were made from class 40A gray cast iron.
59. A tension link is subjected to a repeated, one-direction load of 3000 lb. Specify a suitable material if the link is to be steel and is to have a diameter of 0.50 in.
60. One member of an automatic transfer device in a factory must withstand a repeated tensile load of 800 lb and must not elongate more than 0.010 inch in its 25.0-in length. Specify a suitable steel material and the dimensions for the rod if it has a square cross section.
61. Figure P5–61 shows two designs for a beam to carry a repeated central load of 1200 lb. Which design would have the highest design factor for a given material?

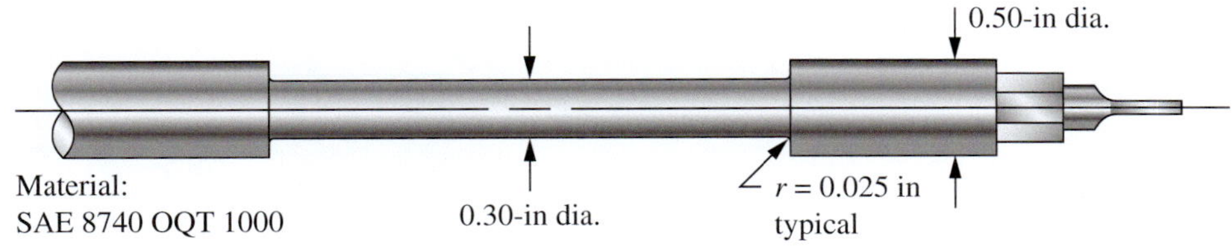

FIGURE P5–57 Screwdriver for Problem 57

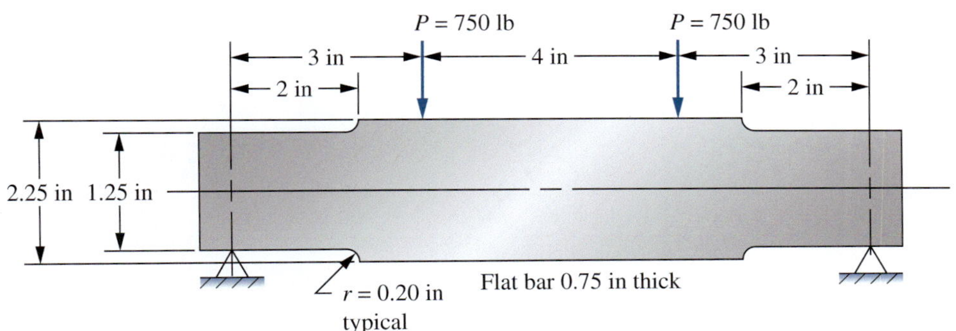

FIGURE P5–58 Beam for Problem 58

62. Refer to Figure P5–61. By reducing the 8.0-in dimension, redesign the beam in Part (b) of the figure so that it has a design factor equal to or higher than that for the design in Part (a).
63. Refer to Figure P5–61. Redesign the beam in Part (b) of the figure by first increasing the fillet radius to 0.40 in and then by reducing the 8.0-in dimension so that the new design has a design factor equal to or higher than that for the design in Part (a).
64. The part shown in Figure P5–64 is made from SAE 1040 HR steel. It is to be subjected to a repeated, one-direction force of 5000 lb applied through two 0.25-in-diameter pins in the holes at each end. Compute the resulting design factor.
65. For the part described in Problem 64, make at least three improvements in the design that will significantly reduce the stress without increasing the weight. The dimensions marked © are critical and cannot be changed. After the redesign, specify a suitable material to achieve a design factor of at least 3.
66. The link shown in Figure P5–66 is subjected to a tensile force that varies from 3.0 to 24.8 kN. Evaluate the design factor if the link is made from SAE 1040 CD steel.

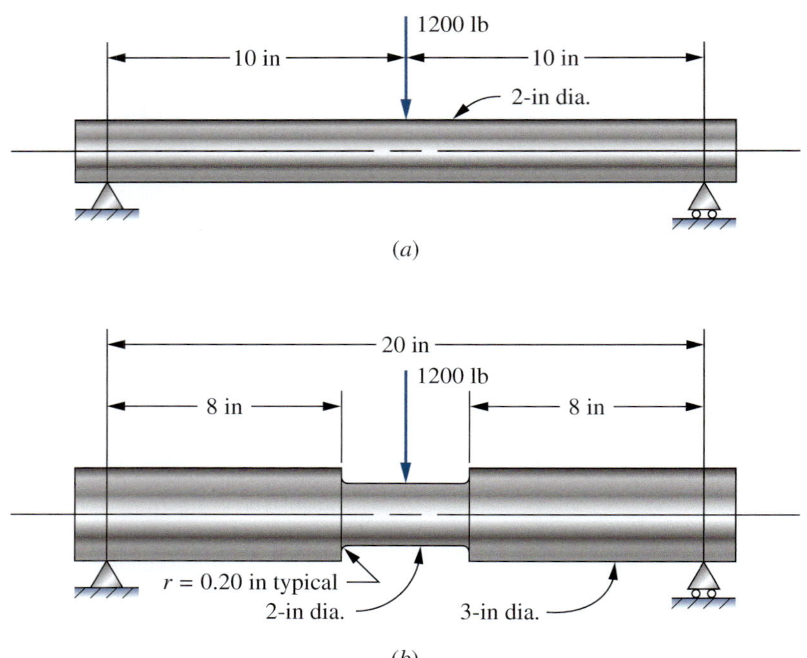

FIGURE P5–61 Beam for Problems 61, 62, and 63

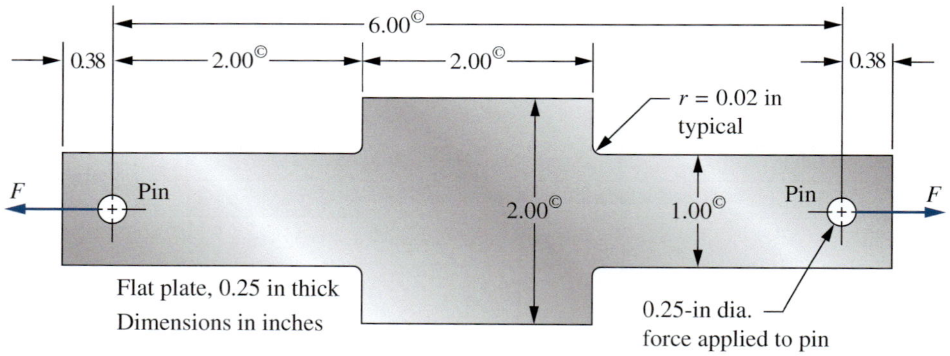

FIGURE P5–64 Beam for Problems 64 and 65

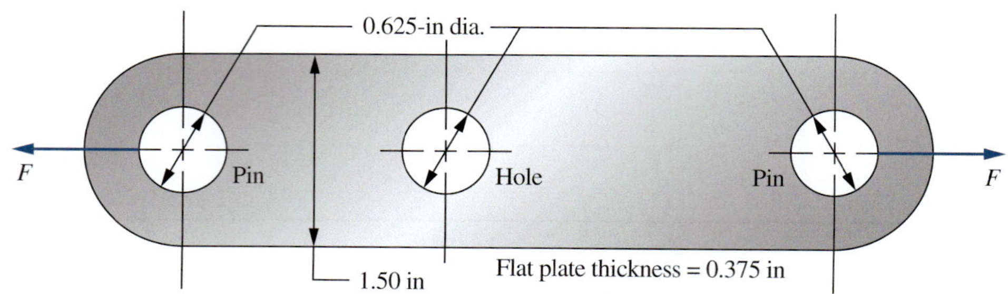

FIGURE P5–66 Link for Problem 66

67. The beam shown in Figure P5–67 carries a repeated, reversed load of 400 N applied at section *C*. Compute the resulting design factor if the beam is made from SAE 1340 OQT 1300.
68. For the beam described in Problem 67, change the steel's tempering temperature to achieve a design factor of at least 2.5.
69. The cantilever shown in Figure P5–69 carries a downward load that varies from 300 to 700 lb. Compute the resulting design factor if the bar is made from SAE 1050 HR steel.
70. For the cantilever described in Problem 69, increase the size of the fillet radius to improve the design factor to at least 3.0 if possible.
71. For the cantilever described in Problem 69, specify a suitable material to achieve a design factor of at least 3.0 without changing the geometry of the beam.
72. Figure P5–72 shows a rotating shaft carrying a steady downward load of 100 lb at *C*. Specify a suitable material.
73. The stepped rod shown in Figure P5–73 is subjected to a direct tensile force that varies from 8500 to 16 000 lb. If the rod is made from SAE 1340 OQT 700 steel, compute the resulting design factor.
74. For the rod described in Problem 73, complete a redesign that will achieve a design factor of at least 3.0.
75. The beam shown in Figure P5–75 carries a repeated, reversed load of 800 lb alternately applied upward and downward. If the beam is made from SAE 1144 OQT 1100, specify the smallest acceptable fillet radius at *A* to ensure a design factor of 3.0.
76. For the beam described in Problem 75, design the section at *B* to achieve a minimum design factor of 3.0. Specify the shape, dimensions, and fillet radius where the smaller part joins the 2.00-by-2.00-in section.

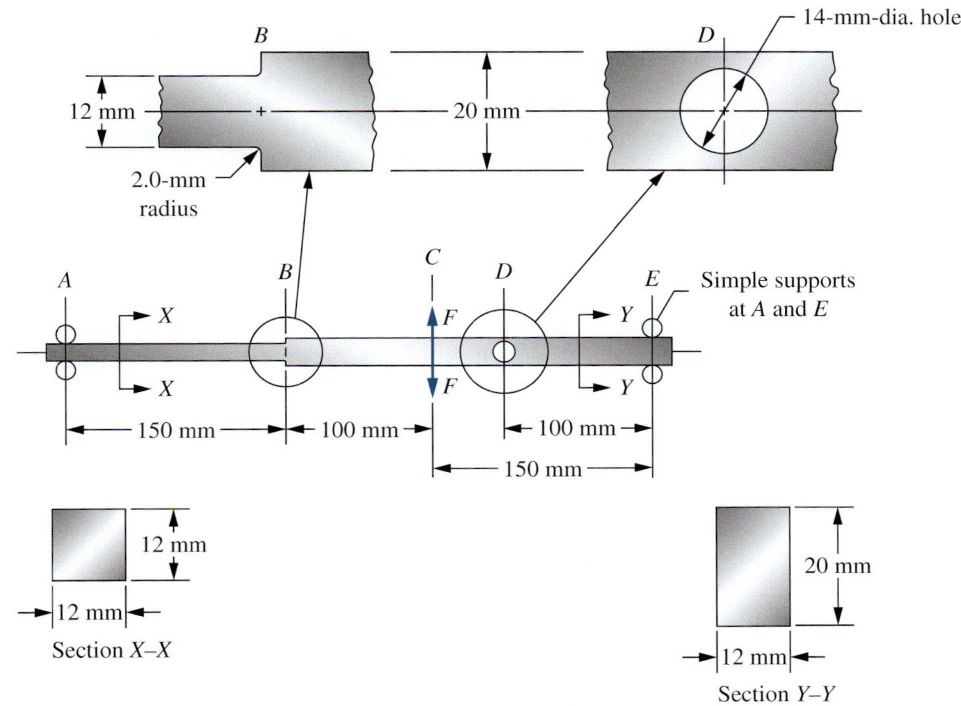

FIGURE P5–67 Beam for Problems 67 and 68

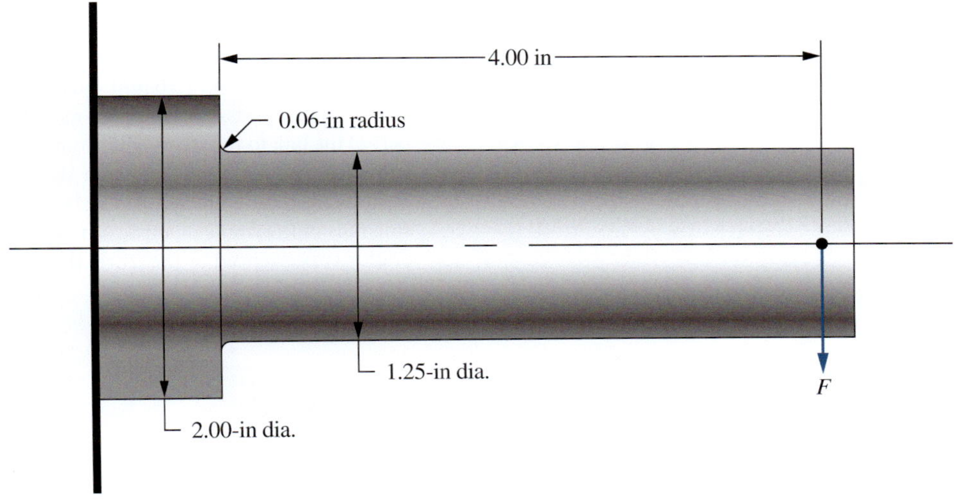

FIGURE P5–69 Cantilever for Problems 69, 70, and 71

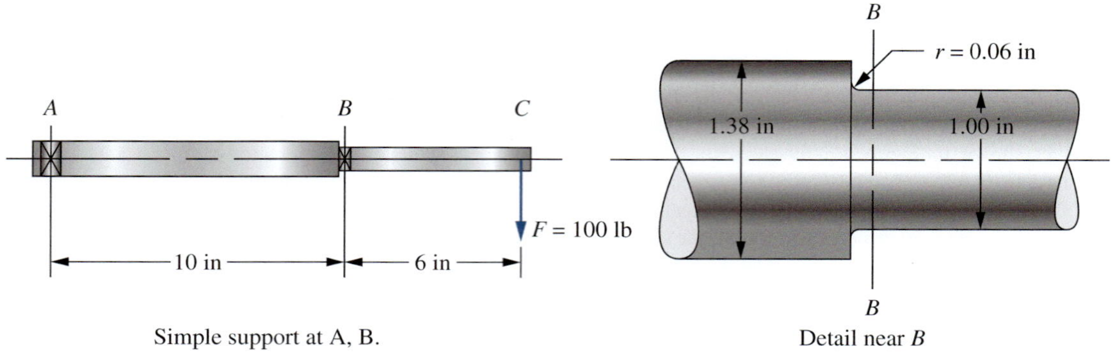

FIGURE P5–72 Shaft for Problem 72

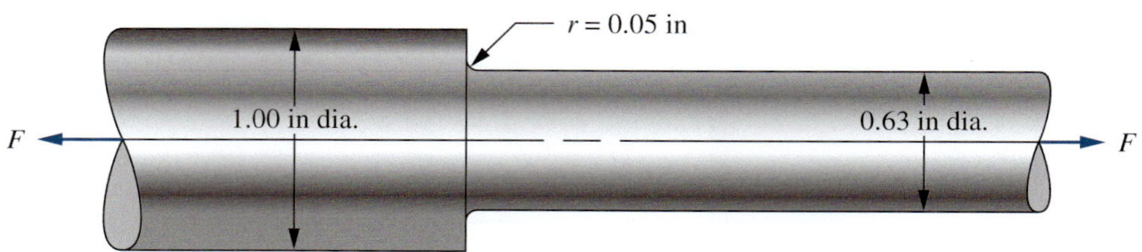

FIGURE P5–73 Rod for Problems 73 and 74

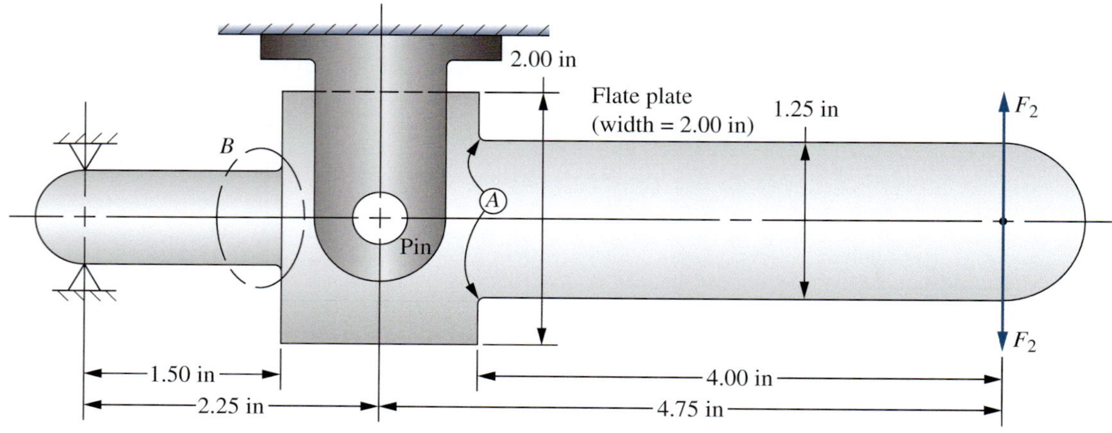

FIGURE P5–75 Beam for Problems 75 and 76

Design Problems

For each of the following problems, complete the requested design to achieve a minimum design factor of 3.0. Specify the shape, the dimensions, and the material for the part to be designed. Work toward an efficient design that will have a low weight.

77. The link shown in Figure P5–77 carries a load of 3000 N that is applied and released many times. The link is machined from a square bar, 12.0 mm on a side, from SAE 1144 OQT 1100 steel. The ends must remain 12.0 mm square to facilitate the connection to mating parts. It is desired to reduce the size of the middle part of the link to reduce weight. Complete the design.

78. Complete the design of the beam shown in Figure P5–78 to carry a large hydraulic motor. The beam is attached to the two side rails of the frame of a truck. Because of the vertical accelerations experienced by the truck, the load on the beam varies

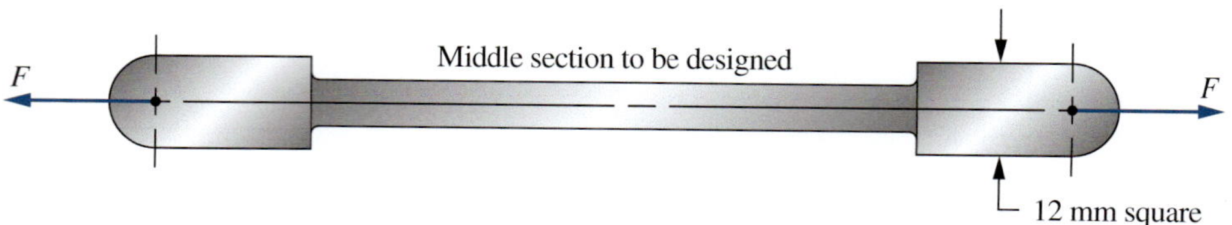

FIGURE P5–77 Link for Problem 77

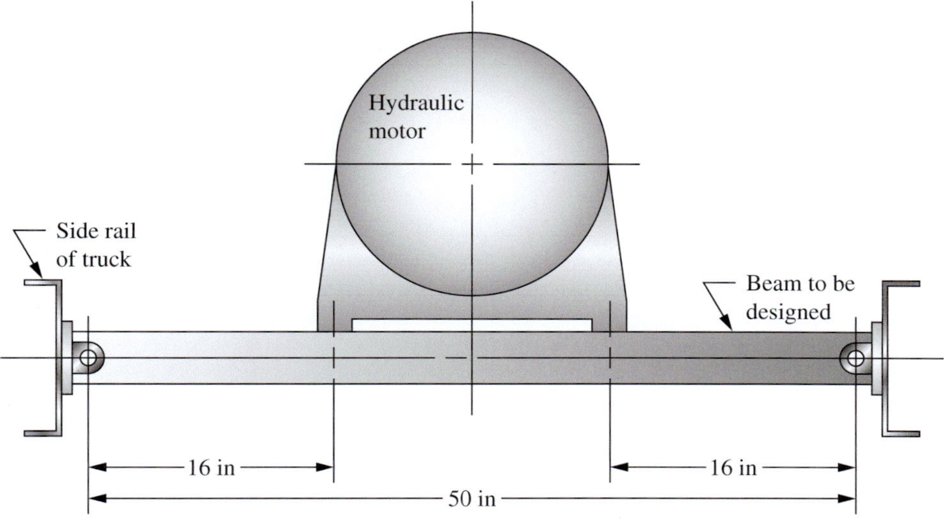

FIGURE P5–78 Beam for Problem 78

from 1200 lb upward to 5000 lb downward. One-half of the load is applied to the beam by each foot of the motor.

79. A tensile member in a truss frame is subjected to a load that varies from 0 to 6500 lb as a traveling crane moves across the frame. Design the tensile member.

80. A hanger for a conveyor system extends outward from two supports as shown in Figure P5–80. The load at the right end varies from 600 to 3800 lb. Design the hanger.

81. Figure P5–81 shows a yoke suspended beneath a crane beam by two rods. Design the yoke if the loads are applied and released many times.

82. For the system shown in Figure P5–81, design the two vertical rods if the loads are applied and released many times.

83. Design the connections between the rods and the yoke and the crane beam shown in Figure P5–81.

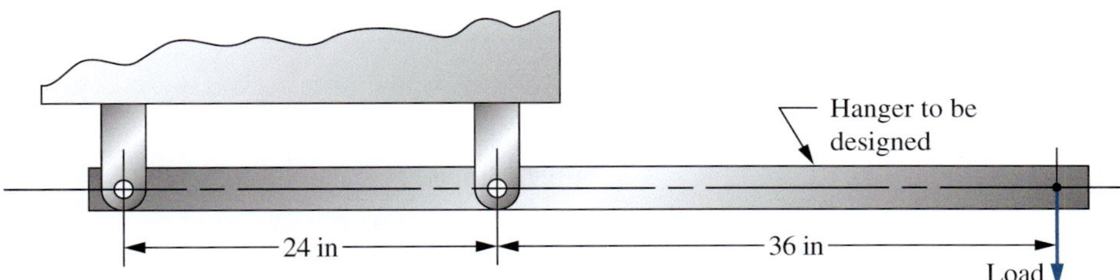

FIGURE P5–80 Hanger for the conveyor system for Problem 80

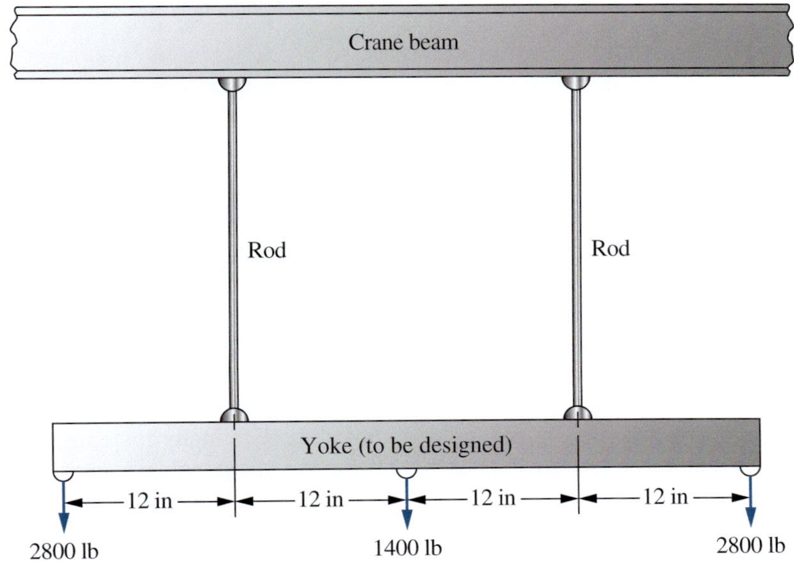

FIGURE P5–81 Yoke and rods for Problems 81, 82, and 83

CHAPTER
SIX

COLUMNS

THE BIG PICTURE Columns

Discussion Map

- A column is a long, slender member that carries an axial compressive load and that fails due to buckling rather than due to failure of the material of the column.

Discover

Find at least 10 examples of columns. Describe them and how they are loaded, and discuss them with your colleagues.

Try to find at least one column that you can load conveniently by hand, and observe the buckling phenomenon.

Discuss with your colleagues the variables that seem to affect how a column fails and how much load it can carry before failing.

This chapter will help you acquire some of the analytical tools necessary to design and analyze columns.

A *column* is a structural member that carries an axial compressive load and that tends to fail by elastic instability, or buckling, rather than by crushing of material. *Elastic instability* is the condition of failure in which the shape of the column is insufficiently rigid to hold it straight under load. At the point of buckling, a radical deflection of the axis of the column occurs suddenly. Then, if the load is not reduced, the column will collapse. Obviously this kind of catastrophic failure must be avoided in structures and machine elements.

Columns are ideally straight and relatively long and slender. If a compression member is so short that it does not tend to buckle, failure analysis must use the methods presented in Chapter 5. This chapter presents several methods of analyzing and designing columns to ensure safety under a variety of loading conditions.

Take a few minutes to visualize examples of column buckling. Find any object that appears to be long and slender—for example, a meter stick, a plastic ruler, a long wooden dowel with a small diameter, a drinking straw, or a thin metal or plastic rod. Carefully apply a downward load on your column while resting the bottom on a desk or the floor. Try to make sure that it does not slide. Gradually increase the load, and observe the behavior of the column until it begins to bend noticeably in the middle. Then hold that level of load. Don't increase it much beyond that level, or the column will likely break!

Now release the load; the column should return to its original shape. The material should not have broken or yielded. But wouldn't you have considered the column to have failed at the point of buckling? Wouldn't it be important to keep the applied load well below the load that caused the buckling to be initiated?

Now look around you. Think of things you are familiar with, or take time to go out and find other examples of columns. Remember, look for relatively long, slender, load-carrying members subjected to compressive loads. Consider parts of furniture, buildings, cars, trucks, toys, play structures, industrial machinery, and construction machinery. Try to find at least 10 examples. Describe their appearance: the material from which they are made, the way they are supported, and the way they are loaded. Do this activity in the classroom or with colleagues; bring the descriptions to class next session for discussion.

Notice that you were asked to find *relatively long, slender,* load-carrying members. How will you know when a member is long and slender? At this point, you should just use your judgment. If the column is available and you are strong enough to load it to buckling, go ahead and try it. Later in this chapter, we will quantify what the terms *long* and *slender* mean.

If the columns you have seen did not actually collapse, what property of the material is highly related to the phenomenon of buckling failure? Remember that the failure was described as *elastic instability*. Then it should seem that the *modulus of elasticity* of the material is a key property, and it is. Review the definition of this property from Chapter 1, and look up representative values in the tables of material properties in Appendices 3–13.

Also note that we specified that the columns are to be initially straight and that loads are to be applied axially. What if these conditions are not met? What if the column is a little crooked before loading? Do you think that it would carry as much compressive loading as one that was straight? Why or why not? What if the column is loaded *eccentrically*, that is, the load is directed off-center, away from the centroidal axis of the column? How will that affect the load-carrying ability? How does the manner of supporting the ends of the column affect its load-carrying ability? What standards exist that guide designers when dealing with columns?

These and other questions will be addressed in this chapter. Any time you are involved in a design in which a compressive load is applied, you should think about analyzing it as a column. The following **You Are the Designer** situation is a good example of such a machine design problem.

YOU ARE THE DESIGNER

You are a member of a team that is designing a commercial compactor to reduce the volume of cardboard and paper waste so that the waste can be transported easily to a processing plant. Figure 6–1 is a sketch of the compaction ram that is driven by a hydraulic cylinder under several thousand pounds of force. The connecting rod between the hydraulic cylinder and the ram must be designed as a column because it is a relatively long, slender compression member. What shape should the cross section of the connecting rod be? From what material should it be made? How is it to be connected to the ram and to the hydraulic cylinder? What are the final dimensions of the rod to be? You, the designer, must specify all of these factors. ■

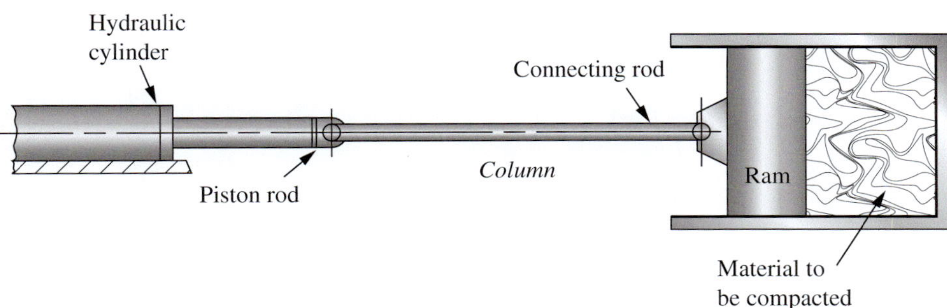

FIGURE 6–1 Waste paper compactor

6–1 OBJECTIVES OF THIS CHAPTER

After completing this chapter, you will be able to:

1. Recognize that any relatively long, slender compression member must be analyzed as a column to prevent buckling.
2. Specify efficient shapes for the cross section of columns.
3. Compute the *radius of gyration* of a column cross section.
4. Specify a suitable value for the *end-fixity factor*, K, and determine the *effective length* of a column.
5. Compute the *slenderness ratio* for columns.
6. Select the proper method of analysis or design for a column based on the manner of loading, the type of support, and the magnitude of the slenderness ratio.
7. Determine whether a column is *long* or *short* based on the value of the slenderness ratio in comparison with the *column constant*.

8. Use the *Euler formula* for the analysis and design of long columns.
9. Use the *J. B. Johnson formula* for the analysis and design of short columns.
10. Analyze crooked columns to determine the allowable load.
11. Analyze columns in which the load is applied with a modest amount of eccentricity to determine the maximum predicted stress and the maximum amount of deflection of the centerline of such columns under load.

6–2 PROPERTIES OF THE CROSS SECTION OF A COLUMN

The tendency for a column to buckle is dependent on the shape and the dimensions of its cross section, along with its length and the manner of attachment to adjacent members or supports. Cross-sectional properties that are important are as follows:

1. The cross-sectional area, A.
2. The moment of inertia of the cross section, I, with respect to the axis about which the value of I is minimum.
3. The least value of the radius of gyration of the cross section, r.

The radius of gyration is computed from

➩ Radius of Gyration

$$r = \sqrt{I/A} \tag{6–1}$$

A column tends to buckle about the axis for which the radius of gyration and the moment of inertia are minimum. Figure 6–2 shows a sketch of a column that has a rectangular cross section. The expected buckling axis is Y-Y because both I and r are much smaller for that axis than for the X-X axis. You can demonstrate this phenomenon by loading a common ruler or meter stick with an axial load of sufficient magnitude to cause buckling. See Appendix 1 for formulas for I and r for common shapes. See Appendix 15 for structural shapes.

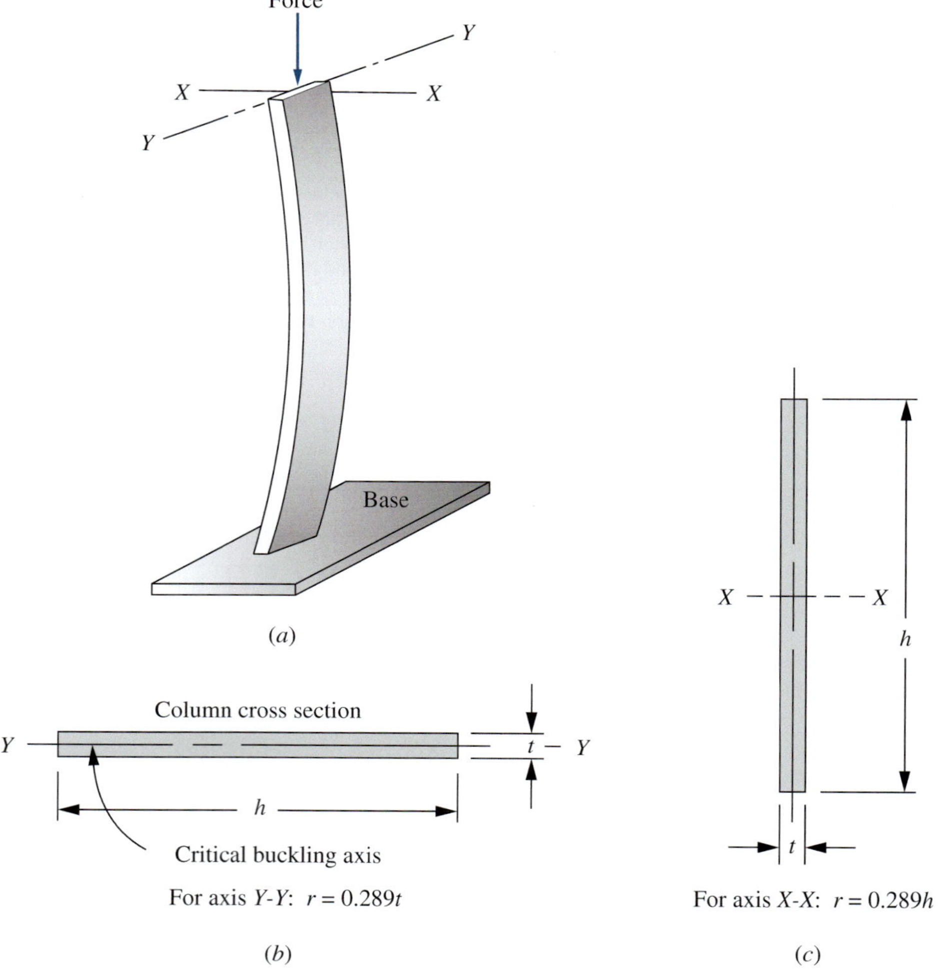

FIGURE 6–2 Buckling of a thin, rectangular column. (a) General appearance of the buckled column. (b) Radius of gyration for Y-Y axis. (c) Radius of gyration for X-X axis

6–3 END FIXITY AND EFFECTIVE LENGTH

The term *end fixity* refers to the manner in which the ends of a column are supported. The most important variable is the amount of restraint offered at the ends of the column to the tendency for rotation. Three forms of end restraint are *pinned*, *fixed*, and *free*.

A *pinned-end* column is guided so that the end cannot sway from side to side, but it offers no resistance to rotation of the end. The best approximation of the pinned end would be a frictionless ball-and-socket joint. A cylindrical pin joint offers little resistance about one axis, but it may restrain the axis perpendicular to the pin axis.

A *fixed end* is one that is held against rotation at the support. An example is a cylindrical column inserted into a tight-fitting sleeve that itself is rigidly supported. The sleeve prohibits any tendency for the fixed end of the column to rotate. A column end securely welded to a rigid base plate is also a good approximation of a fixed-end column.

The *free end* can be illustrated by the example of a flagpole. The top end of a flagpole is unrestrained and unguided, the worst case for column loading.

The manner of support of both ends of the column affects the *effective length* of the column, defined as

➪ Effective Length

$$L_e = KL \qquad \textbf{(6–2)}$$

where L = actual length of the column between supports
K = constant dependent on the end fixity, as illustrated in Figure 6–3

The first values given for K are theoretical values based on the shape of the deflected column. The second values take into account the expected fixity of the column ends in real, practical structures. It is particularly difficult to achieve a true fixed-end column because of the lack of complete rigidity of the support or the means of attachment. Therefore, the higher value of K is recommended.

6–4 SLENDERNESS RATIO

The *slenderness ratio* is the ratio of the effective length of the column to its least radius of gyration. That is,

➪ Slenderness Ratio

$$\text{Slenderness ratio} = L_e/r_{\text{min}} = KL/r_{\text{min}} \qquad \textbf{(6–3)}$$

We will use the slenderness ratio to aid in the selection of the method of performing the analysis of straight, centrally loaded columns.

6–5 TRANSITION SLENDERNESS RATIO

In the following sections, two methods for analyzing straight, centrally loaded columns are presented: (1) the Euler formula for long, slender columns and (2) the J. B. Johnson formula for short columns.

The choice of which method to use depends on the value of the actual slenderness ratio for the column being analyzed in relation to the *transition slenderness ratio*, or *column constant*, C_c, defined as

➪ Column Constant

$$C_c = \sqrt{\frac{2\pi^2 E}{s_y}} \qquad \textbf{(6–4)}$$

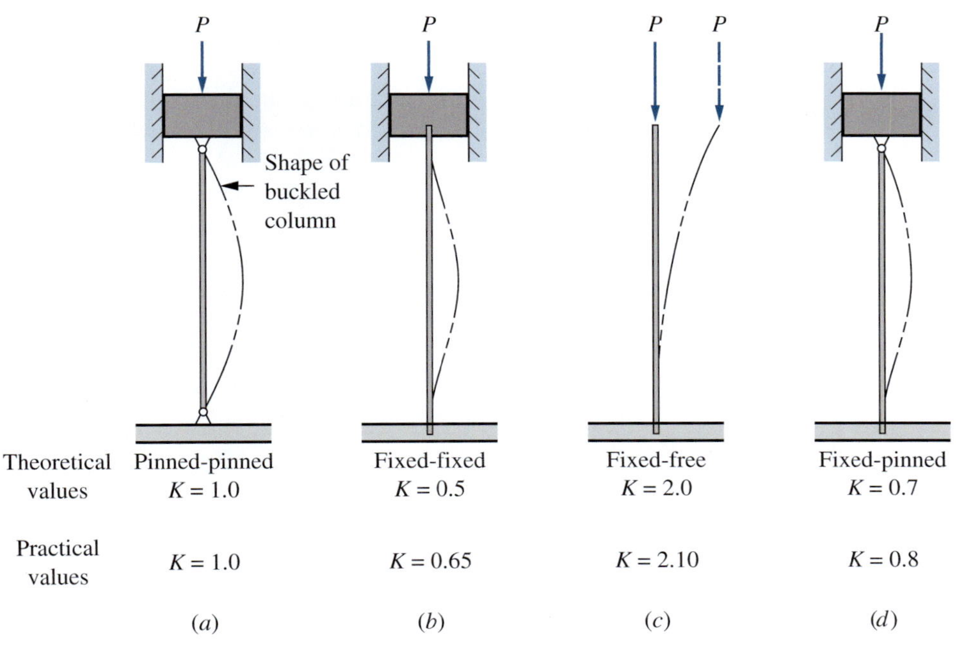

FIGURE 6–3 Values of K for effective length, $L_e = KL$, for different end connections

where E = modulus of elasticity of the material of the column

s_y = yield strength of the material

The use of the column constant is illustrated in the following procedure for analyzing straight, centrally loaded columns.

PROCEDURE FOR ANALYZING STRAIGHT, CENTRALLY LOADED COLUMNS

1. For the given column, compute its actual slenderness ratio.
2. Compute the value of C_c.
3. Compare C_c with KL/r. Because C_c represents the value of the slenderness ratio that separates a long column from a short one, the result of the comparison indicates which type of analysis should be used.
4. If the actual KL/r is greater than C_c, the column is *long*. Use Euler's equation, as described in Section 6–6.
5. If KL/r is less than C_c, the column is *short*. Use the J. B. Johnson formula, described in Section 6–7.

Figure 6–4 is a logical flowchart for this procedure.

The value of the column constant, or transition slenderness ratio, is dependent on the material properties of modulus of elasticity and yield strength. For any given class of material, for example, steel, the modulus of elasticity is nearly constant. Thus, the value of C_c varies inversely as the square root of the yield strength. Figures 6–5 and 6–6 show the resulting values for steel and aluminum, respectively, for the range of yield strengths expected for each material. The figures show that the value of C_c decreases as the yield strength increases. The importance of this observation is discussed in the following section.

6–6 LONG COLUMN ANALYSIS: THE EULER FORMULA

Analysis of a long column employs the Euler formula (see Reference 4):

Euler Formula for Long Columns

$$P_{cr} = \frac{\pi^2 EA}{(KL/r)^2} \tag{6–5}$$

The equation gives the critical load, P_{cr}, at which the column would begin to buckle.

An alternative form of the Euler formula is often desirable. Note that, from Equation (6–5),

$$P_{cr} = \frac{\pi^2 EA}{(KL/r)^2} = \frac{\pi^2 EA}{(KL)^2/r^2} = \frac{\pi^2 EA r^2}{(KL)^2}$$

But, from the definition of the radius of gyration, r,

$$r = \sqrt{I/A}$$

$$r^2 = I/A$$

Then

Alternative Form of Euler Formula

$$P_{cr} = \frac{\pi^2 EAI}{(KL)^2 A} = \frac{\pi^2 EI}{(KL)^2} \tag{6–6}$$

This form of the Euler equation aids in a design problem in which the objective is to specify a size and a shape of a column cross section to carry a certain load. The moment of inertia for the required cross section can be easily determined from Equation (6–6).

Notice that the buckling load is dependent only on the geometry (length and cross section) of the column and the stiffness of the material represented by the modulus of elasticity. The strength of the material is not involved at all. For these reasons, it is often of no benefit to specify a high-strength material in a long column application. A lower-strength material having the same stiffness, E, would perform as well.

Design Factor and Allowable Load

Because failure is predicted to occur at a limiting load, rather than a stress, the concept of a design factor is applied differently than it is for most other load-carrying members. Rather than applying the design factor to the yield strength or the ultimate strength of the material, we apply it to the critical load, from Equation (6–5) or (6-6). For typical machine design applications, a design factor of 3 is used. For stationary columns with well-known loads and end fixity, a lower factor can be used, such as 2.0. A factor of 1.92 is used in some construction applications. Conversely, for very long columns, where there is some uncertainty about the loads or the end fixity, or where special dangers are presented, larger factors are advised. (See References 1 and 2.)

In summary, the objective of column analysis and design is to ensure that the load applied to a column is safe, well below the critical buckling load. The following definitions of terms must be understood:

P_{cr} = critical buckling load

P_a = allowable load

P = actual applied load

N = design factor

Then

Allowable Load

$$P_a = P_{cr}/N$$

The actual applied load, P, must be less than P_a.

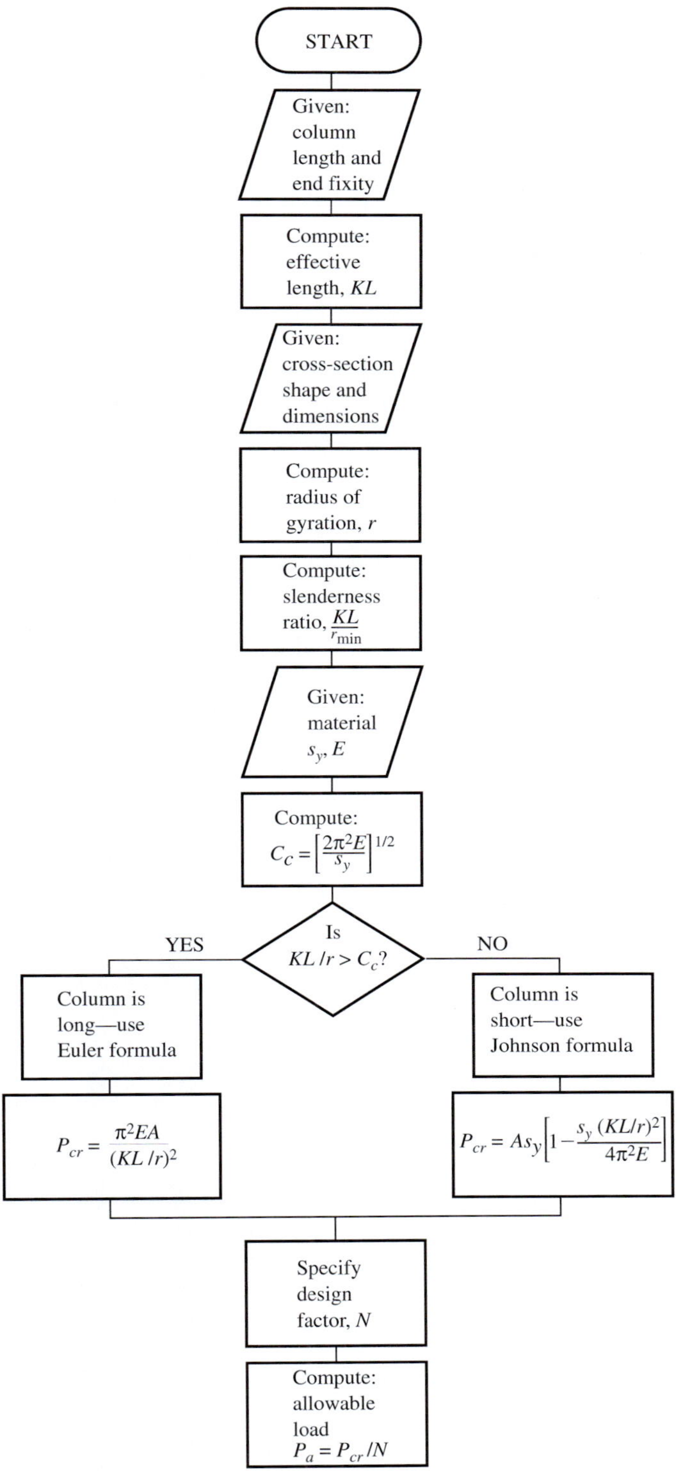

FIGURE 6–4 Analysis of a straight, centrally loaded column

Example Problem 6–1 A column has a solid circular cross section, 1.25 inches in diameter; it has a length of 4.50 ft and is pinned at both ends. If it is made from SAE 1020 cold-drawn steel, what would be a safe column loading?

Solution

Objective: Specify a safe loading for the column.

Given: Solid circular cross section: diameter $= D = 1.25$ in; length $= L = 4.50$ ft.
Both ends of the column are pinned.
Material: SAE 1020 cold-drawn steel.

Analysis Use the procedure in Figure 6–4.

Results **Step 1.** For the pinned-end column, the end-fixity factor is $K = 1.0$. The effective length equals the actual length; $KL = 4.50$ ft $= 54.0$ in.

Step 2. From Appendix 1, for a solid round section,

$$r = D/4 = 1.25/4 = 0.3125 \text{ in}$$

Step 3. Compute the slenderness ratio:

$$\frac{KL}{r} = \frac{1.0(54)}{0.3125} = 173$$

Step 4. Compute the column constant from Equation (6–4). For SAE 1020 cold-drawn steel, the yield strength is 51 000 psi, and the modulus of elasticity is 30×10^6 psi. Then

$$C_c = \sqrt{\frac{2\pi^2 E}{s_y}} = \sqrt{\frac{2\pi^2(30\times10^6)}{51\ 000}} = 108$$

Step 5. Because KL/r is greater than C_c, the column is long, and Euler's formula should be used. The area is

$$A = \frac{\pi D^2}{4} = \frac{\pi(1.25)^2}{4} = 1.23 \text{ in}^2$$

Then the critical load is

$$P_{cr} = \frac{\pi^2 EA}{(KL/r)^2} = \frac{\pi^2(30\times10^6)(1.23)}{(173)^2} = 12\ 200 \text{ lb}$$

At this load, the column should just begin to buckle. A safe load would be a reduced value, found by applying the design factor to the critical load. Let's use $N = 3$ to compute the *allowable load*, $P_a = P_{cr}/N$:

$$P_a = (12\ 200)/3 = 4067 \text{ lb}$$

Comment The safe load on the column is 4067 lb.

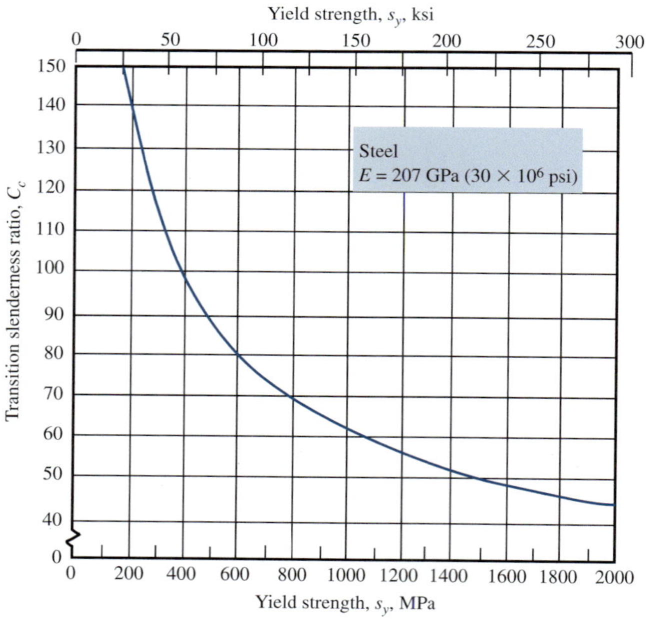

FIGURE 6–5 Transition slenderness ratio C_c vs. yield strength for steel

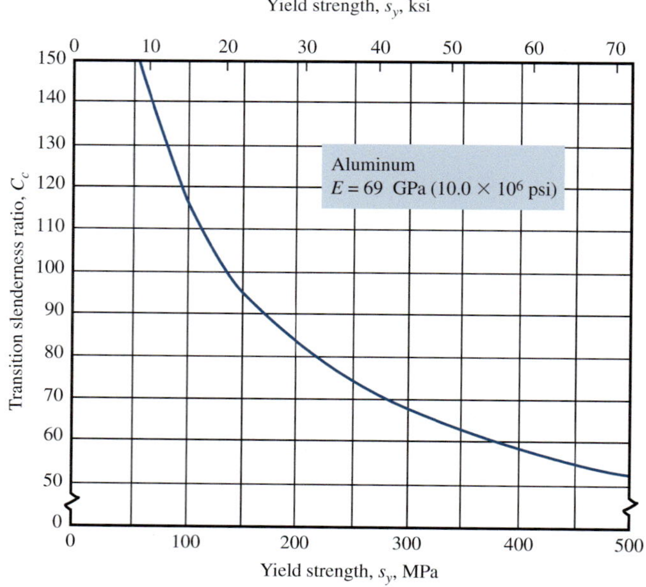

FIGURE 6–6 Transition slenderness ratio C_c vs. yield strength for aluminum

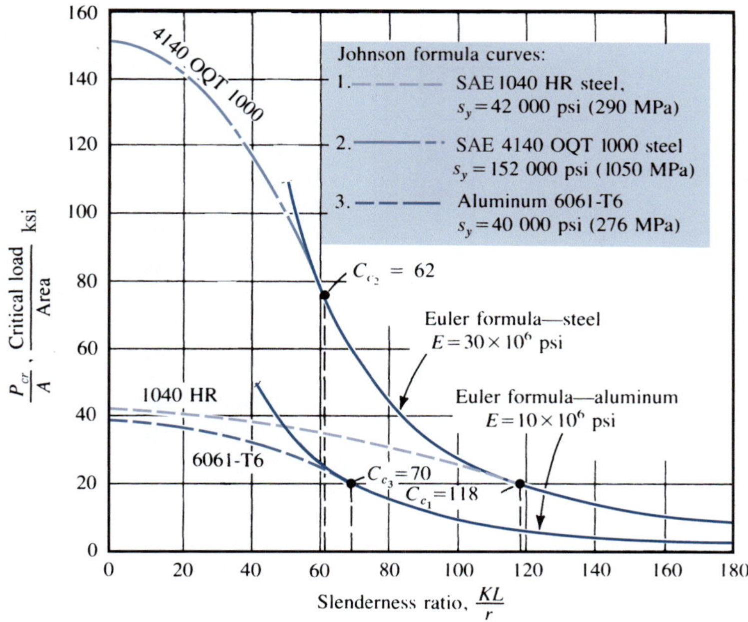

FIGURE 6–7 Johnson formula curves

6–7 SHORT COLUMN ANALYSIS: THE J. B. JOHNSON FORMULA

When the actual slenderness ratio for a column, KL/r, is less than the transition value, C_c, then the column is short, and the J. B. Johnson formula should be used. Use of the Euler formula in this range would predict a critical load greater than it really is.

The J. B. Johnson formula is written as follows:

⇨ J. B. Johnson Formula for Short Columns

$$P_{cr} = As_y\left[1 - \frac{s_y(KL/r)^2}{4\pi^2 E}\right] \quad \textbf{(6–7)}$$

Figure 6–7 shows a plot of the results of this equation as a function of the slenderness ratio, KL/r. Notice that it becomes tangent to the result of the Euler formula at the transition slenderness ratio, the limit of its application. Also, at very low values for the slenderness ratio, the second term of the equation approaches zero, and the critical load approaches the yield load. Curves for three different materials are included in the figure to illustrate the effect of E and s_y on the critical load and the transition slenderness ratio.

The critical load for a short column is affected by the strength of the material in addition to its stiffness, E. As shown in the preceding section, strength is not a factor for a long column when the Euler formula is used.

Note that if a column is to be used as part of a building structure, the methods described in References 1 or 2 must be used.

Example Problem 6–2 Determine the critical load on a steel column having a rectangular cross section, 12 mm by 18 mm, and a length of 280 mm. It is proposed to use SAE 1040 hot-rolled steel. The lower end of the column is inserted into a close-fitting socket and is welded securely. The upper end is pinned (see Figure 6–8).

Solution

Objective Compute the critical load for the column.

Given Solid rectangular cross section: $B = 12$ mm; $H = 18$ mm; $L = 280$ mm.

The bottom of column is fixed; the top is pinned (see Figure 6–8).

Material: SAE 1040 hot-rolled steel.

Analysis Use the procedure in Figure 6–4.

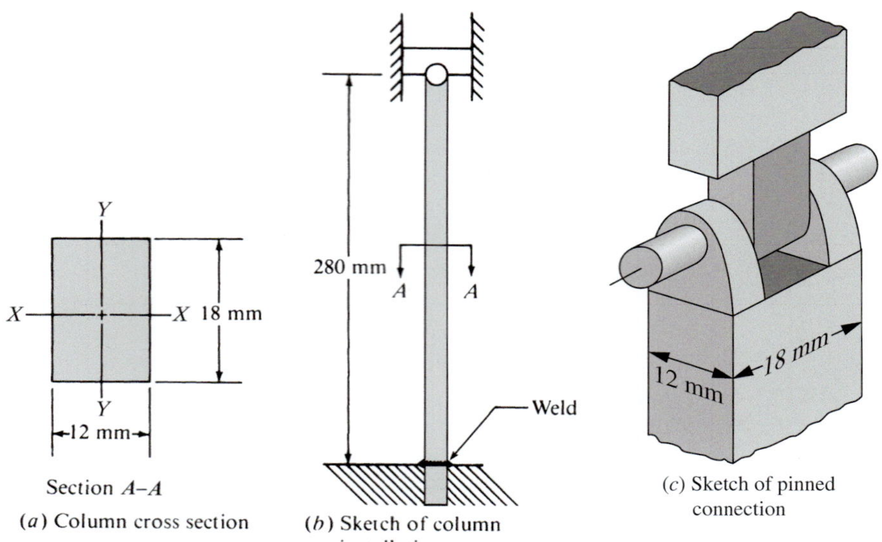

FIGURE 6–8 Steel column

Results **Step 1.** Compute the slenderness ratio. The radius of gyration must be computed about the axis that gives the least value. This is the *Y-Y* axis, for which

$$r = \frac{B}{\sqrt{12}} = \frac{12 \text{ mm}}{\sqrt{12}} = 3.46 \text{ mm}$$

The column has a fixed-pinned end fixity for which $K = 0.8$. Then

$$KL/r = [(0.8)(280)]/3.46 = 64.7$$

Step 2. Compute the transition slenderness ratio. For the SAE 1040 hot-rolled steel, $E = 207$ GPa and $s_y = 290$ MPa. Then, from Equation (6–4),

$$C_c = \sqrt{\frac{2\pi^2(207 \times 10^9 \text{ Pa})}{290 \times 10^6 \text{ Pa}}} = 119$$

Step 3. Then $KL/r < C_c$; thus the column is short. Use the J. B Johnson formula (Eqn. 6–7) to compute the critical load.

$$P_{cr} = As_y\left[1 - \frac{s_y(KL/r)^2}{4\pi^2 E}\right]$$

$$P_{cr} = (216 \text{ mm}^2)(290 \text{ N/mm}^2)\left[1 - \frac{(290 \times 10^6 \text{ Pa})(64.7)^2}{4\pi^2(207 \times 10^9 \text{ Pa})}\right]$$

$$P_{cr} = 53.3 \times 10^3 \text{ N} = 53.3 \text{ kN}$$

Comments This is the critical buckling load. We would have to apply a design factor to determine the allowable load. Specifying $N = 3$ results in $P_a = 17.8$ kN.

6–8 COLUMN ANALYSIS SPREADSHEET

Completing the process described in Figure 6–4 using a calculator, pencil, and paper is tedious. A spreadsheet automates the calculations after you have entered the pertinent data for the particular column to be analyzed. Figure 6–9 shows the output of a spreadsheet used to solve Example Problem 6–1. The layout of the spreadsheet could be done in many ways, and you are encouraged to develop your own style. The following comments describe the features of the given spreadsheet:

1. At the top of the sheet, instructions to the user are given for entering data and for units. This sheet is for U.S. Customary units only. A different sheet would be used if SI metric data were to be used. (See Figure 6–10, which gives the solution for Example Problem 6–2.)
2. On the left side of the sheet are listed the various data that must be provided by the user to run the calculations.

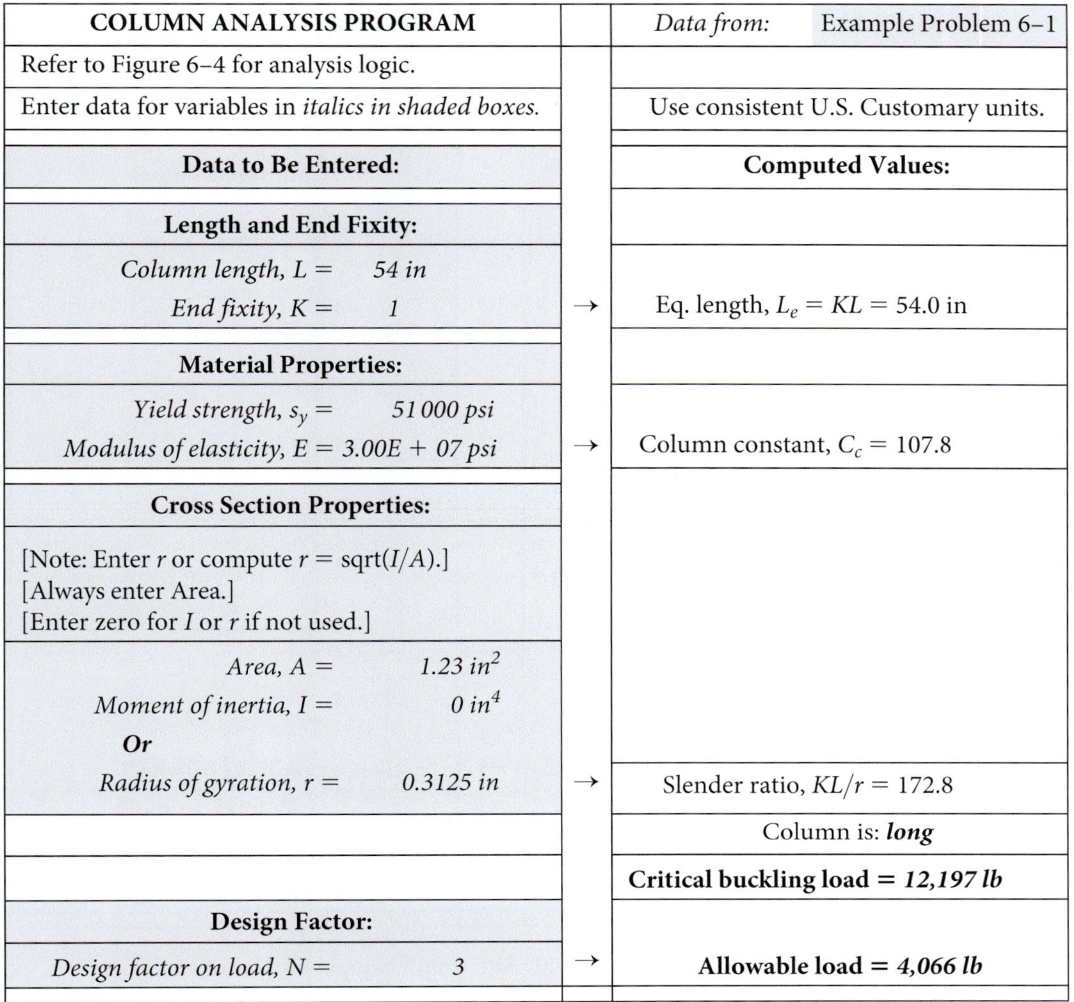

COLUMN ANALYSIS PROGRAM		*Data from:* Example Problem 6–1
Refer to Figure 6–4 for analysis logic.		
Enter data for variables in *italics in shaded boxes.*		Use consistent U.S. Customary units.
Data to Be Entered:		**Computed Values:**
Length and End Fixity:		
Column length, L = *54 in*		
End fixity, K = *1*	→	Eq. length, $L_e = KL = 54.0$ in
Material Properties:		
Yield strength, s_y = *51 000 psi*		
Modulus of elasticity, E = 3.00E + 07 psi	→	Column constant, $C_c = 107.8$
Cross Section Properties:		
[Note: Enter *r* or compute *r* = sqrt(*I*/*A*).] [Always enter Area.] [Enter zero for *I* or *r* if not used.]		
Area, A = *1.23 in*^{2}		
Moment of inertia, I = *0 in*^{4}		
Or		
Radius of gyration, r = *0.3125 in*	→	Slender ratio, $KL/r = 172.8$
		Column is: ***long***
		Critical buckling load = *12,197 lb*
Design Factor:		
Design factor on load, N = *3*	→	**Allowable load = *4,066 lb***

FIGURE 6–9 Spreadsheet for column analysis with data from Example Problem 6–1

On the right are listed the output values. Formulas for computing L_e, C_c, KL/r, and allowable load are written directly into the cell where the computed values show. The output data for the message "Column is: ***long***" and the critical buckling load are produced by *functions* set up within *macros* written in Visual Basic and placed on a separate sheet of the spreadsheet. Figure 6–11 shows the two macros used. The first (*LorS*) carries out the decision process to test whether the column is long or short as indicated by comparison of its slenderness ratio with the column constant. The second (*Pcr*) computes the critical buckling load using either the Euler formula or the J. B. Johnson formula, depending on the result of the *LorS* macro. These functions are called by statements in the cells where "long" and the computed value of the critical buckling load (12 197 lb) are located.

3. Having such a spreadsheet can enable you to analyze several design options quickly. For example, the given problem statement indicated that the ends were pinned, resulting in an end-fixity value of $K = 1$. What would happen if both ends were fixed? Simply changing the value of that one cell to $K = 0.65$ would cause the entire sheet to be recalculated, and the revised value of critical buckling load would be available almost instantly. The result is that $P_{cr} = 28\,868$ lb, an increase of 2.37 times the original value. With that kind of improvement, you, the designer, might be inclined to change the design to produce fixed ends.

6–9 EFFICIENT SHAPES FOR COLUMN CROSS SECTIONS

An *efficient shape* is one that provides good performance with a small amount of material. In the case of columns, the shape of the cross section and its dimensions determine the value of the radius of gyration, *r*. From the definition of the slenderness ratio, KL/r, we can see that as *r* gets larger, the slenderness ratio gets smaller. In the critical load equations, a smaller slenderness ratio results in a larger critical load, the most desirable situation. Therefore, it is desirable to maximize the radius of gyration to design an efficient column cross section.

COLUMN ANALYSIS PROGRAM		*Data from:* Example Problem 6–2
Refer to Figure 6–4 for analysis logic.		
Enter data for variables in *italics in shaded boxes.*		Use consistent SI metric units.
Data to Be Entered:		**Computed Values:**
Length and End Fixity:		
Column length, L = 280 mm		
End fixity, K = 0.8	→	Eq. length, $L_e = KL = 224.0$ mm
Material Properties:		
Yield strength, s_y *= 290 MPa*		
Modulus of elasticity, E = 207 GPa	→	Column constant, $C_c = 118.7$
Cross Section Properties:		
[Note: Enter *r* or compute *r* = sqrt(*I*/*A*).] [Always enter Area.] [Enter zero for *I* or *r* if not used.]		
Area, A = $216\ mm^2$		
Moment of inertia, I = $0\ mm^4$		
Or		
Radius of gyration, r = 3.5 mm	→	Slender ratio, $KL/r = 64.7$
		Column is: ***short***
Design factor:		**Critical buckling load = *53.32 kN***
Design factor on load, N = 3	→	**Allowable load = *17.77 kN***

FIGURE 6–10 Spreadsheet for column analysis with data from Example Problem 6–2

Unless end fixity varies with respect to the axes of the cross section, the column will tend to buckle with respect to the axis with the *least* radius of gyration. So a column with equal values for the radius of gyration in any direction is desirable.

Review again the definition of the radius of gyration:

$$r = \sqrt{I/A}$$

This equation indicates that for a given area of material, we should try to maximize the moment of inertia to maximize the radius of gyration. A shape with a high moment of inertia has its area distributed far away from its centroidal axis.

Shapes that have the desirable characteristics described include circular hollow pipes and tubes, square hollow tubing, and fabricated column sections made from structural shapes placed at the outer boundaries of the section. Solid circular sections and solid square sections are also good, although not as efficient as the hollow sections. Figure 6–12(a–d) illustrates some of these shapes. The built-up section in (e) gives a rigid, boxlike section approximating the hollow square tube in larger sizes. In the case of the section in Figure 6–12(f), the angle sections at the corners provide the greatest contribution to the moment of inertia. The lacing bars merely hold the angles in position. The H-column in (g) has an equal depth and width and relatively heavy flanges and web. The moment of inertia with respect to the *Y-Y* axis is still smaller than for the *X-X* axis, but they are more nearly equal than for most other I-sections designed to be used as beams with bending in only one direction. Thus, this shape would be more desirable for columns.

6–10 THE DESIGN OF COLUMNS

In a design situation, the expected load on the column would be known, along with the length required by the application. The designer would then specify the following:

1. The manner of attaching the ends to the structure that affects the end fixity.
2. The general shape of the column cross section (e.g., round, square, rectangular, and hollow tube).
3. The material for the column.
4. The design factor, considering the application.
5. The final dimensions for the column.

It may be desirable to propose and analyze several different designs to approach an optimum for the application. A computer program or spreadsheet facilitates the process.

```
' Lors Macro
' Determines if column is long or short.
Function LorS(SR, CC)
     If SR > CC Then
        LorS = "long"
     Else
        LorS = "short"
     End if
End Function

' Critical Load Macro
' Uses Euler formula for long columns
' Uses Johnson formula for short columns
Function Pcr(LorS, SR, E, A, Sy)
Const Pi = 3.1415926
     If LorS = "long" Then
        Pcr = Pi ^ 2 * E * A / SR ^ 2
        ' Euler Equation; Eq. (6-4)
     Else
        Pcr = A * Sy(1 - (Sy * SR ^ 2/(4 * Pi ^ 2 * E)))
        ' Johnson Equation; Eq. (6-7)
     End if
End Function
```

FIGURE 6–11 Macros used in the column analysis spreadsheet

It is assumed that items 1 through 4 are specified by the designer for any given trial. For some simple shapes, such as the solid round or square section, the final dimensions are computed from the appropriate formula: the Euler formula, Equation (6–5) or (6–6), or the J. B. Johnson formula, Equation (6–7). If an algebraic solution is not possible, iteration can be done.

In a design situation, the unknown cross-sectional dimensions make computing the radius of gyration and therefore the slenderness ratio, KL/r, impossible. Without the slenderness ratio, we cannot determine whether the column is long (Euler) or short (Johnson). Thus, the proper formula to use is not known.

We overcome this difficulty by making an assumption that the column is either long or short and proceeding with the corresponding formula. Then, after the dimensions are determined for the cross section, the actual value of KL/r will be computed and compared with C_c. This will show whether or not the

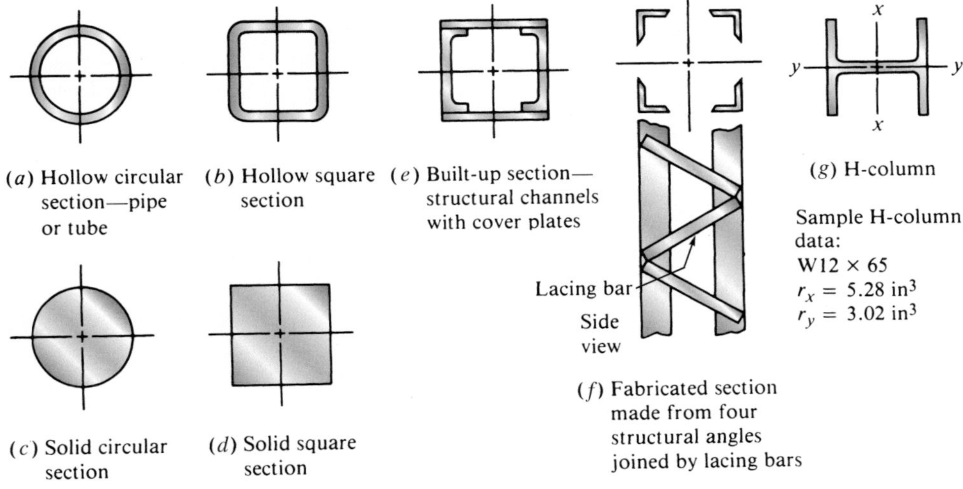

FIGURE 6–12 Column cross sections

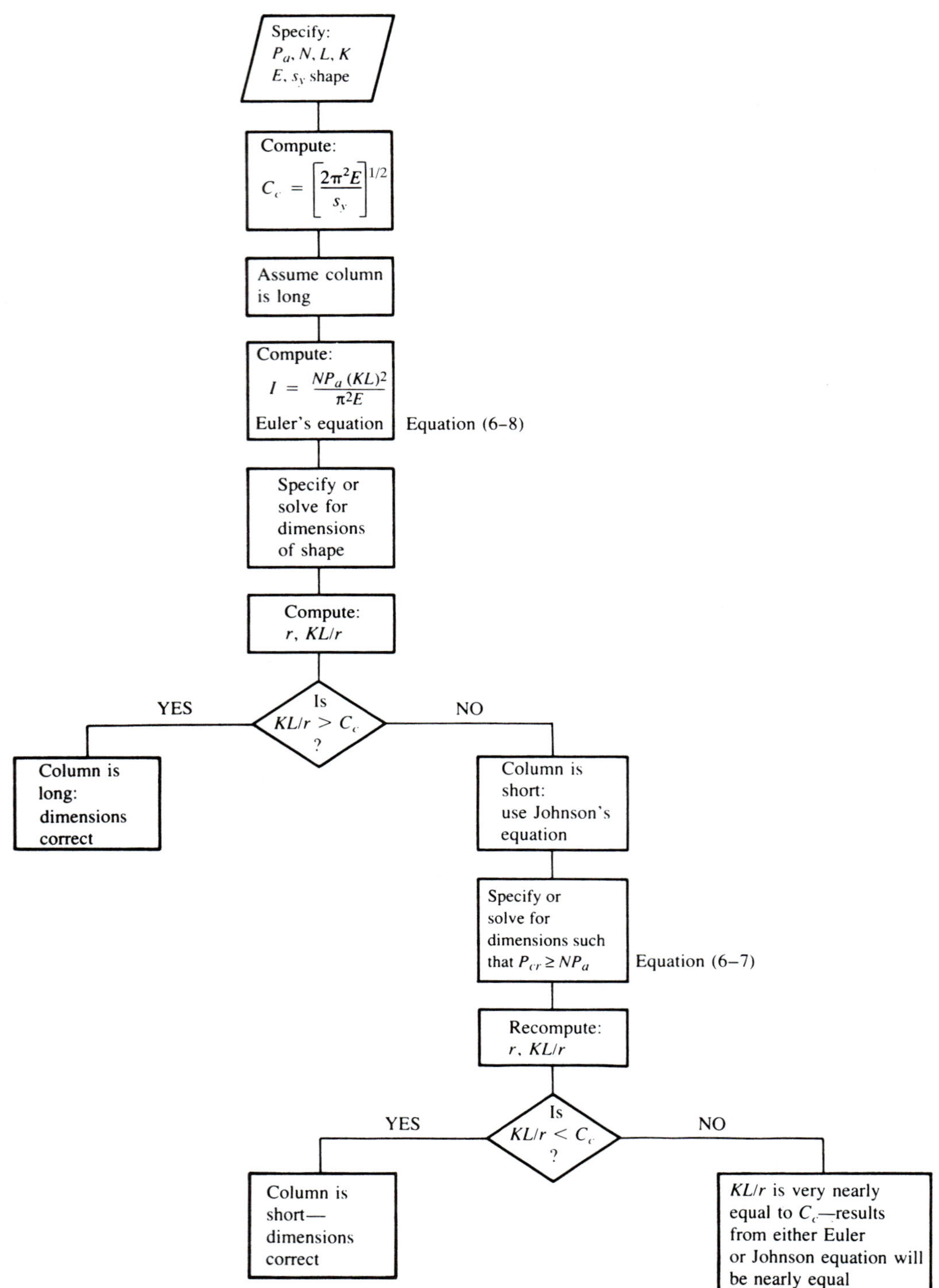

FIGURE 6–13 Design of a straight, centrally loaded column

correct formula has been used. If so, the computed answer is correct. If not, the alternate formula must be used and the computation repeated to determine new dimensions. Figure 6–13 shows a flowchart for the design logic described here.

Design: Assuming a Long Column

Euler's formula is used if the assumption is that the column is long. Equation (6–6) would be the most convenient form because it can be solved for the moment of inertia, I:

⇨ Euler's Formula Solved for Required Value of I

$$I = \frac{P_{cr}(KL)^2}{\pi^2 E} = \frac{NP_a(KL)^2}{\pi^2 E} \tag{6–8}$$

where $P_a =$ allowable load, usually set equal to the actual maximum expected load

Having the required value for I, we can determine the dimensions for the shape by additional computations or by scanning tables of data of the properties of commercially available sections.

The solid circular section is one for which it is possible to derive a final equation for the characteristic dimension, the diameter. The moment of inertia is

$$I = \frac{\pi D^4}{64}$$

Substituting this into Equation (6–8) gives

$$I = \frac{\pi D^4}{64} = \frac{NP_a(KL)^2}{\pi^2 E}$$

Solving for D yields

➩ Required Diameter for a Long, Solid Circular Column

$$D = \left[\frac{64NP_a(KL)^2}{\pi^3 E}\right]^{1/4} \quad \textbf{(6–9)}$$

Design: Assuming a Short Column

The J. B. Johnson formula is used to analyze a short column. It is difficult to derive a convenient form for use in design. In the general case, then, trial and error is used.

For some special cases, including the solid circular section, it is possible to solve the Johnson formula (Eqn. 6–7) for the characteristic dimension, the diameter:

$$P_{cr} = As_y\left[1 - \frac{s_y(KL/r)^2}{4\pi^2 E}\right]$$

But

$$A = \pi D^2/4$$
$$r = D/4 \text{ (from Appendix 1)}$$
$$P_{cr} = NP_a$$

Then

$$NP_a = \frac{\pi D^2}{4}s_y\left[1 - \frac{s_y(KL)^2}{4\pi^2 E(D/4)^2}\right]$$

$$\frac{4NP_a}{\pi s_y} = D^2\left[1 - \frac{s_y(KL)^2(16)}{4\pi^2 ED^2}\right]$$

Solving for D gives

➩ Required Diameter for a Short, Solid Circular Column

$$D = \left[\frac{4NP_a}{\pi s_y} + \frac{4s_y(KL)^2}{\pi^2 E}\right]^{1/2} \quad \textbf{(6–10)}$$

Example Problem 6–3 Specify a suitable diameter of a solid, round cross section for a machine link if it is to carry 9800 lb of axial compressive load. The length will be 25 in, and the ends will be pinned. Use a design factor of 3. Use SAE 1020 hot-rolled steel.

Solution Objective Specify a suitable diameter for the column.

Given Solid circular cross section: $L = 25$ in; use $N = 3$.

Both ends are pinned.

Material: SAE 1020 hot-rolled steel.

Analysis Use the procedure in Figure 6–13. Assume first that the column is long.

Results From Equation (6–9),

$$D = \left[\frac{64NP_a(KL)^2}{\pi^3 E}\right]^{1/4} = \left[\frac{64(3)(9800)(25)^2}{\pi^3(30\times 10^6)}\right]^{1/4}$$

$$D = 1.06 \text{ in}$$

The radius of gyration can now be found:

$$r = D/4 = 1.06/4 = 0.265 \text{ in}$$

The slenderness ratio is

$$KL/r = [(1.0)(25)]/0.265 = 94.3$$

For the SAE 1020 hot-rolled steel, $s_y = 30\,000$ psi. The graph in Figure 6–5 shows C_c to be approximately 138. Thus, the actual KL/r is less than the transition value, and the column must be redesigned as a short column, using Equation (6–10) derived from the Johnson formula:

$$D = \left[\frac{4NP_a}{\pi s_y} + \frac{4s_y(KL)^2}{\pi^2 E}\right]^{1/2}$$

$$D = \left[\frac{4(3)(9800)}{(\pi)(30\,000)} + \frac{4(30\,000)(25)^2}{\pi^2(30\times 10^6)}\right]^{1/2} = 1.23 \text{ in}$$

Checking the slenderness ratio again, we have

$$KL/r = [(1.0)(25)]/(1.23/4) = 81.3$$

Comments This is still less than the transition value, so our analysis is acceptable. A preferred size of $D = 1.25$ in could be specified.

An alternate method of using spreadsheets to design columns is to use an analysis approach similar to that shown in Figure 6–9 but to use it as a convenient "trial and error" tool. You could compute data by hand, or you could look them up in a table for A, I, and r for any desired cross-sectional shape and dimensions and insert the values into the spreadsheet. Then you could compare the computed allowable load with the required value and choose smaller or larger sections to bring the computed value close to the required value. Many iterations could be completed in a short amount of time. For shapes that allow computing r and A fairly simply, you could add a new section to the spreadsheet to calculate these values. An example is shown in Figure 6–14, where the differently shaded box shows the calculations for the properties of a round cross section. The data are from Example Problem 6–3, and the result shown was arrived at with only four iterations.

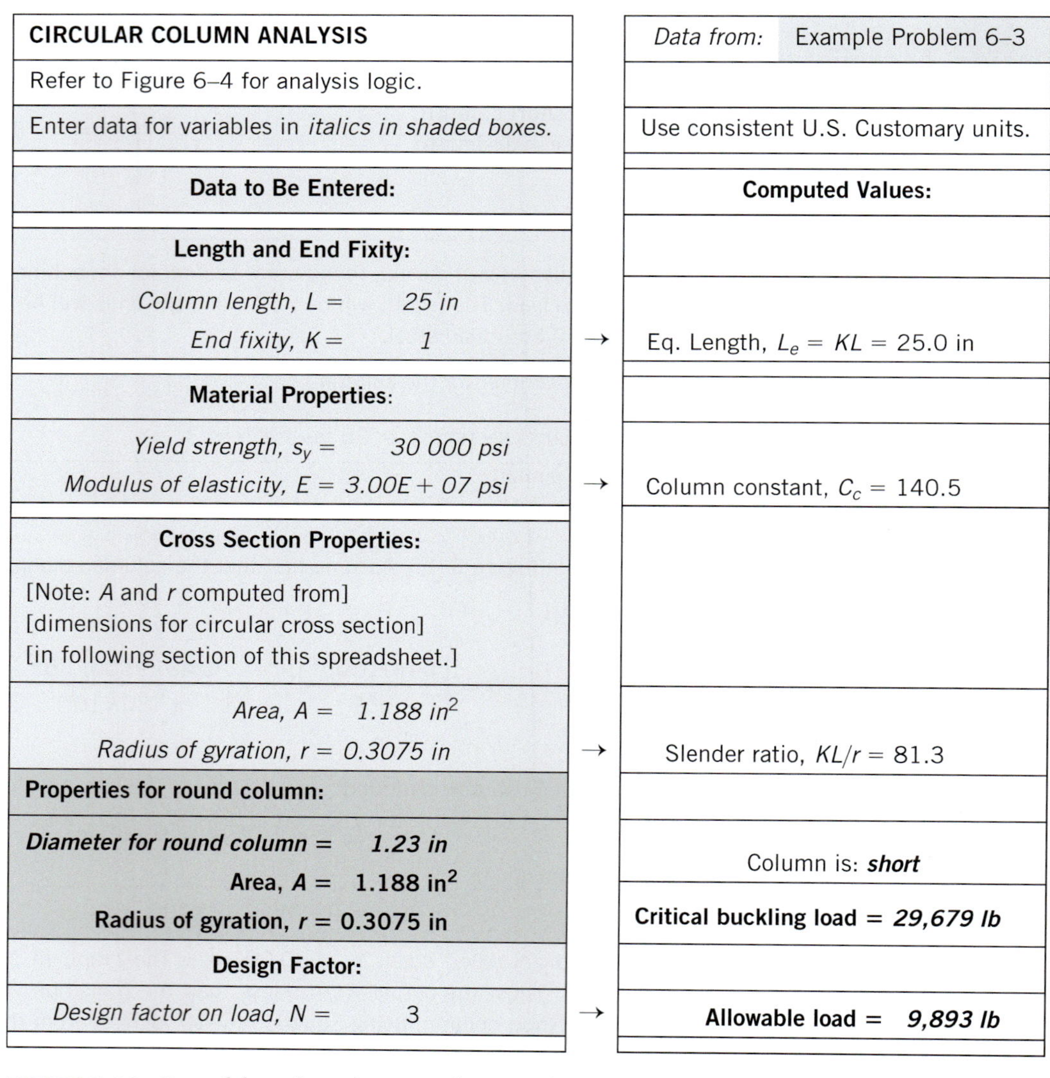

CIRCULAR COLUMN ANALYSIS		*Data from:* Example Problem 6–3
Refer to Figure 6–4 for analysis logic.		
Enter data for variables in *italics in shaded boxes.*		Use consistent U.S. Customary units.
Data to Be Entered:		**Computed Values:**
Length and End Fixity:		
Column length, L = *25 in* *End fixity, K =* *1*	→	Eq. Length, $L_e = KL = 25.0$ in
Material Properties:		
Yield strength, s_y = *30 000 psi* *Modulus of elasticity, E = 3.00E+ 07 psi*	→	Column constant, $C_c = 140.5$
Cross Section Properties:		
[Note: A and r computed from] [dimensions for circular cross section] [in following section of this spreadsheet.]		
Area, A = 1.188 in² *Radius of gyration, r = 0.3075 in*	→	Slender ratio, $KL/r = 81.3$
Properties for round column:		
Diameter for round column = ***1.23 in*** **Area, A = 1.188 in²** **Radius of gyration, r = 0.3075 in**		Column is: ***short*** **Critical buckling load = *29,679 lb***
Design Factor:		
Design factor on load, N = 3	→	**Allowable load = *9,893 lb***

FIGURE 6–14 Spreadsheet for column analysis used as a tool to design a column with a round cross section

6–11 CROOKED COLUMNS

The Euler and Johnson formulas assume that the column is straight and that the load acts in line with the centroid of the cross section of the column. If the column is somewhat crooked, bending occurs in addition to the column action (see Figure 6–15).

The crooked column formula allows an initial crookedness, a, to be considered (see References 6, 7, and 8):

➭ Crooked Column Formula

$$P_a^2 - \frac{1}{N}\left[s_y A + \left(1 + \frac{ac}{r^2}\right)P_{cr}\right]P_a + \frac{s_y A P_{cr}}{N^2} = 0 \quad \textbf{(6–11)}$$

where c = distance from the neutral axis of the cross section about which bending occurs to its outer edge

P_{cr} is defined to be the critical load found from the *Euler formula*.

Although this formula may become increasingly inaccurate for shorter columns, it is not appropriate to switch to the Johnson formula as it is for straight columns.

The crooked column formula is a quadratic with respect to the allowable load P_a. Evaluating all constant terms in Equation (6–11) produces an equation of the form

$$P_a^2 + C_1 P_a + C_2 = 0$$

Then, from the solution for a quadratic equation,

$$P_a = 0.5\left[-C_1 - \sqrt{C_1^2 - 4C_2}\right]$$

The smaller of the two possible solutions is selected.

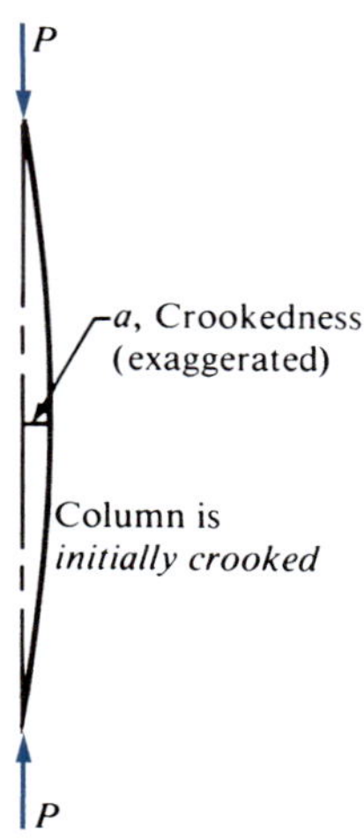

FIGURE 6–15 Illustration of crooked column

Example Problem 6–4 A column has both ends pinned and has a length of 32 in. It has a circular cross section with a diameter of 0.75 in and an initial crookedness of 0.125 in. The material is SAE 1040 hot-rolled steel. Compute the allowable load for a design factor of 3.

Solution

Objective: Specify the allowable load for the column.

Given: Solid circular cross section: $D = 0.75$ in; $L = 32$ in; use $N = 3$.

Both are ends pinned. Initial crookedness = $a = 0.125$ in.

Material: SAE 1040 hot-rolled steel.

Analysis: Use Equation (6–11). First evaluate C_1 and C_2. Then solve the quadratic equation for P_a.

Results

$$s_y = 42\,000 \text{ psi}$$

$$A = \pi D^2/4 = (\pi)(0.75)^2/4 = 0.442 \text{ in}^2$$

$$r = D/4 = 0.75/4 = 0.188 \text{ in}$$

$$c = D/2 = 0.75/2 = 0.375 \text{ in}$$

$$KL/r = [(1.0)(32)]/0.188 = 171$$

$$P_{cr} = \frac{\pi^2 EA}{(KL/r)^2} = \frac{\pi^2(30\,000\,000)(0.442)}{(171)^2} = 4476 \text{ lb}$$

$$C_1 = \frac{-1}{N}\left[s_y A + \left(1 + \frac{ac}{r^2}\right)P_{cr}\right] = -9649$$

$$C_2 = \frac{s_y A P_{cr}}{N^2} = 9.232 \times 10^6$$

The quadratic is therefore

$$P_a^2 - 9649P_a + 9.232 \times 10^6 = 0$$

Comment From this, $P_a = 1077$ lb is the allowable load.

Figure 6–16 shows the solution of Example Problem 6–4 using a spreadsheet. Whereas its appearance is similar to that of the earlier column analysis spreadsheets, the details follow the calculations needed to solve Equation (6–11). On the lower left, two special data values are needed: (1) the crookedness a and (2) the distance c from the neutral axis for buckling to the outer surface of the cross section. In the middle of the right part are listed some intermediate values used in Equation (6–11): C_1 and C_2 as defined in the solution to Example Problem 6–4. The result, the allowable load, P_a, is at the lower right of the spreadsheet. Above that, for comparison, the computed value of the critical buckling load is given for a straight column of the same design. Note that this solution procedure is most accurate for long columns. If the analysis indicates that the column is *short* rather than *long*, the designer should take note of how short it is by comparing the slenderness ratio, KL/r, with the column constant, C_c. If the column is quite short, the designer should not rely on the accuracy of the result from Equation (6–11).

CROOKED COLUMN ANALYSIS			*Data from:* Example Problem 6–4	
Solves Equation 6–11 for allowable load.				
Enter data for variables in *italics in shaded boxes.*			Use consistent U.S. Customary units.	
Data to Be Entered:			**Computed Values:**	
Length and End Fixity:				
Column length, L =	*32 in*			
End fixity, K =	*1*	→	Eq. length, $L_e = KL =$	32.0 in
Material Properties:				
Yield strength, s_y =	*42 000 psi*			
Modulus of elasticity, E =	*3.00E + 07 psi*	→	Column constant, C_c =	118.7
Cross Section Properties:			Euler buckling load =	4,491 lb
[Note: Enter r or compute r = sqrt(I/A).] [Always enter Area.] [Enter zero for *I or r if not used.]*			C_1 in Eq. (6–11) =	−9,678
			C_2 in Eq. (6–11) =	9.259E+06
Area, A =	*0.442 in²*			
Moment of inertia, I =	*0 in²*			
Radius of gyration, r =	*0.188 in*	→	Slender ratio, KL/r =	170.7
Values for Eq. (6–11):			Column is: ***long***	
Initial crookedness = a =	*0.125 in*		***Straight Column***	
Neutral axis to outside = c =	*0.375 in*		**Critical buckling load =**	***4,491 lb***
Design Factor:			***Crooked Column***	
Design factor on load, N =	*3*	→	**Allowable load =**	***1,076 lb***

FIGURE 6–16 Spreadsheet for analysis of crooked columns

6–12 ECCENTRICALLY LOADED COLUMNS

An *eccentric load* is one that is applied away from the centroidal axis of the cross section of the column, as shown in Figure 6–17. The eccentricity, e, is the perpendicular distance from the line of action between the applied end-loads to the centroidal axis of the initially straight column. Such a load exerts bending in addition to the column action that results in the deflected shape shown in the figure. The maximum stress in the deflected column occurs in the outermost fibers of the cross section at the midlength of the column where the maximum deflection, y_{max}, occurs. Let's denote the stress at this point as $\sigma_{L/2}$. Then, for any applied load, P,

➪ Secant Formula for Eccentrically Loaded Columns

$$\sigma_{L/2} = \frac{P}{A}\left[1 + \frac{ec}{r^2}\sec\left(\frac{KL}{2r}\sqrt{\frac{P}{AE}}\right)\right] \quad \textbf{(6–12)}$$

(See References 3, 5, and 9.) Note that this stress is *not* directly proportional to the load. When evaluating the secant in this formula, note that its argument in the parentheses is in *radians*. Also, because most calculators do not have the secant function, recall that the secant is equal to 1/cosine.

For design purposes, we would like to specify a design factor, N, that can be applied to the *failure load* similar to that defined for straight, centrally loaded columns. However, in this case, failure is predicted when the maximum stress in the column exceeds the yield strength of the material. Let's now define a new term, P_y, to be the load applied to the eccentrically loaded column when the maximum stress is equal to the yield strength. Equation (6–12) then becomes

$$s_y = \frac{P_y}{A}\left[1 + \frac{ec}{r^2}\sec\left(\frac{KL}{2r}\sqrt{\frac{P_y}{AE}}\right)\right]$$

Now, if we define the *allowable load* to be

$$P_a = P_y/N$$

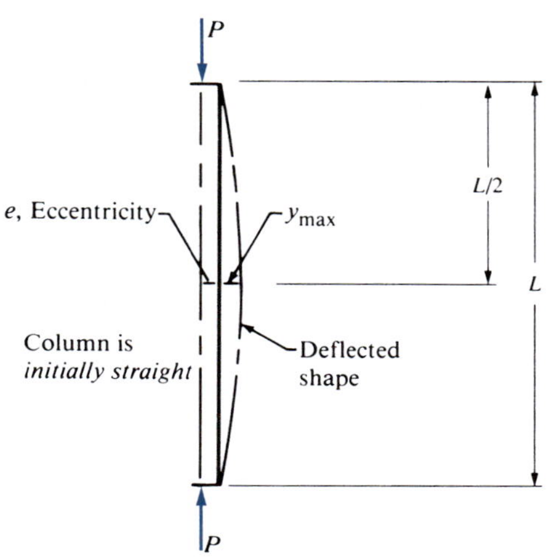

FIGURE 6–17 Illustration of eccentrically loaded columns

or

$$P_y = NP_a$$

this equation becomes

➪ Design Equation for Eccentrically Loaded Columns

$$\text{Required } s_y = \frac{NP_a}{A}\left[1 + \frac{ec}{r^2}\sec\left(\frac{KL}{2r}\sqrt{\frac{NP_a}{AE}}\right)\right] \quad \textbf{(6–13)}$$

This equation cannot be solved for either A or P_a. Therefore, an iterative solution is required, as will be demonstrated in Example Problem 6–6.

Another critical factor may be the amount of deflection of the axis of the column due to the eccentric load:

➪ Maximum Deflection in an Eccentrically Loaded Column

$$y_{max} = e\left[\sec\left(\frac{KL}{2r}\sqrt{\frac{P}{AE}}\right) - 1\right] \quad \textbf{(6–14)}$$

Note that the argument of the secant is the same as that used in Equation (6–12).

Example Problem 6–5 For the column of Example Problem 6–4, compute the maximum stress and deflection if a load of 1075 lb is applied with an eccentricity of 0.75 in. The column is initially straight.

Solution

Objective: Compute the stress and the deflection for the eccentrically loaded column.

Given: Data from Example Problem 6–4, but eccentricity $= e = 0.75$ in.
Solid circular cross section: $D = 0.75$ in; $L = 32$ in; Initially straight
Both ends are pinned; $KL = 32$ in; $r = 0.188$ in; $c = D/2 = 0.375$ in.
Material: SAE 1040 hot-rolled steel; $E = 30\times10^6$ psi, $s_y = 42$ psi

Analysis: Use Equation (6–12) to compute maximum stress. Then use Equation (6–14) to compute maximum deflection.

Results All terms have been evaluated before. Then the maximum stress is found from Equation (6–12):

$$\sigma_{L/2} = \frac{1075}{0.422}\left[1 + \frac{(0.75)(0.375)}{(0.188)^2}\sec\left(\frac{32}{2(0.188)}\sqrt{\frac{1075}{(0.442)(30\times10^6)}}\right)\right]$$

$$\sigma_{L/2} = 29\,300 \text{ psi}$$

The maximum deflection is found from Equation (6–14):

$$y_{max} = 0.75\left[\sec\left(\frac{32}{2(0.188)}\sqrt{\frac{1075}{(0.442)(30\times10^6)}}\right) - 1\right] = 0.293 \text{ in}$$

Comments The maximum stress is 29 300 psi at the midlength of the column. The deflection there is 0.293 in from the original straight central axis of the column.

Example Problem 6–6 The stress in the column found in Example Problem 6–5 seems high for the SAE 1040 hot-rolled steel. Redesign the column to achieve a design factor of at least 3.

Solution Objective Redesign the eccentrically loaded column of Example Problem 6–5 to reduce the stress and achieve a design factor of at least 3.

Given Data from Example Problems 6–4 and 6–5.

Analysis Use a larger diameter. Use Equation (6–13) to compute the required strength. Then compare that with the strength of SAE 1040 hot-rolled steel. Iterate until the stress is satisfactory.

Results Appendix 3 gives the value for the yield strength of SAE 1040 HR to be 42 000 psi. If we choose to retain the same material, the cross-sectional dimensions of the column must be increased to decrease the stress. Equation (6–13) can be used to evaluate a design alternative.

The objective is to find suitable values for A, c, and r for the cross section such that $P_a = 1075$ lb; $N = 3$; $L_e = 32$ in; $e = 0.75$ in; and the value of the entire right side of the equation is less than 42 000 psi. The original design had a circular cross section with a diameter of 0.75 in. Let's try increasing the diameter to $D = 1.00$ in. Then

$$A = \pi D^2/4 = \pi(1.00 \text{ in})^2/4 = 0.785 \text{ in}^2$$

$$r = D/4 = (1.00 \text{ in})/4 = 0.250 \text{ in}$$

$$r^2 = (0.250 \text{ in})^2 = 0.0625 \text{ in}^2$$

$$c = D/2 = (1.00 \text{ in})/2 = 0.50 \text{ in}$$

Now let's call the right side of Equation (6–13) s_y'. Then

$$s_y' = \frac{3(1075)}{0.785}\left[1 + \frac{(0.75)(0.50)}{(0.0625)}\sec\left(\frac{32}{2(0.250)}\sqrt{\frac{(3)(1075)}{(0.785)(30\times10^6)}}\right)\right]$$

$$s_y' = 37\,740 \text{ psi} = \text{required value of } s_y$$

This is a satisfactory result because it is just slightly less than the value of s_y of 42 000 psi for the steel.

Now we can evaluate the expected maximum deflection with the new design using Equation (6–14):

$$y_{max} = 0.75\left[\sec\left(\frac{32}{2(0.250)}\sqrt{\frac{1075}{(0.785)(30\times10^6)}}\right) - 1\right]$$

$$y_{max} = 0.076 \text{ in}$$

Comments The diameter of 1.00 in is satisfactory. The maximum deflection for the column is 0.076 in.

ECCENTRIC COLUMN ANALYSIS		*Data from:* Example Problem 6–6
Solves Equation (6–13) for design stress and Equation (6–14) for maximum deflection.		
Enter data for variables in *italics in shaded boxes.*		Use consistent U.S. Customary units.
Data to Be Entered:		**Computed Values:**
Length and End Fixity:		
Column length, L = *32 in*		
End fixity, K = *1*	→	Eq. length, $L_e = KL = 32.0$ in
Material Properties:		
Yield strength, s_y *=* *42 000 psi*		
Modulus of elasticity, E = 3.00E + 07 psi	→	Column constant, $C_c = 118.7$
Cross Section Properties:		
[Note: Enter *r* or compute $r = \text{sqrt}(I/A)$.]		Argument of sec = 0.749 for strength
[Always enter Area.]		Value of secant = 1.3654
[Enter zero for *I or r if not used.]*		
Area, A = *0.785 in*^{2}		Argument of sec = 0.432 for deflection
Moment of inertia, I = *0 in*^{4}		Value of secant = 1.1014
OR		
Radius of gyration, r = *0.250 in*	→	Slender ratio, $KL/r = 128.0$
Values for Eqs. (6–13) and (6–14)		Column is: ***long***
Eccentricity, e = *0.75 in*		
Neutral axis to outside, c = *0.5 in*		**FINAL RESULTS**
Allowable load, P_a *=* *1,075 lb*		**Req'd yield strength = *37 764 psi***
Design Factor:		***Must be less than actual yield strength:***
Design factor on load, N = *3*		***s_y = 42 000 psi***
		Max deflection, y_{max} = *0.076 in*

FIGURE 6–18 Spreadsheet for analysis of eccentric columns

Figure 6–18 shows the solution of the eccentric column problem of Example Problem 6–6 using a spreadsheet to evaluate Equations (6–13) and (6–14). It is a design aid that facilitates the iteration required to determine an acceptable geometry for a column to carry a specified load with a desired design factor. Note that the data are in U.S. Customary units. At the lower left of the spreadsheet, data required for Equations (6–13) and (6–14) are entered by the designer, along with the other data discussed for earlier column analysis spreadsheets. The **"FINAL RESULTS"** at the lower right show the computed value of the required yield strength of the material for the column and compare it with the given value entered by the designer near the upper left. The designer must ensure that the actual value is greater than the computed value. The last part of the right side of the spreadsheet gives the computed maximum deflection of the column that occurs at its midlength.

REFERENCES

1. Aluminum Association. *Aluminum Design Manual.* Washington, DC: Aluminum Association, 2010.
2. American Institute of Steel Construction. *Steel Construction Manual.* 14th ed. Chicago: American Institute of Steel Construction, 2011.
3. Budynas, Richard G., and Keith J. Nisbett. *Shigley's Mechanical Engineering Design.* 9th ed. New York: McGraw-Hil, 2011.
4. Hibbeler, Russell C. *Mechanics of Materials.* 8th ed. Upper Saddle River, NJ: Pearson Prentice Hall, 2011.
5. Popov, Egor P. *Engineering Mechanics of Solids.* 2nd ed. Upper Saddle River, NJ: Pearson Prentice Hall, 1999.
6. Spotts, Merhyle F., Terry E. Shoup and Lee E. Hornberger. *Design of Machine Elements.* 8th ed. Upper Saddle River, NJ: Pearson Prentice Hall, 2004.
7. Timoshenko, Stephen P. *Strength of Materials.* Vol. 2. 2nd ed. New York: Van Nostrand Reinhold, 1941.

8. Timoshenko, Stephen P., and James. M. Gere. *Theory of Elastic Stability*. 2nd ed. New York: McGraw-Hill, 1961.
9. Young, Warren C., Richard. G. Budynas, and Ali Sadegh. *Roark's Formulas for Stress and Strain*. 8th ed. New York: McGraw-Hill, 2012.

PROBLEMS

1. A column has both ends pinned and has a length of 32 in. It is made of SAE 1040 HR steel and has a circular shape with a diameter of 0.75 in. Determine the critical load.
2. Repeat Problem 1 using a length of 15 in.
3. Repeat Problem 1 with the bar made of aluminum 6061-T4.
4. Repeat Problem 1 assuming both ends fixed.
5. Repeat Problem 1 using a square cross section, 0.65 in on a side, instead of the circular cross section.
6. Repeat Problem 1 with the bar made from high-impact acrylic plastic.
7. A rectangular steel bar has a cross section 0.50 by 1.00 in and is 8.5 in long. The bar has pinned ends and is made of SAE 4150 OQT 1000 steel. Compute the critical load.
8. A steel pipe has an outside diameter of 1.60 in, a wall thickness of 0.109 in, and a length of 6.25 ft. Compute the critical load for each of the end conditions shown in Figure 6–2. Use SAE 1020 HR steel.
9. Compute the required diameter of a circular bar to be used as a column carrying a load of 8500 lb with pinned ends. The length is 50 in. Use SAE 4140 OQT 1000 steel and a design factor of 3.0.
10. Repeat Problem 9 using SAE 1020 HR steel.
11. Repeat Problem 9 with aluminum 2014-T4.
12. In Section 6–10, equations were derived for the design of a solid circular column, either long or short. Perform the derivation for a solid square cross section.
13. Repeat the derivations called for in Problem 12 for a hollow circular tube for any ratio of inside to outside diameter. That is, let the ratio $R = ID/OD$, and solve for the required OD for a given load, material, design factor, and end fixity.
14. Determine the required dimensions of a column with a square cross section to carry an axial compressive load of 6500 lb if its length is 64 in and its ends are fixed. Use a design factor of 3.0. Use aluminum 6061-T6.
15. Repeat Problem 14 for a hollow aluminum tube (6061-T6) with the ratio of $ID/OD = 0.80$. Compare the weight of this column with that of Problem 14.
16. A toggle device is being used to compact scrap steel shavings, as illustrated in Figure P6–16. Design the two links of the toggle to be steel, SAE 5160 OQT 1000, with a circular cross section and pinned ends. The force P required to crush the shavings is 5000 lb. Use $N = 3.50$.
17. Repeat Problem 16, but propose a design that will be lighter than the solid circular cross section.
18. A sling, sketched in Figure P6–18, is to carry 18 000 lb. Design the spreader.
19. For the sling in Problem 18, design the spreader if the angle shown is changed from 30° to 15°.
20. A rod for a certain hydraulic cylinder behaves as a fixed-free column when used to actuate a compactor of industrial waste. Its maximum extended length will be 10.75 ft. If it is to be made of SAE 1144 OQT 1300 steel, determine the required diameter of the rod for a design factor of 2.5 for an axial load of 25 000 lb.
21. Design a column to carry 40 000 lb. One end is pinned, and the other is fixed. The length is 12.75 ft.
22. Repeat Problem 21 using a length of 4.25 ft.
23. Repeat Problem 1 if the column has an initial crookedness of 0.08 in. Determine the allowable load for a design factor of 3.
24. Repeat Problem 7 if the column has an initial crookedness of 0.04 in. Determine the allowable load for a design factor of 3.
25. Repeat Problem 8 if the column has an initial crookedness of 0.15 in. Determine the allowable load for a design factor of 3 and pinned ends only.
26. An aluminum (6063-T4) column is 42 in long and has a square cross section, 1.25 in on a side. If it carries a compressive load of 1250 lb, applied with an eccentricity of 0.60 in, compute the maximum stress in the column and the maximum deflection.
27. A steel (SAE 1020 hot-rolled) column is 3.2 m long and is made from a standard 3-in Schedule 40 steel pipe (see Table A15–17). If a compressive load of 30.5 kN is applied with an eccentricity of 150 mm, compute the maximum stress in the column and the maximum deflection.
28. A link in a mechanism is 14.75 in long and has a square cross section, 0.250 in on a side. It is made from annealed SAE 410 stainless steel. Use $E = 28 \times 10^6$ psi. If it carries a compressive load of 45 lb with an eccentricity of 0.30 in, compute the maximum stress and the maximum deflection.
29. A hollow square steel tube, 40 in long, is proposed for use as a prop to hold up the ram of a punch press during installation of new dies. The ram weighs 75 000 lb. The prop is made from 4 × 4 × 1/4 structural tubing. It is made from steel similar to structural steel, ASTM A500 Grade C. If the load applied by the ram could have an eccentricity of 0.50 in, would the prop be safe?

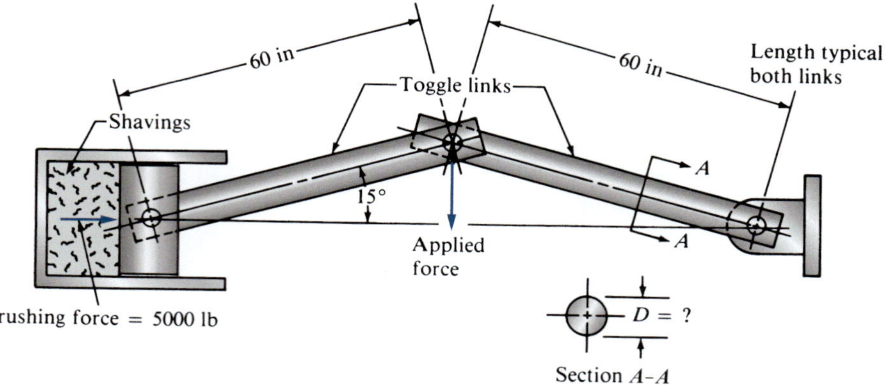

FIGURE P6–16 (Problems 16 and 17)

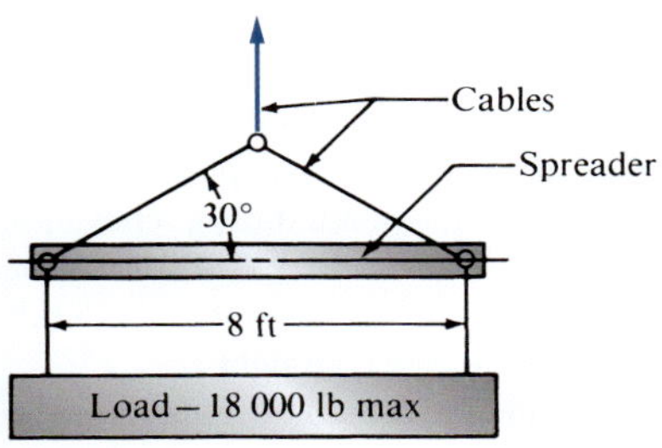

FIGURE P6–18 (Problems 18 and 19)

30. Determine the allowable load on a column 16.0 ft long made from a wide-flange beam shape, W5 × 19. The load will be centrally applied. The end conditions are somewhat between fixed and hinged, say, $K = 0.8$. Use a design factor of 3. Use ASTM A36 structural steel.
31. Determine the allowable load on a fixed-end column having a length of 66 in if it is made from a steel American Standard Beam, S4 × 7.7. The material is ASTM A36 structural steel. Use a design factor of 3.
32. Compute the maximum stress and deflection that can be expected in the steel machine member carrying an eccentric load as shown in Figure P6–32. The load P is 1000 lb. If a design factor of 3 is desired, specify a suitable steel.
33. Specify a suitable steel tube from Table A15–14 to support one side of a platform as shown in Figure P6–33. The material has a yield strength of 36 ksi. The total load on the platform is 55 000 lb, uniformly distributed.
34. Compute the allowable axial load on a steel channel, C5 × 9, made from ASTM A36 structural steel. The channel is 112 in long and can be considered to be pinned at its ends. Use a design factor of 3.
35. Repeat Problem 34 with the ends fixed rather than pinned.
36. Repeat Problem 34, except consider the load to be applied along the outside of the web of the channel instead of being axial.
37. Figure P6–37 shows a 4 × 4 × 1/2 steel column made from ASTM A500 Grade B structural steel. To accommodate a special mounting restriction, the load is applied eccentrically as shown. Determine the amount of load that the column can safely support. The column is supported laterally by the structure.

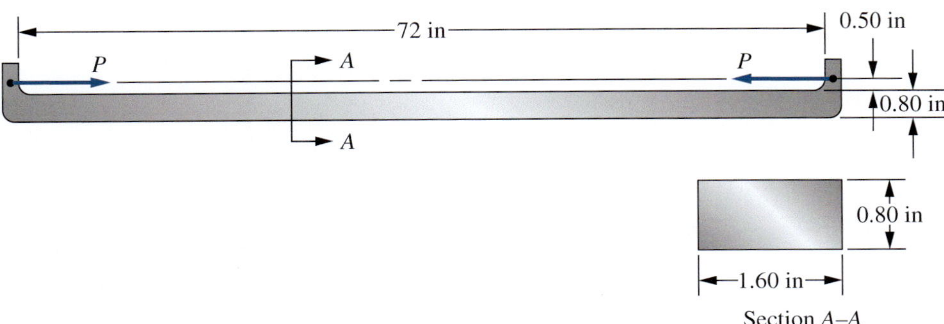

FIGURE P6–32 (Problem 32)

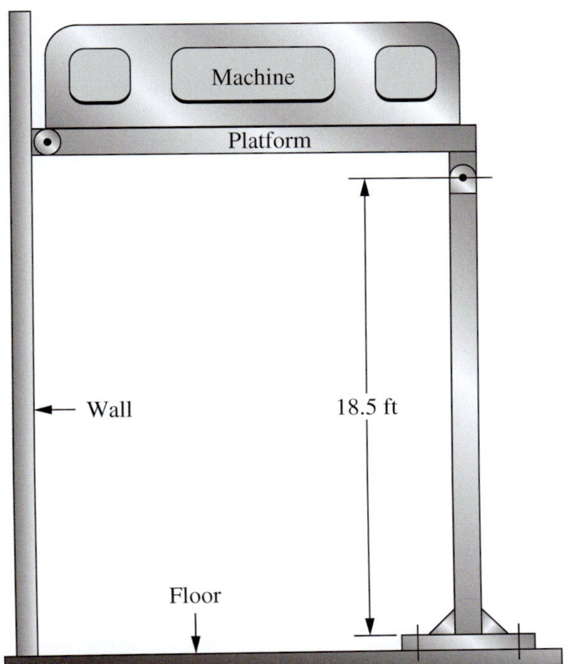

FIGURE P6–33 (Problem 33)

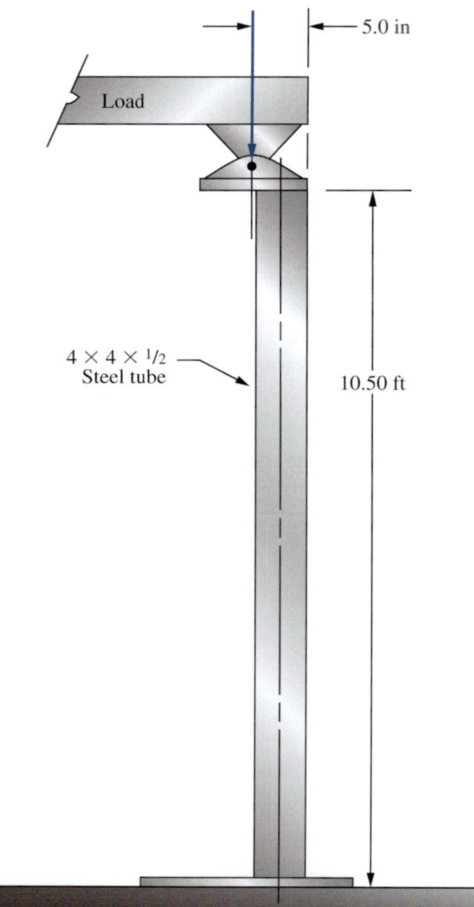

FIGURE P6–37 (Problem 37)

38. The device shown in Figure P6–38 is subjected to opposing forces F. Determine the maximum allowable load to achieve a design factor of 3. The device is made from aluminum 6061-T6.

39. A hydraulic cylinder is capable of exerting a force of 5200 N to move a heavy casting along a conveyor. The design of the pusher causes the load to be applied eccentrically to the piston rod as shown in Figure P6–39. Is the piston rod safe under this loading if it is made from SAE 416 stainless steel in the Q&T 1000 condition? Use $E = 200\,\text{GPa}$.

40. A standard 2-in schedule 40 steel pipe is proposed to be used to support the roof of a porch during renovation. Its length is 13.0 ft. The pipe is made from ASTM A501 structural steel.
   (a) Determine the safe load on the pipe to achieve a design factor of 3 if the pipe is straight.
   (b) Determine the safe load if the pipe has an initial crookedness of 1.25 in.

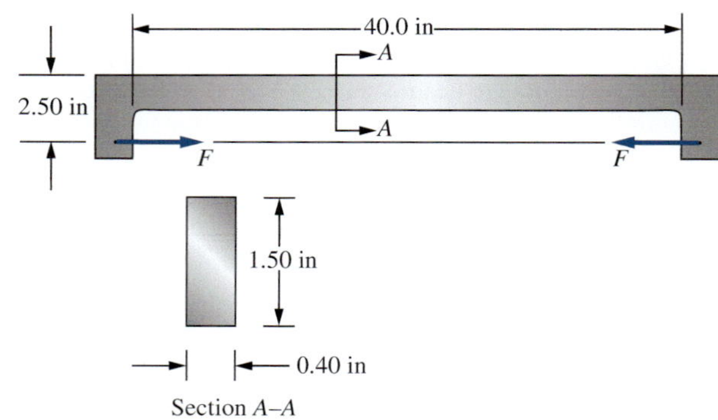

FIGURE P6–38 (Problem 38)

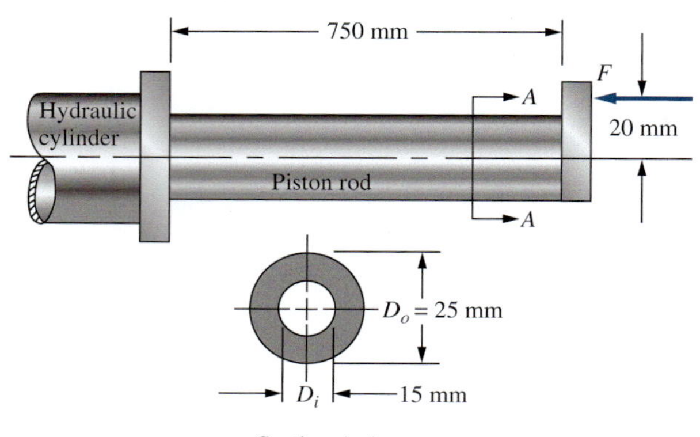

FIGURE P6–39 (Problems 39)

DESIGN OF A MECHANICAL DRIVE

OBJECTIVES AND CONTENT OF PART II

Part II of this book contains nine chapters (Chapters 7–15) that help you gain experience in approaching the design of an important complete device—a *mechanical drive*. The drive, sometimes called a *power transmission*, serves the following functions:

- It receives power from some kind of rotating source such as an electric motor, an internal combustion engine, a gas turbine engine, a hydraulic or pneumatic motor, a steam or water turbine, or even hand rotation provided by the operator.
- The drive typically causes some change in the speed of rotation of the shafts that make up the drive so that the output shaft operates more slowly or faster than the input shaft. Speed reducers are more prevalent than speed increasers.
- The active elements of the drive transmit the power from the input to the output shafts.
- When there is a speed reduction, there is a corresponding increase in the torque transmitted. Conversely, a speed increase causes a reduction in torque at the output compared with the input of the drive.

The chapters of Part II provide the detailed descriptions of the various machine elements that are typically used in power transmissions: *belt drives, chain drives, gears, shafts, bearings, keys, couplings, seals*, and *housings to hold all the elements together*. You will learn the important features of these elements and the methods of analyzing and designing them.

Of equal importance is the information provided on how the various elements interact with each other. You must be sensitive, for example, to how gears are mounted on shafts, how the shafts are supported by bearings, and how the bearings must be mounted securely in a housing that holds the system together. The final completed design must function as an integrated unit.

THE PROCESS OF DESIGNING A MECHANICAL DRIVE

In the design of a power transmission, you would typically know the following:

- ***The nature of the driven machine:*** It might be a machine tool in a factory that cuts metal parts for engines; an electric drill used by professional carpenters or home craft workers; the axle of a farm tractor; the propeller shaft of a turbojet for an airplane; the propeller shaft for a large ship; the wheels of a toy train; a mechanical timing mechanism; or any other of the numerous products that need a controlled-speed drive.
- ***The level of power to be transmitted:*** From the examples just listed, the power demanded may range from thousands of horsepower for a ship, hundreds of horsepower for a large farm tractor or airplane, or a few watts for a timer or a toy.
- ***The rotational speed of the drive motor or other prime mover:*** Typically the prime mover operates at a rather high speed of rotation. The shafts of standard electric motors rotate at about 1200, 1800, or 3600 revolutions per minute (rpm). Actual full-load operating speeds are somewhat less than these speeds, say 1150, 1750, or 3550 rpm. Automotive engines

operate from about 1000 to 6000 rpm. Universal motors in some hand tools (drills, saws, and routers) and household appliances (mixers, blenders, and vacuum cleaners) operate from 3500 to 20 000 rpm. Gas turbine engines for aircraft rotate many thousands of rpm.

- ***The desired output speed of the transmission:*** This is highly dependent on the application. Some gear motors for instruments rotate less than 1.0 rpm. Production machines in factories may run a few hundred rpm. Drives for assembly conveyors may run fewer than 100 rpm. Aircraft propellers may operate at several thousand rpm.

You, the designer, must then do the following:

- Choose the type of power transmission elements to be used: gears, belt drives, chain drives, or other types. In fact, some power transmission systems use two or more types in series to optimize the performance of each.
- Specify how the rotating elements are arranged and how the power transmission elements are mounted on shafts.
- Design the shafts to be safe under the expected torques and bending loads and properly locate the power transmission elements and the bearings. It is likely that the shafts will have several diameters and special features to accommodate keys, couplings, retaining rings, and other details. The dimensions of all features must be specified, along with the tolerances on the dimensions and surface finishes.
- Specify suitable bearings to support the shafts and determine how they will be mounted on the shafts and how they will be held in a housing.
- Specify keys to connect the shaft to the power transmission elements; couplings to connect the shaft from the driver to the input shaft of the transmission or to connect the output shaft to the driven machine; seals to effectively exclude contaminants from entering the transmission; and other accessories.
- Place all of the elements in a suitable housing that provides for the mounting of all elements and for their protection from the environment and their lubrication.

CHAPTERS THAT MAKE UP PART II

To guide you through this process of designing a mechanical drive, Part II includes the following chapters:

Chapter 7: Belt Drives and Chain Drives emphasizes recognizing the variety of commercially available belt and chain drives, the critical design parameters, and the methods used to specify reasonably optimum components of the drive systems.

Chapter 8: Kinematics of Gears describes and defines the important geometric features of gears. Methods of manufacturing gears are discussed, along with the importance of precision in the operation of the gears. The details of how a pair of gears operates are described, and the design and the operation of two or more gear pairs in a gear train are analyzed.

Chapter 9: Spur Gear Design illustrates how to compute forces exerted by one tooth of a gear on its mating teeth. Methods of computing the stresses in the gear teeth are presented, and design procedures for specifying gear-tooth geometry and material are given to produce a safe, long-life gear transmission system.

Chapter 10: Helical Gears, Bevel Gears, and Wormgearing contains approaches similar to those described for spur gears, with special attention to the unique geometry of these types of gears.

Chapter 11: Keys, Couplings, and Seals discusses how to design keys to be safe under the prevailing loads caused by the torque transmitted by them from the shaft to the gears or other elements. Couplings must be specified that accommodate the possible misalignment of connected shafts while transmitting the required torque at operating speeds. Seals must be specified for shafts that project through the sides of the housing and for bearings that must be kept free of contaminants. The maintenance of a reliable supply of clean lubricant for the active elements is essential.

Chapter 12: Shaft Design discusses the fact that in addition to being designed to safely transmit the required torque levels at given speeds, the shafts will probably have several diameters and special features to accommodate keys, couplings, retaining rings, and other details. The dimensions of all features must be specified, along with the tolerances on the dimensions and surface finishes. Completion of these tasks requires some of the skills developed in following chapters. So you will have to come back to this task later.

Chapter 13: Tolerances and Fits is not included in this custom edition.

Chapter 14: Rolling Contact Bearings focuses on commercially available rolling contact bearings such as ball bearings, roller bearings, and tapered roller bearings. You must be able to compute or specify the loads that the bearings will support, their speed of operation, and their expected life. From these data, standard bearings from manufacturers' catalogs will be specified. Then you must review the design process for the shafts as described for Chapter 12 to complete the specification of dimensions and tolerances. It is likely that iteration among the design processes for the power transmission elements, the shafts, and the bearings will be needed to achieve an optimum arrangement.

Chapter 15: Completion of the Design of a Power Transmission merges all of the preceding topics together. You will resolve the details of the design of each element and ensure the compatibility of mating elements. You will review all previous design decisions and assumptions and verify that the design meets specifications. After the individual elements have been analyzed and the iteration among them is complete, they must be packaged in a suitable housing to hold them securely, to protect them from contaminants, and to protect the people who may work around them. The housing must also be designed to be compatible with the driver and the driven machine. That often requires special fastening provisions and means of locating all connected devices relative to one another. Assembly and service must be considered. Then you will present a final set of specifications for the entire power transmission system and document your design with suitable drawings and a written report.

CHAPTER SEVEN

BELT DRIVES AND CHAIN DRIVES

THE BIG PICTURE Belt Drives and Chain Drives

Discussion Map

- Belts and chains are the major types of flexible power transmission elements. Belts operate on sheaves or pulleys, whereas chains operate on toothed wheels called *sprockets*.

Discover

Look around and identify at least one mechanical device having a belt drive and one having a chain drive system.

Describe each system, and make a sketch showing how it receives power from some source and how it transfers power to a driven machine.

Describe the differences between belt drives and chain drives.

In this chapter, you will learn how to select suitable components for belt drives and chain drives from commercially available designs.

Belts and chains represent the major types of flexible power transmission elements. Figure 7–1 shows a typical industrial application of these elements combined with a gear-type speed reducer. This application illustrates where belts, gear drives, and chains are each used to best advantage.

Rotary power is developed by the electric motor, but motors typically operate at too high a speed and deliver too low a torque to be appropriate for the final drive application. Remember, for a given power transmission, the torque is increased in proportion to the amount that rotational speed is reduced. So some speed reduction is often desirable. The high speed of the motor makes belt drives somewhat ideal for that first stage of reduction. A smaller drive pulley is attached to the motor shaft, while a larger diameter pulley is attached to a parallel shaft that operates at a correspondingly lower speed. Pulleys for belt drives are also called *sheaves*.

However, if very large ratios of speed reduction are required in the drive, gear reducers are desirable because they can typically accomplish large reductions in a rather small package. The output shaft of the gear-type speed reducer is generally at low speed and high torque. If both speed and torque are satisfactory for the application, it could be directly coupled to the driven machine.

However, because gear reducers are available only at discrete reduction ratios, the output must often be reduced more before meeting the requirements of the machine. At the low-speed, high-torque condition, chain drives become desirable. The high torque causes high tensile forces to be developed in the chain. The elements of the chain are typically metal, and they are sized to withstand the high forces. The links of chains are engaged in toothed wheels called *sprockets* to provide positive mechanical drive, desirable at the low-speed, high-torque conditions.

In general, belt drives are applied where the rotational speeds are relatively high, as on the first stage of speed reduction from an electric motor or engine. The linear speed of a belt is usually 2500–6500 ft/min, which results in relatively low tensile forces in the belt. At lower speeds, the tension in the belt becomes too large for typical belt cross sections, and slipping may occur between the sides of the belt and the sheave or pulley that carries it.

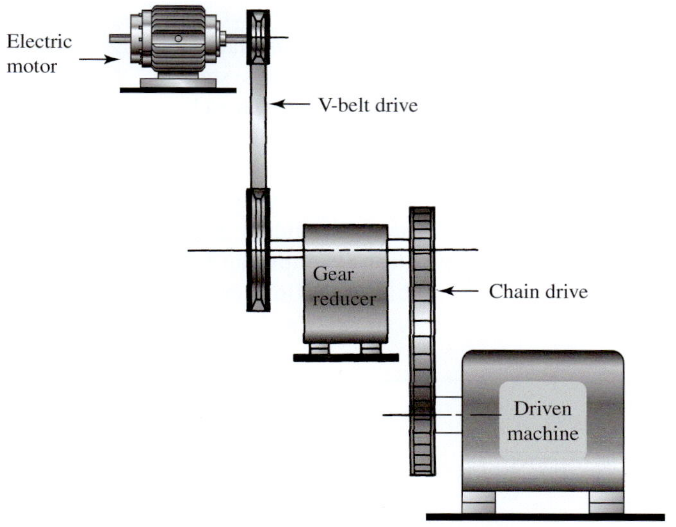

(*a*) Sketch of combination drive

(*b*) Photograph of an actual drive installation. Note that guards have been removed from the belt and chain drives to show detail.

FIGURE 7–1 Combination drive employing V-belts, a gear reducer, and a chain drive
[Source for Part (b): Browning Mfg. Division, Emerson Electric Co., Maysville, KY]

At higher speeds, dynamic effects such as centrifugal forces, belt whip, and vibration reduce the effectiveness of the drive and its life. A speed of 4000 ft/min is generally ideal.

The *synchronous belt drive* employs precision ribs on the inside surface of the belt that engage in matching grooves in the pulleys to enhance their ability to transmit high forces at low speeds. There is also precise timing between the driver and driven pulleys so this type of belt is sometimes called a *timing belt.* Such belt drives often compete with chain drives and gear drives in some applications.

Where have you seen belt drives? Consider mechanical devices around your home or office; vehicles; construction equipment; heating, air-conditioning, and ventilation systems; and industrial machinery. Describe their general appearance. To what was the input pulley attached? Was it operating at a fairly high speed? What was the size of the next pulley? Did it cause the second shaft to rotate at a slower speed? How much slower? Were there more stages of reduction accomplished by belts or by some other reducer? Make a sketch of the layout of the drive system. Make measurements if you can get access to the equipment safely.

Where have you seen chain drives? One obvious place is likely to be the chain on a bicycle where the sprocket attached to the pedal-crank assembly is fairly large and that attached to the rear wheel is smaller. The drive sprocket and/or the driven sprocket assemblies may have several sizes to allow the rider to select many different speed ratios to permit optimum operation under different conditions of speed and hill-climbing demands. Where else have you seen chain drives? Again consider vehicles, construction equipment, and industrial machinery. Describe and sketch at least one chain drive system.

This chapter will help you learn to identify the typical design features of commercially available belt and chain drives. You will be able to specify suitable types and sizes to transmit a given level of power at a certain speed and to accomplish a specified speed ratio between the input and the output of the drive. Installation considerations are also described so that you can put your designs into successful systems.

YOU ARE THE DESIGNER

A plant in Hawaii that produces sugar needs a drive system designed for a machine that chops long pieces of sugar cane into short lengths prior to processing. The machine's drive shaft is to rotate slowly at 30 rpm so that the cane is chopped smoothly and not beaten. The large machine requires a torque of 31 500 lb·in to drive the chopping blades.

Your company is asked to design the drive, and you are given the assignment. What kind of power source should be used? You might consider an electric motor, a gasoline engine, or a hydraulic motor. Most of these run at relatively high speeds, significantly higher than 30 rpm. Therefore, some type of speed reduction is needed. Perhaps you decide to use a drive similar to that shown in Figure 7–1.

Three stages of speed reduction are used. The input sheave of the belt drive rotates at the speed of the motor, while the larger driven sheave rotates at a slower speed and delivers the power to the input of the gear reducer. The larger part of the speed reduction is likely to be accomplished in the gear reducer, with the output shaft rotating slowly and providing a large torque. Remember, as the speed of rotation of a rotating shaft decreases, the torque delivered increases for a given power transmitted. But because there are only a limited number of reducer designs available, the output speed of the reducer will probably not be ideal for the cane chopper input shaft. The chain drive then provides the last stage of reduction.

As the designer, you must decide what type and size of belt drive to use and what the speed ratio between the driving and the driven sheave should be. How is the driving sheave attached to the motor shaft? How is the driven sheave attached to the input shaft of the gear reducer? Where should the motor be mounted in relation to the gear reducer, and what will be the resulting center distance between the two shafts? What speed reduction ratio will the gear reducer provide? What type of gear reducer should be used: helical gears, a worm and worm-gear drive, or bevel gears? How much additional speed reduction must the chain drive provide to deliver the proper speed to the cane-chopper shaft? What size and type of chain should be specified? What is the center distance between the output of the gear reducer and the input to the chopper? Then what length of chain is required? Finally, what motor power is required to drive the entire system at the stated conditions? The information in this chapter will help you answer questions about the design of power transmission systems incorporating belts and chains. Gear reducers are discussed in Chapter 8–10. ■

7–1 OBJECTIVES OF THIS CHAPTER

After completing this chapter, you will be able to:

1. Describe the basic features of a belt drive system.
2. Describe several types of belt drives.
3. Specify suitable types and sizes of belts and sheaves to transmit a given level of power at specified speeds for the input and output sheaves.
4. Specify the primary installation variables for belt drives, including center distance and belt length.
5. Describe the basic features of a chain drive system.
6. Describe several types of chain drives.
7. Specify suitable types and sizes of chains and sprockets to transmit a given level of power at specified speeds for the input and output sprockets.
8. Specify the primary installation variables for chain drives, including center distance between the sheaves, chain length, and lubrication requirements.

7–2 TYPES OF BELT DRIVES

A belt is a flexible power transmission element that seats tightly on a set of pulleys or sheaves. Figure 7–2 shows the basic layout. When the belt is used for speed reduction, the typical case, the smaller sheave is mounted on

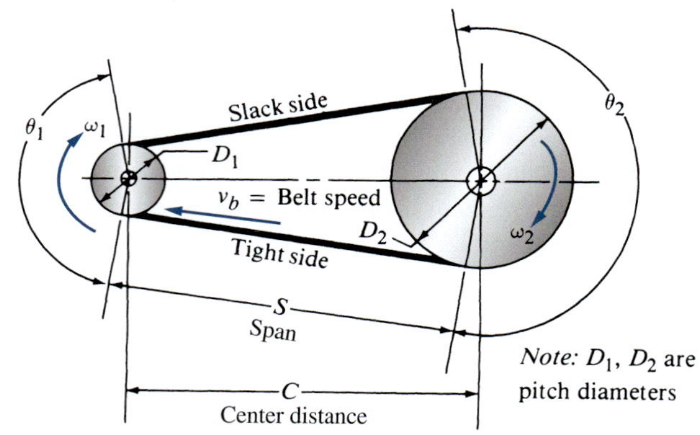

FIGURE 7–2 Basic belt drive geometry

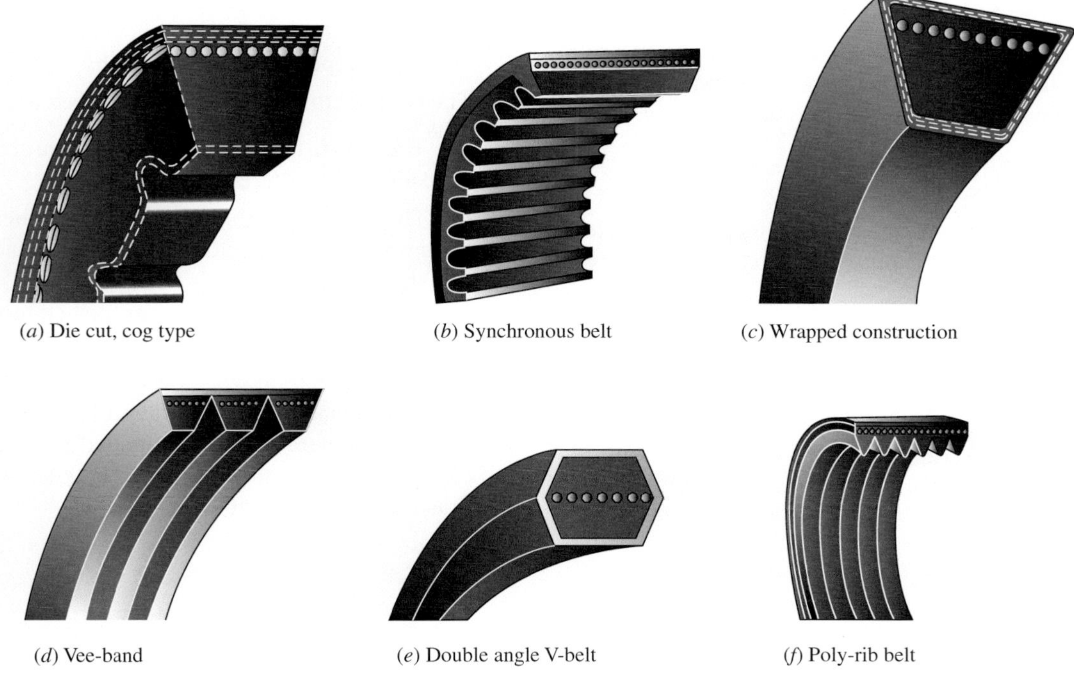

(*a*) Die cut, cog type (*b*) Synchronous belt (*c*) Wrapped construction

(*d*) Vee-band (*e*) Double angle V-belt (*f*) Poly-rib belt

FIGURE 7–3 Examples of belt construction

the high-speed shaft, such as the shaft of an electric motor. The larger sheave is mounted on the driven machine. The belt is designed to ride around the two sheaves without slipping.

The belt is installed by placing it around the two sheaves while the center distance between them is reduced. Then the sheaves are moved apart, placing the belt in a rather high initial tension. When the belt is transmitting power, friction causes the belt to grip the driving sheave, increasing the tension in one side, called the "tight side," of the drive. The tensile force in the belt exerts a tangential force on the driven sheave, and thus a torque is applied to the driven shaft. The opposite side of the belt is still under tension, but at a smaller value. Thus, it is called the "slack side."

Many types of belts are available: flat belts, grooved or cogged belts, standard V-belts, double-angle V-belts, and others. See Figure 7–3 for examples. References 3a–g and 4–8, give more examples and technical data. See also Internet sites 2-14 and 16 for industry data.

The *flat belt* is the simplest type, often made from leather or rubber-coated fabric. The sheave surface is also flat and smooth, and the driving force is therefore limited by the pure friction between the belt and the sheave. Some designers prefer flat belts for delicate machinery because the belt *will* slip if the torque tends to rise to a level high enough to damage the machine.

Synchronous belts, sometimes called *timing belts* [see Figure 7–3(b)], ride on sprockets having mating grooves into which the teeth on the belt seat. This is a positive drive, limited only by the tensile strength of the belt and the shear strength of the teeth. See Section 7–5.

Some cog belts, such as that shown in Figure 7–3(a), are applied to standard V-grooved sheaves. The cogs give the belt greater flexibility and higher efficiency compared with standard belts. They can operate on smaller sheave diameters.

A widely used type of belt, particularly in industrial drives and vehicular applications, is the *V-belt drive*, shown in Figures 7–3(a) and 7–4. The V-shape causes the belt to wedge tightly into the groove, increasing friction and allowing high torques to be transmitted before slipping occurs. Most belts have high-strength cords positioned at the pitch diameter of the belt cross section to increase the tensile strength of the belt. The cords, made from natural fibers, synthetic strands, or steel, are embedded in a firm rubber compound to provide the flexibility needed to allow the belt to pass around the sheave. Often an outer fabric cover is added to give the belt good durability.

The groove angle ranges from 34° to 42° depending on the belt cross-section style and the pitch diameter. Refer to manufacturer's data for sheaves.

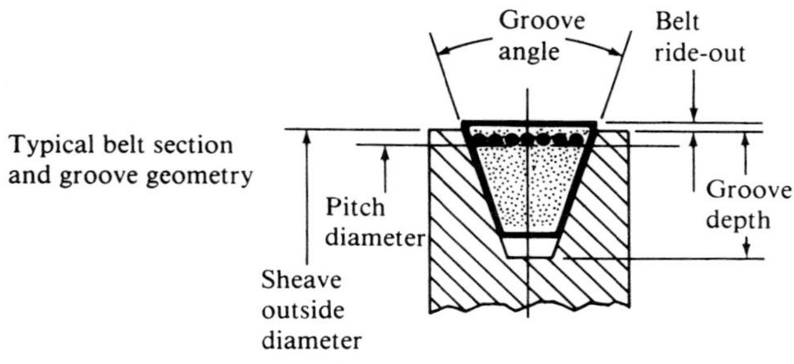

FIGURE 7–4 Cross section of V-belt and sheave groove

7–3 V-BELT DRIVES

The typical arrangement of the elements of a V-belt drive is shown in Figure 7–2. The important observations to be derived from this arrangement are summarized here:

1. The pulley, with a circumferential groove carrying the belt, is called a *sheave* (usually pronounced "shiv").
2. The size of a sheave is indicated by its pitch diameter, slightly smaller than the outside diameter of the sheave.
3. The speed ratio between the driving and the driven sheaves is inversely proportional to the ratio of the sheave pitch diameters. This follows from the observation that there is no slipping (under normal loads). Thus, the linear speed of the pitch line of both sheaves is the same as and equal to the belt speed, v_b. Then

$$v_b = R_1\omega_1 = R_2\omega_2 \qquad (7\text{–}1)$$

But $R_1 = D_1/2$ and $R_2 = D_2/2$. Then

$$v_b = \frac{D_1\omega_1}{2} = \frac{D_2\omega_2}{2} \qquad (7\text{–}1A)$$

The angular velocity ratio is

$$\frac{\omega_1}{\omega_2} = \frac{D_2}{D_1} \qquad (7\text{–}2)$$

4. The relationships between pitch length, L, center distance, C, and the sheave diameters are

$$L = 2C + 1.57(D_2 + D_1) + \frac{(D_2 - D_1)^2}{4C} \qquad (7\text{–}3)$$

$$C = \frac{B + \sqrt{B^2 - 32(D_2 - D_1)^2}}{16} \qquad (7\text{–}4)$$

where $B = 4L - 6.28(D_2 + D_1)$

5. The angle of contact of the belt on each sheave is

$$\theta_1 = 180^\circ - 2\sin^{-1}\left[\frac{D_2 - D_1}{2C}\right] \qquad (7\text{–}5)$$

$$\theta_2 = 180^\circ + 2\sin^{-1}\left[\frac{D_2 - D_1}{2C}\right] \qquad (7\text{–}6)$$

These angles are important because commercially available belts are rated with an assumed contact angle of 180°. This will occur only if the drive ratio is 1 (no speed change). The angle of contact on the smaller of the two sheaves of unequal diameters will always be less than 180°, requiring a lower power rating.

6. The length of the span between the two sheaves, over which the belt is unsupported, is

$$S = \sqrt{C^2 - \left[\frac{D_2 - D_1}{2}\right]^2} \qquad (7\text{–}7)$$

This is important for two reasons: You can check the proper belt tension by measuring the amount of force required to deflect the belt at the middle of the span by a given amount. Also, the tendency for the belt to vibrate or whip is dependent on this length.

7. The contributors to the stress in the belt are as follows:
   (a) The tensile force in the belt, maximum on the tight side of the belt.
   (b) The bending of the belt around the sheaves, maximum as the tight side of the belt bends around the smaller sheave.
   (c) Centrifugal forces created as the belt moves around the sheaves.

   The maximum total stress occurs where the belt enters the smaller sheave, and the bending stress is a major part. Thus, there are recommended minimum sheave diameters for standard belts. Using smaller sheaves drastically reduces belt life.
8. The design value of the ratio of the tight side tension to the slack side tension is 5.0 for V-belt drives. The actual value may range as high as 10.0.

Standard Belt Cross Sections

Commercially available belts are made to one of the standards shown in Figures 7–5 through 7–8. The alignment between the inch sizes and the metric sizes indicates that the paired sizes are actually the same cross section. A "soft conversion" was used to rename the familiar inch sizes with the number for the metric sizes giving the nominal top width in millimeters.

The nominal value of the included angle between the sides of the V-groove ranges from 34° to 42°. The angle on the belt may be slightly different to achieve a tight fit in the groove. Some belts are designed to "ride out" of the groove somewhat.

The designations shown for the various cross sections apply when the belt construction is like that shown in Figure 7–3(c). For the cog-type belt [Figure 7–3(a)] of the same cross section, the letter X is added to the designation. For example, a 5V belt has a smooth inner surface, while a 5VX belt of the cog type. Power transmission ratings are typically higher for the cogged type because they bend more easily around the sheaves with less stress in the belt.

Single automotive V-belts have cross sections ranging across the nine sizes shown in Figure 7–8 and may have either the smooth or cogged (X) type inner surfaces. Many applications employ the vee-band [Figure 7–3(d)] or the poly-rib style [Figure 7–3(f)]. Refer to References 3(a), 3(f), 5, or 8.

7–4 V-BELT DRIVE DESIGN

The factors involved in selection of a V-belt and the driving and driven sheaves and proper installation of the drive are summarized in this section. Abbreviated examples of the data available from suppliers are given for illustration. Catalogs contain extensive data, and step-by-step

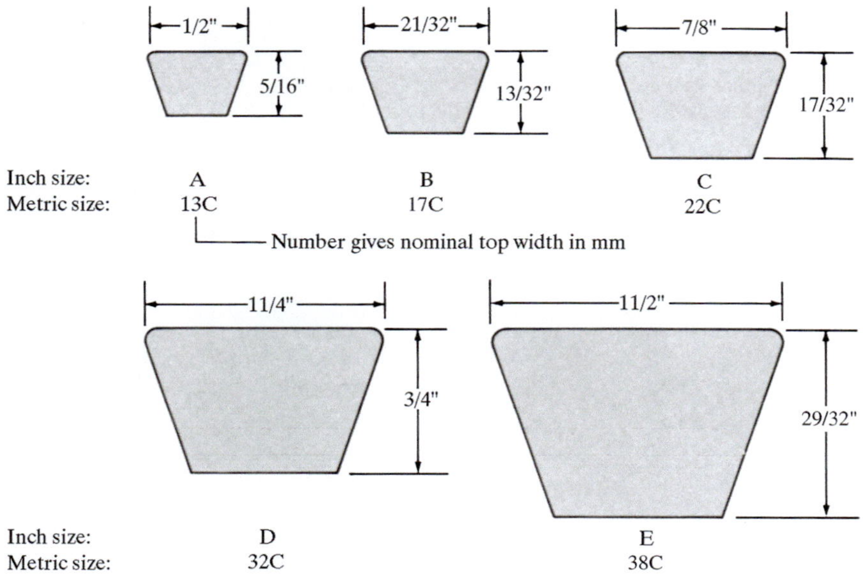

FIGURE 7–5 Heavy-duty industrial V-belts

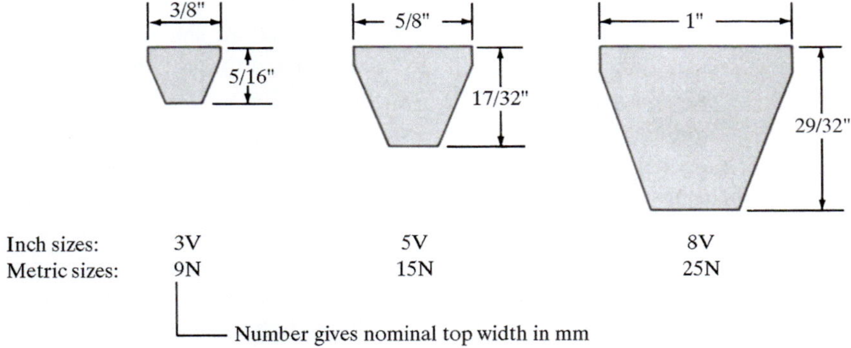

FIGURE 7–6 Industrial narrow-section V-belts

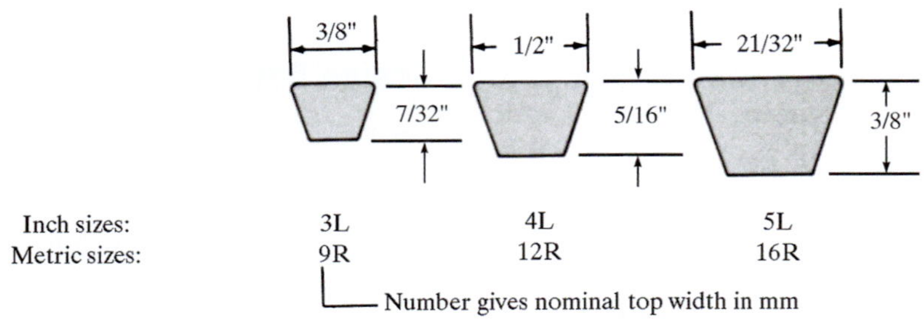

FIGURE 7–7 Light-duty, fractional horsepower (FHP) V-belts

instructions are given for their use. The basic data required for drive selection are the following:

- The rated power of the driving motor or other prime mover
- The service factor based on the type of driver and driven load
- The center distance
- The power rating for one belt as a function of the size and speed of the smaller sheave
- The belt length
- The size of the driving and driven sheaves
- The correction factor for belt length

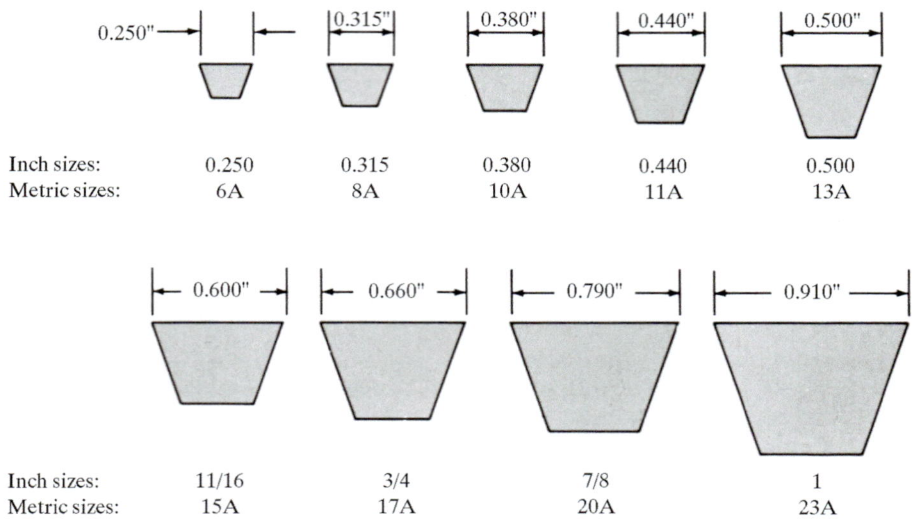

FIGURE 7–8 Automotive V-belts

- The correction factor for the angle of wrap on the smaller sheave
- The number of belts
- The initial tension on the belt

Many design decisions depend on the application and on space limitations. A few guidelines are given here:

- The recommended maximum reduction ratio for a plain V-belt drive is 6:1. For cogged belts it is 7:1. For higher desired ratios use two or more stages of reduction.
- Adjustment for the center distance must be provided in both directions from the nominal value. The center distance must be shortened at the time of installation to enable the belt to be placed in the grooves of the sheaves without force. Provision for increasing the center distance must be made to permit the initial tensioning of the drive and to take up for belt stretch. Manufacturers' catalogs give the data. One convenient way to accomplish the adjustment is the use of a take-up unit, as shown in Figure 14–10(b) and (c).
- If fixed centers are required, idler pulleys should be used. It is best to use a grooved idler on the inside of the belt, close to the large sheave. Adjustable tensioners are commercially available to carry the idler.
- The nominal range of center distances should be

$$D_2 < C < 3(D_2 + D_1) \qquad \textbf{(7–8)}$$

- The angle of wrap on the smaller sheave should be greater than 120°.
- Because of balancing, centrifugal stresses, belt whip, and other dynamic considerations, belt speeds should be under 5000 ft/min or the supplier of the sheaves should be consulted. A recommended maximum belt speed is 6500 ft/min.
- Consider an alternative type of drive, such as a gear type or chain, if the belt speed is less than 1000 ft/min.
- Avoid elevated temperatures around belts.
- Ensure that the shafts carrying mating sheaves are parallel and that the sheaves are in alignment so that the belts track smoothly into the grooves.
- In multibelt installations, matched belts are required. Match numbers are printed on industrial belts, with 50 indicating a belt length very close to nominal. Longer belts carry match numbers above 50; shorter belts below 50.
- Belts must be installed with the initial tension recommended by the manufacturer. Tension should be checked after the first few hours of operation because seating and initial stretch occur.
- Reported power transmission ratings typically are based on belt life of approximately 5000–7000 hours of operation and about 25 000 hours for the sheaves.

Design Data

Catalogs typically give several dozen pages of design data for the various sizes of belts and sheave combinations to ease the job of drive design. The data typically are given in tabular form. Graphical form is used here so that you can get a feel for the variation in performance with design choices. Any design made from the data in this book should be checked against a particular manufacturer's ratings before use.

The data given here are for the narrow-section belts: 3V, 5V, and 8V. These three sizes cover a wide range of power transmission capacities. Note that the cogged versions of these narrow section belts—3VX, 5VX, and 8VX—have higher power ratings and are reported separately in catalogs. Figure 7–9 can be used to choose the basic size for the belt cross section. Note that the power axis is *design power*, the rated power of the prime mover times the service factor from Table 7–1.

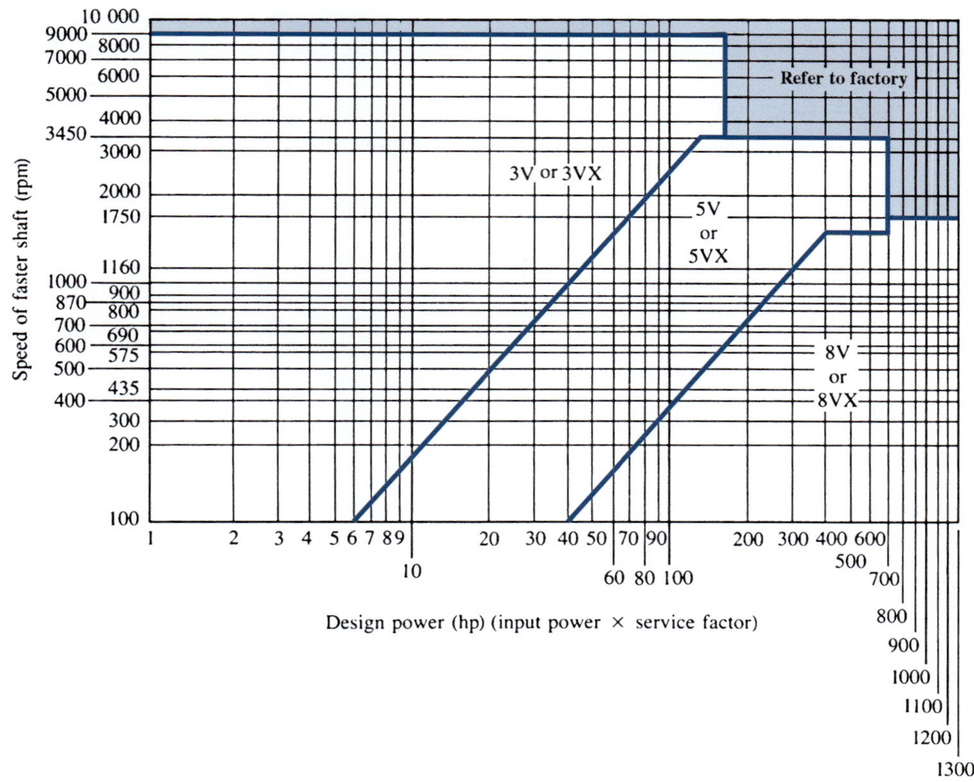

FIGURE 7–9 Selection chart for narrow-section industrial V-belts (Dayco Corp., Dayton, OH)

TABLE 7–1 V-Belt Service Factors[1]

	Driver type					
	AC motors: Normal torque[2] DC motors: Shunt-wound Engines: Multiple-cylinder			AC motors: High torque[3] DC motors: Series-wound, or compound-wound Engines: 4-cylinder or less		
Driven machine type	<6h per day	6–15 h per day	>15h per day	<6h per day	6–15 h per day	>15h per day
Smooth loading	1.0	1.1	1.2	1.1	1.2	1.3
Agitators, light conveyors centrifugal pumps fans and blowers under 10 HP (7.5 kW)						
Light shock loading	1.1	1.2	1.3	1.2	1.3	1.4
Generators, machine tools mixers, fans and blowers over 10 HP (7.5 kW) gravel conveyors						
Moderate shock loading	1.2	1.3	1.4	1.4	1.5	1.6
Bucket elevators, piston pumps textile machinery, hammer mills heavy conveyors, pulverizers						
Heavy shock loading	1.3	1.4	1.5	1.5	1.6	1.8
Crushers, ball mills, hoists rubber mills, and extruders						
Machinery that can choke	2.0	2.0	2.0	2.0	2.0	2.0

[1]Factors given are for speed reducers. For speed increases, multiply listed factors by 1.2.

[2]Synchronous, split-phase, three-phase with starting torque or breakdown torque less than 175% of full-load torque.

[3]Single-phase, three-phase with starting torque or breakdown torque greater than 175% of full-load torque.

Reference 3(c) gives the generic formula for the power rating of a V-belt as,

$$P_{nom} = K(P_b + \Delta P_R + \Delta P_L)$$

Where,

K = factor based on the angle of wrap on the sheave

P_b = Basic power rating for a ratio of 1.0 and a set belt length

ΔP_R = Added power capacity based on speed ratio

ΔP_L = Added power capacity based on belt length

Additional detail is provided in the reference for the individual terms. Manufacturers provide rating data for the particular styles and quality factors for their products. This section gives sample data to demonstrate the process for specifying a particular belt design.

Figures 7–10, 7–11, and 7–12 give the rated power per belt for the three cross sections as a function of the pitch diameter of the smaller sheave and its speed of rotation. The labeled vertical lines in each figure give the standard sheave pitch diameters available.

The basic power rating for a speed ratio of 1.00 is given as the solid curve. A given belt can carry a greater power as the speed ratio increases, up to a ratio of approximately 3.38. Further increases have little effect and may also lead to trouble with the angle of wrap on the smaller sheave. Figure 7–13 is a plot of the data for power to be added to the basic rating as a function of speed ratio for the 5V belt size. The catalog data are given in a stepwise fashion. The maximum power added, for ratios of above 3.38, was used to draw the dashed curves in Figures 7–10, 7–11, and 7–12. In most cases, a rough interpolation between the two curves is satisfactory.

Figure 7–14 gives the value of a correction factor, C_θ, as a function of the angle of wrap of the belt on the small sheave.

Figure 7–15 gives the value of the correction factor, C_L, for belt length. A longer belt is desirable because it reduces the frequency with which a given part of the belt encounters the stress peak as it enters the small sheave. Only certain standard belt lengths are available (Table 7–2).

Example Problem 7–1 illustrates the use of the design data.

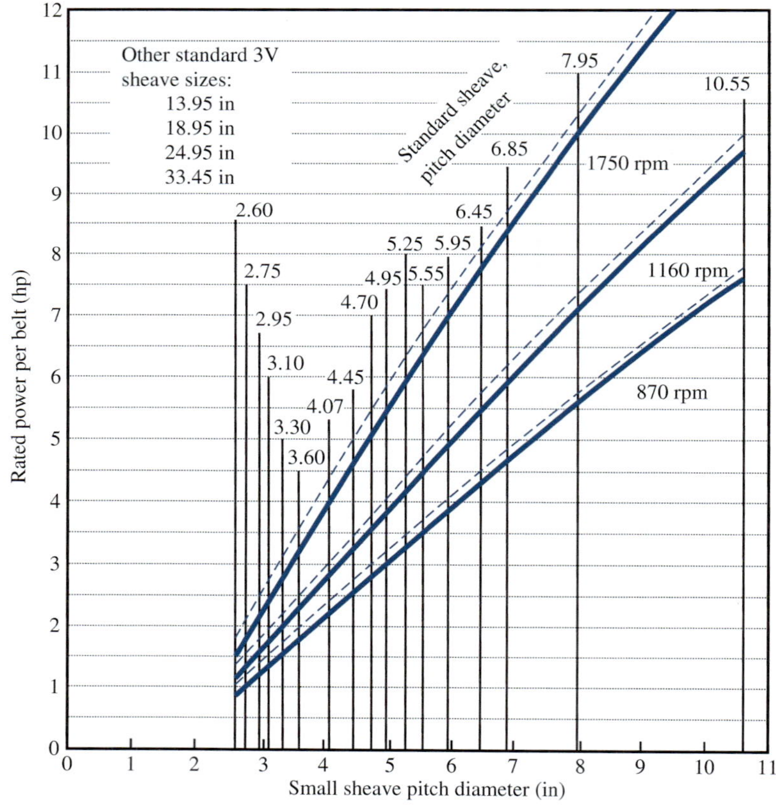

FIGURE 7–10 Power rating: 3V belts

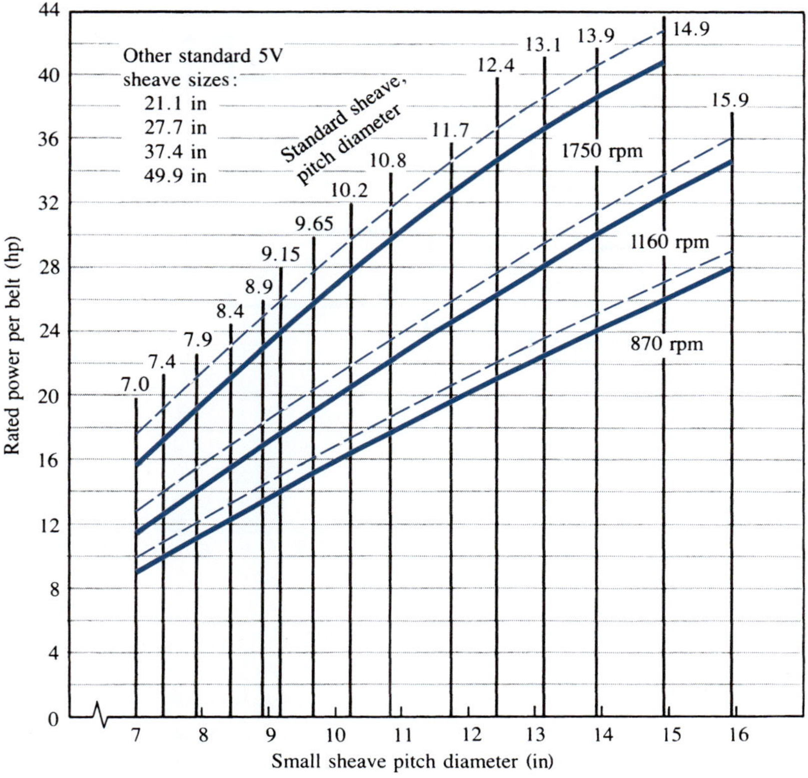

FIGURE 7–11 Power rating: 5V belts

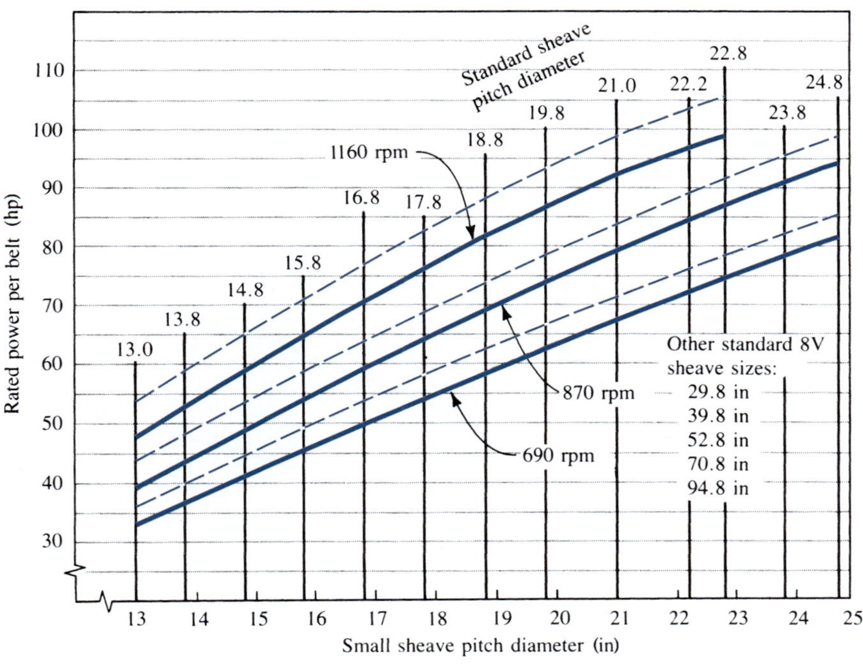

FIGURE 7–12 Power rating: 8V belts

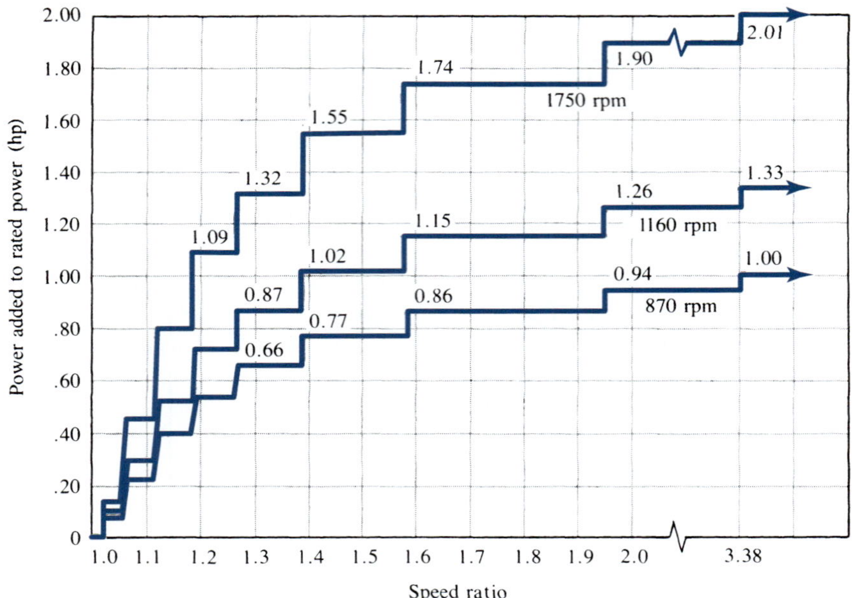

FIGURE 7–13 Power added versus speed ratio: 5V belts

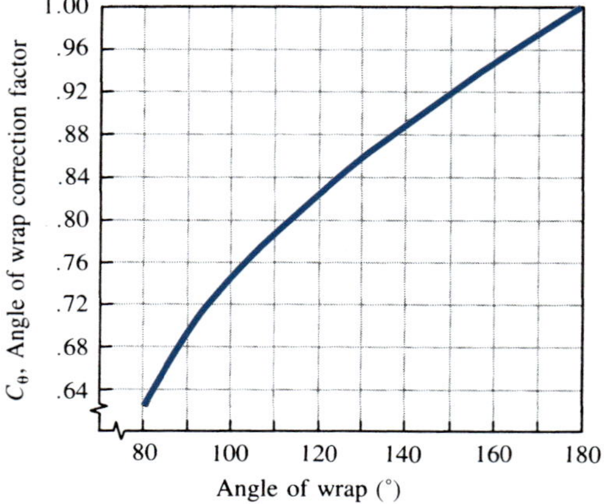

FIGURE 7–14 Angle of wrap correction factor, C_θ

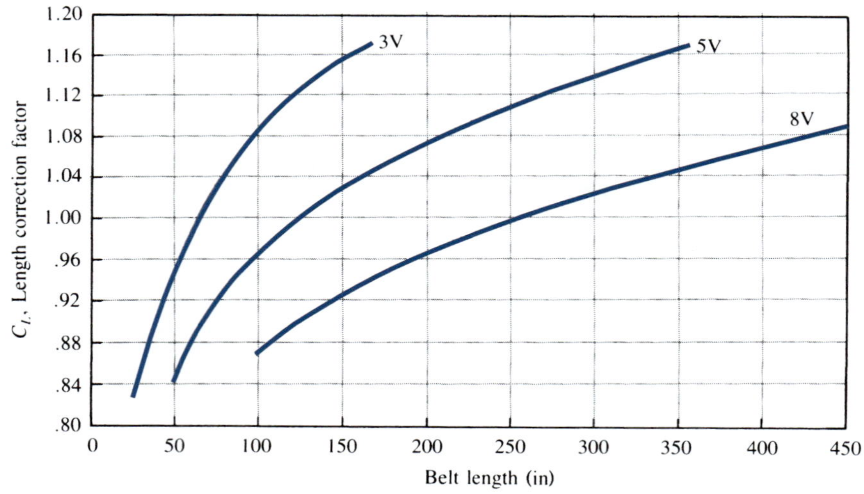

FIGURE 7–15 Belt length correction factor, C_L

TABLE 7–2 Standard Belt Lengths for 3V, 5V, and 8V Belts (in)

3V only	3V and 5V	3V, 5V, and 8V	5V and 8V	8V only
25	50	100	150	375
26.5	53	106	160	400
28	56	112	170	425
30	60	118	180	450
31.5	63	125	190	475
33.5	67	132	200	500
35.5	71	140	212	
37.5	75		224	
40	80		236	
42.5	85		250	
45	90		265	
47.5	95		280	
			300	
165			315	
			335	
			355	

Example Problem 7–1 Design a V-belt drive that has the input sheave on the shaft of an electric motor (normal torque) rated at 50.0 hp at 1160-rpm, full-load speed. The drive is to a bucket elevator in a potash plant that is to be used 12 hours (h) daily at approximately 675 rpm.

Solution Objective Design the V-belt drive.

Given

Power transmitted = 50 hp to bucket elevator

Speed of motor = 1160 rpm; output speed = 675 rpm

Analysis Use the design data presented in this section. The solution procedure is developed within the Results section of the problem solution.

Results ***Step 1.*** Compute the design power. From Table 7–1, for a normal torque electric motor running 12 h daily driving a bucket elevator, the service factor is 1.30. Then the design power is 1.30(50.0 hp) = 65.0 hp.

Step 2. Select the belt section. From Figure 7–9, a 5V belt is recommended for 70.0 hp at 1160-rpm input speed.

Step 3. Compute the nominal speed ratio:

$$\text{Ratio} = 1160/675 = 1.72$$

Step 4. Compute the driving sheave size that would produce a belt speed of 4000 ft/min, as a guide to selecting a standard sheave:

$$\text{Belt speed} = v_b = \frac{\pi D_1 n_1}{12}\ \text{ft/min}$$

Then the required diameter to give $v_b = 4000$ ft/min is

$$D_1 = \frac{12 v_b}{\pi n_1} = \frac{12(4000)}{\pi n_1} = \frac{15\,279}{n_1} = \frac{15\,279}{1160} = 13.17\ \text{in}$$

Step 5. Select trial sizes for the input sheave, and compute the desired size of the output sheave. Select a standard size for the output sheave, and compute the actual ratio and output speed.

TABLE 7–3 Trial Sheave Sizes for Example Problem 7–1

Standard driving sheave size, D_1	Approximate driven sheave size ($1.72D_1$)	Nearest standard sheave, D_2	Actual output speed (rpm)
13.10	22.5	21.1	720
12.4	**21.3**	**21.1**	**682**
11.7	20.1	21.1	643
10.8	18.6	21.1	594
10.2	17.5	15.9	744
9.65	16.6	15.9	704
9.15	**15.7**	**15.9**	**668**
8.9	15.3	14.9	693

For this problem, the trials are given in Table 7–3 (diameters are in inches).

The two trials in **boldface** in Table 7–3 give only about 1% variation from the desired output speed of 675 rpm, and the speed of a bucket elevator is not critical. Because no space limitations were given, let's choose the larger size; $D_1 = 12.4$ in; $D_2 = 21.1$ in.

We can now compute the actual belt speed:

$$v_b = \pi D_1 n_1/12 = \pi(12.4 \text{ in})(1160 \text{ rpm})/12 = 3766 \text{ ft/min}$$

Step 6. Determine the rated power from Figure 7–10, 7–11, or 7–12.

For the 5V belt that we have selected, Figure 7–11 is appropriate. For a 12.4-in sheave at 1160 rpm, the basic rated power is 26.4 hp. Multiple belts will be required. The ratio is relatively high, indicating that some added power rating can be used. This value can be estimated from Figure 7–11 or taken directly from Figure 7–13 for the 5V belt. Power added is 1.15 hp. Then the actual rated power is 26.4 + 1.15 = 27.55 hp.

Step 7. Specify a trial center distance.

We can use Equation (7–8) to determine a nominal acceptable range for C:

$$D_2 < C < 3(D_2 + D_1)$$
$$21.1 < C < 3(21.1 + 12.4)$$
$$21.1 < C < 100.5 \text{ in}$$

In the interest of conserving space, let's try $C = 24.0$ in.

Step 8. Compute the required belt length from Equation (7–3):

$$L = 2C + 1.57(D_2 + D_1) + \frac{(D_2 - D_1)^2}{4C}$$

$$L = 2(24.0) + 1.57(21.1 + 12.4) + \frac{(21.1 - 12.4)^2}{4(24.0)} = 101.4 \text{ in}$$

Step 9. Select a standard belt length from Table 7–2, and compute the resulting actual center distance from Equation (7–4).

In this problem, the nearest standard length is 100.0 in. Then, from Equation (7–4),

$$B = 4L - 6.28(D_2 + D_1) = 4(100) - 6.28(21.1 + 12.4) = 189.6$$

$$C = \frac{189.6 + \sqrt{(189.6)^2 - 32(21.1 - 12.4)^2}}{16} = 23.30 \text{ in}$$

Step 10. Compute the angle of wrap of the belt on the small sheave from Equation (7–5):

$$\theta_1 = 180^\circ - 2\sin^{-1}\left[\frac{D_2 - D_1}{2C}\right] = 180^\circ - 2\sin^{-1}\left[\frac{21.1 - 12.4}{2(23.30)}\right] = 158^\circ$$

Step 11. Determine the correction factors from Figures 7–14 and 7–15. For $\theta = 158°$, $C_\theta = 0.94$. For $L = 100$ in, $C_L = 0.96$.

Step 12. Compute the corrected rated power per belt and the number of belts required to carry the design power:

$$\text{Corrected power} = C_\theta C_L P = (0.94)(0.96)(27.55\text{ hp}) = 24.86\text{ hp}$$

$$\text{Number of belts} = 65.0/24.86 = 2.61\text{ belts (Use 3 belts.)}$$

Comments **Summary of Design**

Input: Electric motor, 50.0 hp at 1160 rpm

Service factor: 1.3

Design power: 65.0 hp

Belt: 5V cross section, 100-in length, 3 belts

Sheaves: Driver, 12.4-in pitch diameter, 3 grooves, 5V. Driven, 21.1-in pitch diameter, 3 grooves, 5V

Actual output speed: 682 rpm

Center distance: 23.30 in

Belt Tension

The initial tension given to a belt is critical because it ensures that the belt will not slip under the design load. At rest, the two sides of the belt have the same tension. As power is being transmitted, the tension in the tight side increases while the tension in the slack side decreases. Without the initial tension, the slack side would go totally loose, and the belt would not seat in the groove; thus, it would slip. Manufacturers' catalogs give data for the proper belt-tensioning procedures.

7–5 SYNCHRONOUS BELT DRIVES

Synchronous belts are constructed with ribs or teeth across the underside of the belt, as shown in Figure 7–3(b). The teeth mate with corresponding grooves in the driving and driven pulleys, called *sprockets*, providing a positive drive without slippage. Therefore, there is a fixed relationship between the speed of the driver and the speed of the driven sprocket. For this reason synchronous belts are often called *timing belts*. In contrast, V-belts can creep or slip with respect to their mating sheaves, especially under heavy loads and varying power demand. Synchronous action is critical to the successful operation of such systems as printing, material handling, packaging, and assembly. Synchronous belt drives are increasingly being considered for applications in which gear drives or chain drives had been used previously.

Figure 7–16 shows a synchronous belt mating with the toothed driving sprocket. Typical driving and driven sprockets are shown in Figure 7–17. At least one of the two sprockets will have side flanges to ensure that the belt does not move axially. Figure 7–18 shows commonly available commercial shapes for synchronous belts. Two series are in use, Metric Sizes (mm) and U.S. customary Sizes (in). The shapes of the cross sections in a part (a) of the figure are drawn full size, showing the pitch lengths from the center of one tooth to the center of the next adjacent tooth. Table 7–4 gives some overall data for the range of belt widths, number of teeth in pulleys, and belt lengths available. Check individual manufacturers' catalogs for available stock sizes. Some manufacturers produce synchronous belts with other shapes for the cogs on the underside of the belt as shown in part (b) of the figure.

Figure 7–3(b) shows detail of the construction of the cross section of a synchronous belt. The tensile strength is provided predominantly by high-strength cords made from fiberglass or similar materials. The cords are encased in a flexible rubber backing material, and the teeth are formed integrally with the backing. Often a fabric covering is used

FIGURE 7–16 Synchronous belt on driving sprocket (Copyright Baldor, used by permission)

FIGURE 7–17 Driving and driven sprockets for synchronous belt drive (Copyright Baldor, used by permission)

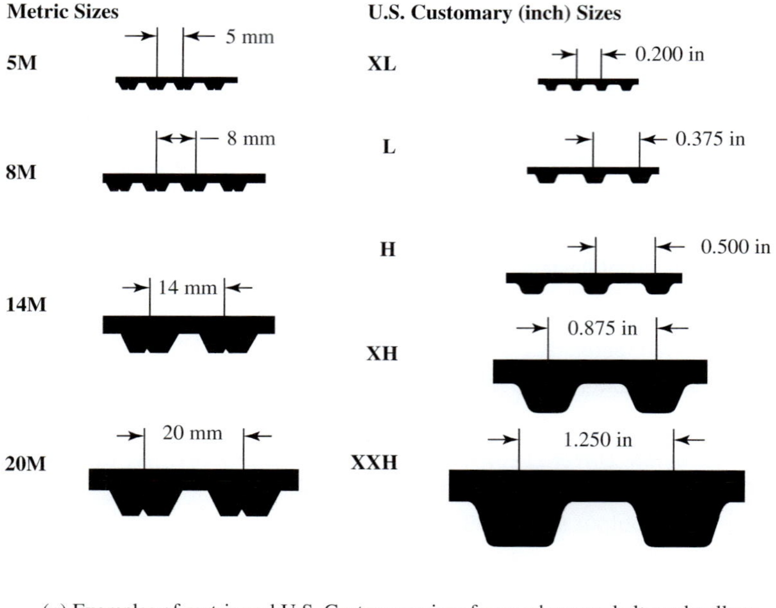

(*a*) Examples of metric and U.S. Customary sizes for synchronous belts and pulleys

(*b*) Alternate shapes for synchronous belt cogs

FIGURE 7–18 Sizes and shapes for synchronous belt cogs

TABLE 7–4 Types and Sizes of Synchronous Belt Drives

Metric Sizes (Dimensions in mm)

Pitch	Belt widths (typical)	Range of number of teeth on pulleys (typical)	Range of belt pitch lengths (mm)	Range of belt pitch lengths No.of teeth
5	15, 25	32–112	350–2000	70–400
8	20, 30, 50, 85	22–192	480–4400	60–550
14	40, 55, 85, 115, 170	28–216	966–6860	69–490
20	115, 170, 230, 290, 340	34–216	2000–6600	100–330

U.S. Customary Sizes (Dimensions in inches)

Pitch	Belt widths (typical)	Range of number of teeth in pulleys	Range of belt pitch lengths (in)	Range of belt pitch lengths No. of teeth
XL–0.200	0.25, 0.375	10–72	5.00–77.00	25–385
L–0.375	0.50, 075, 1.00	10–120	12.38–90.00	33–240
H–0.500	0.75, 1.00, 1.50, 2.00, 3.00	14–156	21.00–170.0	42–340
XH–0.875	2.00, 3.00, 4.00, 5.00, 6.00, 8.00	18–120	50.75–175.0	58–200
XXH–1.250	2.00, 3.00, 4.00, 5.00, 6.00, 8.00	18–90		

Note: Check manufacturers' catalogs for available stock sizes.

on those parts of the belt that contact the sprockets to provide additional wear resistance and higher net shear strength for the teeth. Various widths of the belts are available for each given pitch to provide a wide range of power transmission capacity.

Commercially available sprockets typically employ split-taper bushings in their hubs with a precise bore that provides a clearance of only 0.001 to 0.002 in (0.025 to 0.050 mm) relative to the shaft diameter on which it is to be mounted. Smooth, balanced, concentric operation results.

The process of selecting appropriate components for a synchronous belt drive is similar to that already discussed for V-belt drives. Manufacturers provide selection guides similar to those shown in Figure 7–19 that give the relationship between design power and the rotational speed of the smaller sprocket. These are used to determine the basic belt pitch required. Also, numerous pages of performance data are given showing the power transmission capacity for many combinations of belt width, driving and driven sprocket size, and center distances between the axes of the sprockets for specific belt lengths. In general the selection process involves the following steps. Refer to data and design procedures for specific manufacturers as listed in Internet sites 3–6, 8, 9, 11, 14, and 16. Sites 7 and 13 include standards for synchronous belt drive systems.

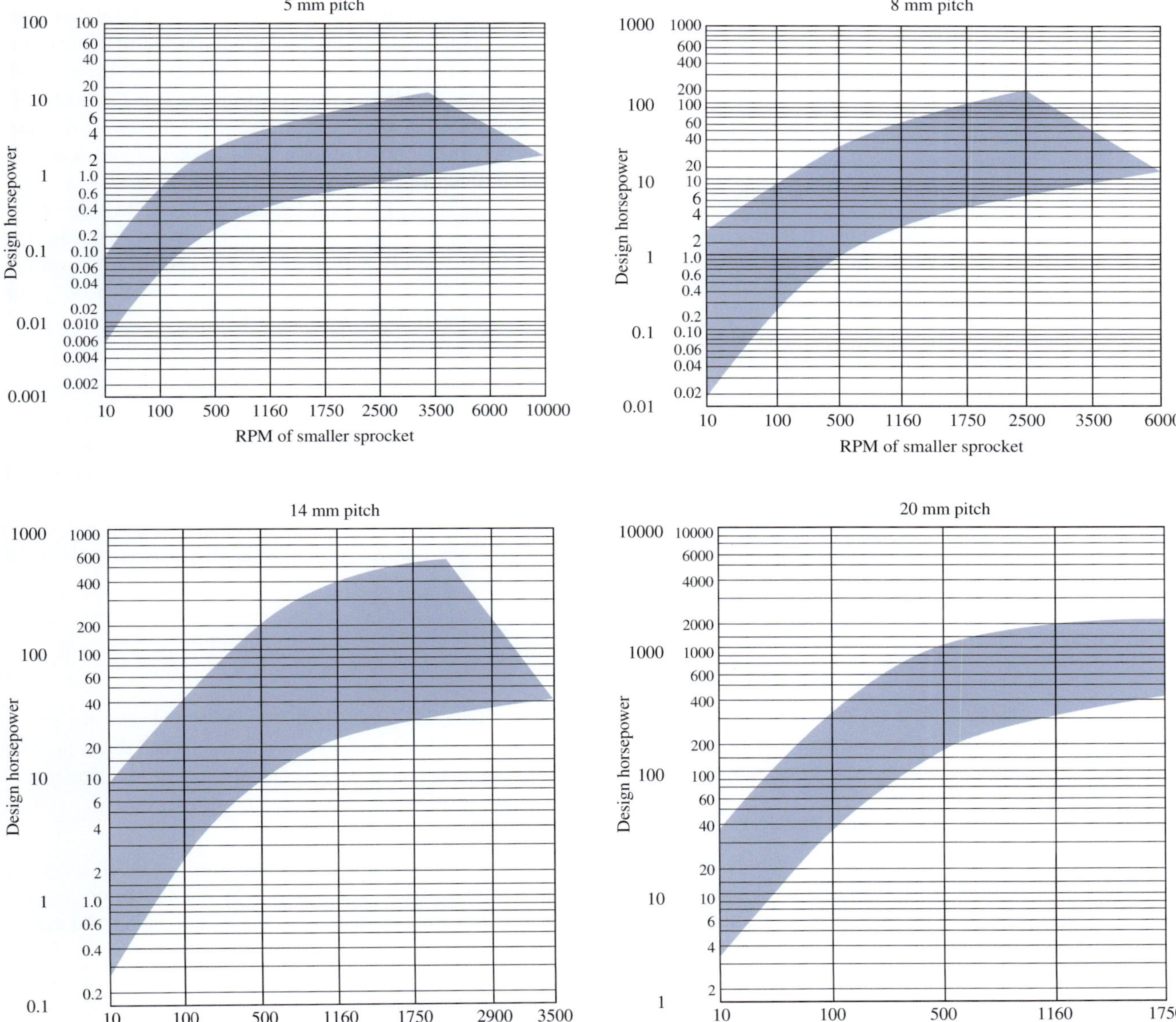

FIGURE 7–19 Belt pitch selection guide for synchronous belts

GENERAL SELECTION PROCEDURE FOR SYNCHRONOUS BELT DRIVES

1. Specify the speed of the driving sprocket (typically a motor or engine) and the desired speed of the driven sprocket.
2. Specify the rated power for the driving motor or engine.
3. Determine a service factor, using manufacturers' recommendations, considering the type of driver and the nature of the driven machine.
4. Calculate the design power by multiplying the driver rated power by the service factor.
5. Determine the required pitch of the belt from a specific manufacturer's data.
6. Calculate the speed ratio between the driver and the driven sprocket.
7. Select several candidate combinations of the number of teeth in the driver sprocket to that in the driven sprocket that will produce the desired ratio.
8. Using the desired range of acceptable center distances, determine a standard belt length that will produce a suitable value.
9. A belt-length correction factor may be required. Catalog data will show factors less than 1.0 for shorter center distances and greater than 1.0 for longer center distances. This reflects the frequency with which a given part of the belt encounters the high-stress area as it enters the smaller sprocket. Apply the factor to the rated power capacity for the belt.
10. Specify the final design details for the sprockets such as flanges, type and size of bushings in the hub, and the bore size to match the mating shafts.
11. Summarize the design, check compatibility with other components of the system, and prepare purchasing documents.

Installation of the sprockets and the belt requires a nominal amount of center distance allowance to enable the belt teeth to slide into the sprocket grooves without force. Subsequently, the center distance will normally have to be adjusted outward to provide a suitable amount of initial tension as defined by the manufacturer. The initial tension is typically less than that required for a V-belt drive. Idlers can be used to take up slack if fixed centers are required between the driver and driven sprockets. However, they may decrease the life of the belt. Consult the manufacturer.

In operation, the final tension in the tight side of the belt is much less than that developed by a V-belt and the slack side tension is virtually zero. The results are lower net forces in the belt, lower side loads on the shafts carrying the sprockets, and reduced bearing loads.

Example Problem 7–2 Design a synchronous belt drive to meet the same requirements as listed for Example Problem 7–1 for which a V-belt drive was specified. Work toward a nominal center distance of 24 in (609 mm) similar to that used before.

Solution

Objective: Design the synchronous belt drive specifying the belt type and size and the numbers of teeth in the driving and driven sprocket.

Given: Power transmitted = P = 50 hp (37.3 kW) to a bucket elevator

Speed of motor = 1160 rpm; target output speed = 675 rpm

Analysis: Use the *General Selection Procedure for Synchronous Belt Drives.* Note that the actual selection procedure used here is from a catalog for a commercially available supplier of synchronous belt drives in order to access their large number of sizes and types of components. Other acceptable designs are possible.

Results:

Steps 1 and 2 are satisfied with the given data.

Step 3. Service factors for synchronous belt drives are similar to those used for V-belt drives as listed in Table 7–1 but they are typically somewhat higher and they vary by manufacturer because of "add-on" factors related to pulley speed and operating conditions. For this application, the manufacturer recommends a service factor of SF = 1.6.

Step 4. Design power = $P \times \text{SF} = 50 \times 1.6 = 80$ hp (59.7 kW)

Step 5. Let's use a metric style belt. From Figure 7–19 the 14 mm belt is recommended for 80 hp and an input speed of 1160 rpm for the sprocket mounted on the motor shaft.

Step 6. Desired ratio = 1160/675 = 1.72

Step 7. The catalog identified ten configurations that met basic design requirements. The one cited here was chosen because of its center distance being closest to the desired value.

Steps 8–11. The selected design is summarized here:

Belt: Pitch = 14 mm (0.551 in); Width = 55 mm (2.17 in); Length = 2100 mm (82.68 in), 150 teeth

Center distance: C = 597 mm (23.5 in)

Driver sprocket: N_1 = 48 teeth; Pitch diameter = 213.9 mm (8.421 in)

Driven sprocket: N_2 = 80 teeth; Pitch diameter = 356.5 mm (14.036 in)

Output speed: 1160 rpm × (N_1/N_2) = 1160(48/80) = 696 rpm

Belt speed: v_b = 2558 ft/min (13.0 m/s)

Belt pull: F_1 = 3727 N (838 lb)

Comments Comparisons of these results with those for the V-belt drive found in Example Problem 7–1 are shown below:

	V-belt drive	Synchronous belt drive
Center distance:	23.3 in	23.5 in
Pitch diameter–Driver:	12.4 in	8.42 in
Pitch diameter–Driven:	21.1 in	14.04 in
Belt width:	3 × 0.625 in(5V) = 1.88 in	2.17 in (55 mm)
Belt speed:	3766 ft/min	2558 ft/min

7–6 CHAIN DRIVES

A chain is a power transmission element made as a series of pin-connected links. The design provides for flexibility while enabling the chain to transmit large tensile forces. See References 1–3 and Internet sites 1, 4, 6–12, and 14–15 for more technical information and manufacturers' data.

When transmitting power between rotating shafts, the chain engages mating toothed wheels, called sprockets. Figure 7–20 shows a typical chain drive.

The most common type of chain is the *roller chain,* in which the roller on each pin provides exceptionally low friction between the chain and the sprockets. Other types include a variety of extended link designs used mostly in conveyor applications (see Figure 7–21).

Roller chain is classified by its *pitch,* the distance between corresponding parts of adjacent links. The pitch is usually illustrated as the distance between the centers of adjacent pins. U.S. Standard roller chain carries a size designation from 40 to 240, as listed in Table 7–5. See Reference 2. The digits (other than the final zero) indicate the pitch of the chain in eighths of an inch, as in the table. For example, the no. 100 chain has a pitch of 10/8 or $1\frac{1}{4}$ in. A series of heavy-duty sizes, with the suffix H on the designation (60H–240H), has the same basic dimensions as

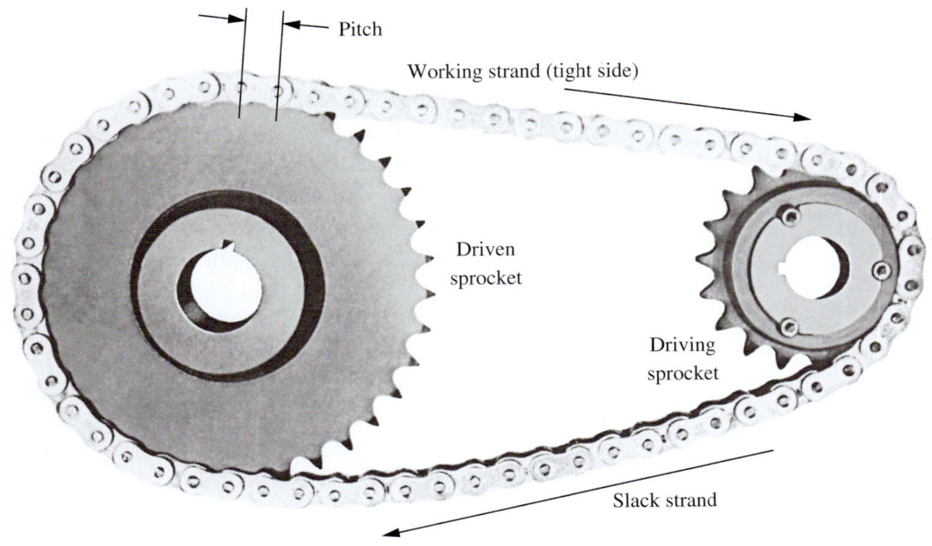

FIGURE 7–20 Roller chain drive (Rexnord Industries, LLC, Milwaukee, WI)

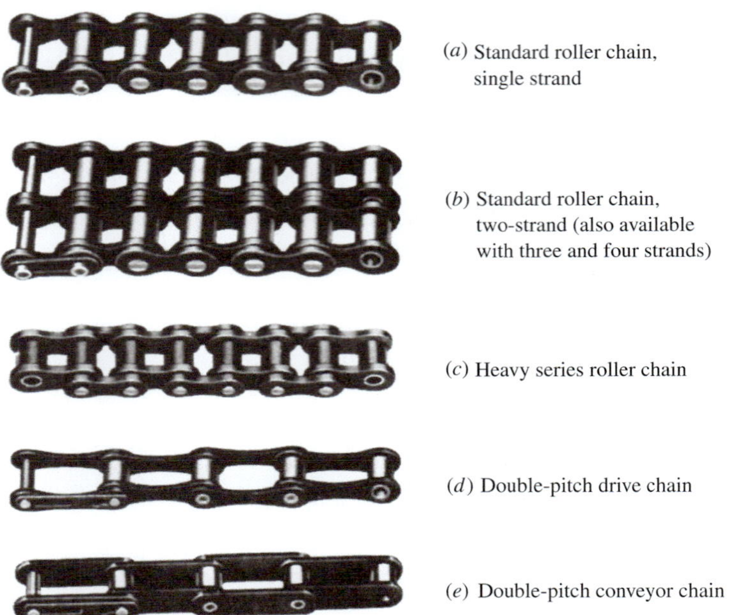

FIGURE 7–21 Some roller chain styles (Rexnord Industries, LLC, Milwaukee, WI)

the standard chain of the same number except for thicker side plates. In addition, there are the smaller and lighter sizes: 25, 35, and 41.

The average tensile strengths of the various chain sizes are also listed in Table 7–5. These data can be used for very low speed drives or for applications in which the function of the chain is to apply a tensile force or to support a load. It is recommended that only 10% of the average tensile strength be used in such applications. For power transmission, the rating of a given chain size as a function of the speed of rotation must be determined, as explained later in this chapter.

TABLE 7–5 U.S. Roller Chain Sizes

Chain number	Pitch (in)	Average tensile strength (lb)
25	1/4	925
35	3/8	2100
41	1/2	2000
40	1/2	3700
50	5/8	6100
60	3/4	8500
80	1	14 500
100	$1\frac{1}{4}$	24 000
120	$1\frac{1}{2}$	34 000
140	$1\frac{3}{4}$	46 000
160	2	58 000
180	$2\frac{1}{4}$	80 000
200	$2\frac{1}{2}$	95 000
240	3	130 000

Reference: ANS1 Standard B29.1.

ISO standards define several different chain types, data for three of which are listed in Table 7–6. One commonly used style from ISO-606 has basically the same design dimensions as for many of the standard U.S. roller chains. Then the pitch and dimensions for sprocket features and bore sizes are listed in the metric unit of mm making it more convenient to integrate familiar chain designs into an all-metric piece of equipment. ISO-3512 includes eight sizes of chain used for heavy-duty power transmission and lifting equipment. Some of the pitches for these chains are also equal to common U.S. sizes. Agricultural equipment such as tractor accessories, planters, harvesters, and mowers employ many chain drives to actuate moving systems. ISO-487 defines eight S-designations that cover a wide range of power transmission and tension pull applications. See Reference 3 and Internet sites 3, 7, 9, 14, and 15 for more information on metric-style chains and for manufacturers' data.

A wide variety of attachments are available to facilitate the application of roller chain to conveying or other material handling uses. Usually in the form of extended plates or tabs with holes provided, the attachments make it easy to connect rods, buckets, parts pushers, part support devices, or conveyor slats to the chain. Figure 7–22 shows some attachment styles.

Figure 7–23 shows a variety of chain types used especially for conveying and similar applications. Such chain typically has a longer pitch than standard roller chain (usually twice the pitch), and the link plates are heavier. The larger sizes have cast link plates.

TABLE 7–6 Metric Roller Chain Sizes and Strength Ratings

General Purpose Power Transmission

ISO-606

Chain Number	Pitch in	Pitch mm	Breaking Strength lb	Breaking Strength kN	Alternate Designation
04B	0.250	6.350	—	—	—
05B	0.315	8.000	989	4.4	—
06B	0.375	9.525	2 001	8.9	Metric 35
08B	0.500	12.700	4 002	17.8	Metric 40
10B	0.625	15.88	4 991	22.2	Metric 50
12B	0.75	19.05	6 497	28.9	Metric 60
16B	1.00	25.40	9 510	42.3	Metric 80
20B	1.25	31.75	14 501	64.5	Metric 100
24B	1.50	38.10	22 010	97.9	Metric 120
28B	1.75	44.45	29 002	129.0	Metric 140
32B	2.00	50.8	37 995	169.0	Metric 160
40B	2.50	63.5	58 993	262.4	Metric 200
48B	3.00	76.2	89 996	400.3	Metric 240
56B	3.50	88.9	122 010	542.7	—
64B	4.00	101.6	160 004	711.7	—
72B	4.50	114.3	202 001	898.5	—

Heavy Duty Power Transmission

ISO-3512

Chain Number	Pitch in	Pitch mm	Breaking Strength lb	Breaking Strength kN
2010	2.500	63.5	58 903	262.0
2512	3.067	77.9	84 982	378
2814	3.500	88.9	116 007	516
3315	4.073	103.5	133 993	596
3618	4.500	114.3	183 004	814
4020	5.000	127.0	236 960	1054
4824	6.000	152.4	341 951	1521
5628	7.000	177.8	464 928	2068

Power Transmission for Agricultural Uses

ISO-487

Chain Number	Pitch in	Pitch mm	Breaking Strength lb	Breaking Strength kN
S32	1.150	29.2	1 799	8.0
S42	1.375	34.9	6 003	26.7
S45	0.843	21.4	4 002	17.8
S52	1.500	38.1	4 002	17.8
S55	1.630	41.4	4 002	17.8
S62	1.650	41.9	6 003	26.7
S77	2.297	58.3	10 004	44.5
S88	2.609	66.3	10 004	44.5

Notes: Not all suppliers offer all sizes.

Breaking strength data must be verified with specific supplier.

(*a*) Slats assembled to attachments to form a flat conveying surface

(*b*) *V* block assembled to attachments to convey round objects of varying diameters

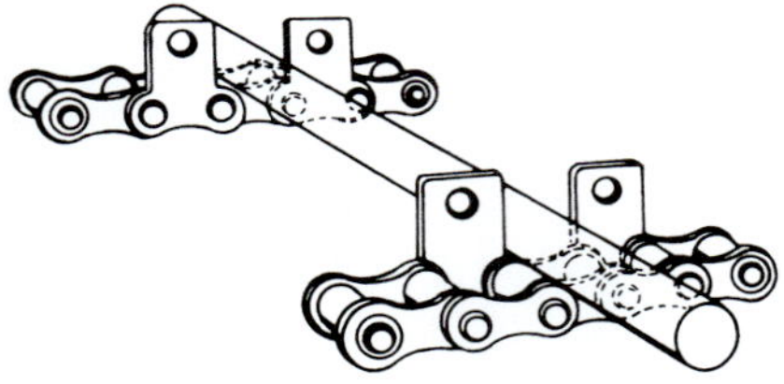

(*c*) Attachments used as spacers to convey and position long objects

FIGURE 7–22 Chain attachments (Rexnord Industries, LLC, Milwaukee, WI)

7–7 DESIGN OF CHAIN DRIVES

The rating of chain for its power transmission capacity considers three modes of failure: (1) fatigue of the link plates due to the repeated application of the tension in the tight side of the chain, (2) impact of the rollers as they engage the sprocket teeth, and (3) galling between the pins of each link and the bushings on the pins.

The ratings are based on empirical data with a smooth driver and a smooth load (service factor = 1.0) and with a rated life of approximately 15 000 h. The important variables are the pitch of the chain and the size and rotational speed of the smaller sprocket. Lubrication is critical to the satisfactory operation of a chain drive. Manufacturers recommend the type of lubrication method for given combinations of chain size, sprocket size, and speed. Details are discussed later.

Tables 7–7, 7–8, and 7–9 list the rated power for three sizes of standard chain: no. 40 (1/2 in), no. 60 (3/4 in), and no. 80 (1.00 in). These are typical of the types of data available for all chain sizes in manufacturers' catalogs. Notice these features of the data:

1. The ratings are based on the speed of the smaller sprocket and an expected life of approximately 15 000 hours.
2. For a given speed, the power capacity increases with the number of teeth on the sprocket. Of course, the larger the number of teeth, the larger the diameter of the sprocket. Note that the use of a chain with a small pitch on a large sprocket produces the quieter drive.
3. For a given sprocket size (a given number of teeth), the power capacity increases with increasing speed up to a point; then it decreases. Fatigue due to the tension in the chain governs at the low to moderate speeds; impact on the sprockets governs at the higher speeds. Each sprocket size has an absolute upper-limit speed due to the onset of galling between the pins and the bushings of the chain. This explains the abrupt drop in power capacity to zero at the limiting speed.
4. The ratings are for a single strand of chain. Although multiple strands do increase the power capacity, they do not provide a direct multiple of the single-strand capacity. Multiply the capacity in the tables by the following factors.

   Two strands: Factor = 1.7

   Three strands: Factor = 2.5

   Four strands: Factor = 3.3
5. The ratings are for a service factor of 1.0. Specify a service factor for a given application according to Table 7–10.

DESIGN GUIDELINES FOR CHAIN DRIVES

The following are general recommendations for designing chain drives:

1. The minimum number of teeth in a sprocket should be 17 unless the drive is operating at a very low speed, under 100 rpm.
2. The maximum speed ratio should be 7.0, although higher ratios are feasible. Two or more stages of reduction can be used to achieve higher ratios.
3. The center distance between the sprocket axes should be approximately 30 to 50 pitches (30–50 times the pitch of the chain).
4. The larger sprocket should normally have no more than 120 teeth.
5. The preferred arrangement for a chain drive is with the centerline of the sprockets horizontal and with the tight side on top.
6. The chain length must be an integral multiple of the pitch, and *an even number of pitches is recommended.* The center distance should be made adjustable to accommodate the chain length and to take up for tolerances and wear. Excessive sag on the slack side should be avoided, especially on drives that are not horizontal. A convenient relation between center distance (C), chain length (L), number of teeth in the small sprocket (N_1), and number of teeth in the large sprocket (N_2), expressed *in pitches,* is

Chain Length in Pitches

$$L = 2C + \frac{N_2 + N_1}{2} + \frac{(N_2 - N_1)^2}{4\pi^2 C} \qquad \textbf{(7–9)}$$

The center distance for a given chain length, again *in pitches,* is

Center Distance in Pitches

$$C = \frac{1}{4}\left[L - \frac{N_2 + N_1}{2} + \sqrt{\left[L - \frac{N_2 + N_1}{2}\right]^2 - \frac{8(N_2 - N_1)^2}{4\pi^2}}\right] \quad \textbf{(7–10)}$$

The computed center distance assumes no sag in either the tight or the slack side of the chain, and thus it is a *maximum*. Negative tolerances or adjustment must be provided. Adjustment for wear must also be provided.

7. The pitch diameter of a sprocket with N teeth for a chain with a pitch of p is

Pitch Diameter of Sprocket

$$D = \frac{p}{\sin(180°/N)} \text{ inches or mm} \quad \textbf{(7–11)}$$

8. The minimum sprocket diameter and therefore the minimum number of teeth in a sprocket are often limited by the size of the shaft on which it is mounted. Check the sprocket catalog.
9. The arc of contact, θ_1, of the chain on the smaller sprocket should be greater than 120°.

Angle or Wrap Smaller Sprocket

$$\theta_1 = 180° - 2\sin^{-1}[(D_2 - D_1)/2C] \quad \textbf{(7–12)}$$

10. For reference, the arc of contact, θ_2, on the larger sprocket is

Angle or Wrap Large Sprocket

$$\theta_2 = 180° + 2\sin^{-1}[(D_2 - D_1)/2C] \quad \textbf{(7–13)}$$

Mill, narrow series
(drive and conveyor sizes)
Offset cast-link chain used primarily in the lumber industry for conveyor applications.

Combination mill
(wide conveyor sizes)
Cast block links and steel sidebar construction for drag conveyor applications.

Heavy-duty drag chain
Cast steel offset block links. Used in ash and clinker conveyors.

Pintle chain
Chain constructed of a series of cast offset links coupled by steel pins or rivets. Suitable for slow-to moderate-speed drive, conveyor and elevator service.

Roller-top transfer
Cast links with top rollers used in several strands to convey material transversely.

Roof-top
Cast roof-shaped links used in several strands on transfer conveyors.

Detachable
Consists of unit links, each with an open-type hook that flexes on the end bar of the adjacent link. Used for slow- to moderate-speed drive and conveyor application.

Drop-forged
Drop-forged inner and outer links coupled by headed pins. Used for trolley, scraper, flight, and similar conveyors.

FIGURE 7–23 Conveyor chains (Rexnord, Industries, LLC, Milwaukee, WI)

TABLE 7–7 Horsepower Ratings—Single Strand Roller Chain No. 40

No. of teeth	0.500 inch pitch										Rotational speed of small sprocket, rev/min												
	10	25	50	100	180	200	300	500	700	900	1000	1200	1400	1600	1800	2100	2500	3000	3500	4000	5000	6000	7000	8000	9000
11	0.06	0.14	0.27	0.52	0.91	1.00	1.48	2.42	3.34	4.25	4.70	5.60	6.49	5.57	4.66	3.70	2.85	2.17	1.72	1.41	1.01	0.77	0.61	0.50	0.00
12	0.06	0.15	0.29	0.56	0.99	1.09	1.61	2.64	3.64	4.64	5.13	6.11	7.09	6.34	5.31	4.22	3.25	2.47	1.96	1.60	1.15	0.87	0.69	0.57	0.00
13	0.07	0.16	0.31	0.61	1.07	1.19	1.75	2.86	3.95	5.02	5.56	6.62	7.68	7.15	5.99	4.76	3.66	2.79	2.21	1.81	1.29	0.98	0.78	0.00	
14	0.07	0.17	0.34	0.66	1.15	1.28	1.88	3.08	4.25	5.41	5.98	7.13	8.27	7.99	6.70	5.31	4.09	3.11	2.47	2.02	1.45	1.10	0.87	0.00	
15	0.08	0.19	0.36	0.70	1.24	1.37	2.02	3.30	4.55	5.80	6.41	7.64	8.86	8.86	7.43	5.89	4.54	3.45	2.74	2.24	1.60	1.22	0.97	0.00	
16	0.08	0.20	0.39	0.75	1.32	1.46	2.15	3.52	4.86	6.18	6.84	8.15	9.45	9.76	8.18	6.49	5.00	3.80	3.02	2.47	1.77	1.34	0.00		
17	0.09	0.21	0.41	0.80	1.40	1.55	2.29	3.74	5.16	6.57	7.27	8.66	10.04	10.69	8.96	7.11	5.48	4.17	3.31	2.71	1.94	1.47	0.00		
18	0.09	0.22	0.43	0.84	1.48	1.64	2.42	3.96	5.46	6.95	7.69	9.17	10.63	11.65	9.76	7.75	5.97	4.54	3.60	2.95	2.11	1.60	0.00		
19	0.10	0.24	0.46	0.89	1.57	1.73	2.56	4.18	5.77	7.34	8.12	9.66	11.22	12.64	10.59	8.40	6.47	4.92	3.91	3.20	2.29	0.09	0.00		
20	0.10	0.25	0.48	0.94	1.65	1.82	2.69	4.39	6.07	7.73	8.55	10.18	11.81	13.42	11.44	9.07	6.99	5.31	4.22	3.45	2.47	0.00			
21	0.11	0.26	0.51	0.98	1.73	1.91	2.83	4.61	6.37	8.11	8.98	10.69	12.40	14.10	12.30	9.76	7.52	5.72	4.54	3.71	2.65	0.00			
22	0.11	0.27	0.53	1.03	1.81	2.01	2.96	4.83	6.68	8.50	9.40	11.20	12.99	14.77	13.19	10.47	8.06	6.13	4.87	3.98	2.85	0.00			
23	0.12	0.28	0.56	1.08	1.90	2.10	3.10	5.05	6.98	8.89	9.83	11.71	13.58	15.44	14.10	11.19	8.62	6.55	5.20	4.26	3.05	0.00			
24	0.12	0.30	0.58	1.12	1.98	2.19	3.23	5.27	7.28	9.27	10.26	12.22	14.17	16.11	15.03	11.93	9.18	6.99	5.54	4.54	0.87	0.00			
25	0.13	0.31	0.60	1.17	2.06	2.28	3.36	5.49	7.59	9.66	10.69	12.73	14.76	16.78	15.98	12.68	9.76	7.43	5.89	4.82	0.00				
26	0.13	0.32	0.63	1.22	2.14	2.37	3.50	5.71	7.89	10.04	11.11	13.24	15.35	17.45	16.95	13.45	10.36	7.88	6.25	5.12	0.00				
28	0.14	0.35	0.67	1.31	2.31	2.55	3.77	6.15	8.50	10.82	11.97	14.26	16.53	18.79	18.94	15.03	11.57	8.80	6.99	5.72	0.00				
30	0.15	0.37	0.72	1.41	2.47	2.74	4.04	6.59	9.11	11.59	12.82	15.28	17.71	20.14	21.01	16.67	12.84	9.76	7.75	6.34	0.00				
32	0.16	0.40	0.77	1.50	2.64	2.92	4.31	7.03	9.71	12.38	13.66	16.30	18.89	21.48	23.14	18.37	14.14	10.76	8.54	1.41					
35	0.18	0.43	0.84	1.64	2.88	3.19	4.71	7.69	10.62	13.52	14.96	17.82	20.67	23.49	26.30	21.01	16.17	12.30	9.76	0.00					
40	0.21	0.50	0.96	1.87	3.30	3.65	5.38	8.79	12.14	15.45	17.10	20.37	23.62	26.85	30.06	25.67	19.76	15.03	0.00						
45	0.23	0.56	1.08	2.11	3.71	4.10	6.06	9.89	13.66	17.39	19.24	22.92	26.57	30.20	33.82	30.63	23.58	5.53	0.00						
	Type A				Type B									Type C											

Type A: Manual or drip lubrication
Type B: Bath or disc lubrication
Type C: Oil stream lubrication

Source: American Chain Association, Naples, FL.

TABLE 7–8 Horsepower Ratings—Single Strand Roller Chain No. 60

No. of teeth	0.750 inch pitch										Rotational speed of small sprocket, rev/min												
	10	25	50	100	120	200	300	400	500	600	800	1000	1200	1400	1600	1800	2000	2500	3000	3500	4000	4500	5000	5500	6000
11	0.19	0.46	0.89	1.72	2.05	3.35	4.95	6.52	8.08	9.63	12.69	15.58	11.85	9.41	7.70	6.45	5.51	3.94	3.00	2.38	1.95	1.63	1.39	1.21	0.00
12	0.21	0.50	0.97	1.88	2.24	3.66	5.40	7.12	8.82	10.51	13.85	17.15	13.51	10.72	8.77	7.35	6.28	4.49	3.42	2.71	2.22	1.86	1.59	1.38	0.00
13	0.22	0.54	1.05	2.04	2.43	3.96	5.85	7.71	9.55	11.38	15.00	18.58	15.23	12.08	9.89	8.29	7.08	5.06	3.85	3.06	2.50	2.10	1.79	0.00	
14	0.24	0.58	1.13	2.19	2.61	4.27	6.30	8.30	10.29	12.26	16.15	20.01	17.02	13.51	11.05	9.26	7.91	5.66	4.31	3.42	2.80	2.34	0.41	0.00	
15	0.26	0.62	1.21	2.35	2.80	4.57	6.75	8.90	11.02	13.13	17.31	21.44	18.87	14.98	12.26	10.27	8.77	6.28	4.77	3.79	3.10	2.60	0.00		
16	0.27	0.66	1.29	2.51	2.99	4.88	7.20	9.49	11.76	14.01	18.46	22.87	20.79	16.50	13.51	11.32	9.66	6.91	5.26	4.17	3.42	1.78	0.00		
17	0.29	0.70	1.37	2.66	3.17	5.18	7.65	10.08	12.49	14.88	19.62	24.30	22.77	18.07	14.79	12.40	10.58	7.57	5.76	4.57	3.74	0.00			
18	0.31	0.75	1.45	2.82	3.36	5.49	8.10	10.68	13.23	15.76	20.77	25.73	24.81	19.69	16.11	13.51	11.53	8.25	6.28	4.98	4.08	0.00			
19	0.33	0.79	1.53	2.98	3.55	5.79	8.55	11.27	13.96	16.63	21.92	27.16	26.91	21.35	17.48	14.65	12.50	8.95	6.81	5.40	0.20	0.00			
20	0.34	0.83	1.61	3.13	3.73	6.10	9.00	11.86	14.70	17.51	23.08	28.59	29.06	23.06	18.87	15.82	13.51	9.66	7.35	5.83	0.00				
21	0.36	0.87	1.69	3.29	3.92	6.40	9.45	12.46	15.43	18.38	24.23	30.02	31.26	24.81	20.31	17.02	14.53	10.40	7.91	6.28	0.00				
22	0.38	0.91	1.77	3.45	4.11	6.71	9.90	13.05	16.17	19.26	25.39	31.45	33.52	26.60	21.77	18.25	15.58	11.15	8.48	0.00					
23	0.40	0.95	1.85	3.61	4.29	7.01	10.35	13.64	16.90	20.13	26.54	32.88	35.84	28.44	23.28	19.51	16.66	11.92	9.07	0.00					
24	0.41	0.99	1.93	3.76	4.48	7.32	10.80	14.24	17.64	21.01	27.69	34.31	38.20	30.31	24.81	20.79	17.75	12.70	9.66	0.00					
25	0.43	1.04	2.01	3.92	4.67	7.62	11.25	14.83	18.37	21.89	28.85	35.74	40.61	32.23	26.38	22.11	18.87	13.51	10.27	0.00					
26	0.45	1.08	2.09	4.08	4.85	7.93	11.70	15.42	19.11	22.76	30.00	37.17	43.07	34.18	27.98	23.44	20.02	14.32	10.90	0.00					
28	0.48	1.16	2.26	4.39	5.23	8.54	12.60	16.61	20.58	24.51	32.31	40.03	47.68	38.20	31.26	26.20	22.37	16.01	0.00						
30	0.52	1.24	2.42	4.70	5.60	9.15	13.50	17.79	22.05	26.26	34.62	42.89	51.09	42.36	34.67	29.06	24.81	17.75	0.00						
32	0.55	1.33	2.58	5.02	5.98	9.76	14.40	18.98	23.52	28.01	36.92	45.75	54.50	46.67	38.20	32.01	27.33	19.56	0.00						
35	0.60	1.45	2.82	5.49	6.54	10.67	15.75	20.76	25.72	30.64	40.39	50.03	59.60	53.38	43.69	36.62	31.26	1.35	0.00						
40	0.69	1.66	3.22	6.27	7.47	12.20	18.00	23.73	29.39	35.02	46.16	57.18	68.12	65.22	53.38	44.74	38.20	0.00							
45	0.77	1.86	3.63	7.05	8.40	13.72	20.25	26.69	33.07	38.39	51.92	64.33	76.63	77.83	63.70	53.38	12.45	0.00							
	Type A				Type B							Type C													

Type A: Manual or drip lubrication
Type B: Bath of disc lubrication
Type C: Oil stream lubrication

Source: American Chain Association, Naples, FL.

TABLE 7–9 Horsepower Ratings—Single Strand Roller Chain No. 80

1.000 inch pitch

No. of teeth	Rotational speed of small sprocket, rev/min																								
	10	25	50	75	88	100	200	300	400	500	600	700	800	900	1000	1200	1400	1600	1800	2000	2500	3000	3500	4000	4500
11	0.44	1.06	2.07	3.05	3.56	4.03	7.83	11.56	15.23	18.87	22.48	26.07	27.41	22.97	19.61	14.92	11.84	9.69	8.12	6.83	4.96	3.77	3.00	2.45	0.00
12	0.48	1.16	2.26	3.33	3.88	4.39	8.54	12.61	16.82	20.59	24.53	28.44	31.23	26.17	22.35	17.00	13.49	11.04	9.25	7.90	5.65	4.30	3.41	2.79	0.00
13	0.52	1.26	2.45	3.61	4.21	4.76	9.26	13.66	18.00	22.31	26.57	30.81	35.02	29.51	25.20	19.17	15.21	12.45	10.43	8.91	6.37	4.85	3.85	3.15	
14	0.56	1.35	2.63	3.89	4.53	5.12	9.97	14.71	19.39	24.02	28.62	33.18	37.72	32.98	28.16	21.42	17.00	13.91	11.66	9.96	7.12	5.42	4.30	3.52	
15	0.60	1.45	2.82	4.16	4.86	5.49	10.68	15.76	20.77	25.74	30.66	35.55	40.41	36.58	31.23	23.76	18.85	15.43	12.93	11.04	7.90	6.01	4.77	0.00	
16	0.64	1.55	3.01	4.44	5.18	5.86	11.39	16.81	22.16	27.45	32.70	37.92	43.11	40.30	34.41	26.17	20.77	17.00	14.25	12.16	8.70	6.62	5.25	0.00	
17	0.68	1.64	3.20	4.72	5.50	6.22	12.10	17.86	23.54	29.17	34.75	40.29	45.80	44.13	37.68	28.66	22.75	18.62	15.60	13.32	9.53	7.25	0.00		
18	0.72	1.74	3.39	5.00	5.83	6.59	12.81	18.91	24.93	30.88	36.79	42.66	48.49	48.08	41.05	31.23	24.78	20.29	17.00	14.51	10.39	7.90	0.00		
19	0.76	1.84	3.57	5.28	6.15	6.95	13.53	19.96	26.31	32.60	38.84	45.03	51.19	52.15	44.52	33.87	26.88	22.00	18.44	15.74	11.26	0.36	0.00		
20	0.80	1.93	3.76	5.55	6.47	7.32	14.24	21.01	27.70	34.32	40.88	47.40	53.88	56.32	48.08	36.58	29.03	23.76	19.91	17.00	12.16	0.00			
21	0.84	2.03	3.95	5.83	6.80	7.69	14.95	22.07	29.08	36.03	42.92	49.77	56.58	60.59	51.73	39.36	31.23	25.56	21.42	18.29	13.09	0.00			
22	0.88	2.13	4.14	6.11	7.12	8.05	15.66	23.12	30.47	37.75	44.97	52.14	59.27	64.97	55.47	42.20	33.49	27.41	22.97	19.61	14.03				
23	0.92	2.22	4.33	6.39	7.45	8.42	16.37	24.17	31.85	39.46	47.01	54.51	61.97	69.38	59.30	45.11	35.80	29.30	24.55	20.97	15.00				
24	0.96	2.32	4.52	6.66	7.77	8.78	17.09	25.22	33.24	41.18	49.06	56.88	64.66	72.40	63.21	48.08	38.16	31.23	26.17	22.35	15.99				
25	1.00	2.42	4.70	6.94	8.09	9.15	17.80	26.27	34.62	42.89	51.10	59.25	67.35	75.42	67.20	51.12	40.57	33.20	27.83	23.76	8.16				
26	1.04	2.51	4.89	7.22	8.42	9.52	18.51	27.32	36.01	44.61	53.14	61.62	70.05	78.43	71.27	54.22	43.02	36.22	29.51	25.20	0.00				
28	1.12	2.71	5.27	7.77	9.06	10.25	19.93	29.42	38.78	48.04	57.23	66.36	75.44	84.47	79.65	60.59	48.08	39.36	32.98	28.16	0.00				
30	1.20	2.90	5.64	8.33	9.71	10.98	21.36	31.52	41.55	51.47	61.32	71.10	80.82	90.50	88.33	67.20	53.33	43.65	36.58	31.23					
32	1.28	3.09	6.02	8.89	10.36	11.71	22.78	33.62	44.32	54.91	65.41	75.84	86.21	96.53	97.31	74.03	58.75	48.08	40.30	5.65					
35	1.40	3.38	6.58	9.72	11.33	12.81	24.92	36.78	48.47	60.05	71.54	82.95	94.29	105.58	111.31	84.68	67.20	55.00	28.15	0.00					
40	1.61	3.87	7.53	11.11	12.95	14.64	28.48	42.03	55.40	68.63	81.76	94.80	107.77	120.67	133.51	103.46	82.10	40.16	0.00						
45	1.81	4.35	8.47	12.49	14.57	16.47	32.04	47.28	62.32	77.21	91.98	106.65	121.24	135.75	150.20	123.45	72.28	0.00							
	Type A				Type B								Type C												

Type A: Manual or drip lubrication
Type B: Bath of disc lubrication
Type C: Oil stream lubrication

Source: American Chain Association, Naples, FL.

TABLE 7–10 Service Factors for Chain Drives

Load type	Type of driver		
	Hydraulic drive	Electric motor or turbine	Internal combustion engine with mechanical drive
Smooth Agitators; fans; generators; grinders; centrifugal pumps; rotary screens; light, uniformly loaded conveyors	1.0	1.0	1.2
Moderate Shock Bucket elevators; machine tools; cranes; heavy conveyors; food mixers and grinders; ball mills; reciprocating pumps; woodworking machinery	1.2	1.3	1.4
Heavy Shock Punch presses; hammer mills; boat propellers; crushers; reciprocating conveyors; rolling mills; logging hoists; dredges; printing presses	1.4	1.5	1.7

Lubrication

It is essential that adequate lubrication be provided for chain drives. There are numerous moving parts within the chain, along with the interaction between the chain and the sprocket teeth. The designer must define the lubricant properties and the method of lubrication.

Lubricant Properties. Petroleum-based lubricating oil similar to engine oil is recommended. Its viscosity must enable the oil to flow readily between chain surfaces that move relative to each other while providing adequate lubrication action. The oil should be kept clean and free of moisture. Table 7–11 gives the recommended lubricant for different ambient temperatures.

TABLE 7–11 Recommended Lubricant for Chain Drives

Ambient temperature		Recommended lubricant
°F	°C	
20 to 40	−7 to 5	SAE 20
40 to 100	5 to 38	SAE 30
100 to 120	38 to 49	SAE 40
120 to 140	49 to 60	SAE 50

Method of Lubrication. The American Chain Association recommends three different types of lubrication depending on the speed of operation and the power being transmitted. See Tables 7–7 to 7–9 or manufacturer's catalogs for recommendations. Refer to the following descriptions of the methods and the illustrations in Figure 7–24.

Type A. *Manual or drip lubrication:* For manual lubrication, oil is applied copiously with a brush or a spout can, at least once every 8 hours of operation. For drip feed lubrication, oil is fed directly onto the link plates of each chain strand.

Type B. *Bath or disc lubrication:* The chain cover provides a sump of oil into which the chain dips continuously. Alternatively, a disc or a slinger can be attached to one of the shafts to lift oil to a trough above the lower strand of chain. The trough then delivers a stream of oil to the chain. The chain itself, then, does not need to dip into the oil.

Type C. *Oil stream lubrication:* An oil pump delivers a continuous stream of oil on the lower part of the chain.

Example Problem 7–3 Design a chain drive for a heavily loaded coal conveyor to be driven by a gasoline engine through a mechanical drive. The input speed will be 900 rpm, and the desired output speed is 230 to 240 rpm. The conveyor requires 15.0 hp.

Solution Objective Design the chain drive.

Given Power transmitted = 15 hp to a coal conveyor

Speed of motor = 900 rpm; output speed range = 230 to 240 rpm

Analysis Use the design data presented in this section. The solution procedure is developed within the Results section of the problem solution.

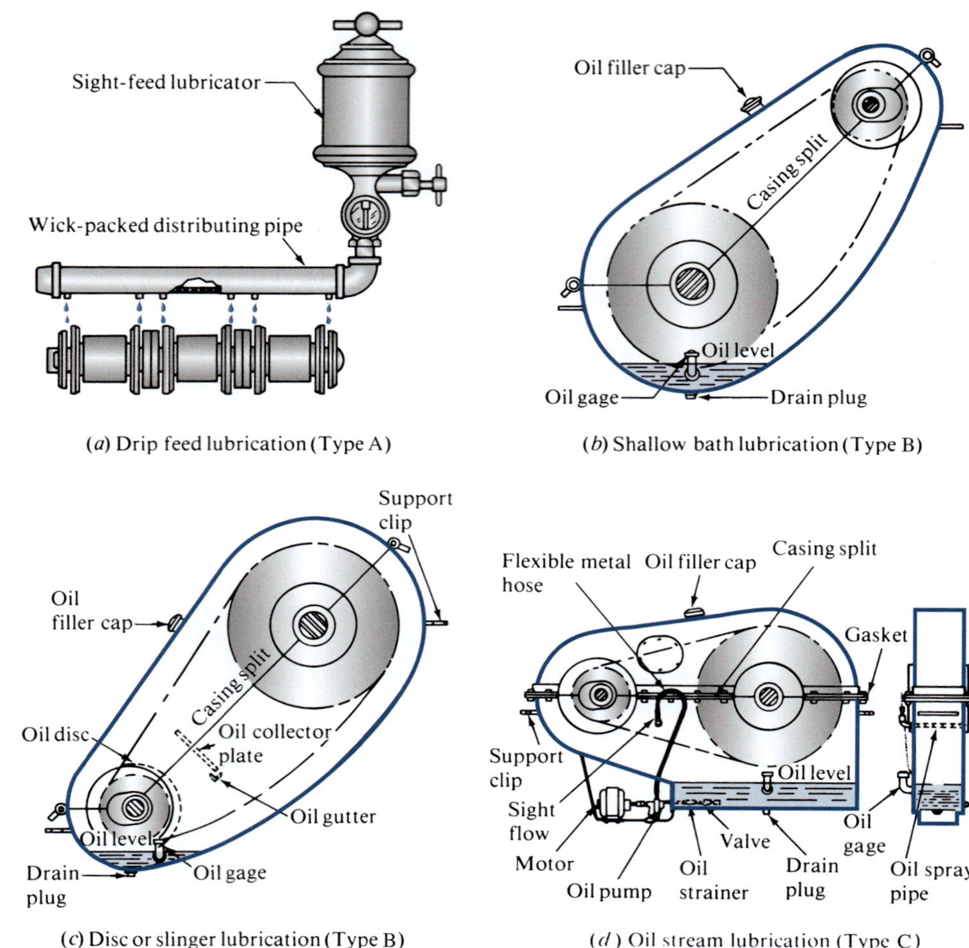

FIGURE 7–24 Lubrication methods (American Chain Association, Naples, FL)

Results

Step 1. Specify a service factor and compute the design power. From Table 7–10, for moderate shock and a gasoline engine drive through a mechanical drive, $SF = 1.4$.

$$\text{Design power} = 1.4(15.0) = 21.0 \text{ hp}$$

Step 2. Compute the desired ratio. Using the middle of the required range of output speeds, we have

$$\text{Ratio} = (900 \text{ rpm})/(235 \text{ rpm}) = 3.83$$

Step 3. Refer to the tables for power capacity (Tables 7–7, 7–8, and 7–9), and select the chain pitch. For a single strand, the no. 60 chain with $p = 3/4$ in seems best. A 17-tooth sprocket is rated at 21.96 hp at 900 rpm by interpolation. At this speed, type B lubrication (oil bath) is required.

Step 4. Compute the required number of teeth on the large sprocket:

$$N_2 = N_1 \times \text{ratio} = 17(3.83) = 65.11$$

Let's use the integer: 65 teeth.

Step 5. Compute the actual expected output speed:

$$n_2 = n_1(N_1/N_2) = 900 \text{ rpm}(17/65) = 235.3 \text{ rpm (Okay!)}$$

Step 6. Compute the pitch diameters of the sprockets using Equation (7–11):

$$D_1 = \frac{p}{\sin(180°/N_1)} = \frac{0.75 \text{ in}}{\sin(180°/17)} = 4.082 \text{ in}$$

$$D_2 = \frac{p}{\sin(180°/N_2)} = \frac{0.75 \text{ in}}{\sin(180°/65)} = 15.524 \text{ in}$$

Step 7. Specify the nominal center distance. Let's use the middle of the recommended range, 40 pitches.

Step 8. Compute the required chain length in pitches from Equation (7–9):

$$L = 2C + \frac{N_2 + N_1}{2} + \frac{(N_2 - N_1)^2}{4\pi^2 C}$$

$$L = 2(40) + \frac{65 + 17}{2} + \frac{(65 - 17)^2}{4\pi^2(40)} = 122.5 \text{ pitches}$$

Step 9. Specify an integral number of pitches for the chain length, and compute the actual theoretical center distance. Let's use 122 pitches, an even number. Then, from Equation (7–10),

$$C = \frac{1}{4}\left[L - \frac{N_2 + N_1}{2} + \sqrt{\left[L - \frac{N_2 + N_1}{2}\right]^2 - \frac{8(N_2 - N_1)^2}{4\pi^2}}\right]$$

$$C = \frac{1}{4}\left[122 - \frac{65 + 17}{2} + \sqrt{\left[122 - \frac{65 + 17}{2}\right]^2 - \frac{8(65 - 17)^2}{4\pi^2}}\right]$$

$$C = 39.766 \text{ pitches} = 39.766(0.75 \text{ in}) = 29.825 \text{ in}$$

Step 10. Compute the angle of wrap of the chain for each sprocket using Equations (7–12) and (7–13). Note that the minimum angle of wrap should be 120 degrees.

For the small sprocket,

$$\theta_1 = 180^\circ - 2\sin^{-1}[(D_2 - D_1)/2C]$$

$$\theta_1 = 180^\circ - 2\sin^{-1}[(15.524 - 4.082)/(2(29.825))] = 158^\circ$$

Because this is greater than 120°, it is acceptable.

For the larger sprocket,

$$\theta_2 = 180^\circ + 2\sin^{-1}[(D_2 - D_1)/2C]$$

$$\theta_2 = 180^\circ + 2\sin^{-1}[(15.524 - 4.082)/(2(29.825))] = 202^\circ$$

Comments **Summary of Design**

Figure 7–25(a) shows a sketch of the design to scale.

Pitch: No. 60 chain, 3/4-in pitch

Length: 122 pitches = 122(0.75) = 91.50 in

Center distance: C = 29.825 in (maximum)

Sprockets: Single-strand, no. 60, 3/4-in pitch

Small: 17 teeth, D = 4.082 in

Large: 65 teeth, D = 15.524 in

Type B lubrication is required. The large sprocket can dip into an oil bath.

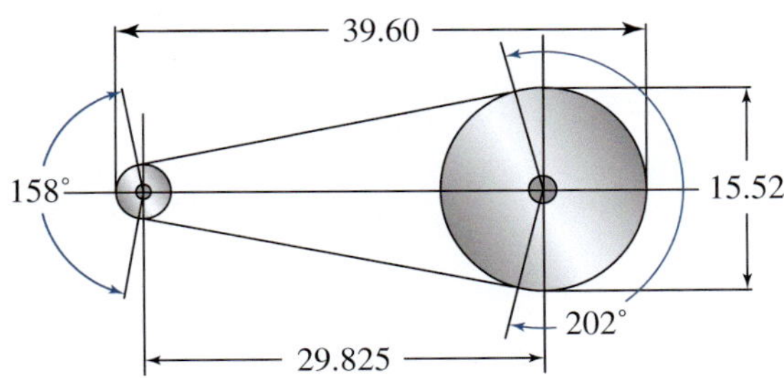

(*a*) Chain drive system for Example Problem 7–3

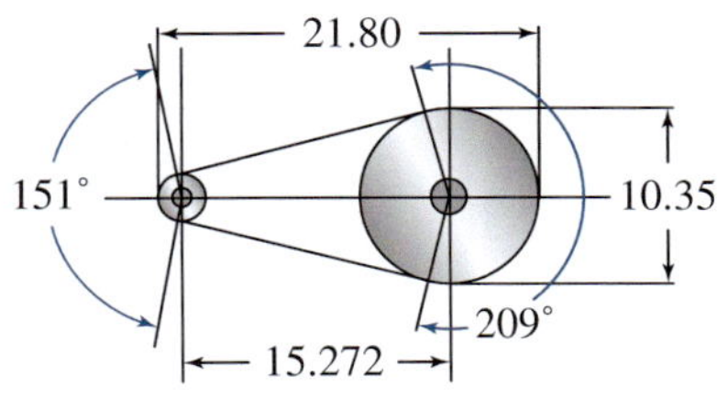

(*b*) Chain drive system for Example Problem 7–4

FIGURE 7–25 Scale drawings of layouts for chain drives for Example Problems 7–3 and 7–4

Example Problem 7–4 Create an alternate design for the conditions of Example Problem 7–3 to produce a smaller drive.

Solution

Objective: Design a smaller chain drive for the application in Example Problem 7–3.

Given: Power transmitted = 15 hp to a conveyor
Speed of motor = 900 rpm; output speed range = 230 to 240 rpm

Analysis: Use a multistrand design to permit a smaller-pitch chain to be used to transmit the same design power (21.0 hp) at the same speed (900 rpm). Use the design data presented in this section. The solution procedure is developed within the Results section of the problem solution.

Results: Let's try a four-strand chain for which the power capacity factor is 3.3. Then the required power per strand is

$$P = 21.0/3.3 = 6.36 \text{ hp}$$

From Table 7–7, we find that a no. 40 chain (1/2-in pitch) with a 17-tooth sprocket will be satisfactory. Type B lubrication, oil bath, can be used.

For the required large sprocket,

$$N_2 = N_1 \times \text{ratio} = 17(3.83) = 65.11$$

Let's use $N_2 = 65$ teeth.

The sprocket diameters are

$$D_1 = \frac{p}{\sin(180^\circ/N_1)} = \frac{0.500 \text{ in}}{\sin(180^\circ/17)} = 2.721 \text{ in}$$

$$D_2 = \frac{p}{\sin(180^\circ/N_2)} = \frac{0.500 \text{ in}}{\sin(180^\circ/65)} = 10.349 \text{ in}$$

For the center distance, let's try the minimum recommended: $C = 30$ pitches.

$$30(0.50 \text{ in}) = 15.0 \text{ in}$$

The chain length is

$$L = 2(30) + \frac{65 + 17}{2} + \frac{(65 - 17)^2}{4\pi^2(30)} = 102.9 \text{ pitches}$$

Specify the integer length, $L = 104$ pitches $= 104(0.50) = 52.0$ in. The actual maximum center distance is

$$C = \frac{1}{4}\left[104 - \frac{65 + 17}{2} + \sqrt{\left[104 - \frac{65 + 17}{2}\right]^2 - \frac{8(65 - 17)^2}{4\pi^2}}\right]$$

$$C = 30.54 \text{ pitches} = 30.54(0.50) = 15.272 \text{ in}$$

Compute the angle of wrap of the chain for each sprocket using Equations (7–12) and (7–13). Note that the minimum angle of wrap should be 120°.

For the small sprocket,

$$\theta_1 = 180^\circ - 2\sin^{-1}[(D_2 - D_1)/2C]$$

$$\theta_1 = 180^\circ - 2\sin^{-1}[(10.349 - 2.721)/(2(15.272))] = 151.1^\circ$$

Because this is greater than 120°, it is acceptable.

For the larger sprocket,

$$\theta_1 = 180^\circ + 2\sin^{-1}[(D_2 - D_1)/2C]$$

$$\theta_2 = 180^\circ + 2\sin^{-1}[(10.349 - 2.721)/(2(15.272))] = 208.9^\circ$$

Comments

Summary

Figure 7–25(b) shows the new design to the same scale as the first design. The space reduction is significant.

Chain: No. 40, 1/2-in pitch, four-strand, 104 pitches, 52.0 in length

Sprockets: No. 40–4 (four strands), $\frac{1}{2}$-in pitch

Small: 17 teeth, $D_1 = 2.721$ in

Large: 65 teeth, $D_2 = 10.349$ in

Maximum center distance: 15.272 in

Type B lubrication (oil bath)

Spreadsheet for Chain Design

Figure 7–26 shows a spreadsheet that assists in the design of chain drives using the procedure developed in this section. The user enters data shown in italics in the gray-shaded cells. Refer to Tables 7–5 to 7–10 for required data. Results for Example Problem 7–4 are shown in the figure.

CHAIN DRIVE DESIGN						
Initial Input Data:	Example Problem 7–4—Multiple strands					
Application:	*Coal conveyor*					
Drive/type:	*Engine-mechanical drive*					
Driven machine:	*Heavily loaded conveyor*					
Power input:	*15* hp					
Service factor:	*1.4*	Table 7–10				
Input speed:	*900* rpm					
Desired output speed:	*235* rpm					
Computed Data:						
Design power:	21 hp					
Speed ratio:	3.83					
Design Decisions—Chain Type and Teeth Numbers:						
Number of strands:	*4*	1	2	3	4	
Strand factor:	*3.3*	1.0	1.7	2.5	3.3	
Required power per strand:	6.36 hp					
Chain number:	*40*	Tables 7–7, 7–8, or 7–9				
Chain pitch:	*0.5* in					
Number of teeth-driver sprocket:	*17*					
Computed no. of teeth-driven sprocket:	65.11					
Enter: Chosen number of teeth:	*65*					
Computed Data:						
Actual output speed:	235.4 rpm					
Pitch diameter-driver sprocket:	2.721 in					
Pitch diameter-driven sprocket:	10.349 in					
Center Distance, Chain Length, and Angle of Wrap:						
Enter: Nominal center distance:	*30* pitches	30 to 50 pitches recommended				
Computed nominal chain length:	102.9 pitches					
Enter: Specified no. of pitches:	*104* pitches	Even number recommended				
Actual chain length:	52.00 in					
Computed actual center distance:	30.545 pitches					
Actual center distance:	15.272 in					
Angle of wrap–driver sprocket:	151.1 degrees	Should be greater than 120 degrees				
Angle of wrap–driven sprocket:	208.9 degrees					

FIGURE 7–26 Spreadsheet for chain design

REFERENCES

1. American Chain Association. *Chains for Power Transmission and Material Handling. 2nd ed.*, Boca Raton, FL: CRC Press Taylor & Francis, 2006.
2. American Society of Mechanical Engineers. *ASME Standard B29.1:1993, Precision Power Transmission Roller Chains, Attachments, and Sprockets.* New York: American Society of Mechanical Engineers, 1993.
3. International Organization for Standardization (ISO). (*Various standards listed below*). Geneva, Switzerland: ISO.
   a. ISO 2790:2004. *Belt Drives—V-belts for the Automotive Industry and Corresponding Pulleys—Dimensions*, 2004.
   b. ISO 5290:2001. *Belt Drives—Grooved Pulleys for Joined Narrow V-belts—Groove Sections 9N/J, 15N/J, and 25N/J*, 2001.
   c. ISO 5292:1995. *Belt Drives—V-belts and V-ribbed Belts—Calculation of Power Ratings*, 1995.
   d. ISO 5296-1:1989. *Synchronous Belt Drives—Belts—Part 1: Pitch Codes MXL, XL, L, H, XH, and XXH—Metric and Inch Dimensions*, 1989.
   e. ISO 8419: 2003. *Belt Drives—Narrow V-belts—Sections 9N/J, 15N/J, and 25N/J (Lengths in the Effective System)*, 2003.
   f. ISO 9981:1998. *Belt Drives—Pulleys and V-ribbed Belts for the Automotive Industry—PK Profile: Dimensions*, 1998.
   g. ISO 9982:1998. *Belt Drives–Pulleys and V-ribbed Belts for the Industrial Applications–PH, PJ, PL and PM Profiles: Dimensions*, 1998.
   h. ISO 487:1998. *Steel Roller Chains, Type S and C*, 1998.
   i. ISO 10823:2004. *Guidelines for the Selection of Roller Chain Drives*, 2004.
4. Society of Automotive Engineers. *SAE Standard J636—V-Belts and Pulleys*. Warrendale, PA: Society of Automotive Engineers, 2001.
5. Society of Automotive Engineers. *SAE Standard J637—Automotive V-Belt Drives*. Warrendale, PA: Society of Automotive Engineers, 2001.
6. Society of Automotive Engineers. *SAE Standard J1278—SI (Metric) Synchronous Belts and Pulleys*. Warrendale, PA: Society of Automotive Engineers, 1993.
7. Society of Automotive Engineers. *SAE Standard J1313—Automotive Synchronous Belt Drives*. Warrendale, PA: Society of Automotive Engineers, 1993.
8. Society of Automotive Engineers. *SAE Standard J1459—V-Ribbed Belts and Pulleys*. Warrendale, PA: Society of Automotive Engineering, 2009.

INTERNET SITES RELATED TO BELT DRIVES AND CHAIN DRIVES

1. **American Chain Association.** www.americanchainassn.org A national trade organization for companies providing products for the chain drive industry. Publishes standards and design aids for designing, applying, and maintaining chain drives and engineering chain conveyor systems. ACA power ratings for the entire range of chain sizes for ANSI Standard roller chains can be downloaded from the site.
2. **Dayco Belt Drives.** www.dayco.com and www.markivauto.com Manufacturer of Dayco industrial belt drive systems under Carlisle Power Transmission Products, Inc., and Dayco automotive belt drive systems under the MarkIV Automotive Company. Complete catalog data for automotive polyrib serpentine belts belts can be downloaded from www.accessdayco.com.
3. **Dodge Power Transmission.** www.dodge-pt.com Manufacturer of numerous power transmission components, including V-belt and synchronous belt drive systems.
4. **Emerson Power Transmission.** www.emerson-ept.com Manufacturer of numerous power transmission components including V-belt drives, synchronous belt drives, and roller chain drives through their Browning and Morse divisions.
5. **Gates Rubber Company.** www.gates.com Rubber products for the automotive and industrial markets including V-belt drives and synchronous belt drives.
6. **Grainger Industrial Supply** www.grainger.com A distributor of a wide array of industrial products including belts, sheaves, chain, and sprockets for power transmission.
7. **International Organization for Standardization** www.iso.org The premier organization for establishing and promulgating technical standards for worldwide implementation. Search on *Standards* and the subject for which you are seeking ISO standards.
8. **Martin Sprocket and Gear Company** www.martinsprocket.com Manufacturer of a wide range of mechanical power transmission products including chain sprockets, V-belt sheaves, synchronous belt sheaves and bushings.
9. **Maryland Metrics** www.mdmetric.com U.S. based company that supplies metric power transmission products rated by the ISO, BS, and DIN standards.
10. **Power Transmission** www.powertransmission.com A comprehensive website for companies providing products for the power transmission industry, many of which supply belt and chain drive systems.
11. **Putnam Precision Molding, Inc.** www.putnamprecision-molding.com Producer of plastic injection molded mechanical drive components, including plastic chain, sprockets, and synchronous belt pulleys.
12. **Rexnord Corporation.** www.rexnord.com Manufacturer of power transmission and conveying components, including roller chain drives and engineered chain drive systems.
13. **SAE International.** www.sae.org The Society of Automotive Engineers, the engineering society for advancing mobility on land or sea, in air or space. Offers standards on V-belts, synchronous belts, pulleys, and drives for automotive applications.
14. **SDP/SI** www.sdp-si.com The Stock Drive Products/Sterling Instruments Company distributes a wide array of mechanical drive components with a heavy emphasis on small, precision mechanical drive components including synchronous belts, pulleys, chain, and sprockets. Both U.S. Customary and Metric styles of components are included.
15. **Wippermann Company** www.wipperman.com Manufacturer of a wide range of chain drive products based in Germany; wide selection of German DIN standard chains and sprockets.
16. **T. B. Wood's Sons Company** www.tbwoods.com Manufacturer of many mechanical drives products, including V-belt drives, synchronous belt drives, and adjustable speed drives.

PROBLEMS

V-Belt Drives

1. Specify the standard 3V belt length (from Table 7–2) that would be applied to two sheaves with pitch diameters of 5.25 in and 13.95 in with a center distance of no more than 24.0 in.
2. For the standard belt specified in Problem 1, compute the actual center distance that would result.
3. For the standard belt specified in Problem 1, compute the angle of wrap on both of the sheaves.
4. Specify the standard 5V belt length (from Table 7–2) that would be applied to two sheaves with pitch diameters of 8.4 in and 27.7 in with a center distance of no more than 60.0 in.
5. For the standard belt specified in Problem 4, compute the actual center distance that would result.
6. For the standard belt specified in Problem 4, compute the angle of wrap on both of the sheaves.
7. Specify the standard 8V belt length (from Table 7–2) that would be applied to two sheaves with pitch diameters of 13.8 in and 94.8 in with a center distance of no more than 144 in.
8. For the standard belt specified in Problem 7, compute the actual center distance that would result.
9. For the standard belt specified in Problem 7, compute the angle of wrap on both of the sheaves.
10. If the small sheave of Problem 1 is rotating at 1750 rpm, compute the linear speed of the belt.
11. If the small sheave of Problem 4 is rotating at 1160 rpm, compute the linear speed of the belt.
12. If the small sheave of Problem 7 is rotating at 870 rpm, compute the linear speed of the belt.
13. For the belt drive from Problems 1 and 10, compute the rated power, considering corrections for speed ratio, belt length, and angle of wrap.
14. For the belt drive from Problems 4 and 11, compute the rated power, considering corrections for speed ratio, belt length, and angle of wrap.
15. For the belt drive from Problems 7 and 12, compute the rated power, considering corrections for speed ratio, belt length, and angle of wrap.
16. Describe a standard 15N belt cross section. To what size belt (inches) would it be closest?
17. Describe a standard 17A belt cross section. To what size belt (inches) would it be closest?

For Problems 18–22 (Table 7–12), design a V-belt drive. Specify the belt size, the sheave sizes, the number of belts, the actual output speed, and the center distance.

Roller Chain

23. Describe a standard roller chain, no. 140.
24. Describe a standard roller chain, no. 60.
25. Specify a suitable standard chain to exert a static pulling force of 1250 lb.
26. Roller chain is used in a hydraulic forklift truck to elevate the forks. If two strands support the load equally, which size would you specify for a design load of 5000 lb?
27. List three typical failure modes of roller chain.
28. Determine the power rating of a no. 60 chain, single-strand, operating on a 20-tooth sprocket at 750 rpm. Describe the preferred method of lubrication. The chain connects a hydraulic drive with a meat grinder.
29. For the data of Problem 28, what would be the rating for three strands?
30. Determine the power rating of a no. 40 chain, single-strand, operating on a 12-tooth sprocket at 860 rpm. Describe the preferred method of lubrication. The small sprocket is applied to the shaft of an electric motor. The output is to a coal conveyor.
31. For the data of Problem 30, what would be the rating for four strands?
32. Determine the power rating of a no. 80 chain, single-strand, operating on a 32-tooth sprocket at 1160 rpm. Describe the preferred method of lubrication. The input is an internal combustion engine, and the output is to a fluid agitator.
33. For the data of Problem 32, what would be the rating for two strands?
34. Specify the required length of no. 60 chain to mount on sprockets having 15 and 50 teeth with a center distance of no more than 36 in.
35. For the chain specified in Problem 34, compute the actual center distance.
36. Specify the required length of no. 40 chain to mount on sprockets having 11 and 45 teeth with a center distance of no more than 24 in.

TABLE 7–12 V-Belt Drive Design Problems

Problem number	Driver type	Driven machine	Service (h/day)	Input speed (rpm)	Input power (hp)	Input power (kW)	Nominal output speed (rpm)
18.	AC motor (HT)	Hammer mill	8	870	25	18.6	310
19.	AC motor (NT)	Fan	22	1750	5	3.73	725
20.	6-cylinder engine	Heavy conveyor	16	1500	40	29.8	550
21.	DC motor (compound)	Milling machine	16	1250	20	14.9	695
22.	AC motor (HT)	Rock crusher	8	870	100	74.6	625

Note: NT indicates a normal-torque electric motor. *HT* indicates a high-torque electric motor.

TABLE 7–13 Chain Drive Design Problems

Problem number	Driver type	Driven machine	Input speed (rpm)	Input power (hp)	Input power (kW)	Nominal output speed (rpm)
38.	AC motor	Hammer mill	310	25	18.6	160
39.	AC motor	Agitator	750	5	3.73	325
40.	6-cylinder engine	Heavy conveyor	500	40	29.8	250
41.	Steam turbine	Centrifugal pump	2200	20	14.9	775
42.	Hydraulic drive	Rock crusher	625	100	74.6	225

37. For the chain specified in Problem 36, compute the actual center distance.

For Problems 38–42 (Table 7–13), design a roller chain drive. Specify the chain size, the sizes and number of teeth in the sprockets, the number of chain pitches, and the center distance.

Synchronous Belts

For any of the sets of problem data in Tables 7–12 and 7–13, use the MDESIGN software or manufacturers' catalogs to specify the belt size and sprocket sizes for synchronous belts. See Internet sites 3-6, 8, 9, 11, 14, and 16 for manufacturers of synchronous belts and sprockets.

CHAPTER
EIGHT

KINEMATICS OF GEARS

THE BIG PICTURE Kinematics of Gears

Discussion Map

- Gears are toothed, cylindrical wheels used for transmitting motion and power from one rotating shaft to another.
- Most gear drives cause a change in the speed of the output gear relative to the input gear.
- Some of the most common types of gears are *spur gears, helical gears, bevel gears, and worm/wormgear sets.*

Discover

Identify at least two machines or devices that employ gears. Describe the operation of the machines or devices and the appearance of the gears.

This chapter will help you learn about the features of different kinds of gears, the kinematics of a pair of gears operating together, and the operation of gear trains having more than two gears.

Gears are toothed, cylindrical wheels used for transmitting motion and power from one rotating shaft to another. The teeth of a driving gear mesh accurately in the spaces between teeth on the driven gear as shown in Figure 8–1. The driving teeth push on the driven teeth, exerting a force perpendicular to the radius of the gear. Thus, a torque is transmitted, and because the gear is rotating, power is also transmitted.

Speed Reduction Ratio. Often gears are employed to produce a change in the speed of rotation of the driven gear relative to the driving gear. In Figure 8–1, if the smaller top gear, called a *pinion*, is driving the larger lower gear, simply called the *gear*, the larger gear will rotate more slowly. The amount of speed reduction is dependent on the ratio of the number of teeth in the pinion to the number of teeth in the gear according to this relationship:

$$n_P/n_G = N_G/N_P \tag{8–1}$$

The basis for this equation will be shown later in this chapter. But to show an example of its application here, consider that the pinion in Figure 8–1 is rotating at 1800 rpm. You can count the number of teeth in the pinion to be 11 and the number of teeth in the gear to be 18. Then we can compute the rotational speed of the gear by solving Equation (8–1) for n_G:

$$n_G = n_P(N_P/N_G) = (1800\text{ rpm})(11/18) = 1100\text{ rpm}$$

When there is a reduction in the speed of rotation of the gear, there is a simultaneous proportional *increase* in the

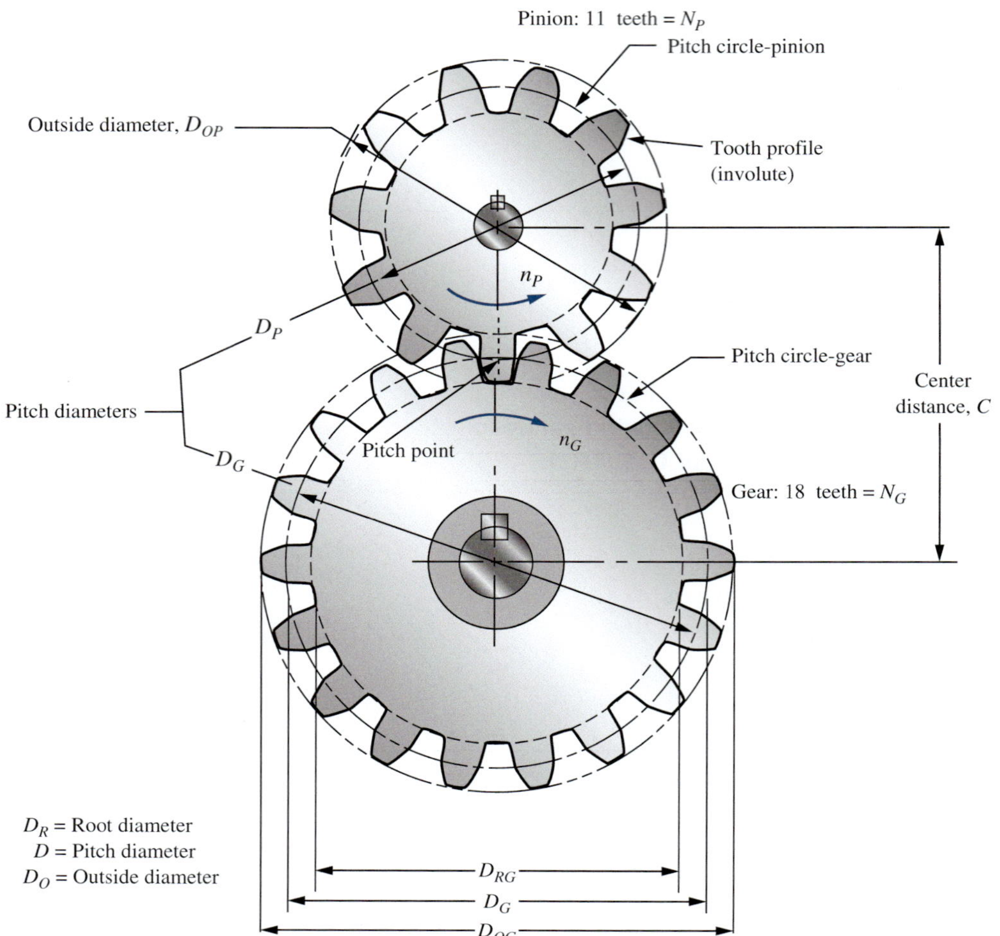

FIGURE 8–1 Pair of spur gears. The pinion drives the gear.

torque transmitted to the shaft carrying the gear. More will be said about this later, also.

Kinds of Gears. Several kinds of gears having different tooth geometries are in common use. To acquaint you with the general appearance of some, their basic descriptions are given here. Later we will describe their geometry more completely.

Figure 8–2 shows a photograph of many kinds of gears. Labels indicate the major types of gears that are discussed in this chapter: *spur gears, helical gears, bevel gears,* and *worm/wormgear sets.* Obviously, the shafts that would carry the gears are not included in this photograph. See References 4, 7, 10, 11, 13, and 19 and Internet sites 1, 4, and 5 for more information on gearing.

Spur gears have teeth that are straight and arranged parallel to the axis of the shaft that carries the gear. The curved shape of the faces of the spur gear teeth have a special geometry called an *involute curve,* described later in this chapter. This shape makes it possible for two gears to operate together with smooth, positive transmission of power. Figure 8–1 also shows the side view of spur gear teeth, and the involute curve shape is evident there. The shafts carrying the gears are parallel.

The teeth of *helical gears* are arranged so that they lie at an angle with respect to the axis of the shaft. The angle, called the *helix angle,* can be virtually any angle. Typical helix angles range from approximately 10° to 30°, but angles up to 45° are practical. The helical teeth operate more smoothly than equivalent spur gear teeth, and stresses are lower. Therefore, a smaller helical gear can be designed for a given power-transmitting capacity as compared with spur gears. One disadvantage of helical gears is that an axial force, called a *thrust force,* is generated in addition to the driving force that acts tangent to the basic cylinder on which the teeth are arranged. The designer must consider the thrust force when selecting bearings that will hold the shaft during operation. Shafts carrying helical gears are typically arranged parallel to each other. However, a special design, called *crossed helical* gears, has 45° helix angles, and their shafts operate 90° to each other.

Bevel gears have teeth that are arranged as elements on the surface of a cone. The teeth of straight bevel gears

FIGURE 8–2 A variety of gear types
(Boston Gear, Quincy, MA)

appear to be similar to spur gear teeth, but they are tapered, being wider at the outside and narrower at the top of the cone. Bevel gears typically operate on shafts that are 90° to each other. Indeed, this is often the reason for specifying bevel gears in a drive system. Specially designed bevel gears can operate on shafts that are at some angle other than 90°. When bevel gears are made with teeth that form a helix angle similar to that in helical gears, they are called *spiral bevel gears*. They operate more smoothly than straight bevel gears and can be made smaller for a given power transmission capacity. When both bevel gears in a pair have the same number of teeth, they are called *miter gears* and are used only to change the axes of the shafts to 90°. No speed change occurs.

Now look closely at Figure 8–3 that shows an example of a large, commercially available reducer with three stages that employs a combination of bevel, helical, and spur gears that were just described. See Internet site 6. Seeing them in one unit can help you appreciate the similarities and differences among them. Follow the flow of power through the reducer as outlined here:

1. The input shaft at the left end carries the spiral bevel pinion for the right angle first stage of reduction.
2. The helical pinion behind the output gear of the bevel gear pair drives the large helical output gear of the second stage of reduction.
3. The output shaft from the helical gear pair carries the spur-type sun gear of a planetary gear train whose output shaft drives the final output shaft projecting from the front of the reducer.

A *rack* is a straight gear that moves linearly instead of rotating. When a circular gear is mated with a rack, as shown toward the right side of Figure 8–2, the combination is called *a rack and pinion drive.* You may have heard that term applied to the steering mechanism of a car or to a part of other machinery. See Section 8–10 for more discussion about a rack.

A *worm and its mating wormgear* operate on shafts that are at 90° to each other. They typically accomplish a rather large speed reduction ratio compared with other types of gears. The worm is the driver, and the wormgear is the driven gear. The teeth on the worm appear similar to screw threads, and, indeed, they are often called *threads* rather than *teeth.* The teeth of the wormgear can be straight like spur gear teeth, or they can be helical. Often the shape of the tip of the wormgear teeth is enlarged to partially wrap around the threads of the worm to improve the power transmission capacity of the set. One disadvantage of the worm/wormgear drive is that it has a somewhat lower mechanical efficiency than most other kinds of gears because there is extensive rubbing contact between the surfaces of the worm threads and the sides of the wormgear teeth.

Where Have You Observed Gears? Think of examples where you have seen gears in actual equipment. Describe the operation of the equipment, particularly the power transmission system. Sometimes, of course, the gears and the shafts are enclosed in a housing, making it difficult for you to observe the actual gears. Perhaps you can find a manual for the equipment that shows the drive system. Or look elsewhere in this chapter and in Chapters 9 and 10 for some photographs of commercially available gear reducers. *(**Note:** If the equipment you are observing is operating, be very careful not to come in contact with any moving parts!)* Try to answer these questions:

- What was the source of the power? An electric motor, a gasoline engine, a steam turbine, a hydraulic motor? Or were the gears operated by hand?

(*a*) External view of reducer

(*b*) Cutaway view showing internal components

FIGURE 8–3 Triple reduction gear reducer employing bevel, helical, and spur gears
(Baldor/Dodge, Greenville, SC)

- How were the gears arranged together, and how were they attached to the driving source and the driven machine?
- Was there a speed change? Can you determine how much of a change?
- Were there more than two gears in the drive system?
- What types of gears were used? (You should refer to Figure 8–2.)
- What materials were the gears made from?
- How were the gears attached to the shafts that supported them?
- Were the shafts for mating gears aligned parallel to each other, or were they perpendicular to one another?
- How were the shafts themselves supported?
- Was the gear transmission system enclosed in a housing? If so, describe it.

This chapter will help you learn the basic geometries and kinematics of gears and pairs of gears operating together. You will also learn how to analyze gear trains having more than two gears so that you can describe the motion of each gear. Then you will learn how to devise a gear train to produce a given speed reduction ratio. In later chapters, you will learn how to analyze gears for their power transmission capacity and to design gear trains to transmit a given amount of power at a specified ratio of the speed of the input shaft to the speed of the output shaft.

YOU ARE THE DESIGNER

A gear-type speed reducer was described in Chapter 1, and a sketch of the layout of the gears within the reducer was shown in Figure 1–12. You are advised to review that discussion now because it will help you understand how the present chapter on *gear geometry* and *kinematics* fits into the design of the complete speed reducer.

Assume that you are responsible for the design of a speed reducer that will take the power from the shaft of an electric motor rotating at 1750 rpm and deliver it to a machine that is to operate at approximately 292 rpm. You have decided to use gears to transmit the power, and you are proposing a double-reduction speed reducer like that shown in Figure 1–12. This chapter will give you the information you need to define the general nature of the gears, including their arrangement and their relative sizes.

The input shaft (shaft 1) is coupled to the motor shaft. The first gear of the gear train is mounted on this shaft and rotates at the same speed as the motor, 1750 rpm. Gear 1 drives the mating gear 2, which is larger, causing the speed of rotation of shaft 2 to be slower than that of shaft 1. But the speed is not yet down to 292 rpm as desired.

The next step is to mount a third gear (gear 3) on shaft 2 and mate it with gear 4 mounted on the output shaft, shaft 3. With proper sizing of all four gears, you should be able to produce an output speed equal or quite close to the desired speed. This process requires knowledge of the concept of *velocity ratio* and the techniques of designing gear trains as presented in this chapter.

But you will also need to specify the appearance of the gears and the geometry of the several features that make up each gear. Whereas the final specification also requires the information from following chapters, you will learn how to recognize common forms of gears and to compute the dimensions of key features. This will be important when completing the design for strength and wear resistance in later chapters.

Let's say that you have chosen to use spur gears in your design. What design decisions must you make to complete the specification of all four gears? The following list gives some of the important parameters for each gear:

- The number of teeth
- The form of the teeth
- The size of the teeth as indicated by the *pitch*
- The width of the face of the teeth
- The style and dimensions of the gear blank into which the gear teeth are to be machined
- The design of the hub for the gear that facilitates its mounting to the shaft
- The degree of precision of the gear teeth and the corresponding method of manufacture that can produce that precision
- The means of attaching the gear to its shaft
- The means of locating the gear axially on the shaft

To make reliable decisions about these parameters, you must understand the special geometry of spur gears as presented first in this chapter. However, there are other forms of gears that you could choose. Later sections give the special geometry of helical gears, bevel gears, and worm/wormgear sets. The methods of analyzing the forces on these various kinds of gears are described in later chapters, including the stress analysis of the gear teeth and recommendations on material selection to ensure safe operation with long life. ■

8–1 OBJECTIVES OF THIS CHAPTER

After completing this chapter, you will be able to:

1. Recognize and describe the main features of *spur gears, helical gears, bevel gears,* and *worm/wormgear sets.*
2. Describe the important operating characteristics of these various types of gears with regard to the similarities and differences among them and their general advantages and disadvantages.
3. Describe the *involute-tooth form* and discuss its relationship to the *law of gearing.*
4. Describe the basic functions of the American Gear Manufacturers Association (AGMA) and identify pertinent standards developed and published by this organization.

5. Define *velocity ratio* as it pertains to two gears operating together.
6. Specify appropriate numbers of teeth for a mating pair of gears to produce a given velocity ratio.
7. Define *train value* as it pertains to the overall speed ratio between the input and output shafts of a gear-type speed reducer (or speed increaser) that uses more than two gears.

8–2 SPUR GEAR STYLES

Figure 8–4 shows several different styles of commercially available spur gears. When gears are large, the spoked design in Part (a) is often used to save weight. The gear teeth are machined into a relatively thin rim that is held by a set of spokes connecting to the hub. The bore of the hub is typically designed to be a close sliding fit with the shaft that carries the gear. A keyway is usually machined into the bore to allow a key to be inserted for positive transmission of torque. The illustration does not include a keyway because this gear is sold as a stock item, and the ultimate user finishes the bore to match a given piece of equipment.

The solid hub design in Figure 8–4(b) is typical of smaller spur gears. Here the finished bore with a keyway is visible. The set screw over the keyway allows the locking of the key in place after assembly.

When spur gear teeth are machined into a straight, flat bar, the assembly is called a rack, as shown in Figure 8–4(c). The rack is essentially a spur gear with an infinite radius. In this form, the teeth become straight-sided, rather than the curved, involute form typical of smaller gears.

Gears with diameters between the small solid form [Part (b)] and the larger spoked form [Part (a)] are often produced with a thinned web as shown in Part (d), again to save weight.

You as a designer may create special designs for gears that you implement into a mechanical device or system. One useful approach is to machine the gear teeth of small pinions directly into the surface of the shaft that carries the gear. This is very often done for the input shaft of gear reducers.

8–3 SPUR GEAR GEOMETRY-INVOLUTE-TOOTH FORM

The most widely used spur gear tooth form is the full-depth involute form. Its characteristic shape is shown in Figure 8–5. See References 10-15 and 18 for more on the kinematics of gearing.

The involute is one of a class of geometric curves called *conjugate curves*. When two such gear teeth are in mesh and rotating, there is a *constant angular velocity* ratio between them: From the moment of initial contact to the moment of disengagement, the speed of the driving gear is in a constant proportion to the speed of the driven gear. The resulting action of the two gears is very smooth. If this were not the case, there would be some speeding up and slowing down during the engagement, with the resulting accelerations causing

(*a*) Spur gear with spoked design

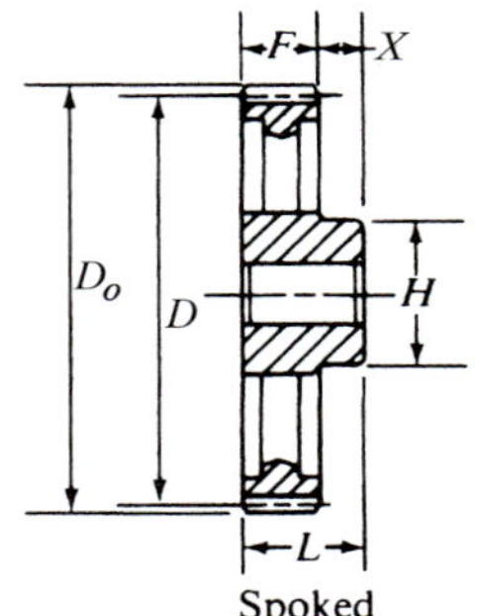

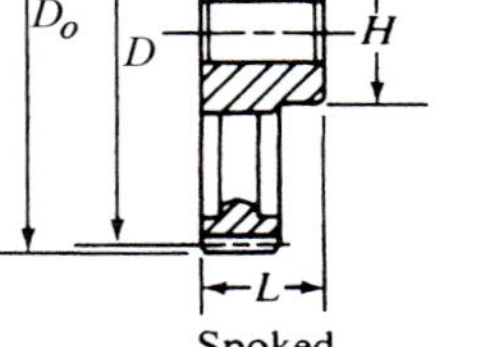

Spoked

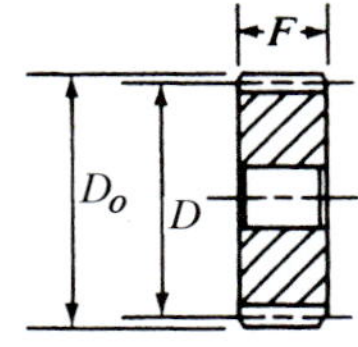

(*b*) Spur gear with solid hub

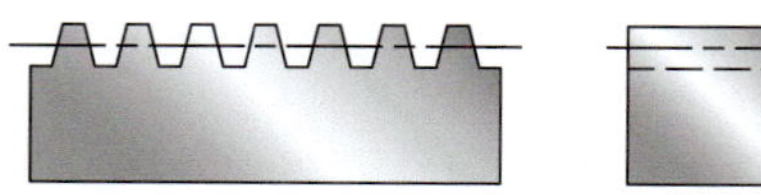

(*c*) Rack-straight spur gear

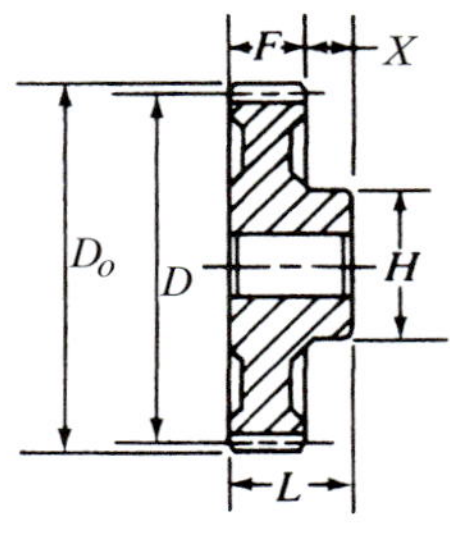

(*d*) Spur gear with thinned web

D_o = outside dia.
D = pitch dia.
F = face width
L = length of hub
X = extension of hub beyond face
H = hub dia.

FIGURE 8–4 Examples of spur gears and a rack
(Power Transmission Solutions, a Business unit of Emerson Industrial Automation)

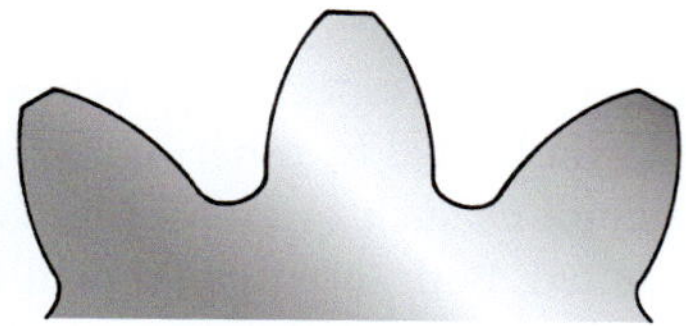

FIGURE 8–5 Involute-tooth form

vibration, noise, and dangerous torsional oscillations in the system.

You can easily visualize an involute curve by taking a cylinder and wrapping a string around its circumference. Tie a pencil to the end of the string. Then start with the pencil tight against the cylinder, and hold the string taut. Move the pencil away from the cylinder while keeping the string taut. The curve that you will draw is an involute. Figure 8–6 is a sketch of the process.

The circle represented by the cylinder is called the *base circle*. Notice that at any position on the curve, the string represents a line tangent to the base circle and, at the same time, perpendicular to the involute. Drawing another base circle along the same centerline in such a position that the resulting involute is tangent to the first one, as shown in Figure 8–7, demonstrates that at the point of contact, the two lines tangent to the base circles are coincident and will stay in the same position as the base circles rotate. This is what happens when two gear teeth are in mesh.

It is a fundamental principle of *kinematics*, the study of motion, that if the line drawn perpendicular to the surfaces of two rotating bodies at their point of contact always crosses the centerline between the two bodies at the same place, the angular velocity ratio of the two bodies will be constant. This is a statement of the *law of gearing*. As demonstrated here, the gear teeth made in the involute-tooth form obey the law.

Of course, only the part of the gear tooth that actually comes into contact with the mating tooth needs to be in the involute form.

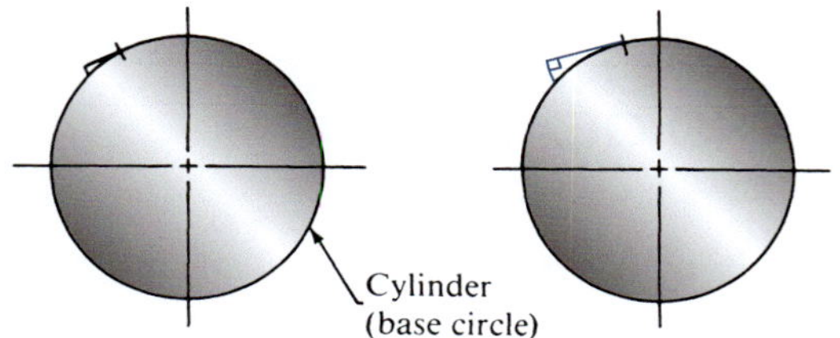

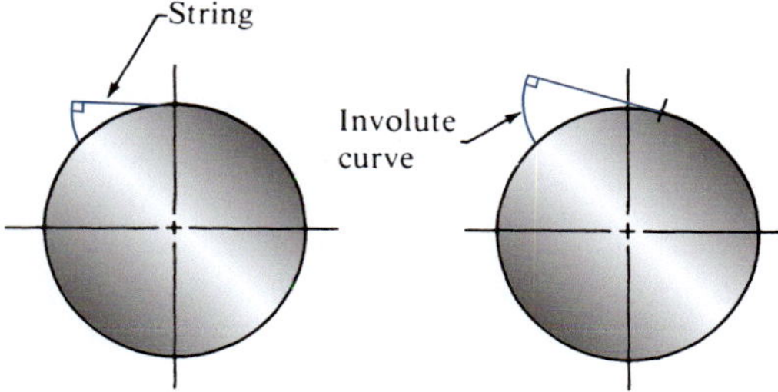

FIGURE 8–6 Graphical presentation of an involute curve

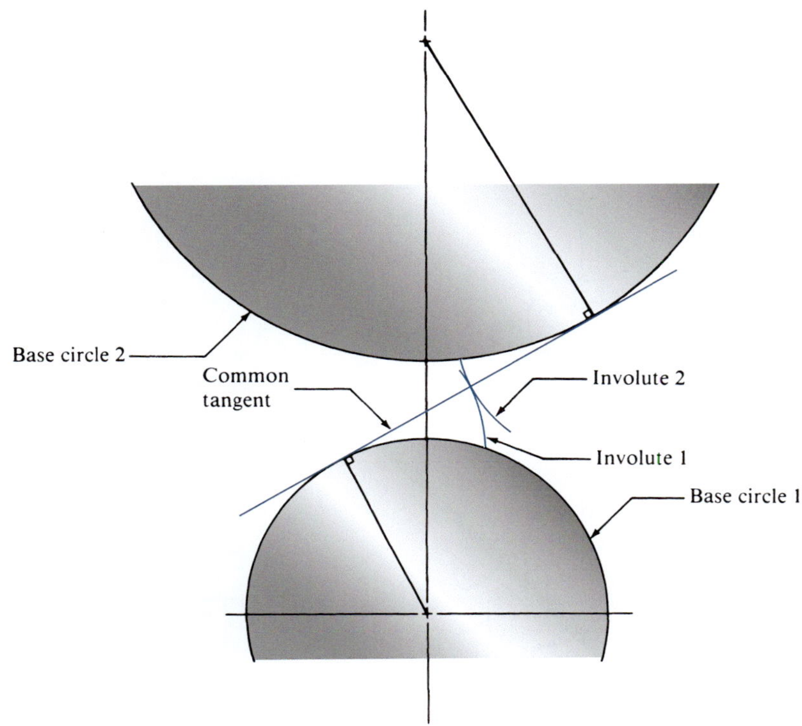

FIGURE 8–7 Mating involutes

8–4 SPUR GEAR NOMENCLATURE AND GEAR-TOOTH FEATURES

This section describes several features of individual spur gear teeth, complete gears, and the basic geometry of two mating gears. Terms and symbols used here conform mostly to American Gear Manufacturers Association (AGMA) standards. Because there is variation among the several applicable standards, the primary reference is AGMA 2001-D04 *Fundamental Rating Factors and Calculation Methods for Involute Spur and Helical Gear Teeth.* This standard is the basis for analytical design methods that are described in Chapters 9 and 10 for spur gears and helical gears, respectively. Where appropriate, the terms and symbols used by other AGMA standards and international standards such as ISO, DIN (Germany), and JIS (Japan) are noted. Both the conventional U.S. system of units, called the *Diametral Pitch System*, and the SI metric system, called the *Metric Module System*, are discussed.

Reference is made to several figures and tables that depict the geometry of interest in the design of gear pairs:

1. Figure 8–1 shows two mating spur gears, indicating the dimensions related to diameters and center distance.
2. Figures 8–8 and 8–9 show details of spur gear teeth and how they engage as the gears rotate. See also Internet sites 7 and 8 for animations of teeth engagement.

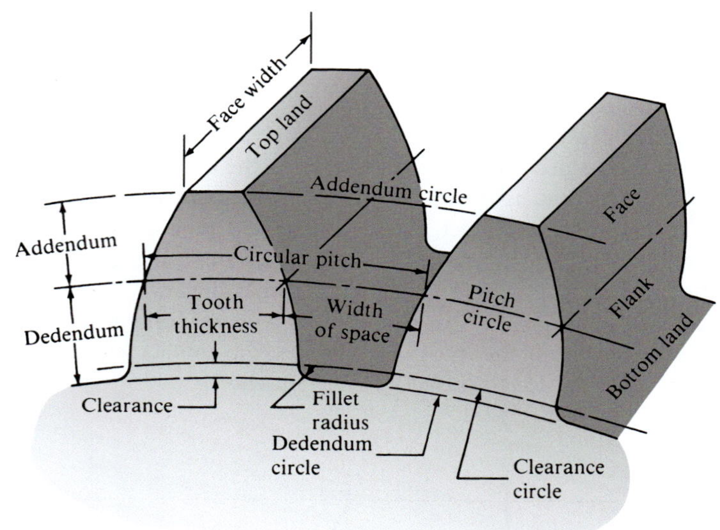

FIGURE 8–8 Spur gear teeth features

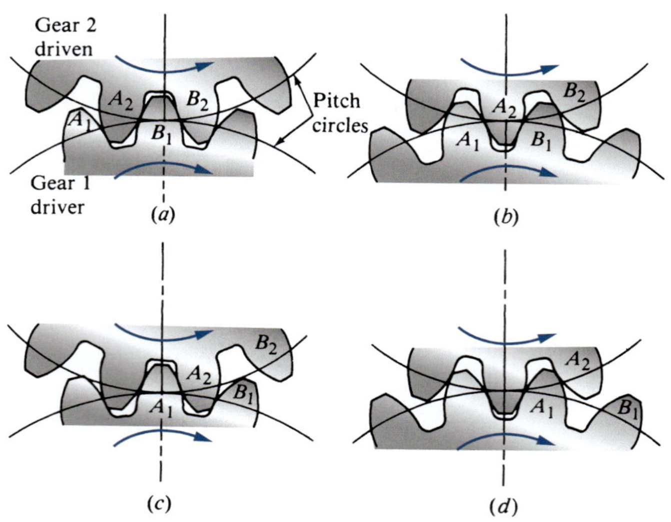

FIGURE 8–9 Cycle of engagement of gear teeth

3. Figures 8–10 and 8–11 show various sizes of gear teeth in both the diametral pitch and metric module systems. Both figures are full size, enabling you to compare physical gears to the drawings to gain an appreciation of gear tooth sizes.
4. Table 8–1 is a composite reference tool for identifying the names, symbols, definitions, units, and formulas related to the several features of gear teeth and mating gears.

A note about accuracy: Gears and gear trains are precision mechanical devices with tolerances on critical dimensions typically in the range of a few ten thousandths of an inch (0.0001 in or about 0.0025 mm). Therefore, it is expected that such dimensions be reported to at least the nearest ten thousandth of an inch (four decimal places) or the nearest 0.001 mm. Some applications require even more precision. See References 8 and 9 for more on accuracy of gearing.

FIGURE 8–10 Gear-tooth size as a function of diametral pitch (Barber-Colman Company, Loves Park, IL)

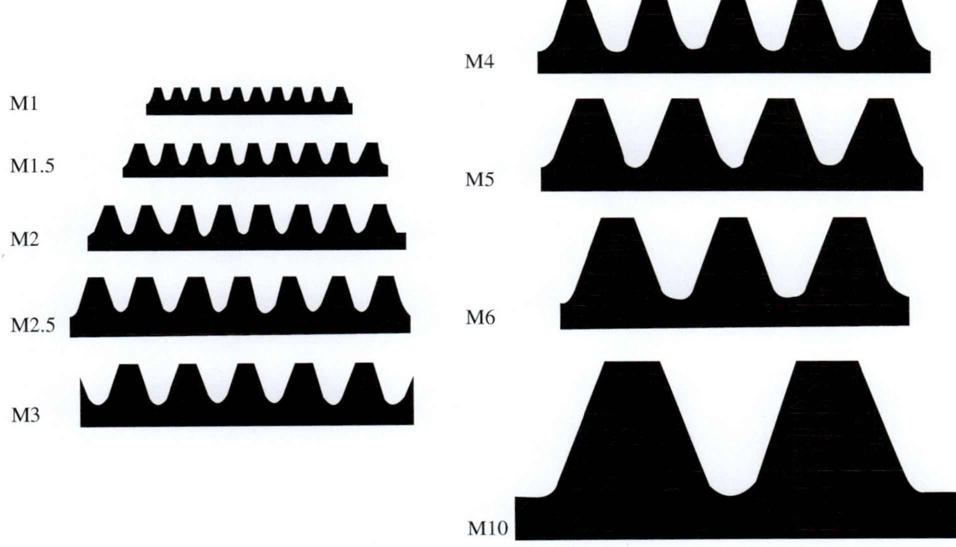

FIGURE 8–11 Selected standard metric modules in rack form—actual size

TABLE 8–1 Gear and Tooth Features, Diameters, Center Distance for a Gear Pair

Pitches	Symbol	Definition	Typical unit	Formulas: General formula	U.S. Full-depth involute system: Coarse pitch $P_d < 20$ (in)	U.S. Full-depth involute system: Fine pitch $P_d \geq 20$ (in)	Metric module system (mm)
Number of teeth	N	Integer count of teeth on a gear					
Circular pitch	p	Arc distance between corresponding points on adjacent teeth	in or mm	$p = \pi D/N$	$p = \pi/P_d$		$p = \pi m$
Diametral pitch	P_d	Number of teeth per inch of pitch diameter	in^{-1}	$P_d = N/D$	$P_d = 25.4/m$		
Module	m	Pitch diameter divided by number of teeth	mm	$m = D/N$			$m = 25.4/P_d$
Diameters							
Pitch diameter	D	Kinematic characteristic diameter for a gear; Diameter of the pitch circle	in or mm		$D = N/P_d$		$D = mN$
Outside diameter	D_o	Diameter to the outside surface of the gear teeth	in or mm		$D_o = (N + 2)/P_d$		$D_o = m(N+2)$
Root diameter	D_R	Diameter to the root circle of the gear at the base of the teeth	in or mm	$D_R = D - 2b$			
Gear Tooth Features							
Addendum	a	Radial distance from pitch circle to outside of tooth	in or mm		$a = 1.00/P_d$		$a = 1.00m$
Dedendum	b	Radial distance from pitch circle to bottom of tooth space	in or mm		$b = 1.25/P_d$	$b = 1.20/P_d + 0.002$	$b = 1.25m^1$
Clearance	c	Radial distance from top of fully engaged tooth of mating gear to bottom of tooth space	in or mm		$c = 0.25/P_d$	$c = 0.20/P_d + 0.002$	$c = 0.25m^1$
Whole depth	h_t	Radial distance from top of a tooth to bottom of tooth space	in or mm	$h_t = a + b$	$h_t = 2.25/P_d$	$h_t = 2.20m + 0.002$	$h_t = 2.25m^1$
Working depth	h_k	Radial distance a gear tooth projects into tooth space of mating gear	in or mm	$h_k = a + a = 2a$	$h_k = 2.00/P_d$	$h_k = 2.00/P_d$	$h_k = 2.00m^1$
Tooth thickness	t	Theoretical arc distance equal to 1/2 of circular pitch	in or mm	$t = p/2$	$t = \pi\ /2P_d$		$t = \pi m/2$
Face width	F	Width of tooth parallel to axis of gear	in or mm	*Design decision*	Approximately $12/P_d$		
Pressure angle	ϕ	Angle between the tangent to the pitch circle and the perpendicular to the gear tooth surface	degrees	*Design decision*	Most common value = 20° Others: 14 1/2°, 25°		
Center Distance	C	Distance from between centerlines of mating gears	in or mm	$C = (D_P + D_G)/2$	$C = (N_P + N_G)/2P_d$		$C = m(N_P + N_G)/2$

Note: [1]Factors in formula for dedendum may vary in metric module system to obtain custom clearance.

Pinion and Gear. For two gears in mesh, the smaller gear is called the *pinion* and the larger is called, simply, the *gear*.

Number of Teeth, *N*. It is essential that there are an integer number of teeth in any gear. This book uses the symbol N for the number of teeth, with N_P for the pinion and N_G for the gear. These subscripts are applied to other gear features as well. Another commonly used symbol for the number of teeth is z, with similar subscripts or simply called z_1 and z_2.

Pitch. Refer to Figures 8–1, 8–8, and 8–9. The *pitch* of a gear, in general, is defined as follows:

> ***The pitch of a gear is the arc distance from a point on a tooth at the pitch circle to the corresponding point on the next adjacent tooth, measured along the pitch circle.***

The pitch circle is defined next. It is important to note that the pitch of both mating gears must be identical to ensure the smooth engagement of the teeth as the gears rotate. Standard pitches are defined in three different systems, described later in this section.

***Pitch Circle* and *Pitch Diameter*.** When two gears are in mesh, they behave as if two smooth rollers are rolling on each other without slipping. The surface of each roller defines the *pitch circle* and its diameter is called the *pitch diameter*. The *pitch diameter*, called D in this book, is used as the characteristic size of the gear for calculations of speeds. Note that the pitch diameter for a gear is a theoretical concept and cannot be measured directly. It falls within the gear teeth and is dependent on which standard system for pitch is specified for a particular gear pair. The commonly used units for D are inches (in) for the U.S. system and millimeters (mm) for the SI metric system.

***Circular Pitch, p*.** The pitch corresponding exactly to the basic definition of pitch given above is called the *circular pitch, p*. Some large gears that are made by casting are made to standard sizes of circular pitch such as those listed in Table 8–2. They represent a very small portion of gears in common use. The formula for p comes from dividing the circumference of the pitch circle of the gear into N parts. That is,

TABLE 8–2 Standard Circular Pitches (in)

10.0	7.5	5.0
9.5	7.0	4.5
9.0	6.5	4.0
8.5	6.0	3.5
8.0	5.5	

➪ **Circular Pitch**

$$p = \pi D/N \tag{8–2}$$

***Diametral Pitch, P_d*.** The most common pitch system in use in the United States at this time is *diametral pitch system*. We use the symbol, P_d, to denote *diametral pitch*. Note that some references use the term *DP*. The definition of P_d is stated here for either the pinion or the gear and both must be identical.

➪ **Diametral Pitch**

$$P_d = N_P/D_P = N_G/D_G \tag{8–3}$$

Analysis of units shows that P_d has the unit of in^{-1}, but the unit is rarely reported. It is necessary to not confuse the terms *diametral pitch*, P_d, and *pitch diameter*, D. Note that designers often refer to gears in this system as, for example, 8-pitch for $P_d = 8$ and 20-pitch for $P_d = 20$.

In this book, we use only those values of P_d listed in Table 8–3 because they are the most readily available as stock gears and most gear manufacturers have tooling for these sizes. Smaller pitches have larger teeth; larger pitches have smaller teeth. Note that pitches under 20 are called *coarse*, while those 20 and higher are called *fine*. Refer to Figure 8–10 that shows actual sizes of teeth with certain diametral pitches. ***Note that not all listed values of P_d are readily available and those such as 7, 9, 11, 22, and 26 should not be specified.***

***Metric Module, m*.** The basic definition of the metric module, m, is given here for both the pinion and the gear and they must be identical.

➪ **Metric Module**

$$m = D_p/N_p = D_G/N_G \tag{8–4}$$

TABLE 8–3 Standard Diametral Pitches (teeth/in)

Coarse pitch ($P_d < 20$)				Fine pitch ($P_d \geq 20$)	
1	2	5	12	20	72
1.25	2.5	6	14	24	80
1.5	3	8	16	32	96
1.75	4	10	18	48	120
				64	

TABLE 8–4 Standard modules

Module (mm)	Equivalent P_d	Closest standard P_d (teeth/in)
0.3	84.667	80
0.4	63.500	64
0.5	50.800	48
0.8	31.750	32
1	25.400	24
1.25	20.320	20
1.5	16.933	16
2	12.700	12
2.5	10.160	10
3	8.466	8
4	6.350	6
5	5.080	5
6	4.233	4
8	3.175	3
10	2.540	2.5
12	2.117	2
16	1.587	1.5
20	1.270	1.25
25	1.016	1

The unit of mm is typically used. Note that smaller values of m denote smaller teeth and vice versa. Some references and vendors' tables use the symbol, M, for module and write M5 for $m = 5$ for example. Figure 8–11 shows actual sizes of nine standard modules, illustrated as the teeth of *racks*, straight gears with infinite diameters. Many more modules are available and Table 8–4 lists a total of 19 values that cover typical applications featured in this book.

Relation between P_d and m. With globally integrated design and marketing of products and systems, it is likely that the need to convert from one system to another will be needed. Note that the definition of P_d is fundamentally the inverse of the definition of m. That is,

$$m = 1/P_d \quad \text{or} \quad P_d = 1/m$$

However, because different units are employed for each term, a conversion factor of 25.4 mm/in is required, resulting in useful forms of the relationship as

➪ Relation between Module and Diametral Pitch

$$m = 25.4/P_d \quad \text{or} \quad P_d = 25.4/m \qquad \textbf{(8–5)}$$

Table 8–4 uses this relationship to compute the equivalent diametral pitch, P_d, for given standard metric modules, m. Note that the conversions do not deliver standard values of P_d such as those listed in Table 8–3. Therefore, we list the nearest standard P_d value in Table 8–4 as an aid to designers who are considering converting a design from one system to the other. We can say, for example, that a module of $m = 1.25$ is a closely similar size to $P_d = 20$. Of course, the design for either system must be completed independently.

Gear Tooth Features. Table 8–1 includes the definitions of several other features of individual teeth that designers must be familiar with. Refer to Figures 8–1 and 8–8 to visualize those features.

Face Width, F. Face width is the width of the gear parallel to the axis of the gear. It is defined by the designer as one of the required *design decisions*. More is said about face width in Chapter 9, where strength of the teeth is considered. For now, we can state that a nominal value for face width is approximately $F \sim 12/P_d$, but a wide range is permitted.

Center Distance, C. One of the most critical dimensions for a gear pair is the *center distance*, defined as the linear distance from the centerline of the pinion to the centerline of the gear as shown in Figure 8–1. The theoretical value is best represented as the sum of the pitch radii of the pinion and the gear. That is,

Pitch radii: $R_P = D_P/2 \quad \text{and} \quad R_G = D_G/2$

Center distance: $C = R_P + R_G = D_P/2 + D_G/2 = (D_P + D_G)/2$ **(8–6)**

Other useful equations for C, recommended for use in this book, are developed here.

Diametral pitch system: From Equation (8–3),

➪ Center Distance in terms of N_G, N_P, and P_d

$$D_P = N_P/P_d \quad \text{and} \quad D_G = N_G/P_d$$
$$C = (D_P + D_G)/2 = (N_P/P_d + N_G/P_d)/2$$
$$= (N_P + N_G)/2P_d \qquad \textbf{(8–7)}$$

Metric module system: From Equation (8–4),

➪ Center Distance in terms of N_G, N_P, and m

$$D_P = mN_P \quad \text{and} \quad D_G = mN_G$$
$$C = (D_P + D_G)/2 = (mN_P + mN_G)/2$$
$$= m(N_P + N_G)/2 \qquad \textbf{(8–8)}$$

An important advantage of the final forms of the center distance Equation (8–7) and (8–8) is that all numbers on the right side are typically integers or exact fractional dimensions such as 1.25, 1.5, or 2.5. Therefore, the highest level of accuracy is obtained from using those forms. Conversely, some values for pitch diameters are irrational numbers. For example, a 12-pitch gear with 65 teeth has a pitch diameter of

$$D = N/P_d = 65/12 = 5.416666\ldots \text{ in} \quad \text{and}$$
$$R = D/2 = 2.708333\ldots \text{ in}$$

Depending on rounding, an inaccurate calculation for center distance could result from using these values.

Comments on Gear Tooth Features. The definitions and formulas given in Table 8–1 yield the theoretical values and it is typical for gear designers and manufacturers to modify some features to produce preferred performance characteristics. It is important to realize that some of these practices change the fundamental geometry of the gears and they may result in weaker tooth shapes, vibration during operation, and/or increased noise. Some examples are given below:

1. *Addendum modification:* The upper part of the flank of a gear tooth is the first to engage its mating tooth and it penetrates most deeply into the tooth space of the mating gear. Some applications benefit from relieving the true surface of the tooth or from shortening the addendum to promote smooth engagement or to avoid damage.
2. *Dedendum modification:* Some designers prefer greater clearance between the bottom of the tooth space and the fully engaged mating tooth to facilitate lubrication. Using a longer dedendum dimension provides this capability. Plastic gears may also employ dedendum modification to accommodate thermal expansion and/or swelling because of moisture absorption.
3. *Center distance modification:* Some combinations of gear tooth geometry result in very small clearances between teeth and possibly interference between the top of the engaging tooth and the lower part of the flank of the mating tooth. These conditions should be avoided by proper design decisions to avoid the interference and this topic is discussed in Chapter 9. However, some designers remove the interference by expanding the center distance slightly from its theoretical dimension.
4. *Tooth thickness modification:* A tooth thickness defined exactly as $\frac{1}{2}$ of the circular pitch will result in a tight fit of a tooth in the tooth space of the mating gear, possibly causing binding or precluding entry of lubrication at the contact point on the teeth. Using a smaller tooth thickness provides space for lubrication and facilitates assembly. The space created is called *backlash* and it is discussed next.

- ***Backlash:*** To provide backlash, the cutter generating the gear teeth can be fed more deeply into the gear blank than the theoretical value on either or both of the mating gears. Alternatively, backlash can be created by adjusting the center distance to a larger value than the theoretical value.

The magnitude of backlash depends on the desired precision of the gear pair and on the size and the pitch of the gears. It is actually a design decision, balancing cost of production with desired performance. The American Gear Manufacturers Association (AGMA) provides recommendations for backlash in their standards. (See Reference 2.) Table 8–5 lists recommended ranges for several values of pitch.

TABLE 8–5 Recommended Minimum Backlash for Coarse Pitch Gears

A. Diametral pitch system (backlash in inches)

	Center distance, C (in)				
P_d	2	4	8	16	32
18	0.005	0.006			
12	0.006	0.007	0.009		
8	0.007	0.008	0.010	0.014	
5		0.010	0.012	0.016	
3		0.014	0.016	0.020	0.028
2			0.021	0.025	0.033
1.25				0.034	0.042

B. Metric module system (backlash in millimeters)

	Center distance, C (mm)				
Module, m	50	100	200	400	800
1.5	0.13	0.16			
2	0.14	0.17	0.22		
3	0.18	0.20	0.25	0.35	
5		0.26	0.31	0.41	
8		0.35	0.40	0.50	0.70
12			0.52	0.62	0.82
18				0.80	1.00

Source: Extracted from AGMA 2002-B88 Standard, *Tooth Thickness Specification and Measurement*, with permission of the publisher, American Gear Manufacturers Association, 1001 North Fairfax Street, 5th floor, Alexandria, VA 22314.

Pressure Angle

> ***The* pressure angle *is the angle between the tangent to the pitch circles and the line drawn normal (perpendicular) to the surface of the gear tooth (see Figure 8–12).***

The normal line is sometimes referred to as the *line of action*. When two gear teeth are in mesh and are transmitting power, the force transferred from the driver to the driven gear tooth acts in a direction along the line of action. Also, the actual shape of the gear tooth depends on the pressure angle, as illustrated in Figure 8–13. The teeth in this figure were drawn according to the proportions for a 20-tooth, 5-pitch gear having a pitch diameter of 4.000 in.

All three teeth have the same tooth thickness because, as stated in Table 8–1 the thickness at the pitch line depends only on the pitch. The difference between the three teeth shown is due to the different pressure angles because the pressure angle determines the size of the base circle. Remember that the base circle is the circle from which the involute is generated. The line of action is always tangent to the base circle. Therefore, the size of the base circle can be found from

➩ Base Circle Diameter

$$D_b = D \cos \phi \tag{8–9}$$

Standard values of the pressure angle are established by gear manufacturers, and the pressure angles of two gears in mesh must be the same. Current standard pressure angles are $14\frac{1}{2}°$, 20°, and 25° as illustrated in Figure 8–13. Actually, the $14\frac{1}{2}°$ tooth form is considered obsolete. Although it is still available, it should be avoided for new designs. The 20° tooth form is the most readily available at this time. The advantages and disadvantages of the different values of pressure angle relate to the strength of the teeth, the occurrence of interference, and the magnitude of forces exerted on the shaft. Interference is discussed in Section 8–5. The other points are discussed in a later chapter.

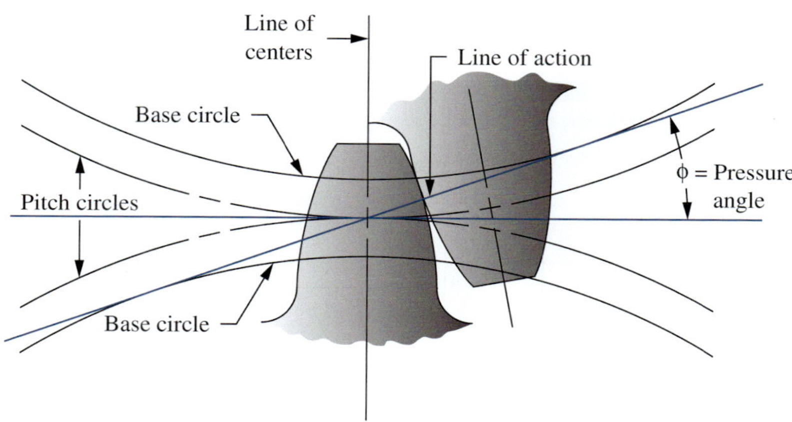

FIGURE 8–12 Pressure angle

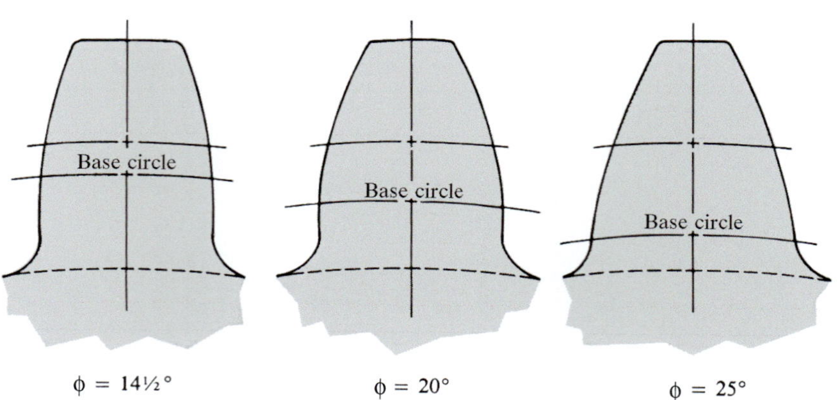

FIGURE 8–13 Full-depth, involute-tooth form for varying pressure angles

TABLE 8–6 Formulas for Use When Implementing Gear Pair Contact Ratio Calculation In U.S. and SI Systems in Terms of Diametral Pitch and Module

	Factors	Diametral Pitch System	Metric Module System
ϕ	Pressure angle		
R_{oP}	Outside radius—pinion	$(N_p + 2)/(2P_d)$	$m(N_p + 2)/2$
R_{bP}	Base circle radius—pinion	$(N_p/2P_d)\cos\phi$	$(mN_p/2)\cos\phi$
R_{oG}	Outside radius—gear	$(N_G + 2)/(2P_d)$	$m(N_G + 2)/2$
R_{bG}	Base circle radius—gear	$(N_G/2P_d)\cos\phi$	$(mN_G/2)\cos\phi$
C	Center distance	$(N_p + N_G)/(2P_d)$	$m(N_p + N_G)/2$
p	Circular pitch	π/P_d	πm

Contact Ratio

When two gears mesh, it is essential for smooth operation that a second tooth begin to make contact before a given tooth disengages. The term *contact ratio* is used to indicate the average number of teeth in contact during the transmission of power. A recommended minimum contact ratio is 1.2 and typical spur gear combinations often have values of 1.5 or higher.

The contact ratio is defined as the ratio of the length of the line-of-action to the base pitch for the gear. The line-of-action is the straight-line path of a tooth from where it encounters the outside diameter of the mating gear to the point where it leaves engagement. The base pitch is the diameter of the base circle divided by the number of teeth in the gear. A convenient formula for computing the contact ratio, m_f, is,

➩ Contact Ratio

$$m_f = \frac{\sqrt{R_{oP}^2 - R_{bP}^2} + \sqrt{R_{oG}^2 - R_{bG}^2} - C\sin\phi}{p\cos\phi} \quad \textbf{(8–10)}$$

where,

ϕ = Pressure angle

R_{oP} = Outside radius of the pinion = $D_{oP}/2$ = $(N_P + 2)/(2P_d)$

R_{bP} = Radius of the base circle for the pinion = $D_{bP}/2 = (D_P/2)\cos\phi = (N_P/2P_d)\cos\phi$

R_{oG} = Outside radius of the gear = $D_{oG}/2$ = $(N_G + 2)/(2P_d)$

R_{bG} = Radius of the base circle for the gear = $D_{bG}/2$ = $(D_G/2)\cos\phi = (N_G/2P_d)\cos\phi$

C = Center distance = $(N_P + N_G)/(2P_d)$

p = Circular pitch = $(\pi D_p/N_p) = \pi/P_d$

For example, consider a pair of gears with the following data:

$$N_P = 18,\ N_G = 64,\ P_d = 8,\ \phi = 20°$$

Then,

$$R_{oP} = (N_P + 2)/(2P_d) = (18 + 2)/[2(8)] = 1.250 \text{ in}$$
$$R_{bP} = (N_P/2P_d)\cos\phi = 18/[2(8)]\cos 20° = 1.05715 \text{ in}$$
$$R_{oG} = (N_G + 2)/(2P_d) = (64 + 2)/[2(8)] = 4.125 \text{ in}$$
$$R_{bG} = (N_G/2P_d)\cos\phi = 64/[2(8)]\cos 20° = 3.75877 \text{ in}$$
$$C = (N_P + N_G)/(2P_d) = (18 + 64)/[2(8)] = 5.125 \text{ in}$$
$$p = \pi/P_d = \pi/8 = 0.392699 \text{ in}$$

Finally, the contact ratio is,

$$m_f = \frac{\sqrt{(1.250)^2 - (1.05715)^2} + \sqrt{(4.125)^2 - (3.75877)^2} - (5.125)\sin 20°}{(0.392699)\cos 20°}$$

$$m_f = 1.66$$

This value is comfortably above the recommended minimum value of 1.20.

Similar developments can be used to determine the factors needed to implement the contact ratio calculation in Equation (8–10) with terms expressed in the Metric Module System. Table 8–6 summarizes the relationships for both systems.

Example Problem 8–1 For the pair of gears shown in Figure 8–1, compute all of the features of the gear teeth described in this section. The gears conform to the standard AGMA form and have a diametral pitch of 12 and a 20° pressure angle.

Solution Given $P_d = 12;\ N_p = 11;\ N_G = 18;\ \phi = 20°$.

Analysis Unless otherwise noted, we use equations from Table 8–1 to compute the features. Refer to the text in this section for explanation of terms.

Note that results are reported to four decimal places as is typical for precise mechanical devices like gears. A similar level of accuracy is expected for problems in this book.

Results *Pitch Diameters*

For the pinion,

$$D_p = N_p/P_d = 11/12 = 0.9167 \text{ in}$$

For the gear,

$$D_G = N_G/P_d = 18/12 = 1.5000 \text{ in}$$

Circular Pitch

Three different approaches could be used.

$$p = \pi/P_d = \pi/12 = 0.2618 \text{ in}$$

Note that data for either the pinion or the gear data may also be used. For the pinion,

$$p = \pi D_P/N_P = \pi(0.9167 \text{ in})/11 = 0.2618 \text{ in}$$

For the gear,

$$p = \pi D_G/N_G = \pi(1.500 \text{ in})/18 = 0.2618 \text{ in}$$

Addendum

$$a = 1/P_d = 1/12 = 0.8333 \text{ in}$$

Dedendum

Note that the 12-pitch gear is considered to be coarse. Thus,

$$b = 1.25/P_d = 1.25/12 = 0.1042 \text{ in}$$

Clearance

$$c = 0.25/P_d = 0.25/12 = 0.0208 \text{ in}$$

Outside Diameters

For the pinion,

$$D_{oP} = (N_P + 2)/P_d = (11 + 2)/12 = 1.0833 \text{ in}$$

For the gear,

$$D_{oG} = (N_G + 2)/P_d = (18 + 2)/12 = 1.6667 \text{ in}$$

Root Diameters

First, for the pinion,

$$D_{RP} = D_P - 2b = 0.9167 \text{ in} - 2(0.1042 \text{ in}) = 0.7083 \text{ in}$$

For the gear,

$$D_{RG} = D_G - 2b = 1.500 \text{ in} - 2(0.1042 \text{ in}) = 1.2917 \text{ in}$$

Whole Depth

$$h_t = a + b = 0.0833 \text{ in} + 0.1042 \text{ in} = 0.1875 \text{ in}$$

Working Depth

$$h_k = 2a = 2(0.0833 \text{ in}) = 0.1667 \text{ in}$$

Tooth Thickness

$$t = \pi/2P_d = \pi/2(12) = 0.1309 \text{ in}$$

Center Distance

$$C = (N_G + N_P)/(2P_d) = (18 + 11)/[2(12)] = 1.2083 \text{ in}$$

Base Circle Diameter

$$D_{bP} = D_P \cos \phi = (0.9167 \text{ in}) \cos (20°) = 0.8614 \text{ in}$$
$$D_{bG} = D_G \cos \phi = (1.5000 \text{ in}) \cos (20°) = 1.4095 \text{ in}$$

8–5 INTERFERENCE BETWEEN MATING SPUR GEAR TEETH

For certain combinations of numbers of teeth in a gear pair, there is interference between the tip of the teeth on the pinion and the fillet or root of the teeth on the gear. Obviously this cannot be tolerated because the gears simply will not mesh. The probability that interference will occur is greatest when a small pinion drives a large gear, with the worst case being a small pinion driving a rack. A *rack* is a gear with a straight pitch line; it can be thought of as a gear with an infinite pitch diameter [see Figure 8–4(c)].

It is the designer's responsibility to ensure that interference does not occur in a given application. The surest way to do this is to control the minimum number of teeth in the pinion to the limiting values shown on the left side of Table 8–7. With this number of teeth or a greater number, there will be no interference with a rack or with any other gear. A designer who desires to use fewer than the listed number of teeth can use a graphical layout to test the combination of pinion and gear for interference. Texts on kinematics provide the necessary procedure. The right side of Table 8–7 indicates the maximum number of gear teeth that you can use for a given number of pinion teeth to avoid interference. (See References 11 and 17.)

Using the information in Table 8–7, we can draw the following conclusions:

1. If a designer wants to be sure that there will not be interference between any two gears when using the $14\frac{1}{2}°$, full-depth, involute system, the pinion of the gear pair must have no fewer than 32 teeth.
2. For the 20°, full-depth, involute system, using no fewer than 18 teeth will ensure that no interference occurs.
3. For the 25°, full-depth, involute system, using no fewer than 12 teeth will ensure that no interference occurs.
4. If a designer desires to use fewer than 18 teeth in a pinion having 20°, full-depth teeth, there is an upper limit to the number of teeth that can be used on the mating gear without interference. For 17 teeth in the pinion, any number of teeth on the gear can be used up to 1309, a very high number. Most gear drive systems use no more than about 200 teeth in any gear. But a 17-tooth

TABLE 8–7 Number of Pinion Teeth to Ensure no Interference

For a pinion meshing with a rack		For a 20°, full-depth pinion meshing with a gear		
Tooth form	**Minimum number of teeth**	**Number of pinion teeth**	**Maximum number of gear teeth**	**Maximum ratio**
$14\frac{1}{2}°$, involute, full-depth	32	17	1309	77.00
20°, involute, full-depth	18	16	101	6.31
25°, involute, full-depth	12	15	45	3.00
		14	26	1.85
		13	16	1.23

pinion *would* have interference with a *rack* which is effectively a gear with an infinite number of teeth or an infinite pitch diameter. Similarly, the following requirements apply for 20° full-depth teeth:

A 16-tooth pinion requires a gear having 101 or fewer teeth, producing a maximum velocity ratio of $N_G/N_P = 101/16 = 6.31$.

A 15-tooth pinion requires a gear having 45 or fewer teeth, producing a maximum velocity ratio of $45/15 = 3.00$.

A 14-tooth pinion requires a gear having 26 or fewer teeth, producing a maximum velocity ratio of $26/14 = 1.85$.

A 13-tooth pinion requires a gear having 16 or fewer teeth, producing a maximum velocity ratio of $16/13 = 1.23$.

As noted earlier, the $14\frac{1}{2}°$ system is considered to be obsolete. The data in Table 8–7 indicate one of the main disadvantages with that system: its potential for causing interference.

Overcoming Interference

If a proposed design encounters interference, there are ways to make it work. But caution should be exercised because the tooth form or the alignment of the mating gears is changed, causing the stress and wear analysis to be inaccurate. With this in mind, the designer can provide for undercutting, modification of the addendum on the pinion or the gear, or modification of the center distance:

Undercutting is the process of cutting away the material at the fillet or root of the gear teeth, thus relieving the interference.

Figure 8–14 shows the result of undercutting. It should be obvious that this process weakens the tooth; this point is discussed further in Chapter 9 in the section on stresses in gear teeth.

To alleviate the problem of interference, increase the addendum of the pinion while decreasing the addendum of the gear. The center distance can remain the same as its theoretical value for the number of teeth in the pair. But the resulting gears are, of course, nonstandard. (See Reference 12.) It is possible to make the pinion of a gear pair larger than standard while keeping the gear standard if the center distance for the pair is enlarged. (See Reference 11.)

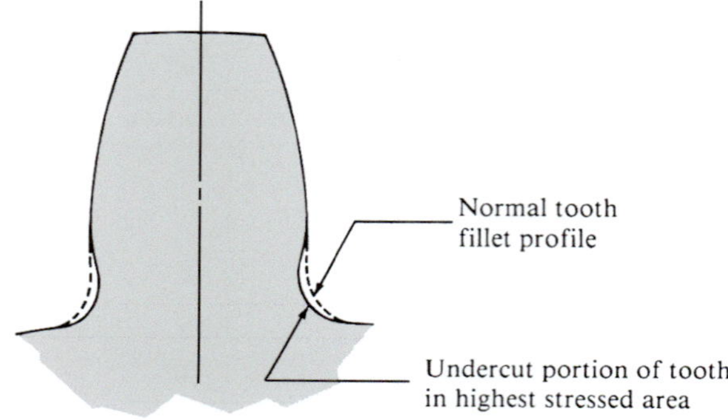

FIGURE 8–14 Undercutting of a gear tooth

8–6 HELICAL GEAR GEOMETRY

Helical and spur gears are distinguished by the orientation of their teeth. On spur gears, the teeth are straight and are aligned with the axis of the gear.

On helical gears, the teeth are inclined at an angle with the axis, that angle being called the *helix angle.* If the gear were very wide, it would appear that the teeth wind around the gear blank in a continuous, helical path. However, practical considerations limit the width of the gears so that the teeth normally appear to be merely inclined with respect to the axis. Figure 8–15 shows two examples of commercially available helical gears.

The forms of helical gear teeth are very similar to those discussed for spur gears. The basic task is to account for the effect of the helix angle.

Helix Angle

The helix for a given gear can be either *left-hand* or *right-hand.* The teeth of a right-hand helical gear would appear to lean to the right when the gear is lying on a flat surface.

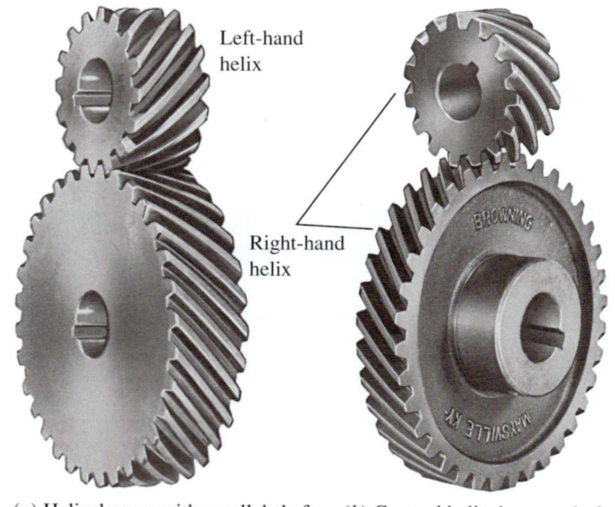

(*a*) Helical gears with parallel shafts (*b*) Crossed helical gears, shafts at right angle

FIGURE 8–15 Helical gears. These gears have a 45° helix angle

(Emerson Power Transmission Corporation, Browning Division, Maysville, KY)

Conversely, the teeth of a left-hand helical gear would lean to the left. In normal installation, helical gears would be mounted on parallel shafts as shown in Figure 8–15(a). To achieve this arrangement, it is required that one gear be of the right-hand design and that the other be left-hand with an equal helix angle. If both gears in mesh are of the same hand, as shown in Figure 8–15(b), the shafts will be at 90° to each other. Such gears are called *crossed helical gears.*

The parallel shaft arrangement for helical gears is preferred because it results in a much higher power-transmitting capacity for a given size of gear than the crossed helical arrangement. In this book, we will assume that the parallel shaft arrangement is being used unless otherwise stated.

Figure 8–16(a) shows the pertinent geometry of helical gear teeth. To simplify the drawing, only the pitch surface of the gear is shown. The pitch surface is the cylinder that passes through the gear teeth at the pitch line. Thus, the diameter of the cylinder is equal to the pitch diameter of the gear. The lines drawn on the pitch surface represent elements of each tooth where the surface would cut into the face of the tooth. These elements are inclined with respect to a line parallel to the axis of the cylinder, and the angle of inclination is the *helix angle,* ψ (the Greek letter *psi*).

The main advantage of helical gears over spur gears is smoother engagement because a given tooth assumes its load gradually instead of suddenly. Contact starts at one end of a tooth near the tip and progresses across the face in a path downward across the pitch line to the lower flank of the tooth, where it leaves engagement. Simultaneously, other teeth are coming into engagement before a given tooth leaves engagement, with the result that a larger average number of teeth are engaged and are sharing the applied loads compared with a spur gear. The lower average load per tooth allows a greater power transmission capacity for a given size of gear, or a smaller gear can be designed to carry the same power.

The main disadvantage of helical gears is that an *axial thrust load* is produced as a natural result of the inclined arrangement of the teeth. The bearings that hold the shaft carrying the helical gear must be capable of reacting against the thrust load.

The helix angle is specified for each given gear design. A balance should be sought to take advantage of the smoother engagement of the gear teeth when the helix angle is high while maintaining a reasonable value of the axial thrust load that increases with increasing helix angle. A typical range of values of helix angles is from 15° to 45°.

Pressure Angles, Primary Planes, and Forces for Helical Gears

To completely describe the geometry of helical gear teeth, we need to define two different pressure angles in addition to the helix angle. The two pressure angles are related to the three primary planes that are illustrated in Figure 8–16: (1) the *tangential plane,* (2) the *transverse plane,* and (3) the *normal plane.* Note that these planes contain the three orthogonal components of the true total normal force that is exerted by a tooth of one gear on a tooth of the mating gear. It may help you to understand the geometry of the teeth, and

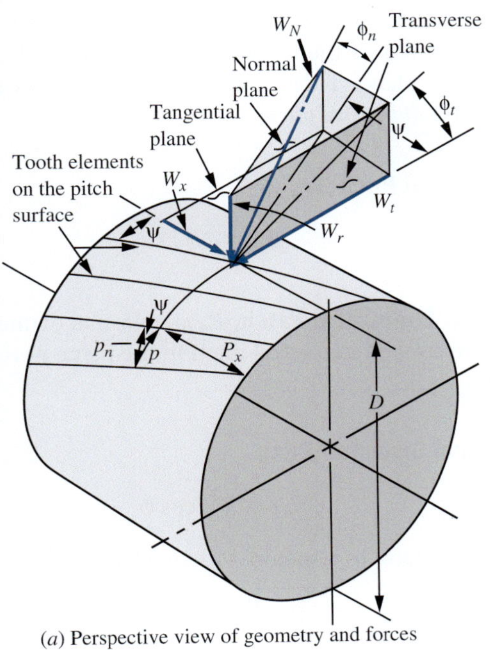

(*a*) Perspective view of geometry and forces

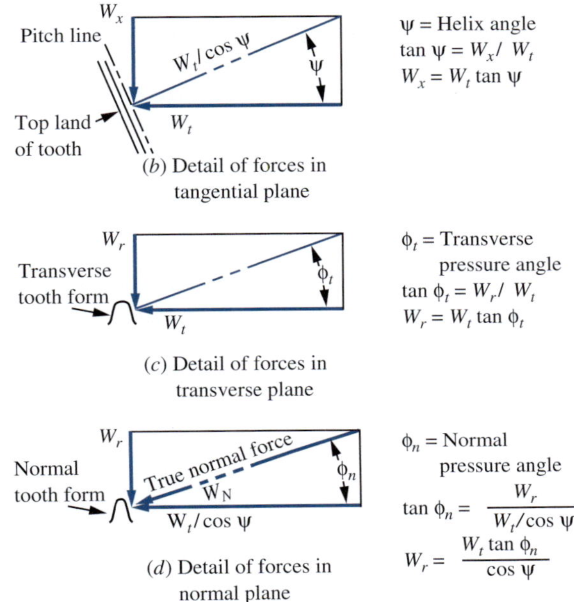

(*b*) Detail of forces in tangential plane

(*c*) Detail of forces in transverse plane

(*d*) Detail of forces in normal plane

FIGURE 8–16 Helical gear geometry and forces

the importance of that geometry, to see how it affects the forces.

We refer first to the *true normal force* as W_N. It acts normal (perpendicular) to the curved surface of the tooth. In reality, we typically do not use the normal force itself in analyzing the performance of the gear. Instead we use its three orthogonal components:

- The *tangential force* (also called the *transmitted force*), W_t, acts tangential to the pitch surface of the gear and perpendicular to the axis of the shaft carrying the gear. This is the force that actually drives the gear. Stress analysis and pitting resistance are both related to the magnitude of the tangential force. It is similar to W_t used in spur gear design and analysis.
- The *radial force*, W_r, acts toward the center of the gear along a radius and tends to separate the two gears in mesh. It is similar to W_r used in spur gear design and analysis.
- The *axial force*, W_x, acts in the tangential plane parallel to the axis of the shaft carrying the gear. Another name for the axial force is the *thrust*. It tends to push the gear along the shaft. The thrust must be reacted by one of the bearings that carry the shaft, so this force is generally undesirable. Spur gears generate no such force because the teeth are straight and are parallel to the axis of the gear.

The plane containing the tangential force, W_t, and the axial force, W_x, is the *tangential plane* [see Figure 8–16(b)]. It is tangential to the pitch surface of the gear and acts through the pitch point at the middle of the face of the tooth being analyzed.

The plane containing the tangential force, W_t, and the radial force, W_r, is the *transverse plane* [see Figure 8–16(c)]. It is perpendicular to the axis of the gear and acts through the pitch point at the middle of the face of the tooth being analyzed. The *transverse pressure angle*, ϕ_t, is defined in this plane as shown in the figure.

The plane containing the true normal force, W_N, and the radial force, W_r, is the *normal plane* [see Figure 8–16(d)]. The angle between the normal plane and the transverse plane is the helix angle, ψ. Within the normal plane we can see that the angle between the tangential plane and the true normal force, W_N, is the *normal pressure angle*, ϕ_n.

In design of a helical gear, there are three angles of interest: (1) the helix angle, ψ; (2) the *normal pressure angle*, ϕ_n; and (3) the *transverse pressure angle*, ϕ_t. Designers must specify the helix angle and one of the two pressure angles. The other pressure angle can be computed from the following relationship:

$$\tan \phi_n = \tan \phi_t \cos \psi \qquad (8\text{–}11)$$

For example, one manufacturer's catalog offers standard helical gears with a normal pressure angle of $14\frac{1}{2}^\circ$ and a 45° helix angle. Then the transverse pressure angle is found from

$$\tan \phi_n = \tan \phi_t \cos \psi$$

$$\tan \phi_t = \tan \phi_n / \cos \psi = \tan(14.5^\circ)/\cos(45^\circ) = 0.3657$$

$$\phi_t = \tan^{-1}(0.3657) = 20.09^\circ$$

Pitches for Helical Gears

To obtain a clear picture of the geometry of helical gears, you must understand the following five different pitches.

Circular Pitch, p. *Circular pitch* is the distance from a point on one tooth to the corresponding point on the next adjacent tooth, measured at the pitch line in the transverse plane. This is the same definition used for spur gears. Then

➩ Circular Pitch

$$p = \pi D/N \qquad (8\text{–}12)$$

Normal Circular Pitch, p_n. *Normal circular pitch* is the distance between corresponding points on adjacent teeth measured on the pitch surface in the normal direction. Pitches p and p_n are related by the following equation:

➩ Normal Circular Pitch

$$p_n = p \cos \psi \qquad (8\text{–}13)$$

Diametral Pitch, P_d. *Diametral pitch* is the ratio of the number of teeth in the gear to the pitch diameter. This is the same definition as the one for spur gears; it applies in considerations of the form of the teeth in the diametral or transverse plane. Thus, this pitch is sometimes called the *transverse diametral pitch:*

➩ Diametral Pitch

$$P_d = N/D \qquad (8\text{–}14)$$

Normal Diametral Pitch, P_{nd}. Normal diametral pitch is the equivalent diametral pitch in the plane normal to the teeth:

➩ Normal Diametral Pitch

$$P_{nd} = P_d/\cos \psi \qquad (8\text{–}15)$$

It is helpful to remember these relationships:

$$P_d p = \pi \qquad (8\text{–}16)$$

$$P_{nd} p_n = \pi \qquad (8\text{–}17)$$

Metric Module, m. As stated for spur gears, the metric module is essentially the inverse of the diametral pitch with the value reported in mm.

➪ Metric Module

$$m = D/N$$

This applies in the transverse plane of the gear.

Normal Metric Module, m_n. This is the inverse of the normal diametral pitch with the value reported in mm. It is the module in the plane normal to the gear tooth.

➪ Normal Metric Module

$$\begin{aligned} m_n &= 1/P_{nd} = 1/(P_d/\cos\psi) = \cos\psi/P_d \\ &= \cos\psi/(N/D) = D\cos\psi/N = m\cos\psi \end{aligned} \quad \textbf{(8–18)}$$

Axial Pitch, P_x. *Axial pitch* is the distance between corresponding points on adjacent teeth, measured on the pitch surface in the axial direction:

➪ Axial Pitch

$$P_x = p/\tan\psi = \pi/(P_d\tan\psi) = \pi m/\tan\psi \quad \textbf{(8–19)}$$

It is necessary to have at least two axial pitches in the face width to have the benefit of full helical action and its smooth transfer of the load from tooth to tooth.

The use of Equations (8–11) through (8–15) and Equation (8–19) is now illustrated in the following example problem.

Example Problem 8–2 A helical gear has a transverse diametral pitch of 12, a transverse pressure angle of $14\frac{1}{2}°$, 28 teeth, a face width of 1.25 in, and a helix angle of 30°. Compute circular pitch, normal circular pitch, normal diametral pitch, axial pitch, pitch diameter, and the normal pressure angle. Compute the number of axial pitches in the face width.

Solution *Circular Pitch*

Use Equation (8–16):

$$p = \pi/P_d = \pi/12 = 0.262 \text{ in}$$

Normal Circular Pitch

Use Equation (8–13):

$$p_n = p\cos\psi = (0.262)\cos(30) = 0.227 \text{ in}$$

Normal Diametral Pitch

Use Equation (8–15):

$$P_{nd} = P_d/\cos\psi = 12/\cos(30) = 13.856$$

Axial Pitch

Use Equation (8–19):

$$P_x = p/\tan\psi = 0.262/\tan(30) = 0.453 \text{ in}$$

Pitch Diameter

Use Equation (8–14):

$$D = N/P_d = 28/12 = 2.333 \text{ in}$$

Normal Pressure Angle

Use Equation (8–11):

$$\phi_n = \tan^{-1}(\tan\phi_t\cos\psi)$$

$$\phi_n = \tan^{-1}[\tan(14\tfrac{1}{2})\cos(30)] = 12.62°$$

Number of Axial Pitches in the Face Width

$$F/P_x = 1.25/0.453 = 2.76 \text{ pitches}$$

Since this is greater than 2.0, there will be full helical action.

8–7 BEVEL GEAR GEOMETRY

Bevel gears are used to transfer motion between nonparallel shafts, usually at 90° to one another. The four primary styles of bevel gears are straight bevel, spiral bevel, zero spiral bevel, and hypoid. Figure 8–17 shows the general appearance of these four types of bevel gear sets. The surface on which bevel gear teeth are machined is inherently a part of a cone. The differences occur in the specific shape of the teeth and in the orientation of the pinion relative to the gear. (See References 3, 5, 14, and 16.)

Straight Bevel Gears

The teeth of a straight bevel gear are straight and lie along an element of the conical surface. Lines along the face of the

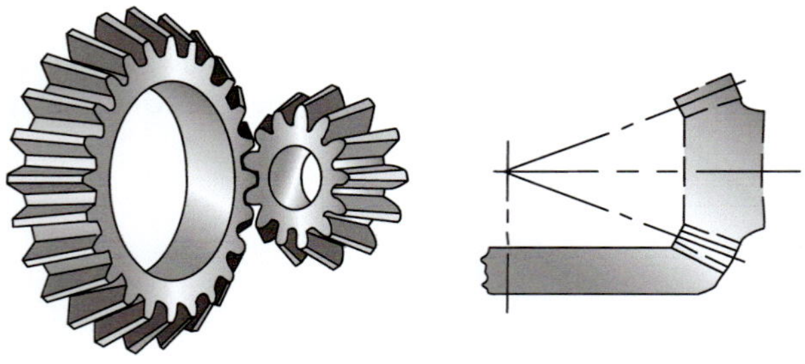

(*a*) Straight bevel

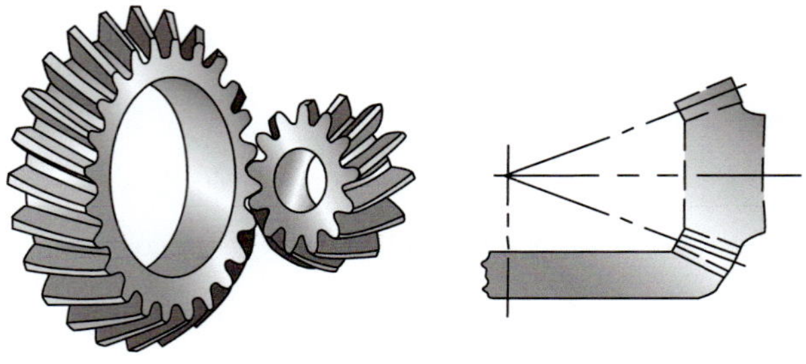

(*b*) Spiral bevel

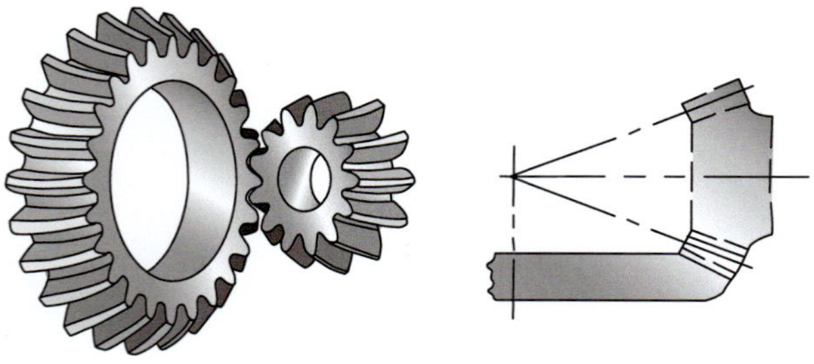

(*c*) Zero spiral bevel

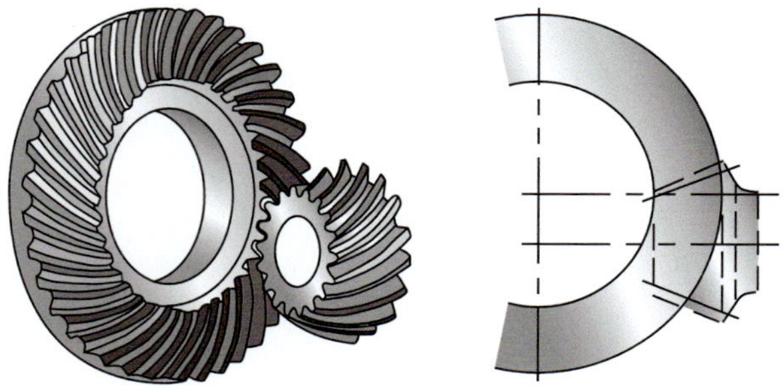

(*d*) Hypoid

(*e*) Photograph of straight bevel gear pair

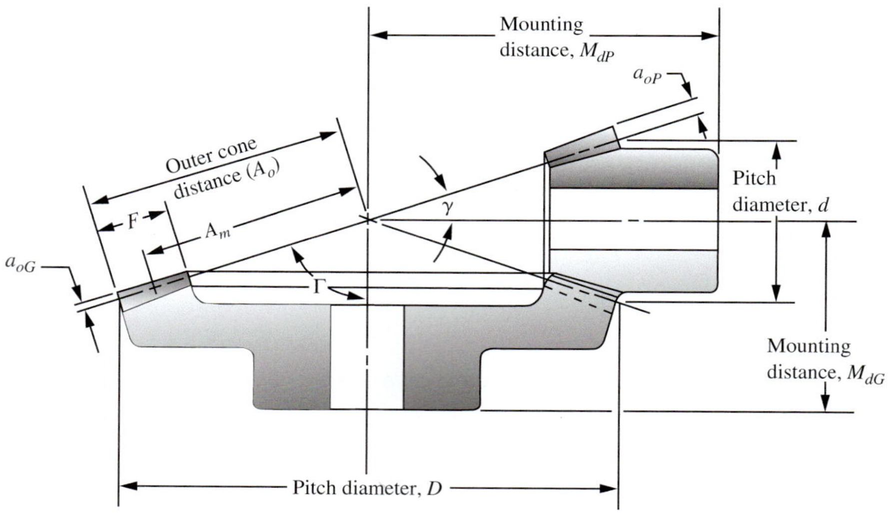

(*f*) Key dimensions of straight bevel gear pair

FIGURE 8–17 Types of bevel gears [Parts (*a*) through (*d*) extracted from ANSI/AGMA 2005-C96, *Design Manual for Bevel Gears,* with the permission of the publisher, the American Gear Manufacturers Association, 1001 North Fairfax Street, 5th floor, Alexandria, VA 22314. Photograph in (*e*) courtesy of Emerson Power Transmission Corporation, Browning Division, Maysville, KY

teeth through the pitch circle meet at the apex of the pitch cone. As shown in Figure 8–17(f), the centerlines of both the pinion and the gear also meet at this apex. In the standard configuration, the teeth are tapered toward the center of the cone.

Key dimensions are specified either at the outer end of the teeth or at the mean, midface position. The relationships that control some of these dimensions are listed in Table 8–8 for the case when the shafts are at the 90° angle. The pitch cone angles for the pinion and the gear are determined by the ratio of the number of teeth, as shown in the table. Note that their sum is 90°. Also, for a pair of bevel gears having a ratio of unity, each has a pitch cone angle of 45°. Such gears, called *miter gears,* are used simply to change the direction of the shafts in a machine drive without affecting the speed of rotation.

You should understand that many more features need to be specified before the gears can be produced. Furthermore, many successful, commercially available gears are made in some nonstandard form. For example, the addendum of the pinion is often made longer than that of the gear. Some manufacturers modify the slope of the root of the teeth to

TABLE 8–8 Geometrical Features of Straight Bevel Gears

Given Diametral pitch $= P_d = N_P/d = N_G/D$ or $m = d/N_P = D/N_G$

where N_P = number of teeth in pinion
N_G = number of teeth in gear

Dimension	**Formula**
Gear ratio	$m_G = N_G/N_P$
Pitch diameters:	
Pinion	$d = N_P/P_d$ or $d = mN_P$
Gear	$D = N_G/P_d$ or $D = mN_G$
Pitch cone angles:	
Pinion	$\gamma = \tan^{-1}(N_P/N_G)$ (lowercase Greek *gamma*)
Gear	$\Gamma = \tan^{-1}(N_G/N_P)$ (uppercase Greek *gamma*)
Outer cone distance	$A_o = 0.5D/\sin(\Gamma)$
Face width must be specified:	$F =$
Nominal face width	$F_{nom} = 0.30A_o$
Maximum face width	$F_{max} = A_o/3$ or $F_{max} = 10/P_d$ or $m/2.54$ (whichever is less)
Mean cone distance	$A_m = A_o - 0.5F$
	(*Note:* A_m is defined for the gear, also called A_{mG}.)
Mean circular pitch	$p_m = (\pi/P_d)(A_m/A_o)$ or $\pi m(A_m/A_o)$
Mean working depth	$h = (2.00/P_d)(A_m/A_o)$ or $2.00m(A_m/A_o)$
Clearance	$c = 0.125h$
Mean whole depth	$h_m = h + c$
Mean addendum factor	$c_1 = 0.210 + 0.290/(m_G)^2$
Gear mean addendum	$a_G = c_1h$
Pinion mean addendum	$a_P = h - a_G$
Gear mean dedendum	$b_G = h_m - a_G$
Pinion mean dedendum	$b_P = h_m - a_P$
Gear dedendum angle	$\delta_G = \tan^{-1}(b_G/A_{mG})$
Pinion dedendum angle	$\delta_P = \tan^{-1}(b_P/A_{mG})$
Gear outer addendum	$a_{oG} = a_G + 0.5F\tan\delta_P$
Pinion outer addendum	$a_{oP} = a_P + 0.5F\tan\delta_G$
Gear outside diameter	$D_o = D + 2a_{oG}\cos\Gamma$
Pinion outside diameter	$d_o = d + 2a_{oP}\cos\gamma$

Source: Extracted from ANSI/AGMA 2005-C96, *Design Manual for Bevel Gears,* with the permission of the publisher, the American Gear Manufacturers Association, 1001 North Fairfax Street, 5th floor, Alexandria, VA 22314.

produce a uniform depth, rather than using the standard, tapered form. Reference 5 gives many more data.

The pressure angle, ϕ, is typically 20°, but 22.5° and 25° are often used to avoid interference. The minimum number of teeth for straight bevel gears is typically 12. More is said about the design of straight bevel gears in Chapter 10.

The mounting of bevel gears is critical if satisfactory performance is to be achieved. Most commercial gears have a defined mounting distance similar to that shown in Figure 8–17(f). It is the distance from some reference surface, typically the back of the hub of the gear, to the apex of the pitch cone. Because the pitch cones of the mating gears have coincident apexes, the mounting distance also locates the axis of the mating gear. If the gear is mounted at a distance smaller than the recommended mounting distance, the teeth will likely bind. If it is mounted at a greater distance, there will be excessive backlash, causing noisy and rough operation.

Spiral Bevel Gears

The teeth of a spiral bevel gear are curved and sloped with respect to the surface of the pitch cone. Spiral angles, ψ, of 20° to 45° are used, with 35° being typical. Contact starts at one end of the teeth and moves along the tooth to its end. For a given tooth form and number of teeth, more teeth are in contact for spiral bevel gears than for straight bevel gears. The gradual transfer of loads and the greater average number of teeth in contact make spiral bevel gears smoother and allow smaller designs than for typical straight bevel gears. Recall that similar advantages were described for a helical gear relative to a spur gear.

The pressure angle, ϕ, is typically 20° for spiral bevel gears, and the minimum number of teeth is typically 12 to avoid interference. But nonstandard spiral gears allow as few as five teeth in the pinion of high-ratio sets if the tips of the teeth are trimmed to avoid interference. The rather high average number of teeth in contact (high contact ratio) for spiral gears makes this approach acceptable and can result in a very compact design. Reference 5 gives the relationships for computing the geometric features of spiral bevel gears that are extensions of those given in Table 8–8.

Zero Spiral Bevel Gears

The teeth of a zero spiral bevel gear are curved somewhat as in a spiral bevel gear, but the spiral angle is zero. These gears can be used in the same mountings as straight bevel gears, but they operate more smoothly. They are sometimes called ZEROL® bevel gears.

Hypoid Gears

The major difference between hypoid gears and the others just described is that the centerline of the pinion for a set of hypoid gears is offset either above or below the centerline of the gear. The teeth are designed specially for each combination of offset distance and spiral angle of the teeth. A major advantage is the more compact design that results, particularly when applied to vehicle drive trains and machine tools. (See References 5, 14, and 16 for more data.)

The hypoid gear geometry is the most general form, and the others are special cases. The hypoid gear has an offset axis for the pinion, and its curved teeth are cut at a spiral angle. Then the spiral bevel gear is a hypoid gear with a zero offset distance. A ZEROL® bevel gear is a hypoid gear with a zero offset and a zero spiral angle. A straight bevel gear is a hypoid gear with a zero offset, a zero spiral angle, and straight teeth.

Example Problem 8–3 Compute the values for the geometrical features listed in Table 8–8 for a pair of straight bevel gears having a diametral pitch of 8, a 20° pressure angle, 16 teeth in the pinion, and 48 teeth in the gear. The shafts are at 90°.

Solution Given $P_d = 8; N_p = 16; N_G = 48.$

Computed Values *Gear Ratio*

$$m_G = N_G/N_P = 48/16 = 3.000$$

Pitch Diameter

For the pinion,

$$d = N_P/P_d = 16/8 = 2.000 \text{ in}$$

For the gear,

$$D = N_G/P_d = 48/8 = 6.000 \text{ in}$$

Pitch Cone Angles

For the pinion,

$$\gamma = \tan^{-1}(N_P/N_G) = \tan^{-1}(16/48) = 18.43°$$

For the gear,

$$\Gamma = \tan^{-1}(N_G/N_P) = \tan^{-1}(48/16) = 71.57°$$

Outer Cone Distance

$$A_o = 0.5D/\sin(\Gamma) = 0.5(6.00\text{ in})/\sin(71.57°) = 3.162\text{ in}$$

Face Width

The face width must be specified:

$$F = 1.000\text{ in}$$

Based on the following guidelines:

Nominal face width:

$$F_{nom} = 0.30A_o = 0.30(3.162\text{ in}) = 0.949\text{ in}$$

Maximum face width:

$$F_{max} = A_o/3 = (3.162\text{ in})/3 = 1.054\text{ in}$$

or

$$F_{max} = 10/P_d = 10/8 = 1.25\text{ in}$$

Mean Cone Distance

$$A_m = A_{mG} = A_o - 0.5F = 3.162\text{ in} - 0.5(1.00\text{ in}) = 2.662\text{ in}$$

Ratio $A_m/A_o = 2.662/3.162 = 0.842$ (This ratio occurs in several following calculations.)

Mean Circular Pitch

$$p_m = (\pi/P_d)(A_m/A_o) = (\pi/8)(0.842) = 0.331\text{ in}$$

Mean Working Depth

$$h = (2.00/P_d)(A_m/A_o) = (2.00/8)(0.842) = 0.210\text{ in}$$

Clearance

$$c = 0.125h = 0.125(0.210\text{ in}) = 0.026\text{ in}$$

Mean Whole Depth

$$h_m = h + c = 0.210\text{ in} + 0.026\text{ in} = 0.236\text{ in}$$

Mean Addendum Factor

$$c_1 = 0.210 + 0.290/(m_G)^2 = 0.210 + 0.290/(3.00)^2 = 0.242$$

Gear Mean Addendum

$$a_G = c_1h = (0.242)(0.210\text{ in}) = 0.051\text{ in}$$

Pinion Mean Addendum

$$a_p = h - a_G = 0.210\text{ in} - 0.051\text{ in} = 0.159\text{ in}$$

Gear Mean Dedendum

$$b_G = h_m - a_G = 0.236\text{ in} - 0.051\text{ in} = 0.185\text{ in}$$

Pinion Mean Dedendum

$$b_P = h_m - a_P = 0.236\text{ in} - 0.159\text{ in} = 0.077\text{ in}$$

Gear Dedendum Angle

$$\delta_G = \tan^{-1}(b_G/A_{mG}) = \tan^{-1}(0.185/2.662) = 3.98°$$

Pinion Dedendum Angle

$$\delta_P = \tan^{-1}(b_P/A_{mG}) = \tan^{-1}(0.077/2.662) = 1.66°$$

Gear Outer Addendum

$$a_oG = a_G + 0.5F\tan\delta_P$$

$$a_{oG} = (0.051\text{ in}) + (0.5)(1.00\text{ in})\tan(1.657°) = 0.0655\text{ in}$$

Pinion Outer Addendum

$$a_oP = a_P + 0.5F\tan\delta_G$$

$$a_oP = (0.159\text{ in}) + (0.5)(1.00\text{ in})\tan(3.975°) = 0.1937\text{ in}$$

Gear Outside Diameter

$$D_o = D + 2a_{oG}\cos\Gamma$$

$$D_o = 6.000\text{ in} + 2(0.0655\text{ in})\cos(71.57°) = 6.041\text{ in}$$

Pinion Outside Diameter

$$d_o = d + 2a_{oP}\cos\gamma$$

$$d_o = 2.000\text{ in} + 2(0.1937\text{ in})\cos(18.43°) = 2.368\text{ in}$$

8–8 TYPES OF WORMGEARING

Wormgearing is used to transmit motion and power between nonintersecting shafts, usually at 90° to each other. The drive consists of a worm on the high-speed shaft which has the general appearance of a power screw thread: a cylindrical, helical thread. The worm drives a wormgear, which has an appearance similar to that of a helical gear. Figures 8–18 shows a typical worm and wormgear set. Sometimes the wormgear is referred to as a *worm wheel* or simply a *wheel* or *gear.* (See Reference 6.) Worms and wormgears can be provided with either right hand or left hand threads on the worm and correspondingly designed teeth on the wormgear affecting the rotational direction of the wormgear.

Several variations of the geometry of wormgear drives are available. The most common one, shown in Figures 8–18 and 8–19, employs a cylindrical worm mating with a wormgear having teeth that are throated, wrapping partially around the worm. This is called a *single-enveloping type* of wormgear drive. The contact between the threads of the worm and wormgear teeth is along a line, and the power transmission capacity is quite good. Many manufacturers offer this type of wormgear set as a stocked item. Installation of the worm is relatively easy because axial alignment is not very critical. However, the wormgear must be carefully aligned radially in order to achieve the benefit of the enveloping action. Figure 8–20 shows a cutaway of a commercial worm-gear reducer.

A simpler form of wormgear drive allows a special cylindrical worm to be used with a standard spur gear or helical gear. Neither the worm nor the gear must be aligned with great accuracy, and the center distance is not critical. However, the contact between the worm threads and the wormgear teeth is theoretically a point, drastically reducing the power transmission capacity of the set. Thus, this type is used mostly for nonprecision positioning applications at low speeds and low power levels.

A third type of wormgear set is the *double-enveloping type* in which the worm is made in an hourglass shape and mates with an enveloping type of wormgear. This results in area contact rather than line or point contact and allows a much smaller system to transmit a given power at a given reduction ratio. However, the worm is more difficult to manufacture, and the alignment of both the worm and the wormgear is very critical.

FIGURE 8–18 Worms and wormgears

(Emerson Power Transmission Corporation, Browning Division, Maysville, KY)

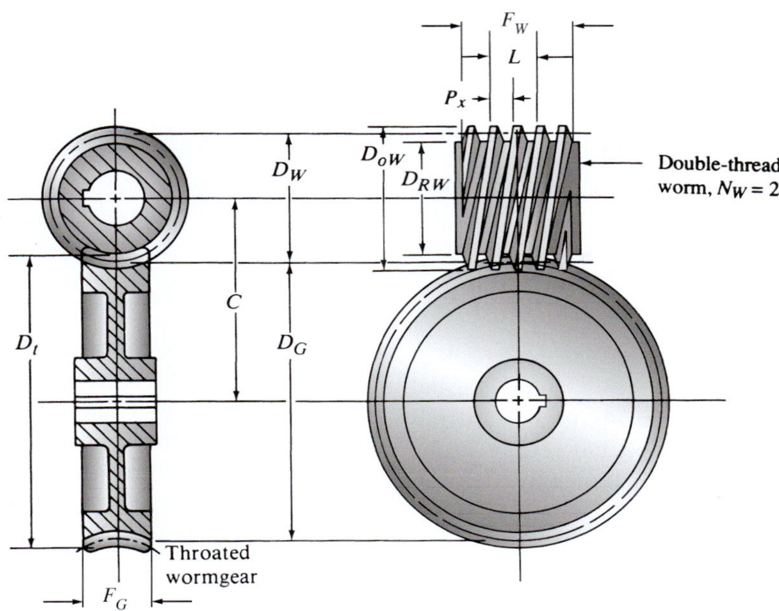

FIGURE 8–19 Single-enveloping wormgear set

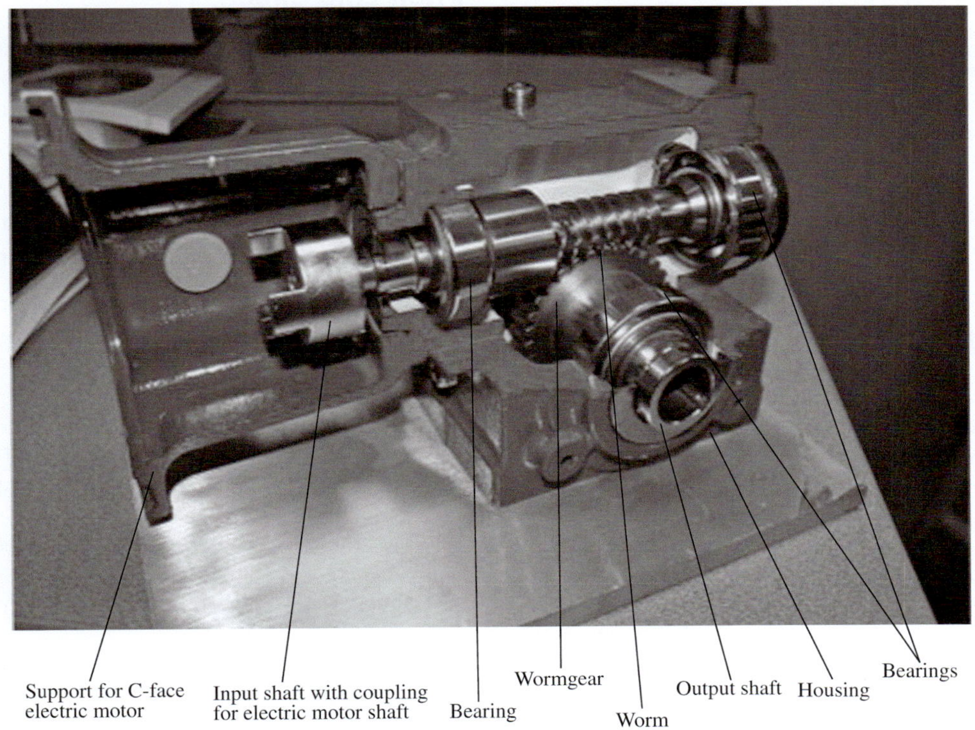

FIGURE 8–20 Cutaway view of a wormgear reducer
(Baldor/Dodge, Greenville, SC)

8–9 GEOMETRY OF WORMS AND WORMGEARS

Pitches, p, P_d, and m

A basic requirement of the worm and wormgear set is that the *axial pitch* of the worm must be equal to the *circular pitch* of the wormgear in order for them to mesh. Figure 8–19 shows the basic geometric features of a single-enveloping worm and wormgear set. *Axial pitch, P_x,* is defined as the distance from a point on the worm thread to the corresponding point on the next adjacent thread, measured axially on the pitch cylinder. As before, the circular pitch is defined for the wormgear as the distance from a point on a tooth on the pitch circle of the gear to the corresponding point on the next adjacent tooth, measured along the pitch circle. Thus, the circular pitch is an arc distance that can be calculated from

➩ Circular Pitch

$$p = \pi D_G/N_G \quad (8\text{–}20)$$

where D_G = pitch diameter of the gear
N_G = number of teeth in the gear

Some wormgears are made according to the circular pitch convention. But, as noted with spur gears, commercially available wormgear sets are usually made to a diametral pitch convention with the following pitches readily available: 48, 32, 24, 16, 12, 10, 8, 6, 5, 4, and 3. See Internet sites 2, 3, 10, and 11. The diametral pitch is defined for the gear as

➩ Diametral Pitch

$$P_d = N_G/D_G \quad (8\text{–}21)$$

The conversion from diametral pitch to circular pitch can be made from the following equation:

$$P_d p = \pi \quad (8\text{–}22)$$

Metric module worms and wormgears are available commercially with modules from 0.50 to 6.00. See Internet sites 9 and 10. As for other types of gears, the module is defined for the wormgear as

$$m = D/N$$

Then the pitch of the wormgear is $p = \pi m$ and this is also the axial pitch of the worm.

Number of Worm Threads, N_W

Worms can have a single thread, as in a typical screw, or multiple threads, usually 2 or 4, but sometimes 3, 5, 6, 8, or more. It is common to refer to the number of threads as N_W and then to treat that number as if it were the number of teeth in the worm. The number of threads in the worm is frequently referred to as the number of *starts;* this is convenient because if you look at the end of a worm, you can count the number of threads that start at the end and wind down the cylindrical worm.

Lead, L

The *lead* of a worm is the axial distance that a point on the worm would move as the worm is rotated one revolution. Lead is related to the axial pitch by

➩ Lead

$$L = N_W P_x \quad (8\text{–}23)$$

Lead Angle, λ

The *lead angle* is the angle between the tangent to the worm thread and the line perpendicular to the axis of the worm. To visualize the method of calculating the lead angle, refer to Figure 8–21, which shows a simple triangle that would be formed if one thread of the worm were unwrapped from the pitch cylinder and laid flat on the paper. The length of the hypotenuse is the length of the thread itself. The vertical side is the lead, L. The horizontal side is the circumference of the pitch cylinder, πD_W, where D_W is the pitch diameter of the worm. Then

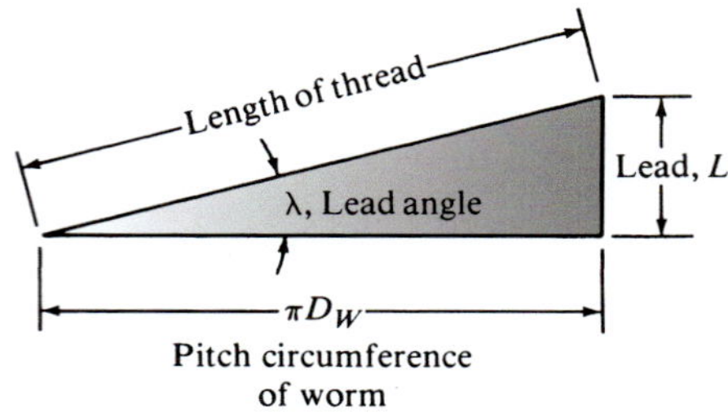

FIGURE 8–21 Lead angle

➩ Lead Angle

$$\tan \lambda = L/\pi D_W \quad (8\text{–}24)$$

Pitch Line Speed, v_t

As before, the pitch line speed is the linear velocity of a point on the pitch line for the worm or the wormgear. For the worm having a pitch diameter D_W in, rotating at n_W rpm,

➩ Pitch Line Speed for worm

$$v_{tW} = \frac{\pi D_W n_W}{12}\ \text{ft/min} \quad \text{or} \quad v_{tW} = \frac{\pi D_W n_W}{60000}\ m/s$$

For the wormgear having a pitch diameter D_G in, rotating at n_G rpm,

➩ Pitch Line Speed for Gear

$$V_{tG} = \frac{\pi D_G n_G}{12}\ \text{ft/min} \quad \text{or} \quad v_{tG} = \frac{\pi D_G n_G}{60000}\ m/s$$

Note that these two values for pitch line speed are *not* equal.

Velocity Ratio, *VR*

It is most convenient to calculate the velocity ratio of a worm and wormgear set from the ratio of the input rotational speed to the output rotational speed:

➩ Velocity Ratio for Worm/Wormgear Set

$$VR = \frac{\text{speed of worm}}{\text{speed of gear}} = \frac{n_W}{n_G} = \frac{N_G}{N_W} \quad (8\text{–}25)$$

Typical commercially available wormgear drives have ratios of 5, 7.5, 10, 12.5, 15, 20, 25, 30, 35, 40, 45, 50, 60, and 70. Drives may have a single worm/wormgear pair or two or more pairs in series. Of course, specially designed drives can have a virtually unlimited array of ratios within limits of size and practicality.

Example Problem 8–4 A wormgear has 52 teeth and a diametral pitch of 6. It mates with a triple-threaded worm that rotates at 1750 rpm. The pitch diameter of the worm is 2.000 in. Compute the circular pitch, the axial pitch, the lead, the lead angle, the pitch diameter of the wormgear, the center distance, the velocity ratio, and the rotational speed of the wormgear.

Solution *Circular Pitch*

$$p = \pi / P_d = \pi/6 = 0.5236 \text{ in}$$

Axial Pitch

$$P_x = p = 0.5236 \text{ in}$$

Lead

$$L = N_W P_x = (3)(0.5236) = 1.5708 \text{ in}$$

Lead Angle

$$\lambda = \tan^{-1}(L/\pi D_W) = \tan^{-1}(1.5708/\pi 2.000)$$

$$\lambda = 14.04°$$

Gear Pitch Diameter

$$D_G = N_G/P_d = 52/6 = 8.667 \text{ in}$$

Center Distance

$$C = (D_W + D_G)/2 = (2.000 + 8.667)/2 = 5.333 \text{ in}$$

Velocity Ratio

$$VR = N_G/N_W = 52/3 = 17.333$$

Gear rpm

$$n_G = n_W/VR = 1750/17.333 = 101 \text{ rpm}$$

Pressure Angle

Most commercially available wormgears are made with pressure angles of $14\frac{1}{2}°$, 20°, 25°, or 30°. The low pressure angles are used with worms having a low lead angle and/or a low diametral pitch. For example, a $14\frac{1}{2}°$ pressure angle may be used for lead angles up to about 17°. For higher lead angles and with higher diametral pitches (smaller teeth), the 20° or 25° pressure angle is used to eliminate interference without excessive undercutting. The 20° pressure angle is the preferred value for lead angles up to 30°. From 30° to 45° of lead angle, the 25° pressure angle is recommended. Either the normal pressure angle, ϕ_n, or the transverse pressure angle, ϕ_t, may be specified. These are related by

➩ Pressure Angle

$$\tan \phi_n = \tan \phi_t \cos \lambda \qquad \textbf{(8–26)}$$

Self-Locking Wormgear Sets

Self-locking is the condition in which the worm drives the wormgear, but if torque is applied to the gear shaft, the worm does not turn. It is locked! The locking action is produced by the friction force between the worm threads and the wormgear teeth, and this is highly dependent on the lead angle. It is recommended that a lead angle no higher than about 5.0° be used in order to ensure that self-locking will occur. This lead angle usually requires the use of a single-threaded worm. Note that the triple-threaded worm in Example Problem 8–4 has a lead angle of 14.04°. It is *not* likely to be self-locking.

Typical Designs of Wormgear Sets

Considerable latitude is permissible in the design of wormgear sets because the worm and wormgear combination is designed as a unit. However, there are some guidelines.

GENERAL GUIDELINES FOR WORM AND WORMGEAR DIMENSIONS

Typical Tooth Dimensions

Table 8–9 shows typical values used for the dimensions of worm threads and gear teeth.

TABLE 8–9 Typical Tooth Dimensions for Worms and Wormgears

Dimension	Formula
Addendum	$a = 0.3183P_x = 1/P_d$
Whole depth	$h_t = 0.6866P_x = 2.157/P_d$
Working depth	$h_k = 2a = 0.6366P_a = 2/P_d$
Dedendum	$b = h_t - a = 0.3683P_x = 1.157/P_d$
Root diameter of worm	$D_{rW} = D_W - 2b$
Outside diameter of worm	$D_{oW} = D_W + 2a = D_W + h_k$
Root diameter of gear	$D_{rG} = D_G - 2b$
Throat diameter of gear	$D_t = D_G + 2a$

Source: Standard AGMA *Design Manual Cylindrical Wormgearing*, with the permission of the publisher, American Gear Manufacturers Association, 1001 North Fairfax Street, 5th floor Alexandria, VA 22314.

Worm Diameter

The diameter of the worm affects the lead angle, which in turn affects the efficiency of the set. For this reason, small diameters are desirable. But for practical reasons and proper proportion with respect to the wormgear, it is recommended that the worm diameter be approximately $C^{0.875}/2.2$, where C is the center distance in inches between the worm and the wormgear. Variation of about 30% is allowed. (See Reference 6.) Thus, the worm diameter should fall in the range

Center Distance-*in* Wormgear Drive

$$1.6 < \frac{C^{0.875}}{D_W} < 3.0 \qquad \textbf{(8–27)}$$

For metric designs with dimensions in mm,

Center Distance-*mm* Wormgear Drive

$$1.07 < \frac{C^{0.875}}{D_W} < 2.0 \qquad \textbf{(8–27M)}$$

The recommended nominal worm diameter is approximately $C^{0.875}/1.54$.

But some commercially available wormgear sets fall outside this range, especially in the smaller sizes. Also, those worms designed to have a through-hole bored in them for installation on a shaft are typically larger than you would find from Equation (8–27). Proper proportion and efficient use of material should be the guide. The worm shaft must also be checked for deflection under operating loads. For worms machined integral with the shaft, the root of the worm threads determines the minimum shaft diameter. For worms having bored holes, sometimes called *shell worms,* care must be exercised to leave sufficient material between the thread root and the keyway in the bore. Figure 8–22 shows the recommended thickness above the keyway to be a minimum of one-half the whole depth of the threads.

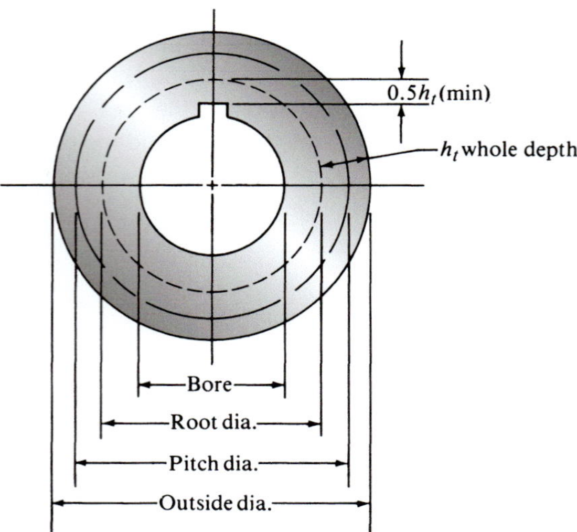

FIGURE 8–22 Shell worm

Wormgear Dimensions

We are concerned here with the single-enveloping type of wormgear, as shown in Figures 8–19 and 8–23. Its addendum, dedendum, and depth dimensions are assumed to be the same as those listed in Table 8–9, measured at the throat of the wormgear teeth. The throat is in line with the vertical centerline of the worm. The recommended face width for the wormgear is

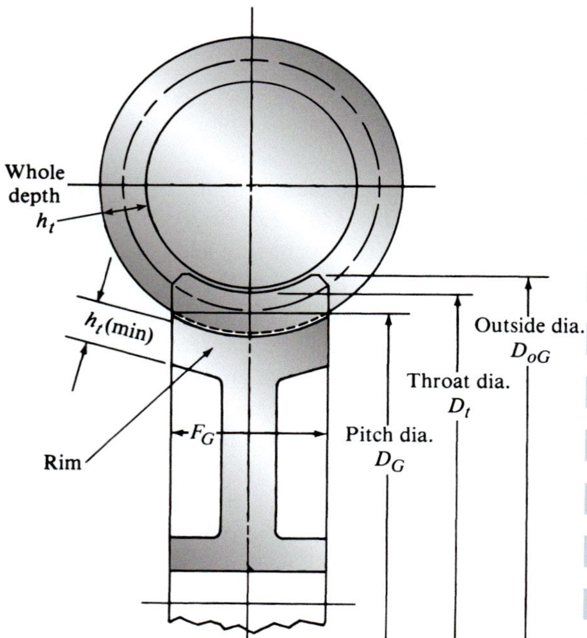

FIGURE 8–23 Wormgear details

Face Width of Wormgear

$$F_G = (D_{oW}^2 - D_W^2)^{1/2} \quad \textbf{(8–28)}$$

This corresponds to the length of the line tangent to the pitch circle of the worm and limited by the outside diameter of the worm. Any face width beyond this value would not be effective in resisting stress or wear, but a convenient value slightly greater than the minimum should be used. The outer edges of the wormgear teeth should be chamfered approximately as shown in Figure 8–23.

Another recommendation, which is convenient for initial design, is that the face width of the gear should be approximately 2.0 times the circular pitch. Because we are working in the diametral pitch system, we will use

$$F_G = 2p = 2\pi/P_d \quad \textbf{(8–29)}$$

However, since this is only approximate and 2π is approximately 6, we will use

$$F_G = 6/P_d \quad \textbf{(8–30)}$$

If the gear web is thinned, a rim thickness at least equal to the whole depth of the teeth should be left.

Face Length of the Worm

For maximum load sharing, the worm face length should extend to at least the point where the outside diameter of the worm intersects the throat diameter of the wormgear. This length is

Face Length of Worm

$$F_w = 2[(D_t/2)^2 - (D_G/2 - a)^2]^{1/2} \quad \textbf{(8–31)}$$

Example Problem 8–5

A worm and wormgear set is to be designed to produce a velocity ratio of 40. It has been proposed that the diametral pitch of the wormgear be 8, based on the torque that must be transmitted. (This will be discussed in Chapter 10.) Using the relationships presented in this section, specify the following:

Worm diameter, D_W
Number of threads in the worm, N_W
Number of teeth in the gear, N_G
Actual center distance, C
Face width of the gear, F_G
Face length of the worm, F_W
Minimum thickness of the rim of the gear

Solution

Many design decisions need to be made, and multiple solutions could satisfy the requirements. Presented here is one solution, along with comparisons with the various guidelines discussed in this section. This type of analysis precedes the stress analysis and the determination of the power-transmitting capacity of the worm and wormgear drive which is discussed in Chapter 10.

Trial Design: Let's specify a double-threaded worm: $N_W = 2$. Then there must be 80 teeth in the wormgear to achieve a velocity ratio of 40. That is,

$$VR = N_G/N_W = 80/2 = 40$$

With the known diametral pitch, $P_d = 8$, the pitch diameter of the wormgear is

$$D_G = N_G/P_d = 80/8 = 10.000 \text{ in}$$

An initial estimate for the magnitude of the center distance is approximately $C = 6.50$ in. We know that it will be greater than 5.00 in, the radius of the wormgear. Using Equation (8–27), the recommended minimum size of the worm is

$$D_W = C^{0.875}/3.0 = 1.71 \text{ in}$$

Similarly, the maximum diameter should be

$$D_W = C^{0.875}/1.6 = 3.21 \text{ in}$$

A small worm diameter is desirable. Let's specify $D_W = 2.25$ in. The actual center distance is

$$C = (D_W + D_G)/2 = 6.125 \text{ in}$$

Worm Outside Diameter

$$D_{oW} = D_W + 2a = 2.25 + 2(1/P_d) = 2.25 + 2(1/8) = 2.50 \text{ in}$$

Whole Depth

$$h_t = 2.157/P_d = 2.157/8 = 0.270 \text{ in}$$

Face Width for Gear

Let's use Equation (8–28):

$$F_G = (D_{oW}^2 - D_W^2)^{1/2} = (2.50^2 - 2.25^2)^{1/2} = 1.090 \text{ in}$$

Let's specify $F_G = 1.25$ in.

Addendum

$$a = 1/P_d = 1/8 = 0.125 \text{ in}$$

Throat Diameter of Wormgear

$$D_t = D_G + 2a = 10.000 + 2(0.125) = 10.250 \text{ in}$$

Recommended Minimum Face Length of Worm

$$F_W = 2[(D_t/2)^2 - (D_G/2 - a)^2]^{1/2} = 3.16 \text{ in}$$

Let's specify $F_W = 3.25$ in.

Minimum Thickness of the Rim of the Gear

The rim thickness should be greater than the whole depth:

$$h_t = 0.270 \text{ in}$$

8–10 VELOCITY RATIO AND GEAR TRAINS

A gear train *is one or more pairs of gears operating together to transmit power.*

Normally there is a speed change from one gear to the next due to the different sizes of the gears in mesh. The fundamental building block of the total speed change ratio in a gear train is the *velocity ratio* between two gears in a single pair.

Velocity Ratio

The* velocity ratio *(VR) is defined as the ratio of the rotational speed of the input gear to that of the output gear for a single pair of gears.

To develop the equation for computing the velocity ratio, it is helpful to view the action of two gears in mesh, as shown in Figure 8–24. The action is equivalent to the action of two smooth wheels rolling on each other without slipping, with the diameters of the two wheels equal to the pitch diameters of the two gears. Remember that when two gears are in mesh, their pitch circles are tangent, obviously, the gear teeth prohibit any slipping.

As shown in Figure 8–24, without slipping there is no relative motion between the two pitch circles at the pitch point, and therefore the linear velocity of a point on either pitch circle is the same. We will use the symbol v_t for this velocity. The linear velocity of a point that is in rotation at a distance R from its center of rotation and rotating with an angular velocity, ω, is found from

⇨ Pitch Line Speed of a Gear

$$v_t = R\omega \quad (8\text{–}32)$$

Using the subscript P for the pinion and G for the gear for two gears in mesh, we have

$$v_t = R_P\omega_P \quad \text{and} \quad v_t = R_G\omega_G$$

This set of equations says that the pitch line speeds of the pinion and the gear are the same. Equating these two and solving for ω_P/ω_G gives our definition for the velocity ratio, VR:

$$VR = \omega_P/\omega_G = R_G/R_P$$

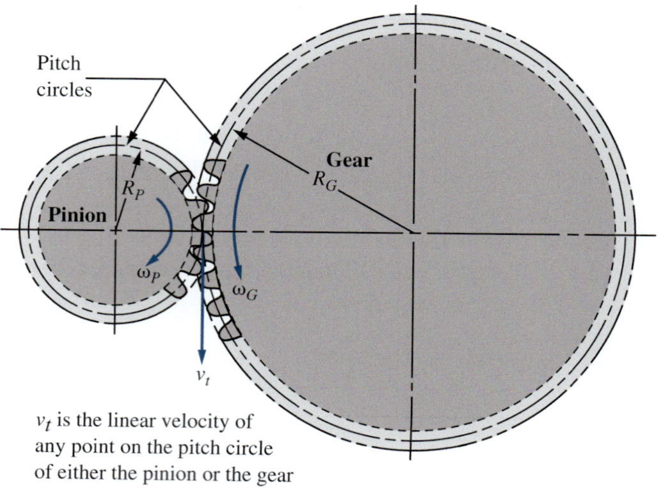

FIGURE 8–24 Two gears in mesh

In general, it is convenient to express the velocity ratio in terms of the pitch diameters, the rotational speeds, or the numbers of teeth of the two gears in mesh. Remember that

$$R_G = D_G/2$$
$$R_P = D_P/2$$
$$D_G = N_G/P_d$$
$$D_P = N_P/P_d$$
$$n_P = \text{rotational speed of the pinion} \quad (\text{in rpm})$$
$$n_G = \text{rotational speed of the gear} \quad (\text{in rpm})$$

The velocity ratio can then be defined in any of the following ways:

➪ Velocity Ratio for Gear Pair

$$VR = \frac{\omega_P}{\omega_G} = \frac{n_P}{n_G} = \frac{R_G}{R_P} = \frac{D_G}{D_P} = \frac{N_G}{N_P} = \frac{\text{speed}_P}{\text{speed}_G} = \frac{\text{size}_G}{\text{size}_P} \tag{8–33}$$

Most gear drives are *speed reducers;* that is, their output speed is lower than their input speed. This results in a velocity ratio greater than 1. If a *speed increaser* is desired, then *VR* is less than 1. Note that not all books and articles use the same definition for velocity ratio. Some define it as the ratio of the output speed to the input speed, the inverse of our definition. It is thought that the use of *VR* greater than 1 for the reducer—that is, the majority of the time—is more convenient.

Train Value

When more than two gears are in mesh, the term* train value *(TV) refers to the ratio of the input speed (for the first gear in the train) to the output speed (for the last gear in the train). By definition the train value is the product of the values of VR for each* gear pair *in the train. In this definition, a gear pair is any set of two gears with a driver and a follower (driven) gear.

Again, *TV* will be greater than 1 for a reducer and less than 1 for an increaser. For example, consider the gear train sketched in Figure 8–25. The input is through the shaft carrying gear *A*. Gear *A* drives gear *B*. Gear *C* is on the same shaft with gear *B* and rotates at the same speed. Gear *C* drives gear *D*, which is connected to the output shaft. Then gears *A* and *B* constitute the first gear pair, and gears *C* and *D* constitute the second pair. The velocity ratios are

$$VR_1 = n_A/n_B$$
$$VR_2 = n_C/n_D$$

The train value is

$$TV = (VR_1)(VR_2) = \frac{n_A n_C}{n_B n_D}$$

But, because they are on the same shaft, $n_B = n_C$, and the preceding equation reduces to

$$TV = n_A/n_D$$

This is the input speed divided by the output speed, the basic definition of the train value. This process can be expanded to any number of stages of reduction in a gear train.

Remember that any of the forms for velocity ratio shown in Equation (8–33) can be used for computing the train value. In design, it is often most convenient to express the velocity ratio in terms of the numbers of teeth in each gear because they must be integers. Then, once the diametral pitch or module is defined, the values of the diameters or radii can be determined.

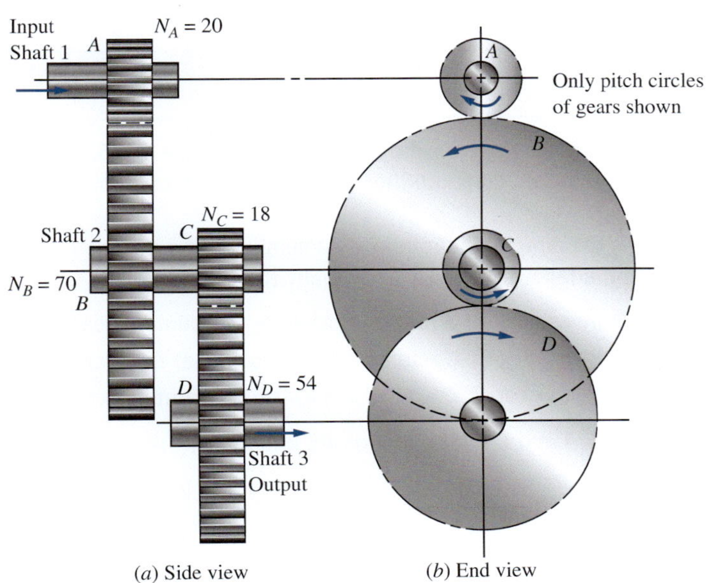

FIGURE 8–25 Double-reduction gear train

The train value of the double-reduction gear train in Figure 8–25 can be expressed in terms of the numbers of teeth in the four gears as follows:

$$VR_1 = N_B/N_A$$

Note that this is the number of teeth in the *driven gear B* divided by the number of teeth in the *driving gear A*. This is the typical format for velocity ratio. Then VR_2 can be found similarly:

$$VR_2 = N_D/N_C$$

Thus, the train value is

$$TV = (VR_1)(VR_2) = (N_B/N_A)(N_D/N_C)$$

This is usually shown in the form

➪ Train Value

$$TV = \frac{N_B N_D}{N_A N_C} \tag{8–34}$$

$$= \frac{\text{product of number of teeth in the driven gears}}{\text{product of number of teeth in the driving gears}}$$

This is the form for train value that we will use most often.

The direction of rotation can be determined by observation, noting that there is a direction reversal for each pair of external gears.

We will use the term* positive train value *to refer to one in which the input and output gears rotate in the same direction. Conversely, if they rotate in the opposite direction, the train value will be negative.

Example Problem 8–5 For the gear train shown in Figure 8–25, if the input shaft rotates at 1750 rpm clockwise, compute the speed of the output shaft and its direction of rotation.

Solution We can find the output speed if we can determine the train value:

$$TV = n_A/n_D = \text{input speed/output speed}$$

Then

$$n_D = n_A/TV$$

But

$$TV = (VR_1)(VR_2) = \frac{N_B}{N_A}\frac{N_D}{N_C} = \frac{70}{20}\frac{54}{18} = \frac{3.5}{1}\frac{3.0}{1} = \frac{10.5}{1} = 10.5$$

Now

$$n_D = n_A/TV = (1750 \text{ rpm})/10.5 = 166.7 \text{ rpm}$$

Gear *A* rotates clockwise; gear *B* rotates counterclockwise.

Gear *C* rotates counterclockwise; gear *D* rotates clockwise.

Thus, the train in Figure 8–25 is a positive train.

Example Problem 8–6 Determine the train value for the train shown in Figure 8–26. If the shaft carrying gear *A* rotates at 1750 rpm clockwise, compute the speed and the direction of the shaft carrying gear *E*.

Solution Look first at the direction of rotation. Remember that a gear pair is defined as any two gears in mesh (a driver and a follower). There are actually three gear pairs:

Gear *A* drives gear *B*: *A* rotates clockwise; *B*, counterclockwise.

Gear *C* drives gear *D*: *C* rotates counterclockwise; *D*, clockwise.

Gear *D* drives gear *E*: *D* rotates clockwise; *E*, counterclockwise.

Because gears *A* and *E* rotate in opposite directions, the train value is negative. Now

$$TV = -(VR_1)(VR_2)(VR_3)$$

In terms of the number of teeth,

$$TV = -\frac{N_B}{N_A}\frac{N_D}{N_C}\frac{N_E}{N_D}$$

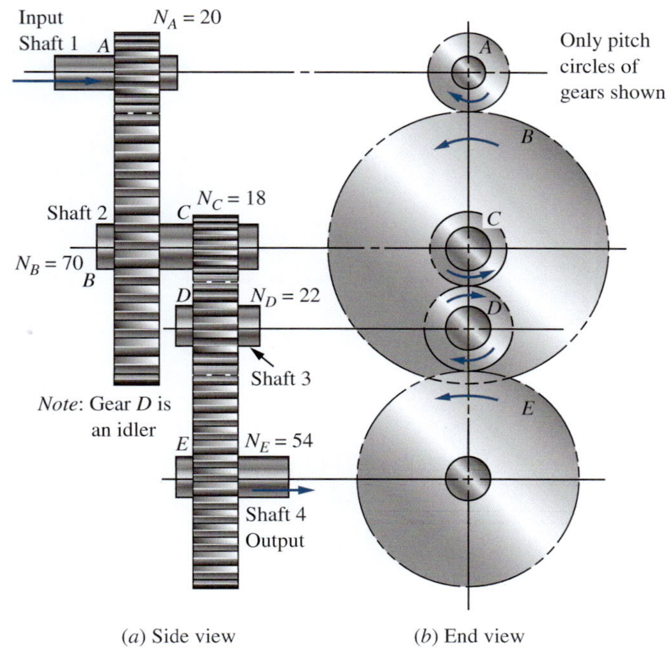

FIGURE 8–26 Double-reduction gear train with an idler. Gear D is an idler

Note that the number of teeth in gear D appears in both the numerator and the denominator and thus can be canceled. The train value then becomes

$$TV = -\frac{N_B}{N_A} \cdot \frac{N_E}{N_C} = -\frac{70}{20} \cdot \frac{50}{18} = -\frac{3.5}{1} \frac{3.0}{1} = -10.5$$

Gear D is called an *idler.* As demonstrated here, it has no effect on the magnitude of the train value, but it does cause a direction reversal. The output speed is then found from

$$TV = n_A/n_E$$

$$n_E = n_A/TV = (1750 \text{ rpm})/(-10.5) = -166.7 \text{ rpm (counterclockwise)}$$

Idler Gear

Example Problem 8–6 introduced the concept of an *idler gear*, defined as follows:

> ***Any gear in a gear train that performs as both a driving gear and a driven gear is called an* idler gear *or simply an* idler.**

The main features of an idler are as follows:

1. An idler does not affect the train value of a gear train because, since it is both a driver and a driven gear, its number of teeth appears in both the numerator and the denominator of the train value equation, Equation (8–34). Thus, any pitch diameter size and any number of teeth may be used for the idler.
2. Placing an idler in a gear train causes a direction reversal of the output gear.
3. An idler gear may be used to fill a space between two gears in a gear train when the desired distance between their centers is greater than the center distance for the two gears alone.

Internal Gear

> **An internal gear *is one for which the teeth are machined on the inside of a ring instead of on the outside of a gear blank.***

An internal gear mating with a standard, external pinion is illustrated at the lower left in Figure 8–2, along with a variety of other kinds of gears.

Figure 8–27 is a sketch of an external pinion driving an internal gear. Note the following:

1. The gear rotates in the *same direction* as the pinion. This is different from the case when an external pinion drives an external gear.

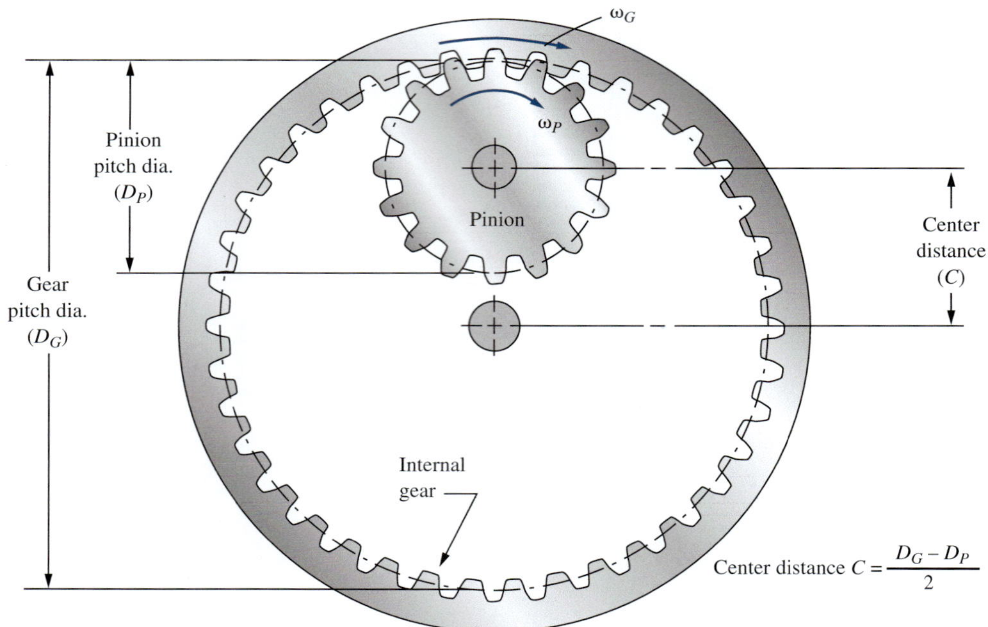

FIGURE 8–27 Internal gear driven by an external pinion

2. The center distance is

➪ Center Distance-Internal Gear

$$C = D_G/2 - D_P/2 = (D_G - D_P)/2 \quad \textbf{(8–35)}$$
$$= (N_G/P_d - N_P/P_d)/2 = (N_G - N_P)/(2P_d)$$

The last form is preferred because its factors are all integers for typical gear trains.

3. The descriptions of most other features of internal gears are the same as those for external gears presented earlier. Exceptions for an internal gear are as follows:

The addendum, a, is the radial distance from the pitch circle to the inside of a tooth.

The inside diameter, D_i, is

$$D_i = D - 2a$$

The root diameter, D_R, is

$$D_R = D + 2b$$

where b = dedendum

Internal gears are used when it is desired to have the same direction of rotation for the input and the output. Also note that less space is taken for an internal gear mating with an external pinion compared with two external gears in mesh.

Velocity of a Rack

Figure 8–28 shows the basic configuration of a *rack-and-pinion* drive. The function of such a drive is to produce a linear motion of the rack from the rotational motion of the driving pinion. The opposite is also true: If the driver

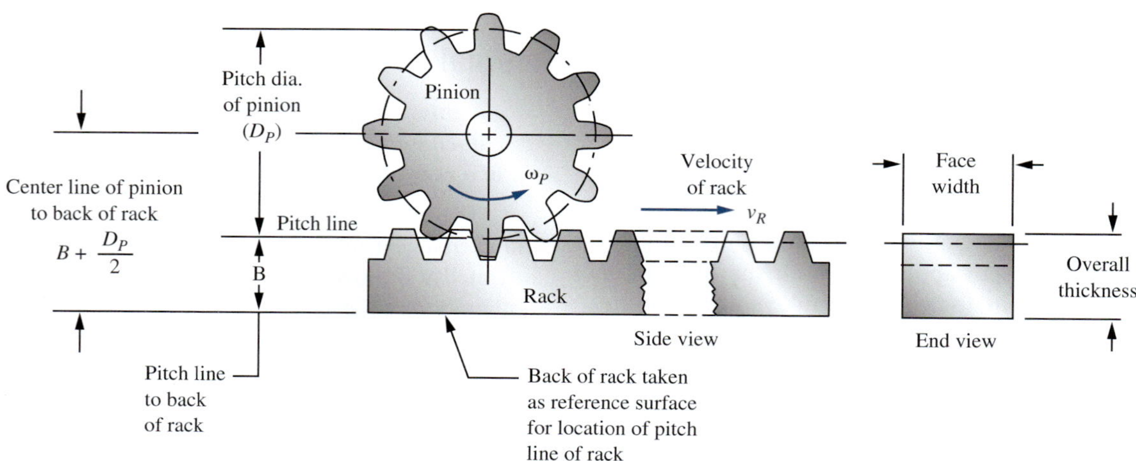

FIGURE 8–28 Rack driven by a pinion

produces the linear motion of the rack, it produces a rotational motion of the pinion.

The linear velocity of the rack, v_R, must be the same as the pitch line velocity of the pinion, v_t, as defined by Equation (8–32), repeated here. Recall that ω_P is the angular velocity of the pinion:

$$v_R = v_t = R_P\omega_P = (D_P/2)\omega_P$$

Note on Units for Pitch Line Speed. In general, any velocity units may be used for pitch line speed at the discretion of the designer as long as careful manipulation of units is exercised. However, there are certain units that are preferred for later calculations in this book and for gear design in particular. The preferred units for pitch line speed are as follows:

U.S. Standard Diametral Pitch System: ft/min or fpm

Metric Module System: m/s

Certain data, design guidelines, and subsequent calculations are keyed to pitch line speed in these units. Because such calculations appear often in the following chapters, we develop here some unit-specific relationships to facilitate calculations in these units. Other assumptions for typical units for related quantities are included in these developments.

U.S. Standard Diametral Pitch System:

Assumptions: Dimensions are in inches; Rotational speed is in rpm

$$v_t = (D/2)\omega = \frac{D(\text{in})}{2}\frac{n(\text{rev})}{(\text{min})}\frac{2\pi(\text{rad})}{(\text{rev})}\frac{1.0(\text{ft})}{12(\text{in})} = \frac{\pi Dn}{12}\text{ ft/min} \quad \textbf{(8–36)}$$

Metric Module System:

Assumptions: Dimensions are in mm; Rotational speed is in rpm

$$v_t = (D/2)\omega = \frac{D\,(\text{mm})}{2}\frac{n\,(\text{rev})}{(\text{min})}\frac{2\pi\,(\text{rad})}{(\text{rev})}\frac{1.0\,(\text{min})}{60\,(\text{s})}\frac{1.0\text{ m}}{1000\text{ mm}} = \frac{\pi Dn}{60\,000}\text{ m/s} \quad \textbf{(8–37)}$$

Provided that data are supplied in the given units, the final form of Equations (8–36) or (8–37) can be used for problems and designs for this book. We refer to this type of equation as *unit specific*, meaning that it is only valid for input data in the proper units.

The concept of center distance does not apply directly for a rack-and-pinion set because the center of the rack is at infinity. But it is critical that the pitch circle of the pinion be tangent to the pitch line of the rack as shown in Figure 8–28. The rack will be machined so that there is a specified dimension between the pitch line and a reference surface, typically the back of the rack. This is dimension B in Figure 8–28. Then the location of the center of the pinion can be computed using the relationships shown in the figure.

Example Problem 8–7 Determine the linear velocity of the rack in Figure 8–28 if the driving pinion rotates at 125 rpm. The pinion has 24 teeth and a diametral pitch of 6.

Solution We will use Equation (8–36). First the pitch diameter of the pinion is computed using Equation (8–3):

$$D_p = N_p/P_d = 24/6 = 4.00\text{ in}$$

Because the units of the given data are now the same as those used to derive Equation (8–36), we can use that equation to compute the pitch line speed of the pinion that is identical to the linear velocity of the rack. For $D = 4.00$ in and $n = 125$ rpm,

$$v_t = v_R = (\pi Dn/12)\text{ ft/min} = [(\pi)(4.00\text{ in})(125\text{ rpm})/12] = 130.9\text{ ft/min}$$

8–11 DEVISING GEAR TRAINS

Now we will show several methods for devising gear trains to produce a desired train value. The result will typically be the specification of the number of teeth in each gear and the general arrangement of the gears relative to each other. The determination of the types of gears will generally not be considered except for how they may affect the direction of rotation or the general alignment of the shafts. Additional details can be specified after completion of the study of the design procedures in later chapters.

Hunting Tooth

Some designers recommend that integer velocity ratios be avoided, if possible, because the same two teeth would come into contact frequently and produce uneven wear patterns on the teeth. For example when using a velocity ratio of exactly 2.0, a given tooth on the pinion would contact the same two teeth on the gear with every two revolutions. In Chapter 9 you will learn that the pinion teeth are often made harder than the gear because the pinion experiences higher stresses. As the gears rotate, the pinion teeth tend to smooth any inherent roughness of the gear teeth, a process sometimes called *wearing in*. Each tooth on the pinion has a slightly different geometry causing unique wear patterns on the few teeth with which it mates.

A more uniform wear pattern will result if the velocity ratio in not an integer. Adding or subtracting one tooth from the number of teeth in the gear produces the result that each pinion tooth would contact a different gear tooth with each revolution and the wear pattern would be more uniform. The added or subtracted tooth is called the *hunting tooth.* Obviously the velocity ratio for the gear pair will be slightly different, but that is often not a concern unless precise timing between the driver and driven gears is required. Consider the following example.

An initial design for a gear pair calls for the pinion to be mounted to the shaft of an electric motor having a nominal speed of 1750 rpm. The pinion has 18 teeth and the gear has 36 teeth, resulting in a velocity ratio of 36/18 or 2.000. The output speed would then be,

Initial design: $n_2 = n_1(N_P/N_G)$

$$= 1750 \text{ rpm } (18/36) = 875 \text{ rpm}$$

Now consider adding or subtracting one tooth from the gear. The output speeds would be,

Modified design: $n_2 = n_1(N_P/N_G)$

$$= 1750 \text{ rpm } (18/35) = 900 \text{ rpm}$$

Modified design: $n_2 = n_1(N_P/N_G)$

$$= 1750 \text{ rpm } (18/37) = 851 \text{ rpm}$$

The output speeds for the modified designs are less than 3.0 percent different from the original design. You would have to decide if that is acceptable in a given design project. However, be aware that the motor speed is typically not exactly 1750 rpm. As discussed in Chapter 21, 1750 rpm is the *full load speed* of a four-pole alternating current electric motor. When operating at a torque less than the full load torque the speed would be greater than 1750 rpm. Conversely, a greater torque would result in a slower speed. When precise speeds are required, a variable speed drive that can be adjusted according to actual loads is recommended.

A few general principles that were discussed earlier in this chapter are reviewed next.

GENERAL PRINCIPLES FOR DEVISING GEAR TRAINS

1. The velocity ratio for any pair of gears can be computed in a variety of ways as indicated in Equation (8–33).
2. The number of teeth in any gear must be an integer.
3. Mating gears must have the same tooth form, pressure angle, and pitch.
4. When external gears mesh, there is a direction reversal of their shafts.
5. When an external pinion meshes with an internal gear, their shafts rotate in the same direction.
6. An idler is a gear that performs as both a driver and a driven gear in the same train. Its size and number of teeth have no effect on the magnitude of the train value, but the direction of rotation is changed.
7. Spur and helical gears operate on parallel shafts.
8. Bevel gears and a worm/wormgear set operate on shafts perpendicular to each other.
9. The number of teeth in the pinion of a gear pair should not be such that it causes interference with the teeth of its mating gear. Refer to Table 8–7.
10. In general, the number of teeth in the gear should not be larger than about 150. This is somewhat arbitrary, but it is typically more desirable to use a double-reduction gear train rather than a very large, single-reduction gear pair.
11. Any given gear train design problem may require one or more pairs of gears depending on the overall train value required, space available, and practical gear size.
12. It is generally desirable to design the gear train with the fewest practical number of gears. Each added gear requires its own shaft that must be supported by bearings.
13. In general, it is expected that the overall gear train be small in size, compact, and arranged in a manner that facilitates assembly.
14. You should determine the required directions of the driver for the gear train and for the output shaft; at least whether the input and output should rotate in the same direction (a positive train) or the opposite direction (a negative train).
15. The use of the "hunting tooth" concept described above for any gear pair is desirable. However, in this book we will consider any practical ratio for a given gear pair to be acceptable.
16. It will be shown in Chapters 9 and 10 that gear pairs operating at slower speeds will be subjected to higher torque and tooth loads as compared with those operating at higher speed. This results in the slower gear teeth having a larger tooth size and circular pitch for a given velocity ratio. For this reason, it is desirable to allocate a higher percentage of the total train value to the higher speed pairs of gears to obtain a more optimal overall design for the train.

Three different design procedures are outlined next and demonstrated in example problems. It is recommended that all three be studied and understood so you can select the method most suitable to any given design situation. Introductory comments are given here on the types of design situations where each procedure should be applied.

1. **Single pair of gears to produce a desired velocity ratio:** This is the fundamental process required to define the number of teeth in the pinion and the gear to produce a desired ratio.
2. **Residual ratio:** This process is used when two or more pairs of gears in a train are required. It calls for specifying all but one of the required ratios to produce an overall train value. Then you will be able to compute the required value of the final ratio. *This is the most general approach and the most frequently used approach in this book.*
3. **Factoring approach:** When two or more pairs of gears in a train are required and an exact ratio for the overall train value is also required, then the velocity ratio of each gear pair must be a factor of the overall train value. Determining the factors of the desired ratio is a necessary skill to apply this method.

Getting started: For any gear train design problem, it is first necessary to determine the minimum number of pairs of gears necessary to produce the overall train value. Here is an outline of the recommended approach.

1. Determine the overall train value, TV, required from the application data.
2. Determine the maximum velocity ratio, VR_{max}, that can be achieved with one pair of gears, considering a reasonable maximum number of teeth for the gear and a number of teeth for the pinion that will not result in interference according to Table 8–7.
3. If the train value is greater than the ratio found from Step 2, divide TV/VR_{max}. This will aid in determining how many pairs of gears are required.
4. Note that the overall train value is the product of the velocity ratio for each pair of gears. That is,

$$TV = VR_1 \times VR_2 \times VR_3 \ldots$$

5. Determine if an idler gear is necessary to achieve the required output shaft direction.
6. Specify each individual velocity ratio using the guidelines listed above.
7. Specify the numbers of teeth in each gear of each pair of gears.
8. Sketch the arrangement of the gears to show how they are placed on shafts in proper relation to each other. At this stage, a schematic diagram similar to Figures 8–25 and 8–26 is adequate.

Design of a Single Pair of Gears to Produce a Desired Velocity Ratio

Example Problem 8–8

Devise a gear train to reduce the speed of rotation of a drive from an electric motor shaft operating at 3450 rpm to approximately 650 rpm. Use $N_{max} = 150$ teeth.

Solution

First we will compute the nominal train value:

$$TV = \text{(input speed)/(output speed)} = 3450/650 = 5.308$$

If a single pair of gears is used then the train value is equal to the velocity ratio for that pair. That is, $TV = VR = N_G/N_P$.

Let's decide that spur gears having 20°, full-depth, involute teeth are to be used. Then we can refer to Table 8–7 and determine that no fewer than 16 teeth should be used for the pinion in order to avoid interference. We can specify the number of teeth in the pinion and use the velocity ratio to compute the number of teeth in the gear:

$$N_G = (VR)(N_P) = (5.308)(N_P)$$

All possible examples are given in Table 8–10.

TABLE 8–10 All Possible Values for N_P and N_G to Produce the Desired Velocity Ratio

N_P	Computed $N_G = (5.308)(N_P)$	Nearest integer N_G	Actual VR: $VR = N_G/N_P$	Actual output speed (rpm): $n_G = n_P/VR = n_P(N_P/N_G)$
16	84.92	85	85/16 = 5.31	649.4
17	90.23	90	90/17 = 5.29	651.7
18	95.54	96	96/18 = 5.33	646.9
19	100.85	101	101/19 = 5.32	649.0
20	106.15	106	106/20 = 5.30	650.9
21	111.46	111	111/21 = 5.29	652.7
22	116.77	117	117/22 = 5.32	648.7
23	122.08	122	122/23 = 5.30	650.4
24	127.38	127	127/24 = 5.29	652.0
25	132.69	133	133/25 = 5.32	648.5
26	138.00	138	138/26 = 5.308	650.0 Exact
27	143.31	143	143/27 = 5.30	651.4
28	148.61	149	149/28 = 5.32	648.3
29	153.92	154 **Too large**		

Conclusion and Comments The combination of $N_P = 26$ and $N_G = 138$ gives the most ideal result for the output speed. But all of the trial values give output speeds reasonably close to the desired value. Only two are more than 2.0 rpm off the desired value. It remains a design decision as to how close the output speed must be to the stated value of 650 rpm. Note that the input speed is given as 3450 rpm, the full load speed of an electric motor. But how accurate is that? The actual speed of the input will vary depending on the load on the motor. Therefore, it is not likely that the ratio must be precise.

Residual Ratio Method

Example Problem 8–9 Devise a gear train for a conveyor drive. The drive motor rotates at 1150 rpm, and it is desired that the output speed for the shaft that drives the conveyor be in the range of 24 to 28 rpm. Use a double-reduction gear train. Power transmission analysis indicates that it would be desirable for the reduction ratio for the first pair of gears to be somewhat greater than that for the second pair.

Solution We use the *getting started* recommendations given earlier to start the solution.

Permissible Train Values

First let's compute the nominal train value that will produce an output speed of 26.0 rpm at the middle of the allowable range:

$$TV_{nom} = (\text{input speed})/(\text{nominal output speed}) = 1150/26 = 44.23$$

Now we can compute the minimum and maximum allowable speed ratio:

$$TV_{min} = (\text{input speed})/(\text{maximum output speed}) = 1150/28 = 41.07$$

$$TV_{max} = (\text{input speed})/(\text{minimum output speed}) = 1150/24 = 47.92$$

Possible Ratio for Single Pair

The maximum ratio that any one pair of gears can produce occurs when the gear has 150 teeth and the pinion has 17 teeth (see Table 8–7). Then

$$VR_{max} = N_G/N_P = 150/17 = 8.82 \text{ (too low)}$$

Possible Train Value for Double-Reduction Train

$$TV = (VR_1)(VR_2)$$

But the maximum value for either VR is 8.82. Then the maximum train value is

$$TV_{max} = (8.82)(8.82) = (8.82)^2 = 77.9$$

A double-reduction train is practical.

Optional Designs

The general layout of the proposed train is shown in Figure 8–29. Its train value is

$$TV = (VR_1)(VR_2) = (N_B/N_A)(N_D/N_C)$$

We need to specify the number of teeth in each of the four gears to achieve a train value within the range just computed. Our approach is to specify two ratios, VR_1 and VR_2, such that their product is within the desired range. If the two ratios were equal, each would be the square root of the target ratio, 44.23. That is,

$$VR_1 = VR_2 = \sqrt{44.23} = 6.65$$

But we want the first ratio to be somewhat larger than the second. Let's specify

$$VR_1 = 8.0 = (N_B/N_A)$$

If we let pinion A have 17 teeth, the number of teeth in gear B must be

$$N_B = (N_A)(8) = (17)(8) = 136$$

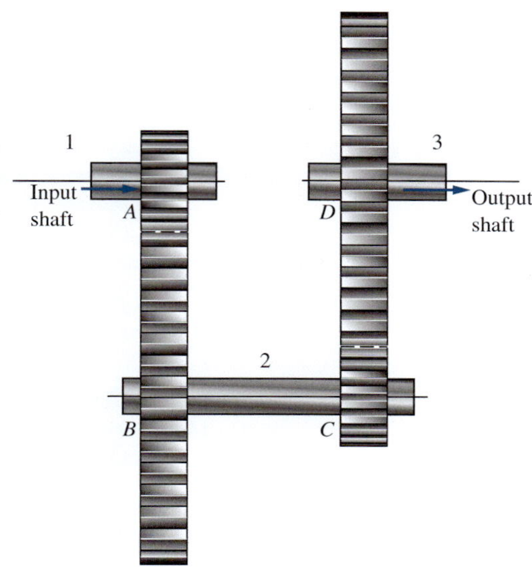

FIGURE 8–29 General layout of proposed gear train

Then the second ratio should be approximately

$$VR_2 = TV/(VR_1) = 44.23/8.0 = 5.53$$

This is the *residual ratio* left after the first ratio has been specified. Now if we specify 17 teeth for pinion *C*, gear *D* must be

$$VR_2 = 5.53 = N_D/N_C = N_D/17$$

$$N_D = (5.53)(17) = 94.01$$

Rounding this off to 94 is likely to produce an acceptable result. Finally,

$$N_A = 17 \qquad N_B = 136 \qquad N_C = 17 \qquad N_D = 94$$

We should check the final design:

$$TV = (136/17)(94/17) = 44.235 = n_A/n_D$$

The actual output speed is

$$n_D = n_A/TV = (1150 \text{ rpm})/44.235 = 26.0 \text{ rpm}$$

This is right in the middle of the desired range.

Factoring Approach for Compound Gear Trains

Example Problem 8–10 Devise a gear train for a recorder for a precision measuring instrument. The input is a shaft that rotates at exactly 3600 rpm. The output speed must be exactly 11.25 rpm. Use 20°, full-depth, involute teeth; no fewer than 17 teeth; and no more than 150 teeth in any gear.

Solution *Target TV*

$$TV_{nom} = 3600/11.25 = 320$$

Maximum Single VR

$$VR_{max} = 150/17 = 8.824$$

Maximum TV for Double Reduction

$$TV_{max} = (8.824)^2 = 77.8 \text{ (too low)}$$

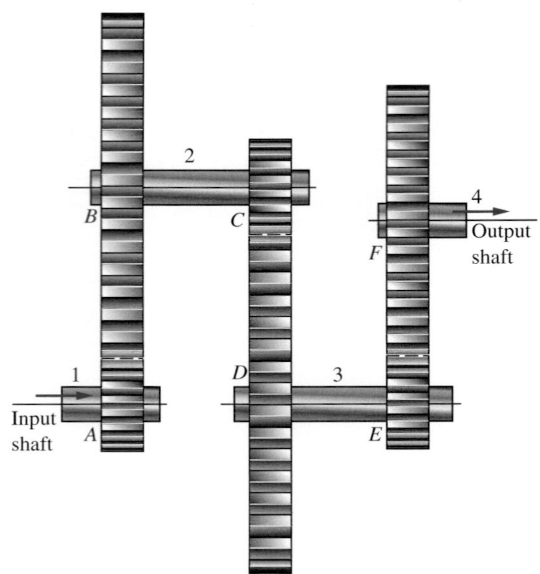

FIGURE 8–30 Triple-reduction gear train

Maximum TV for Triple Reduction

$$TV_{max} = (8.824)^3 = 687 \text{ (okay)}$$

Design a triple-reduction gear train such as that shown in Figure 8–30. The train value is the product of the three individual velocity ratios:

$$TV = (VR_1)(VR_2)(VR_3)$$

If we can find three factors of 320 that are within the range of the possible ratio for a single pair of gears, they can be specified for each velocity ratio.

Factors of 320

One method is to divide by the smallest prime numbers that will divide evenly into the given number, typically 2, 3, 5, or 7. For example,

$$320/2 = 160$$
$$160/2 = 80$$
$$80/2 = 40$$
$$40/2 = 20$$
$$20/2 = 10$$
$$10/2 = 5$$

Then the prime factors of 320 are 2, 2, 2, 2, 2, 2, and 5. We desire a set of three factors, which we can find by combining each set of three "2" factors into their product. That is,

$$(2)(2)(2) = 8$$

Then the three factors of 320 are

$$(8)(8)(5) = 320$$

Now let the number of teeth in the pinion of each pair be 17. The number of teeth in the gears will then be (8)(17) = 136 or (5)(17) = 85. Finally, we can specify

$$N_A = 17 \qquad N_C = 17 \qquad N_E = 17$$
$$N_B = 136 \qquad N_D = 136 \qquad N_F = 85$$

REFERENCES

1. American Gear Manufacturers Association. Standard 1012-G05. *Gear Nomenclature, Definitions of Terms with Symbols.* Alexandria, VA: American Gear Manufacturers Association, 2005.
2. American Gear Manufacturers Association. Standard 2002-B88 (R1996). *Tooth Thickness Specification and Measurement.* Alexandria, VA: American Gear Manufacturers Association, 1996.
3. American Gear Manufacturers Association. Standard 2008-C01. *Assembling Bevel Gears.* Alexandria, VA: American Gear Manufacturers Association, 2001.
4. American Gear Manufacturers Association. Standard 917-B97. *Design Manual for Parallel Shaft Fine-Pitch Gearing.* Alexandria, VA: American Gear Manufacturers Association, 1997.
5. American Gear Manufacturers Association. Standard 2005-D03. *Design Manual for Bevel Gears.* Alexandria, VA: American Gear Manufacturers Association, 2003.
6. American Gear Manufacturers Association. Standard 6022-C93(R2008). *Design Manual for Cylindrical Wormgearing.* Alexandria, VA: American Gear Manufacturers Association, 2008.
7. American Gear Manufacturers Association. Standard 6001-E08. *Design and Selection of Components for Enclosed Gear Drives.* Alexandria, VA: American Gear Manufacturers Association, 2008.
8. American Gear Manufacturers Association. Standard 2000-A88. *Gear Classification and Inspection Handbook—Tolerances and Measuring Methods for Unassembled Spur and Helical Gears (Including Metric Equivalents).* Alexandria, VA: American Gear Manufacturers Association, 1988.
9. American Gear Manufacturers Association. Standard 2015-1-A01. *Accuracy Classification System—Tangential Measurements for Cylindrical Gears.* Alexandria, VA: American Gear Manufacturers Association, 2001.
10. Dooner, David B., and Ali A. Seireg. *The Kinematic Geometry of Gearing: A Concurrent Engineering Approach.* New York: John Wiley & Sons, 1995.
11. Drago, Raymond J. *Fundamentals of Gear Design.* New York: Elsevier Science & Technology Books, 1988.
12. Dudley, Darle W. *Handbook of Practical Gear Design.* Boca Raton, FL: CRC Press, 1994.
13. International Standards Organization. Standard ISO 54:1996. *Cylindrical Gears for General Engineering and for Heavy Engineering—Modules.* Geneva, Switzerland: International Standards Organization, 1996.
14. International Standards Organization. Standard ISO 696:1976. *Straight Bevel Gears for General Engineering and Heavy Engineering.* Geneva, Switzerland: International Standards Organization, 1976.
15. International Standards Organization. Standard ISO 1122:1998. *Vocabulary of Gear Terms—Part 1: Definitions Related to Geometry.* Geneva, Switzerland: International Standards Organization, 1998.
16. International Standards Organization. Standard ISO 23509:2006. *Bevel and Hypoid Gear Geometry.* Geneva, Switzerland: International Standards Organization, 2006.
17. Lipp, Robert. "Avoiding Tooth Interference in Gears." *Machine Design* 54, no. 1 (January 7, 1982).
18. Litvin, Faydor, and Alfonso Fuentes. *Gear Geometry and Applied Theory,* 2nd ed. Cambridge, UK: Cambridge University Press, 2004.
19. Oberg, Erik, Franklin D. Jones, Holbrook L. Horton, and Henry H. Ryffel. *Machinery's Handbook,* 28th ed. New York: Industrial Press, 2008.

INTERNET SITES RELATED TO KINEMATICS OF GEARS

1. **American Gear Manufacturers Association (AGMA).** www.agma.org Develops and publishes voluntary, consensus standards for gears and gear drives. Some standards are jointly published with the American National Standards Institute (ANSI).
2. **Boston Gear Company.** www.bostongear.com A manufacturer of gears and complete gear drives. Part of Altra Industrial Motion, Inc. Data provided for spur, helical, bevel, and worm gearing.
3. **Emerson Power Transmission Corporation.** www.emerson-ept.com The Browning and Morse Divisions produce spur, helical, bevel, and worm gearing and complete gear drives.
4. **Gear Industry Home Page.** www.geartechnology.com Information source for many companies that manufacture or use gears or gearing systems. Includes gear machinery, gear cutting tools, gear materials, gear drives, open gearing, tooling & supplies, software, training and education. Publishes *Gear Technology Magazine, The Journal of Gear Manufacturing.*
5. **Power Transmission Home Page.** www.powertransmission.com Clearinghouse on the Internet for buyers, users, and sellers of power transmission-related products and services. Included are gears, gear drives, and gearmotors.
6. **Baldor/Dodge.** www.dodge-pt.com Manufacturer of many power transmission components, including complete gear-type speed reducers, bearings, and components such as belt drives, chain drives, clutches, brakes, and couplings.
7. **Gear—Wikipedia site** http://en.wikipedia.org/wiki/Gear General discussion of the kinematics and gears, including an animated drawing of meshing gears.
8. **Gear tooth engagement animation** http://school.mech.uwa.edu.au/~dwright/DANotes/gears/meshing/meshing.html#animation An animation of involute gear teeth engaging showing base circles of mating gears and the path of engagement along the line of action as the gears rotate. Copyright: Douglas Wright, University of Western Australia, 2005.
9. **Maryland Metrics Co.** www.mdmetric.com Distributor of a wide variety of metric hardware and power transmission products, including many types of gears.
10. **Stock Drive Products/Sterling Instruments Co.** www.sdp-si.com Distributor of a wide variety of U.S. styles and metric hardware and power transmission products, including many types of gears.
11. **W. M. Berg Co.** www.wmberg.com Distributor of a wide variety of U.S. styles of hardware and power transmission products, including many types of gears.

PROBLEMS

Gear Geometry

1. A gear has 44 teeth of the 20°, full-depth, involute form and a diametral pitch of 12. Compute the following:
   (a) Pitch diameter
   (b) Circular pitch
   (c) Equivalent module
   (d) Nearest standard module
   (e) Addendum
   (f) Dedendum
   (g) Clearance
   (h) Whole depth
   (i) Working depth
   (j) Tooth thickness
   (k) Outside diameter

   Repeat Problem 1 for the following gears:
2. $N = 34; P_d = 24$
3. $N = 45; P_d = 2$
4. $N = 18; P_d = 8$
5. $N = 22; P_d = 1.75$
6. $N = 20; P_d = 64$
7. $N = 180; P_d = 80$
8. $N = 28; P_d = 18$
9. $N = 28; P_d = 20$

For Problems 10–17, repeat Problem 1 for the following gears in the metric module system. Replace Part (c) with equivalent P_d and Part (d) with nearest standard P_d.

10. $N = 34; m = 3$
11. $N = 45; m = 1.25$
12. $N = 18; m = 12$
13. $N = 22; m = 20$
14. $N = 20; m = 1$
15. $N = 180; m = 0.4$
16. $N = 28; m = 1.5$
17. $N = 28; m = 0.8$
18. Define *backlash,* and discuss the methods used to produce it.
19. For the gears of Problems 1 and 12, recommend the amount of backlash.

Velocity Ratio

20. An 8-pitch pinion with 18 teeth mates with a gear having 64 teeth. The pinion rotates at 2450 rpm. Compute the following:
    (a) Center distance
    (b) Velocity ratio
    (c) Speed of gear
    (d) Pitch line speed

    Repeat Problem 20 for the following data:
21. $P_d = 4; N_P = 20; N_G = 92; n_P = 225\,\text{rpm}$
22. $P_d = 20; N_P = 30; N_G = 68; n_P = 850\,\text{rpm}$
23. $P_d = 64; N_P = 40; N_G = 250; n_P = 3450\,\text{rpm}$
24. $P_d = 12; N_P = 24; N_G = 88; n_P = 1750\,\text{rpm}$
25. $m = 2; N_P = 22; N_G = 68; n_P = 1750\,\text{rpm}$
26. $m = 0.8; N_P = 18; N_G = 48; n_P = 1150\,\text{rpm}$
27. $m = 4; N_P = 36; N_G = 45; n_P = 15\,\text{rpm}$
28. $m = 12; N_P = 15; N_G = 36; n_P = 480\,\text{rpm}$

For Problems 29–32, all gears are made in standard 20°, full-depth, involute form. Tell what is wrong with the following statements:

29. An 8-pitch pinion having 24 teeth mates with a 10-pitch gear having 88 teeth. The pinion rotates at 1750 rpm, and the gear at approximately 477 rpm. The center distance is 5.900 in.
30. A 6-pitch pinion having 18 teeth mates with a 6-pitch gear having 82 teeth. The pinion rotates at 1750 rpm, and the gear at approximately 384 rpm. The center distance is 8.3 in.
31. A 20-pitch pinion having 12 teeth mates with a 20-pitch gear having 62 teeth. The pinion rotates at 825 rpm, and the gear at approximately 160 rpm. The center distance is 1.850 in.
32. A 16-pitch pinion having 24 teeth mates with a 16-pitch gear having 45 teeth. The outside diameter of the pinion is 1.625 in. The outside diameter of the gear is 2.938 in. The center distance is 2.281 in.

Housing Dimensions

33. The gear pair described in Problem 20 is to be installed in a rectangular housing. Specify the dimensions X and Y as sketched in Figure P8–33 that would provide a minimum clearance of 0.10 in.

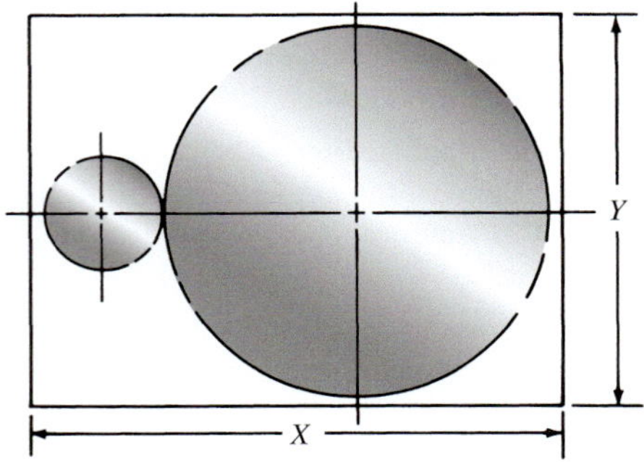

FIGURE P8–33 (Problems 33, 34, 35, and 36)

34. Repeat Problem 33 for the data of Problem 23.
35. Repeat Problem 33 for the data of Problem 26, but make the clearance 2.0 mm.
36. Repeat Problem 33 for the data of Problem 27, but make the clearance 2.0 mm.

Analysis of Simple Gear Trains: Problems 37–40

For the gear trains sketched in the given figures, compute the output speed and the direction of rotation of the output shaft if the input shaft rotates at 1750 rpm clockwise.

37. Use Figure P8–37.
38. Use Figure P8–38.
39. Use Figure P8–39.
40. Use Figure P8–40.

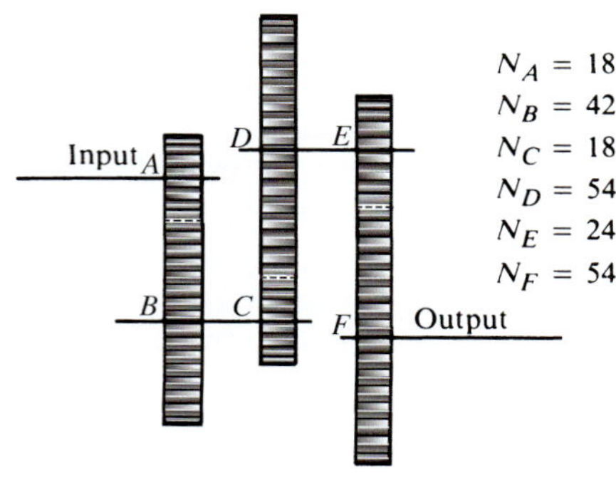

FIGURE P8–37 Gear train layout for Problem 37

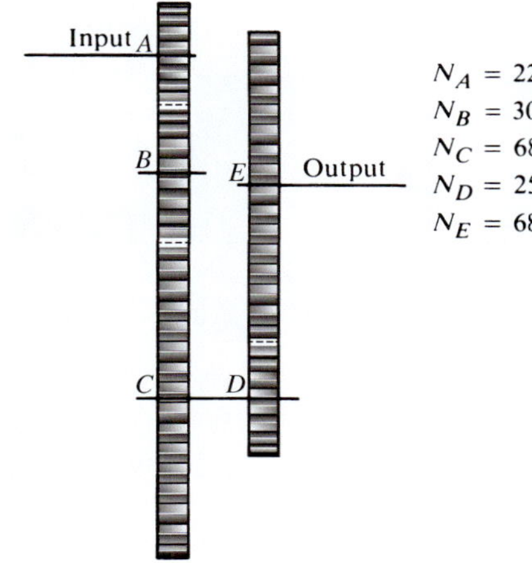

FIGURE P8–38 Gear train layout for Problem 38

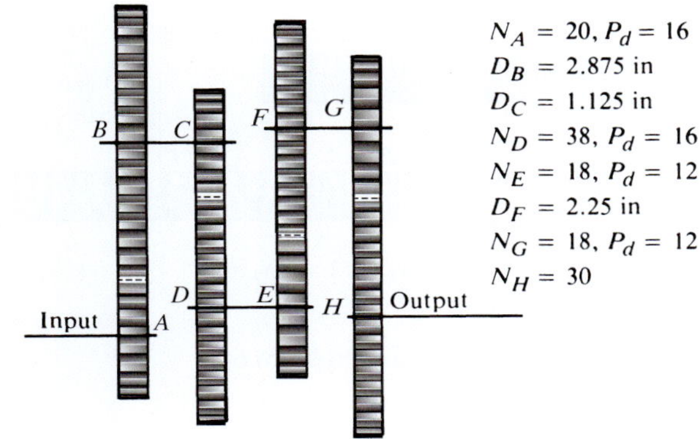

FIGURE P8–39 Gear train layout for Problem 39

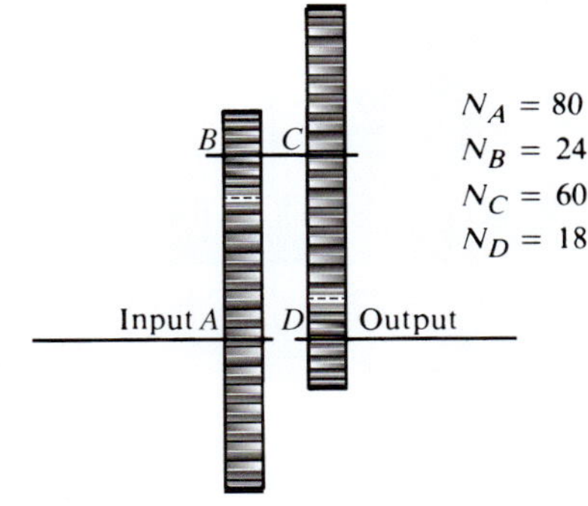

FIGURE P8–40 Gear train layout for Problem 40

Helical Gearing

41. A helical gear has a transverse diametral pitch of 8, a transverse pressure angle of $14\frac{1}{2}°$, 45 teeth, a face width of 2.00 in, and a helix angle of 30°. Compute the circular pitch, normal circular pitch, normal diametral pitch, axial pitch, pitch diameter, and normal pressure angle. Then compute the number of axial pitches in the face width.
42. A helical gear has a normal diametral pitch of 12, a normal pressure angle of 20°, 48 teeth, a face width of 1.50 in, and a helix angle of 45°. Compute the circular pitch, normal circular pitch, transverse diametral pitch, axial pitch, pitch diameter, and transverse pressure angle. Then compute the number of axial pitches in the face width.
43. A helical gear has a transverse diametral pitch of 6, a transverse pressure angle of $14\frac{1}{2}°$, 36 teeth, a face width of 1.00 in, and a helix angle of 45°. Compute the circular pitch, normal circular pitch, normal diametral pitch, axial pitch, pitch diameter, and normal pressure angle. Then compute the number of axial pitches in the face width.
44. A helical gear has a normal diametral pitch of 24, a normal pressure angle of $14\frac{1}{2}°$, 72 teeth, a face width of 0.25 in, and a helix angle of 45°. Compute the circular pitch, normal circular pitch, transverse diametral pitch, axial pitch, pitch diameter, and transverse pressure angle. Then compute the number of axial pitches in the face width.

Bevel Gears

45. A straight bevel gear pair has the following data: $N_P = 15$; $N_G = 45$; $P_d = 6$; 20° pressure angle. Compute all of the geometric features from Table 8–8.
46. Draw the gear pair of Problem 45 to scale. The following additional dimensions are given (refer to Figure 8–17). Mounting distance (M_{dP}) for the pinion = 5.250 in; M_{dG} for the gear = 3.000 in; face width = 1.250 in. Supply any other needed dimensions.
47. A straight bevel gear pair has the following data: $N_P = 25$; $N_G = 50$; $P_d = 10$; 20° pressure angle. Compute all of the geometric features from Table 8–8.
48. Draw the gear pair of Problem 47 to scale. The following additional dimensions are given (refer to Figure 8–17). Mounting distance (M_{dP}) for the pinion = 3.375 in; M_{dG} for the gear = 2.625 in; face width = 0.700 in. Supply any other needed dimensions.
49. A straight bevel gear pair has the following data: $N_P = 18$; $N_G = 72$; $P_d = 12$; 20° pressure angle. Compute all of the geometric features from Table 8–8.
50. A straight bevel gear pair has the following data: $N_P = 16$; $N_G = 64$; $P_d = 32$; 20° pressure angle. Compute all of the geometric features from Table 8–8.
51. A straight bevel gear pair has the following data: $N_P = 12$; $N_G = 36$; $P_d = 48$; 20° pressure angle. Compute all of the geometric features from Table 8–8.

Wormgearing

52. A wormgear set has a single-thread worm with a pitch diameter of 1.250 in, a diametral pitch of 10, and a normal pressure angle of 14.5°. If the worm meshes with a wormgear having 40 teeth and a face width of 0.625 in, compute the lead, axial pitch, circular pitch, lead angle, addendum, dedendum, worm outside diameter, worm root diameter, gear pitch diameter, center distance, and velocity ratio.
53. Three designs are being considered for a wormgear set to produce a velocity ratio of 20 when the wormgear rotates at 90 rpm. All three have a diametral pitch of 12, a worm pitch diameter of 1.000 in, a gear face width of 0.500 in, and a normal pressure angle of 14.5°. One has a single-thread worm and 20 wormgear teeth; the second has a double-thread worm and 40 wormgear teeth; the third has a four-thread worm and 80 wormgear teeth. For each design, compute the lead, axial pitch, circular pitch, lead angle, gear pitch diameter, and center distance.
54. A wormgear set has a double-threaded worm with a normal pressure angle of 20°, a pitch diameter of 0.625 in, and a diametral pitch of 16. Its mating wormgear has 100 teeth and a

face width of 0.3125 in. Compute the lead, axial pitch, circular pitch, lead angle, addendum, dedendum, worm outside diameter, center distance, and velocity ratio.

55. A wormgear set has a four-threaded worm with a normal pressure angle of $14\frac{1}{2}°$, a pitch diameter of 2.000 in, and a diametral pitch of 6. Its mating wormgear has 72 teeth and a face width of 1.000 in. Compute the lead, axial pitch, circular pitch, lead angle, addendum, dedendum, worm outside diameter, center distance, and velocity ratio.

56. A wormgear set has a single-threaded worm with a normal pressure angle of $14\frac{1}{2}°$, a pitch diameter of 4.000 in, and a diametral pitch of 3. Its mating wormgear has 54 teeth and a face width of 2.000 in. Compute the lead, axial pitch, circular pitch, lead angle, addendum, dedendum, worm outside diameter, center distance, and velocity ratio.

57. A wormgear set has a four-threaded worm with a normal pressure angle of 25°, a pitch diameter of 0.333 in, and a diametral pitch of 48. Its mating wormgear has 80 teeth and a face width of 0.156 in. Compute the lead, axial pitch, circular pitch, lead angle, addendum, dedendum, worm outside diameter, center distance, and velocity ratio.

Analysis of Complex Gear Trains

58. The input shaft for the gear train shown in Figure P8–58 rotates at 3450 rpm cw. Compute the rotational speed and direction of the output shaft.

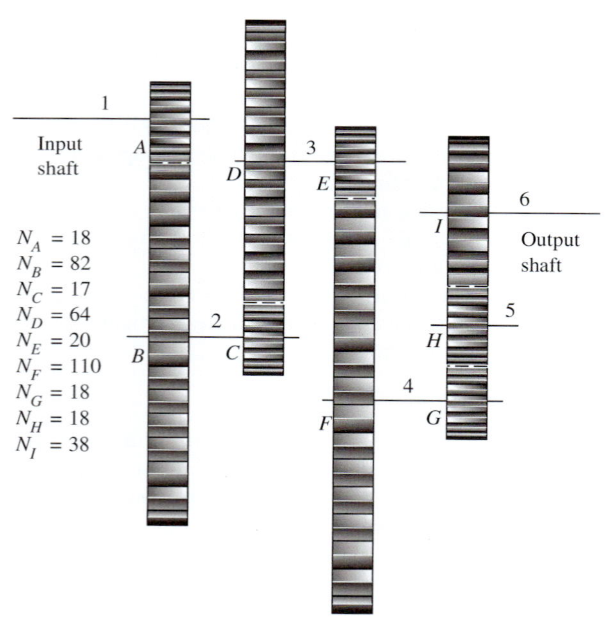

FIGURE P8–58 Gear train for Problem 58

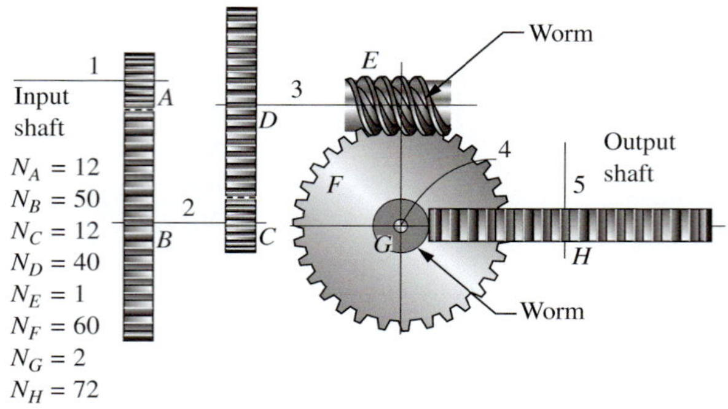

FIGURE P8–59 Gear train for Problem 59

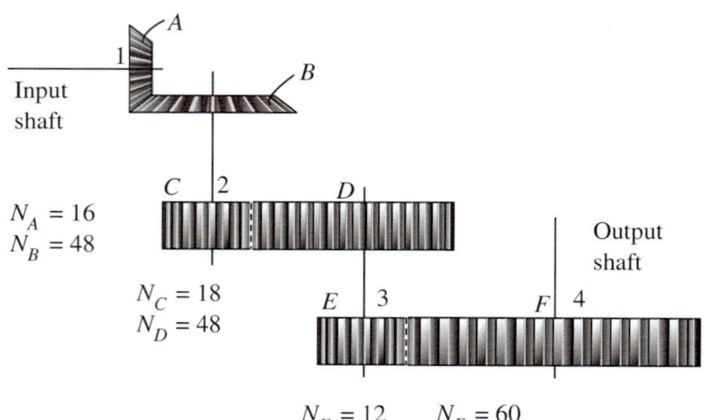

FIGURE P8–60 Gear train for Problem 60

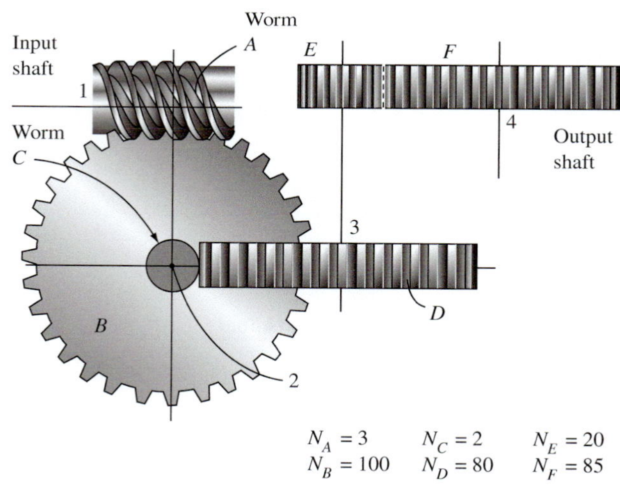

FIGURE P8–61 Gear train for Problem 61

59. The input shaft for the gear train shown in Figure P8–59 rotates at 12 200 rpm. Compute the rotational speed of the output shaft.

60. The input shaft for the gear train shown in Figure P8–60 rotates at 6840 rpm. Compute the rotational speed of the output shaft.

61. The input shaft for the gear train shown in Figure P8–61 rotates at 2875 rpm. Compute the rotational speed of the output shaft.

Kinematic Design of a Single Gear Pair

62. Specify the numbers of teeth for the pinion and gear of a single gear pair to produce a velocity ratio of π as closely as possible. Use no fewer than 16 teeth nor more than 24 teeth in the pinion.

63. Specify the numbers of teeth for the pinion and gear of a single gear pair to produce a velocity ratio of $\sqrt{3}$ as closely as possible. Use no fewer than 16 teeth nor more than 24 teeth in the pinion.

64. Specify the numbers of teeth for the pinion and gear of a single gear pair to produce a velocity ratio of $\sqrt{38}$ as closely as possible. Use no fewer than 18 teeth nor more than 24 teeth in the pinion.
65. Specify the numbers of teeth for the pinion and gear of a single gear pair to produce a velocity ratio of 7.42 as closely as possible. Use no fewer than 18 teeth nor more than 24 teeth in the pinion.

Kinematic Design of Gear Trains

For Problems 66–75, devise a gear train using all external gears on parallel shafts. Use 20° full-depth involute teeth and no more than 150 teeth in any gear. Ensure that there is no interference. Sketch the layout for your design.

Problem no.	Input speed (rpm)	Output speed range (rpm)
66.	1800	2.0 Exactly
67.	1800	21.0 to 22.0
68.	3360	12.0 Exactly
69.	4200	13.0 to 13.5
70.	5500	221 to 225
71.	5500	13.0 to 14.0
72.	1750	146 to 150
73.	850	40.0 to 44.0
74.	3000	548 to 552 Use two pairs
75.	3600	3.0 to 5.0

For Problems 76–80, devise a gear train using any type of gears. Try for the minimum number of gears while avoiding interference and having no more than 150 teeth in any gear. Sketch your design.

Problem no.	Input speed (rpm)	Output speed (rpm)
76.	3600	3.0 to 5.0
77.	1800	8.0 Exactly
78.	3360	12.0 Exactly
79.	4200	13.0 to 13.5
80.	5500	13.0 to 14.0

CHAPTER NINE

SPUR GEAR DESIGN

THE BIG PICTURE — Spur Gear Design

Discussion Map

- A spur gear has involute teeth that are straight and parallel to the axis of the shaft that carries the gear.

Discover

Describe the action of the teeth of the driving gear on those of the driven gear. What kinds of stresses are produced?

How do the geometry of the gear teeth, the materials from which they are made, and the operating conditions affect the stresses and the life of the drive system?

This chapter will help you acquire the skills to perform the necessary analyses and to design safe spur gear drive systems that demonstrate long life.

The goal of this chapter is to help you gain the knowledge and skills necessary to design spur gears to transmit power from a source such as an electric motor, gasoline engine, fluid power motor, turbine, or other prime mover to a driven machine while changing the speed of the input shaft to some other speed at the output shaft. Most gear-type power transmissions are *speed reducers* that deliver the power to the driven machine at a lower speed and a higher torque. Examples of such transmissions are:

1. Electric motor drives through the transmission to a conveyor in a factory
2. Gasoline engine for a vehicle drives through the transmission to the drive wheels
3. Fluid power motor drives through the transmission to a winch on a tractor
4. Water turbine drives through the transmission to an electric generator
5. Gas turbine (jet) engine drives through the transmission to the rotor of a helicopter

Although more rare than speed reducers, transmissions may also be used as *speed increasers*. An important

FIGURE 9–1 Double-reduction spur gear reducer (Bison Gear & Engineering Corporation, St. Charles, IL)

example is a wind turbine that rotates relatively slowly, say 20.0 rpm, and the speed must be increased to 1800 rpm to drive an electric generator.

Figure 9–1 shows a commercially available double-reduction, spur gear type speed reducer driven by an electric motor that can be applied to many kinds of applications. This is an example of the type of reducer addressed in this chapter and about which much of design process discussed in Part II of this book (Chapters 7–15) is directed. Make note of how the gears are positioned and how they are mounted on shafts that are then supported by bearings mounted in a rigid housing.

A *spur gear* is one of the most fundamental types of gears. Its teeth are straight and parallel to the axis of the shaft that carries the gear. The teeth have the involute form described in Chapter 8. So, in general, the action of one tooth on a mating tooth is like that of two convex, curved members in contact: As the driving gear rotates, its teeth exert a force on the mating gear that is tangential to the pitch circles of the two gears. Because this force acts at a distance equal to the pitch radius of the gear, a torque is developed in the shaft that carries the gear. When the two gears rotate, they transmit power that is proportional to the torque. Indeed, that is the primary purpose of the spur gear drive system.

Consider the action described in the preceding paragraph:

- How does that action relate to the design of the gear teeth? Look back at Figure 8–1 as you consider this question and those that follow.
- As the force is exerted by the driving tooth on the driven tooth, what kinds of stresses are produced in the teeth? Consider both the point of contact of one tooth on the other and the whole tooth. Where are stresses a maximum?
- How could the teeth fail under the influence of these stresses?
- What material properties are critical to allow the gears to carry such loads safely and with a reasonable life span?
- What important geometric features affect the level of stress produced in the teeth?
- How does the precision of the tooth geometry affect its operation?
- How does the nature of the application affect the gears? What if the machine that the gears drive is a rock crusher that takes large boulders and reduces them to gravel made up of small stones? How would that loading compare with that of a gear system that drives a fan providing ventilation air to a building?
- What is the influence of the driving machine? Would the design be different if an electric motor were the driver or if a gasoline engine were used?
- The gears are typically mounted on shafts that deliver power from the driver to the input gear of a gear train and that take power from the output gear and transmit it to the driven machine. Describe various ways that the gears can be attached to the shafts and located with respect to each other. How can the shafts be supported?

This chapter contains the kinds of information that you can use to answer such questions and to complete the analysis and design of spur gear power transmission systems.

Chapters 10–15 cover similar topics for helical gears, bevel gears, and wormgearing, along with the design and specification of keys, couplings, seals, shafts, and bearings—all of which are needed to design a complete mechanical drive.

YOU ARE THE DESIGNER

You have already made the design decision that a spur gear type of speed reducer is to be used for a particular application. How do you complete the design of the gears themselves?

This is a continuation of a design scenario that was started in Chapter 1 of this book when the original goals were stated and when an overview of the entire book was given. The introduction to Part II continued this theme by indicating that the arrangement of the chapters is aligned with the design process that you could use to complete the design of the speed reducer.

Then in Chapter 8, you, as the designer, dealt with the kinematics of a gear reducer that would take power from the shaft of an electric motor rotating at 1750 rpm and deliver it to a machine that was to operate at approximately 292 rpm. There you limited your interest to the decisions that affected motion and the basic geometry of the gears. It was decided that you would use a double-reduction gear train to reduce the speed of rotation of the drive system in two stages using two pairs of gears in series. You also learned how to specify the layout of the gear train, along with key design decisions such as the numbers of teeth in all of the gears and the relationships among the diametral pitch, the number of teeth in the gears, the pitch diameters, and the distance between the centers of the shafts that carry those gears. For a chosen diametral pitch, you learned how to compute the dimensions of key features of the gear teeth such as the addendum, dedendum, and tooth width.

But the design is not complete until you have specified the material from which the gears are to be made and until you have verified that the gears will withstand the forces exerted on the gears as they transmit power and the corresponding torque. The teeth must not break, and they must have a sufficiently long life to meet the needs of the customer who uses the reducer.

To complete the design, you need more data: How much power is to be transmitted? To what kind of machine is the power from the output of the reducer being delivered? How does that affect the design of the gears? What is the anticipated duty cycle for the reducer in terms of the number of hours per day, days per week, and years of life expected? What options do you have for materials that are suitable for gears? Which material will you specify, and what will be its heat treatment?

You are the designer. The information in this chapter will help you complete the design. ■

9–1 OBJECTIVES OF THIS CHAPTER

After completing this chapter, you will be able to demonstrate the competencies listed below. They are presented in the order that they are covered in this chapter. The primary objectives are numbers 6, 7, and 8, which involve (a) the calculation of the bending strength and the ability of the gear teeth to resist pitting and (b) the design of gears to be safe with regard to both strength and pitting resistance. The competencies are as follows:

1. Compute the forces exerted on gear teeth as they rotate and transmit power.
2. Describe various methods for manufacturing gears and the levels of precision and quality to which they can be produced.
3. Specify a suitable level of quality for gears according to the use to which they are to be put.
4. Describe suitable metallic materials from which to make the gears, in order to provide adequate performance for both strength and pitting resistance.
5. Use the standards of the American Gear Manufacturers Association (AGMA) as the basis for completing the design of the gears.
6. Use appropriate stress analyses to determine the relationships among the applied forces, the geometry of the gear teeth, the precision of the gear teeth, and other factors specific to a given application, in order to make final decisions about those variables.
7. Perform the analysis of the tendency for the contact stresses exerted on the surfaces of the teeth to cause pitting of the teeth, in order to determine an adequate hardness of the gear material that will provide an acceptable level of pitting resistance for the reducer.
8. Complete the design of the gears, taking into consideration both the stress analysis and the analysis of pitting resistance. The result will be a complete specification of the gear geometry, the material for the gear, and the heat treatment of the material.

Outline of the Chapter: The process of designing spur gear drives contains numerous steps and requires the determination of several quantities that affect the performance of the drives. This chapter builds that process within the following sections, all of which contribute to the final design process described in Section 9–11. The list below gives the primary steps in the process for the design of a drive consisting of two steel gears.

1. Define the goals of the gear drive design: power to be transmitted, speed of the input gear, desired speed of the output gear, the kind of driver that provides the power, the kind of driven machine, and any special design features.
2. Determine the forces applied to the gear (Section 9–3).
3. Define the precision required for the gears by specifying the quality number (Section 9–5).
4. Understand the types of steel alloys typically used for gears and the kinds of heat treatment available (Section 9–7).
5. Propose the geometry of the gears including the pitch, number of teeth in each gear, pitch diameter, form of the teeth, and face width (Section 9–11).
6. Determine the expected stress due to bending in the gear teeth. This step requires specification of several factors that are functions of the manner of use and the manufacturing processes that will be used to produce the gears (Section 9–8).
7. Determine the expected actual contact stress experienced by the face of the teeth (Section 9–9).
8. Using the bending stress from Step 6 and the contact stress from Step 7, determine the strength and hardness required for the materials from which the gears are to be made to ensure adequate safety and life and specify the steel alloy and heat treatment that will meet these requirements (Section 9–10).
9. Summarize the design details.

9–2 CONCEPTS FROM PREVIOUS CHAPTERS

As you study this chapter, it is assumed that you are familiar with the geometry of gear features and the kinematics of one gear driving another as presented in Chapter 8 and illustrated in Figures 8–1, 8–8, 8–12, 8–13, and 8–24. (See also

References 4 and 26.) Key relationships that you should be able to use include the following:

$$\text{Pitch line speed} = v_t = R\omega = (D/2)\omega$$

where R = radius of the pitch circle
D = pitch diameter
ω = angular velocity of the gear

Because the pitch line speed is the same for both the pinion and the gear, values for R, D, and ω can be for either. In the computation of stresses in gear teeth, in the U.S. system, it is usual to express the pitch line speed in the units of ft/min, while the size of the gear is given as its pitch diameter expressed in inches. Speed of rotation is typically given as n rpm—that is, n rev/min. Let's compute the unit-specific equation that gives pitch line speed in ft/min:

➪ **Pitch Line Speed U.S. Units**

$$v_t = (D/2)\omega = \frac{D\text{ in}}{2}\cdot\frac{n\text{ rev}}{\text{min}}\cdot\frac{2\pi\text{ rad}}{\text{rev}}\cdot\frac{1\text{ ft}}{12\text{ in}}$$
$$= (\pi D n/12)\text{ ft/min} \quad \textbf{(9–1)}$$

In the SI metric system, diameter is typically expressed in mm and speed in rpm while the pitch line speed is in m/s. Using a process similar to that used above for Equation 9–1, the following unit-specific expression was developed in Chapter 8.

➪ **Pitch Line Speed SI Units**

$$v_t = \pi D n/60\,000\text{ m/s} \quad \textbf{(9–2)}$$

The velocity ratio can be expressed in many ways. For the particular case of a pinion driving a larger gear,

➪ **Velocity Ratio**

$$\text{Velocity ratio} = VR = \frac{\omega_P}{\omega_G} = \frac{n_P}{n_G} = \frac{R_G}{R_P} = \frac{D_G}{D_P} = \frac{N_G}{N_P} \quad \textbf{(9–3)}$$

A related ratio, m_G, called the *gear ratio*, is often used in analysis of the performance of gears. It is always defined as the ratio of the number of teeth in the larger gear to the number of teeth in the pinion, regardless of which is the driver. Thus, m_G is always greater than or equal to 1.0. When the pinion is the driver, as it is for a speed reducer, m_G is equal to VR. That is,

➪ **Gear Ratio**

$$\text{Gear ratio} = m_G = N_G/N_P \geq 1.0 \quad \textbf{(9–4)}$$

In the U.S. system the diametral pitch, P_d, characterizes the physical size of the teeth of a gear. It is related to the pitch diameter and the number of teeth as follows:

➪ **Diametral Pitch**

$$P_d = N_G/D_G = N_P/D_P \quad \textbf{(9–5)}$$

In the SI metric system, the metric module, m, characterizes the size of the teeth. It is defined as

➪ **Metric Module**

$$m = D_G/N_G = D_P/N_P \quad \textbf{(9–6)}$$

The pressure angle, ϕ, is an important feature that characterizes the form of the involute curve that makes up the active face of the teeth of standard gears. See Figure 8–13. Also notice in Figure 8–12 that the angle between a normal to the involute curve and the tangent to the pitch circle of a gear is equal to the pressure angle.

9–3 FORCES, TORQUE, AND POWER IN GEARING

To understand the method of computing stresses in gear teeth, consider the way power is transmitted by a gear system. For the simple single-reduction gear pair shown in Figure 9–2, power is received from the motor by the input shaft rotating at motor speed. Thus, torque in the shaft can be computed from the following equation:

➪ **Torque**

$$\text{Torque} = \text{power/rotational speed} = P/n \quad \textbf{(9–7)}$$

The input shaft transmits the power from the coupling to the point where the pinion is mounted. The power is transmitted from the shaft to the pinion through the key. The teeth of the pinion drive the teeth of the gear and thus transmit the power to the gear. But again, power transmission actually involves the application of a torque during rotation at a given speed. The torque is the product of the force acting tangent to the pitch circle of the pinion times the pitch radius of the pinion. We will use the symbol W_t to indicate the *tangential force*. As described, W_t is the force exerted *by the pinion teeth on the gear teeth.* But if the gears are rotating at a constant speed and are transmitting a uniform level of power, the system is in equilibrium. Therefore, there must be an equal and opposite tangential force exerted by the gear teeth back on the pinion teeth. This is an application of the principle of action and reaction.

To complete the description of the power flow, the tangential force on the gear teeth produces a torque on the gear equal to the product of W_t times the pitch radius of the gear. Because W_t is the same on the pinion and the gear, but the pitch radius of the gear is larger than that of the pinion, the torque on the gear (the output torque) is greater than the input torque. However, note that the power transmitted is the same or slightly less because of mechanical inefficiencies. The power then flows from the gear through the key to the output shaft and finally to the driven machine.

From this description of power flow, we can see that gears transmit power by exerting a force by the driving teeth on the driven teeth while the reaction force acts back on the teeth of the driving gear. Figure 9–3 shows a single

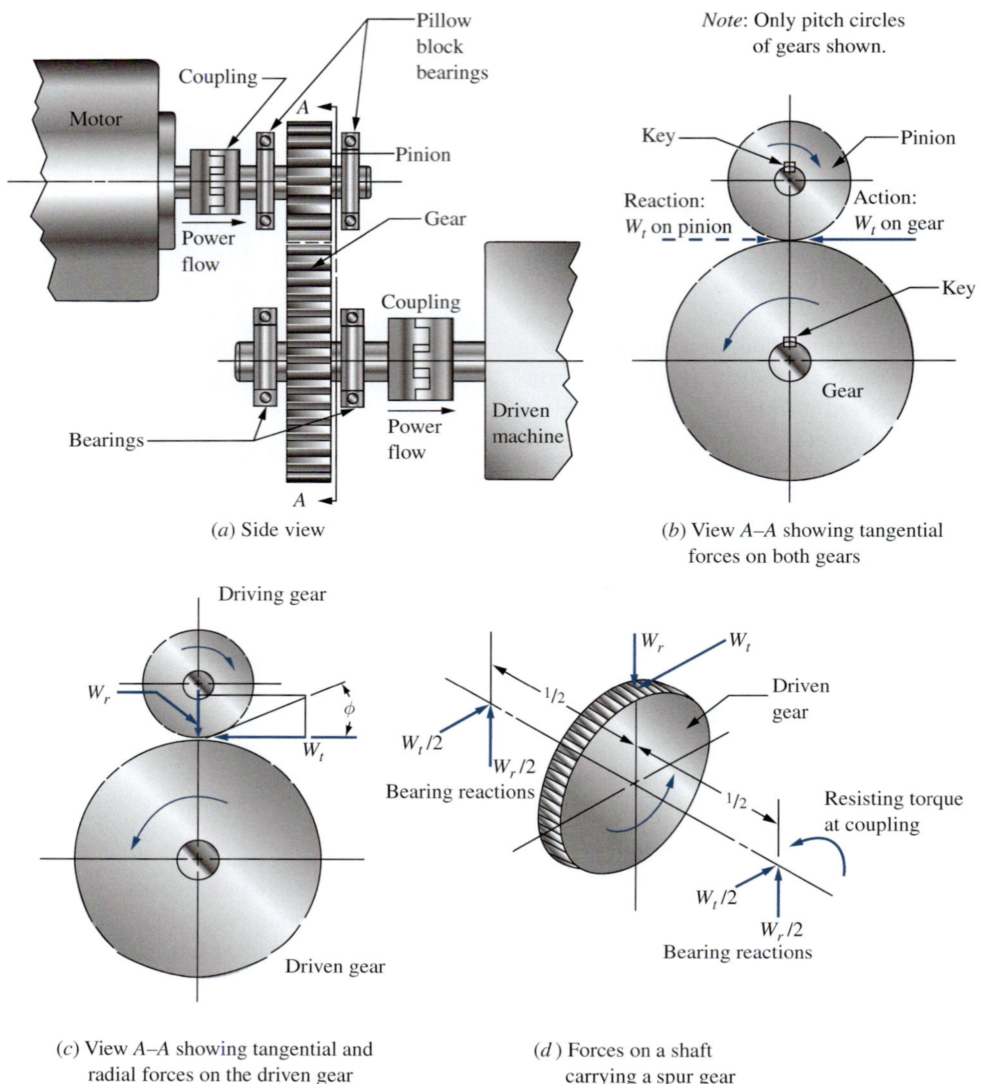

FIGURE 9–2 Power flow through a gear pair

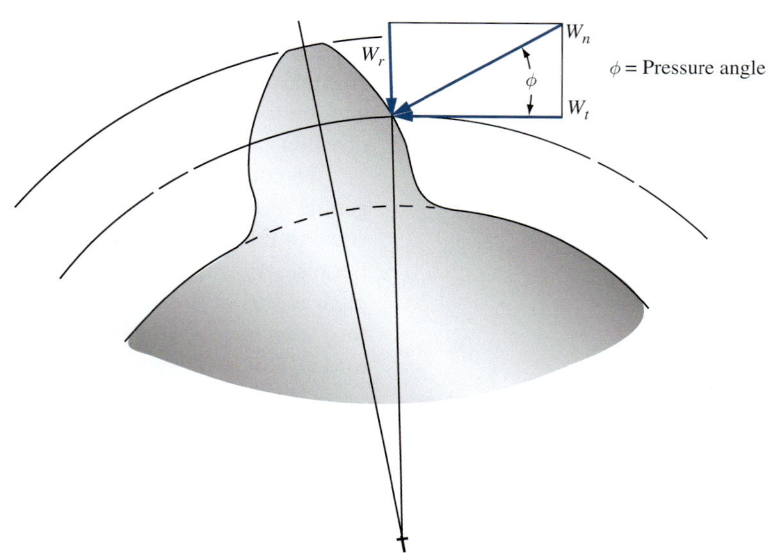

FIGURE 9–3 Forces on gear teeth

gear tooth with the tangential force W_t acting on it. But this is not the total force on the tooth. Because of the involute form of the tooth, the total force transferred from one tooth to the mating tooth acts normal to the involute profile. This action is shown as W_n. The tangential force W_t is actually the horizontal component of the total force. To complete the picture, note that there is a vertical component of the total force acting radially on the gear tooth, indicated by W_r.

The discussion about power flow and forces existing in gears thus far is generic and independent of units. Following discussion is developed primarily for the U.S. unit system based on the diametral pitch, P_d. Later, we adapt the results to the SI metric unit system and module, m.

We will start the computation of forces with the transmitted force, W_t, because its value is based on the given data for power and speed. It is convenient to develop unit-specific equations for W_t because standard practice typically calls for the following units for key quantities pertinent to the analysis of gear sets:

Forces in pounds (lb)

Power in horsepower (hp) (Note that 1.0 hp = 550 lb · ft/s.)

Rotational speed in rpm, that is, rev/min

Pitch line speed in ft/min

Torque in lb · in

The torque exerted on a gear is the product of the transmitted load, W_t, and the pitch radius of the gear. The torque is also equal to the power transmitted divided by the rotational speed. Then

$$T = W_t(R) = W_t(D/2) = P/n$$

Then we can solve for the force, and the units can be adjusted as follows:

➩ Tangential Force

$$W_t = \frac{2P}{Dn} = \frac{2P(\text{hp})}{D(\text{in}) \cdot n(\text{rev/min})} \cdot \frac{550\text{ lb} \cdot \text{ft/s}}{(\text{hp})} \cdot \frac{1.0\text{ rev}}{2\pi\text{ rad}} \cdot \frac{60\text{ s/min}}{} \cdot \frac{12\text{ in}}{\text{ft}}$$

$$W_t = (126\,000)(P)/(nD)\text{ lb} \tag{9–8}$$

Data for either the pinion or the gear can be used in this equation. Other relationships are now developed because they are needed in other parts of the process of analyzing the gears or the shafts that carry them.

Power is also the product of transmitted force, W_t, and the pitch line velocity:

$$P = W_t \cdot v_t$$

Then, solving for the force and adjusting units, we have

➩ Tangential Force

$$W_t = \frac{P}{v_t} = \frac{P(\text{hp})}{v_t(\text{ft/min})} \cdot \frac{550\text{ lb} \cdot \text{ft/s}}{1.0\text{ hp}} \cdot \frac{60\text{ s/min}}{}$$

$$= 33\,000\,(P)/(v_t)\text{ lb} \tag{9–9}$$

We may also need to compute torque in lb · in:

➩ Torque

$$T = \frac{P}{\omega} = \frac{P(\text{hp})}{n(\text{rev/min})} \cdot \frac{550\text{ lb} \cdot \text{ft/s}}{1.0\text{ hp}} \cdot \frac{1.0\text{ rev}}{2\pi\text{ rad}} \cdot \frac{60\text{ s/min}}{} \cdot \frac{12\text{ in}}{\text{ft}}$$

$$T = 63\,000(P)/n\text{ lb} \cdot \text{in} \tag{9–10}$$

These values can be computed for either the pinion or the gear by appropriate substitutions. Remember that the pitch line speed is the same for the pinion and the gear and that the transmitted loads on the pinion and the gear are the same, except that they act in opposite directions.

The normal force, W_n, and the radial force, W_r, can be computed from the known W_t by using the right triangle relations evident in Figure 9–3:

➩ Radial Force

$$W_r = W_t \tan \phi \tag{9–11}$$

➩ Normal Force

$$W_n = W_t/\cos \phi \tag{9–12}$$

where ϕ = pressure angle of the tooth form

In addition to causing the stresses in the gear teeth, these forces act on the shaft. In order to maintain equilibrium, the bearings that support the shaft must provide the reactions. Figure 9–2(d) shows the free-body diagram of the output shaft of the reducer.

Power Flow and Efficiency

The discussion thus far has focused on power, torque, and forces for a single pair of gears. For compound gearing having two or more pairs of gears, the flow of power and the overall efficiency become increasingly important.

Power losses in gear drives made from spur, helical, and bevel gears depend on the action of each tooth on its mating tooth, a combination of rolling and sliding. For accurate, well-lubricated gears, the power loss ranges from 0.5% to 2.0% and is typically taken to be approximately 1.0%. (See Reference 20.) *Because this is quite small, it is customary to neglect it in sizing individual gear pairs; we do that in this book.*

Compound gear drives employ several pairs of gears in series to produce large speed reduction ratios. With 1.0% power loss in each pair, the accumulated power loss for the system can become significant, and it can affect the size of motor to drive the system or the ultimate power and torque available for use at the output. Furthermore, the power loss is transferred to the environment or into the gear lubricant and, for large power transmissions, the management of the heat generated is critical to the overall performance of the unit. The viscosity and load-carrying ability of lubricants is degraded with increasing temperature.

Tracking power flow in a simple or compound gear train is simple, the power is transferred from one gear pair to the next with only a small power loss at each mesh. More complex designs may split the power flow at some point to two or more paths. This is typical of planetary gear trains. In such cases, you should consider the basic relationship among power, torque, and rotational speed shown in Equation (9–7), $P = T \times n$. We can present this in another form. Let the rotational speed, n, that is typically taken to be in the units of rpm, be the more general term *angular velocity*, ω, in the units of rad/s. Now express the torque in terms of the transmitted forces, W_t, and the pitch radius of the gear, R. That is, $T = W_tR$. Equation (9–7) then becomes,

$$P = T \times n = W_tR\omega$$

But $R\,\omega$ is the pitch line velocity for the gears, v_t. Then,

$$P = W_tR\omega = W_tv_t$$

Knowing how the power splits enables the determination of the transmitted load at each mesh.

9–4 GEAR MANUFACTURE

The discussion of gear manufacture will begin with the method of producing the gear blank. Small gears are frequently made from wrought plate or bar, with the hub, web, spokes, and rim machined to final or near-final dimensions before the gear teeth are produced. The face width and the outside diameter of the gear teeth are also produced at this stage. Other gear blanks may be forged, sand cast, or die cast to achieve the basic form prior to machining. A few gears in which only moderate precision is required may be die cast with the teeth in virtually final form.

Large gears are frequently fabricated from components. The rim and the portion into which the teeth are machined may be rolled into a ring shape from a flat bar and then welded. The web or spokes and the hub are then welded inside the ring. Very large gears may be made in segments with the final assembly of the segments by welding or by mechanical fasteners.

The popular methods of machining the gear teeth are form milling, shaping, and hobbing. (See References 5, 18 and 26.)

In *form milling* [Figure 9–4(a)], a milling cutter that has the shape of the tooth space is used, and each space is cut completely before the gear blank is indexed to the position

(*a*) Form milling cutter

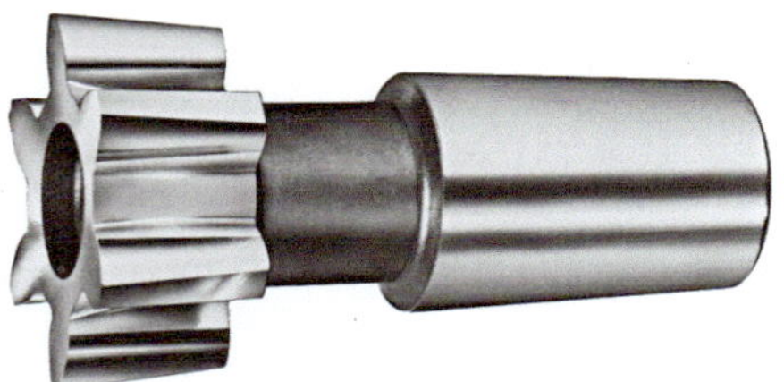

(*b*) Spur gear shaper cutter

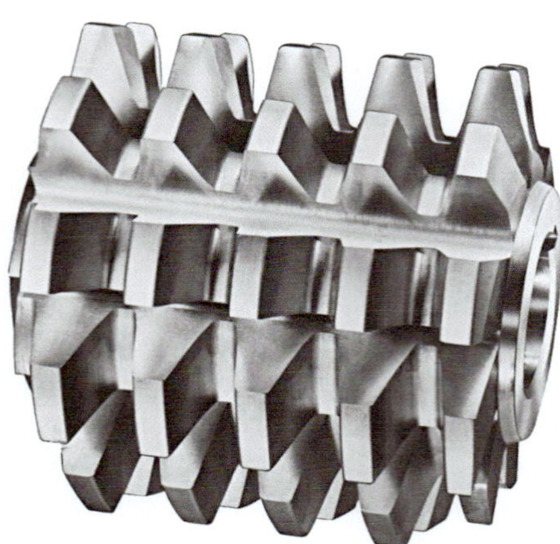

(*c*) Hob for small pitch gears having large teeth

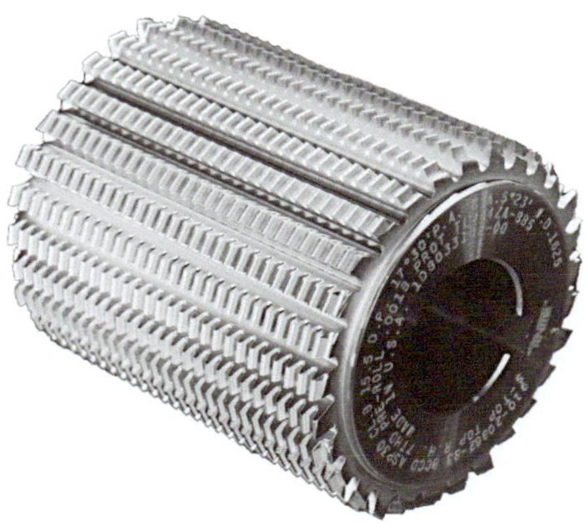

(*d*) Hob for high pitch gears having small teeth

FIGURE 9–4 A variety of gear cutting tools (Gleason Cutting Tools Corporation, Loves Park, IL)

(*a*) Gear being shaped with shaping cutter

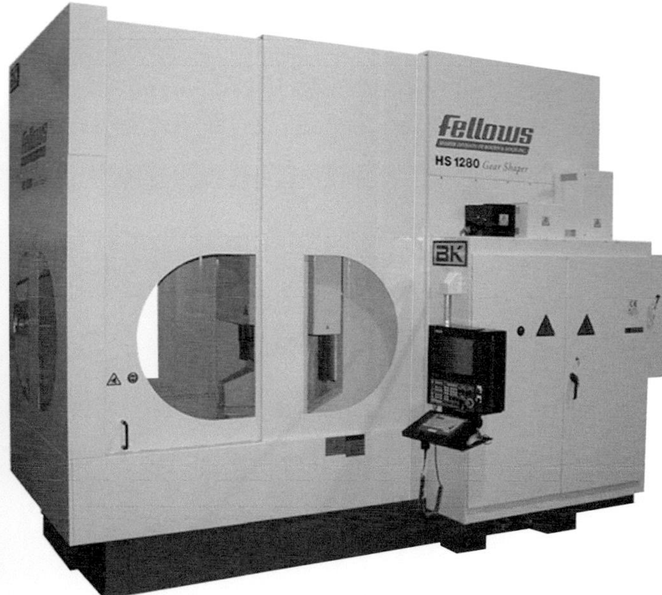

(*b*) Gear shaping machine

FIGURE 9–5 Gear shaping operation and shaping machine (Bourn & Koch, Inc., Rockford, IL USA)

of the next adjacent space. This method is used mostly for large gears, and great care is required to achieve accurate results.

Shaping [Figures 9–4(b) and 9–5] is a process in which the cutter reciprocates, usually on a vertical spindle. The shaping cutter rotates as it reciprocates and is fed into the gear blank. Thus, the involute-tooth form is generated gradually. This process is frequently used for internal gears.

Hobbing [Figures 9–4(c) and (d) and 9–6] is a process similar to milling except that the workpiece (the gear blank) and the cutter (the hob) rotate in a coordinated fashion. Here also, the tooth form is generated gradually as the hob is fed into the blank.

The gear teeth are finished to greater precision after form milling, shaping, or hobbing by the processes of grinding, shaving, and honing. Being products of secondary processes, they are expensive and should be used only where the operation requires high accuracy in the tooth form and spacing. Figure 9–7 shows a gear grinding machine.

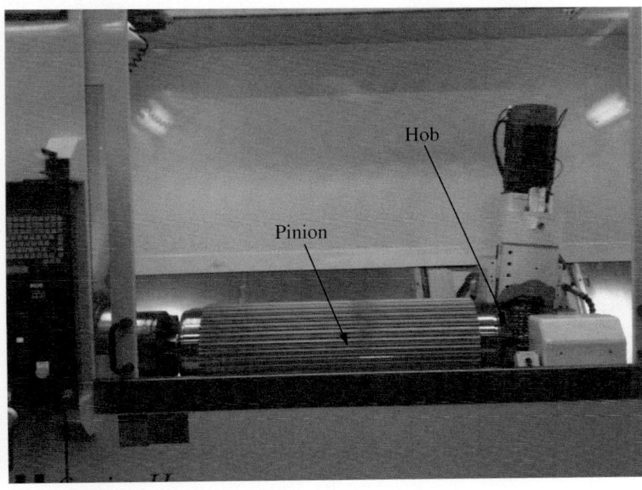

(*a*) Long pinion and a hob in a hobbing machine

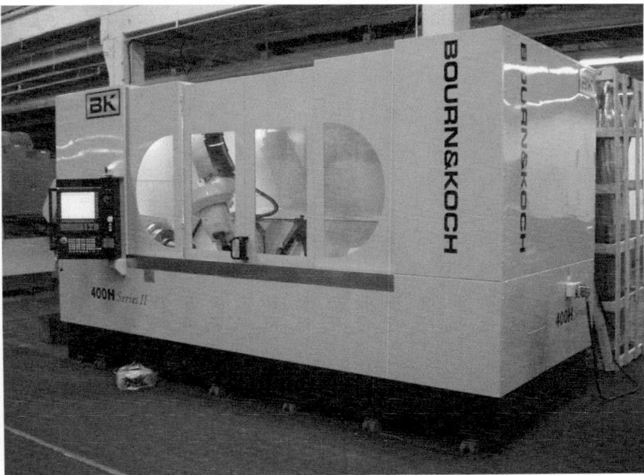

(*b*) Large hobbing machine

FIGURE 9–6 Gear hobbing operation and hobbing machine (Bourn & Koch, Inc., Rockford, IL USA)

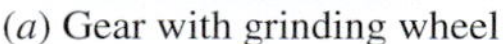

(*a*) Gear with grinding wheel

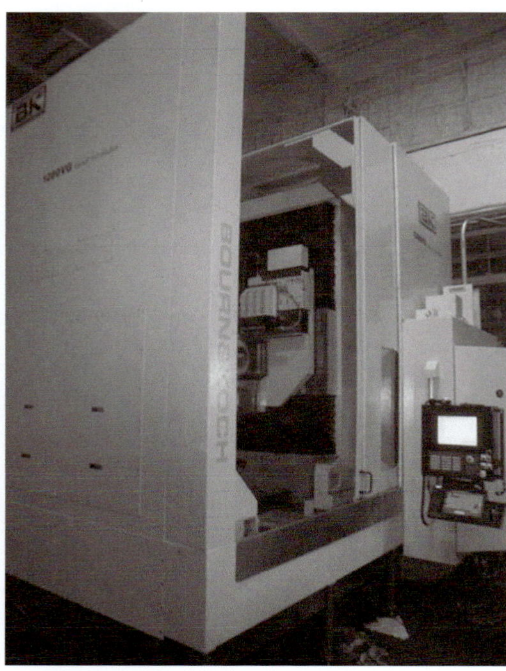

(*b*) Gear grinding machine

FIGURE 9–7 Gear grinding operation and grinding machine (Bourn & Koch, Inc., Rockford, IL USA)

9–5 GEAR QUALITY

Ensuring proper dimensional accuracy of power-transmission gears is essential to their suitability for use in machinery. Stresses in gears, smoothness of operation, life, and noise are affected by the degree of precision obtained in manufacturing. References 5, 7, and 14 provide comprehensive treatments of gear quality. Here, we give an overview of the fundamentals and equipment used to measure gear quality.

Quality in gearing is indicated by either of two methods: (1) the precision of specific features of a single gear, called *analytical measurement* or (2) the composite variation of a product test gear rotating in mesh with a precise master gear, called *functional measurement.* These two methods are described next.

Methods of Gear Measurement

Functional measurement employs a device called a *double flank roll tester* to measure the *total radial composite deviation* and the *tooth-to-tooth radial composite variation.* Figure 9–8 shows a photograph (a) of one commercially available roll tester, a schematic diagram of its construction (b), and a sample of the output produced in the test (c). The primary features of the physical testing device are:

1. Fixed mounting of the axis for the product gear to be tested
2. Mounting for an accurate master gear on a linearly movable slide with means for maintaining full double flank contact between the two gears
3. Measurement device for indicating the movement of the axis for the master gear as the product gear rotates one complete revolution
4. Recorder to display the total excursion of the master gear during the cycle

Note that the total excursion [Part (c) of Figure 9–8] contains a long term component with variations for each tooth's engagement superimposed on it. The total radial composite deviation is measured from the largest excursion of the plot over the complete cycle as shown. To better view the tooth-to-tooth deviation, the long-term component is extracted from the complete plot and shown separately, with the largest deviation for any given tooth reported.

Analytical measurement measures numerous features of individual teeth using a highly sensitive probe moving over the surface of the tooth under the control of a specially designed coordinate measurement machine (CMM). The computer in the complete system guides the probe over specified trajectories while accurately measuring its position. Figure 9–9 shows a commercially available system. Among the measurements that can be made are:

Index variation: The difference between the actual location of a point on the face of a gear tooth at the pitch circle and the corresponding point of a reference tooth measured on the pitch circle. The variation causes inaccuracy in the action of mating gear teeth. Figure 9–10(a) shows a sample output for index variation.

(*a*) Double flank roll testing device

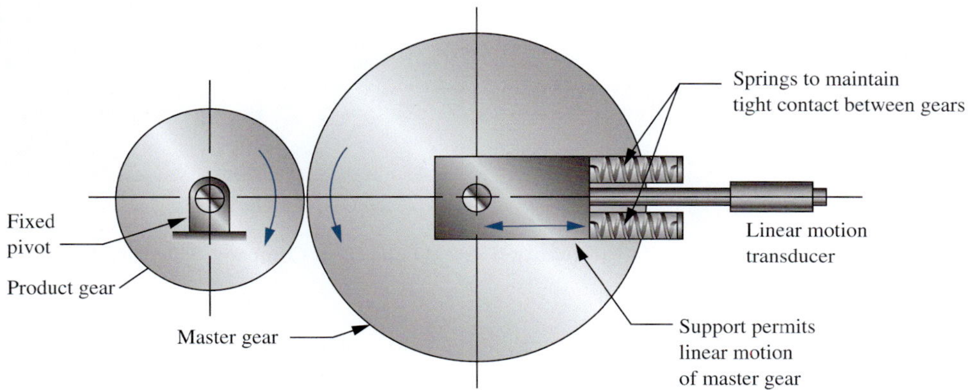

(*b*) Schematic diagram of double flank roll tester

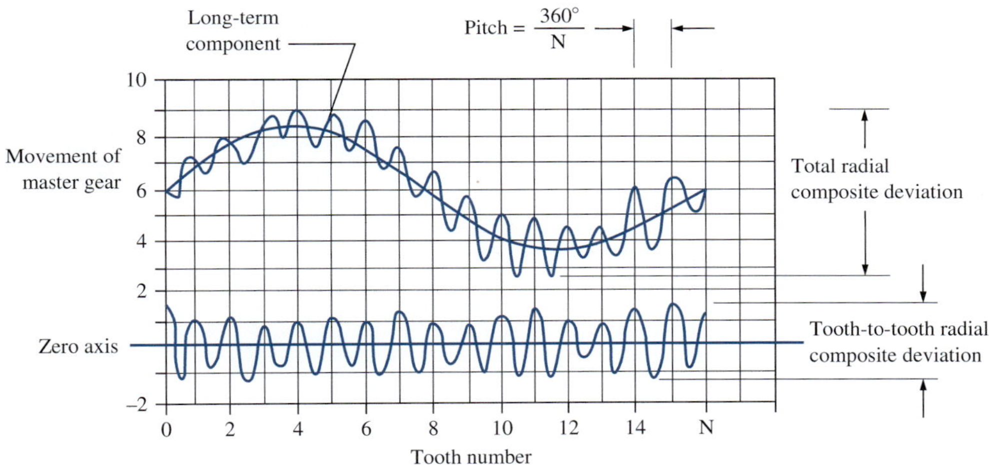

(*c*) Typical chart of roll tester output

FIGURE 9–8 Radial composite deviation testing of gears (Process Equipment Company, Tipp City, Ohio)

Tooth alignment: The deviation of the actual line on the gear tooth surface at the pitch circle from the theoretical line. Measurements are made across the face from one end to the other. For a spur gear, the theoretical line is straight. For a helical gear, it is a part of a helix. Measurement of tooth alignment is sometimes called the *helix* measurement. It is important because excessive misalignment causes nonuniform loading on the gear teeth.

Tooth profile: The measurement of the actual profile of the surface of a gear tooth from the point of the start of

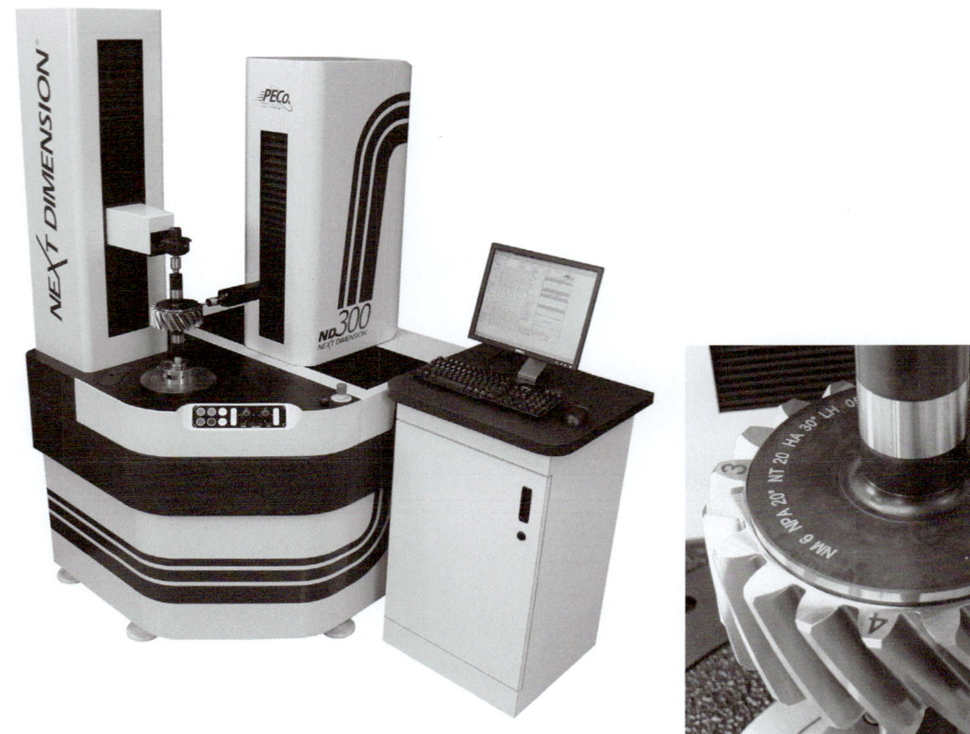

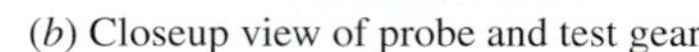

(*a*) Overall view

(*b*) Closeup view of probe and test gear

FIGURE 9–9 Analytical measurement system for gear quality (Process Equipment Company, Tipp City, Ohio)

the active profile to the tip of the tooth. The theoretical profile is a true involute curve. Variations of the actual profile from the theoretical profile cause variations in the instantaneous velocity ratio between the two gears in mesh, affecting the smoothness of the motion and noise. Figure 9–10(b) shows a sample output for profile variation.

Root radius: The radius of the fillet at the base of the tooth. Variations from the theoretical value can affect the meshing of mating gears, creating possible

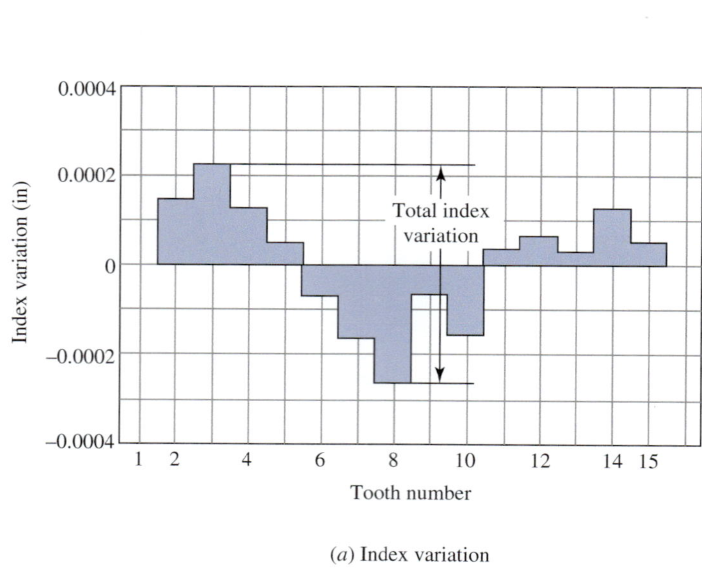

(*a*) Index variation

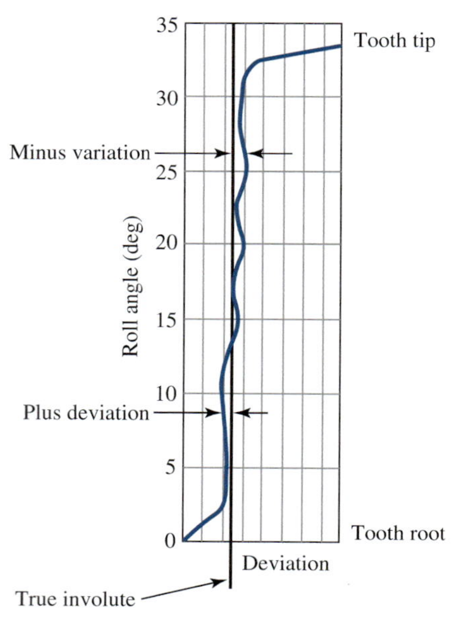

(*b*) Profile variation

FIGURE 9–10 Typical output charts from an analytical measurement system (Process Equipment Company, Tipp City, Ohio)

interference, and the stress concentration factors related to bending stress in the tooth.

Runout: A measure of the eccentricity and out of roundness of a gear. Excessive runout causes the contact point on mating gear teeth to move radially during each revolution.

As the analytical measurement system completes its tests, comparisons are automatically made with the theoretical tooth forms and with tolerance values from the applicable standard (described next) to report the resulting *quality number*. The data are also useful to manufacturing staff for making adjustments to gear cutting and grinding machine control settings to improve the accuracy of the total process.

Using the general capabilities of the analytical measurement system, dimensions of features other than those of the gear teeth may also be determined while the gear is in its fixture. For example, when a gear is machined onto a shaft, diameters, keyseats, shoulder fillets, and other geometric features may be checked for dimensions, perpendicularity, parallelism, and concentricity. Gear segments, composite gears having two or more gears on the same shaft, splines, and cam surfaces can also be measured.

Standards for Gear Quality

Manufacturers, designers, and users of gears must agree on standards used to determine the acceptability of gears produced. Several standard-setting organizations exist, including the AGMA (United States), ISO (International), DIN (German), JIS (Japan), and others. While the standards are similar, they are not identical. Designers and manufacturers should become familiar with the provisions of these standards in order to communicate effectively the intent of their designs, the capability of manufacturing processes, and the acceptability of a given gear.

As indicated in Table 9–1, the current AGMA standard 2015-1-A01 (Reference 15), employs ten accuracy grades from A2 (most precise) to A11 (least precise). An analytical measurement system [Figure 9–9 (a)] performs a total of nine elemental tangential measurements required for high-accuracy gears (A2–A5); five measurements for medium accuracy gears (A6–A9); and three for low accuracy gears (A10–A11).

The AGMA 2015-1-A01 standard replaces standard AGMA 2000-A88 that was used in the previous edition of this book and for which there are many gears in service that were made to that standard. It also replaces ISO 1328-1 that had been an interim standard modeled on the ISO methodology. For reference, Table 9–2 shows the rough relationships among the three standards although the actual values of tolerances are not identical. One notable difference is that the former AGMA 2000 standard employed classifications from Q5 (least precise) to Q15 (most precise); the order is opposite from AGMA 2015. A convenient comparison can be made by noting that the sum of the quality numbers from AGMA 2015 and AGMA 2000 is always 17. For example, Q12 and A5 are closely equivalent.

TABLE 9–1 Accuracy Groups for AGMA 2015 Gear Quality System

Low (L)	A10–A11
Medium (M)	A6–A9
High (H)	A2–A5

TABLE 9–2 General Corelations of AGMA 2000, AGMA 2015, and ISO 1328 Gear Quality Systems

AGMA 2000	AGMA 2015	ISO 1328	AGMA 2000	AGMA 2015	ISO 1328
—Least precise—					
Q5	—	12	Q11	A6	6
Q6	A11	11	Q12	A5	5
Q7	A10	10	Q13	A4	4
Q8	A9	9	Q14	A3	3
Q9	A8	8	Q15	A2	2
Q10	A7	7	—Most precise—		

In this book, we use the AGMA 2015-1-A01 standard. Its classifications of A2 to A11 are keyed to a term called the *dynamic factor, K_v*, introduced later in this chapter. Note that the actual tables of data for tolerances are listed in terms of the metric module system with tolerance values given in micrometers (μm). Be aware that the tolerance values are quite small. For tolerance grade A5, tolerances are in the order of around 6.0 μm (0.00024 in) for single pitch deviation for small gears (100 mm; 3.94 in) with small teeth ($m = 1$ mm; $P_d = 25.4$) to about 22 μm (0.00088 in) for larger gears (800 mm; 31.5 in) with large teeth ($m = 50$ mm; $P_d = 0.51$). Gears are *very precise* mechanical components.

For some applications, often for those of the low accuracy types, the parties involved may agree to use another part of the standard, AGMA 2015-2-A06 for radial measurements based on the total radial composite deviation and the tooth-to-tooth radial composite deviation as shown in Figure 9–8. For this standard, nine classifications are called C4 (most precise) through C12 (least precise). The *double flank roll tester*, described earlier, is used to perform the testing.

Recommended Quality Numbers

All gear-type machinery should be manufactured with good levels of precision reflecting the accuracy with which the gears must operate for good performance, long life, smooth operation, and low noise. The design of the entire system, including shafts, bearings, and the housing must be consistent with the expected degree of precision. Of course, the system should not be made more precise than necessary because

TABLE 9–3 Recommended AGMA Quality Numbers

Application	Quality number	Application	Quality number
Cement mixer drum drive	A11	Small power drill	A9
Cement kiln	A11	Clothes washing machine	A8
Steel mill drives	A11	Printing press	A7
Grain harvester	A10	Computing mechanism	A6
Cranes	A10	Automotive transmission	A6
Punch press	A10	Radar antenna drive	A5
Mining conveyor	A10	Marine propulsion drive	A5
Paper-box-making machine	A9	Aircraft engine drive	A4
Gas meter mechanism	A9	Gyroscope	A2

Machine tool drives and drives for other high-quality mechanical systems

Pitch line speed (fpm)	Quality number	Pitch line speed (m/s)
0–800	A10	0–4
800–2000	A8	4–11
2000–4000	A6	11–22
Over 4000	A4	Over 22

of cost. For this reason, manufacturers recommend quality numbers that will give satisfactory performance at a reasonable cost for a variety of applications. Table 9–3 lists several examples.

Machine tools, such as lathes, machining centers, and grinders, are included in the lower part of Table 9–3 with the recommended quality numbers keyed to the pitch line speed of the gears, as defined in Equations (9–1) and (9–2). Higher speeds require greater accuracy. ***These values should be used for other kinds of precision industrial equipment not listed in the first part of the table.***

9–6 ALLOWABLE STRESS NUMBERS

Later in this chapter, design procedures are presented in which two forms of gear-tooth failure are considered.

A gear tooth acts like a cantilever beam in resisting the force exerted on it by the mating tooth. The point of highest tensile bending stress is at the root of the tooth where the involute curve blends with the fillet. The AGMA has developed a set of *allowable bending stress numbers*, called s_{at}, which are compared to computed bending stress levels in the tooth to rate the acceptability of a design.

A second, independent form of failure is the pitting of the surface of the teeth, usually near the pitch line, where high contact stresses occur. The transfer of force from the driving to the driven tooth theoretically occurs across a line contact because of the action of two convex curves on each other. Repeated application of these high contact stresses can cause a type of fatigue failure of the surface, resulting in local fractures and an actual loss of material. This is called *pitting*. The AGMA has developed a set of *allowable contact stress numbers*, called s_{ac}, which are compared to computed contact stress levels in the tooth to rate the acceptability of a design.

Representative data for s_{at} and s_{ac} are given in the following section for general information and use in problems in this book. More extensive data are given in the AGMA standards listed at the end of the chapter. (See References 6, 8, and 9.) Those documents should be consulted for details beyond the discussion in this book.

9–7 METALLIC GEAR MATERIALS

Gears can be made from a wide variety of materials to achieve properties appropriate to the application. From a mechanical design standpoint, strength and pitting resistance are the most important properties. But, in general, the designer should consider the producibility of the gear, taking into account all of the manufacturing processes involved, from the preparation of the gear blank, through the forming of the gear teeth, to the final assembly of the gear into a machine. Other considerations are weight, appearance, corrosion resistance, noise, and, of course, cost. This section discusses several types of metals used for gears. Plastics are covered in a later section.

TABLE 9–4 Selection of Gear Materials

Heat treatment	Typical alloys (SAE numbers)
Through-hardened or Case-hardened by flame or induction hardening	1045, 4140, 4150, 4340, 4350
Carburizing, case-hardened	1020, 4118, 4320, 4820, 8620, 9310

Many carbon and alloy steels are used for gears and most are given heat treatments to provide controlled hardness and strength (References 8 and 28). Selection of the alloy depends, in part, on the heat-treatment process used to achieve the final properties. Table 9–4 gives some examples.

It is important to recognize that alloys listed for through-hardening, flame, or induction hardening must have good hardenability to achieve the desired hardness levels described in the following section. They must also retain reasonable ductility indicated by the percent elongation in the heat-treated condition. All those listed are medium-carbon steels.

Steels that are to be carburized are typically low carbon steels because the carburization process itself infuses significant amounts of carbon into the case while the steel is at a high temperature. Then the final case hardness is achieved by quenching and tempering, leaving the core of the teeth at a lower strength but with higher ductility.

Steel Gear Materials

Through-Hardened Steels. Gears for machine tool drives and many kinds of medium- to heavy-duty speed reducers and transmissions are typically made from medium-carbon steels. AGMA Standard 2001-D04 gives data for the allowable bending stress number, s_{at}, and the allowable contact stress number, s_{ac}, for steels in the through-hardened condition (Reference 6). Figures 9–11 and 9–12 are graphs relating the stress numbers to the Brinell hardness number, HB, for the teeth. Notice that only knowledge of the hardness is required because of the direct relationship between hardness and the tensile strength of steels. See Appendix 17 for data that correlate the Brinell hardness number, with the tensile strength of steel in ksi. The range of hardnesses covered by the AGMA data is from 180 to 400 HB, corresponding to a tensile strength of approximately 87 to 200 ksi. ***It is not recommended to use through-hardening above 400 HB because of inconsistent performance of the gears in service.*** Typically, case hardening is used when there is a desire to achieve a surface hardness above 400 HB. This is described later in this section.

The hardness measurement for the allowable bending stress number is to be taken at the root of the teeth because that is where the highest bending stress occurs. The allowable contact stress number is related to the surface hardness on the face of the gear teeth where the mating teeth experience high contact stresses.

When selecting a material for gears, the designer must specify one that can be hardened to the desired hardness. Review Chapter 2 for discussions about heat-treatment techniques. Consult Appendices 3 and 4 for representative data. For the higher hardnesses, say, above 250 HB, a medium-carbon-alloy steel with good hardenability is desirable. Examples are, listed in Table 9–4. Ductility is also rather important because of the numerous cycles of stress experienced by gear teeth and the likelihood of occasional overloads, impact, or shock loading. A percent elongation value of 12% or higher is desired.

The curves in Figures 9–11 and 9–12 include two grades of steel: Grade 1 and Grade 2. ***Grade 1 is considered to be the basic standard and will be used for problem solutions in this book.*** Grade 2 requires a higher degree of control of the microstructure, alloy composition, greater cleanliness, prior heat treatment, nondestructive testing performed, core hardness values, and other factors. See AGMA Standard 2001-D04 (Reference 6) for details.

Case-Hardened Steels. Flame hardening, induction hardening, carburizing, and nitriding are processes used to produce a high hardness in the surface layer of gear teeth. See the related discussion in Section 2–6. These processes provide surface hardness values from 50 to 64 HRC (Rockwell C) and correspondingly high values of s_{at} and s_{ac}, as shown in Table 9–5. Special discussions are given below for each of the types of case-hardening processes.

Case-hardened steel gears can be produced to Grades 1, 2, and 3 with Grade 1 referring to typically available steels as discussed for through-hardened steels. Because of the special care required in producing Grades 2 and 3, Table 9–5 shows data for only Grade 1 steels as recommended for use in this book. Furthermore, because nitriding can be done in several ways and it is not used as frequently as carburizing, flame-, or induction hardening, data for design values of bending or pitting resistance stresses are not listed here. Consult Reference 6 for the necessary data.

Flame- and Induction-Hardened Gear Teeth. Recall that these processes involve the local heating of the surface of the gear teeth by high-temperature gas flames or electrical induction coils. By controlling the time and energy input, the manufacturer can control the depth of heating and the depth of the resulting case. It is essential that the heating occur around the entire tooth, producing the hard case on the face of the teeth *and in the fillet and root areas*, in order to use the stress values listed in Table 9–5. This may require a special design for the flame shape or the induction heater. Refer to Reference 6.

The specifications for flame- or induction-hardened steel gear teeth call for a resulting hardness of HRC 50 to 54. Because these processes rely on the inherent hardenability of the steels, you must specify a material that can be hardened to these levels. Normally, medium-carbon-alloy steels (approximately 0.40% to 0.60% carbon) are specified. Table 9–4 and Appendices 3 and 4 list some suitable materials.

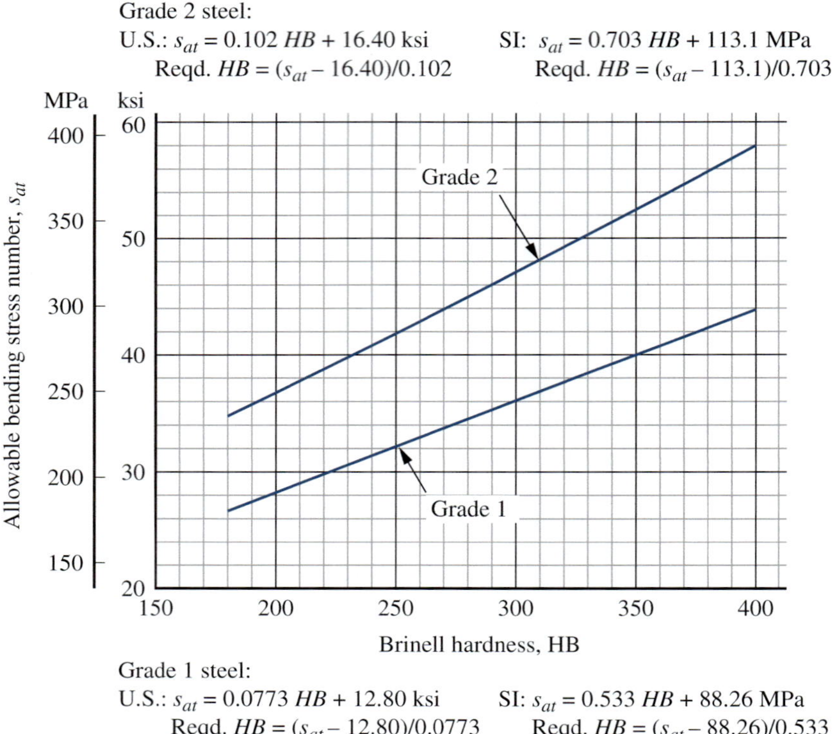

FIGURE 9–11 Allowable bending stress number for through-hardened steel gears, s_{at} (Extracted from AGMA 2001-D04 Standard, *Fundamental Rating Factors and Calculation Methods for Involute Spur and Helical Gear Teeth*, with permission of the publisher, American Gear Manufacturers Association, 1001 North Fairfax Street, 5th floor, Alexandria, VA 22314)

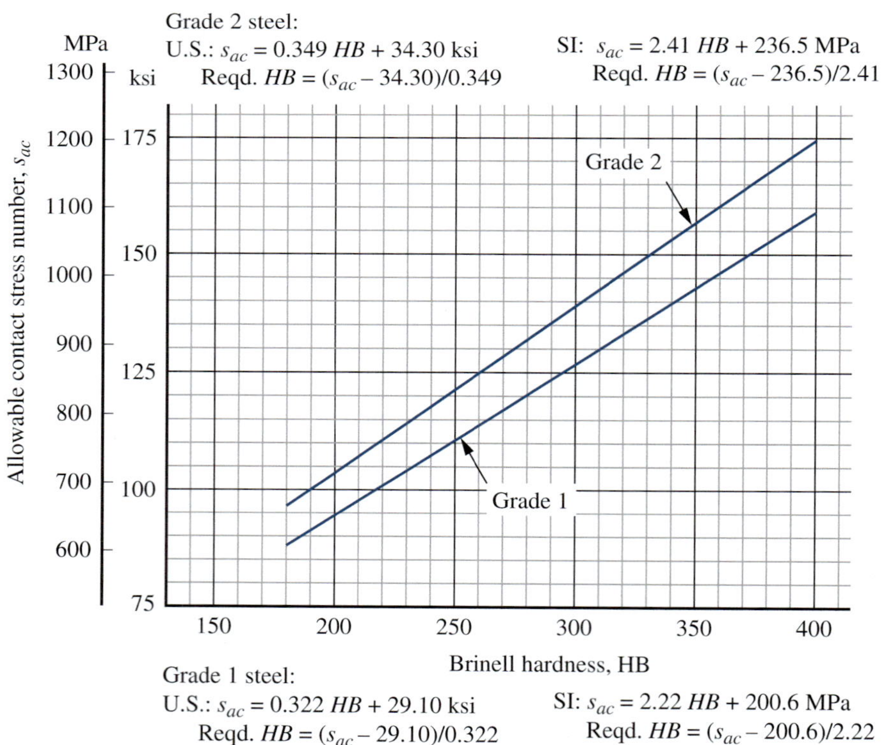

FIGURE 9–12 Allowable contact stress number for through-hardened steel gears, s_{ac} (Extracted from AGMA 2001-D04 Standard, *Fundamental Rating Factors and Calculation Methods for Involute Spur and Helical Gear Teeth*, with permission of the publisher, American Gear Manufacturers Association, 1001 North Fairfax Street, 5th floor, Alexandria, VA 22314)

TABLE 9–5 Allowable Stress Numbers for Case-Hardened Grade 1 Steel Materials

Hardness at surface	Allowable bending stress number, s_{at} (ksi)	(Mpa)	Allowable contact stress number, s_{ac} (ksi)	(Mpa)
Flame- or induction-hardened				
50 HRC	45	310	170	1172
54 HRC	45	310	175	1207
Carburized and case-hardened				
55–64 HRC	55	379	180	1241

Source: Extracted from AGMA Standard 2001-D04, *Fundamental Rating Factors and Calculation Methods for Involute Spur and Helical Gear Teeth*, with the permission of the publisher, American Gear Manufacturers Association, 1001 North Fairfax Street, 5th floor, Alexandria, VA 22314.

Carburizing. Carburizing produces surface hardnesses in the range of 55 to 64 HRC. It results in some of the highest strengths in common use for gears. Special carburizing steels are listed in Appendix 5. Figure 9–13 shows the AGMA recommendation for the thickness of the case for carburized gear teeth. The effective case depth is defined as the depth from the surface to the point where the hardness has reached 50 HRC.

Nitriding. Nitriding produces a very hard *but very thin* case. It is specified for applications in which loads are smooth and well known. Nitriding should be avoided when overloading or shock can be experienced, because the case is not sufficiently strong or well supported to resist such loads. Because of the thin case, the Rockwell 15N scale is used to specify hardness. See Reference 6 for design data for nitrided glass.

Iron and Bronze Gear Materials

Cast Irons. Two types of iron used for gears are *gray cast iron* and *ductile* (sometimes called *nodular*) iron. Table 9–6 gives the common ASTM grades used, with their corresponding allowable bending stress numbers and contact stress numbers. Remember that gray cast iron is brittle, so care should be exercised when shock loading is possible. Also, the higher-strength forms of the other irons have low ductility. Austempered ductile iron (ADI) is being used in some important automotive applications. However, standardized allowable stress numbers have not yet been specified.

Bronzes. Four families of bronzes are typically used for gears: (1) phosphor or tin bronze, (2) manganese bronze, (3) aluminum bronze, and (4) silicon bronze. Yellow brass is also used. Most bronzes are cast, but some are available in

TABLE 9–6 Allowable Stress Numbers for Iron and Bronze Gears

Material designation	Minimum hardness at surface (HB)	Allowable bending stress number, s_{at} (ksi)	(MPa)	Allowable contact stress number, s_{ac} (ksi)	(MPa)
Gray cast iron, ASTM A48, as cast					
Class 20		5	35	50	345
Class 30	174	8.5	59	65	448
Class 40	201	13	90	75	517
Ductile (nodular) iron, ASTM A536					
60-40-18 annealed	140	22	152	77	530
80-55-06 quenched and tempered	179	22	152	77	530
100-70-03 quenched and tempered	229	27	186	92	634
120-90-02 quenched and tempered	269	31	214	103	710
Bronze, sand-cast, $s_{u\ min}$ = 40 ksi (275 MPa)		5.7	39	30	207
Bronze, heat-treated, $s_{u\ min}$ = 90 ksi (620 MPa)		23.6	163	65	448

Source: Extracted from AGMA Standard 2001-D04, *Fundamental Rating Factors and Calculation Methods for Involute Spur and Helical Gear Teeth*, with the permission of the publisher, American Gear Manufacturers Association, 1001 North Fairfax Street, 5th floor, Alexandria, VA 22314.

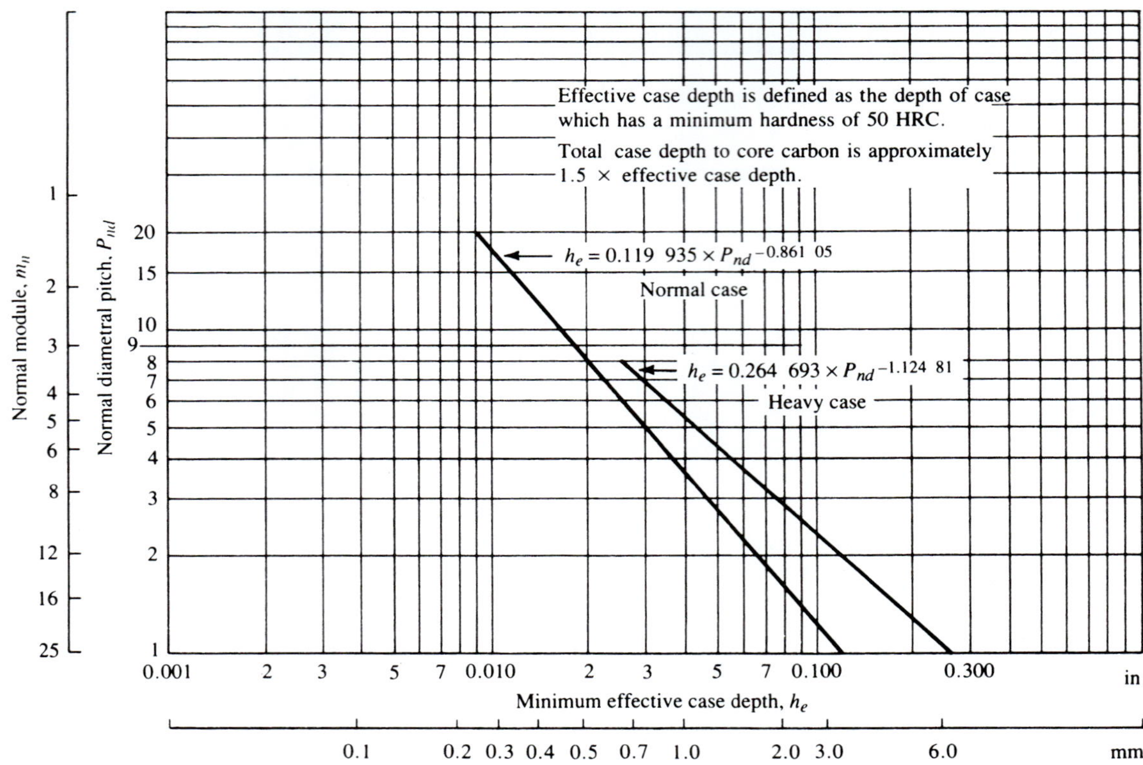

FIGURE 9–13 Effective case depth for carburized gears, h_e (Extracted from AGMA 2001-D04 Standard, *Fundamental Rating Factors and Calculation Methods for Involute Spur and Helical Gear Teeth*, with permission of the publisher, American Gear Manufacturers Association, 1001 North Fairfax Street, 5th floor, Alexandria, VA 22314)

wrought form. Corrosion resistance, good wear properties, and low friction coefficients are some reasons for choosing bronzes for gears. Table 9–6 shows allowable stress numbers for one bronze alloy in two common forms.

9–8 BENDING STRESS IN GEAR TEETH

The stress analysis of gear teeth is facilitated by consideration of the orthogonal force components, W_t and W_r, as shown in Figure 9–3.

The tangential force, W_t, produces a bending moment on the gear tooth similar to that on a cantilever beam. The resulting bending stress is maximum at the base of the tooth in the fillet that joins the involute profile to the bottom of the tooth space. Taking the detailed geometry of the tooth into account, Wilfred Lewis developed the equation for the stress at the base of the involute profile, which is now called the *Lewis equation*:

Lewis Equation for Bending Stress In Gear Teeth

$$\sigma_t = \frac{W_t P_d}{FY} \qquad (9\text{–}13)$$

where W_t = tangential force

P_d = diametral pitch of the tooth

F = face width of the tooth

Y = *Lewis form factor*, which depends on the tooth form, the pressure angle, the diametral pitch, the number of teeth in the gear, and the place where W_t acts

While the theoretical basis for the stress analysis of gear teeth is presented, the Lewis equation must be modified for practical design and analysis. One important limitation is that it does not take into account the stress concentration that exists in the fillet of the tooth. Figure 9–14 is a photograph of a photoelastic stress analysis of a model of a gear tooth. It indicates a stress concentration in the fillet at the root of the tooth as well as high contact stresses at the mating surface (contact stress is discussed in the following section). Comparing the actual stress at the root with that predicted by the Lewis equation enables us to determine the stress concentration factor, K_t, for the fillet area. Placing this into equation (9–13) gives

$$\sigma_t = \frac{W_t P_d K_t}{FY} \qquad (9\text{–}14)$$

The value of the stress concentration factor is dependent on the form of the tooth, the shape and size of the fillet at the root of the tooth, and the point of application of the force on the tooth. Note that the value of the Lewis form factor, Y, also depends on the tooth geometry. Therefore, the two

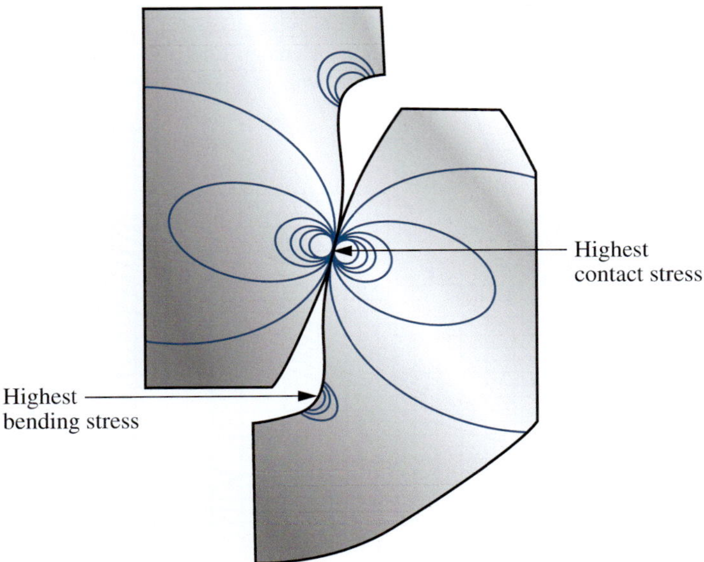

FIGURE 9–14 Photoelastic study of gear teeth under load

factors are combined into one term, the *geometry factor, J,* where $J = Y/K_t$. The value of J also, of course, varies with the location of the point of application of the force on the tooth because Y and K_t vary.

Figure 9–15 shows graphs giving the values for the geometry factor for 20° and 25°, full-depth, involute teeth. The safest value to use is the one for the load applied at the tip of the tooth. However, this value is overly conservative because there is some load sharing by another tooth at the time that the load is initially applied at the tip of a tooth. The critical load on a given tooth occurs when the load is at the highest point of single-tooth contact, when the tooth carries the entire load. The upper curves in Figure 9–15 give the values for J for this condition.

Using the geometry factor, J, in the stress equation gives

$$\sigma_t = \frac{W_t P_d}{FJ} \tag{9–15}$$

The graphs in Figure 9–15 are taken from the former AGMA Standard 218.01 which has been superseded by the two new standards: AGMA 2001-D04, *Fundamental Rating Factors and Calculation Methods for Involute Spur and Helical Gear Teeth,* 1995, and AGMA 908-B89 (R1999), *Geometry Factors for Determining the Pitting Resistance and Bending Strength of Spur, Helical and Herringbone Gear Teeth,* 1999. Standard 908-B89 includes an analytical method for calculating the geometry factor, J. But the values for J are unchanged from those in the former standard. Rather than graphs, the new standard reports values for J for a variety of tooth forms in tables. The graphs from the former standard are shown in Figure 9–15 so that you can visualize the variation of J with the number of teeth in the pinion and the gear.

Note also that J factors for only two tooth forms are included in Figure 9–15 and that the values are valid only for those forms. Designers must ensure that J factors for the tooth form actually used, including the form of the fillet, are included in the stress analysis.

Equation (9–15) can be called the *modified Lewis equation.* Other modifications to the equation are recommended by the AGMA in Standard 2001-D04 for practical design to account for the variety of conditions that can be encountered in service.

The approach used by the AGMA is to apply a series of additional modifying factors to the bending stress from the modified Lewis equation to compute a value called the *bending stress number, s_t.* These factors represent the degree to which the actual loading case differs from the theoretical basis of the Lewis equation. The result is a better estimate of the real level of bending stress that is produced in the teeth of the gear and the pinion.

Then, separately, the allowable bending stress number, s_{at}, is modified by a series of factors that affect that value when the environment is different from the nominal situation assumed when the values for s_{at} are set. The result here is a better estimate of the real level of the bending strength of the material from which the gear or the pinion is made.

The design is completed in a manner that ensures that the bending stress number is less than the modified allowable bending stress number. This process should be completed for both the pinion and the gear of a given pair because materials may be different; the geometry factor, J, is different; and other operating conditions may be different. This is demonstrated in example problems later in this chapter.

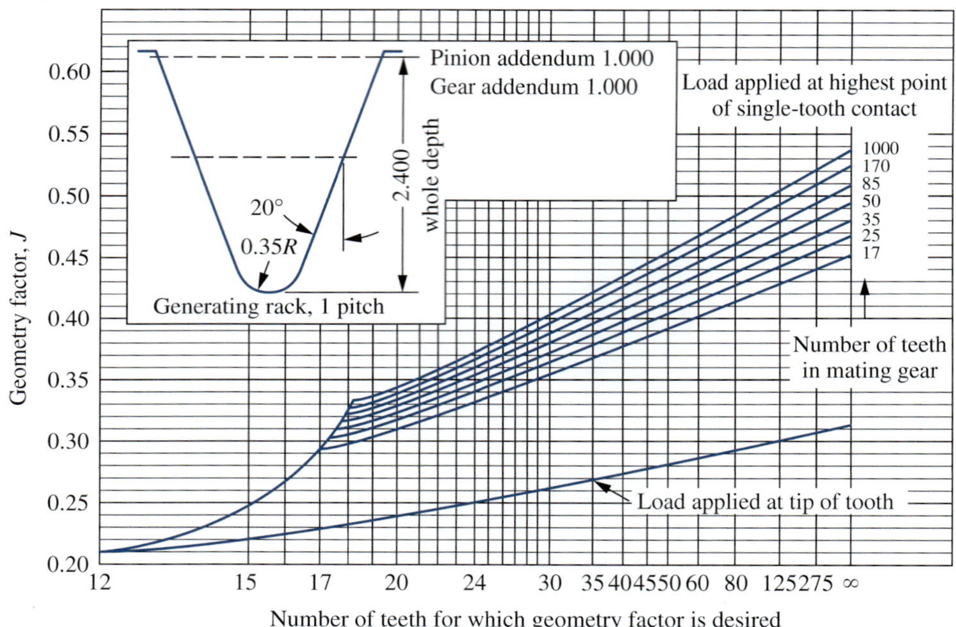

(a) 20° spur gear: standard addendum

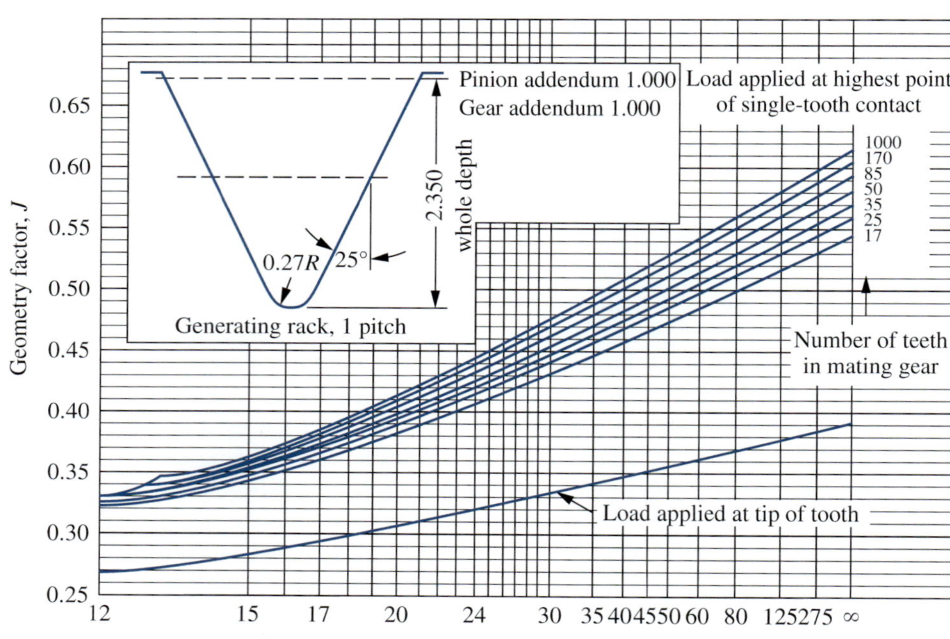

(b) 25° spur gear: standard addendum

FIGURE 9–15 Geometry factor, J (Extracted from AGMA 218.01 Standard, *Rating the Pitting Resistance and Bending Strength of Spur and Helical Involute Gear Teeth,* with the permission of the publisher, American Gear Manufacturers Association, 1001 North Fairfax Street, 5th floor, Alexandria, VA 22314)

Often the major decision to be made is the specification of suitable materials from which to make the pinion and the gear. In such cases, the required basic allowable bending stress number, s_{at}, will be computed. When steel is used, the required hardness of the material is found from the data described in Section 9–6. Finally, the material and its heat treatment are specified to ensure that it will have at least the required hardness.

We proceed now with the discussion of the bending stress number, s_t.

Bending Stress Number, s_t

The design analysis method used here is based primarily on AGMA Standard 2001-D04. However, because values for some of the factors are not included in the standard, data from other sources are added. These data illustrate the kinds of conditions that affect the final design. The designer ultimately has the responsibility for making appropriate design decisions.

The following equation will be used in this book:

Bending Stress Number, s_t

$$s_t = \frac{W_t P_d}{FJ} K_o K_s K_m K_B K_v \qquad (9\text{–}16)$$

where K_o = overload factor for bending strength
K_s = size factor for bending strength
K_m = load-distrubution factor for bending strength
K_B = rim thickness factor
K_v = dynamic factor for bending strength

Methods for specifying values for these factors are discussed below.

Overload Factor, K_o

Overload factors consider the probability that load variations, vibrations, shock, speed changes, and other application-specific conditions may result in peak loads greater than W_t being applied to the gear teeth during operation. A careful analysis of actual conditions should be made, and the AGMA Standard 2001-D04 gives no specific values for K_o. Reference 20 gives some recommended values, and many industries have established suitable values based on experience.

For problem solutions in this book, we will use the values shown in Table 9–7. The primary considerations are the nature of *both* the driving power source and the driven machine. An overload factor of 1.00 would be applied for a perfectly smooth electric motor driving a perfectly smooth generator through a gear-type speed reducer. Any rougher conditions call for a value of K_o greater than 1.00. For power sources we will use the following:

Uniform: Electric motor or constant-speed gas turbine

Light shock: Water turbine, variable-speed drive

Moderate shock: Multicylinder engine

Examples of the roughness of driven machines include the following:

Uniform: Continuous-duty generator, paper and film winders

Light shock: Fans and low-speed centrifugal pumps, liquid agitators, variable-duty generators, uniformly loaded conveyors, rotary positive displacement pumps, metal strip processing

Moderate shock: High-speed centrifugal pumps, reciprocating pumps and compressors, heavy-duty conveyors, machine tool drives, concrete mixers, textile machinery, meat grinders, saws, bucket elevators, freight elevators, escalators, concrete mixers, plastics molding and processing, sewage disposal equipment, winches, cable reels

Heavy shock: Rock crushers, punch press drives, pulverizers, processing mills, tumbling barrels, wood chippers, vibrating screens, railroad car dumpers, log conveyors, lumber handling equipment, metal shears, hammer mills, commercial washers, heavy-duty hoists and cranes, reciprocating feeders, dredges, rubber processing, compactors, plastics extruders

Size Factor, K_s

The AGMA indicates that the size factor can be taken to be 1.00 for most gears. But for gears with large-size teeth or large face widths, a value greater than 1.00 is recommended. Reference 20 recommends a value of 1.00 for diametral pitches of 5 or greater or for a metric module of 5 or smaller. For larger teeth, the values shown in Table 9–8 can be used.

Load-Distribution Factor, K_m

The determination of the load-distribution factor is based on many variables in the design of the gears themselves as well as in the shafts, bearings, housings, and the structure in which the gear drive is installed. Therefore, it is one of the most difficult factors to specify. Much analytical and experimental work is continuing on the determination of values for K_m.

If the intensity of loading on all parts of all teeth in contact at any given time were uniform, the value of K_m would be 1.00. However, this is seldom the case. Any of the following factors can cause misalignment of the teeth on the pinion relative to those on the gear:

1. Inaccurate gear teeth
2. Misalignment of the axes of shafts carrying gears
3. Elastic deformations of the gears, shafts, bearings, housings, and support structures
4. Clearances between the shafts and the gears, the shafts and the bearings, or the bearings and the housing
5. Thermal distortions during operation
6. Crowning or end relief of gear teeth

AGMA Standard 2001-D04 presents extensive discussions of methods of determining values for K_m. One is empirical and considers gears up to 40 in (1000 mm) wide. The other method is analytical and considers the stiffness and mass of individual gears and gear teeth and the total mismatch between mating teeth. We will not provide so much detail. However, rough guidelines are given below.

The designer can minimize the load-distribution factor by specifying the following:

1. Accurate teeth (a low quality number from AGMA 2015)
2. Narrow face widths
3. Gears centered between bearings (straddle mounting)
4. Short shaft spans between bearings
5. Large shaft diameters (high stiffness)

TABLE 9–7 Suggested Overload Factors, K_o

	Driven Machine			
Power source	Uniform	Light shock	Moderate shock	Heavy shock
Uniform	1.00	1.25	1.50	1.75
Light shock	1.20	1.40	1.75	2.25
Moderate shock	1.30	1.70	2.00	2.75

TABLE 9–8 Suggested Size Factors, K_s

Diametral pitch, P_d	Metric module, m	Size factor, K_s
≥ 5	≤ 5	1.00
4	6	1.05
3	8	1.15
2	12	1.25
1.25	20	1.40

6. Rigid, stiff housings
7. High precision and small clearances on all drive components

You are advised to study the details of AGMA Standard 2001-D04 which covers a wide range of physical sizes for gear systems. But the gear designs discussed in this book are of moderate size, typical of power transmissions in light industrial and vehicular applications. A more limited set of data are reported here to illustrate the concepts that must be considered in gear design.

We will use the following equation for computing the value of the load-distribution factor:

$$K_m = 1.0 + C_{pf} + C_{ma} \quad \textbf{(9–17)}$$

where C_{pf} = pinion proportion factor (see Figure 9–16)
C_{ma} = mesh alignement factor (see Figure 9–17)

In this book, we are limiting designs to those with face widths of 15 in or less. Wider face widths call for additional factors. Also, some commercially successful designs employ modifications to the basic tooth form to achieve a more

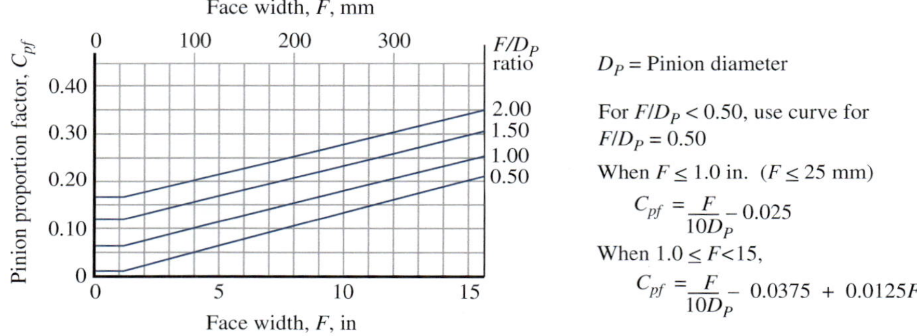

FIGURE 9–16 Pinion proportion factor, C_{pf} (Extracted from AGMA 2001-D04 Standard, *Fundamental Rating Factors and Calculation Methods for Involute Spur and Helical Gear Teeth,* with permission of the publisher, American Gear Manufacturers Association, 1001 North Fairfax Street, 5th floor, Alexandria, VA 22314)

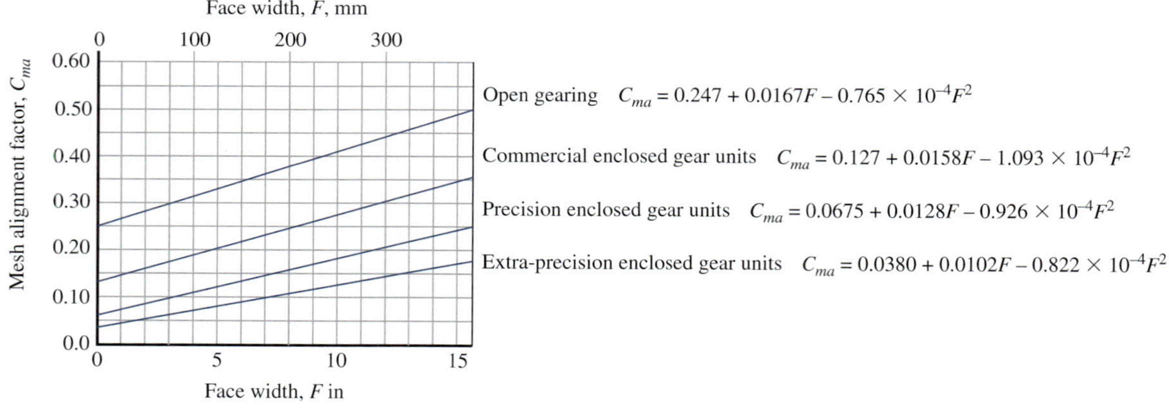

FIGURE 9–17 Mesh alignment factor, C_{ma} (Extracted from AGMA 2001-D04 Standard, *Fundamental Rating Factors and Calculation Methods for Involute Spur and Helical Gear Teeth,* with permission of the publisher, American Gear Manufacturers Association, 1001 North Fairfax Street, 5th floor, Alexandria, VA 22314)

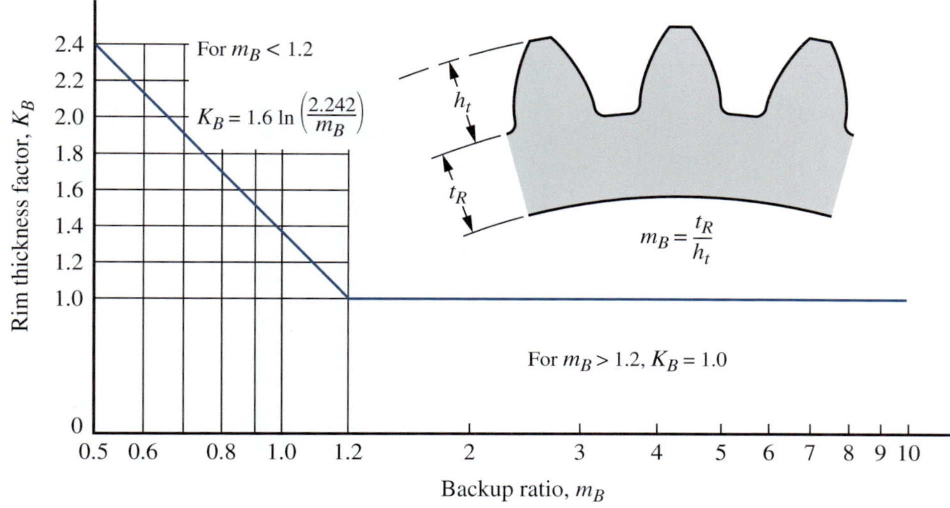

FIGURE 9–18 Rim thickness factor, K_B (Extracted from AGMA 2001-D04 Standard, *Fundamental Rating Factors and Calculation Methods for Involute Spur and Helical Gear Teeth,* with permission of the publisher, American Gear Manufacturers Association, 1001 North Fairfax Street, 5th floor, Alexandria, VA 22314)

uniform meshing of the teeth. Such methods are not discussed in this book.

Figure 9–16 shows that the pinion proportion factor is dependent on the actual face width of the pinion and on the ratio of the face width to the pinion pitch diameter. Figure 9–17 relates the mesh alignment factor to expected accuracy of different methods of applying gears.

- *Open gearing* refers to drive systems in which the shafts are supported in bearings that are mounted on structural elements of the machine with the expectation that relatively large misalignments will result.
- In *commercial-quality enclosed gear units*, the bearings are mounted in a specially designed housing that provides more rigidity than for open gearing, but for which the tolerances on individual dimensions are fairly loose.
- *Precision enclosed gear units* are made to tighter tolerances.
- *Extra-precision enclosed gear* units are made to exacting precision and are often adjusted at assembly to achieve excellent alignment of the gear teeth. Experience with similar units in the field will help you gain better understanding among the different types of designs. Commercial or precision are recommended for this book.

Rim Thickness Factor, K_B

The basic analysis used to develop the Lewis equation assumes that the gear tooth behaves as a cantilever attached to a perfectly rigid support structure at its base. This is true if the gear is made from a solid blank as shown in Figure 8–4(b) or with a blank that has a thinned web as shown in Part (d) of that figure. These are typical of small to medium-sized gears. Larger gears are often made with the spoked design shown in Figure 8–4(a) in order to save material and produce a lighter gear. Commercially made spoked gears can be expected to have a well-supported rim. For these kinds of gears, $K_B = 1.0$ can be used.

However, when designing a spoked gear for a special application, care must be exercised that the rim is sufficiently stiff to support the gear teeth without dangerous stresses created in the rim. Figure 9–18 should be used to estimate the influence of rim thickness. The key geometry parameter is called the *backup ratio*, m_B, where

$$m_B = t_R/h_t$$

$$t_R = \text{rim thickness}$$

$$h_t = \text{whole depth of the gear tooth}$$

For $m_B > 1.2$, the rim is sufficiently strong and stiff to support the tooth, and $K_B = 1.0$.

Special Case for Small Pinions Mounted to a Shaft with a Keyseat. A frequent design option is to mount a small pinion onto a shaft as shown in Figure 9–19, with a key used to transmit the torque from the pinion to the shaft, and requiring a keyway to be machined into the bore of the pinion and into the shaft. Care must be taken to ensure that there is sufficient material above the keyway. It is recommended that the condition of $m_B > 1.2$ be applied in such cases where the t_R, dimension is measured above the top of the keyway. Then the factor $K_B = 1.0$ can be used.

If it is impractical to provide $m_B > 1.2$, it is recommended that *the pinion be machined integral with the shaft,* thus eliminating the need for a key and then using $K_B = 1.0$.

Dynamic Factor, K_v

The dynamic factor accounts for the fact that the load is assumed by a tooth with some degree of impact and

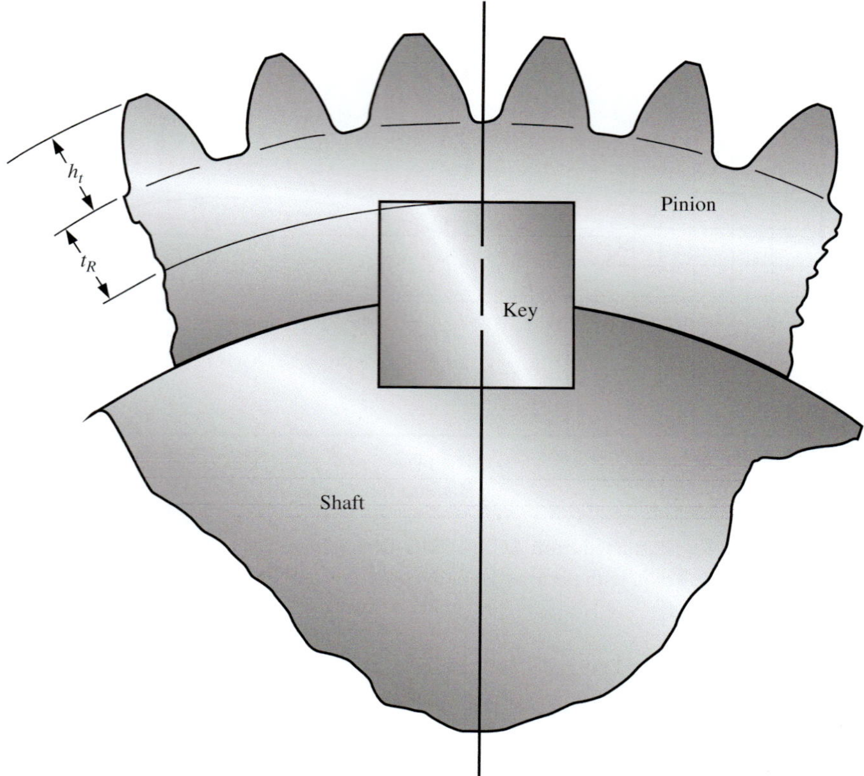

FIGURE 9–19 Pinion mounted on a shaft

that the actual load subjected to the tooth is higher than the transmitted load alone. The value of K_v depends on the accuracy of the tooth profile, the elastic properties of the tooth, and the speed with which the teeth come into contact.

Figure 9–20 shows a graph of the AGMA-recommended values for K_v, where the A_v numbers are the AGMA-quality numbers referred to earlier in Section 9–5. Gears in typical machine design would fall into the classes represented by curves 8 to 11 which are for gears made by hobbing or shaping with average to good tooling. If the teeth are finish-ground or shaved to improve the accuracy of the tooth profile and spacing, curves 6 or 7 should be used. Under special conditions where teeth of high precision are used in applications where there is little chance of developing external dynamic loads, the shaded area can be used ($A_v = 2 - 5$). If the teeth are cut by form milling, curve 12 or higher should be used. Note that the quality 12 gears should not be used at pitch line speeds above 3000 ft/min (15 m/s). Note that the dynamic factors are approximate. For severe applications, especially those operating above 4000 ft/min (20 m/s), approaches taking into account the material properties, the mass and inertia of the gears, and the actual error in the tooth form should be used to predict the dynamic load. (See References 6, 19, 20, 23, 25, and 27.)

TABLE 9–9 Analytical method for computing dynamic factor, K_v

U.S. units		**SI Metric units**	
Given A_v and v_t (ft/min)		Given A_v and v_t (m/s)	
$B = 0.25(A_v - 5.0)^{0.667}$		$B = 0.25(A_v - 5.0)^{0.667}$	
$C = 50 + 56(1.0 - B)$		$C = 3.5637 + 3.9914(1.0 - B)$	
$K_v = \left[\frac{C}{C + \sqrt{v_t}}\right]^{-B}$		$K_v = \left[\frac{C}{C + \sqrt{v_t}}\right]^{-B}$	
Maximum velocity for a given A_v:		Maximum velocity for a given A_v:	
$v_{t\,max} = [C + (14 - A_v)]^2$		$v_{t\,max} = [C + (14 - A_v)]^2$	
A_v	$v_{t\,max}$ (ft/min)	A_v	$v_{t\,max}$ (m/s)
6	10 000	6	50.8
7	8239	7	41.9
8	6867	8	34.9
9	5731	9	29.1
10	4767	10	24.2
11	3937	11	20.0
12	3219	12	16.4

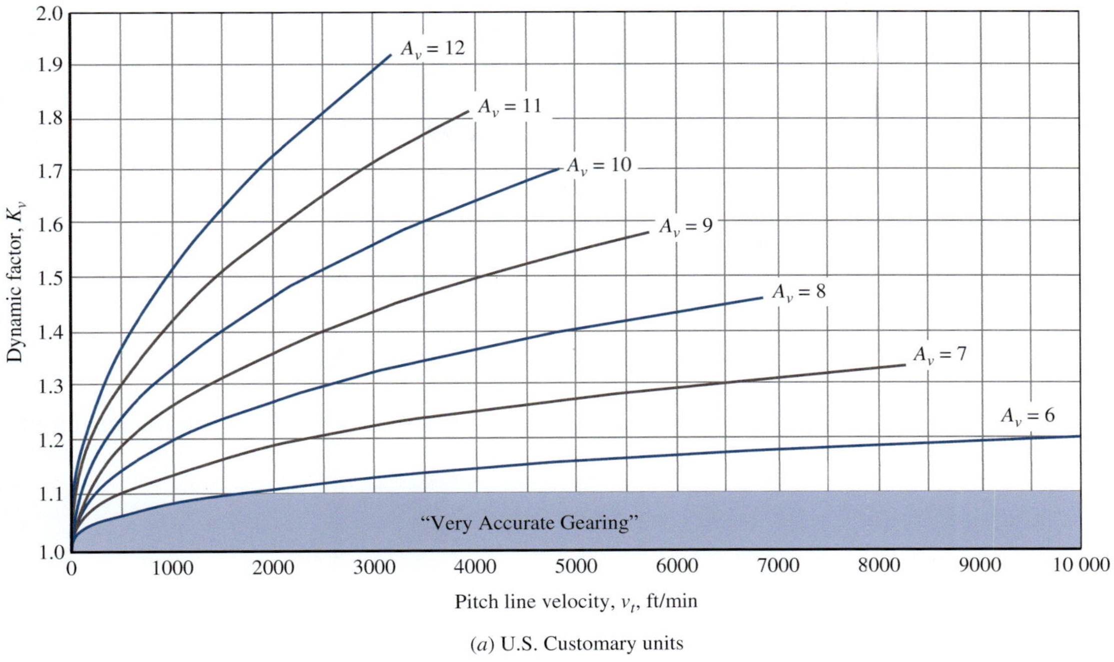

(*a*) U.S. Customary units

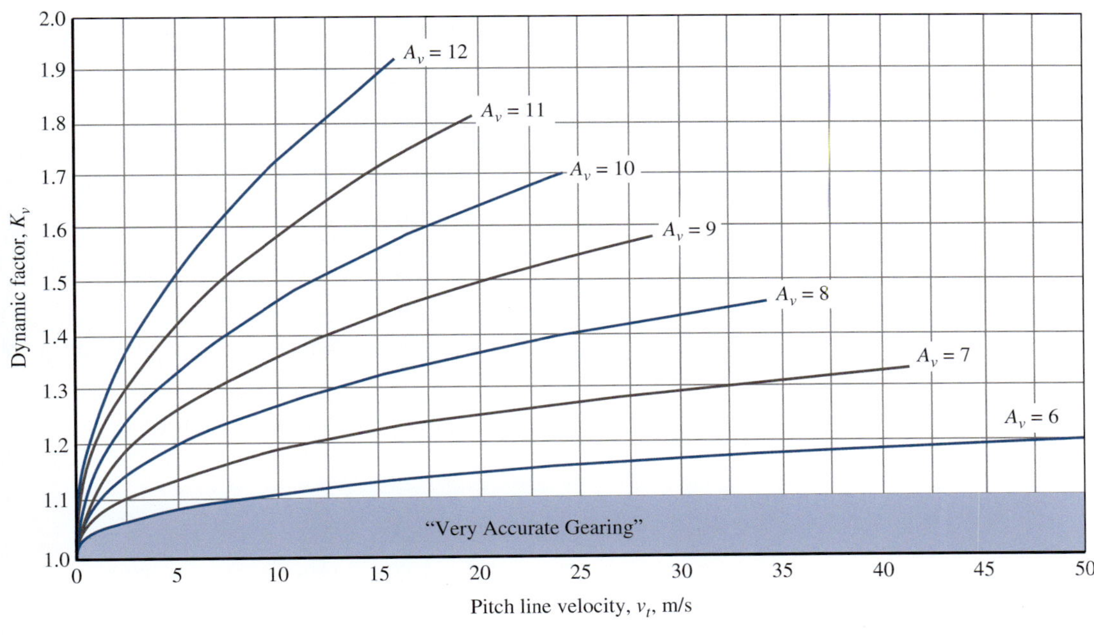

(*b*) SI Metric units

FIGURE 9–20 Dynamic factor, K_v (Adapted from AGMA 2001-D04 Standard, *Fundamental Rating Factors and Calculation Methods for Involute Spur and Helical Gear Teeth,* with the permission of the publisher, American Gear Manufacturers Association, 1001 North Fairfax Street, 5th floor, Alexandria, VA 22314)

Equations for Dynamic Factor, K_v. Reading values for K_v from either part of Figure 9–20 gives adequate accuracy because the charts are approximate and slight differences in reading should not cause difficulty with a gear design. However, it is recommended that an analytical method be used when performing several designs. Later in this chapter, a computer-aided approach using a spreadsheet is shown with the analytical method embedded. Table 9–9 shows equations adapted from Reference 6 and presented in both the U.S. and SI Metric unit systems. Variables involved are the dimensionless quality number (A_v) and the pitch line velocity in ft/min for the U.S. system and m/s for the SI Metric system. Both of these data are specified or computed in early stages of the gear design process. The method involves the calculation of two intermediate terms, B and C, and then computing K_v.

Example Problem 9–1

Compute the bending stress numbers for the pinion and the gear of the pair of gears in Figure 9–2. The pinion rotates at 1750 rpm, driven directly by an electric motor. The driven machine is an industrial saw requiring 25 hp. The gear unit is enclosed and is made to commercial standards. Gears are straddle-mounted between bearings. The following gear data apply:

$$N_P = 20 \quad N_G = 70 \quad P_d = 6 \quad F = 2.25 \text{ in} \quad A_v = 10$$

The gear teeth are 20°, full-depth, involute teeth, and the gear blanks are solid.

Solution

We will use Equation (9–16) to compute the expected stress:

$$s_t = \frac{W_t P_d}{FJ} K_o K_s K_m K_B K_v$$

We can first use the principles from Section 9–3 to compute the transmitted load on the gear teeth:

$$D_P = N_P/P_d = 20/6 = 3.333 \text{ in}$$

$$v_t = \pi D_P n_P/12 = \pi(3.333)(1750)/12 = 1527 \text{ ft/min}$$

$$W_t = 33\,000(P)/v_t = (33\,000)(25)/(1527) = 540 \text{ lb}$$

Other data that are useful to visualize the overall size of the gear pair and that serve as input to later decisions are computed here.

$$D_G = N_G/P_d = 70/6 = 11.667 \text{ in}$$

$$C = (N_P + N_G)/2P_d = (20 + 70)/[(2)(6)] = 7.500 \text{ in}$$

$$n_G = n_P(N_P/N_G) = 1750 \text{ rpm } (20/70) = 500 \text{ rpm}$$

$$\text{Velocity ratio} = VR = \text{Gear ratio} = m_G = N_G/N_P = 70/20 = 3.50$$

From Figure 9–15, we find that $J_P = 0.335$ and $J_G = 0.420$.

The overload factor is found from Table 9–7. For a smooth, uniform electric motor driving an industrial saw generating moderate shock, $K_o = 1.50$ is a reasonable value.

The size factor $K_s = 1.00$ because the gear teeth with $P_d = 6$ are relatively small. See Table 9–8.

The load-distribution factor, K_m, can be found from Equation (9–17) for commercial enclosed gear drives. For this design, $F = 2.25$ in, and

$$F/D_P = 2.25/3.333 = 0.68$$

$$C_{pf} = 0.058 \text{ (Figure 9–16)}$$

$$C_{ma} = 0.162 \text{ (Figure 9–17)}$$

$$K_m = 1.0 + C_{pf} + C_{ma} = 1.0 + 0.058 + 0.162 = 1.22$$

The rim thickness factor, K_B, can be taken as 1.00 because the gears are to be made from solid blanks.

The dynamic factor can be read from Figure 9–20. For $v_t = 1527$ ft/min and $A_v = 10$, $K_v = 1.41$.

The stress can now be computed from Equation (9–16). We will compute the stress in the pinion first:

$$s_{tP} = \frac{(540)(6)}{(2.25)(0.335)}(1.50)(1.0)(1.22)(1.0)(1.41) = 11\,091 \text{ psi}$$

Notice that all factors in the stress equation are the same for the gear except the value of the geometry factor, J. Then the stress in the gear can be computed from

$$s_{tG} = \sigma_{tP}(J_P/J_G) = (11\,091)(0.335/0.420) = 8847 \text{ psi}$$

The stress in the pinion teeth will always be higher than the stress in the gear teeth because the value of J increases as the number of teeth increases.

The magnitudes of the stresses due to bending as found in Example Problem 9–1 are then used to determine the strength and hardness properties of the material from which the gears will be made. Next we show the development of the methods for computing the contact stress that is created at the face of the teeth during power transmission. The bending strength and the resistance to contact stress must both be satisfied by the material selected. It is often found that the contact stress is the determining factor, particularly for gear drives used for long periods. For this reason, material selecion for both types of stress is discussed in Section 9–10.

9–9 CONTACT STRESS IN GEAR TEETH

In addition to being safe from bending, gear teeth must also be capable of operating for the desired life without significant pitting of the tooth form. *Pitting* is the phenomenon in which small particles are removed from the surface of the tooth faces because of the high contact stresses, causing fatigue. Refer again to Figure 9–14 showing the high, localized contact stresses. Prolonged operation after pitting begins causes the teeth to roughen, and eventually the form is deteriorated. Rapid failure follows. Note that both the driving and driven teeth are subjected to these high contact stresses equally. Reference 11 provides a comprehensive treatment of the wear and failure of gear teeth.

The action at the contact point on gear teeth is that of two externally curved surfaces. If the gear materials were infinitely rigid, the contact would be a simple line. Actually, because of the elasticity of the materials, the tooth shape deforms slightly, resulting in the transmitted force acting on a small rectangular area. The resulting stress is called a *contact stress* or *Hertz stress.* Reference 17 gives the following form of the equation for the Hertz stress,

➩ Hertz Contact Stress on Gear Teeth

$$\sigma_c = \sqrt{\frac{W_c}{F}\frac{1}{\pi\{[(1-\nu_1^2)/E_1] + [(1-\nu_2^2)/E_2]\}}\left(\frac{1}{r_1}+\frac{1}{r_2}\right)} \quad \text{(9–18)}$$

where the subscripts 1 and 2 refer to the materials of the two bodies in contact. The tensile modulus of elasticity is E and the Poisson's ratio is ν. W_c is the contact force exerted between the two bodies, and F is the length of the contacting surfaces. The radii of curvature of the two surfaces are called r_1 and r_2.

When Equation (9–18) is applied to gears, F is the face width of the gear teeth and W_c is the normal force delivered by the driving tooth on the driven tooth, found from Equation (9–12) to be,

$$W_N = W_t/\cos\phi \quad \text{(9–19)}$$

The second term in Equation (9–18) (including the square root) can be computed if the elastic properties of the materials for the pinion and gear are known. It is given the name *elastic coefficient, C_P.* That is,

➩ Elastic Coefficient

$$C_P = \sqrt{\frac{1}{\pi\{[(1-\nu_P^2)/E_P] + [(1-\nu_G^2)/E_G]\}}} \quad \text{(9–20)}$$

Table 9–10 gives values for the most common combinations of materials for pinions and gears.

The terms r_1 and r_2 are the radii of curvature of the involute tooth forms of the two mating teeth. These radii change continuously during the meshing cycle as the contact point moves from the top of the tooth through the pitch circle, and onto the lower flank of the tooth before leaving engagement. We can write the following equations for the radius of curvature when contact is at the pitch point,

$$r_1 = (D_P/2)\sin\phi \quad \text{and} \quad r_2 = (D_G/2)\sin\phi \quad \text{(9–21)}$$

However, the AGMA calls for the computation of the contact stress to be made at the lowest point of single tooth contact (LPSTC) because above that point, the load is being shared with other teeth. Computation of the radii of curvature for the LPSTC is somewhat complex. A geometry factor for pitting, I, is defined by the AGMA to include the radii of curvature terms and the cos ϕ term in Equation (9–19) because they all involve the specific geometry of the tooth. The variables required to compute I are the pressure angle ϕ, the gear ratio $m_G = N_G/N_P$, and the number of teeth in the pinion N_p.

Another factor in the contact stress equation is the pinion diameter that is not included in I. The contact stress equation then becomes,

$$\sigma_c = C_P\sqrt{\frac{W_t}{FD_PI}} \quad \text{(9–22)}$$

Values for the geometry factor, I, for a few common cases are graphed in Figure 9–21 and should be used for problem solving in this book. Appendix 18 provides an approach to computing the value for I for spur gears as given in Reference 3.

As with the equation for bending stress in gear teeth, several factors are added to the equation for contact stress as shown below. The resulting quantity is called the *contact stress number, s_c*:

➩ Contact Stress Number

$$s_c = C_P\sqrt{\frac{W_tK_oK_sK_mK_v}{FD_PI}} \quad \text{(9–23)}$$

The values for the overload factor, K_o; the size factor, K_s; the load-distribution factor, K_m; and the dynamic factor, K_v, can be taken to be the same as the corresponding values for the bending stress analysis in the preceding sections.

Equation 9–23 is used to compute the contact stress for both the pinion and the gear; they are equal. It is not correct to use the diameter of the gear, D_G, in this equation. Example Problem 9–2 follows to illustrate the use of this

TABLE 9–10 Elastic Coefficient, C_p

		Gear material and modulus of elasticity, E_G, lb/in² (MPa)					
Pinion material	Modulus of elasticity, E_P, lb/in² (MPa)	Steel 30×10^6 (2×10^5)	Malleable iron 25×10^6 (1.7×10^5)	Nodular iron 24×10^6 (1.7×10^5)	Cast iron 22×10^6 (1.5×10^5)	Aluminum bronze 17.5×10^6 (1.2×10^5)	Tin bronze 16×10^6 (1.1×10^5)
Steel	30×10^6	2300	2180	2160	2100	1950	1900
	(2×10^5)	(191)	(181)	(179)	(174)	(162)	(158)
Mall. iron	25×10^6	2180	2090	2070	2020	1900	1850
	(1.7×10^5)	(181)	(174)	(172)	(168)	(158)	(154)
Nod. iron	24×10^6	2160	2070	2050	2000	1880	1830
	(1.7×10^5)	(179)	(172)	(170)	(166)	(156)	(152)
Cast iron	22×10^6	2100	2020	2000	1960	1850	1800
	(1.5×10^5)	(174)	(168)	(166)	(163)	(154)	(149)
Al. bronze	1.75×10^6	1950	1900	1880	1850	1750	1700
	(1.2×10^5)	(162)	(158)	(156)	(154)	(145)	(141)
Tin bronze	16×10^6	1900	1850	1830	1800	1700	1650
	(1.1×10^5)	(158)	(154)	(152)	(149)	(141)	(137)

Source: Extracted from AGMA Standard 2001-D04, *Fundamental Rating Factors and Calculation Methods for Involute Spur and Helical Gear Teeth,* with the permission of the publisher, American Gear Manufacturers Association, 1001 North Fairfax Street, 5th floor, Alexandria, VA 22314.

Note: Poisson's ratio = 0.30; units for C_p are $(\text{lb/in}^2)^{0.5}$ or $(\text{MPa})^{0.5}$.

equation. Then, we present the method for selecting materials for the pinion and the gear in Section 9–10.

9–10 SELECTION OF GEAR MATERIAL

We now use the values of bending stress found in Section 9–8 and contact stress from Section 9–9 to specify a suitable material and its condition that will withstand those stresses without tooth breakage caused by bending stress and with good resistance to pitting on the face of the teeth caused by contact stress. We develop here an approach for specifying steel for the gears. A similar method can be used for nonferrous materials such as cast iron and bronze. Later in this chapter, we discuss the design of plastic gears. The goal of the process is to ensure that the predicted stresses are less than the allowable strength of the material. Inputs to the decisions are:

1. Estimated bending stress, s_t, as found in Section 9–8, Equation (9–16)
2. Estimated contact stress, s_c, as found in Section 9–9, Equation (9–23)

TABLE 9–11 Reliability Factor, K_R

Realibility	K_R
0.90, one failure in 10	0.85
0.99, one failure in 100	1.00
0.999, one failure in 1000	1.25
0.9999, one failure in 10 000	1.50

TABLE 9–12 Recommended Design Life

Application	Design life (h)
Domestic appliances	1000–2000
Aircraft engines	1000–4000
Automotive	1500–5000
Agricultural equipment	3000–6000
Elevators, industrial fans, multipurpose gearing	8000–15 000
Electric motors, industrial blowers, general industrial machines	20 000–30 000
Pumps and compressors	40 000–60 000
Critical equipment in continuous 24-h operation	100 000–200 000

Source: Eugene A. Avallone and Theodore Baumeister III, eds. *Marks' Standard Handbook for Mechanical Engineers.* 9th ed. New York: McGraw-Hill, 1986.

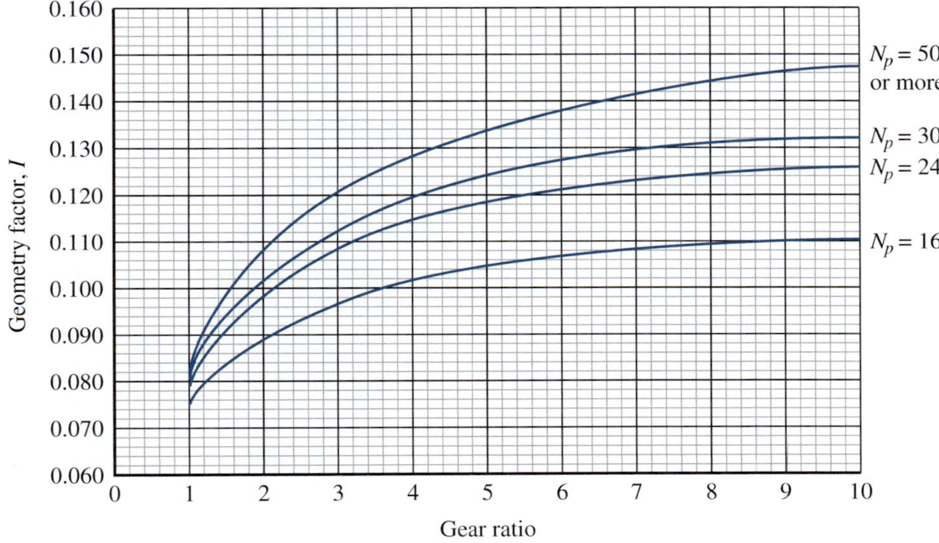

(*a*) 20° pressure angle, full-depth teeth (standard addendum = $1/P_d$)

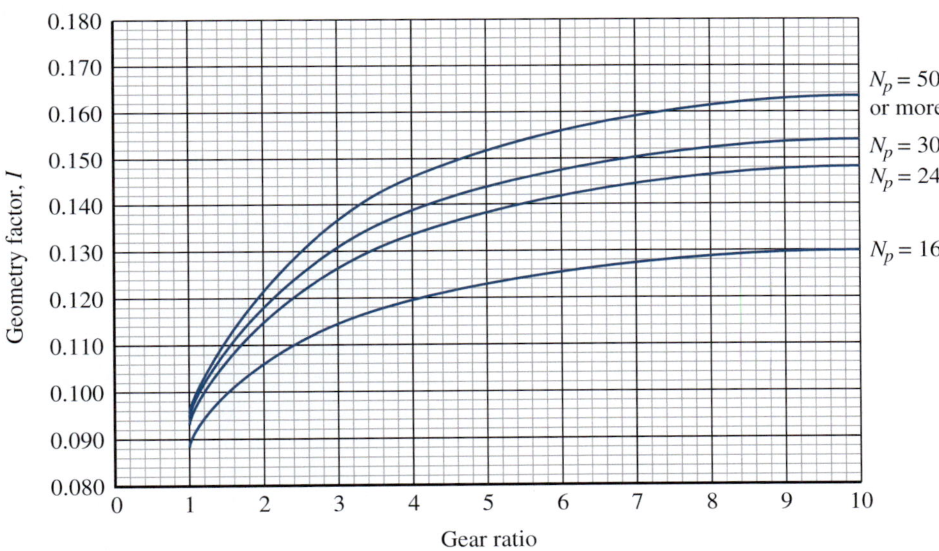

(*b*) 25° pressure angle, full-depth teeth (standard addendum = $1/P_d$)

FIGURE 9–21 External spur pinion geometry factor, *I*, for standard center distances. All curves are for the lowest point of single-tooth contact on the pinion (Extracted from AGMA Standard 218.01, *Rating the Pitting Resistance and Bending Strength of Spur and Helical Involute Gear Teeth,* with the permission of the publisher, American Gear Manufacturers Association, 1001 North Fairfax Street, 5th floor, Alexandria, VA 22314)

3. Desired service life and reliability for the gears, discussed in this section
4. Chosen safety factor, *SF*, discussed in this section
5. Allowable bending stress number, s_{at}, and allowable contact stress number for the material, s_{ac}, discussed in Section 9–7

Note about temperature. The methods developed in this section assume that the operating temperature for the gears is less than 250°F (121°C) and above 32°F (0°C) because we rely on published data for the materials obtained within these limits. Higher temperatures can be considered if data are available on how the material strength is affected. One concern is that many steel gears are heat treated to produce desired strength and ductility and higher operating temperatures can reduce the strength due to tempering. Operating below freezing temperature can be done provided that testing is done to verify impact adequate strength using the Charpy or Izod methods discussed in Chapter 2.

Example Problem 9–2 Compute the contact stress number for the gear pair described in Example Problem 9–1.

Solution Data from Example Problem 9–1 are summarized as follows:

$N_P = 20 \quad N_G = 70, \quad F = 2.25 \text{ in} \quad W_t = 540 \text{ lb} \quad D_P = 3.333 \text{ in}$

$K_o = 1.50 \quad K_s = 1.00 \quad K_m = 1.22 \quad K_v = 1.41 \quad P_d = 6$

The gear teeth are 20°, full-depth, involute teeth. We also need the geometry factor for pitting resistance, I. From Figure 9–21(a), at a gear ratio of $m_G = N_G/N_P = 70/20 = 3.50$ and for $N_P = 20$, we read $I = 0.108$, approximately.

The design analysis for bending strength indicated that two steel gears should be used. Then, from Table 9–10, we find that $C_p = 2300$. Then the contact stress number is

$$s_c = C_p\sqrt{\frac{W_tK_oK_sK_mK_v}{FD_pI}} = 2300\sqrt{\frac{(540)(1.50)(1.0)(1.22)(1.41)}{(2.25)(3.333)(0.108)}}$$

$$s_c = 95\,398 \text{ psi}$$

This value is used for both the pinion and the gear.

General principles for specifying materials. We use the following relationships to guide the process of specifying suitable materials for gears.

$$s_t < s_{at}' \quad \text{and} \quad s_c < s_{ac}' \qquad (9\text{–}24)$$

The "prime" applied to the allowable strengths indicates that adjustments are made to published data for s_{at} and s_{ac} that are presented in Section 9–7. Conditions for those data are:

- Operating temperatures are above 32°F (0°C) and below 250°F (121°C)
- The materials are exposed 10^7 loading cycles
- The expected reliability is 99%, less than one failure in 100
- The safety factor, SF, is 1.00

The adjustments to s_{at} and s_{ac} considered here take the form

$$s_{at}' = s_{at}\frac{Y_N}{(SF)(K_R)} \qquad (9\text{–}25)$$

$$s_{ac}' = s_{ac}\frac{Z_N}{(SF)(K_R)} \qquad (9\text{–}26)$$

where SF = safety factor
K_R = reliability factor
Y_N = bending strength stress cycle factor
Z_N = pitting resistance stress cycle factor

Safety factor: The value of SF is typically taken to be 1.0 because most of the uncertainties involved in computing the bending and contact stresses are included in the equations for s_{at} and s_{ac} by the factors K_o, K_s, K_v, and K_m. A design decision can be made for using $SF > 1.00$ for extra safety or for anticipated undesirable conditions beyond these factors. AGMA recommends

$$1.00 \le SF \le 1.50$$

Reliability factor: Table 9–11 shows typical values for K_R and the choice of which to apply in a given project is a design decision.

Desired life of the gear drive: Designers must evaluate each project to specify what the design life is to be. Table 9–12 shows one set of recommended design life values in hours of operation. The range from 20 000 to 30 000 hours for general industrial machines is a reasonable choice unless known conditions of the types listed exist. The same table for design life is used for rolling contact bearings in Chapter 14. Also needed is the *number of load cycles* that can be computed using this equation:

$$N_c = (60)(L)(n)(q) \qquad (9\text{–}27)$$

where N_c = expected number of cycles of loading
L = design life in hours
n = rotational speed of the gear in rpm
q = number of load applications per revolution

The normal number of load applications per revolution for any given tooth is typically, of course, one. But consider the case of an idler gear that serves as both a driven and a driving gear in a gear train. It receives two cycles of load per revolution: one as it receives power from and one as it delivers power to its mating gears. Also, in certain types of gear trains, one gear may deliver power to two or more gears mating with it. Gears in a planetary gear train often have this characteristic.

As an example of the application of Equation (9–27), consider that the pinion in Example Problems 9–1 and

9–2 is designed to have a life of 20 000 h. When rotating at 1750 rpm. Then

$$N_c = (60)(L)(n)(q) = (60)(20000)(1750)(1)$$
$$= 2.1 \times 10^9 \text{cycles}$$

Because this is higher than 10^7, an adjustment in the allowable bending stress number must be made.

Stress Cycle Factors: Figures 9–22 and 9–23 show recommended values of Y_N for bending stress and Z_N for contact stress for steel gears. Features of these graphs include:

- Each is plotted on log–log scales with the *number of load cycles* on the horizontal axis and the *stress cycle factor* on the vertical axis.

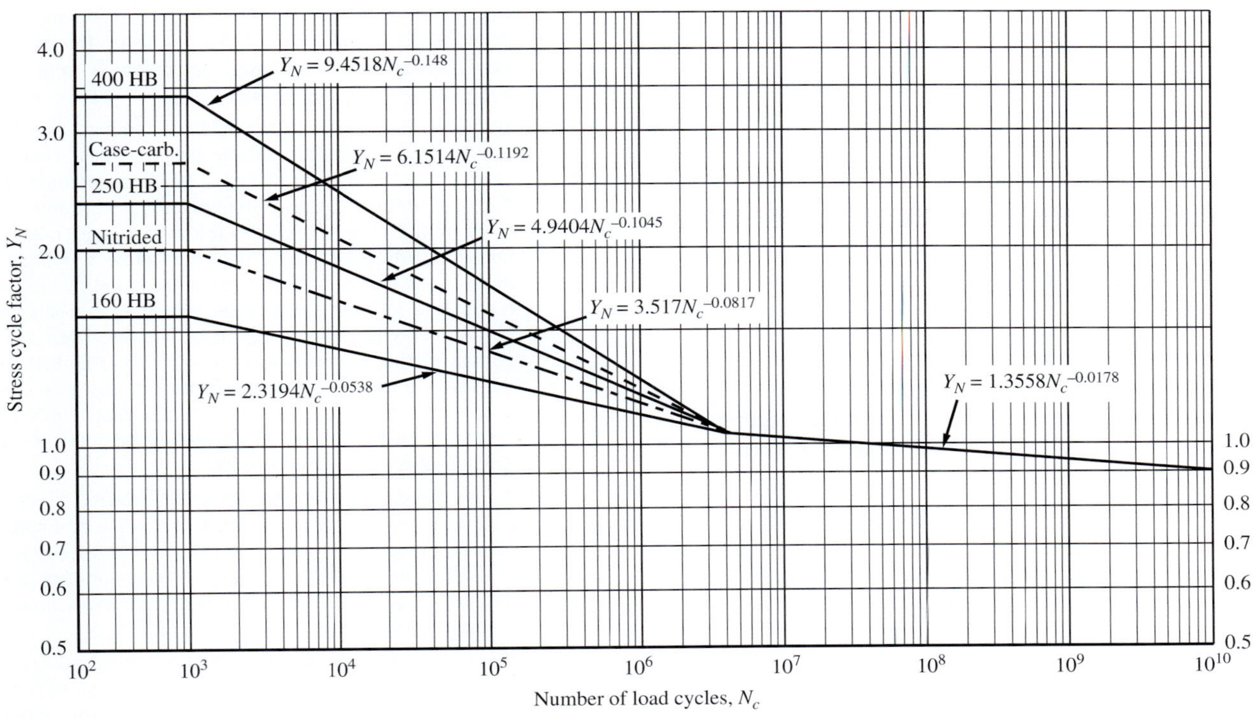

FIGURE 9–22 Bending strength stress cycle factor, Y_N (Adapted from AGMA Standard 2001-D04, *Fundamental Rating Factors and Calculation Methods for Involute Spur and Helical Gear Teeth,* with permission of the publisher, American Gear Manufacturers Association, 1001 North Fairfax Street, 5th floor, Alexandria, VA 22314)

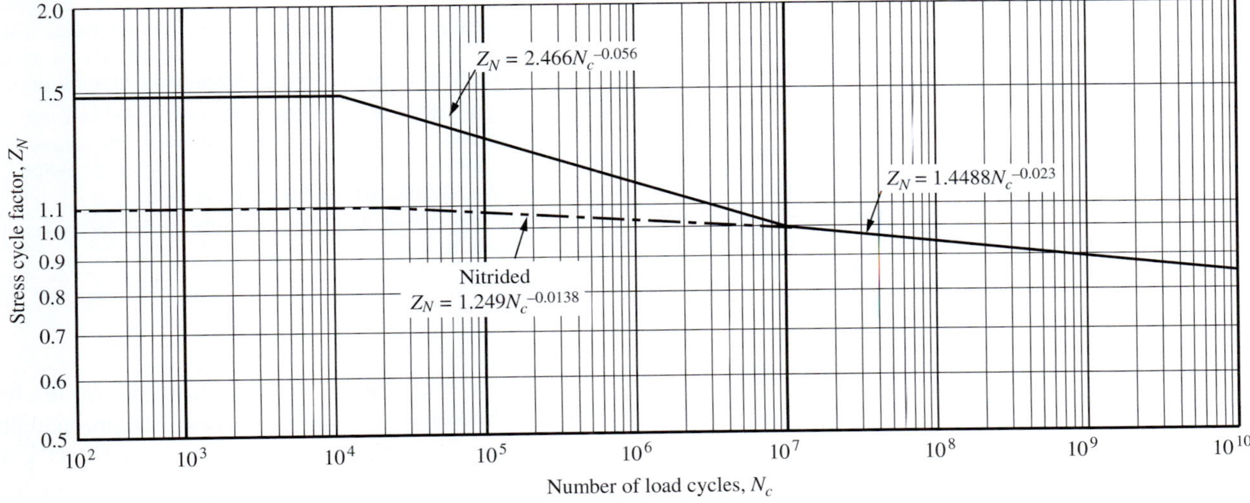

FIGURE 9–23 Pitting resistance stress cycle factor, Z_N (Adapted from AGMA Standard 2001-D04, *Fundamental Rating Factors and Calculation Methods for Involute Spur and Helical Gear Teeth,* with permission of the publisher, American Gear Manufacturers Association, 1001 North Fairfax Street, 5th floor, Alexandria, VA 22314)

- The stress cycle factor is approximately 1.0 for 10^7 load cycles on both graphs.
- The stress cycle factors are less than 1.0 for load cycles higher than 10^7 and the values are independent of the type of steel or its condition. Some designs for critical service applications use values lower than those shown and the AGMA standard provides an added shaded area for use in such conditions.
- The stress cycle factors are greater than 1.0 for load cycles lower than 10^7 and the values are dependent on the type of steel and its condition.
- Equations of an exponential form are given in the graphs to facilitate calculation of the values of the stress cycle factors.
- These charts do not apply to cast iron or to nonferrous materials such as bronze. Values of $Y_N = 1.00$ and $Z_N = 1.00$ can be used while ensuring that the actual strengths of materials specified have a modest margin of safety.

We can now describe the general approach to specifying suitable steels or other metallic materials for gears. The goal is to access the data in Section 9–7, including Figures 9–11 and 9–12 and Tables 9–5, and 9–6, that give accepted values of s_{at} and s_{ac} for through-hardened steels, case-hardened steels using flame or induction hardening, steels that are carburized and case hardened, cast irons, and bronzes. To facilitate this approach, we can combine Equations (9–24) to (9–26) and then solve for s_{at} and s_{ac}.

$$s_t < s'_{at} = s_{at}\frac{Y_N}{(SF)(K_R)}$$

$$s_c < s'_{ac} = s_{ac}\frac{Z_N}{(SF)(K_R)}$$

Then, solving for s_{at} and s_{ac} gives

$$\text{Required } s_{at} > s_t\frac{(SF)(K_R)}{Y_N} \quad \textbf{(9–28)}$$

$$\text{Required } s_{ac} > s_c\frac{(SF)(K_R)}{Z_N} \quad \textbf{(9–29)}$$

Both Equations (9–28) and (9–29) must be satisfied. It should be noted here that for gear drives with relatively long design lives, the design will most often be governed by the pitting resistance and that assumption is made in the following guidelines.

GUIDELINES FOR SPECIFYING METALLIC MATERIALS FOR GEARS

1. Given the bending stress, s_t, the contact stress, s_c, and the rotational speed of both the pinion, n_P, and the gear, n_G from analyses similar to those in Example Problems 9–1 and 9–2. Note that this also required that the material of choice is to be steel, cast iron, or bronze.
2. Decide on the safety factor; typically $SF = 1.00$ unless unusual conditions exist.
3. Decide on the desired reliability and use Table 9–11 to determine K_R; typically use $K_R = 1.00$ for a reliability of 0.99 (one failure in 100).
4. Specify the desired life in hours for the gears using Table 9–12 as a guide,
5. Compute the expected number of load cycles for both the pinion and the gear using Equation (9–27).
6. Use Figure 9–22 to determine Y_N for the pinion and Figure 9–23 to determine Z_N for the gear. The values may be read from the graphs or computed from the equations given in the figures.
7. Use Equation (9–29) to evaluate the required s_{acP} for the pinion because it typically has the most critical value for s_{ac}. Then consult Figure 9–12 to determine if through-hardened Grade 1 steel can be used. Enter the chart from the value of s_{ac} on the vertical axis; project horizontally to the Grade 1 line; then project vertically down to the Brinell hardness axis to read the minimum acceptable HB hardness. Note that the value should be between HB 180 and HB 400. The most desirable portion of this range is about HB 250 to HB 400. If the value is lower, then a more compact gear size can usually be designed. No value greater than HB 400 should be considered for through-hardened steels. Only Grade 1 steels are recommended.
8. If Step 7 produces a reasonable result, proceed to compute s_{acG} for the gear using Equation (9–29) and compute s_{at} for both the pinion and the gear using Equation (9–28). Check that the bending stress numbers are acceptable by consulting Figure 9–11; they will almost always be acceptable.
9. Using the required hardness HB for the pinion computed in Step 7, specify a suitable material and its heat treatment using Appendices 3 and 4. The charts in Appendix 4 are recommended because they give the widest range of tempering temperatures and corresponding hardness values from which to choose. Examine the lower line of data for HB to find a value just larger than the minimum found from Step 7 and specify the tempering temperature that will produce that hardness level. Table 9–4 recommends several steel alloys that are typically used for gears that are through-hardened; SAE 1045 (similar to 1040), 4140, and 4340 are included and charts for these three alloys are shown in Appendix 4.
10. If Step 8 is not successful, either redesign the gears or consider using case-hardened steel with the values for s_{ac} taken from Table 9–5. More guidelines for performing a redesign are given in Section 9–11.
11. If the design is using cast iron or bronze, consult Table 9–6 for values for s_{ac}.

Example Problem 9–3 follows using the results from Example Problems 9–1 and 9–2 as input. The problem illustrates the application of this procedure using the U.S. unit system based on the diametral pitch. Then, in Section 9–11, all of the discussion to this point of the chapter is combined into a more general design procedure. Both the U.S. unit system based on the diametral pitch and the SI Metric unit system based on the module, m, are demonstrated.

A computer-assisted approach to implementing the numerous required calculations is then described in Section 9–12. Gearing design using plastics is discussed in Section 9–13 while other practical considerations for gears and interfaces with other machine elements (shafts, bearings, keys, etc.) are discussed in Section 9–14.

Example Problem 9–3 Specify suitable materials for the pinion and the gear for the drive that is the subject of Example Problems 9–1 and 9–2. Use results from those problems for the computed bending stress and contact stress. Design for a reliability of 0.999, fewer than one failure in 1000. The application is a drive for an industrial saw that will be fully utilized on a normal, one-shift, five-day-per-week operation.

Solution We will use ***Guidelines for specifying metallic materials for gears***, discussed earlier. The intent is to use through-hardened steel gears and we are required to specify the steel alloy and its heat treatment to ensure safe levels for both bending and contact stresses.

Step 1. The critical results from Example Problems 9–1 and 9–2 are:

$$n_P = 1750 \text{ rpm} \qquad n_G = 500 \text{ rpm}$$

$$D_P = 3.333 \text{ in} \qquad F = 2.25 \text{ in}$$

$$s_{tP} = 11091 \text{ psi} \qquad s_{tG} = 8847 \text{ psi}$$

$$s_{cP} = s_{cG} = 95\,398 \text{ psi}$$

Step 2. Design decision: Use a safety factor of $SF = 1.00$ because we know of no unusual conditions not already accounted for in the calculation of the stresses.

Step 3. From Table 9–11, we find $K_R = 1.25$ for the desired reliability of 0.999 as specified in the problem. This is also a design decision.

Step 4. Design decision: Because the saw will be fully utilized in an industrial environment, we choose a life of $L = 20\,000$ hours, using Table 9–12 as a guide.

Step 5. Compute the number of load cycles for the pinion and the gear using Equation (9–27). Each tooth sees one load cycle per revolution, $q = 1$.

$$N_{cP} = (60)(L)(n_P)(q) = (60)(20\,000)(1750)(1) = 2.10 \times 10^9 \text{ cycles}$$

$$N_{cG} = (60)(L)(n_G)(q) = (60)(20\,000)(500)(1) = 6.00 \times 10^8 \text{ cycles}$$

Step 6. From Figure 9–22, we find $Y_{NP} = 0.93$ and $Y_{NG} = 0.95$ for bending stress. From Figure 9–23, we find $Z_{NP} = 0.88$ and $Z_{NG} = 0.91$ for contact stress.

Step 7. Assuming that the contact stress in the pinion will govern the design, use Equation (9–29) for the pinion:

$$\text{Required } s_{acP} > s_c \frac{(SF)(K_R)}{Z_{NP}} = (95\,398 \text{ psi}) \frac{(1.00)(1.25)}{0.88} = 135\,508 \text{ psi}$$

Now we consult Figure 9–12 to evaluate the acceptability of this level of contact stress. For Grade 1 steel, we read that the required surface hardness of a through-hardened steel tooth is HB 330, a highly satisfactory result. We could also compute the required hardness using the equation shown in Step 7.

$$\text{Required HB} = (s_{ac} - 29\,100)/322 = (135\,508 - 29\,100)/322 = 330$$

Step 8. The calculation for required contact stress for the gear gives:

$$\text{Required } s_{acG} > s_c \frac{(SF)(K_R)}{Z_{NG}} = 95\,398 \text{ psi} \frac{(1.00)(1.25)}{0.91} = 131\,041 \text{ psi}$$

It is usual that the contact stress in the gear is lower than in the pinion but that they are quite close in value.

Step 9. Now we compute the required allowable bending stresses for the pinion and the gear:

$$\text{Required } s_{atP} > s_{tP}\frac{(SF)(K_R)}{Y_{NP}} = 11\,091 \text{ psi}\frac{(1.00)(1.25)}{0.93} = 14\,097 \text{ psi}$$

$$\text{Required } s_{atG} > s_{tG}\frac{(SF)(K_R)}{Y_{NG}} = 8847 \text{ psi}\frac{(1.00)(1.25)}{0.95} = 11\,640 \text{ psi}$$

Checking these values with the required hardness of the steel for bending from Figure 9–11 shows that a very low hardness is required, well below HB 180 that is the end point for the Grade 1 line. This verifies that contact stress governs the design.

Step 10. The final step is to specify the material and its heat treatment, in this case based on the contact stress in the pinion for which HB 330 is required. Table 9–4 recommends alloys SAE 1045 (similar to SAE 1040), SAE 4140, and SAE 4340 along with some others. Use of Appendix 4 is recommended for specifying the heat treatment. Let's look at these three possible choices:

SAE 1040 WQT 800, HB 352, $s_y = 90$ ksi, $s_u = 120$ ksi, 20% elongation

SAE 4140 OQT 1000, HB 341, $s_y = 152$ ksi, $s_u = 169$ ksi, 18% elongation

SAE 4340 OQT 1000, HB 363, $s_y = 158$ ksi, $s_u = 170$ ksi, 17% elongation

Any of the three alloys and their heat treatments would be satisfactory and factors such as availability, cost, delivery, and machinability could affect the decision. Perhaps SAE 1040 is preferred because it is a plain carbon steel that is likely less costly and it has the highest percent elongation of the three choices, indicating more ductility.

Summary of the Results for Example Problems 9–1, 9–2, and 9–3

Example Problems 9–1, 9–2, and 9–3 were spread over many pages because the design principles and procedures were developed and demonstrated as those problems were completed. Let's now show the summary of the major design decisions and results of calculations.

Given data:

Application: Industrial saw driven by an electric motor at 1750 rpm

Output speed from the gear: 500 rpm; velocity ratio: 3.50

Production operation: fully utilized; design life = 20 000 hours

Transmitted power: 25 hp

Enclosed gear reducer made to commercial standards

High reliability for gear design: $R = 0.999$; less than one failure in 1000

Design decisions:

Diametral pitch: 6 Pressure angle: $\phi = 20°$ Face width: $F = 2.25$ in

Number of teeth: pinion—20; gear—70

Quality number for gear geometry: $A_v = 10$

Calculated results:

Geometry of gear pair and forces

Pitch diameters: $D_P = 3.333$ in; $D_G = 11.667$ in

Center distance: $C = 7.500$ in Pitch line speed: $v_t = 1527$ ft/min

Transmitted force: $W_t = 540$ lb

Factors in stress analyses:

Overload factor: $K_o = 1.50$ Size factor: $K_s = 1.0$

Load-distribution factor: $K_m = 1.22$

Dynamic factor: $K_v = 1.41$

Stresses in gear teeth:

Bending stress: pinion $s_{tP} = 11091$ psi

gear $s_{tG} = 8847$ psi

Contact stress: Pinion and gear $s_c = 95398$ psi

Required strength of gear material:

The contact stress in the pinion is the determining factor.

Design decisions: Safety factor: $SF = 1.00$ Reliability factor: $K_R = 1.25$

Stress cycle factor for pitting resistance: $Z_{NP} = 0.88$

Required contact stress number: $s_{acP} = 135508$ psi

Required surface hardness for through-hardened steel: $HB = 330$

Specified material—both pinion and gear:

SAE 1040 WQT 800; HB = 352; $s_y = 90$ ksi; $s_u = 120$ ksi; 21% elongation

9–11 DESIGN OF SPUR GEARS

In designs involving gear drives, normally the required speeds of rotation of the pinion and the gear and the amount of power that the drive must transmit are known. These factors are determined from the application. Also, the environment and operating conditions to which the drive will be subjected must be understood. It is particularly important to know the type of driving device and the driven machine, in order to judge the proper value for the overload factor.

The designer must decide the type of gears to use; the arrangement of the gears on their shafts; the materials of the gears, including their heat treatment; and the geometry of the gears: numbers of teeth, diametral pitch, pitch diameters, tooth form, face width, and quality numbers.

This section presents a design procedure that accounts for the bending fatigue strength of the gear teeth and the pitting resistance, called *surface durability*. This procedure makes extensive use of the design equations presented in the preceding sections of the chapter and of the tables of material properties in Appendices 3 through 5, 8, and 12.

You should understand that there is no one best solution to a gear design problem; several good designs are possible. Your judgment and creativity and the specific requirements of the application will greatly affect the final design selected. The purpose here is to provide a means of approaching the problem to create a reasonable design.

Design Objectives

Some overall objectives of a design are listed below. The resulting drive should

- Be compact and small
- Operate smoothly and quietly
- Have long life
- Be low in cost
- Be easy to manufacture
- Be compatible with the other elements in the machine, such as bearings, shafts, the housing, the driver, and the driven machine

The major objective of the design procedure is to define a safe, long-lasting gear drive. General steps and guidelines are outlined here to produce a reasonable initial design. However, because of the numerous variables involved, several iterations are typically made to work toward an optimum design. Details of the procedure are presented in Example Problem 9–4.

Table 9–13 is a combination of a glossary of terms and a reference for finding equations, tables, or figures required to complete a gear design. The procedure given below uses U.S. Customary units and the diametral pitch, P_d, system. Section 9–12 adjusts this procedure using SI Metric units and the metric module, m, system. You should refer back to earlier sections of this chapter and to Chapter 8 for details.

Procedure for Designing a Safe and Long-Lasting Gear Drive

1. From the design requirements, identify the input speed of the pinion, n_P, the desired output speed of the gear, n_G, and the power to be transmitted, P.
2. Choose the type of material for the gears, such as steel, cast iron, or bronze.
3. Considering the type of driver and the driven machine, specify the overload factor, K_o, using Table 9–7. The primary concern is the expected level of shock or impact loading.
4. Specify a trial value for the diametral pitch. When steel gears are used, Figure 9–24 provides initial guidance. The graph of design power transmitted versus the pinion rotational speed was derived for selected pitches and pinion diameters. Design power, $P_{des} = K_oP$. Steel that is through hardened to HB 300 is used. Because of the numerous variables involved, the value of P_d read from the figure is only an initial target value. Subsequent iterations may require considering a different value, either higher or lower.
5. Specify the face width within the following recommended range for general machine drive gears:

Norminal Face Width

$$8/P_d < F < 16/P_d \qquad (9\text{–}30)$$

$$\text{Norminal value of } F = 12/P_d$$

The upper limit given tends to minimize alignment problems and ensure reasonably uniform loading across the face. When the face width is less than the lower limit, it is probable that a more compact design can be achieved with a different pitch. Also, the face width normally is less than twice the pitch diameter of the pinion.

6. Specify or compute the following values:
   - Number of teeth in the pinion (N_P) and the gear (N_G) to achieve the desired output speed of the gear.
   - Compute the actual output speed of the gear and ensure that it is satisfactory.
   - Compute key geometric features: pitch diameters D_P, D_G and center distance, C. Judge that they are acceptable.
   - Compute the pitch line speed, v_t, and the transmitted load, W_t.
   - Determine the geometry factors for bending stress, J_P and J_G, and the geometry factor for contact stress, I.
   - Specify the quality number for the teeth of the gears using Table 9–3 as a guide.
   - Determine values for all of the factors in Equation (9–16) for bending stress and Equation (9–23) for contact stress.
7. Compute the bending stress and the contact stress on the pinion and gear teeth. Judge whether the stresses

TABLE 9–13 Summary and Glossary of Terms Used in Gear Design

Term by cluster	Symbol	Description of use	Reference location
Recommended quality numbers	A_v	Specify value of A_v	Table 9–3
Allowable bending stress number—steel	s_{at}	Specify material	Figure 9–11
Allowable contact stress number—steel	s_{ac}	Specify material	Figure 9–12
Allowable stresses—case-hardened steel	s_{at} and s_{ac}	Specify material	Table 9–5
Allowable stresses—nonferrous	s_{at} and s_{ac}	Specify material	Table 9–6
Geometry factors for bending stress	J_P and J_G	Bending stress	Figure 9–15
Overload factor	K_o	Bending and contact stress	Table 9–7
Size factor	K_s	Bending and contact stress	Table 9–8
Alignment factor: $K_m = 1.0 + C_{pf} + C_{ma}$	K_m	Bending and contact stress	Equation 9–17
Proportion factor:	C_{pf}	Bending and contact stress	Figure 9–16
Mesh alignment factor:	C_{ma}	Bending and contact stress	Figure 9–17
Rim thickness factor	K_B	Bending and contact stress	Figure 9–18
Dynamic factor	K_v	Bending and contact stress	Figure 9–20
Equations for K_v	K_v	Bending and contact stress	Table 9–9
Elastic coefficient	C_P	Contact stress	Table 9–10
Geometry factor for contact stress	I	Contact stress	Figure 9–21
Reliability factor	K_R	Bending and contact stress	Table 9–11
Recommended life in hours	L	Bending and contact stress	Table 9–12
Bending stress cycle factor	Y_N	Bending stress	Figure 9–22
Contact stress cycle factor	Z_N	Contact stress	Figure 9–23
Design pitch selection aid		Specify trial P_d or m	Figure 9–24
		Primary equations used in gear design:	
Bending stress number	s_t	$s_t = \frac{W_t P_d}{FJ} K_o K_s K_m K_B K_v$	Equation 9–16
Required allowable bending stress number	s_{at}	Required $s_{at} > s_t \frac{(SF)(K_R)}{Y_N}$	Equation 9–28
Contact stress number	s_c	$s_c = C_P \sqrt{\frac{W_t K_o K_s K_m K_v}{F D_P I}}$	Equation 9–23
Required allowable contact stress number	s_{ac}	Required $s_{ac} > s_c \frac{(SF)(K_R)}{Z_N}$	Equation 9–29

are reasonable (neither too low nor too high) in terms of being able to specify a suitable material. If not, select a new pitch or revise the number of teeth, pitch diameter, or face width. Typically the contact stress on the pinion is the limiting value for gears designed for a long life.

8. Iterate the design process to seek more optimum designs. It is not unusual to make several trials before settling on a particular design. Using computer aids, such as the spreadsheets described in Section 9–13, can make successive trials quickly.

Guidelines for Adjustments in Successive Iterations

The following relationships should help you determine what changes in your design assumptions you should make after the first set of calculations to achieve a more optimum design:

- Decreasing the numerical value of the diametral pitch results in larger teeth and generally lower stresses. Also, the lower value of the pitch usually means a larger face

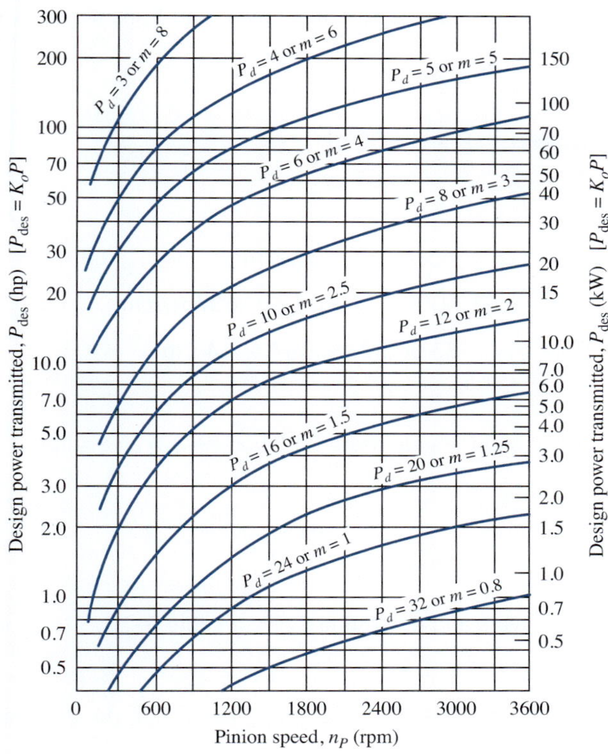

For all curves: 20° full depth teeth;
$N_P = 24$; $N_G = 96$; $m_G = 4.00$; $F = 12/P_d$; $A_v = 11$
Steel gears, HB 300; s_{at} = 36000 psi (250 MPa); s_{ac} = 126000 psi (869 MPa)

FIGURE 9–24 Design power transmitted versus pinion speed for spur gears with different pitches and diameters

width, which decreases stress and increases surface durability.

- Increasing the diameter of the pinion decreases the transmitted load, generally lowers the stresses, and improves the surface durability.
- Increasing the face width lowers the stress and improves the surface durability, but to a generally lesser extent than either the pitch or the pitch diameter changes discussed previously.
- Gears with more and smaller teeth tend to run more smoothly and quietly than gears with fewer and larger teeth.
- Standard values of diametral pitch should be used for ease of manufacture and lower cost (see Table 8–3).
- Using high-alloy steels with high surface hardness results in the most compact system, but the cost is higher.
- Using very accurate gears (with ground or shaved teeth) results in a higher quality number, lower dynamic loads, and consequently lower stresses and improved surface durability, but the cost is higher.
- The number of teeth in the pinion should generally be as small as possible to make the system compact. But the possibility of interference is greater with fewer teeth. Check Table 8–7 to ensure that no interference will occur. (See Reference 24.)

Example Problem 9–4

Design a pair of spur gears to be used as a part of the drive for a chipper to prepare pulpwood for use in a paper mill. Intermittent use is expected. An electric motor transmits 3.0 horsepower to the pinion at 1750 rpm and the gear must rotate between 460 and 465 rpm. A compact design is desired.

Solution and General Design Procedure

Step 1. Considering the transmitted power, P, the pinion speed, n_P, and the application, refer to Figure 9–24 to determine a trial value for the diametral pitch, P_d. The overload factor, K_o, can be determined from Table 9–7, considering both the power source and the driven machine.

For this problem, $P = 3.0$ hp and $n_P = 1750$ rpm, $K_o = 1.75$ (uniform driver, heavy shock driven machine). Then $P_{des} = (1.75)(3.0\text{ hp}) = 5.25$ hp. Try $P_d = 12$ for the initial design.

Step 2. Specify the number of teeth in the pinion. For small size, use 17 to 20 teeth as a start.

For this problem, let's specify $N_P = 18$.

Step 3. Compute the nominal velocity ratio from $VR = n_P/n_G$.

For this problem, use $n_G = 462.5$ rpm at the middle of the acceptable range.

$$VR = n_P/n_G = 1750/462.5 = 3.78$$

Step 4. Compute the approximate number of teeth in the gear from $N_G = N_P(VR)$.

For this problem, $N_G = N_P(VR) = 18(3.78) = 68.04$ Specify $N_G = 68$.

Step 5. Compute the actual velocity ratio from $VR = N_G/N_P$.

For this problem, $VR = N_G/N_P = 68/18 = 3.778$.

Step 6. Compute the actual output speed from $n_G = n_P(N_P/N_G)$.

For this problem, $n_G = n_P(N_P/N_G) = (1750\text{ rpm})(18/68) = 463.2$ rpm. OK.

Step 7. Compute the pitch diameters, center distance, pitch line speed, and transmitted load and judge the general acceptability of the results.

For this problem, the pitch diameters are:

$$D_P = N_P/P_d = 18/12 = 1.500 \text{ in}$$

$$D_G = N_G/P_d = 68/12 = 5.667 \text{ in}$$

Center distance:

$$C = (N_P + N_G)/(2P_d) = (18 + 68)/(24) = 3.583 \text{ in}$$

Pitch line speed: $v_t = \pi D_P n_P/12 = [\pi(1.500)(1.750)]/12 = 687 \text{ ft/min}$

Transmitted load: $W_t = 33\,000(P)/v_t = 33\,000(3.0)/687 = 144 \text{ lb}$

These values seem to be acceptable.

Step 8. Specify the face width of the pinion and the gear using Equation (9–30) as a guide.

For this problem: Lower limit $= 8/P_d = 8/12 = 0.667$ in.

Upper limit $= 16/P_d = 16/12 = 1.333$ in

Nominal value $= 12/P_d = 12/12 = 1.00$ in. Use this value.

Step 9. Specify the type of material for the gears and determine C_p from Table 9–10.

For this problem, specify two steel gears. $C_p = 2300$.

Step 10. Specify the quality number, A_v, using Table 9–3 as a guide. Determine the dynamic factor from Figure 9–20.

For this problem, specify $A_v = 11$ for a wood chipper. $K_v = 1.35$.

Step 11. Specify the tooth form, the bending geometry factors for the pinion and the gear from Figure 9–15 and the pitting geometry factor from Figure 9–21.

For this problem, specify 20° full depth teeth. $J_P = 0.325$, $J_G = 0.410$, $I = 0.104$.

Step 12. Determine the load-distribution factor, K_m, from Equation (9–17) and Figures 9–16 and 9–17. The precision class of the gear system design must be specified. Values may be computed from equations in the figures or read from the graphs.

For this problem: $F = 1.00$ in, $D_P = 1.500$. $F/D_P = 0.667$. Then $C_{pf} = 0.042$.

Specify open gearing for the wood chipper, mounted to the frame. $C_{ma} = 0.264$.

Compute: $K_m = 1.0 + C_{pf} + C_{ma} + 0.042 + 0.264 = 1.31$

Step 13. Specify the size factor, K_s, from Table 9–8.

For this problem, $K_s = 1.00$ for $P_d = 12$.

Step 14. Specify the rim thickness factor, K_B, from Figure 9–18.

For this problem, specify a solid gear blank. $K_B = 1.00$.

Step 15. Specify a service factor, SF, typically from 1.00 to 1.50, based on uncertainty of data.

For this problem, there is no unusual uncertainty. Let $SF = 1.00$.

Step 16. Specify a reliability factor using Table 9–11 as a guideline.

For this problem, specify a reliability of 0.99. $K_R = 1.00$.

Step 17. Specify a design life. Compute the number of loading cycles for the pinion and the gear. Determine the stress cycle factors for bending (Y_N) and pitting (Z_N) for the pinion and the gear.

For this problem, intermittent use is expected. Specify the design life to be 3000 hours, similar to agricultural machinery. The numbers of loading cycles are:

$$N_{cP} = (60)(3000 \text{ hr})(1750 \text{ rpm})(1) = 3.15 \times 10^8 \text{ cycles}$$

$$N_{cG} = (60)(3000 \text{ hr})(463.2 \text{ rpm})(1) = 8.34 \times 10^7 \text{ cycles}$$

Then, from Figure 9–22, $Y_{NP} = 0.96, Y_{NG} = 0.98$. From Figure 9–23, $Z_{NP} = 0.92$, $Z_{NG} = 0.95$.

Step 18. Compute the expected bending stresses in the pinion and the gear using Equation (9–16).

$$s_{tP} = \frac{W_t P_d}{F J_P} K_o K_s K_m K_B K_v = \frac{(144)(12)}{(1.00)(0.325)}(1.75)(1.0)(1.31)(1.0)(1.35) = 16\,455 \text{ psi}$$

$$s_{tG} = s_{tP}(J_P/J_G) = (16\,455)(0.325/0.410) = 13\,044 \text{ psi}$$

Step 19. Adjust the bending stresses using Equation 9–28.

For this problem, for the pinion:

$$s_{atP} > s_{tP}\frac{K_R(SF)}{Y_{(NP)}} = (16\,455)\frac{(1.00)(1.00)}{0.96} = 17\,141 \text{ psi}$$

For the gear:

$$s_{atG} > s_{tG}\frac{K_R(SF)}{Y_{(NG)}} = (13\,044)\frac{(1.00)(1.00)}{0.98} = 13\,310 \text{ psi}$$

Step 20. Compute the expected contact stress in the pinion and the gear from Equation (9–23). Note that this value will be the same for both the pinion and the gear.

$$s_c = C_P\sqrt{\frac{W_t K_o K_s K_m K_v}{F D_P I}} = 2300\sqrt{\frac{(144)(1.75)(1.0)(1.31)(1.35)}{(1.00)(1.50)(0.104)}} = 122\,933 \text{ psi}$$

Step 21. Adjust the contact stresses for the pinion and the gear using Equation (9–29).

$$s_{acP} > s_{cP}\frac{K_R(SF)}{Z_{NP}} = (122\,933)\frac{(1.00)(1.00)}{(0.92)} = 133\,623 \text{ psi}$$

For the gear:

$$s_{acG} > s_{cG}\frac{K_R(SF)}{Z_{NG}} = (122\,933)\frac{(1.00)(1.00)}{(0.95)} = 129\,403 \text{ psi}$$

Step 22. Specify materials for the pinion and the gear that will have suitable through hardening or case hardening to provide allowable bending and contact stresses greater than those required from Steps 19 and 21. Typically the contact stress in the pinion is the controlling factor. Refer to Figures 9–11 and 9–12 and Tables 9–5 and 9–6 for data on required hardness. Refer to Appendices 3 to 5 for properties of steel to specify a particular alloy and heat treatment.

For this problem, the contact stress for the pinion is the controlling factor, as is often the case. A steel must be specified that is rated to handle approximately $s_{ac} = 133.6$ ksi. First check Figure 9–12 to explore whether or not through-hardened steel is practical. We can use the equation for Grade 1 steel in U.S. units to determine the required Brinell hardness number, *HB*.

$$\text{Reqd. } HB = (s_{ac} - 29.10)/0.322 = (133.6 \text{ ksi} - 29.10)/0.322 = 324$$

This value is well within the recommended hardness for through-hardened steels. Using Appendix 4, we can specify SAE 4140 OQT 1000 steel having $HB = 341$ and 18% elongation indicating good ductility. We can also check the required hardness for the gear that has a required $s_{ac} = 129.4$ ksi.

$$\text{Reqd. } HB = (s_{ac} - 29.10)/0.322 = (129.4 \text{ ksi} - 29.10)/0.322 = 311$$

This value can be met by SAE 4140 OQT 1100 steel having $HB = 311$ and 20% elongation. However, because both the pinion and the gear experience nearly the same contact stress, it may be prudent to specify the same heat treatment for both to permit them to be produced by the same process.

9–12 GEAR DESIGN FOR THE METRIC MODULE SYSTEM

Here we take the principles, guidelines, and design methodology that were initially developed for the U.S. Customary unit system based on the diametral pitch, P_d, and adapt them to the SI Metric unit system based on the metric module, m. The primary variables involved are listed below with units in both systems.

Variables	U.S. units	SI units
Length (D, C, pitch)	in	mm
Force (W_t, W_r)	lb	N
Power, P	hp	watts or kW
Pitch line speed, v_t	ft/min	m/s
Stresses	psi or ksi	MPa

Some of the required adjustments to calculations and equations are listed below, to help you to mentally relate the two systems to each other. In general, we recommend that designs be completed in one system or the other with minimal conversions.

- Refer to Table 8–1 gear teeth and gear pair features.
- Charts, tables, and graphs in this chapter contain both sets of units.
- Face width, F, recommended limits:

  U.S. units (in): $8/P_d < F < 16/P_d$ Nominal: $12/P_d$
  SI units (mm): $8m < F < 16m$ Nominal: $12m$
- Pitch line speed, v_t:

  U.S. units (ft/min): $v_t = \pi Dn/12$ ft/min
  [D in inches, n in rpm]

  SI units (m/s): $v_t = \pi Dn/60\,000$ m/s
  [D in mm, n in rpm]
- Transmitted load, W_t:

  U.S. units (lb): $W_t = 33\,000(P)/v_t$ [P in hp, v_t in ft/min]
  SI units (m/s): $W_t = 1000(P)/v_t$ [P in kW, v_t in m/s]

The following example problem uses SI units. The procedure will be virtually the same as that used to design with U.S. Customary units.

Example Problem 9–5 A gear pair is to be designed to transmit 15.0 kilowatts (kW) of power to a large meat grinder in a commercial meat processing plant. The pinion is attached to the shaft of an electric motor rotating at 575 rpm. The gear must operate at 270 to 280 rpm. The gear unit will be enclosed and of commercial quality. Commercially hobbed (quality number A11), 20°, full-depth, involute gears are to be used in the metric module system. The maximum center distance is to be 200 mm. Specify the design of the gears.

Solution The nominal velocity ratio is

$$VR = 575/275 = 2.09$$

Specify an overload factor of $K_o = 1.50$ from Table 9–7 for a uniform power source and moderate shock for the meat grinder. Then compute design power,

$$P_{des} = K_o P = (1.50)(15 \text{ kW}) = 22.5 \text{ kW}$$

From Figure 9–24, $m = 5$ is a reasonable trial module. Then

$$N_P = 18 \quad \text{(design decision)}$$

$$D_P = N_P m = (18)(5) = 90 \text{ mm}$$

$$N_G = N_P(VR) = (18)(2.09) = 37.6 \quad \text{(Use 38.)}$$

$$D_G = N_G m = (38)(5) = 190 \text{ mm}$$

$$\text{Final output speed} = n_G = n_P(N_P/N_G)$$

$$n_G = 575 \text{ rpm} \times (18/38) = 272 \text{ rpm} \quad \text{(OK)}$$

$$\text{Center distance} = C = (N_P + N_G)m/2 \quad \text{[Table 8–1]}$$

$$C = (18 + 38)(5)/2 = 140 \text{ mm} \quad \text{(OK)}$$

In SI units, the pitch line speed in meters per second (m/s) is

$$v_t = \pi D_P n_P/(60\,000) = [(\pi)(90)(575)]/(60\,000) = 2.71 \text{ m/s}$$

In SI units, the transmitted load, W_t, is in newtons (N). If the power, P, is in kW, and v_t is in m/s,

$$W_t = 1000(P)/v_t = (1000)(15)/(2.71) = 5536 \text{ N}$$

Face width, F: Let's specify the nominal $F = 12m = 12(5) = 60$ mm.

Factors in stress analysis:

$K_o = 1.50$ (found earlier)

$K_s = 1.00$ (Table 9–8; $m = 5$) $K_B = 1.00$ (Use solid gear blanks)

$K_R = 1.00$ (Table 9–11; 0.99 reliability) $SF = 1.00$ (No unusual conditions)

$K_v = 1.31$ (Figure 9–20; $A_v = 11$)

$K_m = 1.21$ (Figures 9–16 and 9–17; $F = 60$ mm; $F/D_P = 60/90 = 0.67$)

$J_P = 0.315$; $J_G = 0.380$ (Figure 9–15; $N_P = 18, N_G = 38$)

$C_p = 191$ (Table 9–10) $I = 0.092$ (Figure 9–21)

Pinion contact stress: (Equation 9–23)

$$S_c = C_P\sqrt{\frac{W_t\,K_o\,K_s\,K_m\,K_v}{FD_PI}} = 191\sqrt{\frac{(5536)(1.50)(1.0)(1.21)(1.31)}{(60)(90)(0.092)}} = 983 \text{ MPa}$$

Adjustments for number of cycles, from Figures 9–22 and 9–23:

$$Y_{Np} = 0.94 \quad Z_{NP} = 0.91 \quad Y_{NG} = 0.96 \quad Z_{NG} = 0.92$$

Required $s_{acP} = s_c(SF)(K_R)/Z_{NP} = 983$ MPa $(1.0)(1.0)/0.91 = 1080$ MPa

Using $s_{acP} = 1080$ MPa, Figure 9–12 shows the required hardness = HB 396 for through-hardened Grade 1 steel. This is acceptable but near the upper end of recommended range.

Material specification: From Table A4–5 (other possibilities exist),

SAE 4340 OQT 800; HB 415; $s_y = 1324$ MPa; $s_u = 1448$ MPa; 12% elongation.

Check other stresses: The contact stress for the gear and the bending stress for the pinion and the gear are expected to require less material hardness and strength.

Required $s_{acG} = s_c(SF)(K_R)/Z_{NG} = 983$ MPa$(1.0)(1.0)/0.92 = 1068$ MPa

This is slightly lower than for the pinion (OK)

$$S_{tP} = \frac{W_t\,K_o\,K_s\,K_B\,K_m\,K_v}{FmJ_P} = \frac{(5536)(1.50)(1)(1)(1.21)(1.31)}{(60)(5)(0.315)} = 139 \text{ MPa}$$

Required $s_{atP} = s_{tP}(SF)(K_R)/Y_{NP} = 139$ MPa$(1.0)(1.0)/0.94 = 148$ MPa

Referring to Figure 9–11, it is obvious that bending stress requires far lower hardness for the gear teeth, less than HB 180. The stress in the gear is always less than that in the pinion so it will obviously be safe as well.

Summary of the Design:

$P = 15.0$ kW from an electric motor to a large meat grinder

Pinion speed: $n_P = 575$ rpm Gear speed: $n_G = 272$ rpm

Number of teeth: $N_P = 18$; $N_G = 38$ Center distance: $C = 140.00$ mm

Module: $m = 5$ mm Diameters: $D_P = 90$ mm; $D_G = 190$ mm

Material: SAE 4340 OQT 800

Comment: A redesign may be considered with several possible approaches:

1. Increase the face width, F, to lower the stresses and permit the choice of a material with more moderate required hardness and better ductility. The recommended upper limit of face width is $16m = 16(5) = 80$ mm.
2. Increase the size of pinion and its number of teeth (same module) to lower stresses.

   Possible trial: Module: $m = 5$ mm Number of teeth: $N_P = 22$; $N_G = 46$

   Center distance: $C = 170.00$ mm Diameters: $D_P = 110$ mm; $D_G = 230$ mm
3. Consider case-hardened steel for the initial design, rather than through-hardened steel. A smaller design is possible.

9–13 COMPUTER-AIDED SPUR GEAR DESIGN AND ANALYSIS

This section presents one approach to assisting the gear designer with the many calculations and judgments that must be made to produce an acceptable design. The spreadsheet shown in Figure 9–25 facilitates the completion of a prospective design for a pair of gears in a few minutes by an experienced designer. You must have studied all of the material here and in Chapter 8 in order to understand the data needed in the spreadsheet and to use it effectively.

The recommended use of the spreadsheet is to create a series of design iterations that allow you to progress toward an optimum design in a short amount of time. It follows the process outlined in Section 9–11 up to the point of computing the required allowable bending stress number and the allowable contact stress number for both the pinion and the gear. The designer must use those data to specify suitable materials for the gears and their heat treatments.

Given below is a discussion of the essential features of the spreadsheet. In general, it first calls for the input of basic performance data, allowing a proposed geometry to be specified. The final result is the completion of the stress analyses for bending and pitting resistance for both the pinion and the gear. Equations (9–16) and (9–28) are combined for the bending analysis. The analysis of pitting resistance uses Equations (9–23) and (9–29). The designer must provide data for the several factors in those equations taken from appropriate figures and charts or based on design decisions. Virtually all computations are performed by the spreadsheet, allowing the designer to exercise judgment based on the intermediate results.

The format used for the spreadsheet helps the designer follow the process. After defining the problem at the top of the sheet, the first column at the left calls for several pieces of input data. Any value in italics within a gray-shaded area must be entered by the designer. White areas offer the results of calculations and provide guidance. The upper part of the second column also guides the designer in determining values for the several factors needed to complete the stress analyses for bending and pitting resistance. The area at the lower right of the spreadsheet gives the primary output data on which the design decisions for materials and heat treatments are based.

The data in Figure 9–25 are taken from Example Problem 9–4 which was completed in the traditional manner in Section 9–11.

Discussion of the Use of the Spur Gear Design Spreadsheet

1. **Describing the application:** In the heading of the sheet, the designer is asked to describe the application for identification purposes and to focus on the basic uses for the gears. Use the nature of the prime mover and the driven machine to specify the overload factor, K_o, using Table 9–7 as a guide.
2. **Initial input data:** It is assumed that designers begin with a knowledge of the power transmission requirement, the rotational speed of the pinion of the gear pair, and the desired output speed. Using the nature of the application and the overload factor, K_o, compute the *design power* from

$$P_{\text{des}} = K_o P$$

Then use Figure 9–24 to determine a trial value of the diametral pitch using the design power and the rotational speed of the pinion. The number of teeth in the pinion is a critical design decision because the size of the system depends on this value. Ensure against interference. An initial trial value of $N_P = 17$ to 20 is often a good choice.
3. **Number of gear teeth:** The spreadsheet computes the approximate number of gear teeth to produce the desired output speed from $N_G = N_P(n_G/n_P)$. But, of course, the number of teeth in any gear must be an integer, and the actual value of N_G is entered by the designer.
4. **Computed data:** The seven values reported in the middle of the first column are all determined from the input data, and they allow the designer to evaluate the suitability of the geometry of the proposed design at this point. Changes to the input data can be made at this time if any value is out of the desired range in the judgment of the designer.
5. **Secondary input data:** When a suitable geometry for the gears is obtained, the designer enters the data called for at the lower part of the first column of the spreadsheet. The locations of data in pertinent tables and figures are listed.
6. **Factors in design analysis:** The stress analysis requires many factors to account for the unique situation of the design being pursued. Again, guidance is offered, but the designer must enter the values of the required factors. Many of the factors can have a value of 1.00 for normal conditions.
7. **Alignment factor:** The alignment factor depends on two other factors: the pinion proportion factor and the mesh alignment factor as shown in Figures 9–16 and 9–17. The suggested values in the white areas are computed from the equations given in the figures. Note the listed value of F/D_P. If $F/D_P < 0.50$, use $F/D_P = 0.50$ to find C_{pf}. The designer must decide on the type of gearing to be used (open or closed) and the degree of precision to be designed into the system. The final result is computed from the input data.
8. **Size, and rim thickness factors:** Consult Table 9–8 along with Figure 9–18. Note that the rim thickness factor can be different for the pinion and the gear. Sometimes the smaller pinion is made from a solid blank while the larger gear can use a rim-and-spoke design.

DESIGN OF SPUR GEARS

APPLICATION:	*Example Problem 9–4*		
Wood chipper driven by an electric motor			
Initial Input Data:			
Overload Factor:	$K_o =$	*1.75*	Table 9–7
Transmitted Power:	$P =$	*3*	hp
Design Power	$P_{des} =$	5.25	hp
Diametral Pitch:	$P_d =$	*12*	Fig. 9–24
Input Speed:	$n_P =$	*1750*	rpm
Number of Pinion Teeth:	$N_P =$	*18*	
Desired Output Speed:	$n_G =$	*462.5*	rpm
Computed number of gear teeth:		68.1	
Enter: Chosen No. of Gear Teeth:	$N_G =$	*68*	
Computed Data:			
Actual Output Speed:	$n_G =$	463.2	rpm
Gear Ratio:	$m_G =$	3.78	
Pitch Diameter—Pinion:	$D_P =$	1.500	in
Pitch Diameter—Gear:	$D_G =$	5.667	in
Center Distance:	$C =$	3.583	in
Pitch Line Speed:	$v_t =$	687	ft/min
Transmitted Load:	$W_t =$	144	lb
Secondary Input Data:			
	Min	Nom	Max
Face Width Guidelines (in):	0.667	1.000	1.333
Enter: Face Width:	$F =$	*1.000*	in
Ratio: Face width/pinion diameter:	$F/D_P =$	0.67	
Recommended range of ratio:	$0.50 < F/D_P < 2.00$		
Enter: Elastic Coefficient:	$Cp =$	*2300*	Table 9–10
Enter: Quality Number:	$A_v =$	*11*	Table 9–3
Dynamic Factor:	$K_v =$	1.35	Table 9–9
[Factors for computing K_v:]	$B = 0.826$	$C = 59.75$	
Reference:	$N_P = 18$	$N_G = 68$	
Bending Geometry Factor—Pinion:	$J_P =$	*0.325*	Fig. 9–15
Bending Geometry Factor—Gear:	$J_G =$	*0.410*	Fig. 9–15
Reference:	$m_G =$	3.78	
Enter: Pitting Geometry Factor:	$I =$	*0.104*	Fig. 9–21

Factors in Design Analysis:				
Alignment Factor, $K_m = 1.0 + C_{pf} + C_{ma}$	If $F < 1.0$	If $F > 1.0$		
Pinion Proportion Factor, $C_{pf} =$	0.042	0.042	$[0.50 < F/D_P < 2.00]$	
Enter: $C_{pf} =$	*0.042*	Figure 9–16		
Type of gearing:	Open	Commer.	Precision	Ex. Prec.
Mesh Alignment Factor, $C_{ma} =$	0.264	0.143	0.080	0.048
Enter: $C_{ma} =$	*0.264*	Figure 9–17		
Alignment Factor: $K_m =$	1.31	[Computed]		
Size Factor: $K_s =$	*1.00*	Table 9–8: Use 1.00 if $P_d >= 5$		
Pinion Rim Thickness Factor: $K_{BP} =$	*1.00*	Fig. 9–18: Use 1.00 if solid blank		
Gear Rim Thickness Factor: $K_{BG} =$	*1.00*	Fig. 9–18: Use 1.00 if solid blank		
Service Factor: $SF =$	*1.00*	Use 1.00 if no unusual conditions		
Reliability Factor: $K_R =$	*1.00*	Table 9–11 Use 1.00 for $R = .99$		
Enter: Design Life:	*3000*	hours	See Table 9–12	
		Guidelines: Y_N, Z_N		
		10^7 cycles	$>10^7$	$<10^7$
Pinion—Number of load cycles: $N_P =$	3.2E+08			
Gear—Number of load cycles: $N_G =$	8.3E+07			
Bending Stress Cycle Factor: $Y_{NP} =$	*0.96*	1.00	0.96	Fig. 9–22
Bending Stress Cycle Factor: $Y_{NG} =$	*0.98*	1.00	0.98	Fig. 9–22
Pitting Stress Cycle Factor: $Z_{NP} =$	*0.92*	1.00	0.92	Fig. 9–23
Pitting Stress Cycle Factor: $Z_{NG} =$	*0.95*	1.00	0.95	Fig. 9–23
Stress Analysis: Bending				
Pinion: Required $s_{at} =$	17 102	psi	See Fig. 9–11 or	
Gear: Required $s_{at} =$	13 280	psi	Table 9–5	
Stress Analysis: Pitting				
Pinion: Required $s_{ac} =$	133 471	psi	See Fig. 9–12 or	
Gear: Required $s_{ac} =$	129 256	psi	Table 9–5	
Required hardness of pinion HB:	324	Equations in Fig. 9–12-Grade 1		
Required hardness of gear HB:	311	Equations in Fig. 9–12-Grade 1		
Specify materials, alloy and heat treatment, for most severe requirement.				
One possible material specification:				
Pinion requires HB 324: SAE 4140 OQT 1000; HB 341, 18% Elongation				
Gear requires HB 311: SAE 4140 OQT 1100; HB 311, 20% Elongation				
Comments:				
It would be reasonable to specify the same heat treatment for both the pinion and the gear because their contact stresses are very similar.				

FIGURE 9–25 Spreadsheet solution for spur gear design using data for Example Problem 9–4

9. **Dynamic factor:** The spreadsheet uses the equations included in Table 9–9 to compute the dynamic factor using the quality number and pitch line speed found from data in the first column.
10. **Service factor:** This is a design decision as discussed in Section 9–11. Often a value of 1.00 is used if no unusual conditions are expected that are not already accounted for in other factors. Larger service factors allow for a higher degree of safety or to account for uncertainties.
11. **Reliability factor:** The designer must select a value from Table 9–11 according to the desired level of reliability.
12. **Stress cycle factors:** Here the designer must specify the design life in hours of operation for the gear pair being designed. Table 9–12 provides suggestions according to the use of the system. The number of cycles of stress is then computed for both the pinion and the gear, assuming the normal case of one cycle of one-direction stress per revolution. If the gears operate in a reversing mode, as idlers, or in planetary gear trains, this calculation must be adjusted to account for the multiple cycles of stress experienced in each revolution. Guidelines recommend factors of 1.00 for 10^7 cycles for which the allowable stress numbers are computed. For a larger number of cycles, equations given in Figures 9–22 and 9–23 are used to compute the recommended factors. Because a variety of data are given for the case of fewer than 10^7 cycles, the designer is referred to the figures to determine the factors. In any case, the user of the spreadsheet must enter the selected values.
13. **Stress analyses for bending and pitting resistance:** Finally, the required allowable bending stress number and the required allowable contact stress number are computed using Equations (9–28) and (9–29), adjusted for the special values of factors for the pinion and the gear.
14. **Specification of the materials and their heat treatment:** The final step is left to the designer to use the computed values from the stress analyses and to specify materials that will provide an adequate strength and surface hardness of the gear teeth. Pertinent data are listed in Figures 9–11 and 9–12 and Tables 9–5 and 9–6. The appendices tables for material properties may also be consulted once the required hardnesses of the materials are determined.

9–14 USE OF THE SPUR GEAR DESIGN SPREADSHEET

The spreadsheet developed in Section 9–13 is a useful tool that aids the designer in the process of completing a design for a pair of gears to be safe with regard to bending stresses in the teeth of the gears and for pitting resistance. The use of the spreadsheet was demonstrated for the data in Example Problem 9–4 as shown in Figure 9–25.

An important use for the spreadsheet is to propose and analyze several design alternatives and to work toward a goal of optimizing the design with regard to size, cost, or other parameters important to a particular design objective.

Designers for typical machine and vehicle drives would plan to use Grade 1 steels and standard quenching and tempering heat treatments. Where small size is critical or where cost is not a major concern, case hardening by carburizing, induction or flame hardening, or nitriding can be used. Therefore, it is usually desirable to produce several design alternatives that can be analyzed for cost and manufacturability. Then the final selection can be made with assurance that a reasonably optimum design has been identified.

Successive Iterations. We now continue the design process by making carefully selected changes in design decisions using the *Guidelines for Adjustments in Successive Iterations* from Section 9–11, just before Example Problem 9–4.

The design for the wood chipper drive from Example Problem 9–4 was very satisfactory for the goal of having an efficient design using through-hardened steel. Use of SAE 4140 OQT 1000 steel with a hardness of HB 341 is capable of resisting potential pitting caused by the pinion contact stress of approximately 134 ksi.

Now, how can we improve this design? The answer requires judgment about what constitutes improvement. Generally desirable design objectives for a gear drive, stated at the beginning of Section 9–11, were that the drive should:

Be compact and small	Operate smoothly and quietly
Have long life	Be low in cost
Be easy to manufacture	

Be compatible with other elements in the machine, such as bearings, shafts, the housing, the driver, and the driven machine.

In a given project, some of these objectives may be given higher priority than others. Furthermore, some objectives are counter to others; a highly compact and small drive will not likely be the lowest cost or the easiest to manufacture. However, it may be "worth the cost" to produce a smaller drive. More must be known about the application before making such judgments.

Let's proceed with the premise that the smallest practical gear drive for the wood chipper is desired while cost and ease of manufacture are secondary considerations. Referring to the *Guidelines*, we can conclude that the following changes will facilitate the design of a smaller, yet still safe design:

1. Obviously, each gear must be smaller than the initial design from Example Problem 9–4.
2. Using a higher value of diametral pitch with the same number of teeth will result in smaller gears and a correspondingly smaller center distance.

3. Smaller gears typically result in higher bending and contact stresses in the teeth, requiring higher strength materials.
4. The number of teeth (18) cannot be made much smaller without risking interference.
5. The face width can be used to *fine-tune* the design.
6. More accurate teeth (higher value of A_v) can also be used to *fine-tune* the design.

Design decisions for second trial design: The primary change to achieve a smaller design is to raise the value of diametral pitch. We can try $P_d = 16$ instead of the $P_d = 12$ used in Example Problem 9–4. Now we can show the advantage of using a spreadsheet as a calculation aid. Figure 9–26 shows the final result—a much smaller design that is still safe. ***Reaching this design took only a few minutes.*** The changed data are highlighted within bold boxes. Those that are design decisions are in the gray-shaded boxes while those computed by the spreadsheet have white backgrounds. User-controlled changes are:

1. $P_d = 16$
2. $F = 1.00$ in [At the maximum end of the recommended range; $F_{max} = 16/P_d$]
3. $A_v = 10$ [Modestly more precise than the original value of $A_v = 11$; Note that $K_v = 1.24$ is moderately lower than the value of 1.35 in the initial design.]
4. $K_m = 1.33$ [Modestly higher than the value of 1.31; caused by the change in $C_{pf} = 0.064$ from 0.042 for the initial trial because the ratio F/D_P changed. K_v was computed by the spreadsheet after the computed adjustment to C_{pf} was displayed and selected by the user.]

Comparison of Results:

	Design 1 Example Problem 9–4	Design 2 Example Problem 9–6
a. Pinion diameter: D_P	1.500 in	1.125 in
b. Gear diameter: D_G	5.667 in	4.250 in
c. Center distance: C	3.583 in	2.688 in
d. Pitch line speed: v_t	687 ft/min	515 ft/min
e. Transmitted load: W_t	144 lb	192 lb
f. Required s_{atP}:	17 102 psi	28 496 psi
g. Required s_{atG}:	13 280 psi	22 127 psi
h. Required s_{acP}:	133 471 psi	172 288 psi
i. Required s_{acPG}:	129 256 psi	166 847 psi
j. Material:	SAE 4140 OQT 1000 Through hardened	SAE 6150 OQT 1200, Case hardened by induction hardening

Notes about the changes and their effects:

a. Smaller gears do produce higher stresses.

b. The size of the gears and the overall drive were reduced significantly as can be seen in Figure 9–27 that shows a scale drawing of the two designs. One measure is the set of minimum inside dimensions of the housing to enclose the two gears, shown as x and y.

$$x_1 = 7.333 \text{ in} \quad x_2 = 5.500 \text{ in (25\% smaller)}$$

$$y_1 = 5.833 \text{ in} \quad y_2 = 4.375 \text{ in (25\% smaller)}$$

Where $x = C + D_{oP}/2 + D_{oG}/2$ and $y = D_{oG}$

c. The contact stress in the pinion (the governing stress value) increased by 33% for design #2 as compared with design #1. This was the result of the finer teeth and the smaller pitch diameter for the gears.

d. The value of the contact stress in the pinion of approximately 172 ksi made it impractical to use through-hardened steel. Note that the calculation built into the spreadsheet indicates that a hardness of 445 HB would be required. However, Figure 9–12 indicates that no design should rely on a hardness over 400 HB.

e. Therefore, resisting $s_{acP} = 172$ ksi requires some kind of case hardening, either *induction hardening, flame hardening,* or *carburizing and case hardening* as indicated in Table 9–5.

f. Use of induction hardening was a design decision but induction is a very frequent choice in the gear industry. See Internet site 19 for one provider of commercially available production-oriented induction hardening systems.

g. Specification of SAE 6150 steel for the gears was a design decision based on the need for a highly hardenable steel that can be hardened to 54 HRC minimum (as noted in Table 9–5). Referring to Figure A4–6 indicates that this alloy in the as-quenched condition can be in the range of 627 HB, corresponding to approximately 58 HRC (Appendix 17).

h. Induction hardening is considered a tertiary process for gears because:

   a. Gears are first machined, typically by hobbing.

   b. Then the gears are heat-treated to produce desirable properties in the core of the teeth. In this case, by referring to Figure A4–6, we chose to use the OQT 1200 heat treatment that will produce a core hardness of HB 293 and 20% elongation, a highly ductile condition.

   c. The the gears are induction hardened for a moderate depth to achieve the high pitting resistance on the faces of the teeth. Well-designed processes will also produce case hardening in the root area where the highest bending stresses occur.

i. Because the heat treatment may cause distortion, final grinding or other finishing processes may be needed to produce the final gear quality number; in the case $A_v = 11$.

DESIGN OF SPUR GEARS

APPLICATION:	*Example Problem 9–6*		
Alternate Design: *Wood chipper driven by an electric motor (Ex Prob 9–6)*			
Initial Input Data:			
Overload Factor	$K_o =$	175	Table 9–7
Transmitted Power	$P =$	3	hp
Design Power	$P_{des} =$	5.25	hp
Diametral Pitch	$P_d =$	*16*	Fig. 9–24
Input Speed	$n_P =$	*1750*	rpm
Number of Pinion Teeth	$N_P =$	*18*	
Desired Output Speed	$n_G =$	*462.5*	rpm
Computed number of gear teeth:		68.1	
Enter: Chosen No. of gear Teeth:	$N_G =$	*68*	
Computed Data:			
Actual Output Speed:	$n_G =$	463.2	rpm
Gear Ratio:	$m_G =$	3.78	
Pitch Diameter—Pinion:	$D_p =$	1.125	in
Pitch Diameter—Gear:	$D_G =$	4.250	in
Center Distance:	$C =$	2.688	in
Pitch Line Speed:	$v_t =$	515	ft/min
Transmitted Load:	$W_t =$	192	lb
Secondary Input Data:			
	Min	Norn	Max
Face Width Guidelines (in):	0.500	0.750	1.000
Enter: Face Width:	$F =$	*1.000*	in
Ratio; Face width/pinion diameter: $F/D_P =$		0.89	
Recommended range of ratio: $0.50 < F/D_P < 2.00$			
Enter: Elastic Coefficient:	$C_P =$	*2300*	Table 9–10
Enter: Quality Number:	$A_v =$	10	Table 9–3
Dynamic Factor:	$K_v =$	1.24	Table 9–9
[Factors for computing K_v:]	$B = 0.731$	$C = 65.04$	
Reference:	$N_P = 18$	$N_G = 68$	
Bending Geometry Factor—Pinion:	$J_P = 0.325$		Fig. 9–15
Bending Geometry Factor—Gear:	$J_G = 0.410$		Fig. 9–15
Reference:	$m_G = 3.78$		
Enter: Pitting Geometry Factor	$I = 0.104$		Fig. 9–21

Factors in Design Analysis:				
Alignment Factor, $K_m = 1.0 + C_{pf} + C_{ma}$	If $F < 1.0$	If $F > 1.0$		
Pinion Proportion Factor, $C_{pf} =$	0.064	0.064	[0.50 < F/Dp < 2.00]	
Enter: $C_{pf} =$	*0.064*	Figure 9–16		
Type of gearing:	Open	Commer.	Precision	Ex. Prec.
Mesh Alignment Factor, $C_{ma} =$	0.264	0.143	0.080	0.048
Enter: $C_{ma} =$	*0.264*	Figure 9–17		
Alignment Factor: $K_m =$	1.33	[Computed]		
Size Factor: $K_s =$	*1.00*	Table 9–8: Use 1.00 if $P_d >= 5$		
Pinion Rim Thickness Factor: $K_{BP} =$	*1.00*	Fig. 9–18: Use 1.00 if solid blank		
Gear Rim Thickness Factor; $K_{BG} =$	*1.00*	Fig. 9–18: Use 1.00 if solid blank		
Service Factor: $SF =$	*1.00*	Use 1.00 if no unusual conditions		
Reliability Factor: $K_R =$	*1.00*	Table 9–11 Use 1.00 for $R = .99$		
Enter: Design Life:	*3000*	hours	See Table 9–12	
Pinion-Number of load cycles: $N_P =$	3.2E+08	Guidelines: Y_N, Z_N		
Gear-Number of load cycles: $N_G =$	8.3E+07	10^7 cycles	$> 10^7$	10^7
Bending Stress Cycle Factor: $Y_{NP} =$	*0.96*	1.00	0.96	Fig. 9–22
Bending Stress Cycle Factor: $Y_{NG} =$	*0.98*	1.00	0.98	Fig. 9–22
Pitting Stress Cycle Factor: $Z_{NP} =$	*0.92*	1.00	0.92	Fig. 9–23
Pitting Stress Cycle Factor: $Z_{NG} =$	*0.95*	1.00	0.95	Fig. 9–23
Stress Analysis: Bending				
Pinion: Required $s_{at} =$	28 496	psi	See Fig. 9–11 or	
Gear: Required $s_{at} =$	22 127	psi	Table 9–5	
Stress Analysis: Pitting				
Pinion: Required $s_{ac} =$	172 288	psi	See Fig. 9–12 or	
Gear: Required $s_{ac} =$	166 847	psi	Table 9–5	
Required hardness of pinion HB:	445	Equations in Fig. 9–12-Grade 1		
Required hardness of gear HB:	428	Equations in Fig. 9–12-Grade 1		
Specify materials, alloy and heat treatment, for most severe requirement.				
***Material requirements: Note: Through-hardened steel* cannot be used.**				
Pinion requires case hardening—HRC 54 minimum; induction, flame, or carburizing				
Gear requires case hardening—HRC 50 minimum; induction, flame, or carburizing				
Comments: Possible specification—Induction-hardened SAE 6150 steel				
Pinion and Gear: SAE 6150: Core heat teatment OQT 1200; HB 293				
Induction harden to HRC 54 minimum				

FIGURE 9–26 Spreadsheet solution for Example Problem 9–6; alternate spur gear design using data for Example Problem 9–4

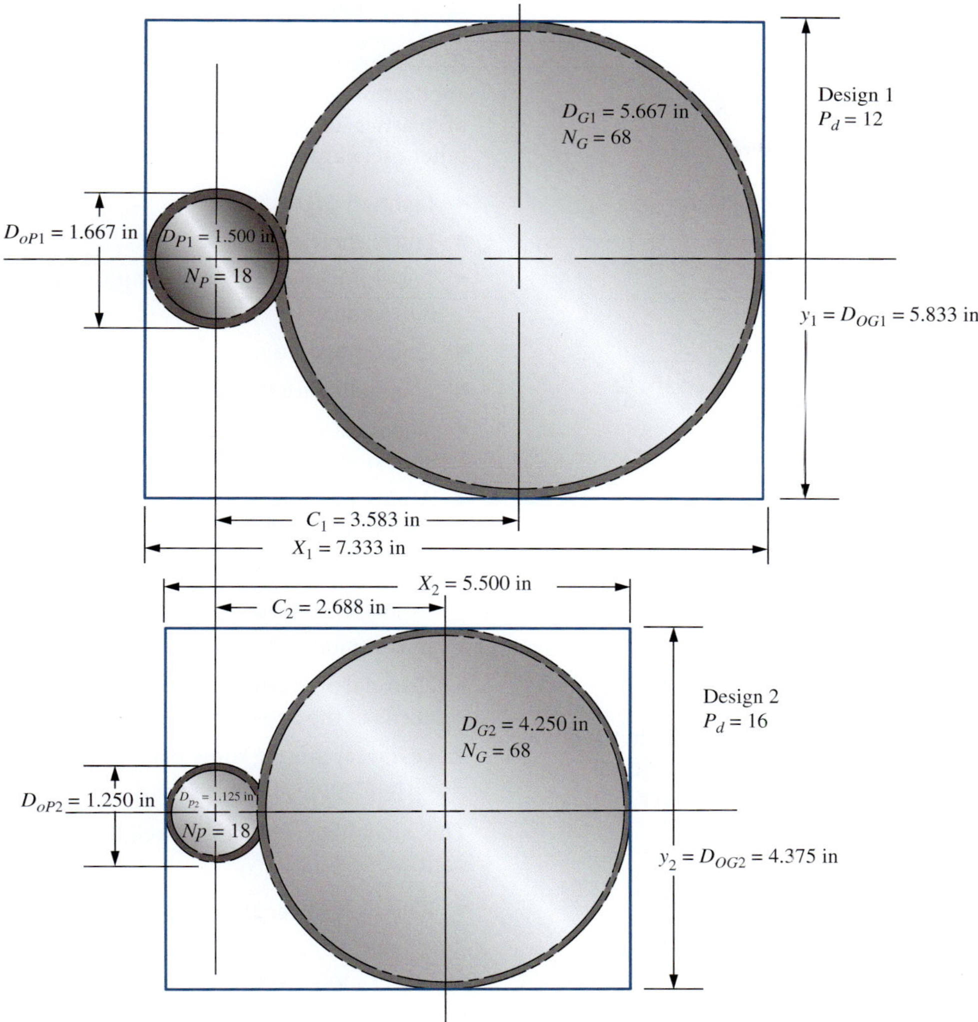

FIGURE 9–27 Comparison of sizes of gear designs for Example Problems 9–4 and 9–6, with minimum inside dimensions for housings

j. This long list of processing steps makes the choice of case hardening by induction hardening much more costly than basic through hardening. The designer must make the judgment that the value of the benefit of the smaller system exceeds the added costs.

It is expected that readers of this book use similar decision-making processes when designing gear drives and specifying materials and heat treatments in their designs.

9–15 POWER-TRANSMITTING CAPACITY

It is sometimes desirable to compute the amount of power that a gear pair can safely transmit after it has been completely defined. The *power-transmitting capacity* P_{cap} is the capacity when the tangential load causes the expected stress to equal the allowable stress number with all of the modifying factors considered. The capacity should be computed for both bending and pitting resistance and for both the pinion and the gear.

When similar materials are used for both the pinion and the gear, it is likely that the pinion will be critical for bending stress. But the most critical condition is usually pitting resistance. The following relationships can be used to compute the power-transmitting capacity. In this analysis, it is assumed that the operating temperature of the gears and their lubricants is 250°F and that gears are produced with the appropriate surface finish.

Bending

We start with Equations (9–16) and (9–28) in which the computed bending stress number is compared with the modified allowable bending stress number for the gear:

$$s_{at} = s_t \frac{(SF)(KR)}{Y_N} = \frac{W_t P_d}{FJ} K_o K_s K_m K_B K_v \times \frac{(SF)(KR)}{Y_N}$$

But solving for W_t gives

$$W_t = \frac{s_{at}Y_N FJ}{(SF)K_R K_o K_s K_m K_B K_v P_d} \tag{9–31}$$

It was shown in Equation (9–8) that

$$W_t = (126\,000)(P)/(n_P D_P)$$

Then substituting into Equation (9–31) and calling the power P_{cap} gives

$$\frac{(126\,000)(P_{cap})}{n_P D_P} = \frac{s_{at}Y_N FJ}{(SF)K_R K_o K_s K_m K_B K_v P_d}$$

Solving for P_{cap}, we have

$$P_{cap} = \frac{s_{at}Y_N FJ n_P D_P}{(126\,000)(P_d)(SF)K_R K_o K_s K_m K_B K_v} \tag{9–32}$$

This equation should be solved for both the pinion and the gear. Most variables will be the same except for s_{at}, Y_N, J, and possibly K_B.

Pitting Resistance

Here we start with Equations (9–23) and (9–29) in which the computed contact stress number is compared with the modified allowable contact stress number for the gear. Equation (9–26) can be expressed in the form

$$s_{ac} = s_c\frac{(SF)(KR)}{Z_N} = C_P\sqrt{\frac{W_t K_o K_s K_m K_v}{D_P FI}} \times \frac{(SF)(KR)}{Z_N}$$

Squaring both sides of this equation and solving for W_t gives

$$\frac{W_t K_o K_s K_m K_v}{D_P FI} = \left[\frac{s_{ac}Z_N}{(SF)K_R C_P}\right]^2$$

$$W_t = \frac{D_P FI}{K_o K_s K_m K_v}\left[\frac{s_{ac}Z_N}{(SF)K_R C_P}\right]^2 \tag{9–33}$$

Now substituting this into Equation (9–8) and solving for the power P_{cap} gives

$$P_{cap} = \frac{W_t D_P n_P}{126\,000} = \frac{D_P n_P D_P FI}{126\,000\,K_o K_s K_m K_v}\left[\frac{s_{ac}Z_N}{(SF)K_R C_P}\right]^2$$

$$P_{cap} = \frac{n_P FI}{126\,000\,K_o K_s K_m K_v}\left[\frac{s_{ac}D_P Z_N}{(SF)K_R C_P}\right]^2 \tag{9–34}$$

Equations (9–32) and (9–34) should be used to compute the power-transmitting capacity for a pair of gears of known design with particular materials. From the material specification along with its condition (typically a heat treatment or case-hardening process), the limiting values for s_{at} and s_{ac} can be found from Figures 9–11 and 9–12 from Tables 9–5 and 9–6. Entering those values along with the known data for the proposed gear design permits the calculation of the power-transmitting capacity, P_{cap}. This value can be shared with colleagues and customers to verify the suitability of a design for a particular application.

9–16 PLASTICS GEARING

Plastics are satisfying an important and growing part of the applications for gearing. Some of the numerous advantages of plastics in gearing systems compared with steels and other metals are:

- Lighter weight
- Lower inertia
- Possibility of running with little or no external lubrication
- Quieter operation
- Low sliding friction, which results in efficient gear meshing
- Chemical resistance and ability to operate in corrosive environments
- Ability to operate well under conditions of moderate vibration, shock, and impact
- Relatively low cost when made in large quantities
- Ability to combine several features into one part
- Accommodation of larger tolerances because of resiliency
- Material properties that can be tailored to meet the needs of the application
- Less wear among some plastics compared to metals in certain applications

The advantages must be weighed against disadvantages such as:

- Relatively lower strength of plastics as compared with metals
- Lower modulus of elasticity
- Higher coefficients of thermal expansion
- Difficulty operating at high temperatures
- Initial high cost for design, development, and mold manufacture
- Dimensional change with moisture absorption that varies with conditions
- Wide range of possible material formulations, which makes design more difficult

Some plastic gears are cut using hobbing or shaping processes similar to those used to cut metallic gears. However, most plastic gears are produced with the injection molding process because of its ability to make large quantities rapidly with low unit cost. Mold design is critical because it must accommodate the shrinking that occurs as the molten plastic solidifies. The typical successful approach accounts for predicted shrinkage by making the die larger than the required finished gear size. However, the allowance is not uniform throughout the gear, and significant amounts of data are required about the material molding properties and the molding process itself to produce plastic gears with high dimensional accuracy. Computer-assisted mold design software that simulates the flow of molten plastic through the mold cavities and the curing process is often used. The gear mold or the gear cutting tools are designed to produce dimensionally accurate gear teeth with tooth thickness controlled to produce a proper amount of backlash during operation. The electrical discharge machining process (EDM) is typically used to produce accurate gear tooth forms in molds made from high-hardness, wear-resistant steels to ensure that large production runs can be made without replacing tooling.

Plastic Materials for Gears

The great variety of plastics available makes material selection difficult, and it is recommended that gear system designers consult with material suppliers, mold designers, and manufacturing staff during the design process. While simulation can aid in reaching a suitable design, it is recommended that testing be done in realistic conditions before committing the design to production. Some of the more popular types of materials used for gears are:

Nylon	Acetal	ABS (acrylonitrile-butadiene-styrene)
Polycarbonate	Polyurethane	Polyester thermoplastic
Polyimide	Phenolic	Polyphenylene sulfide
Polysulfones	Phenylene oxides	Styrene-acrylonitrile (SAN)

Designers must seek a balance of material characteristics appropriate to the application, considering, for example:

- Strength in flexure under fatigue conditions
- High modulus of elasticity for stiffness
- Impact strength and toughness
- Wear and abrasion resistance
- Dimensional stability under expected temperatures
- Dimensional stability due to moisture absorption from liquids and humidity
- Frictional performance and need for lubrication, if any
- Operation in vibration environments
- Chemical resistance and compatibility with the operating environment
- Sensitivity to ultraviolet radiation
- Creep resistance if operated under load for long periods of time
- Flame retarding ability
- Cost
- Ease of processing and molding
- Assembly and disassembly considerations
- Compatibility with mating parts
- Environmental impact during processing, use, recycling, and disposal

The basic plastic materials listed previously are typically modified with fillers and additives to produce optimum as-molded properties. Some of these are:

> ***Reinforcements for strength, toughness, moldability, long-term stability, thermal conductivity, and dimensional stability:*** Long glass fibers, chopped glass fibers, milled glass, woven glass fibers, carbon fibers, glass beads, aluminum flake, mineral, cellulose, rubber modifiers, wood flour, cotton, fabric, mica, talc, and calcium carbonate.
>
> ***Fillers to improve lubricity and overall frictional performance:*** PTFE (polytetrafluoroethylene), silicone, carbon fibers, graphite powders, and molybdenum disulfide (MoS_2).

Refer to Section 2–17 for additional discussion about plastic materials, their properties, and special considerations for selecting plastics. See Internet sites 1, 8, 18, and 20.

Design Strength for Plastic Gear Materials

Data are provided here for typical plastic materials used for gears. They can be applied to problem solving in this book. However, verification of properties for materials to be actually used in a commercial application, with due regard for the operating conditions, should be acquired from the material supplier. The effects of temperature on strength, modulus, toughness, chemical stability, and dimensional precision are particularly important. Manufacturing processes must be controlled to ensure that final properties are consistent with prescribed values.

Table 9–14 lists some selected data for allowable tooth bending stress, s_{at}, in plastic gears. Much additional data for other materials can be found in References 19 and 21. Note the significant increase in allowable strength provided by the glass reinforcement. The combination of glass fibers and the basic plastic matrix performs like a composite material with the amount of reinforcement typically ranging from 20% to 50%.

Material suppliers may be able to provide fatigue data for plastics in charts such as those shown in Figure 9–28, showing allowable bending stress versus number of cycles to failure for DuPont Zytel© nylon resin and Delrin© acetal resin. These data are for molded gears operating at room

TABLE 9–14 Approximate Allowable Tooth Bending Stress, s_{at}, in Plastic Gears

	Approximate allowable bending stress, s_{at}, psi (MPa)	
Material	**Unfilled**	**Glass-filled**
ABS	3000 (21)	6000 (41)
Acetal	5000 (34)	7000 (48)
Nylon	6000 (41)	12 000 (83)
Polycarbonate	6000 (41)	9000 (62)
Polyester	3500 (24)	8000 (55)
Polyurethane	2500 (17)	

Source: Plastics Gearing. Manchester, CT: ABA/PGT Publishing, 1994.

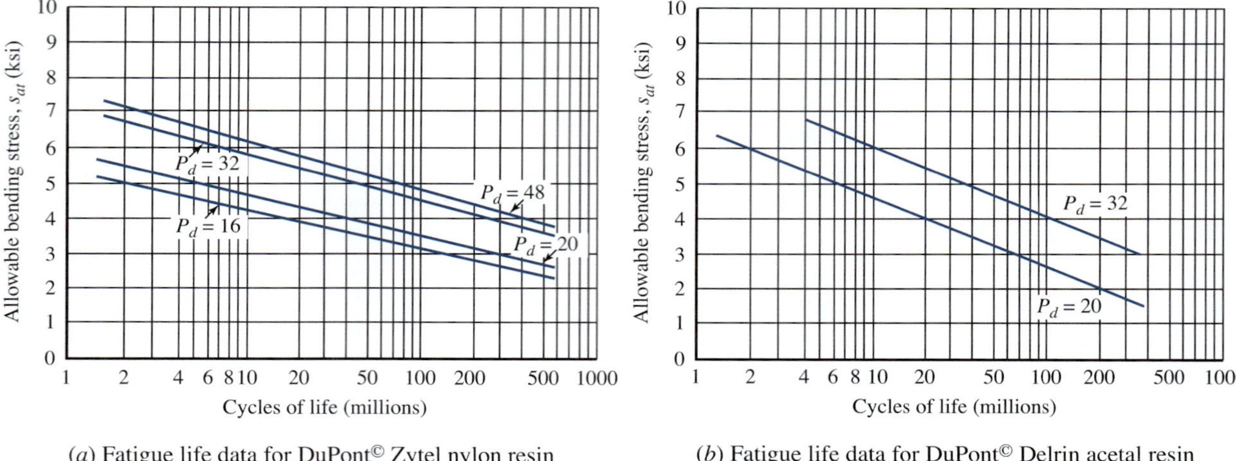

(*a*) Fatigue life data for DuPont© Zytel nylon resin

(*b*) Fatigue life data for DuPont© Delrin acetal resin

FIGURE 9–28 Fatigue life data for two types of plastic materials used for gears (DuPont Polymers, Wilmington, DE)

temperature with diametral pitches shown, pitch line velocity below 4000 ft/min, and continuous lubrication. Reductions should be applied for cut gears, higher temperatures, different pitches, and different lubrication conditions. See Reference 21.

Tooth Geometry

In general, standard tooth geometry for plastic gears conforms to the configurations described in Section 8–4. Standard diametral pitches from Table 8–3 and standard metric modules from Table 8–4 should be used unless there are major advantages to using other values. Suppliers' ability to provide nonstandard pitches should be investigated. Pressure angles of $14\frac{1}{2}°$, 20°, and 25° are used, with 20° usually preferred. Standard formulas for addendum, dedendum, and clearance for full-depth involute teeth are listed in Table 8–1. Gear quality values are set similarly to those for metallic gears as discussed in Section 9–5. The typical AGMA quality number produced by injection molding is in range of A11 To A7.

Designers sometimes use special tooth forms to tailor the strength of plastic gear teeth to the demands of particular applications. The 20° stub tooth system provides a shorter, broader tooth than the standard 20° full-depth tooth system, decreasing tooth-bending stress. The Plastics Gearing Technology unit of the ABA-PGT company has developed another system that is finding favor with some designers. See References 1, 2, 12, and 13.

Many designers of plastic gears prefer to use a longer addendum on the pinion and a shorter addendum on the mating gear to produce more favorable operation because of the greater flexibility of plastics as compared with metals. Tooth thickness is typically thinned on either or both of the pinion and gear to provide acceptable backlash and to ensure that mating gears do not bind. Binding may result from deflection of the teeth under load or from expansions due to increased temperature or moisture absorption from exposure to water or high humidity. Enlarging the center distance is another method employed to adjust for backlash. Designers must specify these feature sizes on drawings and in specifications. Consult AGMA Standard 1006 A97 (R2003) *Tooth Proportions for Plastic Gears* for details. Reference 2 provides useful tables of formulas and data for adjustments to tooth form and center distance. Reference 21 recommends the range of backlash values shown in Figure 9–29.

Shrinkage

During manufacture of plastic gears using injection molding, enlarging the effective diametral pitch and the pitch diameter of the gear teeth cut into the mold accommodates

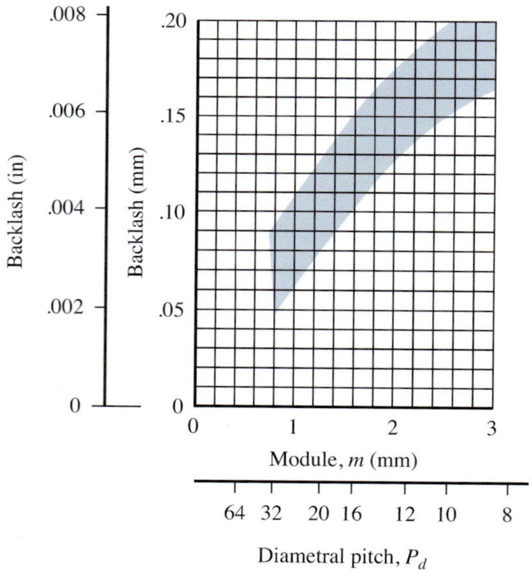

FIGURE 9–29 Recommended backlash for plastic gears

shrinkage. The pressure angle is also adjusted. The nominal corrections are computed as follows:

$$P_{dc} = \frac{P_d}{(1 + S)} \quad \textbf{(9–35)}$$

$$\cos \phi_1 = \frac{\cos \phi}{(1 + S)} \quad \textbf{(9–36)}$$

$$D_c = N/P_{dc} \quad \textbf{(9–37)}$$

where S = shrinkage of material
P_d = standard diametral pitch for the gear
P_{dc} = modified diametral of the teeth in the mold
ϕ = standard pressure angle for the gear
ϕ_1 = modified pressure angle of the teeth in the mold
N = number of teeth
D_c = modified pitch diameter of teeth in the mold

After molding, the teeth should very nearly conform to standard geometry. Additional adjustments are sometimes made, relieving the tips of the teeth for smoother engagement and increasing the tooth width at the base near the point of highest bending stress.

Stress Analysis

Bending stress analysis for plastic gears relies on the basic Lewis formula introduced in Section 9–8, Equation (9–13). The modifying factors called for by the AGMA standards for steel gears are not specified for plastic gears at this time. We can account for uncertainty or shock loading by inserting a safety factor. The overload factor from Table 9–7 can be used as a guide. Testing of the proposed design in realistic conditions should be completed. The bending stress equation then becomes

$$\sigma_t = \frac{W_t P_d (SF)}{FY} \quad \textbf{(9–38)}$$

Values for the Lewis form factor, Y, shown in Table 9–15, describe the geometry of the involute gear teeth acting as a cantilever beam with the load applied near the pitch point. Thus Equation (9–38) gives the bending stress at the root of the tooth. Most plastic gear designs call for a generous fillet radius between the start of the active involute profile on the flank of the tooth and the root, resulting in little, if any, stress concentration.

TABLE 9–15 Lewis Tooth Form Factor, *Y*, for Load Near the Pitch Point

Number Teeth	Tooth Form: 14 1/2° Full Depth	20° Full Depth	20° Stub
14	–	–	0.540
15	–	–	0.566
16	–	–	0.578
17	–	0.512	0.587
18	–	0.521	0.603
19	–	0.534	0.616
20	–	0.544	0.628
22	–	0.559	0.648
24	0.509	0.572	0.664
26	0.522	0.588	0.678
28	0.535	0.597	0.688
30	0.540	0.606	0.698
34	0.553	0.628	0.714
38	0.566	0.651	0.729
43	0.575	0.672	0.739
50	0.588	0.694	0.758
60	0.604	0.713	0.774
75	0.613	0.735	0.792
100	0.622	0.757	0.808
150	0.635	0.779	0.830
300	0.650	0.801	0.855
Rack	0.660	0.823	0.881

Source: DuPont Polymers, Wilmington, DE.

Wear Considerations

Wear of tooth surfaces in plastic gear teeth is a function of the contact stress between mating teeth as it is with metal teeth. Equation (9–18) can be used to compute the contact stress. However, published data are lacking for allowable contact stress values.

In reality, lubrication and the *combination of materials in mating gears* play major roles in the wear life of the pair. Communication with material suppliers and testing of proposed designs are recommended. Presented here are some general guidelines from References 2 and 21.

- Continuously lubricated gearing promotes the longest life.
- With continuous lubrication and light loads, fatigue resistance, not wear, typically determines life.
- Unlubricated gears tend to fail by wear, not fatigue, provided proper design bending stresses are used.
- When continuous lubrication is not practical, initially lubricating the gearing can aid in the run-in process and add life compared with gears that are never lubricated.
- When continuous lubrication is not practical, the combination of a nylon pinion and an acetal gear exhibits low friction and wear.
- Excellent wear performance for relatively high loads and pitch line speeds can be obtained by using a lubricated

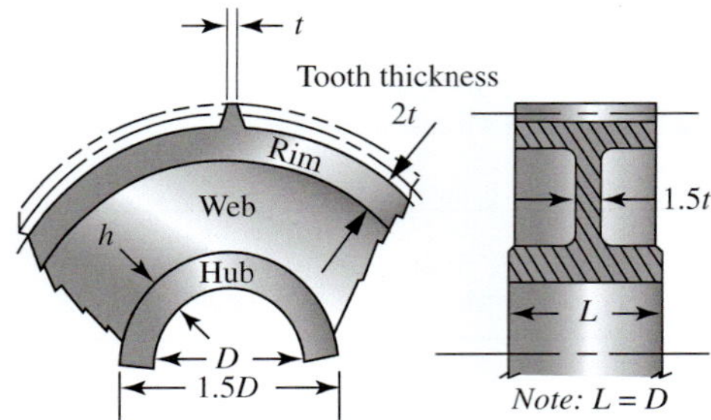

FIGURE 9–30 Suggested plastic gear proportions (DuPont Polymers, Wilmington, DE)

pair of a hardened steel pinion (HRC > 50) mating with a plastic gear made from nylon, acetal, or polyurethane.

- Wear accelerates when operating temperatures rise. Cooling to promote heat dissipation can increase life.

Gear Shapes and Assembly

References 2 and 21 include many recommendations for the geometric design of gears considering strength, inertia, and molding conditions. Many smaller gears are simply made with uniform thickness equal to the face width of the gear teeth. Larger gears often have a rim to support the teeth, a thinned web for lightening and material savings, and a hub to facilitate mounting on a shaft. Figure 9–30 shows recommended proportions. Symmetrical cross sections are preferred, along with balanced section thicknesses to promote good flow of material and to minimize distortion during molding.

Fastening gears to shafts requires careful design. Keys placed in shaft key seats and keyways in the hub of the gear provide reliable transmission of torque. For light torques, setscrews can be used, but slippage and damage of the shaft surface are possible. The bore of the gear hub can be lightly press-fit onto the shaft with care to ensure that a sufficient torque can be transmitted while not overstressing the plastic hub. Knurling the shaft before pressing the gear on increases the torque capability. Some designers prefer to use metal hubs to facilitate the use of keys. Plastic is then molded onto the hub to form the rim and gear teeth.

Design Procedure

Design of plastic gearing should consider a variety of possibilities, and it is likely to be an iterative process. The following procedure outlines the steps for a given trial using U.S. Customary units.

PROCEDURE FOR DESIGNING PLASTIC GEARS

1. Determine the required horsepower, P, to be transmitted and the speed of rotation, n_P, of the pinion in rpm.
2. Specify the number of teeth, N, and select a trial diametral pitch diameter for the pinion.
3. Compute the pinion diameter from $D_P = N_P/P_d$.
4. Compute the transmitted load, W_t (in lb), from Equation (9–8), repeated here.

$$W_t = (126\,000)(P)/(n_P D_P)$$

5. Specify the tooth form and determine the Lewis form factor, Y, from Table 9–15.
6. Specify a safety factor, SF. Refer to Table 9–7 for guidance.
7. Specify the material to be used and determine the allowable stress, s_{at}, from Table 9–15 or Figure 9–33.
8. Solve Equation (9–38) for the face width, F, and compute its value from,

$$F = \frac{W_t P_d (SF)}{s_{at} Y} \quad \textbf{(9–39)}$$

9. Judge the suitability of the computed face width as it relates to the application. Consider its mounting on a shaft, space available in the diametral and axial directions, and whether the general proportions are acceptable for injection molding. See References 2 and 21. No general recommendations are published for the face width of plastic gears and often they are narrower than similar metallic gears.
10. Repeat steps 2 to 9 until a satisfactory design for the pinion is achieved. Specify convenient dimensions for the final value of the face width and other features of the pinion.
11. Considering the desired velocity ratio between the pinion and the gear, compute the required number of teeth in the gear and repeat steps 3 to 9 using the same diametral pitch as the pinion. Using the same face width as for the pinion, the stress in the gear teeth will always be lower than in the pinion because the form factor Y will increase and all other factors will be the same. When the same material is to be used for the gear, it will always be safe. Alternatively, you could compute the bending stress directly from Equation 9–38 and specify a different material for the gear that has a suitable allowable bending stress.

Example Problem 9–7 Design a pair of plastic gears for a paper shredder to transmit 0.25 horsepower at a pinion speed of 1160 rpm. The pinion will be mounted on the shaft of an electric motor that has a diameter of 0.625 in with a keyway for a 3/16 × 3/16 in key. The gear is to rotate approximately 300 rpm.

Given Data $P = 0.25$ hp, $n_P = 1160$ rpm,

Shaft diameter $= D_s = 0.625$ in, Keyway for a 3/16 × 3/16 in key.

Approximate gear speed $= n_G = 300$ rpm

Solution Use the design procedure outlined in this section.

Step 1. Consider the given data.

Step 2. Specify $N_P = 18$ and $P_d = 16$ (Design decisions)

Step 3. $D_P = N_P/P_d = 18/16 = 1.125$ in. This seems reasonable for mounting on the 0.625 in motor shaft.

Step 4. Compute the transmitted load,

$$W_t = (126\,000)(P)/(n_P D_P) = (126\,000)(0.25)/[(1160)(1.125)] = 24.1 \text{ lb}$$

Step 5. Specify 20° full-depth teeth. Then $Y = 0.521$ for 18 teeth from Table 9–15.

Step 6. Specify a safety factor, SF. The shredder will likely experience light shock; the preference is to operate the gears without lubrication. Specify $SF = 1.50$ from Table 9–7.

Step 7. Specify unfilled nylon. From Table 9–14, $s_{at} = 6000$ psi.

Step 8. Compute the required face width using Equation (9–39).

$$F = \frac{W_t P_d(SF)}{s_{at} Y} = \frac{(24.1)(16)(1.50)}{(6000)(0.521)} = 0.185 \text{ in}$$

Step 9. The dimensions seem reasonable.

Step 10. Appendix 2 lists a preferred size for the face width of 0.200 in.

Comment: In summary, the proposed pinion has the following features:

$P_d = 16$, $N_P = 18$ teeth, $D_P = 1.125$ in, $F = 0.200$ in, Bore $= 0.625$ in,

Keyway for a $3/16 \times 3/16$ in key. Unfilled nylon material.

Step 11. Gear design: Specify $F = 0.200$ in, $P_d = 16$. Compute the number of teeth in the gear.

$N_G = N_P(n_P/n_G) = 18(1160/300) = 69.6$ teeth

Specify $N_G = 70$ teeth

Pitch diameter of gear $= D_G = N_G/P_d = 70/16 = 4.375$ in

From Table 9–15, $Y_G = 0.728$ by interpolation.

Stress in gear teeth using Equation 9–38:

$$\sigma_t = \frac{W_t P_d(SF)}{FY} = \frac{(24.1)(16)(1.50)}{(0.200)(0.728)} = 3973 \text{ psi}$$

Comment: This stress level is safe for nylon. The gear could also be made from acetal to achieve better wear performance.

9–17 PRACTICAL CONSIDERATIONS FOR GEARS AND INTERFACES WITH OTHER ELEMENTS

It is important to consider the design of the entire gear system when designing the gears because they must work in harmony with the other elements in the system. This section will briefly discuss some of these practical considerations and will show commercially available speed reducers.

Our discussion so far has been concerned primarily with the gear teeth, including the tooth form, pitch, face width, material selection, and heat treatment. Also to be considered is the type of gear blank. Figures 8–2 and 8–4 show several styles of blanks. Smaller gears and lightly loaded gears are typically made in the plain style. Gears with pitch diameters of approximately 5.0 in through 8.0 in are frequently made with thinned webs between the rim and the hub for lightening, with some having holes bored in the webs for additional lightening. Larger gears, typically with pitch diameters greater than 8.0 in, are made from cast blanks with spokes between the rim and the hub.

In many precision special machines and gear systems produced in large quantities, the gears are machined integral with the shaft carrying the gears. This, of course, eliminates some of the problems associated with mounting and location of the gears, but it may complicate the machining operations.

In general machine design, gears are usually mounted on separate shafts, with the torque transmitted from the shaft to the gear through a key. This setup provides a positive means of transmitting the torque while permitting easy assembly and disassembly. The axial location of the gear must be provided by another means, such as a shoulder on the shaft, a retaining ring, or a spacer (see Chapters 11 and 12).

Other considerations include the forces exerted on the shaft and the bearings that are due to the action of the gears. These subjects are discussed in Section 9–3. The housing design must provide adequate support for the bearings and protection of the interior components. Normally, it must also provide a means of lubricating the gears.

See References 9, 19, 20, 23, and 25–27 for additional practical considerations.

Lubrication

The action of spur gear teeth is a combination of rolling and sliding. Because of the relative motion, and because of the high local forces exerted at the gear faces, adequate lubrication is critical to smoothness of operation and gear life. A continuous supply of oil at the pitch line is desirable for most gears unless they are lightly loaded or operate only intermittently.

In splash-type lubrication, one of the gears in a pair dips into an oil supply sump and carries the oil to the pitch line. At higher speeds, the oil may be thrown onto the inside surfaces of the case; then it flows down, in a controlled fashion, onto the pitch line. Simultaneously, the oil can be directed to the bearings that support the shafts. One difficulty with the splash type of lubrication is that the oil is churned; at high gear speeds, excessive heat can be generated, and foaming can occur.

A positive oil circulation system is used for high-speed and high-capacity systems. A separate pump draws the oil from the sump and delivers it at a controlled rate to the meshing teeth.

The primary functions of gear lubricants are to reduce friction at the mesh and to keep operating temperatures at acceptable levels. It is essential that a continuous film of lubricant be maintained between the mating tooth surfaces of highly loaded gears and that there be a sufficient flow rate and total quantity of oil to maintain cool temperatures. Heat is generated by the meshing gear teeth, by the bearings, and by the churning of the oil. This heat must be dissipated from the oil to the case or to some other external heat-exchange device in order to keep the oil itself below 200°F (approximately 93°C). Above this temperature, the lubricating ability of the oil, as indicated by its viscosity, is severely decreased. Also, chemical changes can be produced in the oil, decreasing its lubricity. Because of the wide variety of lubricants available and the many different conditions under which they must operate, it is recommended that suppliers of lubricants be consulted for proper selection. (See also References 10, 16, 22 and 29.)

The AGMA, in Reference 10, defines several types of lubricants for use in gear drives.

- ***Rust and oxidation inhibited gear oils*** (called R&O) are petroleum based with chemical additives.
- ***Compounded gear lubricants*** (CP) blend 3% to 10% of fatty oils with petroleum oils.
- ***Extreme pressure lubricants*** (EP) include chemical additives that inhibit scuffing of gear tooth faces.
- ***Synthetic gear lubricants*** (S) are special chemical formulations applied mostly in severe operating conditions.

R&O lubricants are supplied in 10 ISO viscosity grades where the lower numbers refer to the lower viscosities. Similar numbers are used for the other types with modified grade designations carrying suffixes *CP*, *EP*, or *S*. The recommended lubricant grade depends on the ambient temperature around the drive and the pitch line velocity of the lowest speed pair of gears in a reducer. See Tables 9–16 and 9–17. Wormgear drives call for higher viscosity grades.

TABLE 9–16 Recommended Lubricant Viscosity Grades for Enclosed Gear Drives

ISO viscosity grade	Midpoint viscosity at 40°C (mm^2/s)	Former AGMA grade equivalent
ISO VG 32	32	0
ISO VG 46	46	1
ISO VG 68	68	2
ISO VG 100	100	3
ISO VG 150	150	4
ISO VG 220	220	5
ISO VG 320	320	6
ISO VG 460	460	7
ISO VG 680	680	8
ISO VG 1000	1000	8A

Notes:

1. Viscostiy unit of mm^2/s is commonly referred to as centistokes (cSt).
2. ISO standard prescribes minimum and maximum kinematic viscosity limits for each grade.

Adapted from AGMA 6013–A06 *Standard for Industrial Enclosed Drives*, with permission of the publisher, American Gear Manufacturers Association, 1001 North Fairfax Street, 5th Floor, Alexandria, VA

TABLE 9–17 Viscosity Grade Guidelines for Enclosed Gear Drives

Approximate temperature range				Approximate pitch line velocity of final stage of the drive, ft/min (m/s)						
Ambient		Oil sump		<1000	1000–2000	2000–3000	3000–4000	4000–5000	5000–6000	6000–7000
°F	°C	°F	°C	(<5)	(5–10)	(10–15)	(15–20)	(20–25)	(25–30)	(30–35)
−40 to −10	−40 to −23.3	<60	<15.6	68S	68S	46S	46S	46S	32S	32S
−10 to +20	−23.3 to −6.7	60 to 90	15.6 to 32.2	100S	100S	68S	68S	46S	32S	32S
20 to 40	−6.7 to 4.5	90 to 150	32.2 to 65.6	150	150	150	68	68	68	46
40 to 80	4.5 to 26.7	150 to 190	65.6 to 87.8	320	220	220	150	100	100	100
80 to 120	26.7 to 48.9	190 to 210	87.8 to 98.9	460	460	320	320	220	150	100
>120	>48.9	>210	>98.9	Not recommended						

Notes: Viscosity grades are for R&O type, unless Synthetic (S) is specified.

See Table 9–16 for listing of ISO viscosity grades.

Adapted from AGMA 6013-A06 *Standard for Industrial Enclosed Drives,* with permission of the publisher, American Gear Manufacturers Association, 1001 North Fairfax Street, 5th Floor, Alexandria, VA

Commercially Available Gear-Type Speed Reducers

By studying the design of commercially available gear-type speed reducers, you should get a better feel for design details and the relationships among the component parts: the gears, the shafts, the bearings, the housing, the means of providing lubrication, and the coupling to the driving and driven machines.

Figure 9–31 shows a double-reduction spur gear speed reducer with an electric motor rigidly attached. Such a unit is often called a *gear motor.* Figure 9–32 shows a triple-reduction unit with spur gears in the final two stages and helical gears in the first stage (as discussed in Chapter 10).

FIGURE 9–31 Double-reduction spur gear reducer (Bison Gear & Engineering Corporation, St. Charles, IL)

FIGURE 9–32 Triple-reduction gear reducer. Helical gears for stage one; spur gears for stages two and three. The pinion for stage three is on the lower left and not visible. (Bison Gear & Engineering Corporation, St. Charles, IL)

The cross-sectional drawing shown with Figure 9–33 gives a clear picture of the several components of a reducer.

The planetary reducer in Figure 9–34 has quite a different design to accommodate the placement of the sun, planet, and ring gears. Figure 9–35 shows the eight-speed transmission from a large farm tractor and illustrates the high degree of complexity that may be involved in the design of transmissions.

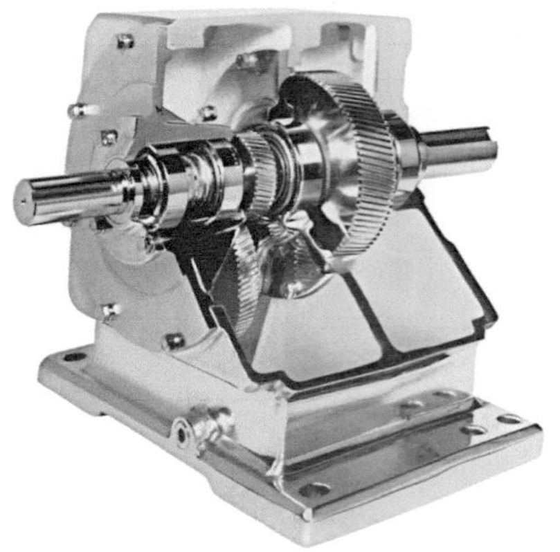

(*a*) Cutaway of a concentric helical gear reducer

(*b*) Complete reducer

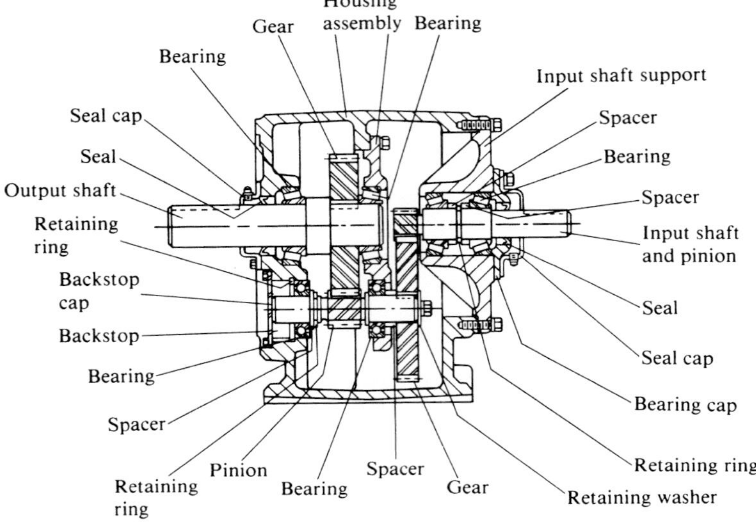

(c) Parts index

FIGURE 9–33 Concentric helical gear reducer (Peerless-Winsmith Subsidiary HBD Industries, Springville, NY)

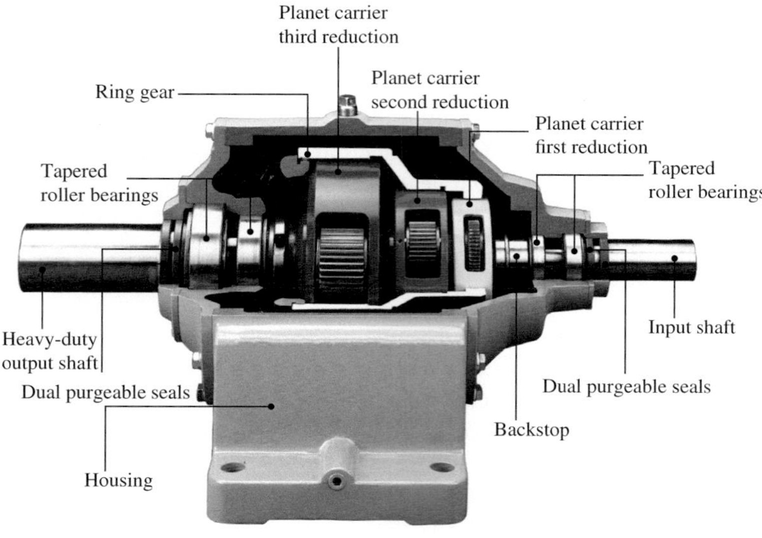

(*a*) Cutaway model with key features labeled

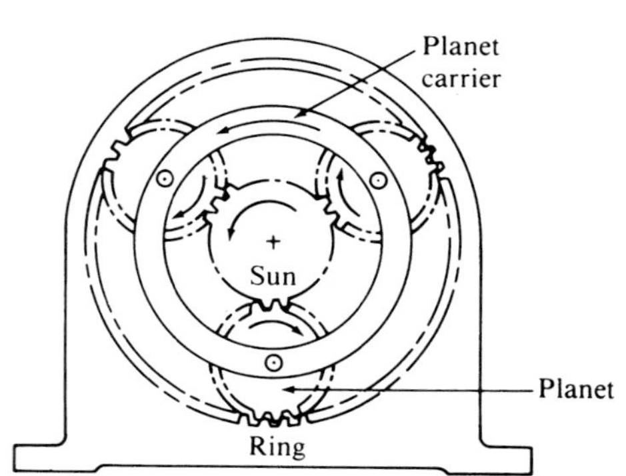

(*b*) Schematic arrangement of planetary gearing

FIGURE 9–34 Cutaway model of a triple reduction planetary gear reducer (Rexnord Industries, LLC, Milwaukee, WI)

FIGURE 9–35 Eight-speed tractor transmission (Case IH, Racine, WI)

REFERENCES

1. ABA-PGT, Inc. *Plastics Gearing.* Manchester, CT:ABA-PGT Publishing, 1994.
2. Adams, Clifford E. *Plastics Gearing: Selection and Application.* Boca Raton, FL: CRC Press, 1986.
3. American Gear Manufacturers Association. Standard 908-B89 (R1999). *Geometry Factors for Determining the Pitting Resistance and Bending Strength of Spur, Helical, and Herringbone Gear Teeth.* Alexandria, VA.: American Gear Manufacturers Association, 1999.
4. American Gear Manufacturers Association. Standard 1012-G05. *Gear Nomenclature, Definitions of Terms with Symbols.* Alexandria, VA: American Gear Manufacturers Association, 2005.
5. Radzevich, Stephen P. *Gear Cutting Tools—Fundamentals of Design and Computation.* Boca Raton, FL: CRC Press, 2010.
6. American Gear Manufacturers Association. Standards 2001-D04 (based on P_d) and 2101-D04 (based on Metric Module, m). *Fundamental Rating Factors and Calculation Methods for Involute Spur and Helical Gear Teeth.* American Gear Manufacturers Association. Alexandria, VA: American Gear Manufacturers Association, 2004.
7. American Gear Manufacturers Association. Standard 2002-B88 (R1996). *Tooth Thickness Specification and Measurement.* Alexandria, VA: American Gear Manufacturers Association, 1996.
8. American Gear Manufacturers Association. Standard 2004-C08 processing. *Gear Materials, Heat Treatment and Processing Manual.* Alexandria, VA: American Gear Manufacturers Association, 2008.
9. American Gear Manufacturers Association. Standard 6013-A06. *Standard for Industrial Enclosed Gear Drives.* Alexandria, VA: American Gear Manufacturers Association, 2006.
10. American Gear Manufacturers Association. Standard 9005-E02. *Industrial Gear Lubrication.* Alexandria, VA: American Gear Manufacturers Association, 2002.
11. American Gear Manufacturers Association. Standard 1010-E95. *Appearance of Gear Teeth-Terminology of Wear and Failure.* Alexandria, VA: American Gear Manufacturers Association, 1995.
12. American Gear Manufacturers Association. Standard 1006-A97 (R2003). *Tooth Proportions for Plastic Gears.* Washington, DC:AGMA, 2003.
13. American Gear Manufacturers Association. Standard 1106-A97 (R2003). *Tooth Proportions for Plastic Gears (Metric).* Washington, DC:AGMA, 2003.
14. American Gear Manufacturers Association. Standard 2000-A88. *Gear Classification and Inspection Handbook—Tolerances and Measuring Methods for Unassembled Spur and Helical Gears (Including Metric Equivalents).* Alexandria, VA: American Gear Manufacturers Association, 1988.
15. American Gear Manufacturers Association. Standard 2015-1-A01. *Accuracy Classification System—Tangential Measurements for Cylindrical Gears.* Alexandria, VA: American Gear Manufacturers Association, 2001.
16. Booser, E. Richard. *Tribology Data Handbook: An Excellent Friction, Lubrication, and Wear Resource.* New York: Taylor & Francis, 1997.
17. Budynas, Richard G., and Keith J. Nisbett. *Shigley's Mechanical Engineering Design.* 9th ed. New York: McGraw-Hill, 2011.
18. Davis, J. R. (Editor). *Gear Materials, Properties and Manufacture.* Materials Park, OH: ASM International, 2005.
19. Drago, Raymond J. *Fundamentals of Gear Design.* New York: Elsevier Science & Technology Books, 1988.
20. Dudley, Darle W. *Handbook of Practical Gear Design.* Boca Raton, FL: CRC Press, 1994.
21. Du Pont Polymers. *Design Handbook for Du Pont Engineering Polymers, Module I—General Design Principles.* Wilmington, DE: Du Pont Polymers, 2000.
22. Errichello, Robert. *Friction, Lubrication, and Wear of Gears.* Materials Park, OH: ASM International, 1992.
23. International Standards Organization. Standard ISO 54:1996. *Cylindrical Gears for General Engineering and for Heavy*

Engineering—Modules. Geneva, Switzerland: International Standards Organization, 1996.

24. Lipp, Robert. "Avoiding Tooth Interference in Gears." *Machine Design* 54, no. 1 (January 7, 1982).
25. Lynwander, Peter. *Gear Drive Systems: Design and Application.* Boca Raton, FL: CRC Press, 1983.
26. Oberg, Erik, Franklin D. Jones, Holbrook L. Horton, and Henry H. Ryffel. *Machinery's Handbook.* 28th ed. New York: Industrial Press, 2008.
27. Radzevich, Stephen P. *Dudley's Handbook of Practical Gear Design and Manufacture.* 2nd ed. Boca Raton, FL: CRC Press, 2012.
28. Rakhit, A. K. *Heat Treatment of Gears: A Practical Guide for Engineers.* Materials Park, OH: ASM International, 2000.
29. Srivastava, S. P. *Advances in Lubrication Additives and Tribology.* Boca Raton, FL: CRC Press, 2009.

INTERNET SITES RELATED TO SPUR GEAR DESIGN

1. **ABA-PGT, Inc.** www.abapgt.com The ABA division produces molds for making plastic gears using injection molding; the PGT division is dedicated to plastic gearing technology.
2. **American Gear Manufacturers Association (AGMA).** www.agma.org Develops and publishes voluntary, consensus standards for gears and gear drives.
3. **Baldor/Dodge.** www.dodge-pt.com Manufacturer of many power transmission components, including complete gear-type speed reducers, bearings, and components such as belt drives, chain drives, clutches, brakes, and couplings.
4. **Bison Gear, Inc.** www.bisongear.com Manufacturer of fractional horsepower gear reducers and gear motors. Site includes several videos showing Bison Gear products.
5. **Boston Gear, Company.** www.bostongear.com Manufacturer of gears and complete gear drives. Part of Altra Industrial Motion, Inc. Data provided for spur, helical, and worm gearing.
6. **Bourn & Koch, Inc.** www.bourn-koch.com Manufacturer of hobbing, grinding, and other types of machines to produce gears, including the Barber-Colman line and Fellows shapers. Also provides remanufacturing services for a wide variety of existing machine tools.
7. **Drivetrain Technology Center.** http://www.arl.psu.edu/centers_dtc.php Research center for gear-type drivetrain technology. Part of the Applied Research Laboratory of Penn State University.
8. **DuPont Polymers.** www.plastics.dupont.com Information and data on plastics and their properties. A database by type of plastic or application.
9. **Emerson Power Transmission Corporation.** www.emerson-ept.com The Browning and Morse Divisions produce spur, helical, bevel, and wormgearing and complete gear drives.
10. **Gear Industry Home Page.** www.geartechnology.com Information source for many companies that manufacture or use gears or gearing systems. Includes gear machinery, gear cutting tools, gear materials, gear drives, open gearing, tooling and supplies, software, training and education. Publishes *Gear Technology Magazine: The Journal of Gear Manufacturing.*
11. **Gleason Corporation.** www.gleason.com Manufacturer of many types of gear cutting machines for hobbing, milling, shaping and grinding. The Gleason Cutting Tools Corporation manufactures a wide variety of milling cutters, hobs, shaper cutters, shaving cutters, and grinding wheels for gear production equipment. Producer of analytical gear inspection systems.
12. **International Organization for Standardization.** www.iso.org Organization that establishes standards for numerous types of products and devices including gearing. Recognized in most parts of the world. Most standards are presented in SI metric system units.
13. **Peerless-Winsmith, Inc.** www.winsmith.com Manufacturer of a wide variety of gear reducers and power transmission products, including wormgearing, planetary gearing, and combined helical/wormgearing. Subsidiary of HBD Industries, Inc.
14. **Power Transmission Engineering.** www.powertransmission.com Clearinghouse on the Internet for buyers, users and sellers of power transmission-related products and services. Included are gears, gear drives, and gear motors. Site includes several videos of power transmission products.
15. **Process Equipment Company.** www.gearinspection.com Producer of innovative machines, products, systems, and services, including gear measurement systems. Provides gear inspection services also.
16. **Quality Transmission Gears.** www.qtcgears.com Supplier of stock metric gears.
17. **Star-SU, Inc.** www.star-su.com Manufacturer of a wide range of gear-production systems including hobbing, grinding, shaping, and shaving, along with cutting tools for the gear industry.
18. **Stock Drive Products—Sterling Instruments.** www.sdp-si.com Manufacturer and distributor of commercial and precision mechanical components, including gear reducers. Site includes an extensive handbook of design and information on metallic and plastic gears.
19. **Eldec induction U.S.A., Inc.** www.eldec-usa.com Producer of complete induction-hardening systems for production operations with special capabilities in gear manufacturing.
20. **Ticona Engineering Polymers.** www.ticona.com Producer of plastic gears made from numerous high-performance polymers for many applications such as automotive, HVAC, industrial power tools, electronics components, medical, and home and garden equipment.

PROBLEMS

Forces on Spur Gear Teeth

1. A pair of spur gears with 20°, full-depth, involute teeth transmits 7.5 hp. The pinion is mounted on the shaft of an electric motor operating at 1750 rpm. The pinion has 20 teeth and a diametral pitch of 12. The gear has 72 teeth. Compute the following:
   a. The rotational speed of the gear
   b. The velocity ratio and the gear ratio for the gear pair
   c. The pitch diameter of the pinion and the gear
   d. The center distance between the shafts carrying the pinion and the gear

e. The pitch line speed for both the pinion and the gear
f. The torque on the pinion shaft and on the gear shaft
g. The tangential force acting on the teeth of each gear
h. The radial force acting on the teeth of each gear
i. The normal force acting on the teeth of each gear

2. A pair of spur gears with 20°, full-depth, involute teeth transmits 50 hp. The pinion is mounted on the shaft of an electric motor operating at 1150 rpm. The pinion has 18 teeth and a diametral pitch of 5. The gear has 68 teeth. Compute the following:
   a. The rotational speed of the gear
   b. The velocity ratio and the gear ratio for the gear pair
   c. The pitch diameter of the pinion and the gear
   d. The center distance between the shafts carrying the pinion and the gear
   e. The pitch line speed for both the pinion and the gear
   f. The torque on the pinion shaft and on the gear shaft
   g. The tangential force acting on the teeth of each gear
   h. The radial force acting on the teeth of each gear
   i. The normal force acting on the teeth of each gear
3. A pair of spur gears with 20°, full-depth, involute teeth transmits 0.75 hp. The pinion is mounted on the shaft of an electric motor operating at 3450 rpm. The pinion has 24 teeth and a diametral pitch of 24. The gear has 110 teeth. Compute the following:
   a. The rotational speed of the gear
   b. The velocity ratio and the gear ratio for the gear pair
   c. The pitch diameter of the pinion and the gear
   d. The center distance between the shafts carrying the pinion and the gear
   e. The pitch line speed for both the pinion and the gear
   f. The torque on the pinion shaft and on the gear shaft
   g. The tangential force acting on the teeth of each gear
   h. The radial force acting on the teeth of each gear
   i. The normal force acting on the teeth of each gear
4. For the data of Problem 1, repeat Parts (g), (h), and (i) if the teeth have 25° full depth instead of 20°.
5. For the data of Problem 2, repeat Parts (g), (h), and (i) if the teeth have 25° full depth instead of 20°.
6. For the data of Problem 3, repeat Parts (g), (h), and (i) if the teeth have 25° full depth instead of 20°.

Gear Manufacture and Quality

7. List three methods for producing gear teeth, and describe each method. Include a description of the cutter for each method along with its motion relative to the gear blank.
8. Specify a suitable quality number for the gears in the drive for a grain harvester.
9. Specify a suitable quality number for the gears in the drive for a high-speed printing press.
10. Specify a suitable quality number for the gears in the drive for an automotive transmission.
11. Specify a suitable quality number for the gears in the drive for a gyroscope used in the guidance system for a spacecraft.
12. List five geometric factors measured by analytical gear quality measurement devices.
13. Identify the AGMA standard that is the basis for gear quality measurements and describe the range of quality numbers it includes. Compare that list with the two most recent predecessor standards that had been in use.
14. Specify a suitable quality number for the gears of Problem 1 if the drive is part of a precision machine tool.
15. Specify a suitable quality number for the gears of Problem 2 if the drive is part of a precision machine tool.
16. Specify a suitable quality number for the gears of Problem 3 if the drive is part of a precision machine tool.

Gear Materials

17. Identify the two major types of stresses that are created in gear teeth as they transmit power. Describe how the stresses are produced and where the maximum values of such stresses are expected to occur.
18. Describe the nature of the data contained in AGMA standards that relate to the ability of a given gear tooth to withstand the major types of stresses that it sees in operation.
19. Describe the general nature of steels that are typically used for gears, and list at least five examples of suitable alloys.
20. Describe the range of hardness that can typically be produced by through-hardening techniques and used successfully in steel gears.
21. Describe the general nature of the differences among steels produced as Grade 1, Grade 2, and Grade 3.
22. Suggest at least three applications in which Grade 2 or Grade 3 steel might be appropriate.
23. Describe three methods of producing gear teeth with strengths greater than can be achieved with through-hardening.
24. What AGMA standard should be consulted for data on the allowable stresses for steels used for gears?
25. In the AGMA standard identified in Problem 24, for what other materials besides steels are strength data provided?
26. Determine the allowable bending stress number and the allowable contact stress number for the following materials:
   a. Through-hardened, Grade 1 steel with a hardness of 200 HB
   b. Through-hardened, Grade 1 steel with a hardness of 300 HB
   c. Through-hardened, Grade 1 steel with a hardness of 400 HB
   d. Through-hardened, Grade 1 steel with a hardness of 450 HB
   e. Through-hardened, Grade 2 steel with a hardness of 200 HB
   f. Through-hardened, Grade 2 steel with a hardness of 300 HB
   g. Through-hardened, Grade 2 steel with a hardness of 400 HB
27. If the design of a steel gear indicates that an allowable bending stress number of 36 000 psi is needed, specify a suitable hardness level for Grade 1 steel. What hardness level would be required for Grade 2 steel?
28. What level of hardness can be expected for gear teeth that are case-hardened by carburizing?
29. Name three typical steels that are used in carburizing.
30. What is the level of hardness that can be expected for gear teeth that are case-hardened by flame or induction hardening?
31. Name three typical steels that are used for flame or induction hardening. What is an important property of such steels?
32. State the minimum hardness level at the surface of gear teeth made from ASTM A536 ductile iron, Grade 80-55-06.
33. Determine the allowable bending stress number and the allowable contact stress number for the following materials:
   a. Flame-hardened SAE 4140 steel, Grade 1, with a surface hardness of 50 HRC
   b. Flame-hardened SAE 4140 steel, Grade 1, with a surface hardness of 54 HRC

c. Carburized and case-hardened SAE 4620 Grade 1 steel, DOQT 300
d. Carburized and case-hardened SAE 4620 Grade 2 steel, DOQT 300
e. Carburized and case-hardened SAE 1118 Grade 1 steel, SWQT 350
f. Gray cast iron, class 20
g. Gray cast iron, class 40
h. Ductile iron, 100-70-03
i. Sand-cast bronze with a minimum tensile strength of 40 ksi (275 MPa)
j. Heat-treated bronze with a minimum tensile strength of 90 ksi (620 MPa)
k. Glass-filled nylon
l. Glass-filled polycarbonate

34. What depth should be specified for the case for a carburized gear tooth having a diametral pitch of 6?
35. What depth should be specified for the case for a carburized gear tooth having a metric module of 6?

Bending Stresses in Gear Teeth

For Problems 36–41, compute the bending stress number, s_t, using Equation (9–16). Assume that the gear blank is solid unless otherwise stated. (*Note that the data in these problems are used in later problems through Problem 59. You are advised to keep solutions to earlier problems accessible so that you can use data and results in later problems. The four problems that are keyed to the same set of data require the analysis of bending stress and contact stress and the corresponding specification of suitable materials based on those stresses. Later design problems, 60–70, use the complete analysis within each problem.*)

36. A pair of gears with 20°, full-depth, involute teeth transmits 10.0 hp while the pinion rotates at 1750 rpm. The diametral pitch is 12, and the quality number is A11. The pinion has 18 teeth, and the gear has 85 teeth. The face width is 1.25 in. The input power is from an electric motor, and the drive is for an industrial conveyor. The drive is a commercial enclosed gear unit.
37. A pair of gears with 20°, full-depth, involute teeth transmits 40 hp while the pinion rotates at 1150 rpm. The diametral pitch is 6, and the quality number is A11. The pinion has 20 teeth, and the gear has 48 teeth. The face width is 2.25 in. The input power is from an electric motor, and the drive is for a cement kiln. The drive is a commercial enclosed gear unit.
38. A pair of gears with 20°, full-depth, involute teeth transmits 0.50 hp while the pinion rotates at 3450 rpm. The diametral pitch is 32, and the quality number is A7. The pinion has 24 teeth, and the gear has 120 teeth. The face width is 0.50 in. The input power is from an electric motor, and the drive is for a small machine tool. The drive is a precision enclosed gear unit.
39. A pair of gears with 25° full-depth, involute teeth transmits 15.0 hp while the pinion rotates at 6500 rpm. The diametral pitch is 10, and the quality number is A5. The pinion has 30 teeth, and the gear has 88 teeth. The face width is 1.50 in. The input power is from a universal electric motor, and the drive is for an actuator on an aircraft. The drive is an extra-precision, enclosed gear unit.
40. A pair of gears with 25°, full-depth, involute teeth transmits 125 hp while the pinion rotates at 2500 rpm. The diametral pitch is 4, and the quality number is A9. The pinion has 32 teeth, and the gear has 76 teeth. The face width is 1.50 in. The input power is from a gasoline engine, and the drive is for a portable industrial water pump. The drive is a commercial enclosed gear unit.
41. A pair of gears with 25°, full-depth, involute teeth transmits 2.50 hp while the pinion rotates at 680 rpm. The diametral pitch is 10, and the quality number is A11. The pinion has 24 teeth, and the gear has 62 teeth. The face width is 1.25 in. The input power is from a vane-type fluid motor, and the drive is for a small lawn and garden tractor. The drive is a commercial enclosed gear unit.

Required Allowable Bending Stress Number

For Problems 42–47, compute the required allowable bending stress number, s_{at}, using Equation (9–29). Assume that no unusual conditions exist unless stated otherwise. That is, use a service factor, *SF*, of 1.00. Then specify a suitable steel and its heat treatment for both the pinion and the gear based on bending stress.

42. Use the data and results from Problem 36. Design for a reliability of 0.99 and a design life of 20 000 h.
43. Use the data and results from Problem 37. Design for a reliability of 0.99 and a design life of 8000 h.
44. Use the data and results from Problem 38. Design for a reliability of 0.9999 and a design life of 12 000 h. Consider that the machine tool is a critical part of a production system calling for a service factor of 1.25 to avoid unexpected down time.
45. Use the data and results from Problem 39. Design for a reliability of 0.9999 and a design life of 4000 h.
46. Use the data and results from Problem 40. Design for a reliability of 0.99 and a design life of 8000 h.
47. Use the data and results from Problem 41. Design for a reliability of 0.90 and a design life of 2000 h. The uncertainty of the actual use of the tractor calls for a service factor of 1.25. Consider using cast iron or bronze if the conditions permit.

Pitting Resistance

For Problems 48–53, compute the expected contact stress number, s_c, using Equation (9–23). Assume that both gears are to be steel unless stated otherwise.

48. Use the data and results from Problems 36 and 42.
49. Use the data and results from Problems 37 and 43.
50. Use the data and results from Problems 38 and 44.
51. Use the data and results from Problems 39 and 45.
52. Use the data and results from Problems 40 and 46.
53. Use the data and results from Problems 41 and 47.

Required Allowable Contact Stress Number

For Problems 54–59, compute the required allowable contact stress number, s_{ac}, using Equation (9–29). Use a service factor, *SF*, of 1.00 unless stated otherwise. Then specify suitable material for the pinion and the gear based on pitting resistance. Use steel unless an earlier decision has been made to use another material. Then evaluate whether the earlier decision is still valid. If not, specify a different material according to the most severe requirement. If no suitable material can be found, consider redesigning the original gears to enable reasonable materials to be used.

54. Use the data and results from Problems 36, 42, and 48.
55. Use the data and results from Problems 37, 43, and 49.
56. Use the data and results from Problems 38, 44, and 50.
57. Use the data and results from Problems 39, 45, and 51.
58. Use the data and results from Problems 40, 46, and 52.
59. Use the data and results from Problems 41, 47, and 53.

Design Problems

Problems 60–70 describe design situations. For each, design a pair of spur gears, specifying (at least) the diametral pitch, the number of teeth in each gear, the pitch diameters of each gear, the center distance, the face width, and the material from which the gears are to be made. Design for recommended life with regard to both strength and pitting resistance. Work toward designs that are compact. Use standard values of diametral pitch, and avoid designs for which interference could occur. See Example Problem 9–4. Assume that the input to the gear pair is from an electric motor unless otherwise stated.

If the data are given in SI units, complete the design in the metric module system with dimensions in millimeters, forces in newtons, and stresses in megapascals. See Example Problem 9–5.

60. A pair of spur gears is to be designed to transmit 5.0 hp while the pinion rotates at 1200 rpm. The gear must rotate between 385 and 390 rpm. The gear drives a reciprocating compressor.
61. A gear pair is to be a part of the drive for a milling machine requiring 20.0 hp with the pinion speed at 550 rpm and the gear speed to be between 180 and 190 rpm.
62. A drive for a punch press requires 50.0 hp with the pinion speed of 900 rpm and the gear speed of 225 to 230 rpm.
63. A single-cylinder gasoline engine has the pinion of a gear pair on its output shaft. The gear is attached to the shaft of a small cement mixer. The mixer requires 2.5 hp while rotating at approximately 75 rpm. The engine is governed to run at approximately 900 rpm.
64. A four-cylinder industrial engine runs at 2200 rpm and delivers 75 hp to the input gear of a drive for a large wood chipper used to prepare pulpwood chips for paper making. The output gear must run between 4500 and 4600 rpm.
65. A small commercial tractor is being designed for chores such as lawn mowing and snow removal. The wheel drive system is to be through a gear pair in which the pinion runs at 600 rpm while the gear, mounted on the hub of the wheel, runs at 170 to 180 rpm. The wheel is 300 mm in diameter. The gasoline engine delivers 3.0 kW of power to the gear pair.
66. A water turbine transmits 75 kW of power to a pair of gears at 4500 rpm. The output of the gear pair must drive an electric power generator at 3600 rpm. The center distance for the gear pair must not exceed 150 mm.
67. A drive system for a large commercial band saw is to be designed to transmit 12.0 hp. The saw will be used to cut steel tubing for automotive exhaust pipes. The pinion rotates at 3450 rpm, while the gear must rotate between 725 and 735 rpm. It has been specified that the gears are to be made from SAE 4340 steel, oil quenched and tempered. Case hardening is *not* to be used.
68. Repeat Problem 67, but consider a case-hardened carburized steel from Appendix 5. Try to achieve the smallest practical design. Compare the result with the design from Problem 67.
69. A gear drive for a special-purpose, dedicated machine tool is being designed to mill a surface on a rough steel casting. The drive must transmit 20 hp with a pinion speed of 650 rpm and an output speed between 110 and 115 rpm. The mill is to be used continuously, two shifts per day, six days per week, for at least five years. Design the drive to be as small as practical to permit its being mounted close to the milling head.
70. A cable drum for a crane is to rotate between 160 and 166 rpm. Design a gear drive for 25 hp in which the input pinion rotates at 925 rpm and the output rotates with the drum. The crane is expected to operate with a 50% duty cycle for 120 hours per week for at least 10 years. The pinion and the gear of the drive must fit within the 24-in inside diameter of the drum, with the gear mounted on the drum shaft.

Power-Transmitting Capacity

71. Determine the power-transmitting capacity for a pair of spur gears having 20°, full-depth teeth, a diametral pitch of 10, a face width of 1.25 in, 25 teeth in the pinion, 60 teeth in the gear, and an AGMA quality class of A9. The pinion is made from SAE 4140 OQT 1000, and the gear is made from SAE 4140 OQT 1100. The pinion will rotate at 1725 rpm on the shaft of an electric motor. The gear will drive a centrifugal pump.
72. Determine the power-transmitting capacity for a pair of spur gears having 20°, full-depth teeth, a diametral pitch of 6, 35 teeth in the pinion, 100 teeth in the gear, a face width of 2.00 in, and an AGMA quality class of A11. A gasoline engine drives the pinion at 1500 rpm. The gear drives a conveyor for crushed rock in a quarry. The pinion is made from SAE 1040 WQT 800. The gear is made from gray cast iron, ASTM A48-83, class 30. Design for 15 000 hr life.
73. It was found that the gear pair described in Problem 72 wore out when driven by a 25-hp engine. Propose a redesign that would be expected to give 15 000 hr life under the conditions described.

Design of Double-Reduction Drives

74. Design a double-reduction gear train that will transmit 10.0 hp from an electric motor running at 1750 rpm to an assembly conveyor whose drive-shaft must rotate between 146 and 150 rpm. Note that this will require the design of two pairs of gears. Sketch the arrangement of the train, and compute the actual output speed.
75. A commercial food waste grinder in which the final shaft rotates at between 40 and 44 rpm is to be designed. The input is from an electric motor running at 850 rpm and delivering 0.50 hp. Design a double-reduction spur gear train for the grinder.
76. A small, powered hand drill is driven by an electric motor running at 3000 rpm. The drill speed is to be approximately 550 rpm. Design the speed reduction for the drill. The power transmitted is 0.25 hp.
77. The output from the drill described in Problem 76 provides the drive for a small bench-scale band saw used in a home shop. The saw blade is to move with a linear velocity of 375 ft/min. The saw blade rides on 9.0-in-diameter wheels. Design a spur gear reduction to drive the band saw. Consider using plastic gears.
78. Design a rack-and-pinion drive to lift a heavy access panel on a furnace. A fluid power motor rotating 1500 rpm will provide 5.0 hp at the input to the drive. The linear speed of the rack is to be at least 2.0 ft/s. The rack moves 6.0 ft each way during the opening and closing of the furnace doors. More than one stage of reduction may be used, but attempt to design with the fewest number of gears. The drive is expected to operate at least six times per hour for three shifts per day, seven days per week, for at least 15 years.
79. Design the gear drive for the wheels of an industrial lift truck. Its top speed is to be 20 mph. It has been decided that the wheels will have a diameter of 12.0 in. A DC motor supplies 20 hp at a speed of 3000 rpm. The design life is 16 hours per day, six days per week, for 20 years.

Plastics Gearing

80. Design a pair of plastic gears to drive a small band saw. The input is from a 0.50 hp electric motor rotating at 860 rpm, and the pinion will be mounted on its 0.75-inch diameter shaft with a keyway for a 0.1875 × 0.1875 in key. The gear is to rotate between 265 and 267 rpm.
81. Design a pair of plastic gears to drive a paper feed roll for an office printer. The pinion rotates at 88 rpm and the gear must rotate between 20 and 22 rpm. The power required is 0.06 hp. Work toward the smallest practical size.
82. Design a pair of plastic gears to drive the wheels of a small remote control car. The gear is mounted on the axle of the wheel and must rotate between 120 and 122 rpm. The pinion rotates at 430 rpm. The power required is 0.025 hp. Work toward the smallest practical size using unfilled nylon.
83. Design a pair of plastic gears to drive a commercial food-chopping machine. The input is from a 0.65 hp electric motor rotating at 1560 rpm, and the pinion will be mounted on its 0.875-inch diameter shaft with a keyway for a 0.1875 × 0.1875 in key. The gear is to rotate between 468 and 470 rpm.

CHAPTER TEN

HELICAL GEARS, BEVEL GEARS, AND WORMGEARING

THE BIG PICTURE Helical Gears, Bevel Gears, and Wormgearing

Discussion Map

- The geometry of helical gears, bevel gears, and wormgearing was described in Chapter 8.
- The principles of stress analysis of gears were discussed in Chapter 9 for spur gears. Much of that information is applicable to the types of gears discussed in this chapter.

Discover

Review Chapters 8 and 9 now.

Recall some of the discussion at the beginning of Chapter 8 about uses for gears that you see in your world. Review that information now, and focus your discussion on helical gears, bevel gears, and wormgearing.

In this chapter, you acquire the skills to perform the necessary analyses to design safe gear drives that use helical gears, bevel gears, and wormgearing and that demonstrate long life.

Much was said in Chapter 8 about the geometry of spur gears, helical gears, bevel gears, and wormgearing and the kinematics of a single pair of gears and trains made from two or more pairs of gears. Also, early parts of Chapter 9 discussed the types of metallic materials commonly used for power transmission gears, methods of manufacturing gears, principles of gear quality, and measurements required to ensure that quality. Then you learned how to analyze the bending stress in the fillet at the base of spur gear teeth and the contact stress along the face of the teeth near the pitch line. This was combined with determining the required bending strength of the gear material to avoid fatigue failure and the required hardness of the face of the gear to provide adequate resistance to surface pitting. The result led to a method of designing spur gear drives to achieve satisfactory performance and life. Section 9–16 applied gear design principles to plastics gearing.

Similar analyses and design approaches are developed in this chapter for helical gears, bevel gears, and wormgearing. You will need to refer back to Chapters 5, 8 and 9 as we proceed. More detailed information can be found in the AGMA standards listed in References 3, 4, 5, 7, and 8.

Figure 10–1 shows a combination of a gear-type speed reducer driven by a close-coupled electric motor; the combination is often called a *gearmotor*. Part (b) of the figure shows the internal arrangement of the three-stage speed reducer. At the right end, a helical pinion shaft is driven directly by the motor and it then drives its mating gear for the first stage of reduction. The same shaft with the output helical gear carries the bevel pinion for the second stage of reduction. The helical pinion for the third stage is placed close to the output bevel gear on the same shaft. The large output helical gear then delivers the power at the greatly reduced speed and at a correspondingly higher torque to the driven machine. The design of the output shaft employs a hollow shaft with taper-lock bushing that facilitates connection to the input shaft of the driven machine.

Figure 10–2 shows a double-reduction, helical gear reducer that receives power from the electric motor mounted above and through the belt drive that produces an initial speed reduction. Take note of the arrangement of the gears, shafts, and bearings in the housing shown to the left. Tapered roller bearings are used on all shafts to carry the combination of radial and axial forces that are inherently produced by helical gears.

Refer back to Figure 8–20 for an example of a worm-gear reducer.

(*a*) Assembly of a gear reducer and a drive motor, called a gearmotor

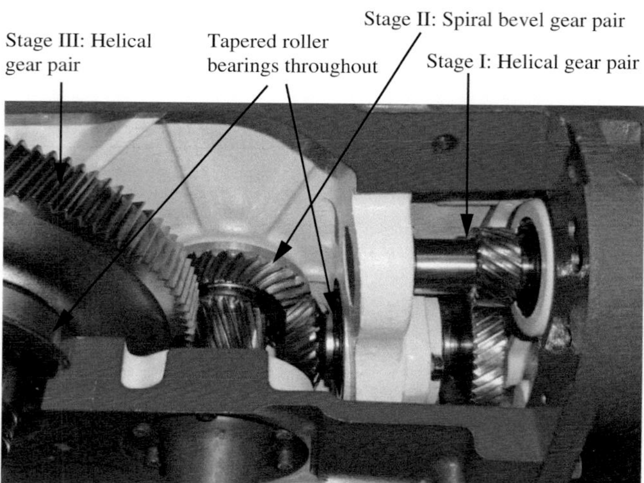

(*b*) Cutaway showing the internal arrangement of the helical-bevel-helical three-stage reduction

FIGURE 10–1 Gearmotor assembly employing a three-stage gear reducer (Baldor/Dodge, Greenville, SC)

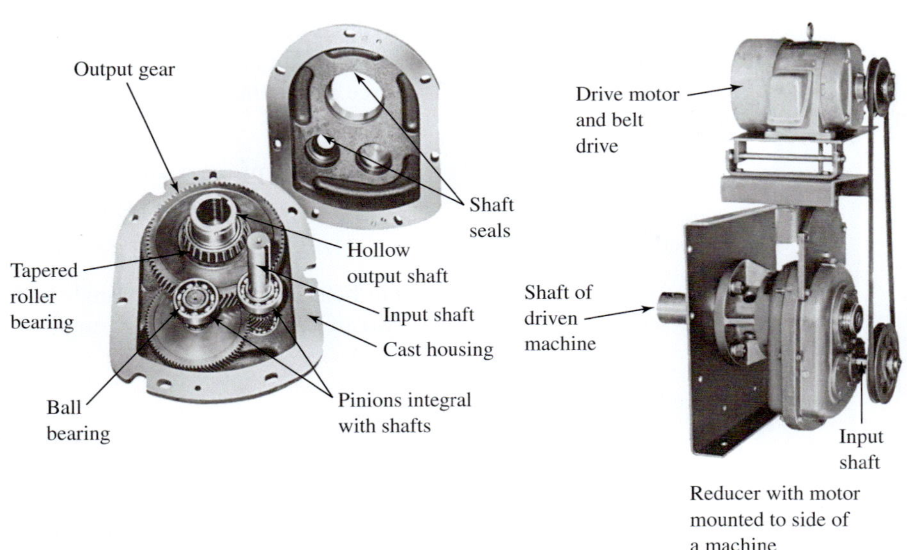

FIGURE 10–2 Helical shaft-mount reducer (Power Transmission Solutions, a business unit of Emerson Industrial Automation)

YOU ARE THE DESIGNER

The gear drives that were designed in Chapter 9 all assumed that spur gears would be used to accomplish the speed reduction or speed increase between the input and the output of the drive. But many other types of gears could have been used. Assume that you are the designer of the drive for the wood chipper described in Example Problem 9–4. How would the design be different if helical gears were used instead of spur gears? What forces would be created and transferred to the shafts carrying the gears and to the bearings carrying the shafts? Would you be able to use smaller gears? How is the geometry of helical gears different from that of spur gears?

Rather than having the input and output shafts parallel as they were in designs up to this time, how can we design drives that deliver power to an output shaft at right angles to the input shaft? What special analysis techniques are applied to bevel gears and wormgearing?

The information in this chapter will help you answer these and other questions. ■

10–1 OBJECTIVES OF THIS CHAPTER

After completing this chapter, you will be able to:

1. Describe the geometry of helical gears and compute the dimensions of key features.
2. Compute the forces exerted by one helical gear on its mating gear.
3. Compute the stress due to bending in helical gear teeth and specify suitable materials to withstand such stresses.
4. Design helical gears for surface durability.
5. Describe the geometry of bevel gears and compute the dimensions of key features.
6. Analyze the forces exerted by one bevel gear on another and show how those forces are transferred to the shafts carrying the gears.
7. Design and analyze bevel gear teeth for strength and surface durability.
8. Describe the geometry of worms and wormgears.
9. Compute the forces created by a wormgear drive system and analyze their effect on the shafts carrying the worm and the wormgear.
10. Compute the efficiency of wormgear drives.
11. Design and analyze wormgear drives to be safe for bending strength and wear.

References at the end of the chapter are recommended for additional information for design and application of helical gears, bevel gears, and wormgearing.

10–2 FORCES ON HELICAL GEAR TEETH

Figure 10–3 shows a photograph of two helical gears in mesh and designed to be mounted on parallel shafts. This is the

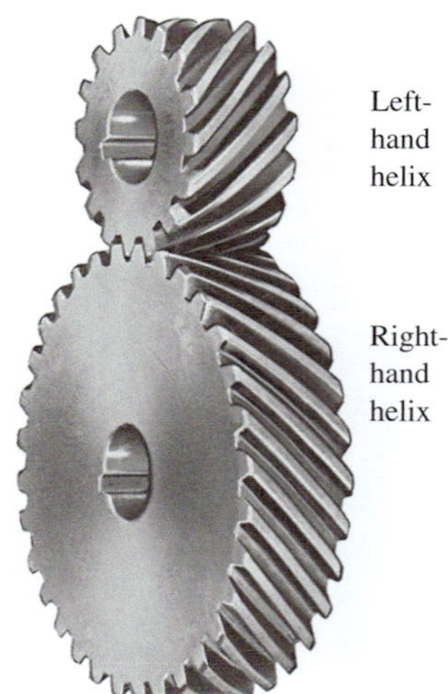

FIGURE 10–3 Helical gears. These gears have a 45° helix angle (Emerson Power Transmission Corporation, Drive and Component Division, Ithaca, NY)

basic configuration that we analyze in this chapter. Refer to Figure 10–4 for a representation of the force system that acts between the teeth of two helical gears in mesh. In Chapter 8, using the same figure, we defined the following forces:

- W_N is the *true normal force* that acts perpendicular to the face of the tooth in the plane normal to the surface of the tooth. The normal plane is shown in Figure 10–4(d). We seldom need to use the value of W_N because its three orthogonal components, defined next, are used in the analyses performed for helical gears. The values for the orthogonal components depend on the following three angles that help define the geometry of the helical gear teeth:

  Normal pressure angle: ϕ_n

  Transverse pressure angle: ϕ_t

  Helix angle: ψ

  For helical gears, the helix angle and one of the other two are specified. The third angle can be computed from

$$\tan \phi_n = \tan \phi_t \cos \psi \tag{10–1}$$

- W_t is the *tangential force* that acts in the transverse plane and tangent to the pitch circle of the helical gear and that causes the torque to be transmitted from the driver to the driven gear. Therefore, this force is often called the *transmitted force.* It is functionally similar to W_t used in the analysis of spur gears in Chapters 8 and 9. We can compute its value from the same equations, as follows:

  If the torque being transmitted (T) and the size of the gear (D) are known,

➪ Tangential Force

$$W_t = T/(D/2) \tag{10–2}$$

If the power being transmitted (P) and the rotational speed (n) are known,

$$T = (P/n) \tag{10–3}$$

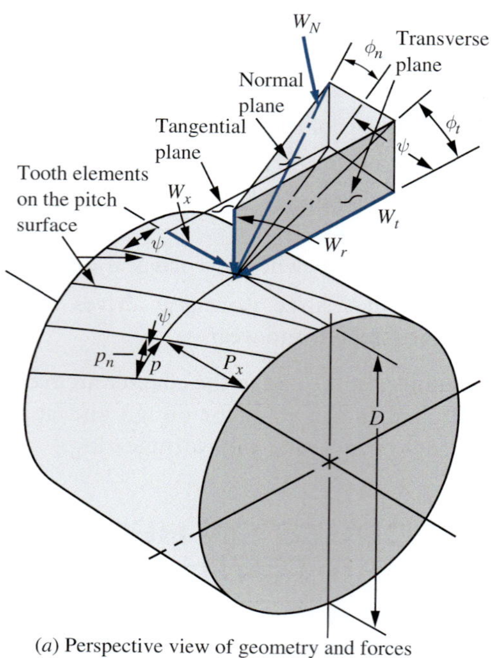

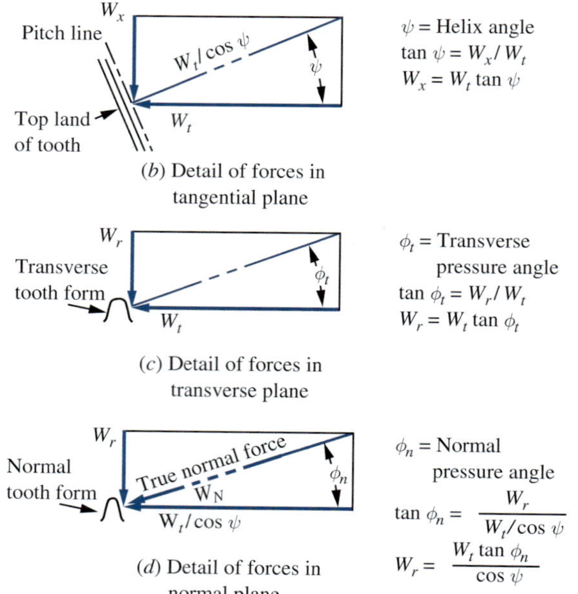

FIGURE 10–4 Helical gear geometry and forces

Power, torque, and forces for the U.S. unit system: Here we bring concepts and equations from Section 9-3 developed for spur gears and apply them for helical gears. *These are unit-specific equations.*

When power, P, is expressed in hp, rotational speed, n, is in rpm, and diameters, D, are in inches:

Torque: $T = 63\,000(P)/n$ lb · in **(10–4)**

Pitch line speed: $v_t = \pi Dn/12$ ft/min **(10–5)**

Tangential force: $W_t = (126\,000)(P)/[(n)(D)]$ lb **(10–6)**

Or: $W_t = (33\,000)(P)/v_t$ lb **(10–7)**

Power, torque, and forces for the SI metric unit system: Again, bringing concepts and equations from Sections 9-3 and 9-12 developed for spur gears, we apply them for helical gears. *These are unit-specific equations.*

When power, P, is expressed in kW, rotational speed, n, is in rpm, and diameters, D, are in mm:

Torque: $T = 9549(P)/n$ N · m **(10–4M)**

Pitch line speed: $v_t = \pi Dn/60\,000$ m/s **(10–5M)**

Tangential force: $W_t = (19\,099)(P)/[(n)(D)]$ N **(10–6M)**

Or: $W_t = (1000)(P)/v_t$ N **(10–7M)**

The value of the tangential load is the most fundamental of the three orthogonal components of the true normal force. The calculation of the bending stress number and the contact stress number of the gear teeth depends on W_t.

- W_r is the *radial force* that acts toward the center of the gear perpendicular to the pitch circle and to the tangential force. It tends to push the two gears apart. As can be seen in Figure 10–4(c),

➪ Radial Force

$$W_r = W_t \tan \phi_t \quad \textbf{(10–8)}$$

where ϕ_t = transverse pressure angle for the helical teeth

- W_x is the *axial force* that acts parallel to the axis of the gear and causes a thrust load that must be resisted by the bearings carrying the shaft. With the tangential force known, the axial force is computed from

➪ Axial Force

$$W_x = W_t \tan \psi \quad \textbf{(10–9)}$$

Example Problem 10–1 A helical gear has a normal diametral pitch of 8, a normal pressure angle of 20°, 32 teeth, a face width of 3.00 in, and a helix angle of 15°. Compute the diametral pitch, the transverse pressure angle, and the pitch diameter. If the gear is rotating at 650 rpm while transmitting 7.50 hp, compute the pitch line speed, the tangential force, the axial force, and the radial force.

Solution *Diametral Pitch [Equation (8–15)]*

$$P_d = P_{nd} \cos \psi = 8 \cos(15) = 7.727$$

Transverse Pressure Angle [Equation (10–1)]

$$\phi_t = \tan^{-1}(\tan \phi_n / \cos \psi)$$

$$\phi_t = \tan^{-1}[\tan(20)/\cos(15)] = 20.65°$$

Pitch Diameter [Equation (8–14)]

$$D = N/P_d = 32/7.727 = 4.141 \text{ in}$$

Pitch Line Speed, v_t [Equation (10–5)]

$$v_t = \pi Dn/12 = \pi(4.141)(650)/12 = 704.7 \text{ ft/min}$$

Tangential Force, W_t [Equation (10–7)]

$$W_t = 33\,000(P)/v_t = 33\,000(7.5)/704.7 = 351 \text{ lb}$$

Axial Force, W_x [Equation (10–9)]

$$W_x = W_t \tan \psi = 351 \tan(15) = 94 \text{ lb}$$

Radial Force, W_r [Equation (10–8)]

$$W_r = W_t \tan \phi_t = 351 \tan(20.65) = 132 \text{ lb}$$

The following example problem illustrates similar calculations for the SI metric system.

Example Problem 10–2 A helical gear has a normal module of 3 mm, a normal pressure angle of 25°, a helix angle of 22°, 32 teeth, and a face width of 75 mm. Compute the transverse module, the transverse pressure angle, and the pitch diameter. Then, if the gear is rotating at 650 rpm while transmitting 5.0 kW of power, compute the pitch line speed, the tangential force, the axial force, and the radial force.

Solution *Transverse Module:* (Equation 8–18)

$$m_n = m \cos \psi$$

$$m = m_n/\cos \psi = 3.00 \text{ mm}/\cos 22° = 3.236 \text{ mm}$$

Transverse Pressure Angle: (Equation 8–11)

$$\phi_t = \text{Tan}^{-1}(\tan \phi_n/\cos \psi) = \text{Tan}^{-1}[\tan(25°)/\cos(22°)] = 23.38°$$

Pitch Diameter: $D = mN = (3.236 \text{ mm})(32) = 103.54 \text{ mm}$

Pitch line speed: (Equation 10–5M)

$$v_t = \pi Dn/(60\,000) = \pi(103.54 \text{ mm})(650 \text{ rpm})/(60\,000) = 3.524 \text{ m/s}$$

Tangential Force: (Equation 10–6M)

$$W_t = (1000)(P)/v_t = (1000)(5.0 \text{ kW})/(3.524 \text{ m/s}) = 1419 \text{ N}$$

Axial Force: (Equation 10–9)

$$W_x = W_t \tan \psi = (1419 \text{ N}) \tan(22°) = 573.3 \text{ N}$$

Radial Force: (Equation 10–8)

$$W_r = W_t \tan \phi_t = (1419 \text{ N}) \tan(23.38°) = 613.5 \text{ N}$$

10–3 STRESSES IN HELICAL GEAR TEETH

We will use the same basic equation for computing the bending stress number for helical gear teeth as we did for spur gear teeth in Chapter 9, given in Equation (9–16) and repeated here:

$$s_t = \frac{W_t P_d}{FJ} K_o K_s K_m K_B K_v$$

Figures 10–5, 10–6, and 10–7 show the values for the geometry factor, J, for helical gear teeth with 15°, 20°, and 22° normal pressure angles, respectively.[1] The K factors are the same as those used for spur gears. See References 9 and 18 and the following locations for values:

K_o = overload factor (Table 9–7)

K_s = size factor (Table 9–8)

K_m = load-distribution factor [Figures 9–16 and 9–17 and Equation (9–17)]

K_B = rim thickness factor (Figure 9–18)

K_v = dynamic factor (Figure 9–20)

For design, a material must be specified that has an allowable bending stress number, s_{at}, greater than the

[1]Figures 10–5, 10–6, and 10–7:
Graphs for the geometry factor, J, for helical gears have been taken from AGMA Standard 218.01-1982, *Standard for Rating the Pitting Resistance and Bending Strength of Spur and Helical Involute Gear Teeth,* with the permission of the publisher, American Gear Manufacturers Association, 1001 North Fairfax Street, 5th floor, Alexandria, VA 22314. This standard has been superseded by two standards: (1) Standard 908-B89 (R 1999), *Geometry Factors for Determining the Pitting Resistance and Bending Strength of Spur, Helical and Herringbone Gear Teeth,* 1999 and (2) Standard 2001-D04, *Fundamental Rating Factors and Calculation Methods for Involute Spur and Helical Gear Teeth,* 2004. The method of calculating the value for J has not been changed. However, the new standards do not contain the graphs. Users are cautioned to ensure that geometry factors for a given design conform to the specific cutter geometry used to manufacture the gears. Standards 908-B89 (R 1999) and 2001-D04 should be consulted for the details of computing the values for J and for rating the performance of the gear teeth.

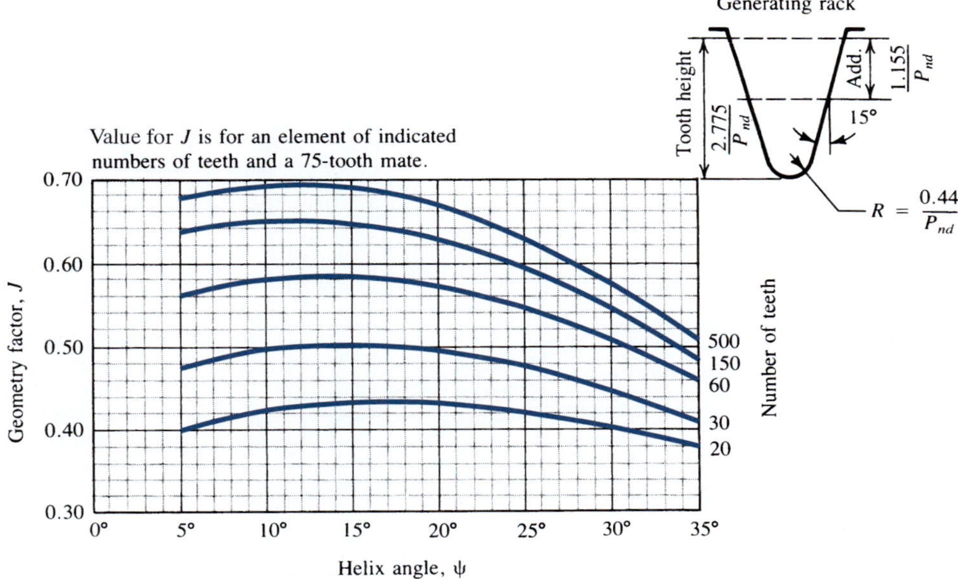

(*a*) Geometry factor (*J*) for 15° normal pressure angle and indicated addendum

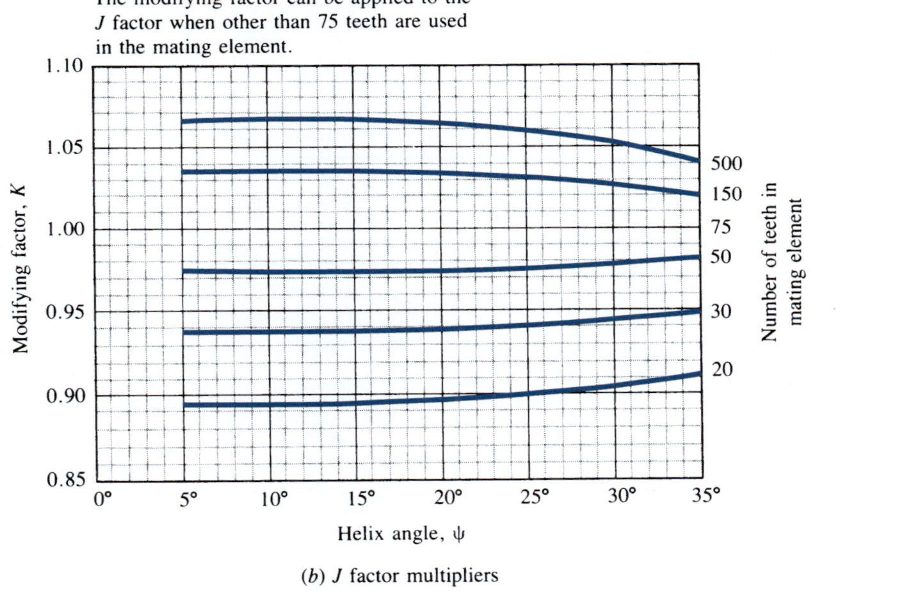

(*b*) *J* factor multipliers

FIGURE 10–5 Geometry factor (*J*) for 15° normal pressure angle

computed bending stress number, s_t. Design values of s_{at} can be found:

Figure 9–11: Steel, through-hardened, Grades 1 and 2
Table 9–5: Case-hardened steels
Table 9–6: Cast iron and bronze

(See also References 11 and 17–23.) The data for steel, iron, and bronze apply to a design life of 10^7 cycles at a reliability of 99% (fewer than one failure in 100). If other values for design life or reliability are desired, the allowable stress can be modified using the procedure described in Section 9–10.

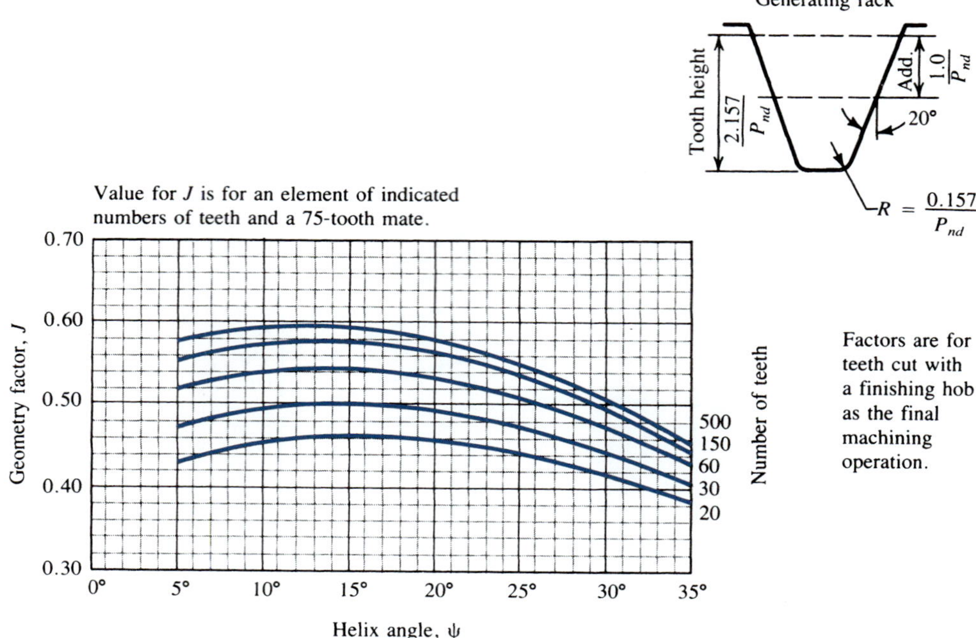

(*a*) Geometry factor (*J*) for 20° normal pressure angle, standard addendum, and finishing hob

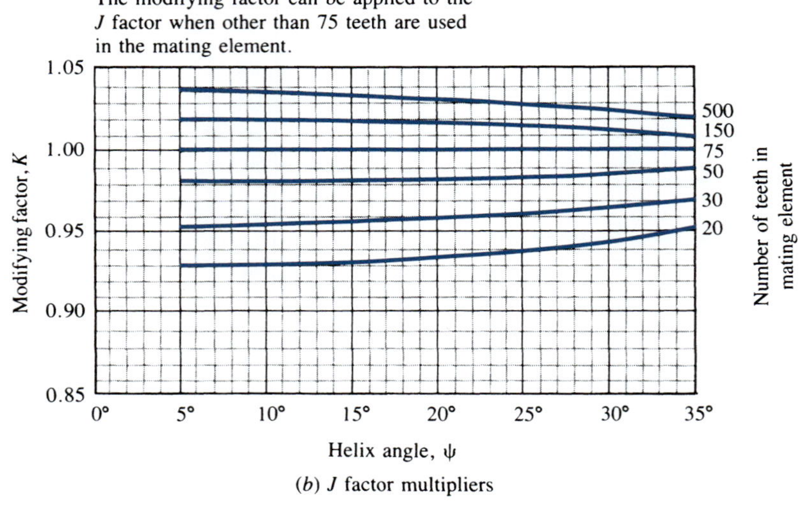

(*b*) *J* factor multipliers

FIGURE 10–6 Geometry factor (*J*) for 20° normal pressure angle

10–4 PITTING RESISTANCE FOR HELICAL GEAR TEETH

Pitting resistance for helical gear teeth is evaluated using the same procedure as that discussed in Chapter 9 for spur gears. Equation (9–23) is repeated here:

$$s_c = C_p\sqrt{\frac{W_tK_oK_sK_mK_v}{FD_pI}} \tag{9–23}$$

All of the factors are the same for helical gears except the geometry factor for pitting resistance, *I*. The values for C_p are found in Table 9–10. Note that the other *K* factors have the same values as the *K* factors discussed and identified in Section 10–3.

Because of the larger variety of geometric features needed to define the form of helical gears, it is not reasonable to reproduce all of the necessary tables of values or the complete formulas for computing *I*. Values change with the gear ratio, the number of teeth in the pinion, the tooth form, the helix angle, and the specific values for addendum,

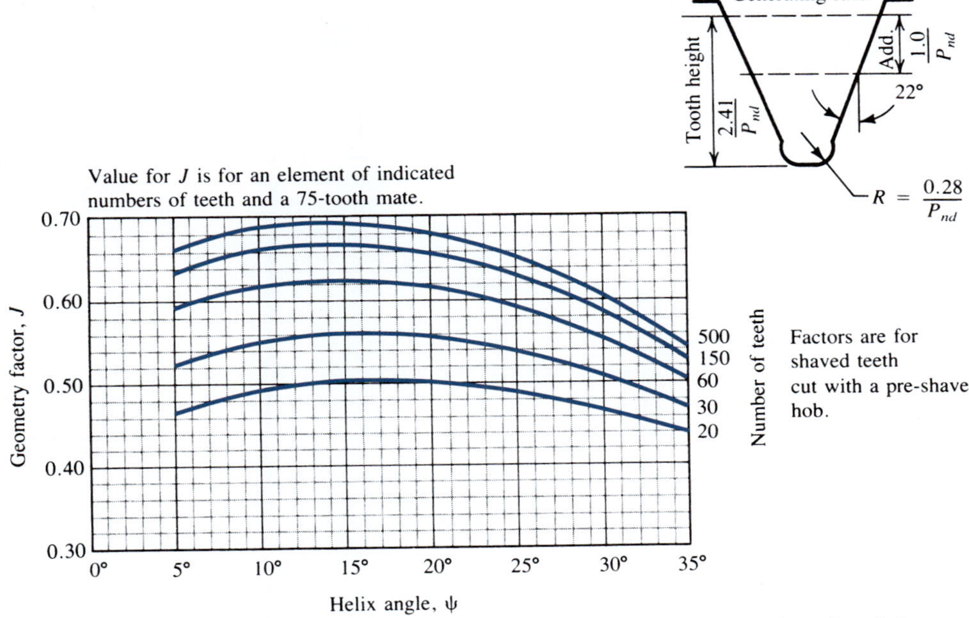

(*a*) Geometry factor (*J*), for 22° normal pressure angle, standard addendum, and pre-shave hob

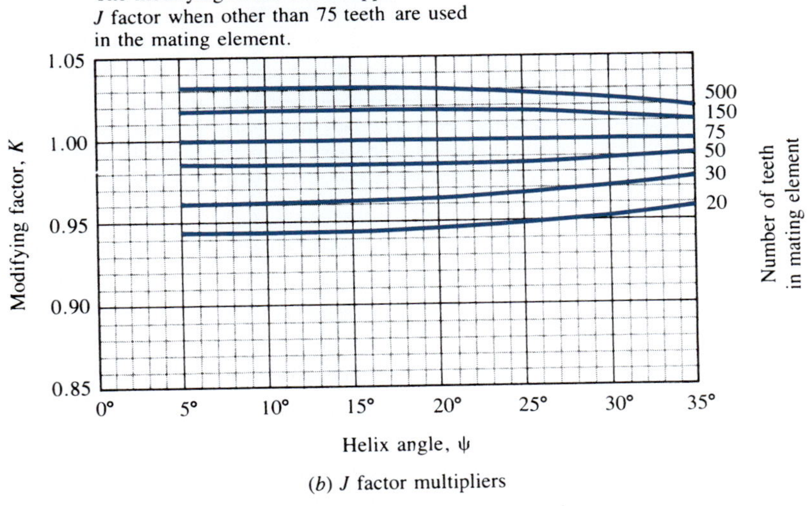

(*b*) *J* factor multipliers

FIGURE 10–7 Geometry factor (*J*) for 22° normal pressure angle

whole depth, and fillet radius. See References 6 and 13 for extensive discussions about the procedures. To facilitate problem solving in this book, Tables 10–1 and 10–2 give a few values for *I*.

For design, when the computed contact stress number is known, a material must be specified that has an allowable contact stress number, s_{ac}, greater than s_c. Design values for s_{ac} can be found from the following:

Figure 9–12: Steel, through-hardened, Grades 1 and 2

Table 9–5: Steel, case-hardened, Grade 1; flame- or induction-hardened, or carburized

Table 9–6: Cast iron and bronze

The data from these sources apply to a design life of 10^7 cycles at a reliability of 99% (fewer than one failure in 100). If other values for design life or reliability are desired, or if a service factor is to be applied, the allowable contact stress number can be modified using the procedure described in Section 9–10.

10–5 DESIGN OF HELICAL GEARS

The example problem that follows illustrates the procedure to design helical gears.

TABLE 10–1 Geometry Factors for Pitting Resistance, *I*, for Helical Gears with 20° Normal Pressure Angle and Standard Addendum

A. Helix angle ψ = 15.0°

Gear teeth	Pinion teeth 17	21	26	35	55
17	0.124				
21	0.139	0.128			
26	0.154	0.143	0.132		
35	0.175	0.165	0.154	0.137	
55	0.204	0.196	0.187	0.171	0.143
135	0.244	0.241	0.237	0.229	0.209

B. Helix angle ψ = 25.0°

Gear teeth	Pinion teeth 14	17	21	26	35	55
14	0.123					
17	0.137	0.126				
21	0.152	0.142	0.130			
26	0.167	0.157	0.146	0.134		
35	0.187	0.178	0.168	0.156	0.138	
55	0.213	0.207	0.199	0.189	0.173	0.144
135	0.248	0.247	0.244	0.239	0.230	0.210

Source: Extracted from AGMA Standard 908-B89 (R 1999), *Geometry Factors for Determining the Pitting Resistance and Bending Strength of Spur, Helical and Herringbone Gear Teeth,* with the permission of the publisher, American Gear Manufacturers Association, 1001 North Fairfax Street, 5th floor, Alexandria, VA 22314.

Example Problem 10–3 A pair of helical gears for a milling machine drive is to transmit 65 hp with a pinion speed of 3450 rpm and a gear speed of 1100 rpm. The power is from an electric motor. Design the gears.

Solution Of course, there are several possible solutions. Here is one. Let's try a normal diametral pitch of 12, 24 teeth in the pinion, a helix angle of 15°, a normal pressure angle of 20°, and a quality number of A9.

Now compute the transverse diametral pitch, the axial pitch, the transverse pressure angle, and the pitch diameter. Then we will choose a face width that will give at least two axial pitches to ensure true helical action.

$$P_d = P_{dn}\cos\psi = 12\cos(15°) = 11.59$$

$$P_x = \frac{\pi}{P_d\tan\psi} = \frac{\pi}{11.59\tan(15°)} = 1.012\text{ in}$$

$$\phi_t = \tan^{-1}(\tan\phi_n/\cos\psi) = \tan^{-1}[\tan(20°)/\cos(15°)] = 20.65°$$

$$d = D_p/P_d = 24/11.59 = 2.071\text{ in}$$

$$F = 2P_x = 2(1.012) = 2.024\text{ in}\quad\text{(nominal face width)}$$

Let's use 2.25 in, a more convenient value. The pitch line speed and the transmitted load are

$$v_t = \pi D_p n/12 = \pi(2.071)(3450)/12 = 1871\text{ ft/min}$$

$$W_t = 33\,000(\text{hp})/v_t = (33\,000)(65)/1871 = 1146\text{ lb}$$

TABLE 10–2 Geometry Factors for Pitting Resistance, I, for Helical Gears with 25° Normal Pressure Angle and Standard Addendum

A. Helix angle ψ = 15.0°

Gear teeth	Pinion teeth 14	17	21	26	35	55
14	0.130					
17	0.144	0.133				
21	0.160	0.149	0.137			
26	0.175	0.165	0.153	0.140		
35	0.195	0.186	0.175	0.163	0.143	
55	0.222	0.215	0.206	0.195	0.178	0.148
135	0.257	0.255	0.251	0.246	0.236	0.214

B. Helix angle ψ = 25.0°

Gear teeth	Pinion teeth 12	14	17	21	26	35	55
12	0.129						
14	0.141	0.132					
17	0.155	0.146	0.135				
21	0.170	0.162	0.151	0.138			
26	0.185	0.177	0.166	0.154	0.141		
35	0.203	0.197	0.188	0.176	0.163	0.144	
55	0.227	0.223	0.216	0.207	0.196	0.178	0.148
135	0.259	0.258	0.255	0.251	0.246	0.235	0.213

Source: Extracted from AGMA Standard 908-B89, *Geometry Factors for Determining the Pitting Resistance and Bending Strength of Spur, Helical and Herringbone Gear Teeth,* with the permission of the publisher, American Gear Manufacturers Association, 1001 North Fairfax Street, 5th floor, Alexandria, VA 22314.

Now we can calculate the number of teeth in the gear:

$$VR = N_G/N_P = n_P/n_G = 3450/1100 = 3.14$$

$$N_G = N_P\,(VR) = 24(3.14) = 75\text{ teeth} \quad \text{(integer value)}$$

The values for the factors in Equation (9–6) must now be determined to enable the calculation of the bending stress. The geometry factor for the pinion is found in Figure 10–6 for 24 teeth in the pinion and 75 teeth in the gear: $J_P = 0.48$. The value of J_G will be greater than the value of J_P, resulting in a lower stress in the gear.

The K factors are

K_o = overload factor = 1.5 (moderate shock)

K_s = size factor = 1.0

K_m = load-distribution factor = 1.26 for $F/D_P = 1.09$ and commercial-quality, enclosed gearing

K_B = rim thickness factor = 1.0 (solid gears)

K_v = dynamic factor = 1.35 for $A_v = 9$ and $v_t = 1871$ ft/min

The bending stress in the pinion can now be computed:

$$s_{tP} = \frac{W_t P_d}{F J_P} K_o K_s K_m K_B K_v$$

$$s_{tP} = \frac{(1146)(11.59)}{(2.25)(0.48)}(1.50)(1.0)(1.26)(1.0)(1.35) = 31\,400\text{ psi}$$

From Figure 9–11, a Grade 1 steel with a hardness of approximately 250 HB would be required. Let's proceed to the design for pitting resistance.

Use Equation (9–23):

$$s_c = C_p\sqrt{\frac{W_t K_o K_s K_m K_v}{F D_p I}}$$

For two steel gears, $C_p = 2300$. Rough interpolation from the data in Table 10–1 for $N_P = 24$ and $N_G = 75$ gives $I = 0.202$. It is recommended that the computational procedure described in the AGMA standards be used to compute a more precise value for critical work. The contact stress is then

$$s_c = 2300\sqrt{\frac{(1146)(1.50)(1.0)(1.26)(1.35)}{(2.25)(2.071)(0.202)}} = 128\,200 \text{ psi}$$

It is obvious that the contact stress governs this design. Let's adjust the solution for a higher reliability and to account for the expected number of cycles of operation. Certain design decisions must be made. For example, consider the following:

Design for a reliability of 0.999 (less than one failure in 1000): $K_R = 1.25$ (Table 9–11).

Design life: Let's design for 10 000 h of life as suggested in Table 9–12 for multipurpose gearing. Then, using Equation (9–27), we can compute the number of cycles of loading. For the pinion rotating at 3450 rpm with one cycle of loading per revolution,

$$N_c = (60)(L)(n)(q) = (60)(10000)(3450)(1.0) = 2.1 \times 10^9 \text{cycles}$$

From Figure 9–24, we find that $Z_N = 0.89$.

No unusual conditions seem to exist in this application beyond those already considered in the various K factors. Therefore, we use a service factor, SF, of 1.00.

We can use Equation (9–27) to apply these factors.

$$\frac{K_R(SF)}{Z_N}s_c = s_{ac} = \frac{(1.25)(1.00)}{(0.89)}(128\,200 \text{ psi}) = 180\,000 \text{ psi}$$

Table 9–5 indicates that Grade 1 steel, case-hardened by carburizing, would be suitable. From Appendix 5, let's specify AISI 4320 SOQT 450, having a case hardness of HRC 59 and a core hardness of 415 HB. This should be satisfactory for both bending and pitting resistance. Both the pinion and the gear should be of this material.

10–6 FORCES ON STRAIGHT BEVEL GEARS

Review Section 8–8 and Figure 8–17 for the geometry of bevel gears. Also see References 1, 10, and 12.

Because of the conical shape of bevel gears and because of the involute-tooth form, a three-component set of forces acts on bevel gear teeth. Using notation similar to that for helical gears, we will compute the tangential force, W_t; the radial force, W_r; and the axial force, W_x. It is assumed that the three forces act concurrently at the midface of the teeth and on the pitch cone (see Figure 10–8). Although the actual point of application of the resultant force is a little displaced from the middle, no serious error results.

The tangential force acts tangential to the pitch cone and is the force that produces the torque on the pinion and the gear. The torque can be computed from the known power transmitted and the rotational speed:

$$T = 63\,000\,P/n$$

Then, using the pinion, for example, the transmitted load is

$$W_{tP} = T/r_m \tag{10–10}$$

where r_m = mean radius of the pinion

The value of r_m can be computed from

$$r_m = d/2 - (F/2)\sin\gamma \tag{10–11}$$

Remember that the pitch diameter, d, is measured to the pitch line of the tooth at its large end. The angle, γ, is the pitch cone angle for the pinion as shown in Figure 10–8(a). The radial load acts toward the center of the pinion, perpendicular to its axis, causing bending of the pinion shaft. Thus,

$$W_{rP} = W_t \tan\phi\,\cos\gamma \tag{10–12}$$

The angle, ϕ, is the pressure angle for the teeth.

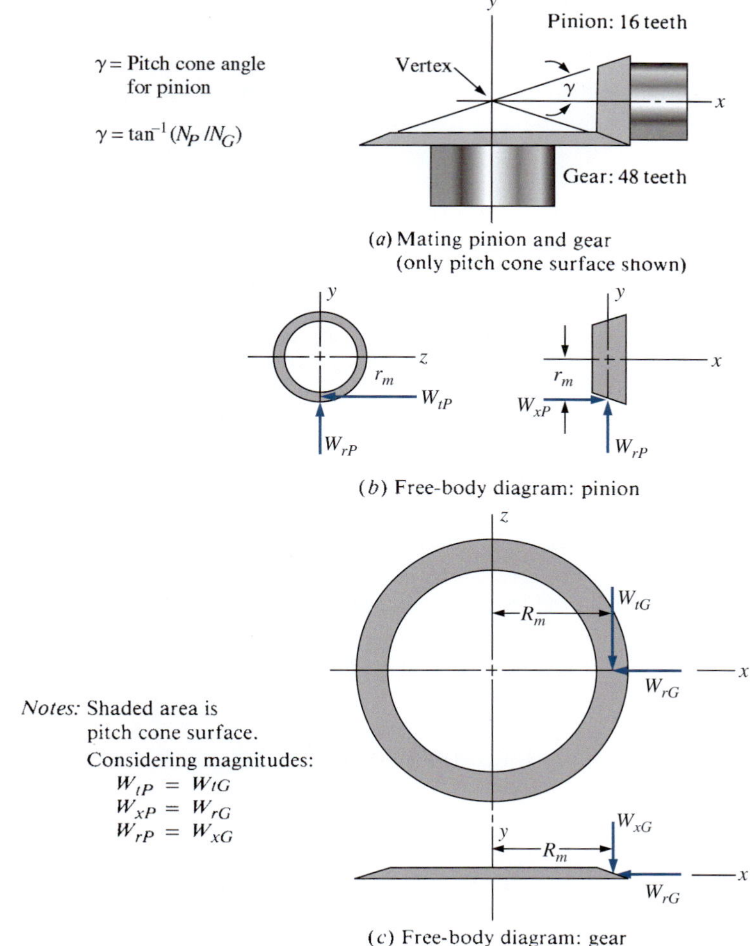

FIGURE 10–8 Forces on bevel gears

The axial load acts parallel to the axis of the pinion, tending to push it away from the mating gear. It causes a thrust load on the shaft bearings. It also produces a bending moment on the shaft because it acts at the distance from the axis equal to the mean radius of the gear. Thus,

$$W_{xP} = W_t \tan \phi \sin \gamma \tag{10–13}$$

The values for the forces on the gear can be calculated by the same equations shown here for the pinion, if the geometry for the gear is substituted for that of the pinion. Refer to Figure 10–8 for the relationships between the forces on the pinion and the gear in both magnitude and direction.

Example Problem 10–4 For the gear pair described in Example Problem 8–3, calculate the forces on the pinion and the gear if they are transmitting 2.50 hp with a pinion speed of 600 rpm. The geometry factors computed in Example Problem 8–3 apply. The data are summarized here.

Summary of Pertinent Results from Example Problem 8–3 and Given Data

Number of teeth in the pinion: $N_P = 16$
Number of teeth in the gear: $N_G = 48$
Diametral pitch: $P_d = 8$
Pitch diameter of pinion: $d = 2.000$ in
Pressure angle: $\phi = 20°$
Pinion pitch cone angle: $\gamma = 18.43°$
Gear pitch cone angle: $\Gamma = 71.57°$
Face width: $F = 1.00$ in

Rotational speed of pinion: $n_P = 600$ rpm
Power transmitted: $P = 2.50$ hp

Solution Forces on the pinion are described by the following equations:

$$W_t = T/r_m$$

But

$$T_p = 63000(P)/n_P = [63000(2.50)]/600 = 263 \text{ lb} \cdot \text{in}$$

$$r_m = d/2 - (F/2)\sin\gamma$$

$$r_m = (2.000/2) - (1.00/2)\sin(18.43°) = 0.84 \text{ in}$$

Then

$$W_t = T_P/r_m = 263 \text{ lb} \cdot \text{in}/0.84 \text{ in} = 313 \text{ lb}$$

$$W_r = W_t \tan\phi \cos\gamma = 313 \text{ lb} \tan(20°)\cos(18.43°) = 108 \text{ lb}$$

$$W_x = W_t \tan\phi \sin\gamma = 313 \text{ lb} \tan(20°)\sin(18.43°) = 36 \text{ lb}$$

To determine the forces on teh gear, first let's calculate the rotational speed of the gear:

$$n_G = n_P(N_P/N_G) = 600 \text{ rpm}(16/48) = 200 \text{ rpm}$$

Then

$$T_G = 63\,000(2.50)/200 = 788 \text{ lb} \cdot \text{in}$$

$$R_m = D/2 - (F/2)\sin\Gamma$$

$$R_m = 6.000/2 - (1.00/2)\sin(71.57°) = 2.53 \text{ in}$$

$$W_t = T_G/R_m = (788 \text{ lb} \cdot \text{in})/(2.53 \text{ in}) = 313 \text{ lb}$$

$$W_r = W_t \tan\phi \cos\Gamma = 313 \text{ lb} \tan(20°)\cos(71.57°) = 36 \text{ lb}$$

$$W_x = W_t \tan\phi \sin\Gamma = 313 \text{ lb} \tan(20°)\sin(71.57°) = 108 \text{ lb}$$

Note from Figure 10–8 that the forces on the pinion and the gear form an *action-reaction pair.* That is, the forces on the gear are equal to those on the pinion, but they act in the opposite direction. Also, because of the 90° orientation of the shafts, the radial force on the pinion becomes the axial thrust load on the gear, and the axial thrust load on the pinion becomes the radial load on the gear.

10–7 BEARING FORCES ON SHAFTS CARRYING BEVEL GEARS

Because of the three-dimensional force system that acts on bevel gears, the calculation of the forces on shaft bearings can be cumbersome. An example is worked out here to show the procedure. In order to obtain numerical data, the arrangement shown in Figure 10–9 is proposed for the bevel gear pair that was the subject of Example Problems 8–3 and 10–4. The locations for the bearings are given with respect to the vertex of the two pitch cones where the shaft axes intersect.

Note that both the pinion and the gear are *straddle-mounted*; that is, each gear is positioned between the supporting bearings. This is the most preferred arrangement because it usually provides the greatest rigidity and maintains the alignment of the teeth during power transmission. Care should be exercised to provide rigid mountings and stiff shafts when using bevel gears.

The arrangement of Figure 10–9 is designed so that the bearing on the right resists the axial thrust load on the pinion, and the lower bearing resists the axial thrust load on the gear.

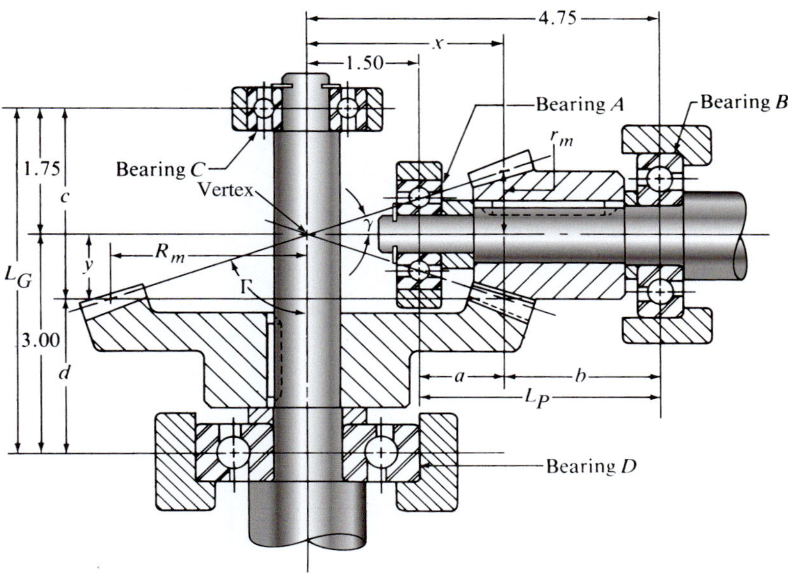

FIGURE 10–9 Layout of bevel gear pair for Example Problem 10–5

Example Problem 10–5 Compute the reaction forces on the bearings that support the shafts carrying the bevel gear pair shown in Figure 10–9. The values of Example Problems 8–3 and 10–4 apply.

Solution Referring to the results of Example Problem 10–4 and Figure 10–8, we have listed the forces acting on the gears:

Force	Pinion	Gear
Tangential	$W_{tP} = 313$ lb	$W_{tG} = 313$ lb
Radial	$W_{rP} = 108$ lb	$W_{rG} = 36$ lb
Axial	$W_{xP} = 36$ lb	$W_{xG} = 108$ lb

It is critical to be able to visualize the directions in which these forces are acting because of the three-dimensional force system. Notice in Figure 10–8 that a rectangular coordinate system has been set up. Figure 10–10 is an isometric sketch of the free-body diagrams of the pinion and the gear, simplified to represent the concurrent forces acting at the pinion/gear interface and at the bearing locations. Although the two free-body diagrams are separated for clarity, notice that you can bring them together by moving the point called *vertex* on each sketch together. This is the point in the actual gear system where the vertices of the two pitch cones lie at the same point. The two pitch points also coincide.

For setting up the equations of static equilibrium needed to solve for the bearing reactions, the distances *a, b, c, d,* L_P, and L_G are needed, as shown in Figure 10–9. These require the two dimensions labeled *x* and *y*. Note from Example Problem 10–4 that

$$x = R_m = 2.53 \text{ in}$$

$$y = r_m = 0.84 \text{ in}$$

Then

$$a = x - 1.50 = 2.53 - 1.50 = 1.03 \text{ in}$$

$$b = 4.75 - x = 4.75 - 2.53 = 2.22 \text{ in}$$

$$c = 1.75 + y = 1.75 + 0.84 = 2.59 \text{ in}$$

$$d = 3.00 - y = 3.00 - 0.84 = 2.16 \text{ in}$$

$$L_P = 4.75 - 1.50 = 3.25 \text{ in}$$

$$L_G = 1.75 + 3.00 = 4.75 \text{ in}$$

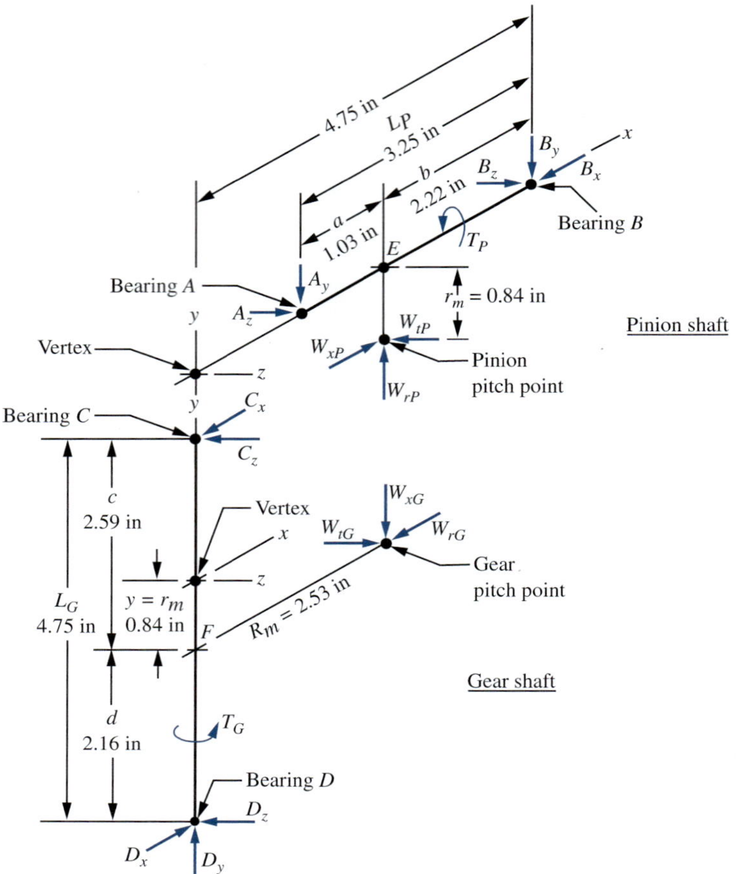

FIGURE 10–10 Free-body diagrams for pinion and gear shafts

These values are shown on Figure 10–10.

To solve for the reactions, we need to consider the horizontal (*x-z*) and the vertical (*x-y*) planes separately. It may help you to look also at Figure 10–11, which breaks out the forces on the pinion shaft in these two planes. Then we can analyze each plane using the fundamental equations of equilibrium.

Bearing Reactions, Pinion Shaft: Bearings A and B

Step 1. To find B_z and A_z: In the *x-z* plane, only W_{tP} acts. Summing moments about *A* yields

$$0 = W_{tP}(a) - B_z(L_P) = 313(1.03) - B_z(3.25)$$

$$B_z = 99.2 \text{ lb}$$

Summing moments about *B* yields

$$0 = W_{tP}(b) - A_z(L_P) = 313(2.22) - A_z(3.25)$$

$$A_z = 214 \text{ lb}$$

Step 2. To find B_y and A_y: In the *x-y* plane, both W_{rP} and W_{xP} act. Summing moments about *A* yields

$$0 = w_{rP}(a) + W_{xP}(r_m) - B_y(L_P)$$

$$0 = 108(1.03) + 36(0.84) - B_y(3.25)$$

$$B_y = 43.5 \text{ lb}$$

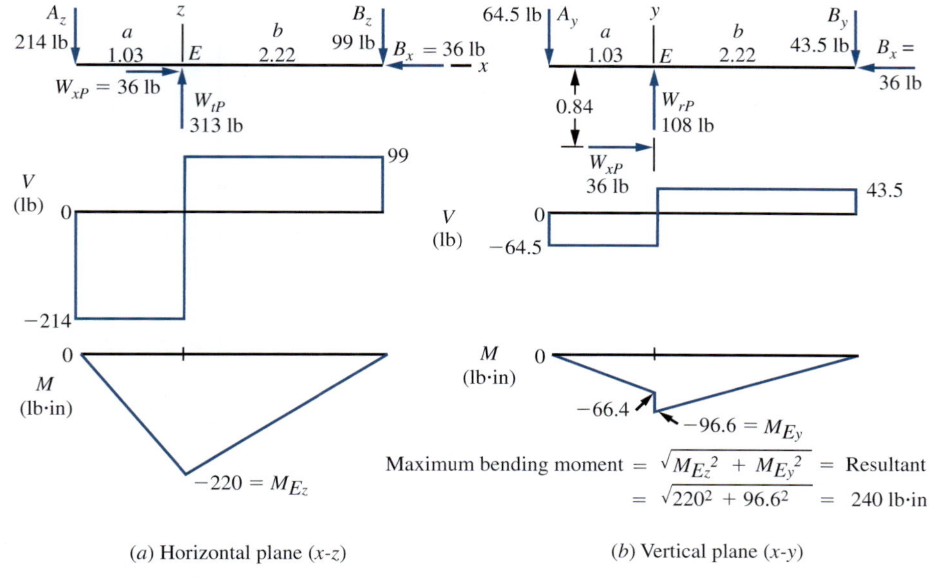

FIGURE 10–11 Pinion shaft bending moments

Summing moments about *B* yields

$$0 = W_{rP}(b) + W_{xP}(r_m) - A_y(L_P)$$

$$0 = 108(2.22) + 36(0.84) - A_y(3.25)$$

$$A_y = 64.5 \text{ lb}$$

Step 3. To find B_x: Summing forces in the *x*-direction yields

$$B_x = W_{xP} = 36 \text{ lb}$$

This is the thrust force on bearing *B*.

Step 4. To find the total radial force on each bearing: Compute the resultant of the *y*- and *z*-components.

$$A = \sqrt{A_y^2 + A_z^2} = \sqrt{64.5^2 + 214^2} = 224 \text{ lb}$$

$$B = \sqrt{B_y^2 + B_z^2} = \sqrt{43.5^2 + 99.2^2} = 108 \text{ lb}$$

Bearing Reactions, Gear Shaft: Bearings C and D

Using similar methods, we can find the forces in Figure 10–12.

$$\left.\begin{aligned} C_z &= 142 \text{ lb} \\ C_x &= 41.1 \text{ lb} \end{aligned}\right\} C = 148 \text{ lb (radial force on C)}$$

$$\left.\begin{aligned} D_z &= 171 \text{ lb} \\ D_x &= 77.1 \text{ lb} \end{aligned}\right\} D = 188 \text{ lb (radial force on } D\text{)}$$

$$D_y = W_{xG} = 108 \text{ lb (thrust force on D)}$$

Summary In selection of the bearings for these shafts, the following capacities are required:

Bearing *A:* 224-lb radial

Bearing *B:* 108-lb radial; 36-lb thrust

Bearing *C:* 148-lb radial

Bearing *D:* 188-lb radial; 108-lb thrust

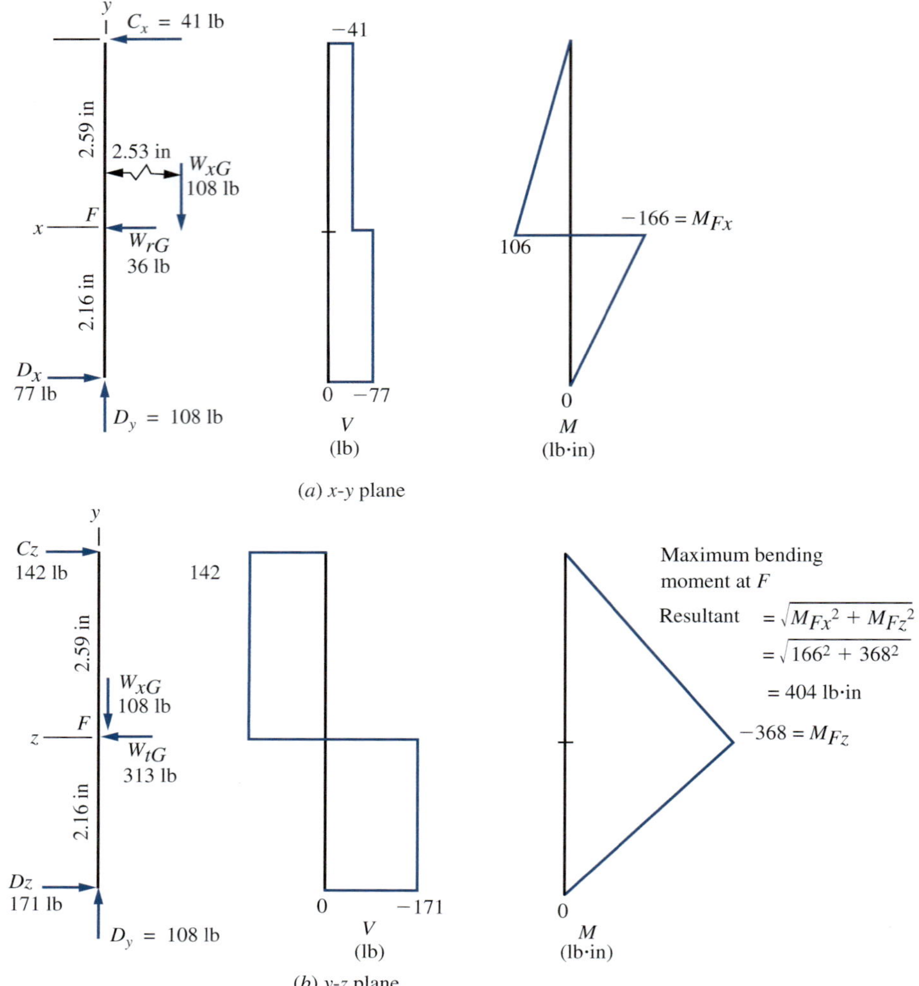

FIGURE 10–12 Gear shaft bending moments

10–8 BENDING MOMENTS ON SHAFTS CARRYING BEVEL GEARS

Because there are forces acting in two planes on bevel gears, as discussed in the preceding section, there is also bending in two planes. The analysis of the shearing force and bending moment diagrams for the shafts must take this into account.

Figures 10–11 and 10–12 show the resulting diagrams for the pinion and the gear shafts, respectively, for the gear pair used for Example Problems 8–3, 10–4, and 10–5. Notice that the axial thrust load on each gear provides a concentrated moment to the shaft equal to the axial load times the distance that it is offset from the axis of the shaft. Also notice that the maximum bending moment for each shaft is the resultant of the moments in the two planes. On the pinion shaft, the maximum moment is 240 lb · in at *E*, where the lines of action for the radial and tangential forces intersect the shaft. Similarly, on the gear shaft, the maximum moment is 404 lb · in at *F*. These data are used in the shaft design (as discussed in Chapter 12).

10–9 STRESSES IN STRAIGHT BEVEL GEAR TEETH

This section generally follows AGMA Standard 2003-C10, *Rating the Pitting Resistance and Bending Strength of Generated Straight Bevel, Zerol Bevel and Spiral Bevel Gear Teeth*, considered to be the primary standard in the United States (Reference 10). However, only straight bevel gears are treated here. The standard presents design analysis in both the U.S. unit system based on diametral pitch, P_d, and the SI Metric unit system based on metric module, m.

This book will maintain similar notations and symbols for gear tooth features, allowable stresses, and modifying factors that were initially presented in Chapter 9 for spur gear design to facilitate the comparison of design approaches. The reader should note that Standard 2003-C10 presents design analysis in SI units using terminology from ISO standards that employ radically different symbol sets. In this book, we maintain similar notations for factors in both systems except for the basic terms, diametral pitch, P_d, and metric module, m.

Furthermore, as introduced in Chapter 9, this book makes the following assumptions:

1. The gears are operating at temperatures between 32°F and 250°F (0°C–120°C) for which the temperature factor $K_T = 1.0$ and, therefore, it is not written in the equations for stress analysis.
2. Gear teeth are made to normal standards without special modifications. This does, however, assume that crowning of the teeth is included in the manufacturing process, typical of the bevel gear industry.
3. Both the pinion and the gear of the bevel gear set are straddle mounted as shown in Figure 10–9, providing the most rigid arrangement. If either gear is not straddle mounted, the overhung arrangement, being generally less stiff, requires the application of additional modifying factors.
4. The hardness of both the pinion and the gear are nearly equal to enable each to withstand the contact stress without pitting, allowing the use of the hardness factor, $C_H = 1.0$, and not including that factor in the equations. The standard contains significant discussion and data to modify allowable strengths when the pinion is significantly harder than the gear, an approach used by some designers to promote *wearing in* of the gear teeth by the harder pinion. Also included in the standard with the hardness factor are adjustments based on surface roughness of the teeth that are not included in this book.
5. When case hardening is used, as is frequently done, the case has sufficient depth and the core hardness and strength are sufficiently high to avoid case crushing or subsurface failure of the core due to either bending or contact stresses. The standard includes much discussion of these factors.
6. Residual stresses in gear teeth are not considered in this section. Note that beneficial compressive residual stresses in the root area from peening processes can significantly enhance the life of gears. Conversely, residual tensile stresses can be detrimental.

Following is the discussion of bending stress and contact stress calculations along with the corresponding parameters and modification factors:

Pitch Diameter, D: An important difference in the analysis of bevel gears is the definition of pitch diameter; it is measured at the large (outer) end of the gear rather than at the middle of the teeth as it was for spur and helical gears. This difference is accommodated in the determination of the geometry factors *J* and *I* that are shown later.

Pitches: The pitch for bevel gears is defined at the outer pitch diameter and is called the *outer transverse pitch.* Calculation is the same as for spur and helical gears with the exception of the definition of pitch diameter given above.

U.S. Units: Outer transverse diametral pitch $= P_d = N/D$ (units are in^{-1}; rarely reported)

SI Metric Units: Outer transverse metric module $= m = D/N$ mm

Tangential Force, W_t: As in Chapter 9, we will use the following unit-specific equations for torque on a gear, pitch line speed, and the resulting tangential force.

U.S. Units: Power, *P*, in hp; rotational speed, *n*, in rpm; diameters, *D*, in inches

$$\text{Pitch line speed} = v_t = \pi Dn/12 \text{ ft/min}$$
$$\text{Torque} = T = 63\,000P/n \text{ lb} \cdot \text{in}$$
$$\text{Tangential force} = W_t = T/(D/2) = 126\,000P/(Dn) \text{ lb}$$
$$\text{Or, Tangential force} = W_t = 33\,000P/v_t \text{ lb}$$

SI Metric Units: Power, *P*, in kW; rotational speed, *n*, in rpm; diameters, *D*, in mm

$$\text{Pitch line speed} = v_t = \pi D\,n/(60\,000) \text{ m/s}$$
$$\text{Torque} = T = 9550\,P/n \text{ N} \cdot \text{m}$$
$$\text{Tangential force} = W_t = T/(D/2)$$
$$= 19.1 \times 10^6 P/(D\,n)\text{N} = 19\,100\,P/(D\,n) \text{ kN}$$
$$\text{Or, Tangential force} = W_t = 1000\,P/v_t \text{ N} = P/v_t \text{ kN}$$

Bending Stress Number, s_t: The maximum bending stress number occurs in the root area of the teeth as it does in spur and helical gears; the equation is:

$$\textbf{U.S. Units:} \quad s_t = \frac{W_t P_d K_o K_s K_m K_v}{FJ} \text{ psi} \quad \textbf{(10–14)}$$

$$\textbf{SI Metric Units:} \quad s_t = \frac{W_t K_o K_s K_m K_v}{FJm} \text{ MPa} \quad \textbf{(10–14M)}$$

Overload Factor, K_o: Use the same values given in Table 9–7.

Size Factor, K_s for Bending Strength: Use Figure 10–13, adapted from AGMA Standard 2003-C10. For $P_d \geq 16 (m \leq 1.6 \text{ mm})$ use $K_s = 0.50$. The equation for the sloped portion of Figure 10–13 is

$$K_s = 0.4867 + 0.2133/P_d \quad \textbf{(10–15)}$$

$$K_s = 0.4867 + 0.008399\,m \quad \textbf{(10–15M)}$$

Load Distribution Factor, K_m: Typical bevel gear teeth are crowned along the profile and from end to end to ensure smooth engagement under foreseeable conditions of tooth accuracy and deflection under load. During the engagement/disengagement cycle, the contact pattern across the teeth should spread the load over the entire area of the faces. When additional misalignment occurs, this pattern is changed and the effects are more prominent with larger face widths and the manner of mounting. Use Figure 10–14 when discrete

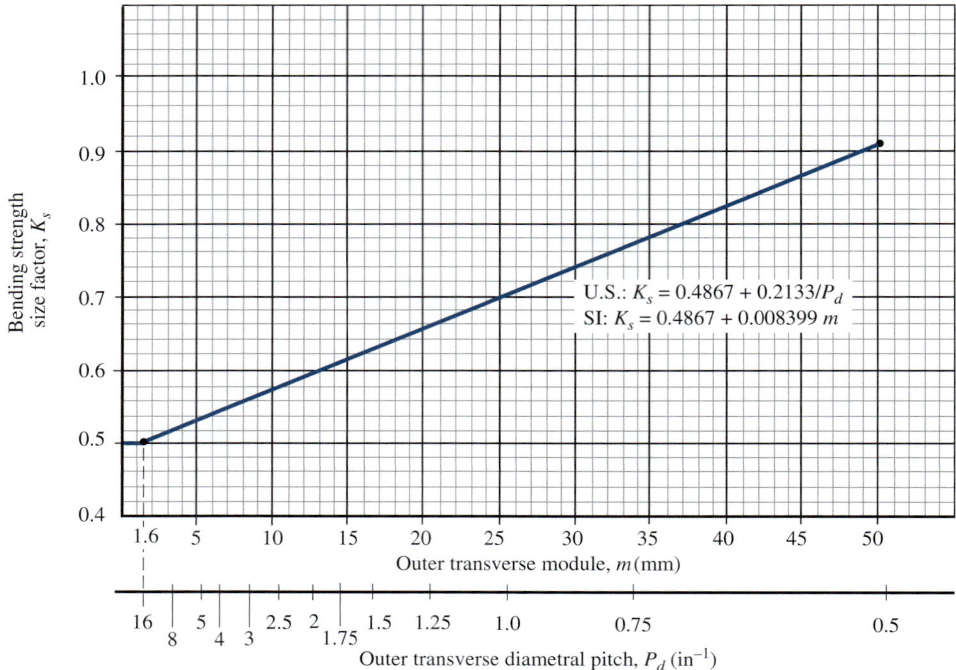

FIGURE 10–13 Size factor for bending stress, K_s, for bevel gears (Adapted from AGMA 2003-C10, *Rating the Pitting Resistance and Bending Strength of Generated Straight Bevel, Zerol Bevel and Spiral Bevel Gear Teeth,* with the permission of the publisher, American Gear Manufacturers Association, 1001 North Fairfax Street, 5th Floor, Alexandria, VA)

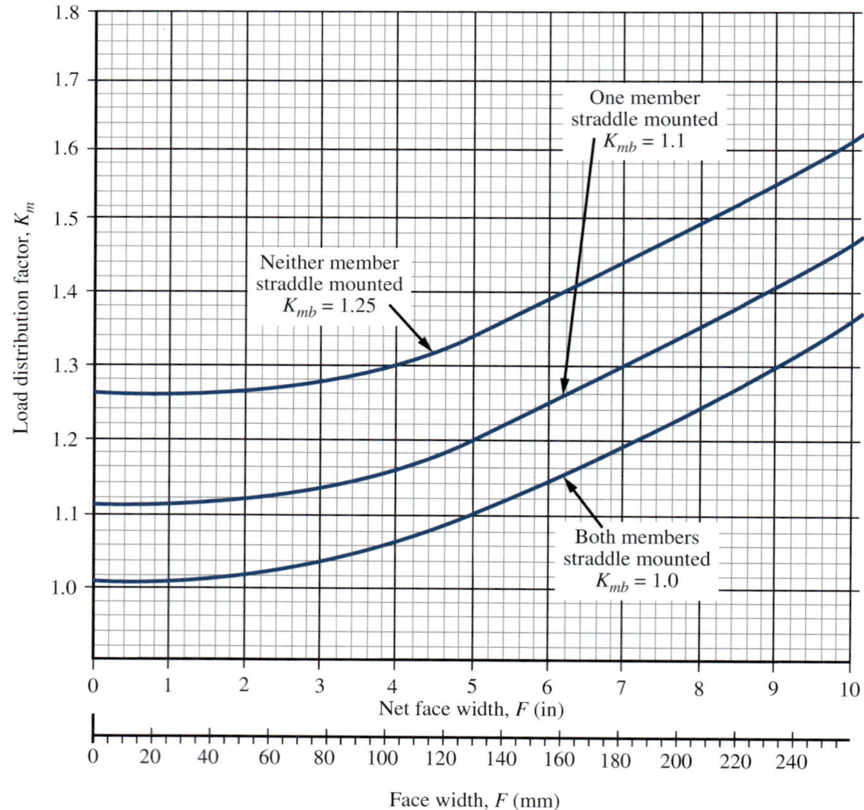

FIGURE 10–14 Load distribution factor, K_m, for crowned bevel gear teeth (Adapted from AGMA 2003-C10, *Rating the Pitting Resistance and Bending Strength of Generated Straight Bevel, Zerol Bevel and Spiral Bevel Gear Teeth,* with the permission of the publisher, American Gear Manufacturers Association, 1001 North Fairfax Street, 5th Floor, Alexandria, VA)

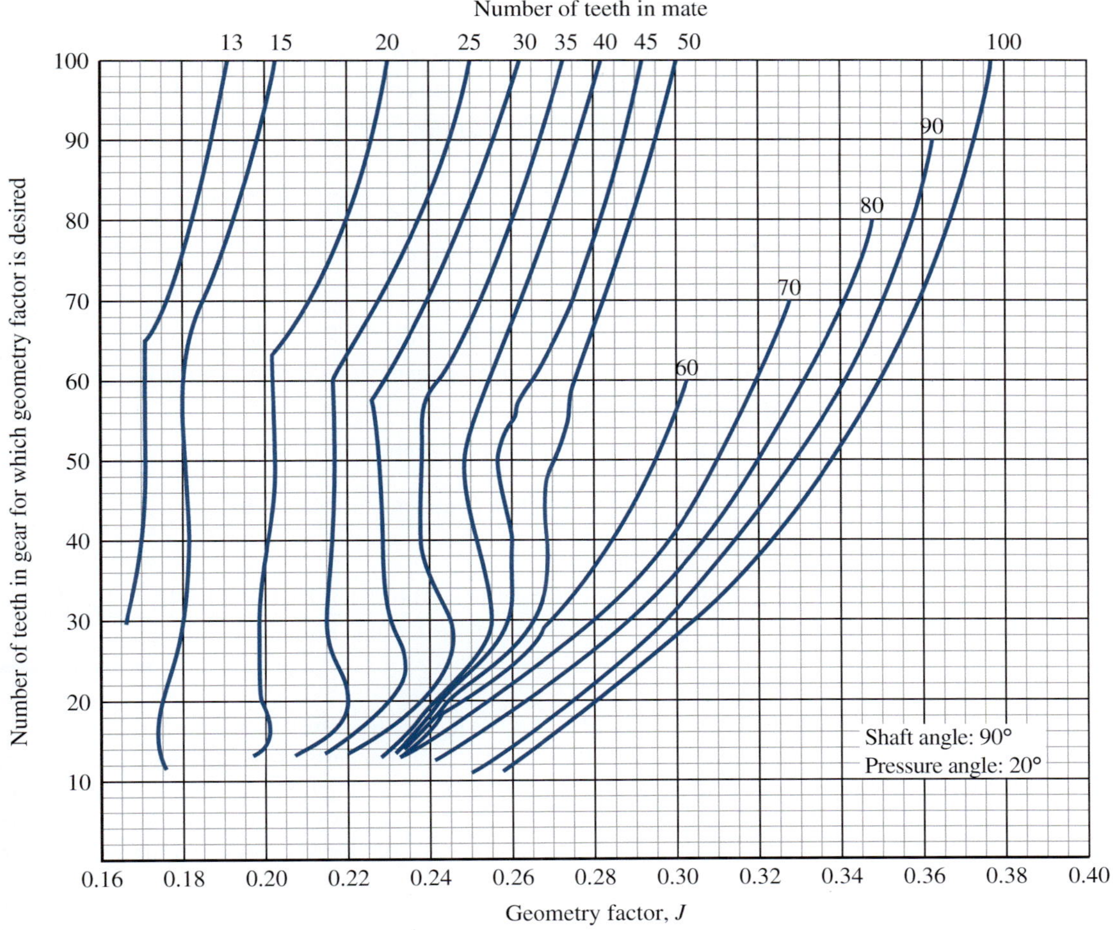

FIGURE 10–15 Geometry factor, *J*, for straight bevel gears with 20° pressure angle and 90° shaft angle (Adapted from AGMA 2003-C10, *Rating the Pitting Resistance and Bending Strength of Generated Straight Bevel, Zerol Bevel and Spiral Bevel Gear Teeth,* with the permission of the publisher, American Gear Manufacturers Association, 1001 North Fairfax Street, 5th Floor, Alexandria, VA)

analysis of deformations is not practical. The three curves conform to the equations:

U.S. Units: $K_m = K_{mb} + 0.0036\,F^2$ **(10–16)**

SI Metric Units: $K_m = K_{mb} + 5.6 \times 10^{-6}\,F^2$ **(10–16M)**

Where,

$K_{mb} = 1.00$ for both gears straddle mounted
$K_{mb} = 1.10$ for one gear straddle mounted
$K_{mb} = 1.25$ for neither gear straddle mounted

Problems in this book can assume that both gears are straddle mounted unless otherwise noted. Designers should take steps to provide a precise and rigid system comprised of the gears, shafts, bearings, and housing.

Dynamic Factor, K_v: Use the same chart shown for spur gears in Figure 9-20, based on the quality system defined in AGMA standards 2015-1-A01 and 2001-D04, in which quality numbers from A11 (least accurate) to A4 (most accurate) are used. Bevel gear standard AGMA 2003-C10 includes the former chart for K_v based on the Q-system of quality from Q5 (least accurate) to Q11 (most accurate). To a reasonable degree of precision for values of K_v in these two charts are compatible provided that the quality number in the A-system plus the quality number in the Q-system add to 17. For example, Q8 is similar to A9 and Q10 is similar to A7.

Geometry Factor for Bending Strength, J: Use Figure 10–15 for problems in this book. This figure is for straight-tooth bevel gears with a 20° pressure angle and a shaft angle between the shafts of the pinion and the gear of 90°. Standard AGMA 2003-C10 gives formulas from which the value for other designs can be computed when detailed geometry for the teeth is known.

Example Problem 10–6 illustrates the calculation for bending stress number for straight bevel gear teeth, using the same data from earlier problems. Then we will introduce allowable stresses and material selection to resist bending stresses.

Example Problem 10–6 Compute the bending stress in the teeth of the bevel pinion shown in Figure 10–9. The data from Example Problem 10–4 apply: $N_P = 16$; $N_G = 48$; $n_P = 600$ rpm; $P = 2.50$ hp; $P_d = 8$; $d = 2.000$ in; $F = 1.00$ in. Assume that the pinion is driven by an electric motor and that the load provides moderate shock. The quality number, A_v, is to be 11.

Solution

$$W_t = \frac{T}{r} = \frac{63\,000(P)}{n_p}\frac{1}{d/2} = \frac{63\,000(2.50)}{600}\frac{1}{2.000/2} = 263 \text{ lb}$$

$$v_t = \pi d n_P/12 = \pi(2.000)(600)/12 = 314 \text{ ft/min}$$

$K_o = 1.50$ (from Table 9–7)

$K_s = 0.4867 + 0.2133/P_d = 0.4867 + 0.2133/8 = 0.513$

$K_m = 1.004$ (both gears straddle-mounted, general commercial-quality)

$J_P = 0.230$ (from Figure 10–5)

$K_v = 1.24$ (Use $A_v = 11$ and $v_t = 314$ ft/min)

(Read from Figure 9–20 or computed from equations in Table 9–9)

Then, from Equation (10–13),

$$s_t = \frac{W_t P_d K_o K_s K_m K_v}{FJ} = \frac{(263)(8)(1.50)(0.5131)(1.004)(1.24)}{(1.00)(0.230)} = 8764 \text{ psi}$$

Allowable Bending Strength Number: The process of finding the required allowable bending strength, s_{at}, for bevel gears is similar to that used for spur and helical gears. The fundamental equation is,

$$s_{wt} = \frac{s_{at} K_L}{(SF)(K_R)} \quad \textbf{(10–17)}$$

Where

s_{wt} = permissible bending stress number considering design life and reliability

K_L = Stress cycle factor for bending
For carburized case-hardened steel bevel gears, use Figure 10–16. Most of the test data for these curves were developed for carburized case-hardened gears so use for through-hardened steel is only approximate. In Chapter 9, a similar factor was called Y_N and that is the term used in ISO standards as well. Note that the curve for life factors above about 3×10^6 is used for most commercial gear drives. The standard permits a lower value for some critical applications.

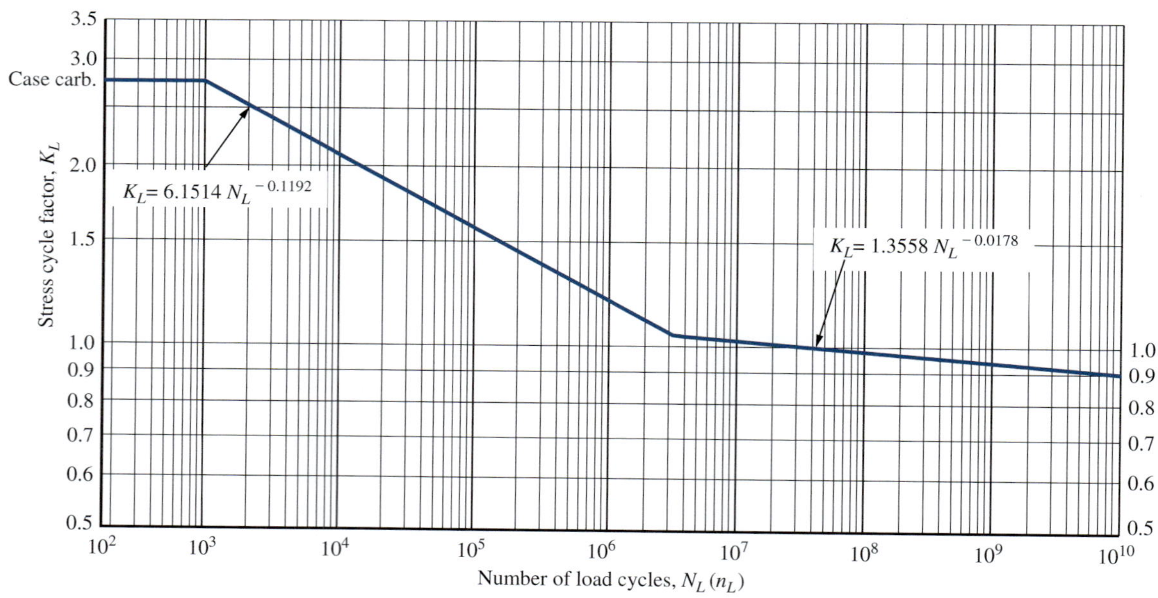

FIGURE 10–16 Stress cycle factor for bending strength, K_L (carburized case-hardened steel bevel gears) (Adapted from AGMA 2003-C10, *Rating the Pitting Resistance and Bending Strength of Generated Straight Bevel, Zerol Bevel and Spiral Bevel Gear Teeth,* with the permission of the publisher, American Gear Manufacturers Association, 1001 North Fairfax Street, 5th Floor, Alexandria, VA)

K_R = Reliability factor. Use Table 10–3. Note that for bending, the reliability factor for bevel gears is the same as for spur and helical gears in Table 9–11.

SF = Safety factor. Typically taken to be 1.00, but values up to about 1.50 can be used for greater uncertainty or for critical systems.

For design where the goal is to specify a suitable gear material, Equation (10–17) can be combined with the equation for s_t from Equation (10–14) and we can solve for the required value of the allowable bending strength, s_{at}. That is,

$$\text{Let } s_t = s_{wt} = \frac{s_{at}K_L}{(SF)(K_R)}$$

Then, the required value of s_{at} is

$$s_{at} = \frac{s_t(SF)(K_R)}{K_L} \quad \textbf{(10–18)}$$

Use Figure 10–17 to determine the required hardness for through-hardened steel, recognizing that HB 400 is the maximum recommended value. Above that, flame, induction, or carburized case-hardened steel should be used up to the limits shown in Table 10–4.

Next, we develop the similar approach for determining the required allowable contact stress number, s_{ac}. Then we will show another example problem in which we identify the critical value on which material selection is based.

TABLE 10–3 Reliability Factors for Allowable Bending and Contact Stresses

Reliability R	Interpretation	Reliability factors: Bending K_R	Reliability factors: Contact C_R
0.9	Fewer than one failure in 10	0.85	0.92
0.99	Fewer than one failure in 100	1.00	1.00
0.999	Fewer than one failure in 1000	1.25	1.12
0.9999	Fewer than one failure in 10 000	1.50	1.22

Source: Adapted from AGMA 2003-C10, *Rating the Pitting Resistance and Bending Strength of Generated Straight Bevel, Zerol Bevel and Spiral Bevel Gear Teeth,* with the permission of the publisher, American Gear Manufacturers Association, 1001 North Fairfax Street, 5th Floor, Alexandria, VA.

Contact Stress Number, s_c: The maximum contact stress number occurs on the face of the teeth as it does in spur and helical gears; the equation is the same for both U.S. and SI units with due attention to units for W_t, F, D_P, and C_p:

U.S. or SI Units:

$$s_c = C_p\sqrt{\frac{W_tK_oK_mK_vC_sC_{xc}}{FD_pI}} \text{ psi or MPa} \quad \textbf{(10–19)}$$

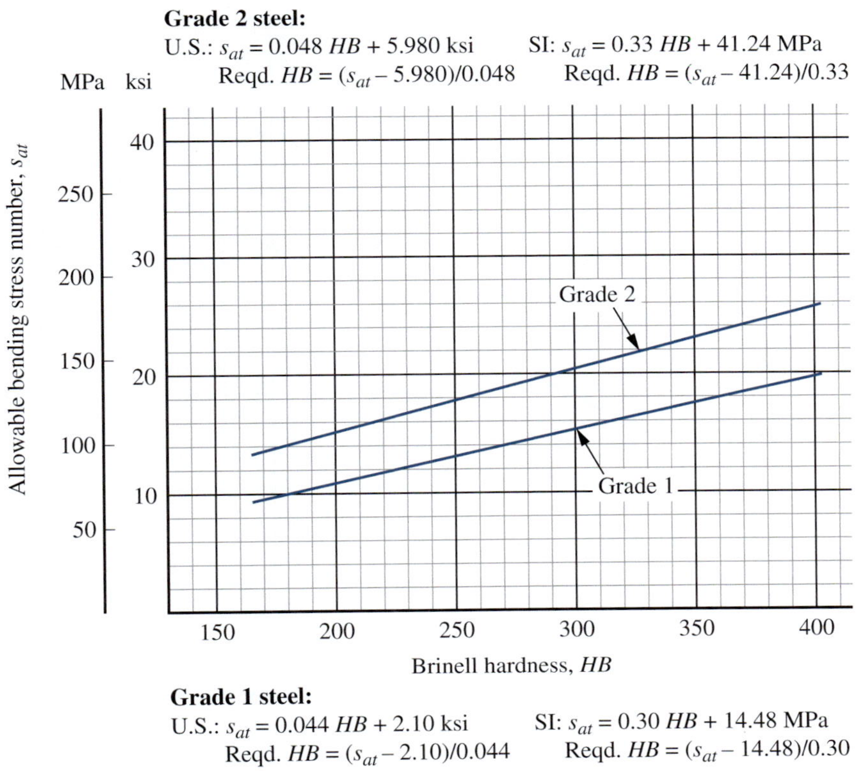

FIGURE 10–17 Allowable bending stress number, s_{at}, for through-hardened steel for bevel gears (Adapted from AGMA 2003-C10, *Rating the Pitting Resistance and Bending Strength of Generated Straight Bevel, Zerol Bevel and Spiral Bevel Gear Teeth,* with the permission of the publisher, American Gear Manufacturers Association, 1001 North Fairfax Street, 5th Floor, Alexandria, VA)

TABLE 10–4 Allowable Stress Numbers for Bevel Gears

Case hardened Grade 1 steel materials				
Hardness at surface	Allowable bending stress number, s_{at} (ksi)	(Mpa)	Allowable contact stress number, s_{ac} (ksi)	(Mpa)
Flame or induction hardened				
50 HRC—Unhardened roots	12.5	86	175	1207
50 HRC—hardened roots	22.5	155	175	1207
Carburized and case hardened				
55–64 HRC	30	207	200	1379

(Adapted from AGMA 2003-C10, *Rating the Pitting Resistance and Bending Strength of Generated Straight Bevel, ZEROL Bevel and Spiral Bevel Gear Teeth,* with the permission of the publisher, American Gear Manufacturers Association, 1001 North Fairfax Street, 5th Floor, Alexandria, VA.)

Note that this equation is used for both the pinion and the gear because the basic contact stress is equal on each. *Do not use the pitch diameter of the gear in this equation.*

All variables in this equation have been discussed in regard to bending stress number except for I, C_p, C_s, and C_{xc}, discussed next.

C_p: The elastic coefficient depends on the modulus of elasticity and Poisson's ratio for the materials of the pinion and the gear. For two steel gears, $C_p = 2300\ \text{psi}^{0.5}(191\ \text{MPa}^{0.5})$. Use the values from Chapter 9 in Table 9–10 for other materials.

C_s: The *size factor for contact stress* is different from the value of K_s used for bending stress. See Figure 10–18. The factor is based on the face width, F. For $F \leq 0.50$ in (12.5 mm), use $C_s = 0.50$. For $F \geq 3.14$ in (80 mm), use $C_s = 0.83$. Between those limits, use the figure or compute the value from the equations given here:

U.S. Units: $$C_s = 0.125F + 0.4375 \quad \textbf{(10–20)}$$

SI Metric Units: $$C_s = 0.00492F + 0.4375 \quad \textbf{(10–20M)}$$

C_{xc}: The *crowning factor for pitting* accounts for the contact pattern between mating teeth. Typical bevel gear production processes create crowning both along the tooth flank and across the full width of the face. This is the preferred approach. However, some bevel gears are not crowned. AGMA Standard 2003-C10 recommends the following factors:

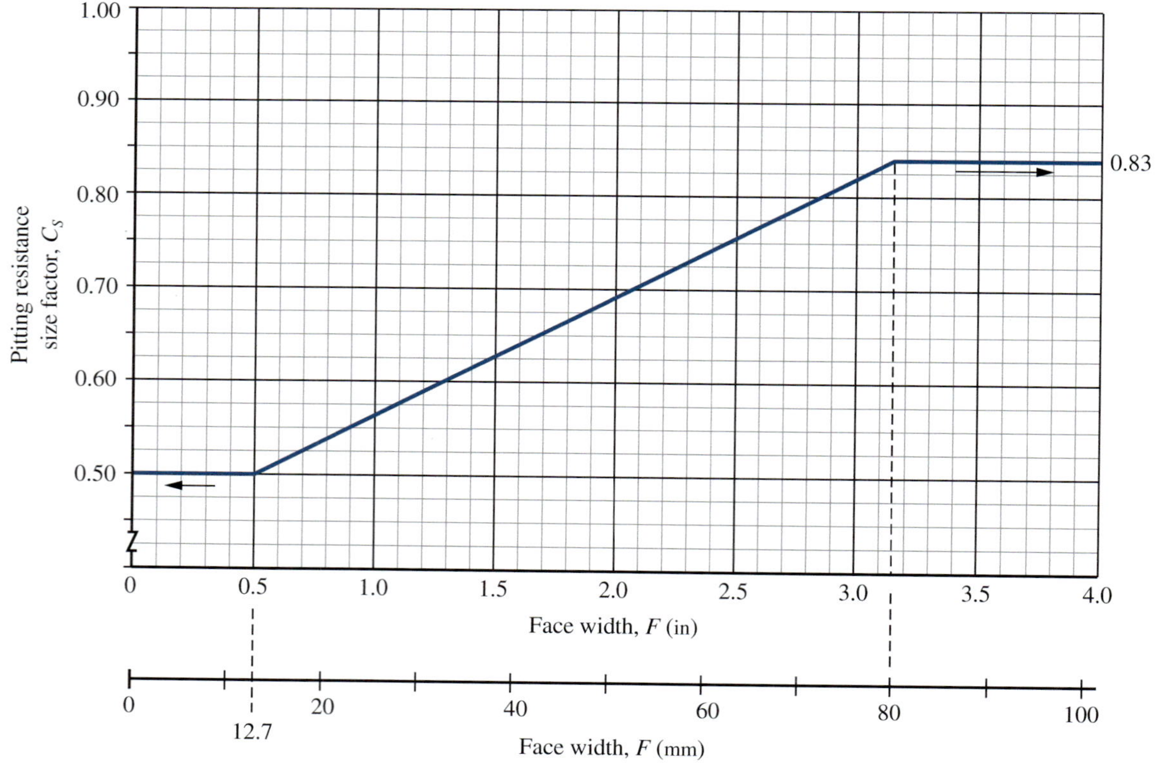

FIGURE 10–18 Pitting resistance size factor, C_s (Adapted from AGMA 2003-C10, *Rating the Pitting Resistance and Bending Strength of Generated Straight Bevel, Zerol Bevel and Spiral Bevel Gear Teeth,* with the permission of the publisher, American Gear Manufacturers Association, 1001 North Fairfax Street, 5th Floor, Alexandria, VA)

$C_{xc} = 1.5$ for properly crowned teeth

$C_{xc} = 2.0$ or larger for non-crowned teeth

I: The *geometry factor for pitting resistance* incorporates the radii of curvature of the pinion and gear teeth and the degree of load sharing between teeth. It is a function of the number of teeth in both the pinion and the gear and, therefore, to the gear ratio. Use Figure 10–19 for values for problem solving in this book. AGMA Standard 2003–C10 provides a formula for I and a detailed procedure for acquiring the necessary data and performing the calculations.

We now revisit the example design problem used throughout this chapter and compute the contact stress.

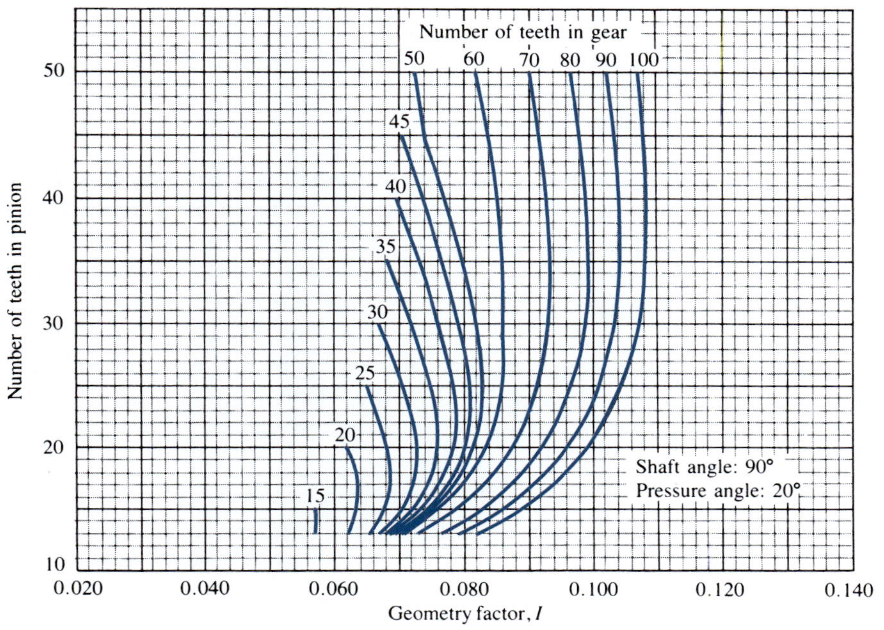

FIGURE 10–19 Geometry factors for straight and ZEROL® bevel gears (Extracted from AGMA 2003-C10, *Rating the Pitting Resistance and Bending Strength of Generated Straight Bevel, ZEROL® Bevel and Spiral Bevel Gear Teeth,* with the permission of the publisher, American Gear Manufacturers Association, 1001 North Fairfax Street, 5th floor, Alexandria, VA 22314)

Example Problem 10–7 Compute the contact stress for the gear pair in Figure 10–9 for the conditions used in Example Problem 10–5: $N_P = 16$; $N_G = 48$; $n_P = 600$ rpm; $P_d = 8$; $F = 1.00$ in; $D_P = 2.000$ in. Both gears are to be steel.

Solution We use Equation (10–19) with U.S. units.

$$s_c = C_p\sqrt{\frac{W_t K_o K_m K_v C_s C_{xc}}{F D_p I}}\ \text{psi}$$

From Example Problem 10–5: $W_t = 263$ lb; $K_o = 1.50$; $K_m = 1.004$; $K_v = 1.24$. Other factors are:

Elastic Coefficient $C_p = 2300\ \text{psi}^{0.5}$ for two steel gears

Size factor $C_s = 0.56$ (Figure 10–18)

Crowning factor $C_{xc} = 1.5$ (Specify properly crowned teeth)

Geometry factor for pitting resistance $I = 0.077$ (Figure 10–19)

Then

$$s_c = 2300\sqrt{\frac{(263)(1.50)(1.004)(1.24)(0.56)(1.5)}{(1.00)(2.000)(0.077)}}\ \text{psi} = 119\,044\ \text{psi}$$

Allowable Contact Stress Number: The process of finding the required allowable contact stress, s_{ac}, for bevel gears is similar to that used for spur and helical gears. The fundamental equation is,

$$s_{wc} = \frac{s_{ac}\,C_L}{(SF)(C_R)} \qquad \textbf{(10–21)}$$

Where

s_{wc} = permissible contact stress number considering design life and reliability.

C_L = Stress cycle factor for contact stress. Use Figure 10–20, developed for carburized case-hardened steel bevel gears. Application to through-hardened gears is approximate.

C_R = Reliability factor. Use Table 10–3. Note that the reliability factor for pitting resistance is equal to the square root of that for bending. The AGMA standard adjusts the allowable bending strengths accordingly.

S_F = Safety factor. Typically taken to be 1.00, but values up to about 1.50 can be used for greater uncertainty or for critical systems.

For design where the goal is to specify a suitable gear material, Equation (10–21) can be combined with the equation for s_c from Equation (10–19) and we can solve for the required value of the allowable bending strength, s_{ac}. That is,

$$\text{Let } s_c = s_{wc} = \frac{s_{ac}C_L}{(SF)(C_R)}$$

Then, the required value of s_{ac} is

$$s_{ac} = \frac{s_c(SF)(C_R)}{C_L} \qquad \textbf{(10–22)}$$

Use Figure 10–21 to determine the required hardness for through-hardened steel, recognizing that HB 400 is the maximum recommended value. Above that, flame, induction, or carburized case-hardened steel should be used up to the limits shown in Table 10–3.

Now we look again at the design analysis example considered before and evaluate both the required value of s_{at} and s_{ac}. Finally, we identify the critical value and specify a suitable material for the two gears.

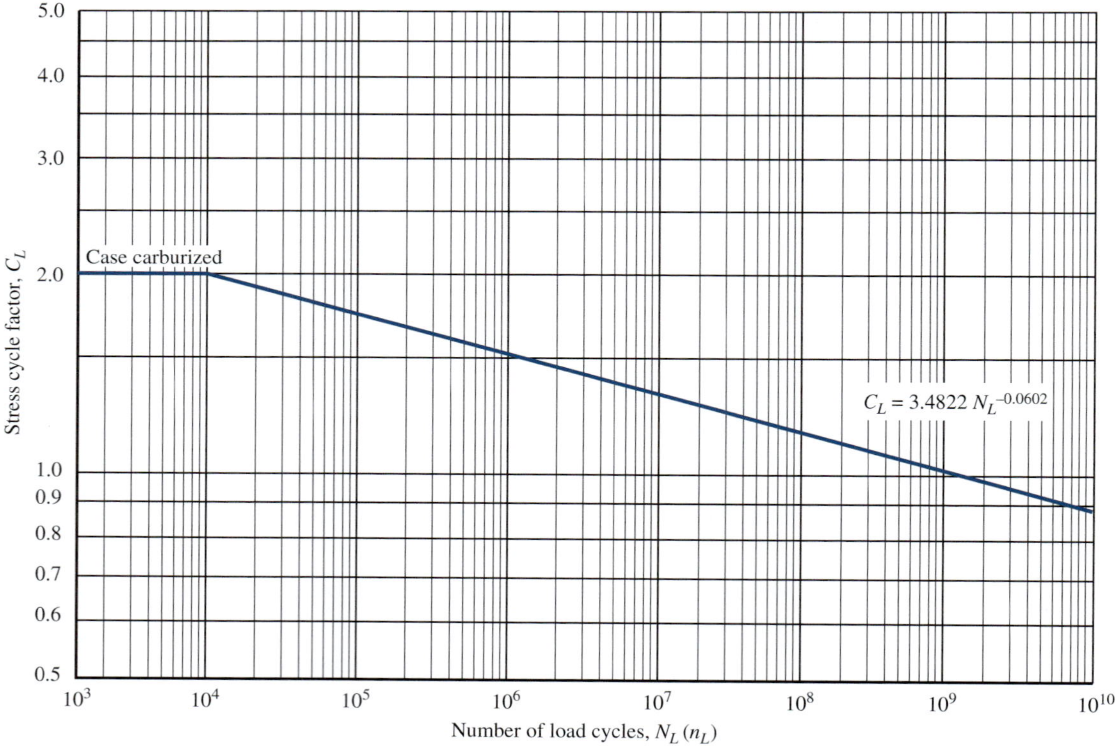

FIGURE 10–20 Stress cycle factor for pitting resistance, C_L (carburized case-hardened steel bevel gears) (Adapted from AGMA 2003-C10, *Rating the Pitting Resistance and Bending Strength of Generated Straight Bevel, Zerol Bevel and Spiral Bevel Gear Teeth,* with permission of the publisher, American Gear Manufacturers Association, 1001 North Fairfax Street, 5th Floor, Alexandria, VA)

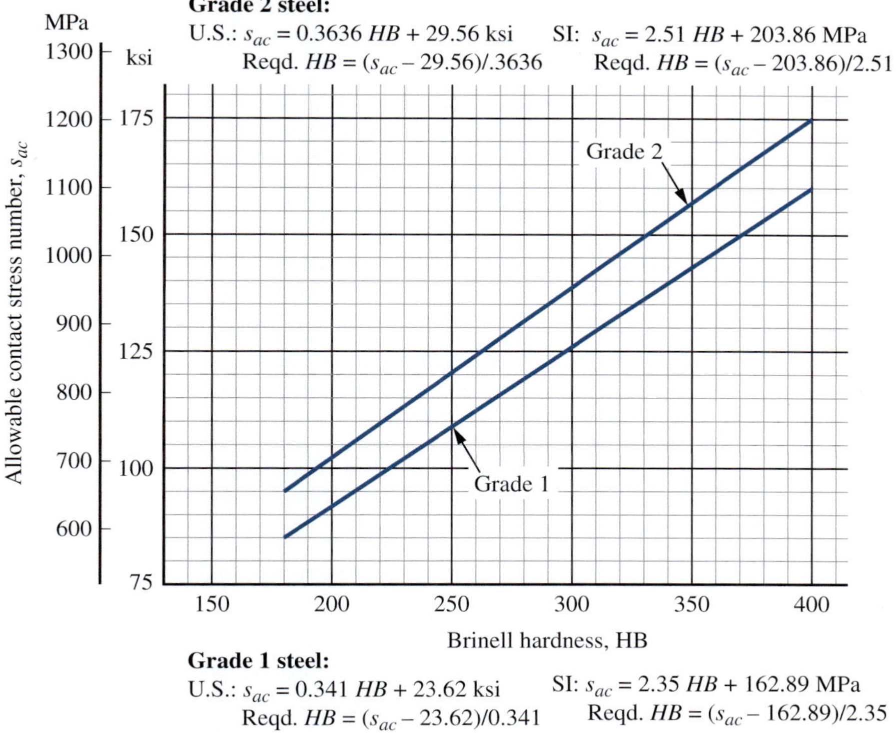

FIGURE 10–21 Allowable contact stress number, S_{ac}, for through-hardened steel for bevel gears (Adapted from AGMA 2003-C10, *Rating the Pitting Resistance and Bending Strength of Generated Straight Bevel, Zerol Bevel and Spiral Bevel Gear Teeth,* with permission of the publisher, American Gear Manufacturers Association, 1001 North Fairfax Street, 5th Floor, Alexandria, VA)

Example Problem 10–8 Specify suitable materials for the level pinion and gear for the data of Example Problems 10–1 to 10–6. Design for a life of 15 000 hours.

Solution From Example Problems 10–5 and 10–6, we find the following data:

Rotational speed of pinion = n_P = 600 rpm

Bending stress number, s_t = 8764 psi

Contact stress number, s_c = 119 044 psi

We apply Equations (10–18) and (10–22).

$$s_{at} = \frac{s_t(SF)(K_R)}{K_L} \tag{10–18}$$

$$s_{ac} = \frac{s_c(SF)(C_R)}{C_L} \tag{10–22}$$

Two of the factors in each equation are design decisions.

Safety factor *SF*:

No additional uncertainties or special requirements are known. Then $SF = 1.00$.

Reliability factors K_R and C_R: Assume a reliability of 0.99, one failure in 100.

Then $K_R = 1.00$ and $C_R = 1.00$.

Life factors K_L and C_L: We use Figures 10–16 and 10–20 that require the expected number of load cycles, N_c. The equation shown below was used in Chapter 9 with q indicating the number of stress cycles per revolution of the gear, typically 1.0. For planetary or split power systems, q can be 2.0 or more.

$$N_c = 60(L)(n_P)(q) = (60)(15\,000)(600)(1) = 5.4 \times 10^8 \text{ load cycles}$$

For bending: $K_L = 0.948$ (computed from equation in Figure 10–16)
For pitting resistance: $C_L = 1.038$ (computed from equation in Figure 10–20)
We can then complete the calculations:

$$s_{at} = \frac{s_t(SF)(K_R)}{K_L} = \frac{(8764)(1.0)(1.0)}{0.948} = 9245 \text{ psi} = 9.245 \text{ ksi}$$

$$s_{ac} = \frac{s_c(SF)(C_R)}{C_L} = \frac{(119\,044)(1.0)(1.0)}{1.038} = 114\,686 \text{ psi} = 114.7 \text{ ksi}$$

Required hardness of steels and material specification:
Use Figure 10–17 for bending: Required $HB = 162$; Low; Use almost any steel.
Use Figure 10–21 for contact stress: Required $HB = 267$.
This value is quite suitable for through-hardened steel.
Use Figure A4–1: Specify SAE 1040 WQT 1000; $HB = 269$
22% elongation; $s_y = 88$ ksi; $s_u = 114$ ksi

Summary and comments: A bevel gear drive has been designed with straight teeth for transmission of power to shafts oriented 90° to each other. Key data were given and used in Chapters 8 and 10 in Example Problems 8–3, and 10–4 to 10–8. Results are summarized as follows:

1. ***Example Problem 8–3:*** Values for geometrical features were calculated for the following input data: $P_d = 8$; 20° pressure angle; $N_P = 16$; $N_G = 48$; 90° shaft angle.
   a. Output data used in subsequent problems include: Gear Ratio = 3.00; $D_P = 2.000$ in; $D_G = 6.000$ in; Pitch cone angle for the pinion $= \gamma = 18.43°$; Pitch cone angle for the gear $= \Gamma = 71.57°$; Face width = F = 1.00 in (design decision); Pinion outside diameter $= D_{oP} = 2.368$ in; Pinion outside diameter $= D_{oP} = 2.368$ in; Gear outside diameter $= D_{oG} = 6.041$ in. Several other detailed gear tooth features were also calculated.
2. ***Example Problem 10–4:*** Forces and torques on the pinion and gear were calculated for a given power transmitted $= P = 2.50$ hp at a pinion speed of 600 rpm.
   a. Results included: Torque on the pinion $= T_P = 263$ lb·in; Torque on the gear $= T_G = 788$ lb·in; Tangential transmitted force at the mean radii $= W_t = 313$ lb; Radial force $= W_r = 108$ lb; Axial force $= W_x = 36$ lb; Mean radius of the pinion teeth $= R_{mP} = 0.84$ in; Mean radius of the gear teeth $= R_{mG} = 2.53$ in;
   b. The forces at the mean radii, torques, and mean radii for the pinion and the gear were used in the analysis of forces on the shafts and bearings in Example Problem 10–5. Note that tangential forces are recalculated in Example Problem 10–6 to conform to conventions for stress analysis of bevel gear teeth.
3. ***Example Problem 10–5:*** See Figure 10–10 for a graphical display of the free-body diagrams of both shafts showing all forces and torques acting on the two shafts carrying the pinion and the gear along with bearing reactions. This required analysis of forces, shearing forces, and bending moments on both shafts in two planes. All four bearings experience radial reaction forces in two planes. Resultants were calculated for each bearing. By a design decision, the thrust forces created by the bevel gears were taken at Bearings *B* and *D*. Results are:
   a. Bearing *A*: Radial load = 224 lb
   b. Bearing *B*: Radial load = 108 lb; Thrust load = 36 lb
   c. Bearing *C*: Radial load = 148 lb
   d. Bearing *D*: Radial load = 188 lb; Thrust load = 108 lb
4. ***Section 10–8:*** The equilibrium analysis to determine forces in Example Problem 10–5 was extended to complete the shearing forces and bending moments in two planes for both shafts. See Figure 10–11 for the pinion shaft and Figure 10–12 for the gear shaft. Resultant bending moments were also calculated that would be needed for completing the design for the shafts, as will be discussed in Chapter 12.
5. ***Example Problem 10–6:*** The bending stress number for the teeth of the pinion was calculated using data from previous problems and additional given data:
   a. The pinion is driven by an electric motor and the load provides moderate shock. The quality number is $A_v = 11$, achievable by general commercial processing.

b. Note that the transmitted forces on the pinion and the gear are recalculated based on the torque transmitted and the outer pitch radii for the pinion and the gear. The differences between these forces and those used in the shaft force analysis are accounted for in the geometry factors J and I. Here we use $W_t = 263$ lb.

c. Bending stress number on pinion teeth $= s_t = 8764$ psi. The stress on the gear teeth will be lower because of the relative value of the geometry factor.

6. ***Example Problem 10–7:*** The contact stress number for the teeth of the pinion was calculated using data from previous problems. A design decision chose to use steel for both the pinion and the gear.

   a. Contact stress number, $s_c = 119\,044$ psi

7. ***Example Problem 10–8:*** The minimum required allowable bending stress number and the minimum allowable contact stress numbers were calculated based on results from Example Problems 10–6 and 10–7 along with the following design decisions:

   a. Reliability $= R = 0.99(<1\,\text{failure in}\,100)$; $K_R = C_R = 1.0$

   b. Safety factor $= SF = 1.0$ assuming no unusual uncertainties beyond other K-factors

   c. Design for a life of 15 000 hours. This is reasonable for general industrial applications, fully utilized.

   d. Results: Bending: $s_{at} = 9245$ psi $= 9.245$ ksi; $s_{ac} = 114\,686$ psi $= 114.7$ ksi

8. ***Required hardness of steels and material specification:***

   Bending: Required HB = 162; Low; Use almost any steel.

   Contact stress–Pitting resistance: Required HB = 267.

   This value controls the material specification and is quite suitable for through-hardened steel.

   Using Figure A4–1, specify SAE 1040 WQT 1000; HB = 269

   22% elongation; $s_y = 88$ ksi; $s_u = 114$ ksi

9. ***Comments on the analysis and design:*** Taken together, the procedures and results summarized here demonstrate a reasonable approach to designing bevel gear pairs with straight teeth. The design is satisfactory as shown. However, alternate designs are practical and other iterations are recommended to explore what improvements can be made. Possible choices for different design decisions are:

   a. A smaller overall drive may be practical by permitting the use of case hardening by flame or induction hardening or carburizing. These would be capable of operating at higher bending and contact stresses produced by smaller gears, but with corresponding higher processing costs.

   b. A more accurate gear quality could be chosen instead of $A_v = 11$ to decrease stresses and improve smoothness of operation and noise, but this also has higher associated costs.

   c. Consider using shot peening or other means of improving the life of the gears.

   d. Spiral bevel gears may permit a more compact design and reference should be made to AGMA Standard 2003–C10 for analysis methodology and additional data required.

Practical Considerations for Bevel Gearing

Factors similar to those discussed for spur and helical gears should be considered in the design of systems using bevel gears. The accuracy of alignment and the accommodation of thrust loads discussed in the example problems are critical.

Figure 10–22 shows the exterior view of a heavy-duty gear reducer for an industrial right angle drive. The input shaft is to the left and the output shaft is the larger one at the right extending through the side of the housing. Figure 10–23 shows a reducer similar to that in Figure 10–22 but with the upper part of the housing removed so the entire three stage reduction can be seen.

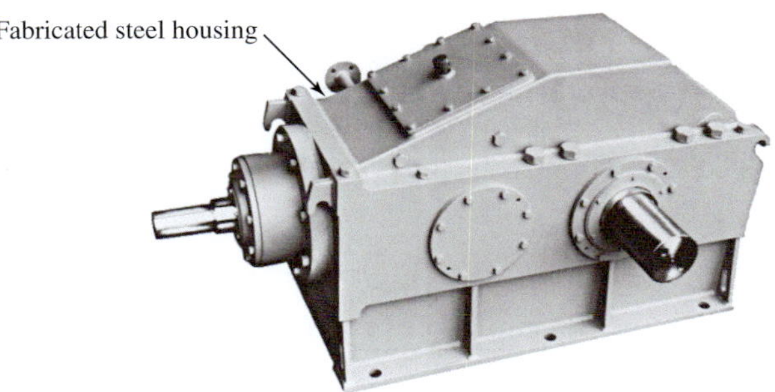

FIGURE 10–22 Heavy-duty right angle gear reducer (Sumitomo Machinery Corporation of America, Teterboro, NJ)

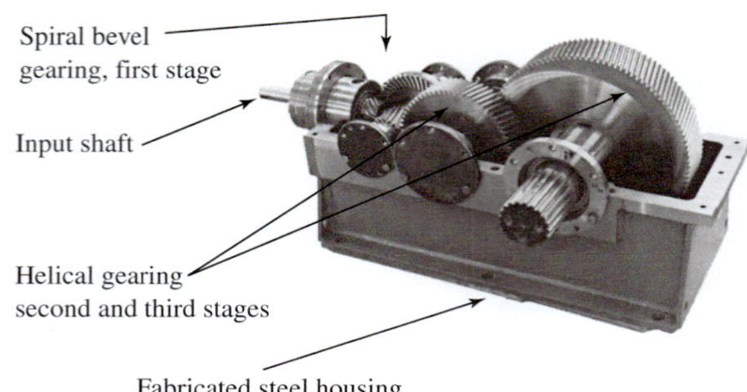

FIGURE 10–23 Three-stage industrial gear reducer employing spiral bevel and helical gears (Sumitomo Machinery Corporation of America, Teterboro, NJ)

The first stage is a spiral bevel gear pair and stages two and three are helical gear pairs. The large speed reduction results in a correspondingly large increase in the torque on the output shaft, requiring its diameter to be larger as shown. Observe, also, the careful placement of the bearings that support all shafts in the housing and how the housing allows for lubrication of the gears and assembly of all components.

10–10 FORCES, FRICTION, AND EFFICIENCY IN WORMGEAR SETS

See Chapter 8 for the geometry of wormgear sets. Also see References 2, 14, 16–18, and 23.

The force system acting on the worm/wormgear set is usually considered to be made of three perpendicular components as was done for helical and bevel gears. There are a tangential force, a radial force, and an axial force acting on the worm and the wormgear. We will use the same notation here as in the bevel gear system.

Figure 10–24 shows two orthogonal views (front and side) of a worm/wormgear pair, showing only the pitch diameters of the gears. The figure shows the separate worm and wormgear with the forces acting on each. Note that because of the 90° orientation of the two shafts,

➭ Forces on Worms and Wormgears

$$\left.\begin{aligned} W_{tG} &= W_{xW} \\ W_{xG} &= W_{tW} \\ W_{rG} &= W_{rW} \end{aligned}\right\} \qquad (10\text{–}23)$$

Of course, the directions of the paired forces are opposite because of the action/reaction principle.

The tangential force on the wormgear is computed first and is based on the required operating conditions of torque, power, and speed at the output shaft.

Pitch Line Speed, v_t

As stated in Chapter 8, the pitch line velocity is the linear velocity of a point on the pitch line for the worm or the wormgear. For the worm having a pitch diameter D_w in, rotating at n_w rpm,

➭ Pitch Line Velocity for Worm

$$v_{tW} = \frac{\pi D_W n_W}{12}\ \text{ft/min} \quad \text{or} \quad v_{tW} = \frac{\pi D_W n_W}{60\,000}\ \text{m/s}$$

For the wormgear having a pitch diameter D_G in, rotating at n_G rpm,

➭ Pitch Line Velocity for Gear

$$V_{tG} = \frac{\pi D_G n_G}{12}\ \text{ft/min} \quad \text{or} \quad v_{tG} = \frac{\pi D_G n_G}{60\,000}\ \text{m/s}$$

Note that these two values for pitch line velocity are *not* equal.

Velocity Ratio, *VR*

It is most convenient to calculate the velocity ratio of a worm and wormgear set from the ratio of the input rotational speed to the output rotational speed:

➭ Velocity Ratio for Worm/Sot

$$VR = \frac{\text{speed of worm}}{\text{speed of gear}} = \frac{n_W}{n_G} = \frac{N_G}{N_W}$$

Coefficient of Friction, μ

Friction plays a major part in the operation of a wormgear set because there is inherently sliding contact between the worm threads and the wormgear teeth. The coefficient of friction is dependent on the materials used, the lubricant, and the sliding velocity. Based on the pitch line speed of the gear, the sliding velocity is

➭ Sliding Velocity for Gear

$$v_s = v_{tG}/\sin\lambda \qquad (10\text{–}24)$$

Based on the pitch line speed of the worm,

➭ Sliding Velocity for Worm

$$v_s = v_{tW}/\cos\lambda \qquad (10\text{–}25)$$

The term λ is the lead angle for the worm thread as defined in Equation (8–24).

The AGMA (see Reference 14) recommends the following formulas to estimate the coefficient of friction for a hardened steel worm (58 HRC minimum), smoothly ground, or polished, or rolled, or with an equivalent finish, operating on a bronze wormgear. The choice of formula depends on the

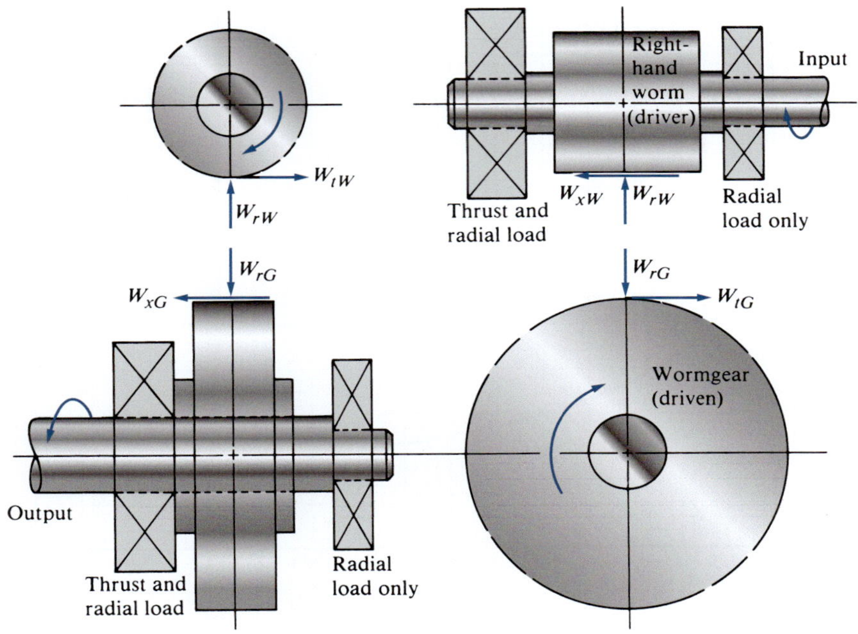

FIGURE 10–24 Forces on a worm and a wormgear

sliding velocity. *Note:* v_s must be in ft/min in the formulas; 1.0 ft/min = 0.0051 m/s.

Static Condition: $v_s = 0$

$$\mu = 0.150$$

Low Speed: $v_s < 10$ ft/min (0.051 m/s)

$$\mu = 0.124e^{(-0.074v_s^{0.645})} \tag{10–26}$$

Higher Speed: $v_s > 10$ ft/min

$$\mu = 0.103e^{(-0.110v_s^{0.450})} + 0.012 \tag{10–27}$$

Figure 10–25 is a plot of the coefficient μ versus the sliding velocity v_s.

Output Torque from Wormgear Drive, T_o

In most design problems for wormgear drives, the output torque and the rotating speed of the output shaft will be known from the requirements of the driven machine. Torque and speed are related to the output power by

➪ Output Torque from Wormgear

$$T_o = \frac{63\,000(P_o)}{n_G} \tag{10–28}$$

By referring to the end view of the wormgear in Figure 10–24, you can see that the output torque is

$$T_o = W_{tG} \cdot r_G = W_{tG}(D_G/2)$$

Then the following procedure can be used to compute the forces acting in a worm/wormgear drive system.

Procedure for Calculating the Forces on a Worm/Wormgear Set

Given:

Output torque, T_o, in lb·in
Output speed, n_G, in rpm
Pitch diameter of the wormgear, D_G, in inches
Lead angle, λ
Normal pressure angle, ϕ_n

Compute:

$$W_{tG} = 2T_o/D_G \tag{10–29}$$

$$W_{xG} = W_{tG}\frac{\cos\phi_n \sin\lambda + \mu\cos\lambda}{[\cos\phi_n \cos\lambda - \mu\sin\lambda]} \tag{10–30}$$

$$W_{rG} = \frac{W_{tG}\sin\phi_n}{\cos\phi_n\cos\lambda - \mu\sin\lambda} \tag{10–31}$$

The forces on the worm can be obtained by observation, using Equation (10–23). Equations (10–30) and (10–31) were derived using the components of both the tangential driving force on the wormgear and the friction force at the location of the meshing worm threads and wormgear teeth. The complete development of the equations is shown in Reference 15.

Friction Force, W_f

The friction force, W_f, acts parallel to the face of the worm threads and the gear teeth and depends on the tangential force on the gear, the coefficient of friction, and the geometry of the teeth:

$$W_f = \frac{\mu W_{tG}}{(\cos\lambda)(\cos\phi_n) - \mu\sin\lambda} \tag{10–32}$$

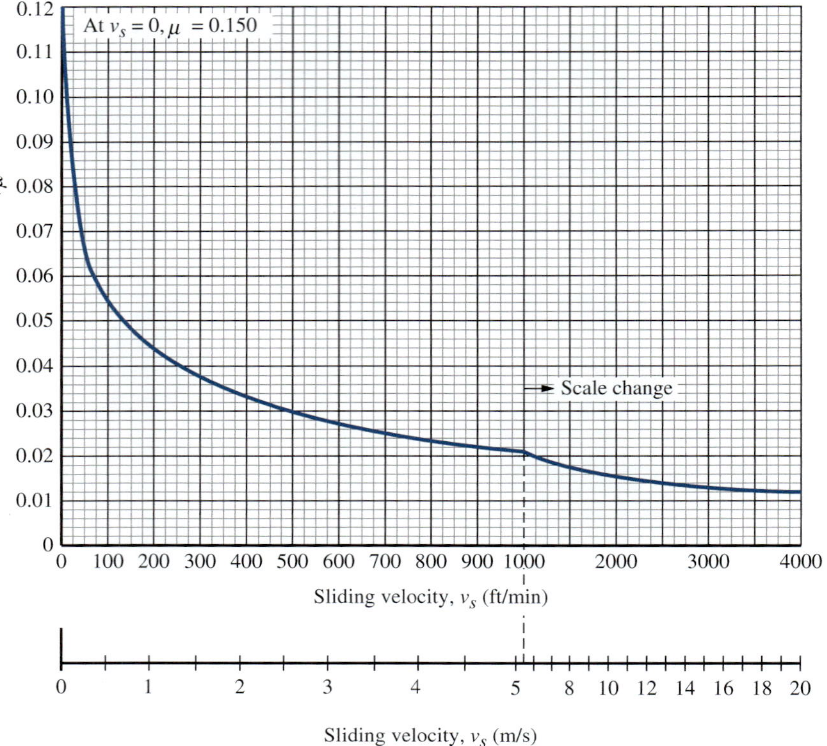

FIGURE 10–25 Coefficient of friction versus sliding velocity for steel worm and bronze wormgear

Power Loss Due to Friction, P_L

Power loss is the product of the friction force and the sliding velocity at the mesh. That is,

$$P_L = \frac{v_s W_f}{33\,000} \quad \textbf{(10–33)}$$

In this equation, the power loss is in hp, v_s is in ft/min, and W_f is in lb.

Input Power, P_i

The input power is the sum of the output power and the power loss due to friction:

$$P_i = P_o + P_L \quad \textbf{(10–34)}$$

Efficiency, η

Efficiency is defined as the ratio of the output power to the input power:

$$\eta = P_o/P_i \quad \textbf{(10–35)}$$

Efficiency for a wormgear drive with the usual case of the input coming through the worm can also be computed directly from the following equation.

$$\eta = \frac{\cos \phi_n - \mu \tan \lambda}{\cos \phi_n + \mu/\tan \lambda} \quad \textbf{(10–36)}$$

Factors Affecting Efficiency

As can be seen in Equation (10–32), the lead angle, the normal pressure angle, and the coefficient of friction all affect the efficiency. The one that has the largest effect, and the one over which the designer has the most control, is the lead angle, λ. The larger the lead angle, the higher the efficiency, up to approximately $\lambda = 45°$. (See Figure 10–26.)

Now, looking back to the definition of the lead angle, note that the number of threads in the worm has a major effect on the lead angle. Therefore, to obtain a high efficiency, use multiple-threaded worms. But there is a disadvantage to this conclusion. More worm threads require more gear teeth to achieve the same ratio, resulting in a larger system overall. The designer is often forced to compromise.

Example Problem: Forces and Efficiency in Wormgearing

Review now the results of Example Problem 8–4, in which the geometry factors for a particular worm and wormgear set were computed. The following example problem extends the analysis to include the forces acting on the system for a given output torque.

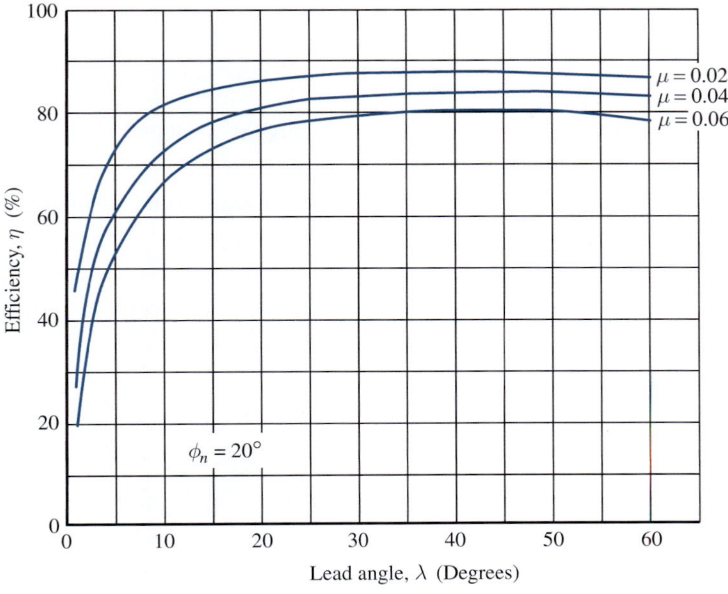

FIGURE 10–26 Efficiency of wormgear drive versus lead angle

Example Problem 10–9 The wormgear drive described in Example Problem 8–4 is transmitting an output torque of 4168 lb·in. The transverse pressure angle is 20°. The worm is made from hardened and ground steel, and the wormgear is bronze. Compute the forces on the worm and the wormgear, the output power, the input power, and the efficiency.

Solution Recall from Example Problem 8–4 that

$$\lambda = 14.04^\circ \qquad D_G = 8.667 \text{ in} \qquad n_G = 101 \text{ rpm}$$

$$n_W = 1750 \text{ rpm} \qquad D_W = 2.000 \text{ in}$$

The normal pressure angle is required. From Equation (8–26),

$$\phi_n = \tan^{-1}(\tan\phi_t \cos\lambda) = \tan^{-1}(\tan 20^\circ \cos 14.04^\circ) = 19.45^\circ$$

Because they recur in several formulas, let's compute the following:

$$\sin\phi_n = \sin 19.45^\circ = 0.333$$

$$\cos\phi_n = \cos 19.45^\circ = 0.943$$

$$\cos\lambda = \cos 14.04^\circ = 0.970$$

$$\sin\lambda = \sin 14.04^\circ = 0.243$$

$$\tan\lambda = \tan 14.04^\circ = 0.250$$

We can now compute the tangential force on the wormgear using Equation (10–29)

$$W_{tG} = \frac{2T_o}{D_G} = \frac{(2)(4168 \text{ lb}\cdot\text{in})}{8.667 \text{ in}} = 962 \text{ lb}$$

The calculations of the axial and radial forces require a value for the coefficient of friction that, in turn, depends on the pitch line speed and the sliding velocity.

Pitch Line Speed of the Gear

$$v_{tG} = \pi D_G n_G/12 = \pi(8.667)(101)/12 = 229 \text{ ft/min}$$

Sliding Velocity [Equation (10–24)]

$$v_s = v_s/\sin\lambda = 229/\sin 14.04^\circ = 944 \text{ ft/min}$$

Coefficient of Friction: From Figure 10–25, at a sliding velocity of 944 ft/min, we can read $\mu = 0.022$.

Now the axial and radial forces on the wormgear can be computed.

Axial Force on the Wormgear [Equation (10–30)]

$$W_{xG} = 962 \text{ lb}\left[\frac{(0.943)(0.243) + (0.022)(0.970)}{(0.943)(0.970) - (0.022)(0.243)}\right] = 265 \text{ lb}$$

Radial Force on the Wormgear [Equation (10–31)]

$$W_{rG} = \left[\frac{(962)(0.333)}{(0.943)(0.970) - (0.022)(0.243)}\right] = 352 \text{ lb}$$

Now the output power, input power, and efficiency can be computed.

Output Power [Equation (10–28)]

$$P_o = \frac{T_o n_G}{63\,000} = \frac{(4168 \text{ lb}\cdot\text{in})(101 \text{ rpm})}{63\,000} = 6.68 \text{ hp}$$

The input power depends on the friction force and the consequent power loss due to friction.

Friction Force [Equation (10–32)]

$$W_f = \frac{\mu W_{tG}}{(\cos\lambda)(\cos\phi_n) - \mu\sin\lambda} = \frac{(0.022)(962 \text{ lb})}{(0.970)(0.943) - (0.022)(0.243)} = 23.3 \text{ lb}$$

Power Loss Due to Friction [Equation (10–33)]

$$P_L = \frac{v_s W_f}{33000} = \frac{(944 \text{ ft/min})(23.3 \text{ lb})}{33\,000} = 0.666 \text{ hp}$$

Input Power [Equation (10–34)]

$$P_i = P_o + P_L = 6.68 + 0.66 = 7.35 \text{hp}$$

Efficiency [Equation (10–35)]

$$\eta = \frac{P_o}{P_i}(100\%) = \frac{6.68 \text{ hp}}{7.35 \text{ hp}}(100\%) = 90.9\%$$

Equation (10–36) could also be used to compute efficiency directly without computing friction power loss.

Self-Locking Wormgear Sets

Self-locking is the condition in which the worm drives the wormgear, but, if torque is applied to the gear shaft, the worm does not turn. It is locked! The locking action is produced by the friction force between the worm threads and the wormgear teeth, and this is highly dependent on the lead angle. It is recommended that a lead angle no higher than about 5.0° be used in order to ensure that self-locking will occur. This lead angle usually requires the use of a single-threaded worm; the low lead angle results in a low efficiency, possibly as low as 60% or 70%.

10–11 STRESS IN WORMGEAR TEETH

We present here an approximate method of computing the bending stress in the teeth of the wormgear. Because the geometry of the teeth is not uniform across the face width, it is not possible to generate an exact solution. However, the method given here should predict the bending stress with sufficient accuracy to check a design because most worm/wormgear systems are limited by pitting, wear, or thermal considerations rather than strength.

The AGMA, in its Standard 6034-B92, does not include a method of analyzing wormgears for strength. The method shown here was adapted from Reference 15. Only the wormgear teeth are analyzed because the worm threads are inherently stronger and are typically made from a stronger material.

The stress in the gear teeth can be computed from

$$\sigma = \frac{W_d}{yFp_n} \qquad \textbf{(10–37)}$$

where W_d = dynamic load on the gear teeth
y = Lewis form factor (see Table 10–5)
F = face width of the gear
p_n = normal circular pitch = $p \cos \lambda = \pi \cos \lambda / P_d$ **(10–38)**

The dynamic load can be estimated from

$$W_d = W_{tG}/K_v \qquad \textbf{(10–39)}$$

and

$$K_v = 1200/(1200 + v_{tG}) \qquad \textbf{(10–40)}$$

$$v_{tG} = \pi D_G n_G/12 = \text{pitch line speed of the gear} \qquad \textbf{(10–41)}$$

Only one value is given for the Lewis form factor for a given pressure angle because the actual value is very difficult to calculate precisely and does not vary much with the number of teeth. The actual face width should be used, up to the limit of two-thirds of the pitch diameter of the worm.

The computed value of tooth bending stress from Equation (10–37) can be compared with the fatigue strength of the material of the gear. For manganese gear bronze, use a fatigue strength of 17 000 psi; for phosphor gear bronze, use 24 000 psi. For cast iron, use approximately 0.35 times the ultimate strength, unless specific data are available for fatigue strength.

TABLE 10–5 Approximate Lewis Form Factor for Wormgear Teeth

ϕ_n	y
$14\frac{1}{2}°$	0.100
20°	0.125
25°	0.150
30°	0.175

10–12 SURFACE DURABILITY OF WORMGEAR DRIVES

AGMA Standard 6034-B92 (see Reference 14) gives a method for rating the surface durability of hardened steel worms operating with bronze gears. The ratings are based on the ability of the gears to operate without significant damage from pitting or wear. All equations from the standard are in the U.S. system. For designs in the SI system, values of parameters should be converted to U.S. units for which the equations were written to determine the corresponding factors. The units are:

W_{tR}—Rated tangential load: lb
Diameters and face width: in
v_s—Sliding velocity: ft/min

Equivalent values in SI units are shown only for reference.

The procedure calls for the calculation of a *rated tangential load,* W_{tR}, from

➩ Rated Tangential Load on Wormgears

$$W_{tR} = C_s D_G^{0.8} F_e C_m C_v \qquad \textbf{(10–42)}$$

where C_s = materials factor (from Figure 10–27)
D_G = pitch diameter of the wormgear, in inches
F_e = effective face width, in inches. Use the actual face width of the wormgear up to a maximum of 0.67 D_W
C_m = ratio correction factor (from Figure 10–28)
C_v = velocity factor (from Figure 10–29)

Conditions on the Use of Equation (10–42)

1. The analysis is valid only for a hardened steel worm (58 HRC minimum) operating with gear bronzes specified in AGMA Standard 6034-B92. The classes of bronzes typically used are tin bronze, phosphor bronze, manganese bronze, and aluminum bronze. The materials factor, C_s, is dependent on the method of casting the bronze, as indicated in Figure 10–27. The values for C_s can be computed from the following formulas.

Sand-Cast Bronzes:

For $D_G > 2.5$ in (64 mm),

$$C_s = 1189.636 - 476.545 \log_{10}(D_G) \qquad \textbf{(10–43)}$$

For $D_G < 2.5$ in (64 mm),

$$C_s = 1000$$

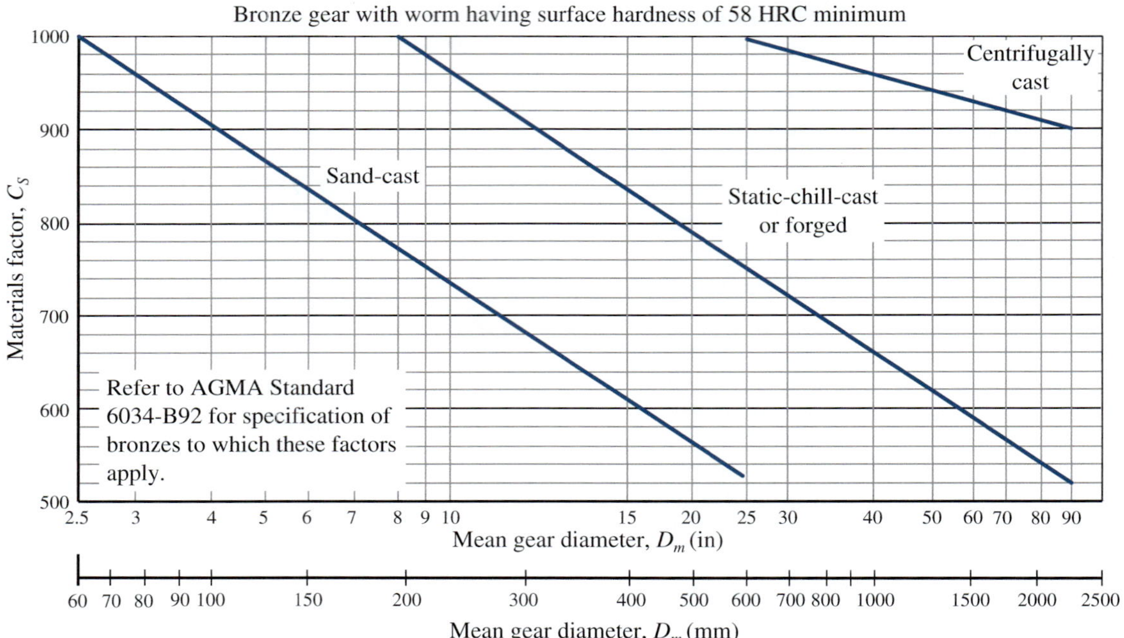

FIGURE 10–27 Materials factor, C_s, for hardened steel worms operating with bronze gears for center distance $C > 3.0$ in (76 mm) (Extracted from AGMA Standard 6034–B92, *Practice for Enclosed Cylindrical Wormgear Speed Reducers and Gearmotors,* with permission of the publisher, American Gear Manufacturers Association, 1001 North Fairfax Street, 5th Floor, Alexandria, VA 22314.)

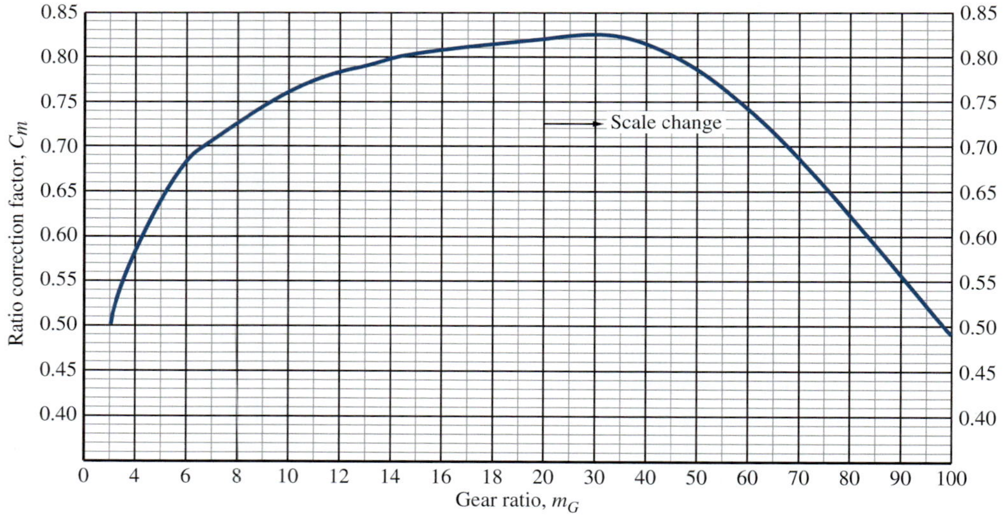

FIGURE 10–28 Ratio correction factor, C_m, versus gear ratio, m_G

Static-Chill-Cast or Forged Bronzes:

For $D_G > 8.0$ in (203 mm),

$$C_s = 1411.651 - 455.825 \log_{10}(D_G) \quad \textbf{(10–44)}$$

For $D_G < 8.0$ in (203 mm)

$$C_s = 1000$$

Centrifugally Cast Bronzes:

For $D_G > 25$ in (635 mm),

$$C_s = 1251.291 - 179.750 \log_{10}(D_G) \quad \textbf{(10–45)}$$

For $D_G < 25$ in (635 mm),

$$C_s = 1000$$

2. The wormgear diameter is the second factor in determining C_s. The *mean diameter* at the midpoint of the working depth of the gear teeth should be used. If standard addendum gears are used, the mean diameter is equal to the pitch diameter.
3. Use the actual face width, F, of the wormgear as F_e if $F < 0.667(D_w)$. For larger face widths, use $F_e = 0.67(D_w)$, because the excess width is not effective.

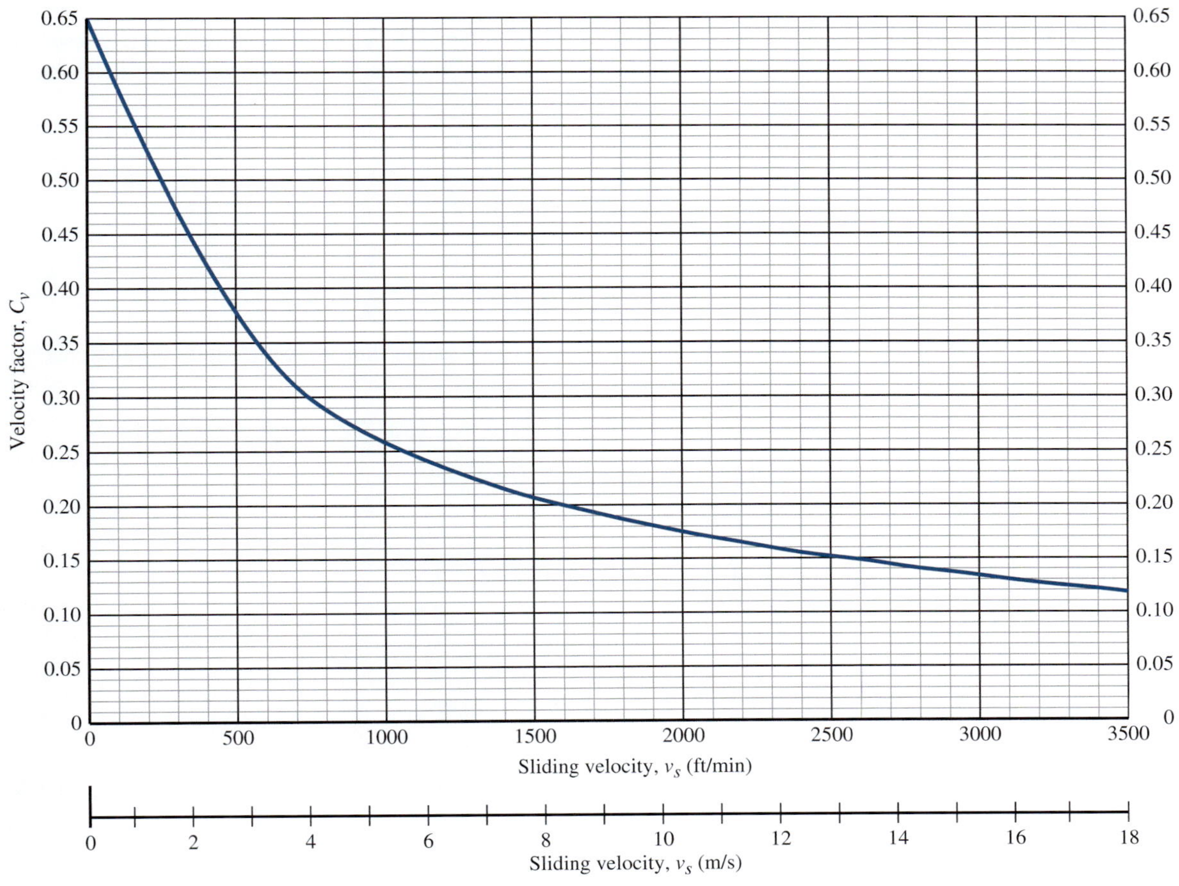

FIGURE 10–29 Velocity factor, C_v, versus sliding velocity

4. The ratio correction factor, C_m, can be computed from the following formulas.

For Gear Ratios, m_G, from 6 to 20

$$C_m = 0.0200(-m_G^2 + 40m_G - 76)^{0.5} + 0.46 \quad \textbf{(10–46)}$$

For Gear Ratios, m_G, from 20 to 76

$$C_m = 0.0107(-m_G^2 + 56m_G + 5145)^{0.5} \quad \textbf{(10–47)}$$

For $m_G > 76$

$$C_m = 1.1483 - 0.00658m_G \quad \textbf{(10–48)}$$

5. The velocity factor depends on the sliding velocity, v_s, computed from Equation (10–24) or (10–25). Values for C_v can be computed from the following formulas.

For v_s from 0 to 700 ft/min (0–3.56 m/s)

$$C_v = 0.659e^{(-0.0011\,v_s)} \quad \textbf{(10–49)}$$

For v_s from 700 to 3000 ft/min (3.56 to 15.24 m/s)

$$C_v = 13.31v_s^{(-0.571)} \quad \textbf{(10–50)}$$

For $v_s > 3000$ ft/min (>15.24 m/s)

$$C_v = 65.52v_s^{(-0.774)} \quad \textbf{(10–51)}$$

6. The proportions of the worm and the wormgear must conform to the following limits defining the maximum and minimum pitch diameters of the worm in relation to the center distance, C, for the gear set. All dimensions are in inches.

$$\text{Maximum}\,D_W = C^{0.875}/1.6 \quad \textbf{(10–52)}$$

$$\text{Minimum}\,D_W = C^{0.875}/3.0 \quad \textbf{(10–53)}$$

7. The shaft carrying the worm must be sufficiently rigid to limit the deflection of the worm at the pitch point to the maximum value of $0.005\sqrt{P_x}$, where P_x is the axial pitch of the worm, which is numerically equal to the circular pitch, p, of the wormgear.
8. When you are analyzing a given wormgear set, the value of the rated tangential load, W_{tR}, must be greater than the actual tangential load, W_t, for satisfactory life.
9. Ratings given in this section are valid only for smooth systems such as fans or centrifugal pumps driven by an electric or hydraulic motor operating under 10 hours per day. More severe conditions, such as shock loading, internal combustion engine drives, or longer hours of operation, require the application of a service factor. Reference 14 lists several such factors based on field experience with specific types of equipment. For problems in this book, factors from Table 9–7 can be used.

Example Problem 10–10 Is the wormgear set described in Example Problem 8–4 satisfactory with regard to strength and wear when operating under the conditions of Example Problem 10–9? The wormgear has a face width of 1.25 in.

Solution From previous problems and solutions,

$$W_{tG} = 962 \text{ lb} \qquad VR = m_G = 17.33$$
$$v_{tG} = 229 \text{ ft/min} \qquad v_s = 944 \text{ ft/min}$$
$$D_G = 8.667 \text{ in} \qquad D_W = 2.000 \text{ in}$$

Assume 58 HRC minimum for the steel worm. Assume that the bronze gear is sand-cast.

Stress

$$K_v = 1200/(1200 + v_{tG}) = 1200/(1200 + 229) = 0.84$$
$$W_d = W_{tG}/K_v = 962/0.84 = 1145 \text{ lb}$$
$$F = 1.25 \text{ in}$$
$$y = 0.125 \text{ (from Table 10–4)}$$
$$p_n = p \cos \lambda = (0.5236)\cos 14.04° = 0.508 \text{ in}$$

Then

$$\sigma = \frac{W_d}{yFp_n} = \frac{1145}{(0.125)(1.25)(0.508)} = 14\,430 \text{ psi}$$

The guidelines in Section 10–12 indicate that this stress level would be adequate for either manganese or phosphor gear bronze.

Surface Durability: Use Equation (10–42):

$$W_{tR} = C_s D_G^{0.8} F_e C_m C_v \tag{10–42}$$

C Factors: The values for the *C* factors can be found from Figures 10–27, 10–28, and 10–29. We find

$$C_s = 740 \text{ for sand-cast bronze and } D_G = 8.667 \text{ in}$$
$$C_m = 0.814 \quad \text{for } m_G = 17.33$$
$$C_v = 0.265 \quad \text{for } v_s = 944 \text{ ft/min}$$

We can use $F_e = F = 1.25$ in if this value is not greater than 0.67 times the worm diameter. For $D_W = 2.000$ in,

$$0.67D_W = (0.67)(2.00 \text{ in}) = 1.333 \text{ in}$$

Therefore, use $F_e = 1.25$ in. Then the rated tangential load is

$$W_{tR} = (740)(8.667)^{0.8}(1.25)(0.814)(0.265) = 1123 \text{ lb}$$

Because this value is greater than the actual tangential load of 962 lb, the design should be satisfactory, provided that the conditions defined for the application of Equation (10–42) are met.

10–13 EMERGING TECHNOLOGY AND SOFTWARE FOR GEAR DESIGN

Comprehending the fundamentals of gear design presented in Chapters 8, 9, and 10 in this book should provide a foundation on which further growth in capabilities to perform reasonable design decisions for gear-type power transmissions can be built. The gear design field is continually adding new technologies and improving upon the methods presented here. Literature in the field often promotes the advantages of gaining additional experience to enable more refined designs and to explore emerging technologies. This section discusses a few emerging technologies and topics beyond the scope of this book. While some technologies mentioned are not truly new and emerging, they are becoming more readily adopted by gear drive design practitioners. Also presented is a brief overview of software available to assist in the gear design process.

Emerging Gearing Technologies

It is strongly recommended that gear design professionals remain aware of the continual upgrading and updating of gear design technology as represented in the extensive suite of standards published by the American Gear Manufacturers Association (AGMA), the International Organization for Standardization (ISO), and other such organizations around the world. With experience, design professionals should also become involved in the standards development process as most of the standards are produced by volunteers representing prominent companies who produce gear drive products and the machinery to make them, or who employ such drives in their products and systems. Industry publications such as *Power Transmission Engineering* and *Gear Technology* provide timely, ongoing commentary and reporting on the state of the art in gear drives and their manufacture. (See Internet sites 10 and 14 in Chapter 9.)

Non-standard Gearing and Gear tooth Forms. To provide a foundation for further study, this book has focused on standard gear tooth forms of the full-depth involute type with standard fillet geometry, addenda, dedenda, center distances, and other features, using methods that were discovered in the late eighteenth century. Major equipment manufacturers provide general-purpose machines to produce standard designs and suppliers of gears and gear drives benefit from standardization in terms of replacement components and are able to specify products from a variety of vendors.

Experience has shown that modifications to some standard features can be beneficial in special applications and in fields where extensive testing can be performed. Examples are automotive, ship propulsion, industrial machinery, construction equipment, and aerospace propulsion. Crowning of teeth and tip relief to promote smooth engagement, geometry modifications that accommodate deformations of teeth under heavy loads, and modified fillets to optimize bending stresses are a few areas of constant exploration. Materials technologies are also areas where fruitful developments are made for steels, other metallic materials, and the widely diverse array of plastic materials. Others pursue radical changes in gear tooth form to enhance strength, stiffness, noise reduction, or pitting resistance.

Internet site 1 describes proprietary designs called Megagear® and Unimegagear® for which increased power density, efficiency, and durability are obtained. The gear tooth profiles are optimized resulting in greater contact area between gear teeth that reduces compressive stress, increases fatigue life, and makes load distribution more uniform.

Internet site 2 describes Direct Gear Design®, an alternative method of analysis and design of involute gears, which separates gear geometry definition from tool selection to achieve the best possible performance for a particular product and application. The result is gear geometry with asymmetric teeth and optimized fillets at the root area. Two involutes of two different base circles are defined that simultaneously increase the contact ratio and operating pressure angle beyond conventional gear limits. Specially designed gear cutting tools are used to produce the gears by hobbing or shaping. Plastic gear molding, die casting, gear forging, and powder metal processes can also be used. The application of Direct Gear Design must be justified by a significant improvement in gear performance.

Internet site 3 describes the patented MGT Frictionless Drive System® that uses the repulsive forces of magnets to provide a link between a drive shaft and the driven one while allowing both shafts to rotate completely independently of each other. Consequently there is no friction generated between them and the efficiency and life of the equipment are improved. The system can be applied either as gearing or couplings. Safety is improved because if the system is overloaded by a certain amount, the driver slips with respect to the driven member. Coupler faces do not touch and the gap between them is typically a few millimeters.

Internet site 4 and Reference 24 describe the field of non-circular gears and their manufacture. Forms can be elliptical, square, or designed to a wide variety of special functions. Sometimes used as cam substitutes in applications where torque or radial loads must be handled. Applications include speed matching on assembly lines, variable cutoff cycles for cutting products from continuous webs, linear motion with quick return, positive and negative rotation mechanisms, flowmeters, bicycle sprockets, variable speed windshield wipers, and stop and dwell motion devices.

Peening. We have emphasized that successful gear design must demonstrate that the bending stress and contact stress in gears must be controlled to limits of the fatigue strength and pitting resistance of the material from which the gear is made. It stands to reason, therefore, that enhancing the fatigue strength and pitting resistance of the material can lead to safer, lighter, and longer lasting gears. One method of accomplishing that is called *peening*, a process of bombarding the high-stress surfaces of gears with hard materials, called *shot*, at high velocity in a controlled manner. Each impact of a piece of shot permanently dents the target material producing the final condition of compression. Shot media come in a variety of shapes such as:

- Cast spheres (steel, stainless steel, cast iron, aluminum oxide, zinc, titanium, and others)
- cut wire (steel, aluminum, zinc, copper, and others)
- beads (ceramic, glass, aluminosilicate, plastic)
- Crushed glass, coal slag, hard rock, garnet

The process leaves the surface with a desirable residual compressive stress and a high hardness that extends the load-carrying ability and life of the treated component. One major objective for using shot peening is to selectively treat areas of a part that have inherently high tensile stress due to operating conditions or those for which tensile stresses at the surface are created by previous processing steps such as grinding, welding, and aggressive machining. Peening counteracts the residual tensile stresses resulting in lower net final stresses under load. Applications are found in numerous types of products from aerospace (landing gear; turbine components), automotive (gears, engine components, structural elements), engines (crankshafts, camshafts, connecting rods, shafts, leaf springs, coil springs), construction equipment (high-wear surfaces, actuator arms, shafts), medical devices (replacement knees), energy production and transmission, and recreation equipment.

Internet sites 5 and 6 describe the types of equipment used for shot peening and the materials and forms of shot available. Two common types are:

- Wheel blasting—Shot is introduced at the center of a rapidly spinning wheel and centrifugally accelerated as it passes to the periphery and is then flung at high velocity toward the target area.
- Air blasting—Pressurized air passes through a nozzle into which the shot is injected and then blown at high velocity toward the target.

Related products commonly called sand blasters are used to clean surfaces of scale and to produce a textured surface finish without necessarily producing the residual compressive stress in the surface. Vibratory finishing employs hard media of many materials and shapes into which parts are discharged to remove scale and deburr machined edges and surfaces, by repetitive rubbing of the media against the parts.

Software for Gear Design

As may be obvious from working through the topics and design methodologies covered in this chapter and Chapter 9, the subject of gear design is computationally demanding. This book includes a few focused calculation aids in the form of spreadsheets that give a hint to the advantages of using such computer-aided engineering approaches. Developers of several commercially available software packages with far greater capabilities for gear design covering a wide range of gear types are listed as Internet sites 7–13. Some of the developers also offer consulting services to aid in planning and implementing complex gear drives.

REFERENCES

1. American Gear Manufacturers Association. Standard 2008-C01. *Assembling Bevel Gears.* Alexandria, VA: American Gear Manufacturers Association, 2001.
2. American Gear Manufacturers Association. Standard 6022-C93 (R2008). *Design Manual for Cylindrical Wormgearing.* Alexandria, VA: American Gear Manufacturers Association, 2008.
3. American Gear Manufacturers Association. Standard AGMA 917-B97. *Design Manual for Parallel Shaft Fine-Pitch Gearing.* Alexandria, VA: American Gear Manufacturers Association, 1997.
4. American Gear Manufacturers Association. Standard 6001-E08. *Design and Selection of Components for Enclosed Gear Drives.* Alexandria, VA: American Gear Manufacturers Association, 2008.
5. American Gear Manufacturers Association. *Standard 2015-1-A01. Accuracy Classification System—Tangential Measurements for Cylindrical Gears.* Alexandria, VA: American Gear Manufacturers Association, 2001.
6. American Gear Manufacturers Association. Standard AGMA 908-B89 (R1999). *Geometry Factors for Determining the Pitting Resistance and Bending Strength of Spur, Helical, and Herringbone Gear Teeth.* Alexandria, VA: American Gear Manufacturers Association, 1999.
7. American Gear Manufacturers Association. Standard AGMA 1012-G05. *Gear Nomenclature, Definitions of Terms with Symbols.* Alexandria, VA: American Gear Manufacturers Association, 2005.
8. American Gear Manufacturers Association. Standard AGMA 2000-A88. *Gear Classification and Inspection Handbook—Tolerances and Measuring Methods for Unassembled Spur and Helical Gears (Including Metric Equivalents).* Alexandria, VA: American Gear Manufacturers Association, 1988. Partial replacement of AGMA 390.03.
9. American Gear Manufacturers Association. Standards 2001-D04 (based on P_d) and 2101-D04 (based on Metric Module, m). *Fundamental Rating Factors and Calculation Methods for Involute Spur and Helical Gear Teeth.* Alexandria, VA: American Gear Manufacturers Association, 2004.
10. American Gear Manufacturers Association. Standard AGMA 2003-C10. *Rating the Pitting Resistance and Bending Strength of Generated Straight Bevel, ZEROL® Bevel, and Spiral Bevel Gear Teeth.* Alexandria, VA: American Gear Manufacturers Association, 2010.
11. American Gear Manufacturers Association. Standard AGMA 2004-C08. *Gear Materials, Heat Treatment and Processing Manual.* Alexandria, VA: American Gear Manufacturers Association, 2008.
12. American Gear Manufacturers Association. Standard AGMA 2005-D03. *Design Manual for Bevel Gears.* Alexandria, VA: American Gear Manufacturers Association, 2003.
13. American Gear Manufacturers Association. Standard AGMA 6013-A06. *Standard for Industrial Enclosed Gear Drives.* Alexandria, VA: American Gear Manufacturers Association, 2006.
14. American Gear Manufacturers Association. Standard AGMA 6034-B92 (R1999). *Practice for Enclosed Cylindrical Wormgear Speed Reducers and Gearmotors.* Alexandria, VA: American Gear Manufacturers Association, 1999.
15. Budynas, R. G., and K. J. Nisbett. *Shigley's Mechanical Engineering Design.* 9th ed. New York: McGraw-Hill, 2011.
16. Crosher, W. P. *Design and Application of the Worm Gear.* New York: American Society of Mechanical Engineers, 2002.
17. Drago, Raymond J. *Fundamentals of Gear Design.* New York: Elsevier Science & Technology Books, 1988.

18. Dudley, Darle W. *Handbook of Practical Gear Design.* Boca Raton, FL: CRC Press, 1994.
19. International Standards Organization. Standard ISO 54:1996. *Cylindrical Gears for General Engineering and for Heavy Engineering—Modules.* Geneva, Switzerland: International Standards Organization, 1996.
20. Kimotho, J., and Kihiu, J. *Design Optimization of Multistage Gear Trains.* Saarbruken, Germany: VDM Verlag Dr. Muller GMBH & Co., 2010.
21. Lynwander, Peter. *Gear Drive Systems: Design and Application.* Boca Raton, FL: CRC Press, 1983.
22. Oberg, Erik, Franklin D. Jones, Holbrook L. Horton, and Henry H. Ryffel. *Machinery's Handbook.* 28th ed. New York: Industrial Press, 2008.
23. Radzevich, S. P. *Handbook of Practical Gear Design.* Boca Raton, FL: CRC Press, 1994.
24. Litvin, F. L., A. Fuentes, I. Gonzalez, and K Hayasaka. *Noncircular Gears: Design and Generation.* Cambridge, UK: Cambridge University Press, 2010.

INTERNET SITES RELATED TO HELICAL GEARS, BEVEL GEARS, AND WORMGEARING

Refer to the list of Internet sites at the end of Chapter 9, Spur Gears. Virtually all sites listed there are also relevant to the design of helical gears, bevel gears, and wormgearing.

1. **Power Engineering and Manufacturing, Ltd.** www.prmltd.com Manufacturer of custom speed reducers and creator of the Megagear® gear tooth form.
2. **AKGears, LLC.** www.akgears.com Creator of the Direct Gear Design® gear tooth form and developer of its application to a variety of products and mechanical systems.
3. **Magnetic Gearing and Turbine Corporation.** www.mgt.com.au Developer and producer of the MTG Frictionless Drive System® that uses the repulsive forces of magnets to provide torque and power transmission without physical contact between the mating parts.
4. **Cunningham Industries, Inc.** www.cunningham-ind.com Manufacturer of elliptical and other non-circular gears.
5. **The Shot Peener.** www.shotpeener.com A site dedicated to providers of shot peening services or machinery to perform shot peening. The site contains valuable information about the technology and its applications.
6. **Wheelabrator, Inc.** www.wheelabratorgroup.com Manufacturer of equipment and media for shot peening including wheelblast and airblast processes along with vibratory finishing equipment.
7. **TEDATA.** www.tedata.com Developer of the MDESIGN calculation and analysis software for numerous mechanical components including gears. Developer of the special version of the MDESIGN software included with this book.
8. **Smart Manufacturing Technology, Ltd.** www.smartmt.com Developer of the MASTA gear design and analysis software and provider of design services for gear-type transmissions.
9. **Universal Technical Systems, Inc.** www.uts.com Developer of the versatile UTS technical analysis software that includes the Integrated Gear Software (IGS) for gear design.
10. **Drive System Technology, Inc.** www.geardoc.com Mechanical power transmission consultants and developer of the PowerGear gear design software. Company founded by Raymond Drago, Chief Engineer of Drive Systems Technology, Inc. (see Reference 17).
11. **K2 Plastics, Inc.** www.k2plasticsinc.com Engineering consultants for the design and production of plastic gears, utilizing their own custom plastic gear design software. The website's *Engineering and Design* page includes an animated, dynamic, full-color FEA image of gear teeth in contact that displays bending and contact stresses during power transmission.
12. **KISSsoft, U.S.A., LLC.** www.kisssoft.ag Developer of the KISSsoft gear design software for several common gear styles along with related machine components and complete power transmissions.
13. **AGMA Software Products.** www.agma.org/publications.software-products/ Developer of a set of software tied directly to the AGMA standards. Included are Bevel Gears Rating Suite, Gear Rating Suite (for spur and helical gears), AGMA 2015-1-A01 Calculator (Accuracy Classification System–Tangential Measurements for Cylindrical Gears).

PROBLEMS

Helical Gearing

1. A helical gear has a transverse diametral pitch of 8, a transverse pressure angle of $14\frac{1}{2}°$, 45 teeth, a face width of 2.00 in, and a helix angle of 30°.
   (a) If the gear transmits 5.0 hp at a speed of 1250 rpm, compute the tangential force, the axial force, and the radial force.
   (b) If the gear operates with a pinion having 15 teeth, compute the bending stress in the pinion teeth. The power comes from an electric motor, and the drive is to a reciprocating pump. Specify a quality number for the teeth.
   (c) Specify a suitable material for the pinion and the gear considering both strength and pitting resistance.

2. A helical gear has a normal diametral pitch of 12, a normal pressure angle of 20°, 48 teeth, a face width of 1.50 in, and a helix angle of 45°.
   (a) If the gear transmits 2.50 hp at a speed of 1750 rpm, compute the tangential force, the axial force, and the radial force.
   (b) If the gear operates with a pinion having 16 teeth, compute the bending stress in the pinion teeth. The power comes from an electric motor, and the drive is to a centrifugal blower. Specify a quality number for the teeth.
   (c) Specify a suitable material for the pinion and the gear considering both strength and pitting resistance.

3. A helical gear has a transverse diametral pitch of 6, a transverse pressure angle of $14\frac{1}{2}°$, 36 teeth, a face width of 1.00 in, and a helix angle of 45°.
   (a) If the gear transmits 15.0 hp at a speed of 2200 rpm, compute the tangential force, the axial force, and the radial force.

(b) If the gear operates with a pinion having 12 teeth, compute the bending stress in the pinion teeth. The power comes from a six-cylinder gasoline engine, and the drive is to a concrete mixer. Specify a quality number for the teeth.

(c) Specify a suitable material for the pinion and the gear considering both strength and pitting resistance.

4. A helical gear has a normal diametral pitch of 24, a normal pressure angle of $14\frac{1}{2}°$, 72 teeth, a face width of 0.25 in, and a helix angle of 45°.

   (a) If the gear transmits 0.50 hp at a speed of 3450 rpm, compute the tangential force, the axial force, and the radial force.

   (b) If the gear operates with a pinion having 16 teeth, compute the bending stress in the pinion teeth. The power comes from an electric motor, and the drive is to a winch that will experience moderate shock. Specify a quality number for the teeth.

   (c) Specify a suitable material for the pinion and the gear considering both strength and pitting resistance.

For Problems 5–11, complete the design of a pair of helical gears to operate under the stated conditions. Specify the geometry of the gears and the material and its heat treatment. Assume that the drive is from an electric motor unless otherwise specified. Consider both strength and pitting resistance.

5. A pair of helical gears is to be designed to transmit 5.0 hp while the pinion rotates at 1200 rpm. The gear drives a reciprocating compressor and must rotate between 385 and 390 rpm.

6. A helical gear pair is to be a part of the drive for a milling machine requiring 20.0 hp with the pinion speed at 550 rpm and the gear speed to be between 180 and 190 rpm.

7. A helical gear drive for a punch press requires 50.0 hp with the pinion rotating at 900 rpm and the gear speed at 225 to 230 rpm.

8. A single-cylinder gasoline engine has the pinion of a helical gear pair on its output shaft. The gear is attached to the shaft of a small cement mixer. The mixer requires 2.5 hp while rotating at approximately 75 rpm. The engine is governed to run at approximately 900 rpm.

9. A four-cylinder industrial engine runs at 2200 rpm and delivers 75 hp to the input gear of a helical gear drive for a large wood chipper used to prepare pulpwood chips for paper making. The output gear must run between 4500 and 4600 rpm.

10. A small commercial tractor is being designed for chores such as lawn mowing and snow removal. The wheel drive system is to be through a helical gear pair in which the pinion runs at 450 rpm while the gear, mounted on the hub of the wheel, runs at 75 to 80 rpm. The wheel has an 18-in diameter. The two-cylinder gasoline engine delivers 20.0 hp to the wheels.

11. A water turbine transmits 15.0 hp to a pair of helical gears at 4500 rpm. The output of the gear pair must drive an electric power generator at 3600 rpm. The center distance for the gear pair must not exceed 4.00 in.

12. Determine the power-transmitting capacity of a pair of helical gears having a normal pressure angle of 20°, a helix angle of 15°, a normal diametral pitch of 10, 20 teeth in the pinion, 75 teeth in the gear, and a face width of 2.50 in, if they are made from SAE 4140 OQT 1000 steel. They are of typical commercial quality. The pinion will rotate at 1725 rpm on the shaft of an electric motor. The gear will drive a centrifugal pump.

13. Repeat Problem 12 with the gears made from SAE 4620 DOQT 300 carburized, case-hardened steel. Then compute the axial and radial forces on the gears.

Bevel Gears

14. A straight bevel gear pair has the following data: $N_P = 15$; $N_G = 45$; $P_d = 6$; 20° pressure angle. If the gear pair is transmitting 3.0 hp, compute the forces on both the pinion and the gear. The pinion speed is 300 rpm. The face width is 1.25 in. Compute the bending stress and the contact stress for the teeth, and specify a suitable material and heat treatment. The gears are driven by a gasoline engine, and the load is a concrete mixer providing moderate shock. Assume that neither gear is straddle-mounted.

15. A straight bevel gear pair has the following data: $N_P = 25$; $N_G = 50$; $P_d = 10$; 20° pressure angle. If the gear pair is transmitting 3.5 hp, compute the forces on both the pinion and the gear. The pinion speed is 1250 rpm. The face width is 0.70 in. Compute the bending stress and the contact stress for the teeth, and specify a suitable material and heat treatment. The gears are driven by a gasoline engine, and the load is a conveyor providing moderate shock. Assume that neither gear is straddle-mounted.

16. Design a pair of straight bevel gears to transmit 5.0 hp at a pinion speed of 850 rpm. The gear speed should be approximately 300 rpm. Consider both strength and pitting resistance. The driver is a gasoline engine, and the driven machine is a heavy-duty conveyor.

17. Design a pair of straight bevel gears to transmit 0.75 hp at a pinion speed of 1800 rpm. The gear speed should be approximately 475 rpm. Consider both strength and pitting resistance. The driver is an electric motor, and the driven machine is a reciprocating saw.

Wormgearing

18. A wormgear set has a single-thread worm with a pitch diameter of 1.250 in, a diametral pitch of 10, and a normal pressure angle of 14.5°. If the worm meshes with a wormgear having 40 teeth and a face width of 0.625 in, compute the gear pitch diameter, the center distance, and the velocity ratio. If the wormgear set is transmitting 924 lb · in of torque at its output shaft, which is rotating at 30 rpm, compute forces on the gear, efficiency, input speed, input power, and stress on the gear teeth. If the worm is hardened steel and the gear is chilled bronze, evaluate the rated load, and determine whether the design is satisfactory for pitting resistance.

19. Three designs are being considered for a wormgear set to produce a velocity ratio of 20 when the worm gear rotates at 90 rpm. All three have a diametral pitch of 12, a worm pitch diameter of 1.000 in, a gear face width of 0.500 in, and a normal pressure angle of 14.5°. One has a single-thread worm and 20 wormgear teeth; the second has a double-thread worm and 40 wormgear teeth; the third has a four-thread worm and 80 wormgear teeth. For each design, compute the rated output torque, considering both strength and pitting resistance. The worms are hardened steel, and the wormgears are chilled bronze.

20. For each of the three designs proposed in Problem 19, compute the efficiency.

The data for Problems 21, 22, and 23 are given in Table 10–6. Design a wormgear set to produce the desired velocity ratio when transmitting the given torque at the output shaft for the given output rotational speed.

TABLE 10–6 Data for Problems 21–23

Problem	*VR*	Torque (lb · in)	Output speed (rpm)
21.	7.5	984	80
22.	3	52.5	600
23.	40	4200	45

24. Compare the two designs described in Table 10–7 when each is transmitting 1200 lb · in in of torque at its output shaft, which rotates at 20 rpm. Compute the forces on the worm and the wormgear, the efficiency, and the input power required.

TABLE 10–7 Designs for Problem 24

Design	P_d	N_t	N_G	D_w	F_G	Pressure angle
A	6	1	30	2.000	1.000	14.5°
B	10	2	60	1.250	0.625	14.5°

CHAPTER ELEVEN

KEYS, COUPLINGS, AND SEALS

THE BIG PICTURE Keys, Couplings, and Seals

Discussion Map

- Keys and couplings connect functional parts of mechanisms and machines, allowing moving parts to transmit power or to locate parts relative to each other.
- Retaining rings hold assemblies together or hold parts on shafts, such as keeping a sprocket in position or holding a wheel on an axle.
- Seals protect critical components by excluding contaminants or by retaining fluids inside the housing of a machine.

Discover

Look around you and identify several examples of the use of keys, couplings, retaining rings, and seals in automobiles, trucks, home appliances, shop tools, gardening equipment, or bicycles.

This chapter will help you understand the functions and design requirements of such devices. In addition, you will learn to recognize commercially available designs and apply them properly.

Think of how two or more parts of a machine can be connected for the purpose of locating one part with respect to another. Now think about how that connection must be designed if the parts are moving and if power must be transmitted between them.

This chapter presents information on commercially available products that accomplish these functions. The generic categories of keys, couplings, and seals actually encompass numerous different designs.

A *key* is used to connect a drive member such as a belt pulley, chain sprocket, or gear to the shaft that carries it. (See Figure 11–1.) Torque and power are transmitted across the key to or from the shaft. But how does the power get into or out of the shaft? One way might be that the output from the shaft of a motor or engine is connected to the input shaft of a transmission through a *flexible coupling* that reliably transmits power but allows for some misalignment between the shafts during operation because of the flexing of frame members or through progressive misalignment due to wear.

Seals may be difficult to see because they are typically encased in a housing or are covered in some way. Their function is to protect critical elements of a machine from contamination due to dust, dirt, or water or other fluids while allowing rotating or translating machine elements to move to accomplish their desired functions. Seals exclude undesirable materials from the inside of a mechanism, or they hold critical lubrication or cooling fluids inside the housing of the mechanism.

Look at machines that you interact with each day, and identify parts that fit the descriptions given here. Look in the engine compartment of a car or a truck. How are drive pulleys, linkages, latches, the hinges for the hood, the fan, the windshield wipers, and any other moving part connected to something else—the frame of the car, a rotating shaft, or some other moving part? If you are familiar with the inner workings of the engine and transmission, describe how those parts are connected. Look at the steering and suspension systems, the water pump, the fuel pump, the brake fluid reservoir, and the suspension struts or shock absorbers. Try to see where seals are used. Can you see the universal joints, sometimes called the *constant-velocity* (CV) joints, in the drive train? They should be connecting the output shaft from the transmission to the final parts of the drive train as the power is delivered to the wheels.

Find a small lawn or garden tractor at home or at a local home store. Typically their mechanisms are accessible, although protected from casual contact for safety reasons. Trace how power is transmitted from the engine, through a transmission, through a drive chain or belt, and all the way to the wheels or to the blade of a mower. How are functional parts connected together?

Look at home appliances, power tools in a home shop, and gardening equipment. Can you see parts that are held in place with *retaining rings?* These are typically thin, flat rings that are pressed onto shafts or inserted into grooves to hold a wheel on a shaft, to hold a gear or a pulley in position along the length of a shaft, or to simply hold some part of the device in place.

How are the keys, couplings, seals, retaining rings, and other connecting devices made? What materials are used? How are they installed? Can they be removed? What kinds of forces must they resist? How does their special geometry accomplish the desired function? How could they fail?

This chapter will help you become familiar with such mechanical components, with some of the manufacturers that offer commercial versions, and with correct application methods.

YOU ARE THE DESIGNER

In the first part of Chapter 8, you were the designer of a gear-type speed reducer whose conceptual design is shown in Figure 8–3. It has four gears, three shafts, and six bearings, all contained within a housing. How are the gears attached to the shafts? One way is to use keys at the interface between the hub of the gears and the shaft. You should be able to design the keys. How is the input shaft connected to the motor or engine that delivers the power? How is the output shaft connected to the driven machine? One way is to use flexible couplings. You should be able to specify commercially available couplings and apply them properly, considering the amount of torque they must transmit and how much misalignment they should permit.

How are the gears located axially along the shafts? Part of this function may be provided by shoulders machined on the shaft. But that works only on one side. On the other side, one type of locating means is a retaining ring that is installed into a groove in the shaft after the gear is in place. Rings or spacers may be used at the left of gears *A* and *B* and at the right of gears *C* and *D*. Notice that the input and output shafts extend outside the housing. How can you keep contaminants from the outside from getting inside? How can you keep lubricating oil inside? Shaft seals can accomplish that function. Seals may also be provided on the bearings to retain lubricant inside and in full contact with the rotating balls or rollers of the bearing.

You should be familiar with the kinds of materials used for seals and with their special geometries. These concepts are discussed in this chapter. ■

11–1 OBJECTIVES OF THIS CHAPTER

After completing this chapter, you will be able to:

1. Describe several kinds of *keys.*
2. Specify a suitable size key for a given size shaft.
3. Specify suitable materials for keys.
4. Complete the design of keys and the corresponding keyways and keyseats, giving the complete geometries.
5. Describe *splines* and determine their torque capacity.
6. Describe several alternate methods of fastening machine elements to shafts.
7. Describe *rigid couplings* and *flexible couplings.*
8. Describe several types of flexible couplings.
9. Describe *universal joints.*
10. Describe *retaining rings* and other means of locating elements on shafts.
11. Specify suitable seals for shafts and other types of machine elements.

11–2 KEYS

A *key* is a machinery component placed at the interface between a shaft and the hub of a power-transmitting element for the purpose of transmitting torque [see Figure 11–1(a)]. The key is demountable to facilitate assembly and disassembly of the shaft system. It is installed in an axial groove machined into the shaft, called a *keyseat.* A similar groove in the hub of the power-transmitting element is usually called a *keyway,* but it is more properly also a keyseat. The key is typically installed into the shaft keyseat first; then the hub keyseat is aligned with the key, and the hub is slid into position.

Square and Rectangular Parallel Keys

The most common type of key for shafts up to $6\frac{1}{2}$ in in diameter is the square key, as illustrated in Figure 11–1(b). The rectangular key, Figure 11–1(c), is recommended for larger shafts and is used for smaller shafts where the shorter height can be tolerated. Both the square and the rectangular keys are referred to as *parallel keys* because the top and bottom and the sides of the key are parallel. (See Internet sites 1 and 20.)

Table 11–1 gives the preferred dimensions for parallel keys as a function of shaft diameter for both U.S. sizes and SI Metric sizes. The width is approximately one-quarter of the diameter of the shaft. See References 7 and 9 for more detailed dimensions and tolerances.

The keyseats in the shaft and the hub are designed so that exactly one-half of the height of the key is bearing on the side of the shaft keyseat and the other half on the side of the hub keyseat. Figure 11–2 shows the resulting geometry. The distance Y is the radial distance from the theoretical top of the shaft, before the keyseat is machined, to the top edge of the finished keyseat to produce a keyseat depth of exactly $H/2$. To assist in machining and inspecting the shaft or the hub, the dimensions S and T can be computed and shown on the part drawings. The equations are given in Figure 11–2. Tabulated values of Y, S, and T are available in References 7 and 9.

As discussed later in Chapter 12, keyseats in shafts are usually machined with either an end mill or a circular milling cutter, producing the profile or sled runner keyseat, respectively (refer to Figure 12–6). In general practice, the keyseats and keys are left with essentially square ends and edges. But radiused keyseats and chamfered keys can be used to reduce the stress concentrations. Table 11–2 shows suggested values from ANSI Standard B17.1.

As alternates to the use of parallel keys, taper keys, gib head keys, pin keys, and Woodruff keys can be used to provide special features of installation or operation. Figure 11–3 shows the general geometry of these types of keys.

Taper Keys and Gib Head Keys

Taper keys are designed to be inserted from the end of the shaft after the hub is in position rather than installing the key first and then sliding the hub over the key as with parallel keys. The taper extends over at least the length of the hub, and the height, H, measured at the end of the hub, is the same as for the parallel key. The taper is typically 1/8 in per foot. Note that this design gives a smaller bearing area on the sides of the key, and the bearing stress must be checked.

The *gib head key* [Figure 11–3(c)] has a tapered geometry inside the hub that is the same as that of the plain taper key. But the extended head provides the means of extracting the key from the same end at which it was installed. This is very desirable if the opposite end is not accessible to drive the key out.

Pin Keys

The *pin key,* shown in Figure 11–3(d), is a cylindrical pin placed in a cylindrical groove in the shaft and hub. Lower stress concentration factors result from this design as compared with parallel or taper keys. A close fit between the pin

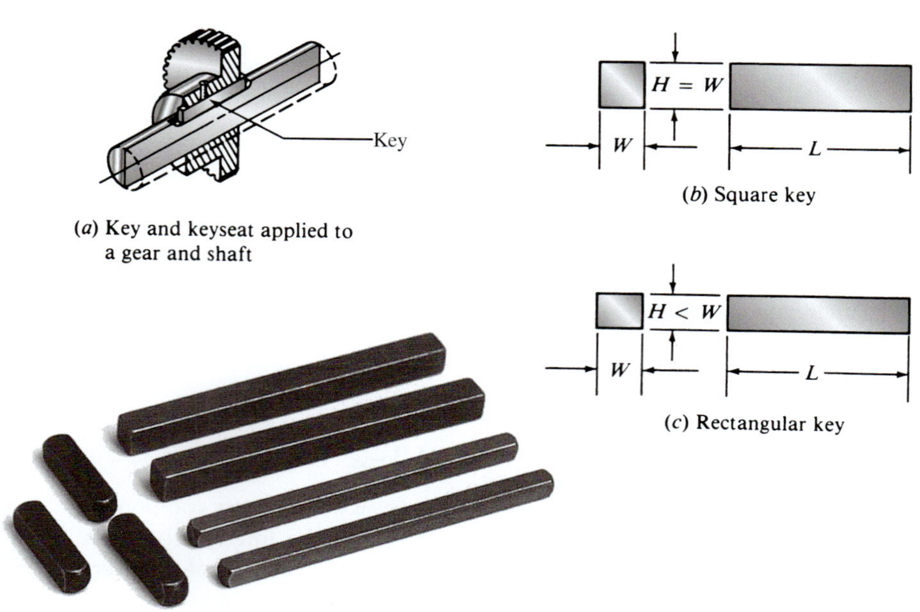

(*a*) Key and keyseat applied to a gear and shaft

(*b*) Square key

(*c*) Rectangular key

(*d*) Commercially available keys

FIGURE 11–1 Parallel keys. (Source for (d): Driv-Lok, Inc., Sycamore, IL)

TABLE 11–1 Key Size vs. Shaft Diameter

U.S. inch sizes				SI metric sizes			
Nominal shaft diameter		Key dimensions		Nominal shaft diameter		Key dimensions	
Over (in)	to-including (in)	Width, *W* (in)	Height, *H* (in)	Over (mm)	to-including (mm)	Width, *W* (mm)	Height, *H* (mm)
0.3125	0.4375	0.09375	0.09375	6	8	2	2
0.4375	0.5625	0.1250	0.1250	8	10	3	3
0.5625	0.875	0.1875	0.1875	10	12	4	4
0.875	1.250	0.2500	0.2500	12	17	5	5
1.250	1.375	0.3125	0.3125	17	22	6	6
1.375	1.75	0.375	0.375	22	30	8	7
1.75	2.25	0.500	0.500	30	38	10	8
2.25	2.75	0.625	0.625	38	44	12	8
2.75	3.25	0.750	0.750	44	50	14	9
3.25	3.75	0.875	0.875	50	58	16	10
3.75	4.50	1.00	1.00	58	65	18	11
4.50	5.50	1.25	1.25	65	75	20	12
5.50	6.50	1.50	1.50	75	85	22	14
6.50	7.50	1.75	1.50	85	95	25	14
7.50	9.00	2.00	1.50	95	110	28	16
9.00	11.00	2.50	1.75	110	130	32	18
11.00	13.00	3.00	2.00	130	150	36	20
13.00	15.00	3.50	2.50	150	170	40	22
15.00	18.00	4.00	3.00	170	200	45	25
18.00	22.00	5.00	3.50	200	230	50	28
22.00	26.00	6.00	4.00	230	260	56	32
26.00	30.00	7.00	5.00	260	290	63	32
				290	330	70	36
				330	380	80	40
				380	440	90	45
				440	500	100	50

Source-U.S. Sizes: Reprinted from ANSI Standard B17.1–1967 by permission of the American Society of Mechanical Engineers. All rights reserved.

Note: Key sizes above the horizontal line are square; others are rectangular.

and the groove is required to ensure that the pin does not move and that the bearing is uniform along the length of the pin.

Woodruff Keys

Where light loading and relatively easy assembly and disassembly are desired, the *Woodruff key* should be considered. Figure 11–3(e) shows the standard configuration. The circular groove in the shaft holds the key in position while the mating part is slid over the key. The stress analysis for this type of key proceeds in the manner discussed below for the parallel key, taking into consideration the particular geometry of the Woodruff key. ANSI Standard B17.2–1967 lists the dimensions for a large number of standard Woodruff keys and their mating keyseats. (See Reference 8) Table 11–3 provides a sampling. Note that the *key number* indicates the nominal key dimensions. The last two digits give the nominal diameter, *B*, in eighths of an inch, and the digits preceding the last two give the nominal width, *W*, in thirty-seconds of an inch. For example, key number 1210 has a diameter of 10/8 in ($1\frac{1}{4}$ in), and a width of 12/32 in (3/8 in). The actual size of the key is slightly smaller than half of the full circle, as shown in dimensions *C* and *F* in Table 11–3.

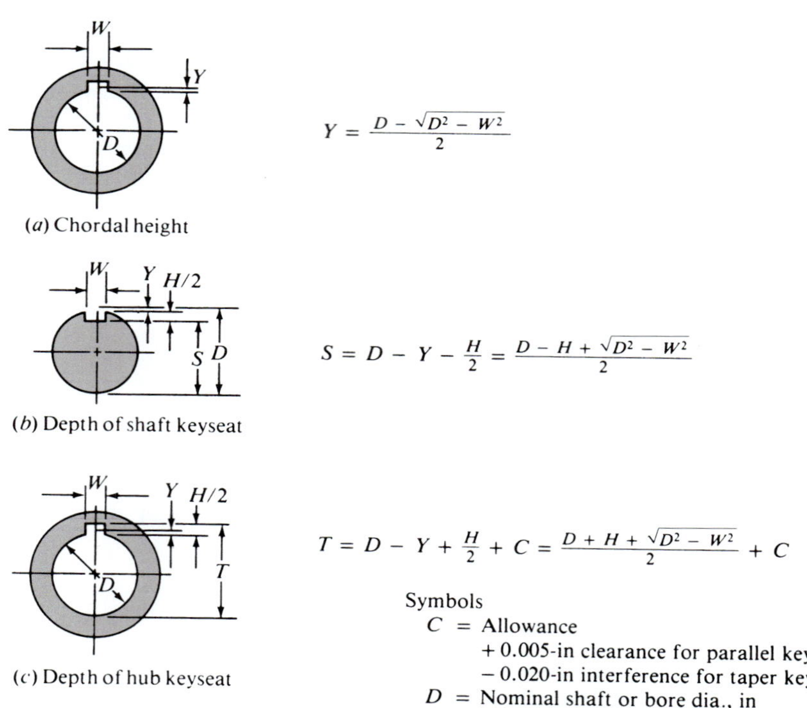

FIGURE 11–2 Dimensions for parallel keyseats

TABLE 11–2 Suggested Fillet Radii and Key Chamfers

$H/2$, keyseat depth			
Over	**To (incl.)**	**Fillet radius**	**45° chamfer**
1/8	1/4	1/32	3/64
1/4	1/2	1/16	5/64
1/2	7/8	1/8	5/32
7/8	$1\frac{1}{4}$	3/16	7/32
$1\frac{1}{4}$	$1\frac{3}{4}$	1/4	9/32
$1\frac{3}{4}$	$2\frac{1}{2}$	3/8	13/32

Source: Reprinted from ASME B17.1–1967 by permission of the American Society of Mechanical Engineers. All rights reserved.
Note: All dimensions are given in inches.

Selection and Installation of Keys and Keyseats

The key and the keyseat for a particular application are usually designed after the shaft diameter is specified by the methods of Chapter 12. Then, with the shaft diameter as a guide, the size of the key is selected from Table 11–1. The only remaining variables are the length of the key and its material. One of these can be specified, and the requirements for the other can then be computed.

Typically the length of a key is specified to be a substantial portion of the hub length of the element in which it is installed to provide for good alignment and stable operation. But if the keyseat in the shaft is to be in the vicinity of other geometric changes, such as shoulder fillets and ring grooves, it is important to provide some axial clearance between them so that the effects of the stress concentrations are not compounded.

The key can be cut off square at its ends or provided with a radius at each end when installed in a profile keyseat

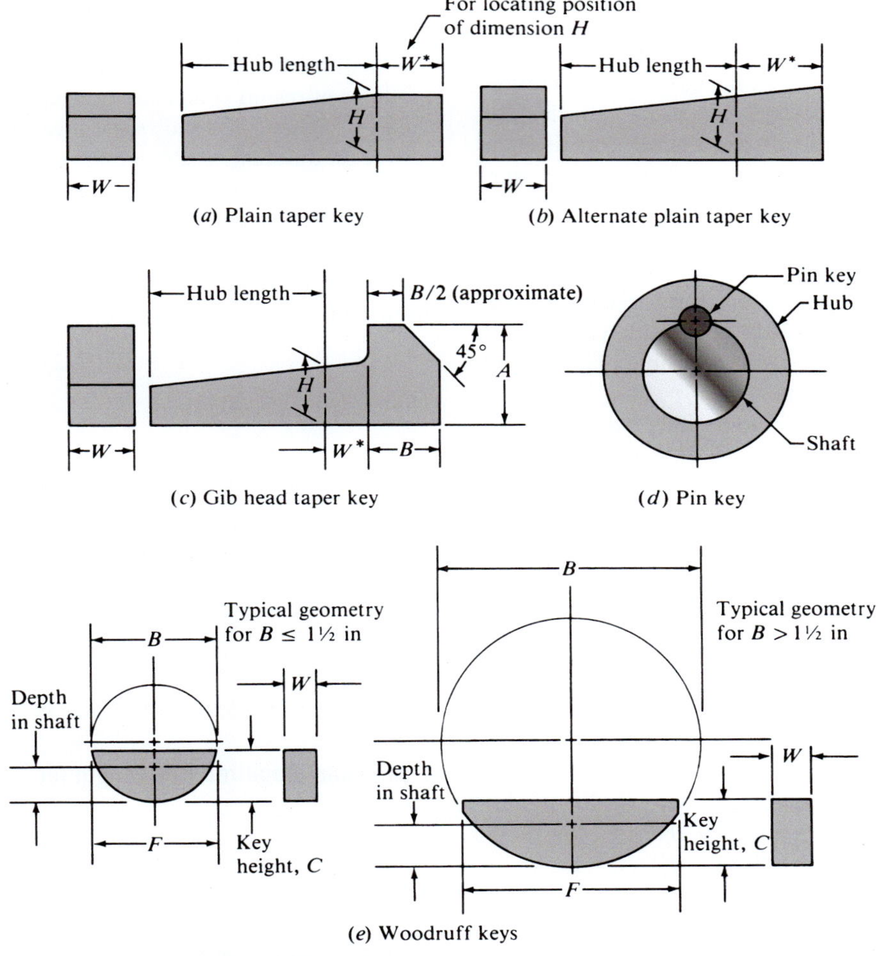

FIGURE 11–3 Key types

TABLE 11–3 Woodruff Key Dimensions

Key number	Nominal key size, $W \times B$	Actual length, F	Height of key, C	Shaft keyseat depth	Hub keyseat depth
202	$1/16 \times 1/4$	0.248	0.104	0.0728	0.0372
204	$1/16 \times 1/2$	0.491	0.200	0.1668	0.0372
406	$1/8 \times 3/4$	0.740	0.310	0.2455	0.0685
608	$3/16 \times 1$	0.992	0.435	0.3393	0.0997
810	$1/4 \times 1\frac{1}{4}$	1.240	0.544	0.4170	0.1310
1210	$3/8 \times 1\frac{1}{4}$	1.240	0.544	0.3545	0.1935
1628	$1/2 \times 3\frac{1}{2}$	2.880	0.935	0.6830	0.2560
2428	$3/4 \times 3\frac{1}{2}$	2.880	0.935	0.5580	0.3810

Source: Reprinted from ASME B17.2–1967, by permission of the American Society of Mechanical Engineers. All rights reserved.
Note: All dimensions are given in inches.

to improve location. Square cut keys are usually used with the sled-runner-type keyseat.

The key is sometimes held in position with a set screw in the hub over the key. However, the reliability of this approach is questionable because of the possibility of the set screw's backing out with vibration of the assembly. Axial location of the assembly should be provided by more positive means, such as shoulders, retaining rings, or spacers.

11–3 MATERIALS FOR KEYS

Keys are made from plain carbon steel, alloy steels, stainless steels, and some non-ferrous metals. Even plastics are used for small devices under low loads. This book will focus primarily on steels for general industrial applications. Table 11–4 lists a variety of steels and one aluminum alloy to illustrate the types of materials available. (See Internet site 20.)

For problem solutions in this book, it is recommended that the low-carbon steel SAE 1018 be considered for most applications. It is a low-cost, readily available material and its strength is generally adequate.

If higher strength is necessary to produce a design with a reasonable length, medium-carbon SAE 1035 or 1045, or alloy steels SAE 4140 or 8630 are recommended. The high-carbon steel SAE 1095 may be used but it may have low ductility.

Where corrosion resistance is necessary, the stainless steels listed in Table 11–4 may be considered. Aluminum 6061 is less frequently used for keys but may be desirable for material compatibility reasons.

11–4 STRESS ANALYSIS TO DETERMINE KEY LENGTH

There are two basic modes of potential failure for keys transmitting power: (1) shear across the shaft/hub interface and (2) compression failure due to the bearing action between the sides of the key and the shaft or hub material. The analysis for either failure mode requires an understanding of the forces that act on the key. Figure 11–4 shows the idealized case in which the torque on the shaft creates a force on the left side of the key. The key in turn exerts a force on the right side of the hub keyseat. The reaction force of the hub back on the key then produces a set of opposing forces that place the key in direct shear over its cross section, $W \times L$. The magnitude of the shearing force can be found from

$$F = T/(D/2)$$

The shearing stress is then

$$\tau = \frac{F}{A_s} = \frac{T}{(D/2)(WL)} = \frac{2T}{DWL} \qquad \textbf{(11–1)}$$

In design, we can set the shearing stress equal to a design stress in shear for the maximum shear stress theory of failure:

$$\tau_d = 0.5s_y/N$$

Then the required length of the key is

➩ Minimum Required Key Length for Shear

$$L_{\min} = \frac{2T}{\tau_d DW} \qquad \textbf{(11–2)}$$

TABLE 11–4 Selection of Materials Used for Keys

Material designation	Tensile strength s_u (ksi)	Tensile strength s_u (MPa)	Yield strength s_y (ksi)	Yield strength s_y (MPa)
Carbon steels (SAE)				
1018	64	441	54	372
1035	72	496	39.5	272
1045	91	627	77	531
1095	140	965	83	572
Alloy steels (SAE)				
4140	102	703	90	621
8630	100	690	95	655
Stainless steels (SAE)				
303	90	621	35	241
304	85	586	35	241
316	85	586	35	241
416	75	517	40	276
Aluminum				
6061	18	124	12	83

Source: Adapted from Internet site 20.
Note: Strength properties typical, not guaranteed.

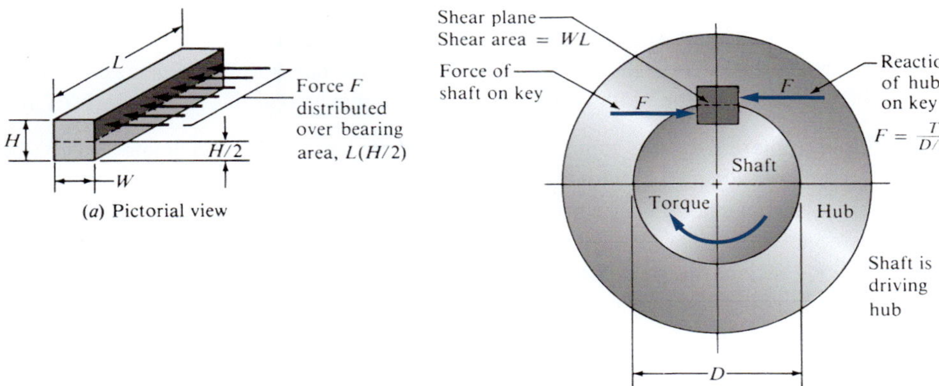

FIGURE 11–4 Forces on a key.

The failure in bearing is related to the compressive stress on the side of the key, the side of the shaft keyseat, or the side of the hub keyseat. The area in compression is the same for either of these zones, $L \times (H/2)$. Thus, the failure occurs on the surface with the lowest compressive yield strength. Let's define a *design stress for compression* as

$$\sigma_d = s_y/N$$

Then the compressive stress is

$$\sigma = \frac{F}{A_c} = \frac{T}{(D/2)(L)(H/2)} = \frac{4T}{DLH} \quad \textbf{(11–3)}$$

Letting this stress equal the design compressive stress allows the computation of the required length of the key for this mode of failure:

➩ Minimum Required Key Length for Compression

$$L_{\min} = \frac{4T}{\sigma_d DH} \quad \textbf{(11–4)}$$

For the design of a square key in which the strength of the key material is lower than that of the shaft or the hub, Equations (11–2) and (11–4) produce the same result. Substituting the design stress into either equation would give

➩ Minimum Required Key Length if Key Material is Weakest

$$L_{\min} = \frac{4TN}{DWs_y} \quad \textbf{(11–5)}$$

But be sure to evaluate the length from Equation (11–4) if either the shaft or the hub has a lower yield strength than the key.

DESIGN PROCEDURE FOR PARALLEL KEYS

1. Complete the design of the shaft into which the key will be installed, and specify the actual diameter at the location of the keyseat.
2. Select the size of the key from Table 11–1.
3. Specify a suitable design factor, N. In typical industrial applications, $N = 3$ is adequate to accommodate accidental overloads and shock.
4. Specify the material for the key, usually SAE 1018 steel. A higher-strength material can be used.
5. Determine the yield strength of the materials for the key, the shaft, and the hub.
6. If a square key is used and the key material has the lowest strength, use Equation (11–5) to compute the minimum required length of the key. This length will be satisfactory for both shear and bearing stress.
7. If a rectangular key is used, or if either the shaft or the hub has a lower strength than the key, use Equation (11–4) to compute the minimum required length of the key based on bearing stress. Also, use Equation (11–2) or (11–5) to compute the minimum required length based on shear of the key. The larger of the two computed lengths governs the design. Check to be sure that the computed length is shorter than the hub length. If not, a higher-strength material must be selected and the design process repeated. Alternatively, two keys or a spline can be used instead of a single key.
8. Specify the actual length of the key to be equal to or longer than the computed minimum length. A convenient standard size should be specified using the preferred basic sizes shown in Appendix A2–1. ***The key should extend over all or a substantial part of the length of the hub. But the keyseat should not run into other stress raisers such as shoulders or grooves.***
9. Complete the design of the keyseat in the shaft and the keyway in the hub using the equations in Figure 11–2. ANSI Standard B17.1 should be consulted for standard tolerances on dimensions for the key and the keyseats.
10. See also Chapter 15 for additional details concerning tolerancing and chamfering.

Example Problem 11–1 A portion of a shaft where a gear is to be mounted has a diameter of 2.00 in. The gear transmits 2965 lb · in of torque. The shaft is to be made of SAE 1040 cold-drawn steel. The gear is made from SAE 8650 OQT 1000 steel. The width of the hub of the gear mounted at this location is 1.75 in. Design the key.

Solution From Table 11–1, the standard key dimension for a 2.00-in-diameter shaft would be 0.500 in square. See Figure 11–5 for the proposed design.

Material selection is a design decision. Let's choose SAE 1018 steel with $s_y = 54\ 000$ psi as listed in Table 11–4.

A check of the yield strengths of the three materials in the key, the shaft, and the hub indicates that the key is the weakest material. Then Equation (11–5) can be used to compute the minimum required length of the key:

$$L = \frac{4TN}{DWs_y} = \frac{4(2965)(3)}{(2.00)(0.500)(54\ 000)} = 0.659 \text{ in}$$

This length is well below the width of the hub of the gear. Notice that the design of the shaft includes retaining rings on both sides of the gear. It is desirable to keep the keyseat well clear of the ring grooves. Therefore, let's specify the length of the key to be 1.50 in.

Summary In summary, the key has the following characteristics:

Material: SAE 1018 steel keystock

Width: 0.500 in

Height: 0.500 in

Length: 1.50 in.

Figure 11–5 shows some details of the completed design. A profile keyseat in the shaft is shown.

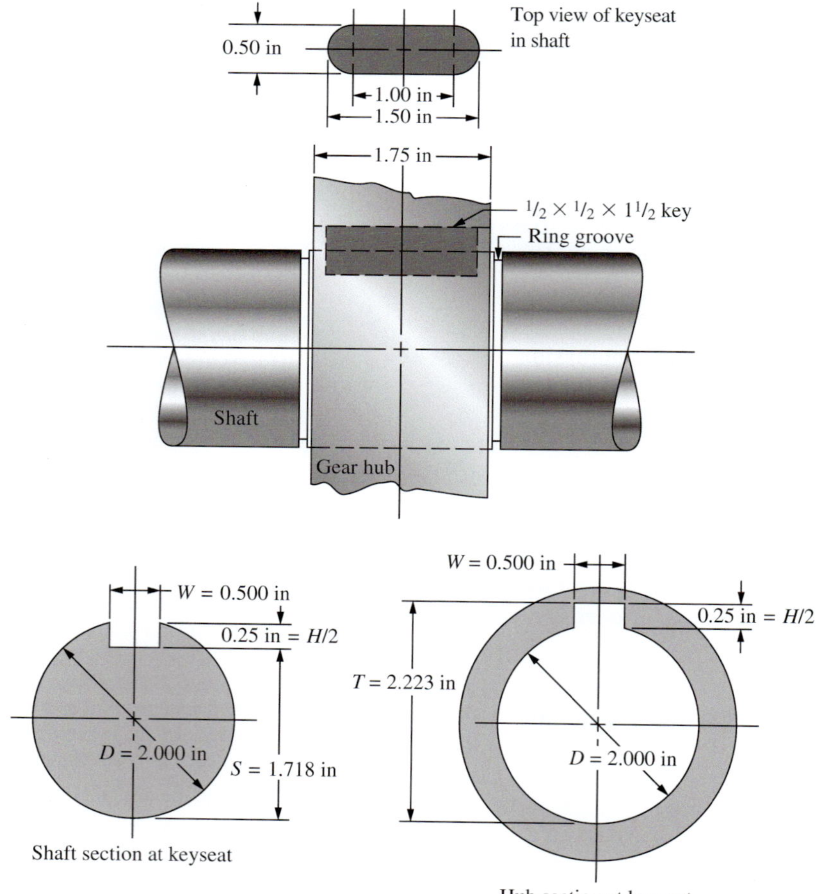

FIGURE 11–5 Details for proposed design of key and keyseats

Shear and Bearing Areas for Woodruff Keys

The geometry of Woodruff keys makes it more difficult to determine the shear area and the bearing area for use in stress analyses. Figure 11–3(e) shows that the bearing area on the side of the key in the keyseat is a segment of a circle. The shear area is the product of the chord of that segment times the thickness of the key. The following equations describe the geometry:

Given

B = nominal diameter of the cylinder of which the key is a part

W = width (thickness) of the key

C = full height of the key

d_s = depth of the keyseat in the shaft

Results

$$\text{Shear area} = A_s = 2W\sqrt{d_s(B - d_s)} \qquad \textbf{(11–6)}$$

To define the equations for the bearing areas on the side of the key in the shaft and in the hub, we first define three geometric variables, G, L, and J, as follows:

$$G = (\pi/180)B\cos^{-1}\{2[(B/2) - d_s]/B\}$$
$$L = 2\sqrt{d_s(B - d_s)}$$
$$J = (\pi/180)B\cos^{-1}\{2[(B/2) - C]/B\}$$

Bearing Area in Shaft

$$A_{c\,\text{shaft}} = 0.5\{G(B/2) - L[(B/2) - d_s]\} \qquad \textbf{(11–7)}$$

$$A_{c\,\text{hub}} = 0.5\{J(B/2) - F[(B/2) - C]\} - A_{c\,\text{shaft}} \qquad \textbf{(11–8)}$$

11–5 SPLINES

A *spline* can be described as a series of axial keys machined into a shaft, with corresponding grooves machined into the bore of the mating part (gear, sheave, sprocket, and so on; see Figure 11–6). The splines perform the same function as a key in transmitting torque from the shaft to the mating element. The advantages of splines over keys are many. Because usually four or more splines are used, as compared with one or two keys, a more uniform transfer of the torque and a lower loading on a given part of the shaft/hub interface result. The splines are integral with the shaft, so no relative motion can occur as between a key and the shaft. Splines are accurately machined to provide a controlled fit between the mating internal and external splines. The surface of the spline is often hardened to resist wear and to facilitate its use in applications in which axial motion of the mating element is desired. Sliding motion between a standard parallel key and the mating element should not be permitted. Because of the multiple splines on the shaft, the mating element can be indexed to various positions.

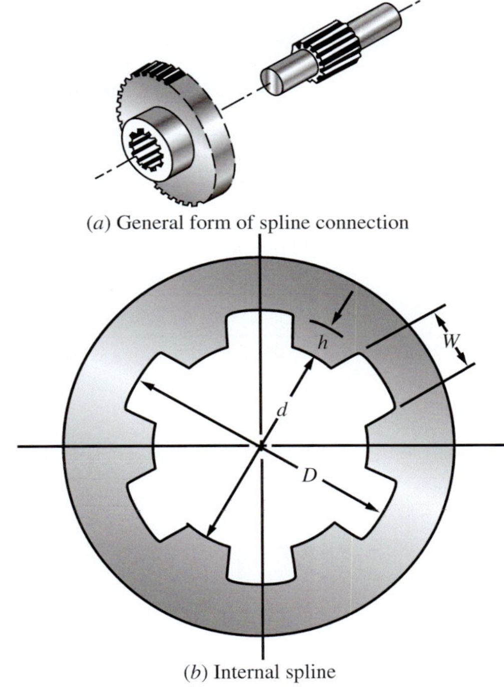

(*a*) General form of spline connection

(*b*) Internal spline

FIGURE 11–6 Straight-sided spline

Splines can be either straight-sided or involute. The involute form is preferred because it provides for self-centering of the mating element and because it can be machined with standard hobs used to cut gear teeth.

Straight-Sided Splines

Straight splines are made according to the specifications of the Society of Automotive Engineers (SAE) and usually contain 4, 6, 10, or 16 splines. Figure 11–6 shows the six-spline version, from which you can see the basic design parameters of D (major diameter), d (minor diameter), W (spline width), and h (spline depth). The dimensions for d, W, and h are related to the nominal major diameter D by the formulas given in Table 11–5. Note that the values of h and d differ according to the use of the spline. The permanent fit, A, is used when the mating part is not to be moved after installation. The B fit is used if the mating part will be moved along the shaft without a torque load. When the mating part must be moved under load, the C fit is used.

The torque capacity for SAE splines is based on the limit of 1000-psi bearing stress on the sides of the splines, from which the following formula is derived:

➭ Torque Capacity for a Spline

$$T = 1000NRh \qquad \textbf{(11–9)}$$

where N = number of splines

R = mean radius of the splines

h = depth of the splines (from Table 11–5)

TABLE 11–5 Formulas for SAE Straight Splines

Number of splines	W, for all fits	A: Permanent fit		B: To slide without load		C: To slide under load	
		h	*d*	*h*	*d*	*h*	*d*
Four	0.241D	0.075D	0.850D	0.125D	0.750D		
Six	0.250D	0.050D	0.900D	0.075D	0.850D	0.100D	0.800D
Ten	0.156D	0.045D	0.910D	0.070D	0.860D	0.095D	0.810D
Sixteen	0.098D	0.045D	0.910D	0.070D	0.860D	0.095D	0.810D

Note: These formulas give the maximum dimensions for *W*, *h*, and *d*.

The torque capacity is per inch of length of the spline. But note that

$$R = \frac{1}{2}\left[\frac{D}{2} + \frac{d}{2}\right] = \frac{D + d}{4}$$

$$h = \frac{1}{2}(D - d)$$

Then

$$T = 1000N\frac{(D + d)}{4}\frac{(D - d)}{2} = 1000N\frac{(D^2 - d^2)}{8} \quad \textbf{(11–10)}$$

This equation can be further refined for each of the types of splines in Table 11–5 by substitution of the appropriate relationships for N and d. For example, for the six-spline version and the B fit, $N = 6$, $d = 0.850D$, and $d^2 = 0.7225D^2$. Then

$$T = 1000(6)\frac{[D^2 - 0.7225D^2]}{8} = 208D^2$$

Thus, the required diameter to transmit a given torque would be

➪ Required Spline Diameter for a given Torque

$$D = \sqrt{T/208}$$

In these formulas, dimensions are in inches and the torque is in pound-inches. We use this same approach to find the torque capacities and required diameters for the other versions of straight splines (Table 11–6).

The graphs in Figure 11–7 enable you to choose an acceptable diameter for a spline to carry a given torque, depending on the desired fit, *A*, *B*, or *C*. The data were taken from Table 11–6.

TABLE 11–6 Torque Capacity for Straight Splines per inch of Spline Length

Number of splines	Fit	Torque capacity	Required diameter
4	*A*	$139D^2$	$\sqrt{T/139}$
4	*B*	$219D^2$	$\sqrt{T/219}$
6	*A*	$143D^2$	$\sqrt{T/143}$
6	*B*	$208D^2$	$\sqrt{T/208}$
6	*C*	$270D^2$	$\sqrt{T/270}$
10	*A*	$215D^2$	$\sqrt{T/215}$
10	*B*	$326D^2$	$\sqrt{T/326}$
10	*C*	$430D^2$	$\sqrt{T/430}$
16	*A*	$344D^2$	$\sqrt{T/344}$
16	*B*	$521D^2$	$\sqrt{T/521}$
16	*C*	$688D^2$	$\sqrt{T/688}$

Involute Splines

Involute splines are typically made with pressure angles of 30°, 37.5°, or 45°. The 30° form is illustrated in Figure 11–8, showing the two types of fit that can be specified. The *major diameter fit* produces accurate concentricity between the shaft and the mating element. In the *side fit,* contact occurs only on the sides of the teeth, but the involute form tends to center the shaft in the mating splined hub.

Figure 11–8 also gives some of the basic formulas for key features of involute splines in the U.S. Customary Unit System with dimensions in inches. (See Reference 5.) The terms are similar to those for involute spur gears, which are discussed more completely in Chapter 8. The basic spline size is governed by its *diametral pitch, P:*

$$P = N/D \quad \textbf{(11–11)}$$

where N = number of spline teeth

D = pitch diameter

The diametral pitch, then, is *the number of teeth per inch of pitch diameter.* Only even numbers of teeth from 6 to 60 are typically used. Up to 100 teeth are used on some 45° splines. Note that the pitch diameter lies *within* the tooth and is related to the major and minor diameters by the relationships shown in Figure 11–8.

The *circular pitch, p,* is the distance from one point on a tooth to the corresponding point on the next adjacent tooth,

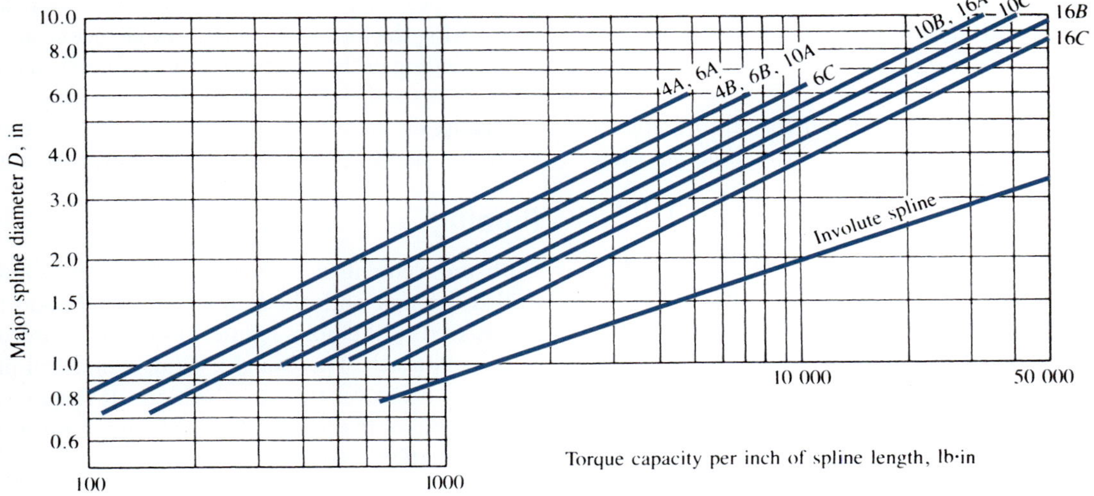

FIGURE 11–7 Torque capacity per inch of spline length, lb · in.

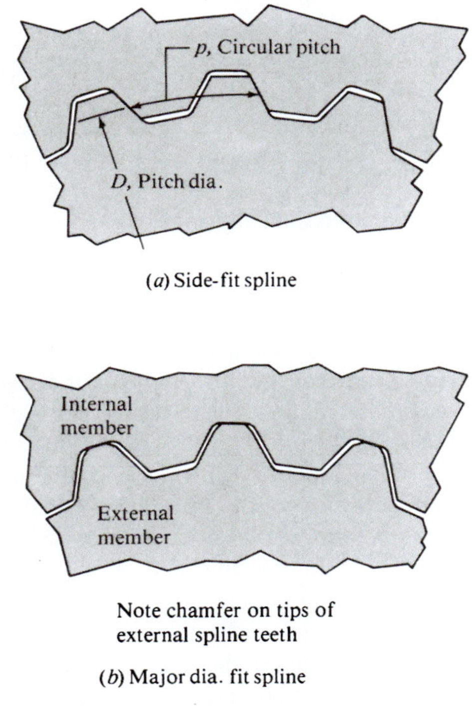

N = Number of spline teeth
P = Diametral pitch
$D = N/P$ = Pitch dia.
$p = \pi/P$ = Circular pitch

Minor dia.:
Internal: $\frac{N-1}{P}$
External: $\frac{N-1.35}{P}$

Major dia.:
Internal: $\frac{N+1.35}{P}$ side fit
$\frac{N+1}{P}$ major dia. fit
External: $\frac{N+1}{P}$

FIGURE 11–8 30° involute spline.

measured along the pitch circle. To find the nominal value of p, divide the circumference of the pitch circle by the number of spline teeth. That is,

$$p = \pi D/N \quad \textbf{(11–12)}$$

But because $P = N/D$, we can also say

$$p = \pi/P \quad \textbf{(11–13)}$$

The *tooth thickness*, t, is the thickness of the tooth measured along the pitch circle. Then the theoretical value is

$$t = p/2 = \pi/2P$$

The nominal value of the width of the tooth space is equal to t.

Standard Diametral Pitches. The following are the 17 standard diametral pitches in common use:

2.5	3	4	5	6	8	10	12	16
20	24	32	40	48	64	80	128	

The common designation for an involute spline is given as a fraction, P/P_s, where P_s is called the *stub pitch* and is always equal to $2P$. Thus, if a spline had a diametral pitch of 4, it would be called a 4/8 pitch spline. For convenience, we will use only the diametral pitch.

Length of Splines. Common designs use spline lengths from $0.75D$ to $1.25D$, where D is the pitch diameter of the spline. If these standards are used, the shear strength of the splines will exceed that of the shaft on which they are machined.

Metric Module Splines. The dimensions of splines made to metric standards are related to the *module, m,* where

$$m = D/N \tag{11–14}$$

and both D and m are in millimeters. (See Reference 6.) Note that the symbol Z is used in place of N for the number of teeth in standards describing metric splines. Other features of metric splines can be found from the following formulas:

$$\text{Pitch diameter} = D = mN \tag{11–15}$$

$$\text{Circular pitch} = p = \pi m \tag{11–16}$$

$$\text{Basic tooth thickness} = t = \pi m/2 \tag{11–17}$$

Standard Modules. There are 15 standard modules:

0.25	0.50	0.75	1.00	1.25	1.50	1.75	2.00
2.50	3	4	5	6	8	10	

Design Aids

Refer to References 9, 10, and 11 for additional design guidelines. Reference 9 includes extensive tables of data for splines and offers application and design information. Reference 10 gives information on the application, operation, dimensioning, and manufacture of involute splines as applied to automotive applications. Data are included on allowable shear stress, allowable compressive stress, wear life factor, spline overload factor, and fatigue life factor. Reference 11 focuses on metric module splines.

11–6 OTHER METHODS OF FASTENING ELEMENTS TO SHAFTS

The following discussion will acquaint you with some of the ways in which power-transmitting elements can be attached to shafts without keys or splines. In most cases the designs have not been standardized, and analysis of individual cases, considering the forces exerted on the elements and the manner of loading of the fastening means, is necessary. In several of the designs, the analysis of shear and bearing will follow a procedure similar to that shown for keys. If a satisfactory analysis is not possible, testing of the assembly is recommended.

Pinning

With the element in position on the shaft, a hole can be drilled through both the hub and the shaft, and a pin can be inserted in the hole. Figure 11–9 shows three examples of this approach. The straight, solid, cylindrical pin is subjected to shear over two cross sections. If there is a force, F, on each end of the pin at the shaft/hub interface, and if the shaft diameter is D, then

$$T = 2F(D/2) = FD$$

or $F = T/D$. With the symbol d representing the pin diameter, the shear stress in the pin is

$$\tau = \frac{F}{A_s} = \frac{T}{D(\pi d^2/4)} = \frac{4T}{D(\pi d^2)} \tag{11–18}$$

Letting the shear stress equal the design stress in shear as before, solving for d gives the required pin diameter:

➪ Required Diameter for a Pin

$$d = \sqrt{\frac{4T}{D(\pi)(\tau_d)}} \tag{11–19}$$

Sometimes the diameter of the pin is purposely made small to ensure that the pin will break if a moderate overload is encountered, in order to protect critical parts of a mechanism. Such a pin is called a *shear pin.*

One problem with a cylindrical pin is that fitting it adequately to provide precise location of the hub and to prevent the pin from falling out is difficult. The *taper pin* overcomes some of these problems, as does the *split spring pin* shown in Figure 11–9(c). For the split spring pin, the hole is made slightly smaller than the pin diameter so that a light force is required to assemble the pin in the hole. The spring force retains the pin in the hole and holds the assembly in position. But, of course,

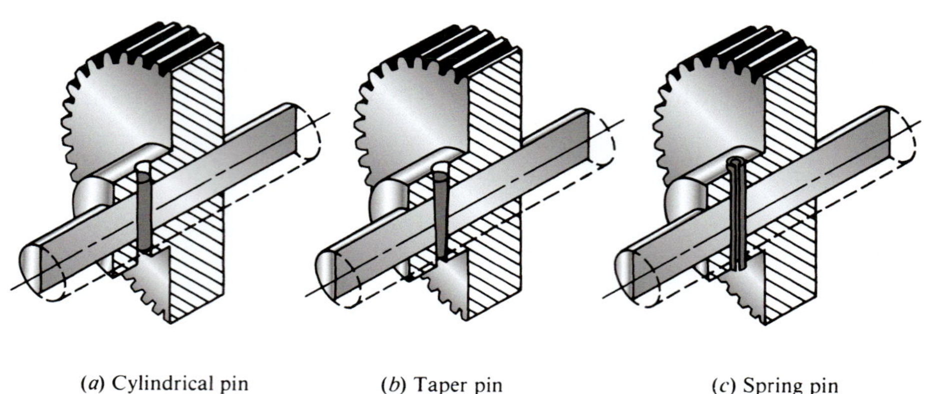

(*a*) Cylindrical pin (*b*) Taper pin (*c*) Spring pin

FIGURE 11–9 Pinning.

the presence of any of the pin-type connections produces stress concentrations in the shaft. (See Internet site 1.)

Keyless Hub to Shaft Connections

Using a steel ring compressed tightly around a smooth shaft allows torque to be transmitted between the hub of a power-transmitting element and a shaft without having a key between the two elements. Figure 11–10 shows a commercially available product called a *Locking Assembly*™ from Ringfeder® Corporation that employs this principle.

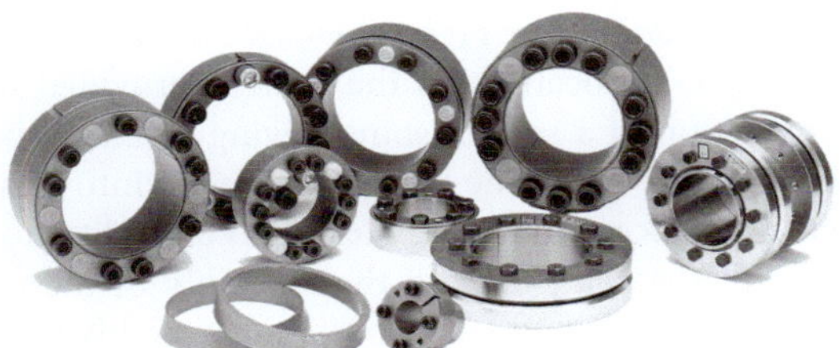

(*a*) A variety of styles

(*b*) Locking assembly applied to a gear

FIGURE 11–10 Ringfeder® Locking Assemblies. (Ringfeder® Corporation, Westwood, NJ)

The Locking Assembly™ employs steel rings with opposing mating tapers held together with a series of fasteners. With the Locking Assembly™ placed completely within a counter bore of the hub of virtually any kind of power-transmitting element such as a gear, sprocket, fan wheel, cam, coupling, or turbine rotor, the fasteners can then be tightened. Initially there is a small clearance between the inside diameter of the locking device and the shaft as well as the hub bore. This clearance facilitates easy assembly and positioning of the hub. After the hub is positioned in the desired location on the shaft, the fasteners are tightened to a specified torque in a specific sequence. As the bolts are tightened, they draw the opposing tapered rings together, generating a radial movement of the inner ring toward the shaft and a simultaneous outward movement of the outer ring toward the ID of the hub. Once the initial clearances are eliminated, further tightening of the bolts results in a high pressure against the shaft and the hub. When the bolts are properly torqued using a torque wrench, the final contact pressure combined with friction allows for the transmission of a predetermined and predictable amount of torque between the hub and the shaft.

The connection can transmit axial forces in the form of thrust loads as well as torque as with helical gears, for example. Table 11–7 lists examples of torque and axial force capacities for selected sizes of one model of the Ringfeder Locking Assemblies®. All the metric data are for catalog metric sizes with capacities given for torque in kN·m and axial force in kN. Data in the upper nonshaded part for the inch-sized models are from catalog data with capacities given for torque in lb·ft and axial force in lb. The shaft sizes listed are fairly close to the metric sizes in the left part of the table to illustrate comparisons. The lower, shaded part shows data in U.S. units converted from the metric data for

TABLE 11–7 Capacities of Selected Sizes of Ringfeder Locking Assemblies

Metric sizes			Inch sizes		
Shaft size (mm)	Transmissible torque (kN·m)	Transmissible axial force (kN)	Shaft size (in)	Transmissible torque (lb·ft)	Transmissible axial force (lb)
25	0.460	30.0	1.000	337	8088
50	2.07	80.0	2.000	1808	21 696
75	5.81	160	3.000	4332	34 656
100	11.1	220	4.000	8489	50 934
150	28.0	380	6.000	22 762	91 048
200	63.5	640	7.875	46 707	142 345
300	183	1220	11.8	134 987	274 281
400	384	1920	15.7	283 252	431 655
600	896	2989	23.6	66 0921	671 987
800	1550	3876	31.5	1 143 334	871 402
1000	2375	4749	39.4	1 751 145	1 067 670
			Converted from metric data		

Note: All unshaded data are from catalog listings.

the larger sizes because the maximum standard inch-size model is for a shaft size of 7.875 in.

Unlike a thermal or pressure fit connection, the locking device can be easily removed since it is a *mechanical shrink fit*. The pressures generated within the locking device itself allow the stresses to remain within the elastic limits of the materials. Removal is simply done by carefully loosening the screws, thus allowing the rings to slide apart and return the Locking Assembly™ to its original relaxed condition. The element can then be repositioned or removed at any time. Some locking devices have self-releasing taper angles, which allow them to self-release when the fasteners are loosened. Other locking devices have self-locking tapers that require the gentle pressing apart of the locking device parts. Different applications dictate which type of device is better suited.

Advantages of the keyless connection are the elimination of keys, keyways, or splines, and the cost of machining them; tight fit of the driving element around the shaft; the ability to transmit reversing or dynamically changing loads; and easy assembly, disassembly, and adjustment of the elements. General engineering information is provided on the Internet site 2 for Ringfeder® Corporation for installation dimensions, tolerances, lubrication, and surface finishes required. The rated torque values are for use on solid shafts; hollow shafts require additional analysis. Special considerations are necessary for the hub design to ensure that stresses remain below the yield strength of the hub material.

Polygon Hub to Shaft Connection

Figure 11–11 shows a shaft to hub connection that employs special mating polygon shapes to transmit torque without keys or splines. See Internet site 17 for available sizes and application information. German standards DIN 32711 and 32712 describe the forms. They can be produced on shaft sizes from 0.188 in (4.76 mm) to 8.00 in (203 mm). The three-sided configuration is called the P3 profile, and the four-sided design is called the PC4 profile. CNC turning and grinding can be used to produce the external form while broaching typically produces the internal form. Torque is transmitted by distributing the load on each side of the polygon, eliminating the shearing action inherent with keys or splines. Dimensions can be controlled for tight or press fits for precision, backlash-free location or with a sliding fit for ease of assembly.

Split Taper Bushing

A *split taper bushing* (see Figure 11–12) uses a key to transmit torque. Axial location on the shaft is provided by the clamping action of a split bushing having a small taper on its outer surface. When the bushing is pulled into a mating hub with a set of capscrews, the bushing is brought into tight contact with the shaft to hold the assembly in the proper axial position. The small taper locks the assembly together. Removal of the bushing is accomplished by removing the capscrews and using them in push-off holes to force the hub off the taper. The assembly can then be easily disassembled.

Set Screws

A *set screw* is a threaded fastener driven radially through a hub to bear on the outer surface of a shaft (see Figure 11–13). The point of the set screw is flat, oval, cone-shaped, cupped, or any of several proprietary forms. The point bears on the shaft or digs slightly into its surface. Thus, the set screw transmits torque by the friction between the point and the shaft or by the resistance of the material in shear. The capacity for torque transmission is somewhat variable, depending on the hardness of the shaft material and the clamping force created when the screw is installed. Furthermore, the screw may loosen during operation because of vibration. For these

Polygon profile produced in a variety of products

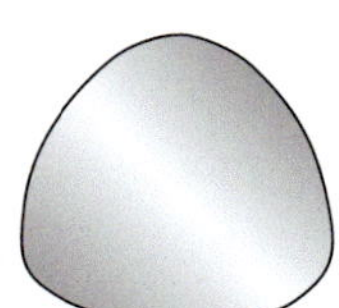

Three-sided P3 external profile

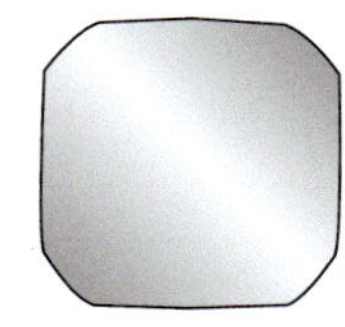

Four-sided PC4 external profile

FIGURE 11–11 Polygon hub to shaft connections (General Polygon Systems, Inc., Millville, NJ)

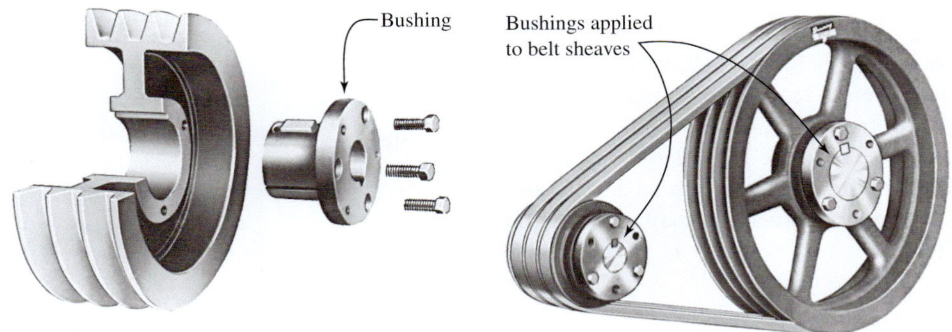

FIGURE 11–12 V-belt sheaves having three grooves with a split taper bushing for shaft mounting (Power Transmission Solutions, a business unit of Emerson Industrial Automation)

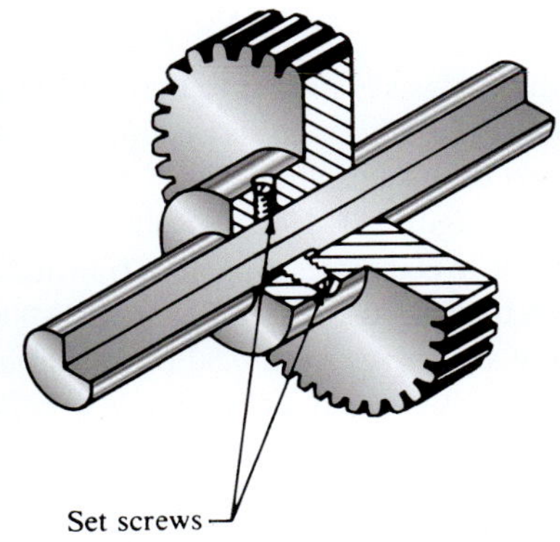

FIGURE 11–13 Set screws

reasons, set screws should be used with care. Some manufacturers provide set screws with plastic inserts in the side among the threads. When the set screw is screwed into a tapped hole, the plastic is deformed by the threads and holds the screw securely, resisting vibration. Using a liquid adhesive also helps resist loosening. (See Internet site 21.)

Another problem with using set screws is that the shaft surface is damaged by the point; this damage may make disassembly difficult. Machining a flat on the surface of the shaft may help reduce the problem and also produce a more consistent assembly.

When set screws are properly assembled on typical industrial shafting, their force capability is approximately as follows (see Reference 9):

Screw Diameter	*Holding Force*
1/4 in	100 lb
3/8 in	250 lb
1/2 in	500 lb
3/4 in	1300 lb
1 in	2500 lb

Taper and Screw

The power-transmitting element (gear, sheave, sprocket, or other) that is to be mounted at the end of a shaft can be secured with a screw and a washer in the manner shown in Figure 11–14(a). The taper provides good concentricity and moderate torque transmission capacity. Because of the

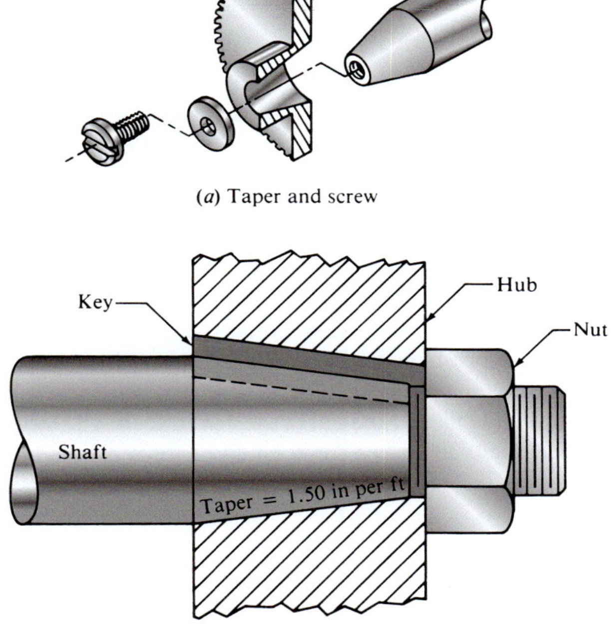

FIGURE 11–14 Tapered shaft for fastening machine elements to shafts

machining required, the connection is fairly costly. A modified form uses the tapered shaft with a threaded end for the application of a nut, as shown in Figure 11–14(b). The inclusion of a key lying in a keyseat machined parallel to the taper increases the torque-transmitting capacity greatly and ensures positive alignment.

Press Fit

Making the diameter of the shaft greater than the bore diameter of the mating element results in an interference fit. The resulting pressure between the shaft and the hub permits the transmission of torque at fairly high levels, depending on the degree of interference. This is discussed in more detail in Chapter 13. Sometimes the press fit is combined with a key, with the key providing the positive drive and the press fit ensuring concentricity and holding the part in position axially.

Molding

Plastic and die cast gears can be molded directly to their shafts. Often the gear is applied to a location that is knurled to improve the ability to transmit torque. A modification of this procedure is to take a separate gear blank with a prepared hub, locate it over the proper position on a shaft, and then cast zinc into the space between the shaft and the hub to lock them together.

11–7 COUPLINGS

The term *coupling* refers to a device used to connect two shafts together at their ends for the purpose of transmitting power. There are two general types of couplings: rigid and flexible.

Rigid Couplings

Rigid couplings are designed to draw two shafts together tightly so that no relative motion can occur between them. This design is desirable for certain kinds of equipment in which precise alignment of two shafts is required and can be provided. In such cases, the coupling must be designed to be capable of transmitting the torque in the shafts.

Figure 11–15 shows three styles of rigid couplings. (See Internet site 3.) For the style in part (a), the two flanged parts are installed on the shafts to be coupled and the bolts in the flange are drawn together by a series of bolts. The load path is then from the driving shaft to its flange, through the bolts, into the mating flange, and out to the driven shaft. The torque places the bolts in shear. The total shear force on the bolts depends on the radius of the bolt circle, $D_{bc}/2$, and the torque, T. That is,

$$F = T/(D_{bc}/2) = 2T/D_{bc}$$

Letting N be the number of bolts, the shear stress in each bolt is

$$\tau = \frac{F}{A_s} = \frac{F}{N(\pi d^2/4)} = \frac{2T}{D_{bc}N(\pi d^2/4)} \qquad \textbf{(11–20)}$$

Letting the stress equal the design stress in shear and solving for the bolt diameter, we have

➩ Required Bolt Diameter for Rigid Coupling

$$d = \sqrt{\frac{8T}{D_{bc}N\pi\tau_d}} \qquad \textbf{(11–21)}$$

Notice that this analysis is similar to that for pinned connections in Section 11–6. The analysis assumes that the bolts are the weakest part of the coupling.

The style in Figure 11–15(b) is called a *taper-lock rigid coupling,* available in eight sizes capable of transmitting torques from 5050 lb·in to 254 500 lb·in (655 N·m to 33 000 N·m) and accommodating shaft sizes from 0.50 in to 6.00 in (12.7 mm to 152 mm). Part (c) is called a *ribbed rigid coupling* available in bore sizes up to 7.00 in (178 mm) and can transmit up to 254 400 lb·in (33 000 N·m) of torque. (See Internet site 3.)

Rigid couplings should be used only when the alignment of the two shafts can be maintained very accurately, not only at the time of installation but also during operation of the machines. If significant angular, radial, or axial misalignment occurs, stresses that are difficult to predict and that may lead to early failure due to fatigue will be induced in the shafts. These difficulties can be overcome by the use of flexible couplings.

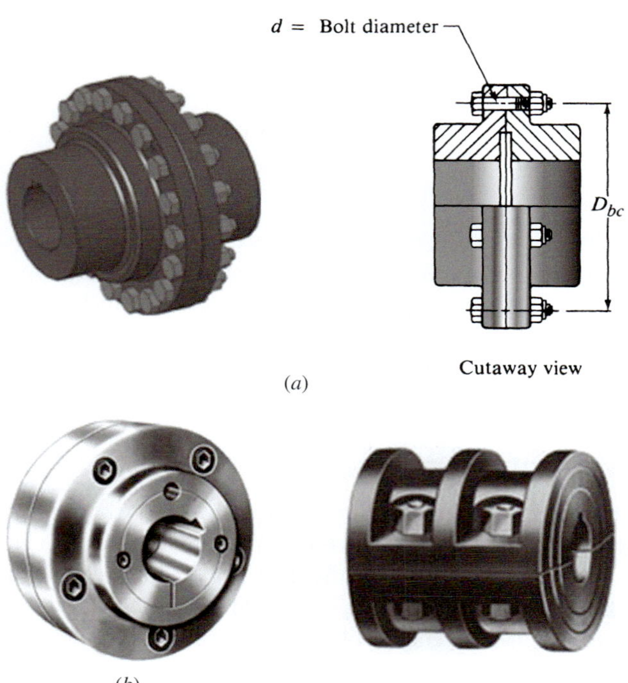

FIGURE 11–15 Rigid coupling (Baldor/Dodge, Greenville, SC)

Flexible Couplings

Flexible couplings are designed to transmit torque smoothly while permitting some axial, radial, and angular misalignment. The flexibility is such that when misalignment does occur, parts of the coupling move with little or no resistance. Thus, no significant axial or bending stresses are developed in the shaft.

Many types of flexible couplings are available commercially, as shown in Figures 11–16 through 11–25. Each is designed to transmit a given limiting torque. The manufacturer's catalog lists the design data from which you can choose a suitable coupling. Remember that torque equals power divided by rotational speed. So for a given size of coupling, as the speed of rotation increases, the amount of power that the coupling can transmit also increases, although not always in direct proportion. Of course, centrifugal effects determine the upper limit of speed.

The degree of misalignment that can be accommodated by a given coupling should be obtained from the manufacturer's catalog data, with values varying with the size and design of the coupling. Small couplings may be limited to parallel misalignment of 0.005 in, although larger couplings may allow 0.030 in

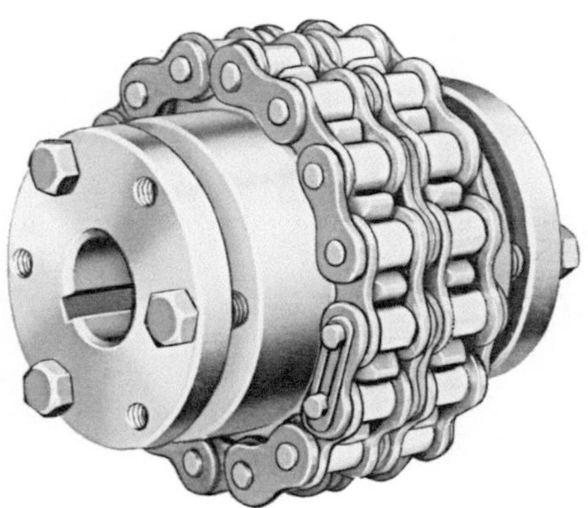

FIGURE 11–16 Chain coupling. Torque is transmitted through a double roller chain. Clearances between the chain and the sprocket teeth on the two coupling halves accommodate misalignment (Emerson Power Transmission Corporation, Ithaca, NY)

FIGURE 11–17 Ever-Flex coupling. The features of this coupling are that it (1) generally minimizes torsional vibration; (2) cushions shock loads; (3) compensates for parallel misalignment up to 1/32 in; (4) accommodates angular misalignment of ±3°; and (5) provides adequate end float, ±1/32 in (Emerson Power Transmission Corporation, Ithaca, NY)

FIGURE 11–18 Grid-Flex coupling. Torque is transmitted through a flexible spring steel grid. Flexing of the grid permits misalignment and makes it torsionally resilient to resist shock loads (Emerson Power Transmission Corporation, Ithaca, NY)

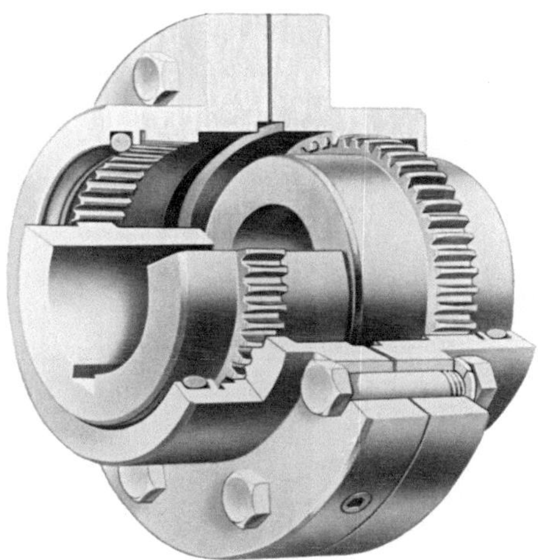

FIGURE 11–19 Gear coupling. Torque is transmitted between crown-hobbed teeth from the coupling half to the sleeve. The crown shape on the gear teeth permits misalignment (Emerson Power Transmission Corporation, Ithaca, NY)

FIGURE 11–20 Bellows coupling. The inherent flexibility of the bellows accommodates the misalignment (Stock Drive Products, New Hyde Park, NY)

FIGURE 11–21 PARA-FLEX® coupling. Using an elastomeric element permits misalignment and cushions shocks (Baldor/Dodge, Greenville, SC)

FIGURE 11–22 D-Flex coupling used mainly for connecting pumps and motors (Baldor/Dodge, Greenville, SC)

FIGURE 11–23 Dynaflex® coupling. Torque is transmitted through elastomeric material that flexes to permit misalignment and to attenuate shock loads (Lord Corporation, Erie, PA)

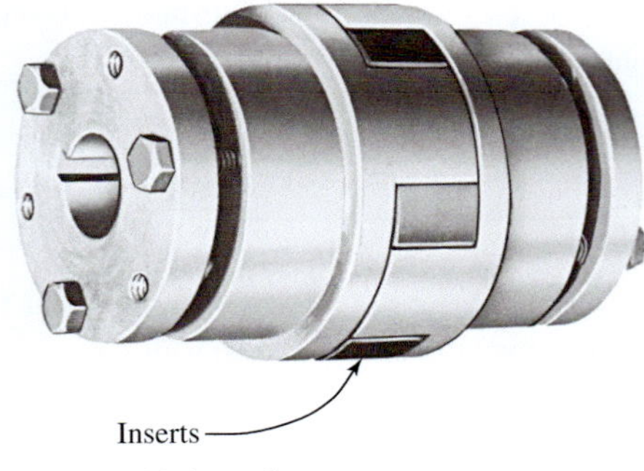

(*a*) Assembled coupling

(*b*) Types of inserts

FIGURE 11–24 Jaw-type coupling (Emerson Power Transmission Corporation, Ithaca, NY)

FIGURE 11–25 FORM-FLEX® coupling. Torque is transmitted from hubs through laminated flexible elements to the spacer (T. B. Wood's Incorporated, Chambersberg, PA)

or more. Typical allowable angular misalignment is ±3°. Axial movement allowed, sometimes called *end float,* is up to 0.030 in for many types of couplings. (See References 1–4 and the Internet sites 3–8, 17, and 22, for the manufacturers.)

11-8 UNIVERSAL JOINTS

When an application calls for accommodating misalignment between mating shafts that is greater than the three degrees typically provided by flexible couplings, a *universal joint* is often employed. Figures 11–26 to 11–29 show some of the styles that are available. See Internet sites 5, 6, 9, 10, and 22 for commercially available universal joints. Angular misalignments of up to 45 degrees are possible at low rotational speeds with single universal joints like that shown in Figure 11–26(a), consisting of two yokes, a center bearing block, and two pins that pass through the block at right angles. Approximately 20 to 30 degrees is more reasonable for speeds above 10 rpm. Single universal joints have the disadvantage that the rotational speed of the output shaft is nonuniform in relation to the input shaft.

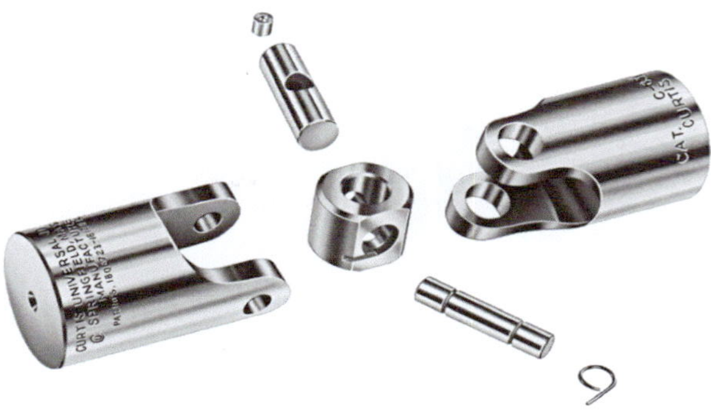

(*a*) Single universal joint components

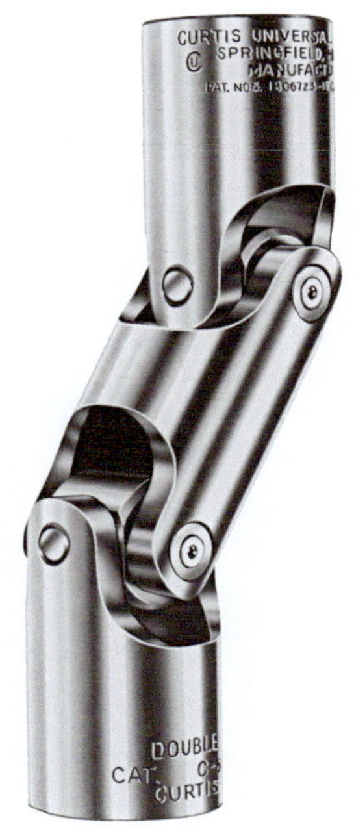

(*b*) Double universal joint

FIGURE 11–26 Single and double universal joints (Curtis Universal Joint Co., Inc., Springfield, MA)

A double universal joint as shown in Figure 11–26(b) allows the connected shafts to be parallel and offset by large amounts. Furthermore, the second joint cancels the nonuniform oscillation of the first joint so the input and output rotate at the same uniform speed. See Internet sites 5, 6, 9, 10, 19, and 22.

Figure 11–27 shows a vehicular universal joint connecting an engine or transmission to the drive wheels used in some rear-wheel drive cars, light and heavy trucks, agricultural equipment, and construction vehicles. The spider assembly contains needle-bearing rollers on each arm. The right end shows a ball stud yoke, a flange yoke, and a center coupling yoke that make up a *double Cardan universal joint.*

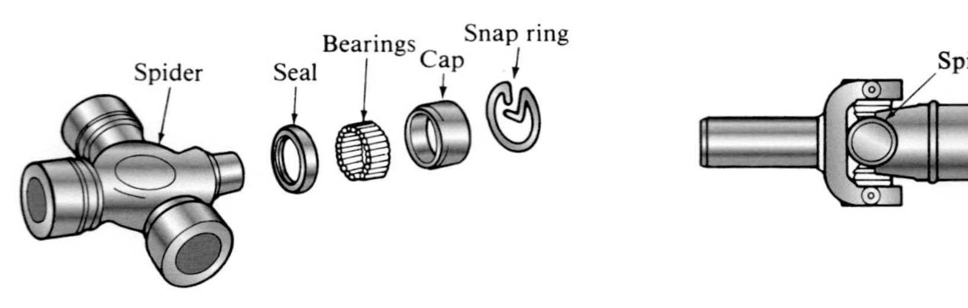

(*a*) Automotive universal joint components

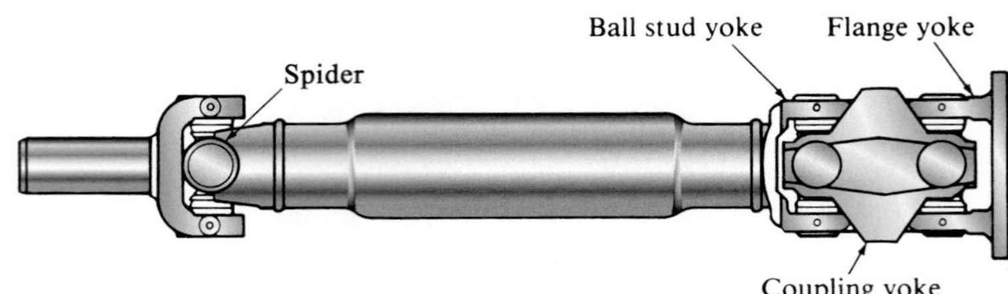

(*b*) Automotive propeller shaft with double Cardan universal joint

FIGURE 11–27 Automotive universal joints

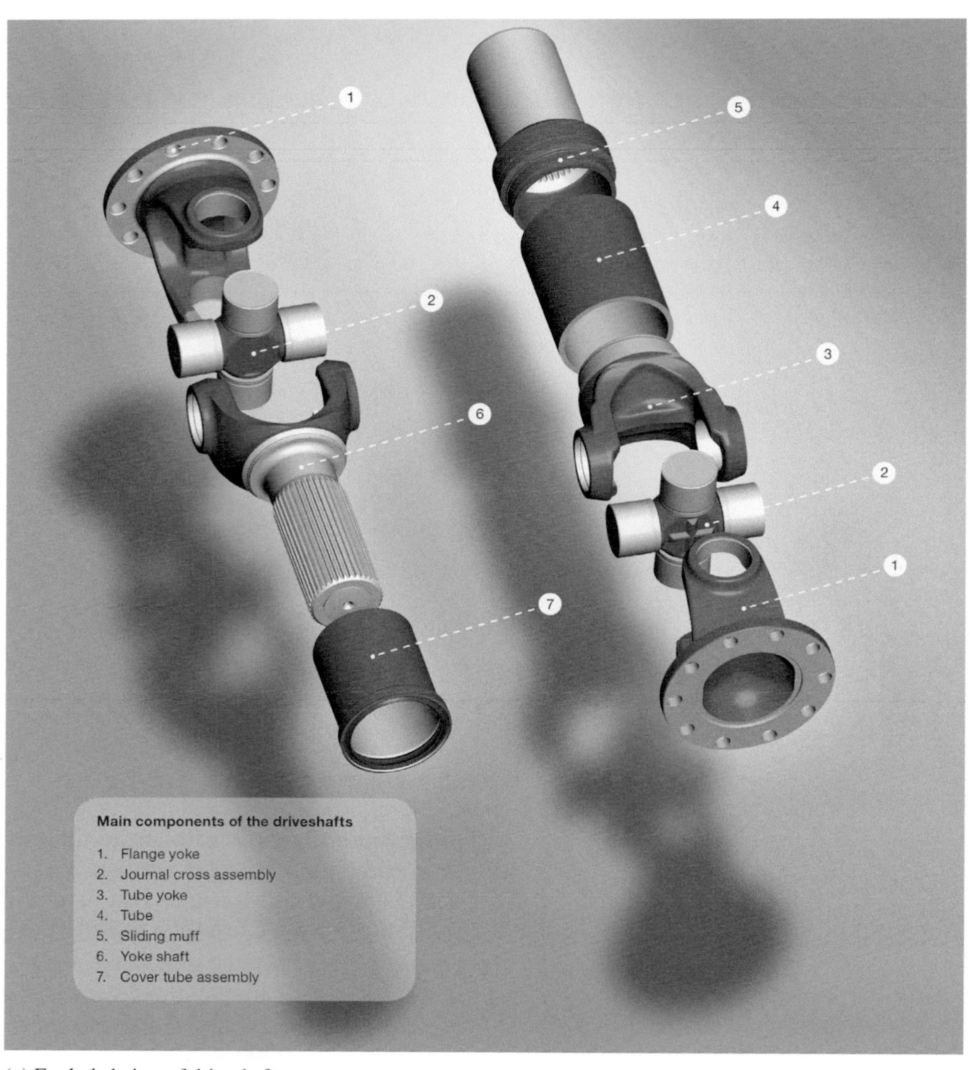

(*a*) Exploded view of driveshaft components

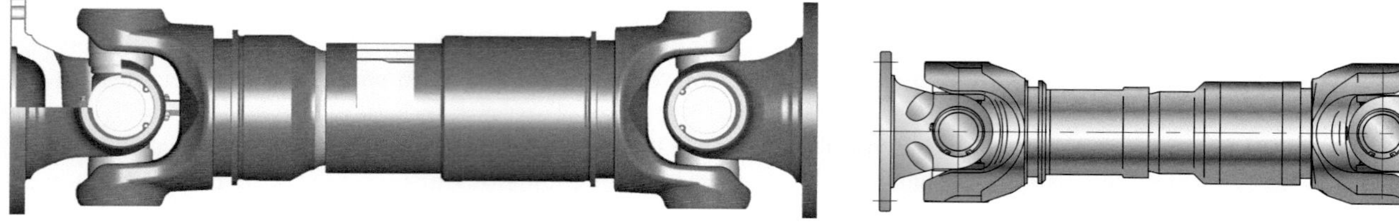

(*b*) Side view of assembly with partial cutaway sections

(*c*) Drawing of a driveshaft assembly

FIGURE 11–28 Industrial drive shaft incorporating a pair of universal joints (GWB–A Dana Brand, Essen, Germany and Toledo, OH)

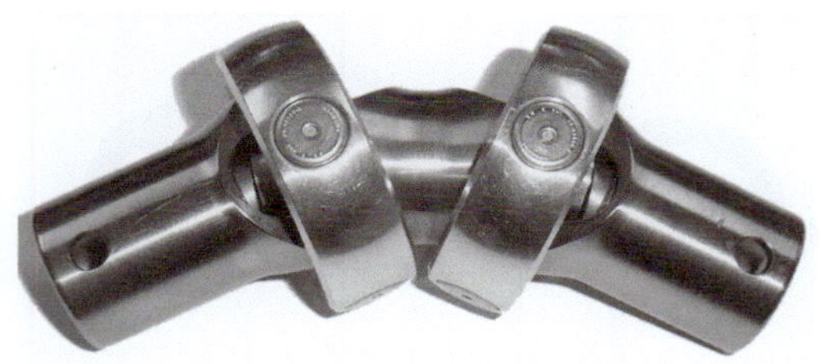

FIGURE 11–29 Cornay™ universal joint (Drive Technologies, Inc., Longmont, CO)

Another style, called a *constant velocity joint*, or simply a *CV joint*, is often used as a key component of front wheel drive and all wheel drive vehicle drivelines.

Figure 11–28 shows a heavy-duty industrial type double universal joint. Some of this type will have a two-part connecting tube that is splined to allow for sizable changes in axial position as well as accommodating the angular or parallel misalignment.

Figure 11–29 shows a novel design called the Cornay™ universal joint that produces true constant velocity of the output shaft through all drive angles up to 90 degrees. Compared with standard universal joint designs, the Cornay™ joint can operate at higher speeds, carry higher torque levels, and produce less vibration.

Manufacturers' literature should be consulted for specifying a suitable size of universal joint for a given application. The primary variables involved are the torque to be transmitted, the rotational speed, and the angle at which the joint will be operating. A *speed/angle factor* is computed as the product of the rotational speed and the operating angle. From this value, an *operating use factor* is determined that is applied to the basic torque load to compute the required rated torque for the joint. All manufacturers provide such data in their catalogs.

11–9 RETAINING RINGS AND OTHER MEANS OF AXIAL LOCATION

The preceding sections of this chapter have focused on means of connecting machine elements to shafts for the purpose of transmitting power. Therefore, they emphasized the ability of the elements to withstand a given torque at a given speed of rotation. It must be recognized that the axial location of the machine elements must also be ensured by the designer.

The choice of the means for axial location depends heavily on whether or not axial thrust is transmitted by the element. Spur gears, V-belt sheaves, and chain sprockets produce no significant thrust loads. Therefore, the need for axial location affects only incidental forces due to vibration, handling, and shipping. Although not severe, these forces should not be taken lightly. Movement of an element in relation to its mating element in an axial direction can cause noise, excessive wear, vibration, or complete disconnection of the drive. Any bicycle rider who has experienced the loss of a chain can appreciate the consequences of misalignment. Recall that for spur gears, the strength of the gear teeth and the wear resistance are both directly proportional to the face width of the gear. Axial misalignment decreases the effective face width.

Some of the methods discussed in Section 11–6 for fastening elements to shafts for the purpose of transmitting power also provide some degree of axial location. Refer to the discussions of pinning, keyless hub to shaft connections, split taper bushings, set screws, the taper and screw, the taper and nut, the press fit, molding, and the several methods of mechanically locking the elements to the shaft.

Among the wide variety of other means available for axial location, we will discuss the following:

- Retaining rings
- Collars
- Shoulders
- Spacers
- Locknuts

Retaining Rings

Retaining rings are placed on a shaft, in grooves cut into the shaft, or in internal recesses to prevent the axial movement of a machine element or to hold internal components in place. Figure 11–30 shows sketches of common generic forms of retaining rings. The basic external design [Figure 11–30(*a*)] is most often used for retaining gears and other power transmission elements in axial position on shafts. This is discussed further in Chapter 12 on shaft design. There are numerous styles available from several vendors, including those listed in Internet sites 7, 11, and 21, from which catalog data may be found along with design guidance for preparing a shaft or housing to accept the rings. In general, retaining rings are classified as either *external* or *internal*. Some install in carefully sized grooves, while others hold their position with gripping force between the ring and the mating element. The amount of axial thrust capacity and the shoulder height provided by the different ring styles vary and catalog data should be consulted. Rings are made to either U.S. Customary dimensions or SI metric sizes.

Collars

A *collar* is a ring slid over the shaft and positioned adjacent to a machine element for the purpose of axial location. It is held in position, typically, by set screws. Its advantage is that axial location can be set virtually anywhere along the shaft to allow adjustment of the position at the time of assembly. The disadvantages are chiefly those related to the use of set screws themselves (Section 11–6).

Shoulders

A *shoulder* is the vertical surface produced when a diameter change occurs on a shaft. Such a design is an excellent method for providing for the axial location of a machine element, at least on one side. Several of the shafts illustrated

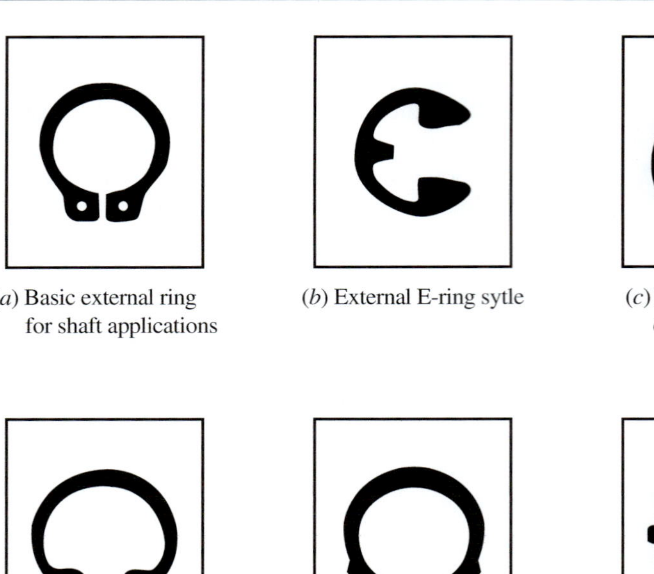

(*a*) Basic external ring for shaft applications

(*b*) External E-ring sytle

(*c*) Radially installed external ring

(*d*) Pressed-on external ring

(*e*) Basic internal ring with inward projections

(*f*) Internal ring with outward projections

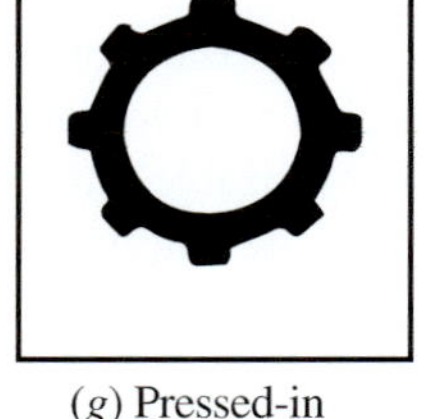

(*g*) Pressed-in internal ring

FIGURE 11–30 Examples of retaining ring styles

in Chapter 12 incorporate shoulders. The main design considerations are providing (1) a sufficiently large shoulder to locate the element effectively and (2) a fillet at the base of the shoulder that produces an acceptable stress concentration factor and that is compatible with the geometry of the bore of the mating element (see Figures 12–2 and 12–7).

Spacers

A *spacer* is similar to a collar in that it is slid over the shaft against the machine element that is to be located. The primary difference is that set screws and the like are not necessary because the spacer is positioned *between* two elements and thus controls only the relative position between them. Typically, one of the elements is positively located by some other means, such as a shoulder or a retaining ring. Consider the shaft in Figure 11–31 on which two spacers locate each gear with respect to its adjacent bearing. The bearings are in turn located on one side by the housing supporting their outer races. The gears also seat against shoulders on one side. The retaining rings on the shaft secure the subassembly prior to installing it into the housing.

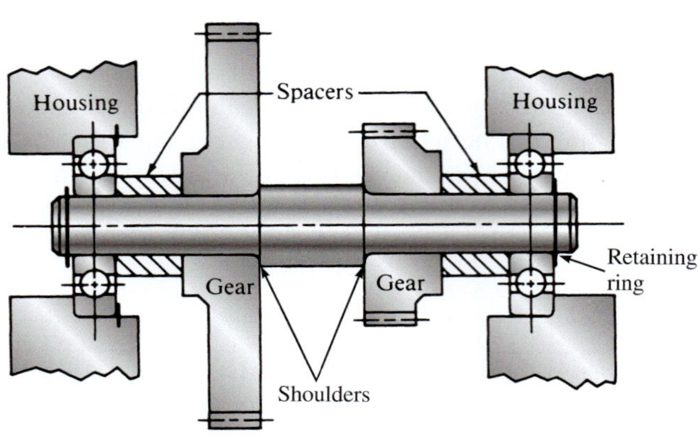

FIGURE 11–31 Application of spacers and shoulders

Locknuts

When an element is located at the end of a shaft, a *locknut* can be used to retain it on one side. Figure 11–32 shows a bearing retainer type of locknut. These are available as stock items from bearing suppliers.

Caution against Overconstraint

One practical consideration in the matter of axial location of machine elements is exercising care that elements are not *overconstrained.* Under certain conditions of differential thermal expansion or with an unfavorable tolerance stack, the elements may be forced together so tightly as to cause dangerous axial stresses. At times it may be desirable to locate only one bearing positively on a shaft and to permit the other to float slightly in the axial direction. The floating element may be held lightly with an axial spring force accommodating the thermal expansion without creating dangerous forces.

11–10 TYPES OF SEALS

Seals are an important part of machine design in situations where the following conditions apply:

1. Contaminants must be excluded from critical areas of a machine.
2. Lubricants must be contained within a space.

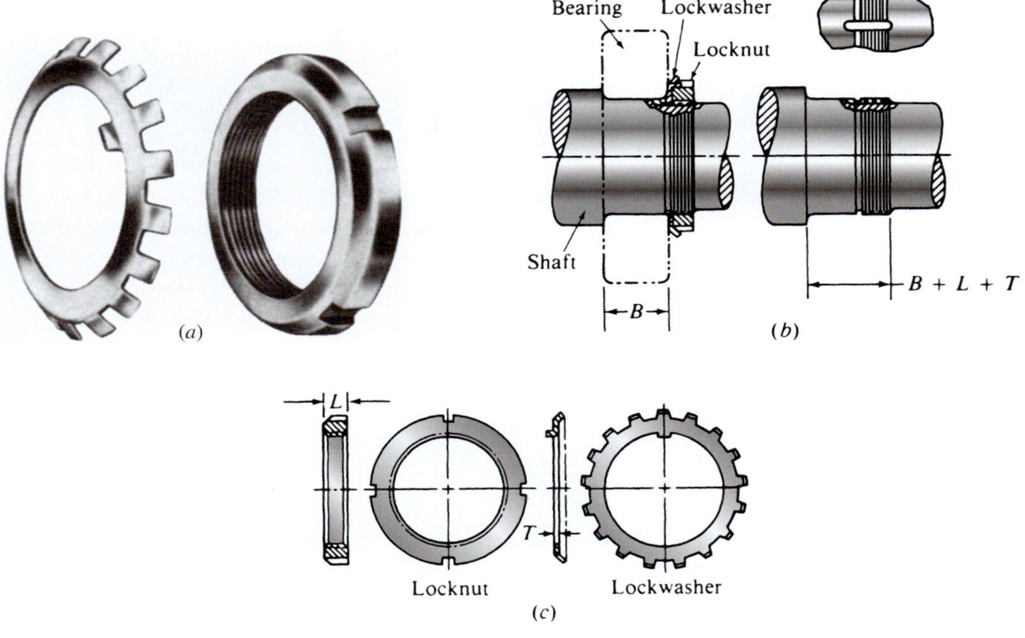

FIGURE 11–32 Locknut and lockwasher for retaining bearings. Dimensions are given in manufacturer's catalog; part sizes are compatible with standard bearing sizes (SKF Industries, Inc., Norristown, PA)

3. Pressurized fluids must be contained within a component such as a valve or a hydraulic cylinder.

Some of the parameters affecting the choice of the type of sealing system, the materials used, and the details of its design are as follows:

1. The nature of the fluids to be contained or excluded.
2. Pressures on both sides of the seal.
3. The nature of any relative motion between the seal and the mating components.
4. Temperatures on all parts of the sealing system.
5. The degree of sealing required: Is some small amount of leakage permissible?
6. The life expectancy of the system.
7. The nature of the solid materials against which the seal must act: corrosion potential, smoothness, hardness, wear resistance.
8. Ease of service for replacement of worn sealing elements.

The number of designs for sealing systems is virtually limitless, and only a brief overview will be presented here. More comprehensive coverage can be found in References 12–15. Often, designers rely on technical information provided by manufacturers of complete sealing systems or specific sealing elements. Also, in critical or unusual situations, testing of a proposed design is advised. See Internet sites 12–15, 22, and 23.

The choice of a type of sealing system depends on the function that it must perform. Common conditions in which seals must operate are listed here, along with some of the types of seals used.

1. Static conditions such as sealing a closure on a pressurized container: elastomeric O-rings; T-rings; hollow metal O-rings; and sealants such as epoxies, silicones, and butyl caulking (Figure 11–33).
2. Sealing a closed container while allowing relative movement of some part, such as diaphragms, bellows, and boots (Figure 11–34).
3. Sealing around a continuously reciprocating rod or piston, such as in a hydraulic cylinder or a spool valve in a hydraulic system: lip seal; U-cup seal; V-packing; and split ring seals, sometimes called *piston rings* (Figure 11–35).
4. Sealing around a rotating shaft such as the input or output shafts of a speed reducer, transmission, or engine: lip seal; wipers and scrapers; and face seals (Figure 11–36).
5. Protection of rolling-element bearings supporting shafts to keep contaminants from the balls and rollers (Figure 11–37).
6. Sealing the active elements of a pump or compressor to retain the pumped fluid: face seals and V-packing.
7. Sealing infrequently moved elements such as a valve stem of a fluid-flow control valve: compression packings and V-packings.
8. Sealing between hard, rigid surfaces such as between a cylinder head and the block of an engine: resilient gaskets. See Internet site 15.
9. Circumferential seals, such as at the tips of turbine blades, and on large, high-speed rotating elements: labyrinth seals; abradable seals; and hydrostatic seals.

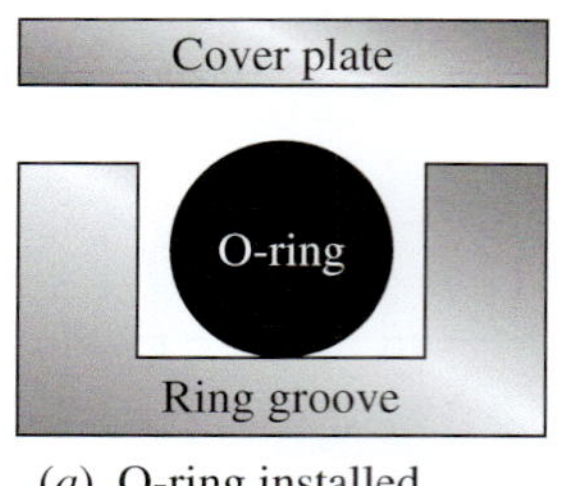

(*a*) O-ring installed without compression.

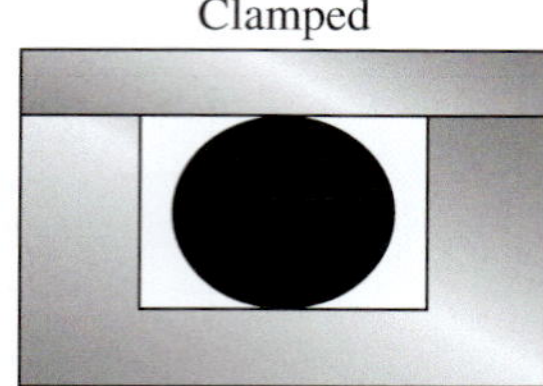

(*b*) Compressed O-ring. Approximately 10% squeeze recommended.

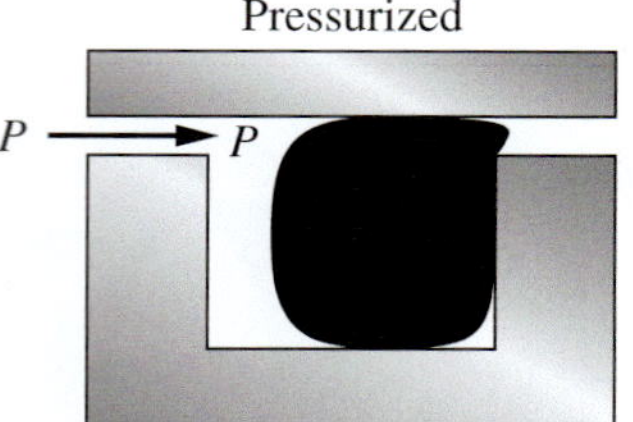

(*c*) O-ring installed with clearance permits extrusion into space. Damage to ring can occur.

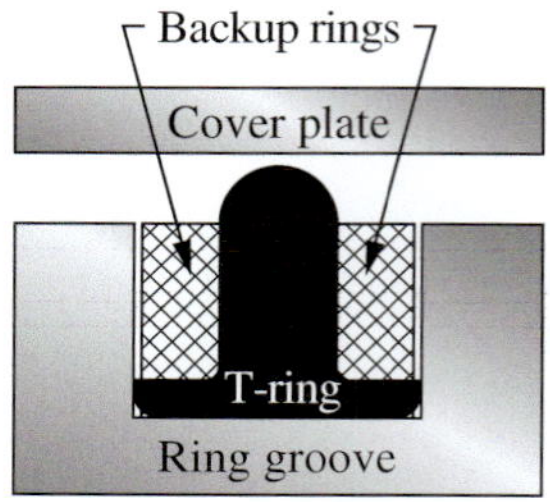

(*a*) Initial installation of T-ring and backup rings.

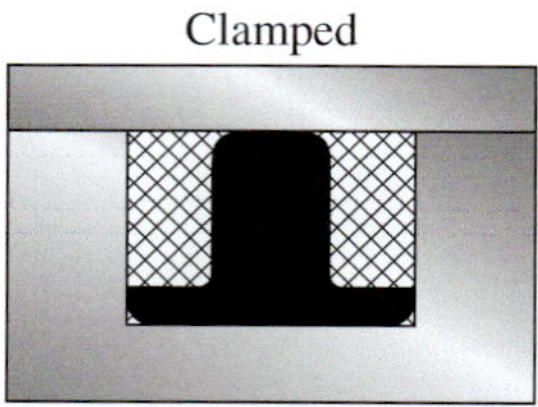

(*b*) Backup rings squeeze bottom and sides of T-rings.

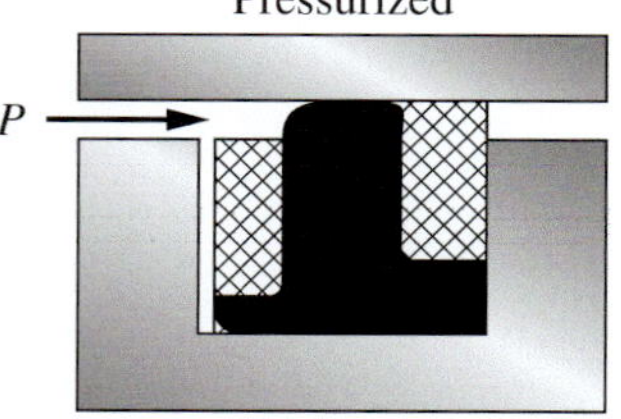

(*c*) Pressure causes T-ring to move backup ring into gap. Harder backup ring does not extrude.

FIGURE 11–33 O-rings and T-rings used as static seals

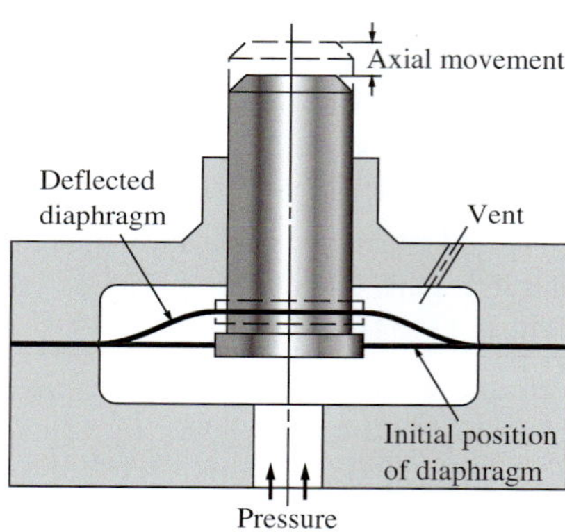

FIGURE 11–34 Application of a diaphragm seal

11–11 SEAL MATERIALS

Most seal materials are resilient to permit the sealing points to follow minor variations in the geometry of mating surfaces. Flexing of parts of the seal cross section also occurs in some designs, calling for resiliency in the materials. Alternatively, as in the case of hollow metal O-rings, the shape of the seal allows the flexing of hard materials to occur. Face seals require rigid, hard materials that can withstand constant sliding motion and that can be produced with fine accuracy, flatness, and smoothness.

Elastomers

Resilient seals such as O-rings, T-rings, and lip seals are often made from synthetic elastomers such as the following:

Neoprene	Butyl	Nitrile (Buna-*N*)
Fluorocarbon	Silicone	Fluorosilicone
Butadiene	Polyester	Ethylene propylene
Polysulfide	Polyurethane	Epichlorohydrin
Polyacrylate	PNF (Phosphonitrilic fluoroelastomer)	

Many proprietary formulations within these general classifications are available under trade names from seal producers and plastics manufacturers.

Properties needed in a given installation will limit the selection of candidate materials. The following list gives some of the more prevalent requirements for seals and some of the materials that meet those requirements:

Weather resistance: Silicone, fluorosilicone, fluorocarbon, ethylene propylene, polyurethane, polysulfide, polyester, neoprene, epichlorohydrin, and PNF.

Petroleum fluid resistance: Polyacrylate, polyester, PNF, nitrile, polysulfide, polyurethane, fluorocarbon, and epichlorohydrin.

Acid resistance: Fluorocarbon.

High-temperature operation: Ethylene propylene, fluorocarbon, polyacrylate, silicone, and PNF.

Cold-temperature operation: Silicone, fluorosilicone, ethylene propylene, and PNF.

Tensile strength: Butadiene, polyester, and polyurethane.

Abrasion resistance: Butadiene, polyester, and polyurethane.

Impermeability: Butyl, polyacrylate, polysulfide, and polyurethane.

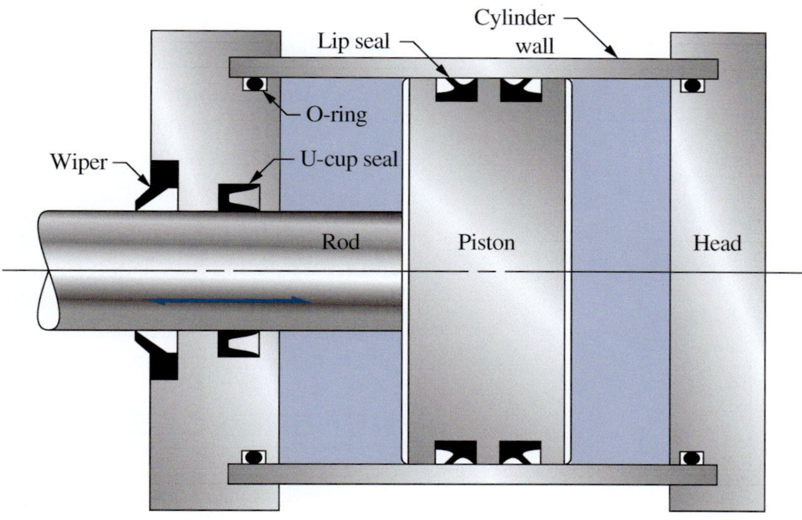

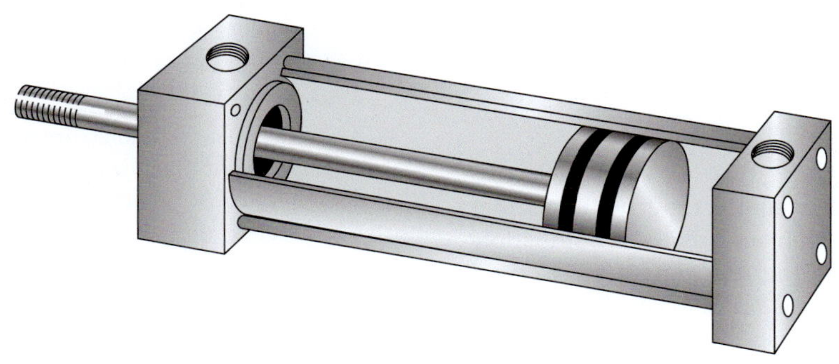

FIGURE 11–35 Lip seals, U-cup seal, wiper, and O-rings applied to a hydraulic actuator

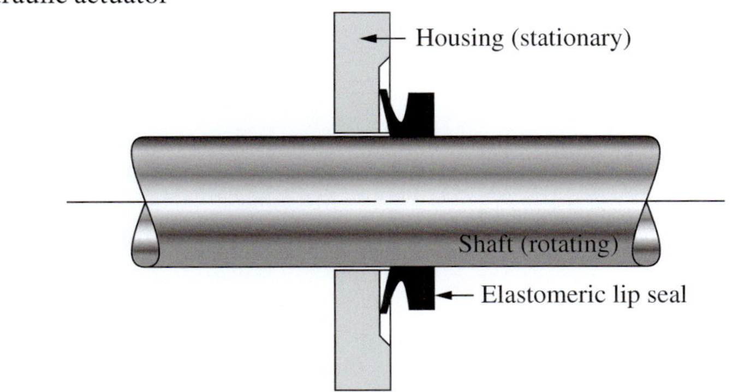

(*a*) Lip-type face seal

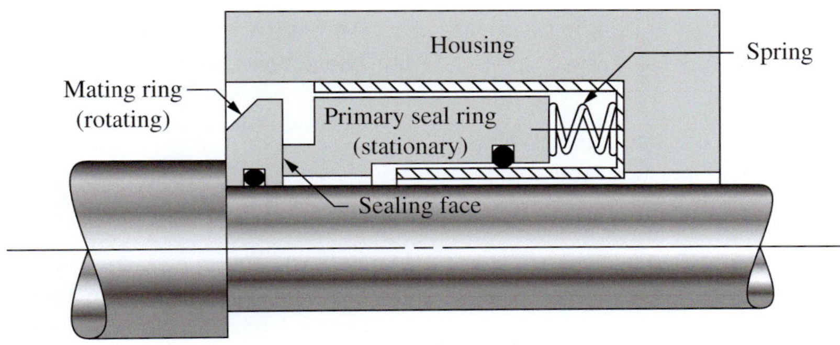

(*b*) Mechanical face seal

FIGURE 11–36 Face seals

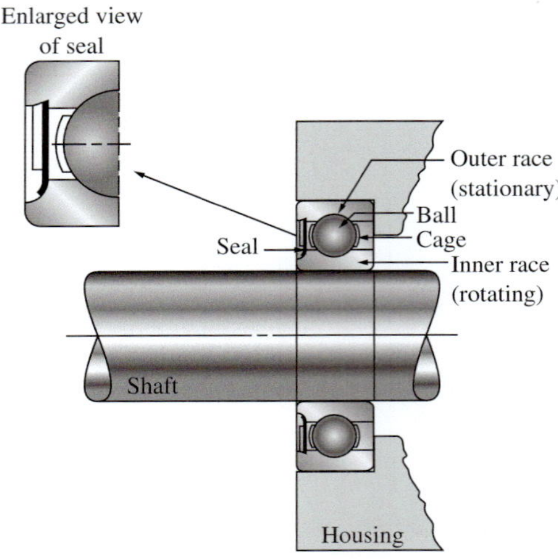

FIGURE 11–37 Seal for ball bearing

Rigid Materials

Face seals, and the parts of other types of sealing systems against which elastomers seal, require rigid materials that can withstand the sliding action and that are compatible with the environment around the seal. Some typical rigid materials used in sealing systems are described in the following list:

Metals: Carbon steel, stainless steel, cast iron, nickel alloys, bronze, and tool steels.

Plastics: Nylon, filled polytetrafluoroethylene (PTFE), and polyimide.

Carbon, ceramics, tungsten-carbide.

Plating: Chromium, cadmium, tin, nickel, and silver.

Flame-sprayed compounds.

Packings

Packings for sealing shafts, rods, valve stems, and similar applications are made from a variety of materials, including leather, cotton, flax, several types of plastics, braided or twisted wire made from copper or aluminum, laminated cloth and elastomeric materials, and flexible graphite.

Gaskets

Common gasket materials are cork, cork and rubber compounds, filled rubber, paper, resilient plastics, and foams. See Internet site 15.

Shafts

When radial lip seals are required around shafts, the shafts are typically steel. They should be hardened to HRC 30 to resist scoring of the surface. Tolerance on the diameter of the shaft on which the seal bears should conform to the following recommendations to ensure that the seal lip can follow the variations:

Shaft Diameter (in)	*Tolerance (in)*
$D \leq 4.000$	± 0.003
$4.000 < D \leq 6.000$	± 0.004
$D > 6.000$	± 0.005

The surface of the shaft and any areas over which the seal must pass during installation should be free of burrs to protect against tearing the seal. A surface finish of 10 to 20 μin is recommended, with adequate lubrication to ensure full contact and to reduce friction between the seal and the shaft surface.

REFERENCES

1. American Gear Manufacturers Association. AGMA 9009-D02. *Nomenclature for Flexible Couplings.* Alexandria, VA: American Gear Manufacturers Association, 2002.
2. American Gear Manufacturers Association. ANSI/AGMA 9002-B04. *Bores and Keyways for Flexible Couplings (Inch Series).* Alexandria, VA: American Gear Manufacturers Association, 2004.
3. American Gear Manufacturers Association. AGME 9112-A04. *Bores and Keyways for Flexible Couplings (Metric Series).* Alexandria, VA: American Gear Manufacturers Association, 2004.
4. American Gear Manufacturers Association. ANSI/AGMA 9001-B97 (R2003). *Lubrication of Flexible Couplings.* Alexandria, VA: American Gear Manufacturers Association, 1997.
5. American National Standards Institute. ANSI B92.1–1986. *Involute Splines.* New York: American National Standards Institute, 1996.
6. American National Standards Institute. ANSI B92.2–1980 R1989. *Metric Module Involute Splines.* New York: American National Standards Institute, 1989.
7. American Society of Mechanical Engineers. ANSI B17.1–67. R98. *Keys and Keyseats.* New York: American Society of Mechanical Engineers, 1998.
8. American Society of Mechanical Engineers. ANSI B17.2–67. R98. *Woodruff Keys and Keyseats.* New York: American Society of Mechanical Engineers, 1998.
9. Oberg, Erik, et al. *Machinery's Handbook.* 28th ed. New York: Industrial Press, 2008.
10. Society of Automotive Engineers. *Standard AS-84,* (R2004) *Splines, Involute (Full Fillet).* Warrendale, PA: Society of Automotive Engineers, 2004.
11. International Standards Organization. *ISO Standard 4156-1:2005, Straight cylindrical involute splines–Metric Module.* Geneva, Switzerland: International Standards Organization, 2005.
12. Flitney, R. K. *Seals and Sealing Handbook.* 5th ed. New York: Elsevier Science, 2008.
13. Summers-Smith, J. D. (Editor). *Mechanical Seal Practice for Improved Performance.* New York: John Wiley & Sons, 2005.
14. Bickford, John. *Gaskets and Gasketed Joints.* Boca Raton, FL: CRC Press, 1998.
15. Czernik, Daniel. *Gaskets: Design, Selection, and Testing.* New York: McGraw-Hill, 1996.

INTERNET SITES FOR KEYS, COUPLINGS, AND SEALS

1. Driv-Loc, Inc. www.driv-lok.com Manufacturer of a wide variety of parallel keys and press-fit fastening devices such as grooved pins, dowel pins, spring pins, and studs. For keys, select *Products,* then *Additional Solutions*.
2. Ringfeder Corporation. www.ringfeder.com Manufacturer of Ringfeder keyless hub to shaft locking devices.
3. Baldor/Dodge. www.dodge-pt.com Manufacturer of a wide variety of power transmission components including flexible couplings, gearing, belt drives, clutches and brakes, and bearings. For couplings, select *Products*, then *PT Components.*
4. Emerson Power Transmission, Inc. www.emerson-ept.com Manufacturer of a wide variety of power transmission equipment including flexible couplings and universal joints under the Kop-Flex, Browning, and Morse brands. Other brands include McGill, Rollway, Sealmaster, and Van Gorp for gearing, belt drives, chain drives, bearings, conveyor pulleys, and clutches.
5. Dana Corporation—Spicer Drivetrain Components. www.spicerparts.com Manufacturer of flexible couplings, universal joints, and driveshafts for vehicular and industrial applications using the Dana and Spicer brand names.
6. GWB—A Dana Brand www.gwb-essen.de/ Manufacturer of heavy-duty driveshafts for industrial equipment, locomotives, and similar applications.
7. Stock Drive Products/Sterling Instrument. www.sdp-si.com Manufacturer and distributor of precision machine components and assemblies, including flexible couplings, gearing, clutches and brakes, and fasteners.
8. T. B. Wood's Sons Company. www.tbwoods.com Manufacturer of mechanical, electrical, and electronic industrial power transmission products including flexible couplings, synchronous belt drives, V-belt drives, gearmotors, and gearing.
9. Curtis Universal Joint Company. www.curtisuniversal.com Manufacturer of universal joints for the industrial and aerospace markets.
10. Cooper Power Tools/Apex Tool Group. www.apexuniversal.com Manufacturer of universal joints for the military, aerospace, performance racing, and industrial power transmission markets.
11. Rotor Clip Company, Inc. www.rotorclip.com Manufacturer of a wide variety of retaining rings for industrial, commercial, military, and consumer products.
12. Federal Mogul Corporation/National Seals. www.federal-mogul.com/national Manufacturer of seals for engines, transmissions, wheels, differentials, and industrial applications. Select Automotive OEM products technology, then powertrain sealing.
13. American Seal & Packing Company. www.americansealand-packing.com Manufacturer of O-rings, backups, square rings, mechanical seals, custom molded parts, and related products. Division of Steadman & Associates, Inc.
14. Mechanical Seals.net. www.mechanicalseals.net Part of the site for American Seal and Packing Company providing technical information about the design of various types of seals, how they work, how they are maintained, and the materials used.
15. Industrial Gasket & Shim Company. www.igscorp.com Manufacturer of gaskets, shims, and custom fabricated seals, expansion joints, and industrial sealants.
16. American High Performance Seals, Inc. www.ahpseals.com Manufacturer of a wide variety of wipers and seals for rods, pistons, and rotary applications. Site includes numerous graphic representations of seal cross sections and tables of materials and their properties.
17. Lord Corporation—Vibration & Motion Control Products. www.lord.com Manufacturer of flexible couplings, vibration mounts, and shock isolation mounts using elastomeric materials bonded to metals. Select Products & Solutions, then Vibration & Motion Control.
18. General Polygon Systems, Inc. www.generalpolygon.com Providers of mechanical shaft to hub connections using the General Polygon System.
19. Drive Technologies, LLC www.cornay.com Developer and manufacturer of the Cornay™ universal joint, drive shafts, power take offs, CV joints, and complete drive systems.
20. G. L. Huyett Co. www.huyett.com Supplier of numerous machine components including square and rectangular keys and keystock, bars, shafts, shim and shimstock in both U.S. and SI sizes and in several grades of material. From the home page, select *Power Transmission.*
21. Loctite Corporation www.loctiteproducts.com/threadlockers.shtml Manufacturer of numerous types of adhesives, including Loctite® Threadlocker® that inhibits loosening of screws and other threaded products.
22. Grainger www.grainger.com Distributor of a huge array of products for machinery components, hardware, motors, power transmission drive products, bearings, flexible couplings, universal joints, fasteners, retaining rings, keys and key stock, raw materials, welding products, adhesives, seals and sealants, threadlocking materials, linear motion devices, and many more.
23. Federal Mogul Corporation www.federalmogul.com Manufacturer of automotive components and products for industrial uses, energy, and transport, including seals for wheels, crankshafts, camshafts, and oil control rings on pistons.

PROBLEMS

For Problems 1–4 and 7, determine the required key geometry: length, width, and height. Use SAE 1018 steel for the keys if a satisfactory design can be achieved. If not, use a higher-strength material from Table 11–4. Unless otherwise stated, assume that the key material is weakest when compared with the shaft material or the mating elements.

1. Specify a key for a gear to be mounted on a shaft with a 2.00-in diameter. The gear transmits 21 000 lb·in of torque and has a hub length of 4.00 in.
2. Specify a key for a gear carrying 21 000 lb·in of torque if it is mounted on a 3.60-in-diameter shaft. The hub length of the gear is 4.00 in.
3. A V-belt sheave transmits 1112 lb·in of torque to a 1.75-in-diameter shaft. The sheave is made from ASTM class 20 cast iron and has a hub length of 1.75 in.
4. A chain sprocket delivers 110 hp to a shaft at a rotational speed of 1700 rpm. The bore of the sprocket is 2.50 inches in diameter. The hub length is 3.25 in.

5. Specify a suitable spline having a *B* fit for each of the applications in Problems 1–4.
6. Design a cylindrical pin to transmit the power, as in Problem 4. But design it so that it will fail in shear if the power exceeds 220 hp.
7. Specify a key for both the sprocket and the wormgear from Example Problem 12–3. Note the specifications for the final shaft diameters at the end of the problem.
8. Describe a Woodruff key no. 204.
9. Describe a Woodruff key no. 1628.
10. Make a detailed drawing of a Woodruff key connection between a shaft and the hub of a gear. The shaft has a diameter of 1.500 in. Use a no. 1210 Woodruff key. Dimension the keyseat in the shaft and the hub.
11. Repeat Problem 10, using a no. 406 Woodruff key in a shaft having a 0.500-in diameter.
12. Repeat Problem 10, using a no. 2428 Woodruff key in a shaft having a 3.250-in diameter.
13. Compute the torque that could be transmitted by the key of Problem 10 on the basis of shear and bearing if the key is made from SAE 1018 cold-drawn steel with a design factor of $N = 3$.
14. Repeat Problem 13 for the key of Problem 11.
15. Repeat Problem 13 for the key of Problem 12.
16. Make a drawing of a four-spline connection having a major diameter of 1.500 in and an *A* fit. Show critical dimensions.
17. Make a drawing of a 10-spline connection having a major diameter of 3.500 in and a *B* fit. Show critical dimensions.
18. Make a drawing of a 16-spline connection having a major diameter of 2.500 in and a *C* fit. Show critical dimensions.
19. Determine the torque capacity of the splines in Problems 16–18.
20. Describe the manner in which a set screw transmits torque if it is used in place of a key. Discuss the disadvantages of such an arrangement.
21. Describe a press fit as it would be used to secure a power transmission element to a shaft.
22. Describe the main differences between rigid and flexible couplings as they affect the stresses in the shafts that they connect.
23. Discuss a major disadvantage of using a single universal joint to connect two shafts with angular misalignment.
24. Describe five ways to locate power transmission elements positively on a shaft axially.
25. Describe three situations in which seals are applied in machine design.
26. List eight parameters that should be considered when selecting a type of seal and specifying a particular design.
27. Name three means of sealing a pressurized container under static conditions.
28. Name three methods of sealing a closed container while allowing relative movement of some part.
29. Name four types of seals used on reciprocating rods or pistons.
30. Name three types of seals applied to rotating shafts.
31. Describe the method of sealing a ball bearing from contaminants.
32. Describe an O-ring seal, and sketch its installation.
33. Describe a T-ring seal, and sketch its installation.
34. Describe some advantages of T-ring seals over O-rings.
35. Describe a diaphragm seal and the type of situation in which it is used.
36. Describe suitable methods of sealing the sides of a piston against the inner walls of the cylinder of a hydraulic actuator.
37. Describe the function of a scraper or wiper on a cylinder rod.
38. Describe the essential elements of a mechanical face seal.
39. Name at least six kinds of elastomers commonly used for seals.
40. Name at least three kinds of elastomers that are recommended for use when exposed to weather.
41. Name at least three kinds of elastomers that are recommended for use when exposed to petroleum-based fluids.
42. Name at least three kinds of elastomers that are recommended for use when exposed to cold-temperature operation.
43. Name at least three kinds of elastomers that are recommended for use when exposed to high-temperature operation.
44. A sealing application involves the following conditions: exposure to high-temperature petroleum fluids; impermeability. Specify a suitable elastomer for the seal.
45. A sealing application involves the following conditions: exposure to high temperature and weather; impermeability; and high strength and abrasion resistance. Specify a suitable elastomer for the seal.
46. Describe suitable shaft design details where elastomeric seals contact the shaft.

CHAPTER TWELVE

SHAFT DESIGN

THE BIG PICTURE Shaft Design

Discussion Map

- A shaft is a rotating machine component that transmits power.

Discover

Identify examples of mechanical systems that incorporate power-transmitting shafts. Describe their geometry and the forces and torques that are exerted on them.

What kinds of stresses are produced in the shaft?

How are other elements mounted on the shaft? How does the shaft geometry accommodate them? How is the shaft supported? What kinds of bearings are used?

This chapter provides approaches that you can use to design shafts that are safe for their intended use. But you have the final responsibility for the design.

A *shaft* is the component of a mechanical device that transmits rotational motion and power. It is integral to any mechanical system in which power is transmitted from a prime mover, such as an electric motor or an engine, to other rotating parts of the system. Can you identify some kinds of mechanical systems incorporating rotating elements that transmit power?

Here are some examples: gear-type speed reducers, belt or chain drives, conveyors, pumps, fans, agitators, and many types of automation equipment. What others can you think of? Consider household appliances, lawn maintenance equipment, parts of a car, power tools, and machines around an office or in your workplace. Describe them and discuss how shafts are used. From what source is power delivered into the shaft? What kind of power-transmitting element, if any, is on the shaft itself? Or is the shaft simply transmitting the rotational motion and torque to some other element? If so, how is the shaft connected to that element?

Visualize the forces, torques, and bending moments that are created in the shaft during operation. In the process of transmitting power at a given rotational speed, the shaft is inherently subjected to a torsional moment, or *torque*. Thus, torsional shear stress is developed in the shaft. Also, a shaft usually carries power-transmitting components, such as gears, belt sheaves, or chain sprockets, which exert forces on the shaft in the transverse direction (perpendicular to its axis). These transverse forces cause bending moments to be developed in the shaft, requiring analysis of the stress due to bending. In fact, most shafts must be analyzed for combined stress.

Describe the specific geometry of shafts from some types of equipment that you can examine. Make a sketch of any variations in geometry that may occur, such as changes in diameter, to produce shoulders, grooves, keyseats, or holes. How are any power-transmitting elements held in position along the length of the shaft? How are the shafts supported? Typically, bearings are used to support the shaft while permitting rotation relative to the housing of the machine. What kinds of bearings are used? Do they have rolling elements such as ball bearings? Or are they smooth-surfaced bearings? What materials are used?

It is likely that you will find much diversity in the design of the shafts in different kinds of equipment. You should see that the functions of a shaft have a large influence on its design. Shaft geometry is greatly affected by the mating elements such as bearings, couplings, gears, chain sprockets, or other kinds of power-transmitting elements.

This chapter provides approaches that you can use to design shafts that are safe for their intended use. But the final responsibility for the design is yours because it is impractical to predict in a book all conditions to which a given shaft will be subjected.

YOU ARE THE DESIGNER

Consider the gear-type speed reducer shown in Figures 1–12 and 8–3. As you continue the design, three shafts must be designed. The input shaft carries the first gear in the gear train and rotates at the speed of the prime mover, typically an electric motor or an engine. The middle shaft carries two gears and rotates more slowly than the input shaft because of the first stage of speed reduction. The final gear in the train is carried by the third shaft, which also transmits the power to the driven machine. From what material should each shaft be made? What torque is being transmitted by each shaft, and over what part of the shaft is it acting? How are the gears to be located on the shafts? How is the power to be transmitted from the gears to the shafts, or vice versa? What forces are exerted on the shaft by the gears, and what bending moments result? What forces must the bearings that support each shaft resist? What are the minimum acceptable diameters for the shafts at all sections to ensure safe operation? What should be the final dimensional specifications for the many features of the shafts, and what should be the tolerances on those dimensions? The material in this chapter will help you make these and other shaft design decisions. ■

12–1 OBJECTIVES OF THIS CHAPTER

After completing this chapter, you will be able to:

1. Propose reasonable geometries for shafts to carry a variety of types of power-transmitting elements, providing for the secure location of each element and the reliable transmission of power.
2. Compute the forces exerted on shafts by gears, belt sheaves, and chain sprockets.
3. Determine the torque distribution on shafts.
4. Prepare shearing force and bending moment diagrams for shafts in two planes.
5. Account for stress concentration factors commonly encountered in shaft design.
6. Specify appropriate design stresses for shafts.
7. Apply the shaft design procedure shown in this chapter, to determine the required diameter of shafts at any section to resist the combination of torsional shear stress and bending stress.
8. Specify reasonable final dimensions for shafts that satisfy strength requirements and installation considerations and that are compatible with the elements mounted on the shafts.
9. Consider the influence of shaft rigidity on its dynamic performance.

12–2 SHAFT DESIGN PROCEDURE

Because of the simultaneous occurrence of torsional shear stresses and normal stresses due to bending, the stress analysis of a shaft virtually always involves the use of a combined stress approach. The recommended approach for shaft design and analysis is the *distortion energy theory of failure*. This theory was introduced in Chapter 5 and will be discussed more fully in Section 12–5. Vertical shear stresses and direct normal stresses due to axial loads may also occur. On very short shafts or on portions of shafts where no bending or torsion occurs, such stresses may be dominant. The discussions in Chapters 3–5 explain the appropriate analysis.

The specific tasks to be performed in the design and analysis of a shaft depend on the shaft's proposed design in addition to the manner of loading and support. With this in

PROCEDURE FOR DESIGN OF A SHAFT

1. Determine the rotational speed of the shaft.
2. Determine the power or the torque to be transmitted by the shaft.
3. Determine the design of the power-transmitting components or other devices that will be mounted on the shaft, and specify the required location of each device.
4. Specify the location of bearings to support the shaft. Normally two and only two bearings are used to support a shaft. The reactions on bearings supporting radial loads are assumed to act at the midpoint of the bearings. For example, if a single-row ball bearing is used, the load path is assumed to pass directly through the balls. If thrust (axial) loads exist in the shaft, you must specify which bearing is to be designed to react against the thrust load. Then the bearing that does not resist the thrust should be permitted to move slightly in the axial direction to ensure that no unexpected and undesirable thrust load is exerted on that bearing.

   Bearings should be placed on either side of the power-transmitting elements if possible to provide stable support for the shaft and to produce reasonably well-balanced loading of the bearings. The bearings should be placed close to the power-transmitting elements to minimize bending moments. Also, the overall length of the shaft should be kept small to keep deflections at reasonable levels.
5. Propose the general form of the geometry for the shaft, considering how each element on the shaft will be held in position axially and how power transmission from each element to the shaft is to take place. For example, consider the shaft in Figure 12–1, which is to carry two gears as the intermediate shaft in a double-reduction, spur gear-type speed reducer. Gear *A* accepts power from gear *P* by way of the input shaft. The power is transmitted from gear *A* to the shaft through the key at the interface between the gear hub and the shaft. The power is then transmitted down the shaft to point *C*, where it passes through another key into gear *C*. Gear *C* then transmits the power to gear *Q* and thus to the output shaft. The locations of the gears and bearings are dictated by the overall configuration of the reducer.

   It is now decided that the bearings will be placed at points *B* and *D* on the shaft to be designed. But how will the bearings and the gears be located so as to ensure that they stay in position during operation, handling, shipping, and so forth? Of course, there are many ways to do this. One way is proposed in Figure 12–2. Shoulders are to be machined in the shaft to provide surfaces against which to seat the bearings and the gears on one side of each element. The gears are restrained on the other side by retaining rings snapped into grooves in the shaft. The bearings will be held in position by the housing acting on the outer races of the bearings. Keyseats will be machined in the shaft at the location of each gear. This proposed geometry provides for positive location of each element.
6. Determine the magnitude of torque that the shaft sees at all points. It is recommended that a torque diagram be prepared, as will be shown later.
7. Determine the forces that are exerted on the shaft, both radially and axially.
8. Resolve the radial forces into components in perpendicular directions, usually vertically and horizontally.
9. Solve for the reactions on all support bearings in each plane.

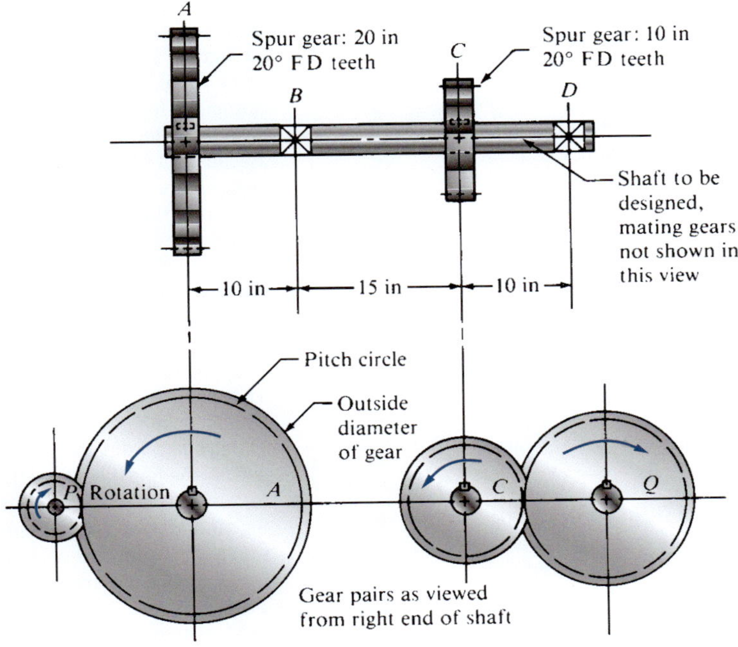

FIGURE 12–1 Intermediate shaft for a double-reduction, spur gear-type speed reducer

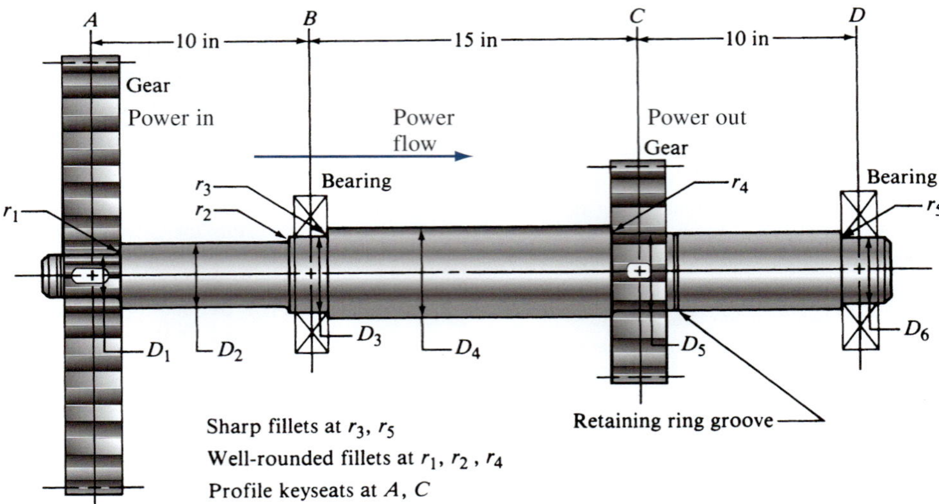

FIGURE 12–2 Proposed geometry for the shaft in Figure 12–1. Sharp fillets at r_3, r_5; well-rounded fillets at r_1, r_2, r_4; profile keyseats at A, C.

10. Produce the complete shearing force and bending moment diagrams to determine the distribution of bending moments in the shaft.
11. Select the material from which the shaft will be made, and specify its condition: cold-drawn, heat-treated, and so on. As indicated in Table 2–9, suggested steel materials for shafts are plain carbon or alloy steels with medium carbon content, such as SAE 1040, 4140, 4340, 4640, 5150, 6150, and 8650. Good ductility with percent elongation above about 12% is recommended. Determine the ultimate strength, yield strength, and percent elongation of the selected material.
12. Determine an appropriate design stress, considering the manner of loading (smooth, shock, repeated and reversed, or other).
13. Analyze each critical point of the shaft to determine the minimum acceptable diameter of the shaft at that point in order to ensure safety under the loading at that point. In general, the critical points are several and include those where a change of diameter takes place, where the higher values of torque and bending moment occur, and where stress concentrations occur.
14. Specify the final dimensions, surface finishes, tolerances, geometric dimensioning details, fillet radii, shoulder heights, keyseat dimensions, retaining ring groove geometry, and other details for each part of the shaft, ensuring that the minimum diameter dimensions from Step 13 are satisfied. Some considerations are listed here while additional information and guidelines are presented in Chapters 13–15.

    a. Power-transmission elements such as gears, belt sheaves, and chain sprockets are typically placed on shafts with small clearances to facilitate assembly while retaining reliable operation. Chapter 13 describes fits between elements that should be applied to final tolerancing decisions. For high-precision devices and those operating at high speeds, the elements are often applied with a light interference fit to minimize vibration and to maintain precise alignment of mating members. The chosen fit determines the final limit dimensions for the shaft. See Reference 12.
    b. The inner races of rolling contact bearings are typically mounted on shafts with a light interference fit as specified by the bearing manufacturer. It is essential that bearing manufacturer's specifications are shown on the shaft drawing. Chapters 13–15 give more information and some example data. Table 15–5 shows that for a ball bearing the specified total tolerance for the shaft diameter at a bearing seat is typically in the range of 0.0003 in to 0.0011 in [approximately 7.6 micrometers (μm) to 28.0 μm]. See examples of the bearing seat dimensioning in Figures 15–6 and 15–7, *typically shown as limit dimensions*. For the purposes of this chapter, it is sufficient to state that the specified bearing inner race diameter determines the final shaft dimension.
    c. Because of the close tolerances at bearing seats, these parts of the shaft are typically ground. Other parts may be produced by turning with fine-finishing passes to achieve the desired fit and satisfactory fatigue performance.
    d. Keyseats in the shaft should be dimensioned as shown in Figure 11–2(b), giving the *S* dimension for ease of manufacture. See also Figures 15–6 and 15–7 for examples.
    e. This chapter presents a method of accounting for stress concentrations due to shoulder fillets that is initially approximate. The designer is expected to specify the final fillet radii that result in safe stresses.
    f. Retaining ring manufacturers specify the dimensioning and tolerancing required for the grooves into which the rings are seated. Designers must show these details on the shaft drawings. See Figures 15–6 and 15–7 for examples.
    g. Axial dimensions for shaft features are typically shown relative to some critical datum feature such as a shoulder against which a gear or bearing is mounted.

mind, the following is a recommended general approach for the design of a shaft.

This process will be demonstrated after the concepts of force and stress analysis are presented.

12–3 FORCES EXERTED ON SHAFTS BY MACHINE ELEMENTS

Gears, belt sheaves, chain sprockets, and other elements typically carried by shafts exert forces on the shaft that cause bending moments. The following is a discussion of the methods for computing these forces for some cases. In general, you will have to use the principles of statics and dynamics to determine the forces for any particular element.

Spur Gears

The force exerted on a gear tooth during power transmission acts normal (perpendicular) to the involute-tooth profile, as discussed in Chapter 9 and shown in Figure 12–3. It is convenient for the analysis of shafts to consider the rectangular components of this force acting in the radial and tangential directions. It is most convenient to compute the tangential force, W_t, directly from the known torque being transmitted by the gear. For U.S. Customary units,

Torque

$$T = 63\,000\,(P)/n \tag{12–1}$$

Tangential Force

$$W_t = T/(D/2) \tag{12–2}$$

where P = power being transmitted in hp

n = rotational speed in rpm

T = torque on the gear in lb · in

D = pitch diameter of the gear in inches

The angle between the total force and the tangential component is equal to the pressure angle, ϕ, of the tooth form. Thus, if the tangential force is known, the radial force can be computed directly from

Radial Forces

$$W_r = W_t \tan \phi \tag{12–3}$$

and there is no need to compute the normal force at all. For gears, the pressure angle is typically $14\frac{1}{2}°$, 20°, or 25°.

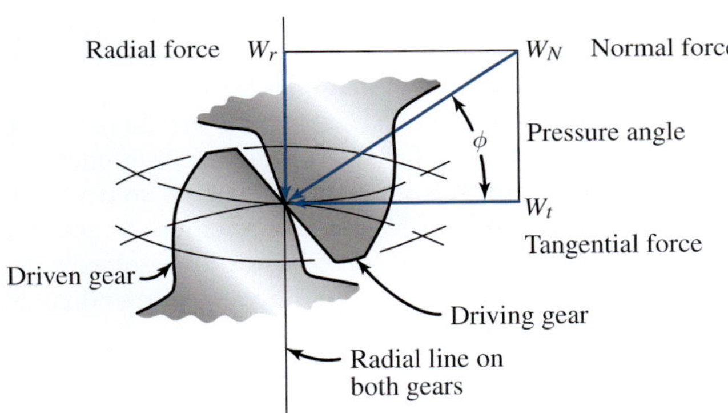

FIGURE 12–3 Forces on teeth of a driven gear

Directions for Forces on Mating Spur Gears

Representing the forces on gears in their correct directions is essential to an accurate analysis of forces and stresses in the shafts that carry the gears. The force system shown in Figure 12–4(*a*) represents the action of the driving gear *A* on the driven gear *B*. The tangential force, W_t, pushes perpendicular to the radial line causing the driven gear to rotate. The radial force, W_r, exerted by the driving gear *A*, acts along the radial line tending to push the driven gear *B* away.

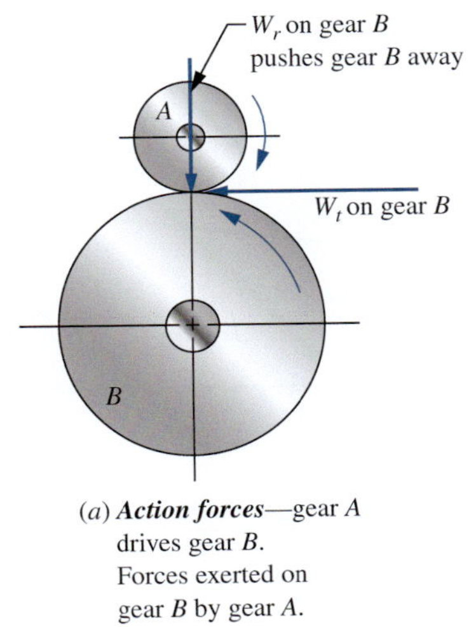

(*a*) ***Action forces***—gear *A* drives gear *B*. Forces exerted on gear *B* by gear *A*.

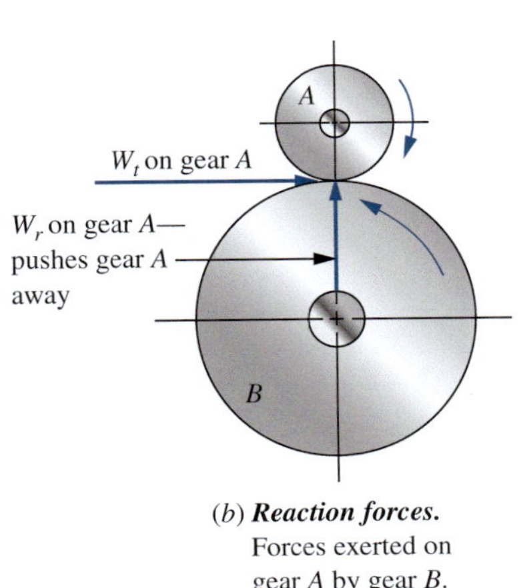

(*b*) ***Reaction forces.*** Forces exerted on gear *A* by gear *B*.

FIGURE 12–4 Directions for forces on mating spur gears

An important principle of mechanics states that for each action force, there is an equal and opposite reaction force. Therefore, as shown in Figure 12–4, the driven gear pushes back on the driving gear with a tangential force opposing that of the driving gear and a radial force that tends to push the driving gear away. For the orientation of the gears shown in Figure 12–4, note the following directions for forces:

Action: *Driver pushes on driven gear*	***Reaction:*** *Driven gear pushes back on driver*
W_t: Acts to the left	W_t: Acts to the right
W_r: Acts downward	W_r: Acts upward

In summary, whenever you need to determine the direction of forces acting on a given gear, first determine whether it is a driver or a driven gear. Then visualize the action forces of the driver. If the gear of interest is the driven gear, these are the forces on it. If the gear of interest is the driver gear, the forces on it act in the opposite directions to the action forces.

Helical Gears

In addition to the tangential and radial forces encountered with spur gears, helical gears produce an axial force (as discussed in Chapter 10). First compute the tangential force from Equations (12–1) and (12–2). Then, if the helix angle of the gear is ψ, and if the normal pressure angle is ϕ_n, the radial force can be computed from

➪ Radial Force

$$W_r = W_t \tan \phi_n / \cos \psi \qquad (12\text{–}4)$$

The axial force is

➪ Axial Force

$$W_x = W_t \tan \psi \qquad (12\text{–}5)$$

Bevel Gears

Refer to Chapter 10 to review the formulas for the three components of the total force on bevel gear teeth in the tangential, radial, and axial directions. Example Problem 10–5 gives a comprehensive analysis of the forces, torques, and bending moments on shafts carrying bevel gears.

Worms and Wormgears

Chapter 10 also gives the formulas for computing the forces on worms and wormgears in the tangential, radial, and axial directions. See Example Problem 10–9.

Chain Sprockets

Figure 12–5 shows a pair of chain sprockets transmitting power. The upper part of the chain is in tension and produces the torque on either sprocket. The lower part of the chain, referred to as the *slack side*, exerts no force on either sprocket. Therefore, the total bending force on the shaft carrying the sprocket is equal to the tension in the tight side of the chain. If the torque on a certain sprocket is known,

➪ Force in Chain

$$F_c = T/(D/2) \qquad (12\text{–}6)$$

where D = pitch diameter of that sprocket

Notice that the force, F_c, acts along the direction of the tight side of the belt. Because of the size difference between the two sprockets, that direction is at some angle from the centerline between the shaft centers. A precise analysis would call for the force, F_c, to be resolved into components parallel to the centerline and perpendicular to it. That is,

$$F_{cx} = F_c \cos \theta \quad \text{and} \quad F_{cy} = F_c \sin \theta$$

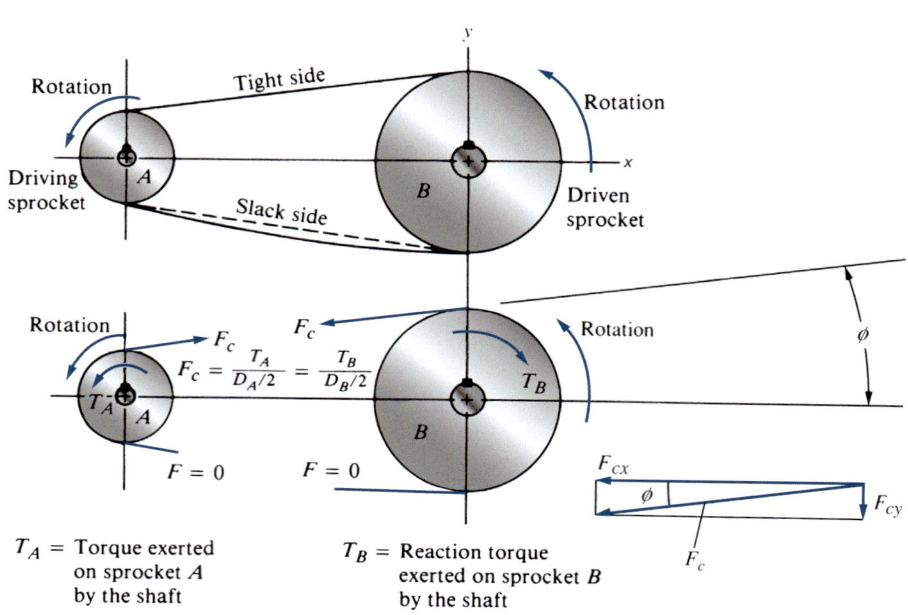

FIGURE 12–5 Forces on chain sprockets

where the x-direction is parallel to the centerline
the y-direction is perpendicular to it
the angle θ is the angle of inclination of the tight side of the chain with respect to the x-direction

These two components of the force would cause bending in both the x-direction and the y-direction. Alternatively, the analysis could be carried out in the direction of the force, F_c, in which single plane bending occurs.

If the angle θ is small, little error will result from the assumption that the entire force, F_c, acts along the x-direction. *Unless stated otherwise, this book will use this assumption.*

V-Belt Sheaves

The general appearance of the V-belt drive system looks similar to the chain drive system. But there is one important difference: Both sides of the V-belt are in tension, as indicated in Figure 12–6. The tight side tension, F_1, is greater than the "slack side" tension, F_2, and thus there is a net driving force on the sheaves equal to

➩ Net Driving Force

$$F_N = F_1 - F_2 \tag{12–7}$$

The magnitude of the net driving force can be computed from the torque transmitted:

➩ Net Driving Force

$$F_N = T/(D/2) \tag{12–8}$$

But notice that the bending force on the shaft carrying the sheave is dependent on the *sum*, $F_1 + F_2 = F_B$. To be more precise, the components of F_1 and F_2 parallel to the line of centers of the two sprockets should be used. But unless the two sprockets are radically different in diameter, little error will result from $F_B = F_1 + F_2$.

To determine the bending force, F_B, a second equation involving the two forces F_1 and F_2 is needed. This is provided by assuming a ratio of the tight side tension to the slack side tension. For V-belt drives, the ratio is normally taken to be

$$F_1/F_2 = 5 \tag{12–9}$$

It is convenient to derive a relationship between F_N and F_B of the form

$$F_B = CF_N \tag{12–10}$$

where C = constant to be determined

$$C = \frac{F_B}{F_N} = \frac{F_1 + F_2}{F_1 - F_2} \tag{12–11}$$

But from Equation (12–9), $F_1 = 5F_2$. Then

$$C = \frac{F_1 + F_2}{F_1 - F_2} = \frac{5F_2 + F_2}{5F_2 - F_2} = \frac{6F_2}{4F_2} = 1.5$$

Equation (12–10) then becomes, for V-belt drives,

➩ Bending Force on Shaft for V-Belt Drive

$$F_B = 1.5\,F_N = 1.5T/(D/2) \tag{12–12}$$

It is customary to consider the bending force, F_B, to act as a single force in the direction along the line of centers of the two sheaves as shown in Figure 12–6.

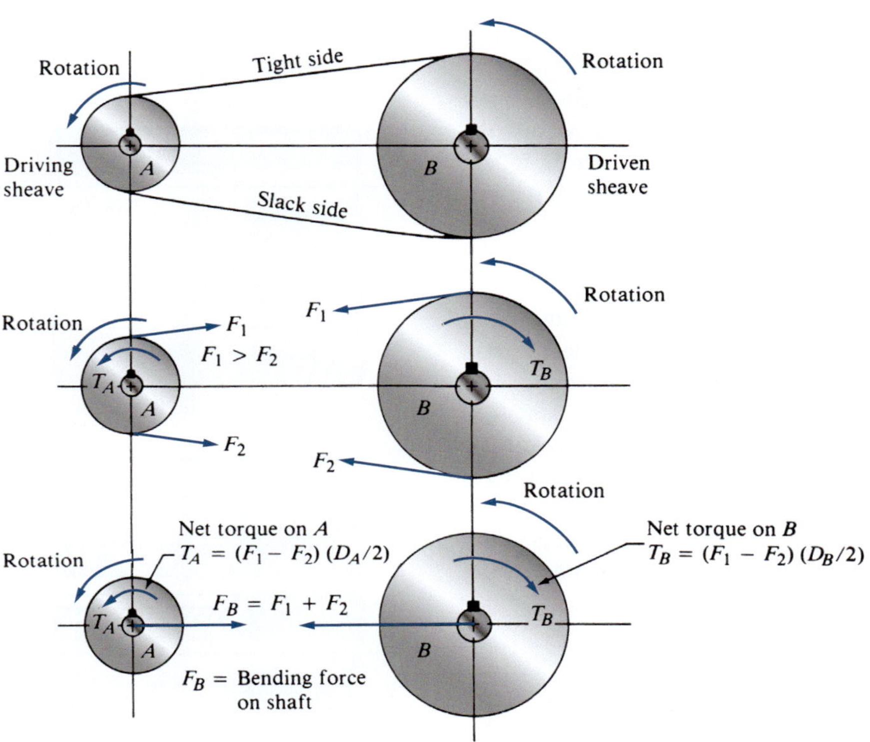

FIGURE 12–6 Forces on belt sheaves or pulleys

Flat-Belt Pulleys

The analysis of the bending force exerted on shafts by flat-belt pulleys is identical to that for V-belt sheaves except that the ratio of the tight side to the slack side tension is typically taken to be 3 instead of 5. Using the same logic as with V-belt sheaves, we can compute the constant C to be 2.0. Then, for flat-belt drives,

Bending Force on Shaft for Flat-Belt Drive

$$F_B = 2.0\,F_N = 2.0T/(D/2) \qquad \textbf{(12–13)}$$

Flexible Couplings

More detailed discussion of flexible couplings was presented in Chapter 11, but it is important to observe here how the use of a flexible coupling affects the design of a shaft.

A flexible coupling is used to transmit power between shafts while accommodating minor misalignments in the radial, angular, or axial directions. Thus, the shafts adjacent to the couplings are subjected to torsion, but the misalignments cause no axial or bending loads.

12–4 STRESS CONCENTRATIONS IN SHAFTS

In order to mount and locate the several types of machine elements on shafts properly, a final design typically contains several diameters, keyseats, ring grooves, and other geometric discontinuities that create stress concentrations. The shaft design proposed in Figure 12–2 is an example of this observation.

These stress concentrations must be taken into account during the design analysis. But a problem exists because the true design values of the stress concentration factors, K_t, are unknown at the start of the design process. Most of the values are dependent on the diameters of the shaft and on the fillet and groove geometries, and these are the objectives of the design.

You can overcome this dilemma by establishing a set of preliminary design values for commonly encountered stress concentration factors, which can be used to produce initial estimates for the minimum acceptable shaft diameters. Then, after the refined dimensions are selected, you can analyze the final geometry to determine the real values for stress concentration factors. Comparing the final values with the preliminary values will enable you to judge the acceptability of the design. The final values of K_t can be determined using Internet site 3 in Chapter 3.

Preliminary Design Values for K_t

Considered here are the types of geometric discontinuities most often found in power-transmitting shafts: keyseats, shoulder fillets, and retaining ring grooves. In each case, a suggested design value is relatively high in order to produce a conservative result for the first approximation to the design. Again it is emphasized that the final design should be checked for safety. That is, if the final value is less than the original design value, the design is still safe. Conversely, if the final value is higher, the stress analysis for the design must be rechecked.

Keyseats. A *keyseat* is a longitudinal groove cut into a shaft for the mounting of a key, permitting the transfer of torque from the shaft to a power-transmitting element, or vice versa. The detail design of keys was covered in Chapter 11.

Two types of keyseats are most frequently used: profile and sled runner (see Figure 12–7). The profile keyseat is milled into the shaft, using an end mill having a diameter equal to the width of the key. The resulting groove is flat-bottomed and has a sharp, square corner at its end. The sled runner keyseat is produced by a circular milling cutter having a width equal to the width of the key. As the cutter begins or ends the keyseat, it produces a smooth radius. For this reason, the stress concentration factor for the sled runner keyseat is lower than that for the profile keyseat. Normally used design values are

$$K_t = 2.0 \quad \text{(profile)}$$

$$K_t = 1.6 \quad \text{(sled runner)}$$

Each of these is to be applied to the bending stress calculation for the shaft, using the full diameter of the shaft. The factors take into account both the reduction in cross section and the effect of the discontinuity. Consult the references listed for more detail about stress concentration factors for keyseats. (See Reference 6.) If the torsional shear stress is

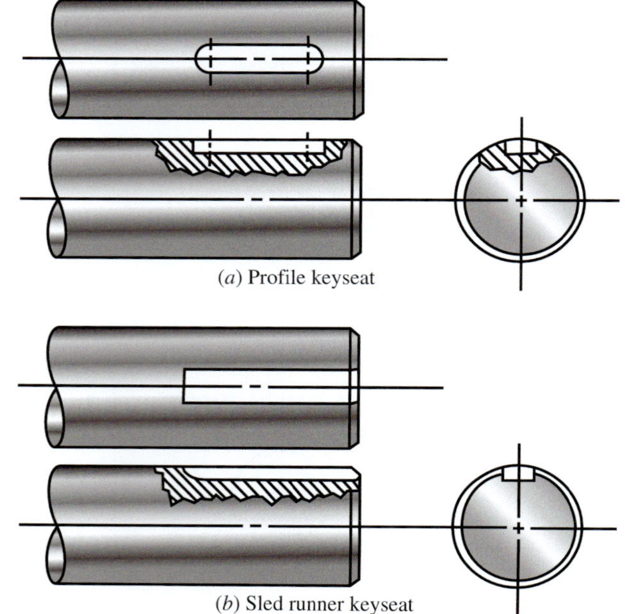

FIGURE 12–7 Keyseats

fluctuating rather than steady, the stress concentration factor is also applied to that.

Shoulder Fillets. When a change in diameter occurs in a shaft to create a shoulder against which to locate a machine element, a stress concentration dependent on the ratio of the two diameters and on the radius in the fillet is produced (see Figure 12–8). It is recommended that the fillet radius be as large as possible to minimize the stress concentration, but at times the design of the gear, bearing, or other element affects the radius that can be used. For the purpose of design, we will classify fillets into two categories: sharp and well-rounded.

The term *sharp* here does not mean truly sharp, without any fillet radius at all. Such a shoulder configuration would have a very high stress concentration factor and should be avoided. Instead, *sharp* describes a shoulder with a relatively small fillet radius. One situation in which this is likely to occur is where a ball or roller bearing is to be located. The inner race of the bearing has a factory-produced radius, but it is small. The fillet radius on the shaft must be smaller yet in order for the bearing to be seated properly against the shoulder. When an element with a large chamfer on its bore is located against the shoulder, or when nothing at all seats against the shoulder, the fillet radius can be much larger (*well-rounded*), and

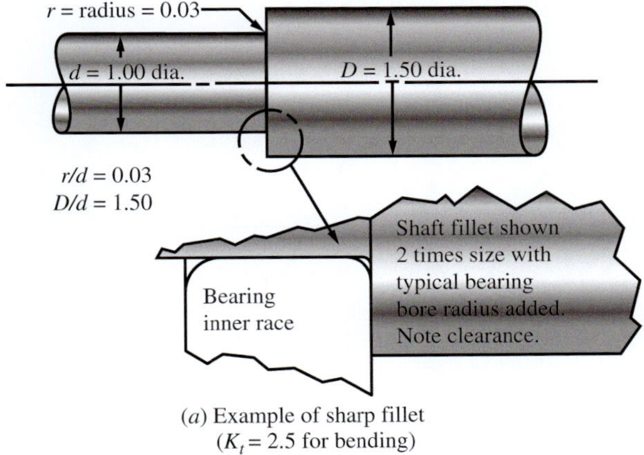

(a) Example of sharp fillet ($K_t = 2.5$ for bending)

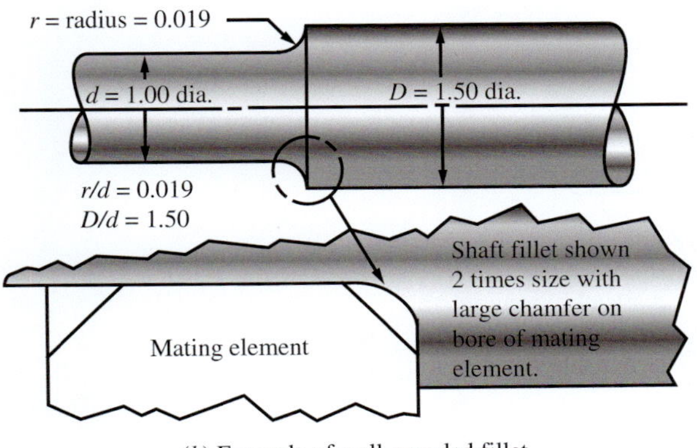

(b) Example of well-rounded fillet ($K_t = 1.5$ for bending)

FIGURE 12–8 Fillets on shafts

the corresponding stress concentration factor is smaller. We will use the following values for design for bending:

$$K_t = 2.5 \quad \text{(sharp fillet)}$$

$$K_t = 1.5 \quad \text{(well-rounded fillet)}$$

Referring to Internet site 3 in Chapter 3, you can see that these values correspond to ratios of r/d of approximately 0.03 for the sharp fillet case and 0.19 for the well-rounded fillet for a D/d ratio of 1.50.

Retaining Ring Grooves. Retaining rings are used for many types of locating tasks in shaft applications. The rings are installed in grooves in the shaft after the element to be retained is in place. The geometry of the groove is dictated by the ring manufacturer. Its usual configuration is a shallow groove with straight side walls and bottom and a small fillet at the base of the groove. The behavior of the shaft in the vicinity of the groove can be approximated by considering two sharp-filleted shoulders positioned close together. Thus, the stress concentration factor for a groove is fairly high.

For preliminary design, we will apply $K_t = 3.0$ to the bending stress at a retaining ring groove to account for the rather sharp fillet radii. The stress concentration factor is not applied to the torsional shear stress if it is steady in one direction.

The computed estimate for the minimum required diameter at a ring groove is at the base of the groove. You should increase this value by approximately 6% to account for the typical groove depth to determine the nominal size for the shaft. Apply a ring groove factor of 1.06 to the computed required diameter.

12–5 DESIGN STRESSES FOR SHAFTS

In a given shaft, several different stress conditions can exist at the same time. For any part of the shaft that transmits power, there will be a torsional shear stress, while bending stress is usually present on the same parts. Only bending stresses may occur on other parts. Some points may not be subjected to either bending or torsion but will experience vertical shearing stress. Axial tensile or compressive stresses may be superimposed on the other stresses. Then there may be some points where no significant stresses at all are created.

Thus, the decision of what design stress to use depends on the particular situation at the point of interest. In many shaft design and analysis projects, computations must be done at several points to account completely for the variety of loading and geometry conditions that exist.

Several cases discussed in Chapter 5 for computing design factors, N, are useful for determining design stresses for shaft design. The bending stresses will be assumed to be completely reversed and repeated because of the rotation of the shaft. Because ductile materials perform better under such loads, it will be assumed that the material for the shaft is ductile. It will also be assumed that the torsional loading is relatively constant and acting in one direction. If other situations exist, consult the appropriate case from Chapter 5.

The symbol τ_d will be used for the design stress when a shear stress is the basis for the design. The symbol σ_d will be used when a normal stress is the basis.

Design Shear Stress—Steady Torque

It was stated in Chapter 5 that the best predictor of failure in ductile materials due to a steady shear stress was the distortion energy theory in which the design shear stress is computed from

$$\tau_d = s_y/(N\sqrt{3}) = (0.577s_y)/N \qquad \textbf{(12–14)}$$

We will use this value for steady torsional shear stress, vertical shear stress, or direct shear stress in a shaft.

Design Shear Stress—Reversed Vertical Shear

Points on a shaft where no torque is applied and where the bending moments are zero or very low are often subjected to significant vertical shearing forces which then govern the design analysis. This typically occurs where a bearing supports an end of a shaft and where no torque is transmitted in that part of the shaft.

Figure 12–9(a) shows the distribution of vertical shearing stresses on such a circular cross section. Note that the maximum shearing stress is at the neutral axis of the shaft, that is, at the diameter. The stress decreases in a roughly parabolic manner to zero at the outer surface of the shaft.

The diagram in Figure 12–9(b) can be explained by visualizing an element of the shaft, beginning at the top, with a shearing force acting downward on the cross section. Then follow that element as it rotates one revolution.

1. Look straight at the circular cross section from the end of the shaft [see Figure 12–9(c)].
2. An element at the top of the shaft sees zero shearing stress as indicated in Figure 12–9(a).
3. After rotating 90 degrees counterclockwise, placing it on the left side of the horizontal diameter, that same element sees the maximum shearing stress, directed downward.
4. That same element sees zero shearing stress again when it rotates 90 degrees more, placing it at the bottom of the shaft.
5. Now as the same element rotates 90 degrees more to a place on the right side of the horizontal diameter, note that the element is upside down from its orientation when it was on the left side, while the shearing force still acts downward.
6. However, in relation to the element, the stress is acting in the *opposite direction* as compared with the direction in Step 3, and the stress on the element has reversed from the value seen in Step 3.

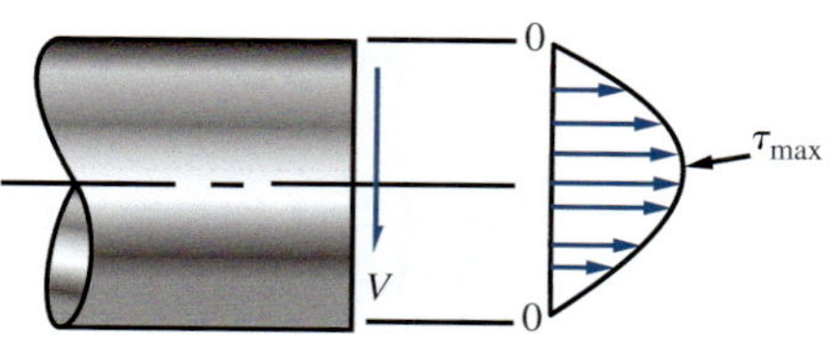

(*a*) Distribution of shearing stress on a circular cross section of a shaft

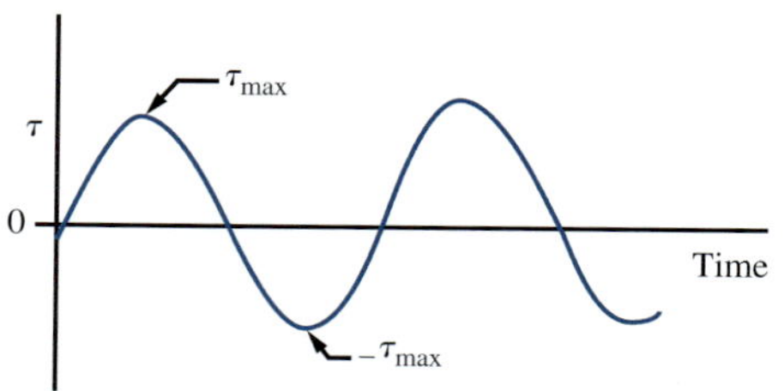

(*b*) Variation of shearing stress on a given element at the surface of a circular rotating shaft

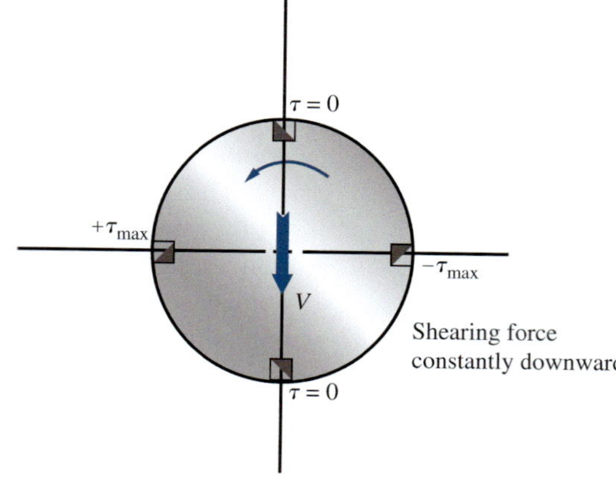

(*c*) Visualization of orientation of element for one revolution

FIGURE 12–9 Shearing stress in a rotating shaft due to vertical shearing force, *V*

7. Rotating another 90 degrees back to the top of the shaft, the element sees zero shearing stress again.

Therefore, we can conclude that any particle of the shaft on its outside surface sees a reversal of the direction of the shearing stress during each rotation as shown in Figure 12–9(b).

Recall from strength of materials that the maximum vertical shearing stress for the special case of a solid circular cross section can be computed from

$$\tau_{max} = 4V/3A$$

where V = vertical shearing force

A = area of the cross section

Where stress concentration factors are to be considered,

$$\tau_{max} = K_t(4V/3A)$$

Also note, as shown in Figure 12–9, that the rotation of the shaft causes any point at the outer part of the cross section to experience a reversing shearing stress that varies from $+\tau_{max}$ to zero to $-\tau_{max}$ to zero in each revolution. The design equations applied to this type of loading are developed next:

$$N = s'_{sn}/\tau_{max}$$

We recommend using the distortion energy theory. Then the endurance strength in shear is

$$s'_{sn} = 0.577s'_n$$

Thus, the above equation can be written in the form

$$N = 0.577s'_n/\tau_{max}$$

Expressed as a design stress, this is

$$\tau_d = 0.577s'_n/N$$

Now letting $\tau_{max} = \tau_d = K_t(4V)/3A$ gives

$$\frac{K_t(4V)}{3A} = \frac{0.577s'_n}{N}$$

Solving for N gives

$$N = \frac{0.577s'_n(3A)}{K_t(4V)} = \frac{0.433s'_n(A)}{K_t(V)} \tag{12–15}$$

Equation (12–15) is useful if the goal is to evaluate the design factor for a given magnitude of loading, a given geometry for the shaft, and given material properties.

Now, solving for the required area gives

$$A = \frac{K_t(V)N}{0.433s'_n} = \frac{2.31K_t(V)N}{s'_n}$$

But our usual objective is the design of the shaft to determine the required diameter. By substituting

$$A = \pi D^2/4$$

we can solve for D:

➭ Required Shaft Diameter

$$D = \sqrt{2.94\, K_t(V)N/s'_n} \tag{12–16}$$

This equation should be used to compute the required diameter for a shaft where a vertical shearing force V is the only significant loading present. In most shafts, the resulting diameter will be much smaller than that required at other parts of the shaft where significant values of torque and bending moment occur. Also, practical considerations may require that the shaft be somewhat larger than the computed minimum to accommodate a reasonable bearing at the place where the shearing force is equal to the radial load on the bearing.

Implementation of Equations (12–15) and (12–16) has the complication that values for the stress concentration factor under conditions of vertical shearing stress are not well known. Published data such as that in Reference 6 in this chapter and Internet site 3 in Chapter 3 report values for stress concentration factors for axial normal stress, bending normal stress, and torsional shear stress. But values for vertical shearing stress are rarely reported. As an approximation, we will use the values for K_t for torsional shear stress when using these equations.

Design Normal Stress—Fatigue Loading

For the repeated, reversed bending in a shaft caused by transverse loads applied to the rotating shaft, the design stress is related to the endurance strength of the shaft material. The actual conditions under which the shaft is *manufactured* and *operated* should be considered when specifying the design stress.

Refer to the discussion in Section 5–4 for the method of computing the estimated actual endurance strength, s'_n, for use in shaft design. The process starts with using the graph in Figure 5–8 to determine the endurance strength as a function of the ultimate tensile strength of the material, adjusted for the surface finish. Equation (5–6) adjusts this value by applying four factors for the type of material, the type of stress, reliability, and the size of the cross section. When rotating steel shafts are being designed, the values for the material factor and the type of stress factor are both equal to 1.0. Use Table 5–2 for the reliability factor. Use Figure 5–9 or the equations in Table 5–3 to determine the size factor.

Stress concentration factors will be accounted for in the design equation to be developed later.

The design stress used here for parts of the shaft subjected to reversed bending stress due to the rotation of the shaft is

$$\sigma_d = s'_n/N \tag{12–17}$$

Note: Other factors not considered here could have adverse effect on the endurance strength of the shaft material, and therefore on the design stress. Examples are:

- Variation in peak stress levels above the nominal endurance strength for some periods of time, even quite short periods for high overstressing

- Temperatures above approximately 400°F (204°C)
- Vibration that induces variations in stress not included in the analysis
- Residual stresses: dangerous when tensile, possibly beneficial when compressive
- Case hardening of the shaft material, resulting in nonuniform strength distribution near the surface and changing the net endurance strength
- Interference fits of mating elements that introduce additional local stresses and stress concentrations
- Corrosion that may roughen the surface and decrease the effective endurance strength
- Thermal cycling over and above the applied stresses, changing the span of stress cycles
- Axial tensile or compressive stresses. (A later section addresses this case.)

Testing of actual components under realistic operating conditions is recommended when any of these conditions exist.

Design Factor, *N*

Refer to Section 5–7 for discussion of factors that affect the choice of design factor. We will typically use $N = 2.5$ to 3.0 in this book, indicating a moderate level of uncertainty about actual material strengths, loading conditions, and long-term environmental factors. Higher values are sometimes justified for critical designs without clear knowledge of actual conditions. Conversely, lower values can be used when extensive, reliable data are available.

12–6 SHAFTS IN BENDING AND TORSION ONLY

Examples of shafts subjected to bending and torsion only are those carrying spur gears, V-belt sheaves, or chain sprockets. The power being transmitted causes the torsion, and the transverse forces on the elements cause bending. In the general case, the transverse forces do not all act in the same plane. In such cases, the bending moment diagrams for two perpendicular planes are prepared first. Then the resultant bending moment at each point of interest is determined. The process will be illustrated in Example Problem 12–1.

A design equation is now developed based on the assumption that the bending stress in the shaft is repeated and reversed as the shaft rotates, but that the torsional shear stress is nearly uniform. The design equation is based on the principle shown graphically in Figure 12–10 in which the vertical axis is the ratio of the reversed bending stress to the endurance strength of the material. (See Reference 8.) The horizontal axis is the ratio of the torsional shear stress to the yield strength of the material in shear. The points having the value of 1.0 on these axes indicate impending failure in pure bending or pure torsion, respectively. Experimental data show that failure under combinations of bending and torsion roughly follows the

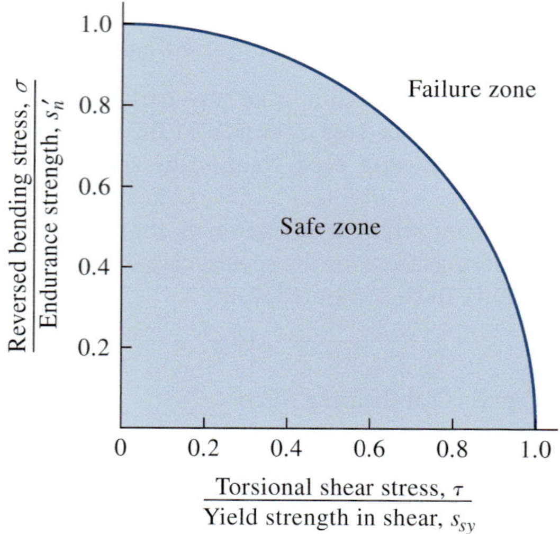

FIGURE 12–10 Basis for shaft design equation for repeated, reversed bending stress and steady torsional shear stress

curve connecting these two points, which obeys the following equation:

$$(\sigma/s_n')^2 + (\tau/s_{ys})^2 = 1 \qquad (12\text{–}18)$$

We will use $s_{ys} = s_y/\sqrt{3}$ for the distortion energy theory. Also, a design factor can be introduced to each term on the left side of the equation to yield an expression based on *design stresses:*

$$(N\sigma/s_n')^2 + (N\tau\sqrt{3}/s_y)^2 = 1$$

Now we can introduce a stress concentration factor for bending in the first term only, because this stress is repeated. No factor is needed for the torsional shear stress term because it is assumed to be steady, and stress concentrations have little or no effect on the failure potential. Then

$$(K_t N\sigma/s_n')^2 + (N\tau\sqrt{3}/s_y)^2 = 1 \qquad (12\text{–}19)$$

For rotating, solid, circular shafts, the bending stress due to a bending moment, M, is

$$\sigma = M/S \qquad (12\text{–}20)$$

where $S = \pi D^3/32$ is the rectangular section modulus. The torsional shear stress is

$$\tau = T/Z_p \qquad (12\text{–}21)$$

where $Z_p = \pi D^3/16$ is the polar section modulus.

Note that $Z_p = 2S$ and that, therefore,

$$\tau = T/(2S)$$

Substituting these relationships into Equation (12–19) gives

$$\left[\frac{K_t NM}{Ss_n'}\right]^2 + \left[\frac{NT\sqrt{3}}{2Ss_y}\right]^2 = 1 \qquad (12\text{–}22)$$

Now the terms N and S can be factored out, and the terms $\sqrt{3}$ and 2 can be brought outside the bracket in the torsion term:

$$\left[\frac{N}{S}\right]^2\left[\left[\frac{K_t M}{s_n'}\right]^2 + \frac{3}{4}\left[\frac{T}{s_y}\right]^2\right] = 1$$

We now take the square root of the entire equation:

$$\frac{N}{S}\sqrt{\left[\frac{K_t M}{s_n'}\right]^2 + \frac{3}{4}\left[\frac{T}{s_y}\right]^2} = 1$$

Let $S = \pi D^3/32$ for a solid circular shaft.

$$\frac{32N}{\pi D^3}\sqrt{\left[\frac{K_t M}{s_n'}\right]^2 + \frac{3}{4}\left[\frac{T}{s_y}\right]^2} = 1 \qquad \textbf{(12–23)}$$

Now we can solve for the diameter D:

Design Equation for Shaft Design

$$D = \left[\frac{32\,N}{\pi}\sqrt{\left[\frac{K_t M}{s_n'}\right]^2 + \frac{3}{4}\left[\frac{T}{s_y}\right]^2}\right]^{1/3} \qquad \textbf{(12–24)}$$

Equation (12–24) is used for shaft design in this book. It is compatible with the standard ANSI B106.1M-1985. (See Reference 1.) Note that Equation (12–24) can also be used for pure bending or pure torsion.

12–7 SHAFT DESIGN EXAMPLES—BENDING AND TORSION ONLY

Next we show two design examples for shaft design using the General Design Procedure given in Section 12–2. The general process must be adapted to the specific nature of each problem. Example Problems 12–1 and 12–2 carry some combinations of spur gears, belt sheaves, and chain sprockets that exert forces normal to the shaft to cause bending. The forces may act in any direction around the shaft depending on the placement of mating elements. Notice that none of these types of elements exerts forces parallel to the axis of the shaft tending to cause direct axial tension or compression.

Section 12–8 will modify this procedure for cases in which axial loads do exist, as with shafts carrying helical gears or wormgears.

Recommended Basic Sizes for Shafts

When mounting a commercially available element, of course, follow the manufacturer's recommendation for the basic size of the shaft and the tolerance.

In the U.S. Customary unit system, diameters are usually specified to be common fractions or their decimal equivalents. Appendix 2 lists the preferred basic sizes that you can use for dimensions over which you have control in decimal-inch, fractional-inch, and metric units. (See Reference 2.)

When commercially available unmounted bearings are to be used on a shaft, it is likely that their bores will be in metric dimensions. Typical sizes available and their decimal equivalents are listed in Table 14–3.

Design Example 12–1

Design the shaft shown in Figures 12–1 and 12–2. It is to be machined from SAE 1144 OQT 1000 steel. The shaft is part of the drive for a large blower system supplying air to a furnace. Gear A receives 200 hp from gear P. Gear C delivers the power to gear Q. The shaft rotates at 600 rpm. Use $D_A = 20.00$ in. and $D_C = 10.00$ in. Pressure angle $= \phi = 20°$

Solution

First determine the properties of the steel for the shaft. From Figure A4–2, $s_y = 83\,000$ psi, $s_u = 118\,000$ psi, and the percent elongation is 19%. Thus, the material has good ductility. Using Figure 5–8, we can estimate $s_n = 42\,000$ psi. For reversed bending, let $K_m = K_{st} = 1.0$.

A size factor should be applied to the endurance strength because the shaft will be quite large to be able to carry 200 hp. Although we do not know the actual size at this time, we might select $C_s = 0.75$ from Figure 5–9 as an estimate.

A reliability factor should also be specified. This is a design decision. For this problem, let's design for a reliability of 0.99 and use $C_R = 0.81$. Now we can compute the estimated actual endurance strength:

$$s_n' = s_n C_s C_R = (42\,000)(0.75)(0.81) = 25\,500 \text{ psi}$$

The design factor is taken to be $N = 2$. The blower is not expected to present any unusual shock or impact.

Now we can compute the torque in the shaft from Equation (12–1):

$$T = 63\,000(P)/n = 63\,000(200)/600 = 21\,000 \text{ lb}\cdot\text{in}$$

Note that only that part of the shaft from A to C is subjected to this torque. There is zero torque from the right of gear C over to bearing D. See Figure 12–11(d).

Forces on the Gears: Figure 12–11 shows the two pairs of gears with the forces acting *on gears* A *and* C *shown.* Observe that gear A is driven by gear P, and gear C drives gear Q. It is very important

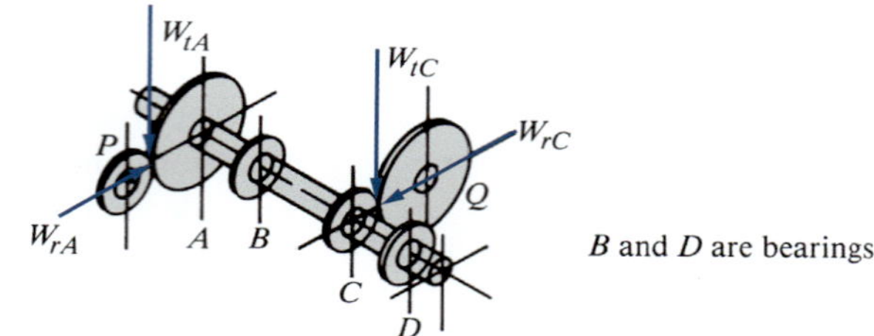

(*a*) Pictorial view of forces on gears *A* and *C*

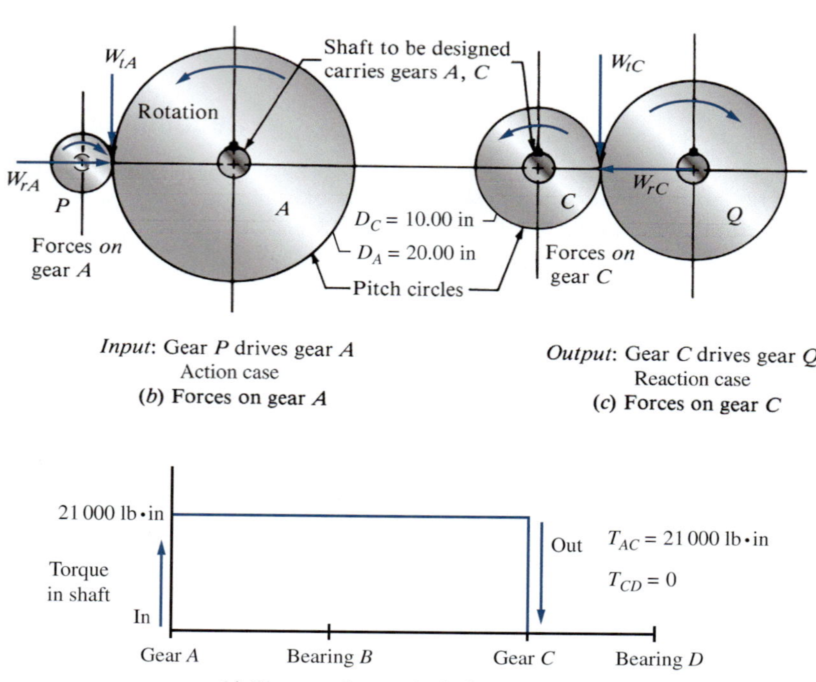

FIGURE 12–11 Forces on gears *A* and *C*

for the directions of these forces to be correct. The values of the forces are found from Equations (12–2) and (12–3).

$$W_{tA} = T_A/(D_A/2) = 21\ 000/(20/2) = 2100 \text{ lb} \downarrow$$

$$W_{rA} = W_{tA} \tan(\phi) = 2100 \tan(20°) = 764 \text{ lb} \rightarrow$$

$$W_{tC} = T_C/(D_C/2) = 21\ 000/(10/2) = 4200 \text{ lb} \downarrow$$

$$W_{rC} = W_{tC} \tan(\phi) = 4200 \tan(20°) = 1529 \text{ lb} \leftarrow$$

Forces on the Shaft: The next step is to show these forces on the shaft in their proper planes of action and in the proper direction. The reactions at the bearings are computed, and the shearing force and bending moment diagrams are prepared. The results are shown in Figure 12–12.

We continue the design by computing the minimum acceptable diameter of the shaft at several points along the shaft. We use the proposed design from Figure 12–2. At each point, we will observe the magnitude of torque and the bending moment that exist at the point, and we will estimate the value of any stress concentration factors. If more than one stress concentration exist in the vicinity of the point of interest, the larger value is used for design. This assumes that the geometric discontinuities themselves do not interact, which is good practice. For example, at point *A*, the keyseat should end well before the shoulder fillet begins.

1. ***Point A:*** Gear *A* produces torsion in the shaft from *A* and to the right. To the left of *A*, where there is a retaining ring, there are no forces, moments, or torques.

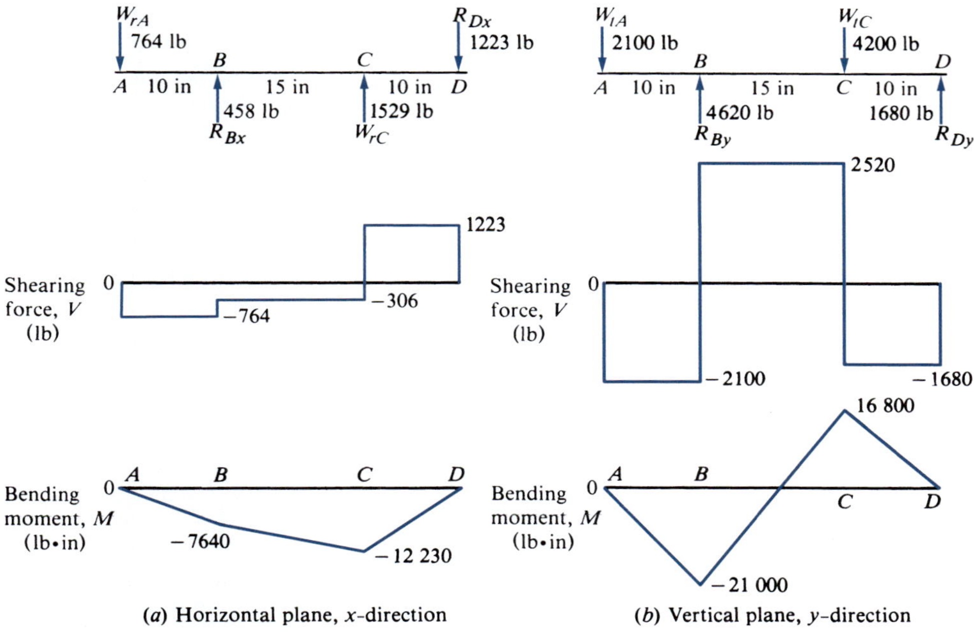

FIGURE 12-12 Load, shear, and moment diagrams for the shaft in Figure 12–10

The moment at A is zero because it is a free end of the shaft. Now we can use Equation (12–24) to compute the required diameter for the shaft at A, using only the torsion term.

$$D_1 = \left[\frac{32\ N}{\pi}\sqrt{\frac{3}{4}\left(\frac{T}{s_y}\right)^2}\right]^{1/3}$$

$$D_1 = \left[\frac{32(2)}{\pi}\sqrt{\frac{3}{4}\left(\frac{21\ 000}{83\ 000}\right)^2}\right]^{1/3} = 1.65\ \text{in}$$

2. ***Point B:*** Point B is the location of a bearing with a sharp fillet to the right of B and a well-rounded fillet to the left. It is desirable to make D_2 at least slightly smaller than D_3 at the bearing seat to permit the bearing to be slid easily onto the shaft up to the place where it is pressed to its final position. There is usually a light press fit between the bearing bore and the shaft seat. The torque in the shaft at B is

$$T_B = 21\ 000\ \text{lb}\cdot\text{in}$$

The bending moment at B is the resultant of the moment in the x- and y-planes from Figure 12–12:

$$M_B = \sqrt{M_{Bx}^2 + M_{By}^2} = \sqrt{(7640)^2 + (21\ 000)^2} = 22\ 350\ \text{lb}\cdot\text{in}$$

To the left of B (diameter D_2),

$$K_t = 1.5\ \text{(well-rounded fillet)}$$

Using Equation (12–24) because of the combined stress condition,

$$D_2 = \left[\left(\frac{32N}{\pi}\right)\sqrt{\left(\frac{K_t\ M}{s_n'}\right)^2 + \frac{3}{4}\left(\frac{T}{s_y}\right)^2}\right]^{1/3}$$

$$D_2 = \left[\frac{32(2)}{\pi}\sqrt{\left[\frac{1.5(22\ 350)}{25\ 500}\right]^2 + \frac{3}{4}\left[\frac{21\ 000}{83\ 000}\right]^2}\right]^{1/3} = 3.30\ \text{in} \qquad \textbf{(12-24a)}$$

At B and to the right of B (diameter D_3), everything is the same, except the value of $K_t = 2.5$ for the sharp fillet. Then

$$D_3 = \left[\frac{32(2)}{\pi}\sqrt{\left[\frac{2.5(22\ 350)}{25\ 500}\right]^2 + \frac{3}{4}\left[\frac{21\ 000}{83\ 000}\right]^2}\right]^{1/3} = 3.55\ \text{in}$$

Notice that D_4 will be larger than D_3 in order to provide a shoulder for the bearing. Therefore, it will be safe. Its actual diameter will be specified after we have completed the stress analysis and selected the bearing at B. The bearing manufacturer's catalog will specify the minimum acceptable diameter to the right of the bearing to provide a suitable shoulder against which to seat the bearing.

3. ***Point C:*** Point C is the location of gear C with a well-rounded fillet to the left, a profile keyseat at the gear, and a retaining ring groove to the right. The use of a well-rounded fillet at this point is actually a design decision that requires that the design of the bore of the gear accommodate a large fillet. Usually this means that a chamfer is produced at the ends of the bore. The bending moment at C is

$$M_C = \sqrt{M_{Cx}^2 + M_{Cy}^2} = \sqrt{(12\ 230)^2 + (16\ 800)^2} = 20\ 780\ \text{lb}\cdot\text{in}$$

To the left of C the torque of 21 000 lb·in exists with the profile keyseat giving $K_t = 2.0$. Then

$$D_5 = \left[\frac{32(2)}{\pi}\sqrt{\left[\frac{2.0(20\ 780)}{25\ 500}\right]^2 + \frac{3}{4}\left[\frac{21\ 000}{83\ 000}\right]^2}\right]^{1/3} = 3.22\ \text{in}$$

To the right of C there is no torque, but the ring groove suggests $K_t = 3.0$ for design, and there is reversed bending. We can use Equation (12–24) with $K_t = 3.0$, $M = 20\ 780$ lb·in and $T = 0$.

$$D_5 = \left[\frac{32(2)}{\pi}\sqrt{\left(\frac{(3.0)(20\ 780)}{25\ 500}\right)^2}\right]^{1/3} = 3.68\ \text{in}$$

Applying the ring groove factor of 1.06 raises the diameter to 3.90 in.

This value is higher than that computed for the left of C, so it governs the design at point C.

4. ***Point D:*** Point D is the seat for bearing D, and there is no torque or bending moment here. However, there is a vertical shearing force equal to the reaction at the bearing. Using the resultant of the x- and y-plane reactions, the shearing force is

$$V_D = \sqrt{(1223)^2 + (1680)^2} = 2078\ \text{lb}$$

We can use Equation (12–16) to compute the required diameter for the shaft at this point:

$$D = \sqrt{2.94\ K_t(V)N/s_n'} \qquad \textbf{(12–16)}$$

Referring to Figure 12–2, we see a sharp fillet near this point on the shaft. Then a stress concentration factor of 2.5 should be used:

$$D_6 = \sqrt{\frac{2.94(2.5)(2078)(2)}{25\ 500}} = 1.094\ \text{in}$$

This is very small compared to the other computed diameters, and it will usually be so. In reality, the diameter at D will probably be made much larger than this computed value because of the size of a reasonable bearing to carry the radial load of 2078 lb.

Summary The computed minimum required diameters for the various parts of the shaft in Figure 12–2 are as follows:

$$D_1 = 1.65\ \text{in}$$

$$D_2 = 3.30\ \text{in}$$

$$D_3 = 3.55\ \text{in}$$

$$D_5 = 3.90\ \text{in}$$

$$D_6 = 1.094\ \text{in}$$

Also, D_4 must be somewhat greater than 3.90 in in order to provide adequate shoulders for gear C and bearing B.

We now specify convenient decimal-inch dimensions for the six diameters. Choose the bearing seat dimensions from Table 14–3. Choose all other dimensions from Appendix 2. Table 12–1 shows one possible set of recommended diameters.

TABLE 12–1 Recommended Diameters

Mating part	Diameter number	Minimum diameter	Specified diameter (basic size)
	(from Design Example 12–1 and Figure 12–2)		
Gear	D_1	1.65 in	1.800 in
Nothing	D_2	3.30 in	3.400 in
Bearing	D_3	3.55 in	3.7402 in (95 mm)
Nothing	D_4	$>D_3$ or D_5	4.400 in
Gear	D_5	3.90 in	4.000 in
Bearing	D_6	1.094 in	1.7717 in (45 mm)

Diameters D_3 and D_6 are the decimal equivalents of the metric diameters of the inner races of bearings from Table 14–3. The procedures in Chapter 14 would have to be used to determine whether bearings having those diameters are suitable to carry the given radial loads. Also, D_4 would have to be checked to see whether it provides a sufficiently high shoulder against which to seat the bearing mounted at point B on the shaft. Then detailed specifications for fillet radii, lengths, keyseats, and retaining ring grooves would have to be defined. The actual values for stress concentration factors and the size factor should then be determined. Finally, the stress analysis should be repeated to ensure that the resulting design factor is acceptable. Equation (12–23) can be solved for N and evaluated for actual conditions.

Design Example 12–2

The shaft shown in Figure 12–13 receives 110 hp from a water turbine through a chain sprocket at point C. The gear pair at E delivers 80 hp to an electrical generator. The V-belt sheave at A delivers 30 hp to a bucket elevator that carries grain to an elevated hopper. The shaft rotates at 1700 rpm. The sprocket, sheave, and gear are located axially by retaining rings. The sheave and gear are keyed with sled runner keyseats, and there is a profile keyseat at the sprocket. Use SAE 1040 cold-drawn steel for the shaft. Compute the minimum acceptable diameters D_1 through D_7 as defined in Figure 12–13.

Solution

First, the material properties for the SAE 1040 cold-drawn steel are found from Appendix 3:

$$s_y = 71\,000 \text{ psi} \quad s_u = 80\,000 \text{ psi}$$

Then from Figure 5–8, $s_n = 30\,000$ psi. Let's design for a reliability of 0.99 and use $C_R = 0.81$. The shaft size should be moderately large, so we can assume $C_s = 0.85$ as a reasonable estimate. Then the modified endurance strength is

$$s'_n = s_n C_s C_R = (30\,000)(0.85)(0.81) = 20\,650 \text{ psi}$$

This application is fairly smooth: a turbine drive and a generator and a conveyor at the output points. A design factor of $N = 2$ should be satisfactory.

Torque Distribution in the Shaft: Recalling that all of the power comes into the shaft at C, we can then observe that 30 hp is delivered down the shaft from C to the sheave at A. Also, 80 hp is delivered down the shaft from C to the gear at E. From these observations, the torque in the shaft can be computed:

$$T_A = T_{AC} = 63\,000(30)/1700 = 1112 \text{ lb}\cdot\text{in} \qquad \text{from } A \text{ to } C \text{ in shaft}$$

$$T_E = T_{CE} = 63\,000(80)/1700 = 2965 \text{ lb}\cdot\text{in} \qquad \text{from } C \text{ to } E \text{ in shaft}$$

Figure 12–14 shows a plot of the torque distribution *in the shaft* superimposed on the sketch of the shaft. When designing the shaft at C, we will use 2965 lb·in *at C and to the right*, but we can use 1112 lb·in *to the left of C*. Notice that no part of the shaft is subjected to the full 110 hp that comes into the sprocket at C. The power splits into two parts as it enters the shaft. When analyzing the sprocket itself, we must use the full 110 hp and the corresponding torque:

$$T_C = 63\,000(110)/1700 = 4076 \text{ lb}\cdot\text{in} \qquad \text{(torque on the sprocket)}$$

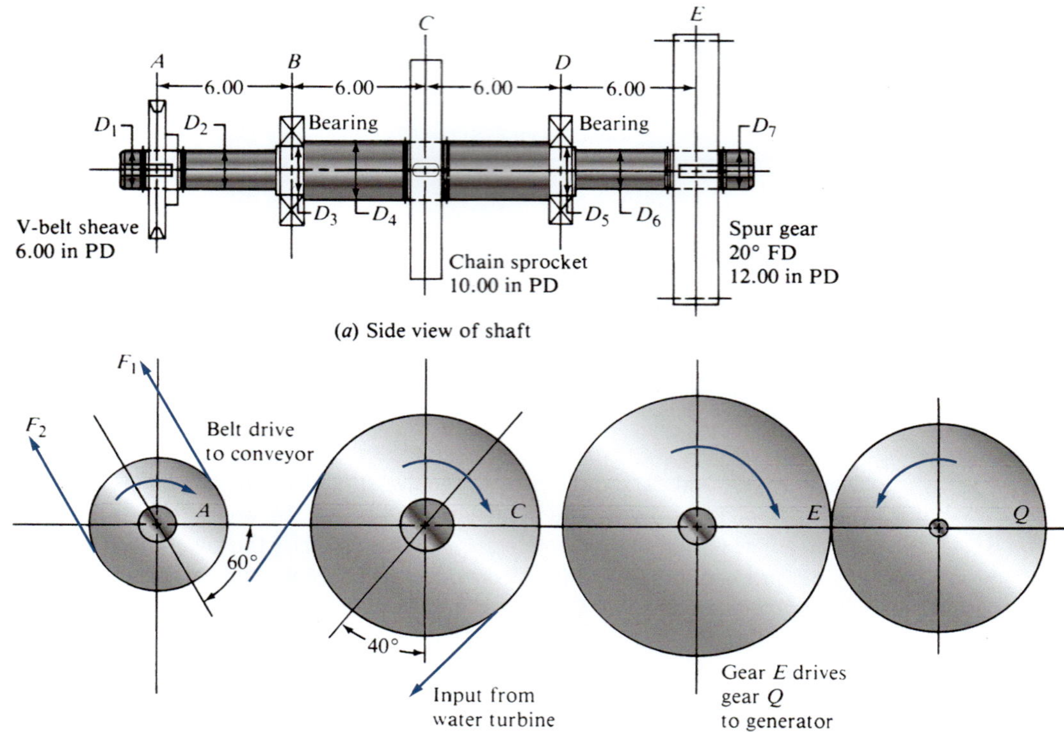

FIGURE 12–13 Shaft design for example problem 12–2

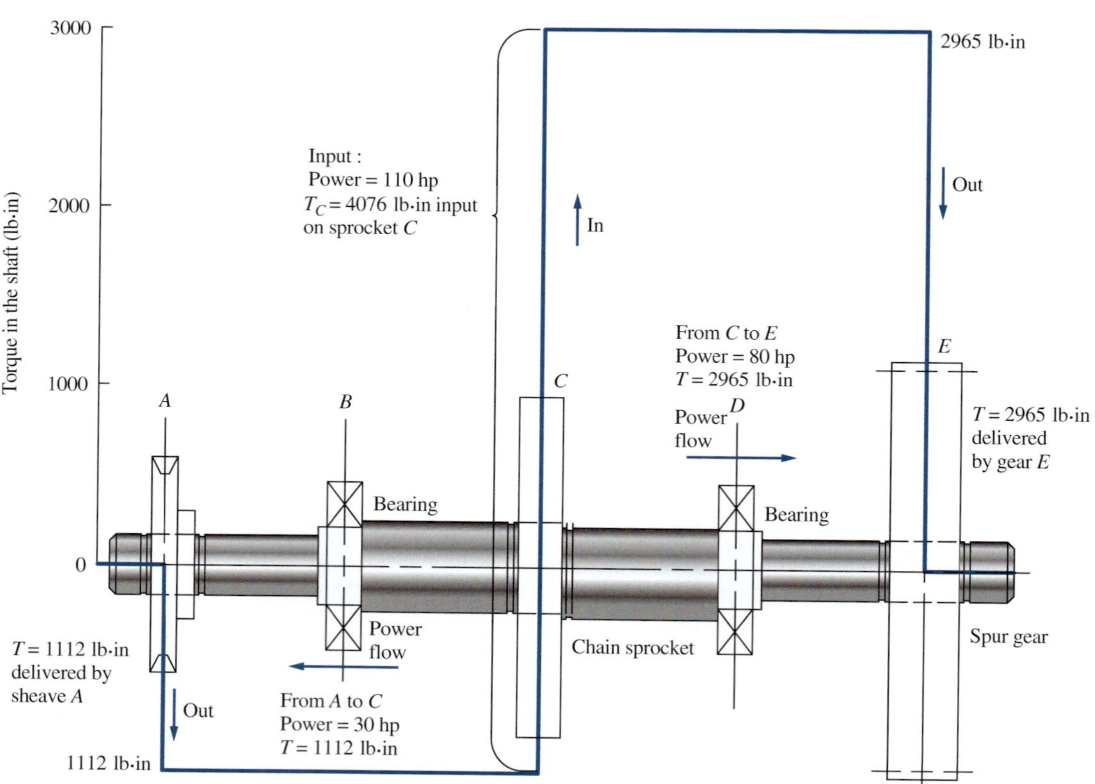

FIGURE 12–14 Torque distribution in the shaft

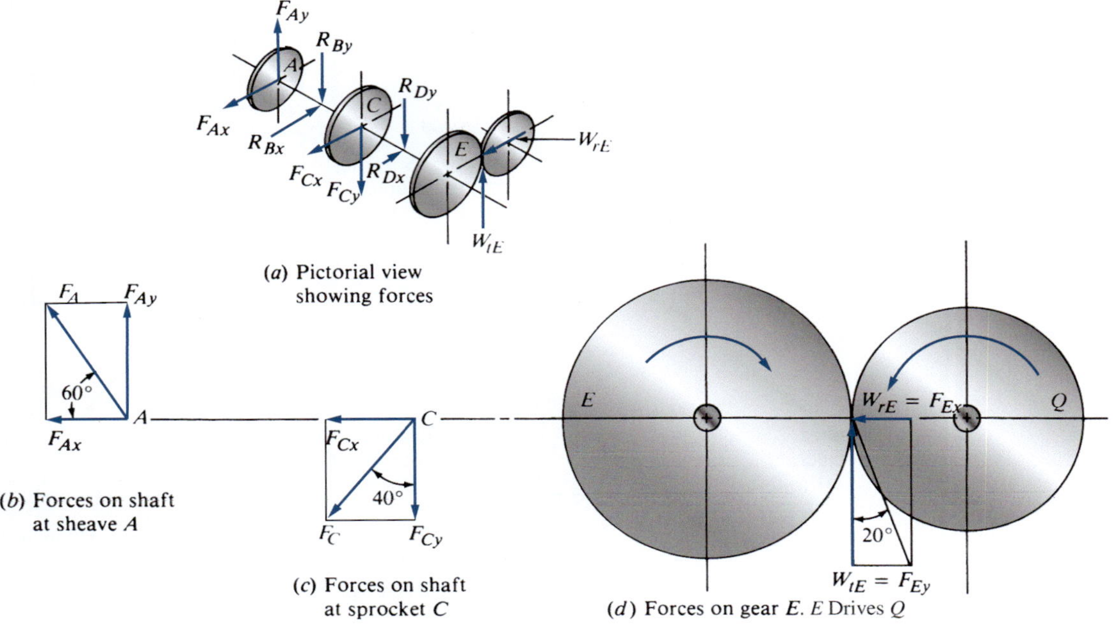

FIGURE 12–15 Forces resolved into x- and y-components

Forces: We will compute the forces at each element separately and show the component forces that act in the vertical and horizontal planes, as in Design Example 12–1. Figure 12–15 shows the directions of the applied forces and their components for each element.

1. ***Forces on sheave A:*** Use Equations (12–7), (12–8), and (12–12):

$$F_N = F_1 - F_2 = T_A/(D_A/2) = 1112/3.0 = 371 \text{ lb} \quad \text{(net driving force)}$$

$$F_A = 1.5\ F_N = 1.5(371) = 556 \text{ lb} \quad \text{(bending force)}$$

The bending force acts upward and to the left at an angle of 60° from the horizontal. As shown in Figure 12–15, the components of the bending force are

$$F_{Ax} = F_A\cos(60°) = (556)\cos(60°) = 278 \text{ lb} \leftarrow \text{(toward the left)}$$

$$F_{Ay} = F_A\sin(60°) = (556)\sin(60°) = 482 \text{ lb} \uparrow \text{(upward)}$$

2. ***Forces on sprocket C:*** Use Equation (12–6):

$$F_C = T_C/(D_C/2) = 4076/5.0 = 815 \text{ lb}$$

This is the bending load on the shaft. The components are

$$F_{Cx} = F_C\sin(40°) = (815)\sin(40°) = 524 \text{ lb} \leftarrow \text{(to the left)}$$

$$F_{Cy} = F_C\cos(40°) = (815)\cos(40°) = 624 \text{ lb} \downarrow \text{(downward)}$$

3. ***Forces on gear E:*** The transmitted load is found from Equation (12–2), and the radial load from Equation (12–3). The directions are shown in Figure 12–15.

$$F_{Ey} = W_{tE} = T_E/(D_E/2) = 2965/6.0 = 494 \text{ lb} \uparrow \text{(upward)}$$

$$F_{Ex} = W_{rE} = W_{tE}\tan(\phi) = (494)\tan(20°) = 180 \text{ lb} \leftarrow \text{(to the left)}$$

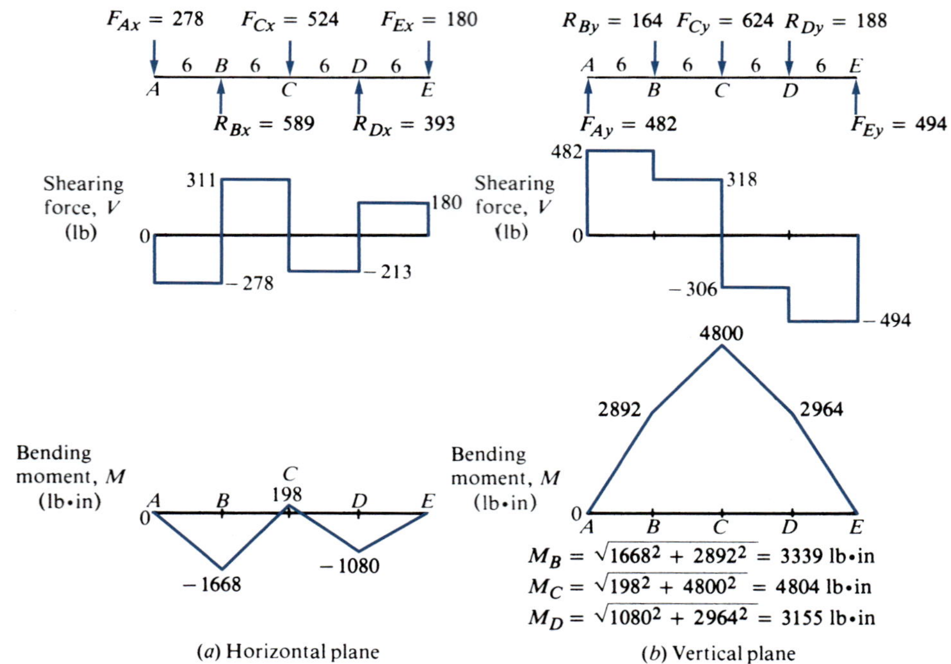

FIGURE 12–16 Load, shear, and moment diagrams

Load, Shear, and Moment Diagrams: Figure 12–16 shows the forces acting on the shaft at each element, the reactions at the bearings, and the shearing force and bending moment diagrams for both the horizontal (x-) and vertical (y-) planes. In the figure, the computations of the resultant bending moment at points B, C, and D are also shown.

Design of the Shaft: We will use Equation (12–24) to determine the minimum acceptable diameter of the shaft at each point of interest. Because the equation requires a fairly large number of individual operations, and because we will be using it at least seven times, it may be desirable to write a computer program just for that operation. Or the use of a spreadsheet would be nearly ideal. See Section 12–9. Note that we can use Equation (12–24) even though there is only torsion or only bending by entering zero for the missing value.

Equation (12–24) is repeated here for reference. In the solution below, the data used for each design point are listed. You may want to verify the calculations for the required minimum diameters. The design factor of $N = 2$ has been used.

$$D = \left[\frac{32N}{\pi}\sqrt{\left(\frac{K_t M}{s_n'}\right)^2 + \frac{3}{4}\left(\frac{T}{s_y}\right)^2}\right]^{1/3}$$

1. ***Point A:*** Torque = 1112 lb · in; moment = 0. The sheave is located with retaining rings. Because the torque is steady, we will not use a stress concentration factor in this calculation, as discussed in Section 12–4. But then we will find the nominal diameter at the groove to the right of A by increasing the computed result by about 6%. The result should be conservative for typical groove geometries.

   Using Equation (12–24), $D_1 = 0.65$ in. Increasing this by 6% gives $D_1 = 0.69$ in.

2. ***To the left of point B:*** This is the relief diameter leading up to the bearing seat. A well-rounded fillet radius will be specified for the place where D_2 joins D_3. Thus,

   $$\text{Torque} = 1112 \text{ lb}\cdot\text{in} \quad \text{Moment} = 3339 \text{ lb}\cdot\text{in} \quad K_t = 1.5$$

   Then $D_2 = 1.70$ in.

3. ***At point B and to the right:*** This is the bearing seat with a shoulder fillet at the right, requiring a fairly sharp fillet:

   $$\text{Torque} = 1112 \text{ lb}\cdot\text{in} \quad \text{Moment} = 3339 \text{ lb}\cdot\text{in} \quad K_t = 2.5$$

   Then $D_3 = 2.02$ in.

TABLE 12–2 Specification of Values

Mating part	Diameter number	Minimum diameter	Specified diameter	
			Fraction	Decimal
Sheave	D_1	0.69	$1\frac{3}{4}$	1.750
Nothing	D_2	1.70	$1\frac{3}{4}$	1.750
Bearing	D_3	2.02	$2\frac{1}{4}$	2.250
Sprocket	D_4	2.57	$2\frac{3}{4}$	2.750
Bearing	D_5	1.98	2	2.000
Nothing	D_6	1.68	$1\frac{3}{4}$	1.750
Gear	D_7	0.96	$1\frac{3}{4}$	1.750

4. ***At point C:*** It is planned that the diameter be the same all the way from the right of bearing *B* to the left of bearing *D*. The worst condition is at the right of *C*, where there is a ring groove and the larger torque value is

$$\text{Torque} = 2965 \text{ lb}\cdot\text{in} \quad \text{Moment} = 4804 \text{ lb}\cdot\text{in} \quad K_t = 3.0$$

Then $D_4 = 2.57$ in after applying the ring groove factor of 1.06.

5. ***At point D and to the left:*** This is a bearing seat similar to that at *B*:

$$\text{Torque} = 2965 \text{ lb}\cdot\text{in} \quad \text{Moment} = 3155 \text{ lb}\cdot\text{in} \quad K_t = 2.5$$

Then $D_5 = 1.98$ in.

6. ***To the right of point D:*** This is a relief diameter similar to D_2:

$$\text{Torque} = 2965 \text{ lb}\cdot\text{in} \quad \text{Moment} = 3155 \text{ lb}\cdot\text{in} \quad K_t = 1.5$$

Then $D_6 = 1.68$ in.

7. ***At point E:*** The gear is mounted with retaining rings on each side:

$$\text{Torque} = 2965 \text{ lb}\cdot\text{in} \quad \text{Moment} = 0 \quad K_t = 3.0$$

Then $D_7 = 0.96 \text{lb}\cdot\text{in}$ after applying the ring groove factor of 1.06.

Summary with Convenient Values Specified

Using Appendix 2, we specify convenient fractions at all places, including bearing seats (see Table 12–2). It is assumed that inch bearings of the pillow block type will be used.

We decided to make the diameters D_1, D_2, D_6, and D_7 the same to minimize machining and to provide a little extra safety factor at the ring grooves. Again the bearing bore sizes would have to be checked against the load rating of the bearings. The size of D_4 would have to be checked to see that it provides a sufficient shoulder for the bearings at *B* and *D*.

The size factor and the stress concentration factors must also be checked.

12–8 SHAFT DESIGN EXAMPLE—BENDING AND TORSION WITH AXIAL FORCES

Example Problem 12–3 follows that shows a shaft carrying a chain sprocket along with a wormgear. One of the components of the forces acting on the wormgear acts parallel to the axis of the shaft, creating a direct axial stress component in the shaft to the left of point D. We describe one method for handling this situation within the development of the stress analysis for that part of the shaft in the problem solution.

Except for the presence of the axial stress in the shaft, the design procedure is the same as that for Design Examples 12–1 and 12–2. For this reason, much of the detailed manipulation of formulas is omitted to condense the presentation of the analysis.

Example Problem 12–3 A wormgear is mounted at the right end of the shaft as shown in Figure 12–17. The gear has the same design as that discussed in Example Problem 10–7 and delivers 6.68 hp to the shaft at a speed of 101 rpm. The magnitudes and directions of the forces on the gear are given in the figure. Notice that there is a system of three orthogonal forces acting on the gear. The power is transmitted by a chain sprocket at B to drive a conveyor removing cast iron chips from a machining system. Design the shaft.

Solution The torque on the shaft from the wormgear at point D to the chain sprocket at B is

$$T_B = T_D = T_{BD} = W_{tG}(D_G/2) = 962(4.333) = 4168\ \text{lb}\cdot\text{in}$$

also, $T_{AB} = 0$. The torque diagram is drawn in Figure 12–17.

The force on the chain sprocket is

$$F_c = T_B/(D_s/2) = 4168/(6.71/2) = 1242\ \text{lb}$$

This force acts horizontally toward the right as viewed from the end of the shaft.

Bending Moment Diagrams: Figure 12–18 shows the forces acting on the shaft in both the vertical and the horizontal planes and the corresponding shearing force and bending moment diagrams. You should review these diagrams, especially that for the vertical plane, to grasp the effect of the axial force of 265 lb. Notice that because it acts above the shaft, it creates a bending moment at the end of the shaft of

$$M_D = W_{XG} \cdot D_{G/2} = (265\ \text{lb})(8.667\ \text{in}/2)$$

$$M_D = 1148\ \text{lb.in}$$

It also affects the reactions at the bearings. The resultant bending moments at B, C, and D are also shown in the figure.

In the design of the entire system, we must decide which bearing will resist the axial force. For this problem, let's specify that the bearing at C will transfer the axial thrust force to the housing. This decision places a compressive stress in the shaft from C to D and requires that means be provided to transmit the axial force from the wormgear to the bearing. The geometry proposed in Figure 12–17 accomplishes this, and it will be adopted for the following stress analysis. The

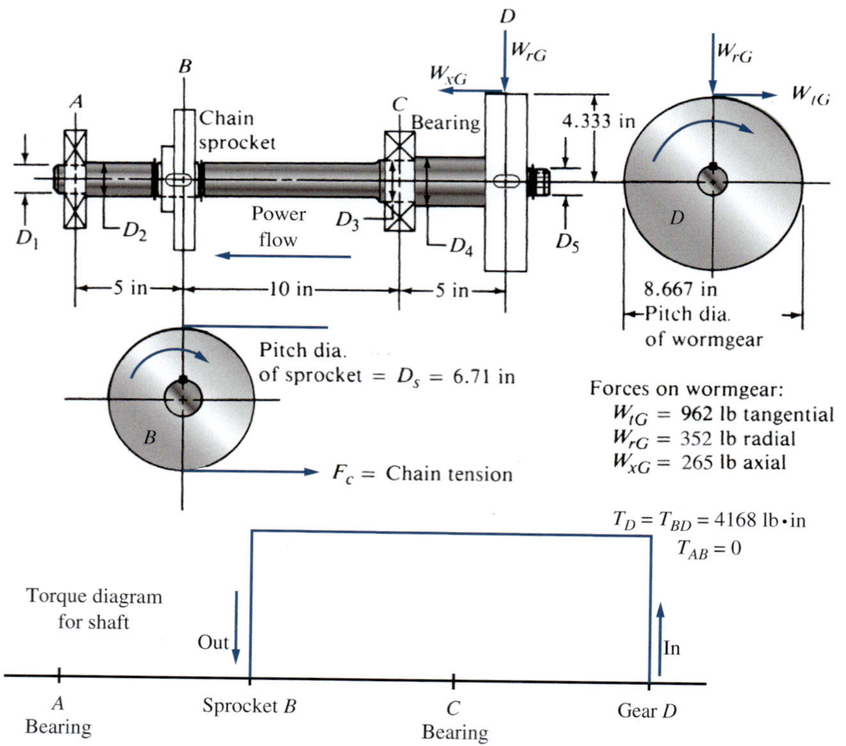

FIGURE 12–17 Shaft design

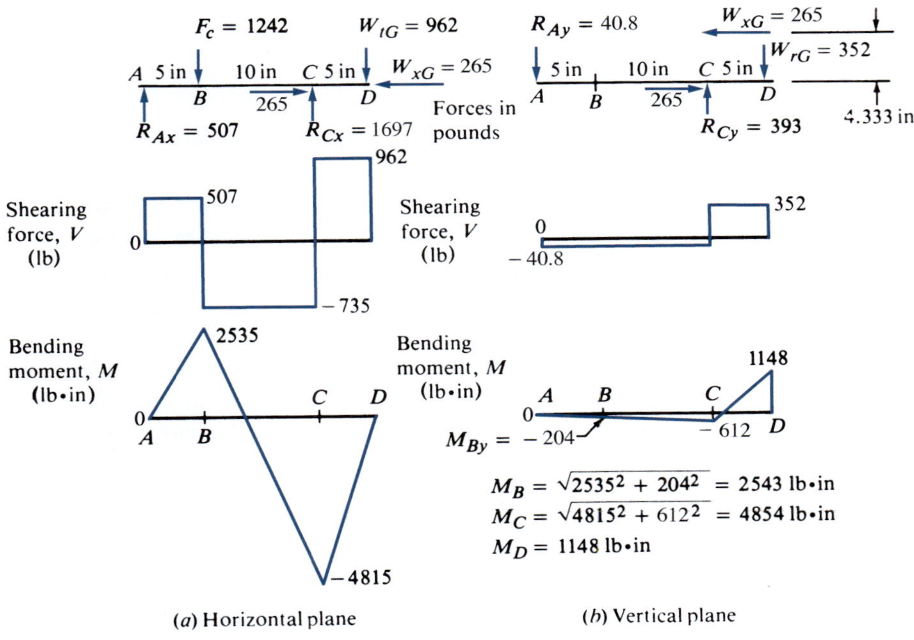

FIGURE 12–18 Load, shear, and moment diagrams for the shaft in Figure 12–16

procedures are the same as those used in Design Examples 12–1 and 12–2 and only summary results will be shown. The consideration of the axial compressive stress is discussed, along with the computations at point *C* on the shaft.

Material Selection and Design Strengths: A medium-carbon steel with good ductility and a fairly high strength is desired for this demanding application. We will use SAE 1340 OQT 1000 (Appendix 3), having an ultimate strength of 144 000 psi, a yield strength of 132 000 psi, and a 17% elongation. From Figure 5–8 we estimate s_n = 50 000 psi. Let's use an initial size factor of 0.80 and a reliability factor of 0.81. Then

$$s'_n = (50\,000\text{ psi})(0.80)(0.81) = 32\,400\text{ psi}$$

Because the use of the conveyor is expected to be rough, we will use a design factor of $N = 3$, higher than average.

Except at point *A*, where only a vertical shear stress exists, the computation of the minimum required diameter is done using Equation (12–24).

1. ***Point A:*** The left bearing mounts at point *A*, carrying the radial reaction force only, which acts as a vertical shearing force in the shaft. There is no torque or bending moment here.

   The vertical shearing force is

   $$V = \sqrt{R_{Ax}^2 + R_{Ay}^2} = \sqrt{(507)^2 + (40.8)^2} = 509\text{ lb}$$

   We can use Equation (12–16) to compute the required diameter for the shaft at this point:

   $$D = \sqrt{2.94\,K_t(V)N/s'_n} \qquad \textbf{(12–16)}$$

   Referring to Figure 12–17, we see a sharp fillet near this point on the shaft. Then a stress concentration factor of 2.5 should be used:

   $$D = \sqrt{\frac{2.94(2.5)(509)(3)}{32\,400}} = 0.588\text{ in}$$

   As seen before, this is quite small, and the final specified diameter will probably be larger, depending on the bearing selected.

2. ***Point B:*** The chain sprocket mounts at point *B*, and it is located axially by retaining rings on both sides. The critical point is at the right of the sprocket at the ring groove, where T = 4168 lb · in, M = 2543 lb · in, and K_t = 3.0 for bending.

The computed minimum diameter required is $D_2 = 1.93$ in at the base of the groove. We should increase this by approximately 6%, as discussed in Section 12–3. Then

$$D_2 = 1.06(1.93 \text{ in}) = 2.05 \text{ in}$$

3. ***To the left of point C:*** This is the relief diameter for the bearing seat. The diameter here will be specified to be the same as that at *B*, but different conditions occur: Torque = 4168 lb·in, $M = 4854$ lb·in, and $K_t = 1.5$ for the well-rounded fillet for bending only. The required diameter is 1.91 in. Because this is smaller than that at *B,* the previous calculation will govern.
4. ***At point C and to the right:*** The bearing will seat here, and it is assumed that the fillet will be rather sharp. Thus, $T = 4168$ lb·in, $M = 4854$ lb·in, and $K_t = 2.5$ for bending only. The required diameter is $D_3 = 2.26$ in.

   The axial thrust load acts between points *C* and *D*. The inclusion of this load in the computations would greatly complicate the solution for the required diameters. In most cases, the axial normal stress is relatively small compared with the bending stress. Also, the fact that the stress is compressive improves the fatigue performance of the shaft, because fatigue failures normally initiate at points of tensile stress. For these reasons, the axial stress is ignored in these calculations. The computed diameters are also interpreted as nominal minimum diameters, and the final selected diameter is larger than the minimum. This, too, tends to make the shaft safe even when there is an added axial load. When in doubt, or when a relatively high axial tensile stress is encountered, the methods of Chapter 5 should be applied. Long shafts in compression should also be checked for buckling.
5. ***Point D:*** The wormgear mounts at point *D*. We will specify that a well-rounded fillet will be placed to the left of *D* and that there will be a sled runner keyseat. Thus, $T = 4168$ lb·in, $M = 1148$ lb·in, and $K_t = 1.6$ for bending only. The computed required diameter is $D_5 = 1.24$ in. Notice that D_4 must be greater than either D_3 or D_5 because it provides the means to transfer the thrust load from the wormgear to the inner race of the bearing at *C*.

Summary and Selection of Convenient Diameters Table 12–3 presents a summary of the required diameters and the specified diameters for all parts of the shaft in this design example. See Figure 12–17 for the locations of the five diameters. For this application we have chosen to use fractional-inch dimensions from Appendix 2 except at the bearing seats, where the use of metric bearing bores from Table 14–3 are selected.

Comment As a partial check on the validity of the design decision to complete the calculations for shaft diameters while avoiding the axial stress between points *C* and *D* due to the axial force from the wormgear, let's now compute the magnitude of the axial compressive stress due to the 265 lb force. The shaft size between points *C* and *D* is $D_4 = 3.00$ in. Then,

$$\text{Axial stress} = \sigma_{CD} = -W_{xG}/A_4$$

$$A_4 = \pi D_4^2/4 = \pi(3.00 \text{ in})^2/4 = 7.069 \text{ in}^2$$

$$\sigma_{CD} = -265 \text{ lb}/7.069 \text{ in}^2 = -37.5 \text{ psi Compression}$$

Because of this rather low stress level, it seems reasonable to conclude that avoiding the axial compressive stress in the initial design analysis was warranted.

TABLE 12–3 Summary of Shaft Diameters

Mating part	Diameter number	Minimum diameter	Specified diameter	
			Fraction (metric)	Decimal
Bearing *A*	D_1	0.59 in	(35 mm)	1.3780 in
Sprocket *B*	D_2	2.05 in	$2\frac{1}{4}$ in	2.250 in
Bearing *C*	D_3	2.26 in	(65 mm)	2.5591 in
(Shoulder)	D_4	$>D_3$	3 in	3.000 in
Wormgear *D*	D_5	1.24 in	$1\frac{1}{2}$ in	1.500 in

12–9 SPREADSHEET AID FOR SHAFT DESIGN

A spreadsheet is useful to organize the data required to compute the minimum required shaft diameter at various points along a shaft and to complete the calculation using Equations (12–16) and (12–24). Note that Equation (12–24) can be used for bending only, torsion only, or the combination of bending and torsion.

Figure 12–19 shows a typical example, using U.S. Customary units, for data from Design Example 12–1. Describe the application in the upper panel for future reference. Then complete the following steps:

- Enter the shaft material specification along with its ultimate and yield strength properties found from tables in the appendices.
- Find the basic endurance strength from Figure 5–8 considering the ultimate tensile strength and the manner of production (ground, machined, etc.).
- Enter values for the size factor and the reliability factor. The spreadsheet then computes the modified endurance strength, s'_n.
- Enter the stress concentration factor for the point of interest.
- Enter the design factor.
- After analysis such as that shown in Design Example 12–1, enter the torque and the components of the bending moment in the x- and y-planes that exist at the point of interest along the shaft. The spreadsheet computes the combined bending moment.
- Enter the components of the vertical shearing force in the x- and y-planes. The spreadsheet computes the combined shearing force.

The minimum acceptable shaft diameters from both Equation (12–16) (vertical shear only) and Equation (12–24) (torsion and/or bending) are computed. You must observe which required diameter is larger.

DESIGN OF SHAFTS			
Application:	*Design Example 12–1, Drive for a Blower System* *Diameter D_3—To right of point B—Bending and torsion*		
This design aid computes the minimum acceptable diameter using Equation (12–24) for shafts subjected to steady torsion and/or rotating bending. Equation (12–16) is used when only vertical shear stress is present.			
Input Data	***(Insert values in italics.)***		
Shaft material specification:	*SAE 1144 OQT 1000 Steel*		
Tensile strength:	$s_u =$	*118 000 psi*	
Yield strength:	$s_y =$	*83 000 psi*	
Basic endurance strength:	$s_n =$	*42 000 psi*	From Figure 5–8
Size factor:	$C_s =$	*0.75*	From Figure 5–9
Reliability factor:	$C_R =$	*0.81*	From Table 5–2
Modified endurance strength:	$s'_n =$	25 515 psi	***Computed***
Stress concentration factor:	$K_t =$	*2.5 Sharp fillet*	
Design factor:	$N =$	*2* Nominal N=2	
Shaft Loading Data: Bending and Torsion			
Bending moment components:	$M_x =$	*21 000 lb·in*	*M_y=7640 lb·in*
Combined bending moment:	$M =$	22 347 lb·in	***Computed***
Torque:	$T =$	*21 000 lb·in*	
Minimum shaft diameter:	$D =$	**3.55 in**	***Computed from Eq. (12–24)***
Shaft Loading Data: Vertical Shearing Force Only			
Shearing force components:	$V_x =$	*764 lb*	*V_y=2520 lb*
Combined shearing force:	$V =$	2633 lb	***Computed***
Minimum shaft diameter:	$D =$	**1.232 in**	***Computed from Eq. (12–16)***

FIGURE 12–19 Spreadsheet aid for shaft design

12–10 SHAFT RIGIDITY AND DYNAMIC CONSIDERATIONS

The design processes outlined in this chapter thus far have concentrated on stress analysis and ensuring that the shaft is safe with regard to the bending and torsional shear stresses imposed on the shaft by power-transmitting elements. Rigidity of the shaft is also a major concern for several reasons:

1. Excessive radial deflection of the shaft may cause active elements to become misaligned, resulting in poor performance or accelerated wear. For example, the center distance between shafts carrying precision gears should not vary more than approximately 0.005 in (0.13 mm) from the theoretical dimension. Improper meshing of the gear teeth would occur and actual bending and contact stresses could be significantly higher than those predicted in the design for the gears.
2. Guidelines are reported in Section 20–2 for the recommended limits for bending and torsional deflection of a shaft according to its intended precision.
3. Shaft deflection is also an important contributor to a tendency for shafts to vibrate as they rotate. A flexible shaft will oscillate in both bending and torsional modes, causing movements that are greater than the static deflections due just to gravity and applied loads and torques. A long slender shaft tends to whip and whirl with relatively large deformations from its theoretical straight axis.
4. The shaft itself and the elements mounted on it should be balanced. Any amount of unbalance causes centrifugal forces to be created that rotate with the shaft. Large unbalances and high rotating speeds may create unacceptable force levels and shaking of the rotating system. An example with which you may be familiar is an automobile wheel that is "out of balance." You can actually feel the vibration through the steering wheel as you drive. Having the tire and wheel balanced reduces the vibration to acceptable levels.
5. The dynamic behavior of the shaft may become dangerously destructive if it is operated near its *critical speed.* At critical speed, the system is in resonance, the deflection of the shaft continues to increase virtually without bound, and it will eventually self-destruct.

Critical speeds of typical machinery shaft designs are several thousand revolutions per minute. However, many variables are involved. Long shafts used for vehicular driveshafts, power screws, or agitators may have much lower critical speeds and they must be checked. It is typical to keep operating speeds to at least 20% below or above the critical speed. When operating above the critical speed, it is necessary to accelerate the shaft rapidly through the critical speed because it takes time for dangerous oscillations to develop.

Analysis to determine the critical speed is complex, and computer programs are available to assist in the calculations. See Internet site 1. The objective is to determine the natural frequency of the shaft while carrying the static weight of elements such as gears, sprockets, and pulleys. The stiffness of bearings is also a factor. One fundamental expression for natural frequency, ω_n, is,

$$\omega_n = \sqrt{k/m}$$

where k is the stiffness of the shaft and m is its mass. It is desirable to have a high critical speed, well above operating speed, so stiffness should be high and mass low. The primary variables over which a designer has control are the material and its modulus of elasticity, E, its density, ρ, the shaft diameter, D, and the shaft length, L. The following functional relationship can help to understand the influence of each of these variables:

$$\omega_n \propto (D/L^2)\sqrt{E/\rho}$$

where the symbol $\propto$ indicates proportionality among the variables. Using this function as a guide, the following actions can reduce potential problems with deflections or critical speeds.

1. Making the shaft more rigid can prevent undesirable dynamic behavior.
2. Larger shaft diameters add rigidity.
3. Shorter shaft lengths reduce deflections and lower critical speeds.
4. Placing active elements on the shaft close to support bearings is recommended.
5. Reducing the weight of elements carried by the shaft reduces static deflection and lowers critical speed.
6. Selection of a material for the shaft with a high ratio of E/ρ is desirable. While most metals have similar ratios, that for composite materials is typically high. For example, long driveshafts for vehicles that must operate at high speeds are frequently made from hollow tubular composite materials using carbon fibers. See Internet sites 2 and 3.
7. Bearings should have a high stiffness in terms of radial deflection as a function of load.
8. Mountings for bearings and housings should be designed with high rigidity.
9. References 3, 4, 5, and 7 provide additional information and analytical methods for estimating critical speeds. References 9–11 discuss machinery vibration.

Internet site 4 includes a section on critical speeds that can be used for calculating approximate values for several different support conditions and loading patterns. Internet sites 5 and 6 describe two software packages that contain beam analysis programs that can be used to calculate beam deflections. Included with this book, the MDESIGN software also contains a shaft analysis software module.

12–11 FLEXIBLE SHAFTS

At times it is desirable to transmit rotational motion and power between two points that are not aligned with each other. Flexible shafts can be used in such situations to couple a driver, such as a motor, to a driven device along a curved or dynamically moving path. The flexibility allows the driven

TABLE 12–4 Sampling of Operating Torque Capacity for Flexible Shafts

Data for shafts of medium stiffness operating at 12 in (305 mm) radius

Shaft Diameter		Torque capacity	
(in)	(mm)	(lb·in)	(N·m)
0.127	3.23	1.7	0.19
0.183	4.65	4.3	0.49
0.245	6.22	20	2.26
0.304	7.72	33	3.73
0.495	12.57	86	9.72
0.740	18.80	265	29.95
0.990	25.15	386	43.62

Data adapted from Internet site 7.

point to be offset from the driving point in either parallel or angular fashion. Unidirectional shafts are used for power transmission in applications such as automation systems, industrial machinery, agricultural equipment, aircraft actuators, seat adjusters, medical devices, dentistry, speedometers, woodworking, and jeweler's tools. Bidirectional flexible shafts are used in remote control, valve actuation, and safety devices. They can often provide more design flexibility than other design options such as right-angle gear drives or universal joints. The offset between the driver and driven machine can be as much as 180° provided that the bend radius is above a specified minimum. Torque capacity increases with increasing bend radius. Efficiencies are in the range from 90% to 95%.

The construction consists of a flexible cable-like core connected to fittings on each end. The fittings facilitate connection with the driving and driven machines. The casing protects nearby equipment or people from contact with the rotating core. Designs are available for transmitting power in either direction, clockwise or counterclockwise. Some will operate bidirectionally.

See Internet sites 7–9 for additional information about commercially available flexible shafts. Table 12–4 gives a sampling of sizes and ratings from Internet site 7. Many more sizes are available.

REFERENCES

1. American Society of Mechanical Engineers. ANSI Standard B106.1M-1985. *Design of Transmission Shafting*. New York: American Society of Mechanical Engineers, 1985.
2. American Society of Mechanical Engineers. ANSI Standard B4.1-67. *Preferred Limits and Fits for Cylindrical Parts*. New York: American Society of Mechanical Engineers, 1994.
3. Avallone, E., and T. Baumeister, Ali Sadegh. *Marks' Standard Handbook for Mechanical Engineers*. 11th ed. New York: McGraw-Hill, 2007.
4. Piersol, Allan G., and Thomas Paez (Editors). *Harris' Shock and Vibration Handbook*. 6th ed. New York: McGraw-Hill, 2010.
5. Oberg, Erik, et al. *Machinery's Handbook*. 28th ed. New York: Industrial Press, 2008.
6. Pilkey, Walter D., and Deborah F. Pilkey, *Peterson's Stress Concentration Factors*. 3rd ed. New York: John Wiley & Sons, 2008.
7. Budynas, R. G., and K. J. Nisbett, *Shigley's Mechanical Engineering Design*, 9th ed. New York: McGraw-Hill, 2011.
8. Soderberg, C. R. "Working Stresses." *Journal of Applied Mechanics* 57 (1935): A–106.
9. Wowk, Victor. *Machinery Vibration: Alignment*. New York: McGraw-Hill, 2000.
10. Wowk, Victor. *Machinery Vibration: Balancing*. New York: McGraw-Hill, 1998.
11. Wowk, Victor. *Machinery Vibration: Measurement and Analysis*. New York: McGraw-Hill, 1991.
12. American Society of Mechanical Engineers. ANSI Standard B4.2-1978. *Preferred Metric Limits and Fits*. New York: American Society of Mechanical Engineers, 1978.

INTERNET SITES FOR SHAFT DESIGN

1. **Hexagon Software.** www.hexagon.de Computer software called WL1+ for shaft calculations. Can handle up to 100 cylindrical or conical shaft segments and up to 50 individual forces. Computes shearing force and bending moment diagrams, slope, deflection, bending stresses, and shearing stresses. Critical speeds for flexure and torsional vibration are calculated.
2. **Spinning Composites**. www.spinning-composites.com Manufacturer of composite shafts for vehicular and industrial applications. Includes discussion of critical speeds.
3. **Advanced Composite Products and Technology, Inc**. www.acpt.com Manufacturer of composite products for aerospace, military, commercial and industrial markets, including driveshafts and high-speed rotors.
4. **RoyMech.** www.roymech.co.uk An Internet site for useful information, tables, and formulas related to mechanical engineering; developed by Roy Beardmore. One section deals with critical speeds of shafts: www.roymech.co.uk/Useful_Tables/Drive/Shaft_Critical_Speed.html.
5. **Orand Systems, Inc.** www.orandsystems.com Developer and distributor of *Beam 2D Stress Analysis* software. Analyzes beams with numerous loads, cross sections, and support conditions. Computes load, shearing force, bending moment, slope, and deflection outputs in graphical and numerical data formats. Can be used for the bending calculations for shafts.
6. **MDSolids.** www.mdsolids.com Educational software for stress analysis that includes several modules including beams, torsion, trusses, and columns.
7. **S. S. White Technologies, Inc.** www.sswt.com Manufacturer of flexible shafts. Site includes rating charts for torque capacity versus the minimum radius of the shaft during operation.
8. **Elliott Manufacturing Co.** www.elliottmfg.com Manufacturer of flexible shafts.
9. **Otto Suhner AG.** www.suhner-transmission-expert.com Manufacturer of flexible shafts, spiral bevel gears, and small, high speed air motors and electric motors. Site includes rating charts for power and torque capacity of flexible shafts.

PROBLEMS

Problems 1 through 30 relate to one of Figures P12–1 through P12–17 showing shafts carrying a variety of combinations of gears, belt sheaves, chain sprockets, and a few other items such as a flywheel and a propeller-type fan. There are multiple ways in which the problems may be assigned, as some deal with only torque levels and radial forces in the shaftdue to selected power-transmitting elements while others are comprehensive design problems for the shaft that carries several elements. The following table may help instructors decide how to assign the problems for student solution and may help students comprehend how the sets of problems lead to the more general shaft design. Any combination of problems may be chosen.

Torques and Forces Acting Radial to Shaft	Comprehensive
Figure P12–1: P1—Gear *B*; P14—Sheave *D*	P22
Figure P12–2: P2—Gear *C*; P12—Sprocket *D*; P13—Pulley *A*	P23
Figure P12–3: P3—Gear *B*; P15—Sprocket *C*; P16—Sheaves *D*, *E*	P24
Figure P12–4: P4—Gear *A*; P19—Sprockets *C*, *D*	P25
Figure P12–5: P5—Gear *D*; P20—Sheave *A*; P21—Sprocket *E*	P26
Figure P12–6: P6—Gear *E*; (No separate analysis of Sheave *A*)	P27 (Includes Sheave *A*)
Figure P12–7: P7—Gear *C*; P8—Gear *A*	P28 (Includes Sheaves *D*, *E*)
Figure P12–9: P9—Gear *C*; P10—Gear *D*; P11—Gear *F*	P29 (Includes Sheave *B*)
Figure P12–17: P17—Sheave *C*; P18—Pulley *D*	P30 (Includes Fan *A*)

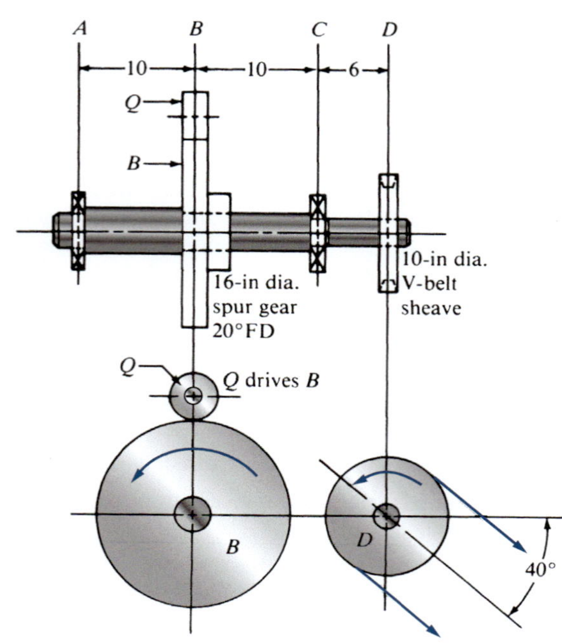

FIGURE P12–1 (Problems 1, 14, and 22)

Torques and Forces Acting Radial to Shaft

1. See Figure P12–1. The shaft rotating at 550 rpm carries a spur gear *B* having 96 teeth with a diametral pitch of 6. The teeth are of the 20°, full-depth, involute form. The gear receives 30 hp from a pinion directly above it. Compute the torque delivered to the shaft and the tangential and radial forces exerted on the shaft by the gear.
2. See Figure P12–2. The shaft rotating at 200 rpm carries a spur gear *C* having 80 teeth with a diametral pitch of 8. The teeth are of the 20°, full-depth, involute form. The gear delivers 6 hp to a pinion directly below it. Compute the torque delivered by the shaft to gear *C* and the tangential and radial forces exerted on the shaft by the gear.
3. See Figure P12–3. The shaft rotating at 480 rpm carries a spur pinion *B* having 24 teeth with a diametral pitch of 8. The teeth are of the 20°, full-depth, involute form. The pinion delivers 5 hp to a gear directly below it. Compute the torque delivered

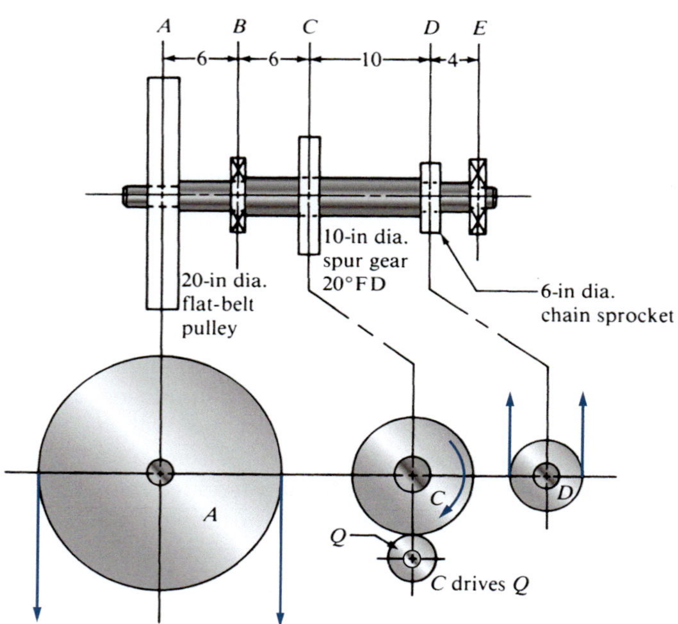

FIGURE P12–2 (Problems 2, 12, 13, and 23)

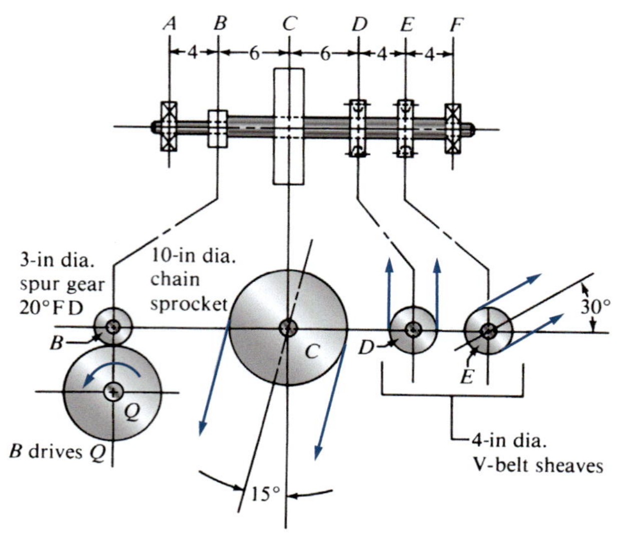

FIGURE P12–3 (Problems 3, 15, 16, and 24)

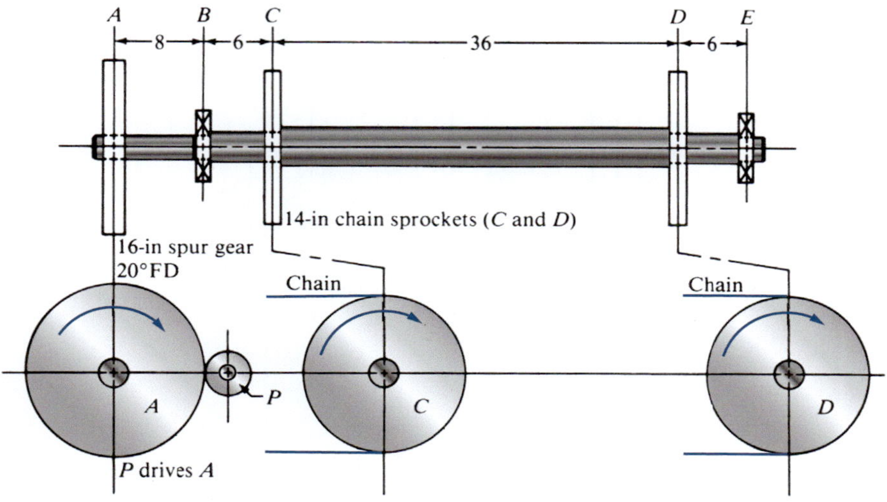

FIGURE P12–4 (Problems 4, 19, and 25)

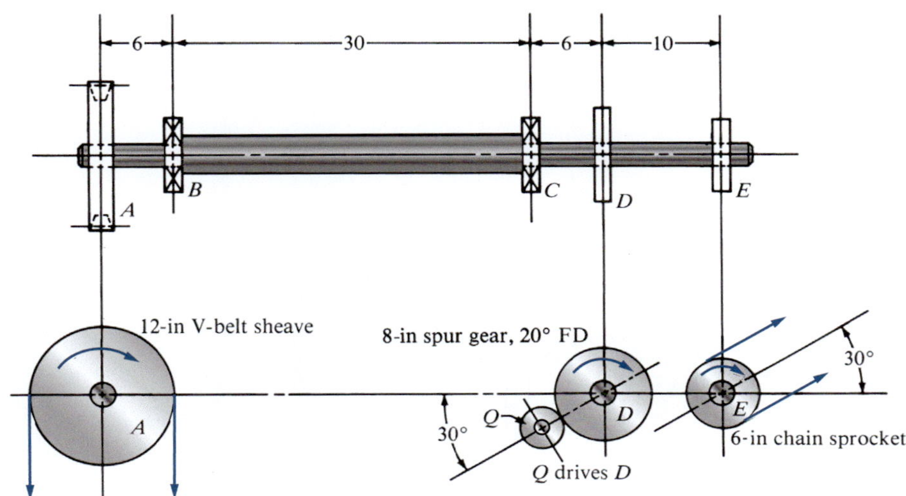

FIGURE P12–5 (Problems 5, 20, 21, and 26)

by the shaft to pinion *B* and the tangential and radial forces exerted on the shaft by the pinion.

4. See Figure P12–4. The shaft rotating at 120 rpm carries a spur gear *A* having 80 teeth with a diametral pitch of 5. The teeth are of the 20°, full-depth, involute form. The gear receives 40 hp from a pinion to the right as shown. Compute the torque delivered to the shaft and the tangential and radial forces exerted on the shaft by the gear.

5. See Figure P12–5. The shaft rotating at 240 rpm carries a spur gear *D* having 48 teeth with a diametral pitch of 6. The teeth are of the 20°, full-depth, involute form. The gear receives 15 hp from pinion *Q* located as shown. Compute the torque delivered to the shaft and the tangential and radial forces exerted on the shaft by the gear. Resolve the tangential and radial forces into their horizontal and vertical components, and determine the net forces acting on the shaft at *D* in the horizontal and vertical directions.

6. See Figure P12–6. The shaft rotating at 310 rpm carries a spur pinion *E* having 36 teeth with a diametral pitch of 6. The teeth are of the 20°, full-depth, involute form. The pinion delivers 20 hp to a gear to the left as shown. Compute the torque delivered by the shaft to pinion *E* and the tangential and radial forces exerted on the shaft by the pinion. Include the weight of the pinion.

7. See Figure P12–7. The shaft rotating at 480 rpm carries a spur gear *C* having 50 teeth with a diametral pitch of 5. The teeth are of the 20°, full-depth, involute form. The gear receives 50 hp from a pinion directly below it. Compute the torque delivered to the shaft and the tangential and radial forces exerted on the shaft by the gear.

8. See Figure P12–7. The shaft rotating at 480 rpm carries a spur pinion *A* having 30 teeth with a diametral pitch of 6. The teeth are of the 20°, full-depth, involute form. The pinion delivers 30 hp to a gear to the left as shown. Compute the torque delivered by the shaft to pinion *A* and the tangential and radial forces exerted on the shaft by the pinion.

9. See Figure P12–9. The shaft rotating at 220 rpm carries a spur pinion *C* having 60 teeth with a diametral pitch of 10. The teeth are of the 20°, full-depth, involute form. The pinion delivers 5 hp to a gear directly above it. Compute the torque delivered by the shaft to pinion *C* and the tangential and radial forces exerted on the shaft by the pinion.

10. See Figure P12–9. The shaft rotating at 220 rpm carries a spur gear *D* having 96 teeth with a diametral pitch of 8. The teeth

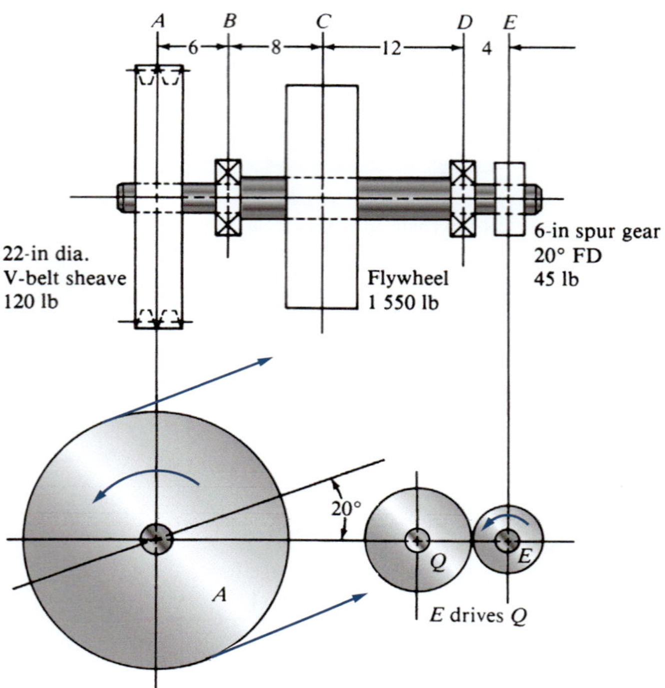

FIGURE P12–6 (Problems 6 and 27)

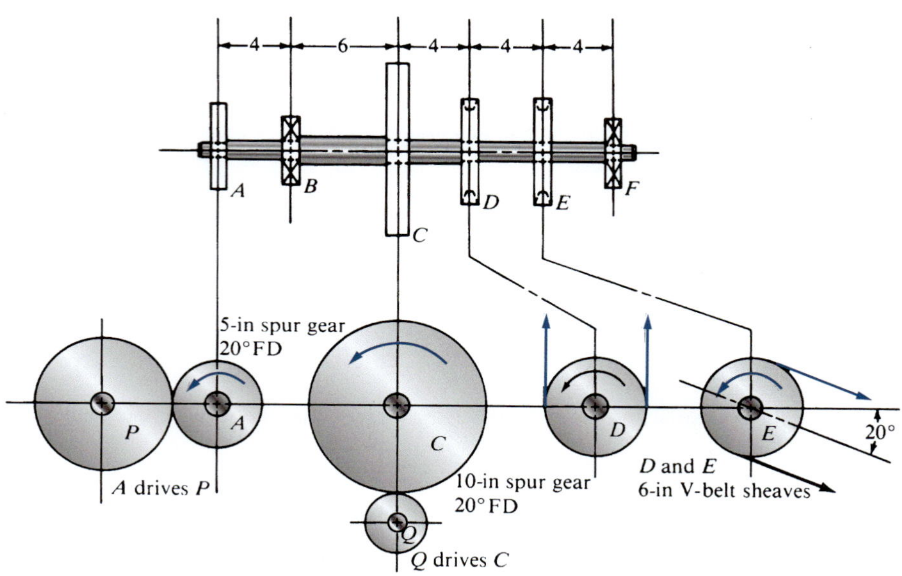

FIGURE P12–7 (Problems 7, 8, and 28)

are of the 20°, full-depth, involute form. The gear receives 12.5 hp from a pinion directly below it. Compute the torque delivered to the shaft and the tangential and radial forces exerted on the shaft by the gear.

11. See Figure P12–9. The shaft rotating at 220 rpm carries a spur pinion *F* having 60 teeth with a diametral pitch of 10. The teeth are of the 20°, full-depth, involute form. The pinion delivers 5 hp to a gear located as shown. Compute the torque delivered by the shaft to pinion *F* and the tangential and radial forces exerted on the shaft by the pinion. Resolve the forces into their horizontal and vertical components, and determine the net forces acting on the shaft at *F* in the horizontal and vertical directions.

12. See Figure P12–2. The shaft rotating at 200 rpm carries a 6-in-diameter chain sprocket *D* that delivers 4 hp to a mating sprocket above. Compute the torque delivered by the shaft to sprocket *D* and the force exerted on the shaft by the sprocket.

13. See Figure P12–2. The shaft rotating at 200 rpm carries a 20-in-diameter flat-belt pulley at *A* that receives 10 hp from below. Compute the torque delivered by the pulley to the shaft and the force exerted on the shaft by the pulley.

14. See Figure P12–1. The shaft rotating at 550 rpm carries a 10-in-diameter V-belt sheave at *D* that delivers 30 hp to a mating sheave as shown. Compute the torque delivered by the shaft to the sheave and the total force exerted on the shaft by the sheave. Resolve the force into its horizontal and vertical

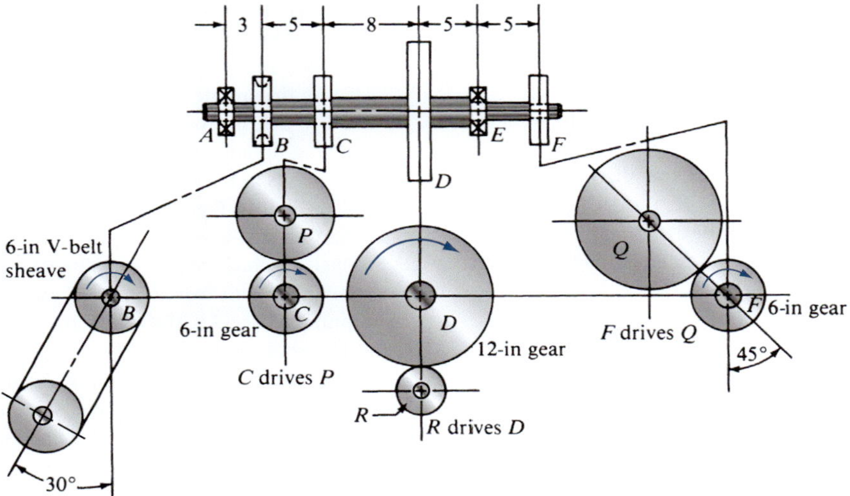

FIGURE P12–9 (Problems 9, 10, 11, and 29)

components, and show the net forces acting on the shaft at *D* in the horizontal and vertical directions.

15. See Figure 12–3. The shaft rotating at 480 rpm carries a 10-in-diameter chain sprocket at *C* that receives 11 hp from a mating sprocket below and to the left as shown. Compute the torque delivered to the shaft by the sprocket and the total force exerted on the shaft by the sprocket. Resolve the force into its horizontal and vertical components, and show the net forces acting on the shaft at *C* in the horizontal and vertical directions.

16. See Figure 12–3. The shaft rotating at 480 rpm carries two 4-in-diameter sheaves at *D* and *E* that each deliver 3 hp to mating sheaves as shown. Compute the torque delivered by the shaft to each sheave and the total force exerted on the shaft by each sheave. Resolve the force at *E* into its horizontal and vertical components, and show the net forces acting on the shaft at *E* in the horizontal and vertical directions.

17. See Figure P12–17. The shaft rotating at 475 rpm carries a 10-in V-belt sheave at *C* that receives 15.5 hp from a mating sheave to the left as shown. Compute the torque delivered to the shaft at *C* by the sheave and the total force exerted on the shaft at *C* by the sheave.

18. See Figure P12–17. The shaft rotating at 475 rpm carries a 6-in-diameter flat-belt pulley at *D* that delivers 3.5 hp to a mating pulley above and to the right as shown. Compute the torque delivered to the shaft by the pulley and the total force exerted on the shaft by the pulley. Resolve the force into its horizontal and vertical components, and show the net forces acting on the shaft at *D* in the horizontal and vertical directions.

19. See Figure P12–4. The shaft rotating at 120 rpm carries two identical 14-in-diameter chain sprockets at *C* and *D*. Each sprocket delivers 20 hp to mating sprockets to the left as shown. Compute the torque delivered by the shaft to each sprocket and the total force exerted on the shaft by each sprocket.

20. See Figure P12–5. The shaft rotating at 240 rpm carries a 12-in-diameter V-belt sheave at *A* that delivers 10 hp to a mating sheave directly below. Compute the torque delivered by the shaft to the sheave and the total force exerted on the shaft at *A* by the sheave.

21. See Figure P12–5. The shaft rotating at 240 rpm carries a 6-in-diameter chain sprocket at *E* that delivers 5.0 hp to a mating sprocket to the right and above as shown. Compute the torque delivered by the shaft to the sprocket and the total force exerted on the shaft by the sprocket. Resolve the force into its horizontal and vertical components, and show the net forces acting on the shaft at *E* in the horizontal and vertical directions.

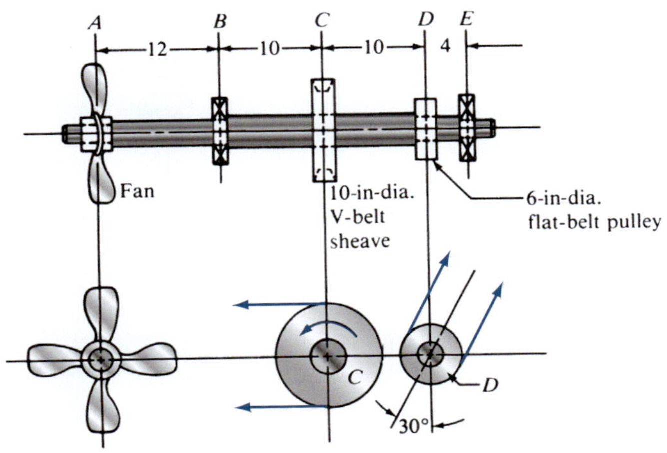

FIGURE P12–17 (Problems 17, 18, and 30)

Comprehensive Shaft Design Problems

For each of the following problems 22 to 30, it will be necessary to do the following:

(a) Determine the magnitude of the torque in the shaft at all points
(b) Compute the forces acting on the shaft at all power-transmitting elements
(c) Compute the reactions at the bearings
(d) Draw the complete load, shear, and bending moment diagrams

Neglect the weight of the elements on the shafts, unless otherwise noted.

The objective of any problem, at the discretion of the instructor, could be any of the following:

- Design the complete shaft, including the specification of the overall geometry and the consideration of stress concentration factors. The analysis would show the minimum acceptable diameter at each point on the shaft to be safe from the standpoint of strength.
- Given a suggested geometry of one part of the shaft, specify the minimum acceptable diameter for the shaft at that point.
- Specify the required geometry at any selected element on the shaft: a gear, sheave, bearing, or other.
- Make a working drawing of the design for the shaft, following the appropriate stress analysis, and specify the final dimensions.
- Suggest how the given shaft can be redesigned by moving or reorienting the elements on the shaft to improve the design to produce lower stresses, smaller shaft size, more convenient assembly, and so on.
- Incorporate the given shaft into a more comprehensive machine, and complete the design of the entire machine. In most problems, the type of machine for which the shaft is being designed is suggested.

22. The shaft in Figure P12–1 is part of a drive for an automated transfer system in a metal stamping plant. Gear Q delivers 30 hp to gear B. Sheave D delivers the power to its mating sheave as shown. The shaft carrying B and D rotates at 550 rpm. Use SAE 1040 cold-drawn steel.
23. The shaft in Figure P12–2 rotates at 200 rpm. Pulley A receives 10 hp from below. Gear C delivers 6 hp to the mating gear below it. Chain sprocket D delivers 4 hp to a shaft above. Use SAE 1117 cold-drawn steel.
24. The shaft in Figure P12–3 is part of a special machine designed to reclaim scrap aluminum cans. The gear at B delivers 5 hp to a chopper that cuts the cans into small pieces. The V-belt sheave at D delivers 3 hp to a blower that draws air through the chopper. V-belt sheave E delivers 3 hp to a conveyor that raises the shredded aluminum to an elevated hopper. The shaft rotates at 480 rpm. All power comes into the shaft through the chain sprocket at C. Use SAE 1137 OQT 1300 steel for the shaft. The elements at B, C, D, and E are held in position with retaining rings and keys in profile keyseats.
25. The shaft in Figure P12–4 is the drive shaft for a large, bulk material conveyor. The gear receives 40 hp and rotates at 120 rpm. Each chain sprocket delivers 20 hp to one side of the conveyor. Use SAE 1020 cold-drawn steel.
26. The shaft in Figure P12–5 is part of a conveyor drive system that feeds crushed rock into a railroad car. The shaft rotates at 240 rpm and is subjected to moderate shock during operation. All power is input to the gear at D. The V-belt sheave at A delivers 10.0 hp vertically downward. Chain sprocket E delivers 5.0 hp. Note the position of the gear Q, which drives gear D.
27. Figure P12–6 illustrates an intermediate shaft of a punch press that rotates at 310 rpm while transmitting 20 hp from the V-belt sheave to the gear. The flywheel is not absorbing or giving up any energy at this time. Consider the weight of all elements in the analysis.
28. The shaft in Figure P12–7 is part of a material-handling system aboard a ship. All power comes into the shaft through gear C, which rotates at 480 rpm. Gear A delivers 30 hp to a hoist. V-belt sheaves D and E each deliver 10 hp to hydraulic pumps. Use SAE 3140 OQT 1000 steel.
29. The shaft in Figure P12–9 is part of an automatic machining system. All power is input through gear D. Gears C and F drive two separate tool feed devices requiring 5.0 hp each. The V-belt sheave at B requires 2.5 hp to drive a coolant pump. The shaft rotates at 220 rpm. All gears are spur gears with 20°, full-depth teeth. Use SAE 1020 cold-drawn steel for the shaft.
30. The shaft in Figure P12–17 is part of a grain-drying system. At A is a propeller-type fan that requires 12 hp when rotating at 475 rpm. The fan weighs 34 lb, and its weight should be included in the analysis. The flat-belt pulley at D delivers 3.5 hp to a screw conveyor handling the grain. All power comes into the shaft through the V-belt sheave at C. Use SAE 1144 cold-drawn steel.

Shifts with both Radial and Axial Loads

Problems P31 to P34 deal with shafts carrying helical gears and wormgears that produce forces directed axially in addition to radial forces.

Torques and Forces Acting Radial and Axial to Shaft	Comprehensive
Figure P12–31: P31—Helical Gear *B*	P32
Figure P12–33: P33—Wormgear *C*	P34 (Includes Sheave *A*)

Forces and Torque on Shafts—Helical Gears

31. See Figure P12–31. The shaft is rotating at 650 rpm, and it receives 7.5 hp through a flexible coupling. The power is delivered to an adjacent shaft through a single helical gear B having

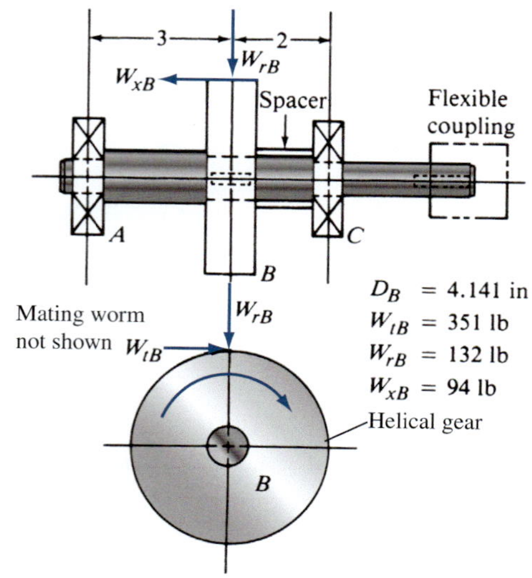

FIGURE P12–31 (Problems 31 and 32)

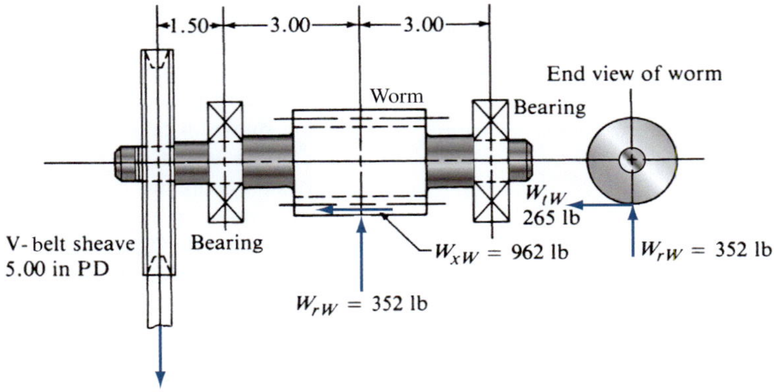

FIGURE P12–33 (Problems 33 and 34)

a normal pressure angle of 20° and a helix angle of 15°. The pitch diameter for the gear is 4.141 in. Review the discussion of forces on helical gears in Chapter 10, and use Equations (12–1), (12–2), (12–4), and (12–5) in this chapter to verify the values for the forces given in the figure. Draw the complete free-body diagrams for the shaft in both the vertical plane and the horizontal plane. Then draw the complete shearing force and bending moment diagrams for the shaft in both planes.

32. Figure P12–31 shows a helical gear mounted on a shaft that rotates at 650 rpm while transmitting 7.5 hp. The gear is also analyzed in Example Problem 10–1, and the tangential, radial, and axial forces on it are shown in the figure. The pitch diameter of the gear is 4.141 in. The power is delivered from the shaft through a flexible coupling at its right end. A spacer is used to position the gear relative to bearing *C*. The thrust load is taken at bearing *A*.

Forces and Torque on Shafts—Wormgears

33. See Figure P12–33. The shaft is rotating at 1750 rpm, and it receives 7.5 hp through a 5.00-in V-belt drive from a mating belt sheave directly below. The power is delivered by a worm with a pitch diameter of 2.00 in. The forces on the worm were used in Design Example 12–3 and are shown in the figure. Review the description of these forces in Chapter 10. Draw the complete free-body diagrams for the shaft in both the vertical plane and the horizontal plane. Then draw the complete shearing force and bending moment diagrams for the shaft in both planes.

34. The shaft shown in Figure P12–33 is the input shaft of a wormgear drive. The V-belt sheave receives 7.5 hp from directly downward. The worm rotates at 1750 rpm and has a pitch diameter of 2.000 in. This is the driving worm for the wormgear described in Design Example 12–3. The tangential, radial, and axial forces are shown in the figure. The worm is to be machined integrally with the shaft, and it has a root diameter of 1.614 in. Assume that the geometry of the root area presents a stress concentration factor of 1.5 for bending. Analyze the stress in the area of the worm thread root, and specify a suitable material for the shaft.

Other Comprehensive Design Problems

Problems 35 to 41 contain a variety of loading situations for which the general solution procedure must be adapted. Some of the problems involve more than one shaft, considering shafts for mating gears and multiple reductions.

Figure P12-35: P35—Double reduction helical drive
Figure P10-8 in Chapter 10: P36—Bevel gear drive
Figure P12-37: P37—Bevel gear drive with two chain sprockets
Figure P12-38: P38—Double reduction spur gear drive; design three shafts.
Figure P12-39: P39—Drive system consisting of an electric motor, a V-belt drive, a double reduction spur gear-type reducer, and a chain drive.
Figure P12-40: P40—Shaft with three spur gears
Figure P12-41: P41—Shaft for windshield wiper mechanism with two levers

35. The double-reduction, helical gear reducer shown in Figure P12–35 transmits 5.0 hp. Shaft 1 is the input, rotating at 1800 rpm and receiving power directly from an electric

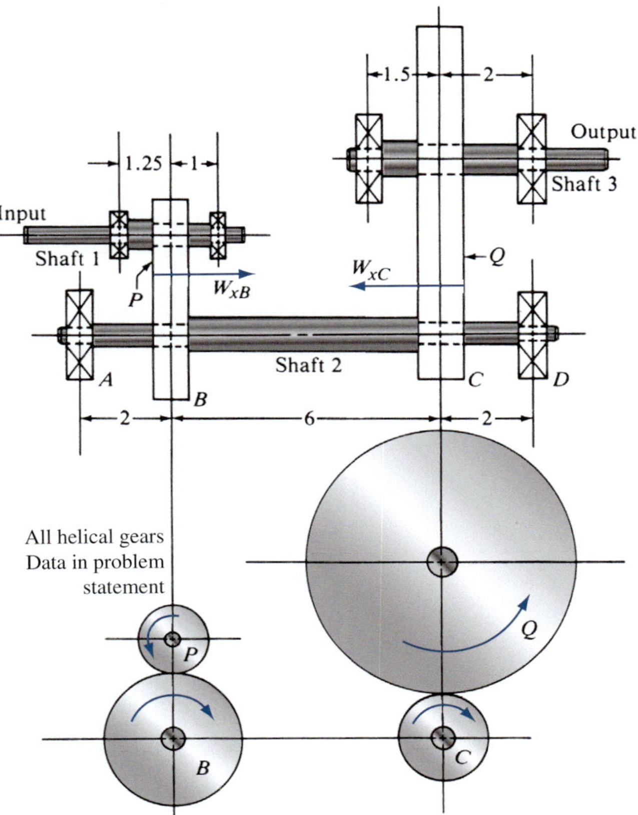

FIGURE P12–35 (Problem 35)

TABLE 12–5

Gear	Diametral pitch	Pitch diameter	Number of teeth	Face width
P	8	1.500 in	12	0.75 in
B	8	3.000 in	24	0.75 in
C	6	2.000 in	12	1.00 in
Q	6	6.000 in	36	1.00 in

motor through a flexible coupling. Shaft 2 rotates at 900 rpm. Shaft 3 is the output, rotating at 300 rpm. A chain sprocket is mounted on the output shaft as shown and delivers the power upward. The data for the gears are given in Table 12–5. Each gear has a 14½° normal pressure angle and a 45° helix angle. The combinations of left- and right-hand helixes are arranged so that the axial forces oppose each other on shaft 2 as shown. Use SAE 4140 OQT 1200 for the shafts.

36. Complete the design of the shafts carrying the bevel gear and pinion shown in Figure 10–8. The forces on the gears, the bearing reactions, and the bending moment diagrams are developed in Example Problems 10–3 and 10–4 and are shown in Figures 10–8 through 10–12. Assume that the 2.50 hp enters the pinion shaft from the right through a flexible coupling. The power is delivered by way of the lower extension of the gear shaft through another flexible coupling. Use SAE 1040 OQT 1200 for the shafts.
37. The vertical shaft shown in Figure P12–37 is driven at a speed of 600 rpm with 4.0 hp entering through the bevel gear. Each of the two chain sprockets delivers 2.0 hp to the side to drive mixer blades in a chemical reactor vessel. The bevel gear has a diametral pitch of 5, a pitch diameter of 9.000 in, a face width of 1.31 in, and a pressure angle of 20°. Use SAE 4140 OQT 1000 steel for the shaft. See Chapter 10 for the methods for computing the forces on the bevel gear.
38. Figure P12–38 shows a sketch for a double-reduction spur gear train. Shaft 1 rotates at 1725 rpm, driven directly by an electric motor that transmits 15.0 hp to the reducer. The gears

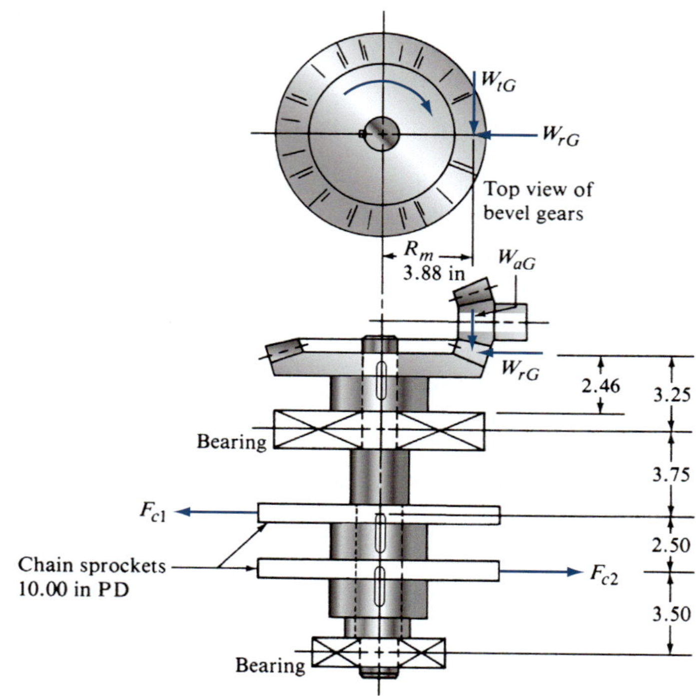

FIGURE P12–37 Bevel gears shown in section for Problem 37. See Chapter 10

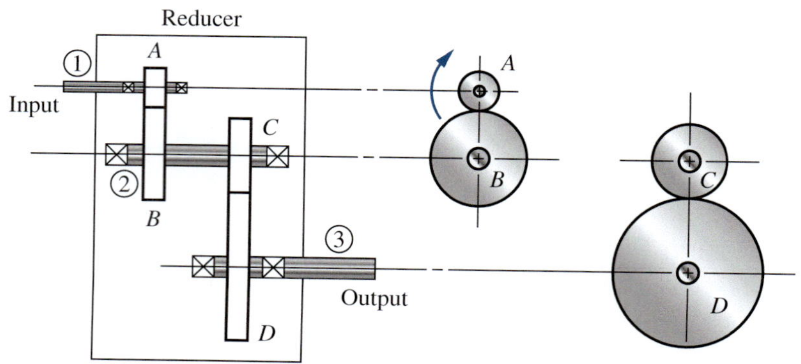

FIGURE P12–38 Gear reducer for Problems 38 and 39

in the train all have 20°, full-depth teeth and the following numbers of teeth and pitch diameters:

Gear *A*	Gear *B*
18 teeth	54 teeth
1.80-in dia.	5.40-in dia.
Gear *C*	**Gear *D***
24 teeth	48 teeth
4.00-in dia.	8.00-in dia.

Note that the speed reduction for each gear pair is proportional to the ratio of the number of teeth. Therefore, shaft 2 rotates at 575 rpm, and shaft 3 rotates at 287.5 rpm. Assume that all shafts carry 15.0 hp. The distance from the middle of each bearing to the middle of the face of the nearest gear is 3.00 in. Shaft 2 has a length of 5.00 in between the two gears, making the total distance between the middle of the two bearings equal to 11.00 in. The extensions of the input and the output shafts carry torque, but no bending loads are exerted on them. Complete the design of all three shafts. Provide for profile keyseats for each gear and sled-runner keyseats at the outboard ends of the input and output shafts. Provide for the location of each gear and bearing on the shaft.

39. Figure P12–39 shows a speed reducer with a V-belt drive delivering power to the input shaft and a chain drive taking power from the output shaft and delivering it to a conveyor. The drive motor supplies 12.0 hp and rotates 1150 rpm.

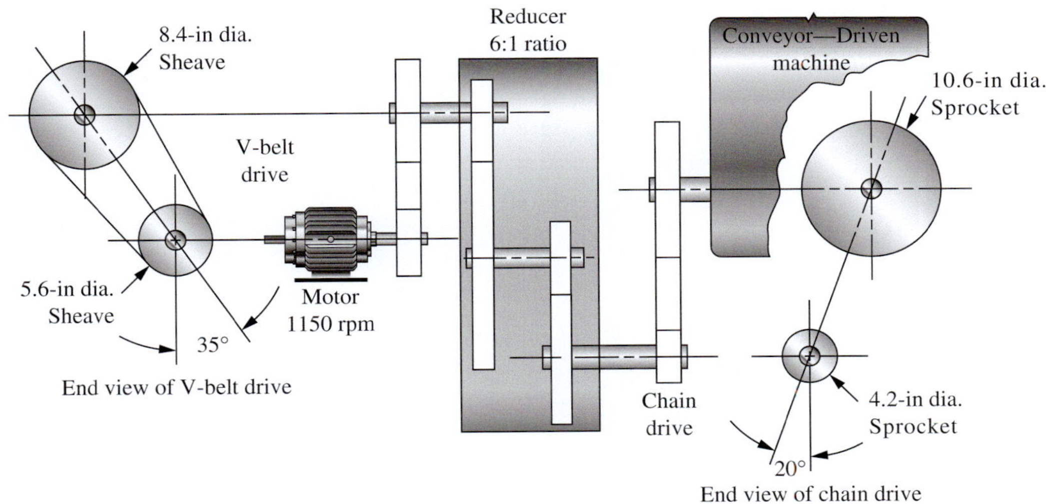

FIGURE P12–39 Drive system for Problem 39

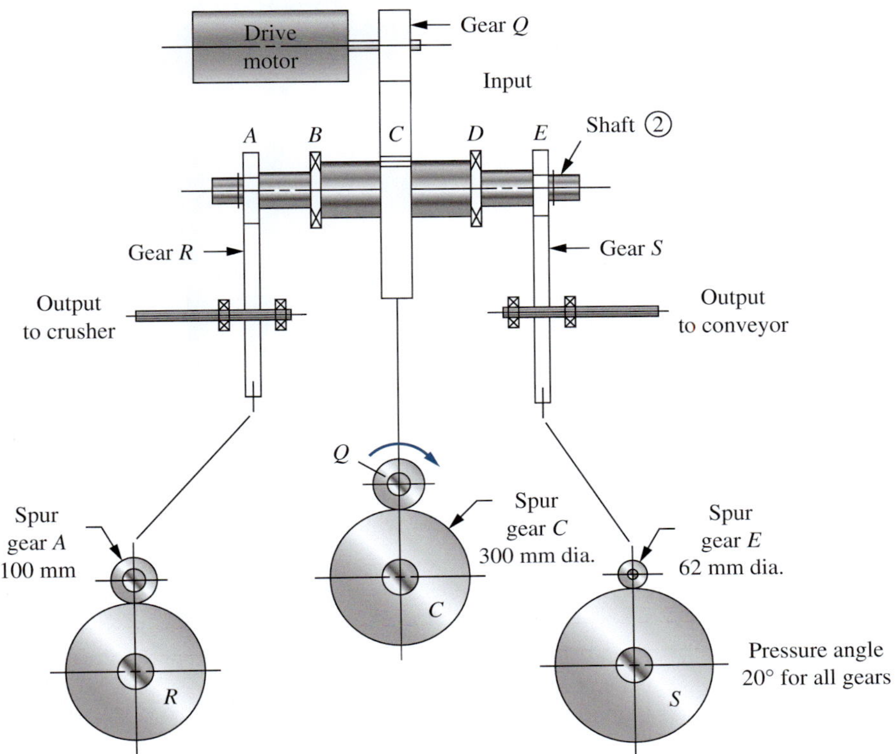

FIGURE P12–40 Drive system for Problem 40

The speed reductions of the V-belt drive and the chain drive are proportional to the ratio of the diameter of the driving and driven sheaves or sprockets. The arrangement of the gears in the reducer is the same as that described in Figure P12–38 and Problem 38. Determine the forces applied to the motor shaft, to each of the three shafts of the reducer, and to the conveyor driveshaft. Then complete the design of the three shafts of the reducer, assuming that all shafts transmit 12.0 hp.

40. Figure P12–40 shows a drive for a system to crush coal and deliver it by conveyor to a railroad car. Gear *A* delivers 15 kW to the crusher, and gear *E* delivers 7.5 kW to the conveyor. All power enters the shaft through gear *C*. The shaft carrying gears *A, C,* and *E* rotates at 480 rpm. Design that shaft. The distance from the middle of each bearing to the middle of the face of the nearest gear is 100 mm.

41. A shaft is to be designed for a linkage of a windshield wiper mechanism for a truck (see Figure P12–41). A 20-N force is applied to lever 1 by an adjacent link. The reaction force, F_2, on lever 2 is transmitted to another link. The distance, *d*, between elements is 20.0 mm. Bearings *A* and *C* are straight, cylindrical, bronze bushings, 10.0 mm long. Design the shaft and levers 1 and 2.

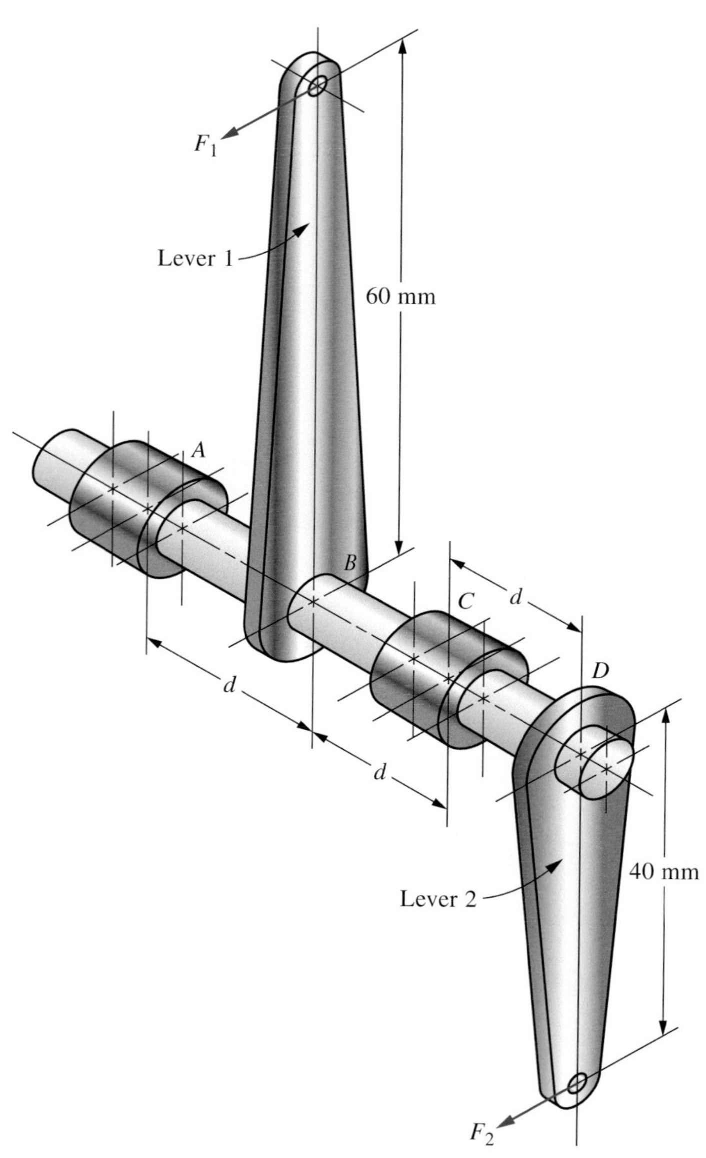

FIGURE P12–41 Shaft and levers for windshield wiper system for Problem 41

CHAPTER FOURTEEN

ROLLING CONTACT BEARINGS

THE BIG PICTURE Rolling Contact Bearings

Discussion Map

- *Bearings* are used to support a load while permitting relative motion between two elements of a machine.
- Some bearings use rolling elements, such as spherical balls or cylindrical or tapered rollers. This results in a very low coefficient of friction.

Discover

Look for examples of bearings on machines, cars, trucks, bicycles, and consumer products.

Describe the bearings, including how they are installed and what kinds of forces are exerted on them.

This chapter presents information about such bearings and gives methods for analyzing bearings and for selecting commercially available bearings.

The purpose of a bearing is to support a load while permitting relative motion between two elements of a machine. The term *rolling contact bearings* refers to the wide variety of bearings that use spherical balls or some other type of roller between the stationary and the moving elements. The most common type of bearing supports a rotating shaft, resisting purely radial loads or a combination of radial and axial (thrust) loads. Some bearings are designed to carry only thrust loads. Most bearings are used in applications involving rotation, but some are used in linear motion applications.

The components of a typical rolling contact bearing are the inner race, the outer race, and the rolling elements. Figure 14–1 shows the common single-row, deep-groove ball bearing. Usually the outer race is stationary and is held by the housing of the machine. The inner race is pressed onto the rotating shaft and thus rotates with it. Then the balls roll between the outer and inner races. The load path is from the shaft, to the inner race, to the balls, to the outer race, and finally to the housing. The presence of the balls allows a very smooth, low-friction rotation of the shaft. The typical coefficient of friction for a rolling

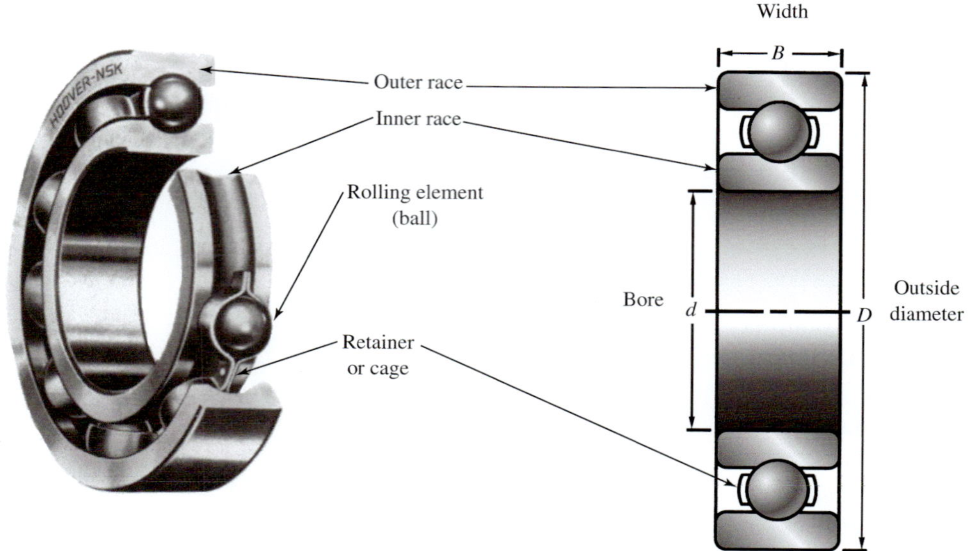

FIGURE 14–1 Single-row, deep-groove ball bearing (NSK Corporation, Ann Arbor, MI)

contact bearing is approximately 0.001 to 0.005. These values reflect only the rolling elements themselves and the means of retaining them in the bearing. The presence of seals, excessive lubricant, or unusual loading increases these values.

Look for consumer products, industrial machinery, or transportation equipment (cars, trucks, bicycles, and so on), and identify any uses of rolling contact bearings. Be sure that the machine is turned off, and then try to gain access to the mechanical drive shafts used to transmit power from a motor or an engine to some moving parts of the machine. Are these shafts supported in ball or roller bearings? Or are they supported in plain surface bearings in which the shaft passes through cylindrical members called *bushings* or *bearings,* typically with lubricants present between the rotating shaft and the stationary bearing? Plain surface bearings are discussed in Chapter 16.

For the shafts that are supported in ball or roller bearings, describe the bearing. How is it mounted on the shaft? How is it mounted in the housing of the machine? Can you identify what kinds of forces are acting on the bearing and in what directions they act? Are the forces directed radially toward the centerline of the shaft? Is there any force that acts parallel to the axis of the shaft? Compare the bearings that you find with the photographs that are in this chapter. Which varieties of bearings have you found? Measure or estimate the physical size of the bearings, particularly the diameter of the bore that is in contact with the shaft, the outside diameter, and the width. Can you see the rolling elements—balls or rollers? If so, sketch them and estimate their diameters and/or lengths. Are any of the rolling elements tapered rollers like those shown later in this chapter in Figure 14–7?

When you complete this chapter, you will be able to identify several kinds of rolling contact bearings and specify suitable bearings to carry specified loads. You will also be able to apply such bearings properly, planning for their installation on shafts and in housings.

YOU ARE THE DESIGNER

In Chapter 12, you were the designer of a shaft that was rotating at 600 rpm carrying two gears as a part of a power transmission system. Figures 12–1 and 12–2 showed the basic layout that was proposed. The shaft was designed to be supported on two bearings at points B and D. Then, in Example Problem 12–1, we completed the force analysis by computing the forces applied to the shaft by the gears and then computing the *reactions at the bearings.* Figure 12–11 showed the results, summarized here:

$$R_{Bx} = 458 \text{ lb} \qquad R_{By} = 4620 \text{ lb}$$
$$R_{Dx} = 1223 \text{ lb} \qquad R_{Dy} = 1680 \text{ lb}$$

where x refers to the horizontal direction and y refers to the vertical direction. All forces on the bearings are in the radial direction. Your task now is to specify suitable rolling contact bearings for the shaft to withstand those forces and transfer them from the shaft to the housing of the speed reducer.

What kind of bearing should be selected? How do the forces just identified affect the choice? What life expectancy is reasonable for the bearings, and how does that affect the selection of the bearings? What size should be specified? How are the bearings to be installed on the shaft, and how does that affect the detailed design of the shaft? What limit dimensions should be defined for the bearing seats on the shaft? How is the bearing to be located axially on the shaft? How is it to be installed in the housing and located there? How is lubrication to be provided for the bearings? Is there a need for shields and seals to keep contaminants out of the bearings? The information in this chapter will help you make these and other design decisions. ■

14–1 OBJECTIVES OF THIS CHAPTER

After completing this chapter, you will be able to:

1. Identify the types of rolling contact bearings that are commercially available, and select the appropriate type for a given application, considering the manner of loading and installation conditions.
2. Use the relationship between forces on bearings and the life expectancy for the bearings to determine critical bearing selection factors.
3. Use manufacturers' data for the performance of ball bearings to specify suitable bearings for a given application.
4. Recommend appropriate values for the design life of bearings.
5. Compute the *equivalent load* on a bearing corresponding to combinations of radial and thrust loads applied to it.
6. Specify mounting details for bearings that affect the design of the shaft onto which the bearing is to be seated and the housing into which it is to be installed.
7. Compute the equivalent loads on tapered roller bearings.
8. Describe the special design of thrust bearings.
9. Describe several types of commercially available mounted bearings and their application to machine design.
10. Understand certain practical considerations involved in the application of bearings, including lubrication, sealing, limiting speeds, bearing tolerance classes, and standards related to the manufacture and application of bearings.
11. Consider the effects of varying loads on the life expectancy and specification of bearings.

14–2 TYPES OF ROLLING CONTACT BEARINGS

Here we will discuss seven different types of rolling contact bearings and the applications in which each is typically used. Many variations on the designs shown are available. As each is discussed, refer to Table 14–1 for a comparison of the performance relative to the others.

Radial loads act toward the center of the bearing along a radius. Such loads are typical of those created by power transmission elements on shafts such as spur gears, V-belt drives, and chain drives. *Thrust loads* are those that act parallel to the axis of the shaft. The axial components of the forces on helical gears, worms and wormgears, and bevel gears are thrust loads. Also, bearings supporting shafts with vertical axes are subjected to thrust loads due to the weight of the shaft and the elements on the shaft as well as from axial operating forces. *Misalignment* refers to the angular deviation of the axis of the shaft at the bearing from the true axis of the bearing itself. An excellent rating for misalignment in Table 14–1 indicates that the bearing can accommodate up to 4.0° of angular deviation. A bearing with a fair rating can withstand up to 0.15°, while a poor rating indicates that rigid shafts with less than 0.05° of misalignment are required. Manufacturers' catalogs should be consulted for specific data. Check Internet sites 1–8.

Single-Row, Deep-Groove Ball Bearing

Sometimes called *Conrad bearings*, the single-row, deep-groove ball bearing (Figure 14–1) is what most people think of when the term *ball bearing* is used. The inner race is typically pressed on the shaft at the bearing seat with a slight interference fit to ensure that it rotates with the shaft. The spherical rolling elements, or balls, roll in a deep groove in both the inner and the outer races. The spacing of the balls is maintained by retainers or "cages." Although designed primarily for radial load-carrying capacity, the deep groove allows it to carry a fairly sizable thrust load. The thrust load would be applied to one side of the inner race by a shoulder on the shaft. The load would pass across the side of the groove, through the ball, to the opposite side of the outer race, and then to the housing. The radius of the ball is slightly smaller than the radius of the groove to allow free rolling of the balls. The contact between a ball and the race is theoretically at a point, but it is actually a small circular area because of the deformation of the elements. Because the load is carried on a small area, very high local contact stresses occur. To increase the capacity of a single-row bearing, a bearing with a greater number of balls, or larger balls operating in larger-diameter races, should be used.

TABLE 14–1 Comparison of Bearing Types

Bearing type	Radial load capacity	Thrust load capacity	Misalignment capability
Single-row, deep-groove ball	Good	Fair	Fair
Double-row, deep-groove ball	Excellent	Good	Fair
Angular contact	Good	Excellent	Poor
Cylindrical roller	Excellent	Poor	Fair
Needle	Excellent	Poor	Poor
Spherical roller	Excellent	Fair/good	Excellent
Tapered roller	Excellent	Excellent	Poor

FIGURE 14–2 Double-row, deep-groove ball bearing (NSK Corporation, Ann Arbor, MI)

Double-Row, Deep-Groove Ball Bearing

Adding a second row of balls (Figure 14–2) increases the radial load-carrying capacity of the deep-groove type of bearing compared with the single-row design because more balls share the load. Thus, a greater load can be carried in the same space, or a given load can be carried in a smaller space. The greater width of double-row bearings often adversely affects the misalignment capability.

Angular Contact Ball Bearing

One side of each race in an angular contact bearing is higher to allow the accommodation of greater thrust loads compared with the standard single-row, deep-groove bearing. The sketch in Figure 14–3 shows the preferred angle of the resultant force (radial and thrust loads combined), with commercially available bearings having angles of 15° to 40°.

Cylindrical Roller Bearing

Replacing the spherical balls with cylindrical rollers (Figure 14–4), with corresponding changes in the design of the races, gives a greater radial load capacity. The pattern of contact between a roller and its race is theoretically a line, and it becomes a rectangular shape as the members deform under load. The resulting contact stress levels are lower than for equivalent-size ball bearings, allowing smaller bearings to carry a given load or a given-size bearing to carry a higher load. Thrust load capacity is poor because any thrust load would be applied to the side of the rollers, causing rubbing, not true rolling motion. It is recommended that *no* thrust load be applied. Roller bearings are often fairly wide, giving them only fair ability to accommodate angular misalignment.

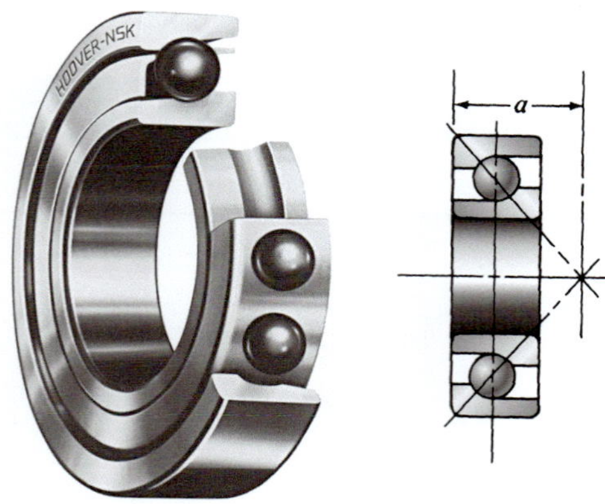

FIGURE 14–3 Angular contact ball bearing (NSK Corporation, Ann Arbor, MI)

FIGURE 14–4 Cylindrical roller bearing (NSK Corporation, Ann Arbor, MI)

Needle Bearing

Needle bearings (Figure 14–5) are actually roller bearings, but they have much smaller-diameter rollers, as you can see by comparing Figures 14–4 and 14–5. A smaller radial space is typically required for needle bearings to carry a given load than for any other type of rolling contact bearing. This makes it easier to design them into many types of equipment and components such as pumps, universal joints, precision instruments, and household appliances. The cam follower shown in Figure 14–5(b) is another example in which the antifriction operation of needle bearings can be built in with little radial space required. As with other roller bearings, thrust and misalignment capabilities are poor.

Spherical Roller Bearing

The spherical roller bearing (Figure 14–6) is one form of *self-aligning bearing*, so called because there is actual relative rotation of the outer race relative to the rollers and the inner race when angular misalignments occur. This gives the excellent rating for misalignment capability while retaining virtually the same ratings on radial load capacity.

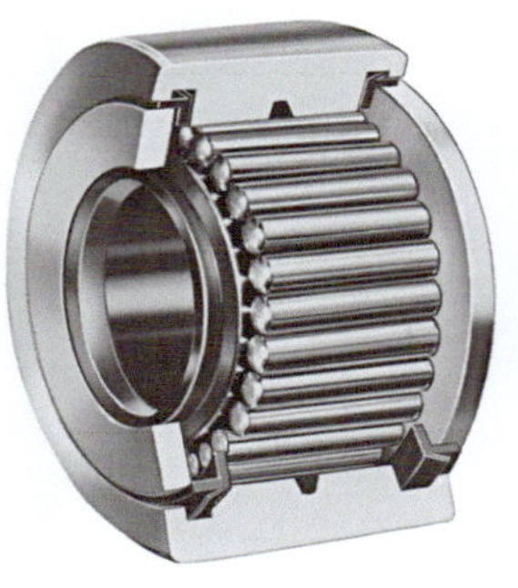

(*a*) Single- and double-row needle bearings

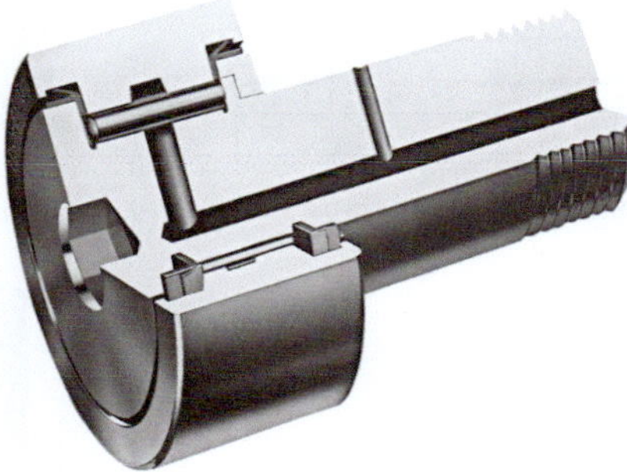

FIGURE 14–5 Needle bearings (McGill Manufacturing Co., Inc., Bearing Division, Valparaiso, IN)

(*b*) Needle bearings adapted to cam followers

FIGURE 14–6 Spherical roller bearing (NSK Corporation, Ann Arbor, MI)

FIGURE 14–7 Tapered roller bearing (NSK Corporation, Ann Arbor, MI)

Tapered Roller Bearing

Tapered roller bearings (Figure 14–7) are designed to take substantial thrust loads along with high radial loads, resulting in excellent ratings on both. They are often used in wheel bearings for vehicles and mobile equipment and in heavy-duty machinery having inherently high thrust loads. Section 14–13 gives additional information about their application. Figures 8–20, 9–33, 10–1, and 10–2 show tapered roller bearings applied in gear-type speed reducers.

14–3 THRUST BEARINGS

The bearings discussed so far in this chapter have been designed to carry radial loads or a combination of radial and thrust loads. Many machine design projects demand a bearing that resists only thrust loads, and several types of standard thrust bearings are commercially available. The same types of rolling elements are used: spherical balls, cylindrical rollers, and tapered rollers (see Figure 14–8).

Most thrust bearings can take little or no radial load. Then the design and the selection of such bearings are dependent only on the magnitude of the thrust load and the design life. The data for basic dynamic load rating and basic static load rating are reported in manufacturers' catalogs in the same way as they are for radial bearings.

14–4 MOUNTED BEARINGS

In many types of heavy machines and special machines produced in small quantities, mounted bearings rather than unmounted bearings are selected. The mounted bearings,

(*a*) Example of ball thrust bearings

(*c*) Examples of roller thrust bearings

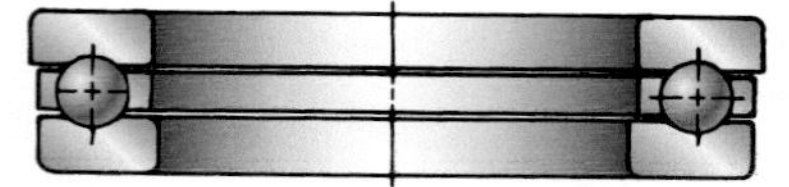

(*b*) Typical cross section of ball thrust bearing

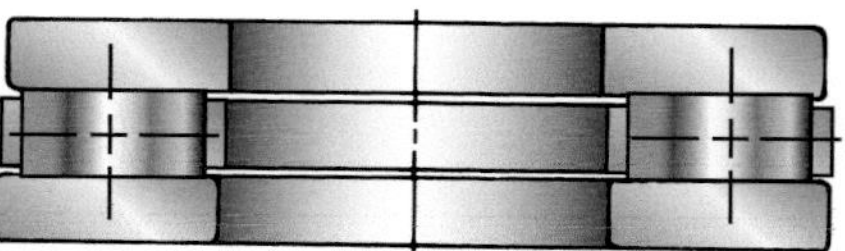

Standard cylindrical roller-thrust bearing

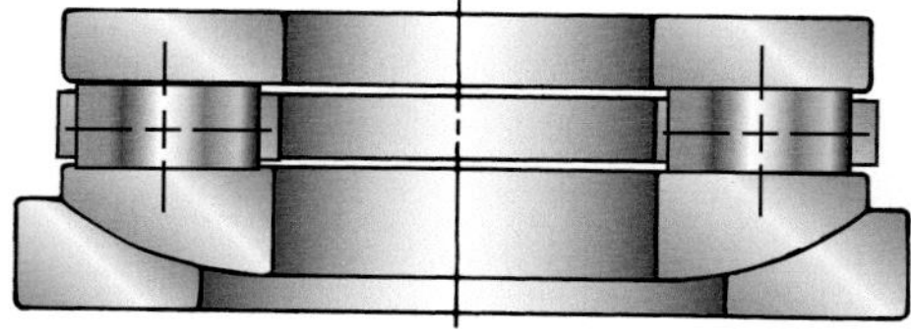

Self-aligning thrust bearing

FIGURE 14–8 Thrust bearings (Andrews Bearing Corp., Spartanburg, SC)

sometimes called housed bearings, provide a means of attaching the bearing unit directly to the frame of the machine with bolts rather than inserting it into a machined recess in a housing as is required in unmounted bearings.

Figure 14–9 shows the most common configuration for a mounted bearing: the *pillow block*. The housing is made from formed steel, cast iron, or cast steel, with holes or slots provided for attachment during assembly of the machine, at which time alignment of the bearing unit is adjusted. The bearings themselves can be of virtually any of the types discussed in the preceding sections; ball, tapered roller, or spherical roller. Misalignment capability is an important application consideration because of the conditions of use of such bearings. This capability is provided either in the bearing construction itself or in the housing as shown in Figure 14–9.

Because the bearing itself is similar to those already discussed, the selection process is also similar. Most catalogs provide extensive data listing the load-carrying capacity at specified rated life values. Internet site 7 includes an example.

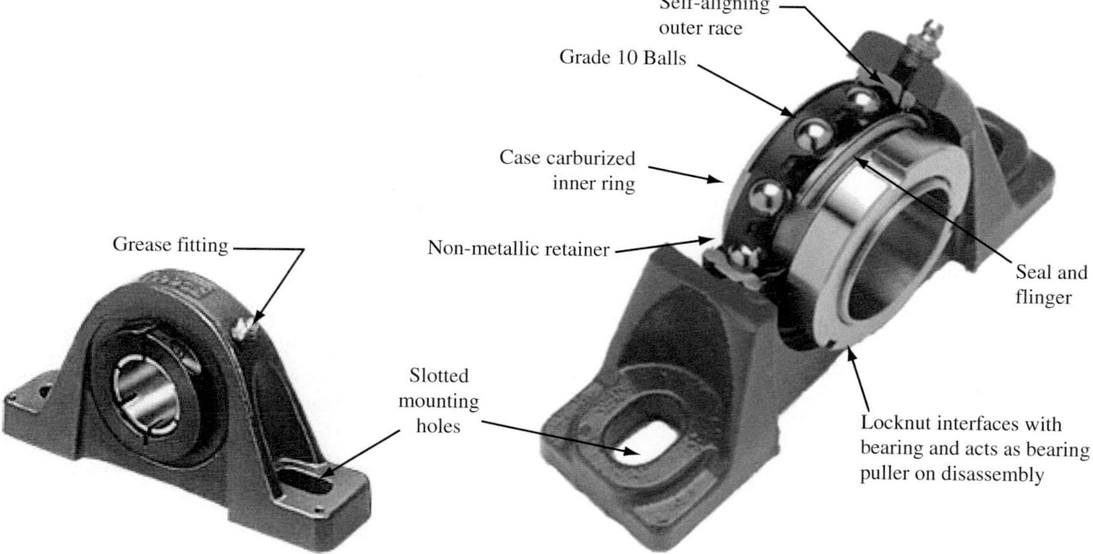

FIGURE 14–9 Ball bearing pillow block (Baldor/Dodge, Greenville, SC)

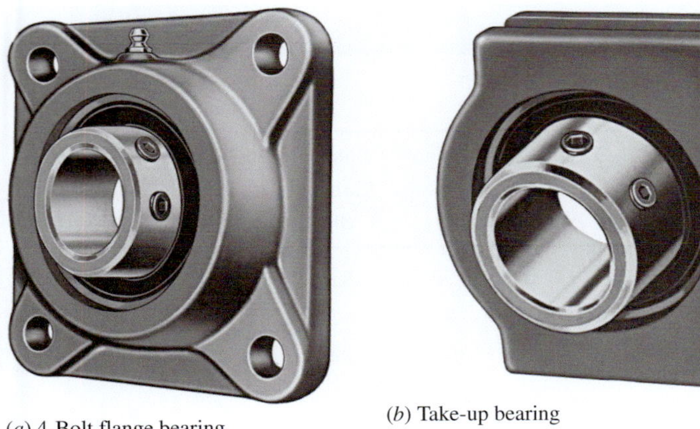

(*a*) 4-Bolt flange bearing

(*b*) Take-up bearing

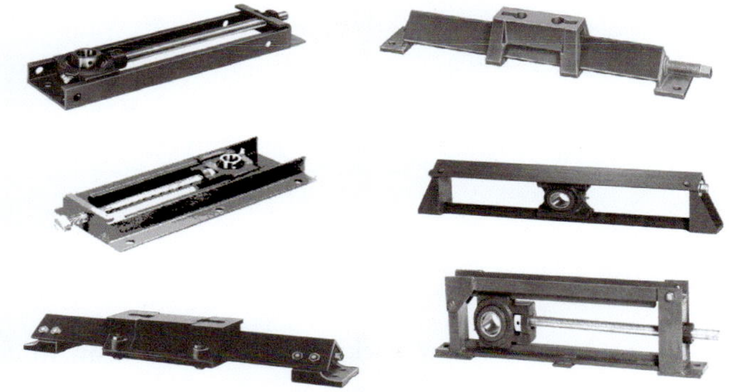

(*c*) Selected styles of take-up frames

FIGURE 14–10 Forms of mounted bearings (Baldor/Dodge, Greenville, SC)

Other forms of mounted bearings are shown in Figure 14–10. The *flange units* are designed to be mounted on the vertical side frames of machines, holding horizontal shafts. Again, several bearing types and sizes are available. The term *take-up unit* refers to a bearing mounted in a housing, which in turn is mounted in a frame that allows movement of the bearing with the shaft in place. Used on conveyors, chain drives, belt drives, and similar applications, the take-up unit permits adjustment of the center distance of the drive components at the time of installation and during operation to accommodate wear or stretch of parts of the assembly.

14–5 BEARING MATERIALS

The load applied to a rolling contact bearing passes through the outer race to the rolling elements (balls or rollers) and then to the inner race, resulting in forces exerted on very small areas of contact. Contact stresses of approximately 300 000 psi (2070 MPa) are not uncommon. To withstand such high stresses, the balls, rollers, and races are typically made from very hard, high-strength steel or ceramic. Other metallic materials used include some titanium/nickel alloys and Monel metal (alloys of primarily nickel, copper, and cobalt). Where lighter loads are encountered or when resistance to corrosion by specific materials is needed, plastic bearings can be used. See Table 14–2 for properties of selected materials.

The most widely used bearing material is SAE 52100 steel having a very high carbon content of 0.95% to 1.10% along with 1.3% to 1.6% chromium, 0.25% to 0.45% manganese, 0.2% to 0.35% silicon, and other alloying elements in low, but controlled, amounts. Impurities are carefully minimized. The steel is through hardened to the range of 58 to 65 HRC to give it the ability to resist high contact stress. Some tool steels, particularly M1 and M50, are also used. Steel alloys such as SAE 3310, 4620, and 8620, case hardened by carburizing to achieve high surface hardness and a tough core, are sometimes used. Careful control of case depth is required to withstand critical subsurface stresses. Some more lightly loaded bearings and those exposed to corrosive environments are made from SAE 440C stainless steel.

Rolling elements, races, and other components can also be made from ceramic materials such as silicon nitride (Si_3N_4), zirconia oxide (ZrO_2), alumina oxide (Al_2O_3), and silicon carbide (SiC). Compared with typical bearing steels, these materials have much lower density, higher hardness, and higher modulus of elasticity values leading to lighter weight, lower wear, and higher stiffness. They typically have better corrosion resistance and can operate at much higher temperatures as well. Their applications include machine tools, railroad equipment, semiconductor manufacturing, aerospace devices, and any equipment operating at high temperatures. Uses of ceramics can be for all bearing elements (balls, rollers, races, cages) or hybrid designs consisting of ceramic rolling elements in steel races with steel or polymeric cages. Figure 14–11 shows examples of full ceramic bearings in radial ball type and in spherical ball type. Note that silicon

TABLE 14–2 Comparison of Bearing Materials

		Steels		Titanium/ Nickel	Cermics				Monel	Plastic
Property	Unit	Beariing Steel 52 100	Stainless Steel 440C	Nitinol 60NiTi	Silicon Nitride Si_3N_4	Zirconia Oxide ZrO_2	Alumina Oxide Al_2O_3	Silicon Carbide SiC	K-500 Metal	Polyamide (Nylon 66)
Density	kg/m^3	7680	7750	6700	3230	6050	3920	3120	8434	1360
	lbm/ft^3	480	484	418	202	378	245	195	527	85
Modulus of elasticity	GPa	207	200	114	300	210	340	440	179	4.2
	ksi	30 000	29 000	16 500	43 500	30 500	49 300	63 800	25 950	610
Hardness	Vickers	700	700	650	1500	1200	1650	2800	263	—
Flexural strength	MPa	2240	1930	900	450	210	230	380	965	82
	ksi	325	280	131	65	31	33	55	140	11.9
Maximum use temperature	°C	300	350	400	1050	750	1500	1700	315	130
	°F	570	660	750	1920	1380	2730	3100	600	270

Note: Data taken from a variety of sources and are representative only. Actual properties highly dependent on specific composition, processing, thermal treatment, and form.

(*a*) Full ceramic—radial ball bearings

(*b*) Full ceramic—spherical ball bearings

FIGURE 14–11 Ceramic bearings in radial ball and spherical ball forms (Boca Bearings Company, Delray Beach, FL)

nitride is shiny black, zirconia oxide is shiny white, alumina oxide is dull white, and silicon carbide is dull black. Internet site 1 shows ceramic bearings with bores as small as 3.0 mm (0.1181 in) and up to 50 mm (1.9685 in). Inch bore sizes are as small as 0.0937 in (2.38 mm) up to 1.25 in (31.75 mm).

Alloys of titanium and nickel, sometimes called Nitinol, are applied where their anti-corrosive properties are desirable. The lower density compared with steel is also advantageous. One alloy is listed in Table 14–2. Development continues for other titanium alloys.

Monel metal is an alloy of primarily nickel, copper, and cobalt, with smaller amounts of iron, silicon, manganese, carbon, aluminum, titanium, and sulfur. Its resistance to fresh water, sea water, most acids, calcium chloride, and petroleum oils make it attractive to marine applications, chemical plants, oil and gas industries, and water and wastewater treatment facilities. One alloy is listed in Table 14–2. (See Internet site 12.)

Bearings made from plastics are typically used for lightly loaded, moderate precision applications requiring good corrosion resistance and light weight. Several types of plastic formulations can be used such as acetal, polyamide (PA, nylon) and thermoplastic polyester (PBT). Hybrid designs using plastic housing elements with stainless steel rolling balls or rollers are often used. One glass-reinforced polyamide material is listed in Table 14–2. (See Internet sites 13 and 14 for some suppliers.)

14-6 LOAD/LIFE RELATIONSHIP

Despite using steels with very high strength, all bearings have a finite life and will eventually fail due to fatigue because of the high contact stresses. But, obviously, the lighter the load, the longer the life, and vice versa. The relationship between load, P, and life, L, for rolling contact bearings can be stated as

Relationship between Bearing Load and Life

$$\frac{L_2}{L_1} = \left(\frac{P_1}{P_2}\right)^k \tag{14–1}$$

where $k = 3.00$ for ball bearings
$k = 3.33$ for roller bearings

14–7 BEARING MANUFACTURERS' DATA

The selection of a rolling contact bearing from a manufacturer's catalog involves considerations of load-carrying capacity and the geometry of the bearing. Table 14–3 shows data for three classes of single-row, deep-groove ball bearings that are similar to those in manufactures' catalogs.

Standard bearings are available in several classes, typically extra-light, light, medium, and heavy classes. The designs differ in the size and number of load-carrying elements (balls or rollers) in the bearing. The bearing number usually indicates the class and the size of the bore of the bearing. Most bearings are manufactured with nominal dimensions in metric units, and the last two digits of the bearing number indicate the nominal bore size. The bore-size convention can be seen from the data in Table 14–3. Note that for bore sizes 04 and above, the nominal bore dimension in millimeters is five times the last two digits in the bearing number.

The number preceding the last two digits indicates the class. Typically, the 100 series is for extra-light, 200 for light, 300 for medium, and 400 for heavy-duty. Figure 14–12 shows the relative size of the different classes of bearings. The three digits may be preceded by others to indicate a design code for the type of bearing. In Table 14–3, the number 6 indicates *single-row, deep-groove ball bearings*.

Inch-type bearings are available with bores ranging from 0.125 through 15.000 in.

Considering load-carrying capacity first, the data reported for each bearing design will include a basic dynamic load rating, C, and a basic static load rating, C_o.

The *basic static load rating*, C_o, is the load that the bearing can withstand without permanent deformation of any component. If this load is exceeded, the most probable result would be the indentation of one of the bearing races by the rolling elements. The deformation would be similar to that produced in the Brinell hardness test, and the failure is sometimes referred to as *brinelling*. Operation of the bearing after brinelling would be very noisy, and the impact loads on the indented area would produce rapid wear and progressive failure of the bearing.

To understand the *basic dynamic load rating*, C, it is necessary first to discuss the concept of the rated life of a bearing. Fatigue occurs over a large number of cycles of loading; for a bearing, that would be a large number of revolutions. Also, fatigue is a statistical phenomenon with considerable spread of the actual life of a group of bearings of a given design. The rated life is the standard means of reporting the results of many tests of bearings of a given design. It represents the life that 90% of the bearings would achieve successfully at a rated load. Note that it also represents the life that 10% of the bearings would not achieve. The rated life is thus typically referred to as the L_{10} *life* at the rated load.

Now the *basic dynamic load rating* C, can be defined as that load to which the bearings can be subjected while achieving a rated life (L_{10}) of 1 million revolutions (rev). Thus, the manufacturer supplies you with one set of data relating load and life. Equation (14–1) can be used to compute the expected life at any other load.

You should be aware that different manufacturers use other bases for the rated life. For example, some use 90 million cycles as the rated life and determine the rated load for that life. Also, some will report *average life*, for which 50% of the bearings will not survive. Thus, average life can be called the L_{50} life, not the L_{10}. Note that the average life is approximately five times as long as the L_{10} life. (See Internet site 2.) Be sure that you understand the basis for the ratings in any given catalog. (See References 2, 5, 6, 7, 10, and 12 for additional analysis of roller bearing performance.)

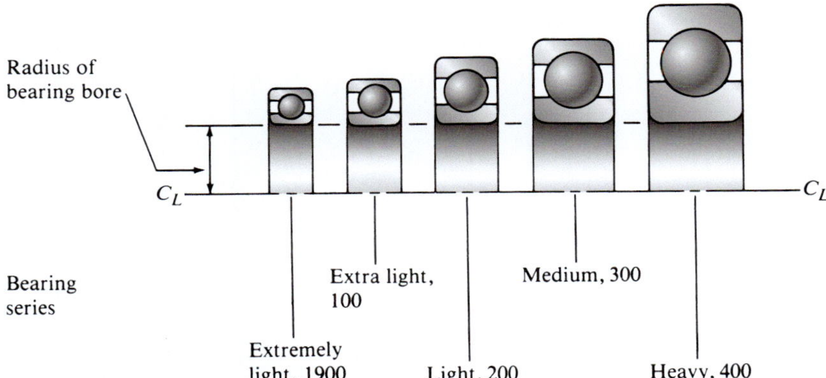

FIGURE 14–12 Relative sizes of bearing series

TABLE 14-3 Dimensions for Single Row, Deep-groove Ball Bearings

Bearing number	Nominal bearing dimensions						Basic load ratings				Maximum fillet radius r_{max}[1]		Minimum shaft shoulder diameter, S		Maximum housing shoulder diameter, H		Bearing mass	
	Bore, d		Outside dia., D		Width, B		Static, C_o		Dynamic, C									
	mm	in	mm	in	mm	in	kN	lb_f	kN	lb_f	mm	in	mm	in	mm	in	kg	lb_m
6000	**10**	0.3937	26	1.0236	8	0.3150	1.96	441	4.62	1039	0.3	0.012	12	0.472	24	0.945	0.019	0.042
6200	**10**	0.3937	30	1.1811	9	0.3543	2.36	531	5.07	1140	0.6	0.024	14	0.551	26	1.024	0.032	0.071
6300	**10**	0.3937	35	1.3780	11	0.4331	8.06	1812	3.40	764	0.6	0.024	14	0.551	31	1.220	0.053	0.117
6001	**12**	0.4724	28	1.1024	8	0.3150	2.36	531	5.07	1140	0.3	0.012	14	0.551	26	1.024	0.022	0.049
6201	**12**	0.4724	32	1.2598	10	0.3937	3.10	697	6.89	1549	0.6	0.024	16	0.630	28	1.102	0.037	0.082
6301	**12**	0.4724	37	1.4567	12	0.4724	4.15	933	9.75	2192	1.0	0.039	17	0.669	32	1.260	0.060	0.132
6002	**15**	0.5906	32	1.2598	9	0.3543	2.85	641	5.59	1257	0.3	0.012	17	0.669	30	1.181	0.030	0.066
6202	**15**	0.5906	35	1.3780	11	0.4331	3.75	843	7.80	1754	0.6	0.024	19	0.748	31	1.220	0.045	0.099
6302	**15**	0.5906	42	1.6535	13	0.5118	5.40	1214	11.40	2563	1.0	0.039	20	0.787	37	1.457	0.082	0.181
6003	**17**	0.6693	35	1.3780	10	0.3937	3.25	731	6.05	1360	0.3	0.012	19	0.748	33	1.299	0.039	0.086
6203	**17**	0.6693	40	1.5748	12	0.4724	4.75	1068	9.56	2149	0.6	0.024	21	0.827	36	1.417	0.065	0.143
6303	**17**	0.6693	47	1.8504	14	0.5512	6.55	1473	13.50	3035	1.0	0.039	22	0.866	42	1.654	0.120	0.265
6004	**20**	0.7874	42	1.6535	12	0.4724	5.00	1124	9.36	2104	0.6	0.024	24	0.945	38	1.496	0.069	0.152
6204	**20**	0.7874	47	1.8504	14	0.5512	6.55	1473	12.70	2855	1.0	0.039	25	0.984	42	1.654	0.110	0.243
6304	**20**	0.7874	52	2.0472	15	0.5906	7.80	1754	15.90	3575	1.0	0.039	27	1.063	45	1.772	0.140	0.309
6005	**25**	0.9843	47	1.8504	12	0.4724	6.55	1473	11.20	2518	0.6	0.024	29	1.142	43	1.693	0.080	0.176
6205	**25**	0.9843	52	2.0472	15	0.5906	7.80	1754	14.00	3147	1.0	0.039	30	1.181	47	1.850	0.130	0.287
6305	**25**	0.9843	62	2.4409	17	0.6693	11.60	2608	22.50	5058	1.0	0.039	32	1.260	55	2.165	0.230	0.507
6006	**30**	1.1811	55	2.1654	13	0.5118	8.30	1866	13.30	2990	1.0	0.039	35	1.378	50	1.969	0.160	0.353
6206	**30**	1.1811	62	2.4409	16	0.6299	11.2	2518	19.5	4384	1.0	0.039	35	1.378	57	2.244	0.200	0.441
6306	**30**	1.1811	72	2.8346	19	0.7480	16.0	3597	28.1	6317	1.0	0.039	37	1.457	65	2.559	0.350	0.772
6007	**35**	1.3780	62	2.4409	14	0.5512	10.2	2293	15.9	3575	1.0	0.039	40	1.575	57	2.244	0.160	0.353
6207	**35**	1.3780	72	2.8346	17	0.6693	15.3	3440	25.5	5733	1.0	0.039	42	1.654	65	2.559	0.290	0.639
6307	**35**	1.3780	80	3.1496	21	0.8268	19.0	4272	33.2	7464	1.5	0.059	43	1.693	72	2.835	0.460	1.014
6008	**40**	1.5748	68	2.6772	15	0.5906	11.6	2608	16.8	3777	1.0	0.039	45	1.772	63	2.480	0.190	0.419
6208	**40**	1.5748	80	3.1496	18	0.7087	19.0	4272	30.7	6902	1.0	0.039	47	1.850	73	2.874	0.370	0.816
6308	**40**	1.5748	90	3.5433	23	0.9055	24.0	5396	41.0	9218	1.5	0.059	48	1.890	82	3.228	0.630	1.389

Bearing number	Nominal bearing dimensions						Basic load ratings				Maximum fillet radius r_{max}[1]		Minimum shaft shoulder diameter, S		Maximum housing shoulder diameter, H		Bearing mass	
	Bore, d		Outside dia., D		Width, B		Static, C_o		Dynamic, C									
	mm	in	mm	in	mm	in	kN	lb_f	kN	lb_f	mm	in	mm	in	mm	in	kg	lb_m
6009	**45**	1.7717	75	2.9528	16	0.6299	14.6	3282	20.8	4676	1.0	0.039	50	1.969	70	2.756	0.250	0.551
6209	**45**	1.7717	85	3.3465	19	0.7480	21.6	4856	33.2	7464	1.0	0.039	52	2.047	78	3.071	0.410	0.904
6309	**45**	1.7717	100	3.9370	25	0.9843	31.5	7082	52.7	11 848	1.5	0.059	53	2.087	92	3.622	0.830	1.830
6010	**50**	1.9685	80	3.1496	16	0.6299	16.0	3597	21.6	4856	1.0	0.039	55	2.165	75	2.953	0.260	0.573
6210	**50**	1.9685	90	3.5433	20	0.7874	23.2	5216	35.1	7891	1.0	0.039	57	2.244	83	3.268	0.460	1.014
6310	**50**	1.9685	110	4.3307	27	1.0630	38.0	8543	61.8	13 894	2.0	0.079	59	2.323	101	3.976	1.050	2.315
6011	**55**	2.1654	90	3.5433	18	0.7087	21.2	4766	28.1	6317	1.0	0.039	62	2.441	83	3.268	0.390	0.860
6211	**55**	2.1654	100	3.9370	21	0.8268	29.0	6520	43.6	9802	1.5	0.059	63	2.480	92	3.622	0.610	1.345
6311	**55**	2.1654	120	4.7244	29	1.1417	45.0	10 117	71.5	16 075	2.0	0.079	64	2.520	111	4.370	1.350	2.977
6012	**60**	2.3622	95	3.7402	18	0.7087	23.2	5216	29.6	6655	1.0	0.039	67	2.638	88	3.465	0.420	0.926
6212	**60**	2.3622	110	4.3307	22	0.8661	32.5	7307	47.5	10 679	1.5	0.059	68	2.677	102	4.016	0.780	1.720
6312	**60**	2.3622	130	5.1181	31	1.2205	52.0	11 691	81.9	18 413	2.0	0.079	71	2.795	119	4.685	1.700	3.749
6013	**65**	2.5591	100	3.9370	18	0.7087	25.0	5621	30.7	6902	1.0	0.039	72	2.835	93	3.661	0.440	0.970
6213	**65**	2.5591	120	4.7244	23	0.9055	40.5	9105	55.9	12 567	1.5	0.059	73	2.874	112	4.409	0.990	2.183
6313	**65**	2.5591	140	5.5118	33	1.2992	60.0	13 489	92.3	20 751	2.0	0.079	76	2.992	129	5.079	2.100	4.631
6014	**70**	2.7559	110	4.3307	20	0.7874	31.0	6969	37.7	8476	1.0	0.039	77	3.031	103	4.055	0.600	1.323
6214	**70**	2.7559	125	4.9213	24	0.9449	45.0	10 117	60.5	13 602	1.5	0.059	78	3.071	117	4.606	1.050	2.315
6314	**70**	2.7559	150	5.9055	35	1.3780	68.0	15 288	104.0	23 381	2.0	0.079	81	3.189	139	5.472	2.500	5.513
6015	**75**	2.9528	115	4.5276	20	0.7874	33.5	7531	39.7	8925	1.0	0.039	82	3.228	108	4.252	0.640	1.411
6215	**75**	2.9528	130	5.1181	25	0.9843	49.0	11 016	66.3	14 906	1.5	0.059	83	3.268	122	4.803	1.200	2.646
6315	**75**	2.9528	160	6.2992	37	1.4567	76.5	17 199	114.0	25 629	2.0	0.079	86	3.386	149	5.866	3.000	6.615
6016	**80**	3.1496	125	4.9213	22	0.8661	40.0	8993	47.5	10 679	1.0	0.039	87	3.425	118	4.646	0.850	1.874
6216	**80**	3.1496	140	5.5118	26	1.0236	55.0	12 365	70.2	15 782	2.0	0.079	89	3.504	131	5.157	1.400	3.087
6316	**80**	3.1496	170	6.6929	39	1.5354	86.5	19 447	124.0	27 878	2.0	0.079	91	3.583	159	6.260	3.600	7.938

(continued)

TABLE 14-3 Dimensions for Single Row, Deep-groove Ball Bearings (*continued*)

Bearing number	Nominal bearing dimensions						Basic load ratings				Maximum fillet radius r_{max}[1]		Minimum shaft shoulder diameter, *S*		Maximum housing shoulder diameter, *H*		Bearing mass	
	Bore, *d*		Outside dia., *D*		Width, *B*		Static, C_o		Dynamic, *C*									
	mm	in	mm	in	mm	in	kN	lb_f	kN	lb_f	mm	in	mm	in	mm	in	kg	lb_m
6017	**85**	3.3465	130	5.1181	22	0.8661	43.0	9667	49.4	11 106	1.0	0.039	92	3.622	123	4.843	0.890	1.962
6217	**85**	3.3465	150	5.9055	28	1.1024	64.0	14 388	83.2	18 705	2.0	0.079	94	3.701	141	5.551	1.800	3.969
6317	**85**	3.3465	180	7.0866	41	1.6142	96.5	21 695	133.0	29 901	2.5	0.098	98	3.858	167	6.575	4.250	9.371
6018	**90**	3.5433	140	5.5118	24	0.9449	50.0	11 241	58.5	13 152	1.5	0.059	98	3.858	132	5.197	1.150	2.536
6218	**90**	3.5433	160	6.2992	30	1.1811	73.5	16 524	95.6	21 493	2.0	0.079	99	3.898	151	5.945	2.150	4.741
6318	**90**	3.5433	190	7.4803	43	1.6929	108.0	24 281	143.0	32 149	2.5	0.098	103	4.055	177	6.969	4.900	10.805
6019	**95**	3.7402	145	5.7087	24	0.9449	54.0	12 140	60.5	13 602	1.5	0.059	103	4.055	137	5.394	1.200	2.646
6219	**95**	3.7402	170	6.6929	32	1.2598	81.5	18 323	108.0	24 281	2.0	0.079	106	4.173	159	6.260	2.600	5.733
6319	**95**	3.7402	200	7.8740	45	1.7717	118.0	26 529	153.0	34 397	2.5	0.098	108	4.252	187	7.362	5.650	12.458
6020	**100**	3.9370	150	5.9055	24	0.9449	54.0	12 140	60.5	13 602	1.5	0.059	108	4.252	142	5.591	1.250	2.756
6220	**100**	3.9370	180	7.0866	34	1.3386	93.0	20 908	124.0	27 878	2.0	0.079	111	4.370	169	6.654	3.150	6.946
6320	**100**	3.9370	215	8.4646	47	1.8504	140.0	31 475	174.0	39 119	2.5	0.098	113	4.449	202	7.953	7.000	15.435
6021	**105**	4.1339	160	6.2992	26	1.0236	65.5	14 726	72.8	16 367	2.0	0.079	114	4.488	151	5.945	1.600	3.528
6221	**105**	4.1339	190	7.4803	36	1.4173	104.0	23 381	133.0	29 901	2.0	0.079	116	4.567	179	7.047	3.700	8.159
6321	**105**	4.1339	225	8.8583	49	1.9291	153.0	34 397	182.0	40 917	2.5	0.098	118	4.646	212	8.346	8.250	18.191
6022	**110**	4.3307	170	6.6929	28	1.1024	73.5	16 524	81.9	18 413	2.0	0.079	119	4.685	161	6.339	1.950	4.300
6222	**110**	4.3307	200	7.8740	38	1.4961	118.0	26 529	143.0	32 149	2.0	0.079	121	4.764	189	7.441	4.350	9.592
6322	**110**	4.3307	240	9.4488	50	1.9685	180.0	40 468	203.0	45 638	2.5	0.098	123	4.843	227	8.937	9.550	21.058
6024	**120**	4.7244	180	7.0866	28	1.1024	80.0	17 986	85.2	19 155	2.0	0.079	129	5.079	171	6.732	2.050	4.520
6224	**120**	4.7244	215	8.4646	40	1.5748	118.0	26 529	146.0	32 824	2.0	0.079	131	5.157	204	8.031	5.150	11.356
6324	**120**	4.7244	260	10.2362	55	2.1654	186.0	41 817	208.0	46 763	2.5	0.098	133	5.236	247	9.724	14.500	31.973
6026	**130**	5.1181	200	7.8740	33	1.2992	100.0	22 482	106.0	23 831	2.0	0.079	139	5.472	191	7.520	3.150	6.946
6226	**130**	5.1181	230	9.0551	40	1.5748	132.0	29 676	156.0	35 072	2.5	0.098	143	5.630	217	8.543	5.800	12.789
6326	**130**	5.1181	280	11.0236	58	2.2835	216.0	48 561	229.0	51 484	3.0	0.118	146	5.748	264	10.394	18.000	39.690

[1] Maximum fillet on shaft shoulder that will clear radius on bearing race

Example Problem 14–1 A catalog lists the basic dynamic load rating for a ball bearing to be 7050 lb for a rated life of 1 million rev. What would be the expected L_{10} life of the bearing if it were subjected to a load of 3500 lb?

Solution In Equation (14–1),

$$P_1 = C = 7050 \text{ lb} \qquad \text{(basic dynamic load rating)}$$
$$P_2 = P_d = 3500 \text{ lb} \qquad \text{(design load)}$$
$$L_1 = 10^6 \text{ rev} \qquad (L_{10} \text{ life at load } C)$$
$$k = 3 \qquad \text{(ball bearing)}$$

Then letting the life, L_2, be called the *design life,* L_d, at the design load,

$$L_2 = L_d = L_1\left(\frac{P_1}{P_2}\right)^k = 10^6\left(\frac{7050}{3500}\right)^{3.00} = 8.17 \times 10^6 \text{ rev}$$

This must be interpreted as the L_{10} life at a load of 3500 lb.

14–8 DESIGN LIFE

We will use the method developed in Example Problem 14–1 to refine the procedure for computing the required basic dynamic load rating C for a given design load P_d and a given design life L_d. If the reported load data in the manufacturer's literature is for 10^6 revolutions, Equation (14–1) can be written as

➩ Design Life

$$L_d = (C/P_d)^k(10^6) \qquad \textbf{(14–2)}$$

The required C for a given design load and life would be

➩ Basic Dynamic Load Rating

$$C = P_d(L_d/10^6)^{1/k} \qquad \textbf{(14–3)}$$

Most people do not think in terms of the number of revolutions that a shaft makes. Rather, they consider the speed of rotation of the shaft, usually in rpm, and the design life of the machine, usually in hours of operation. The design life is specified by the designer, considering the application. As a guide, Table 14–4 can be used. Now, for a specified design life in hours, and a known speed of rotation in rpm, the number of design revolutions for the bearing would be

$$L_d = (h)(\text{rpm})(60 \text{ min/h})$$

Example Problem 14–2 Compute the required basic dynamic load rating, C, for a ball bearing to carry a radial load of 650 lb from a shaft rotating at 600 rpm that is part of an assembly conveyor in a manufacturing plant.

Solution From Table 14–4, let's select a design life of 30 000 hours. Then L_d is

$$L_d = (30000 \text{ h})(600 \text{ rpm})(60 \text{ min/h}) = 1.08 \times 10^9 \text{ rev}$$

From Equation (14–3),

$$C = 650(1.08 \times 10^9/10^6)^{1/3} = 6670 \text{ lb}$$

TABLE 14–4 Recommended Design Life for Bearings

Application	Design life L_{10}, h
Domestic appliances, instruments, medical apparatus	1000–2000
Aircraft engines	1000–4000
Automotive	1500–5000
Agricultural equipment, hoists, construction machines	3000–6000
Elevators, industrial fans, multipurpose gearing, rotary crushers, cranes	8000–15 000
Electric motors, industrial blowers, general industrial machines, conveyors	20 000–30 000
Pumps and compressors, textile machinery, rolling mill drives	40 000–60 000
Critical equipment in continuous, 24-h operation; power plants, ship drives	100 000–200 000

Source: Eugene A. Avallone and Theodore Baumeister III, eds., *Marks' Standard Handbook for Mechanical Engineers,* 9th ed. New York: McGraw-Hill, 1986.

14–9 BEARING SELECTION: RADIAL LOADS ONLY

The selection of a bearing takes into consideration the load capacity, as discussed, and the geometry of the bearing ensuring that it can be installed conveniently in the machine. We will first consider unmounted bearings carrying radial loads only. Then we will consider unmounted bearings carrying a combination of radial and thrust loads. The term *unmounted* refers to the case in which the designer must provide for the proper application of the bearing onto the shaft and into a housing.

PROCEDURE FOR SELECTING A BEARING—RADIAL LOAD ONLY

1. The bearing is normally selected after the shaft design has progressed to the point where the required minimum diameter of the shaft has been determined, using the techniques presented in Chapter 12. The radial loads are also known, along with the orientation of the bearings with respect to other elements in the machine.
2. Specify the design load on the bearing, usually called *equivalent load*. The method of determining the equivalent load when only a radial load, R, is applied takes into account whether the inner or the outer race rotates.

Equivalent Load, Radial Load Only

$$P = VR \tag{14–4}$$

The factor V is called a *rotation factor* and takes the value of 1.0 if the inner race of the bearing rotates, which is usually the case. Use $V = 1.2$ if the outer race rotates.

3. Determine the minimum acceptable diameter of the shaft that will limit the bore size of the bearing.
4. Select the type of bearing, using Table 14–1 as a guide.
5. Specify the design life of the bearing, using Table 14–4.
6. Compute the required basic dynamic load rating, C, from Equation (14–3).
7. Identify a set of candidate bearings that have the required basic dynamic load rating.
8. Select the bearing having the most convenient geometry, also considering its cost and availability.
9. Determine mounting conditions, such as shaft seat diameter and tolerance, housing bore diameter and tolerance, means of locating the bearing axially, and special needs such as seals or shields.

Example Problem 14–3 Select a single-row, deep-groove ball bearing to carry 650 lb of pure radial load from a shaft that rotates at 600 rpm. The design life is to be 30 000 h. The bearing is to be mounted on a shaft with a minimum acceptable diameter of 1.48 in.

Solution Note that this is a pure radial load and the inner race is to be pressed onto the shaft and rotate with it. Therefore, the factor $V = 1.0$ in Equation (14–4), and the design load is equal to the radial load. These are the same data used in Example Problem 14–2, where we found the required basic dynamic load rating, C, to be 6670 lb. From Table 14–3, giving design data for three classes of bearings, we find that we could use a bearing 6208 C having a bore of 40 mm (1.5748 in).

Summary of Data for the Selected Bearing

Bearing number: 6208, single-row, deep-groove ball bearing
Bore: $d = 40$ mm (1.5748 in)
Outside diameter: $D = 80$ mm (3.1496 in)
Width: $B = 18$ mm (0.7087 in)
Maximum fillet radius: $r = 1.0$ mm (0.039 in)
Basic dynamic load rating: $C = 6902$ lb.

14–10 BEARING SELECTION: RADIAL AND THRUST LOADS COMBINED

When both radial and thrust loads are exerted on a bearing, the equivalent load is the constant radial load that would produce the same rated life for the bearing as the combined loading. The method of computing the equivalent load, P, for such cases is presented in the manufacturer's catalog and takes the form

Equivalent Load with Radial and Thrust Loads

$$P = VXR + YT \tag{14–5}$$

where P = equivalent load
V = rotation factor (as defined)
R = applied radial load
T = applied thrust load
X = radial factor
Y = thrust factor

TABLE 14–5 Radial and Thrust Factors for Single-row, Deep-groove Ball Bearings

e	T/C_o	Y	e	T/C_o	Y
0.19	0.014	2.30	0.34	0.170	1.31
0.22	0.028	1.99	0.38	0.280	1.15
0.26	0.056	1.71	0.42	0.420	1.04
0.28	0.084	1.55	0.44	0.560	1.00
0.30	0.110	1.45			

Note: $X = 0.56$ for all values of Y.

The values of X and Y vary with the specific design of the bearing and with the magnitude of the thrust load relative to the radial load. For relatively small thrust loads, $X = 1$ and $Y = 0$, so the equivalent load equation reverts to the form of Equation (14–4) for pure radial loads. To indicate the limiting thrust load for which this is the case, manufacturers list a factor called e. If the ratio $T/R > e$, Equation (14–5) must be used to compute P. If $T/R < e$, Equation (14–4) is used. Table 14–5 shows one set of data for a single-row, deep-groove ball bearing, for which the value of $X = 0.56$ for all values of Y. Note that both e and Y depend on the ratio T/C_o, where C_o is the static load rating of a particular bearing. This presents a difficulty in bearing selection because the value of C_o is not known until the bearing has been selected. Therefore, a simple trial-and-error method is applied. If a significant thrust load is applied to a bearing along with a radial load, perform the following steps:

PROCEDURE FOR SELECTING A BEARING—RADIAL AND THRUST LOAD

1. Assume a value of Y from Table 14–5. The value $Y = 1.50$ is reasonable, being at about the middle of the range of possible values.
2. Compute $P = VXR + YT$.
3. Compute the required basic dynamic load rating C from Equation (14–3).
4. Select a candidate bearing having a value of C at least equal to the required value.
5. For the selected bearing, determine C_o.
6. Compute T/C_o.
7. From Table 14–5, determine e.
8. If $T/R > e$, then determine Y from Table 14–5.
9. If the new value of Y is different from that assumed in Step 1, repeat the process.
10. If $T/R < e$, use Equation (14–4) to compute P, and proceed as for a pure radial load.

Example Problem 14–4 Select a single-row, deep-groove ball bearing from Table 14–3 to carry a radial load of 1850 lb and a thrust load of 675 lb. The shaft is to rotate at 1150 rpm, and a design life of 20 000 h is desired. The minimum acceptable diameter for the shaft is 3.10 in.

Solution Use the procedure outlined above.

Step 1. Use $X = 0.56$ and assume $Y = 1.50$.

Step 2. $P = VXR + YT = (1.0)(0.56)(1850) + (1.50)(675) = 2049$ lb.

Step 3. The required basic dynamic load rating C is found from Equation (14–3).

$$L_d = (\text{h})(\text{rpm})(60\ \text{min/h}) = (20000\ \text{h})(1150\ \text{rev/min})(60\ \text{min/h}) = 1.38 \times 10^9\ \text{rev}$$

$$C = P_d(L_d/10^6)^{1/k} = 2049\ \text{lb}(1.38 \times 10^9/10^6)^{1/3} = 22812\ \text{lb}$$

Step 4. From Table 14–3, we can use bearing number 6316 having a bore of 80 mm (3.1496 in), just larger than the stated $D_{min} = 3.10$ in.

Step 5. For bearing number 6316, $C = 27\,878$ lb and $C_o = 19\,447$ lb.

Step 6. $T/C_o = 675\ \text{lb}/(19\,447\ \text{lb}) = 0.0347$.

Step 7. From Table 14–5, $e = 0.25$ approximately.

Step 8. $T/R = 675 \text{ lb}/1850 \text{ lb} = 0.365$. Because $T/R > e$, we can find $Y = 1.78$ from Table 14–5 by interpolation based on $T/C_o = 0.0347$.

Step 9. Recompute $P = (1.0)(0.56)(1850 \text{ lb}) + (1.78)(675 \text{ lb}) = 2238 \text{ lb}$

$$C = P_d(L_d/10^6)^{1/k} = 2238 \text{ lb}(1.38 \times 10^9/10^6)^{1/3} = 24910 \text{ lb}$$

The value of $C = 27\,878$ lb for bearing 6316 is still satisfactory. Therefore, we specify that bearing for this application.

Adjustment of Life Rating for Reliability

Thus far we have used the basic L_{10} life for selecting rolling contact bearings. This is the general industrial practice and the basis for data published by most bearing manufacturers. Recall that L_{10} life indicates a 90% probability that the selected bearing would carry its rated dynamic load for the specified number of design hours. That leaves a 10% probability that any given bearing would have a lower life.

Certain applications call for greater reliability. Examples can be found in the aerospace, military, instrumentation, and medical fields. It is then desirable to be able to adjust the expected life of a bearing for higher reliability. The following equation provides a method.

$$L_{aR} = C_R L_{10} \quad \textbf{(14–6)}$$

where

L_{10} = Life in millions of revolutions for 90% reliability
L_{aR} = Life adjusted for reliability
C_R = Adjustment factor for reliability

Table 14–6 gives values of C_R for reliabilities between 90% and 99%.

It should be noted that one result of designing for higher reliability is that the bearings would be larger and more expensive.

14–11 BEARING SELECTION FROM MDESIGN SOFTWARE AND MANUFACTURERS' CATALOGS

Manual methods of selection of single-row, deep-groove ball bearings in previous sections demonstrate the basic fundamentals and use the basic rating tables in Figure 14–3. The required basic dynamic load rating C was found from Equation (14–3) after determining the values of the design load P_d and the design life L_d. Adjustment of the life rating for reliability other than the typical L_{10} life (90% reliability) was added in Equation (14–6). All of these data were determined using standard methods of rating bearings developed by industry standards such as that described in Reference 13.

Most bearing manufacturers produce large numbers of styles and sizes of bearings and report performance data in comprehensive catalogs. Furthermore, other factors beyond those listed in this book are included. Bearing standards permit manufacturers to define additional adjustment factors for special bearing properties such as:

- The quality of steel or other materials used in the bearing
- Special manufacturing processes
- Unique design features of the bearings
- Lubrication effectiveness represented by the nature of the oil film between the rolling elements and the races
- Temperature of operation and the resulting viscosity of the oil.

It is impractical to report all these factors in a textbook and some are likely different for each particular manufacturer. Therefore, it is recommended that either the MDESIGN software provided with this book or online bearing selection software provided by particular manufacturers be used in selecting bearings for problems in this book and for design projects.

Here we describe the basic process for using the Ball and Roller Bearings module from the MDESIGN software, noting that the module includes comprehensive data from the catalog of the SKF Bearing Company (Internet site 2). The process for other manufacturers should be similar and may be accessed from Internet sites 2–6 or others.

Different levels of detail for bearing life analysis can be selected within MDESIGN: *Normal* and *Extended*. The Normal level requires only basic design decisions and produces results similar to those already demonstrated in this book. The Extended level requires design decisions for

TABLE 14–6 Life Adjustment Factors for Reliability, C_R

Reliability (%)	C_R	Life designation
90	1.0	L_{10}
95	0.62	L_5
96	0.53	L_4
97	0.44	L_3
98	0.33	L_2
99	0.21	L_1

bearing lubricant viscosity grade, temperature of the surrounding environment, volume flow rate of oil through the bearing, cooling factor for external heat removal, and contamination factor for the cleanliness of the oil. The resulting life value for the Extended level of analysis is typically far higher than the Normal level.

For problems in this book, the following steps outline the input data required for using MDESIGN and the interpretation of output data. Before entering data, choose *Tools* from the top menu bar and then *Measures system* at the left end; either *US system, Metric system,* or *All systems.* This controls the default units for each input and output data value. Note that each value has a means of changing the selected unit within the selected system.

Input Data:

1. *Dynamic loading* should be selected for the type of loading for typical problems in this book. The bearings are to be shaft-mounted with the inner race rotating and the outer race stationary. The objective of the analysis is to calculate bearing life. *Static loading* can be selected for bearings that rotate slowly or simply oscillate, where the static load limit is calculated.
2. *Normal level of computation* is selected for problems in this book unless otherwise specified, where loads and life are analyzed but thermal conditions are not.
3. *In detail* is selected for the level of results reported so the user can see all performance data.
4. *SKF* is the only manufacturer's catalog available in this version of MDESIGN.
5. *New begin* is selected initially for Bearing selection because no prior data exist.

Bearing requirements:

1. Only *no* for the *design loading cases* is available for this version of MDESIGN. The full version allows different loading levels to be entered for different numbers of revolutions to determine the life using a process similar to that outlined in Section 14–15 of the book.
2. Next enter the radial load, F_r, and the axial load, F_a, applied to the bearing.
3. Enter the *Operating speed* in rpm.
4. Use Table 14–4 to decide on the *Required fatigue life* in hours.
5. The typical value for *Required reliability* is 90%, for the L_{10} life. Refer to Table 14–6 for other choices.
6. *Required stress index,* S_o, is the ratio of the actual maximum load on the bearing to the rated static load. A typical design value for S_o is 2.0.
7. Values for *Min. bore diameter, Max. outside diameter,* and *Max. bearing width* are all design decisions. The minimum acceptable shaft diameter computed for strength is typically chosen for the *Min. bore diameter.* Others are typically set quite large to produce the maximum number of candidate bearing sizes generated by the program. If space limitations exist for the design of bearing mounts, they may be used.

Bearing parameters:

1. *Bearing type* is selected from the list of 14 available types listed on the graphic aid on the Input page.
2. All other choices are typically left blank to produce the maximum number of candidate bearing sizes generated by the program. If the user wants to evaluate a particular bearing type and style, its designation information can be entered.

Calculate:

1. Two common methods of starting the *Calculate* process are:
   a. Select *Calculate* from the left end of the *Calculation* tab on the top menu.
   b. Select *F10* on the keyboard.
2. After a few seconds of searching, MDESIGN shows a screen of candidate bearings starting with those having the smallest bore that is larger than the value entered for *Min. bore diameter.*
   a. Note that all bearing choices are in metric sizes, regardless of what unit system is selected because the SKF catalog is metric.
   b. Typically select one of the smallest sizes to start by highlighting your choice and clicking *OK.* Note that several candidate bearings have the same numerical designation; e.g. 6208, but some have modifying letters. All such bearings are basically the same except for seals, cage materials, shields, or other available accessories.
   c. After a few seconds of processing time, the Output page is posted. Under the *View* tab, select *Show secondary view* to view both the Input and Output data.
3. Scan the output data for the geometric features of the selected bearing. Note that the selected bearing shows drawn to scale in a graphic pane in the lower right of the viewing screen.
   a. *Be careful that an adequate level of precision is reported for feature dimensions. Most data are integer values of metric dimensions and conversion to US units may not produce sufficient decimal places of precision. Hand calculations may be needed. Bore size and outside diameter must be known to four decimal places in inches (ten-thousandths of an inch) for proper fit of the bearing on the shaft.*
4. Check that the *Static stress index* is greater than 2.0 or the specified input value. If so, it will be printed in green.
5. Check the *Nominal fatigue life* in both revolutions (rev) and hours (L_h). This value should exceed the *Required fatigue life* entered on the Input page. If so, it will be printed in green. But it should not be excessively higher than the required value. If it is, then a smaller or lighter series bearing should be available.

6. Check the *Result evaluation* area at the end of the data set. If some aspect of performance does not meet design goals, it will be printed in red. Otherwise, it will be in green.
7. Consider whether there could be a more optimum design. To try a different choice of bearing size, simply select *F10* (*Calculate*) again and the selection screen will reappear.
8. Continue trying alternative design decisions until an optimum bearing is found.
9. Print both the input and output pages by using *Ctrl P* or select *Print* from the left pull-down main menu.
10. If it is desired to try a different style of bearing, return to the Input screen and make whatever adjustments needed to the data and then repeat these steps.

14–12 MOUNTING OF BEARINGS

Up to this point, we have considered the load-carrying capacity of the bearings and the bore size in selecting a bearing for a given application. Although these are the most critical parameters, the successful application of a bearing must consider its proper mounting. Bearings are precision machine elements. Great care must be exercised in their handling, mounting, installation, and lubrication.

The primary considerations in the mounting of a bearing are as follows:

- The shaft seat diameter and tolerances
- The housing internal bore and tolerances
- The shaft shoulder diameter against which the inner race of the bearing will be located
- The housing shoulder diameter provided for locating the outer race
- The radius of the fillets at the base of the shaft and housing shoulders
- The means of retaining the bearing in position

In a typical installation, the bore of the bearing makes a light interference fit on the shaft, and the outside diameter of the outer race makes a close clearance fit in the housing bore. To ensure proper operation and life, the mounting dimensions must be controlled to a total tolerance of only *a few ten-thousandths of an inch.* Most catalogs specify the limit dimensions for both the shaft seat diameter and the housing bore diameter.

Likewise, the catalog will specify the desirable shoulder diameters for the shaft and the housing that will provide a secure surface against which to locate the bearing while ensuring that the shaft shoulder contacts only the inner race and the housing shoulder contacts only the outer race. Table 14–3 includes these values.

The fillet radius specified in the catalog (see r in Table 14–3) is the maximum permissible radius *on the shaft and in the housing* that will clear the external radius on the bearing races. Using too large a radius would not permit the bearing to seat tightly against the shoulder. Of course, the actual fillet radius should be made as large as possible up to the maximum to minimize the stress concentration at the shoulder.

Bearings can be retained in the axial direction by many of the means described in Chapter 11. Three popular methods are retaining rings, end caps, and locknuts. Figure 14–13 shows one possible arrangement. Note that for the left bearing, the shaft diameter is slightly smaller to the left of the bearing seat. This allows the bearing to be slid easily over the shaft up to the place where it must be pressed on. The retaining ring for the outer race could be supplied as a part of the outer race instead of as a separate piece.

The right bearing is held on the shaft with a locknut threaded on the end of the shaft. See Figure 14–14 for the design of standard locknuts. The internal tab on the lockwasher engages a groove in the shaft, and one of the external

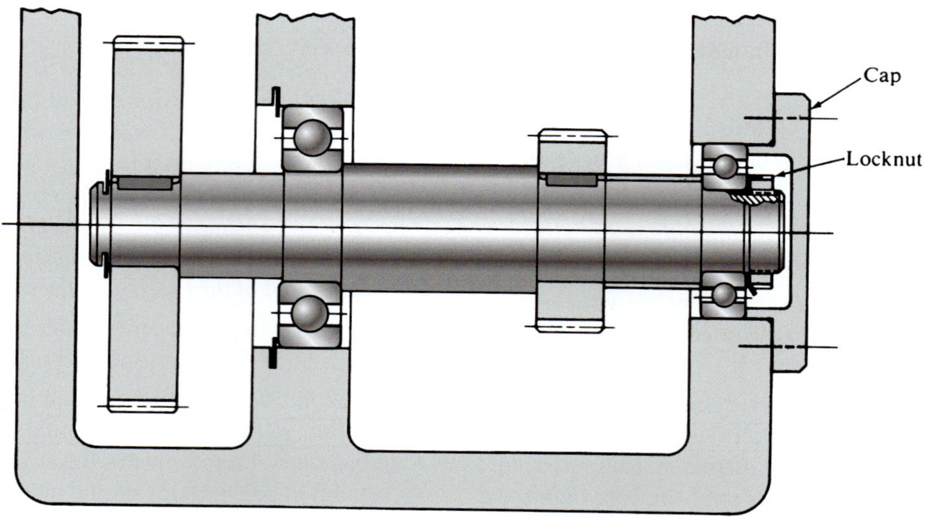

FIGURE 14–13 Bearing mounting illustration

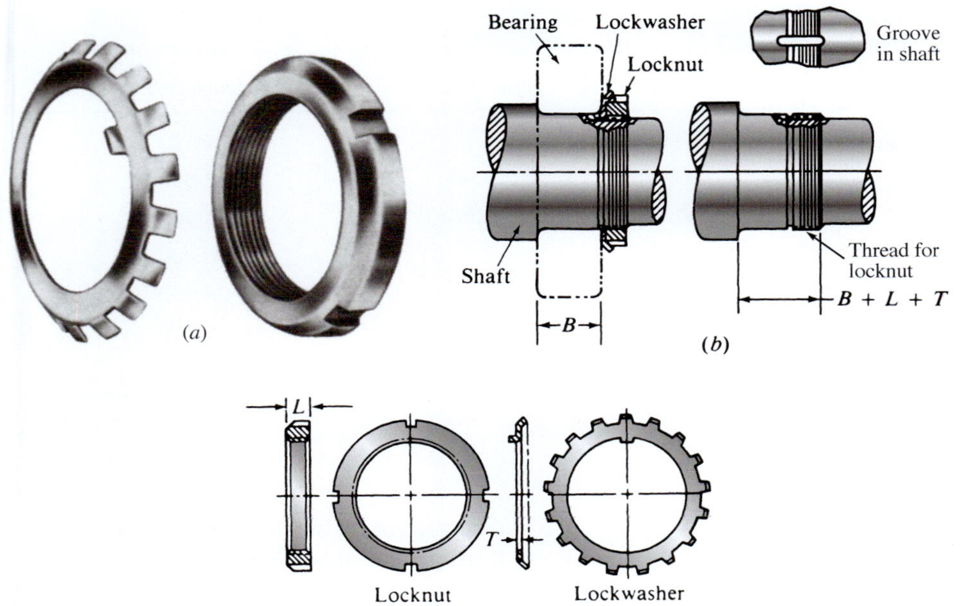

FIGURE 14–14 Locknut and lockwasher for retaining bearings (SKF USA, Inc., Norristown, PA)

tabs is bent into a groove on the nut after it is seated to keep the nut from backing off. The external cap not only protects the bearing but also retains the outer race in position.

Care must be exercised to ensure that the bearings are not overly constrained. If both bearings are held tightly, any changes in dimensions due to thermal expansion or unfavorable tolerance stackup would cause binding and could lead to dangerous unexpected loads on the bearings. It is desirable to give one bearing complete location while allowing the other bearing to float axially.

14–13 TAPERED ROLLER BEARINGS

The taper on the rollers of tapered roller bearings, evident in Figure 14–7, results in a load path different from that for the bearings discussed thus far. Figure 14–15 shows two tapered roller bearings supporting a shaft with a combination of a radial load and a thrust load. The design of the shaft is such that the thrust load is resisted by the left bearing. But a peculiar feature of this type of bearing is that a radial load on one of the bearings creates a thrust on the opposing bearing, also; this feature must be considered in analysis of the bearing.

The location of the radial reaction must also be determined with care. Part (b) of Figure 14–15 shows a dimension a that is found by the intersection of a line perpendicular to the axis of the roller and the centerline of the shaft. The radial reaction at the bearing acts through this point. The distance a is reported in the tables of data for the bearings.

The American Bearings Manufacturers' Association (ABMA) recommends the following approach in computing the equivalent loads on a tapered roller bearing:

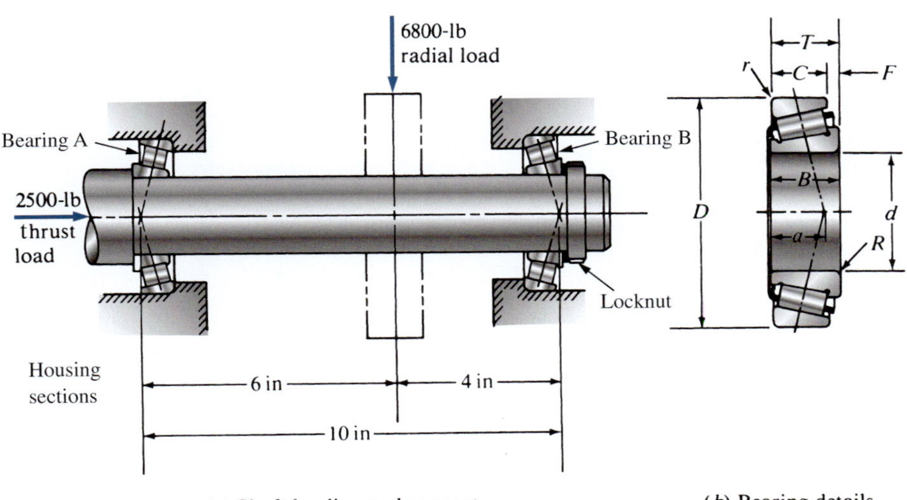

FIGURE 14–15 Example of tapered roller bearing installation

TABLE 14–7 Tapered Roller Bearing Data

Bore, d	Outside diameter, D	Width, T	a	Thrust factor, Y	Basic dynamic load rating, C
1.0000	2.5000	0.8125	0.583	1.71	8370
1.5000	3.0000	0.9375	0.690	1.98	12 800
1.7500	4.0000	1.2500	0.970	1.50	21 400
2.0000	4.3750	1.5000	0.975	2.02	26 200
2.5000	5.0000	1.4375	1.100	1.65	29 300
3.0000	6.0000	1.6250	1.320	1.47	39 700
3.5000	6.3750	1.8750	1.430	1.76	47 700

Note: Dimensions are in inches. Load C is in pounds for an L_{10} life of 1 million rev.

Equivalent Load for Tapered Roller Bearing

$$P_A = 0.4F_{rA} + 0.5\frac{Y_A}{Y_B}F_{rB} + Y_AT_A \quad (14\text{–}7)$$

$$P_B = F_{rB} \quad (14\text{–}8)$$

where P_A = equivalent radial load on bearing A
P_B = equivalent radial load on bearing B
F_{rA} = applied radial load on bearing A
F_{rB} = applied radial load on bearing B
T_A = thrust load on bearing A
Y_A = thrust factor for bearing A from tables
Y_B = thrust factor for bearing B from tables

Table 14–7 shows an abbreviated set of data from a catalog to illustrate the method of computing equivalent loads.

For the several hundred designs of standard tapered roller bearings available commercially, the value of the thrust factor varies from as small as 1.07 to as high as 2.26. In design problems, a trial-and-error procedure is usually necessary. Example Problem 14–5 illustrates one approach.

Example Problem 14–5 The shaft shown in Figure 14–15 carries a transverse load of 6800 lb and a thrust load of 2500 lb. The thrust is resisted by bearing A. The shaft rotates at 350 rpm and is to be used in a piece of agricultural equipment. Specify suitable tapered roller bearings for the shaft.

Solution The radial loads on the bearings are

$$F_{rA} = 6800(4\text{ in}/10\text{ in}) = 2720\text{ lb}$$
$$F_{rB} = 6800(6\text{ in}/10\text{ in}) = 4080\text{ lb}$$
$$T_A = 2500\text{ lb}$$

To use Equation (14–7), we must assume values of Y_A and Y_B. Let's use $Y_A = Y_B = 1.75$. Then

$$P_A = 0.40(2720) + 0.5\frac{1.75}{1.75}4080 + 1.75(2500) = 7503\text{ lb}$$
$$P_B = F_{rB} = 4080\text{ lb}$$

Using Table 14–4 as a guide, let's select 4000 h as a design life. Then the number of revolutions would be

$$L_d = (4000\text{ h})(350\text{ rpm})(60\text{ min/h}) = 8.4 \times 10^7\text{ rev}$$

The required basic dynamic load rating can now be calculated from Equation (14–3), using $k = 3.33$:

$$C_A = P_A(L_d/10^6)^{1/k}$$
$$C_A = 7503(8.4 \times 10^7/10^6)^{0.30} = 28\,400\text{ lb}$$

Similarly,

$$C_B = 4080(8.4 \times 10^7/10^6)^{0.30} = 15400\text{ lb}$$

From Table 14–7, we can choose the following bearings.

Bearing A

$d = 2.5000$ in $D = 5.0000$ in

$C = 29\,300$ lb $Y_A = 1.65$

Bearing B

$d = 1.7500$ in $D = 4.0000$ in

$C = 21\,400$ lb $Y_B = 1.50$

We can now recompute the equivalent loads:

$$P_A = 0.40(2720) + 0.5\frac{1.65}{1.50}4080 + 1.65(2500) = 7457 \text{ lb}$$

$$P_B = F_{rB} = 4080 \text{ lb}$$

From these, the new values of $C_A = 28\,200$ lb and $C_B = 15\,400$ lb are still satisfactory for the selected bearings.

One caution must be observed in using the equations for equivalent loads for tapered roller bearings. If, from Equation (14–7), the equivalent load on bearing A is less than the applied radial load, the following equation should be used.

If $P_A < F_{rA}$, then let $P_A = F_{rA}$ and compute P_B.

$$P_B = 0.4F_{rB} + 0.5\frac{Y_B}{Y_A}F_{rA} - Y_B T_A \quad \textbf{(14–9)}$$

A similar analysis is used for angular contact ball bearings in which the design of the races results in a load path similar to that for tapered roller bearings. Figure 14–3 shows an angular contact bearing and the angle through the pressure center. This is equivalent to the line perpendicular to the axis of the tapered roller bearing. The radial reaction on the bearing acts through the intersection of this line and the axis of the shaft. Also, a radial load on one bearing induces a thrust load on the opposing bearing, requiring the application of the equivalent load formulas of the type used in Equations (14–7) and (14–9). The angle of the load line in commercially available angular contact bearings ranges from 15° to 40°.

The Ball and Roller Bearing module in MDESIGN also contains an option to design a tapered roller bearing, accessing the sizeable database of sizes from the SKF catalog.

14–14 PRACTICAL CONSIDERATIONS IN THE APPLICATION OF BEARINGS

This section discusses lubrication of bearings, installation, preloading, stiffness, operation under varying loads, sealing, limiting speeds, and standards and bearing tolerance classes related to the manufacture and application of bearings.

Lubrication

The functions of lubrication in a bearing unit are as follows:

1. To provide a low-friction film between the rolling elements and the races of the bearing and at points of contact with cages, guiding surfaces, retainers, and so on.
2. To protect the bearing components from corrosion.
3. To help dissipate heat from the bearing unit.
4. To carry heat away from the bearing unit.
5. To help dispel contaminants and moisture from the bearing.

Rolling contact bearings are usually lubricated with either grease or oil. Under normal ambient temperatures (approximately 70°F) and relatively slow speeds (under 500 rpm), grease is satisfactory. At higher speeds or higher ambient temperatures, oil lubrication applied in a continuous flow is required, possibly with external cooling of the oil.

Oils used in bearing lubrication are usually clean, stable mineral oils. Under lighter loads and lower speeds, light oil is used. Heavier loads and/or higher speeds require heavier oils up to SAE 30. A recommended upper limit for lubricant temperature is 160°F. The choice of the correct oil or grease depends on many factors, so each application should be discussed with the bearing manufacturer. In general, a kinematic viscosity of 13 to 21 centistokes should be maintained at the operating temperature of the lubricant in the bearing. Manufacturer's recommendations should be sought.

In some critical applications such as bearings in jet engines and high-speed devices, lubricating oil is pumped under pressure to an enclosed housing for the bearing where the oil is directed at the rolling elements themselves. A controlled return path is also provided. The temperature of the oil in the sump is monitored and controlled with heat exchangers or refrigeration to maintain oil viscosity within acceptable limits. Such systems provide reliable lubrication and ensure the removal of heat from the bearing.

See Section 14–15 for additional discussion about the importance of oil film thickness in bearings. References 1, 3, 5, 6, and 11 contain an extensive amount of information on this subject.

Greases used in bearings are mixtures of lubricating oils and thickening agents, usually soaps such as lithium or barium. The soaps act as carriers for the oil which is drawn out at the point of need within the bearing. Additives to resist corrosion or oxidation of the oil itself are sometimes added. Classifications of greases specify the operating temperatures to which the greases will be exposed, as defined by the American Bearing Manufacturers' Association (ABMA), and outlined in Table 14–8.

Internet site 1 lists many styles of greases, oils, and solid lubricants for ceramic bearings, metal bearings, hybrids, and other precision equipment. Grease thickeners include barium complex soap, polyurea, PTFE, and lithium. Also listed is an Ultra-Dry Lube DL-5 made from a modified tungsten disulfide (WS_2) in lamellar form that is chemically inert, non-toxic, and corrosion resistant. It can operate over a temperature range of −188°C to +538°C (−350°F to +1000°F). The nominal coefficient of friction is 0.030. Some full ceramic bearings can operate well without lubrication, an important feature for applications in food processing, semiconductor production, aerospace devices, and high vacuum installations.

Installation

It has already been stated that most bearings should be installed with a light interference fit between the bore of the bearing and the shaft to preclude the possibility of rotation of the inner race of the bearing with respect to the shaft. Such a condition would result in uneven wear of the bearing elements and early failure. To install the bearing then requires rather heavy forces applied axially. Care must be exercised so that the bearing is not damaged during installation. The installation force should be applied directly to the inner race of the bearing.

If the force were applied through the outer race, the load would be transferred through the rolling elements to the inner race. Because of the small contact area, it is likely that such transfer of forces would overstress some element, exceeding the static load capacity. Brinelling would result, along with the noise and rapid wear that accompany this condition. For large bearings it may be necessary to heat the bearing to expand its diameter in order to keep the installation forces within reason. Removal of bearings intended for reuse requires similar precautions. Bearing pullers are available to facilitate this task.

TABLE 14–8 Types of Greases Used for Bearing Lubrication

Group	Type of grease	Operation temperature range (°F)
I	General-purpose	−40 to 250
II	High-temperature	0 to 300
III	Medium-temperature	32 to 200
IV	Low-temperature	−67 to 225
V	Extremely high-temperature	Up to 450

Preloading

Some bearings are made with internal clearances that must be taken up in a particular direction to ensure satisfactory operation. In such cases, preloading must be provided, usually in the axial direction. On horizontal shafts, springs are typically used, with axial adjustment of the spring deflection sometimes provided to adjust the amount of preload. When space is limited, the use of Belleville washers is desirable because they provide high forces with small deflections. Shims can be used to adjust the actual deflection and preload obtained (see Chapter 18). On vertical shafts the weight of the shaft assembly itself may be sufficient to provide the required preload.

Bearing Stiffness

Stiffness is the deflection that a given bearing undergoes when carrying a given load. Usually the radial stiffness is most important because the dynamic behavior of the rotating shaft system is affected. The critical speed and the mode of vibration are both functions of the bearing stiffness. Generally, the softer the bearing (low stiffness), the lower the critical speed of the shaft assembly. Stiffness is measured in the units used for springs, such as pounds per inch or newtons per millimeter. Of course, the stiffness values are quite high, with values of 500 000 to 1 000 000 lb/in reasonable (87.5 to 175 MN/m). The manufacturer should be consulted when such information is needed, because it is rarely included in standard catalogs.

Operation under Varying Loads

The load/life relationships used thus far assume that the load is reasonably constant in magnitude and direction. If the load varies considerably, an effective mean load must be used for determining the expected life of the bearing. (See References 4, 6, and 12.) Oscillating loads also require special analysis because only a few of the rolling elements share the load. See Section 14–16 for additional information about life prediction under varying loads.

Sealing

When the bearing is to operate in dirty or moist environments, special shields and seals are usually specified. They can be provided on either or both sides of the rolling elements. Shields are typically metal and are fixed to the stationary race, but they remain clear of the rotating race. Seals are made of elastomeric materials and do contact the rotating race. Bearings fitted with both seals and shields and precharged at the factory with grease are sometimes called *permanently lubricated.* Although such bearings are likely to give many years of satisfactory service, extreme conditions can produce a degradation of the lubricating properties of the grease. The presence of seals also increases the friction in a bearing. Sealing can be provided outside the bearing in the housing or at the shaft/housing interface. On high-speed shafts, a *labyrinth seal,* consisting of a non-contacting ring around the shaft with a few thousandths of an inch radial clearance, is frequently used. Grooves, sometimes in the form of a thread, are machined in the ring; the relative motion of the shaft with respect to the ring creates the sealing action.

Limiting Speeds

Most catalogs list limiting speeds for each bearing. Exceeding these limits may result in excessively high operating temperatures due to friction between the cages supporting the rolling elements. Generally, the limiting speed is lower for larger bearings than for smaller bearings. Also, a given bearing will have a lower limiting speed as loads increase. With special care, either in the fabrication of the bearing cage or in the lubrication of the bearing, bearings can be operated at higher speeds than those listed in the catalog. The manufacturer should be consulted in such applications. The use of ceramic rolling elements with their lower mass can result in higher limiting speeds.

Standards

Several groups are involved in standard setting for the bearing industry. Some of them are as follows:

American Bearing Manufacturers Association	(ABMA) (Internet site 11)
Annular Bearing Engineers Committee	(ABEC)
American National Standards Institute	(ANSI)
International Standards Organization	(ISO)

Many standards are listed by the ANSI and ABMA organizations including:

Load Ratings and Fatigue Life for Ball Bearings, ABMA 9

Load Ratings and Fatigue Life for Roller Bearings, ABMA 11

Tolerances

Several different tolerance classes are recognized in the bearing industry to accommodate the needs of the wide variety of equipment using rolling contact bearings. In general, of course, all bearings are precision machine elements and should be treated as such. As noted before, the general range of tolerances is of the order of a few ten-thousandths of an inch. The standard tolerance classes are defined by ABEC, as identified below.

ABEC 1:	Standard radial ball and roller bearings
ABEC 3:	Semiprecision instrument ball bearings
ABEC 5:	Precision radial ball and roller bearings
ABEC 5P:	Precision instrument ball bearings
ABEC 7:	High-precision radial ball bearings
ABEC 7P:	High-precision instrument ball bearings

Most machine applications would use ABEC 1 tolerances, the data for which are usually provided in the catalogs. Machine tool spindles requiring extra smooth and accurate running would use ABEC 5 or ABEC 7 classes.

14–15 IMPORTANCE OF OIL FILM THICKNESS IN BEARINGS

In bearings operated at heavy loads or high speeds, it is critical to maintain a film of lubricating oil at the surface of the rolling elements. A steady supply of clean lubricant with an adequate viscosity is required. With careful analysis of the geometry of the bearing, the speed of rotation, and the properties of the lubricant, it is possible to estimate the thickness of the oil film between the rolling elements and the races. Although the film thickness may be only a few microinches, it has been shown that oil starvation of the contact area is a chief cause of premature failure of rolling contact bearings. Conversely, if a film thickness significantly greater than the height of surface roughness features can be maintained, the expected life will be several times greater than published data from catalogs. (See References 4, 6, and 12.)

The nature of the lubrication at the interface between rolling elements and the races is called *elastohydrodynamic lubrication* because it depends on the specific elastic deformation of the mating surfaces under the influence of high contact stresses and the creation of a pressurized film of lubricant by the dynamic action of the rolling elements.

The data needed to evaluate film thickness for ball bearings are given in the following lists:

Bearing Geometry Factors

Ball diameter

Number of balls

Radius of curvature of the inner race groove in both the circumferential and the axial directions

Pitch diameter of the bearing; the average of the bore diameter and the outside diameter

Surface roughness of the balls and the races

Contact angle for angular contact bearings

Bearing Material Factors

Modulus of elasticity of balls and races

Poisson's ratio

Lubricant Factors

Dynamic viscosity at the operating temperature *within* the bearing

Pressure coefficient of viscosity; the change of viscosity with pressure

Operational Factors

Rotational speed of both the inner and the outer races

Radial load

Thrust load

The details of the analysis are given in Reference 6. The results of the analysis include the following:

The minimum film thickness of the lubricant, h_o

The composite roughness of the balls and the races, S

The ratio $\Lambda = h_o/S$

The service life of the bearing is dependent on the value of Λ:

- If $\Lambda < 0.90$, a life less than the manufacturer's rated life is to be expected because of surface damage due to inadequate lubricant film.
- If Λ is in the range of 0.90 to 1.50, rated service life can be expected.
- If Λ is in the range of 1.50 to 3.0, an increase of life of up to three times the rated life is possible.
- If Λ is greater than 3.0, a life of up to six times the rated life is possible.

General Recommendations for Achieving Long Life for Bearings

1. Choose a bearing with an adequate rated life using the procedures outlined in this chapter.
2. Ensure that the bearing has a fine surface finish and is not damaged by rough handling, poor installation practices, corrosion, vibration, or exposure to electrical current flow.
3. Ensure that operating loads are within the design values.
4. Supply the bearing with a copious flow of clean lubricant at an adequate viscosity at the operating temperature within the bearing according to the bearing manufacturer's recommendations. Provide external cooling for the lubricant if necessary. For an existing system, this is the factor over which you have the most control without a significant redesign of the system itself.
5. If redesign is possible, design the system to operate at as low a speed as possible.

14–16 LIFE PREDICTION UNDER VARYING LOADS

The design and analysis procedures used so far in this book assumed that the bearing would operate at a single design load throughout its life. It is possible to predict the life of the bearing under such conditions fairly accurately using manufacturers' data published in catalogs. If loads vary with time, a modified procedure is required.

One procedure that is recommended by bearing manufacturers is called the *Palmgren-Miner rule,* or sometimes simply *Miner's rule*. References 9 and 10 describe their work, and Reference 8 discusses a modified approach more tailored to bearings.

The basis of Miner's rule is that if a given bearing is subjected to a series of loads of different magnitudes for known lengths of time, each load contributes to the eventual failure of the bearing in proportion to the ratio of the load to the expected life that the bearing would have under that load. Then the cumulative effect of the series of loads must account for all such contributions to failure.

A similar approach, described in Reference 6, introduces the concept of *mean effective load, F_m*:

➩ Mean Effective Load under Varying Loads

$$F_m = \left(\frac{\Sigma_i (F_i)^p N_i}{N}\right)^{1/p} \tag{14–10}$$

where F_i = individual load among a series of i loads
N_i = number of revolutions at which F_i operates
N = total number of revolutions in a complete cycle
p = exponent on the load/life relationship; $p = 3$ for ball bearings, and $p = 10/3$ for rollers

Alternatively, if the bearing is rotating at a constant speed, and because the number of revolutions is proportional to the time of operation, N_i can be the number of minutes of operation at F_i, and N is the sum of the number of minutes in the total cycle. That is,

$$N = N_1 + N_2 + \cdots + N_i$$

Then the total expected life, in millions of revolutions of the bearing, would be

$$L = \left(\frac{C}{F_m}\right)^p \tag{14–11}$$

Example Problem 14–6 A single-row, deep-groove ball bearing number 6208 is subjected to the following set of loads for the given times:

Condition	F_i	Time
1	650 lb	30 min
2	750 lb	10 min
3	250 lb	20 min

This cycle of 60 min is repeated continuously throughout the life of the bearing. The shaft carried by the bearing rotates at 600 rpm. Estimate the total life of the bearing.

Solution Using Equation (14–10), we have

$$F_m = \left(\frac{\Sigma_i (F_i)^p N_i}{N}\right)^{1/p} \quad \textbf{(14–10a)}$$

$$F_m = \left(\frac{30(650)^3 + 10(750)^3 + 20(250)^3}{30 + 10 + 20}\right)^{1/3} = 597 \text{ lb}$$

Now use Equation (14–11):

$$L = \left(\frac{C}{F_m}\right)^p \quad \textbf{(14–11a)}$$

From Table 14–3, for the 6208 bearing, we find that $C = 6902$ lb. Then

$$L = \left(\frac{6902}{597}\right)^3 = 1545 \text{ million rev}$$

At a rotational speed of 600 rpm, the number of hours of life would be

$$L = \frac{1545 \times 10^6 \text{ rev}}{1} \cdot \frac{\text{min}}{600 \text{ rev}} \cdot \frac{\text{h}}{60 \text{ min}} = 42\,917 \text{ h}$$

Note that this is the same bearing used in Example Problem 14–3 which was selected to operate at least 30 000 hr when carrying a steady load of 650 lb.

14–17 BEARING DESIGNATION SERIES

An example of a bearing designation system has been shown in Table 14–3, using the 6000 series for single-row, deep-groove ball bearings. The following list shows the series designations for other types of bearings. Note that different manufacturers may use modifications of these series designations.

Series	Bearing type	Example for 30 mm bore, light (206)
1000	Double row self-aligning ball bearing	1206
2000	Tapered roller bearing	2206
6000	Single row, deep groove ball bearing	6206
7000	Single row angular contact ball bearing	7206
21 000	Spherical roller bearing	21 206
51 000	Ball thrust bearing	51 206
81 000	Cylindrical roller thrust bearing	81 206
NU	Single row cylindrical roller	NU206
Needle bearings use a modified system in which the last two digits indicate the bore size in mm.		
NK	Cylindrical needle roller bearing	NK30

REFERENCES

1. Association for Iron and Steel Technology-AIST. *Lubrication Engineers Manual*. 4th ed. Warrendale, PA: AIST, 2010.
2. Avallone, Eugene, Theodore Baumeister III, and Ali Sadegh. *Mark's Standard Handbook for Mechanical Engineers*. 11th ed. New York: McGraw-Hill, 2007.
3. Bloch, Heinz P. *Practical Lubrication for Industrial Facilities*. 2nd ed. Lilburn, GA: Fairmont Press, 2010.
4. Brandlein, Johannes, Paul Eschmann, Ludwig Hasbargen, and Karl Weigand. *Ball and Roller Bearings: Design and Application*. 3rd ed. New York: John Wiley & Sons, 2000.
5. Gresham, Robert M., and George E. Totten, editors. *Lubrication and Maintenance of Industrial Machinery: Best Practices and Reliability*. Boca Raton, FL: CRC Press, 2009.
6. Harris, Tedric A., and Michael N. Kotzalas. *Rolling Bearing Analysis*. 5th ed. Boca Raton, FL: CRC Press, 2007.
7. Juvinall, Robert C., and Kurt M. Marshek. *Fundamentals of Machine Component Design*. 5th ed. New York: John Wiley & Sons, 2011.
8. Kauzlarich, James J. "The Palmgren-Miner Rule Derived." *Proceedings of the 15th Leeds/Lyon Symposium of Tribology*. (Leeds, United Kingdom, September 6–9, 1988).
9. Miner, M. A. "Cumulative Damage in Fatigue." *Journal of Applied Mechanics* 67 (1945): A159–A164.
10. Palmgren, Arvid. *Ball and Roller Bearing Engineering*. 3rd ed. Philadelphia, PA: Burbank, 1959.
11. Pirro, D. M., and A. A. Wessol, editors. *Lubrication Fundamentals*. 2nd ed. Boca Raton, FL: CRC Press, 2002.
12. Changsen, Wan. *Analysis of Rolling Contact Bearings*. New York: John Wiley & Sons, 2006.
13. ASME Tribology Division. *Standard ISO 281/2—Life Ratings for Modern Rolling Bearings*. New York: ASME Press, 2003.

INTERNET SITES RELATED TO ROLLING CONTACT BEARINGS

1. **Boca Bearings Company.** www.bocabearings.com manufacturer of many styles of rolling contact bearings made from steel, ceramic, and ceramic hybrids in both inch and metric sizes. Special capabilities in miniature bearings as small as 0.040 in and 0.60 mm bore. Applications to precision industrial machinery, semiconductor industry, flowmeters, medical devices, robotics, optics, cryogenic pumps, food processing, bicycles, RC cars, and planes. Also markets specialized lubricants for bearings.
2. **SKFUSA, Inc.** www.skf.com Manufacturer of rolling contact bearings under the SKF brand. Online catalog.
3. **FAG Bearings.** www.fag.com Manufacturer of rolling contact bearings under the FAG and IMA brands. Online catalog. A subsidiary of the Schaeffler Group.
4. **RBC Bearings.** www.rbcbearings.com Manufacturer of many types of rolling contact bearings, cam followers, and specialized integrated assemblies containing bearings for the automotive, agricultural, lawn equipment, semiconductor, medical, material handling, and other similar markets.
5. **NSK Corporation.** www.nsk.com Manufacturer of rolling contact bearings under the NSK brand. Online catalog.
6. **Timken Corporation.** www.timken.com Manufacturer of rolling contact bearings under the Timken brand. Online catalog.
7. **Baldor/Dodge.** www.dodge-pt.com Manufacturer of numerous power transmission products, including mounted rolling contact bearings along with many other types of bearings. Online catalog.
8. **Emerson Power Transmission.** www.emerson-ept.com Manufacturer of numerous power transmission products, including bearings from their Browning, McGill, Rollway, and Sealmaster units. Online catalog.
9. **PowerTransmission.com.** www.powertransmission.com A site that lists numerous manufacturers of power transmission components including rolling contact bearings, gears, gear drives, clutches, couplings, motors, and more. Descriptive information about the listed companies is provided along with links to their websites.
10. **Machinery Lubrication.** www.machinerylubrication.com On online service of *Machinery Lubrication Magazine* providing technical articles and information on lubrication of industrial machinery, including bearings.
11. **American Bearing Manufacturers Association.** www.americanbearings.org A non-profit association of manufacturers of anti-friction bearings. ABMA defines national standards for bearings.
12. **High Temp Metals, Inc.** www.hightempmetals.com Distributor of high temperature, corrosion resistant grades of nickel and cobalt alloys.
13. **Igus, Inc.** www.igus.com Producer of plastic ball bearings using their proprietary polymers xiros® and iglide®.
14. **KMS Bearings, Inc.** www.kmsbearings.com Producer of mounted bearings made from combinations of plastic housings and stainless steel balls.

PROBLEMS

1. A radial ball bearing has a basic dynamic load rating of 2350 lb for a rated (L_{10}) life of 1 million rev. What would its L_{10} life be when operating at a load of 1675 lb?
2. Determine the required basic dynamic load rating for a bearing to carry 1250 lb from a shaft rotating at 880 rpm if the design life is to be 20 000 h.
3. A catalog lists the basic dynamic load rating for a ball bearing to be 3150 lb for a rated life of 1 million rev. What would be the L_{10} life of the bearing if it were subjected to a load of (a) 2200 lb and (b) 4500 lb?
4. Compute the required basic dynamic load rating, *C*, for a ball bearing to carry a radial load of 1450 lb at a shaft speed of 1150 rpm for an industrial fan.
5. Specify suitable bearings for the shaft of Example Problem 12–1. Note the data contained in Figures 12–1, 12–2, 12–11, and 12–12.
6. Specify suitable bearings for the shaft of Example Problem 12–2. Note the data contained in Figures 12–13, 12–14, 12–15, and 12–16.
7. Specify suitable bearings for the shaft of Example Problem 12–3. Note the data contained in Figures 12–17 and 12–18.
8. For any of the bearings specified in Problem 2–7, make a scale drawing of the shaft, the bearings, and the part of the housing to support the outer races of the bearings. Be sure to consider fillet radii and axial location of the bearings.
9. A bearing is to carry a radial load of 455 lb and no thrust load. Specify a suitable bearing from Table 14–3 if the shaft rotates at 1150 rpm and the design life is 20 000 hr.

For each of the problems in Table 14–9, repeat Problem 9 with the new data.

TABLE 14–9

Problem number	Radial load	Thrust load	rpm	Design life, h
10.	857 lb	0	450	30 000
11.	1265 lb	645 lb	210	5000
12.	235 lb	88 lb	1750	20 000
13.	2875 lb	1350 lb	600	15 000
14.	3.8 kN	0	3450	15 000
15.	5.6 kN	2.8 kN	450	2000
16.	10.5 kN	0	1150	20 000
17.	1.2 kN	0.85 kN	860	20 000

18. In Chapter 12, Figures P12–1 through P12–40 showed shaft design exercises related to the problems at the end of the chapter. For each bearing of each shaft, specify a suitable bearing from Table 14–3. If the shaft design has already been completed to the point that the minimum acceptable diameter of the shaft at the bearing seat is known, consider that diameter when specifying the bearing. Note the Chapter 12 problem statements for the speed of the shaft and the loading data.
19. Bearing number 6324 from Table 14–3 is carrying the set of loads in Table P14–19 while rotating at 600 rpm. Compute the

expected L_{10} life for the bearing under these conditions if the cycle repeats continuously.

TABLE P14–19

Condition	Load, F_i	Time, N_i
1	4500 lb	25 min
2	2500 lb	15 min

20. Bearing number 6314 from Table 14–3 is carrying the set of loads in Table P14–20 while rotating at 600 rpm. Compute the expected L_{10} life for the bearing under these conditions if the cycle repeats continuously.

TABLE P14–20

Condition	Load, F_i	Time, N_i
1	2500 lb	25 min
2	1500 lb	15 min

21. Bearing number 6209 from Table 14–3 is carrying the set of loads in Table P14–21 while rotating at 1700 rpm. Compute the expected L_{10} life for the bearing under these conditions if the cycle repeats continuously.

TABLE P14–21

Condition	Load, F_i	Time, N_i
1	600 lb	480 min
2	200 lb	115 min
3	100 lb	45 min

22. Bearing number 6209 from Table 14–3 is carrying the set of loads in Table P14–22 while rotating at 1700 rpm. Compute the expected L_{10} life for the bearing under these conditions if the cycle repeats continuously.

TABLE P14–22

Condition	Load, F_i	Time, N_i
1	450 lb	480 min
2	180 lb	115 min
3	50 lb	45 min

23. Bearing number 6205 from Table 14–3 is carrying the set of loads in Table P14–23 while rotating at 101 rpm for one 8-h shift. Compute the expected L_{10} life for the bearing under these conditions if the cycle repeats continuously. If the machine operates two shifts per day, six days per week, in how many weeks would you expect to have to replace the bearing?

TABLE P14–23

Condition	Load, F_i	Time, N_i
1	500 lb	6.75 h
2	800 lb	0.40 h
3	100 lb	0.85 h

24. Bearing number 6211 from Table 14–3 is carrying the set of loads in Table P14–24 while rotating at 101 rpm for one 8-h shift. Compute the expected L_{10} life for the bearing under these conditions if the cycle repeats continuously. If the machine operates two shifts per day, six days per week, in how many weeks would you expect to have to replace the bearing?

TABLE P14–24

Condition	Radial load	Thrust load	Time, N_i
1	1750 lb	350 lb	6.75 h
2	600 lb	250 lb	0.40 h
3	280 lb	110 lb	0.85 h

25. Compute the basic dynamic load rating, *C*, for a ball bearing to carry a radial load of 1450 lb at a shaft speed of 1150 rpm for 15 000 hours. Use a reliability of 95%.

26. Compute the basic dynamic load rating, *C*, for a ball bearing to carry a radial load of 509 lb at a shaft speed of 101 rpm for 20 000 hours. Use a reliability of 99%.

27. Compute the basic dynamic load rating, *C*, for a ball bearing to carry a radial load of 436 lb at a shaft speed of 1700 rpm for 5000 hours. Use a reliability of 97%.

28. Compute the basic dynamic load rating, *C*, for a ball bearing to carry a radial load of 1250 lb at a shaft speed of 880 rpm for a design life of 20 000 hours. Use a reliability of 95%.

CHAPTER FIFTEEN

COMPLETION OF THE DESIGN OF A POWER TRANSMISSION

THE BIG PICTURE — Completion of the Design of a Power Transmission

Discussion Map

- Now we bring together the concepts and design procedures from the last eight chapters to complete the design of the power transmission.

Discover

Consider how all of the machine elements that you have studied in Chapters 7–14 fit together.

Think also about the entire life cycle of the transmission, from its design to its disposal.

This chapter presents a summary of the steps that you should take to complete the design of the power transmission. Some procedures are quite detailed. You should be able to apply this experience to any design you are responsible for in the future.

This is where all of the work of Part II of this book comes together. In Chapters 7–14, you learned important concepts and design procedures for many kinds of machine elements that could all be part of a given power transmission. In each case, we mentioned how the elements need to work together. Now we show the approach to completing the design of a power transmission, which demonstrates an integrated approach and illustrates the title of this book, ***Machine Elements in Mechanical Design.*** The emphasis is on the whole design.

The lesson of this chapter is that you as a designer must constantly keep in mind how the part you are currently working on fits with other parts and how its design can affect the design of other parts. You must also consider how the part is to be manufactured, how it may be serviced and repaired as necessary, and how it will eventually be taken out of service. What will happen to the materials in the product when they have served their useful life as part of your current project?

Although we are using a power transmission in this example, the skills and insights that you will acquire should be transferable to the design of almost any other mechanical device or system.

15–1 OBJECTIVES OF THIS CHAPTER

After completing this chapter, you will be able to:

1. Bring together the individual components of a mechanical, gear-type power transmission into a unified, complete system.
2. Resolve the interface questions where two components fit together.
3. Establish reasonable tolerances and limit dimensions on key dimensions of components, especially where assembly and operation of the components are critical.
4. Verify that the final design is safe and suitable for its intended purpose.
5. Add details to some of the components that were not considered in earlier analyses.

15–2 DESCRIPTION OF THE POWER TRANSMISSION TO BE DESIGNED

The project to be completed in this chapter is the design of a single-reduction speed reducer that uses spur gears. We will use the data from Example Problem 9–1 in which the gears for the drive for an industrial saw were designed. You should review that problem now and note that it was carried forward through Example Problems 9–1 through 9–3.

We will also use elements of the mechanical design process that were first outlined in Sections 1–4 and 1–5. We will state the functions and design requirements for the power transmission, establish a set of criteria for evaluating design decisions, and implement the design tasks that were outlined in Section 1–5. References 3, 8, and 10–13 in this chapter provide additional approaches that you may find useful for other design projects.

Basic Statement of the Problem

We will design a power transmission for an industrial saw that will be used to cut tubing for vehicle exhaust pipes to length prior to the forming processes. The saw will receive 25 hp from the shaft of an electric motor rotating at 1750 rpm. The drive shaft for the saw should rotate at approximately 500 rpm.

Functions, Design Requirements, and Selection Criteria for the Power Transmission

Functions. The functions of the power transmission are as follows:

1. To receive power from an electric motor through a rotating shaft.
2. To transmit the power through machine elements that reduce the rotational speed to a desired value.
3. To deliver the power at the lower speed to an output shaft, which ultimately drives the saw.

Design Requirements. Additional information is presented here for the specific case of the industrial saw. You would normally be responsible for acquiring the necessary information and for making design decisions at this point in the design process. You would be involved in the design of the saw and would be able to discuss its desirable features with colleagues in marketing, sales, manufacturing planning, and production management and field service, and, perhaps, with customers. The kinds of information that you should seek are provided in the following list:

1. The reducer must transmit 25 hp.
2. The input is from an electric motor whose shaft rotates at a full-load speed of 1750 rpm. It has been proposed to use a NEMA frame 284T motor having a shaft diameter of 1.875 in and a keyway to accommodate a $1/2 \times 1/2$ in key. See Figure 21–18 and Table 21–3, for more data on the dimensions of the motor. Sections 21–3 to 21–8 describe industrial AC motors.
3. The output of the reducer delivers power to the saw through a shaft that rotates at 500 rpm. The speed reduction ratio will then be 3.50.
4. A mechanical efficiency of greater than 95% is desirable.
5. The minimum torque delivered to the saw should be 2950 lb · in.
6. The saw is a band saw. The cutting operation is generally smooth, but moderate shock may be encountered as the saw blade engages the tubes and if there is any binding of the blade in the cut.
7. The speed reducer will be mounted on a rigid plate that is part of the base of the saw. The means of mounting the reducer should be specified.
8. The design for the shaft for the band-saw drive has not yet been completed. It is likely that its diameter will be the same as that of the output shaft of the reducer.
9. Whereas a small, compact size for the reducer is desirable, space in the machine base should be able to accommodate most reasonable designs.
10. The saw is expected to operate 16 hours per day, 5 days per week, with a design life of 5 years. This is approximately 20 000 hours of operation.
11. The machine base will be enclosed and will prevent any casual contact with the reducer. However, the functional components of the reducer should be enclosed in their own rigid housing to protect them from contaminants and to provide for the safety of those who work with the equipment.
12. The saw will operate in a factory environment and should be capable of operation in the temperature range of 50°F to 100°F.
13. The saw is expected to be produced in quantities of 5000 units per year.
14. A moderate cost is critical to the marketing success of the saw.

Selection Criteria. The list of criteria should be produced by an interdisciplinary team composed of people having broad experience with the market for and use of such equipment. The details will vary according to the specific design. As an illustration of the process, the following criteria are suggested for the present design:

1. ***Safety:*** The speed reducer should operate safely and provide a safe environment for people near the machine.
2. ***Cost:*** Low cost is desirable so that the saw appeals to a large set of customers.
3. ***Small size.***
4. ***High reliability.***
5. ***Low maintenance.***
6. ***Smooth operation; low noise; low vibration.***

15–3 DESIGN ALTERNATIVES AND SELECTION OF THE DESIGN APPROACH

There are many ways that the speed reduction for the saw can be accomplished. Figure 15–1 shows four possibilities: (a) belt drive, (b) chain drive, (c) gear-type drive connected through flexible couplings, and (d) gear-type drive with a belt drive on the input side and connected to the saw with a flexible coupling.

Reference 9 includes a more extensive set of alternatives and a more detailed analysis. Internet site 2 provides numerous examples of commercially available speed reducers.

Selection of the Basic Design Approach

Table 15–1 shows an example of the rating that could be done to select the type of design to be produced for the speed reducer for the saw. A 10-point scale is used, with 10 being the highest rating. Of course, with more information about the actual application, a different design approach could be selected. Also, it may be desirable to proceed with more than one design to determine more of the details, thus allowing a more rational decision. A modification of the design decision matrix calls for weighting factors to be assigned to each criterion to reflect its relative importance. See References 5 and 10 and Internet site 1 for extensive discussions of rational decision analysis techniques.

On the basis of this decision analysis, let's proceed with (c), the design of a *gear-type speed reducer using flexible couplings* to connect with the drive motor and the driven

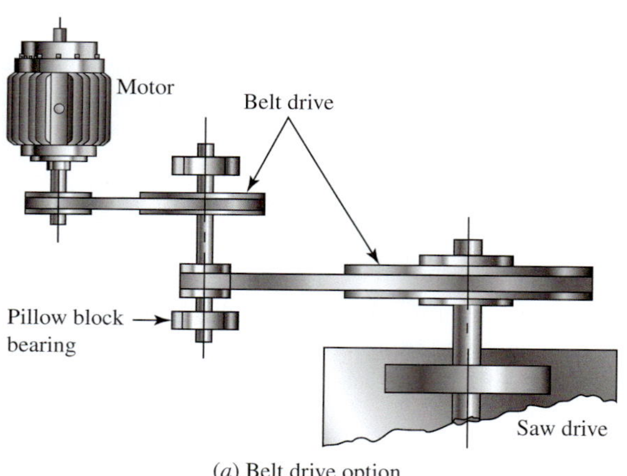

(*a*) Belt drive option

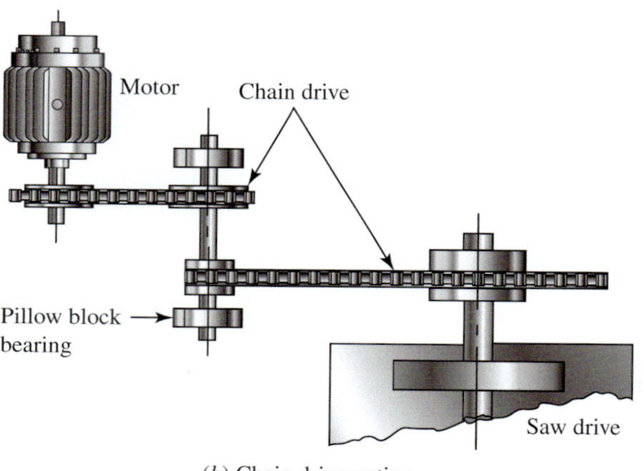

(*b*) Chain drive option

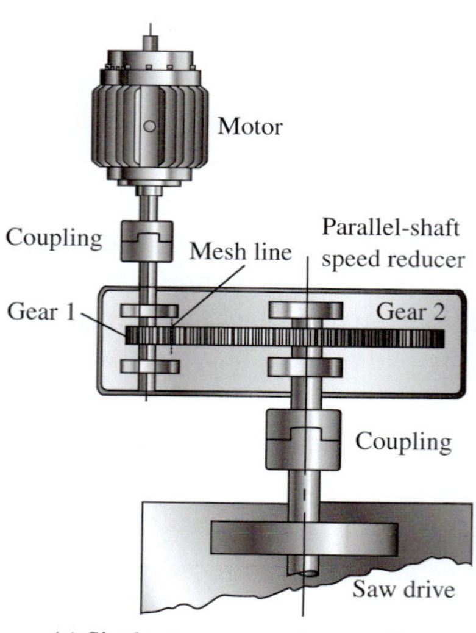

(*c*) Single-stage gear-reducer option

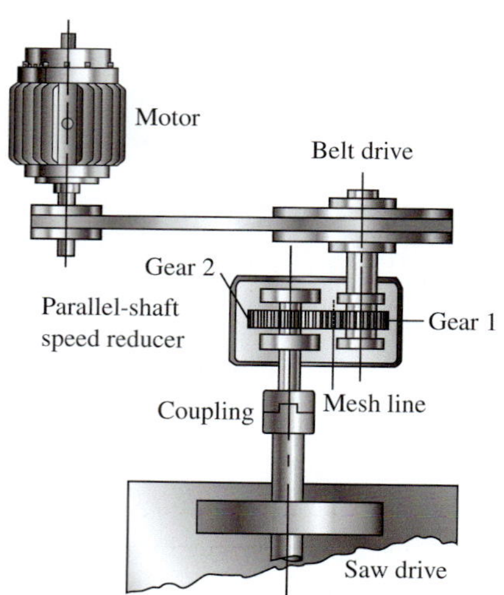

(*d*) Belt drive with single-stage gear-reducer option

FIGURE 15–1 Options for speed reduction for saw drive design project

TABLE 15–1 Decision Analysis Chart

	Alternatives			
Criteria	(a) Belt	(b) Chain	(c) Gear with flexible couplings	(d) Gear with input belt drive
1. Safety	6	6	9	7
2. Cost	9	8	7	6
3. Size	5	6	9	6
4. Reliability	7	6	10	7
5. Maintenance	6	5	9	6
6. Smoothness	8	6	9	8
Total:	41	31	53	40

shaft of the saw. It is considered to have a higher level of safety for operators and maintenance people because its rotating components are enclosed. The input and output shafts and the couplings can be covered at the time of installation. Reliability is expected to be higher because precision metallic parts are used and the drive is enclosed in a sealed housing. The flexing of belts and the significant number of moving parts in a chain drive are judged to provide lower reliability. Initial cost may be higher than for belt or chain drives. However, it is expected that maintenance will be somewhat less, leading to lower cost overall. The space taken by the design should be small, simplifying the design of other parts of the saw. Design alternative (d) is attractive if there is some interest in providing a variable speed operation in the future. By using different belt drive ratios, we can achieve different cutting speeds for the saw. A further alternative would be to consider a variable-speed electric drive motor, either to replace the need for a reducer at all or to be used in conjunction with the gear-type reducer.

15–4 DESIGN ALTERNATIVES FOR THE GEAR-TYPE REDUCER

Now that we have selected the gear-type reducer, we need to decide which type to use. Here are some alternatives:

1. ***Single-reduction spur gears:*** The nominal ratio of 3.50 : 1 is reasonable for a single pair of gears. Spur gears produce only radial loads which simplify selection of the bearings that support the shafts. Efficiency should be greater than 95% with reasonable precision of the gears, bearings, and seals. Spur gears are relatively inexpensive to produce. Shafts would be parallel and should be fairly easy to align with the motor and the drive shaft for the saw.
2. ***Single-reduction helical gears:*** These gears are equally practical as spur gears. Shaft alignment is similar. A smaller size should be possible because of the greater capacity of helical gears. However, axial thrust loads would be created which must be accommodated by the bearings and the housing. Cost is likely to be somewhat higher.
3. ***Bevel gears:*** These gears produce a right-angle drive which may be desirable, but not necessary in the present design. They are also somewhat more difficult to design and assemble to achieve adequate precision.
4. ***Worm and wormgear drive:*** This drive also produces a right-angle drive. It is typically used to achieve a higher reduction ratio than 3.50 : 1. Efficiency is usually much lower than the 95% called for in the design requirements. Heat generation could be a problem with 25 hp and the lower efficiency. A larger motor could be required to overcome the loss of power and still provide the required torque at the output shaft.

Design Decision for the Gear Type

For the present design, we choose the *single-reduction spur gear reducer*. Its simplicity is desirable, and the final cost is likely to be lower than that of the other proposed designs. The smaller size of the helical reducer is not considered to be of high priority.

15–5 GENERAL LAYOUT AND DESIGN DETAILS OF THE REDUCER

Figure 15–2 shows the proposed arrangement of the components for the single-reduction spur gear–type speed reducer. Note that the illustration in Part (b) is the top view. The design involves the following tasks:

1. Design a pinion and gear to transmit 25 hp with a pinion speed of 1750 rpm and a gear speed of 500 rpm. The ratio is 3.50 : 1. Design for both strength and pitting resistance to achieve approximately 20 000 hours of life and a reliability of at least 0.999.
2. Design two shafts, one for the pinion and one for the gear. Provide positive axial location for the gears on the shaft. The input shaft must be designed to extend

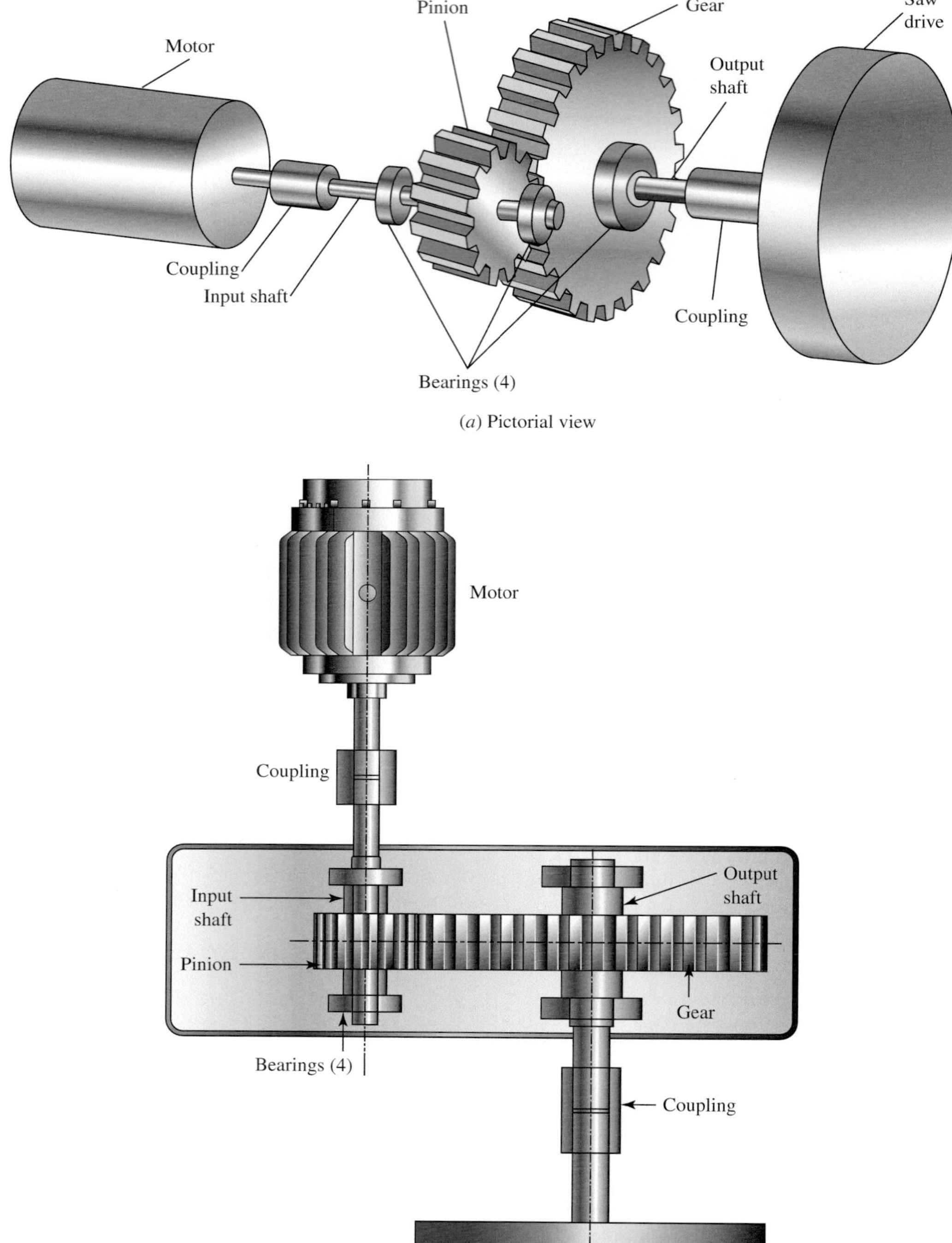

FIGURE 15–2 General layout of saw drive with single-stage gear reducer

beyond the housing to enable the motor shaft to be coupled to it. The output shaft must accommodate a coupling that mates with the drive shaft of the saw. Use a design reliability of 0.999.

3. Design six keys: one for each gear; one for the motor; one for the input shaft at the coupling; one for the output shaft at the coupling; and one for the drive shaft for the saw.
4. Specify two flexible couplings: one for the input shaft and one for the output shaft.
5. Specify four commercially available rolling contact bearings, two for each shaft. The L_{10} design life should be 20 000 hours.
6. Design a housing to enclose the gears and the bearings and to support them rigidly.

7. Provide a means of lubricating the gears within the housing.
8. Provide seals for the input and output shafts at the place where they pass through the housing. We will not specify the particular seals because of lack of data in this book. However, refer to Chapter 11 for suggestions for the types of seals that may be suitable.

Gear Design

The conditions described above are similar to those in Example Problems 9–1, 9–2, and 9–3. We propose modest changes to design decisions to achieve a gear pair that would operate more smoothly and quietly with more and smaller teeth. Namely, we select:

- Diametral pitch: $P_d = 8$ instead of 6 as was used before
- Number of pinion teeth = 28; Number of gear teeth = 98 instead of 20 and 70 before

The modified design is shown in Figure 15–3 in the form of the output from the gear design spreadsheet developed in Chapter 9. The primary results that affect the completion of the design for the reducer are summarized below.

- Diametral pitch: $P_d = 8$; 20°, full-depth, involute teeth
- Number of teeth in the pinion: $N_P = 28$
- Number of teeth in the gear: $N_G = 98$
- Diameter of the pinion: $D_P = 3.500$ in
- Diameter of the gear: $D_G = 12.250$ in
- Center distance: $C = 7.875$ in
- Face width: $F = 2.00$ in
- Quality number: $A_{vee} = 9$
- Tangential force: $W_t = 514$ lb
- Required bending stress number for pinion: $s_{at} = 20\,915$ psi
- Required contact stress number for pinion: $s_{ac} = 153\,363$ psi; requires 386 HB steel
- Material specified: SAE 4140 OQT 800; 429 HB; $s_u = 210$ ksi; 16% elongation

Shaft Design

1. ***Forces:*** Figure 15–4(a) shows the proposed configuration for the input shaft that carries the pinion and connects to the motor shaft through a flexible coupling. Figure 15–4(b) shows the output shaft, similarly configured. The only active forces on the shafts are the tangential force and the radial force from the gear teeth. The flexible couplings at the ends of the shafts allow torque transmission, but no radial or axial forces are transmitted when the alignment of the shafts is within the recommended limits for the coupling. (See Chapter 11.) Without the flexible couplings, it is very likely that significant radial loads would be produced, requiring somewhat larger diameters for the shaft and larger bearings. (See Chapter 12.)

   The spreadsheet analysis for the gears gives the tangential force as $W_t = 514$ lb. It acts downward in the vertical plane on the pinion and upward on the gear. The radial force is

$$W_r = W_r \tan\phi = (514\ \text{lb})\tan(20°) = 187\ \text{lb}$$

   The radial force acts horizontally toward the left on the pinion, tending to separate the pinion from the gear. It acts toward the right on the gear.

2. ***Torque values:*** The torque on the input shaft is

$$T_1 = (63\,000)(P)/n_p = (63\,000)(25)/1750 = 900\ \text{lb}\cdot\text{in}$$

   This value acts from the coupling at the left end of the shaft to the pinion where the power is delivered through the key to the pinion and thus to the mating gear.

   The torque on the output shaft is computed next, assuming that no power is lost. The resulting value of torque is conservative for use in the shaft design:

$$T_2 = (63\,000)(P)/n_G = (63\,000)(25)/500 = 3150\ \text{lb}\cdot\text{in}$$

   The torque acts in the output shaft from the gear to the coupling at the right end of the shaft. Assuming that the system is 95% efficient, the actual output torque is approximately

$$T_o = T_2(0.95) = 2992\ \text{lb}\cdot\text{in}$$

   This value is within the required range, as indicated in the design requirements, item 5.

3. ***Shearing force and bending moment diagrams:*** Figure 15–4 shows the shearing force and bending moment diagrams for the two shafts. Because the active loading occurs only at the gears, the form of each diagram is the same in the vertical and horizontal directions. The first number given is the value for load, the shearing force, or the bending moment in the vertical plane. The second number in parentheses is the value in the horizontal plane. The maximum bending moment in each shaft occurs where the gears are mounted. The values are

$$M_y = 643\ \text{lb}\cdot\text{in} \qquad M_x = 234\ \text{lb}\cdot\text{in}$$

   The resultant moment is $M_{max} = 684$ lb · in.

   The bending moment is zero at the bearings and in the extensions for the input and output shafts.

4. ***Support reactions—bearing forces:*** The reactions at all bearings are the same for this example because of the simplicity of the loading pattern and the symmetry of the design. The horizontal and vertical components are

$$F_y = 257\ \text{lb} \qquad F_x = 93.5\ \text{lb}$$

   The resultant force is the radial force that must be carried by the bearings: $F_r = 274$ lb. This value also produces vertical shearing stress in the shaft at the bearings.

5. ***Material selection for shafts:*** Each shaft will have a series of different diameters, shoulders with fillets, keyseats, and a ring groove as shown in Figure 15–4. Thus, much machining will be required. The shafts will be subjected to a combination of steady torque and reversed, repeated bending during normal use as the saw cuts steel tubing for vehicular exhaust systems. Occasional moderate shock loading is expected as the saw engages the tubing

APPLICATION:	Industrial saw + electirc motor		
Data for design project in Chapter 15			
Initial Input Data:			
Overload Factor:	K_o =	1.50 Table 9–7	
Input Power:	P =	*25* hp	
Design Power:	P_{des} =	37.5	
Input Speed:	n_P =	*1750* rpm	
Diametral Pitch:	P_d =	*8*	
Number of Pinion Teeth:	N_P =	*28*	
Desired Output Speed:	n_G =	*500* rpm	
Computed Number of Gear Teeth:		98.0	
Enter: Chosen No. of Gear Teeth:	N_G =	*98*	
Computed Data:			
Actual Output Speed:	n_G =	500.0 rpm	
Gear Ratio:	m_G =	3.50	
Pitch Diameter—Pinion:	D_P =	3.500 in	
Pitch Diameter—Gear:	D_G =	12.250 in	
Center Distance:	C =	7.875 in	
Pitch Line Speed:	v_t =	1604 ft/min	
Transmitted Load:	W_t =	514 lb	
Secondary Input Data:			
	Min	Nom	Max
Face Width Guidelines (in):	1.000	1.500	2.000
Enter: Face Width:	F =	*2.000* in	
Ratio: Face Width/Pinion Diameter:	F/D_P =	0.57	
Recommended Range of Ratio:	$0.50 < F/D_P < 2.00$		
Enter: Elastic Coefficient:	C_p = *2300*	Table 9–10	
Enter: Quality Number:	A_v = *9*	Table 9–3	
Dynamic Factor:	K_v = 1.33	Table 9–9	
[Factors for computing K_v :]	B = 0.630	C = 70.71	
Reference: N_P = 28		N_G = 98	
Bending Geometry Factor–Pinion:	J_P = *0.380*	Figure 9–15	
Bending Geometry Factor–Gear:	J_G = *0.440*	Figure 9–15	
Reference:	m_G = 3.50		
Enter: Pitting Geometry Factor:	I = *0.115*	Figure 9–21	

Factors in Design Analysis:				
Alignment Factor, $K_m = 1.0 + C_{pf} + C_{ma}$	If $F < 1.0$	If $F > 1.0$		
Pinion Proportion Factor, C_{pf} =	0.032	0.045	$[0.50 < F/D_P < 2.00]$	
Enter: C_{pf} =	*0.045*	Figure 9–16		
Type of Gearing:	Open	Commer.	Precision	Ex. Prec.
Mesh Alignment Factor, C_{ma} =	0.280	0.158	0.093	0.058
Enter: C_{ma} =	*0.158*	Figure 9–17		
Alignment Factor: K_m =	1.20	[Computed]		
Size Factor: K_s =	*1.00*	Table 9–7: Use 1.00 if $P_d >= 5$		
Pinion Rim Thickness Factor: K_{BP} =	*1.00*	Figure 9–18: Use 1.00 if solid blank		
Gear Rim Thickness Factor: K_{BG} =	*1.00*	Figure 9–18: Use 1.00 if solid blank		
Service Factor: SF =	*1.00*	Use 1.00 if no unusual conditions		
Reliability Factor: K_R =	*1.50*	Table 9–9: Use 1.00 for $R = 0.99$		
Enter: Design Life:	*20000* hours	See Table 9–8		
Pinion— Number of Load Cycles: N_P =	2.1E + 09	Guidelines: Y_N, Z_N		
Gear—Number of Load Cycles: N_G =	6.0E + 08	10^7 cycles	$>10^7$	$<10^7$
Bending Stress Cycle Factor: Y_{NP} =	*0.93*	1.00	0.93	Figure 9–20
Bending Stress Cycle Factor: Y_{NG} =	*0.95*	1.00	0.95	Figure 9–20
Pitting Stress Cycle Factor: Z_{NP} =	*0.88*	1.00	0.88	Figure 9–22
Pitting Stress Cycle Factor: Z_{NG} =	*0.91*	1.00	0.91	Figure 9–22
Stress Analysis: Bending				
Pinion: Required s_{at} =	20 912 psi	See Figure 9–11 or		
Gear: Required s_{at} =	17 680 psi	Table 9–5 or 9–6		
Stress Analysis: Pitting				
Pinion: Required s_{ac} =	153 353 psi	See Figure 9–12 or		
Gear: Required s_{ac} =	148 298 psi	Table 9–5 or 9–6		
Required Hardness of Pinion HB:	386	Equations in Figure 9–12-Grade 1		
Required Hardness of Gear HB:	370	Equations in Figure 9–12-Grade 1		
Specify materials, alloy and heat treatment, for most severe requirement.				
One possible material specification:				
Pinion requires HB 386: SAE 4140 OQT 800; HB 429, 16% Elongation				
Gear requires HB 370: SAE 4140 OQT 800; HB 429, 16% Elongation				
Comments:				
It is reasonable to specify the same material for both the pinion and the gear because the required hardness is simialr.				

FIGURE 15–3 Spreadsheet data used for final design project for power transmission in Chapter 15. Input data from Example Problems 9–1 to 9–3

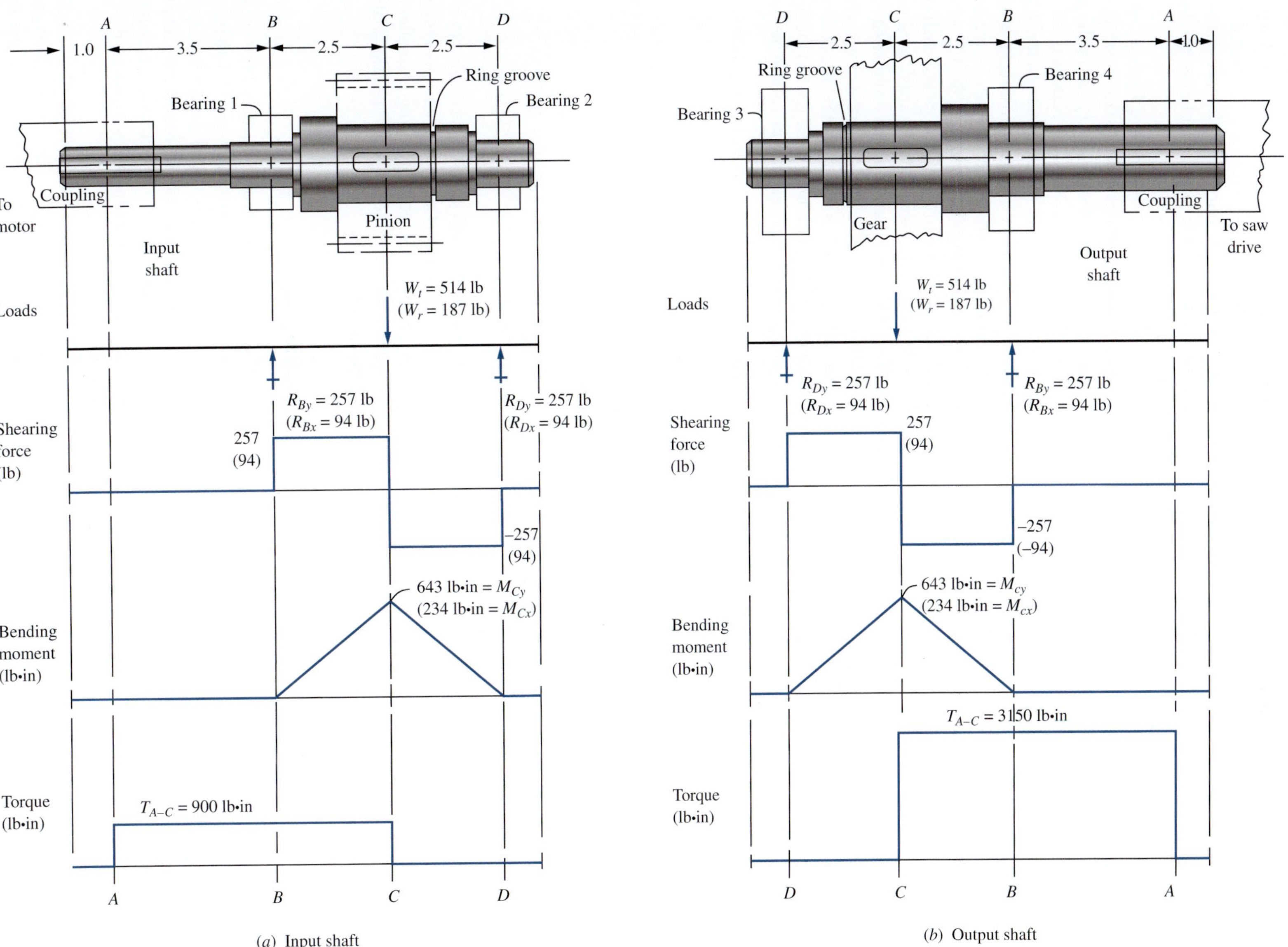

(*a*) Input shaft

(*b*) Output shaft

FIGURE 15–4 Diagrams of shaft configurations, shearing force, bending moment, and torque

and when binding occurs in the cut due to dullness of the blade or unusually hard steel in the tubing.

These conditions call for a steel that has moderately high strength, good fatigue resistance, good ductility, and good machinability. Such shafts are typically made from a medium-carbon-alloy steel (0.30% to 0.60% carbon) in either the cold-drawn or the oil-quenched and tempered condition. Good machinability is obtained from a steel having moderately high sulfur content, a characteristic of the 1100 series. Where good hardenability is also desired, higher manganese content is used.

An example of such an alloy is AISI 1144, which has 0.40% to 0.48% carbon, 1.35% to 1.65% manganese, and 0.24% to 0.33% sulfur. It is called a *resulfurized, free-machining* grade of steel. Figure A4–2 shows the range of properties available for this material when it is oil-quenched and tempered. We select a tempering temperature of 1000°F, which produces good balance between strength and ductility.

In summary, the material specified is,

AISI 1144 OQT 1000 steel: $s_u = 118\,000$ psi;
$s_y = 83\,000$ psi;
20% elongation

The endurance strength of the material can be estimated using the method outlined in Chapters 5 and 12:

Basic endurance strength: $s_n = 43\,000$ psi (from Figure 5–8 for a machined surface)

Size factor: $C_s = 0.81$ (from Figure 5–9 with an estimate of a 2.0-in diameter)

Reliability factor: $C_R = 0.75$ (desired reliability of 0.999)

Modified endurance strength:
$s'_n = s_n(C_s)(C_R) = (43\,000 \text{ psi})(0.81)(0.75) = 26\,100 \text{psi}$

6. ***Design factor, N:*** The choice of a design factor, N, should consider the many factors discussed in Chapter 5 where a nominal value of $N = 2$ was suggested for general machine design. With the expectation of moderate shock and impact loading, let's specify $N = 4$ for extra safety.
7. ***Minimum allowable shaft diameters:*** The minimum allowable shaft diameters are now computed at several sections along the shaft using Equation (12–24) if there is any combination of torsion or bending loads at the section of interest. For those sections subjected only to vertical shearing loads, such as at the bearings labeled D in Figure 15–4, Equation (12–16) is used. Table 15–2 summarizes the data used in these equations for each section and reports the computed minimum diameter. The spreadsheet from Section 12–9 was used to complete the analysis.

The last column in Table 15–2 also lists some *preliminary* design decisions for convenient diameters at the given locations. These will be reevaluated and refined as the design is completed. The diameters suggested for the shaft extensions at A on both the input and the output shafts have been set to standard values available for the bores of flexible couplings. The actual flexible couplings are discussed later.

Note that the diameters for the bearing seats at sections B and D have not been given. The reason is that the next task in the design project is to specify commercially available rolling contact bearings to carry the radial loads with suitable life. The diameters of the shafts must be specified according to the limit dimensions recommended by the bearing manufacturer. Therefore, we will leave Table 15–2 as it is for now and revisit it after completing the bearing selection process.

Another check advisable at this stage of the design process is at section C where the pinion is mounted on the shaft to ensure that adequate material is above the keyseat in the pinion hub to the root of the pinion teeth. This issue was discussed in Chapter 9 in relation to the *rim thickness factor,* K_B, and shown in Figures 9–18 and 9–19. The dimension, t_R, is defined as the radial distance from the top of the key to the root of the teeth and it is recommended that $t_R/h_t > 1.2$, where h_t is the whole depth of the teeth. For the proposed shaft design with $D_C = 1.75$ in, a 3/8 in square key is used, as shown in Table 11–1. It can be shown that the radial distance to the top of the key is 1.044 in. We need to calculate the pinion root diameter and then the root radius. Recalling that the dedendum, b, is the distance from the pitch line of the gear to the bottom of the tooth space, we find the radius to the root to be:

$$\begin{aligned}\text{Root radius} &= R_R = D/2 - b = D/2 - 1.25/P_d \\ &= 3.500/2 - 1.25/8 = 1.594 \text{ in}\end{aligned}$$

Then,

$$t_R = R_R - 1.044 \text{ in} = 1.594 - 1.044 = 0.550 \text{ in}$$

The whole depth of the gear teeth is:

$$\begin{aligned}h_t &= a + b = 1/P_d + 1.25/P_d \\ &= 2.25/P_d = 2.25/8 = 0.281 \text{ in}\end{aligned}$$

$$t_R/h_t = 0.550/0.281 = 1.95 > 1.2$$

Therefore, the space is adequate and it is acceptable to mount the pinion on the shaft. If this check failed, it is recommended that the design be changed—possibly using a smaller shaft diameter or machining the pinion directly into the shaft and eliminating the need for a key.

Bearing Selection

We will use the method outlined in Section 14–9 to select commercially available, single-row, deep-groove ball bearings from the data given in Table 14–3. The design load is equal to the radial load, and the value can be found from the shaft analysis shown in Figure 15–4. The reactions at the supports for each shaft are, in fact, the radial loads to which the bearings are subjected.

Because of the symmetry of the design of this system, and because there are no radial loads on the shaft except those produced by the action of the gear teeth, the radial loads on each of the four bearings in this design are the same. Earlier in this chapter, in the shaft design section, we determined that the bearing load is 274 lb.

TABLE 15–2 Summary of the Results of Shaft Diameter Calculations for the Preliminary Design Sizing for the Shaft

A. Input shaft

			Bending moments		Shearing forces				Diameter (in)	
Section	**Diameter (and related component)**	**Torque (lb · in)**	M_x **(lb · in)**	M_y **(lb · in)**	V_x **(lb)**	V_y **(lb)**	K_t	**Feature**	**Minimum**	**Design**
A	D_1 *(coupling)*	900	0	0	0	0	1.60	S.R. keyseat	0.73	0.875
B (to right)	D_2 *(bearing)*	900	0	0	0	0	2.50	Sharp fillet	0.73	*
Note: D_3 *must be greater than* D_2 *or* D_4 *to provide shoulders for bearing and gear.*										2.000
C	D_4 *(gear)*	900	234	643	94	257	2.00	Prof. keyseat	1.29	1.750
C (to right)	D_4 *(gear)*	0	234	643	94	257	3.00	Ring groove	1.47	1.750
D	D_5 *(bearing)*	0	0	0	94	257	2.50	Sharp fillet	0.56	*

B. Output shaft

			Bending moments		Shearing forces				Diameter (in)	
Section	**Diameter (and related component)**	**Torque (lb · in)**	M_x **(lb · in)**	M_y **(lb · in)**	V_x **(lb)**	V_y **(lb)**	K_t	**Feature**	**Minimum**	**Design**
A	D_1 *(coupling)*	3150	0	0	0	0	1.60	S.R. keyseat	1.10	1.250
B (to left)	D_2 *(bearing)*	3150	0	0	0	0	2.50	Sharp fillet	1.10	*
Note: D_3 *must be greater than* D_2 *or* D_4 *to provide shoulders for bearing and gear.*										2.000
C	D_4 *(gear)*	3150	234	643	94	257	2.00	Prof. keyseat	1.36	1.750
C (to left)	D_4 *(gear)*	0	234	643	94	257	3.00	Ring groove	1.47	1.750
D	D_5 *(bearing)*	0	0	0	94	257	2.50	Sharp fillet	0.56	*

Note: Bearing seat diameters denoted by * are to be specified. S.R. = sled runner; Prof. = profile.

Recall that design life for the bearings, L_d, is the total number of revolutions expected in service. Therefore, it is dependent on both the rotational speed of the shaft and the design life in hours. We are using a design life of 20 000 hours for all bearings. Shaft 1, the input shaft, rotates at 1750 rpm, resulting in a total number of revolutions of

$$L_d = (20\,000 \text{ h})(1750 \text{ rev/min})(60 \text{ min/h}) = 2.10 \times 10^9 \text{ rev}$$

Shaft 2, the output shaft, rotates at 500 rpm. Then its design life is

$$L_d = (20000 \text{ h})(500 \text{ rev/min})(60 \text{ min/h}) = 6.0 \times 10^8 \text{ rev}$$

Data for the bearings in Table 14–3 are for a life of 1.0 million rev (10^6 rev).

Now we will use Equation (14–3) with $k = 3$ to compute the required basic dynamic load rating C for each ball bearing. For the bearings on shaft 1,

$$C = P_d(L_d/10^6)^{1/k} = (274 \text{ lb})(2.10 \times 10^9/10^6)^{1/3} = 3510 \text{ lb}$$

Similarly, for the bearings on shaft 2,

$$C = P_d(L_d/10^6)^{1/k} = (274 \text{lb})(6.0 \times 10^8/10^6)^{1/3} = 2310 \text{ lb}$$

Listed in Table 15–3 are candidate bearings for each shaft having basic dynamic load ratings at least as high as those just computed. We also need to refer to Table 15–2 to determine the minimum acceptable diameters for the shafts at each bearing seat to ensure that the inner race diameter for the bearing is compatible.

We have selected the smallest bearing for each location on shaft 1 that had a suitable value for the basic dynamic load rating. For shaft 2, we decided that the diameter of the shaft extension should be 1.25 in and that the bearing bore must be larger. Bearing 6207 provides a suitable bore and an additional safety factor on the load rating.

Note that the dimensions for the bearings are actually listed in mm by the manufacturer as indicated in Table 14–3. The decimal-inch equivalents are somewhat

TABLE 15–3 Candidate Bearings for Shafts 1 and 2

A. For shaft 1: Minimum bore = 0.73 in at section *B*; 0.56 in at section *D*; C_{min} = 3516 lb

Bearing no.	*C* (lb)	*d* (in)	*D* (in)	*B* (in)	r_{max} (in)	Comment
6206	4384	1.1811	2.4409	0.6299	0.039	Large for both *B* and *D*
6305	5058	0.9843	2.4409	0.6693	0.039	Specify for both *B* and *D*

B. For shaft 2: Minimum bore = 1.10 in at section *B*; 0.56 in at section *D*; C_{min} = 2310 lb

Bearing no.	*C* (lb)	*d* (in)	*D* (in)	*B* (in)	r_{max} (in)	Comment
6205	3147	0.9843	2.0472	0.5906	0.039	Small for *B;* specify for *D*
6207	5733	1.3780	2.8346	0.6693	0.039	Specify for *B*

inconvenient, but they must be used. The dimensions in mm are listed below:

Inch dimension	mm	Inch dimension	mm
0.5906	15	1.3780	35
0.6299	16	2.0472	52
0.6693	17	2.4409	62
0.9843	25	2.8346	72
1.1811	30	0.039	1.00-mm (fillet radius)

Bearing Mounting on the Shafts and in the Housing

With the specifications for the bearings, we can finalize the basic dimensions for the shaft diameters. Table 15–4 is an update of the data in Table 15–2 with the bearing bore dimensions given. A few other changes are included as well, which is typical of the iterative nature of design. For example, the diameter of the input shaft at the coupling (section *A*) was made slightly smaller than the bearing seat diameter. This permits the bearing to be slid onto the shaft easily to the point where it is then pressed into position on its seat at section *B* and against the shoulder. Another check needs to be made at section *D* for both shafts where a 1.750-in diameter steps down to the bearing seat diameter of 0.984 in. There is the possibility that the step is too large and that it may interfere with the outer race of the bearing. That will be checked as we complete the details of mounting the bearings. If interference does occur, it should be a simple matter to provide another small step to make the bearing shoulder an acceptable height.

Mounting of ball and roller bearings on shafts and into housings requires very careful consideration of limit dimensions on all mating parts to ensure proper fits as defined by the bearing manufacturer. The total tolerances on shaft diameters are only a few ten-thousandths of an inch in sizes up to about 6.00 in. Total tolerances on housing bore diameters range from about 0.001 to 0.004 in for sizes from about 1.00 in to over 16.0 in. Violation of the recommended fits will likely cause unsatisfactory performance and possibly early failure of the bearing.

The bore of a bearing is typically pressed onto the shaft seat with a light interference fit to ensure that the inner race rotates with the shaft. The *OD* of the bearing is typically a close sliding fit in the housing, with the minimum clearance being zero. This facilitates installation and allows some slight movement of the bearing as thermal deformation occurs during operation. Tighter fits than those recommended by the manufacturer may cause the rolling elements to bind between the inner and outer races, resulting in higher loads and higher friction. Looser fits may permit the outer race to rotate relative to the housing, a very undesirable situation.

Only one of the two bearings on a shaft should be located and held fixed axially in the housing to provide proper alignment of functional components such as the gears in this design. The second bearing should be installed in a way that allows some small axial movement during operation. If the second bearing is held fixed also, it is likely that extra axial loads will be developed for which the bearing has not been designed.

We first discuss the specification of shaft limit dimensions at the bearing seats.

Bearing Seat Diameters. Because most commercially available bearings are produced to metric dimensions, the fits are specified according to the tolerance system of the International Standards Organization (ISO). Only a sampling of the data are listed here to illustrate the process of specifying limit dimensions for shafts and housings to accommodate bearings. Manufacturers' catalogs include much more extensive data.

For bearings carrying moderate to heavy loads such as those in this example design, the following tolerance grades are recommended for the bearing seats on shafts and housing bore fits with the outer race:

Bearing bore diameter range	Tolerance grade
10–18 mm	j5
20–100 mm	k5
105–140 mm	m5
150–200 mm	m6
Housing bore (any)	H8

TABLE 15–4 Summary of the Results of the Shaft Diameter Calculations for the Revised Design Sizing for the Shaft

A. Input shaft

			Bending moments		Shearing forces				Diameter (in)	
Section	**Diameter (and related component)**	**Torque (lb · in)**	**M_x (lb · in)**	**M_y (lb · in)**	**V_x (lb)**	**V_y (lb)**	**K_t**	**Feature**	**Minimum**	**Design**
A	*D_1 (coupling)*	900	0	0	0	0	1.60	S.R. keyseat	0.73	0.875
B (to right)	*D_2 (bearing)*	900	0	0	0	0	2.50	Sharp fillet	0.73	0.984
Note: D_3 must be greater than D_2 or D_4 to provide shoulders for bearing and gear.										2.000
C	*D_4 (gear)*	900	234	643	94	257	2.00	Prof. keyseat	1.29	1.750
C (to right)	*D_4 (gear)*	0	234	643	94	257	3.00	Ring groove	1.47	1.750
D	*D_5 (bearing)*	0	0	0	94	257	2.50	Sharp fillet	0.56	0.984

B. Output shaft

			Bending moments		Shearing forces				Diameter (in)	
Section	**Diameter (and related component)**	**Torque (lb · in)**	**M_x (lb · in)**	**M_y (lb · in)**	**V_x (lb)**	**V_y (lb)**	**K_t**	**Feature**	**Minimum**	**Design**
A	*D_1 (coupling)*	3150	0	0	0	0	1.60	S.R. keyseat	1.10	1.250
B (to left)	*D_2 (bearing)*	3150	0	0	0	0	2.50	Sharp fillet	1.10	1.378
Note: D_3 must be greater than D_2 or D_4 to provide shoulders for bearing and gear.										2.000
C	*D_4 (gear)*	3150	234	643	94	257	2.00	Prof. keyseat	1.36	1.750
C (to left)	*D_4 (gear)*	0	234	643	94	257	3.00	Ring groove	1.47	1.750
D	*D_5 (bearing)*	0	0	0	94	257	2.50	Sharp fillet	0.56	0.984

Note: S.R. = sled runner; Prof. = profile.

Table 15–5 shows representative data for the actual limit dimensions for these grades over the size ranges included for the bearings listed in Table 14–3. Note that the bearing bore and the bearing *OD* dimensions are those expected from the bearing manufacturer. You must control the shaft diameter and the housing bore to the specified minimum and maximum dimensions. The table also lists the minimum and maximum fits that result. The symbol L indicates that there is a net clearance (loose) fit; T indicates an interference (tight) fit. So bearings must be pressed onto the shaft seat. Sometimes heating of the bearing and cooling of the shaft are used to produce a clearance to facilitate assembly. When the parts return to normal temperatures, the final fit is produced.

We now show the determination of the limit dimensions for the shaft at each bearing seat.

Shaft 1: Input Shaft. Both bearings 1 and 2 are number 6305.

Nominal bore = 25 mm (0.9843 in)

From Table 15–5: k5 ISO tolerance grade on the shaft seat; limits of 0.9847–0.9844 in

Resulting fit between bearing bore and shaft seat: 0.0001 in tight to 0.0008 in tight

OD of the outer race = 62 mm (2.4409 in)

From Table 15–5: H8 ISO tolerance grade on the housing bore; limits of 2.4409–2.4427 in

Resulting fit between outer race and housing bore: 0.0 to 0.0023 in loose

Shaft 2: Output Shaft. Bearing 3 at *D* is number 6205.

Nominal bore = 25 mm (0.9843 in)

From Table 15–5: k5 ISO tolerance grade on the shaft seat; limits of 0.9847–0.9844 in

Resulting fit between bearing bore and shaft seat: 0.0001 in tight to 0.0008 in tight

OD of the outer race = 52 mm (2.0472 in)

From Table 15–5: H8 ISO tolerance grade on the housing bore; limits of 2.0472–2.0490 in

Resulting fit between outer race and housing bore: 0.0 to 0.0023 in loose

TABLE 15–5 Shaft and Housing Fits for Bearings

A. Shaft fits

Bearing bore			ISO tolerance grade	Shaft diameter		Limits of fit	
Nominal (mm)	Maximum (in)	Minimum (in)		Maximum (in)	Minimum (in)	Minimum (in)	Maximum (in)
10	0.3937	0.3934	j5	0.3939	0.3936	0.0001L	0.0005T
12	0.4724	0.4721	j5	0.4726	0.4723	0.0001L	0.0005T
15	0.5906	0.5903	j5	0.5908	0.5905	0.0001L	0.0005T
17	0.6693	0.6690	j5	0.6695	0.6692	0.0001L	0.0005T
20	0.7874	0.7870	k5	0.7878	0.7875	0.0001T	0.0008T
25	0.9843	0.9839	k5	0.9847	0.9844	0.0001T	0.0008T
30	1.1811	1.1807	k5	1.1815	1.1812	0.0001T	0.0008T
35	1.3780	1.3775	k5	1.3785	1.3781	0.0001T	0.0010T
40	1.5748	1.5743	k5	1.5753	1.5749	0.0001T	0.0010T
45	1.7717	1.7712	k5	1.7722	1.7718	0.0001T	0.0010T
50	1.9685	1.9680	k5	1.9690	1.9686	0.0001T	0.0010T
55	2.1654	2.1648	k5	2.1660	2.1655	0.0001T	0.0012T
60	2.3622	2.3616	k5	2.3628	2.3623	0.0001T	0.0012T
65	2.5591	2.5585	k5	2.5597	2.5592	0.0001T	0.0012T
70	2.7559	2.7553	k5	2.7565	2.7560	0.0001T	0.0012T
75	2.9528	2.9522	k5	2.9534	2.9529	0.0001T	0.0012T
80	3.1496	3.1490	k5	3.1502	3.1497	0.0001T	0.0012T
85	3.3465	3.3457	k5	3.3472	3.3466	0.0001T	0.0015T
90	3.5433	3.5425	k5	3.5440	3.5434	0.0001T	0.0015T
95	3.7402	3.7394	k5	3.7409	3.7403	0.0001T	0.0015T
100	3.9370	3.9362	k5	3.9377	3.9371	0.0001T	0.0015T
105	4.1339	4.1331	m5	4.1350	4.1344	0.0005T	0.0019T
110	4.3307	4.3299	m5	4.3318	4.3312	0.0005T	0.0019T
115	4.5276	4.5268	m5	4.5287	4.5281	0.0005T	0.0019T
120	4.7244	4.7236	m5	4.7255	4.7249	0.0005T	0.0019T
125	4.9213	4.9203	m5	4.9226	4.9219	0.0006T	0.0023T
130	5.1181	5.1171	m5	5.1194	5.1187	0.0006T	0.0023T
140	5.5118	5.5108	m5	5.5131	5.5124	0.0006T	0.0023T
150	5.9055	5.9045	m6	5.9071	5.9061	0.0006T	0.0026T
160	6.2992	6.2982	m6	6.3008	6.2998	0.0006T	0.0026T
170	6.6929	6.6919	m6	6.6945	6.6935	0.0006T	0.0026T
180	7.0866	7.0856	m6	7.0882	7.0872	0.0006T	0.0026T
190	7.4803	7.4791	m6	7.4821	7.4810	0.0007T	0.0030T
200	7.8740	7.8728	m6	7.8758	7.8747	0.0007T	0.0030T

B. Housing fits

Bearing *OD*			ISO tolerance grade	Housing bore		Limits of fit	
Nominal (mm)	Maximum (in)	Minimum (in)		Maximum (in)	Minimum (in)	Minimum (in)	Maximum (in)
30	1.1811	1.1807	H8	1.1811	1.1824	0	0.0017L
32	1.2598	1.2594	H8	1.2598	1.2613	0	0.0019L
35	1.3780	1.3776	H8	1.3780	1.3795	0	0.0019L

TABLE 15–5 (*continued*)

B. Housing fits

	Bearing *OD*		ISO	Housing bore		Limits of fit	
Nominal (mm)	Maximum (in)	Minimum (in)	tolerance grade	Maximum (in)	Minimum (in)	Minimum (in)	Maximum (in)
37	1.4567	1.4563	H8	1.4567	1.4582	0	0.0019L
40	1.5748	1.5744	H8	1.5748	1.5763	0	0.0019L
42	1.6535	1.6531	H8	1.6535	1.6550	0	0.0019L
47	1.8504	1.8500	H8	1.8504	1.8519	0	0.0019L
52	2.0472	2.0467	H8	2.0472	2.0490	0	0.0023L
62	2.4409	2.4404	H8	2.4409	2.4427	0	0.0023L
72	2.8346	2.8341	H8	2.8346	2.8364	0	0.0023L
80	3.1496	3.1491	H8	3.1496	3.1514	0	0.0023L
85	3.3465	3.3459	H8	3.3465	3.3486	0	0.0027L
90	3.5433	3.5427	H8	3.5433	3.5454	0	0.0027L
100	3.9370	3.9364	H8	3.9370	3.9391	0	0.0027L
110	4.3307	4.3301	H8	4.3307	4.3328	0	0.0027L
120	4.7244	4.7238	H8	4.7244	4.7265	0	0.0027L
125	4.9213	4.9206	H8	4.9213	4.9238	0	0.0032L
130	5.1181	5.1174	H8	5.1181	5.1206	0	0.0032L
140	5.5118	5.5111	H8	5.5118	5.5143	0	0.0032L
150	5.9055	5.9048	H8	5.9055	5.9080	0	0.0032L
160	6.2992	6.2982	H8	6.2992	6.3017	0	0.0035L
170	6.6929	6.6919	H8	6.6929	6.6954	0	0.0035L
180	7.0866	7.0856	H8	7.0866	7.0891	0	0.0035L
190	7.4803	7.4791	H8	7.4803	7.4831	0	0.0040L
200	7.8740	7.8728	H8	7.8740	7.8768	0	0.0040L
215	8.4646	8.4634	H8	8.4646	8.4674	0	0.0040L
225	8.8583	8.8571	H8	8.8583	8.8611	0	0.0040L
230	9.0551	9.0539	H8	9.0551	9.0579	0	0.0040L
240	9.4488	9.4476	H8	9.4488	9.4516	0	0.0040L
250	9.8425	9.8413	H8	9.8425	9.8453	0	0.0040L
260	10.2362	10.2348	H8	10.2362	10.2394	0	0.0046L
270	10.6299	10.6285	H8	10.6299	10.6331	0	0.0046L
280	11.0236	11.0222	H8	11.0236	11.0268	0	0.0046L
290	11.4173	11.4159	H8	11.4173	11.4205	0	0.0046L
300	11.8110	11.8096	H8	11.8110	11.8142	0	0.0046L
310	12.2047	12.2033	H8	12.2047	12.2079	0	0.0046L
320	12.5984	12.5968	H8	12.5984	12.6019	0	0.0051L
340	13.3858	13.3842	H8	13.3858	13.3893	0	0.0051L
360	14.1732	14.1716	H8	14.1732	14.1767	0	0.0051L
380	14.9606	14.9590	H8	14.9606	14.9641	0	0.0051L
400	15.7480	15.7464	H8	15.7480	15.7515	0	0.0051L
420	16.5354	16.5336	H8	16.5354	16.5392	0	0.0056L

Note: L = loose; T = tight.

Shaft 2: Output Shaft. Bearing 4 at B is number 6207.

Nominal bore = 35 mm (1.3780 in)

From Table 15–5: k5 ISO tolerance grade on the shaft seat; limits of 1.3785–1.3781 in

Resulting fit between bearing bore and shaft seat: 0.0001 in tight to 0.0010 in tight

OD of the outer race = 72 mm (2.8346 in)

From Table 15–5: H8 ISO tolerance grade on the housing bore; limits of 2.8346–2.8364 in

Resulting fit between outer race and housing bore: 0.0 to 0.0023 in loose.

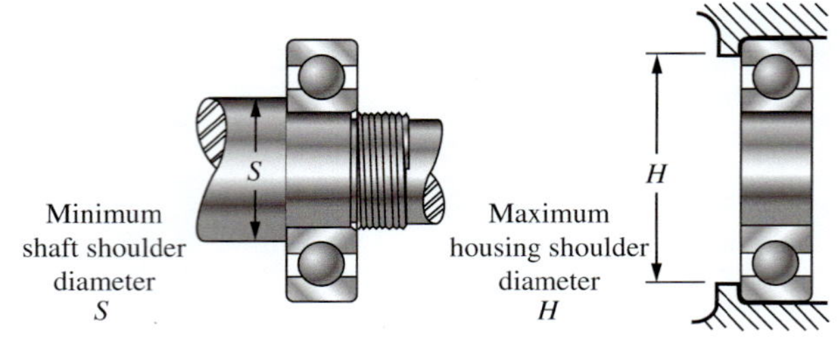

Notes:
S = minimum shaft shoulder diameter
Maximum diameter should not exceed the mean diameter of the bearing at the middle of the balls.
H = maximum housing shoulder diameter
Minimum diameter should not be less than the mean diameter of the bearing at the middle of the balls.
Refer to Table 14–3 in chapter 14 for values for S and H.

FIGURE 15–5 Shaft and housing shoulder diameters

Shaft and Housing Shoulder Diameters. Each of the bearings in this design is to be seated against a shoulder on one side of the bearing. The shaft shoulder must be sufficiently large to provide a solid, flat surface against which to seat the side of the inner race. But the shoulder must not be so high that it contacts the outer race because the inner race rotates at shaft speed and the outer race is stationary.

Similarly, a shoulder in the housing must provide for the solid location of the outer race but not be such that it contacts the inner race.

Bearing manufacturers' catalogs provide data such as those shown in Table 14–3 to guide you in specifying suitable shoulder heights. As shown in Figure 15–5, the value of S is the minimum shaft shoulder diameter. The nominal maximum diameter is the mean diameter for the bearing at the middle of the balls. The value of H is the maximum housing shoulder diameter, with the nominal minimum diameter being the mean diameter for the bearing.

For example, in the present design, the minimum shoulder diameter at each bearing on shaft 1 should be 1.260 in as indicated for the bearing number 305 in Table 14–3. (Note that the bearing number 6305 specified for the shaft is of the same *series* as the number 305, indicating that it would have similar dimensions.) The maximum housing shoulder diameter for the number 305 bearing is 2.165 in, where the outer race is to seat against a shoulder.

On shaft 2, the shoulder for bearing 6205 should also be at least 1.181 in, and the shoulder for bearing 6207 should be 1.654 in minimum. The maximum housing shoulder diameter for the 6205 bearing on shaft 2 is 1.850. For the 6207 bearing, the maximum housing shoulder diameter is 2.559 in.

Table 15–6 shows the pertinent data used to decide on the values for the shoulder diameters and, in the last two columns, gives the specified values. Where the specified shoulder diameter is less than the preliminary value shown in Table 15–4, another step in the shaft will be used to provide the proper shoulder for the bearing and for the gear. This can be seen in the drawings of the shafts given at the end of this chapter.

The use of a 1.75-in diameter for the shoulder at D on shaft 1 was specified because that is the diameter of the shaft chosen earlier. It is a bit higher than the mean diameter of the bearing, but it should still be lower than that of the outer race. More complete data in a manufacturer's catalog indicates the diameter of the inner surface of the outer race to be 2.00 in, so the 1.75-in diameter is acceptable.

TABLE 15–6 Shaft and Housing Shoulder Diameters for Bearings in the Present Design

A. Shaft 1

Bearing no.	Bore	*OD*	Mean diameter	*S* Minimum	*H* Maximum	Specified shaft shoulder	Specified housing shoulder
6305 at *B*	0.9843	2.4409	1.713	1.260	2.165	1.50	2.00
6305 at *D*	0.9843	2.4409	1.713	1.260	2.165	1.75	2.00

B. Shaft 2

Bearing no.	Bore	*OD*	Mean diameter	*S* Minimum	*H* Maximum	Specified shaft shoulder	Specified housing shoulder
6207 at *B*	1.3780	2.8346	2.106	1.654	2.559	2.00	2.25
6205 at *D*	0.9843	2.0472	1.516	1.181	1.850	1.50	1.75

Fillet Radii. Each of the bearings specified for the reducer calls for the maximum fillet radius at the shoulder that locates the bearing to be 0.039 in. (See Table 14–3.) Let's specify the limits on the radius to be 0.039 to 0.035. Before committing to the design, we will check the stress concentration factor at each shoulder.

Flexible Couplings. The use of flexible couplings on both the input and the output shafts has been taken into account in the shaft design and analysis. They allow the transmission of torque between two shafts but do not exert significant radial or axial forces on the shaft. In the present design, the use of flexible couplings made the shaft design simpler and decreased the loads on bearings compared with having a device such as a belt sheave or a chain sprocket on the shaft.

Now we specify suitable couplings for the input and output shafts. Chapter 11 showed many examples for such couplings, and you should review them now. It is impractical to reproduce data for all couplings in this book. As you read this section, it would be good for you to seek a copy of the catalog of one of the manufacturers of couplings and study their recommended selection procedures. (See the Internet sites for Chapter 11.)

We have selected couplings of the type pictured in Figure 11–16, called the *Browning Ever-flex Coupling* from Emerson Power Transmission, a division of the Emerson Electric Company. Rubber flex members are permanently bonded to steel hubs, and the flexing of the rubber accommodates parallel misalignment of the mating shafts up to 0.032 in, angular misalignment of $\pm 3°$, and axial end float of the shafts of up to ± 0.032 in. It is important for you to design the drive system for the saw to provide this alignment of the input shaft to the drive motor and of the output shaft to the drive shaft of the saw.

The selection of a suitable coupling relies on the power transmission rating of the various sizes available. But the power rating must be correlated to the speed of rotation because the real variable is the torque to which the coupling is subjected. Both the input and the output couplings transmit nominally 25 hp in this design for the drive for the saw. But the input shaft rotates at 1750 rpm, and the output shaft rotates at 500 rpm. Because torque is inversely proportional to speed of rotation, the torque experienced by the coupling on the output shaft is approximately 3.5 times higher than that on the input shaft. The coupling catalog data also call for the use of a *service factor* based on the kind of machine being driven, and some suggested values are included in the catalog. We judge that a service factor of 1.5 is suitable for the saw which will see mostly smooth power transmission with occasional moderate shock loading.

The service factor is applied to the nominal power being transmitted to compute a value for the *normal rating* for the couplings. Then,

$$\begin{aligned}\text{Normal rating} &= \text{power input} \times \text{service factor}\\ &= 25\ \text{hp}(1.5) = 37.5\ \text{hp}\end{aligned}$$

The catalog tables list coupling number CFR6 with a suitable normal rating at 1750 rpm for the input shaft and number CFR9 for the output shaft at 500 rpm.

We specify hubs for the couplings that have machined bores and keyways with a range of bores allowed. Each coupling half can have a different bore according to the shaft size on which it is to be mounted. For the input shaft, we have specified the diameter to be 0.875 in (7/8 in), and this will be the specification for the bore of that half for the CFR6 coupling. The keyway is $3/16 \times 3/32$ to accept a 3/16-in square key. The nominal maximum length of shaft inside each half of the coupling is 2.56 in.

The other half of the CFR6 coupling mounts on the motor shaft. Recall that the design requirements listed at the beginning of this design process specified a 25-hp motor with a NEMA Frame 284T. Table 21–3 gives the shaft diameter for this motor to be 1.875 in (1 7/8 in) with a $1/2 \times 1/4$ in keyway to accept a 1/2-in square key. This will be the bore specified for the motor half of the CFR6 coupling. The reason for the large difference in the sizes of the shafts for the motor and for our reducer is that the general-purpose motor must be designed to carry a significant side load, and our shaft does not.

The output shaft of the reducer at the coupling has a diameter of 1.250 in, and the input shaft for the saw will have the same size. Therefore, both halves of the CFR9 coupling will have that bore with a $1/4 \times 1/8$ in keyway to accept a 1/4-in square key. The nominal maximum shaft length inside each half of the coupling is 3.125 in.

Keys and Keyseats. A total of six keys need to be specified: two for each half of the flexible couplings on the input and output shafts, and one for each gear in the reducer. The methods of Chapter 11 are used to verify the suitability of the keys and to specify the required length using Equation (11–5). We will use standard key sizes made from SAE 1018 CD steel having a yield strength of 54 000 psi.

1. ***Keys for CFR6 coupling on the input shaft:*** First let's check the keys inside the couplings because their sizes have already been specified by the coupling manufacturer. The coupling half that mounts on the input shaft is critical because its bore diameter of 0.875 in is smaller, resulting in larger forces on the key when transmitting the torque of 900 lb · in which was computed earlier during the shaft design. The key is 3/16 in square (0.188 in). We use a design factor N of 4 as we did in the shaft design. Then, from Equation (11–5),

$$L = \frac{4TN}{DWs_y} = \frac{4(900\ \text{lb}\cdot\text{in})(4)}{(0.875\ \text{in})(0.188\ \text{in})(54000\ \text{psi})} = 1.62\ \text{in}$$

We can specify a key length of 2.50 in for extra safety and to match the length of the hub of the CFR6 coupling. The 1/2-in key for the motor shaft should be made to match the length of the coupling hub also, and it should be very safe because of the larger key size and the larger shaft size carrying the same torque.

2. ***Keys for the CFR9 coupling on the output shaft:*** For the output shaft and the drive shaft for the saw,

$T = 3150 \text{ lb} \cdot \text{in}$

$D = 1.25 \text{ in}$

$W = 0.250 \text{ in}$ (key width)

$$L = \frac{4TN}{DWs_y} = \frac{4(3150 \text{ lb} \cdot \text{in})(4)}{(1.25 \text{ in})(0.250 \text{ in})(54\,000 \text{ psi})} = 2.99 \text{ in}$$

We will make the key length 3.125 in (3 1/8 in), the full length of the hub of the CFR9 coupling. The conservative design factor of 4 should make this length acceptable.

3. ***Key for the pinion on shaft 1:*** The bore of the pinion is to be nominally 1.75 in as determined in the shaft design and shown in Table 15–4. The key size for this diameter should be 3/8 in square according to Table 11–1. The torque being transmitted is 900 lb · in. Then Equation (11–5) gives

$$L = \frac{4TN}{DWs_y} = \frac{4(900 \text{ lb} \cdot \text{in})(4)}{(1.75 \text{ in})(0.375 \text{ in})(54\,000 \text{ psi})} = 0.406 \text{ in}$$

The face width of the gear is 2.00 in. Let's use a key length of 1.50 in and center the profile keyseat at section *C* on the shaft so that the keyseat does not significantly interact with the ring groove to the right or with the shoulder fillet to the left.

4. ***Key for the gear on shaft 2:*** The bore of the gear is to be nominally 1.75 in as determined in the shaft design and shown in Table 15–4. The key size for this diameter should be 3/8 in square according to Table 11–1. The torque being transmitted is 3150 lb · in. Then Equation (11–5) gives

$$L = \frac{4TN}{DWs_y} = \frac{4(3150 \text{ lb} \cdot \text{in})(4)}{(1.75 \text{ in})(0.375 \text{ in})(54\,000 \text{ psi})} = 1.42 \text{ in}$$

The face width of the gear is 2.00 in. Let's use a key length of 1.50 in for this key also.

The key designs are summarized in the following list:

Summary of Key Designs

Motor shaft: 1/2-in square key × 2.50 in long

Input shaft of reducer at coupling: 3/16-in square key × 2.50 in long; sled runner keyseat

Input shaft at pinion: 3/8-in square key × 1.50 in long; profile shaft keyseat

Output shaft at gear: 3/8-in square key × 1.50 in long; profile shaft keyseat

Output shaft at coupling: 1/4-in square key × 3.125 in long; sled runner keyseat

Drive shaft for saw at coupling: 1/4-in square key × 3.125 in long; sled runner keyseat

The tolerances for the keys and keyseats are summarized as follows: Standard square bar stock in SAE 1018 steel is available to be used for keys. Typical tolerances are given in Table 15–7. Also given are recommended tolerances on the keyseat width dimension and the resulting fit between the key and the keyseat. A small clearance fit is desirable to permit easy assembly while not allowing the key to rock noticeably when installed.

Tolerances on Other Shaft Dimensions. You should review the discussion in Sections 13–5 and 13–9, on fits and tolerancing, respectively. See also References 1, 4, and 7 in Chapter 15, along with other comprehensive texts on technical drawing and interpretation of engineering drawings.

The fit of the bore of the gears on the shafts or the fit of the *ID* of the couplings on the ends of the shafts must be specified. Either a close sliding fit or a close locational fit is recommended for such components. Data are given in Table 13–3 for the RC2, RC5, and RC8 fits. More complete data are available in References 1, 2, 3, 5, 6, and 8 in Chapter 13. We will apply the RC5 fit to accurate mating parts that must assemble easily but where little perceptible play between the parts is desired. The RC fits use the *basic hole system,* as illustrated in Chapter 13.

TABLE 15–7 Tolerances and Fits for Keys and Keyseats

Key size (width in inches)	Tolerance on key (all +0.000)	Tolerance on keyseat (all −0.000)	Fit range
Up to 1/2	−0.002	+0.002	0.000–0.004
Over 1/2 to 3/4	−0.002	+0.003	0.000–0.005
Over 3/4 to 1	−0.003	+0.003	0.000–0.006
Over 1 to $1\frac{1}{2}$	−0.003	+0.004	0.000–0.007
Over $1\frac{1}{2}$ to $2\frac{1}{2}$	−0.004	+0.004	0.000–0.008

15-6 FINAL DESIGN DETAILS FOR THE SHAFTS

Figures 15–6 and 15–7 show the final design for the input and output shafts. Data from throughout this chapter have been used to specify pertinent dimensions. Where fillets are specified, a final check on the stress condition has been made to ensure that the estimated stress concentration factors used in the earlier design analysis are satisfactory and that the final stress levels are safe.

Keyseat details have been shown in sections below the main dimensioned shaft drawings. See Chapter 11 for computing the vertical dimension from the bottom of the shaft to the bottom of the keyseat.

The retaining ring grooves are drawn to the dimensions specified for a basic external ring for a 1.75-in-diameter shaft as pictured in Figure 11–30(a). See Internet sites 11 and 21 in Chapter 11 for providers of retaining rings and application details.

Four shaft diameters are sized to the RC5 fit in Figures 15–6 and 15–7. On shaft 1 they are at the extension where the coupling mounts and at the pinion. On shaft 2 they are at the gear and at the coupling location on the output extension. Table 15–8 summarizes the data for the fits. You should verify these using the procedure shown in Chapter 13. Note that the limit dimensions for both the shaft diameter and the bore of the mating element are given

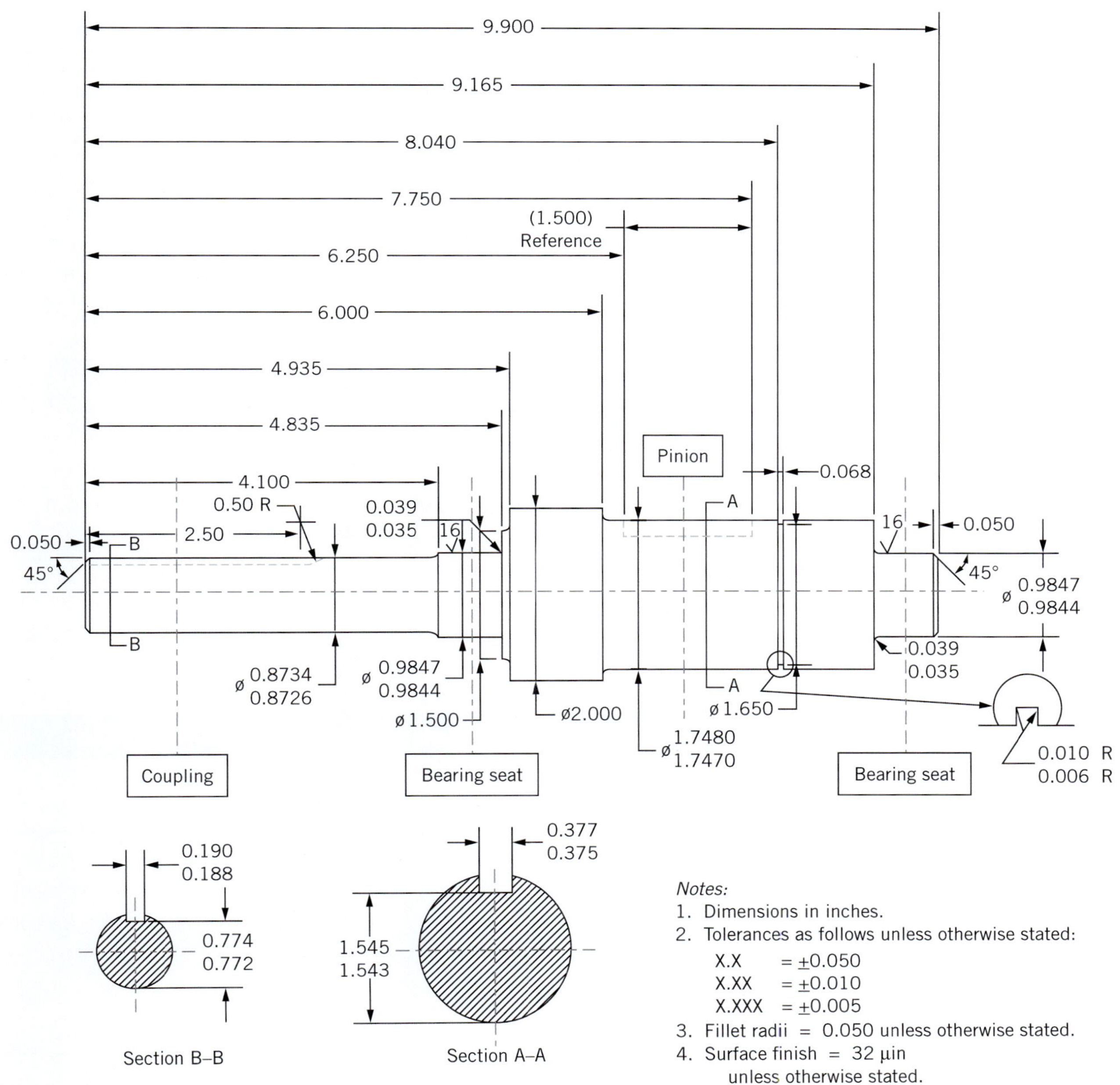

FIGURE 15–6 Final design of the input shaft

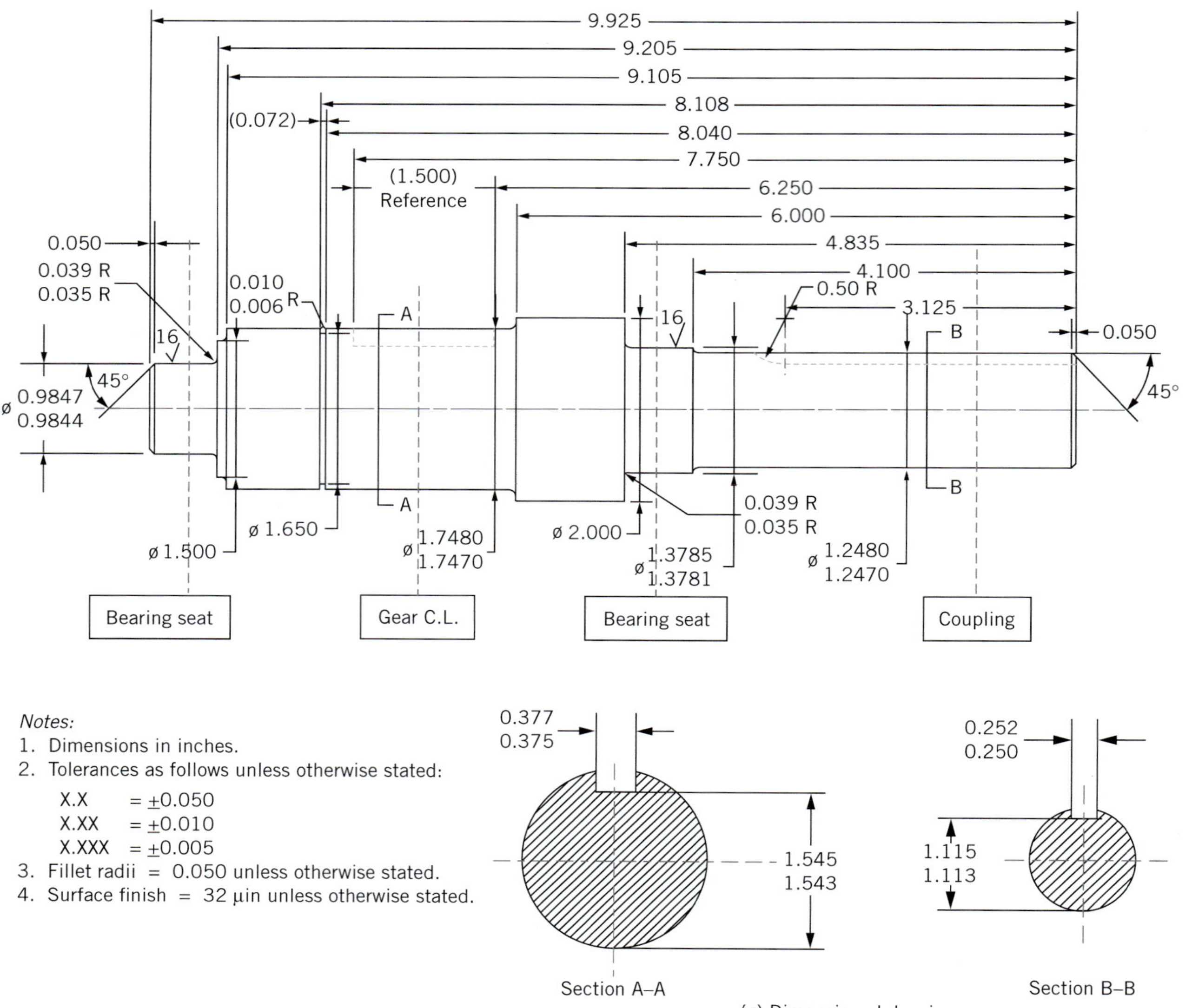

FIGURE 15–7 Final design of the output shaft

TABLE 15–8 Calculations and Specifications for Fits of Elements on the Shafts of the Transmission

Location	Nominal diameter	Bore limit dimensions for external element	*OD* limit dimensions for shaft	Fit
Shaft 1:				
Input at coupling	0.8750	0.8762/0.8750	0.8734/0.8726	+0.0026/+0.0036
Pinion	1.7500	1.7516/1.7500	1.7480/1.7470	+0.0020/+0.0046
Shaft 2:				
Gear	1.7500	1.7516/1.7500	1.7480/1.7470	+0.0020/+0.0046
Output at coupling	1.2500	1.2516/1.2500	1.2480/1.2470	+0.0020/+0.0046

Note: RC5 fit is used for all locations; dimensions are in inches.

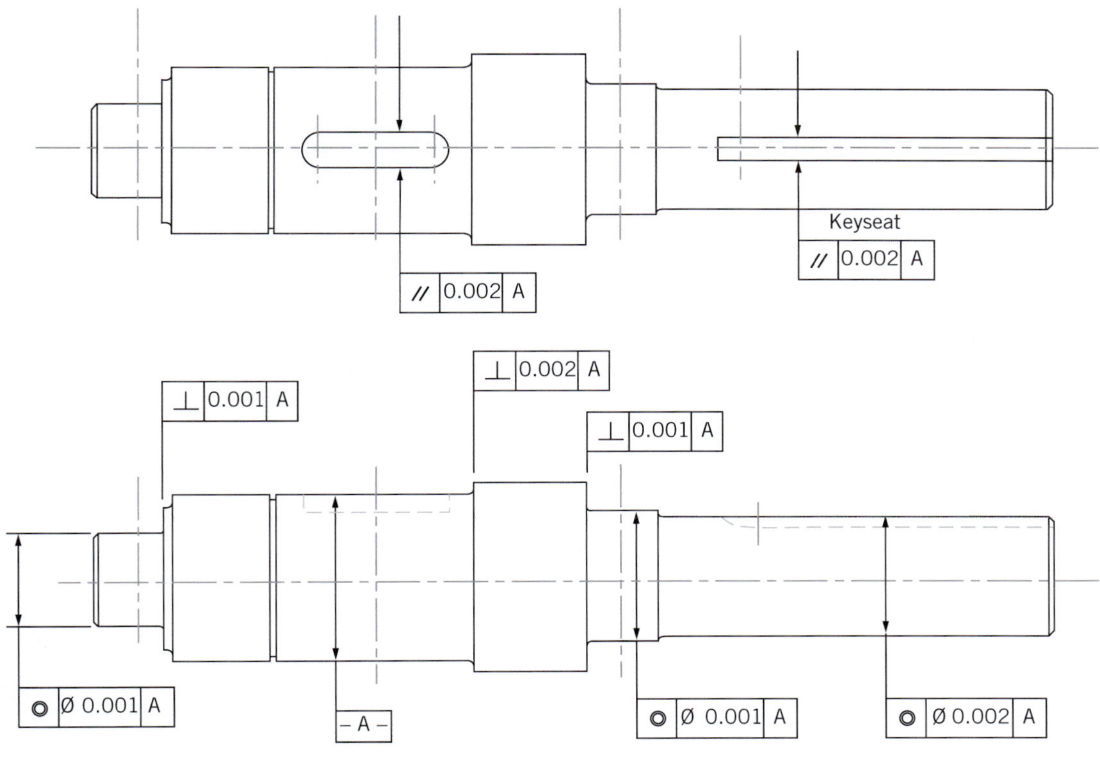

(*b*) Geometric tolerances

FIGURE 15–7 (*continued*)

and that the total tolerance on each dimension is small, less than 0.002 in for any dimension. Note also the small variation of the clearance on mating parts as indicated in the last column called "Fit."

For the shafts in this project, we have specified geometric tolerances for concentricity of four critical diameters for each shaft. Figure 15–7(b) documents the approach for the output shaft. Because of the similarity of the two shafts, the nature of the callouts would be the same for the input shaft. The datum or reference diameter is specified at the gear. Then the diameters at the two bearing seats and at the end of the shaft where the coupling mounts are controlled with concentricity feature control blocks. Shoulders for locating bearings and gears are controlled for perpendicularity to the axis of the shaft as represented by the gear diameter. The keyseat is controlled for parallelism to the axis of the shaft.

15–7 ASSEMBLY DRAWING

Figure 15–8 is an assembly drawing for the reducer with all features drawn to scale. The housing has been shown as a rectangular box shape for simplicity. Bearings are held in bearing retainers which are then fastened to the housing walls. Special attention to the alignment of the retainers is required and would be an important task for the detailing of the housing, not shown here.

The assembly of all components into the housing is facilitated by having the right side removable. Again, alignment of the cover piece with the main housing is critical.

Seals have been shown in the bearing retainers where the shafts penetrate the side walls of the housing. See Chapter 11 for more information on seals.

Critique of the Design

Figures 15–6 to 15–8 present drawings that meet the basic design requirements established at the beginning of this chapter. It is likely that refinements could be made if more detail were available about the saw for which the reducer is being designed.

It appears that the length of the shafts could be made somewhat shorter. The distances between the center of the gears and the bearings was arbitrarily set at 2.50 in at the start of the design process when dimensions for any components were unknown. Now that the nominal size of the gears, bearings, and couplings are known, further iterations on the design could result in a smaller package.

You should look at other commercially available gear-type speed reducers for other features that might be built into this design. Note particularly Figures 9–31, 9–32, and 9–33 in Chapter 9 and Figures 10–1, 10–2, 10–22, and 10-23 in Chapter 10.

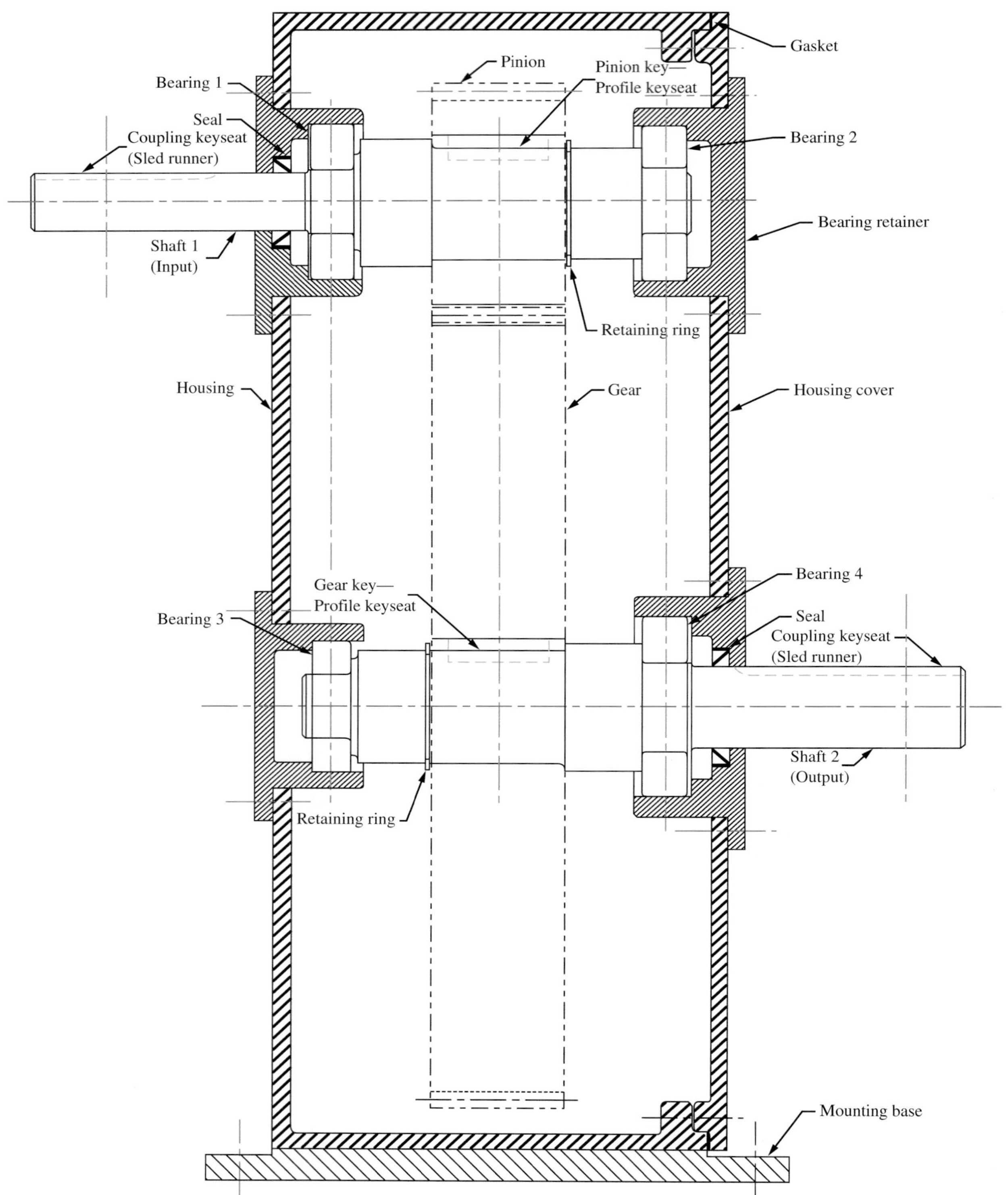

FIGURE 15–8 Assembly drawing for the reducer

REFERENCES

1. American Society of Mechanical Engineers. ASME Standard Y14.5. *Dimensioning and Tolerancing*. New York: American Society of Mechanical Engineers, 2009.
2. Bertoline, G., E. Wiebe, N. Hartman, and W. Ross. *Fundamentals of Graphic Communication*. 6th ed. New York: McGraw-Hill, 2010.
3. Cross, Nigel. *Engineering Design Methods: Strategies for Product Design*. 4th ed. New York: John Wiley & Sons, 2008.
4. Earle, James H. *Engineering Design Graphics*, 12th ed. Upper Saddle River, NJ: Pearson Prentice Hall, 2007.
5. Kepner, Charles H., and Benjamin B. Tregoe. *The New Rational Manager: An Updated Edition for a New World*. Princeton, NJ: Princeton Research Press, 1997.
6. National Research Council. *Improving Engineering Design: Designing for Competitive Advantage*. Washington, DC: National Academy Press, 1991.
7. Oberg, Erik, et al. *Machinery's Handbook*. 28th ed. New York: Industrial Press, 2008.
8. Pahl, G., W. Beitz, J. Feldhusen, and K. Grote. *Engineering Design: A Systematic Approach*. 3rd ed. London: Springer, 2007.
9. Peerless-Winsmith, Inc. *The Speed Reducer Book: A Practical Guide to Enclosed Gear Drives*. Springville, NY: Peerless-Winsmith, 1980.
10. Pugh, Stuart. *Total Design: Integrated Methods for Successful Product Engineering*. Reading, MA: Addison-Wesley, 1991.
11. Pugh, Stuart, and Don Clausing. *Creating Innovative Products Using Total Design: The Living Legacy of Stuart Pugh*. Reading, MA: Addison-Wesley, 1996.
12. Ullman, David G. *The Mechanical Design Process*. 4th ed. New York: McGraw-Hill, 2010.
13. Ulrich, K., and S. Eppinger. *Product Design and Development*. 4th ed. New York: McGraw-Hill/Irwin, 2007.

INTERNET SITES RELATED TO TRANSMISSION DESIGN

1. **Kepner-Tregoe, Inc.** www.kepner-tregoe.com A management consulting and training firm specializing in strategic and operational decision making. The firm's work evolves from the popular book listed as Reference 5. The Kepner-Tregoe approach is applicable to problem solving and decision making in product development and manufacturing operations. An online self-study learning mode called *eLearning Problem Solving and Decision Making* is now offered.
2. **Peerless-Winsmith, Inc.** www.winsmith.com Manufacturer of a broad line of speed reducers. This company produced the book listed in Reference 9.

DESIGN DETAILS AND OTHER MACHINE ELEMENTS

OBJECTIVES AND CONTENT OF PART III

Chapters 16–23 present methods of analysis and design of several important machine elements that were not particularly pertinent to the design of a power transmission as presented in Part II of this book. These chapters can be covered in any order or can be used as reference material for general design projects.

Chapter 16: Plain Surface Bearings discusses plain surface bearings, sometimes called *journal bearings*. These bearings employ smooth-surfaced elements to support loads from shafts or other devices in which relative motion occurs. You will learn how to specify materials for the parts of the bearing, the geometry of the bearing, and the lubricant. Full-film hydrodynamic lubrication and boundary lubrication are discussed.

Chapter 17: Linear Motion Elements discusses devices that convert rotary motion to linear motion, or vice versa. They are often used in machine tools, automation equipment, and parts of construction machinery. You will learn about their geometry and how to analyze their performance.

Chapter 18: Springs is not included in this custom edition.

Chapter 19: Fasteners describes machine screws, bolts, nuts, and set screws. You will learn about the kinds of materials used for fasteners and how to design them for safe, reliable performance. Also discussed briefly are rivets, quick-operating fasteners, welding, brazing, soldering, and adhesive bonding.

Chapter 20: Machine Frames, Bolted Connections, and Welded Joints presents the important skill of designing a frame for rigidity as well as strength. You will gain experience with analyzing the forces and stresses in bolted and welded connections that hold load-carrying members together. You will also learn how to analyze eccentric loads on connections.

Chapter 21: Electric Motors and Controls discusses the many types of AC and DC motors that are commercially available. Because a great many mechanical design projects involve the use of an electric motor as the prime mover, you will learn how to match their performance characteristics to the needs of the machine being designed and to specify motor controls.

Chapter 22: Motion Control: Clutches and Brakes discusses the many types of clutches and brakes that can be applied. Clutches provide the path to connect power from a prime mover to a driven machine. Brakes bring the moving equipment to a stop or slow it down. You will learn how to analyze their performance and to either design them or specify commercially available units.

Chapter 23: Design Projects presents several projects that you can complete yourself.

CHAPTER SIXTEEN

PLAIN SURFACE BEARINGS

THE BIG PICTURE Plain Surface Bearings

Discussion Map

- The purpose of a bearing is to support a load while permitting relative motion between two elements of a machine. This chapter discusses *plain surface* bearings where the two parts that move relative to each other do not have rolling elements between them.

Discover

Look around your home and in your car for products that have plain surface bearings. Find some that experience rotational motion and some that have linear sliding contact. Consider simple items such as hinges, door locks, latches, or the wheels of a lawn mower. Most of these experience boundary lubrication.

Now see if you can find information about the crankshaft bearings in the engine of your car. Such bearings usually employ full-film hydrodynamic lubrication. *What can you discover about them?*

How do you think a large telescope or an astronomical radio antenna is supported to permit it to be moved easily and positioned precisely? One method is called hydrostatic bearings. *What can you discover about such bearings?*

This chapter will help you explore all of these kinds of bearings and complete the basic design analyses required to ensure satisfactory operation.

The purpose of a bearing is to support a load while permitting relative motion between two elements of a machine. This chapter discusses *plain surface* bearings where the two parts that move relative to each other do not have rolling elements between them. Note that rolling contact bearings were discussed in Chapter 14.

Plain surface bearings for rotating parts are, of course, cylindrical with a typical arrangement as shown in Figure 16–1. The inner member, called the *journal*, is usually that part of a shaft where any radial reaction forces are transferred to the base of the machine. The stationary member that mates with the journal is called simply the *bearing*. Other names used for plain surface bearings are *journal bearings* and *sleeve bearings*.

Where have you seen such plain surface bearings in operation? As you did in Chapter 14 on rolling contact bearings, look for consumer products, industrial machinery, or transportation equipment (cars, trucks, bicycles, and so on—any device with rotating shafts). If the bearings are not visible from the outside (and that is often the case), you may need to partially disassemble the product to see inside.

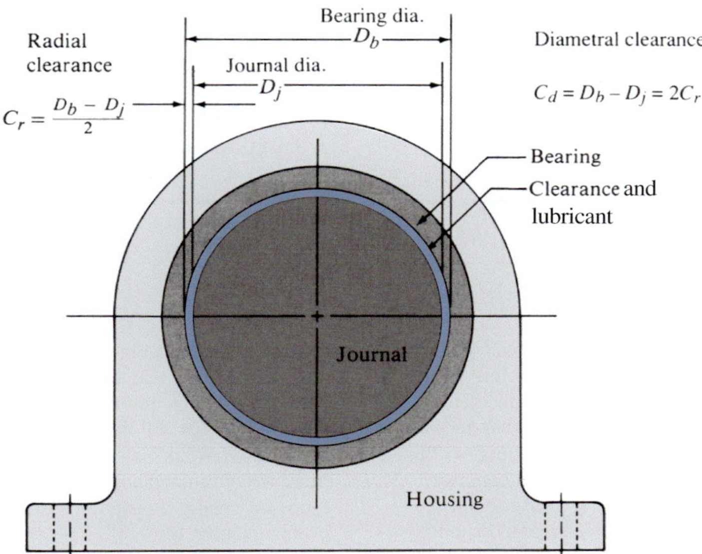

FIGURE 16–1 Bearing geometry

But there are some obvious examples that should be accessible to you. The wheels of a household lawn mower, wheelbarrow, or simple cart are typically mounted directly on axles to form the plain surface bearing. Hand tools such as grass and hedge clippers, pruning shears, pliers, and adjustable or ratchet wrenches use plain surface bearings at points where one part must rotate relative to another.

Look at almost any hinge for a door. It incorporates a solid pin riding inside a cylindrical outer member. That is an example of a plain surface bearing. Garage doors often have rollers that ride in channels, but the shafts that carry the rollers are operating in plain cylindrical bearings, often with very large clearances. The cable system that connects to the counterbalance springs runs over pulleys having plain surface bearings rotating on stationary shafts. If the door has an electric door opener, several of its components probably use plain surface bearings. Get a ladder and climb up to inspect them. *But be careful of the moving parts!*

Besides the rotational motion of the kinds of applications described previously, some plain surface bearings have linear sliding motion. Look inside a printer for a computer of the ink-jet or dot-matrix-pin type. Identify the part that delivers the actual image to the paper as it traverses across the page. It typically slides on a polished rod with great precision. Many linkages contain parts that slide. Look at locking mechanisms, latches, staplers, switches, car seat adjusters, gear shifters for cars, parts of window-lift devices in cars, and coin-operated machines. Can you identify parts that have linear sliding motion?

If you visit a factory, you will likely see numerous examples of both rotational and linear sliding action. Transfer devices, reciprocating machinery, machine tool slides, and packaging equipment employ many plain surface bearings.

Most of the examples just mentioned have intermittent and relatively slow motion between the mating parts, whether the motion is rotational or linear. In such applications, there is often a lubricant between the moving parts. Or the materials are carefully chosen to produce low friction and reliable, smooth motion. This kind of bearing experiences *boundary lubrication* as discussed in this chapter.

You may be familiar with the bearings on the crankshaft of an internal combustion engine. They are often called *journal bearings* because they have the classic configuration shown in Figure 16–1. But consider their general operation. After the engine has been started, the crank rotates at several thousand revolutions per minute, typically from about 1500 to 6000 rpm. This is one of the most important applications of plain surface bearings using *full-film hydrodynamic lubrication*. In such bearings, there is a continuous film of lubricant, usually oil, that actually lifts the rotating shaft journal off the stationary bearing. Thus, there is no metal-to-metal contact between the members. Much of this chapter is devoted to this kind of bearing and the analysis of the conditions required to maintain the supporting oil film. Engine bearings are some of the most highly engineered mechanical devices.

Can you identify other examples of bearings that use full-film hydrodynamic lubrication?

Another example of a class of plain surface bearings is called *hydrostatic*. Pumps create high pressure in a fluid such as oil that is delivered by carefully shaped pads where the pressure lifts the part to be moved. Then it can be moved easily, either slowly or rapidly. Think about a huge telescope or radio antenna that must be positioned accurately and moved with ease. Such applications are rare but important, justifying the large engineering and technical effort of designing, installing, and maintaining them.

This chapter will help you understand boundary lubrication, full-film hydrodynamic lubrication, and hydrostatic lubrication. Also discussed are more general aspects of friction, lubrication, wear, and the lubricants that are used to minimize friction.

YOU ARE THE DESIGNER

Your company is designing a conveyor system to transport products in a shipping and receiving system of a major manufacturer. The design has progressed to the point where it has been decided to use a flexible belt conveyor driven by flat pulleys at either end. The shafts that carry the pulleys must be supported in bearings in the side frames of the conveyor. Your task is to design the bearings.

First you must decide what type of bearing to use: plain surface or journal bearings as discussed in this chapter, or rolling contact bearings, which are discussed in Chapter 14. For the plain surface bearings, you must determine whether full-film hydrodynamic lubrication can be achieved with its advantages of low friction and long life. Or will the shaft operate in the bearing with boundary lubrication? What materials will the bearing and the journal be made from? What dimensions will be specified for all components? What lubricant should be used? These and other questions are discussed in this chapter. ■

16–1 OBJECTIVES OF THIS CHAPTER

After completing this chapter, you will be able to:

1. Describe the three modes of operation of a plain surface bearing (boundary, mixed-film, and full-film hydrodynamic lubrication), and discuss the conditions under which each will normally occur.
2. Discuss the significance of the bearing parameter, $\mu n/p$.
3. List the decisions that a bearing designer must make to completely define a plain surface bearing system.
4. List the materials often used for journals and bearings, and describe their important properties.
5. Define the *pV factor* and use it in the design of boundary-lubricated bearings.
6. Describe the operation of full-film hydrodynamically lubricated bearings.
7. Complete the design of full-film bearings, defining the size of the journal and bearing, the diametral clearance, the bearing length, the minimum film thickness, the surface finish, the lubricant, and the resulting frictional performance of the bearing system.
8. Describe a hydrostatic bearing system and complete the basic design of such bearings.
9. Define *tribology* and discuss the essential characteristics of friction, lubrication, and wear as applied to machinery.
10. Describe the general nature of oils and greases and their effects on lubrication and wear.

16–2 THE BEARING DESIGN TASK

The term *plain surface bearing* refers to the kind of bearing in which two surfaces move relative to each other without the benefit of rolling contact. Thus, there is sliding contact. The actual shape of the surfaces can be anything that permits the relative motion. The most common shapes are flat surfaces and concentric cylinders. Figure 16–1 shows the basic geometry of a cylindrical plain surface bearing.

A given bearing system can operate with any of three types of lubrication:

Boundary lubrication: There is actual contact between the solid surfaces of the moving and the stationary parts of the bearing system, although a film of lubricant is present.

Mixed-film lubrication: There is a transition zone between boundary and full-film lubrication.

Full-film lubrication: The moving and stationary parts of the bearing system are separated by a complete film of lubricant that supports the load. The term *hydrodynamic lubrication* is often used to describe this type.

All of these types of lubrication can be encountered in a bearing without external pressurization of the bearing. If lubricant under pressure is supplied to the bearing, it is called a *hydrostatic bearing*, which is discussed separately. Running dry surfaces together is not recommended unless there is inherently good lubricity between the mating materials. Some plastics are used dry, as discussed in Sections 16–4 and 16–5.

Bearing design involves so many design decisions that it is not possible to develop a procedure that produces the single best design. Thus, several feasible designs could be proposed, and the designer must make judgments based on knowledge of the application and on the principles of bearing operation to define the final design. The following lists identify the information needed to design a bearing system and the types of design decisions that must be made. (See References 6 and 13.) In this discussion, it is assumed that the bearing will be cylindrical, like that used to support a rotating shaft. Modified lists could be made for linear sliding surfaces or some other geometry.

Bearing Requirements

Magnitude, direction, and degree of variation of the radial load

Magnitude and direction of the thrust load, if any

Rotational speed of the journal (shaft)

Frequency of starts and stops, and duration of idle periods

Magnitude of the load when the system is stopped and when it is started

Life expectancy of the bearing system

Environment in which the bearing will operate

Design Decisions

Materials for the journal and the bearing

Diameters, including tolerances, of the journal and the bearing

Nominal value and range of journal clearance in the bearing

Surface finish for the journal and the bearing

Length of the bearing

Method of manufacturing the bearing system

Type of lubricant to be used and the means of supplying it

Operating temperature of the bearing system and of the lubricant

Method of maintaining the lubricant cleanliness and temperature

Analyses Required

Type of lubrication: boundary, mixed-film, full-film

Coefficient of friction

Frictional power loss

Minimum film thickness

Thermal expansion

Heat dissipation required and the means of accomplishing it

Shaft stiffness and slope of the shaft in the bearing

16–3 BEARING PARAMETER, $\mu n/p$

The performance of a bearing differs radically, depending on which type of lubrication occurs. There is a marked decrease in the coefficient of friction when the operation changes from boundary to full-film lubrication. Wear also decreases with full-film lubrication. Thus, it is desirable to understand the conditions under which one or the other type of lubrication occurs.

The creation of full-film lubrication, the most desirable type, is encouraged by light loads, high relative speed between the moving and stationary parts, and presence of a high-viscosity lubricant in the bearing in copious supply. For a rotating journal bearing, the combined effect of these three factors, as it relates to the friction in the bearing, can be evaluated by computing the *bearing parameter*, $\mu n/p$. The viscosity of the lubricant is indicated by μ, the rotational speed by n, and the bearing load by the pressure, p. To compute the pressure, divide the applied radial load on the bearing by the *projected area* of the bearing, that is, the product of the length times the diameter.

The bearing parameter $\mu n/p$ is dimensionless when each term is expressed in consistent units. Some unit systems that can be used are listed here:

Unit System	Viscosity, μ	Rotational Speed, n	Pressure, p
SI metric	N·s/m² or Pa·s	rev/s	N/m² or Pa
English	lb·s/in² or reyn	rev/s	lb/in²
Old metric (obsolete)	dyne·s/cm² or poise	rev/s	dynes/cm²

The effect of the bearing parameter is shown in Figure 16–2, sometimes called the *Stribeck curve*, which is a plot of the coefficient of friction, f, for the bearing versus the value of $\mu n/p$. At low values of $\mu n/p$, boundary lubrication occurs, and the coefficient of friction is high. For example, with a steel shaft sliding slowly in a lubricated bronze bearing (boundary lubrication), the value of f would be approximately 0.08 to 0.14. At high values of $\mu n/p$, the full hydrodynamic film is created, and the value of f is normally in the range of 0.001 to 0.005. Note that this compares favorably with precision rolling contact bearings. Between boundary and full-film lubrication, the *mixed-film* type, which is some combination of the other two, occurs. The dashed curve shows the general nature of the variation in film thickness in the bearing.

It is recommended that designers avoid the mixed-film zone because it is virtually impossible to predict the performance of the bearing system. Also notice that the curve is

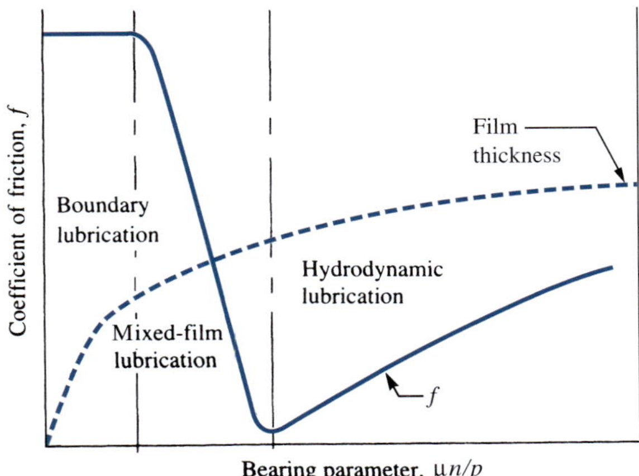

FIGURE 16–2 Bearing performance and types of lubrication related to the bearing parameter, $\mu n/p$. The Stribeck curve

very steep in this zone. Thus, a small change in any of the three factors, μ, n, or p, produces a large change in f, resulting in erratic performance of the machine.

The value of $\mu n/p$ at which full-film lubrication is produced is difficult to predict. Besides the individual factors of speed, pressure (load), and viscosity (a function of the type of lubricant and its temperature), the variables affecting the production of the film include the quantity of lubricant supplied, the adhesion of the lubricant to the surfaces, the materials of the journal and the bearing, the structural rigidity of the journal and the bearing, and the surface roughness of the journal and the bearing. After completing the design process presented later in this chapter, you are advised to test the design.

In general, boundary lubrication should be expected for slow-speed operation with a surface speed less than about 10 ft/min (0.05 m/s). Reciprocating or oscillating motion or a combination of a light lubricant and high pressure would also produce boundary lubrication.

The design of bearings to produce full-film lubrication is described in Section 16–6. In general, it requires a surface speed of greater than 25 ft/min (0.13 m/s) continuously in one direction with an adequate supply of oil at a proper viscosity.

16–4 BEARING MATERIALS

In rotational applications, the journal on the shaft is frequently steel. The stationary bearing may be made of any of a wide variety of materials, including the following:

Bronze
Babbitt
Aluminum
Zinc
Porous metals
Plastics (nylon, TFE, PTFE, phenolic, acetal, polycarbonate, filled polyimide)

The properties desirable for materials used for plain bearings are unique, and compromises must frequently be made. The following list discusses these properties.

1. ***Strength:*** The function of the bearing is to carry the applied load and deliver it to the supporting structure. At times, the loads are varied, requiring fatigue strength as well as static strength.
2. ***Embeddability:*** The property of embeddability relates to the ability of the material to hold contaminants in the bearing without causing damage to the rotating journal. Thus, a relatively soft material is desirable.
3. ***Corrosion resistance:*** The total environment of the bearing must be considered, including journal material, lubricant, temperature, air-borne particulates, and corrosive gases or vapors.
4. ***Cost:*** Always an important factor, cost includes not only the material cost but also processing and installation costs.

A brief discussion of the performance of some of the bearing materials follows.

Cast Bronze

The name *bronze* refers to several alloys of copper with tin, lead, zinc, or aluminum, either singly or in combination. The leaded bronzes contain 25% to 35% lead, which gives them good embeddability and resistance to seizing under conditions of boundary lubrication. However, their strength is relatively low. The cast bearing bronze, SAE CA932, has 83% copper, 7% tin, 7% lead, and 3% zinc. It possesses a good combination of properties for such uses as pumps, machinery, and appliances. Tin and aluminum bronzes have higher strength and hardness and can carry greater loads, particularly in impact situations. But they rate lower on embeddability. See Internet sites 1, 6, 7, 9, and 10.

Babbitt

Babbitts may be lead-based or tin-based, having nominally 80% of the parent metal. Various alloy compositions of copper and antimony (as well as lead and tin) can tailor the properties to suit a particular application. Because of their softness, babbitts have outstanding embeddability and resistance to seizure, important properties in applications in which boundary lubrication occurs. They have rather low strength, however, and are often applied as liners in steel or cast iron housings. See Internet site 9.

Aluminum

With the highest strength of the commonly used bearing materials, aluminum is used in severe applications in engines, pumps, and aircraft. The high hardness of aluminum bearings results in poor embeddability, requiring clean lubricants.

Zinc

Bearings made from zinc alloys offer good protection to running without a continuous supply of lubricant, although they perform best if lubricated. Standard bearing greases are often used. When operating on steel journals, a thin film of the softer zinc material is transferred to the steel to protect it from wear and damage. It performs well in most atmospheric conditions except for continuously wet environments and exposure to seawater. See Internet site 8.

Porous Metals

Products of the powder metal industry, porous metals are sintered from powders of bronze, iron, and aluminum; some are mixed with lead or copper. The sintering leaves a large number of voids in the bearing material into which lubricating oil is forced. Then, during operation, the oil migrates out of the pores, supplying the bearing. Such bearings are particularly good for slow-speed, reciprocating, or oscillating motions. See Internet sites 6, 7, and 17.

Plastics

Generally referred to as *self-lubricating materials,* plastics used in bearing applications have inherently low friction characteristics. They can be operated dry, but most improve in performance with a lubricant present. Embeddability is usually good, as is resistance to seizure. But many have low strength, limiting their load-carrying capacity. Backing with metal sleeves is frequently done to improve load-carrying capacity. Major advantages are corrosion resistance and, when operated dry, freedom from contamination. These properties are particularly important in the processing of foodstuffs and chemical products. See Internet sites 2, 3, 4, 5, 7, and 17.

Because of the complex chemical names for plastic materials and the virtually infinite combinations of base materials, reinforcements, and fillers used, it is difficult to characterize bearing plastics. Most are composites of several components. The group known as *fluoropolymers* are popular because of the very low coefficient of friction (0.05 to 0.15) and good wear resistance. Phenolics, polycarbonates, acetals, nylons, and many other plastics are also used for bearings. Among the chemical names and abbreviations found in this field are the following:

PTFE: Polytetrafluoroethylene

PA: Polyamide

PPS: Polyphenylene sulfide

PVDF: Polyvinylidene fluoride

PEEK: Polyetheretherketone

PEI: Polyetherimide

PES: Polyethersulfone

PFA: Perfluoroalkoxy-modified tetrafluoroethylene

Reinforcements and fillers used with plastic bearing materials include glass fibers, milled glass, carbon fibers, bronze powders, PTFE, PPS, and some solid lubricants, such as graphite and molybdenum disulfide.

16–5 DESIGN OF BOUNDARY-LUBRICATED BEARINGS

A boundary-lubricated bearing is typically a smooth hollow cylinder through which a shaft passes that rotates slowly, oscillates, or slides linearly. Figure 16–3 shows two styles of commercially available bearings. Part (a) shows a flange-mounted self-lubricating bearing that can operate in corrosive atmospheres or under water or other light liquids. It can be exposed to temperatures as low as −200°F up to +1000°F (−129°C to 538°C). Part (b) shows a bearing lined with Babbitt [for up to 130°F (54°C)] or bronze [for up to 300°F (149°C)] that is lubricated periodically with grease.

The factors to be considered when selecting materials for bearings and specifying the design details include the following:

Coefficient of friction: Both static and dynamic conditions should be considered.

Load capacity, p: Radial load divided by the projected area of the bearing (lb/in^2 or Pa).

Speed of operation, V: The relative speed between the moving and stationary components, usually in ft/min or m/s.

Temperature at operating conditions.

Wear limitations.

Producibility: Machining, molding, fastening, assembly, and service.

Besides specifying the material for the bearing, the designer must determine the combination of the inside diameter, D, and the length, L, for the bearing.

pV Factor

In addition to the individual consideration of load capacity, p, and speed of operation, V, the product pV is an important performance parameter for bearing design when boundary lubrication occurs. The pV value is a measure of the ability of the bearing material to accommodate the frictional energy generated in the bearing. At the limiting pV value, the bearing will not achieve a stable temperature limit, and rapid failure will occur. A practical design value for pV is one-half of the limiting pV value, given in Table 16–1.

(*a*) Self-lubricated flange-mounted bearing

(*b*) Grease-lubricated journal bearing

FIGURE 16–3 Two styles of boundary-lubricated bearings (Baldor/Dodge, Greenville, SC)

TABLE 16–1 Typical Performance Parameters for Bearing Materials in Boundary Lubrication at Room Temperature

Material	*pV* psi-fpm	*pV* MPa-m/s	
Vespel® SP-21 polyimide	300 000	10.500	Trademark of DuPont Co.
DP11™, dry	286 000	10.02	GGB Bearing Technology
Manganese bronze (C86200)	150 000	5.250	Also called SAE 430A
Aluminum bronze (C95400)	125 000	4.375	Also called SAE 68A
DX®10, dry or oiled	80 000	2.800	GGB Bearing Technology
Leaded tin bronze (C93800)	75 000	2.625	Also called SAE 660
DU®	51 400	1.800	GGB Bearing Technology
BU dry lubricant bearing	51 000	1.785	See note 1
Porous bronze/oil impregnated	50 000	1.750	
Babbitt: high tin content (89%)	30 000	1.050	
DP11™, dry	28 600	1.000	GGB Bearing Technology
Babbitt: low tin content (10%)	18 000	0.630	
Graphite/Metallized	15 000	0.525	Graphite Metallizing Corp.
Rulon® PTFE: 641	10 000	0.350	Food and drug applications (see note 2)
Rulon® PTFE: J	7500	0.263	Filled PTFE
Polyurethane: UHMW	4000	0.140	Ultra-high molecular weight
Nylon® 101	3000	0.105	Trademark of DuPont Co.

Source: Internet sites 2–7, 17, and 18

Notes:

[1]BU Bearings consist of bonded layers of a steel backing and a porous bronze matrix overlaid with PTFE/lead bearing material. A film from the bearing material is transferred to the journal during operation.

[2]Rulon® is a registered trademark of Saint-Gobain Performance Plastics Company. Bearings are made from Rulon® PTFE (polytetrafluoroethylene) material in a variety of formulations and physical constructions.

Units for pV. The nominal units for pV are simply the product of the units for pressure and the units for velocity. In the U.S. Customary System, this is, considering only units,

$$p = F/LD = \text{lb/in}^2 = \text{psi}$$
$$V = \pi Dn/12 = \text{ft/min} = \text{fpm}$$
for continuous rotation in rpm
$$pV = (\text{lb/in}^2)(\text{ft/min}) = \text{psi-fpm}$$

Another way of looking at these units is to rearrange them into the form

$$pV = (\text{ft}\cdot\text{lb/min})/\text{in}^2$$

The numerator represents one unit for power or energy transfer per unit time. The denominator represents area. Therefore, pV can be thought of as the rate of energy input to the bearing per unit of projected area of the bearing if the coefficient of friction is 1.0. Of course, the actual coefficient of friction is normally much less than one. Then you can think of pV as a comparative measure of the bearing's ability to absorb energy without overheating.

In SI units, force, F, is in newtons (N) and bearing dimensions are in mm. The pressure is then

➩ Bearing Pressure

$$p = F/LD = \text{N/mm}^2 = \text{MPa} \quad \textbf{(16–1)}$$

The linear velocity of the surface of the journal is typically computed from

$$v = \pi Dn/(60\,000)\ \text{m/s}$$

with D in mm and n in rpm. Then the units for pV are

$$pV = (\text{MPa})(\text{m/s})$$

A useful conversion to the U.S. Customary System is

$$1.0\ \text{psi-fpm} = 3.50 \times 10^{-5}\ \text{MPa}\cdot\text{m/s}$$

Again, the units for pV can be reformatted to reflect the rate of energy transfer per unit area:

$$pV = \text{MPa}\cdot\text{m/s} = \frac{10^6\,\text{N}}{\text{m}^2}\cdot\frac{\text{m}}{\text{s}} = \frac{10^6\,\text{W}}{\text{m}^2}$$

where $1\ \text{W} = 1\ \text{N}\cdot\text{m/s}$

Operating Temperature

Most plastics are limited to approximately 200°F (93°C). However, PTFE can operate at 500°F (260°C). Babbitt is limited to 300°F (150°C), while tin-bronze and aluminum can operate at 500°F (260°C). A major advantage of carbon-graphite bearings is their ability to operate at up to 750°F (400°C).

Design Procedure for Continuously Rotating Shafts in Bearings

The following is one method of completing the preliminary design of boundary-lubricated plain surface bearings.

PROCEDURE FOR DESIGNING BOUNDARY-LUBRICATED PLAIN SURFACE BEARINGS

Given information: Radial load on the bearing, F (lb or N); speed of rotation, n (rpm); nominal minimum shaft diameter, D_{min} (in or mm) (based on stress or deflection analysis).

Objectives of the design process: To specify the nominal diameter and length of the bearing and a material that will have a safe value of pV.

1. Specify a trial diameter, D, for the journal and the bearing.
2. Specify a ratio of bearing length to diameter, L/D, typically in the range of 0.5 to 2.0. For nonlubricated (dry-rubbing) or oil-impregnated porous bearings, $L/D = 1$ is recommended. (See Reference 6.) For carbon-graphite bearings, $L/D = 1.5$ is recommended.
3. Compute $L = D(L/D)$ nominal length of the bearing.
4. Specify a convenient value for L.
5. Compute the bearing pressure (lb/in^2 or Pa):
$$p = F/LD$$
6. Compute the linear speed of the journal surface:

   U.S. Customary units: $V = \pi Dn/12$ ft/min or fpm

   SI metric units: $V = \pi Dn$ (60 000) m/s

   Also note: 1.0 ft/min = 0.005 08 m/s

   1.0 m/s = 197 ft/min
7. Compute pV (psi-fpm or MPa·m/s).
8. Multiply $2(pV)$ to obtain a design value for pV.
9. Specify a material from Table 16–1 with a rated value of pV equal to or greater than the design value.
10. Complete the design of the bearing system considering diametral clearance, lubricant selection, lubricant supply, surface finish specification, thermal control, and mounting considerations. Often the supplier of the bearing material provides recommendations for many of these design decisions.
11. Nominal diametral clearance: Many factors affect the final specification for clearance, such as need for precision, thermal expansion of all parts of the bearing system, load variations, shaft deflection expected, means of providing lubricant, and manufacturing capability. One long-used rule of thumb is to provide 0.001 in of clearance per inch of journal diameter. Figure 16–4

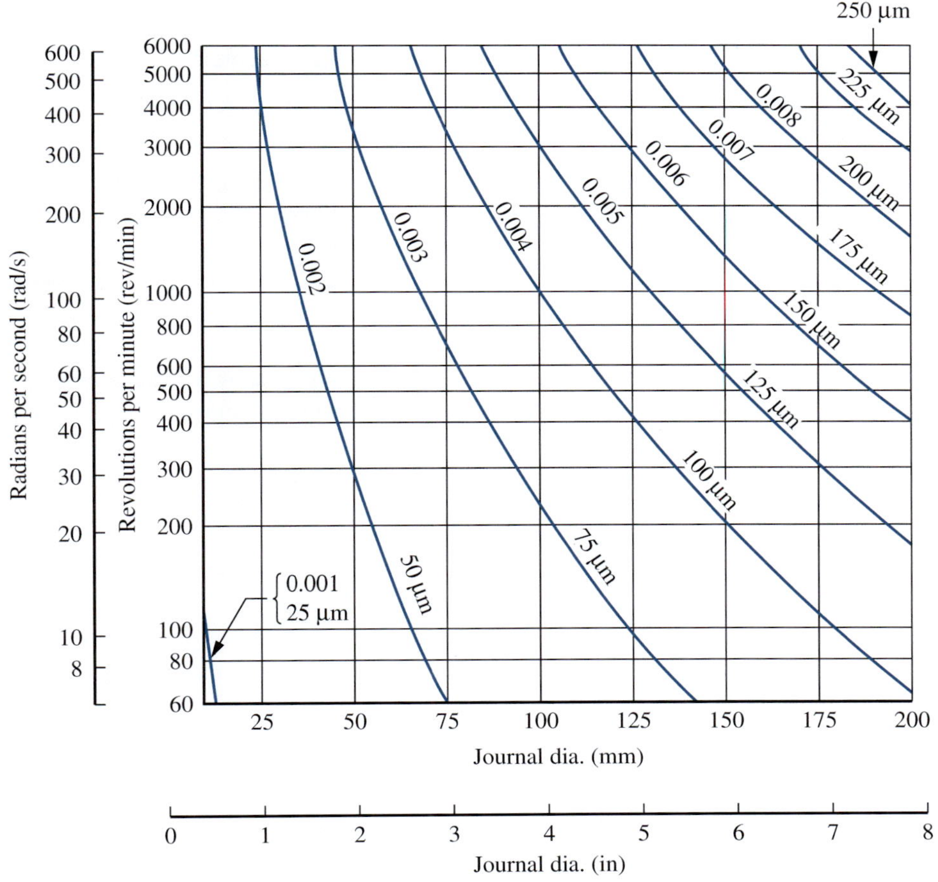

FIGURE 16–4 Minimum recommended diametral clearance for bearings considering journal diameter and rotational speed (Taken from R. J. Welsh. *Plain Bearing Design Handbook*. London: Butterworths, 1983)

shows recommended minimum values for clearance based on journal diameter and rotational speed under steady loads. These values apply to the smallest clearance under any combination of tolerances on the dimensions of the bearing system to avoid problems of bearing heating and eventual seizure. The operating clearance will then be higher than these values because of manufacturing tolerances. Performance over the entire range of clearances should be evaluated, preferably by testing.

Example Problem 16–1 A bearing is to be designed to carry a radial load of 150 lb from a shaft having a minimum acceptable diameter of 1.50 in and rotating at 500 rpm. Design the bearing to operate under boundary-lubrication conditions.

Solution We will use the design procedure previously outlined.

Step 1. Trial diameter: Try $D = D_{min} = 1.50$ in.

Steps 2–4. Try $L/D = 1.0$. Then $L = D = 1.50$ in.

Step 5. Bearing pressure:

$$p = F/LD = (150 \text{ lb})/(1.50 \text{ in})(1.50 \text{ in}) = 66.7 \text{ psi}$$

Step 6. Journal speed:

$$V = \pi Dn/12 = \pi(1.50)(500)/12 = 196 \text{ ft/min}$$

Step 7. pV factor:

$$pV = (66.7 \text{ psi})(196 \text{ fpm}) = 13\,100 \text{ psi-fpm}$$

Step 8. Design value of $pV = 2(13\,100) = 26\,200$ psi-fpm.

Step 9. From Table 16–1, we could use a bearing made from high tin babbitt having a rated value of pV of 30 000 psi-fpm.

Steps 10–11. Nominal diametral clearance: From Figure 16–4, we can recommend a minimum $C_d = 0.002$ in based on $D = 1.50$ in and $n = 500$ rpm. Other design details are dependent on the system into which the bearing will be placed.

Example Problem 16–2 Design a boundary-lubricated plain surface bearing to carry a radial load of 2.50 kN from a shaft rotating at 1150 rpm. The nominal minimum diameter of the journal is 65 mm.

Solution We will use the design procedure previously outlined.

Step 1. Trial diameter: Try $D = 75$ mm.

Steps 2–4. Try $L/D = 1.0$. Then $L = D = 75$ mm.

Step 5. Bearing pressure:

$$p = F/LD = (2500 \text{ N})/(75 \text{ mm})(75 \text{ mm}) = 0.444 \text{ N/mm}^2 = 0.444 \text{ MPa}$$

Step 6. Journal speed:

$$V = \pi Dn/(60\,000) = \pi(75)(1150)/(60\,000) = 4.52 \text{ m/s}$$

Step 7. pV factor:

$$pV = (0.444 \text{ MPa})(4.52 \text{ m/s}) = 2.008 \text{ MPa}\cdot\text{m/s}$$

Step 8. Design value for $pV = 2(2.008) = 4.016$ MPa·m/s.

Step 9. From Table 16–1, we can specify aluminum bronze (C95400) having a pV rating of 4.375 MPa·m/s.

Steps 10–11. From Figure 16–4, we can recommend a minimum $C_d = 100$ μm (0.100 mm or 0.004 in) based on $D = 75$ mm and 1150 rpm.

Alternate design: The pV factor for the initial design, while satisfactory, is somewhat high and may require careful lubrication. Consider the following alternate design having a larger bearing diameter.

Step 1. Try $D = 150$ mm.

Step 2. Let $L/D = 1.25$.

Step 3. Then

$$L = D(L/D) = (150 \text{ mm})(1.25) = 187.5 \text{ mm}$$

Step 4. Let's use the more convenient value of 175 mm for L.

Step 5. Bearing pressure:

$$p = F/LD = (2500 \text{ N})/(175 \text{ mm})(150 \text{ mm}) = 0.095 \text{ N/mm}^2 = 0.095 \text{ MPa}$$

Step 6. Journal speed:

$$V = \pi Dn/(60\,000) = \pi(150)(1150)/(60\,000) = 9.03 \text{ m/s}$$

Step 7. pV factor:

$$pV = (0.095 \text{ MPa})(9.03 \text{ m/s}) = 0.860 \text{ MPa}\cdot\text{m/s}$$

Step 8. Design value of $pV = 2(0.860) = 1.720$ MPa·m/s.

Step 9. From Table 16–1, we can specify an oil-impregnated porous bronze bearing having a pV rating of 1.750 MPa·m/s or a BU dry-lubricated bearing having a pV rating of 1.785 MPa·m/s.

Steps 10–11. From Figure 16–4, we can recommend a minimum $C_d = 150$ μm (0.150 mm or 0.006 in) based on $D = 150$ mm and 1150 rpm. Other design details are dependent on the system into which the bearing will be placed.

Design Procedure for Boundary-Lubricated Bearings under Oscillating Loading

Many bearings of the boundary-lubricated type oscillate on stationary shafts rather than having continuously rotating shafts within the stationary bearings. Examples are bearings on steering shafts, suspension elements for vehicles, brake and accelerator pedals, seat position adjusters, shift mechanisms, door hinges, fitness exercising machines, hospital beds, hydraulic cylinder actuators, and hatch covers.

The design procedure is modified under these conditions. We can describe the oscillation by the following data:

φ = Angle of the oscillation in one direction, degrees

n_o = Number of complete oscillation cycles in both directions per minute, cycles/min

D = Nominal inside diameter of the bearing, in or mm

L = Length of the bearing, in or mm

F = Applied radial load, lb or N

Now we can define an equivalent number of revolutions/min for the bearing motion:

$$n_{eq} = n_o(2\varphi)/360 \text{ equivalent revolutions/min} \quad (16\text{–}2)$$

We can then substitute this value in previously developed equations for the relative velocity between the bearing and the shaft as follows.

U.S. Customary units; D in inches:

$$V = \pi Dn_{eq}/12 \text{ ft/min} \quad (16\text{–}3)$$

SI units; D in mm: $V = \pi Dn_{eq}/(60\,000)$ m/s **(16–4)**

The pV value is then calculated in the same manner as for rotating shafts in bearings.

Example Problem 16–3 A design for an industrial oven door has a 15 mm diameter horizontal bar over the oven opening on which the door hangs. Two smooth cylindrical door hinges are lined with the graphite metalized bearing material listed in Table 16–1. The door weighs 20.40 kN and rotates 110° from vertical to 20° above horizontal as the door is opened to insert or withdraw parts to be heated. It then returns to the original position. The door is opened and closed five times per minute. Determine the required length of the hinges to produce a pV value no more than 25% of the limiting value.

Solution From the given data:

$\varphi = 110°$

$n_o = 5.0$ cycles/min

$D = 15$ mm

$L =$ Length of the bearing, in or mm

$F = (20.40\text{ kN})/2 = 10.2\text{ kN} = 10\,200\text{ N}$ on each hinge

pV limit $= 0.525$ MPa·m/s

The allowable pV value is

$$(pV)_{all} = pV\text{ limit} \times 0.25 = 0.525\text{ MPa}\cdot\text{m/s} \times 0.25 = 0.1313\text{ MPa}\cdot\text{m/s}$$

We can now compute the equivalent number of revolutions per minute from:

$$n_{eq} = n_o(2\varphi)/360 = (5.0\text{ cycles/min})(2)(110°)/360° = 3.06$$

Now the equivalent sliding velocity is

$$V_e = \pi D n_{eq}/(60\,000)\text{ m/s} = V = \pi(15)(3.06)/(60\,000)\text{ m/s} = 0.00240\text{ m/s}$$

We can now solve for the limiting bearing pressure:

$$p_{all} = (pV)_{all}/V_e = (0.1313\text{ MPa}\cdot\text{m/s})/(0.00240\text{ m/s}) = 54.69\text{ MPa} = 54.69\text{ N/mm}^2$$

Let $p = F/LD$, we can solve for the required length, L.

$$L = F/pD = (10\,200\text{ N})/(54.69\text{ N/mm}^2)(15\text{ mm}) = 12.43\text{ mm}$$

A preferred value of $L = 16$ mm would be reasonable.

Wear Considerations for Plain Boundary-Lubricated Bearings

Wear is an important consideration in boundary-lubricated bearings because of the inherent sliding action between the shaft and the bearing. Each manufacturer provides wear data unique to its materials. See Section 16–10 for additional discussion of wear.

16–6 FULL-FILM HYDRODYNAMIC BEARINGS

In the *full-film hydrodynamic bearing*, the load on the bearing is supported on a continuous film of lubricant, usually oil, so that no contact between the bearing and the rotating journal occurs. A pressure must be developed in the oil in order to carry the load. With proper design, the motion of the journal inside the bearing creates the necessary pressure.

Figure 16–5 shows the progressive action in a plain surface bearing from start-up to the steady-state hydrodynamic operation. Observe that boundary lubrication and mixed-film lubrication precede the establishment of full-film hydrodynamic lubrication. At start-up, the radial load applied through the journal to the bearing forces the journal off center in the direction of the load, taking up all clearance [Figure 16–5(a)]. At the initial slow rotational speeds, the friction between the journal and the bearing causes the journal to climb up the bearing wall, somewhat as shown in Figure 16–5(b). Because of the viscous shear stresses developed in the oil, the moving journal draws oil into the converging, wedge-shaped area above the region of contact. The resulting pumping action produces a pressure in the oil film; when the pressure is sufficiently high, the journal is lifted from the bearing. The frictional forces are greatly reduced under this operating condition, and the journal moves eventually to the steady-state position shown in Figure 16–5(c). Note that the journal is offset from the direction of the load; that there is a certain eccentricity, e, between the geometric center of the bearing and the center of the journal; and that there is a point of minimum film thickness, h_o, at the nose of the wedge-shaped pressurized zone.

Figure 16–6 illustrates the general form of the pressure distribution within a full-film hydrodynamically lubricated

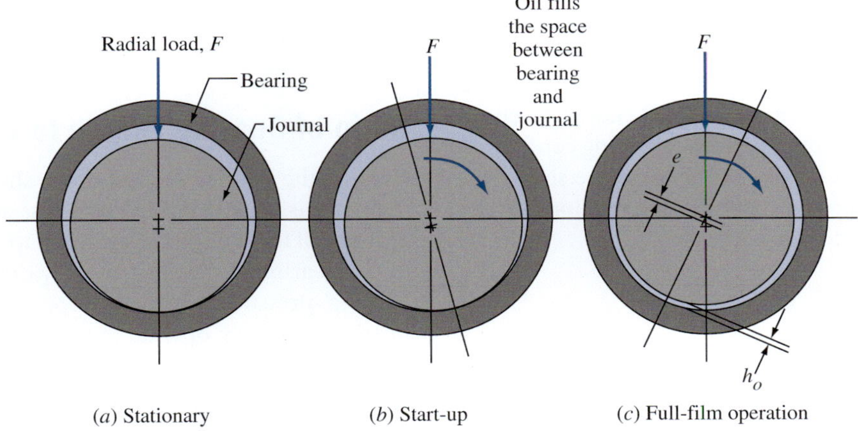

FIGURE 16–5 Position of the journal relative to the bearing as a function of mode of operation

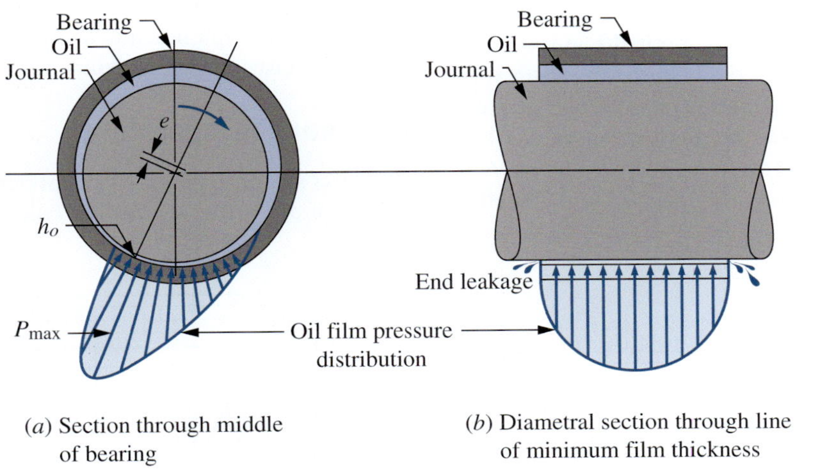

FIGURE 16–6 Pressure distribution in the oil film for hydrodynamic lubrication

bearing. The clearance between the bearing and the journal is highly exaggerated. Part (a) of the figure shows the rise of pressure as the rotating shaft draws oil into the converging wedge, approaching the point of minimum film thickness. The maximum pressure occurs there and then falls rapidly to nearly zero as the space between the journal and the bearing diverges again. The integrated effect of the pressure distribution is a force sufficient to support the shaft on a film of oil without metal-to-metal contact.

Figure 16–6(b) shows the pressure distribution axially along the shaft through the line of minimum film thickness or maximum pressure. The highest pressure value occurs in the middle of the bearing length, and it falls rapidly as the ends are approached because the pressure outside the bearing is ambient pressure, typically atmospheric pressure. There is continual leakage flow from both ends of the bearing. This illustrates the importance of providing a means of continually supplying oil to the bearing to maintain the full-film operation. Without a steady, adequate supply of oil, the system would not be able to create the pressurized film to carry the shaft, and boundary lubrication would result. The significantly higher frictional forces thus created would cause rapid heating of the bearing/journal interface, and seizure would likely occur very rapidly.

16–7 DESIGN OF FULL-FILM HYDRO-DYNAMICALLY LUBRICATED BEARINGS

The following discussion presents several guidelines for bearing design for typical industrial applications. The design procedure is based primarily on information in References 1–5.

Surface Roughness

A ground journal with a surface roughness average of 16–32 microinches (μin), or 0.40 to 0.80 μm, is recommended for bearings of good quality. The bearing should be equally smooth or made from one of the softer materials so that "wearing in" can smooth the high spots, creating a good fit between the journal and the bearing. In high-precision equipment, polishing or lapping can be used to produce a surface finish of the order of 8 to 16 μin (0.20 to 0.40 μm).

Minimum Film Thickness

The limiting acceptable value of the minimum film thickness depends on the surface roughness of the journal and the bearing because the film must be thick enough to eliminate any solid contact during expected operating conditions. The suggested design value also depends on the size of the journal. For ground journals, the following relationship can be used to estimate the design value:

$$h_o = 0.00025D \tag{16–5}$$

where D = diameter of the bearing

Diametral Clearance

The clearance between the journal and the bearing depends on the nominal diameter of the bearing, the precision of the machine for which the bearing is being designed, the speed of rotation, and the surface roughness of the journal. The coefficient of thermal expansion for both the journal and the bearing must also be considered to ensure a satisfactory clearance under all expected operating conditions. An overall guideline of making the clearance in the range of 0.001 to 0.002 times the bearing diameter can be used. Figure 16–4 shows a graph of the recommended minimum diametral clearance as a function of the bearing diameter and the rotational speed. Some variation above the values from the curve is permissible.

Ratio of Bearing Length to Diameter

Because the journal is a part of the shaft itself, its minimum diameter is usually limited by stress and deflection considerations of the type discussed in Chapter 12. Then the length of the bearing is specified to provide for a suitable level of bearing pressure. General-purpose industrial machinery bearings typically operate at approximately 200 to 500 psi (1.4 to 3.4 MPa) bearing pressure, based on the projected area of the bearing [p = load/(LD)]. The pressure may range from as low as 50 psi (0.34 MPa) for light-duty equipment to 2000 psi (13.4 MPa) for heavy machinery under varying loads, such as in internal combustion engines. The leakage of oil from the bearing is also affected by the bearing length. The typical range of length to diameter ratio (L/D) for full-film hydrodynamic bearings is from 0.35 to 1.5. But many successful bearings operate outside this range.

Lubricant Temperature

The viscosity of the oil is a critical parameter in the performance of a bearing. Figure 16–7 shows the great variation

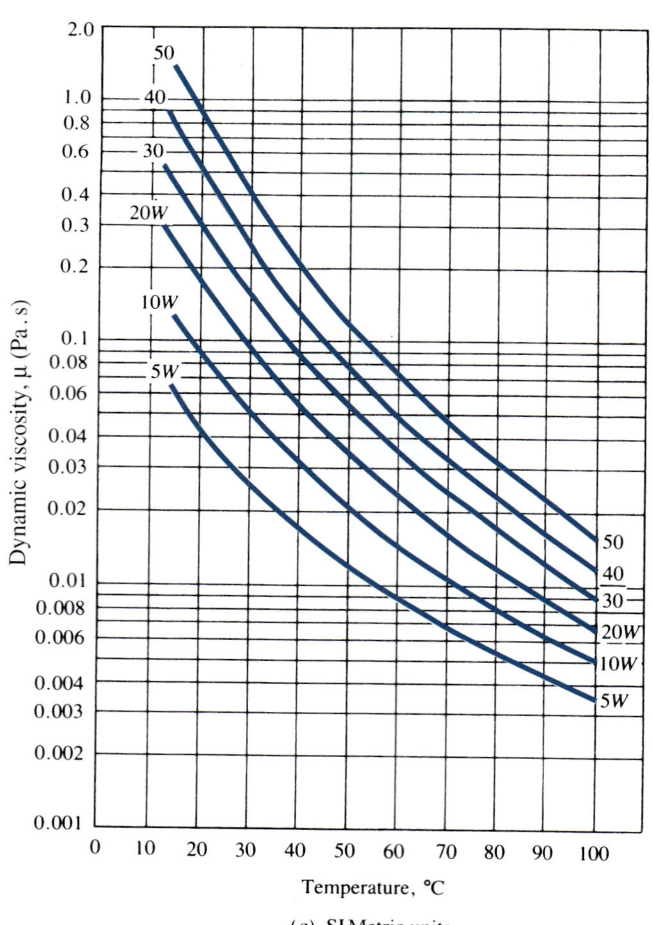

(*a*) SI Metric units

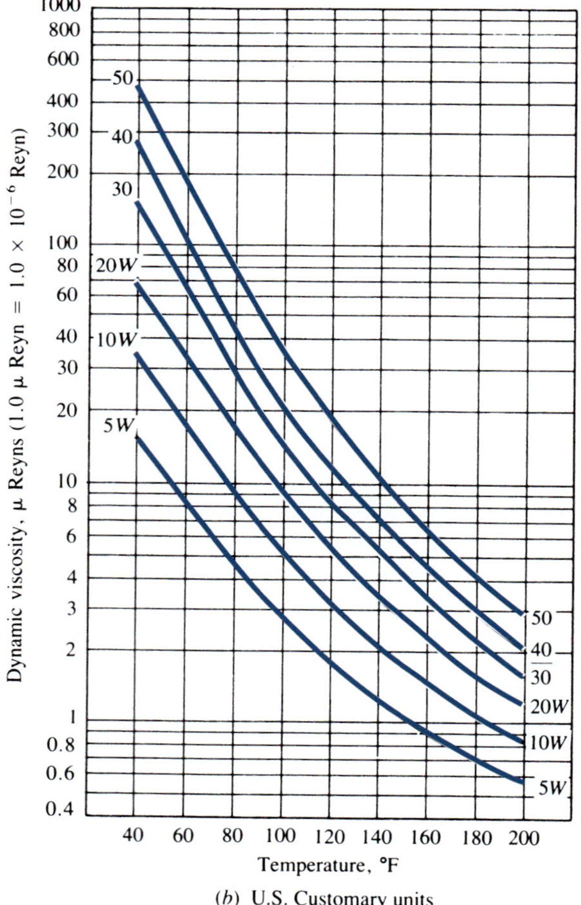

(*b*) U.S. Customary units

FIGURE 16–7 Viscosity versus temperature for SAE oils

of viscosity with temperature, which indicates that temperature control is advisable. Also, most petroleum lubricating oils should be limited to approximately 160°F (70°C) in order to retard oxidation. The temperature of interest is, of course, that inside the bearing. Frictional energy or thermal energy from the equipment itself can increase the oil temperature above that in the supply reservoir. In design examples we will select the lubricant that will ensure a satisfactory viscosity at 160°F, unless otherwise noted. If the actual operating temperature is lower, the resulting film thickness will be greater than the design value—a conservative result. It is incumbent on the designer to ensure that the limiting temperature is not exceeded, using forced cooling if necessary.

Lubricant Viscosity

The specification of the lubricant for the bearing is one of the final decisions to be made in the design procedure that follows. In the calculations, it is the dynamic viscosity, μ, which is used. In the U.S. Customary Unit (lb-in-s) System, the dynamic viscosity is expressed in lb·s/in^2, which is given the name *reyn* in honor of Osbourne Reynolds, who performed much significant work in fluid flow. In SI units, the standard unit is N·s/m^2 or Pa·s. Some prefer the unit of poise or centipoise, derived from the dyne-cm-s metric system. Some useful conversions are

$$1.0 \text{ reyn} = 6895 \text{ Pa}\cdot\text{s}$$

$$1.0 \text{ Pa}\cdot\text{s} = 1000 \text{ centipoise}$$

Other viscosity conversions may also be useful. (See Reference 18.) Figure 16–7 shows graphs of dynamic viscosity versus temperature for both SI and U.S. Customary units. Note in Figure 16–7(b) that common values in U.S. Customary units are very small. Values from the scale must be multiplied by 10^{-6}.

Sommerfeld Number

The combined effect of many of the variables involved in the operation of a bearing under hydrodynamic lubrication can be characterized by the dimensionless number, S, called the *Sommerfeld number*. Some refer to this as the *bearing characteristic number*, defined as follows:

➪ Bearing Characteristic Number

$$S = \frac{\mu n_s (R/C_r)^2}{p} \tag{16–6}$$

Note that S is similar to the bearing parameter, $\mu n/p$, discussed in Section 16–3, in that it involves the combined effect of viscosity, rotational speed, and bearing pressure. In order for S to be dimensionless, the following units should be used for the factors:

	U.S. Customary Units	SI Units
μ	lb·s/in^2 (reyns)	Pa·s (N·s/m^2)
n_s	rev/s	rev/s
p	lb/in^2 (psi)	Pa (N/m^2)
R, C_r	in	m or mm

Any consistent units can be used. Figure 16–8, adapted from Reference 4, shows the relationship between the Sommerfeld number and the film thickness ratio, h_o/C_r.

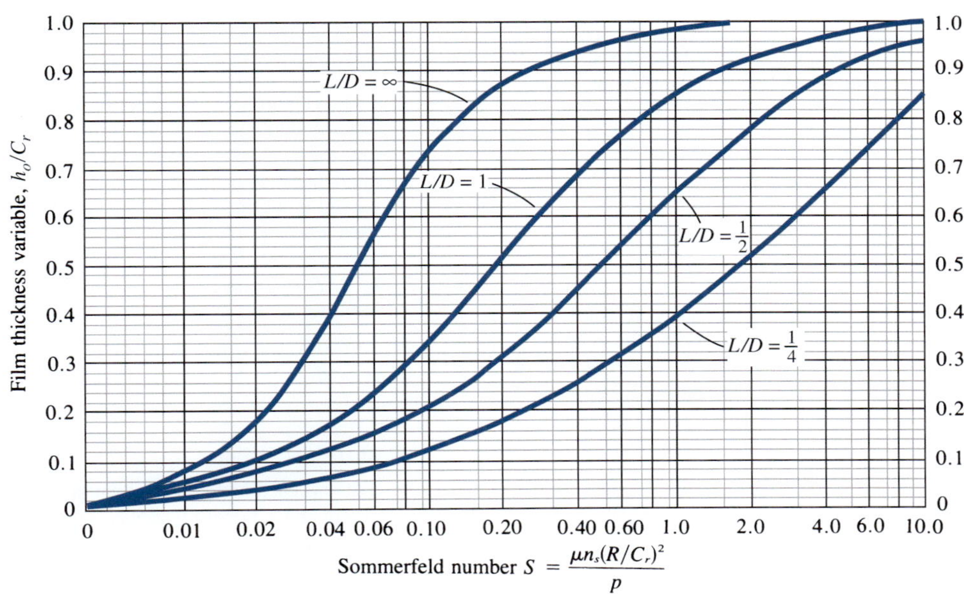

FIGURE 16–8 Film thickness variable, h_o/C_r, versus Sommerfeld number, S (Adapted from John Boyd and Albert A. Raimondi, "A Solution for the Finite Journal Bearing and Its Application to Analysis and Design," Parts I and II. *Transactions of the American Society of Lubrication Engineers*, Vol. 1, No. 1, 1958)

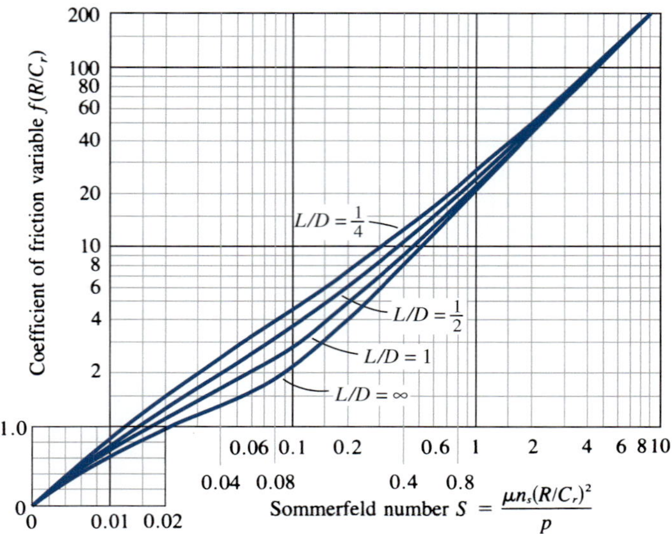

FIGURE 16–9 Coefficient of friction variable, $f(R/C_r)$, versus Sommerfeld number, S

Figure 16–9 shows the relationship between S and the coefficient of friction variable, $f(R/C_r)$. These values are used in the design procedure that follows. Because many design decisions are required, several acceptable solutions are possible.

Design Procedure

PROCEDURE FOR DESIGNING FULL-FILM HYDRODYNAMICALLY LUBRICATED BEARINGS

Because the bearing design is usually done after the stress analysis of the shaft is completed, the following items are typically known:

Radial load on the bearing, F, usually in lb or N

Rotational speed, n, usually in rpm

Nominal shaft diameter at the journal, sometimes specified as a minimum acceptable diameter based on strength and stiffness

The results of the design procedure produce values for the actual journal diameter, the bearing length, the diametral clearance, the minimum film thickness of the lubricant during operation, the surface finish for the journal, the lubricant and its maximum operating temperature, the coefficient of friction, the friction torque, and the power dissipated because of friction.

1. Specify a trial value for the journal diameter, D, and the radius, $R = D/2$.
2. Specify a nominal operating bearing pressure, usually 200 to 500 psi (1.4 to 3.4 MPa), where $p = F/LD$. Solve for L:

$$L = F/pD$$

Then compute L/D. It might be desirable to redefine L/D to be a convenient value from 0.25 to 1.5 in order to use the design charts available. Finally, specify the actual design value of L/D and L, and compute the actual $p = F/LD$.

3. Referring to Figure 16–4, specify the diametral clearance, C_d, based on the values for D and n. Then compute $C_r = C_d/2$ and the ratio R/C_r.
4. Specify the desired surface finish for the journal and bearing, also based on the application. a typical value is 16 to 32 μin average (0.40 to 0.80 μm).
5. Compute the nominal minimum film thickness from Equation (16–5),

$$h_o = 0.00025D$$

6. Compute h_o/C_r, the film thickness ratio.
7. From Figure 16–8, determine the value of the Sommerfeld number for the selected film thickness ratio and the L/D ratio. Exercise care in interpolating in this chart because of the logarithmic axis and the nonlinear spread between curves. For $L/D > 1$, only approximate data can be obtained. For $L/D = 1.5$, interpolate approximately one-fourth of the way between the curves for $L/D = 1$ and $L/D =$ infinity. For $L/D = 2$, go about halfway.
8. Compute the rotational speed n_s in revolutions per second:

$$n_s = n/60$$

where n is in rpm

9. Because each factor of the Sommerfeld number is known except the lubricant viscosity, μ, solve for the required minimum viscosity that will produce the desired minimum film thickness:

Required Minimum Lubricant Viscosity

$$\mu = \frac{Sp}{n_s(R/C_r)^2} \quad (16\text{–}7)$$

10. Specify a maximum acceptable lubricant temperature, usually 160°F or 70°C. Select a lubricant from Figure 16–7 that will have at least the required viscosity at the operating temperature. If the selected lubricant has a viscosity greater than that computed in Step 9, recompute S for the new value of viscosity. The resulting value for the minimum film thickness will now be somewhat greater than the design value, a generally desirable result. You can consult Figure 16–8 again to determine the new value of the minimum film thickness, if desired.
11. From Figure 16–9, obtain the coefficient of friction variable, $f(R/C_r)$.
12. Compute $f = f(R/C_r)/(R/C_r)$ = coefficient of friction.
13. Compute the friction torque. The product of the coefficient of friction and the load F gives the frictional force at the surface of the journal. That value multiplied by the radius gives the torque:

Friction Torque

$$T_f = F_f R = fFR \quad \textbf{(16–8)}$$

14. Compute the power dissipated in the bearing from the relationship between power, torque, and speed that we have used many times:

Frictional Power Loss

$$P_f = T_f n/63\,000 \text{ hp} \quad \textbf{(16–9)}$$

This frictional power loss represents the rate of energy input to the lubricant within the bearing that can cause the temperature to rise. It is a part of the energy that must be removed from the bearing to maintain satisfactory lubricant viscosity.

Example Problem 16–4 will illustrate the procedure.

Example Problem 16–4 Design a plain surface journal bearing to carry a steady radial load of 1500 lb while the shaft rotates at 850 rpm. The shaft stress analysis determines that the minimum acceptable diameter at the journal is 2.10 in. The shaft is part of a machine requiring good precision.

Solution

Step 1. Let's select $D = 2.50$ in. Then $R = 1.25$ in.

Step 2. For $p = 200$ psi, L must be

$$L = F/pD = 1500/(200)(2.50) = 3.00 \text{ in}$$

For this value of L, $L/D = 3.00/2.50 = 1.20$. In order to use one of the standard design charts, let's change L to 2.50 in so that $L/D = 1.0$. This is not essential, but it eliminates interpolation. The actual pressure is then

$$p = F/LD = 1500/(2.50)(2.50) = 240 \text{ psi}$$

This is an acceptable pressure.

Step 3. From Figure 16–4, $C_d = 0.003$ in is suitable for the diametral clearance, based on $D = 2.50$ in and $n = 850$ rpm, and $C_r = C_d/2 = 0.0015$ in. Also,

$$R/C_r = 1.25/0.0015 = 833$$

This value is used in later calculations.

Step 4. For the precision desired in this machine, use a surface finish of 16 to 32 μin, requiring a ground journal.

Step 5. Minimum film thickness (design value):

$$h_o = 0.00025D = 0.00025(2.50) = 0.0006 \text{ in (approx)}$$

Step 6. Film thickness variable:

$$h_o/C_r = 0.0006/0.0015 = 0.40$$

Step 7. From Figure 16–8, for $h_o/C_r = 0.40$ and $L/D = 1$, we can read $S = 0.13$.

Step 8. Rotational speed in revolutions per second:

$$n_s = n/60 = 850/60 = 14.2 \text{ rev/s}$$

Step 9. Solve for the viscosity from the Sommerfeld number, S:

$$\mu = \frac{Sp}{n_s(R/C_r)^2} = \frac{(0.13)(240)}{(14.2)(833)^2} = 3.17 \times 10^{-6}\ \text{reyn}$$

Step 10. From the viscosity chart, Figure 16–7, SAE 30 oil is required to ensure a sufficient viscosity at 160°F. The actual expected viscosity of SAE 30 at 160°F is approximately 3.3×10^{-6} reyn.

Step 11. For the actual viscosity, the Sommerfeld number would be

$$S = \frac{\mu n_s(R/C_r)^2}{p} = \frac{(3.3 \times 10^{-6})(14.2)(833)^2}{240} = 0.135$$

Step 12. Coefficient of friction (from Figure 16–9): $f(R/C_r) = 3.5$ for $S = 0.135$ and $L/D = 1$. Now, because $R/C_r = 833$,

$$f = 3.5/833 = 0.0042$$

Step 13. Friction torque:

$$T_f = fFR = (0.0042)(1500)(1.25) = 7.88\ \text{lb}\cdot\text{in}$$

Step 14. Frictional power:

$$P_f = T_f n/63\,000 = (7.88)(850)/63\,000 = 0.106\ \text{hp}$$

Comment A qualitative evaluation of the result would require more knowledge about the application. But note that a coefficient of friction of 0.0042 is quite low. It is likely that a machine requiring such a large shaft and with such high bearing forces also requires a large power to drive it. Then the 0.106-hp friction power would appear small.

Thermal considerations are also important to determine how much energy must be dissipated from the bearing. Converting the frictional power loss to thermal power gives

$$P_f = 0.106\ \text{hp}\,\frac{745.7\ \text{W}}{\text{hp}} = 79.0\ \text{W}$$

Expressing this in U.S. Customary units gives

$$P_f = 79.0\ \text{W}\,\frac{1\ \text{Btu/hr}}{0.293\ \text{W}} = 270\ \text{Btu/hr}$$

16–8 PRACTICAL CONSIDERATIONS FOR PLAIN SURFACE BEARINGS

The design of the bearing system must consider the method of delivery of the lubricant to the bearing, the distribution of the lubricant within the bearing, the quantity of lubricant required, the amount of heat generated in the bearing and its effect on the temperature of the lubricant, the dissipation of heat from the bearing, the maintenance of the lubricant in a clean condition, and the performance of the bearing over the complete range of operating conditions that the bearing is likely to experience.

Many of these factors are simply design details that must be worked out along with the other aspects of the machine design. But some guidelines and general recommendations will be presented here.

The lubricant can be delivered to the bearing by a pump, perhaps driven from the same source that drives the entire machine. In some gear drives, one of the gears is designed to dip into an oil sump and carry oil up to the gear mesh and to the bearings. An external oil cup can be used to supply oil by gravity if the lubricant quantity required is small.

Methods of estimating the quantity of oil required, considering the leakage of oil from the ends of the bearing, are available. (See References 1, 4, 6, 7, 9, 12, 13, and 20.)

The delivery of oil to the bearing should always be in an area opposite the location of the hydrodynamic pressure that supports the load. Otherwise, the oil delivery hole would destroy the pressurization of the film.

Grooving is frequently used to distribute the oil along the length of the bearing. Oil would be delivered through a radial hole in the bearing at the midlength point. The groove would extend axially in both directions from the hole, but it would terminate somewhat before the end of the bearing in order to keep the oil from leaking to the side. The rotation of the journal then carries the oil around to the area where the hydrodynamic film is generated. Figure 16–10 shows several styles of grooving in use.

Cooling of the bearing itself, or of the oil in the sump that supplies the oil, must always be considered. Natural convection may be sufficient to transfer heat away and maintain an acceptable bearing temperature. If not, forced convection can be used. In severe cases of heat generation, especially where the bearing system operates in a hot area such as a furnace, liquid coolant can be pumped through a jacket around the bearing. Some commercially available bearings provide this feature. Placing a heat exchanger in the oil sump or pumping the oil through an external heat exchanger can also be done. Figure 16–11 shows one commercially available style of a plain bearing with coolant pipes to allow use of water, air, or oil as the internal coolant.

The lubricant can be cleaned by passing it through filters as it is pumped to the bearing. Magnetic plugs within the sump can be effective in attracting and holding metallic particles that can score the bearing if allowed to enter the clearance space between the journal and the bearing. Of course, frequent changing of the oil is also desirable.

The design procedure used in the preceding section was completed for one set of conditions: a given temperature, diametral clearance, load, and rotational speed. If any of these factors varies during the operation of the machine, the performance of the bearing must be evaluated under the new conditions. The testing of a prototype under a variety of conditions is also desirable. References 1–7, 10–17, and

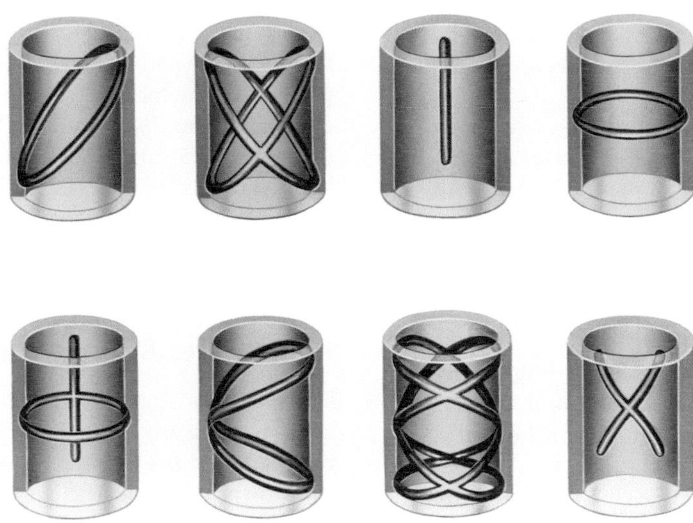

FIGURE 16–10 Examples of styles of grooving for plain bearings (Bunting Bearings Corp., Holland, OH)

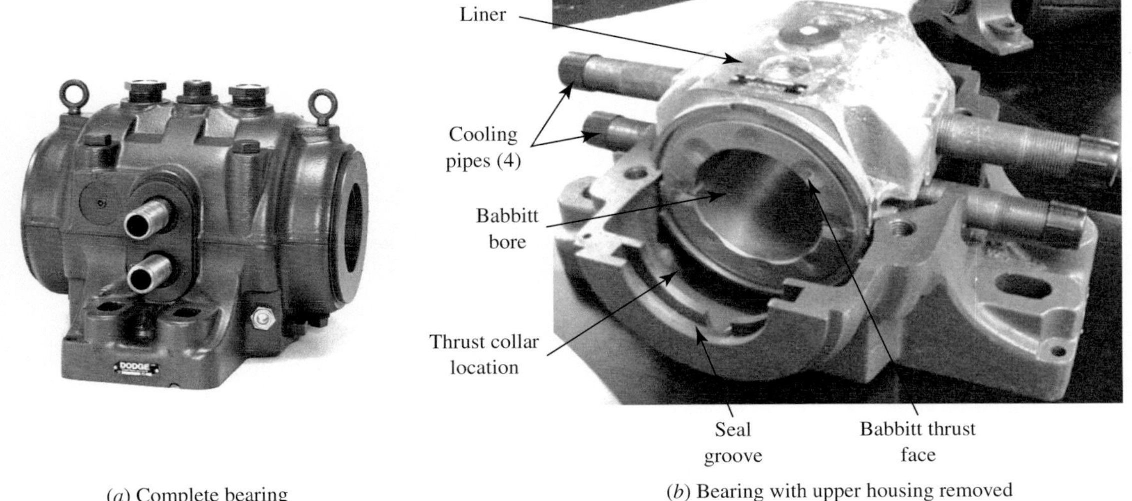

(*a*) Complete bearing

(*b*) Bearing with upper housing removed

FIGURE 16–11 Sleevoil hydrodynamically lubricated bearing with cooling pipes (Baldor/Dodge, Greenville, SC)

19–21 discuss other practical considerations for the design of plain surface bearings.

16–9 HYDROSTATIC BEARINGS

Recall that hydrodynamic lubrication results from the creation of a pressurized film of oil sufficient to carry the load on the bearing, with the film being generated by the motion of the journal itself within the bearing. It was noted that a steady relative motion between the journal and the bearing is required to generate and maintain the film.

In some types of equipment, the conditions are such that a hydrodynamic film cannot be developed. Reciprocating or oscillating devices or very-slow-moving machines are examples. If the load on the bearing is very high, it may not be possible to generate a sufficiently high pressure in the film to support the load. Even in cases in which hydrodynamic lubrication can be developed during normal operation of the machine, there is still mixed-film or boundary lubrication during the start-up and shut-down cycles. This may be unacceptable.

Consider the design of the mount for a telescope or an antenna system in which rotation of the base is required to be at very slow speed and to have smooth motion. Also, low friction is desired in order to keep the drive system small and to provide fast response and accurate positioning. The mount is essentially a thrust bearing supporting the weight of the system.

In this type of application, *hydrostatic lubrication* is desirable. Lubricant is supplied to the bearing under high pressure, several hundred psi or higher, and the pressure acting on the area of the bearing literally lifts the load off the bearing, even with the equipment stationary.

Figure 16–12 shows the main elements of a hydrostatic bearing system. A positive displacement pump draws oil from a reservoir and delivers it under pressure to a supply manifold from which several bearing pads can be supplied. At each bearing pad, the oil passes through a control element that permits the balancing of the system. The control element may be a flow-control valve, a length of small-diameter tubing, or an orifice, any of which offers a resistance to the flow of oil and permits the several bearing pads to operate at a sufficiently high pressure to lift the load on that pad. When the system is in operation, the oil enters a recess within the bearing pad. For example, Figure 16–12(b) shows a circular pad with a circular recess in its center supplied with oil through a central hole. The load initially rests on the land area, sealing the recess. As the pressure in the recess reaches the level where the product of the pressure times the recess area equals the applied load, the load is lifted from the pad. Immediately there is a flow of oil across the land area under the lifted load, and the pressure decreases to atmospheric pressure at the outside of the pad. The flow of oil must be maintained at a level that matches the outflow from the pad. When equilibrium occurs, the integrated product of the local pressure times the area lifts the load a certain distance, h, usually in the range of 0.001 to 0.010 in (0.025 to 0.25 mm). The film thickness, h, must be large enough to ensure no solid contact over the range of operating conditions, but it should be kept as small as possible to minimize the flow of oil through each bearing and the pump power required to drive the system.

Hydrostatic Bearing Performance

Three factors characterizing the performance of a hydrostatic bearing are its load-carrying capacity, the flow of oil required, and the pumping power required, as indicated by the dimensionless coefficients a_f, q_f, and H_f. The magnitudes of the coefficients depend on the design of the pad:

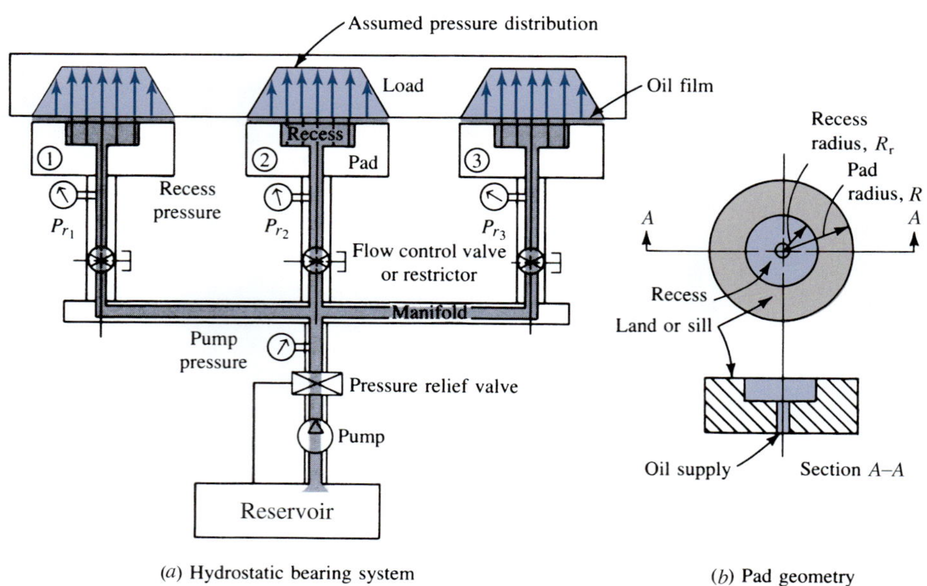

(*a*) Hydrostatic bearing system

(*b*) Pad geometry

FIGURE 16–12 Main elements of a hydrostatic bearing system

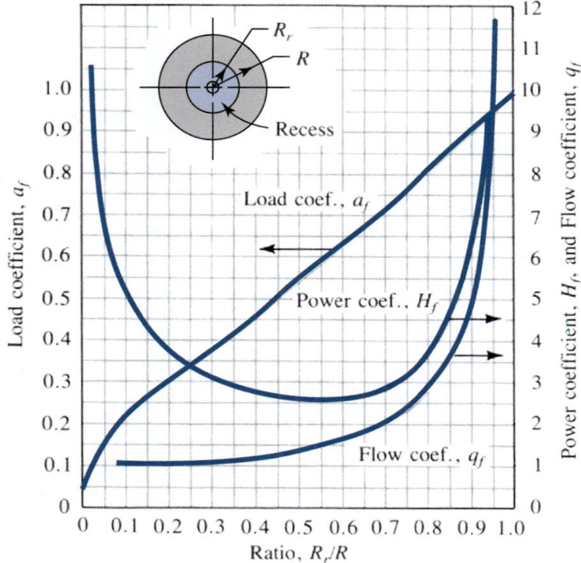

FIGURE 16–13 Dimensionless performance coefficients for circular pad hydrostatic bearing (Cast Bronze Institute. *Cast Bronze Hydrostatic Bearing Design Manual.* New York: Copper Development Association, 1975)

⇨ Load-Carrying Capacity

$$F = a_f A_p p_r \tag{16–10}$$

⇨ Oil Flow Required

$$Q = q_f \frac{F}{A_p}\frac{h^3}{\mu} \tag{16–11}$$

⇨ Pumping Power Required

$$P = p_r Q = H_f \left(\frac{F}{A_p}\right)^2 \frac{h^3}{\mu} \tag{16–12}$$

where F = load on the bearing, lb or N

Q = volume flow rate of oil, in^3/s or m^3/s

P = pumping power, lb·in/s or N·m/s (watts)

a_f = pad load coefficient, dimensionless

q_f = pad flow coefficient, dimensionless

H_f = pad power coefficient, dimensionless (*Note:* $H_f = q_f/a_f$)

A_p = pad area, in^2 or m^2

p_r = oil pressure in the recess of the pad, psi or Pa

h = film thickness, in or m

μ = dynamic viscosity of the oil, lb·s/in² (reyn) or Pa·s

Figure 16–13 shows the typical variation of the dimensionless coefficients as a function of the pad geometry for a circular pad with a circular recess. As the size of the recess, R_r/R, increases, the load-carrying capacity increases, as indicated by a_f. But at the same time, the flow through the bearing increases, as indicated by q_f. The increase is gradual up to a value of R_r/R of approximately 0.7, and then rapid for higher ratios. This higher flow rate requires much higher pumping power, as indicated by the rapidly increasing power coefficient. At very low ratios of R_r/R, the load coefficient decreases rapidly. The pressure in the recess would have to increase to compensate in order to lift the load. The higher pressure requires more pumping power. Therefore, the power coefficient is high either at very small ratios of R_r/R or at high ratios. The minimum power is required for ratios between 0.4 and 0.6.

These general characteristics of hydrostatic bearing pad performance are typical for many different geometries of pads. Extensive data for the performance of different pad shapes are published. (See Reference 8.)

Example Problem 16–5 illustrates the basic design procedure for hydrostatic bearings.

Example Problem 16–5 A large antenna mount weighing 12 000 lb is to be supported on three hydrostatic bearings such that each bearing pad carries 4000 lb. A positive displacement pump will be used to deliver oil at a pressure up to 500 psi. Design the hydrostatic bearings.

Solution We will choose the circular pad design for which the performance coefficients are available in Figure 16–13. The results of the design will specify the dimensions of the pads, the required oil pressure in the recess of each pad, the type of oil required and its temperature, the thickness of the film of oil when the bearings are supporting the load, the flow rate of oil required, and the pumping power required.

Step 1. From Figure 16–13, the minimum power required for a circular pad bearing would occur with the ratio R_r/R of approximately 0.50. For that ratio, the value of the load coefficient is $a_f = 0.55$. The pressure at the bearing recess will be somewhat below the maximum available of 500 psi because of the pressure drop in the restrictor placed between the supply manifold and the pad. Let's design for a recess pressure of approximately 400 psi. Then, from Equation (16–10),

$$A_p = \frac{F}{a_f p_r} = \frac{4000\,\text{lb}}{0.55(400\,\text{lb/in}^2)} = 18.2\ \text{in}^2$$

But $A_p = \pi D^2/4$. Then the required pad diameter is

$$D = \sqrt{4A_p/\pi} = \sqrt{4(18.2)/\pi} = 4.81\ \text{in}$$

For convenience, let's specify $D = 5.00$ in. Then the actual pad area will be

$$A_p = \pi D^2/4 = (\pi)(5.00\ \text{in})^2/4 = 19.6\ \text{in}^2$$

The required recess pressure is then

$$p_r = \frac{F}{a_f A_p} = \frac{4000\ \text{lb}}{0.55(19.6\ \text{in}^2)} = 370\ \text{lb/in}^2$$

Also,

$$R = D/2 = 5.00\ \text{in}/2 = 2.50\ \text{in}$$

$$R_r = 0.50R = 0.50(2.50\ \text{in}) = 1.25\ \text{in}$$

Step 2. Specify the design value of the film thickness, h. It is recommended that h be between 0.001 and 0.010 in. Let's use $h = 0.005$ in.

Step 3. Specify the lubricant and the operating temperature. Let's select SAE 30 oil and assume that the maximum oil temperature in the oil film will be 120°F. A method of estimating the actual film temperature during operation may be consulted. (See Reference 8.) From the viscosity/temperature curves, Figure 16–7, the viscosity is approximately 8.3×10^{-6} reyn (lb·s/in²).

Step 4. Compute the oil flow through the bearing from Equation (16–11). The value of $q_f = 1.4$ can be found from Figure 16–13:

$$Q = q_f \frac{F}{A_p}\frac{h^3}{\mu} = (1.4)\frac{4000\ \text{lb}}{19.6\ \text{in}^2}\frac{(0.005\ \text{in})^3}{8.3 \times 10^{-6}\ \text{lb}\cdot\text{s/in}^2}$$

$$Q = 4.30\ \text{in}^3/\text{s}$$

Step 5. Compute the pumping power required from Equation (16–12). The value of $H_f = 2.6$ can be found from Figure 16–13:

$$P = p_r Q = H_f\left(\frac{F}{A_p}\right)^2\frac{h^3}{\mu} = 2.6\left(\frac{4000}{19.6}\right)^2\frac{(0.005\ \text{in})^3}{8.3 \times 10^{-6}} = 1631\ \text{lb}\cdot\text{in/s}$$

For convenience, we can convert this to horsepower:

$$P = \frac{1631\ \text{lb}\cdot\text{in}}{\text{s}}\,\frac{1.0\ \text{ft}}{12\ \text{in}}\,\frac{1.0\ \text{hp}}{550\ \text{lb}\cdot\text{ft/s}} = 0.247\ \text{hp}$$

16–10 TRIBOLOGY: FRICTION, LUBRICATION, AND WEAR

The study of friction, lubrication, and wear, called *tribology*, involves many disciplines, such as solid mechanics, fluid mechanics, material science, and chemistry. It is appropriate to engage several specialists on design teams where critical bearing or general lubrication designs are completed. For study beyond the scope of this book, see References 1–3, 9, 12–17, and 19–22.

This section presents some general principles of lubrication that can apply to a variety of design situations where relative motion occurs between mating machine elements. The goal is to help you recognize the many parameters that must be considered in designing machinery and in analyzing failures or unsatisfactory operation of existing machines. Much of the discussion relates to minimizing or controlling *friction*, defined generally as the resistance to parallel motion of mating surfaces. *Lubrication* is used to minimize friction by introducing a film of material that inherently reduces the force required to move one component relative to its mating component. Some materials have inherently low friction coefficients and can operate satisfactorily without external lubrication. When the relative motion results in physical contact between the surfaces of the mating components, some of the surface material can be removed, resulting in *wear*.

Friction

Not all friction is undesirable. Consider the need for drive wheels to use friction to develop propulsive force against floors, rails, or roads. Clutches and brakes employ friction to engage machinery, accelerate it, decelerate it, bring it to a stop, or hold it in position. See Chapter 22. Clamps and collets use friction to hold work pieces during manufacturing operations. In such applications, large, consistent frictional forces are desirable.

Most other applications where sliding contact occurs between mating components call for minimizing friction in order to minimize the forces, torques, and power required to drive the system. We are most concerned with these applications in this section.

The primary phenomena involved in creating friction are adhesion, elastic effects such as rolling resistance, viscoelastic effects, and hydrodynamic resistance. *Adhesion* is the bonding between dissimilar materials. The strength of adhesion depends on the structure and chemistry of the mating materials. Surface characteristics also contribute, such as the height of roughness peaks and valleys, called *asperities*. Sometimes asperities on mating parts are deformed or fractured during relative motion, while for other conditions, the motion is resisted as the asperities ride up and down on one another. *Rolling resistance* is caused by the elastic deformation of the moving body or the surface on which it moves. The geometry of the members in rolling contact, the magnitude of applied forces, and the elasticity of the contacting materials all play a part in determining the amount of resistance. *Viscoelastic effects* relate to the forces caused by deformation of flexible materials, such as elastomers, during contact. *Hydrodynamic resistance*, also called the *viscous effect*, is caused by the relative motion among the molecules of fluid lubricants between the moving mating components. This is the primary form of resistance in full-film hydrodynamically lubricated bearings. All or many of these forms of friction exist simultaneously in most practical machines.

Lubricants

Some of the important functions of lubricants are to reduce friction, remove heat from bearings and other machine elements where friction occurs, and to suspend contaminants. Several properties contribute to satisfactory lubricant performance:

- Good lubricity or oiliness to promote low friction
- Adequate viscosity for the application
- Low volatility under operating conditions
- Satisfactory flow characteristics at the temperatures encountered in use
- Appropriate thermal conductivity and specific heat to perform the heat transfer function
- Good chemical and thermal stability and the ability to maintain desirable characteristics for a reasonable period of use
- Compatibility with other materials in the system such as bearings, seals, and machine parts, particularly with regard to corrosion protection and degradation
- Environmental friendliness

Here we discuss the basic nature of oils, greases, and solid lubricants.

Oils. Vendors of lubricating oils offer a huge variety of grades. One general classification is between refined natural petroleum oil and synthetic lubricants. Typically, natural oils are lower cost and can provide satisfactory service for general-purpose lubrication. Additives are frequently compounded with the natural oil to enhance viscosity, improve viscosity index (decrease the variation of viscosity with temperature), reduce corrosion potential, retard oxidation or other forms of chemical breakdown, or increase the ability to withstand localized high pressure. Synthetic lubricants are specially designed chemical formulations that can be tailored to specific applications. While performance is better than natural oils, cost is typically higher.

Oil lubricants are offered in categories such as engine oils, gear oils, compressor oils, turbine oils, general-purpose lubricating oils, chain lubricants, bearing oils, food machinery lubricants, automatic transmission fluids, hydraulic oils, and metalworking fluids. Internet sites 11–13 and 16 provide good examples of the types of lubricants available. Properties of lubricants that can affect selection are viscosity grade, viscosity index, and corrosion protection.

Viscosity grades are typically reported using the ISO (International Standards Organization) rating system. The grade number is the kinematic viscosity of the oil in centistokes (mm^2/s) measured at 40°C (104°F). Common ISO grades for lubricants are 32, 46, 68, 100, 150, 220, 320, 460, 680, and 1000. Gear oils are typically grades 150 to 680, depending on environmental temperature and degree of loading. Often the viscosity will also be reported at 100°C (212°F) as an indication of the variation of viscosity with temperature. This is not a part of the ISO classification and is influenced primarily by the property of viscosity index, discussed later in this section.

The American Gear Manufacturers Association (AGMA) formerly defined lubricant numbers from 0 through 15, with numbers 3 to 8 being the most common for general power-transmission uses. The basic viscosities for the former AGMA grades 0 to 8 correlate with ISO viscosity grades as follows.

AGMA	ISO	AGMA	ISO	AGMA	ISO
0	32	3	100	6	320
1	46	4	150	7	460
2	68	5	220	8	680

Suffixes are added to the AGMA grade designations: EP for extreme pressure, CP for compounded oils, and S for synthetic oils. See References 10, 16, 22, and 29 for Chapter 9. See also Section 9–17.

SAE viscosity grades are also used to report oil viscosity. Similar to the grades used in automotive engines, SAE oils are also good for general lubrication and for gear-type power transmissions. Common grades are SAE 20, 30, 40, 50, 60, 85, 90, 140, and 250. These grades must conform to limits of kinematic viscosity in centistokes measured at 100°C (212°F). The W-grades of SAE oils, such as SAE 20W, must have viscosities less than specified limits at low temperatures ranging from −5°C to −55°C (−3°F to −67°F). See Reference 18. This ensures that the lubricant can flow to critical surfaces in cold environments, particularly during equipment startup. Gear oils are typically SAE grades 80 to 250.

For details of SAE viscosity grades, see:

- SAE J300 *Engine Oil Viscosity Classifications*
- SAE J306 *Automotive Gear Lubricant Viscosity Classifications*

Always use the latest version.

Viscosity index (VI) is a measure of how greatly the viscosity of a fluid changes with temperature. VI is determined by measuring the viscosity of the sample fluid at 40°C (104°F) and 100°C (212°F) and comparing these values with those of certain reference fluids that were assigned VI values of 0 and 100.

A fluid with a high VI exhibits a small change in viscosity with temperature.

A fluid with a low VI exhibits a large change in viscosity with temperature.

For most lubricants, a high VI is desirable because it would provide more reliable protection and exhibit a more uniform performance as the temperature varies. Commercially available lubricants report VI values from approximately 90 to 250. Gear lubricants typically have a VI of approximately 150. Additives, typically organic polymers, are used to improve viscosity and adjust viscosity index.

Corrosion protection can be tailored for specific applications by blending various additives with the base oil. Rust inhibitors for ferrous materials or copper and bronze corrosion protection are common. Oxidation inhibitors are used to prolong the life of oils. Extreme pressure additives help prevent scuffing in heavily loaded applications. Foam control additives prevent foaming where gears or other machine elements churn in an oil bath.

References 3, 6, 16, 19, and 22 describe recent developments in the field of lubricant technology and tribology. Reference 9 describes recent work in using composite materials containing natural fiber reinforcements and the resulting tribology for such materials. Reference 22 emphasizes the influences of the rapidly growing field of micro- and nanoscale phenomena in tribology. Major lubricant suppliers (Internet sites 11–13 and others) are developing new additives and synthetic formulations to serve the alternative energy fields of wind and solar power, with special attention to the wide range of environmental conditions in which they must operate. Some are applied at temperatures as low as −55°C (−67°F) and as high as 50°C (122°F) with related changes in humidity and exposure to corrosive elements.

Greases. Grease is a two-phase lubricant composed of a thickener dispersed in a base fluid, typically an oil. When applied to the interface between moving components, the grease tends to remain in place and adhere to the surfaces. The oil provides lubrication in a manner similar to that already discussed. As long as there is a sufficient amount of grease at the interface, continuous lubrication is provided. Some mechanisms are said to be lubricated for life. However, the designer must take great care to ensure that grease is not displaced from the critical areas that need it or to design the system for periodic reapplication of the grease. Some components such as bearings are provided with grease fittings for replenishing the supply and for discharging contaminated or oxidized grease.

Several kinds of thickeners are used in combination with natural or synthetic oils. See Internet sites 14 and 15 for examples. Thickeners are soaps formed by the reaction of animal or vegetable fats with alkaline substances such as lithium, calcium, an aluminum complex, clay, polyurea, and others. Lithium 12-hydroxy stearate is the most often used form. Soaps have a smooth buttery consistency and hold the oil in suspension until it is drawn out in the zone to be lubricated. Additives provide extreme pressure (EP) capability, rust protection, oxidation stability, and improved pumpability of the grease.

The National Lubricating Grease Institute (NLGI) defines nine ranges of consistency, labeled 000 through 6, from semifluid to soft, medium hard, hard, and rigid block. Grade #2 is good for general industrial applications.

Solid Lubricants. Some applications cannot employ oils or greases because of contamination of other parts of the system, exposure to foods, excessively high or low temperatures, operation in a vacuum, or other environmental considerations. In such cases, the designer may specify solid materials that have inherently good lubricating properties or add solid lubricants on critical surfaces. The section on boundary lubrication has already discussed various formulations of PTFE (polytetrafluoroethylene) as examples of materials with good lubricity.

A solid lubricant is a thin solid film that reduces friction and wear. Some are applied in powder form by brushing, spraying, or dipping, and they then adhere to mating surfaces. Binders are often blended with the base material to facilitate application and to promote adhesion. Curing in air or by baking is required.

Molybdenum disulfide (MoS_2) and graphite are two often-used types of solid lubricants. Others are lead iodide (PbI_2), silver sulfate ($AgSO_4$), tungsten disulfide, and stearic acid. An example of their effectiveness is the reduction of the coefficient of sliding friction for steel-on-steel from approximately 0.50 for dry clean surfaces to the range from 0.03 to 0.06.

Many solid lubricant developments are designed to enhance the efficiency and improve wear resistance of the many bearing surfaces in automotive and truck engines. Internet site 19 describes new IROX™ bearings that have an overlay of polyamide-imide (PAI) polymer resin binder containing a number of additives dispersed throughout the matrix. These additives provide enhanced wear resistance, strength, thermal conductivity, and embeddability.

Wear

Wear is the gradual removal of surface material from sliding surfaces. It is a complex process with numerous variables. Only testing under real conditions of service can predict actual wear in a given system. Several types of wear occur:

- Pitting, galling, scuffing, or scoring that typically result from high contact stresses and fatigue of surface material during rolling or sliding contact.
- Abrasive wear, mechanical scraping, cutting, or scratching such as by hard contaminants in the interface between mating parts.
- Fretting, the cyclic sliding of very small amplitude that displaces surface material. The subsequent accumulation of the debris tends to accelerate the process. Continued operation produces a surface appearance similar to corrosion and can generate small cracks from which ultimate fatigue failure can occur. It often occurs when tight-fitting parts are subjected to oscillating loads or vibration.
- Impingement wear caused by the eroding of material due to hard moving materials hitting the surface, perhaps carried by air or fluids. High velocity fluids, such as the discharge from high-pressure washers, can themselves cause such wear.

While it is not possible to prescribe specific approaches to reduce wear, the following are the things a designer may try. Again, testing is the only way to ensure satisfactory operation.

1. Keep contact force low between sliding surfaces.
2. Maintain low temperature at the mating surfaces.
3. Use hard contacting surfaces.
4. Produce smooth mating surfaces.
5. Maintain continuous lubrication to reduce friction.
6. Keep the relative velocity between mating surfaces low.
7. Specify materials that have inherently good wear properties.

Many material suppliers will report wear properties of their materials when operating against a similar or dissimilar material. These data are acquired by testing under carefully controlled laboratory conditions. Typically one member of a pair of materials is moved at a known velocity such as by rotation. The mating material is held stationary and under a known load. Careful measurements are made of the original weight and dimensions of the specimens of mating materials. After a sizable time of operation, the specimens are again weighed and measured to determine how much material has been removed. The results are reported as wear, computed from an equation such as,

$$K = W/FVT \qquad \textbf{(16–13)}$$

where K = wear factor for the materials

W = wear measured as loss of weight or volume

F = applied load

V = linear relative velocity between the sliding members

T = time of operation

Comparing the K factors for a variety of materials being considered can aid the designer in material selection.

REFERENCES

1. Avraham, Harnoy. *Bearing Design in Machinery: Engineering Tribology and Lubrication.* Boca Raton, FL: CRC Press, 2003.
2. Bayer, Raymond G. *Mechanical Wear Fundamentals and Testing.* Boca Raton, FL: CRC Press, 2004.
3. Bloch, Heinz P. *Practical Lubrication for Industrial Facilities.* 2nd ed. Lilburn, GA: Fairmont Press, 2010.
4. Boyd, John, and Albert A. Raimondi. "A Solution for the Finite Journal Bearing and Its Application to Analysis and Design." Parts I, II, and III. *Transactions of the American Society of Lubrication Engineers.* 1, No. 1 (1958): 159–209.
5. Boyd, John, and Albert A. Raimondi. "Applying Bearing Theory to the Analysis and Design of Journal Bearings." Parts I and II. *Journal of Applied Mechanics* 73 (1951): 298–316.
6. Bruce, Robert W., editor. *Handbook of Lubrication and Tribology, Volume II: Theory and Design.* 2nd ed. Boca Raton, FL: CRC Press, 2012.

7. Cast Bronze Institute. *Cast Bronze Bearing Design Manual.* New York: Copper Development Association, 1979.
8. Cast Bronze Institute. *Cast Bronze Hydrostatic Bearing Design Manual.* New York: Copper Development Association, 1975.
9. Chand, Neven. *Tribology of Natural Fiber Polymer Composites.* Boca Raton, FL: CRC Press, 2009.
10. Copper Development Association. *Cast Copper Alloy Sleeve Bearings—Selection Guide.* New York: Copper Development Association, 1997.
11. Juvinall, Robert C., and Kurt M. Marshek. *Fundamentals of Machine Component Design.* 5th ed. New York: John Wiley & Sons, 2011.
12. Hamrock, Bernard J. *Fundamentals of Fluid Film Lubrication.* Boca Raton, FL: CRC Press, 2004.
13. Khonsari, Michael M., and E. Richard Booser. *Applied Tribology: Bearing Design and Lubrication* 2nd ed . New York: John Wiley & Sons, 2008.
14. Ludema, Kenneth C. *Friction, Wear, Lubrication: A Textbook in Tribology.* Boca Raton, FL: CRC Press, 1996.
15. Mang, Theo, and Wilfried Dresel. *Lubricants and Lubrications* 2nd ed. New York: Wiley-VCH, 2007.
16. Mang, Theo, Kirsten Bobzin, and Bartels Thorsten. *Industrial Tribology: Tribosystems, Friction, Wear, and Surface Engineering, Lubrication.* New York: John Wiley & Sons, 2011.
17. Miyoshi, Kazuhisa. *Solid Lubrication Fundamentals and Applications.* Boca Raton, FL: CRC Press, 2002.
18. Mott, Robert L. *Applied Fluid Mechanics.* 6th ed. Upper Saddle River, NJ: Prentice Hall, 2006.
19. Srivastava, Som Prakash. *Advances in Lubricant Additives and Tribology.* Boca Raton, FL: CRC Press, 2009.
20. Totten, George E. editor. *Handbook of Lubrication and Tribology: Volume I Application and Maintenance.* 2nd ed. Boca Raton, FL: CRC Press, 2006.
21. Totten, George E., and Hong Liang. *Mechanical Tribology: Materials, Characterization, and Applications.* Boca Raton, FL: CRC Press, 2004.
22. Yip-Wah Chung. *Micro- and Nanoscale Phenomena in Tribology.* Boca Raton, FL: CRC Press, 2012.

INTERNET SITES RELATED TO PLAIN BEARINGS AND LUBRICATION

1. **Copper Development Association (CDA).** www.copper.org Industry association of companies and organizations involved in the production and use of copper, including cast bronze bearings and porous bronze bushings. Site includes some technical information on the design of plain bronze bearings. Also publishes the useful *Cast Bronze Bearing Design Manual.*
2. **Thomson Engineering & Polymers.** www.nyliner.com Manufacturer of plain plastic bearings under the brand names Nyliner® and Nyliner Plus®.
3. **Saint-Gobain Performance Plastics.** www.saint-gobain-northamerica.com Manufacturer of plain plastic bearings under the brand name Rulon®, available in 15 formulations.
4. **GGB Bearing Technology.** www.ggbearings.com Manufacturer of a variety of plain bearings made from a composite of PTFE plastic, porous bronze, and steel or fiberglass under the brand names DU®, DX®, Gar-Fil®, and Gar-Max® and others.
5. **Graphite Metallizing Corporation.** www.graphalloy.com Manufacturer of plain bearings under the Graphalloy® brand name. Graphalloy® is a graphite-metal alloy formed from molten metal and graphite to make a uniform, solid, self-lubricating bushing material, available in over 100 grades.
6. **Beemer Precision, Inc.** www.oilite.com Manufacturer of plain bearings made from cast bronze and porous sintered bronze impregnated with oil under the Oilite® and Excelite brands.
7. **Bunting Bearings Corporation.** www.buntingbearings.com Manufacturer of plain bearings made from cast bronze, porous sintered bronze impregnated with oil, and plastic bearings made from Nylon, PTFE, and Vespel®. Also makes the BU bearing consisting of a steel backing strip, a porous bronze matrix impregnated and overlaid with a PTFE/lead lining for low friction coefficient.
8. **Zincaloy, Inc.** www.zincaloy.com Manufacturer of continuous cast ZA-12 zinc-aluminum alloy (Zincaloy™) used for bearings in mining, construction, forestry, and off-road vehicle industries. Team Tube Ltd. is an authorized distributor of Zincaloy bearing stock. www.teamtube.com.
9. **Baldor/Dodge.** www.dodge-pt.com Manufacturer of mounted plain self-lubricating sleeve bearings under the Solidlube® brand used in general industrial, material handling, and bulk materials industries. Also available are Bronzoil® bearings with oil-impregnated porous bronze, babbitted and bronze bushed bearings, and flange-mounted M Series Hydrodynamic bearings specifically designed for large motors and generators.
10. **Waukesha Bearings Corporation.** www.waukbearing.com Manufacturer of fluid film hydrodynamic sleeve bearings, tilting pad thrust and journal bearings, magnetic bearings and housings for the bearings. These bearings are used for turbines, compressors, generators, gearboxes, and other types of rotating equipment.
11. **Exxon-Mobol, Inc.** www.mobil.com Producer of a wide variety of lubricants for general industry, automotive, turbine, compressor, and engine applications. Data sheets provided under the *Industrial lubricants* page.
12. **Castrol Industrial.** www.castrol.com Manufacturer of lubricants for vehicles and industrial applications such as bearings, gears, chains, and general lubrication applications. Select *Industrial Lubricants* from the home page.
13. **Shell Oil Company.** www.shell.us/lubricants.com Producer of a wide variety of lubricants for general industry, automotive, turbine, compressor, and engine applications. Data sheets provided under the *Lubricants* page.
14. **National Lubricating Grease Institute.** www.nlgi.org An association of companies and research organizations that develop, manufacture, distribute, and use greases. Establishes standards and provides publications and technical articles on grease.
15. **Lubrizol Corporation.** www.lubrizol.com Manufacturer of a large variety of greases and additives for the lubrication industry.
16. **LubeLink.com** www.lubelink.com Lists live links to companies providing lubrication products and services, testing laboratories, original equipment manufacturers that are heavy

users of lubricants for vehicular and industrial applications, and industry associations related to the lubrication field.

17. **RBC Bearings.** www.rbcbearings.com Manufacturer of a wide array of bearings including spherical plain bearings, rod end bearings, and self-lubricated bearings. Products are used wherever pivoting, high load bearing applications are found in applications such as in construction equipment, agricultural machinery, hydraulic cylinder connections, vehicle suspensions, and articulated joints.
18. **Plastics International.** www.plasticsintl.com Supplier of a large number of types and grades of plastics. Selecting the *Material Properties* tab reveals a list of over 100 plastics with mechanical properties. Values of *pV factor* are given in the selectable PDF file for materials commonly used for bearings, such as Delrin, PEEK, and Rulon.
19. **Federal Mogul Corporation.** www.federalmogul.com Manufacturer of numerous components and systems for automotive applications including engine bearings. Select *Automotive OEM Technology and Products → Powertrain Sealing and Bearings → Bearings, Bushings and Washers.*

PROBLEMS

For Problems 1–8 and the data listed in Table 16–2, design a plain surface bearing, using the boundary-lubricated approach of Section 16–5. Use an L/D ratio for the bearing in the range of 0.50 to 1.50. Compute the pV factor, and specify a material from Table 16–1.

TABLE 16–2

Problem number	Radial load (lb)	Shaft diameter (in)	Shaft speed (rpm)
1.	225	3.00	1750
2.	100	1.50	1150
3.	200	1.25	850
4.	75	0.50	600
5.	850	4.50	625
6.	500	3.75	450
7.	800	3.00	350
8.	60	0.75	750

For Problems 9–18 and the data listed in Table 16–3, design a hydrodynamically lubricated bearing, using the method outlined in Section 16–7. Specify the journal nominal diameter, the bearing length, the diametral clearance, the minimum film thickness of the lubricant during operation, the surface finish of the journal and bearing, the lubricant, and its maximum operating temperature. For your design, compute the coefficient of friction, the friction torque, and the power dissipated as the result of friction.

For Problems 19–28 and the data listed in Table 16–4, design a hydrostatic bearing of circular shape. Specify the pad diameter, the recess diameter, the recess pressure, the film thickness, the lubricant and its temperature, the oil flow rate, and the pumping power. The load specified is for a single bearing. You may choose to use multiple bearings. (The supply pressure is the maximum available at the pump.)

TABLE 16–3

Problem number	Radial load	Min. shaft diameter	Shaft speed (rpm)	Application
9.	1250 lb	2.60 in	1750	Electric motor
10.	2250 lb	3.50 in	850	Conveyor
11.	1875 lb	2.25 in	1150	Air compressor
12.	1250 lb	1.75 in	600	Precision spindle
13.	500 lb	1.15 in	2500	Precision spindle
14.	850 lb	1.45 in	1200	Idler sheave
15.	4200 lb	4.30 in	450	Shaft for chain drive
16.	18.7 kN	100 mm	500	Conveyor
17.	2.25 kN	25 mm	2200	Machine tool
18.	5.75 kN	65 mm	1750	Printer

TABLE 16–4

Problem number	Load	Supply pressure
19.	1250 lb	300 psi
20.	5000 lb	300 psi
21.	3500 lb	500 psi
22.	750 lb	500 psi
23.	250 lb	150 psi
24.	500 lb	150 psi
25.	22.5 kN	2.0 MPa
26.	1.20 kN	750 kPa
27.	8.25 kN	1.5 MPa
28.	12.5 kN	1.5 MPa

CHAPTER SEVENTEEN

LINEAR MOTION ELEMENTS

THE BIG PICTURE Linear Motion Elements

Discussion Map

- Many kinds of mechanical devices produce linear motion for machines such as automation equipment, packaging systems, and machine tools.
- *Power screws, jacks,* and *ball screws* are designed to convert rotary motion to linear motion and to exert the necessary force to move a machine element along a desired path. They use the principle of a screw thread and its mating nut.

Discover

Visit a machine shop and see if you can identify power screws, ball screws, or other linear motion devices. Look in particular at the lathes and the milling machines. Manual machines likely use power screws. Those using computer numerical control should have ball screws. Describe the shape of the threads. How are the screws powered? How are they fastened to other machine parts?

Can you find other equipment that uses linear motion devices? Look in laboratories where materials are tested or where large forces must be generated.

This chapter will help you learn to analyze power screws and ball screw drives and to specify suitable sizes for a given application.

A common requirement in mechanical design is to move components in a straight line. Elevators move vertically up or down. Machine tools move cutting tools or parts to be machined in straight lines, either horizontally or vertically, to shape metal into desired forms. A precision measuring device moves a probe in straight lines to determine electronically the dimensions of a part. Assembly machines require many straight-line motions to insert components and to fasten them together. A packaging machine moves products into cartons, closes flaps, and seals the cartons.

Some examples of components and systems that facilitate linear motion are:

Power screws	Jacks
Linear actuators	Ball bushings
Linear solenoids	X-Y-Z tables
Ball screws	Fluid power cylinders
Linear slides	Rack-and-pinion sets
Positioning stages	Gantry tables

Figure 17–1(a) shows a cutaway view of a jack that employs a power screw to produce linear motion. Power is delivered to the input shaft by an electric motor. The worm, machined integral with the input shaft, drives the wormgear, resulting in a reduction in the speed of rotation. The inside of the wormgear has machined threads that engage the external threads of the power screw, driving it vertically. Figure 17–1(b) shows the screw jack mated with an external gear reducer and motor to provide a complete linear motion system. Limit switches, position sensors, and programmable logic controllers can be used to control the motion cycle. These and other types of linear actuators can be seen on Internet sites 1–13.

Rack-and-pinion sets are discussed in Chapter 8. Fluid power actuators employ oil hydraulic or pneumatic fluid pressure to extend or retract a piston rod within a cylinder as discussed in textbooks on fluid power. Positioning stages, X-Y-Z tables, and gantry tables typically are driven by precision stepper motors or servomechanisms that allow precision location of components

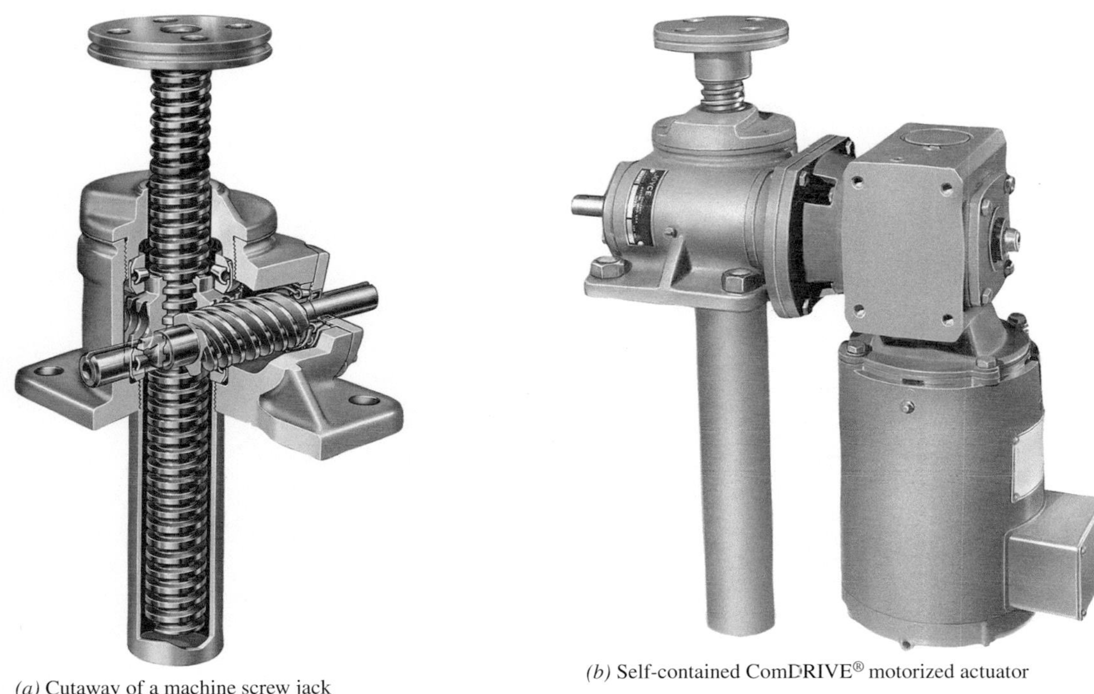

(*a*) Cutaway of a machine screw jack

(*b*) Self-contained ComDRIVE® motorized actuator

FIGURE 17–1 Examples of linear motion machine elements (Joyce/Dayton Corporation, Dayton, OH)

anywhere within their control volume. Linear solenoids are devices that cause a rodlike core to be extended or retracted as power is supplied to an electrical coil, producing rapid movement over small distances. Applications are seen in office equipment, automation devices, and packaging systems. See Internet site 11.

Linear slides and ball bushings are designed to guide mechanical components along a precise linear track. Low friction materials or rolling contact elements are used to produce smooth motion with low power required. See Internet sites 1, 3, 5, and 7–9.

Power screws and ball screws are designed to convert rotary motion to linear motion and to exert the necessary force to move a machine element along a desired path. Power screws operate on the classic principle of the screw thread and its mating nut. If the screw is supported in bearings and rotated while the nut is restrained from rotating, the nut will translate along the screw. If the nut is made an integral part of a machine, for example, the tool holder for a lathe, the screw will drive the tool holder along the bed of the machine to take a cut. Conversely, if the nut is supported while it is rotating, the screw can be made to translate. The screw jack uses this approach.

A ball screw is similar in function to a power screw, but the configuration is different. The nut contains many small, spherical balls that make rolling contact with the threads of the screw, giving low friction and high efficiencies when compared with power screws. Modern machine tools, automation equipment, vehicle steering systems, and actuators on aircraft use ball screws for high precision, fast response, and smooth operation.

Visit a machine shop where there are metal-cutting machine tools. Look for examples of power screws that convert rotary motion to linear motion. They are likely to be on manual lathes moving the tool holder. Or look at the table drive for a milling machine. Inspect the form of the threads of the power screw. Are they of a form similar to that of a screw thread with sloped sides? Or are the sides of the threads straight? Compare the screw threads with those shown in Figure 17–2 for square, Acme, and buttress forms.

While in the shop, do you see any type of material-testing equipment or a device called an *arbor press* that exerts large axial forces? Such machines often employ square-thread power screws to produce the axial force and motion from rotational input, through either a hand crank or an electric motor drive. If they are not in the machine shop, look for them in the metallurgy lab or another room where materials testing is done.

Now look further in the machine shop. Are there machines that use digital readouts to indicate position of the table or the tool? Are there computer numerical control machine tools? Any of these types of machines should have ball screws rather than the traditional power screws because ball screws require significantly less power and torque to drive them against a given load.

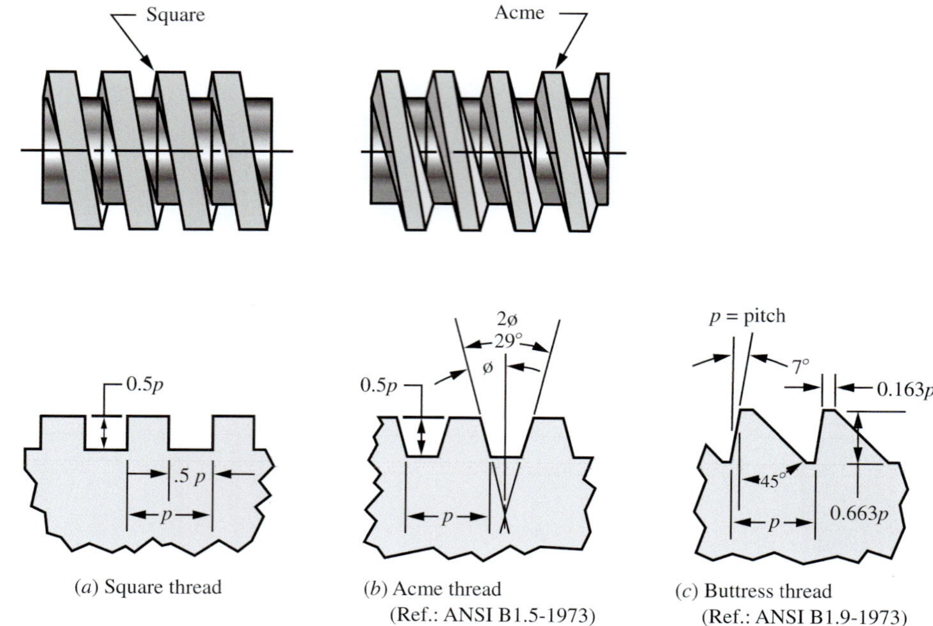

FIGURE 17–2 Forms of power screw threads [(b), ANSI/ASME Standard B1.5-1973; (c), ANSI Standard B1.9-1973]

They can also be moved faster and positioned more accurately than power screws. You may or may not be able to see the recirculating balls in the nut of the power screw, as illustrated in Figure 17–3. But you should be able to see the different-shaped threads looking like grooves with circular bottoms in which the spherical balls roll.

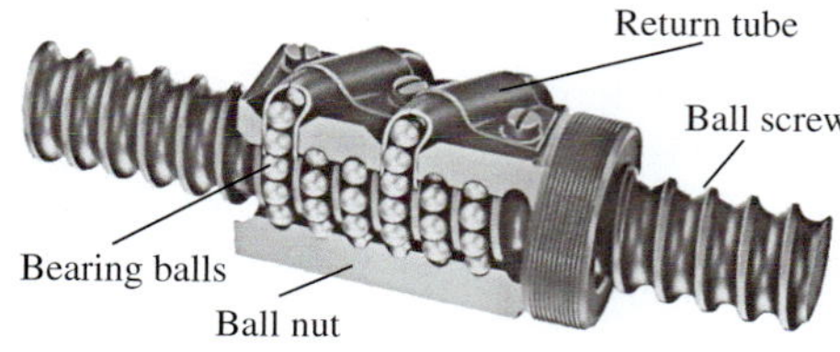

FIGURE 17–3 Ball bearing screw (Thomson Industries, Inc., Port Washington, NY)

Have you seen such power screws or ball screws anywhere outside a machine shop? Some garage door openers employ a screw drive, but others use chain drives. Perhaps your home has a screw jack or a scissors jack for raising the car to change a tire. Both use power screws. Have you ever sat in a seat on an airplane where you can see the mechanisms that actuate the flaps on the rear edge of the wings? Try it sometime, and observe the actuators during takeoff or landing. It is likely that you will see a ball screw in action.

This chapter will help you learn the methods of analyzing the performance of power screws and ball screws and to specify the proper size for a given application.

YOU ARE THE DESIGNER

You are a member of a plant engineering team for a large steel processing plant. One of the furnaces in the plant in which steel is heated prior to final heat treatment is installed beneath the floor, and the large ingots are lowered vertically into it. While the ingots are soaking in the furnace, a large, heavy hatch is placed over the opening to minimize the escape of heat and provide more uniform temperature. The hatch weighs 25 000 lb.

You are asked to design a system that will permit the hatch to be raised at least 15 in above the floor within 12.0 s and to lower it again within 12.0 s.

What design concept would you propose? Of course, there are many feasible concepts, but suppose that you proposed a system like that sketched in Figure 17–4. An overhead support structure is suggested on which a worm/wormgear drive system would be mounted. One shaft would be driven directly by the gear drive while a second shaft would be driven simultaneously by a chain drive. The shafts are power screws, supported in bearings at the top and bottom. A yoke is connected to the hatch and is mounted on the screws with the nuts that mate with the screw integral with the yoke. Therefore, as the screw rotates, the nuts carry the yoke and the hatch vertically upward or downward.

As the designer of the hatch lift system, you must make several decisions. What size screw is required to ensure that it can safely raise the 25 000-lb hatch? Notice that the screws are placed in tension as

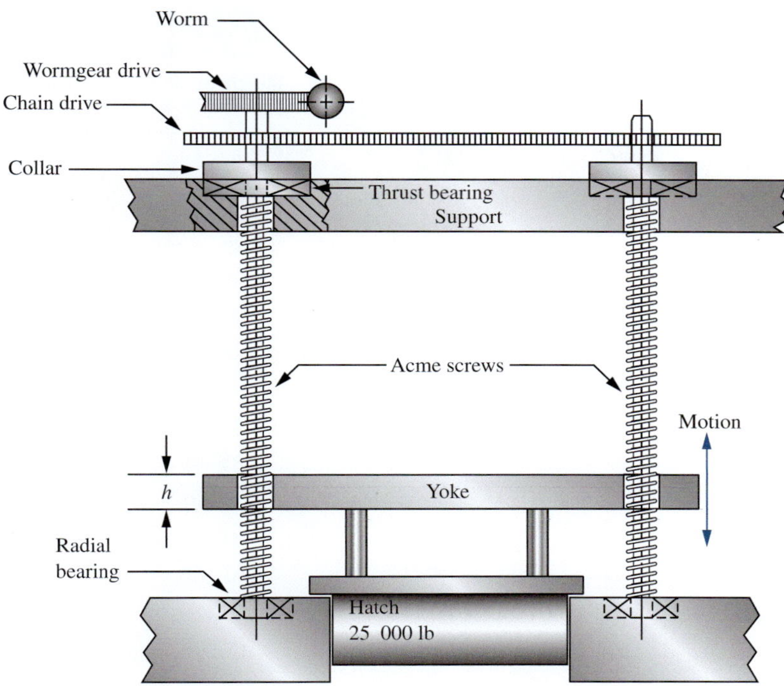

FIGURE 17–4 An Acme screw-driven system for raising a hatch

they are supported on the collars on the upper support system. What diameter, thread type, and thread size should be used? The sketch suggests the Acme thread form. What other styles are available? At what speed must the screws be rotated to raise the hatch in 12.0 s or less? How much power is required to drive the screws? What safety concerns exist while the system is handling this heavy load? What advantage would there be to using a ball screw rather than a power screw?

The material in this chapter will help you make these decisions, along with providing methods of computing stresses, torques, and efficiencies. ■

17–1 OBJECTIVES OF THIS CHAPTER

After completing this chapter, you will be able to:

1. Describe the operation of a power screw and the general form of *square threads, Acme threads,* and *buttress threads* as they are applied to power screws.
2. Compute the torque that must be applied to a power screw to raise or lower a load.
3. Compute the efficiency of power screws.
4. Compute the power required to drive a power screw.
5. Describe the design of a ball screw and its mating nut.
6. Specify suitable ball screws for a given set of requirements of load, speed, and life.
7. Compute the torque required to drive a ball screw, and compute its efficiency.

17–2 POWER SCREWS

Figure 17–2 shows three types of power screw threads: the square thread, the Acme thread, and the buttress thread. Of these, the square and buttress threads are the most efficient. That is, they require the least torque to move a given load along the screw. However, the Acme thread is not greatly less efficient, and it is easier to machine. The buttress thread is desirable when force is to be transmitted in only one direction.

Table 17–1 gives the preferred combinations of basic major diameter, D, and number of threads per inch, n, for Acme screw threads in U.S. units. The pitch, p, is the distance from a point on one thread to the corresponding point on the adjacent thread, and $p = 1/n$. More data for power screws can be found in References 1 and 4.

Other pertinent dimensions listed in Table 17–1 include the minimum minor diameter and the minimum pitch diameter of a screw with an external thread. When you are performing stress analyses on the screw, the safest approach is to compute the area corresponding to the minor diameter for tensile or compressive stresses. However, a more accurate stress computation results from using the *tensile stress area* (listed in Table 17–1), found from

➭ Tensile Stress Area for Screw Threads

$$A_t = \frac{\pi}{4}\left[\frac{D_r + D_p}{2}\right]^2 \qquad (17\text{–}1)$$

This is the area corresponding to the average of the minor (or root) diameter, D_r, and the pitch diameter, D_p. The data

TABLE 17–1 Preferred Acme Screw Threads

Nominal major diameter, D (in)	Threads per in, n	Pitch, $p = 1/n$ (in)	Minimum minor diameter, D_r (in)	Minimum pitch diameter, D_p (in)	Tensile stress area, A_t (in^2)	Shear stress area, A_s (in^2)[a]
1/4	16	0.0625	0.1618	0.2043	0.026 32	0.3355
5/16	14	0.0714	0.2140	0.2614	0.044 38	0.4344
3/8	12	0.0833	0.2632	0.3161	0.065 89	0.5276
7/16	12	0.0833	0.3253	0.3783	0.097 20	0.6396
1/2	10	0.1000	0.3594	0.4306	0.1225	0.7278
5/8	8	0.1250	0.4570	0.5408	0.1955	0.9180
3/4	6	0.1667	0.5371	0.6424	0.2732	1.084
7/8	6	0.1667	0.6615	0.7663	0.4003	1.313
1	5	0.2000	0.7509	0.8726	0.5175	1.493
$1\frac{1}{8}$	5	0.2000	0.8753	0.9967	0.6881	1.722
$1\frac{1}{4}$	5	0.2000	0.9998	1.1210	0.8831	1.952
$1\frac{3}{8}$	4	0.2500	1.0719	1.2188	1.030	2.110
$1\frac{1}{2}$	4	0.2500	1.1965	1.3429	1.266	2.341
$1\frac{3}{4}$	4	0.2500	1.4456	1.5916	1.811	2.803
2	4	0.2500	1.6948	1.8402	2.454	3.262
$2\frac{1}{4}$	3	0.3333	1.8572	2.0450	2.982	3.610
$2\frac{1}{2}$	3	0.3333	2.1065	2.2939	3.802	4.075
$2\frac{3}{4}$	3	0.3333	2.3558	2.5427	4.711	4.538
3	2	0.5000	2.4326	2.7044	5.181	4.757
$3\frac{1}{2}$	2	0.5000	2.9314	3.2026	7.388	5.700
4	2	0.5000	3.4302	3.7008	9.985	6.640
$4\frac{1}{2}$	2	0.5000	3.9291	4.1991	12.972	7.577
5	2	0.5000	4.4281	4.6973	16.351	8.511

[a]Per inch of length of engagement.

reflect the minimums for commercially available screws according to recommended tolerances.

Another failure mode for a power screw is the shearing of the threads in the axial direction, cutting them away from the main shaft near the pitch diameter. The shear stress is computed from the direct stress formula,

$$\tau = F/A_s$$

The shear stress area, A_s, listed in Table 17–1 is also found in published data and represents the area in shear approximately at the pitch line of the threads for a 1.0-in length of engagement. Other lengths would require that the area be modified by the ratio of the actual length to 1.0 in.

Metric Power Screws

Metric power screws use a trapezoidal thread system that is similar to the Acme thread, but with slightly different thread form and made to metric dimensions. One notable difference between the Acme and ISO trapezoidal metric thread is that the angle $\phi = 15°$ instead of 14½° shown for the Acme design in Figure 17–2(b). Table 17–1M shows a selected list of sizes from the larger list of ISO trapezoidal screws. Each of these is a single- thread design. Many more diameters and pitches are available, along with many other styles with two or more threads that produce longer leads that are multiples of the pitch.

Torque Required to Move a Load

When using a power screw to exert a force, as with a jack raising a load, you need to know how much torque must be applied to the nut of the screw to move the load. The parameters involved are the force to be moved, F; the size of the screw, as indicated by its pitch diameter, D_p; the lead of the screw, L; and the coefficient of friction, f. Note that the *lead* is defined as the axial distance that the screw would move in one complete revolution. For the usual case of a single-threaded screw, the lead is equal to the pitch and can be read from Table 17–1 or computed from $L = p = 1/n$ for U.S. Acme Designs.

TABLE 17–1M Examples of Power Screws with Metric Trapezoidal Screw Thread

ISO thread system—External threads

Major diameter, D (mm)	Pitch, p (mm)	Pitch diameter, D_p (mm)	Minor diameter, D_r (mm)	Tensile stress area (mm^2)
8	1.5	7.25	6.2	35.52
10	2	9.0	7.5	53.46
12	3	10.5	8.5	70.88
14	3	12.5	10.5	103.9
16	3	14.5	12.5	143.1
20	4	18.0	15.5	220.4
22	5	19.5	16.5	254.5
24	5	21.5	18.5	314.2
28	5	25.5	22.5	452.4
30	6	27.0	23.0	490.9
32	6	29.0	33.0	754.8
36	6	33.0	29.0	754.8
40	7	36.5	32.0	921.3
46	8	42.0	37.0	1225
50	8	46.0	41.0	1486
55	9	50.5	45.0	1791
60	9	55.5	50.0	2185
70	10	65.0	59.0	3019
80	10	75.0	69.0	4072
90	12	84.0	77.0	5090
100	12	94.0	87.0	6433
120	14	113.0	104.0	9246
125	14	122.0	109.0	10 477

In the development of Equation (17–2) for the torque required to turn the screw, Figure 17–5(a), which depicts a load being pushed up an inclined plane against a friction force, is used. This is a reasonable representation for a square thread if you think of the thread as being unwrapped from the screw and laid flat. The torque for an Acme thread is slightly different from this due to the thread angle. The revised equation for the Acme thread will be shown later.

The torque computed from Equation (17–2) is called T_u, implying that the force is applied to move a load up the plane, that is, to raise the load. This observation is

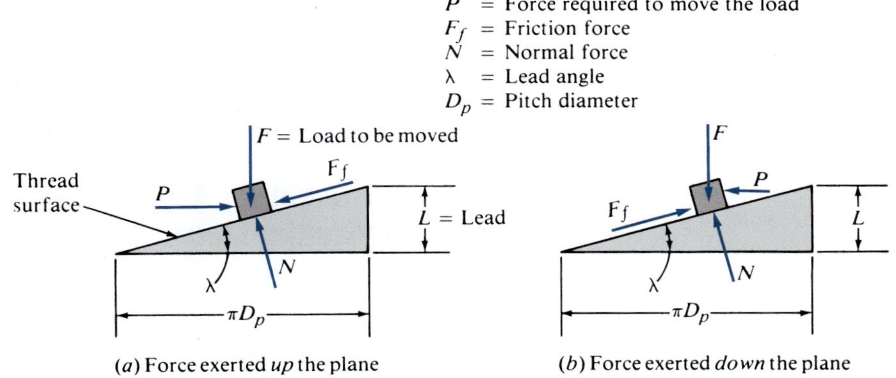

FIGURE 17–5 Screw thread force analysis

completely appropriate if the load is raised vertically, as with a jack. If, however, the load is horizontal or at some angle, Equation (17–2) is still valid if the load is to be advanced along the screw "up the thread." Equation (17–4) shows the required torque, T_d, to lower a load or move a load "down the thread."

The torque to move a load up the thread is

➪ Torque Required to Move a Load up a Power Screw with a Square Thread

$$T_u = \frac{FD_p}{2}\left[\frac{L + \pi f D_p}{\pi D_p - fL}\right] \tag{17–2}$$

This equation accounts for the force required to overcome friction between the screw and the nut in addition to the force required just to move the load. If the screw or the nut bears against a stationary surface while rotating, there will be an additional friction torque developed at that surface. For this reason, many jacks and similar devices incorporate antifriction bearings at such points.

The coefficient of friction for use in Equation (17–2) depends on the materials used and the manner of lubricating the screw. For well-lubricated steel screws acting in steel nuts, $f = 0.15$ should be conservative.

An important factor in the analysis for torque is the angle of inclination of the plane. In a screw thread, the angle of inclination is referred to as the *lead angle*, λ. It is the angle between the tangent to the helix of the thread and the plane transverse to the axis of the screw. It can be seen from Figure 17–5 that

$$\tan\lambda = L/(\pi D_p) \tag{17–3}$$

where πD_p = circumference of the pitch line of the screw

Then if the rotation of the screw tends to raise the load (move it up the incline), the friction force opposes the motion and acts down the plane.

Conversely, if the rotation of the screw tends to lower the load, the friction force will act up the plane, as shown in Figure 17–5(b). The torque analysis changes, producing Equation (17–4):

➪ Torque Required to Lower a Load Down a Power Screw with a Square Thread

$$T_d = \frac{FD_p}{2}\left[\frac{\pi f D_p - L}{\pi D_p + fL}\right] \tag{17–4}$$

If the screw thread is steep (i.e., if it has a high lead angle), the friction force may not be able to overcome the tendency for the load to "slide" down the plane, and the load will fall due to gravity. In most cases for power screws with single threads, however, the lead angle is rather small, and the friction force is large enough to oppose the load and keep it from sliding down the plane. Such a screw is called *self-locking*, a desirable characteristic for jacks and similar devices. Quantitatively, the condition that must be met for self-locking is

$$f > \tan\lambda \tag{17–5}$$

The coefficient of friction must be greater than the tangent of the lead angle. For $f = 0.15$, the corresponding value of the lead angle is 8.5°. For $f = 0.1$, for very smooth, well-lubricated surfaces, the lead angle for self-locking is 5.7°. The lead angles for the screw designs listed in Table 17–1 range from 1.94° to 5.57°. Thus, it is expected that all would be self-locking. However, operation under vibration should be avoided, as this may still cause movement of the screw.

Efficiency of a Power Screw

Efficiency for the transmission of a force by a power screw can be expressed as the ratio of the torque required to move the load without friction to that with friction. Equation (17–2) gives the torque required with friction, T_u. Letting $f = 0$, the torque required without friction, T', is

$$T' = \frac{FD_p}{2}\frac{L}{\pi D_p} = \frac{FL}{2\pi} \tag{17–6}$$

Then the efficiency, e, is

➪ Efficiency of a Power Screw

$$e = \frac{T'}{T_u} = \frac{FL}{2\pi T_u} \tag{17–7}$$

Alternate Forms of the Torque Equations

Equations (17–2) and (17–4) can be expressed in terms of the lead angle, rather than the lead and the pitch diameter, by noting the relationship in Equation (17–3). With this substitution, the torque required to move the load would be

➪ Torque to Raise a Load with a Square Thread

$$T_u = \frac{FD_p}{2}\left[\frac{(\tan\lambda + f)}{(1 - f\tan\lambda)}\right] \tag{17–8}$$

and the torque required to lower the load is

➪ Torque to Lower a Load with a Square Thread

$$T_d = \frac{FD_p}{2}\left[\frac{(f - \tan\lambda)}{(1 + f\tan\lambda)}\right] \tag{17–9}$$

Adjustment for Acme Threads and Trapezoidal Metric Threads

The difference between Acme threads and square threads is the presence of the thread angle, ϕ. Note from Figure 17–1 that $2\phi = 29°$, and therefore $\phi = 14.5°$. For the trapezoidal metric thread, $\phi = 15°$. This changes the direction of action of the forces on the thread from that depicted in

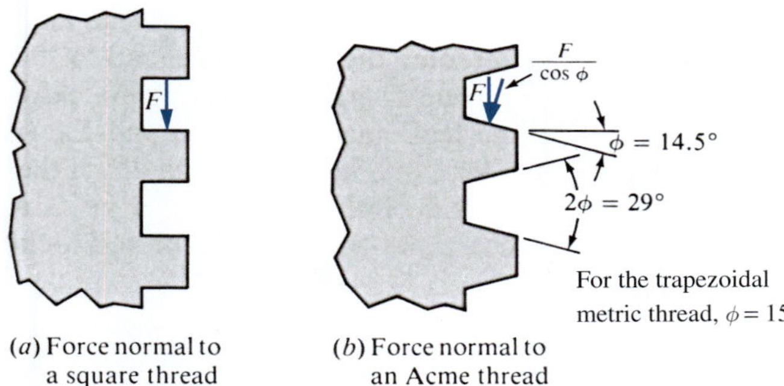

FIGURE 17–6 Force on an Acme thread

Figure 17–5. Figure 17–6 shows that F would have to be replaced by $F/\cos\phi$. Carrying this through, the analysis for torque would give the following modified forms of Equations (17–8) and (17–9). The torque to move the load up the thread is

➩ Torque to Raise a Load with an Acme Thread

$$T_u = \frac{FD_p}{2}\left[\frac{(\cos\phi\tan\lambda + f)}{(\cos\phi - f\tan\lambda)}\right] \quad \textbf{(17–10)}$$

and the torque to move the load down the thread is

➩ Torque to Lower a Load with an Acme Thread

$$T_d = \frac{FD_p}{2}\left[\frac{(f - \cos\phi\tan\lambda)}{(\cos\phi + f\tan\lambda)}\right] \quad \textbf{(17–11)}$$

Power Required to Drive a Power Screw

If the torque required to rotate the screw is applied at a constant rotational speed, n, then the power in horsepower to drive the screw is

$$P = \frac{Tn}{63\,000}$$

Example Problem 17–1 Two Acme-threaded power screws are to be used to raise a heavy access hatch, as sketched in Figure 17–4. The total weight of the hatch is 25 000 lb, divided equally between the two screws. Select a satisfactory screw from Table 17–1 on the basis of tensile strength, limiting the tensile stress to 10 000 psi. Then determine the required thickness of the yoke that acts as the nut on the screw to limit the shear stress in the threads to 5000 psi. For the screw thus designed, compute the lead angle, the torque required to raise the load, the efficiency of the screw, and the torque required to lower the load. Use a coefficient of friction of 0.15.

Solution The load to be lifted places each screw in direct tension. Therefore, the required tensile stress area is

$$A_t \frac{F}{\sigma_d} = \frac{12\,500\text{ lb}}{10000\text{ lb/in}^2} = 1.25\text{ in}^2$$

From Table 17–1, a $1\frac{1}{2}$-in-diameter Acme thread screw with four threads per inch would provide a tensile stress area of 1.266 in^2.

For this screw, each inch of length of a nut would provide 2.341 in^2 of shear stress area in the threads. The required shear area is then

$$A_s = \frac{F}{\tau_d} = \frac{12\,500\text{ lb}}{5000\text{ lb/in}^2} = 2.50\text{ in}^2$$

Then the required length of the yoke would be

$$h = 2.5\text{ in}^2\left[\frac{1.0\text{ in}}{2.341\text{ in}^2}\right] = 1.07\text{ in}$$

For convenience, let's specify $h = 1.25$ in.

The lead angle is (remember that $L = p = 1/n = 1/4 = 0.250$ in)

$$\lambda = \tan^{-1}\frac{L}{\pi D_p} = \tan^{-1}\frac{0.250}{\pi(1.3429)} = 3.39^\circ$$

The torque required to raise the load can be computed from Equation (17–10):

$$T_u = \frac{FD_p}{2}\left[\frac{(\cos\phi\tan\lambda + f)}{(\cos\phi - f\tan\lambda)}\right] \quad \textbf{(17–10)}$$

Using $\cos\phi = \cos(14.5°) = 0.968$, and $\tan\lambda = \tan(3.39°) = 0.0592$, we have

$$T_u = \frac{(12\,500\text{ lb})(1.3429\text{ in})}{2}\frac{[(0.968)(0.0592) + 0.15]}{[0.968 - (0.15)(0.0592)]} = 1809\text{ lb}\cdot\text{in}$$

The efficiency can be computed from Equation (17–7):

$$e = \frac{FL}{2\pi T_u} = \frac{(12\,500\text{ lb})(0.250\text{ in})}{2(\pi)(1809\text{ lb}\cdot\text{in})} = 0.275 \text{ or } 27.5\%$$

The torque required to lower the load can be computed from Equation (17–11):

$$T_d = \frac{FD_p}{2}\left[\frac{(f - \cos\phi\tan\lambda)}{(\cos\phi + f\tan\lambda)}\right] \qquad \textbf{(17–11)}$$

$$T_d = \frac{(12\,500\text{ lb})(1.3429\text{ in})}{2}\frac{[0.15 - (0.968)(0.0592)]}{[0.968 + (0.15)(0.0592)]} = 796\text{ lb}\cdot\text{in}$$

Example Problem 17–2 It is desired to raise the hatch in Figure 17–4 a total of 15.0 inches in no more than 12.0 s. Compute the required rotational speed for the screws and the power required.

Solution The screw selected in the solution for Example Problem 17–1 was a $1\frac{1}{2}$-in A cme-threaded screw with four threads per inch. Thus, the load would be moved 1/4 in with each revolution. The linear speed required is

$$V = \frac{15.0\text{ in}}{12.0\text{ s}} = 1.25\text{ in/s}$$

The required rotational speed would be

$$n = \frac{1.25\text{ in}}{\text{s}}\,\frac{1\text{ rev}}{0.25\text{ in}}\,\frac{60\text{ s}}{\text{min}} = 300\text{ rpm}$$

Then the power required to drive each screw would be

$$P = \frac{Tn}{63\,000} = \frac{(1809\text{ lb}\cdot\text{in})(300\text{ rpm})}{63\,000} = 8.61\text{ hp}$$

Multiple Start Thread Forms for Power Screws

The relatively low efficiency of standard single-thread Acme screws (approximately 30% or less) can be a strong disadvantage. Higher efficiencies can be achieved using high lead, multiple thread designs. The higher lead angle produces efficiencies in the 30% to 70% range. It should be understood that some mechanical advantage is lost so that higher torques are required to move a particular load as compared with single-thread screws. (See Internet sites 3, 5, 7, 10, 12, and 13.)

17–3 BALL SCREWS

The basic action of using screws to produce linear motion from rotation was described in Section 17–2 on power screws. A special adaptation of this action that minimizes the friction between the screw threads and the mating nut is the *ball screw*.

Figure 17–3 shows a cutaway view of a commercially available ball screw. It replaces the sliding friction of the conventional power screw with the rolling friction of bearing balls. The bearing balls circulate in hardened steel races formed by concave helical grooves in the screw and the nut. All reactive loads between the screw and the nut are carried by the bearing balls which provide the only physical contact between these members. As the screw and the nut rotate relative to each other, the bearing balls are diverted from one end and are carried by the ball-guide return tubes to the opposite end of the ball nut. This recirculation permits unrestricted travel of the nut in relation to the screw. (See Internet sites 3, 5, 7, 10, and 12.)

Applications of ball screws occur in automotive steering systems, machine tool tables, linear actuators, jacking and positioning mechanisms, aircraft controls such as flap

FIGURE 17–7 Application of a ball screw (Thomson Industries, Inc., Port Washington, NY)

actuating devices, packaging equipment, and instruments. Figure 17–7 shows a machine with a ball screw installed to move a component along the ways of the bed.

The application parameters to be considered in selecting a ball screw include the following:

The axial load to be exerted by the screw during rotation

The rotational speed of the screw

The maximum static load on the screw

The direction of the load

The manner of supporting the ends of the screw

The length of the screw

The expected life

The environmental conditions

Load-Life Relationship

When transmitting a load, a ball screw experiences stresses similar to those on a ball bearing, as discussed in Chapter 14. The load is transferred from the screw to the balls, from the balls to the nut, and from the nut to the driven device. The contact stress between the balls and the races in which they roll eventually causes fatigue failure, indicated by pitting of the balls or the races.

Thus, the rating of ball screws gives the load capacity of the screw for a given life which 90% of the screws of a given design will survive. This is similar to the L_{10} life of ball bearings. Because ball screws are typically used as linear actuators, the most pertinent life parameter is the distance traveled by the nut relative to the screw.

Manufacturers usually report the rated load that a given screw can exert for 1 million in (25.4 km) of cumulative travel. The relationship between load, P, and life, L, is also similar to that for a ball bearing:

➪ Relationship between Bearing Load and Life

$$\frac{L_2}{L_1} = \left(\frac{P_1}{P_2}\right)^3 \tag{17–12}$$

Thus, if the load on a ball screw is doubled, the life is reduced to one-eighth of the original life. If the load is cut in half, the life is increased by eight times. Figure 17–8 shows the nominal performance of ball screws of a small variety of sizes. Many more sizes and styles are available.

Torque and Efficiency

The efficiency of a ball bearing screw is typically taken to be 90%. This far exceeds the efficiency for power screws

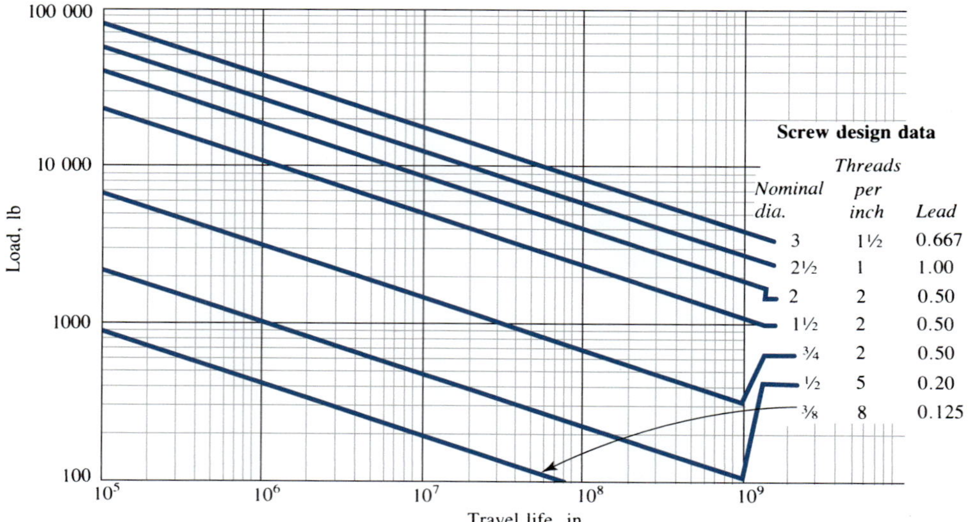

FIGURE 17–8 Ball screw performance

without rolling contact that are typically in the range of 20% to 30%. Thus, far less torque is required to exert a given load with a given size of screw. Power is correspondingly reduced. The computation of the torque to turn is adapted from Equation (17–7), relating efficiency to torque:

Efficiency of a Ball Screw

$$e = \frac{FL}{2\pi T_u} \tag{17–7}$$

Then, using $e = 0.90$,

Torque to Drive a Ball Screw

$$T_u = \frac{FL}{2\pi e} = 0.177FL \tag{17–13}$$

Because of the low friction, ball screws are virtually never self-locking. In fact, designers also use this property to advantage by purposely using the applied load on the nut to rotate the screw. This is called *backdriving*; backdriving torque can be computed from

Backdriving Torque for a Ball Screw

$$T_b = \frac{FLe}{2\pi} = 0.143FL \tag{17–14}$$

Example Problem 17–3 Select suitable ball screws for the application described in Example Problems 17–1 and 17–2 and illustrated in Figure 17–4. The hatch must be lifted 15.0 in to open it eight times per day, and then it must be closed. The design life is 10 years. The lifting or lowering is to be completed in no more than 12.0 s.

For the screw selected, compute the torque to turn the screw, the power required, and the actual expected life.

Solution The data required to select a screw from Figure 17–8 are the load and the travel of the nut on the screw over the desired life. The load is 12 500 lb on each screw:

$$\text{Travel} = \frac{15.0 \text{ in}}{\text{stroke}}\,\frac{2 \text{ strokes}}{\text{cycle}}\,\frac{8 \text{ cycles}}{\text{day}}\,\frac{365 \text{ days}}{\text{year}}\,\frac{10 \text{ years}}{} = 8.76 \times 10^5 \text{ in}$$

From Figure 17–8, the 2-in screw with two threads per inch and a lead of 0.50 in is satisfactory.

The torque required to turn the screw is

$$T_u = 0.177FL = 0.177(12\,500)(0.50) = 1106 \text{ lb}\cdot\text{in}$$

The rotational speed required is

$$n = \frac{1 \text{ rev}}{0.50 \text{ in}}\,\frac{15.0 \text{ in}}{12.0 \text{ s}}\,\frac{60 \text{ s}}{\text{min}} = 150 \text{ rpm}$$

The power required for each screw is

$$P = \frac{Tn}{63\,000} = \frac{(1106)(150)}{63\,000} = 2.63 \text{ hp}$$

Compare this with the 8.61 hp required for the Acme screw in Example Problem 17–2.

The actual travel life expected for this screw at a 12 500-lb load would be approximately 3.2×10^6 in, using Figure 17–8. This is 3.65 times longer than required.

17–4 APPLICATION CONSIDERATIONS FOR POWER SCREWS AND BALL SCREWS

This section discusses additional application considerations that apply to both power screws and ball screws. Details may change according to the specific geometry and manufacturing processes. Supplier data should be consulted.

Critical Speed

The proper application of ball screws must take into account their vibration tendencies, particularly when operating at relatively high speeds. Long slender screws may exhibit the phenomenon of *critical speed* at which the screw would tend to vibrate or whirl about its axis, possibly reaching dangerous amplitudes. Therefore, it is recommended that the operating speed of the screw be below 0.80 times the critical speed. An estimate for the critical speed, offered by Roton Products, Inc. (Internet site 10), is:

$$n_c = \frac{4.76 \times 10^6 dK_s}{(SF)L^2} \tag{17–15}$$

where,

d = minor diameter of the screw (in)
K_s = end fixity factor
L = length between supports (in)
SF = safety factor

The end fixity factor, K_s, depends on the manner of supporting the ends of the screw with the possibilities being:

1. Simply supported at each end by one bearing: $K_s = 1.00$
2. Fixed at each end by two bearings that prevent rotation at the support: $K_s = 2.24$
3. Fixed at one end and simply supported at the other: $K_s = 1.55$
4. Fixed at one end and free at the other: $K_s = 0.32$

The value of the safety factor is a design decision, often taken to be in the range from 1.25 to 3.0. Note that the screw length is squared in the denominator, indicating that a relatively long screw would have a low critical speed. The best designs would employ a short length, rigid fixed supports, and large diameter. See References 2 and 3 for additional information on critical speeds.

Column Buckling

Ball screws that carry axial compressive loads must be checked for column buckling. The parameters, similar to those discussed in Chapter 6, are the material from which the screw is made, the end fixity, the diameter, and the length. Long screws should be analyzed using the Euler formula, Equation (6–5) or (6–6), while the J. B. Johnson formula, Equation (6–7), is used for shorter screws. End fixity depends on the rigidity of supports similar to that described above for critical speed. However, the factors are different for column loading.

1. Simply supported at each end by one bearing: $K_s = 1.00$
2. Fixed at each end by two bearings that prevent rotation at the support: $K_s = 4.00$
3. Fixed at one end and simply supported at the other: $K_s = 2.00$
4. Fixed at one end and free at the other: $K_s = 0.25$

Suppliers of commercially available ball screws include data for allowable compressive load in their catalogs. (See Internet sites 3, 5, 7, 10, and 12.)

Material for Screws

Ball screws are typically made from carbon or alloy steels using thread-rolling technology. After the threads are formed, induction heating improves the hardness and strength of the surfaces on which the circulating balls roll for wear resistance and long life. Ball screw nuts are made from alloy steel that is case hardened by carburizing.

Power screws are typically produced from carbon or alloy steels such as SAE 1018, 1045, 1060, 4130, 4140, 4340, 4620, 6150, 8620, and others. For corrosive environments or where high temperatures are experienced, stainless steels are used, such as SAE 304, 305, 316, 384, 430, 431, or 440. Some are made from aluminum alloys 1100, 2014, or 3003.

Power screw nuts are made from steels for moderate loads and when operating at relatively low speeds. Grease lubrication is recommended. Higher speeds and loads call for lubricated bronze nuts that have superior wear performance. Applications requiring lighter loads can use plastic nuts that have inherently good lubricity without external lubrication. Examples of such applications are food processing equipment, medical devices, and clean manufacturing operations.

REFERENCES

1. Avallone, E., T. Baumeister III, and A. Sadegh. *Mark's Standard Handbook for Mechanical Engineers.* 11th ed. New York: McGraw-Hill, 2007.
2. Budynas, R. G., and K. J. Nisbett. *Shigley's Mechanical Engineering Design.* 9th ed. New York: McGraw-Hill, 2011.
3. Juvinall, R. C., and K. M. Marshek. *Fundamentals of Machine Component Design.* 5th ed. New York: John Wiley & Sons, 2011.
4. Oberg, E., et al. *Machinery's Handbook.* 28th ed. New York: Industrial Press, 2008.

INTERNET SITES FOR LINEAR MOTION ELEMENTS

1. **PowerTransmission.com.** www.powertransmission.com Online listing of numerous manufacturers and suppliers of linear motion devices, including ball screws, lead (power) screws, linear actuators, screw jacks, and linear slides.
2. **Ball-screws.net.** www.ball-screws.net Site lists numerous manufacturers of ball screws, some with online catalogs.
3. **Thomson Industries, Inc.** www.thomsonlinear.com Manufacturer of ball screws, ball bushings, linear motion guides, clutches, brakes, actuators, and other motion control elements. Site contains selection software for U.S. and metric dimensions. Thomson is part of the Danaher Motion Group of Danaher Corporation.
4. **Danaher Motion Group.** www.danahermotion.net Part of Danaher Corporation. Manufacturers of Thomson motion control devices, motors, drives, controls, air bearing rotary stages, and high precision X- Y- Z stages.
5. **THK Linear Motion Systems.** www.thk.com/us Manufacturer of ball screws, ball splines, linear motion guides, linear bushings, linear actuators, and other motion control products.
6. **Joyce/Dayton Company.** www.joycedayton.com Manufacturer of a wide variety of jacks for commercial and industrial applications. Included are power screw and ball screw types with integral gear drives and complete motorized actuator systems.
7. **SKF Linear Motion.** www.linearmotion.skf.com Manufacturer of high efficiency metric screws, linear guiding systems, actuators, and controls.
8. **Specialty Motions, Inc. (SMI).** www.smi4motion.com Manufacturer and supplier of linear motion systems and components from miniature to large sizes. Ball screws, positioning stages, linear bearings and bushings, ball slides and others.
9. **Techno, Inc.** www.techno-isel.com Manufacturer of a wide variety of linear motion products including slides, X-Y and X-Y-Z tables, gantry tables, and controls.
10. **Roton Products, Inc.** www.roton.com Manufacturer of a large variety of power screws, ball screws, and worms in U.S. and metric dimensions.
11. **Ledex.** www.ledex.com Manufacturer of small linear and rotary solenoid actuators and related products under the Ledex and Dormeyer® brands. A unit of Johnson Electric.
12. **Nook Industries.** www.nookindustries.com Manufacturer of power screws and screw jacks in U.S. and metric dimensions and thread forms.
13. **Power Jacks Group.** www.powerjacks.com Manufacturer of metric screw jacks and actuators.

PROBLEMS

1. Name four types of threads used for power screws.
2. Make a scale drawing of an Acme thread having a major diameter of $1\frac{1}{2}$ in and four threads per inch. Draw a section 2.0 in long.
3. Repeat Problem 2 for a buttress thread.
4. Repeat Problem 2 for a square thread.
5. If an Acme-thread power screw is loaded in tension with a force of 30 000 lb, what size screw from Table 17–1 should be used to maintain a tensile stress below 10 000 psi?
6. For the screw chosen in Problem 5, what would be the required axial length of the nut on the screw that transfers the load to the frame of the machine if the shear stress in the threads must be less than 6000 psi?
7. Compute the torque required to raise the load of 30 000 lb with the Acme screw selected in Problem 5. Use a coefficient of friction of 0.15.
8. Compute the torque required to lower the load with the screw from Problem 5.
9. If a square-thread screw having a major diameter of 3/4 in and six threads per inch is used to lift a load of 4000 lb, compute the torque required to rotate the screw. Use a coefficient of friction of 0.15.
10. For the screw of Problem 9, compute the torque required to rotate the screw when lowering the load.
11. Compute the lead angle for the screw of Problem 9. Is it self-locking?
12. Compute the efficiency of the screw of Problem 9.
13. If the load of 4000 lb is lifted by the screw described in Problem 9 at the rate of 0.5 in/s, compute the required rotational speed of the screw and the power required to drive it.
14. A ball screw for a machine table drive is to be selected. The axial force to be transmitted by the screw is 600 lb. The table moves 24 in per cycle, and it is expected to cycle 10 times per hour for a design life of 10 years. Select an appropriate screw.
15. For the screw selected in Problem 14, compute the torque required to drive the screw.
16. For the screw selected in Problem 14, the normal travel speed of the table is 10.0 in/min. Compute the power required to drive the screw.
17. If the cycle time for the machine in Problem 14 were decreased to obtain 20 cycles/h instead of 10, what would be the expected life in years of the screw originally selected?
18. Specify a suitable size for a metric trapezoidal power screw that is subjected to a tensile load of 125 kN while keeping the tensile stress below 75 MPa.
19. Compute the torque required to raise the load vertically for the screw selected in Problem 18. Use a coefficient of friction of 0.15.
20. Compute the torque required to lower the load with the screw specified in Problem 18.
21. For the screw specified in Problem 18, compute the power required to raise the load a total of 4250 mm in 7.5 s.
22. Compute the lead angle for the screw specified in Problem 18. Will it be self-locking?
23. Compute the efficiency for the screw specified in Problem 18 when raising the load.
24. Specify a suitable size for a metric trapezoidal power screw that is subjected to a tensile load of 8500 N while keeping the tensile stress below 110 MPa.
25. Compute the torque required to raise the load vertically for the screw selected in Problem 24. Use a coefficient of friction of 0.15.
26. Compute the torque required to lower the load with the screw specified in Problem 24.
27. For the screw specified in Problem 24, compute the power required to raise the load a total of 240 mm in 3.5 s.
28. Compute the lead angle for the screw specified in Problem 24. Will it be self-locking?
29. Compute the efficiency for the screw specified in Problem 24 when raising the load.
30. Compute the critical speed for the ball screw specified in Problem 14. The total length of the screw between single bearings at each end is 28.0 in.

CHAPTER NINETEEN

FASTENERS

THE BIG PICTURE Fasteners

Discussion Map

- *Fasteners* connect or join two or more components. Common types are *bolts* and *screws* such as those illustrated in Figures 19–1 through 19–4.

Discover

Look for examples of bolts and screws. List how many types you have found. For what functions were they being used? What kinds of forces are the fasteners subjected to? What materials are used for the fasteners?

In this chapter, you will learn to analyze the performance of fasteners and to select suitable types and sizes.

A *fastener* is any device used to connect or join two or more components. Literally hundreds of fastener types and variations are available. The most common are threaded fasteners referred to by many names, among them bolts, screws, nuts, studs, lag screws, and set screws.

A *bolt* is a threaded fastener designed to pass through holes in the mating members and to be secured by tightening a nut from the end opposite the head of the bolt. See Figure 19–1(a), called a *hex head bolt*. Several other types of bolts are shown in Figure 19–2.

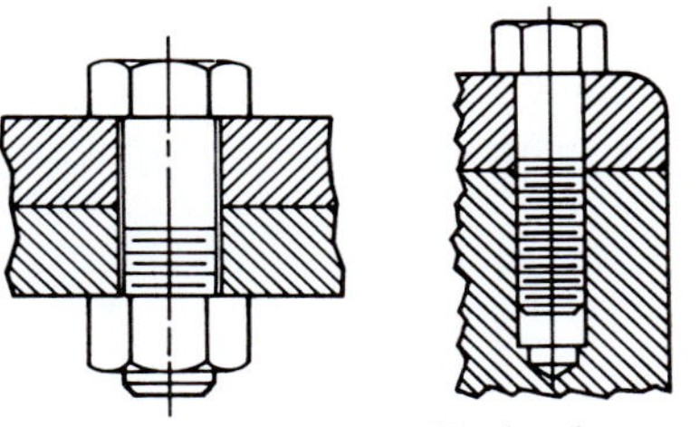

FIGURE 19–1 Comparison of a bolt with a screw (Reference 9)

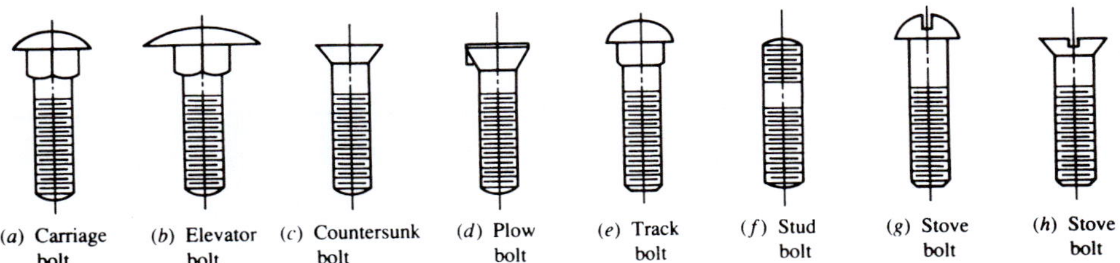

FIGURE 19–2 Bolt styles. See also the hex head bolt in Figure 19–1 (Reference 9)

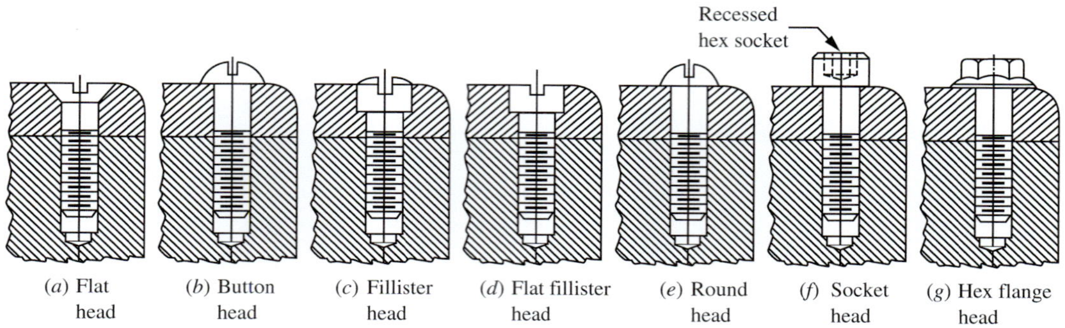

FIGURE 19–3 Cap screws or machine screws. See also the hex head cap screw in Figure 19–1 (Reference 9)

A *screw* is a threaded fastener designed to be inserted through a hole in one member to be joined and into a threaded hole in the mating member. See Figure 19–1(b). The threaded hole may have been preformed, for example, by tapping, or it may be formed by the screw itself as it is forced into the material. *Machine screws,* also called *cap screws,* are precision fasteners with straight-threaded bodies that are turned into tapped holes (see Figure 19–3). A popular type of machine screw is the socket head cap screw. The usual configuration, shown in Figure 19–3(f), has a cylindrical head with a recessed hex socket. Also readily available are flat head styles for countersinking to produce a flush surface, button head styles for a low profile appearance, and shoulder screws providing a precision bearing surface for location or pivoting. See Internet sites 9 and 11. *Sheet-metal screws, lag screws, self-tapping screws,* and *wood screws* usually form their own threads. Figure 19–4 shows a few styles.

Search for examples where the kinds of fasteners illustrated in Figures 19–1 through 19–4 are used. How many can you find? Make a list using the names for the fasteners in the figures. Describe the application. What function is the fastener performing? What kinds of forces are exerted on each fastener during service? How large is the fastener? Measure as many dimensions as you can. What material is each fastener made from?

Look in your car, particularly under the hood in the engine compartment. If you can, also look under the chassis to see where fasteners are used to hold different components onto the frame or some other structural member.

Look also at bicycles, lawn and garden equipment, grocery carts, display units in a store, hand tools, kitchen appliances, toys, exercise equipment, and furniture. If you have access to a factory, you should be able to identify hundreds or thousands of examples. Try to get some insight about where certain types of fasteners are used and for what purposes.

In this chapter, you will learn about many of the types of fasteners that you will encounter, including how to analyze their performance.

References 1-7 and Internet sites 1, 2, 14, and 15 provide extensive coverage of the science and technology of fasteners and are recommended as sources to supplement the treatment in this book.

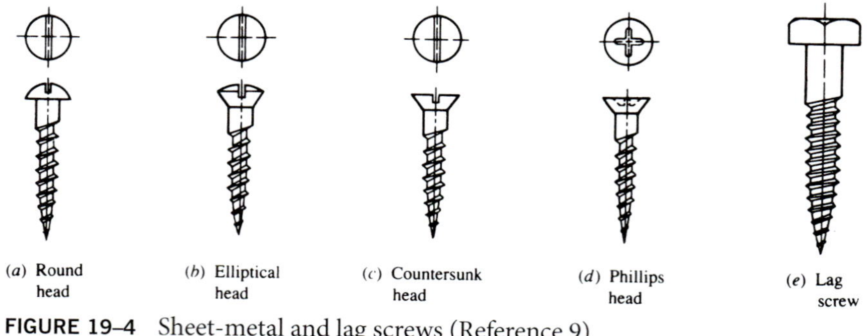

FIGURE 19–4 Sheet-metal and lag screws (Reference 9)

YOU ARE THE DESIGNER

Review Figure 15–6 which shows the assembly of the gear-type power transmission that was designed in that chapter. Fasteners are called for in several places on the housing for the transmission, but they were not specified in that chapter. The four *bearing retainers* are to be fastened to the housing and the cover by threaded fasteners. The cover itself is to be attached to the housing by fasteners. Finally, the mounting base has provisions for using fasteners to hold the entire transmission to a support structure.

You are the designer. What kinds of fasteners would you consider for these applications? What material should be used to make them? What strength should the material have? If threaded fasteners are used, what size should the threads be, and how long must they be? What head style would you specify? How much torque should be applied to the fastener to ensure that there is sufficient clamping force between the joined members? How does the design of the gasket between the cover and the housing affect the choice of the fasteners and the specification of the tightening torque for them? What alternatives are there to the use of threaded fasteners to hold the components together and to still allow disassembly?

This chapter presents information that you can use to make such design decisions. The references at the end of the chapter give other valuable sources of information from the large body of knowledge about fasteners. ■

19–1 OBJECTIVES OF THIS CHAPTER

After completing this chapter, you will be able to:

1. Describe a bolt in comparison with a machine screw.
2. Name and describe nine styles of heads for bolts.
3. Name and describe six styles of heads for machine screws.
4. Describe sheet-metal screws and lag screws.
5. Describe six styles of set screws and their application.
6. Describe nine types of locking devices that restrain a nut from becoming loose on a bolt.
7. Use tables of data for various grades of steel materials used for bolts as published by the Society of Automotive Engineers (SAE) and the American Society for Testing and Materials (ASTM), and for standard metric grades.
8. List at least 10 materials other than steel that are used for fasteners.
9. Use tables of data for standard screw threads in the American Standard and metric systems for dimensions and stress analysis.
10. Define *proof load, clamping load,* and *tightening torque* as applied to bolts and screws, and compute design values.
11. Compute the effect of adding an externally applied force on a bolted joint, including the final force on the bolts and the clamped members.
12. List and describe 16 different coating and finishing techniques that are used for metal fasteners.
13. Describe rivets, quick-operating fasteners, welding, brazing, and adhesives, and contrast them with bolts and screws for fastening applications.

19–2 BOLT MATERIALS AND STRENGTH

In machine design, most fasteners are made from steel because of its high strength, high stiffness, good ductility, and good machinability and formability. But varying compositions and conditions of steel are used. The strength of steels used for bolts and screws is used to determine its *grade*, according to one of several standards. Three strength ratings are frequently available: the familiar tensile strength and yield strength plus the proof strength. The *proof strength*, similar to the elastic limit, is defined as the stress at which the bolt or the screw would undergo permanent deformation. It usually ranges between 0.90 and 0.95 times the yield strength.

Three major standards-setting organizations relevant to fasteners in the United States are listed in References 1, 2, 11, and 17–28. See Internet sites 1, 2, 14, and 15. Note that some of the listed standards are for metric fasteners whereas most others focus on the U.S. inch-pound system and a few address both units. The fastener-producing industry is represented by the Industrial Fasteners Institute that publishes a comprehensive document listed in Reference 10. There is much cross-referencing among these organizations. It is essential that designers are familiar with the entire content of the latest versions of applicable standards that present far more information than is mentioned in this book.

You should become familiar with a variety of vendors of commercially available screws, bolts, and other types of fasteners. As a sample, check out Internet sites 3–13.

Table 19–1 lists selected basic data for four SAE grades of fasteners from Reference 18 with the tensile strength generally increasing with higher grade numbers. Take note that strengths are related to specified ranges of sizes of fasteners. Some grades are made with special marking on the heads to assist users in the field to ensure that the proper grade is being used.

Table 19–2 lists data extracted from six different ASTM standards, each focused on a particular grade of steel or a type of fastener. The full designations for these grades are listed in References 20–24 and 27. Reference 25 is the specification for nuts most applicable to the listed bolt grades. Numerous grades are described. Reference 10 contains tables showing the commercially available grades of nuts for each style of bolt.

Examples of metric grades of steels used for bolts are shown in Table 19–3, with data extracted from a variety of sources including References 1, 2, 19, and 28. Note the significantly different system for denoting the grades.

TABLE 19–1 SAE Grades of Steels for Fasteners

Grade number	Bolt size (in)	Tensile strength (ksi)	Yield strength (ksi)	Proof strength (ksi)	Head marking
1	1/4–$1\frac{1}{2}$	60	36	33	None
2	1/4–3/4	74	57	55	None
	>3/4–$1\frac{1}{2}$	60	36	33	
5	1/4–1	120	92	85	[3 radial lines]
	>1–$1\frac{1}{2}$	105	81	74	
8	1/4–$1\frac{1}{2}$	150	130	120	[6 radial lines]

TABLE 19–2 ASTM Grades for Bolt Steels

ASTM grade	Bolt size (in)	Tensile strength (ksi)	Yield strength (ksi)	Proof strength (ksi)	Head marking
A307	1/4–4	60	(Not reported)		None
A325	1/2–1	120	92	85	A 325
	$>1–1\frac{1}{2}$	105	81	74	
A354-BC	$1/4–2\frac{1}{2}$	125	109	105	BC
A354-BD	$1/4–2\frac{1}{2}$	150	130	120	
A449	1/4–1	120	92	85	
	$>1–1\frac{1}{2}$	105	81	74	
	$>1\frac{1}{2}–3$	90	58	55	
A490	1/2–1/2	150	130	120	A 490
A574	0.060–1/2	180		140	(Socket head cap screws)
	5/8–4	170		135	

TABLE 19–3 Metric Grades of Steels for Bolts

Grade	Bolt size	Tensile strength (MPa)	Yield strength (MPa)	Proof strength (MPa)
4.6	M5–M36	400	240	225
4.8	M1.6–M16	420	340[a]	310
5.8	M5–M24	520	415[a]	380
8.8	M17–M36	830	660	600
9.8	MI.6–M16	900	720[a]	650
10.9	M6–M36	1040	940	830
12.9	M1.6–M36	1220	1100	970

[a]Yield strengths are approximate and are not included in the standard.

Metric bolts and screws use a numerical code system ranging from 4.6 to 12.9, with higher numbers indicating higher strengths. The numbers before the decimal point are approximately 0.01 times the tensile strength of the material in MPa. The last digit with the decimal point is the approximate ratio of the yield strength of the material to the tensile strength.

Approximate Equivalencies among SAE, ASTM, and Metric Grades of Bolt Steels. The following list shows approximate equivalents that may be useful when comparing designs for which specifications include combinations of SAE, ASTM, and metric grades of bolt steels. The individual standards should be consulted for specific strength data.

SAE Grade	ASTM Grade	Metric Grade
J429 Grade 1	A307 Grade A	Grade 4.6
J429 Grade 2	______	Grade 5.8
J429 Grade 5	A449, A325	Grade 8.8
J429 Grade 8	A354 Grade BD, A490	Grade 10.9
	A574	Grade 12.9

Listed below are some of the more commonly used bolt steels according to Reference 10.

U.S. units	SI units
SAE J429, grades 5 and higher	SAE J1199 and ISO, grades 8.8, 10.9, and 12.9
ASTM A325, A384, A449, A490	ASTM A325M, A490M, A574M

The company listed in Internet site 9 provides high-strength heat-treated alloy steel socket head cap screws having the following strengths:

Size Range	Tensile Strength (ksi)	Tensile Strength (MPa)	Yield Strength (ksi)	Yield Strength (MPa)
0–5/8	190	1310	170	1172
3/4–3	180	1241	155	1069

Roughly equivalent performance is obtained from metric socket head cap screws made to the metric strength grade 12.9. The same geometry is available in corrosion-resistant stainless steel, typically 19–8, at somewhat lower strength levels. See Internet site 11.

Aluminum is used for its corrosion resistance, light weight, and fair strength level. Its good thermal and electrical conductivity may also be desirable. The most widely used alloys are 2024-T4, 6061-T6, and 7075-T73. Properties of these materials are listed in Appendix 9.

Brass, copper, and *bronze* are also used for their corrosion resistance. Ease of machining and an attractive appearance are also advantages. Certain alloys are particularly good for resistance to corrosion in marine applications. See Appendix 12.

Nickel and its alloys, such as *Monel* and *Inconel* (from the International Nickel Company), provide good performance at elevated temperatures while also having good corrosion resistance, toughness at low temperatures, and an attractive appearance. See Appendix 11–1.

Stainless steels are used primarily for their corrosion resistance. Alloys used for fasteners include 19–8, 410, 416, 430, and 431. In addition, stainless steels in the 300 series are nonmagnetic. See Appendix 6 for properties.

A high strength-to-weight ratio is the chief advantage of *titanium* alloys used for fasteners in aerospace applications. Appendix 11–2 gives a list of properties of several alloys.

Plastics are used widely because of their light weight, corrosion resistance, insulating ability, and ease of manufacture. Nylon 6/6 is the most frequently used material, but others include ABS, acetal, TFE fluorocarbons, polycarbonate, polyethylene, polypropylene, and polyvinylchloride. Appendix 13 lists several plastics and their properties. In addition to being used in screws and bolts, plastics are used extensively where the fastener is designed specially for the particular application. See also Reference 13.

Coatings and *finishes* are provided for metallic fasteners to improve appearance or corrosion resistance. Some also lower the coefficient of friction for more consistent results relating tightening torque to clamping force. Steel fasteners can be finished with black oxide, bluing, bright nickel, phosphate, and hot-dip zinc. Plating can be used to deposit cadmium, copper, chromium, nickel, silver, tin, and zinc. Various paints, lacquers, and chromate finishes are also used. Aluminum is usually anodized. Check environmental hazards for coatings and finishes.

19–3 THREAD DESIGNATIONS AND STRESS AREA

Table 19–4 shows pertinent dimensions for threads in the American Standard styles, and Table 19–5 gives SI metric styles. For consideration of strength and size, the designer must

TABLE 19–4 American Standard Screw Threads

A. American Standard thread dimensions, numbered sizes

Basic major diameter, D (in)	Coarse threads: UNC		Fine threads: UNF	
	Size-Threads per inch, n	Tensile stress area (in^2)	Size-Threads per inch, n	Tensile stress area (in^2)
0.0600	—	—	0–80	0.001 80
0.0730	1–64	0.002 63	1–72	0.002 78
0.0860	2–56	0.003 70	2–64	0.003 94
0.0990	3–48	0.004 87	3–56	0.005 23
0.1120	4–40	0.006 04	4–48	0.006 61
0.1250	5–40	0.007 96	5–44	0.008 30
0.1380	6–32	0.009 09	6–40	0.010 15
0.1640	8–32	0.0140	8–36	0.014 74
0.1900	10–24	0.0175	10–32	0.0200
0.2160	12–24	0.0242	12–28	0.0258

B. American Standard thread dimensions, fractional sizes

Basic major diameter, D (in)	Coarse threads: UNC		Fine threads: UNF	
	Size-Threads per inch, n	Tensile stress area (in^2)	Size-Threads per inch, n	Tensile stress area (in^2)
0.2500	1/4–20	0.0318	1/4–28	0.0364
0.3125	5/16–18	0.0524	5/16–24	0.0580
0.3750	3/8–16	0.0775	3/8–24	0.0878
0.4375	7/16–14	0.1063	7/16–20	0.1187
0.5000	1/2–13	0.1419	1/2–20	0.1599
0.5625	9/16–12	0.182	9/16–18	0.203
0.6250	5/8–11	0.226	5/8–18	0.256
0.7500	3/4–10	0.334	3/4–16	0.373
0.8750	7/8–9	0.462	7/8–14	0.509
1.000	1–8	0.606	1–12	0.663
1.125	$1\frac{1}{8}$–7	0.763	$1\frac{1}{8}$–12	0.856
1.250	$1\frac{1}{4}$–7	0.969	$1\frac{1}{4}$–12	1.073
1.375	$1\frac{3}{8}$–6	1.155	$1\frac{3}{8}$–12	1.315
1.500	$1\frac{1}{2}$–6	1.405	$1\frac{1}{2}$–12	1.581
1.750	$1\frac{3}{4}$–5	1.90		
2.000	2–$4\frac{1}{2}$	2.50		

TABLE 19–5 Metric Sizes of Screw Threads

	Coarse threads		Fine threads	
Basic major diameter, *D* (mm)	Basic thread designation *MD* × Pitch (mm) (mm)	Tensile stress area (mm^2)	*MD* × Pitch (mm) (mm)	Tensile stress area (mm^2)
1	M1 × 0.25	0.460	—	—
1.6	M1.6 × 0.35	1.27	M16 × 0.20	1.57
2	M2 × 0.4	2.07	M2 × 0.25	2.45
2.5	M2.5 × 0.45	3.39	M25 × 0.35	3.70
3	M3 × 0.5	5.03	M3 × 0.35	5.61
4	M4 × 0.7	8.78	M4 × 0.5	9.79
5	M5 × 0.8	14.2	M5 × 0.5	16.1
6	M6 × 1	20.1	M6 × 0.75	22.0
8	M8 × 1.25	36.6	M8 × 1	39.2
10	M10 × 1.5	58.0	M10 × 1.25	61.2
12	M12 × 1.75	84.3	M12 × 1.25	92.1
16	M16 × 2	157	M16 × 1.5	167
20	M20 × 2.5	245	M20 × 1.5	272
24	M24 × 3	353	M24 × 2	384
30	M30 × 3.5	561	M30 × 2	621
36	M36 × 4	817	M36 × 3	865
42	M42 × 4.25	1121		
48	M48 × 5	1473		

know the basic major diameter, the pitch of the threads, and the area available to resist tensile loads. Note that the pitch is equal to $1/n$, where n is the number of threads per inch in the American Standard system. In the SI, the pitch in millimeters is designated directly. The tensile stress area listed in Tables 19–4 and 19–5 takes into account the actual area cut by a transverse plane. Because of the helical path of the thread on the screw, such a plane will cut near the root on one side of the screw but will cut near the major diameter on the other. The equation for the tensile stress area for American Standard threads is

Tensile Stress Area for UNC or UNF Threads

$$A_t = (0.7854)[D - (0.9743)p]^2 \qquad \textbf{(19–1)}$$

where D = major diameter
p = pitch of the thread

For metric threads, the tensile stress area is

Tensile Stress Area for Metric Threads

$$A_t = (0.7854)[D - (0.9382)p]^2 \qquad \textbf{(19–2)}$$

For most standard screw thread sizes, at least two pitches are available: the *coarse* series and the *fine thread* series. Both are included in Tables 19–4 and 19–5.

The smaller American Standard threads use a number designation from 0 to 12. The corresponding major diameter is listed in Table 19–4(A). The larger sizes use fractional-inch designations. The decimal equivalent for the major diameter is shown in Table 19–4(B). Metric threads list the major diameter and the pitch in millimeters, as shown in Table 19–5. Samples of the standard designations for a thread are given next.

American Standard: Basic size followed by number of threads per inch and the thread series designation.

10–24 UNC	10–32 UNF
1/2–13 UNC	1/2–20 UNF
$1\frac{1}{2}$–6 UNC	$1\frac{1}{2}$–12 UNF

Metric: M (for "metric"), followed by the basic major diameter and then the pitch in millimeters.

M3 × 0.5 M3 × 0.35 M10 × 1.5

19–4 CLAMPING LOAD AND TIGHTENING OF BOLTED JOINTS

Clamping Load

When a bolt or a screw is used to clamp two parts, the force exerted between the parts is the *clamping load*. The designer is responsible for specifying the clamping load and for ensuring that the fastener is capable of withstanding the load. The maximum clamping load is often taken to be 0.75 times the proof load, where the proof load is the product of the proof stress times the tensile stress area of the bolt or screw.

Tightening Torque

The clamping load is created in the bolt or the screw by exerting a tightening torque on the nut or on the head of the screw. An approximate relationship between the torque and the axial tensile force in the bolt or screw (the clamping force) is

➩ **Tightening Torque**

$$T = KDP \tag{19–3}$$

where T = torque, lb · in
D = nominal outside diameter of threads, in
P = clamping load, lb
K = constant dependent on the lubrication present

For average commercial conditions, use $K = 0.15$ if any lubrication at all is present. Even cutting fluids or other residual deposits on the threads will produce conditions consistent with $K = 0.15$. If the threads are well cleaned and dried, $K = 0.20$ is better. Of course, these values are approximate, and variations among seemingly identical assemblies should be expected. Testing and statistical analysis of the results are recommended.

Example Problem 19–1 A set of three bolts is to be used to provide a clamping force of 12 000 lb between two components of a machine. The load is shared equally among the three bolts. Specify suitable bolts, including the grade of the material, if each is to be stressed to 75% of its proof strength. Then compute the required tightening torque.

Solution The load on each screw is to be 4000 lb. Let's specify a bolt made from SAE Grade 5 steel, having a proof strength of 85 000 psi. Then the allowable stress is

$$\sigma_a = 0.75(85000\,\text{psi}) = 63750\,\text{psi}$$

The required tensile stress area for the bolt is then

$$A_t = \frac{\text{load}}{\sigma_a} = \frac{4000\ \text{lb}}{63\ 750\ \text{lb/in}^2} = 0.0627\,\text{in}^2$$

From Table 19–4(B), we find that the 3/8–16 UNC thread has the required tensile stress area. The required tightening torque will be

$$T = KDP = 0.15(0.375\ \text{in})(4000\ \text{lb}) = 225\ \text{lb}\cdot\text{in}$$

Equation (19–3) is adequate for general mechanical design. A more complete analysis of the torque to create a given clamping force requires more information about the joint design. There are three contributors to the torque. One, which we will refer to as T_1, is the torque required to develop the tensile load in the bolt, P_t, using the inclined plane nature of the thread.

$$T_1 = \frac{P_t\, l}{2\pi} = \frac{P_t}{2\pi n} \tag{19–4}$$

where l is the lead of the bolt thread and $l = p = 1/n$.

The second component of the torque, T_2, is that required to overcome friction between the mating threads, computed from,

$$T_2 = \frac{d_p\, \mu_1\, P_t}{2\cos\alpha} \tag{19–5}$$

where

d_p = pitch diameter of the thread
μ_1 = coefficient of friction between the thread surfaces
α = 1/2 of the thread angle, typically 30°

The third component of the torque, T_3, is the friction between the underside of the head of the bolt or nut and the clamped surface. This friction force is assumed to act at the middle of the friction surface and is computed from,

$$T_3 = \frac{(d + b)\, \mu_2\, P_t}{4} \tag{19–6}$$

where

d = major diameter of the bolt
b = outside diameter of the friction surface on the underside of the bolt
μ_2 = coefficient of friction between the bolt head and the clamped surface

The total torque is then,

$$T_{tot} = T_1 + T_2 + T_3 \tag{19–7}$$

See References 4–8 and 12–16 for additional discussion on torque for bolts. It is important to note that many variables are involved in the contributors to the relationship between the applied torque and the tensile preload given to the bolt. Accurate prediction of the coefficients of friction is difficult. The accuracy with which the specified torque is applied is affected by the precision of the measurement device used, such as a torque wrench, pneumatic nut runner, or hydraulic nut driver, as well as the operator's skill. Reference 4 has an extensive discussion of the large variety of torque wrenches available.

Other Methods of Bolt Tightening

Measuring the torque applied to the bolt, screw, or nut during installation is convenient. However, because of the many variables involved, the actual clamping force created may vary significantly. Fastening of critical connections often uses other methods of bolt tightening that more directly relate to the clamping force. Situations where these methods may be used are structural steel connections, flanges for high-pressure systems, nuclear power plant components, cylinder head and connecting rod bolts for engines, aerospace structures, turbine engine components, propulsion systems, and military equipment.

Turn of the Nut Method. The bolt is first tightened to a snug fit to bring all of the parts of the joint into intimate contact. Then the nut is given an additional turn with a wrench of between one-third and one full turn, depending on the size of the bolt. One full turn would produce a stretch in the bolt equal to the lead of the thread, where $l = p = 1/n$. The elastic behavior of the bolt determines the amount of the resulting clamping force. References 4, 8, and 17 give more details.

Tension Control Bolting Products. Special bolts are available that include a carefully sized neck on one end connected to a splined section. The spline is held fixed as the nut is turned. When a predetermined torque is applied to the nut, the neck section breaks and the tightening stops. Consistent connection performance results.

Another form of tension control bolt employs a tool that exerts direct axial tension on the bolt, swages a collar into annular grooves or the threads of the fastener, and then breaks a small-diameter part of the bolt at a predetermined force. The result is a predictable amount of clamping force in the joint. See Internet site 13.

Wavy Flanged Bolt. The underside of the head of this bolt is formed in a wavy pattern during manufacture. When torque is applied to the joint, the wavy surface is deformed to become flat against the clamped surface when the proper amount of tension has been created in the body of the bolt. See Internet site 9.

Direct Tension Indicator (DTI) Washers. The DTI washer has several raised areas on its upper surface. A regular washer is then placed over the DTI washer and a nut tightens the assembly until the raised areas are flattened to a specified degree, creating a predictable tension in the bolt. See Internet site 10.

Ultrasonic Tension Measurement and Control. Recent developments have resulted in the availability of equipment that imparts ultrasonic acoustic waves to bolts as they are tightened with the timing of the reflected waves being correlated to the amount of stretch and tension in the bolt. See Reference 4.

Tighten to Yield Method. Most fasteners are provided with guaranteed yield strength; therefore, it takes a predictable amount of tensile force to cause the bolt to yield. Some automatic systems use this principle by sensing the relationship between applied torque and the rotation of the nut and stopping the process when the bolt begins to yield. During the elastic part of the stress-strain curve for the bolt, a linear change in torque versus rotation occurs. At yield, a dramatic increase in rotation with little or no increase in torque signifies yielding. A variation on this method, called the *logarithmic rate method (LRM)*, determines the peak of the curve of the logarithm of the rate of torque versus turn data and then applies a preset amount of additional turn to the nut. See Internet site 16.

Internet site 12 describes a testing system that aids in analyzing the effectiveness of bolt-tightened joints. A pressure sensitive thin film is placed at the interface between the surfaces to be clamped, somewhat like a gasket. Then the bolted connection is made with bolts tightened to specified values. Subsequent removal of the bolts reveals quantitative and qualitative data about the distribution of clamping pressure on the critical surfaces. The company also offers consulting services on joint design and production methods.

19–5 EXTERNALLY APPLIED FORCE ON A BOLTED JOINT

The analysis shown in Example Problem 19–1 considers the stress in the bolt only under static conditions and only for the clamping load. It was recommended that the tension on the bolt be very high, approximately 75% of the proof load for the bolt. Such a load will use the available strength of the bolt efficiently and will prevent the separation of the connected members.

When a load is applied to a bolted joint over and above the clamping load, special consideration must be given to the behavior of the joint. Initially, the force on the bolt (in tension) is equal to the force on the clamped members (in compression). Then some of the additional load will act to stretch the bolt beyond its length assumed after the clamping load was applied. Another portion will result in a *decrease* in the compressive force in the clamped member. Thus, only part of the applied force is carried by the bolt. The amount is dependent on the relative stiffness of the bolt and the clamped members.

If a stiff bolt is clamping a flexible member, such as a resilient gasket, most of the added force will be taken by the bolt because it takes little force to change the compression in the gasket. In this case, the bolt design must take into account not only the initial clamping force but also the added force.

Conversely, if the bolt is relatively flexible compared with the clamped members, virtually all of the externally applied load will initially go to decreasing the clamping force until the members actually separate, a condition usually interpreted as failure of the joint. Then the bolt will carry the full amount of the external load.

In practical joint design, a situation between the extremes previously described would normally occur. In typical "hard" joints (without a soft gasket), the stiffness of the clamped members is approximately three times that of the bolt. The externally applied load is then shared by the bolt and the clamped members according to their relative stiffnesses as follows:

$$F_b = P + \frac{k_b}{k_b + k_c}F_e \quad \textbf{(19–8)}$$

$$F_c = P - \frac{k_c}{k_b + k_c}F_e \quad \textbf{(19–9)}$$

where F_e = externally applied load
P = initial clamping load [as used in Equation (19–3)]
F_b = final force in bolt
F_c = final force on clamped members
k_b = stiffness of bolt
k_c = stiffness of clamped members

Example Problem 19–2

Assume that the joint described in Example Problem 19–1 was subjected to an additional external load of 3000 lb after the initial clamping load of 4000 lb was applied. Also assume that the stiffness of the clamped members is three times that of the bolt. Compute the force in the bolt, the force in the clamped members, and the final stress in the bolt after the external load is applied.

Solution

We will first use Equations (19–8) and (19–9) with $P = 4000$ lb, $F_e = 3000$ lb, and $k_c = 3k_b$:

$$F_b = P + \frac{k_b}{k_b + k_c}F_e = P + \frac{k_b}{k_b + 3k_b}F_e = P + \frac{k_b}{4k_b}F_e$$

$$F_b = P + F_e/4 = 4000 + 3000/4 = 4750\ lb$$

$$F_c = P - \frac{k_c}{k_b + k_c}F_e = P - \frac{3k_b}{k_b + 3k_b}F_e = P - \frac{3k_b}{4k_b}F_e$$

$$F_c = P - 3F_e/4 = 4000 - 3(3000)/4 = 1750\ lb$$

Because F_c is still greater than zero, the joint is still tight. Now the stress in the bolt can be found. For the 3/8–16 bolt, the tensile stress area is 0.0775 in^2. Thus,

$$\sigma = \frac{P}{A_t} = \frac{4750\,\text{lb}}{0.0775\,\text{in}^2} = 61\,300 \text{ psi}$$

The proof strength of the Grade 5 material is 85 000 psi, and this stress is approximately 72% of the proof strength. Therefore, the selected bolt is still safe. But consider what would happen with a relatively "soft" joint, discussed in Example Problem 19–3.

Example Problem 19–3

Solve Example Problem 19–2 again, but assume that the joint has a flexible elastomeric gasket separating the clamping members and that the stiffness of the bolt is then 10 times that of the joint.

Solution

The procedure will be the same as that used previously, but now $k_b = 10k_c$. Thus,

$$F_b = P + \frac{k_b}{k_b + k_c}F_e = P + \frac{10k_c}{10k_c + k_c}F_e = P + \frac{10k_c}{11k_c}F_e$$

$$F_b = P + 10F_e/11 = 4000 + 10(3000)/11 = 6727 \text{ lb}$$

$$F_c = P - \frac{k_c}{k_b + k_c}F_e = P - \frac{k_c}{10k_c + k_c}F_e = P - \frac{k_c}{11k_c}F_e$$

$$F_c = P - F_e/11 = 4000 - 3000/11 = 3727 \text{ lb}$$

The stress in the bolt would be

$$\sigma = \frac{6727 \text{ lb}}{0.0775 \text{ in}^2} = 86\,800 \text{ psi}$$

This exceeds the proof strength of the Grade 5 material and is dangerously close to the yield strength.

19–6 THREAD STRIPPING STRENGTH

In addition to sizing a bolt on the basis of axial tensile stress, the threads must be checked to ensure that they will not be stripped off by shearing. The variables involved in the shear strength of the threads are the materials of the bolt, the nut, or the internal threads of a tapped hole, the length of engagement, L_e, and the size of the threads. The details of analysis depend on the relative strength of the materials.

Internal Thread Material Stronger than Bolt Material. For this case, the strength of the threads of the bolt will control the design. Here, we present an equation for the required length of engagement, L_e, of the bolt threads that will have at least the same strength in shear as the bolt itself does in tension.

$$L_e = \frac{2\,A_{tB}}{\pi\,(ID_{\mathrm{Nmax}})\,[0.5 + 0.57735\,n(PD_{\mathrm{Bmin}} - ID_{\mathrm{Nmax}})]} \tag{19–10}$$

where A_{tB} = tensile stress area of bolt
ID_{Nmax} = maximum inside (root) diameter of nut threads
n = number of threads per inch
PD_{Bmin} = minimum pitch diameter of bolt threads

The subscripts B and N refer to the bolt and nut, respectively. The subscripts min and max refer to the minimum and maximum values, respectively, considering the tolerances on thread dimensions. Reference 14 gives data for tolerances as a function of the class of thread specified.

For a given length of engagement, the resulting shear area for the bolt threads is

$$A_{sB} = \pi\,L_e\,ID_{\mathrm{Nmax}}[0.5 + 0.57735\,n(PD_{\mathrm{Bmin}} - ID_{\mathrm{Nmax}})] \tag{19–11}$$

Nut Material Weaker than Bolt Material. This is particularly applicable when the bolt is inserted into a tapped hole in cast iron, aluminum, or some other material with relatively low strength. The required length of engagement to develop at least the full strength of the bolt is

$$L_e = \frac{s_{utB}\,(2\,A_{tB})}{s_{utN}\,\pi\,OD_{\mathrm{Bmin}}\,[0.5 + 0.57735\,n\,(OD_{\mathrm{Bmin}} - PD_{\mathrm{Nmax}})]} \tag{19–12}$$

where

s_{utB} = ultimate tensile strength of the bolt material
s_{utN} = ultimate tensile strength of the nut material
OD_{Bmin} = minimum outside diameter of the bolt threads
PD_{Nmax} = maximum pitch diameter of the nut threads

The shear area of the root of the threads of the nut is

$$A_{sN} = \pi L_e OD_{\mathrm{Bmin}}[0.5 + 0.57735\,n(OD_{\mathrm{Bmin}} - PD_{\mathrm{Nmax}})] \tag{19–13}$$

Equal Strength for Nut and Bolt Material. For this case failure is predicted as shear of either part at the nominal pitch diameter, PD_{nom}. The required length of engagement to develop at least the full strength of the bolt is

$$L_e = \frac{4\,A_{tB}}{\pi\,PD_{\mathrm{nom}}} \tag{19–14}$$

The shear stress area for the nut or bolt threads is

$$A_s = \pi PD_{\mathrm{nom}}\,L_e/2 \tag{19–15}$$

19–7 OTHER TYPES OF FASTENERS AND ACCESSORIES

Most bolts and screws have enlarged heads that bear down on the part to be clamped and thus exert the clamping force. *Set screws* are headless, are inserted into tapped holes, and are designed to bear directly on the mating part, locking it into place. Figure 19–5 shows several styles of points and drive means for set screws. Caution must be used with set screws, as with any threaded fastener, so that vibration does not loosen the screw.

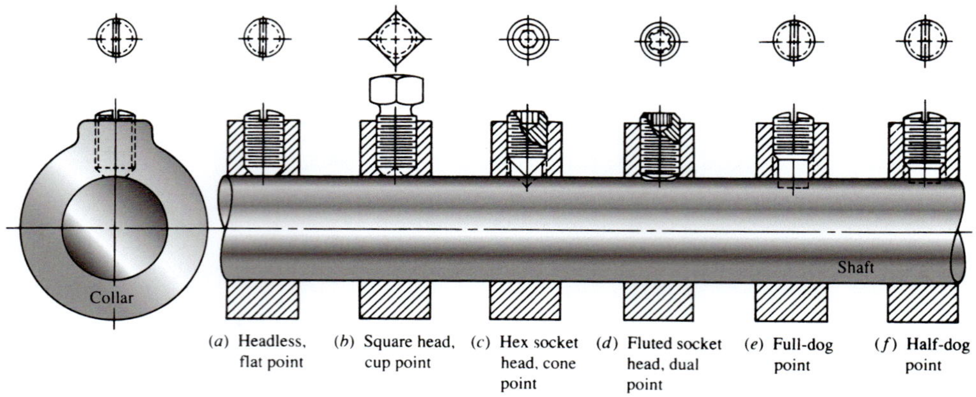

FIGURE 19–5 Set screws with different head and point styles applied to hold a collar on a shaft (Reference 9)

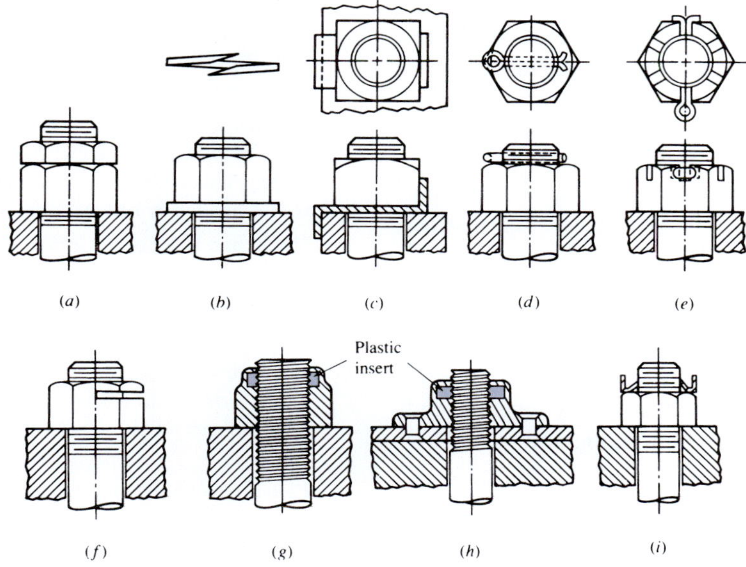

FIGURE 19–6 Locking devices (Reference 9)

A *washer* may be used under either or both the bolt head and the nut to distribute the clamping load over a wide area and to provide a bearing surface for the relative rotation of the nut. The basic type of washer is the plain flat washer, a flat disc with a hole in it through which the bolt or screw passes. Other styles, called *lockwashers*, have axial deformations or projections that produce axial forces on the fastener when compressed. These forces keep the threads of the mating parts in intimate contact and decrease the probability that the fastener will loosen in service.

Figure 19–6 shows several means of using washers and other types of locking devices. Part (a) is a jam nut tightened against the regular nut. Part (b) is the standard lockwasher. Part (c) is a locking tab that keeps the nut from turning. Part (d) is a cotter inserted through a hole drilled through the bolt. Part (e) uses a cotter, but it also passes through slots in the nut. Part (f) is one of several types of thread-deformation techniques used. Part (g), an *elastic stop nut*, uses a plastic insert to keep the threads of the nut in tight contact with the bolt. This may be used on machine screws as well. In part (h), the elastic stop nut is riveted to a thin plate, allowing a mating part to be bolted from the opposite side. The thin metal device in (i) bears against the top of the nut and grips the threads, preventing axial motion of the nut.

A *stud* is like a stationary bolt attached permanently to a part of one member to be joined. The mating member is then placed over the stud, and a nut is tightened to clamp the parts together.

Additional variations occur when these types of fasteners are combined with different head styles. Several of these are shown in the figures already discussed. Others are listed next:

Square	Hex	Heavy hex	Hex jam
Hex castle	Hex flat	Hex slotted	12-point
High crown	Low crown	Round	T-head
Pan	Truss	Hex washer	Flat countersunk
Plow	Cross recess	Fillister	Oval countersunk
Hex socket	Spline socket	Button	Binding

Additional combinations are created by consideration of the American National Standards or British Standard (metric); material grades; finishes; thread sizes; lengths; class (tolerance grade); manner of forming heads (machining, forging, cold heading); and the manner of forming threads (machining, die cutting, tapping, rolling, and plastic molding).

Thus, you can see that comprehensive treatment of threaded fasteners encompasses extensive data. See the References and Internet sites listed at the end of the chapter.

19–8 OTHER MEANS OF FASTENING AND JOINING

Thus far, this chapter has focused on screws and bolts because of their wide applications. Other types of fastening means will now be discussed.

Rivets are nonthreaded fasteners, usually made of steel or aluminum. They are originally made with one head, and the opposite end is formed after the rivet is inserted through holes in the parts to be joined. Steel rivets are formed hot, whereas aluminum can be formed at room temperatures. Of course, riveted joints are not designed to be assembled more than once. (See Internet sites 3 and 4.)

A large variety of *quick-operating fasteners* is available. Many are of the quarter-turn type, requiring just a 90° rotation to connect or disconnect the fastener. Access panels, hatches, covers, and brackets for removable equipment are attached with such fasteners. Similarly, many varieties of *latches* are available to provide quick action with, perhaps, added holding power. (See Internet sites 3 and 4.)

Welding involves the metallurgical bonding of metals, usually by the application of heat with an electric arc, a gas flame, or electrical resistance heating under heavy pressure. Welding is discussed in Chapter 20.

Brazing and *soldering* use heat to melt a bonding agent that flows into the space between parts to be joined, adhering to both parts and then solidifying as it cools. *Brazing* uses relatively high temperatures, above 840°F (450°C), using alloys of copper, silver, aluminum, silicon, or zinc. Of course, the metals to be joined must have a significantly higher melting temperature. Metals successfully brazed include plain carbon and alloy steels, stainless steels, nickel alloys, copper, aluminum, and magnesium. *Soldering* is similar to brazing, except that it is performed at lower temperatures, less than 840°F (450°C). Several soldering alloys of lead-tin, tin-zinc, tin-silver, lead-silver, zinc-cadmium, zinc-aluminum, and others are used. Brazed joints are generally stronger than soldered joints due to the inherently higher strength of the brazing alloys. Most soldered joints are fabricated with interlocking lap joints to provide mechanical strength, and then the solder is used to hold the assembly together and possibly to provide sealing. Joints in piping and tubing are frequently soldered.

Adhesives are seeing wide use. Versatility and ease of application are strong advantages of adhesives used in an array of products from toys and household appliances to automotive and aerospace structures. (See Internet sites 3 and 17.) Some types include the following:

Acrylics: Used for many metals and plastics.

Cyanoacrylates: Very fast curing; flow easily between well-mated surfaces.

Epoxies: Good structural strength; joint is usually rigid. Some require two-part formulations. A large variety of formulations and properties are available.

Anaerobics: Used for securing nuts and bolts and other joints with small clearances; cures in the absence of oxygen.

Silicones: Flexible adhesive with good high-temperature performance (400°F, 200°C).

Polyester hot melt: Good structural adhesive; easy to apply with special equipment.

Polyurethane: Good bonding; provides a flexible joint.

REFERENCES

1. ASTM International. *ISO Standards Handbooks—Fasteners and Screw Threads—Volumes 1&2.* West Conshohocken, PA: ASTM International, 2001.
2. ASTM International. *Publication STP 587, Metric Mechanical Fasteners.* West Conshohocken, PA: ASTM International, originally published 1975, updated 2011.
3. Barrett, R. T., and National Aeronautics and Space Administration (NASA). *Fastener Design Manual.* Seattle, WA: Create Space/Amazon.com, 2011.
4. Bickford, J. H. *Introduction to the Design and Behavior of Bolted Joints: Non-Gasketed Joints.* 4th ed. Boca Raton, FL: CRC Press, 2008. [See also Payne, Item 16.]
5. Bickford, J. H., and Sayed Nassar (eds). *Handbook of Bolts and Bolted Joints.* New York: Marcel Dekker, 1998.
6. Blake, Alexander. *Design of Mechanical Joints.* Boca Raton, FL: CRC Press, 1985. [Note: This is a reference book recommended by International Fasteners Institute.]
7. Blake, Alexander. *What Every Engineer Should Know About Threaded Fasteners—Materials and Design.* Boca Raton, FL: CRC Press, 1986. [Note: This is a reference book recommended by International Fasteners Institute.]
8. Fastener Technology International. *Torque Tensioning: A Ten Part Compilation.* Stow, OH: Fastener Technology International/ Initial Publications, Inc., 1990. [Note: This is a reference book recommended by International Fasteners Institute.]
9. Hoelscher, R. P., et al. *Graphics for Engineers.* New York: John Wiley & Sons, 1968.
10. Industrial Fasteners Institute. *IFI Inch Fastener Standards Book.* 8th ed. Independence, OH: Industrial Fasteners Institute, 2011.
11. International Organization for Standardization. *ISO Metric Screw Thread and Fastener Handbook.* 5th ed. Independence, OH: Industrial Fasteners Institute, 2001; Updated 2010.
12. Kulak, G. L., J. W. Fisher, and J. H. A. Struik. *Guide to Design Criteria for Bolted and Riveted Joints.* 2nd ed. New York: John Wiley & Sons, 1987.
13. Lincoln, Brayton, K. J. Gomes, and J. F. Branden. *Mechanical Fastening of Plastics: An Engineering Handbook.* Boca Raton, FL: CRC Press, 1984. [Note: This is a reference book recommended by International Fasteners Institute.]
14. Oberg, E., F. D. Jones, H. L. Horton, and H. H. Ryffel. *Machinery's Handbook.* 28th ed. New York: Industrial Press, 2008.
15. Parmley, R. O. *Standard Handbook of Fastening and Joining.* 3rd ed. New York: McGraw-Hill, 1997.
16. Payne, James R. *Introduction to the Design and Behavior of Bolted Joints: Gasketed Bolted Joints.* 4th ed. Boca Raton, FL: CRC Press, 2010. [See also Bickford, Item 4. Now separate books, originally published though 3rd ed. as one book by John H. Bickford.]
17. Research Council on Structural Connections. *Specification for Structural Joints Using High-Strength Bolts.* Chicago, IL: Research Council on Structural Connections, 2010.
18. SAE International. *SAE Standard J429 Mechanical and Material Requirements for Externally Threaded Fasteners.* Warrendale, PA: SAE International, 2011.
19. SAE International. *SAE Standard J1199 Mechanical and Material Requirements for Metric Externally Threaded Fasteners.* Warrendale, PA: SAE International, 2001.
20. ASTM International. *ASTM Standard A307, 2010 Standard Specifications for Carbon Steel Bolts and Studs, 60,000 psi Tensile Strength.* West Conshohocken, PA: ASTM International, 2010, DOI 10.1520/A0307-10.
21. ASTM International. *ASTM Standard A325, 2010 Standard Specifications Structural Bolts, Steel, 120/105 Minimum Tensile Strength.* West Conshohocken, PA: ASTM International, 2010, DOI 10.1520/A0325-10.
22. ASTM International. *ASTM Standard A354, 2007 Standard Specifications for Quenched and Tempered Alloy Steel Bolts, Studs and Other Externally Threaded Fasteners.* West Conshohocken, PA: ASTM International, 2007, DOI 10.1520/A0354-07A.
23. ASTM International. *ASTM Standard A449, 2010 Standard Specifications for Hex Cap Screws, Bolts and Studs, Steel, Heat Treated, 120/105/90 Minimum Tensile Strength, General Use.* West Conshohocken, PA: ASTM International, 2010, DOI 10.1520/A0449-10.
24. ASTM International. *ASTM Standard A490, 2010 Standard Specifications for Structural Bolts, Alloy Steel, Heat Treated,*

150 ksi Minimum Tensile Strength. West Conshohocken, PA: ASTM International, 2010, DOI 10.1520/A0490-10AE01.

25. ASTM International. *ASTM Standard A563, 2007 Standard Specifications for Carbon and Alloy Steel Nuts.* West Conshohocken, PA: ASTM International, 2007, DOI 10.1520/A0563-07A.
26. ASTM International. *ASTM Standard A563M, 2007 Standard Specifications for Carbon and Alloy Steel Nuts (Metric).* West Conshohocken, PA: ASTM International, 2007, DOI 10.1520/A0563M-07.
27. ASTM International. *ASTM Standard A574, 2011 Standard Specifications for Alloy Steel Socket Head Cap Screws.* West Conshohocken, PA: ASTM International, 2011, DOI 10.1520/A0574-11.
28. ASTM International. *ASTM Standard A574M, 2011 Standard Specifications for Alloy Steel Socket Head Cap Screws (Metric).* West Conshohocken, PA: ASTM International, 2011, DOI 10.1520/A0574M-11. [Metric companion to Specification A574.]

INTERNET SITES RELATED TO FASTENERS

1. **Industrial Fasteners Institute (IFI).** www.industrial-fasteners.org An association of manufacturers and suppliers of bolts, nuts, screws, rivets, and special formed parts and the materials and equipment to make them. IFI develops standards, organizes research, and conducts education programs related to the fasteners industry.
2. **Research Council on Structural Connections (RCSC).** www.boltcouncil.org An organization that stimulates and supports research on structural connections, prepares and publishes standards, and conducts educational programs. See Reference 12.
3. **Accurate Fasteners, Inc.** www.actfast.com A supplier of bolts, cap screws, nuts, rivets, and numerous other types of fasteners for general industry uses. Also provides adhesives for structural and nonstructural applications, thread locking adhesives, and tapes.
4. **The Fastener Group.** www.fastenergroup.com A supplier of bolts, cap screws, nuts, rivets, and numerous other types of fasteners for military and general industry uses.
5. **Haydon Bolts, Inc**. www.haydonbolts.com Manufacturer of bolts, nuts, and numerous other types of fasteners for the construction industry.
6. **Nucor Fastener Division.** www.nucor-fastener.com Manufacturer of hex head cap screws in SAE, ASTM, and metric grades, hex nuts, and structural bolts, nuts and washers.
7. **Nylok Fastener Corporation.** www.nylok.com Manufacturer of Nylok® self-locking fasteners for automotive, aerospace, consumer products, agricultural, industrial, furniture, and many other applications.
8. **Phillips Screw Company.** www.phillips-screw.com Developer of the Phillips® screwdriver. Manufacturer of related fasteners for the aerospace, automotive, and industrial markets.
9. **Unbrako Group.** www.unbrako.com Manufacturer of high-strength socket head cap screws and related products for industrial, aerospace, automotive, and other applications under the Unbrako and Durlok brands. Part of Deepak Fasteners, Ltd. and Interfast (USA). Engineering guide available on the website.
10. **St. Louis Screw & Bolt Company.** www.stlouisscrewbolt.com Manufacturer of bolts, nuts, and washers to ASTM standards for the construction industry. Offers direct tension indicator (DTI) washers.
11. **Acument® Global Technologies.** www.acumentnorthamerica.com Provider of fasteners and related products to the automotive, electronics, and other manufacturing industries under the Camcar® and Ring Screw brands. Products include cap screws, thread forming self-tapping screws, the Torx Plus® drive system, and other related products.
12. **Sensor Products, Inc.** www.sensorprod.com Provider of tactile thin film surface pressure and force sensors useful in designing and verifying the effectiveness of bolted joints, gaskets, seals, and related applications. Also offers consulting services to clients.
13. **Alcoa Fastening Systems/Huck Fasteners.** www.afshuck.net Manufacturer of Huck LockBolts and structural blind fasteners for the truck, trailer, agricultural, HVAC, aerospace, industrial equipment, and other markets, often replacing welding or traditional nut/bolt fastening methods.
14. **ASTM International.** www.astm.org/standards A standards-setting organization with numerous standards for engineering materials, fasteners, and associated products.
15. **SAE International.** www.sae.org/standards SAE International is a global body of scientists, engineers, and practitioners that advances self-propelled vehicle and system knowledge in a neutral forum for the benefit of society. The mission involves standards-setting activities mostly related to aerospace, automotive, and commercial vehicles.
16. **Hytorc Company.** www.hytorc.com Manufacturer of hydraulic and pneumatic automatic torque tensioners, mechanical tensioners, and *Stretch-to-Load* technology.
17. **Loctite Adhesives.** www.loctiteproducts.com Provider of adhesives for industry, construction, commercial, and household use, including epoxies, cyanoacrylates, contact cements, construction adhesives and the Loctite threadlocker. Part of Henkel Corporation that offers several dozen additional adhesive brands serving a wide range of uses. www.henkelna.com.

PROBLEMS

1. Describe the difference between a screw and a bolt.
2. Define the term *proof strength.*
3. Define the term *clamping load.*
4. Specify suitable machine screws to be installed in a pattern of four, equally spaced around a flange, if the clamping force between the flange and the mating structure is to be 6000 lb. Then recommend a suitable tightening torque for each screw.
5. What would be the tensile force in a machine screw having an 8–32 thread if it is made from SAE Grade 5 steel and is stressed to its proof strength?
6. What would be the tensile proof force in newtons (N) in a machine screw having a major diameter of 4 mm with standard fine threads if it is made from steel having a metric strength grade of 8.6?
7. What would be the nearest standard metric screw thread size to the American Standard 7/8–14 thread? By how much do their major diameters differ?

8. A machine screw is found with no information given as to its size. The following data are found by using a standard micrometer caliper: The major diameter is 0.196 in; the axial length for 20 full threads is 0.630 in. Identify the thread.
9. A threaded fastener is made from nylon 6/6 with an M10 × 1.5 thread. Compute the maximum tensile force that can be permitted in the fastener if it is to be stressed to 75% of the tensile strength of the nylon 66 dry. See Appendix 13.
10. Compare the tensile force that can be carried by a 1/4–20 screw if it is to be stressed to 50% of its tensile strength and if it is made from each of the following materials:
    - **(a)** Steel, SAE Grade 2
    - **(b)** Steel, SAE Grade 5
    - **(c)** Steel, SAE Grade 8
    - **(d)** Steel, ASTM Grade A307
    - **(e)** Steel, ASTM Grade A574
    - **(f)** Steel, metric Grade 8.8
    - **(g)** Aluminum 2024-T4
    - **(h)** SAE 430 annealed
    - **(i)** Ti–6A1–4V annealed
    - **(j)** Nylon 66 dry
    - **(k)** Polycarbonate
    - **(l)** High-impact ABS
11. Describe the differences among welding, brazing, and soldering.
12. What types of metals are typically brazed?
13. What are some common brazing alloys?
14. What materials make up commonly used solders?
15. Name five common adhesives, and give the typical properties of each.
16. The label of a common household adhesive describes it as a *cyanoacrylate*. What would you expect its properties to be?
17. Find three commercially available adhesives from your home, a laboratory, a machine shop, or your workplace. Try to identify the generic nature of the adhesive, and compare it with the list presented in this chapter.

CHAPTER TWENTY

MACHINE FRAMES, BOLTED CONNECTIONS, AND WELDED JOINTS

THE BIG PICTURE — Machine Frames, Bolted Connections, and Welded Joints

Discussion Map

- As you develop the design of machine elements (as you have throughout this book), you must also design the housing, frame, or structure that supports them and protects them from the elements.

Discover

Select a variety of products, machines, vehicles, and even toys. Observe how they are built. What is the basic shape of the structure that holds everything together? Why was that shape chosen by the designer? What functions are performed by the frame?

What kinds of forces, bending moments, and torsional moments (torques) are produced when the product is operating? How are they managed and controlled? What is the load path that delivers them to the ultimate structure?

This chapter will help you identify some efficient approaches to the design of structures and frames and to analyze the performance of fasteners and welded joints loaded in many different ways.

Thus far in this book, you have studied individual machine elements while simultaneously considering how those elements must work together in a more comprehensive machine. As the design progresses, there comes a time when *you must put it all together.* But then you are faced with the questions, "What do I put it in? How do I hold all of the functional components safely, allowing assembly and service while providing a secure, rigid structure?"

It is impractical to generate a completely general approach to the design of a structure or a frame for a machine, a vehicle, a consumer product, or even a toy. Each is different with regard to its functions; the number, size, and type of components in the product; the intended use; and the expected demand for an aesthetic design. For example, toys often exhibit clever design approaches because the manufacturer wants to provide a safe, functional toy while minimizing the material used and the amount of personnel time required to produce the toy.

In this chapter you will explore some basic concepts for creating a satisfactory frame design, considering the shape of structural components, material properties, the use of fasteners such as bolts, and the fabrication of welded assemblies. You will learn some of the techniques for analyzing and designing bolted assemblies to consider loads on the bolts in several directions. The design of welded joints to be safe and rigid is discussed.

Chapter 19 covered part of the story, that dealing with bolts loaded in pure tension, as in a clamping function. This chapter extends that chapter to consider eccentrically loaded joints: those that must resist a combination of direct shear and a bending moment on a bolt pattern.

The ability of a welded joint to carry a variety of loads is discussed with the objective of designing the weld. Here both uniformly loaded and eccentrically loaded joints are treated.

To appreciate the value of such study, select a variety of products, machines, and vehicles, and observe how they are built. What is the basic shape of the structure that holds everything together? Where are forces, bending moments, and torsional moments (torques) generated? What kinds of stress do they create? Consider the *load path* by following a force from the place where it is generated through all of the means by which that force or its effects are passed to a series of members and to the point where it is supported by a basic frame of the machine or where it is delivered out of the system of interest. Considering the load paths for all forces that exist in a mechanical device should give you an understanding of the desirable characteristics of the structure and should help you know how to produce a design that optimizes the management of the forces exerted on it.

Take apart a variety of mechanical devices to observe their structure. How did shape contribute to the rigidity and safety of the device? Is it rather stiff? Or is it more flexible? Are there ribs or thickened sections to provide special strength or rigidity for certain parts?

This chapter will help you identify some efficient approaches to the design of structures and frames and to analyze the performance of fasteners and welded joints loaded in many ways.

The subject of machine frames and structures is quite complex. It is discussed from the standpoint of general principles and guidelines, rather than specific design techniques. Critical frames are typically designed with computerized finite-element analysis. Also, experimental stress analysis techniques are often used to verify designs.

YOU ARE THE DESIGNER

In Chapter 16, you were the designer of bearings for conveyor systems for a large product distribution center. Now, how do you design the frame and the structure of the conveyor system? What general form is desirable? What materials and shape should be used for the structural elements? Are the elements loaded in tension, compression, bending, shear, torsion, or some combination of these types of stress? How do the manner of loading and the nature of the stress in the structure affect these decisions? Should the frame be fabricated from standard structural shapes and bolted together? Or should it be made from steel plate and welded? How about using aluminum? Or should it be cast from cast iron or cast steel? Can it be molded from plastic? Can composites be used? How would the weight of the structure be affected? How much rigidity is desirable for this kind of structure? What shapes of load-carrying elements contribute to a stiff, rigid structure, as well as one that is safe in resisting applied stresses?

If the structure involves bolting or welding, how are the joints to be designed? What forces, both magnitude and direction, must be carried by the fasteners or the welds?

The material in this chapter will help you make some of these design decisions. Much of the information is general in nature rather than giving you specific design procedures. You must exercise judgment and creativity not only in the design of the conveyor frame but also in the analysis of its components. Because the frame design could evolve into a form too complex for analysis using traditional techniques of stress analysis, you may have to employ finite-element modeling to determine whether the design is adequate or, perhaps, overdesigned. Perhaps one or more prototypes should be built for testing. ■

20–1 OBJECTIVES OF THIS CHAPTER

After completing this chapter, you will be able to:

1. Apply the principles of stress and deflection analysis to propose a reasonable and efficient shape for a structure or frame and for the components involved.
2. Specify materials that are well suited to the demands of a given design, given certain conditions of load, environment, fabrication requirements, safety, and aesthetics.
3. Analyze eccentrically loaded bolted joints.
4. Design welded joints to carry many types of loading patterns.

20–2 MACHINE FRAMES AND STRUCTURES

The design of machine frames and structures is largely art in that the components of the machine must be accommodated. The designer is often restricted in where supports can be placed in order not to interfere with the operation of the machine or in order to provide access for assembly or service.

But, of course, technical requirements must be met as well for the structure itself. Some of the more important design parameters include the following:

Strength	Stiffness
Appearance	Cost to manufacture
Corrosion resistance	Weight
Size	Noise reduction
Vibration limitation	Life

Because of the virtually infinite possibilities for design details for frames and structures, this section will concentrate on general guidelines. The implementation of the guidelines would depend on the specific application. Factors to consider in starting a design project for a frame are now summarized:

- Forces exerted by the components of the machine through mounting points such as bearings, pivots, brackets, and feet of other machine elements
- Manner of support of the frame itself
- Precision of the system: allowable deflection of components
- Environment in which the unit will operate
- Quantity of production and facilities available
- Availability of analytical tools such as computerized stress analysis, past experience with similar products, and experimental stress analysis
- Relationship to other machines, walls, and so on

Again, many of these factors require judgment by the designer. The parameters over which the designer has the most control are material selection, the geometry of load-carrying parts of the frame, and manufacturing processes. A review of some possibilities follows.

Materials

As with machine elements discussed throughout this book, the material properties of strength and stiffness are of prime importance. Chapter 2 presented an extensive amount of information about materials, and the Appendices contain much useful information. In general, steel ranks high in strength compared with competing materials for frames. But it is often better to consider more than just yield strength, ultimate tensile strength, or endurance strength alone. The complete design can be executed in several candidate materials to evaluate the overall performance. Considering the *ratio of strength to density,* sometimes referred to as the *strength-to-weight ratio* or *specific strength,* may lead to a different material selection. Indeed, this is one reason for the use of aluminum, titanium, and composite materials in aircraft, aerospace vehicles, and transportation equipment.

Rigidity of a structure or a frame is frequently the determining factor in the design, rather than strength. In these cases, the stiffness of the material, indicated by its modulus of elasticity, is the most important factor. Again, the *ratio of stiffness to density,* called *specific stiffness,* may need to be evaluated. See Table 2–17 and Figures 2–23, 2–24, 2–31, and 2–32 for data.

Recommended Deflection Limits

Actually, only intimate knowledge of the application of a machine member or a frame can give a value for an acceptable deflection. But some guidelines are available to give you a place to start. (See Reference 7.)

Deflection Due to Bending

General machine part: 0.0005 to 0.003 in/in of beam length

Moderate precision: 0.00001 to 0.0005 in/in

High precision: 0.000 001 to 0.000 01 in/in

Deflection (Rotation) Due to Torsion

General machine part: 0.001° to 0.01°/in of length

Moderate precision: 0.000 02° to 0.0004°/in

High precision: 0.000 001° to 0.000 02°/in

Suggestions for Design to Resist Bending

Scrutiny of a table of deflection formulas for beams in bending such as those in Appendix 14 would yield the following form for the deflection:

$$\Delta = \frac{PL^3}{KEI} \tag{20–1}$$

where P = load
L = length between supports
E = modulus of elasticity of the material in the beam
I = moment of inertia of the cross section of the beam
K = a factor depending on the manner of loading and support

Some obvious conclusions from Equation (20–1) are that the load and the length should be kept small, and the values of E and I should be large. Note the cubic function of the length. This means, for example, that reducing the length by a factor of 2.0 would reduce the deflection by a factor of 8.0, obviously a desirable effect.

Figure 20–1 shows the comparison of four types of beam systems to carry a load, P, at a distance, a, from a rigid support. A beam simply supported at each end is taken as the "basic case." Using standard beam formulas, we computed the value of the bending moment and the deflection in terms of P and a, and these values were arbitrarily normalized to be 1.0. Then the values for the three other cases were computed and ratios determined relative to the basic case. The data show that a fixed-end beam gives both the lowest bending moment and the lowest deflection, while the cantilever gives the highest values for both.

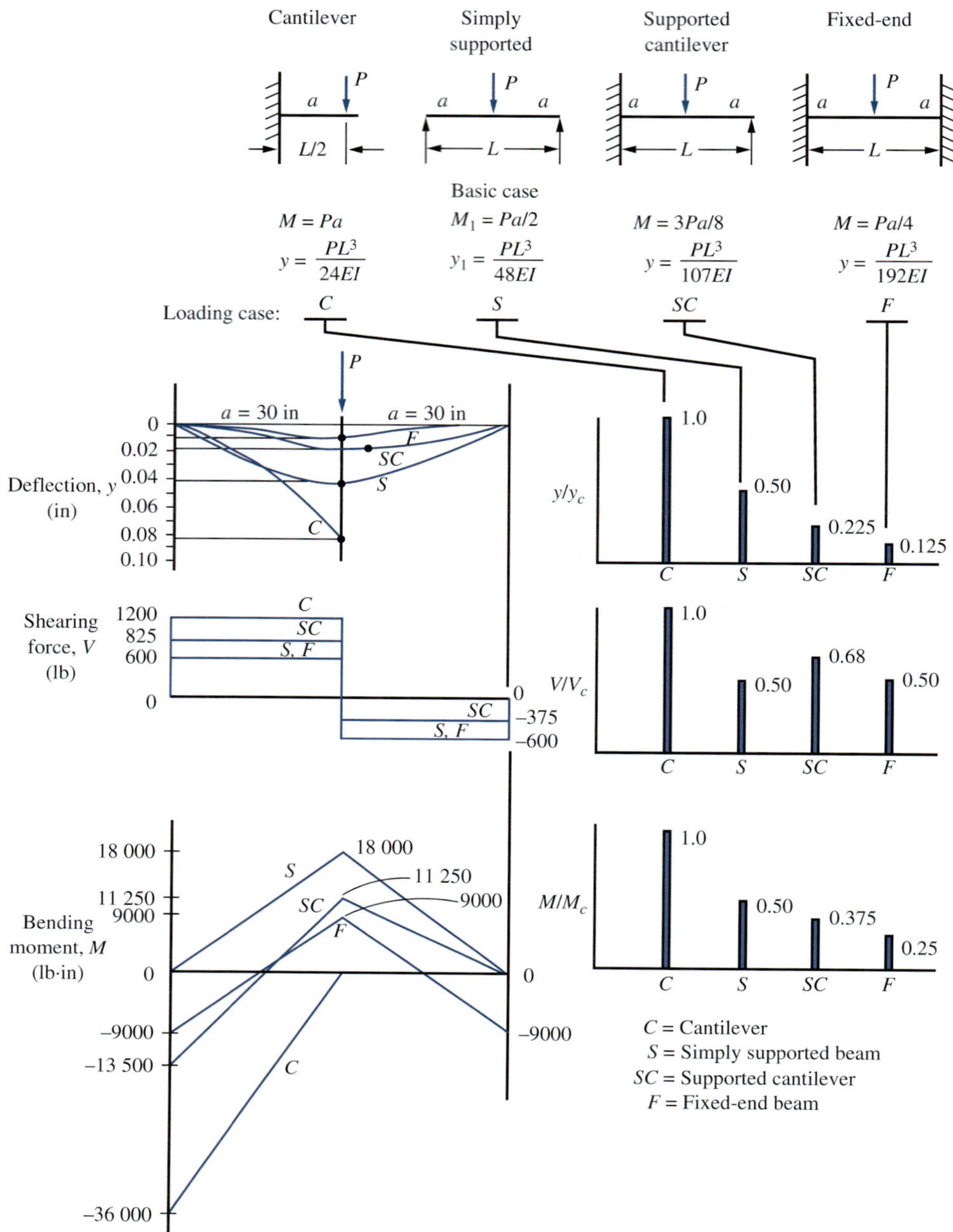

FIGURE 20–1 Comparison of methods of supporting a load on a beam (Reference 17)

In summary, the following suggestions are made for designing to resist bending:

1. Keep the length of the beam as short as possible, and place loads close to the supports.
2. Maximize the moment of inertia of the cross section in the direction of bending. In general, you can do so by placing as much of the material as far away from the neutral axis of bending as possible, as in a wide-flange beam or a hollow rectangular section.
3. Use a material with a high modulus of elasticity.
4. Use fixed ends for the beam where possible.
5. Consider lateral deflection in addition to deflection in the primary load direction. Such loads could be encountered during fabrication, handling, shipping, careless use, or casual bumping.
6. Be sure to evaluate the final design with regard to both strength and rigidity. Some approaches to improving rigidity (increasing I) can actually increase the stress in the beam because the section modulus is decreased.
7. Provide rigid corner bracing in open frames.
8. Cover an open frame section with a sheet material to resist distortion. This process is sometimes called *panel stiffening*.

9. Consider a truss-type construction to obtain structural stiffness with lightweight members.
10. When designing an open space frame, use diagonal bracing to break sections into triangular parts, an inherently rigid shape.
11. Consider stiffeners for large panels to reduce vibration and noise. See Figure 2–29.
12. Add bracing and gussets to areas where loads are applied or at supports to help transfer the forces into adjoining members.
13. Beware of load-carrying members with thin, extended flanges that may be placed in compression. Local buckling, sometimes called *crippling* or *wrinkling,* could occur.
14. Place connections at points of low stress if possible.

See also References 1, 4, 6, 7, and 10 for additional design and analysis techniques.

Suggestions for Design of Members to Resist Torsion

Torsion can be created in a machine frame member in a variety of ways: A support surface may be uneven; a machine or a motor may transmit a reaction torque to the frame; a load acting to the side of the axis of the beam (or any place away from the flexural center of the beam) would produce twisting.

In general, the torsional deflection of a member is computed from

$$\theta = \frac{TL}{GR} \tag{20–2}$$

where T = applied torque or twisting moment
L = length over which torque acts
G = shear modulus of elasticity of the material
R = torsional rigidity constant

The designer must choose the shape of the torsion member carefully to obtain a rigid structure. The following suggestions are made:

1. Use closed sections wherever possible. Examples are solid bars with large cross section, hollow pipe and tubing, closed or hollow rectangular or square tubing, and special closed shapes that approximate a tube.
2. Conversely, avoid open sections made from thin materials. Figure 20–2 shows a dramatic illustration.
3. For wide frames, brackets, tables, bases, and so on, use diagonal braces placed at 45° to the sides of the frame (see Figure 20–3).
4. Use rigid connections, such as by welding members together.

Most of the suggestions made in this section can be implemented regardless of the specific type of frame designed:

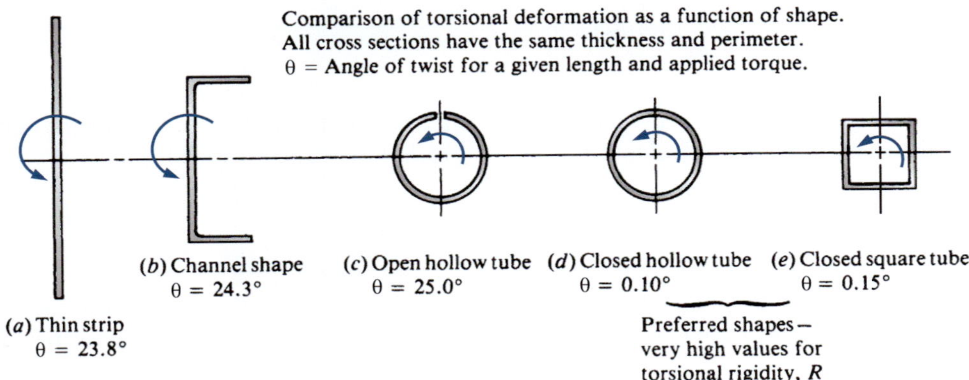

FIGURE 20–2 Comparison of torsional deformation as a function of shape

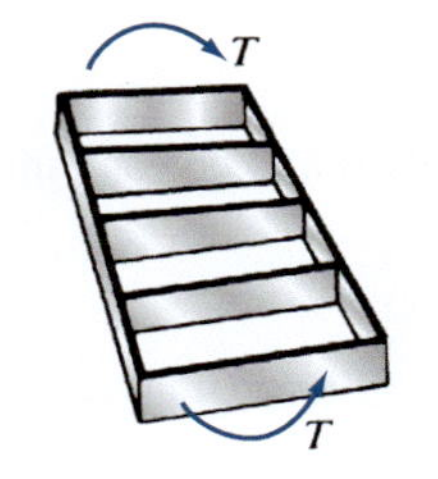

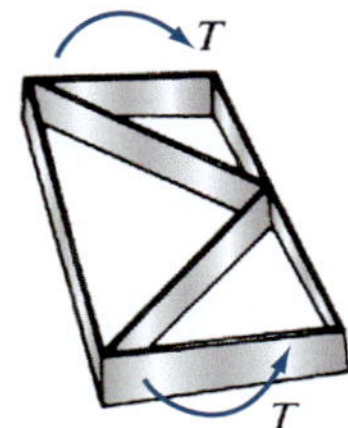

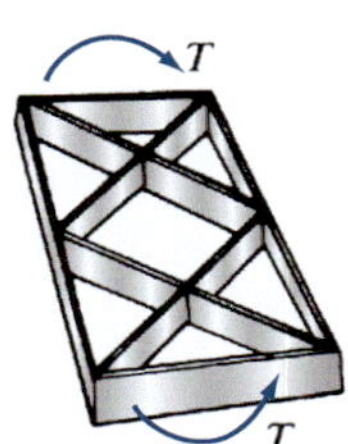

(*a*) Conventional cross bracing
θ = 10.8°

(*b*) Single diagonal bracing
θ = 0.30°

(*c*) Double diagonal bracing
θ = 0.10°

FIGURE 20–3 Comparison of torsional angle of twist, θ, for boxlike frames. Each has the same basic dimensions and applied torque

castings made from cast iron, cast steel, aluminum, zinc, or magnesium; weldments made from steel or aluminum plate; formed housings from sheet metal or plate; or plastic moldings. The references in this chapter provide valuable additional guidance as you complete the design of frames, structures, and housings. See References 1, 4, 6, 7, 10–12, and 18. See also Internet sites 1–9.

20–3 ECCENTRICALLY LOADED BOLTED JOINTS

Figure 20–4 shows an example of an eccentrically loaded bolted joint. The motor on the extended bracket places the bolts in shear because its weight acts directly downward. But there also exists a moment equal to $P \times a$ that must be resisted. The moment tends to rotate the bracket and thus to shear the bolts.

The basic approach to the analysis and design of eccentrically loaded joints is to determine the forces that act on each bolt because of all the applied loads. Then, by a process of superposition, the loads are combined vectorially to determine which bolt carries the greatest load. That bolt is then sized. The method will be illustrated in Example Problem 20–1.

The American Institute of Steel Construction (AISC) lists allowable stresses for bolts made from ASTM grade steels, as shown in Table 20–1. These data are for bolts used in standard-sized holes, 1/16 in larger than the bolt. Also, a *friction-type connection,* in which the clamping force is sufficiently large that the friction between the mating parts helps carry some of the shear load, is assumed. (See References 1–3, 14, 19, 20, and 22.)

In the design of bolted joints, you should ensure that there are no threads in the plane where shear occurs. The body of the bolt will then have a diameter equal to the major diameter of the thread. You can use the tables in Chapter 19 to select the standard size for a bolt.

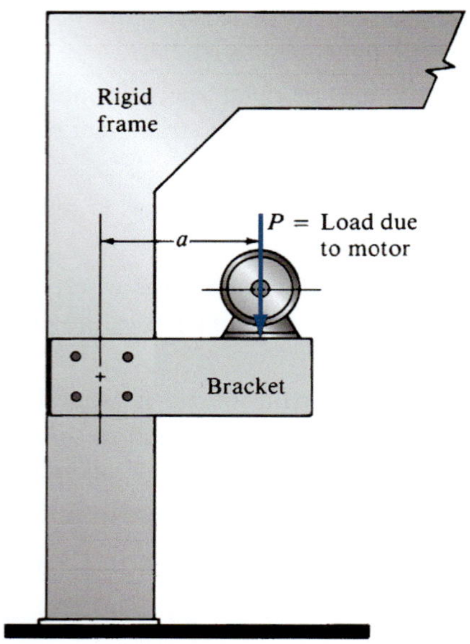

FIGURE 20–4 Eccentrically loaded bolted joint

TABLE 20–1 Allowable stresses for bolts

ASTM grade	Allowable shear stress	Allowable tensile stress
A307	10 ksi (69 MPa)	20 ksi (138 MPa)
A325 and A449	17.5 ksi (121 MPa)	44 ksi (303 MPa)
A490	22 ksi (152 MPa)	54 ksi (372 MPa)

Example Problem 20–1 For the bracket in Figure 20–4, assume that the total force P is 3500 lb and the distance a is 12 in. Design the bolted joint, including the location and number of bolts, the material, and the diameter.

Solution The solution shown is an outline of a procedure that can be used to analyze similar joints. The data of this problem illustrate the procedure.

Step 1. Propose the number of bolts and the pattern. This is a design decision, based on your judgment and the geometry of the connected parts. In this problem, let's try a pattern of four bolts placed as shown in Figure 20–5.

Step 2. Determine the direct shear force on the bolt pattern and on each individual bolt, assuming that all bolts share the shear load equally:

$$\text{Shear load} = P = 3500 \text{ lb}$$
$$\text{Load per bolt} = F_s = P/4 = 3500 \text{ lb}/4 = 875 \text{ lb/bolt}$$

The shear force acts directly downward on each bolt.

Step 3. Compute the **moment** to be resisted by the bolt pattern: the product of the overhanging load and the distance to the **centroid** of the bolt pattern. In this problem, $M = P \times a = (3500 \text{ lb})(12 \text{ in}) = 42\ 000 \text{ lb} \cdot \text{in}$.

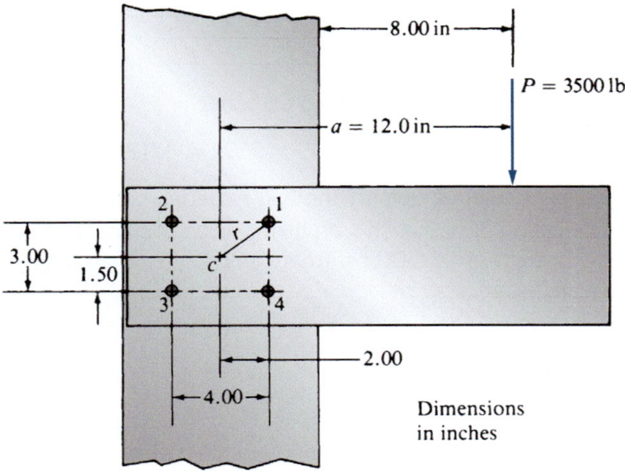

(a) Proposed bolt pattern

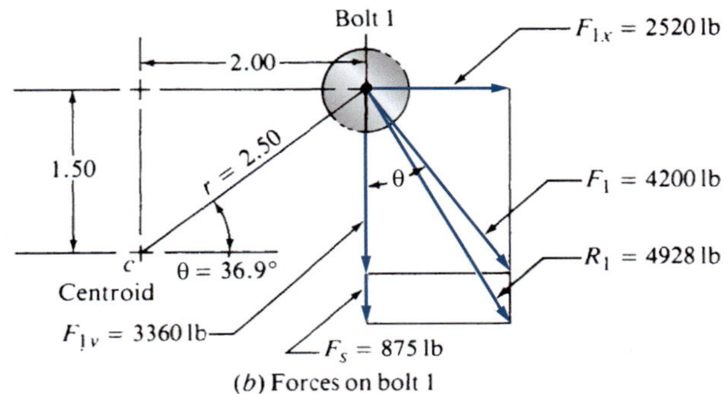

(b) Forces on bolt 1

FIGURE 20–5 Geometry of bolted joint and forces on bolt 1

Step 4. Compute the radial distance from the centroid of the bolt pattern to the center of each bolt. In this problem, each bolt has a radial distance of

$$r = \sqrt{(1.50 \text{ in})^2 + (2.00 \text{ in})^2} = 2.50 \text{ in}$$

Step 5. Compute the sum of the *squares* of all radial distances to all bolts. In this problem, all four bolts have the same r. Then

$$\Sigma r^2 = 4(2.50 \text{ in})^2 = 25.0 \text{ in}^2$$

Step 6. Compute the force on each bolt required to resist the bending moment from the relation

$$F_i = \frac{Mr_i}{\Sigma r^2} \tag{20–3}$$

where r_i = radial distance from the centroid of the bolt pattern to the ith bolt
F_i = force on the ith bolt due to the moment. The force acts perpendicular to the radius.

In this problem, all such forces are equal. For example, for bolt 1,

$$F_1 = \frac{Mr_1}{\Sigma r^2} = \frac{(42\,000 \text{ lb} \cdot \text{in})(2.50 \text{ in})}{25.0 \text{ in}^2} = 4200 \text{ lb}$$

Step 7. Determine the resultant of all forces acting on each bolt. A vector summation can be performed either analytically or graphically, or each force can be resolved into horizontal and vertical components. The components can be summed and then the resultant can be computed.

Let's use the latter approach for this problem. The shear force acts directly downward, in the y-direction. The x- and y-components of F_1 are

$$F_{1x} = F_1 \sin\theta = (4200 \text{ lb})\sin(36.9°) = 2520 \text{ lb}$$
$$F_{1y} = F_1 \cos\theta = (4200)\cos(36.9°) = 3360 \text{ lb}$$

The total force in the y-direction is then

$$F_{1y} + F_s = 3360 + 875 = 4235 \text{ lb}$$

Then the resultant force on bolt 1 is

$$R_1 = \sqrt{(2520)^2 + (4235)^2} = 4928 \text{ lb}$$

Step 8. Specify the bolt material; compute the required area for the bolt; and select an appropriate size. For this problem, let's specify ASTM A325 steel for the bolts having an allowable shear stress of 17 500 psi from Table 20–1. Then the required area for the bolt is

$$A_s = \frac{R_1}{\tau_a} = \frac{4928 \text{ lb}}{17\,500 \text{ lb/in}^2} = 0.282 \text{ in}^2$$

The required diameter would be

$$D = \sqrt{\frac{4A_s}{\pi}} = \sqrt{\frac{4(0.282 \text{ in}^2)}{\pi}} = 0.599 \text{ in}$$

Let's specify a 5/8-in bolt having a diameter of 0.625 in.

20–4 WELDED JOINTS

The design of welded joints requires consideration of the manner of loading on the joint, the types of materials in the weld and in the members to be joined, and the geometry of the joint itself. The load may be either uniformly distributed over the weld such that all parts of the weld are stressed to the same level, or the load may be eccentrically applied. Both are discussed in this section.

The materials of the weld and the parent members determine the allowable stresses. Table 20–2 lists several

TABLE 20–2 Allowable shear stresses on fillet welds for steel and aluminum

A. Steel

Electrode type	Typical metals joined (ASTM grade)	Allowable shear stress
E60	A36, A500	18 ksi (124 MPa)
E70	A242, A441	21 ksi (145 MPa)
E80	A572, Grade 65	24 ksi (165 MPa)
E90		27 ksi (186 MPa)
E100		30 ksi (207 MPa)
E110		33 ksi (228 MPa)

B. Aluminum

	Filler Alloy							
	1100		4043		5356		5556	
	Allowable Shear Stress							
Metal joined	**ksi**	**MPa**	**ksi**	**MPa**	**ksi**	**MPa**	**ksi**	**MPa**
1100	3.2	22	4.8	33				
3003	3.2	22	5.0	34				
6061			5.0	34	7.0	48	8.5	59
6063			5.0	34	6.5	45	6.5	45

examples for steel and aluminum. The allowables listed are for shear on fillet welds. For steel, welded by the electric arc method, the type of electrode is an indication of the tensile strength of the filler metal. For example, the E70 electrode has a minimum tensile strength of 70 ksi (483 MPa). Additional data are available in publications of the American Welding Society (AWS), the American Institute for Steel Construction (AISC), and the Aluminum Association (AA). See Reference 1 and Internet sites 3, 7, and 10.

Types of Joints

Joint type refers to the relationship between mating parts, as illustrated in Figure 20–6. The butt weld allows a joint to be the same nominal thickness as the mating parts and is usually loaded in tension. If the joint is properly made with the appropriate weld metal, the joint will be stronger than the parent metal. Thus, no special analysis of the joint is required if the joined members themselves are shown to be safe. Caution is advised, however, when the materials to be joined are adversely affected by the heat of the welding process. Heat-treated steels and many aluminum alloys are examples. The other types of joints in Figure 20–6 are assumed to place the weld in shear.

Types of Welds

Figure 20–7 shows several types of welds named for the geometry of the edges of the parts to be joined. Note the special edge preparation required, especially for thick plates, to permit the welding rod to enter the joint and build a continuous weld bead.

Size of Weld

The five types of groove-type welds in Figure 20–7 are made as complete penetration welds. Then, as indicated before for

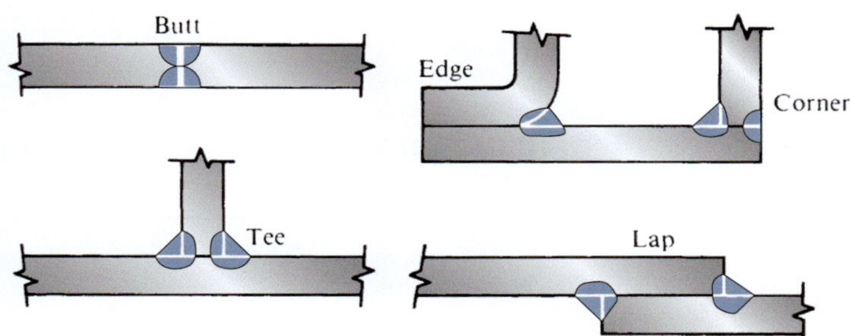

FIGURE 20–6 Types of weld joints

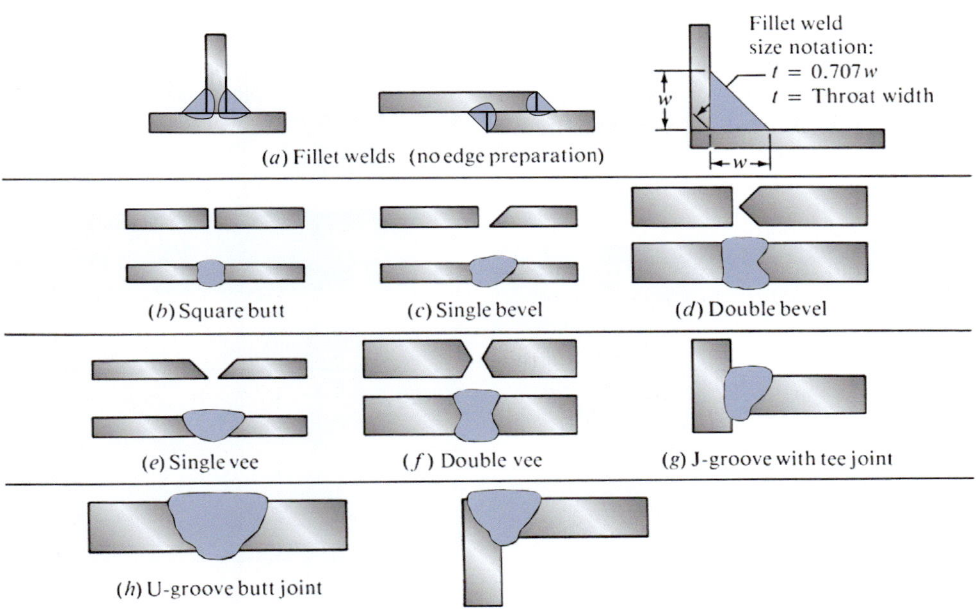

FIGURE 20–7 Some types of welds showing edge preparation

butt welds, the weld is stronger than the parent metals, and no further analysis is required.

Fillet welds are typically made as equal-leg right triangles, with the size of the weld indicated by the length of the leg. A fillet weld loaded in shear would tend to fail along the shortest dimension of the weld that is the line from the root of the weld to the theoretical face of the weld and normal to the face. The length of this line is found from simple trigonometry to be $0.707w$, where w is the leg dimension.

The objectives of the design of a fillet welded joint are to specify the length of the legs of the fillet; the pattern of the weld; and the length of the weld. Presented here is the method that treats the weld as a line having no thickness. The method involves determining the *maximum force per inch* of weld leg length. Comparing the actual force with an allowable force allows the calculation of the required leg length.

Table 20–3 gives data for the allowable shear stress and the allowable force per inch for some combinations of base metal and welding electrode. In general, the allowables for building-type structures are for steady loads. The values for bridge-type loading accounts for the cyclic effects. For true fatigue-type repeated loading, refer to the literature. (See References 1, 6–10, 13, 15, 16, 21, and 23.)

Method of Treating Weld as a Line

Four different types of loading are considered here: (1) direct tension or compression, (2) direct vertical shear, (3) bending, and (4) twisting. The method allows the designer to perform calculations in a manner very similar to that used to design the load-carrying members themselves. In general, the weld is analyzed separately for each type of loading to determine the force per inch of weld size due to each load. The loads are then combined vectorially to determine the maximum force. This maximum force is compared with the allowables from Table 20–3 to determine the size of the weld required. See References 6 and 7.

TABLE 20–3 Allowable shear stresses and forces on welds

Base metal ASTM grade	Electrode	Allowable shear stress	Allowable force per inch of leg
Building-type structures:			
A36, A441	E60	13 600 psi	9600 lb/in
A36, A441	E70	15 800 psi	11 200 lb/in
Bridge-type structures:			
A36	E60	12 400 psi	8800 lb/in
A441, A242	E70	14 700 psi	10 400 lb/in

The relationships used are summarized next:

Type of loading	Formula (and equation number) for force per inch of weld	
Direct tension or compression	$f = P/A_w$	(20–4)
Direct vertical shear	$f = V/A_w$	(20–5)
Bending	$f = M/S_w$	(20–6)
Twisting	$f = Tc/J_w$	(20–7)

In these formulas, the geometry of the weld is used to evaluate the terms A_w, S_w, and J_w, using the relationships shown in Figure 20–8. Note the similarity between these formulas and those used to perform the stress analysis. Note, also, the similarity between the geometry factors for welds and the properties of areas used for the stress analysis. Because the weld is treated as a line having no thickness, the units for the geometry factors are different from those of the area properties, as indicated in Figure 20–8.

The use of this method of weld analysis will be demonstrated with example problems. In general, the method requires the steps in the *General procedure for Designing Welded Joints.*

GENERAL PROCEDURE FOR DESIGNING WELDED JOINTS

1. Propose the geometry of the joint and the design of the members to be joined.
2. Identify the types of stresses to which the joint is subjected (bending, twisting, vertical shear, direct tension, or compression).
3. Analyze the joint to determine the magnitude and the direction of the force on the weld due to each type of load.
4. Combine the forces vectorially at the point or points of the weld where the forces appear to be maximum.
5. Divide the maximum force on the weld by the allowable force from Table 20–3 to determine the required leg size for the weld. Note that when thick plates are welded, there are minimum acceptable sizes for the welds as listed in Table 20–4.

TABLE 20–4 Minimum weld sizes for thick plates

Plate thickness (in)	Minimum leg size for fillet weld (in)
≤1/2	3/16
>1/2–3/4	1/4
>3/4–$1\frac{1}{2}$	5/16
>$1\frac{1}{2}$–$2\frac{1}{4}$	3/8
>$2\frac{1}{4}$–6	1/2
>6	5/8

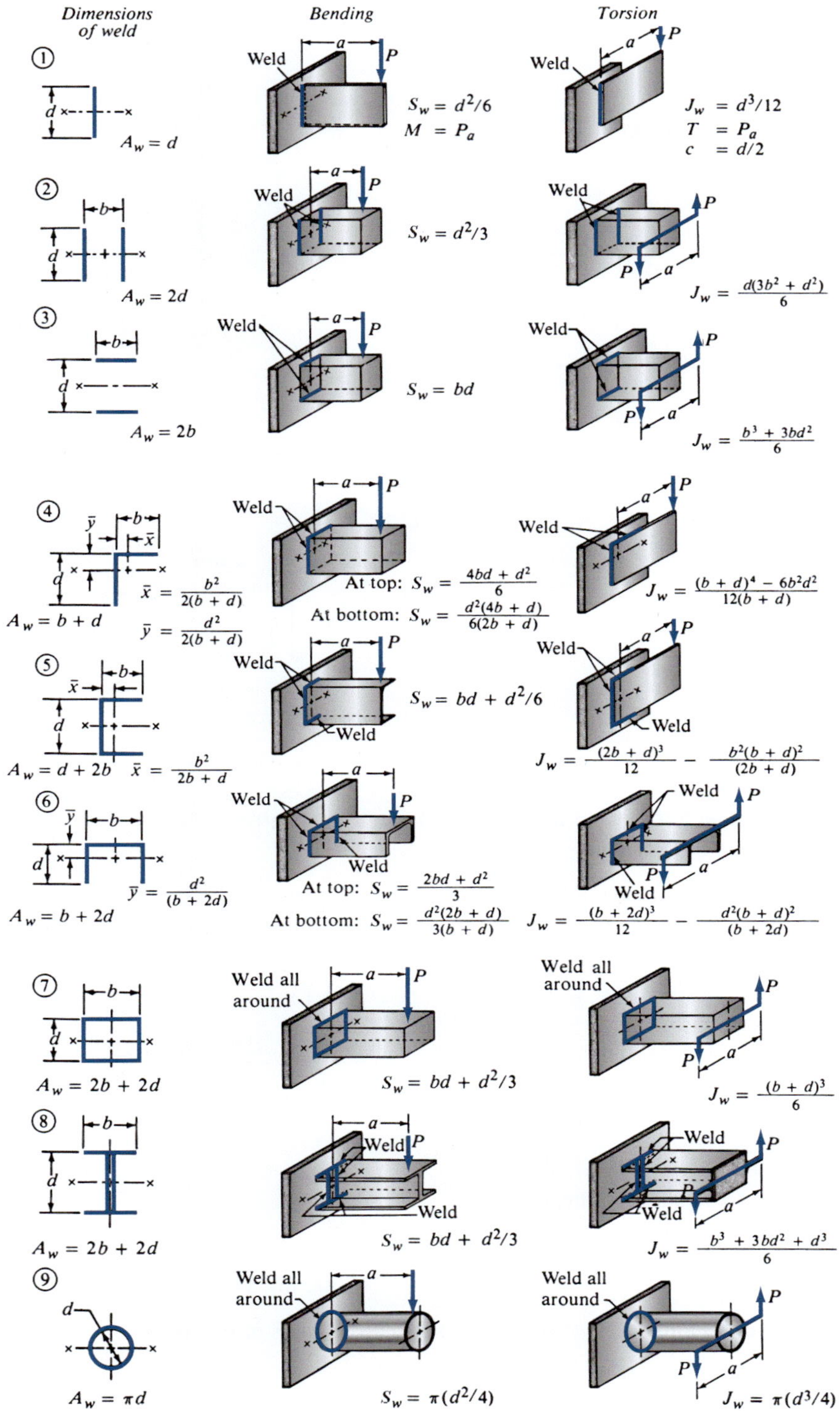

FIGURE 20–8 Geometry factors for weld analysis

Example Problem 20–2

Design a bracket similar to that in Figure 20–4, but use welding to attach the bracket to the column. The bracket is 6.00 in high and is made from ASTM A36 steel having a thickness of 1/2 in. The column is also made from A36 steel and is 8.00 in wide.

Solution

Step 1. The proposed geometry is a design decision and may have to be subjected to some iteration to achieve an optimum design. For a first trial, let's use the C-shaped weld pattern shown in Figure 20–9.

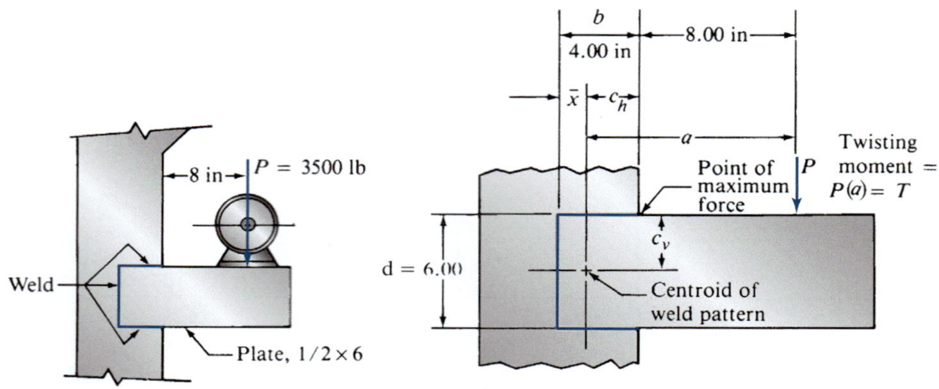

(a) Basic design of bracket

(b) Dimensions of bracket

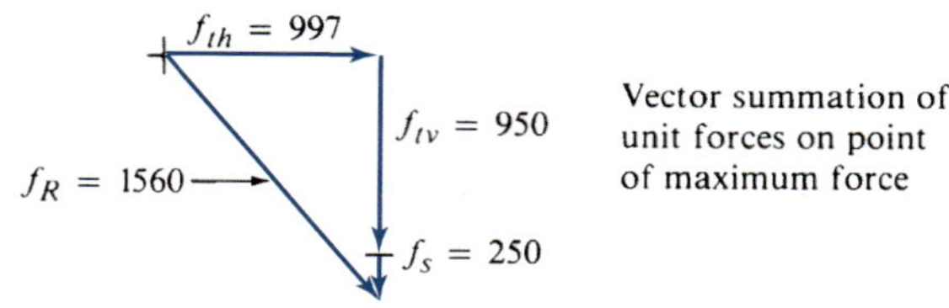

(c) Analysis of forces

FIGURE 20–9 C-shaped weld bracket

Step 2. The weld will be subjected to direct vertical shear and twisting caused by the 3500-lb load on the bracket.

Step 3. To compute the forces on the weld, we must know the geometry factors A_w and J_w. Also, the location of the centroid of the weld pattern must be computed [see Figure 20–9(b)]. Use Case 5 in Figure 20–8.

$$A_w = 2b + d = 2(4) + 6 = 14 \text{ in}$$

$$J_w = \frac{(2b + d)^3}{12} - \frac{b^2(b + d)^2}{(2b + d)} = \frac{(14)^3}{12} - \frac{16(10)^2}{14} = 114.4 \text{ in}^3$$

$$\bar{x} = \frac{b^2}{2b + d} = \frac{16}{14} = 1.14 \text{ in}$$

Force Due to Vertical Shear

$$V = P = 3500 \text{ lb}$$

$$f_s = P/A_w = (3500 \text{ lb})/14 \text{ in} = 250 \text{ lb/in}$$

This force acts vertically downward on all parts of the weld.

Forces Due to the Twisting Moment

$$T = P[8.00 + (b - \bar{x})] = 3500[8.00 + (4.00 - 1.14)]$$

$$T = 3500(10.86) = 38\,010 \text{ lb}\cdot\text{in}$$

The twisting moment causes a force to be exerted on the weld that is perpendicular to a radial line from the centroid of the weld pattern to the point of interest. In this case, the end of the weld to the upper right experiences the greatest force. It is most convenient to break the force down into horizontal and vertical components and then subsequently recombine all such components to compute the resultant force:

$$f_{th} = \frac{Tc_v}{J_w} = \frac{(38\,010)(3.00)}{114.4} = 997 \text{ lb/in}$$

$$f_{tv} = \frac{Tc_h}{J_w} = \frac{(38\,010)(2.86)}{114.4} = 950 \text{ lb/in}$$

Step 4. The vectorial combination of the forces on the weld is shown in Figure 20–9(c). Thus, the maximum force is 1560 lb/in.

Step 5. Selecting an E60 electrode for the welding, we find that the allowable force per inch of weld leg size is 9600 lb/in (Table 20–3). Then the required weld leg size is

$$w = \frac{1560 \text{ lb/in}}{9600 \text{ lb/in per in of leg}} = 0.163 \text{ in}$$

Table 20–4 shows that the minimum size weld for a 1/2-in plate is 3/16 in (0.188 in). That size should be specified.

Example Problem 20–3

A steel strap, 1/4 in thick, is to be welded to a rigid frame to carry a dead load of 12 500 lb, as shown in Figure 20–10. Design the strap and its weld.

Solution

The basic objectives of the design are to specify a suitable material for the strap, the welding electrode, the size of the weld, and the dimensions W and h, as shown in Figure 20–10.

Let's specify that the strap is to be made from ASTM A441 structural steel and that it is to be welded with an E70 electrode, using the minimum size weld, 3/16 in. Appendix 7 gives the yield strength of the A441 steel as 42 000 psi. Using a design factor of 2, we can compute an allowable stress of

$$\sigma_a = 42000/2 = 21000 \text{psi}$$

Then the required area of the strap is

$$A = \frac{P}{\sigma_a} = \frac{12\,500 \text{ lb}}{21\,000 \text{ lb/in}^2} = 0.595 \text{ in}^2$$

But the area is $W \times t$, where $t = 0.25$ in. Then the required width W is

$$W = A/t = 0.595/0.25 = 2.38 \text{ in}$$

Let's specify that $W = 2.50$ in

To compute the required length of the weld h, we need the allowable force on the 3/16-in weld. Table 20–3 indicates the allowable force on the A441 steel welded with an E70 electrode to be 11 200 lb/in per in of leg size. Then

$$f_a = \frac{11\,200 \text{ lb/in}}{1.0 \text{ in leg}} \times 0.188 \text{ in leg} = 2100 \text{ lb/in}$$

The actual force on the weld is

$$f_a = P/A_w = P/2h$$

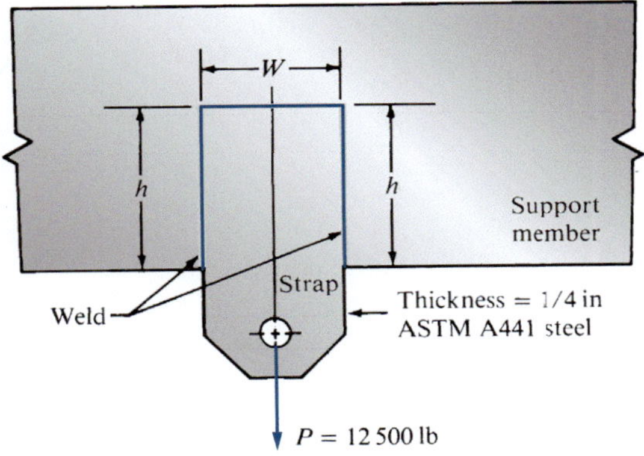

FIGURE 20–10 Steel strap

Then solving for h gives

$$h = \frac{P}{2(f_a)} = \frac{12\ 500\ \text{lb}}{2(2100\ \text{lb/in})} = 2.98\ \text{in}$$

Let's specify $h = 3.00$ in.

Example Problem 20–4

Evaluate the design shown in Figure 20–11 with regard to stress in the welds. All parts of the assembly are made of ASTM A36 structural steel and are welded with an E60 electrode. The 2500-lb load is a dead load.

Solution

The critical point would be the weld at the top of the tube where it is joined to the vertical surface. At this point, there is a three-dimensional force system acting on the weld as shown in Figure 20–12. The offset location of the load causes a twisting on the weld that produces a force f_t on the weld toward the left in the y-direction. The bending produces a force f_b acting outward along the x-axis. The vertical shear force f_s acts downward along the z-axis.

From statics, the resultant of the three force components would be

$$f_R = \sqrt{f_t^2 + f_b^2 + f_s^2}$$

Now each component force on the weld will be computed.

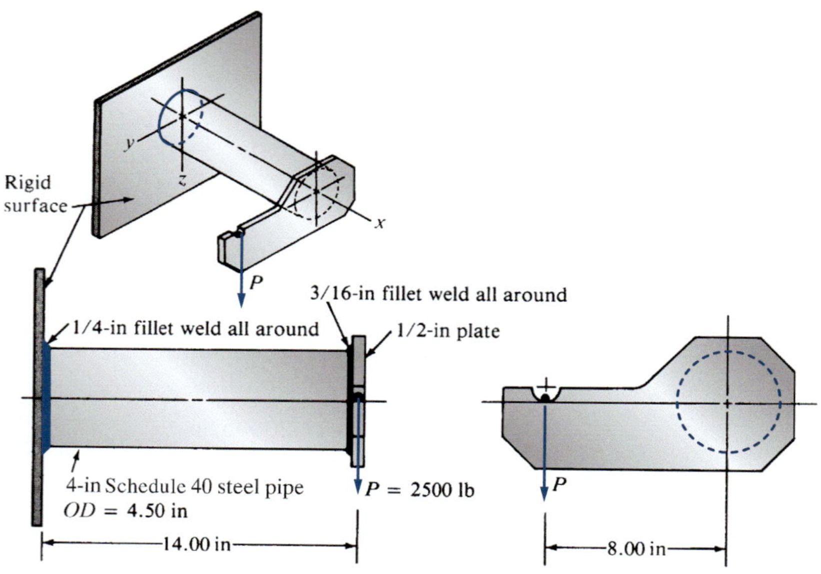

FIGURE 20–11 Bracket assembly

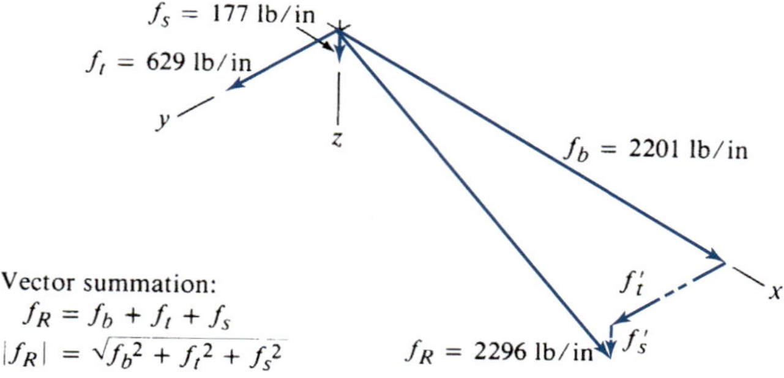

FIGURE 20–12 Force vectors

Twisting Force, f_t

$$f_t = \frac{Tc}{J_w}$$

$$T = (2500 \text{ lb})(8.00 \text{ in}) = 20\,000 \text{ lb}\cdot\text{in}$$

$$c = OD/2 = 4.500/2 = 2.25 \text{ in}$$

$$J_w = (\pi)(OD)^3/4 = (\pi)(4.500)^3/4 = 71.57 \text{ in}^3$$

Then

$$f_t = \frac{Tc}{J_w} = \frac{(20\,000)(2.25)}{71.57} = 629 \text{ lb/in}$$

Bending Force, f_b

$$f_b = \frac{M}{S_w}$$

$$M = (2500 \text{ lb})(14.00 \text{ in}) = 35\,000 \text{ lb}\cdot\text{in}$$

$$S_w = (\pi)(OD)^2/4 = (\pi)(4.500)^2/4 = 15.90 \text{ in}^2$$

Then

$$f_b = \frac{M}{S_w} = \frac{35\,000}{15.90} = 2201 \text{ lb/in}$$

Vertical Shear Force, f_s

$$f_s = \frac{P}{A_w}$$

$$A_w = (\pi)(OD) = (\pi)(4.500 \text{ in}) = 14.14 \text{ in}$$

$$f_s = \frac{P}{A_w} = \frac{2500}{14.14} = 177 \text{ lb/in}$$

Now the resultant can be computed:

$$f_R = \sqrt{f_t^2 + f_b^2 + f_s^2}$$

$$f_R = \sqrt{629^2 + 2201^2 + 177^2} = 2296 \text{ lb/in}$$

Comparing this with the allowable force on a 1.0-in weld gives

$$w = \frac{2296 \text{ lb/in}}{9600 \text{ lb/in per in of leg size}} = 0.239 \text{ in}$$

The 1/4 in fillet specified in Figure 20–11 is satisfactory.

REFERENCES

1. American Institute of Steel Construction. *Steel Construction Manual.* 14th ed. Chicago, IL: American Institute of Steel Construction, 2011.
2. Bickford, J. H. *Introduction to the Design and Behavior of Bolted Joints: Non-Gasketed Joints.* 4th ed. Boca Raton, FL: CRC Press, 2008.
3. Bickford, J. H., and Sayed Nassar, editors. *Handbook of Bolts and Bolted Joints.* New York: Marcel Dekker, 1978.
4. Blair, M., T. L. Stevens, and B. Linskey, editors. *Steel Castings Handbook.* 6th ed. Materials Park, OH: ASM International, 1996.
5. Blake, Alexander. *Design of Mechanical Joints.* Boca Raton, FL: CRC Press, 1985.
6. Blodgett, O. W. *Design of Welded Structures.* Cleveland, OH: James F. Lincoln Arc Welding Foundation, 1976.
7. Blodgett, O. W. *Design of Weldments.* Cleveland, OH: James F. Lincoln Arc Welding Foundation, 1998.
8. Cary, H. B., and S. Helzer. *Modern Welding Technology.* 6th ed. Upper Saddle River, NJ: Pearson/Prentice Hall, 2004.
9. Geary, D., and R. Miller. *Welding.* 2nd ed. New York: McGraw-Hill, 2011.
10. Hicks, J. *Welded Design: Theory and Practice.* Cambridge, UK and Philadelphia, PA: Woodhead Publishing, 2001.

11. Humpston, G., and Jacobson, D. M. *Principles of Soldering.* Materials Park, OH: ASM International, 2004.
12. Jacobson, D. M., and G. Humpston. *Principles of Brazing.* Materials Park, OH: ASM International, 2005.
13. Jeffus, L. *Welding Principles and Applications.* 7th ed. Florence, KY: Delmar Cengage Learning, 2011.
14. Kulak, G. L., J. W. Fisher, and J. H. A. Struik. *Guide to Design Criteria for Bolted and Riveted Joints.* 2nd ed. New York: John Wiley & Sons, 1987.
15. Lohwasser, D., and A. W. Chen. *Friction Stir Welding: From Basics to Application.* Cambridge, UK and Philadelphia, PA: Woodhead Publishing, 2009.
16. Mathers, Gene. *The Welding of Aluminum and Its Alloys.* Boca Raton, FL: CRC Press, 2002.
17. Mott, R. L. *Applied Strength of Materials.* 5th ed. Upper Saddle River, NJ: Pearson/Prentice Hall, 2008.
18. North American Die Casting Association. *Cast View Design Software.* Rosemont, IL: North American Die Casting Association, 2002.
19. Oberg, E., F. D. Jones, H. L. Horton, and H. H. Ryffel. *Machinery's Handbook.* 28th ed. New York: Industrial Press, 2008.
20. Parmley, R. O. *Standard Handbook of Fastening and Joining.* 3rd ed. New York: McGraw-Hill, 1997.
21. Raj, B., V. Shankar, and A. K. Bhaduri. *Welding Technology for Engineers.* Materials Park, OH: ASM International, 2006. (Jointly published by Narosa Publishing House, New Delhi, India).
22. Research Council on Structural Connections. *Specification for Structural Joints Using High-Strength Bolts.* Chicago, IL: Research Council on Structural Connections, 2010.
23. Weman, K. *Welding Processes Handbook.* 2nd ed. Cambridge, UK and Philadelphia, PA: Woodhead Publishing, 2012.

INTERNET SITES FOR MACHINE FRAMES, BOLTED CONNECTIONS, AND WELDED JOINTS

Also refer to the Internet sites from Chapter 19 on Fasteners that includes many related to bolted connections.

1. **Steel Founders' Society of America.** www.sfsa.org An association of companies providing foundry services. Site includes useful casting information, a glossary of foundry terms, and a framework for designing.
2. **American Foundry Society.** www.afsinc.org A professional society that promotes research and technology for the foundry industry. Site includes a section on *How to Design Castings*, numerous design tutorials, and videos of casting operations.
3. **American Welding Society**. www.aws.org A professional society that develops standards for the welding industry, including AWS D1.1 Structural Welding Code-Steel (2010), AWS D1.2 Structural Welding Code-Aluminum (2008), and many others. Site offers ways to access standards and its American Welding Online learning system. Also provides links to SENSE (School Excelling through National Skill Standards Education) and guidelines for welder training.
4. **James F. Lincoln Foundation.** www.jflf.org An organization that promotes education and training in welding technology. The site includes technical papers and general information about welding processes, joint design, and guides to welded steel construction. The site includes an Education Center with PowerPoint slides and other training materials.
5. **Miller Electric Company.** www.millerwelds.com A manufacturer of a wide variety of welding equipment and accessories for the professional welding community and occasional welders. The site includes a Training/Education section that can be searched for information about numerous welding processes. Site includes a Resources section containing Tech Tips, welding guidelines, and welding videos.
6. **Lincoln Electric Company.** www.lincolnelectric.com A manufacturer of a wide variety of welding equipment and accessories for the welding industry. The site includes a Knowledge section that provides technical articles about welding technology.
7. **Hobart Brothers Company.** www.hobartbrothers.com A manufacturer of numerous types of filler metal products, electrodes, and steel wire for the welding industry. Site offers a Support tab containing downloadable posters and charts on welding.
8. **Hobart Institute of Welding Technology.** www.welding.org An educational organization that provides instruction in the performance of welding techniques.
9. **Metal Casting Design.** www.metalcastingdesign.com An online resource for metal casting buyers and design engineers, sponsored by the American Foundry Society. Includes a metal casting process selector, casting tutorials, and training for buyers and designers. The casting tutorials follow a casting project from design through manufacturing issues and terminology involved in the selection, design, purchase, production, and use of metal castings.
10. **Aluminum Association.** www.aluminum.org The Association promotes the use of aluminum and represents U.S. and foreign-based primary producers of aluminum, aluminum recyclers, and producers of fabricated products, as well as industry suppliers. Publishes *Aluminum Standards and Data* in both U.S. and metric units, *Aluminum Design Manual,* and *Aluminum Structural Welding Code.*

PROBLEMS

For Problems 1–6, design a bolted joint to join the two members shown in the appropriate figure. Specify the number of bolts, the pattern, the bolt grade, and the bolt size.

1. Figure P20–1

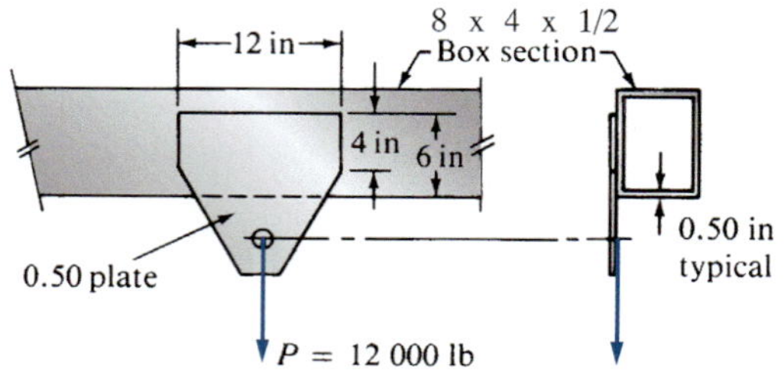

FIGURE P20–1 (Problems 1, 7, and 13)

2. Figure P20–2

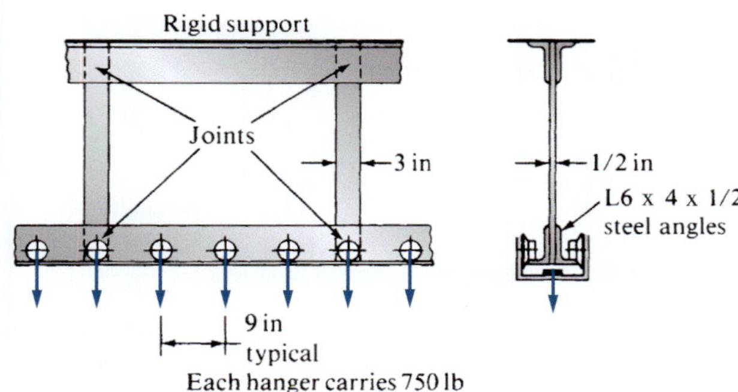

FIGURE P20–2

3. Figure P20–3

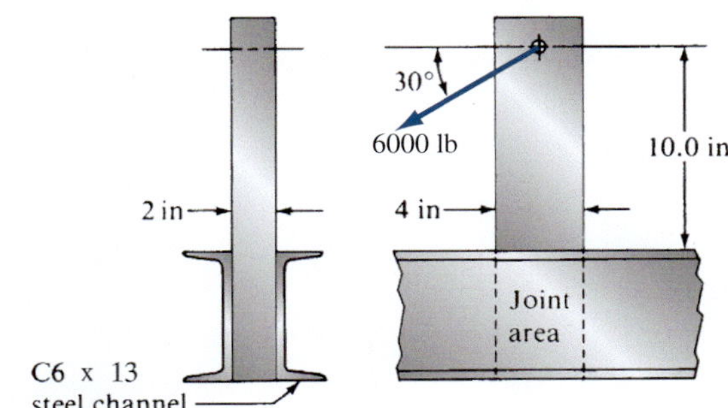

FIGURE P20–3

4. Figure P20–4

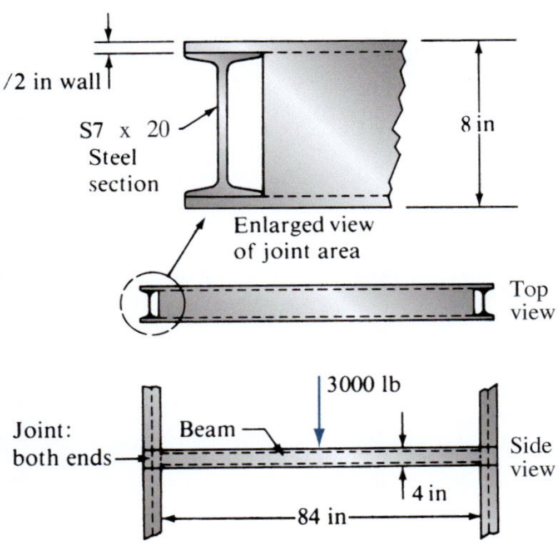

FIGURE P20–4 (Problems 4 and 8)

5. Figure P20–5

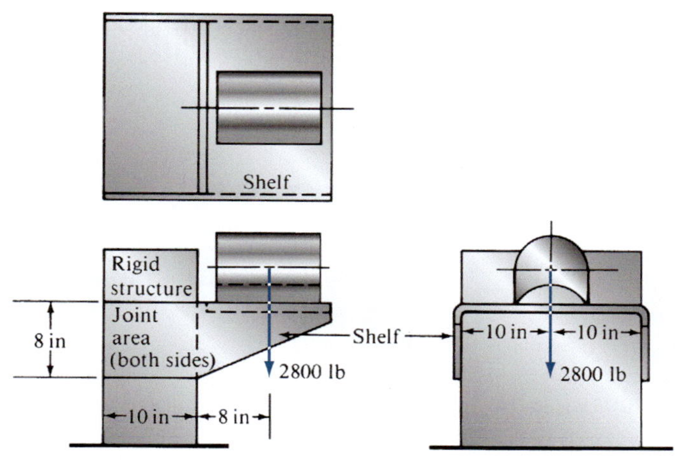

FIGURE P20–5 (Problems 5 and 9)

6. Figure P20–6

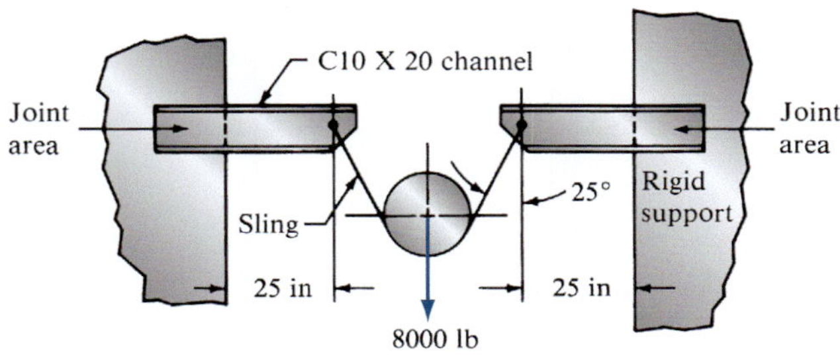

Load shared equally by *four* brackets (only two shown)

FIGURE P20–6 (Problems 6 and 10)

For Problems 7–12, design a welded joint to join the two members shown in the appropriate figure. Specify the weld pattern, the type of electrode to be used, and the size of weld. In Problems 7–9, the members are ASTM A36 steel. In Problems 10–12, the members are ASTM A441 steel. Use the method of treating the joint as a line, and use the allowable forces per inch of leg for building-type structures from Table 20–3.

7. Figure P20–1
8. Figure P20–4
9. Figure P20–5
10. Figure P20–6
11. Figure P20–11

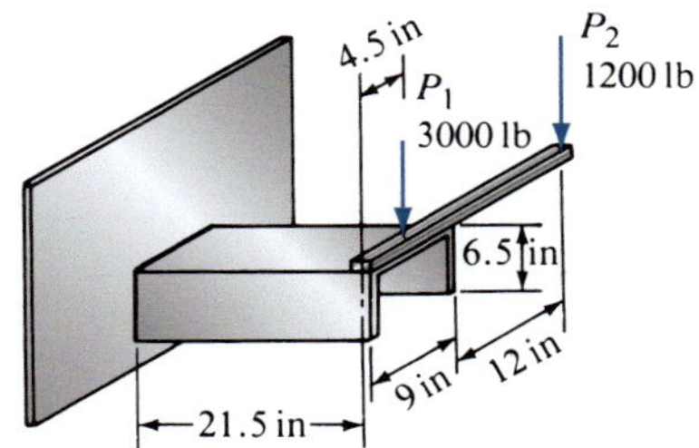

FIGURE P20–11 (Problems 11 and 12)

12. Figure P20–11 (but $P_2 = 0$)

For Problems 13–16, design a welded joint to join the two aluminum members shown in the appropriate figure. Specify the weld pattern, the type of filler alloy, and the size of weld. The types of materials joined are listed in the problems.

13. Figure P20–1: 6061: alloy (but $P = 4000$ lb)
14. Figure P20–14: 6061 alloy

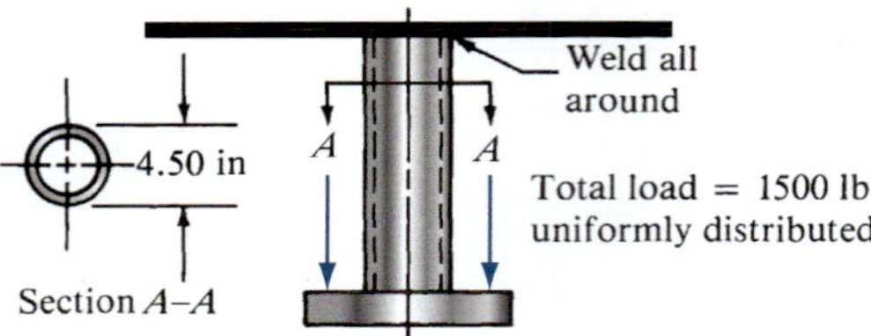

FIGURE P20–14

15. Figure P20–15: 6063 alloy

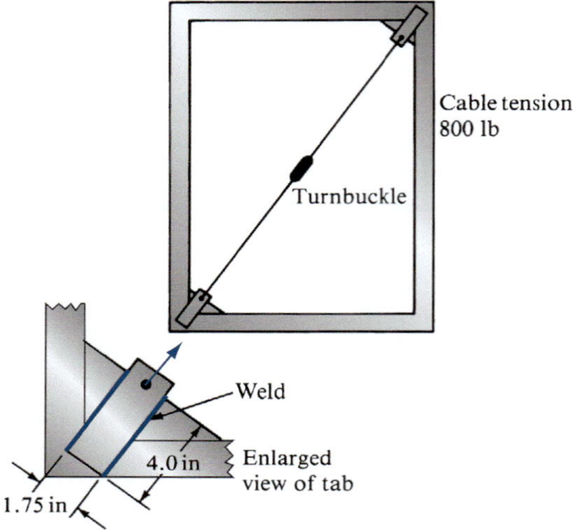

FIGURE P20–15

16. Figure P20–16: 3003 alloy

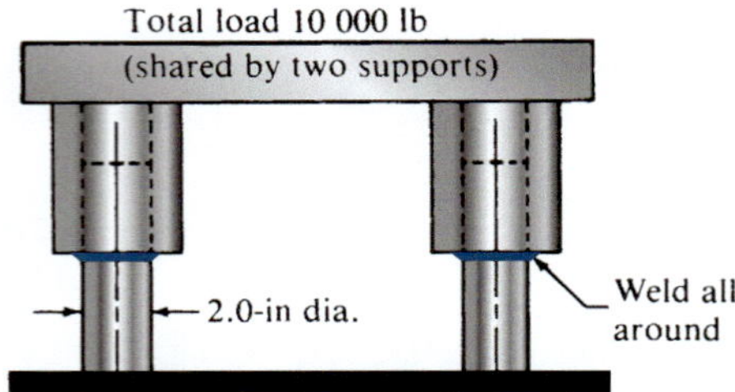

FIGURE P20–16

17. Compare the weight of a tensile rod carrying a dead load of 4800 lb if it is made from (a) SAE 1020 HR steel; (b) SAE 5160 OQT 1300 steel; (c) aluminum 2014-T6; (d) aluminum 7075-T6; (e) titanium 6A1-4V, annealed; and (f) titanium 3A1-13V-11Cr, aged. Use $N = 2$ based on yield strength.

CHAPTER TWENTY ONE

ELECTRIC MOTORS AND CONTROLS

THE BIG PICTURE Electric Motors and Controls

Discussion Map

- Electric motors provide the power for a huge array of products in homes, factories, schools, commercial facilities, transportation equipment, and many portable devices.
- The two major classifications of motors are *alternating current (AC)* and *direct current (DC)*. Some can operate on either type of power.

Discover

Look for a variety of machines and products that are powered by electric motors. Choose both large and small devices; some portable and some that plug into standard electrical outlets. Look around your home, at work, and in a factory.

Try to find the nameplate for each motor, and copy down as much information as you can. How are the nameplate data related to the performance characteristics of the motor? Is it an AC motor or a DC motor? How fast does it run? What is its electrical rating in terms of voltage and current?

Some motors that will operate well in the United States will run differently or will not run at all in other countries. Why? What are the electric voltage and frequency standards in various parts of the world?

This chapter will help you identify many kinds of motors, understand the general performance characteristics of each, and apply them properly.

The electric motor is widely used for providing the prime power to industrial machinery, consumer products, and business equipment. This chapter will describe the various types of motors and will discuss their performance characteristics. The objective is to provide you with the background needed to specify motors and to communicate with vendors to acquire the proper motor for a given application.

Types of motors discussed in this chapter are direct current (DC), alternating current (AC, both single-phase and three-phase), universal motors, stepper motors, and variable-speed AC motors. Motor controls, batteries, and other means of providing DC power are also discussed.

While the appearance of different types of motors and the design of component parts vary significantly, it

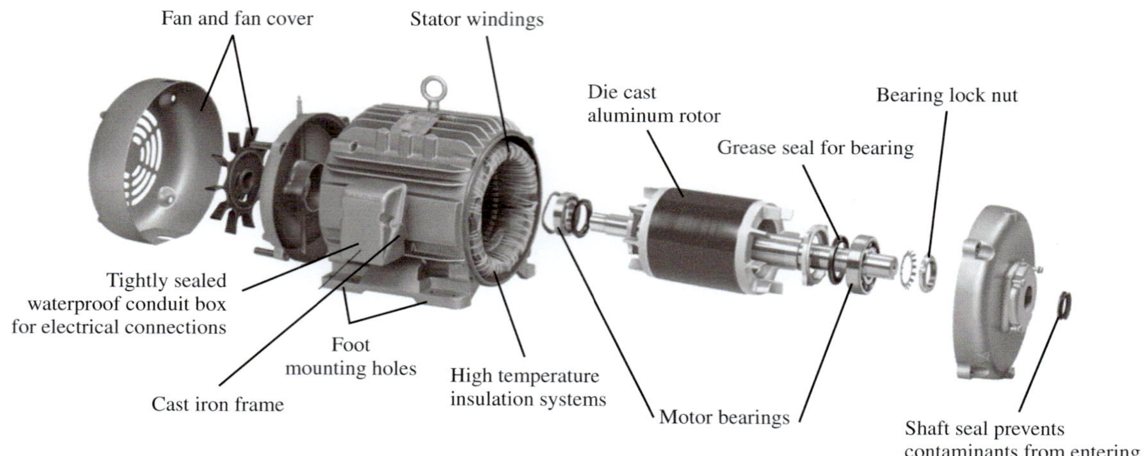

FIGURE 21–1 Exploded view of components for an AC electric induction motor (Baldor Eletric Company, Greenville, SC)

should be helpful at this time to study Figure 21–1 that shows a typical industrial-sized AC electric motor. The exploded view shows the primary components and how they are positioned relative to each other. The delivery of output power from the motor is through the *shaft* that typically runs through from the rear through the front of the *frame* extending out to enable connection to the driven machine or to mount a power-transmission element such as a gear, belt sheave, or chain sprocket. The shaft is supported by two precision *bearings,* typically rolling contact type, that accurately locate the axis concentric with that of the *housing.* The *rotor* of the motor is fixed to the shaft between the two bearings. The rotor reacts electromagnetically with the *windings in the stator* that is fitted inside the stationary housing. This particular motor style is called a *totally enclosed, fan cooled AC electric induction motor, TEFC* for short. Note the *fan* on the left side that rotates with the motor shaft and draws air over the housing and out the rear cover to remove heat generated by the induction process.

Now you should look for a variety of machines that are powered by electric motors: some large and some small; some portable, some that plug into standard electrical outlets, or others that are hardwired directly to an electrical circuit. Look around your home, in your car, at work, in gas stations, in shops or movie theaters, or in a factory if possible.

Try to find the nameplate for each motor and copy down as much information as you can. What does each piece of information mean? Are some electrical wiring diagrams included? How are the nameplate data related to the performance characteristics of the motor? Is it an AC motor or a DC motor? How fast does it run? What is its electrical rating in terms of voltage and current? For the motors in portable devices, what kind of power supply is used? What kinds of batteries are required? How many? How are they connected? In series? In parallel? How do those factors relate to the voltage rating of the motor? What is the relationship between the kind of work done by the motor and the time between charging sessions?

Compare some of the larger motors with the photographs in Section 21–8 of this chapter. Can you identify the frame type? What company manufactured the motor?

See if you can find more information about the motors or their manufacturers from the Internet. Look up some of the company names in this chapter to learn more about the lines of motors that they offer. Do a broader search to find as many different types as you can.

What is the rated power for each motor? What units are used for the power rating? Convert all of the power ratings to watts. (See Appendix 17.) Then convert them all to horsepower to gain an appreciation for the physical size of various motors and to visualize the comparison between the SI unit of watts and the older U.S. Customary unit of horsepower.

Separate the motors you find into those that are AC and those that are DC. Do you see any significant differences between the two classes of motors? Within each class, what variations do you see? Do any of the nameplates describe the type of motor, such as *synchronous, universal, split-phase, NEMA design C,* or other designations?

What kinds of controls are connected to the motors? Switches? Starters? Speed controls? Protection devices?

This chapter will help you identify many kinds of motors, understand the general performance characteristics of each, and apply them properly.

YOU ARE THE DESIGNER

Consider a conveyor system that you are to design. One possibility for driving the conveyor system is to use an electric motor. What type should be used? At what speed will it operate? What type of electric power is available to supply the motor? What power is required? What type of housing and mounting style should be specified? What are the dimensions of the motor? How is the motor connected to the drive pulley of the conveyor system? The information in this chapter will help you to answer these and other questions. ■

21–1 OBJECTIVES OF THIS CHAPTER

After completing this chapter, you will be able to:

1. Describe the factors that must be specified to select a suitable motor.
2. Describe the principles of operation of AC motors.
3. Identify the typical classifications of AC electric motors according to rated power.
4. Identify the common voltages and frequencies of AC power and the speed of operation of AC motors operating on those systems.
5. Describe single-phase and three-phase AC power.
6. Describe the typical frame designs, sizes, and enclosure styles of AC motors.
7. Describe the general form of a motor performance curve.
8. Describe the comparative performance of *shaded-pole, permanent-split capacitor, split-phase,* and *capacitor-start,* single-phase AC motors.
9. Describe three-phase, squirrel-cage AC motors.
10. Describe the comparative performance of *NEMA design B, NEMA design C, NEMA design D,* and *wound-rotor* three-phase AC motors.
11. Describe *synchronous motors.*
12. Describe *universal motors.*
13. Describe three means of producing DC power and the common voltages produced.
14. Describe the advantages and disadvantages of DC motors compared with AC motors.
15. Describe four basic designs of DC motors—*shunt-wound, series-wound, compound-wound,* and *permanent magnet*—and describe their performance curves.
16. Describe torque motors, servomotors, stepper motors, brushless DC motors, and printed circuit motors.
17. Describe motor control systems for system protection, speed control, starting, stopping, and overload protection.
18. Describe speed control of AC motors.
19. Describe the control of DC motors.

21–2 MOTOR SELECTION FACTORS

As a minimum, the following items must be specified for motors:

- Motor type: DC, AC, single-phase, three-phase, and so on
- Power rating and speed
- Operating voltage and frequency
- Type of enclosure
- Frame size
- Mounting details

In addition, there may be many special requirements that must be communicated to the vendor. The primary factors to be considered in selecting a motor include the following:

- Operating torque, operating speed, and power rating. Note that these are related by the equation

$$\text{Power} = \text{torque} \times \text{speed}$$

- Starting torque.
- Load variations expected and corresponding speed variations that can be tolerated.
- Current limitations during the running and starting phases of operation.
- Duty cycle: how frequently the motor is to be started and stopped.
- Environmental factors: temperature, presence of corrosive or explosive atmospheres, exposure to weather or to liquids, availability of cooling air, and so on.
- Voltage variations expected: Most motors will tolerate up to $\pm 10\%$ variation from the rated voltage. Beyond this, special designs are required.
- Shaft loading, particularly side loads and thrust loads that can affect the life of shaft bearings.

Motor Size

A rough classification of motors by size is used to group motors of similar design. Horsepower (hp) is currently used most frequently, with the metric unit of watts or kilowatts also used at times. The conversion is

$$1.0\text{ hp} = 0.746\text{ kW} = 746\text{ W}$$

The classifications are as follows:

- ***Subfractional horsepower:*** 1 to 40 millihorsepower (mhp), where 1 mhp = 0.001 hp. Thus, this range includes 0.001 to 0.040 hp (0.75 to 30 W, approximately).
- ***Fractional horsepower:*** 1/20 to 1.0 hp (37 to 746 W, approximately).
- ***Integral horsepower:*** 1.0 hp (0.75 kW) and larger.

References 1–8 provide additional information about the selection and application of electric motors. See also Internet sites 1–4, 14, and 15.

21–3 AC POWER AND GENERAL INFORMATION ABOUT AC MOTORS

Alternating current (AC) power is produced by the electric utility and delivered to the industrial, commercial, or residential consumer in a variety of forms. In the United States, AC power has a frequency of 60 hertz (Hz) or 60 cycles/s. In many other countries, 50 Hz is used. Some aircraft use 400-Hz power from an onboard generator.

AC power is also classified as single-phase or three-phase. Most residential units and light commercial installations have only single-phase power, carried by two conductors plus ground. The waveform of the power would appear like that in Figure 21–2, a single continuous sine wave at the system frequency whose amplitude is the rated voltage of the power. Three-phase power is carried on a three-wire system and is composed of three distinct waves of the same amplitude and frequency, with each phase offset from the next by 120°, as illustrated in Figure 21–3. Industrial and large commercial installations use three-phase power for the larger electrical loads because smaller motors are possible and there are economies of operation.

AC Voltages

Some of the more popular voltage ratings available in AC power are listed in Table 21–1. Given are the nominal system voltage and the typical motor voltage rating for that system in both single-phase and three-phase. In most cases, the highest voltage available should be used because the current flow for a given power is smaller. This allows smaller conductors to be used.

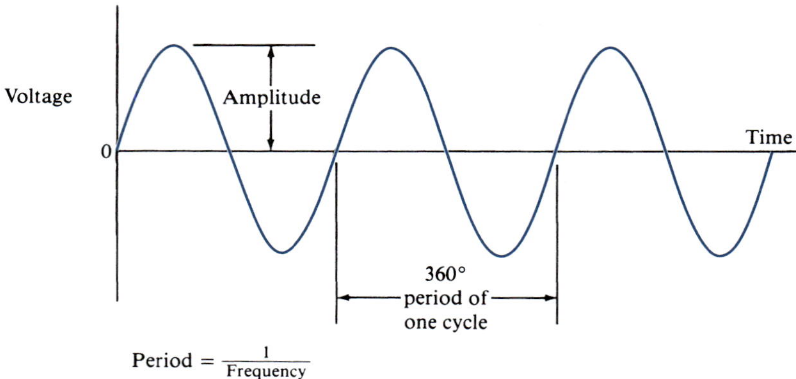

FIGURE 21–2 Single-phase AC power

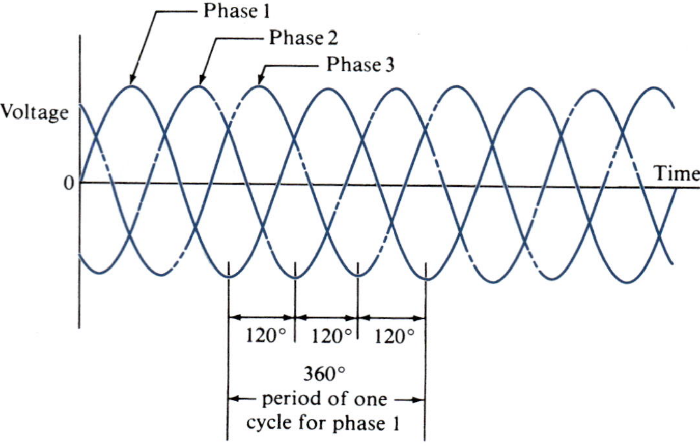

FIGURE 21–3 Three-phase AC power

TABLE 21–1 AC Motor Voltages

System voltage	Motor voltage ratings: Single-phase	Three-phase
120	115	115
120/208	115	200
240	230	230
480		460
600		575

TABLE 21–2 AC Motor Speeds for 60 Hertz Power

Number of poles	Synchronous speed (rpm)	Full-load[a] speed (rpm)
2	3600	3450
4	1800	1725
6	1200	1140
8	900	850
10	720	690
12	600	575

[a]Approximately 95% of synchronous speed (normal slip).

Speeds of AC Motors

An AC motor at zero load would tend to operate at or near its *synchronous speed, n_s,* which is related to the frequency, f, of the AC power and to the number of electrical poles, p, wound into the motor, according to the equation

➯ Synchronous Speed

$$n_s = \frac{120f}{p} \text{rev/min} \qquad \textbf{(21–1)}$$

Motors have an even number of poles, usually from 2 to 12, resulting in the synchronous speeds listed in Table 21–2 for 60-Hz power. But the induction motor, the most widely used type, operates at a speed progressively slower than its synchronous speed as the load (torque) demand increases. When the motor is delivering its rated torque, it will be operating near its rated or full-load speed, also listed in Table 21–2. Note that the full-load speed is not precise and that those listed are for motors with normal slip of approximately 5%. Some motors described later are "high-slip" motors having lower full-load speeds. Some four-pole motors are rated at 1750 rpm at full load, indicating only about 3% slip. *Synchronous motors* operate precisely at the synchronous speed with no slip.

21–4 PRINCIPLES OF OPERATION OF AC INDUCTION MOTORS

Later in this chapter, we will discuss the particular details of several different types of AC motors, but the most common of these is the *induction motor.* Refer again to Figure 21–1 that shows the components in an exploded view. Also look ahead to Figures 21–13 to 21–17 for photos of complete motors. The two active parts of an induction motor are the *stator,* or stationary element, and the *rotor,* or rotating element. Figure 21–4 shows a longitudinal cross section of an induction motor showing the stator, in the form of a hollow cylinder, fixed in the housing. The rotor is positioned inside the stator and is carried on the shaft. The shaft, in turn, is supported by bearings in the housing.

The stator is constructed of many thin, flat discs of steel, called *laminations,* stacked together and insulated from one another. Figure 21–5 shows the shape of the laminations, including a series of slots around the inside. These slots are aligned as the stator laminations are stacked, thus forming channels along the length of the stator core. Several layers of copper wire are passed through the channels and are looped around to form a set of continuous coils, called *windings.* The pattern of the coils

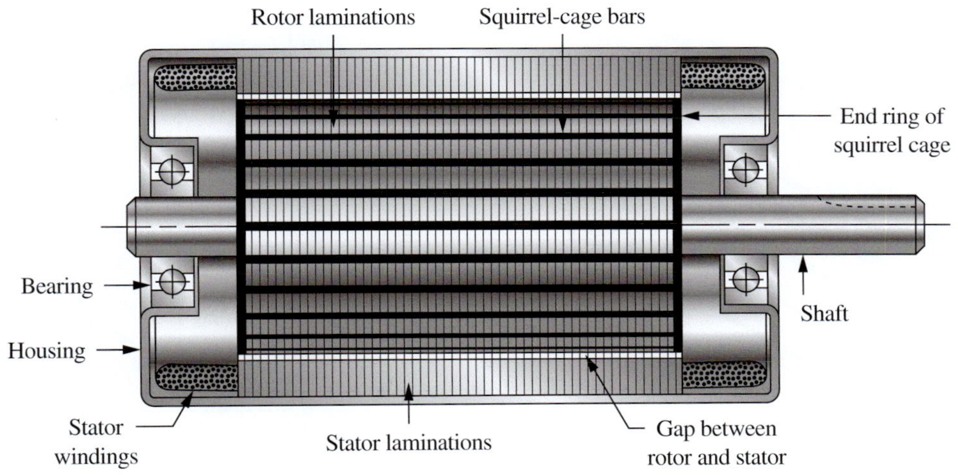

FIGURE 21–4 Longitudinal section through an induction motor

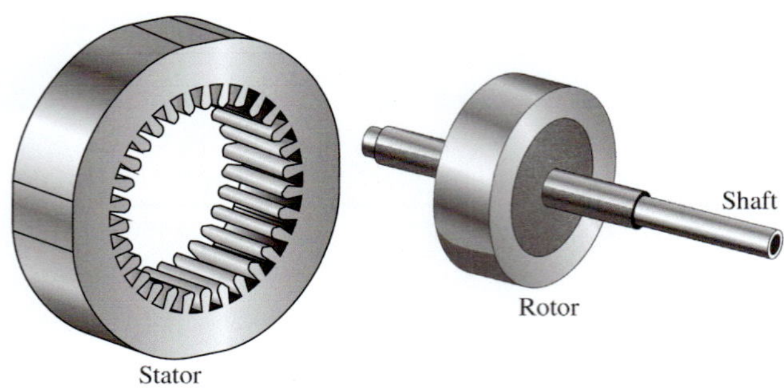

FIGURE 21–5 Induction motor laminations

in the stator determines the number of poles for the motor, usually 2, 4, 6, 8, 10, or 12. Table 21–2 shows that the rotational speed of the motor depends on the number of poles.

The rotor also has a lamination stack with channels along the length. The channels are typically filled with solid bars made from a good electrical conductor such as copper or aluminum, with the ends of all bars connected to continuous rings at each end. In some smaller rotors, the complete set of bars and end rings are cast from aluminum as a unit. As shown in Figure 21–6, if this casting were viewed without the laminations, it would appear similar to a squirrel cage. Thus, induction motors are often called squirrel-cage motors. The combination of the squirrel cage and the laminations is fixed onto the motor shaft with good precision to assure concentric alignment with the stator and good dynamic balance while rotating. When the rotor is installed in the supporting bearings and is inserted inside the stator, there is a small gap of approximately 0.020 in (0.50 mm) between the outer surface of the rotor and the inner surface of the stator.

Three-Phase Motors

The principles of operation of AC motors is first discussed for three-phase induction motors. Single-phase motor designs are discussed later. The three-phase electrical power, shown schematically in Figure 21–3, is connected to the stator windings. As the current flows in the windings, electromagnetic fields are created that are exposed to the conductors in the rotor. Because the three phases of power are displaced from each other in time, the effect created is a set of fields rotating around the stator. A conductor placed in a moving magnetic field has a current induced in it, and a force is exerted perpendicular to the conductor. The force acts near the periphery of the rotor, thus creating a torque to rotate the rotor.

It is the production of the induced current in the rotor that leads to calling such motors *induction motors.* Note also that there are no direct electrical connections to the rotor, thus greatly simplifying the design and construction of the motor and contributing to its high reliability.

21–5 AC MOTOR PERFORMANCE

The performance of electric motors is usually displayed on a graph of speed versus torque, as shown in Figure 21–7. The *vertical axis* is the rotational speed of the motor as a percentage of synchronous speed. The *horizontal axis* is the torque developed by the motor as a percentage of the full-load or rated torque. When exerting its full-load torque, the motor operates at its full-load speed and delivers the rated power. See Table 21–2 for a listing of synchronous speeds and full-load speeds.

The torque at the bottom of the curve where the speed is zero is called the *starting torque* or *locked-rotor torque.* It is the torque available to initially get the load moving and begin its acceleration. This is one of the most important selection parameters for motors, as will be discussed in the descriptions of the individual types of motors.

The "knee" of the curve, called the *breakdown torque,* is the maximum torque developed by the motor during acceleration. The slope of the speed/torque curve in the vicinity of the full-load operating point is an indication of *speed regulation.* A flat curve (a low slope) indicates good speed regulation with little variation in speed as load varies. Conversely, a steep curve (a high slope) indicates poor speed regulation, and the motor will exhibit wide swings in speed as load varies. Such motors produce a "soft" acceleration of a load which may be an advantage in some applications. But where fairly constant speed is desired, a motor with good speed regulation should be selected.

FIGURE 21–6 Squirrel cage rotor without laminations

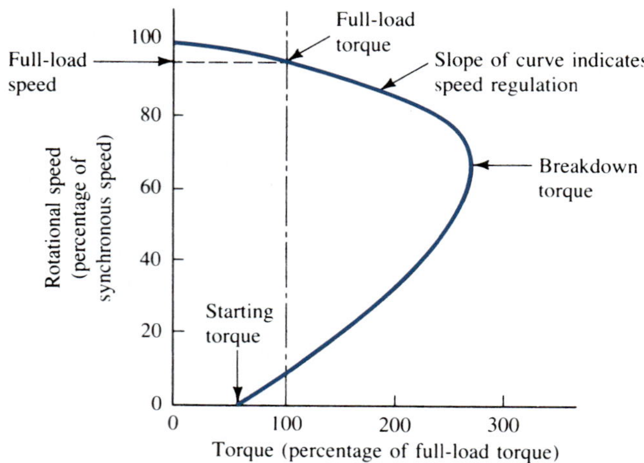

FIGURE 21–7 General form of motor performance curve

21–6 THREE-PHASE, SQUIRREL-CAGE INDUCTION MOTORS

Three of the most commonly used three-phase AC motors are simply designated as designs B, C, and D by the National Electrical Manufacturers Association (NEMA). They differ primarily in the value of starting torque and in the speed regulation near full load. Figure 21–8 shows the performance curves for these three designs for comparison. Each of these designs employs the solid squirrel-cage type of rotor, and thus there is no electrical connection to the rotor.

The four-pole design with a synchronous speed of 1800 rpm is the most common and is available in virtually all power ratings from 1/4 hp to 500 hp. Certain sizes are available in 2-pole (3600 rpm), 6-pole (1200 rpm), 8-pole (900 rpm), 10-pole (720 rpm), and 12-pole (600 rpm) designs.

NEMA Design B

The performance of the three-phase design B motor is similar to that of the single-phase split-phase motor described later. It has a moderate starting torque (about 150% of full-load torque) and good speed regulation. The breakdown torque is high, usually 200% of full-load torque or more. Starting current is fairly high, at approximately six times full-load current. The starting circuit must be selected to be able to handle this current for the short time required to bring the motor up to speed.

Typical uses for the design B motor are centrifugal pumps, fans, blowers, and machine tools such as grinders and lathes.

NEMA Design C

High starting torque is the main advantage of the design C motor. Loads requiring 200% to 300% of full-load torque to start can be driven. Starting current is typically lower than for the design B motor for the same starting torque. Speed regulation is good and is about the same as for the design B motor. Reciprocating compressors, refrigeration systems, heavily loaded conveyors, and ball-and-rod mills are typical uses.

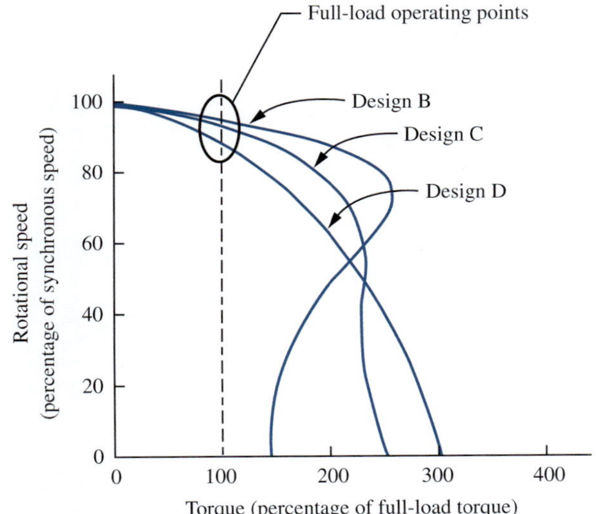

FIGURE 21–8 Performance curves for three-phase motors: designs B, C, and D

NEMA Design D

The design D motor has a high starting torque, about 300% of full-load torque. But it also has poor speed regulation, which results in large speed changes with varying loads. Sometimes called a *high-slip motor,* it operates at 5% to 13% slip at full load, whereas the designs B and C operate at 3% to 5% slip. Thus, the full-load speed will be lower for the design D motor.

The poor speed regulation is considered an advantage in some applications and is the main reason for selecting the design D motor for such uses as punch presses, shears, sheet-metal press brakes, cranes, elevators, and oil well pumps. Allowing the motor to slow down significantly when loads increase gives the system a "soft" response, reducing shock and jerk felt by the drive system and the driven machine. Consider an elevator: When a heavily loaded elevator cab is started, the acceleration should be smooth and soft, and the cruising speed should be approached without excessive jerk. This comment also applies to a crane. If a large jerk occurs when the crane hook is heavily loaded, the peak acceleration will be high. The resulting high inertia force may break the cable.

Wound-Rotor Motors

As the name implies, the rotor of the *wound-rotor motor* has electrical windings that are connected through slip rings to the external power circuit. The selective insertion of resistance in the rotor circuit allows the performance of the motor to be tailored to the needs of the system and to be changed with relative ease to accommodate system changes or to actually vary the speed of the motor.

Figure 21–9 shows the results obtained by changing the resistance in the rotor circuit. Note that the four curves are all for the same motor, with curve 0 giving the performance with zero external resistance. This is similar to design B. Curves 1, 2, and 3 show the performance with progressively higher levels of resistance in the rotor circuit. Thus, the starting torque and the speed regulation (softness) can be tuned to the load. Speed adjustment under a given load up to approximately 50% of full-load speed can be obtained.

The wound-rotor design is used in such applications as printing presses, crushing equipment, conveyors, and hoists.

Synchronous Motors

Entirely different from the squirrel-cage induction motor or the wound-rotor motor, the *synchronous motor* operates precisely at the synchronous speed with no slip. Such motors are available in sizes from subfractional, used for timers and instruments, to several hundred horsepower to drive large air compressors, pumps, or blowers.

The synchronous motor must be started and accelerated by a means separate from the synchronous motor components themselves because they provide very little torque at zero speed.

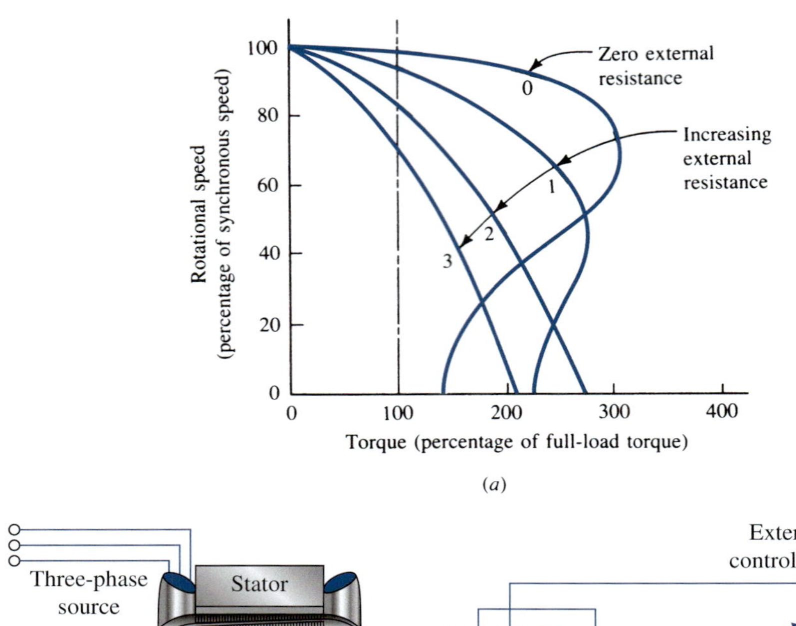

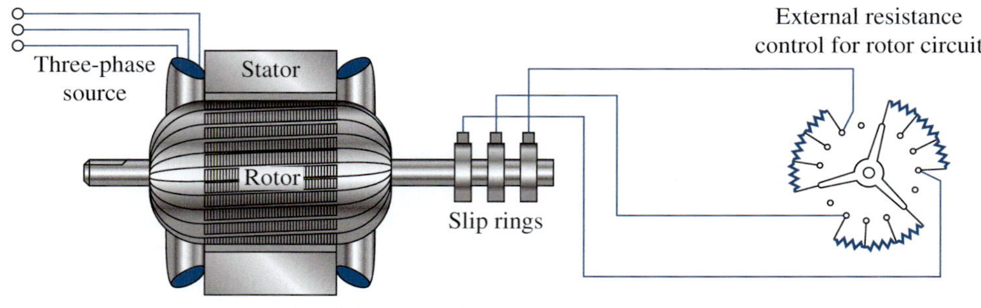

FIGURE 21–9 (a) Performance curves for a three-phase, wound-rotor motor with varying external resistance in the rotor circuit. (b) Schematic of wound-rotor motor with external resistance control

Typically, there will be a separate squirrel-cage type of winding within the normal rotor which initially accelerates the motor shaft. When the speed of the rotor is within a few percent of the synchronous speed, the field poles of the motor are excited, and the rotor is pulled into synchronism. At that point the squirrel cage becomes ineffective, and the motor continues to run at speed regardless of load variations, up to a limit called the *pull-out torque.* A load above the pull-out torque will pull the motor out of synchronism and cause it to stop.

Universal Motors

Universal motors operate on either AC or direct current (DC) power. Their construction is similar to that of a series-wound DC motor, which is described later. The rotor has electrical coils in it that are connected to the external circuit through a commutator on the shaft, a kind of slip ring assembly made of several copper segments on which stationary carbon brushes ride. Contact is maintained with light spring pressure.

Universal motors usually run at high speeds, 3500 to 20 000 rpm. This results in a high power-to-weight and high power-to-size ratio for this type of motor, making it desirable for handheld tools such as drills, saws, and food mixers. Vacuum cleaners and sewing machines also frequently use universal motors. Figure 21–10 shows a typical set of speed/torque curves for a high-speed version of the universal motor, showing the performance for 60-Hz and 25-Hz AC power and DC power. Note that performance near the rated load is similar, regardless of the nature of the input power. Note also that these motors have poor speed regulation; that is, the speed varies greatly with load.

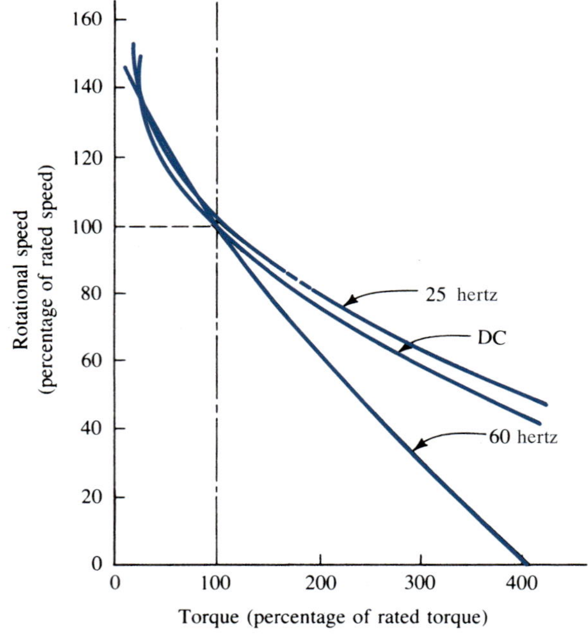

FIGURE 21–10 Performance curves for a universal motor

21–7 SINGLE-PHASE MOTORS

The four most common types of single-phase motors are the *split-phase, capacitor-start, permanent-split capacitor,* and *shaded-pole.* Each is unique in its physical construction and in the manner in which the electrical components are connected to provide for starting and running of the motor. The emphasis here is not on how to design the motors, but rather on the performance, so that a suitable motor can be selected.

Figure 21–11 shows the performance characteristics of these four types of motors so that they can be compared. The special features of the performance curves for the four types of motors are discussed in later sections.

In general, the construction of single-phase motors is similar to that for three-phase motors, consisting of a fixed stator, a solid rotor, and a shaft carried on bearings. The induction principle discussed earlier applies also to single-phase motors. Differences occur because single-phase power does not inherently rotate around the stator to create a moving field. Each type uses a different scheme for initially starting the motor. See Figure 21–12.

Single-phase motors are usually in the subfractional or fractional horsepower range from 1/50 hp (15 W) to 1.0 hp (750 W), although some are available up to 10 hp (7.5 kW).

Split-Phase Motors

The stator of the split-phase motor [Figure 21–12(b)] has two windings: the *main winding,* which is continuously connected to the power line, and the *starting winding,* which is connected only during the starting of the motor. The starting winding creates a slight phase shift that creates the initial torque to start and accelerate the rotor. After the rotor reaches approximately 75% of its synchronous speed, the starting winding is cut out by a centrifugal switch, and the rotor continues to run on the main winding.

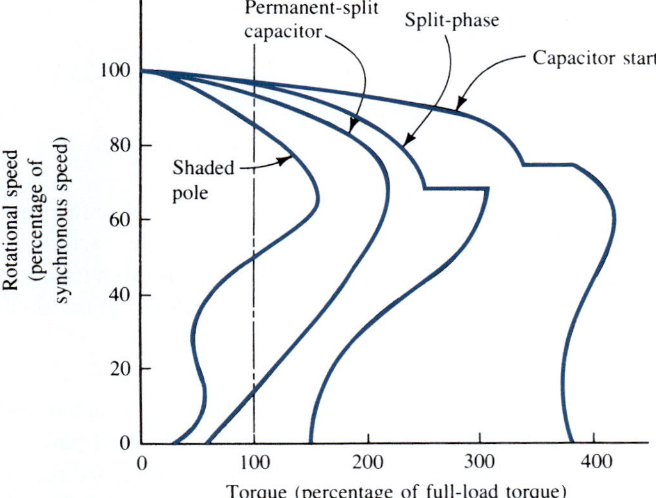

FIGURE 21–11 Performance curves of four types of single-phase electric motors

The performance curve for the split-phase motor is shown in Figure 21–11. It has moderate starting torque, approximately 150% of full-load torque. It has good efficiency and is designed for continuous operation. Speed regulation is good. One of the disadvantages is that it requires a centrifugal switch to cut out the starting winding. The step in the speed/torque curve indicates this cutout.

These characteristics make the split-phase motor one of the most popular types, used in business machines, machine tools, centrifugal pumps, AC electric lawn mowers, and similar applications.

Capacitor-Start Motors

Like the split-phase motor, the capacitor-start motor [Figure 21–12(c)] also has two windings: a *main* or *running winding* and a *starting winding.* But in it, a capacitor is connected in series with the starting winding, giving a much higher starting torque than with the split-phase motor. A starting torque of 250% of full load or higher is common. Again a centrifugal switch is used to cut out the starting winding and the capacitor. The running characteristics of the motor are then very similar to those of the split-phase motor: good speed regulation and good efficiency for continuous operation.

Disadvantages include the switch and the relatively bulky capacitor. Frequently the capacitor is mounted conspicuously right on top of the motor. It may also be integrated into a package containing the starting switch, a relay, or other control elements.

Uses for the capacitor-start motor include the many types of machines that need the high starting torque. Examples include heavily loaded conveyors, refrigeration compressors, and pumps and agitators for heavy fluids.

Permanent-Split Capacitor Motors

A capacitor is connected in series with the starting winding at all times. The starting torque of the permanent-split capacitor motor [Figure 21–12(d)] is typically quite low, approximately 40% or less of full-load torque. Thus, only low-inertia loads such as fans and blowers are usually used. An advantage is that you can tailor the running performance and the speed regulation to match the load by selecting the appropriate capacitor value. Also, no centrifugal switch is required.

Shaded-Pole Motors

The shaded-pole motor [Figure 21–12(e)] has only one winding, the *main* or *running winding.* The starting reaction is created by the presence of a copper band around one side of each pole. The low-resistance band "shades" the pole to produce a rotating magnetic field to start the motor.

The shaded-pole motor is simple and inexpensive, but it has a low efficiency and a very low starting torque. Speed regulation is poor, and it must be fan-cooled during normal operation. Thus, its primary use is in shaft-mounted fans and blowers where the air is drawn over the motor. Some small pumps, toys, and intermittently used household items also employ shaded-pole motors because of their low cost.

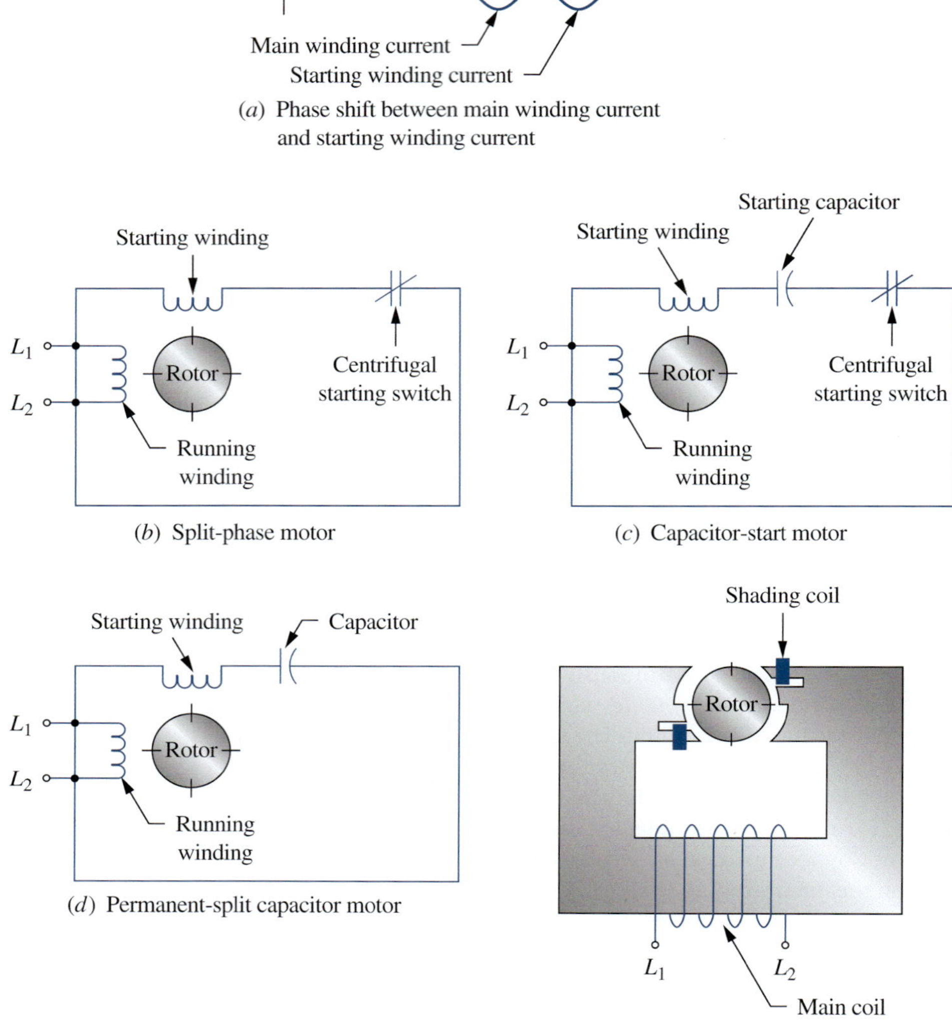

(a) Phase shift between main winding current and starting winding current

(b) Split-phase motor

(c) Capacitor-start motor

(d) Permanent-split capacitor motor

(e) Shaded-pole motor

FIGURE 21–12 Schematic diagrams of single-phase motors

21–8 AC MOTOR FRAME TYPES AND ENCLOSURES

Frame Types

The design of the equipment in which the motor is to be mounted determines the type of frame required. Some types are described below.

Foot-mounted. The most widely used type for industrial machinery, the foot-mounted frame has integral feet with a standard hole pattern for bolting the motor to the machine (see Figure 21–13).

Cushion Base. A foot mounting is provided with resilient isolation of the motor from the frame of the machine to reduce vibration and noise. This frame type is often used for fans and appliances where the motor is enclosed in the shell or housing of the product.

C-Face Mounting. A machined face is provided on the shaft end of the motor which has a standard pattern of tapped holes. Driven equipment is then bolted directly to the motor. The design of the face is standardized by the National Electrical Manufacturers Association (NEMA). See Figure 21–14.

(*a*) Open, protected, dripproof

(*b*) Totally enclosed, nonventilated lint-proof

(*c*) Totally enclosed, fan-cooled

(*d*) Explosion proof, dust ignition proof

FIGURE 21–13 Foot-mounted motors with various enclosure types (Baldor/Reliance Electric, Greenville, SC)

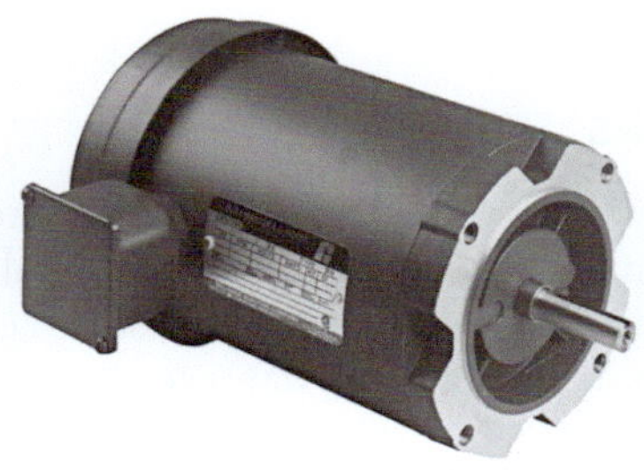

FIGURE 21–14 C-face motor. See Figure 8–20 for an example of a reducer designed to mate with a C-face motor (Baldor/Reliance Electric, Greenville, SC)

D-Flange Mounting. A machined flange is provided on the shaft end of the motor with a standard pattern of through clearance holes for bolts for attaching the motor to the driven equipment. The flange design is controlled by NEMA.

Vertical Mounting. Vertical mounting is a special design because of the effects of the vertical orientation on the bearings of the motor. Attachment to the driven equipment is usually through C-face or D-flange bolt holes as previously described.

Unmounted. Some equipment manufacturers purchase only the bare rotor and stator from the motor manufacturer and then build them into their machine. Compressors for refrigeration equipment are usually built in this manner.

Special-Purpose Mountings. Many special designs are made for fans, pumps, oil burners, and so on.

Enclosures

The housings around the motor that support the active parts and protect them vary with the degree of protection required. Some enclosure types are described next.

Open. Typically a light-gage sheet-metal housing is provided around the stator with end plates to support the shaft bearings. The housing contains several holes or slots that permit cooling air to enter the motor. Such a motor must be protected by the housing of the machine itself.

Protected. Sometimes called *open dripproof*, ventilating openings are provided only on the lower part of the housing so that liquids dripping on the motor from above cannot enter the motor. This is probably the most widely used type [see Figure 21–13(a)].

Totally Enclosed Nonventilated (TENV). No openings at all are provided in the housing, and no special provisions are made for cooling the motor except for fins cast into the frame to promote convective cooling. The design protects the motor from harmful atmospheres [see Figure 21–13(b)].

Totally Enclosed Fan-cooled (TEFC). The totally enclosed fan-cooled TEFC design is similar to the TENV design, except a fan is mounted to one end of the shaft to draw air over the finned housing [see Figure 21–13(c)].

TEFC-XP. The TEFC-XP (explosion-proof) design is similar to the TEFC housing, except special protection is provided for electrical connections to prohibit fire or explosion in hazardous environments [see Figure 21–13(d)].

IEEE 841-2001 Severe Duty Motors. See Figure 21–15. This special type of housing is designed for demanding environments such as in steel mills, petrochemical plants, and pulp and paper mills where corrosive environments, high temperatures, and high humidity often occur. Extra fin area, some stainless steel components, and corrosion resistant paints and seals are employed.

Washdown Duty Motors. See Figure 21–16. When motors are used in food processing areas, clean rooms, medical facilities, and similar locations, this special type of housing is designed to be washed down with sanitary cleaning fluid and to withstand direct hosing.

Brake Motors. See Figure 21–17. Note the extension to the left side of the motor that houses a braking system that can be tied into a control system to stop the motor shaft quickly in case of safety issues or to accommodate process cycling. See Chapter 22 for more discussion of brake designs.

FIGURE 21–15 IEEE 841-2001 severe duty motor (Baldor Electric, Greenville, SC)

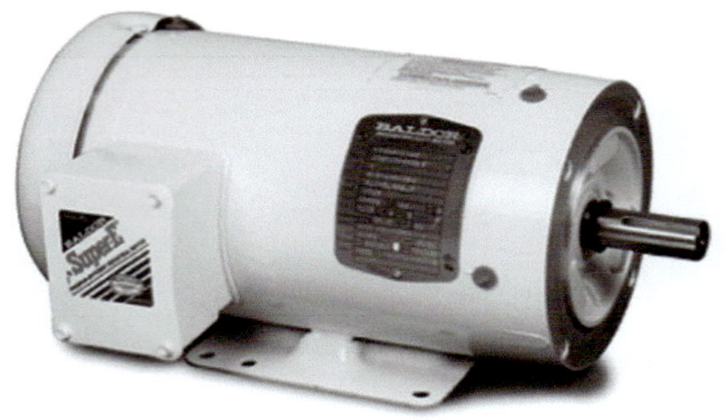

FIGURE 21–16 Washdown duty motor (Baldor Electric, Greenville, SC)

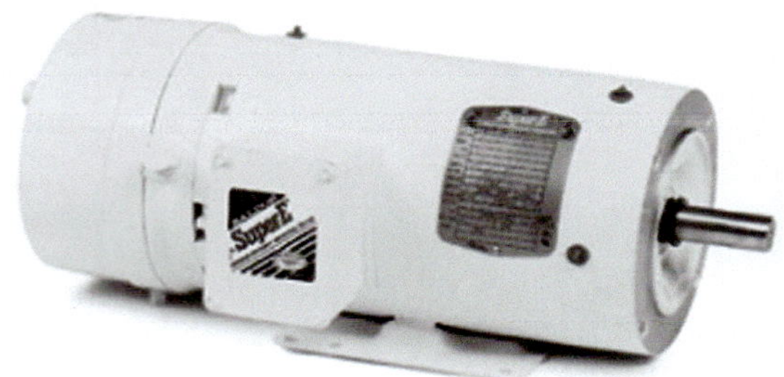

FIGURE 21–17 Brake motor (Baldor Electric, Greenville, SC)

Frame Sizes

The critical dimensions of motor frames are controlled by NEMA frame sizes. Included are the overall height and width; the height from the base to the shaft centerline; the shaft diameter, length, and keyway size; and mounting hole pattern dimensions. A few selected motor frame sizes for 1725-rpm, three-phase, induction, foot-mounted dripproof motors are listed in Table 21–3. For the descriptions of the dimensions, refer to Figure 21–18.

TABLE 21–3 Motor Frame Sizes

		Dimensions (in)								
hp	Frame size	*A*	*C*	*D*	*E*	*F*	*O*	*U*	*V*	Keyway
1/4	48	5.63	9.44	3.00	2.13	1.38	5.88	0.500	1.50	0.05 flat
1/2	56	6.50	10.07	3.50	2.44	1.50	6.75	0.625	1.88	3/16 × 3/32
1	143T	7.00	10.69	3.50	2.75	2.00	7.00	0.875	2.00	3/16 × 3/32
2	145T	7.00	11.69	3.50	2.75	2.50	7.00	0.875	2.00	3/16 × 3/32
5	184T	9.00	13.69	4.50	3.75	2.50	9.00	1.125	2.50	1/4 × 1/8
10	215T	10.50	17.25	5.25	4.25	3.50	10.56	1.375	3.13	5/16 × 5/32
15	254T	12.50	22.25	6.25	5.00	4.13	12.50	1.625	3.75	3/8 × 3/16
20	256T	12.50	22.25	6.25	5.00	5.00	12.50	1.625	3.75	3/8 × 3/16
25	284T	14.00	23.38	7.00	5.50	4.75	14.00	1.875	4.38	1/2 × 1/4
30	286T	14.00	24.88	7.00	5.50	5.50	14.00	1.875	4.38	1/2 × 1/4
40	324T	16.00	26.00	8.00	6.25	5.25	16.00	2.125	5.00	1/2 × 1/4
50	326T	16.00	27.50	8.00	6.25	6.00	16.00	2.125	5.00	1/2 × 1/4

Note: All motors are four-pole, three-phase, 60-Hz, AC induction motors. Refer to Figure 21–18 for description of dimensions.

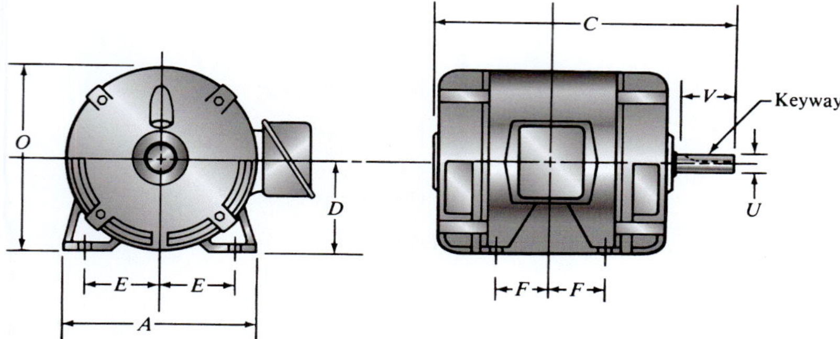

FIGURE 21–18 Key for NEMA standard motor frame dimensions listed in Table 21–3

21–9 CONTROLS FOR AC MOTORS

Motor controls must perform several functions, as outlined in Figure 21–19. The complexity of the control depends on the size and the type of motor involved. Small fractional or subfractional motors may sometimes be started with a simple switch that connects the motor directly to the full line voltage. Larger motors, and some smaller motors on critical equipment, require greater protection. See Internet sites 1, 3, and 5–9 and References 2–8.

The functions of motor controls are as follows:

1. To start and stop the motor
2. To protect the motor from overloads that would cause the motor to draw dangerously high current levels
3. To protect the motor from overheating
4. To protect personnel from contact with hazardous parts of the electrical system
5. To protect the controls from the environment
6. To prohibit the controls from causing a fire or an explosion
7. To provide controlled torque, acceleration, speed, or deceleration of the motor
8. To provide for the sequential starting of a series of motors or other devices
9. To provide for the coordinated operation of different parts of a system, possibly including several motors
10. To protect the conductors of the branch circuit in which the motor is connected

The proper selection of a motor control system requires knowledge of at least the following factors:

1. The type of electrical service: voltage and frequency; single- or three-phase; current limitations
2. The type and size of motor: power and speed ratings; full-load current rating; locked-rotor current rating
3. Operation desired: duty cycle (continuous, start/stop, or intermittent); single or multiple discrete speeds, or variable-speed operation; one-direction or reversing
4. Environment: temperature; water (rain, snow, sleet, sprayed, or splashed water); dust and dirt; corrosive gases or liquids; explosive vapors or dusts; oils or lubricants
5. Space limitations
6. Accessibility of controls
7. Noise or appearance factors

Starters

There are several classifications of motor starters: manual or magnetic; one-direction or reversing; two-wire or three-wire control; full-voltage or reduced-voltage starting; single-speed or multiple-speed; normal stopping, braking, or plug stopping. All of these typically include some form of overload protection, which will be discussed later.

Manual and Magnetic, Full-Voltage, One-Direction Starting

Figure 21–20 shows the schematic connection diagram for manual starters for single-phase and three-phase motors.

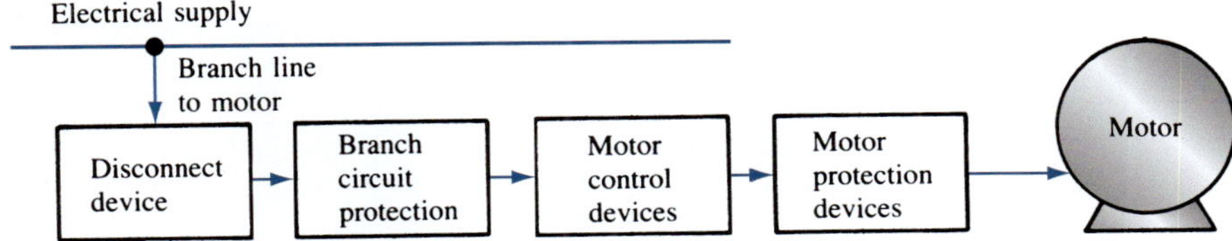

FIGURE 21–19 Motor control block diagram

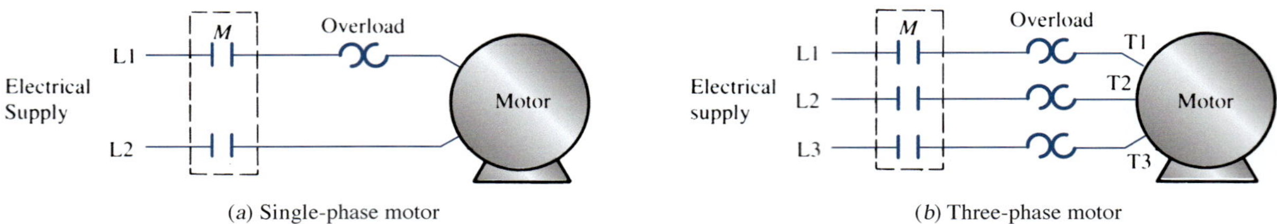

FIGURE 21–20 Manual starters. M = normally open contactors. All actuate together

The symbol M indicates a normally open contactor (switch) that is actuated manually, for example, by throwing a lever. The contactors are rated according to the motor power that they can safely handle. The power rating indirectly relates to the current drawn by the motor, and the contactor design must (1) safely make contact during the start-up of the motor, considering the high starting current; (2) carry the expected range of operating current without overheating; and (3) break contact without excessive arcing that could burn the contacts. The ratings are established by NEMA. Tables 21–4 and 21–5 show the ratings for selected NEMA starter sizes.

Note in Figure 21–20 that overload protection is required in all three lines for three-phase motors but in only one line of the single-phase motor.

Figure 21–21 shows the schematic connection diagrams for magnetic starters using two-wire and three-wire controls. The "start" button in the three-wire control is a momentary contact type. As it is actuated manually, the coil in parallel with the switch is energized, and it magnetically closes the line contactors marked M. The contacts remain closed until the stop button is pushed or until the line voltage drops to a set low value. (Remember, a low line voltage causes the motor to draw excessive current.) Either case causes the magnetic contactors to open, stopping the motor. The start button must be manually pushed again to restart the motor.

The two-wire control has a manually operated start button that stays engaged after the motor starts. As a safety feature, the switch will open when a low-voltage condition occurs. But when the voltage rises again to an acceptable level, the contacts close, restarting the motor. You must ensure that this is a safe operating mode.

TABLE 21–4 Ratings of AC Full-Voltage Starters for Single-Phase Power

NEMA size number	Current rating (amperes)	Power rating at given voltages: 110 V (hp)	110 V (kW)	220 V (hp)	220 V (kW)	440 and 550 V (hp)	440 and 550 V (kW)
00		1/2	0.37	3/4	0.56		
0	15	1	0.75	$1\frac{1}{2}$	1.12	$1\frac{1}{2}$	1.12
1	25	$1\frac{1}{2}$	1.12	3	2.24	5	3.73
2[a]	50	3	2.24	$7\frac{1}{2}$	5.60	10	7.46
3[a]	100	$7\frac{1}{2}$	5.60	15	11.19	25	18.65

[a]Applies to magnetically operated starters only.

TABLE 21–5 Ratings of AC Full-Voltage Starters for Three-Phase Power

NEMA size number	Current rating (amperes)	Power rating at given voltages: 110 V (hp)	110 V (kW)	220 V (hp)	220 V (kW)	440 and 550 V (hp)	440 and 550 V (kW)
00		3/4	0.56	1	0.75	1	0.75
0	15	$1\frac{1}{2}$	1.12	2	1.49	2	1.49
1	25	3	2.24	5	3.73	$7\frac{1}{2}$	5.60
2	50	$7\frac{1}{2}$	5.60	15	11.19	25	18.65
3	100	15	11.19	30	22.38	50	37.30

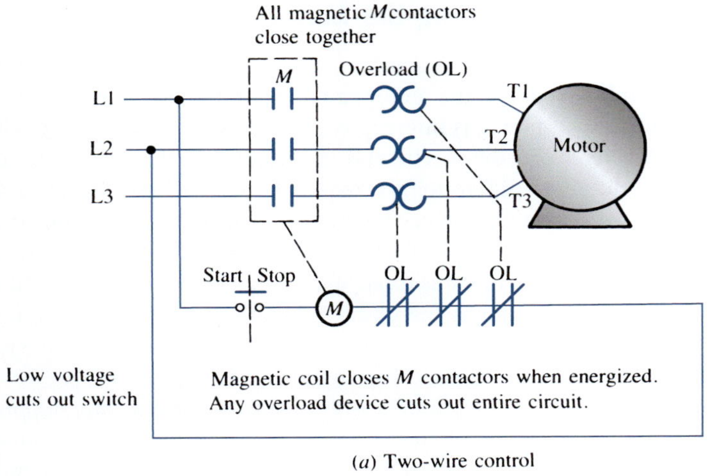

(*a*) Two-wire control

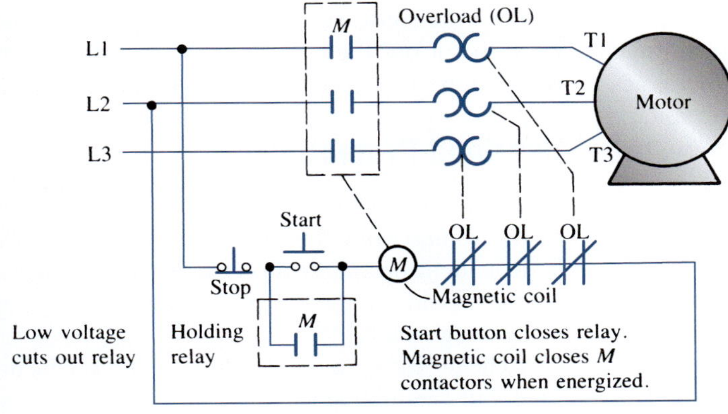

(*b*) Three-wire control

FIGURE 21–21 Magnetic starters for three-phase motors

Reversing Starters

Figure 21–22 shows the connection for a reversing starter for a three-phase motor. You can reverse the direction of rotation of a three-phase motor by interchanging any two of the three power lines. The *F* contactors are used for the forward direction. The *R* contactors would interchange L1 and L3 to reverse the direction. The *Forward* and *Reverse* pushbuttons actuate only one of the sets of contactors.

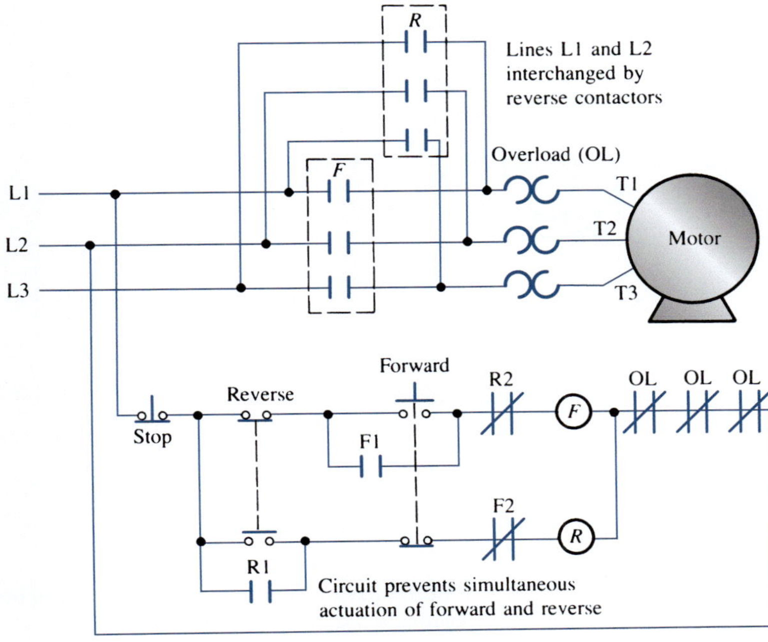

FIGURE 21–22 Reversing control for a three-phase motor

Reduced-Voltage Starting

The motors discussed in the previous sections and the circuits shown in Figures 21–20 through 21–22 employ full-voltage starting. That is, when the system is actuated, the full line voltage is applied to the motor terminals. This will give the maximum starting effort, but in some cases it is not desirable. To limit jerk, to control the acceleration of a load, and to limit the starting current, reduced-voltage starting is sometimes used. This gentle start is used on some conveyors, hoists, pumps, and similar loads.

Figure 21–23 shows one method of providing a reduced voltage to the motor when starting. The first action is the closing of the contactors marked *A*. Thus, the power to the motor passes through a set of resistors that reduces the voltage at each motor terminal. A typical reduction would be to approximately 65% of normal line voltage. The peak line current would be reduced to 65% of normal locked-rotor current, and the starting torque would be 42% of normal locked-rotor torque. (See Reference 4.) After the motor is accelerated, the main contactors *M* are closed, and full voltage is applied to the motor. A timer is typically used to control the sequencing of the *A* and *M* contactors.

Dual-Speed Motor Starting

A dual-speed motor with two separate windings to produce the different speeds can be started with the circuit shown in Figure 21–24. The operator selectively closes either the *F* (fast) or *S* (slow) contacts to obtain the desired speed. The other features of starting circuits discussed earlier can also be applied to this circuit.

Stopping the Motor

Where no special conditions exist when the system is shut down, the motor can be permitted to coast to a stop after the power is interrupted. The time required to stop will depend on the inertia and the friction in the system. If controlled, rapid stopping is required, external brakes can be used. *Brake motors*, which have a brake integral with the motor, are available. Typically, the design is of the "fail-safe" nature, in which the brake is disengaged by an electromagnetic coil when the motor is energized. When the motor is de-energized, either on purpose or because of power failure, the brake is actuated by mechanical spring force. See Figure 21–17.

On circuits with reversing starters, *plug stopping* can be used. When it is desired to stop the motor running in the forward direction, the control can be switched immediately to reverse. There would then be a decelerating torque applied to the rotor, stopping it quickly. Care must be exercised to cut out the reversing circuit when the motor is at rest to prevent it from continuing in the reverse direction.

Overload Protection

The chief cause of failure in electric motors is overheating of the wound coils due to excessive current. The current is dependent on the load on the motor. A short circuit, of course, would cause a virtually instantaneously high current of a damaging level.

The protection against a short circuit can be provided by *fuses*, but careful application of fuses to motors is essential. A fuse contains an element that literally melts when a particular level of current flows through it. Thus, the circuit is

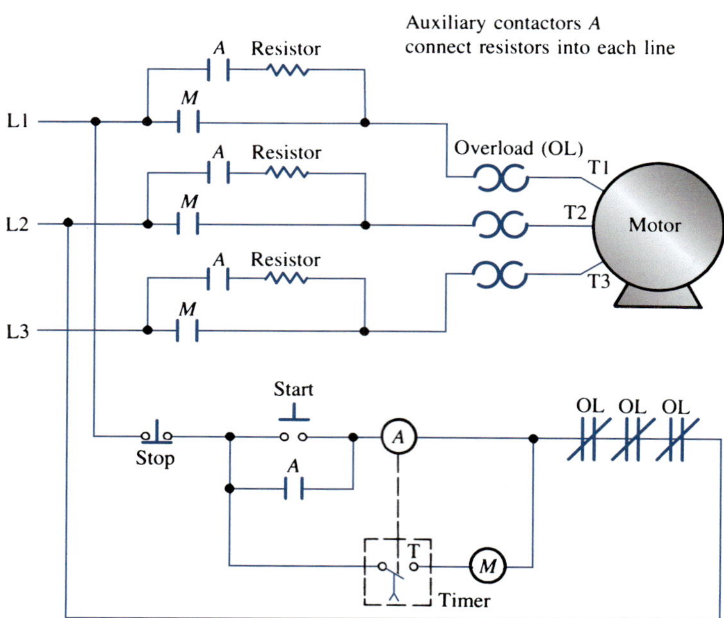

FIGURE 21–23 Reduced-voltage starting by primary resistor method

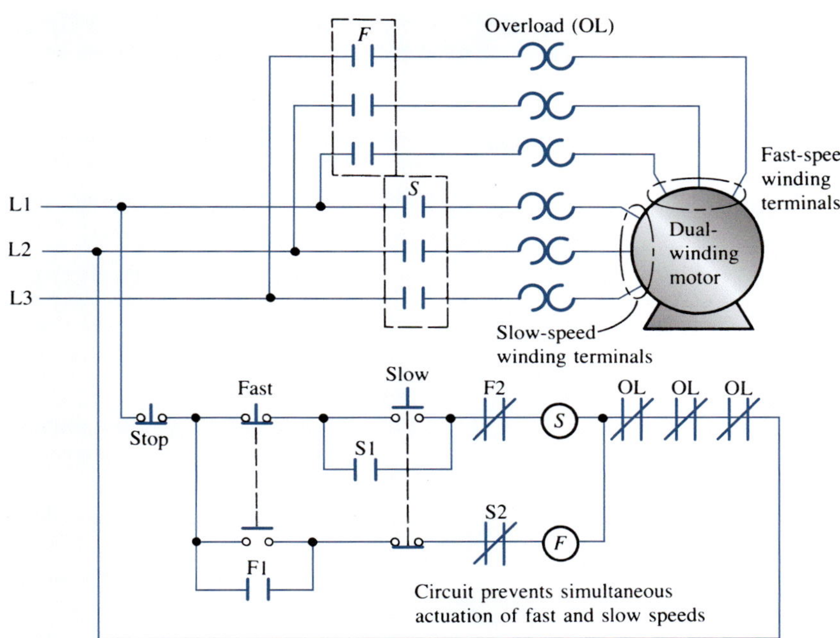

FIGURE 21–24 Speed control for a dual-winding, three-phase motor

opened. Reactivating the circuit would require replacing the fuse. Time-delay fuses, or "slow-blowing" fuses, are needed for motor circuits to prevent the fuses from blowing when the motor starts, drawing the relatively high starting current that is normal and not damaging. After the motor starts, the fuse will blow at a set value of overcurrent.

Fuses are inadequate for larger or more critical motors because they provide protection at only one level of overcurrent. Each motor design has a characteristic *overheating curve*, as shown in Figure 21–25. This indicates that the motor could withstand different levels of overcurrent for different periods of time. For example, for the motor heating curve of Figure 21–25, a current twice as high as the full-load current (200%) could exist for up to 9 min before a damaging temperature was produced in the windings. But a 400% overload would cause damage in less than 2 min. An ideal overload protection device would parallel the overheating curve of the given motor, always cutting out the motor at a safe current level, as shown in Figure 21–25. Devices are available commercially to

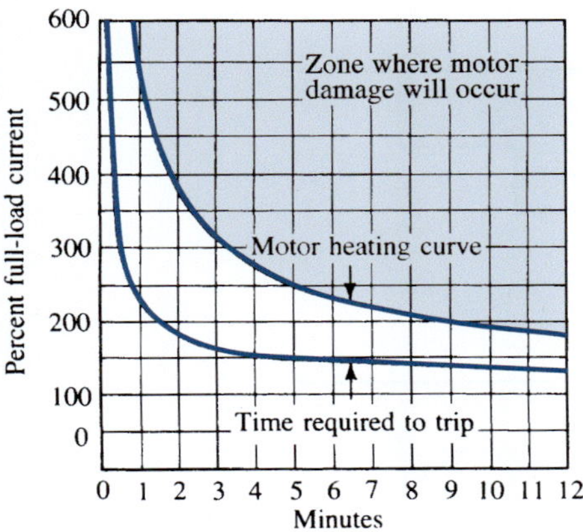

FIGURE 21–25 Motor heating curve and response curve of a typical overload protector (Square D by Schneider Electric Company, Palatine, IL)

provide this protection. Some use special melting alloys, bimetallic strips similar to a thermostat, or magnetic coils that are sensitive to the current flowing in them. Most large motor starters include overload protection integral with the starter.

Another type of overload protection uses a temperature-sensitive device inserted in the windings of the motor when it is made. Then it opens the motor circuit when the windings reach a dangerous temperature, regardless of the reason.

Solid-State Overload Relay

Difficulties with thermal element or bimetallic overload devices can be overcome by using a solid-state overload relay. Thermal devices that use a melting element require the replacement of that element after tripping, resulting in extra cost for maintenance supplies and personnel. Both thermal element and bimetallic overload devices are affected by varying ambient temperatures that can change the actual current flow protection level. Temperature compensation devices are available, but they require careful setting and knowledge of expected conditions. Solid-state overload relays overcome these difficulties because only sensed current flow level is used to produce the tripping action. They are inherently insensitive to ambient temperature swings. Furthermore, they can sense the current flow in each of the three windings of three-phase motors and provide protection if any one of the phases experiences a failure or a given increase in current. This provides protection not only for the motor but also for associated equipment that may be damaged if the motor fails suddenly. See Internet site 5 for additional information.

Enclosures for Motor Controls

As stated before, one of the functions of a motor control system is to protect personnel from contact with dangerous parts of the electrical system. Also, protection of the system from the environment must be provided. These functions are accomplished by the enclosure.

NEMA has established standards for enclosures for the variety of environments encountered by motor controls. The most frequent types are described in Table 21–6.

AC Variable-Speed Drives

Standard AC motors operate at a fixed speed for a given load if powered by AC power at a fixed frequency, for example, 60 Hz. Variable-speed operation can be obtained with a control system that produces a variable frequency power. Two such control types are in common use: the *six-step method* and the *pulse-width modulation (PWM) method.* Either system takes the 60-Hz line voltage and first rectifies it to a DC voltage. The six-step method then uses an inverter to produce a series of square waves that provides a voltage to the motor winding which varies both voltage and frequency in six steps per cycle. In the PWM system, the DC voltage is input to an inverter that produces a series of pulses of variable width. The rate of polarity reversals determines the frequency applied to the motor. See Figures 21–26 and 21–27.

TABLE 21–6 Motor Control Enclosures

NEMA design number	Description
1	General-purpose: indoor use; not dusttight
3	Dusttight, raintight: outdoor-weather-resistant
3R	Dusttight, rainproof, sleet-resistant
4	Watertight: can withstand a direct spray of water from a hose; used on ships and in food processing plants where washdown is required
4X	Watertight, corrosion-resistant
7	Hazardous locations, class I: can operate in areas where flammable gases or vapors are present
9	Hazardous locations, class II: combustible dust areas
12	Industrial use: resistant to dust, lint, oil, and coolants
13	Oiltight, dusttight

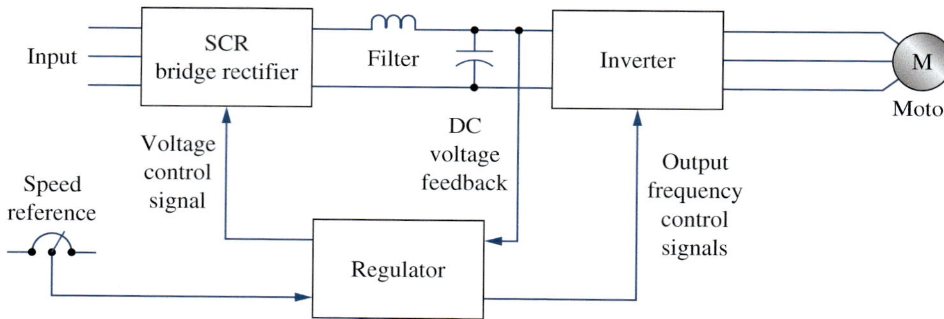

(*a*) Schematic diagram of variable voltage inverter

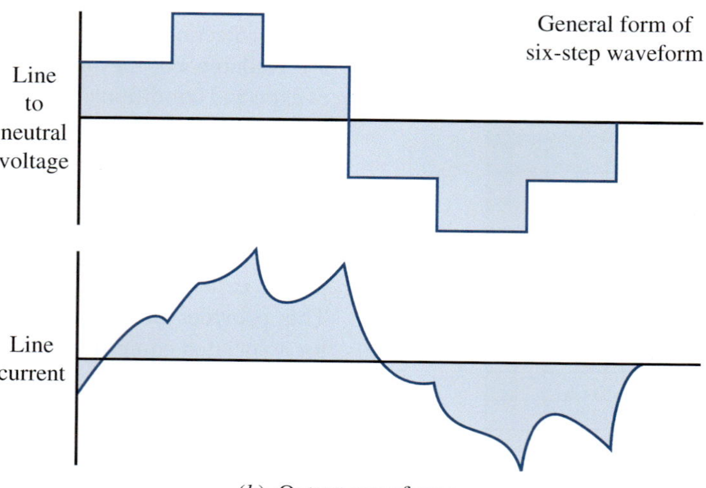

(*b*) Output waveforms

FIGURE 21–26 Six-step method of variable-speed AC motor control (Rockwell Automation/Allen-Bradley, Milwaukee, WI)

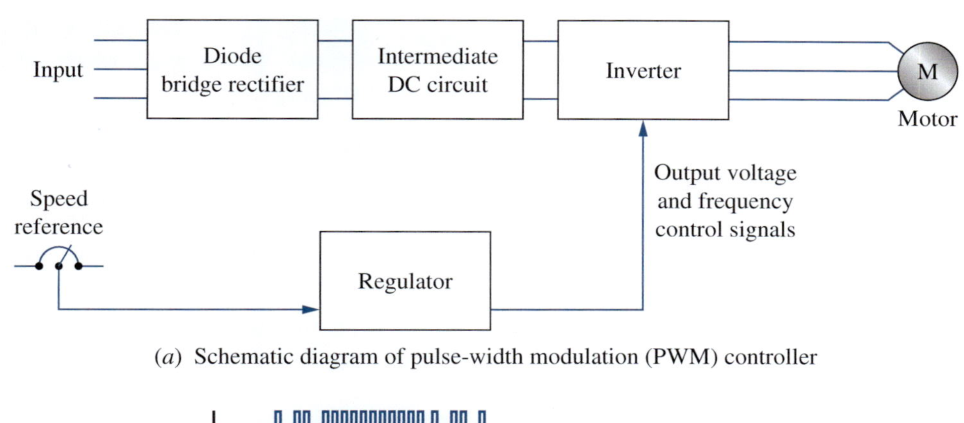

(*a*) Schematic diagram of pulse-width modulation (PWM) controller

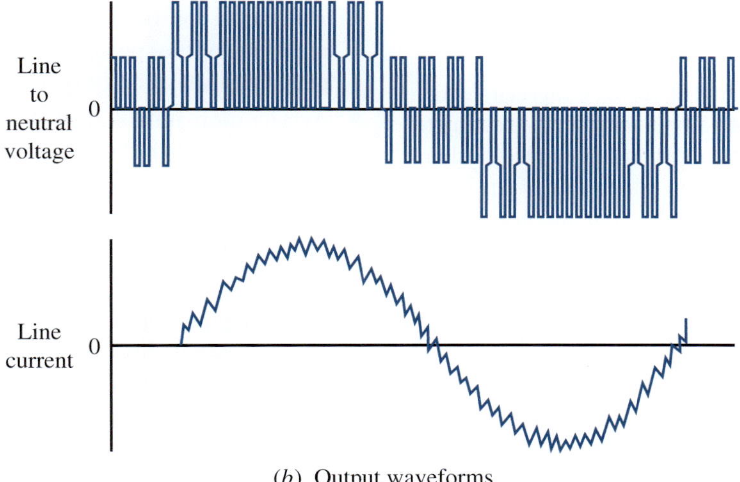

(*b*) Output waveforms

FIGURE 21–27 Pulse-width modulation method of variable-speed AC motor control (Rockwell Automation/Allen-Bradley, Milwaukee, WI)

Reasons for Applying Variable-Speed Drives

It is often desirable to vary the speed of mechanical systems to obtain operating characteristics that are more nearly optimum for the application. For example:

1. Conveyor speed can be varied to match the demand of production.
2. The delivery of bulk materials to a process can be varied continuously.
3. Automatic control can provide synchronization of two or more system components.
4. Dynamic control of system operation can be used during start-up and stopping sequences, for torque control, or for in-process acceleration and deceleration control, often necessary when processing continuous webs such as paper or plastic film.
5. Spindle speeds of machine tools can be varied to produce optimum cutting for given materials, depth of cut, feeds, or cutting tools.
6. The speeds of fans, compressors, and liquid pumps can be varied in response to needs for cooling or for product delivery.

All of these situations allow for more flexible and better process control. Cost savings are also obtained, particularly for item 6. The difference in the power required to operate a pump at two speeds is proportional to the cube of the ratio of the speeds. For example, if the speed of the motor is reduced to one-half of its original speed, the power required to operate the pump is reduced to $1/8$ of the original power. Matching the pump speed to the required delivery of the fluid can accumulate significant savings. Similar savings result for fans and compressors.

21–10 DC POWER

DC motors have several inherent advantages over AC motors, as discussed in the next section. A disadvantage of DC motors is that a source of DC power must be available. Most residential, commercial, and industrial locations have only AC power provided by the local utility. Four approaches are used to provide DC power: batteries, hydrogen fuel cells, generators, and rectifiers. See Internet sites 5–9 for additional information on the types of power and control, and sites 10–13 for extended information about batteries.

1. ***Batteries:*** Typically available battery voltages and their uses are:
   a. Alkaline and rechargeable nickel metal hydride: 1.5, 3.6, 4.8, 9.0 V
      - Flashlights, portable phones, sensors, small kitchen appliances, small power tools

b. Nickel cadmium: 9.6, 12.0, 14.4, 18.0, 24.0 V, power tools
c. Sealed lead acid (SLA), flooded valve regulated lead acid (VRLA), absorbed glass mats (AGM): 2.0, 4.0, 6.0, 8.0, 12.0, 24.0, 36.0 V
   - Automotive, marine, wheelchairs, small scooters, forklift trucks, automatic guided vehicles, golf carts, lawn mowers, garden tractors, floor scrubbers, utility switches, network power units, uninterruptable power supplies (UPS), small solar power storage units, telecommunications equipment.
d. Lithium ion: 12.0, 18.0, 24.0, 28.0, 36.0, 48.0 V
   - Commercial-grade power tools, large capacity standby power, hybrid vehicles, plug-in electric vehicles

2. ***Hydrogen fuel cells:*** In a single hydrogen fuel cell, hydrogen passing through a plate-type anode, an electrolyte, and a cathode converts chemical energy to electrical energy between two electrodes on either side of the cell at approximately 1.0 V per cell. Only water and heat are discharged from the cells. Multiple cells are stacked together to produce sufficient voltage and power capacity for a given application. Potential uses include small handheld devices and laptop computers, telecommunication equipment, auxiliary power for residential, commercial, or industrial applications, portable power for remote installations, and all types of transportation. Power capacities range from about 50 W to several hundred kW and higher. Systems rated in the MW range are being considered. Still considered an emerging technology, rapid advancements are being made.
3. ***Generators:*** Powered by AC electric motors, internal combustion engines, turbine engines, wind devices, water turbines, and so on; DC generators produce pure DC. The usual voltages are 115 and 230 V. Some industries maintain such generators to provide DC power throughout the plant.
4. ***Rectifiers:*** *Rectification* is the process of converting AC power with its sinusoidal variation of voltage with time to DC power, which ideally is nonvarying. A readily available device is the *silicon-controlled rectifier (SCR)*. One difficulty with rectification of AC power to produce DC power is that there is always some amount of "ripple," a small variation of voltage as a function of time. Excessive ripple can cause overheating of the DC motor. Most commercially available SCR devices produce DC power with an acceptably low ripple. Table 21–7 lists the commonly used DC voltage ratings for motors powered by rectified AC power as defined by NEMA.

21–11 DC MOTORS

The advantages of direct current motors are summarized here:

- The speed is adjustable by use of a simple rheostat to adjust the voltage applied to the motor.
- The direction of rotation is reversible by switching the polarity of the voltage applied to the motor.

TABLE 21–7 DC Motor Voltage Ratings

Input AC voltage	DC motor rating	NEMA code
115 V AC, one-phase	90 V DC	K
230 V AC, one-phase	180 V DC	K
230 V AC, three-phase	240 V DC	C or D
460 V AC, three-phase	500 V DC or	C or D
	550 V DC	
460 V AC, three-phase	240 V DC	E

- Automatic control of speed is simple to provide for matching of the speeds of two or more motors or to program a variation of speed as a function of time.
- Acceleration and deceleration can be controlled to provide the desired response time or to decrease jerk.
- Torque can be controlled by varying the current applied to the motor. This is desirable in tension control applications, such as the winding of film on a spool.
- Dynamic braking can be obtained by reversing the polarity of the power while the motor is rotating. The reversed effective torque slows the motor without the need for mechanical braking.
- DC motors typically have quick response, accelerating quickly when voltage is changed, because they have a small rotor diameter, giving them a high ratio of torque to inertia.

DC motors have electric windings in the rotor, and each coil has two connections to the commutator on the shaft. The commutator is a series of copper segments through which the electric power is transferred to the rotor. The current path from the stationary part of the motor to the commutator is through a pair of brushes, usually made of carbon, which are held against the commutator by light coil or leaf springs. Maintenance of the brushes is one of the disadvantages of DC motors.

DC Motor Types

Four commonly used DC motor types are the *shunt-wound, series-wound, compound-wound,* and *permanent magnet* motors. They are described in terms of their speed/torque curves in a manner similar to that used for AC motors. One difference here is that the speed axis is expressed in percentage of *full-load rated speed,* rather than a percentage of synchronous speed, as that term does not apply to DC motors.

Shunt-Wound DC Motor. The electromagnetic field is connected in parallel with the rotating armature, as sketched in Figure 21–28. The speed/torque curve shows relatively good speed regulation up to approximately two times full-load torque, with a rapid drop in speed after that point. The speed at no load is only slightly higher than the full-load speed. Contrast this with the series-wound motor described next. Shunt-wound motors are used mainly for small fans and blowers.

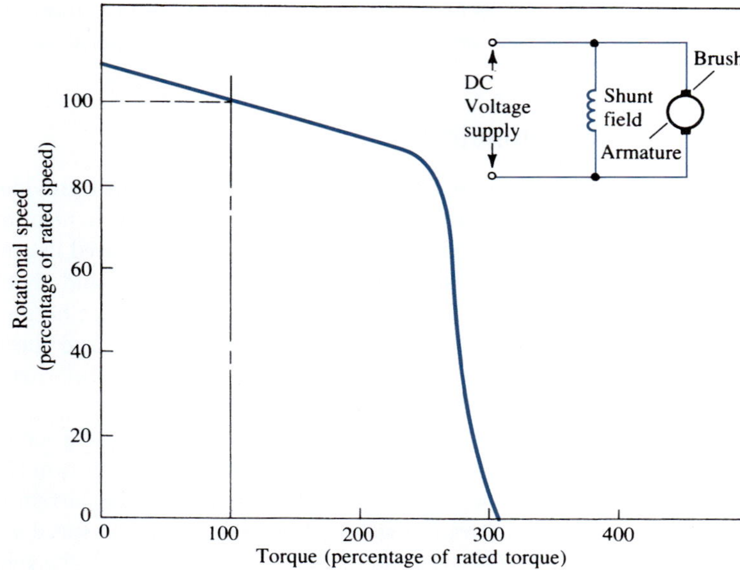

FIGURE 21–28 Shunt-wound DC motor performance curve

Series-Wound DC Motor. The electromagnetic field is connected in series with the rotating armature, as shown in Figure 21–29. The speed/torque curve is steep, giving the motor a soft performance that is desirable in cranes, hoists, and traction drives for vehicles. The starting torque is very high, as much as 800% of full-load rated torque. A major difficulty, however, with series-wound motors is that the no-load speed is theoretically unlimited. The motor could reach a dangerous speed if the load were to be accidentally disconnected. Safety devices, such as overspeed detectors that shut off the power, should be used.

Compound-Wound DC Motors. The compound-wound DC motor employs both a series field and a shunt field, as sketched in Figure 21–30. It has a performance somewhat between that of the series-wound and the shunt-wound motors. It has fairly high starting torque and a soft speed characteristic, but it has an inherently controlled no-load speed. This makes it good for cranes, which may suddenly lose their loads. The motor would run slowly when heavily loaded for safety and control, and fast when lightly loaded to improve productivity.

Permanent Magnet DC Motors. Instead of using electromagnets, the permanent magnet DC motor uses permanent magnets to provide the field for the armature. The direct current passes through the armature, as shown in Figure 21–31. The field is nearly constant at all times and results in a linear speed/torque curve. Current draw also varies linearly with torque. Applications include fans

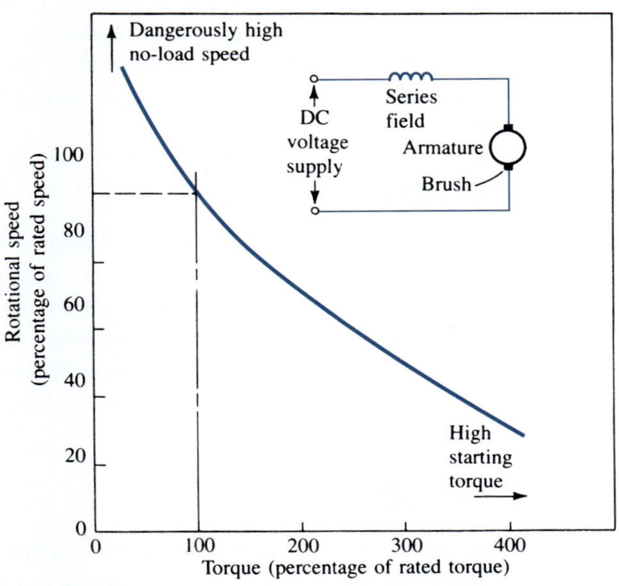

FIGURE 21–29 Series-wound DC motor performance curve

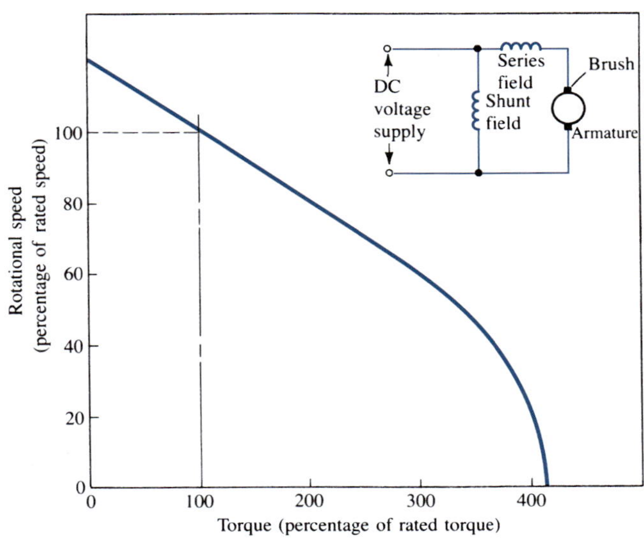

FIGURE 21–30 Compound-wound DC motor performance curve

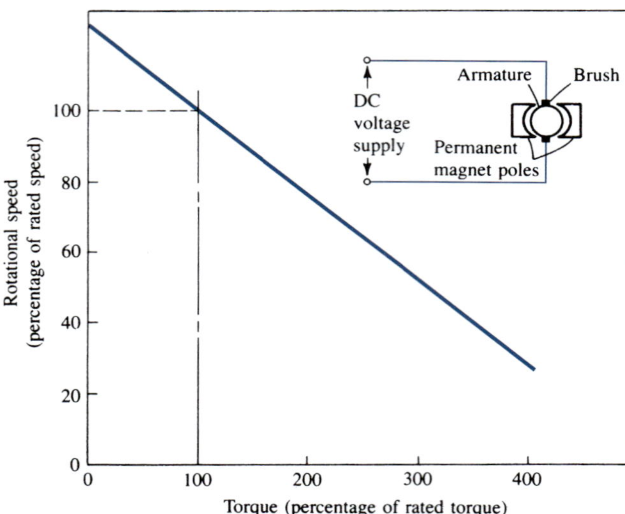

FIGURE 21–31 Permanent magnet DC motor performance curve

and blowers to cool electronics packages in aircraft, small actuators for control in aircraft, automotive power assists for windows and seats, and fans in automobiles for heating and air conditioning. These motors frequently have built-in gear-type speed reducers to produce a low-speed, high-torque output.

21–12 DC MOTOR CONTROL

Starting DC motors presents essentially the same problems as discussed for AC motors in terms of limiting the starting current and the provision of switching devices and holding relays of sufficient capacity to handle the operating loads. The situation is made somewhat more severe, however, by the presence of the commutators in the rotor circuit which are more sensitive to overcurrent.

Speed control is provided by variation of the resistance in the lines containing the armature or the field of the motor. The details depend on whether the motor is a series, shunt, or compound type. The variable-resistance device, sometimes called a *rheostat*, can provide either stepwise variation in resistance or continuously varying resistance. Figure 21–32 shows the schematic diagrams for several types of DC motor speed controls. See Internet sites 1, 3, and 5–9 and References 2–8.

21–13 OTHER TYPES OF MOTORS

Torque Motors

As the name implies, *torque motors* are selected for their ability to exert a certain torque rather than for a rated power. Frequently this type of motor is operated at a stalled condition to maintain a set tension on a load. The continuous operation at slow speed or at zero speed causes heat generation to be a potential problem. In severe cases, external cooling fans may be required.

By special design, several of the AC and DC motor types discussed elsewhere in this chapter can be used as torque motors.

Servomotors

Either AC or DC *servomotors* are available to provide automatic control of position or speed of a mechanism in response to a control signal. Such motors are used in aircraft actuators, instruments, computer printers, and machine tools. Most have rapid response characteristics because of the low inertia of the rotating components and the relatively high torque exerted by the motor. See Internet sites 3, 7–9, and 15.

Figure 21–33 shows a schematic diagram of a servomotor controller system. Three control loops are shown: (1) the position loop, (2) the velocity loop, and (3) the current loop. Speed control is effected by sensing the motor speed with a tachometer and feeding the signal back through the velocity loop to the controller. Position is sensed by an optical encoder or a similar device on the driven load, with the signal fed back through the position loop to the controller. The controller sums the inputs, compares them with the desired value set by the control program, and generates a signal to control the motor. Thus, the system is a closed-loop servo-control. Typical uses for this type of control would be for numerical control machine tools, special-purpose assembly machines, and aircraft control surface actuators.

Stepping Motors

A stream of electronic pulses is delivered to a *stepping motor*, which then responds with a fixed rotation (step) for each pulse. Thus, a very precise angular position can be obtained by counting and controlling the number of pulses delivered to the motor. Several step angles are available in commercially provided motors, such as 1.8°, 3.6°, 7.5°, 15°, 30°, 45°, and 90°. When the pulses are stopped, the motor stops automatically and is held in position. Because many of these motors are connected through a gear-type speed reducer to the load, very precise positioning is possible to a small fraction of a step. Also, the reducer provides a torque increase. See Internet site 14.

Brushless Motors

The typical DC motor requires brushes to make contact with the rotating commutator on the shaft of the motor. This is a major failure mode of such motors. In the *brushless DC motor*, the switching of the rotor coils is accomplished by solid-state electronic devices, resulting in very long life. The emission of electromagnetic interference is likewise reduced compared with brush-type DC motors.

Printed Circuit Motors

The rotor of the *printed circuit motor* is a flat disc that operates between two permanent magnets. The resulting design has a relatively large diameter and small axial length; sometimes it

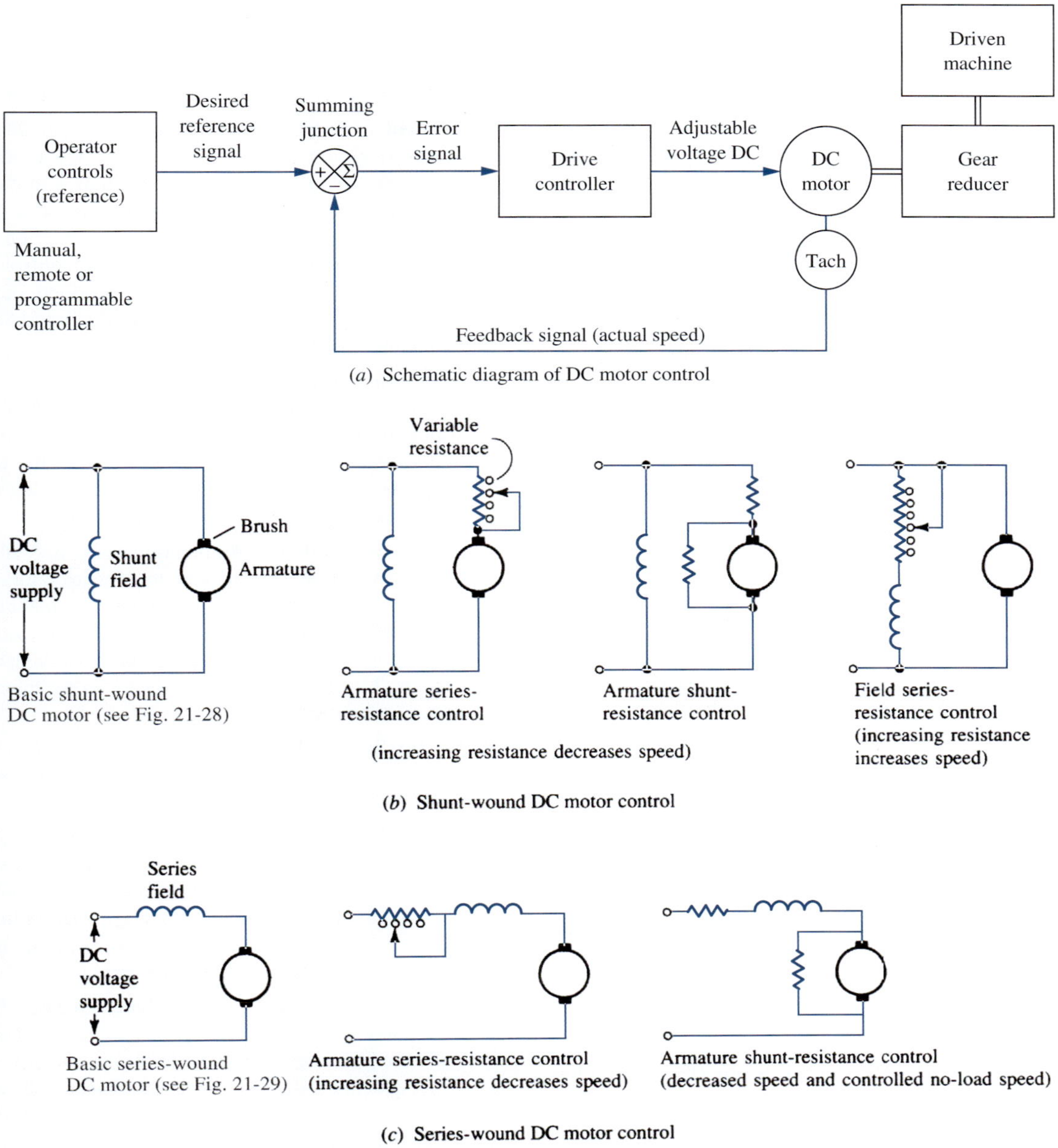

Driven machine
Operator controls (reference)
Desired reference signal
Summing junction
Error signal
Drive controller
Adjustable voltage DC
DC motor
Gear reducer
Tach
Manual, remote or programmable controller
Feedback signal (actual speed)
(a) Schematic diagram of DC motor control
Variable resistance
DC voltage supply
Shunt field
Brush
Armature
Basic shunt-wound DC motor (see Fig. 21-28)
Armature series-resistance control
Armature shunt-resistance control
(increasing resistance decreases speed)
Field series-resistance control (increasing resistance increases speed)
(b) Shunt-wound DC motor control
Series field
DC voltage supply
Basic series-wound DC motor (see Fig. 21-29)
Armature series-resistance control (increasing resistance decreases speed)
Armature shunt-resistance control (decreased speed and controlled no-load speed)
(c) Series-wound DC motor control

FIGURE 21–32 DC motor control [(*a*), Rockwell Automation/Allen-Bradley, Milwaukee, WI]

Position loop
Velocity loop
Current loop
Tach
Encoder
Position controller
Motor
Transmission
Load
Servo controller

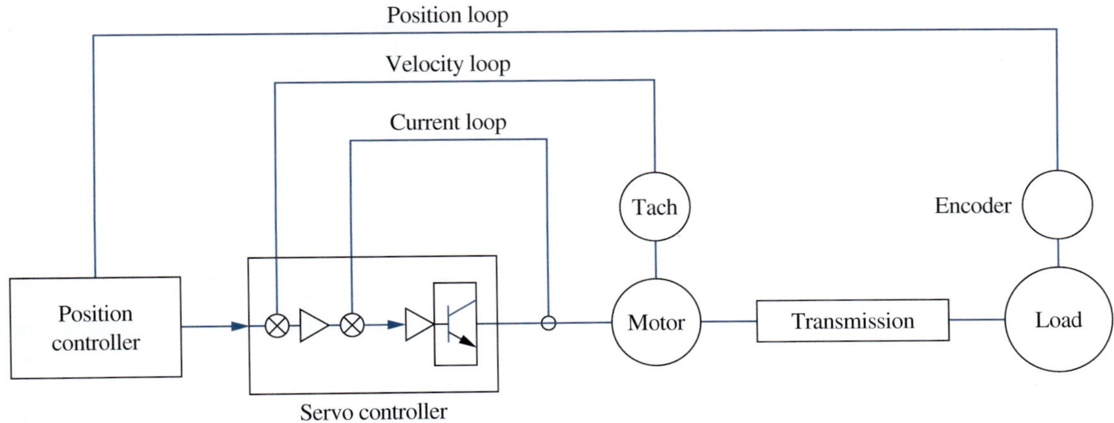

FIGURE 21–33 Servomotor controller system (Rockwell Automation/Allen-Bradley, Milwaukee, WI)

is called a *pancake motor.* The rotor has a very low inertia, so high acceleration rates are possible.

Linear Motors

Linear motors are electrically similar to rotary motors, except the components, stator and rotor, are laid flat instead of being formed into a cylindrical shape. Types include brush-type DC motors, brushless DC motors, stepping motors, and the single-phase type of motor. Capacity is measured in terms of the force that the motor can exert, ranging typically from a few pounds to 2500 lb. Speeds range from 100 in/s (1.02 to 2.54 m/s). See Internet sites 9 and 15.

REFERENCES

1. Avallone, Eugene P., Theodore Baumeister III., and Ali Sadegh. *Marks' Handbook for Mechanical Engineers.* 11th ed. New York: McGraw-Hill, 2011.
2. Hubert, Charles I. I. *Electric Machines: Theory, Operating Applications, and Controls.* 2nd ed. Upper Saddle River, NJ: Prentice Hall, 2002.
3. Skvarenina, Timothy L., and William E. DeWitt. *Electrical Power and Controls.* 2nd ed. Upper Saddle River, NJ: Prentice Hall, 2005.
4. Wildi, Theodore. *Electrical Machines, Drives, and Power Systems.* 6th ed. Upper Saddle River, NJ: Prentice Hall, 2006.
5. Hughes, Austin. *Electrical Motors and Drives: Fundamentals, Types and Applications.* Amsterdam, Holland: Elsevier Science & Technology Books, 2006.
6. Polka, David. *Motors and Drives: A Practical Technology Guide.* Research Triangle Park, NC: ISA: The Instrumentation, Systems, and Automation Society, 2003.
7. Herman, Stephen. *Industrial Motor Control.* 6th ed. Florence, KY: Delmar Cengage Learning, 2010.
8. Miller, Rex, and Miller, Mark. *Industrial Electricity and Motor Controls.* New York: McGraw-Hill, 2008.

INTERNET SITES FOR ELECTRIC MOTORS AND CONTROLS

1. **Reliance Electric/Baldor Electric.** www.reliance.com Manufacturer of AC and DC motors and the associated controls. The site provides a Motor Technical Reference resource giving motor performance, applications guidelines, motor construction information, and an online catalog.
2. **Regal Beloit EPC, Inc.** www.regal-beloit.com/rbcelectricmotors Manufacturer of electric motors for driving pumps, conveyors, food processing machinery, machine tools, office equipment, and other commercial and industrial applications. Sizes range from 1/12 to 1250 hp.
3. **Baldor Electric Company.** www.baldor.com Manufacturer of AC and DC motors, gear motors, servomotors, linear motors, generators, linear motion products, and controls for a wide range of industrial and commercial applications. An online catalog, CAD drawings, and performance charts are included.
4. **U.S. Electric Motors Company.** www.usmotors.com Manufacturer of a wide variety of AC and DC electric motors from $\frac{1}{4}$ to 4000 hp for general and specific applications. An online catalog is included. Part of Nidec Motor Corporation.
5. **Schneider Electric.** www.schneider-electric.us Manufacturer of electrical control devices and systems for industry, energy and infrastructure, buildings, residential, and data centers/networks fields. The industry market includes process automation, machine control and monitoring, power supply and distribution, energy monitoring and control, and many others. Some motor control-related products such as starters and contactors are offered under the brand Square D by Schneider Electric.
6. **Eaton/Electrical-USA.** www.eaton.com/electrical/USA Manufacturer of a wide range of electrical control and power distribution products for industrial, commercial, and residential applications. Select "Product and services" for motor control centers, circuit breakers, power conditioning equipment, switchgear, brakes, and many other products.
7. **Allen-Bradley/Rockwell Automation.** www.rockwellautomation.com Manufacturer of a wide variety of controls for automation, including motor contactors, AC motor drives, motor control centers, servomotors, programmable logic controllers, sensors, relays, circuit protection devices, network communication, and control systems. Use www.ab.com/catalogs to access information on motor controls and related products.
8. **GE Intelligent Platforms.** www.ge-ip.com Manufacturer of an extensive line of programmable logic controllers, motion controls, and process solutions.
9. **Parker Automation.** www.parkermotion.com Manufacturer of a wide range of automation devices and systems for industrial applications. Includes the Compumotor line of servomotors and controllers and the Trilogy Systems line of linear electric motors and linear positioners for industrial automation. Part of Parker Hannifin Corporation.
10. **Deka Batteries—East Penn Manufacturing Co.** www.dekabatteries.com Manufacturer of a wide variety of sizes of lead-acid batteries for power tools, garden equipment, network electric power systems, industrial mobile equipment, golf carts, wheelchairs, railroad diesel engine starting systems, and many others.
11. **Exide Technologies.** www.exide.com Manufacturer of a full line of batteries for automotive, truck, lawn and garden, commercial, marine, industrial forklifts, automated guided vehicles, telecommunications, computer backup and security systems, emergency lighting, and various military equipment.
12. **U.S. Battery.** www.usbattery.com Manufacturer of 6 V, 8 V, and 12 V batteries for golf carts, industrial sweepers and scrubbers, marine, and multipurpose applications.
13. **Battery Council International.** www.batterycouncil.org An organization of manufacturers and users of batteries and battery-powered equipment for training, recommended test procedures, charging, and safety.
14. **Oriental Motor U.S.A. Corporation.** www.orientalmotor.com Manufacturer of stepping motors with either AC or DC input power and low speed synchronous motors for a variety of industrial automation and commercial product applications.
15. Beckhoff Drive Technology. www.beckhoff.com Manufacturer of brushless synchronous servomotors and synchronous linear motors for a variety of industrial automation and commercial product applications.

PROBLEMS

1. List six items that must be specified for electric motors.
2. List eight factors to be considered in the selection of an electric motor.
3. Define *duty cycle.*
4. How much variation in voltage will most AC motors tolerate?
5. State the relationship among torque, power, and speed.
6. What does the abbreviation AC stand for?
7. Describe and sketch the form of single-phase AC power.
8. Describe and sketch the form of three-phase AC power.
9. What is the standard frequency of AC power in the United States?
10. What is the standard frequency of AC power in Europe?
11. What type of electrical power is available in a typical American residence?
12. How many conductors are required to carry single-phase power? How many for three-phase power?
13. Assume that you are selecting an electric motor for a machine to be used in an industrial plant. The following types of AC power are available: 120-V, single-phase; 240-V, single-phase; 240-V, three-phase; 480-V, three-phase. In general, for which type of power would you specify your motor?
14. Define *synchronous speed* for an AC motor.
15. Define *full-load speed* for an AC motor.
16. What is the synchronous speed of a four-pole AC motor when running in the United States? In France?
17. A motor nameplate lists the full-load speed to be 3450 rpm. How many poles does the motor have? What would its approximate speed be at zero load?
18. If a four-pole motor operates on 400-Hz AC power, what will its synchronous speed be?
19. If an AC motor is listed as a normal-slip, four-pole/six-pole motor, what will its approximate full-load speeds be?
20. What type of control would you use to make an AC motor run at variable speeds?
21. Describe a C-face motor.
22. Describe a D-flange motor.
23. What does the abbreviation NEMA stand for?
24. Describe a protected motor.
25. Describe a TEFC motor.
26. Describe a TENV motor.
27. What type of motor enclosure would you specify to be used in a plant that manufactures baking flour?
28. What type of motor would you specify for a meat grinder if the motor is to be exposed?
29. Figure P21–29 shows a machine that is to be driven by a 5-hp, protected, foot-mounted AC motor having a 184T frame. The motor is to be aligned with the shaft of the driven machine. Make a complete dimensioned drawing, showing standard side and top views of the machine and the motor. Design a suitable mounting base for the motor showing the motor mounting holes.
30. Define *locked-rotor torque.* What is another term used for this parameter?
31. What is meant if one motor has a poorer speed regulation than another?
32. Define *breakdown torque.*
33. Name the four most common types of single-phase AC motors.
34. Refer to the AC motor performance curve in Figure P21–34.
    a. What type of motor is the curve likely to represent?
    b. If the motor is a six-pole type, rated at 0.75 hp, how much torque can it exert at the rated load?
    c. How much torque can the motor develop to start a load?
    d. What is the breakdown torque for the motor?
35. Repeat Parts (b), (c), and (d) of Problem 34 if the motor is a two-pole type, rated at 1.50 kW.
36. A cooling fan for a computer is to operate at 1725 rpm, direct-driven by an electric motor. The speed/torque curve for the fan is shown in Figure P21–36. Specify a suitable motor, giving the type of motor, horsepower, and number of poles.
37. Figure P21–37 shows the speed/torque curve for a household refrigeration compressor, designed to operate at 3450 rpm. Specify a suitable motor, giving the type, power rating in watts, and number of poles.
38. How is speed adjusted for a wound-rotor motor?
39. What is the full-load speed of a 10-pole synchronous motor?

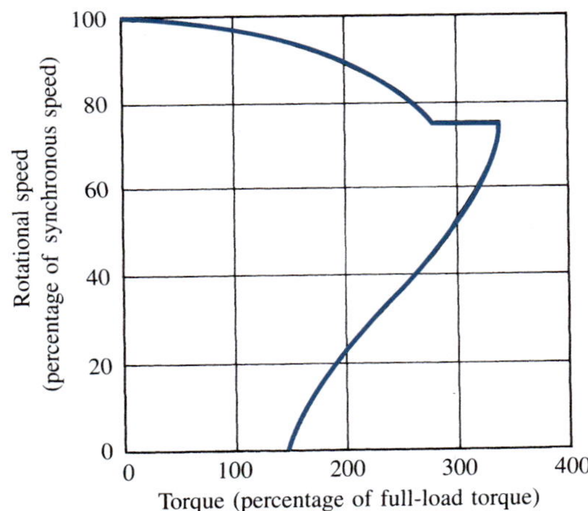

FIGURE P21–34 (Problems 34 and 35)

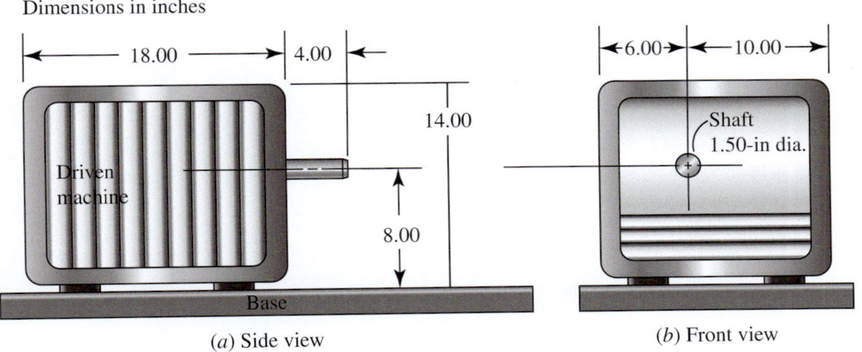

FIGURE P21–29

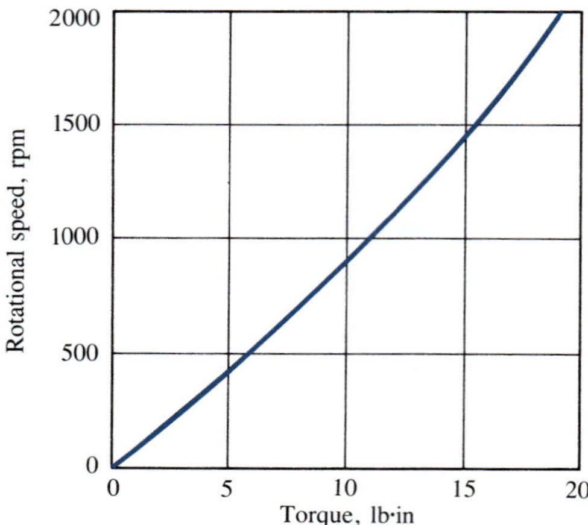

FIGURE P21–36

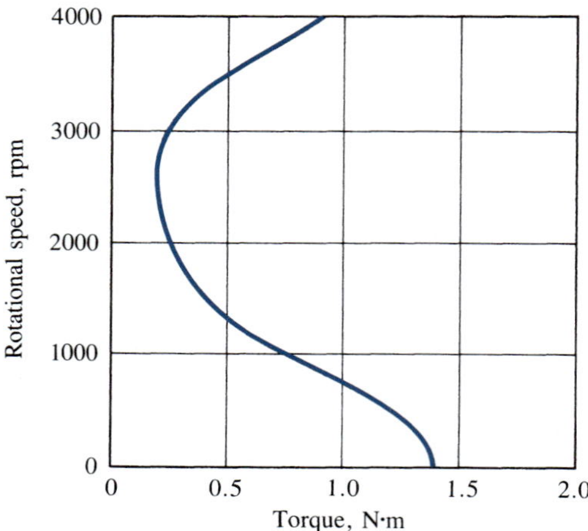

FIGURE P21–37

40. What is meant by the term *pull-out torque,* as applied to a synchronous motor?
41. Discuss the reasons that universal motors are frequently used for handheld tools and small appliances.
42. Why is the adjective *universal* used to describe a universal motor?
43. Name four ways to produce DC power.
44. List 12 common DC voltages.
45. What is an SCR control? For what is it used?
46. If a DC motor drive advertises that it produces *low ripple,* what does the term mean?
47. If you want to use a DC motor in your home, and your home has only standard 115 V AC, single-phase power, what will you need? What type of motor should you get?
48. List seven advantages of DC motors with respect to AC motors.
49. Discuss two disadvantages of DC motors.
50. Name four types of DC motors.
51. What happens to a series-wound DC motor if the load on the motor falls to nearly zero?
52. Assume that a permanent magnet DC motor can exert a torque of 15.0 N · m when operating at 3000 rpm. What torque could it exert at 2200 rpm?
53. List 10 functions of a motor control.
54. What size motor starter is required for a 10-hp, three-phase motor, operating on 220 V?
55. A single-phase, 110-V AC motor has a nameplate rating of 1.00 kW. What size motor starter is required?
56. What does the term *plug stopping* mean, and how is it accomplished?
57. Why is a fuse an inadequate protection device for an industrial motor?
58. What type of motor control enclosure would you specify for use in a car wash?
59. What could you do to the control circuit for a standard series DC motor to give it a controlled no-load speed?
60. What happens if you connect a resistance in series with the armature of a shunt-wound DC motor?
61. What happens if you connect a resistance in series with the shunt field of a shunt-wound DC motor?

CHAPTER TWENTY TWO

MOTION CONTROL: CLUTCHES AND BRAKES

THE BIG PICTURE — Motion Control: Clutches and Brakes

Discussion Map

- A *brake* is a device used to bring a moving system to rest, to slow its speed, or to control its speed to a certain value under varying conditions.
- A *clutch* is a device used to connect or disconnect a driven component from the prime mover of the system.

Discover

Where do you use brakes?

Consider the brakes for a car or a bicycle. Describe the components and the actuation cycle in as much detail as you can. Discuss your findings with your colleagues.

What kinds of equipment other than vehicles use clutches or brakes? Describe some scenarios.

Describe the physics of the operation of a clutch and a brake, considering energy and inertia effects.

This chapter helps you explore all of these issues, and many convenient design and analysis equations are developed. Many kinds of commercially available clutches and brakes are pictured and described.

Machine systems require control whenever the speed or the direction of the motion of one or more components is to be changed. When a device is started initially, it must be accelerated from rest to the operating speed. As a function is completed, the system must frequently be brought to rest. In continuously operating systems, changing speeds to adjust to different operating conditions is often necessary. Safety sometimes dictates motion control, as with a load being lowered by a hoist or an elevator.

In this chapter we will be concerned mostly with control of rotary motion in systems driven by electric motors, engines, turbines, and the like. Ultimately, linear motion may be involved through linkages, conveyors, or other mechanisms.

The machine elements most frequently used for motion control are the clutch and the brake, defined as follows:

- A *clutch* is a device used to connect or disconnect a driven component from the prime mover of the system. For example, in a machine that must cycle frequently, the driving motor is left running continuously, and a clutch is interposed between the motor and the driven machine. Then the clutch is cycled on and off to connect and disconnect the load. This permits the motor to operate at an efficient speed, and it also permits the system to cycle more rapidly because there is no need to accelerate the heavy motor rotor with each cycle.
- A *brake* is a device used to bring a moving system to rest, to slow its speed, or to control its speed to a certain value under varying conditions.

Where do you use brakes? An obvious answer is in a car or a bicycle where safe operation requires fast, smooth stopping when emergency conditions occur or simply when you need to stop at a stop sign or in your driveway. And you do not always need to bring the car or bicycle to a full stop. Slowing down to conform to prevailing traffic flow or to round a curve requires braking to decrease the speed from a higher to a lower speed.

What really happens when you apply the brakes of a bicycle? Can you describe the essential elements of the braking system? With hand brakes, your actuation of the brake lever pulls on a cable that, in turn, pulls on a linkage at the brake assembly mounted above the rim of a wheel. The linkage causes the friction pad to press against the rim. When you pull harder on the lever, a higher force is developed between the pad and the rim. This is called the *normal force*. Recall from physics and statics that a frictional force is created between surfaces moving relative to each other when a normal force presses them together. The frictional force acts in a direction opposite to the relative motion, and thus it tends to slow the motion. With a sufficiently large frictional force applied for a sufficient time, the wheel stops. Notice also that the frictional force acts at a fairly large radius from the center of the wheel. Thus, the force causes a frictional *torque* to be developed, and what is really happening is the deceleration of the angular velocity of the wheel. But since this is directly proportional to the linear speed of the bicycle, you experience the stopping action as a linear deceleration.

But that is not all! Have you ever felt the brake pad after a hard stop? The fact that it gets warm is an indication that the brake absorbs energy during the stopping action. Where does the energy come from? Can you compute the amount of energy that must be absorbed? What are the parameters involved in this calculation? Compare the amount of energy that must be absorbed in stopping a bicycle that you are riding with that involved in stopping a large airliner landing at 120 mph with a full load of people and baggage in addition to the huge weight of the airplane itself. Imagine what those brakes look like in comparison with the bicycle brake!

What other kinds of equipment besides transportation vehicles require brakes? Consider elevators, escalators, hoists, and winches, which must stop and hold a load after lifting it. Machine tools, conveyors, and other manufacturing equipment must often be brought to a safe, quick stop.

But that equipment must also be *accelerated* to start a new cycle of operation. How can that be achieved? One way is to start and stop the motor or the engine that drives the equipment. However, that is inconvenient and time-consuming, and it may lead to early failure of the system.

What if you had to stop the engine of your car each time you came to a stop sign and then had to restart it to move on? How does the automotive system allow you to stop and then continue driving without shutting off the engine? Have you ever driven a standard shift car? A "stick shift"? A *clutch* is used to engage and disengage the drive train from the engine. In cars with automatic transmissions clutches are also built into the transmission.

What other kinds of equipment use clutches? Describe some examples from your personal experience, or create a scenario in which it would be desirable to employ a clutch.

Now, consider what the task of a clutch is. Some parts of a machine are rotating continuously, while others are temporarily stationary. Then you engage the clutch. What happens? Consider the physics of the situation. The stationary parts must be accelerated from zero speed to the desired speed according to the design of the drive system. Inertia must be overcome—sometimes rotational inertia; sometimes linear inertia; and sometimes both kinds in the same machine. How long do you want it to take to accelerate the load to its operating speed? You must realize that a shorter time requires a higher value of accelerating torque to be developed by the clutch, and it increases the technical demands on the system in terms of strength of its components, smoothness, and wear life of the frictional materials that actually accomplish the engagement.

This chapter helps you explore all of these issues, and many convenient design and analysis equations are developed. Several different styles of brakes and clutches are discussed, and photographs or detailed drawings of several commercially available designs are shown. It might be desirable to keep this book handy for future projects so that you can look up the details of different clutch and brake designs.

YOU ARE THE DESIGNER

Your company manufactures conveyor systems for warehouses and truck terminals. The conveyors deliver cartons of materials to any of several stations where trucks are to be loaded. To save energy and to decrease the wear on operating parts of the conveyor system, it is decided to operate only the parts of the system that have a demand to deliver a carton. The system is to be operated automatically through a series of sensors, switches, programmable controllers, and an overall supervisory computer control. Your assignment is to recommend the type and the size of clutch and brake units to start and stop the various conveyors.

Some of the design decisions you will have to make are as follows:

1. How much time should be allowed to get the conveyors up to speed after the initial command to start has been given?
2. How fast should the conveyors come to a stop?
3. How many cycles per hour are expected?
4. How much room is available to install the clutch and brake units?
5. What means are available to actuate the clutch and brake units: electric power, compressed air, hydraulic pressure, or some other means?
6. What general type of clutch and brake should be used?
7. What size and model of units should be specified?

Along with these decisions, you will need information about the conveyor system itself, such as the following:

1. How much load will be on the conveyors when they are started and stopped?
2. What is the design of the conveyor system itself, and what are the weights, shapes, and dimensions of its components?
3. How is the conveyor driven: electric motor, hydraulic motor, or another means?
4. Are the products moved all on one level, or are there elevation changes in the system?

The information in this chapter will help you design such a system. ■

22–1 OBJECTIVES OF THIS CHAPTER

After completing this chapter, you will be able to:

1. Define the terms *clutch* and *brake*.
2. Differentiate between a clutch and a *clutch coupling*.
3. Describe a fail-safe brake.
4. Describe a clutch-brake module.
5. Specify the required capacity of a clutch or a brake to drive a given system reliably.
6. Compute the time required to accelerate a system or to bring it to a stop with the application of a given torque.
7. Define the inertia of a system in terms of its Wk^2 value.
8. Compute the energy-dissipation requirements of a clutch or a brake.
9. Determine the response time for a clutch-brake system.
10. Describe at least five types of clutches and brakes.
11. Name six actuation means used for clutches and brakes.
12. Perform the design and analysis of plate-type, caliper disc, cone, drum and shoe, and band brakes and clutches.
13. Name nine other types of clutches and brakes.

22–2 DESCRIPTIONS OF CLUTCHES AND BRAKES

Several arrangements of clutches and brakes are shown in Figure 22–1. By convention, the term *clutch* is reserved for the application in which the connection is made to a shaft parallel to the motor shaft, as illustrated in Figure 22–1(a). If the connection is to a shaft in-line with the motor, the term *clutch coupling* is used, as shown in Figure 22–1(b).

Also by convention, a brake [Figure 22–1(c)] is actuated by some overt action: the application of a fluid pressure, the switching on of an electric current, or the moving of a lever by hand. A brake that is spring-applied automatically in the absence of an overt action is called a *fail-safe brake* [Figure 22– 1(d)]: When the power goes off, the brake goes on.

When the functions of both a clutch and a brake are required in a system, they are frequently arranged in the same unit, the *clutch-brake module*. When the clutch is activated, the brake is deactivated, and vice versa [see Figure 22–1(e)].

A *slip clutch* is a clutch that, by design, transmits only a limited torque; thus, it will slip at any higher torque. It is used to provide a controlled acceleration to a load that is smooth and that requires a smaller motor power. It is also used as a safety device, protecting expensive or sensitive components in the event that a system is jammed.

Most of the discussion in this chapter will concern clutches and brakes that transmit motion through friction at the interface between two rotating parts moving at different speeds. Other types are discussed briefly in the last section.

References 1–7 and Internet sites 1–9 provide additional information on the design, selection, and application of motion control systems, clutches, and brakes.

22–3 TYPES OF FRICTION CLUTCHES AND BRAKES

Clutches and brakes that use friction surfaces as the means of transmitting the torque to start or stop a mechanism can be classified according to the general geometry of the friction surfaces and by the means used to actuate them. In some cases, the same basic geometry can be used as either a clutch or a brake by selectively attaching the friction elements to the driver, the driven machine, or the stationary frame of the machine.

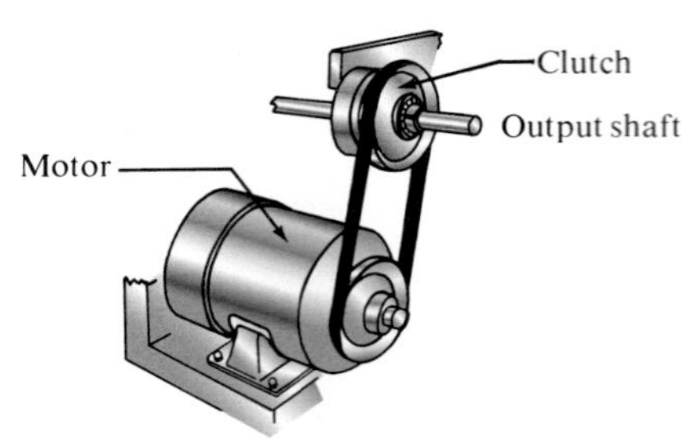

(*a*) Clutch: Transmits rotary motion to a parallel shaft only when coil is energized, by using sheaves, sprockets, gears, or timing pulleys.

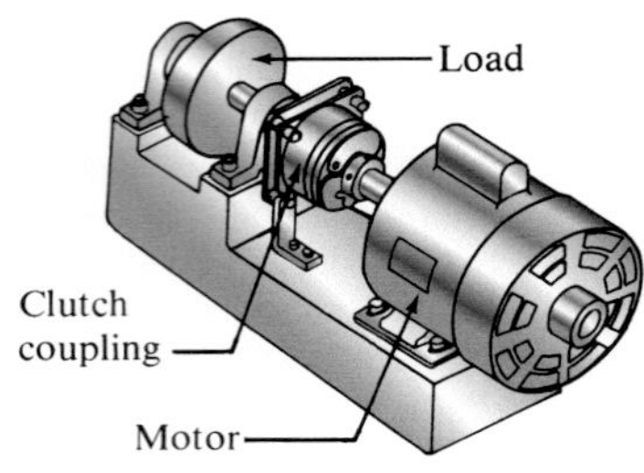

(*b*) Clutch coupling: Transmits rotary motion to an in-line shaft only when coil is energized. Split shaft applications.

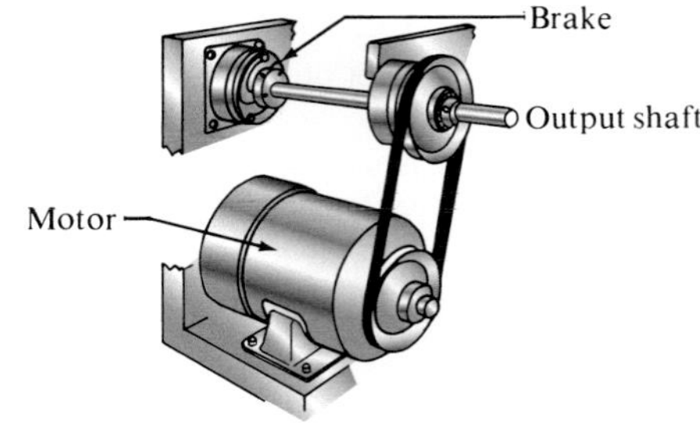

(*c*) Brake: Stops (brakes) load when coil is energized. Panel shows clutch imparting rotary motion to the output (load) shaft while brake is de-energized. Conversely, de-energizing clutch coil and energizing brake coil causes load to stop.

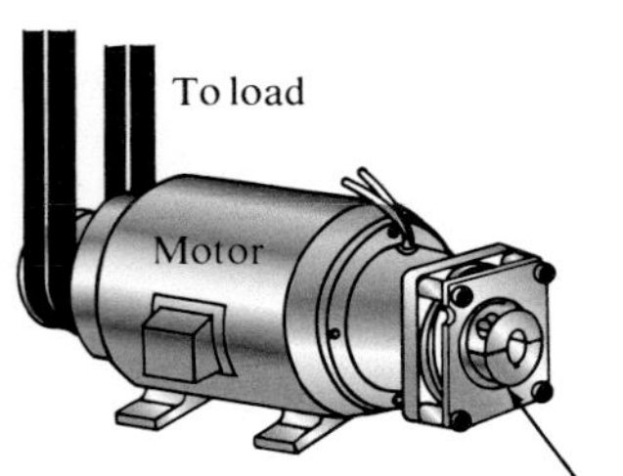

(*d*) Fail-safe brake: Stops load by de-energization of coil; power off—brake on.

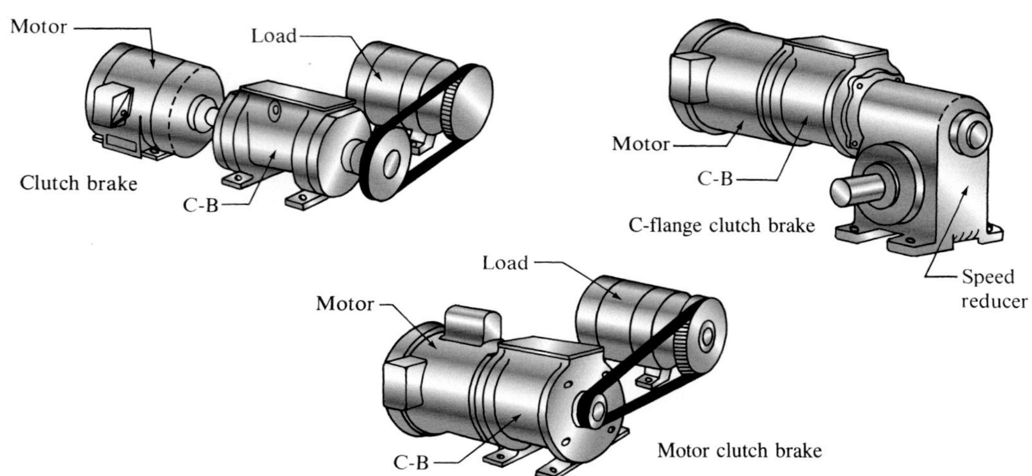

(*e*) Three types of mountings for clutch-brake modules: *Clutch brake* combines functions of clutch and brake in a complete preassembled package with input and output shafts. *C-flange clutch brake* performs same function but for use between NEMA "C" flange motor and speed reducer. *Motor clutch brake* is a preassembled module for mounting to a NEMA "C" flange motor and presenting an output shaft for connection to load.

FIGURE 22–1 Typical applications of clutches and brakes (Electroid Company, Springfield, NJ)

The following types of clutches and brakes are sketched in Figure 22–2:

1. ***Plate clutch or brake:*** Each friction surface is in the shape of an annulus on a flat plate. One or more friction plates move axially to contact a mating smooth plate, usually made of steel, to which the friction torque is transmitted.
2. ***Caliper disc brake:*** A disc-shaped rotor is attached to the machine to be controlled. Friction pads covering only a small portion of the disc are contained in a fixed

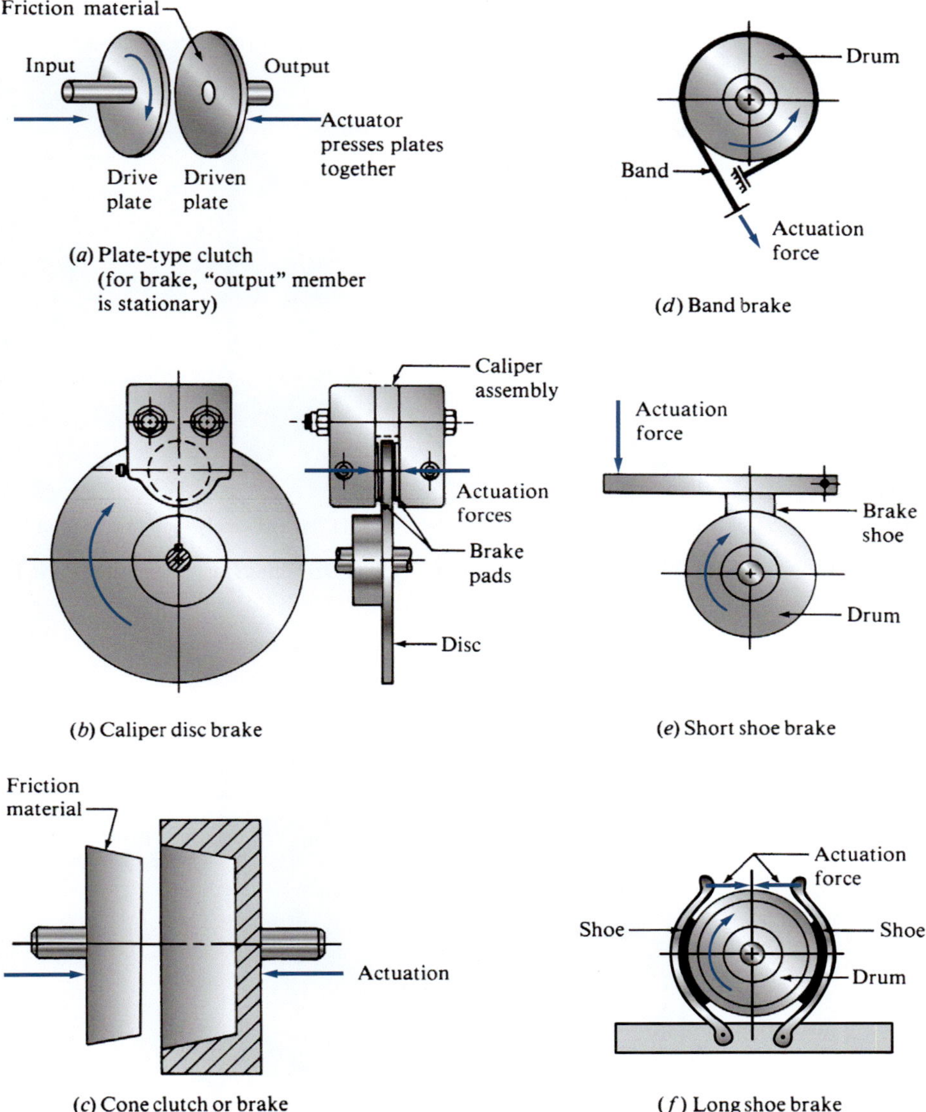

FIGURE 22–2 Types of friction clutches and brakes [(b), Tol-O-Matic, Hamel, MN]

assembly called a *caliper* and are forced against the disc by air pressure or hydraulic pressure.

3. ***Cone clutch or brake:*** A cone clutch or brake is similar to a plate clutch or brake except that the mating surfaces are on a portion of a cone instead of on a flat plate.
4. ***Band brake:*** Used only as a brake, the friction material is on a flexible band that nearly surrounds a cylindrical drum attached to the machine to be controlled. When braking is desired, the band is tightened on the drum, exerting a tangential force to stop the load.
5. ***Block or shoe brake:*** Curved, rigid pads faced with the friction material are forced against the surface of a drum, from either the outside or the inside, exerting a tangential force to stop the load.

Actuation

The following are the means used to actuate clutches or brakes. Each may be applied to several of the types described. Figures 22–3 through 22–9 show a variety of commercially available designs.

Manual. The operator provides the force, usually through a lever arrangement to achieve force multiplication. Figure 22–3 shows a power take-off for agricultural or construction equipment where a manually operated clutch engages and disengages the driven shaft from the engine. The driven shaft drives accessory equipment such as a fertilizer spreader, concrete mixer, or wood chipper.

(*a*) Components of power take-off

(*b*) Clutch assembly for power take-off

FIGURE 22–3 Power take-off with manual clutch (North American Cluth & Driveline, Inc., Machesney Park, IL)

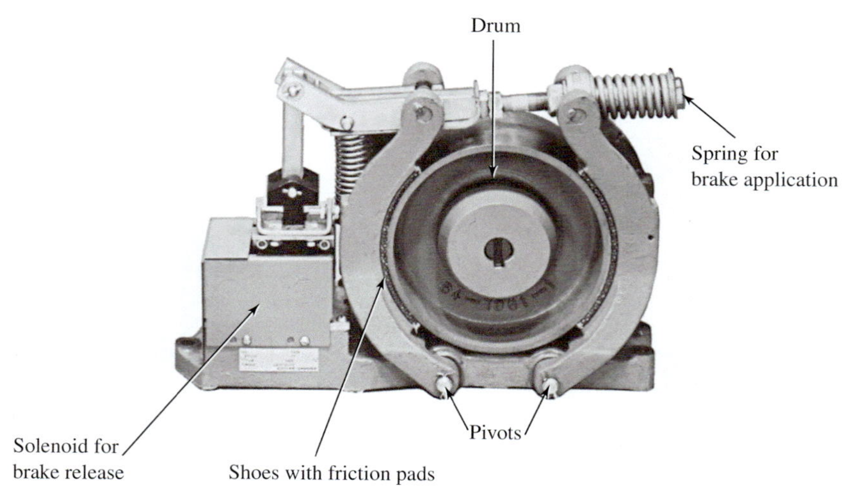

FIGURE 22–4 Spring-applied, electrically released, long shoe brake (Eaton Corp., Cutler-Hammer Products, Milwaukee, WI)

Spring-applied. Sometimes called a *fail-safe* design when applied to a brake, the brake is applied automatically by springs unless there is some opposing force present. Thus, if power fails, or if air pressure or hydraulic pressure is lost, or if the operator is unable to perform the function, the springs apply the brake and stop the load. The concept may also be used to engage or to disengage a clutch.

Centrifugal. A centrifugal clutch is sometimes employed to permit the driving system to accelerate without a connected load. Then, at a preselected speed, centrifugal force moves the clutch elements into contact to connect the load. As the system slows, the load will be automatically disconnected.

Pneumatic. Compressed air is introduced into a cylinder or some other chamber. The force produced by the pressure on a piston or a diaphragm brings the friction surfaces into contact with the members connected to the load.

Hydraulic. Hydraulic brakes are similar to the pneumatic type, except that they use oil hydraulic fluids instead of air. The hydraulic actuator is usually applied where high actuation forces are required.

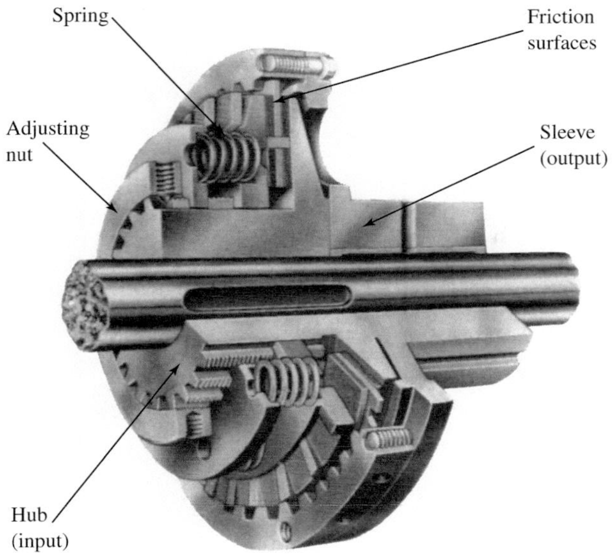

FIGURE 22–5 Slip clutch. Springs apply normal pressure on the friction plates. The spring force is adjustable to vary the torque level at which clutch will slip (The Hilliard Corp., Elmira, NY)

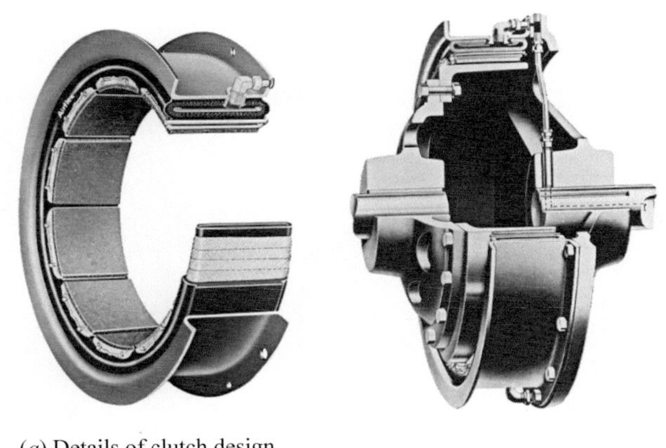

(*a*) Details of clutch design

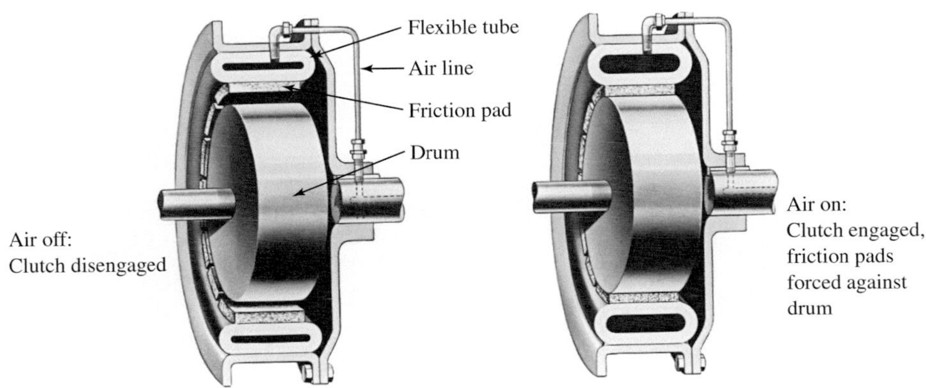

(*b*) Clutch activation cycle

FIGURE 22–6 Air-actuated clutch or brake (Eaton Corp., Airflex Division, Cleveland, OH)

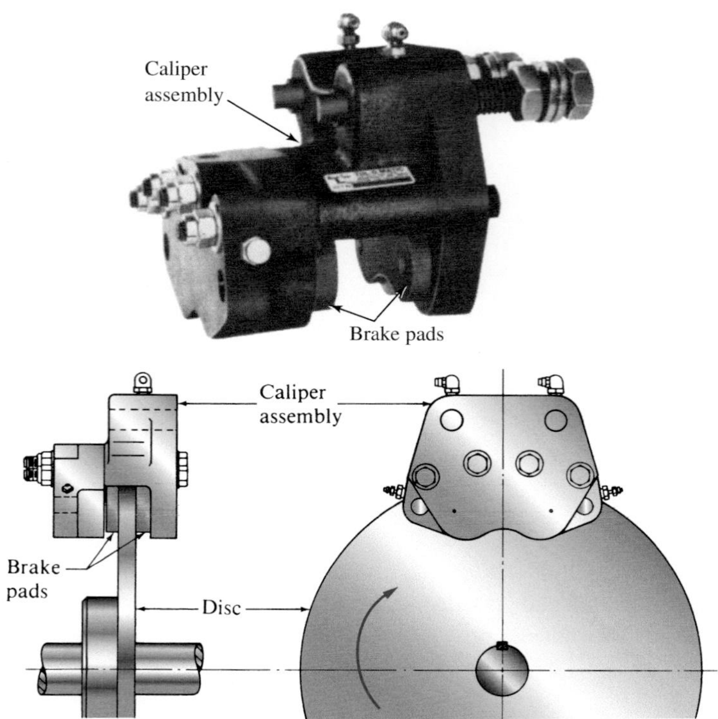

FIGURE 22–7 Hydraulically actuated disc brake assembly (Tol-O-Matic, Hamel, MN)

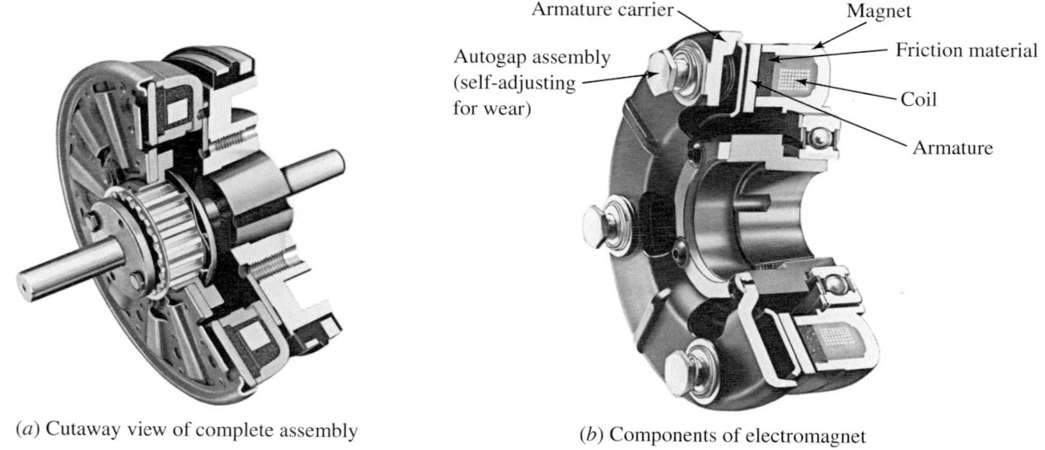

(*a*) Cutaway view of complete assembly

(*b*) Components of electromagnet

FIGURE 22–8 Electrically actuated plate-type clutch or brake (Warner Electric, Inc., South Beloit, IL)

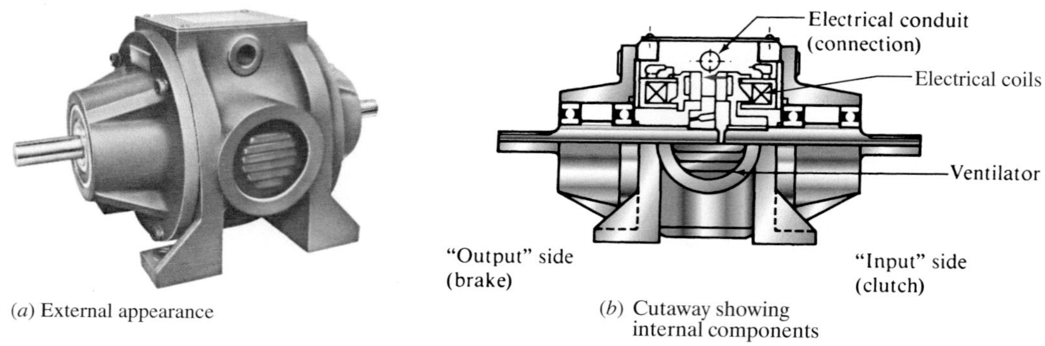

(*a*) External appearance

(*b*) Cutaway showing internal components

FIGURE 22–9 Electrically operated clutch-brake module (Electroid Company, Springfield, NJ)

Electromagnetic. An electric current is applied to a coil, creating an electromagnetic flux. The magnetic force then attracts an armature attached to the machine that is to be controlled. The armature is usually of the plate type.

22–4 PERFORMANCE PARAMETERS

The principles of physics tell us that whenever the speed or the direction of the motion of a body is changed, there must be a force exerted on the body. If the body is in rotation, a torque must be applied to the system to speed it up or to slow it down. When a change in speed occurs, a corresponding change in the kinetic energy of the system occurs. Thus, motion control inherently involves the control of energy, either adding energy to accelerate a system or absorbing energy to decelerate it.

The parameters involved in the rating of clutches and brakes are as follows:

1. Torque required to accelerate or decelerate the system
2. Time required to accomplish the speed change
3. The cycling rate: number of on/off cycles per unit time
4. The inertia of the rotating or translating parts
5. The environment of the system: temperature, cooling effects, and so on
6. Energy-dissipation capability of the clutch or the brake
7. Physical size and configuration
8. Actuation means
9. Life and reliability of the system
10. Cost and availability

Two basic methods are used to determine the torque capacity required of a clutch or a brake. One relates the capacity to the power of the motor driving the system. Recall that, in general, power = torque × rotational speed ($P = Tn$). The required torque capacity is then usually expressed in the form

⇨ Required Torque Capacity of Clutch or Brake

$$T = \frac{CPK}{n} \tag{22–1}$$

where C = conversion factor for units

K = service factor based on the application

More will be said about these later.

Note that the torque required is inversely proportional to the rotational speed. For this reason, it is advisable to locate the clutch or the brake on the highest-speed shaft in the system so that the required torque is a minimum. The size, cost, and response time are all typically lower when the torque is lower. One disadvantage is that the shaft to be accelerated or decelerated must undergo a larger change in speed, and the amount of slipping may be greater. This effect may generate more frictional heat, leading to thermal problems. However, it is offset by the increased cooling effect because of the faster motion of the clutch or brake parts.

The value for the K factor in the torque equation is largely a design decision. Some of the typical guidelines follow:

1. For brakes under average conditions, use $K = 1.0$.
2. For clutches in light duty where the output shaft does not assume its normal load until after it is up to speed, use $K = 1.5$.
3. For clutches in heavy duty where large attached loads must be accelerated, use $K = 3.0$.
4. For clutches in systems having varying loads, use a K factor at least equal to the factor by which the motor breakdown torque exceeds the full-load torque. This was discussed in Chapter 21, but for a typical industrial motor (design B), use $K = 2.75$. For a high starting torque motor (design C or capacitor-start motor), $K = 4.0$ may be required. This ensures that the clutch will be able to transmit at least as much torque as the motor and that it will not slip after getting up to speed.
5. For clutches in systems driven by gasoline engines, diesel engines, or other prime movers, consider the peak torque capability of the driver; $K = 5.0$ may be required.

The following list relates the value for C to typically used units for torque, power, and rotational speed. For example, if power is in hp and speed is in rpm, then to obtain torque in lb · ft, use $T = 5252(P/n)$.

Torque	Power	Speed	C
lb · ft	hp	rpm	5252
lb · in	hp	rpm	63 025
N · m	W	rad/s	1
N · m	W	rpm	9.549
N · m	kW	rpm	9549

Although the torque computation method of Equation (22–1) will produce generally satisfactory performance in typical applications, it does not provide a means of estimating the actual time required to accelerate the load with a clutch or to decelerate the load with a brake. The method described next should be used for systems with large inertia such as conveyors or presses having flywheels.

22–5 TIME REQUIRED TO ACCELERATE A LOAD

The basic principle involved is taken from dynamics:

$$T = I\alpha$$

where I = mass moment of inertia of the components being accelerated

α (alpha) = angular acceleration—that is, the rate of change of angular velocity

The usual objective of such an analysis is to determine the torque required to produce a change in rotational speed, Δn, of a system in a given amount of time, t. But $\Delta n/t = \alpha$.

Radius of gyration:

$$k^2 = \frac{1}{2}(R_1^2 + R_2^2)$$

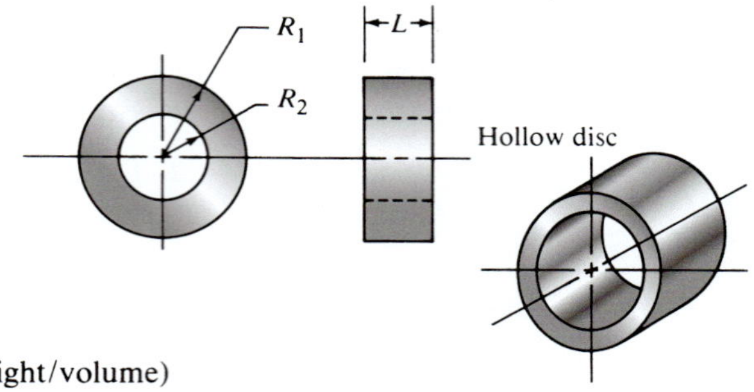

Volume:

$$V = \pi(R_1^2 - R_2^2)L$$

Weight:

$$W = \delta_W V$$

δ_W = Weight density (weight/volume)

Inertia (Wk^2):

$$Wk^2 = \delta_W Vk^2 = \delta_W \pi(R_1^2 - R_2^2)L(R_1^2 + R_2^2)/2$$

$$Wk^2 = \frac{\pi \delta_W L}{2}(R_1^4 - R_2^4)$$

Typical units: L, R_1, R_2 in inches

δ_W in lb/in^3

Wk^2 in lb·ft^2

$$Wk^2 = \frac{\pi}{2} \times \delta_W \frac{\text{lb}}{\text{in}^3} \times L(\text{in}) \times (R_1^4 - R_2^4)\ \text{in}^4 \times \frac{1\ \text{ft}^2}{144\ \text{in}^2}$$

$$Wk^2 = \frac{\delta_W L(R_1^4 - R_2^4)}{91.67}\ \text{lb·ft}^2$$

Special case for steel: $\delta_W = 0.283$ lb/in^3

$$Wk^2 = \frac{L(R_1^4 - R_2^4)}{323.9}\ \text{lb·ft}^2$$

FIGURE 22–10 Inertia properties of a hollow disc

Also, it is more convenient to express the mass moment of inertia in terms of the *radius of gyration, k*. By definition,

$$k = \sqrt{I/m} \quad \text{or} \quad k^2 = I/m$$

where m = mass

$m = W/g$

Then

$$I = mk^2 = Wk^2/g$$

The equation for torque then becomes

➪ Torque Required to Accelerate an Inertia Load

$$T = I\alpha = \frac{Wk^2}{g}\frac{(\Delta n)}{t} \qquad \text{(22–2)}$$

The term Wk^2 is frequently called simply the *inertia* of the load, although that designation is not strictly correct. A large proportion of the components in a machine system to be accelerated are in the form of cylinders or discs. Figure 22–10 gives the relationships for the radius of gyration and Wk^2 for hollow discs. Solid discs are simply a special case with an inside radius of zero. We can analyze more complex objects by considering them to be made from a set of simpler discs. Example Problem 22–1 illustrates the process.

Now the torque required to accelerate the pulley can be computed. Equation (22–2) can be put into a more convenient form by noting that T is usually expressed in lb · ft, Wk^2 in lb · ft^2, n in rpm, and t in s. Using g = 32.2 ft/s^2 and converting units gives

$$T = \frac{Wk^2(\Delta n)}{308t}\ \text{lb·ft} \qquad \text{(22–3)}$$

Example Problem 22–1 Compute the value of Wk^2 for the steel flat-belt pulley shown in Figure 22–11.

Solution The pulley can be considered to be made up of three components, each of which is a hollow disc. The Wk^2 for the total pulley is the sum of that for each component.

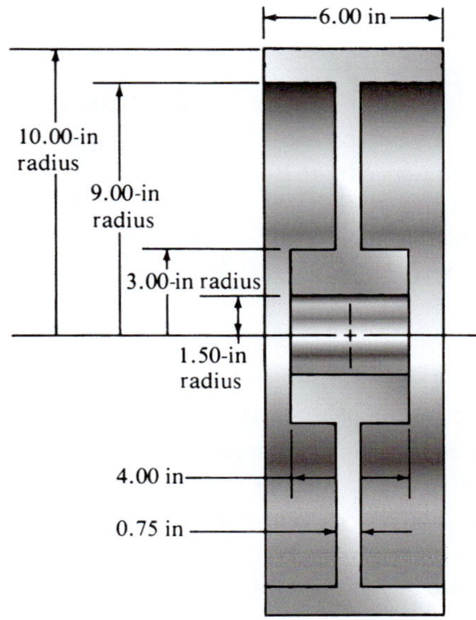

FIGURE 22–11 Steel flat-belt pulley

Part 1. Using the formula from Figure 22–10 for a steel disc, we have

$$Wk^2 = \frac{(R_1{}^4 - R_2{}^4)(L)}{323.9}\ \text{lb}\cdot\text{ft}^2 = \frac{[(10.0)^4 - (9.0)^4](6.0)}{323.9}$$

$$Wk^2 = 63.70\ \text{lb}\cdot\text{ft}^2$$

Part 2.

$$Wk^2 = \frac{[(9.0)^4 - (3.0)^4](0.75)}{323.9} = 15.00\ \text{lb}\cdot\text{ft}^2$$

Part 3.

$$Wk^2 = \frac{[(3.0)^4 - (1.5)^4]\ (4.0)}{323.9} = 0.94\ \text{lb}\cdot\text{ft}^2$$

$$\text{Total } Wk^2 = 63.70 + 15.00 + 0.94 = 79.64\ \text{lb}\cdot\text{ft}^2$$

Example Problem 22–2 Compute the torque that a clutch must transmit to accelerate the pulley of Figure 22–11 from rest to 550 rpm in 2.50 s. From Example Problem 22–1, $Wk^2 = 79.64\ \text{lb}\cdot\text{ft}^2$.

Solution Use Equation (22–3):

$$T = \frac{(79.64)(550)}{308(2.5)} = 56.9\ \text{lb}\cdot\text{ft}$$

In summary, if a clutch that is capable of exerting at least 56.9 lb·ft of torque engages a shaft carrying the pulley shown in Figure 22–11, the pulley can be accelerated from rest to 550 rpm in 2.50 s or less.

22–6 INERTIA OF A SYSTEM REFERRED TO THE CLUTCH SHAFT SPEED

In many practical machine systems, there are several elements on several shafts, operating at differing speeds. It is required that the effective inertia of the entire system *as it affects the clutch* be determined. The effective inertia of a connected load operating at a rotational speed different from that of the clutch is proportional to the square of the ratio of the speeds. That is,

Effective Inertia

$$Wk_e^2 = Wk^2\left(\frac{n}{n_c}\right)^2 \tag{22–4}$$

where n = speed of the load of interest

n_c = speed of the clutch

Example Problem 22–3 Compute the total inertia of the system in Figure 22–12 as seen by the clutch. Then compute the time required to accelerate the system from rest to a motor speed of 550 rpm if the clutch exerts a torque of 240 lb·ft. The Wk^2 for the armature of the clutch, which must also be accelerated, is 0.22 lb·ft^2, including the 1.25-in shaft.

Solution The clutch and gear A will be rotating at 550 rpm, but because of the gear reduction, gear B, its shaft, and the pulley rotate at

$$n_2 = 550 \text{ rpm } (24/66) = 200 \text{ rpm}$$

Now compute the inertia for each element referred to the clutch speed. Assume that the gears are discs having outside diameters equal to the pitch diameter of the gear and inside diameters equal to the shaft diameter. The equation in Figure 22–10 for a steel disc will be used to compute Wk^2.

Gear *A*

$$Wk^2 = [(2.00)^4 - (0.625)^4](2.50)/323.9 = 0.122 \text{ lb}\cdot\text{ft}^2$$

Gear *B*

$$Wk^2 = [(5.50)^4 - (1.50)^4](2.50)/323.9 = 7.02 \text{ lb}\cdot\text{ft}^2$$

But, because of the speed difference, the effective inertia is

$$Wk_e^2 = 7.02(200/550)^2 = 0.93 \text{ lb}\cdot\text{ft}^2$$

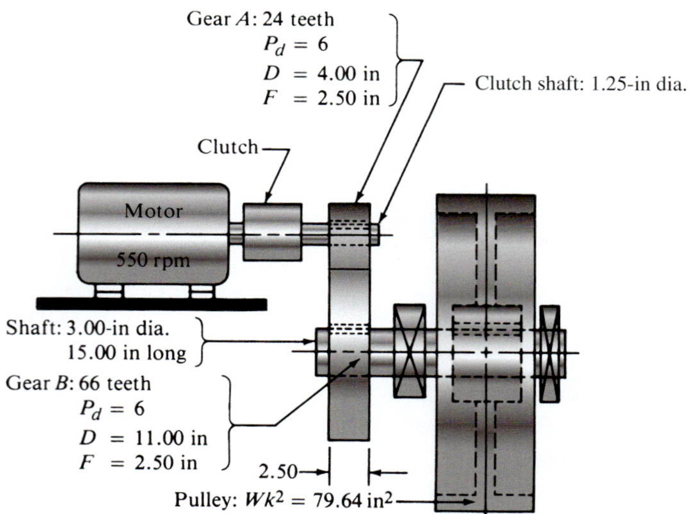

FIGURE 22–12 Given system

Pulley

From Example Problem 22–1, $Wk^2 = 79.64\ \text{lb}\cdot\text{ft}^2$. The effective inertia is

$$Wk_e^2 = 79.64(200/550)^2 = 10.53\ \text{lb}\cdot\text{ft}^2$$

Shaft

$$Wk^2 = (1.50)^4(15.0)/323.9 = 0.234\ \text{lb}\cdot\text{ft}^2$$

The effective inertia is

$$Wk_e^2 = 0.234(200/550)^2 = 0.03\ \text{lb}\cdot\text{ft}^2$$

The total effective inertia as seen by the clutch is

$$Wk_e^2 = 0.22 + 0.12 + 0.93 + 10.53 + 0.03 = 11.83\ \text{lb}\cdot\text{ft}^2$$

Solving Equation (22–3) for the time gives

$$t = \frac{Wk_e^2(\Delta n)}{308T} = \frac{(11.83)(550)}{308(240)} = 0.088\ s$$

22–7 EFFECTIVE INERTIA FOR BODIES MOVING LINEARLY

To this point, we have dealt only with components that rotate. Many systems include linear devices such as conveyors, hoist cables and their loads, or reciprocating racks driven by gears that also have inertia and must be accelerated. It would be convenient to represent these devices with an effective inertia measured by Wk^2 as we have for rotating bodies. We can accomplish this by relating the equations for kinetic energy for linear and rotary motion. The actual kinetic energy of a translating body is

$$KE = \frac{1}{2}mv^2 = \frac{1}{2}\frac{W}{g}v^2 = \frac{Wv^2}{2g}$$

where v = linear velocity of the body

We will use ft/min for the units of velocity. For a rotating body,

$$KE = \frac{1}{2}I\omega^2 = \frac{1}{2}\frac{Wk^2}{g}\omega^2 = \frac{Wk^2\omega^2}{2g}$$

Letting Wk^2 be the effective inertia, and equating these two formulas, gives

$$Wk_e^2 = W\left(\frac{v}{\omega}\right)^2$$

where ω must be in rad/min to be consistent

Using n rpm rather than ω rad/min, we must substitute $\omega = 2\pi n$. Then

➪ Effective Inertia for a Load Moving Linearly

$$Wk_e^2 = W\left(\frac{v}{2\pi n}\right)^2 \qquad \textbf{(22–5)}$$

Example Problem 22–4 The conveyor in Figure 22–13 moves at 80 ft/min. The combined weight of the belt and the parts on it is 140 lb. Compute the equivalent Wk^2 inertia for the conveyor referred to the shaft driving the belt.

Solution The rotational speed of the shaft is

$$\omega = \frac{V}{R} = \frac{80\ \text{ft}}{\text{min}}\ \frac{1}{5.0\ \text{in}}\ \frac{12\ \text{in}}{\text{ft}} = 192\ \text{rad/min}$$

Then the equivalent Wk^2 is

$$Wk_e^2 = W\left(\frac{V}{\omega}\right)^2 = (140\ \text{lb})\left(\frac{80\ \text{ft/min}}{192\ \text{rad/min}}\right)^2 = 24.3\ \text{lb}\cdot\text{ft}^2$$

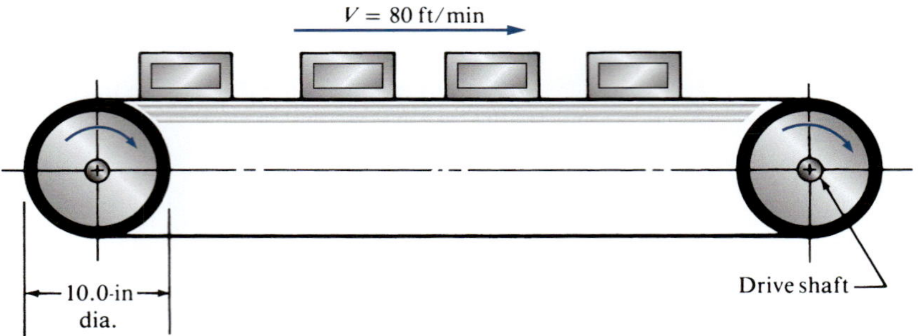

FIGURE 22–13 Conveyor moving at 80 ft/min

22–8 ENERGY ABSORPTION: HEAT-DISSIPATION REQUIREMENTS

When you are using a brake to stop a rotating object or using a clutch to accelerate it, the clutch or brake must transmit energy through the friction surfaces as they slip in relation to each other. Heat is generated at these surfaces, tending to increase the temperature of the unit. Of course, heat is then dissipated from the unit, and for a given set of operating conditions an equilibrium temperature is achieved. That temperature must be sufficiently low to ensure a long life of the friction elements and other operating parts of the unit, such as electric coils, springs, and bearings.

The energy to be absorbed or dissipated by the unit per cycle is equal to the change in kinetic energy of the components being accelerated or stopped. That is,

➩ Energy Absorption by a Brake

$$E = \Delta KE = \frac{1}{2} I\omega^2 = \frac{1}{2} mk^2\omega^2 = \frac{Wk^2\omega^2}{2g}$$

For typical units in the U.S. Customary System ($\omega = n$ rpm; Wk^2 in lb · ft; and $g = 32.2$ ft/s^2), we get

$$E = \frac{Wk^2(\text{lb}\cdot\text{ft}^2)}{2(32.2\ \text{ft/s}^2)}\frac{n^2\ \text{rev}^2}{\text{min}^2}\frac{(2\pi)^2\ \text{rad}}{\text{rev}^2}\frac{1\ \text{min}^2}{60^2\ \text{s}^2}$$

➩ Energy Absorption in U.S. Customary Units

$$E = 1.7 \times 10^{-4}\ Wk^2n^2\ \text{lb}\cdot\text{ft} \qquad (22\text{–}6)$$

In SI units, mass is in kilograms (kg), radius of gyration is in meters (m), and angular velocity is in radians per second (rad/s). Then

$$E = \frac{1}{2}I\omega^2 = \frac{1}{2}mk^2\omega^2(\text{kg}\cdot\text{m}^2/\text{s}^2)$$

But the newton unit is equal to the kg · m/s^2. Then

➩ Energy Absorption in SI Units

$$E = \frac{1}{2}mk^2\omega^2\ \text{N}\cdot\text{m} \qquad (22\text{–}7)$$

No further conversion factors are required.

If there are repetitive cycles of operation, the energy from Equation (22–6) or (22–7) must be multiplied by the cycling rate, usually in cycles/min in U.S. Customary units and cycles/s for SI units. The result would be the energy generation per unit time, which must be compared with the heat-dissipation capacity of the clutch or brake being considered for the application.

When the clutch-brake cycles on and off, part of its operation is at the full operating speed of the system, and part is at rest. The combined heat-dissipation capacity is the average of the capacity at each speed weighted by the proportion of the cycle at each speed (see Example Problem 22–5).

22–9 RESPONSE TIME

The term *response time* refers to the time required for the unit (clutch or brake) to accomplish its function after action is initiated by the application of an electric current, air pressure, spring force, or manual force. Figure 22–14 shows a complete cycle using a clutch-brake module. The straight-line curve is idealized while the curved line gives the general form of system motion. The actual response time will vary, even for a given unit, with variations in the load, environment, or other operating conditions.

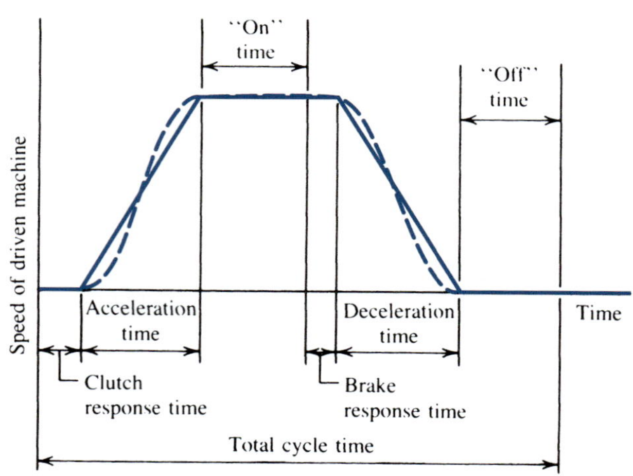

FIGURE 22–14 Typical cycle of engagement and disengagement for a clutch

TABLE 22–1 Clutch-Brake Sample Performance Data

Unit size	Torque capacity (lb · ft)	Inertia, Wk^2 (lb · ft^2)	Heat dissipation (ft · lb/min) At rest	Heat dissipation (ft · lb/min) 1800 rpm	Response time (s) Clutch	Response time (s) Brake
A	0.42	0.000 17	750	800	0.022	0.019
B	1.25	0.0014	800	1200	0.032	0.024
C	6.25	0.021	1050	2250	0.042	0.040
D	20.0	0.108	2000	6000	0.090	0.089
E	50.0	0.420	3000	13 000	0.110	0.105
F	150.0	1.17	9000	62 000	0.250	0.243
G	240.0	2.29	18 000	52 000	0.235	0.235
H	465.0	5.54	20 000	90 000	0.350	0.350
I	700.0	13.82	26 000	190 000	0.512	0.512

Note: Torque ratings are static. The torque capacity decreases as the speed difference between the parts being engaged increases. Interpolation may be used on heat-dissipation data.

Commercially available clutches and brakes for typical machine applications have response times from a few milliseconds (1/1000 s) for a small device, such as a paper transport, to approximately 1.0 s for larger machines, such as an assembly conveyor. Manufacturers' literature should be consulted. To give you a feel for the capabilities of commercially available clutches and brakes, Table 22–1 gives sample data for electrically powered units.

Example Problem 22–5 For the system in Figure 22–12, and using the data from Example Problem 22–3, estimate the total cycle time if the system is controlled by unit *G* from Table 22–1 and the system must stay on (at steady speed) for 1.50 s and off (at rest) for 0.75 s. Also estimate the response time of the clutch-brake and the acceleration and deceleration times. If the system cycles continuously, compute the heat-dissipation rate, and compare it with the capacity of the unit.

Solution Figure 22–15 shows the estimated total cycle time for the system to be 2.896 s. From Table 22–1, we find that the clutch-brake exerts 240 lb · ft of torque and has a response time of 0.235 s for both the clutch and the brake.

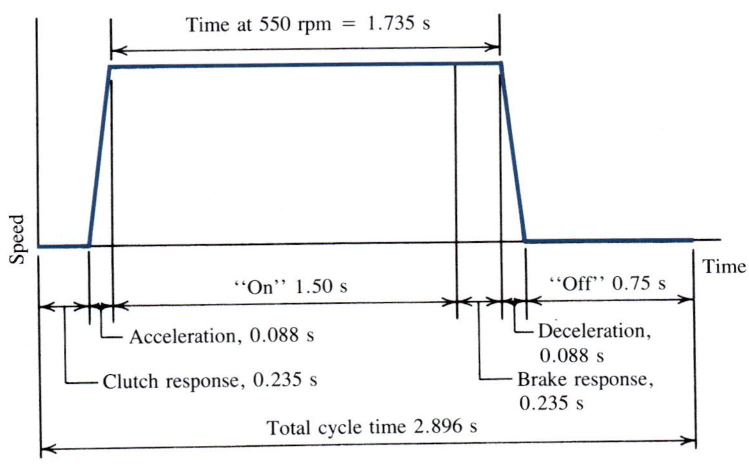

FIGURE 22–15 Estimated total cycle time

Acceleration and Deceleration Time [Equation (22–3)]

$$t = \frac{Wk_e^2(\Delta n)}{308T} = \frac{(11.83)(550)}{308(240)} = 0.088\ \text{s}$$

Cycling Rate and Heat Dissipation [Equation (22–6)]

For a total cycle time of 2.896 s, the number of cycles per minute would be

$$C = \frac{1.0\ \text{cycle}}{2.896\ \text{s}}\frac{60\ \text{s}}{\text{min}} = 20.7\ \text{cycles/min}$$

The energy generated with each engagement of either the clutch or the brake is

$$E = 1.7 \times 10^{-4}Wk^2n^2 = 1.7 \times 10^{-4}(11.83)(550)^2 = 608\ \text{lb}\cdot\text{ft}$$

The energy generation per minute is

$$E_t = 2EC = (2)(608\ \text{lb}\cdot\text{ft/cycle})(20.7\ \text{cycles/min}) = 25\,200\ \text{lb}\cdot\text{ft/min}$$

This is greater than the heat-dissipation capacity of unit G at rest (18 000 lb·ft/min). Then let's compute a weighted average capacity for this cycle. First, referring to Figure 22–15, approximately 1.735 s is "at speed," or 550 rpm. The balance of the cycle, 1.161 s, is at rest. From Table 22–1, and interpolating between zero speed and 1800 rpm, the heat-dissipation rate at 550 rpm is approximately 28 400 lb·ft/min. Then the weighted average capacity for unit G is

$$E_{avg} = \frac{t_0}{t_t}E_0 + \frac{t_{550}}{t_t}E_{550}$$

where t_t = total cycle time
t_0 = time at rest (0 rpm)
t_{550} = time at 550 rpm
E_0 = heat-dissipation capacity at rest
E_{550} = heat-dissipation capacity at 550 rpm

Then

$$E_{avg} = \frac{1.161}{2.896}(18\,000) + \frac{1.735}{2.896}(28\,400) = 24\,230\ \text{lb}\cdot\text{ft/min}$$

This is a little lower than required, and the design would be marginal. Fewer cycles per minute should be specified.

22–10 FRICTION MATERIALS AND COEFFICIENT OF FRICTION

Many of the types of clutches and brakes discussed in this chapter employ mating surfaces driven through friction materials. The function of these materials is to develop a substantial friction force when a normal force is created by the brake actuation means. The friction force produces a force or a torque that retards existing motion if applied as a brake or that accelerates a member at rest or moving at a slower speed if applied as a clutch.

The desirable properties of friction materials are as follows:

1. They should have a relatively high coefficient of friction when operating against the mating materials in the system. The highest coefficient of friction is not always the best choice, because smooth engagement is often aided by a more moderate frictional force or torque.
2. The coefficient of friction should be relatively constant over the range of operating pressures and temperatures so that reliable, predictable performance can be expected.
3. The materials should have good resistance to wear.
4. The materials must be chemically compatible with mating components.
5. Environmental hazards must be minimized.

Several different materials are used for friction elements in clutches and brakes, and many are proprietary to a particular manufacturer. Common in the past were various asbestos-based compounds having coefficients of friction in

TABLE 22–2 Coefficients of Friction

Friction material	Dynamic friction coefficient		Pressure range	
	Dry	In oil	(psi)	(kPa)
Molded compounds	0.25–0.45	0.06–0.10	150–300	1035–2070
Woven materials	0.25–0.45	0.08–0.10	50–100	345–690
Sintered metal	0.15–0.45	0.05–0.08	150–300	1035–2070
Cork	0.30–0.50	0.15–0.25	8–15	55–100
Wood	0.20–0.45	0.12–0.16	50–90	345–620
Cast iron	0.15–0.25	0.03–0.06	100–250	690–1725
Paper-based		0.10–0.15		
Graphite/resin		0.10–0.14		

the range of 0.35 to 0.50. Asbestos has been shown to be a health hazard and is being replaced by molded compounds of polymers and rubber. Where flexibility is required, as in band brakes, the base material is woven into a fabric, sometimes strengthened with metal wire, saturated with a resin, and cured. Cork and wood are also sometimes used. Paper-based materials are used in some oil-filled clutches. In severe environments, cast iron, sintered iron or other metals, or graphite materials are employed. Table 22–2 gives approximate ranges of coefficients of friction and the pressure that the materials can withstand.

For automotive applications, standards are set by the Society of Automotive Engineers. In SAE J866 (Reference 7), a set of codes is defined to classify friction materials according to the coefficient of friction, regardless of the material used. Table 22–3 lists these codes.

In problems in this book that require a coefficient of friction, we will use a value of 0.25 unless otherwise stated. Based on the values reported here, this is a relatively low value that should yield conservative designs.

Materials for Discs and Drums

A variety of metals are applied in the manufacture of brake and clutch discs and drums. The material must have a sufficient strength, ductility, and stiffness to withstand the applied forces while maintaining a precise geometry. It must also be able to absorb heat from the friction surface and dissipate it to the environment.

TABLE 22–3 Coefficient of Friction Classification Codes of the Society of Automotive Engineers

Code letter	Coefficient of friction
C	Not over 0.15
D	Over 0.15 but not over 0.25
E	Over 0.25 but not over 0.35
F	Over 0.35 but not over 0.45
G	Over 0.45 but not over 0.55
H	Over 0.55
Z	Unclassified

Some of the popular choices are gray cast iron, ductile iron, carbon steel, and copper alloys. Many discs and drums are cast for cost reasons and to obtain near net shape of the parts that require little machining after casting. Cast iron has low cost and high thermal conductivity as compared with ductile iron. However, ductile iron is better able to withstand shock or impact loading. Copper alloys have far higher thermal conductivity than the other materials, but they have poorer wear performance.

22–11 PLATE-TYPE CLUTCH OR BRAKE

Figure 22–2(a) shows a simple sketch of a plate-type clutch, and Figure 22–8 shows a cutaway view of a commercially available, electrically actuated clutch or brake. In Figure 22–8, an electromagnet exerts an axial force that draws the friction surfaces together. When two bodies are brought into contact with a normal force between them, a friction force that tends to resist relative motion is created. This is the principle on which the plate-type clutch or brake is based.

Friction Torque

As the friction plate rotates in relation to the mating plate with an axial force pressing them together, the friction force acts in a tangential direction, producing the brake or clutch torque. At any point, the local pressure times the differential area at the point is the *normal force*. The normal force times the coefficient of friction is the *friction force*. The friction force times the radius of the point is the *torque* produced at this point. The *total torque* is the sum of all of the torques over the entire area of the plate. We can find the sum by integrating over the area.

There is usually some variation of pressure over the surface of the friction plate, and some assumption about the nature of the variation must be made before the total torque can be computed. A conservative assumption that yields a useful result is that the friction surface will wear uniformly over the entire area as the brake or clutch operates. This assumption implies that the product of the local pressure, p, times the linear relative speed, v, between the plates is constant. Wear has been found to be approximately proportional to the product of p times v.

Taking all these factors into account and completing the analysis produces the following result for friction torque:

$$T_f = fN(R_o + R_i)/2$$

But the last part of this relation is the mean radius, R_m, of the annular plate. Then

➪ Friction Torque on Annular Plate

$$T_f = fNR_m \tag{22–8}$$

As stated before, this is a conservative result, meaning that the actual torque produced would be slightly larger than predicted.

Wear Rating

Note that the torque is proportional to the mean radius, but that no area relationship is involved in Equation (22–8). Thus, the completion of the design for final dimensions requires some other parameter. The missing factor in Equation (22–8) is the expected wear rate of the friction material. It should be obvious that even with the same mean radius, a brake with a larger area would wear less than one with a smaller area.

Manufacturers of friction materials can assist in the final determination of the relationship between wear and the area of the friction surface. However, the following guidelines allow the estimation of the physical size of brakes and will be used for problem solutions in this book.

The wear rating, WR, will be based on the frictional power, P_f, absorbed by the brake per unit area, A, where

➪ Frictional Power

$$P_f = T_f\omega \tag{22–9}$$

and ω is the angular velocity of the disc. In SI units, with torque in N · m and ω in rad/s, the frictional power is in N · m/s or watts. In the U.S. Customary System, with torque in lb · in and angular velocity expressed as n rpm, the frictional power is in hp, computed from

$$P_f = \frac{T_f n}{63\,000}\text{ hp} \tag{22–10}$$

For industrial applications, we will use

➪ Wear Rating

$$WR = P_f/A \tag{22–11}$$

where $WR = 0.04\text{ hp/in}^2$ for frequent applications, a conservative rating

$WR = 0.10\text{ hp/in}^2$ for average service

$WR = 0.40\text{ hp/in}^2$ for infrequently used brakes allowed to cool somewhat between applications

Example Problem 22–6 Compute the dimensions of an annular plate-type brake to produce a braking torque of 300 lb · in. Springs will provide a normal force of 320 lb between the friction surfaces. The coefficient of friction is 0.25. The brake will be used in average industrial service, stopping a load from 750 rpm.

Solution ***Step 1.*** Compute the required mean radius. From Equation (22–8),

$$R_m = \frac{T_f}{fN} = \frac{300\text{ lb}\cdot\text{in}}{(0.25)(320\text{ lb})} = 3.75\text{ in}$$

Step 2. Specify a desired ratio of R_o/R_i, and solve for the dimensions. A reasonable value for the ratio is approximately 1.50. The range can be from 1.2 to about 2.5, at the designer's choice. Using 1.50, $R_o = 1.50R_i$, and

$$R_m = (R_o + R_i)/2 = (1.5R_i + R_i)/2 = 1.25R_i$$

Then

$$R_i = R_m/1.25 = (3.75\text{ in})/1.25 = 3.00\text{ in}$$

$$R_o = 1.50R_i = 1.50(3.00) = 4.50\text{ in}$$

Step 3. Compute the area of the friction surface:

$$A = \pi(R_o^2 - R_i^2) = \pi[(4.50)^2 - (3.00)^2] = 35.3\text{ in}^2$$

Step 4. Compute the frictional power absorbed:

$$P_f = \frac{T_f n}{63\,000} = \frac{(300)(750)}{63\,000} = 3.57 \text{ hp}$$

Step 5. Compute the wear ratio:

$$WR = \frac{P_f}{A} = \frac{3.57 \text{ hp}}{35.3 \text{ in}^2} = 0.101 \text{ hp/in}^2$$

Step 6. Judge the suitability of WR. If WR is too high, return to Step 2 and increase the ratio. If WR is too low, decrease the ratio. In this example, WR is acceptable.

A more compact unit can be designed if more than one friction plate is used. We multiply the friction torque for one plate by the number of plates to determine the total friction torque. One disadvantage of this approach is that the heat dissipation is relatively poorer than for the single plate.

Improving Wear Performance of Brakes

Actual wear in a given application depends on a combination of many variables. Friction materials are relatively softer and weaker than the metallic materials used for discs and drums. Wear is often characterized as adhesion. As the surface of the friction material rubs over the high spots of the metal, plastic deformation occurs at the surface and particles are sheared off, breaking the bond between particles or dislodging filler materials from the polymer bonding agents. This process accelerates when surface temperatures rise as the brake absorbs the energy required to stop the rotating system. The thermal behavior of the system is critical to good life. If temperatures rise above about 400°F (200°C), wear rate increases significantly and the coefficient of friction decreases, leading to poorer braking performance called *fade*.

It is difficult to predict the life of a given brake system analytically, and testing under real operating conditions is recommended for new designs. The following list gives general principles of improving wear performance:

- Specify friction materials that have relatively low adhesion when in contact with the disc or drum material.
- Specify friction materials that have high bonding strength between constituent particles.
- Provide high hardness on the surface of the disc or drum by heat treatment.
- Keep the pressure between the friction material and the material of the disc or drum as low as practical.
- Maintain the surface temperature at the interface between the friction material and the material of the disc or drum as low as practical by promoting heat transfer away from the system by conduction, convection, and radiation. Forced airflow or cooling with water is often applied in critical situations.
- Provide a smooth surface finish on the discs and drums.
- Provide lubricants such as oil or graphite at the friction interface.
- Exclude abrasive contaminants from the friction interface.
- Minimize slipping between the clutch or brake elements by promoting lockup of the engaging elements.

22–12 CALIPER DISC BRAKES

The disc brake pads are brought into contact with the rotating disc by fluid pressure acting on a piston in the caliper. The pads are either round or in a short crescent shape to cover more of the disc surface [see Figures 22–2(b) and 22–7]. However, one advantage of the disc brake is that the disc is exposed to the atmosphere, promoting the dissipation of heat. Because the disc rotates with the machine to be controlled, heat dissipation is further enhanced. The cooling effect improves the fade resistance of this type of brake relative to the shoe-type brake.

Designs for the friction torque and for the wear rating are similar to those explained for plate-type brakes.

22–13 CONE CLUTCH OR BRAKE

The angle of inclination of the conical surface of the cone clutch or brake is typically 12°. A lower angle could be used with care, but there is a tendency for the friction surfaces to engage suddenly with a jerk. As the angle increases, the amount of axial force required to produce a given friction torque increases. Thus, 12° is a reasonable compromise.

Referring to Figure 22–16, we see that as an axial force F_a is applied by a spring, manually or by fluid pressure, a normal force N is created between the mating friction surfaces, all around the periphery of the cone. The desired friction force F_f is produced in the tangential direction, where $F_f = fN$. It is assumed that the friction force acts at the mean radius of the cone so that the friction torque is

$$T_f = F_f R_m = fNR_m \quad \textbf{(22–8)}$$

In addition to the tangentially directed friction force, a friction force develops along the surface of the cone and

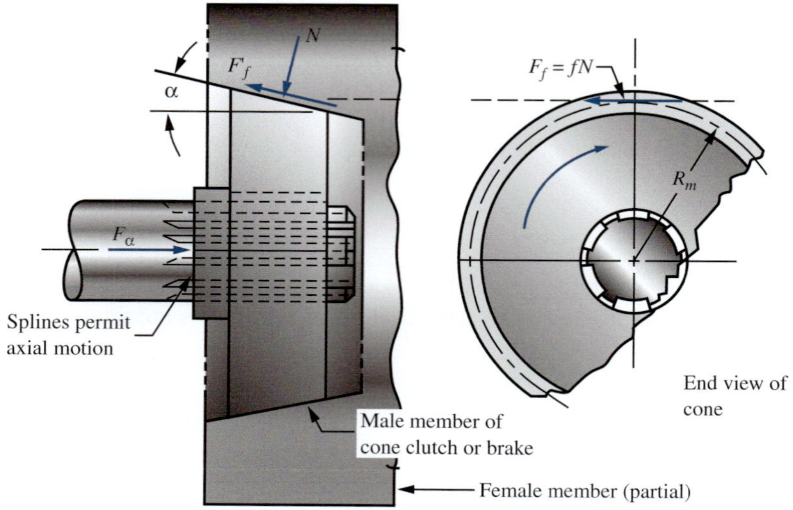

FIGURE 22–16 Cone clutch or brake

opposes the tendency for the member having the internal cone surface to move axially away from the external cone. We will call this force F'_f and compute it also from

$$F'_f = fN$$

For the equilibrium condition of the external cone, the sum of the horizontal forces must be zero. Then

$$F_a = N\sin\alpha + F'_f\cos\alpha = N\sin\alpha + fN\cos\alpha$$
$$= N(\sin\alpha + f\cos\alpha)$$

or

$$N = \frac{F_a}{\sin\alpha + f\cos\alpha} \quad \text{(22–12)}$$

Substituting this into Equation (22–8) gives

➪ **Frictional Torque on a Cone Clutch or Brake**

$$T_f = \frac{fR_mF_a}{\sin\alpha + f\cos\alpha} \quad \text{(22–13)}$$

Example Problem 22–7 Compute the axial force required for a cone brake if it is to exert a braking torque of 50 lb·ft. The mean radius of the cone is 5.0 in. Use $f = 0.25$. Try cone angles of 10°, 12°, and 15°.

Solution We can solve Equation (22–13) for the axial force F_a:

$$F_a = \frac{T_f(\sin\alpha + f\cos\alpha)}{fR_m} = \frac{(50\text{ lb}\cdot\text{ft})(\sin\alpha + 0.25\cos\alpha)}{(0.25)(5.0/12)\text{ ft}}$$

$$F_a = 480(\sin\alpha + 0.25\cos\alpha)\text{ lb}$$

Then the values of F_a as a function of the cone angle are as follows:

For $\alpha = 10°$
$F_a = 202$ lb

For $\alpha = 12°$
$F_a = 217$ lb

For $\alpha = 15°$
$F_a = 240$ lb

22–14 DRUM BRAKES

Short Shoe Drum Brakes

Figure 22–17 shows a sketch of a drum brake in which the actuating force W acts on the lever that pivots on pin A. This creates a normal force between the shoe and the rotating drum. The resulting friction force is assumed to act tangential to the drum if the shoe is short. The friction force times the radius of the drum is the friction torque that slows the drum.

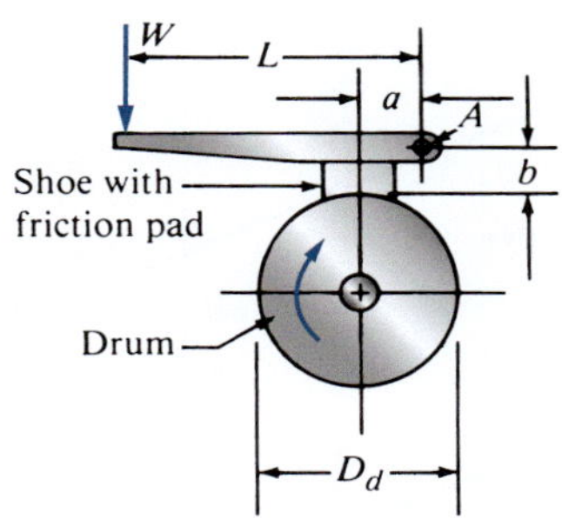

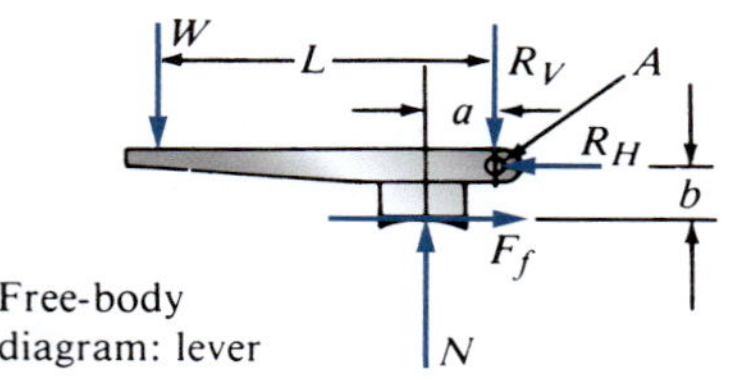

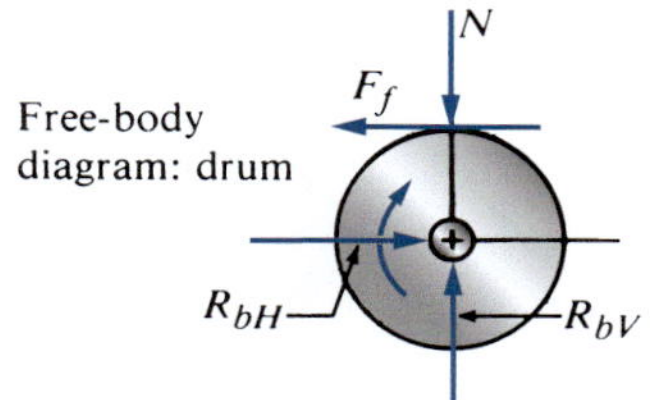

(*a*) Lever design 1

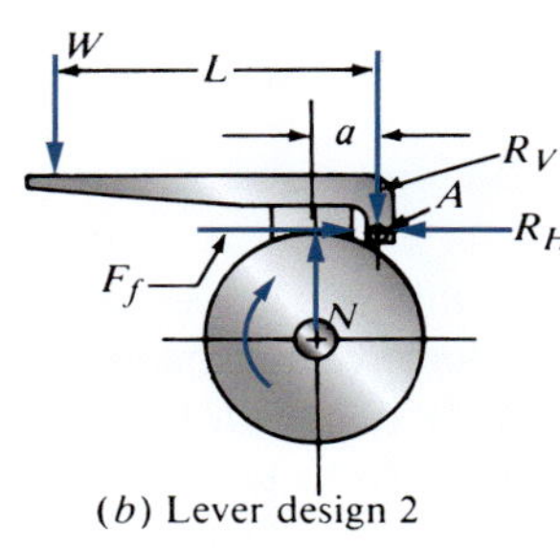

(*b*) Lever design 2

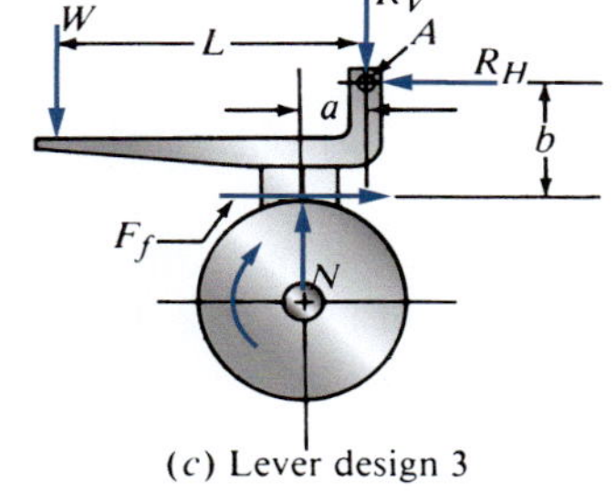

(*c*) Lever design 3

FIGURE 22–17 Short shoe drum brake

The objectives of the analysis are to determine the relationship between the applied load and the friction force and to be able to evaluate the effect of design decisions such as the size of the drum, the lever dimensions, and the placement of the pivot *A*. The free-body diagrams in Figure 22–17(a) support this analysis. For the lever, we can sum moments about the pivot *A*:

$$\Sigma M_A = 0 = WL - Na + F_f b \qquad \textbf{(22–14)}$$

But note that $F_f = fN$ or $N = F_f/f$; where $f =$ coefficient of friction. Then

$$0 = WL - F_f a/f + F_f b = WL - F_f(a/f - b)$$

Solving for *W* gives

$$W = \frac{F_f(a/f - b)}{L} \qquad \textbf{(22–15)}$$

Solving for F_f gives

➪ Friction Force on Drum Brake

$$F_f = \frac{WL}{(a/f - b)} \qquad \textbf{(22–16)}$$

We can use these equations for the friction torque by noting

➪ Friction Torque

$$T_f = F_f D_d/2 \qquad \textbf{(22–17)}$$

where $D_d =$ diameter of the drum

Note the alternate positions of the pivot in Parts (b) and (c) of Figure 22–17. In (b), the dimension $b = 0$.

Example Problem 22–8 Compute the actuation force required for the short shoe drum brake of Figure 22–17(a) to produce a friction torque of 50 lb · ft. Use a drum diameter of 10.0 in, $a = 3.0$ in, and $L = 15.0$ in. Consider values of f of 0.25, 0.50, and 0.75, and different points of location of pivot *A* such that *b* ranges from 0 to 6.0 in.

Solution The required friction force can be found from Equation (22–17):

$$F_f = 2T_f/D_d = (2)(50\ \text{lb}\cdot\text{ft})/(10/12\ \text{ft}) = 120\ \text{lb}$$

In Equation (22–15), we can substitute for *a*, *L*, and F_f:

$$W = \frac{F_f(a/f - b)}{L} = \frac{120\ \text{lb}\,[(3.0\ \text{in})/f - b]}{15.0\ \text{in}} = 8(3/f - b)\ \text{lb}$$

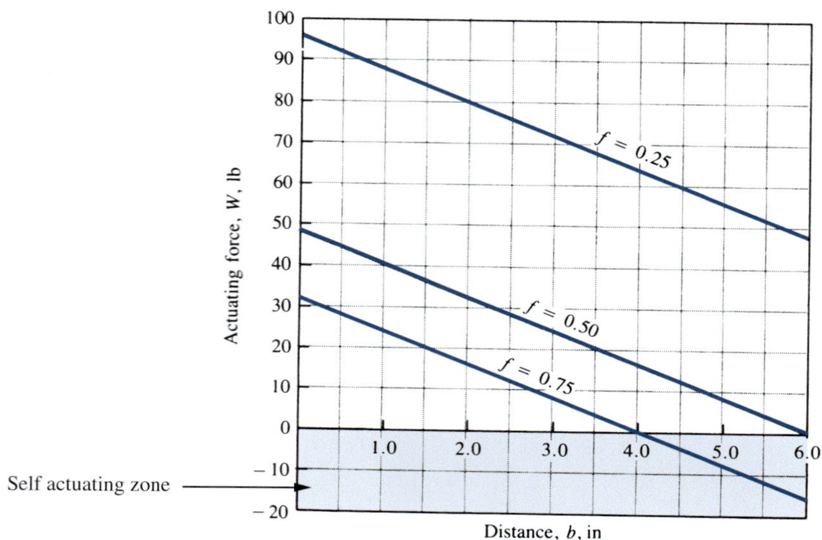

FIGURE 22–18 Results: actuating load force versus distance b

We can substitute the varying values of f and b into this last equation to compute the data for the curves of Figure 22–18, showing the actuating force versus the distance b for different values of f. Note that for some combinations, the value of W is *negative*. This means that the brake is *self-actuating* and that an upward force on the lever would be required to release the brake.

Long Shoe Drum Brakes

The assumption used for short shoe brakes, that the resultant friction force acted at the middle of the shoe, cannot be used in the case of shoes covering more than about 45° of the drum. In such cases, the pressure between the friction lining and the drum is very nonuniform, as is the moment of the friction force and of the normal force with respect to the pivot of the shoe.

The following equations govern the performance of a long shoe brake, using the terminology from Figure 22–19. (See Reference 4.)

1. ***Friction torque on drum:***

$$T_f = r^2 f w p_{\max}(\cos\theta_1 - \cos\theta_2) \quad \textbf{(22–18)}$$

2. ***Actuation force:***

$$W = (M_N + M_f)/L \quad \textbf{(22–19)}$$

where M_N = moment of normal force with respect to the hinge pin

$$M_N = 0.25 p_{\max} w r C[2(\theta_2 - \theta_1) - \sin 2\theta_2 + \sin 2\theta_1] \quad \textbf{(22–20)}$$

M_f = moment of friction force with respect to the hinge pin

$$M_f = f p_{\max} w r[r(\cos\theta_1 - \cos\theta_2) + 0.25C(\cos 2\theta_2 - \cos 2\theta_1)] \quad \textbf{(22–21)}$$

The sign of M_f is negative (−) if the drum surface is moving away from the pivot and positive (+) if it is moving toward the pivot.

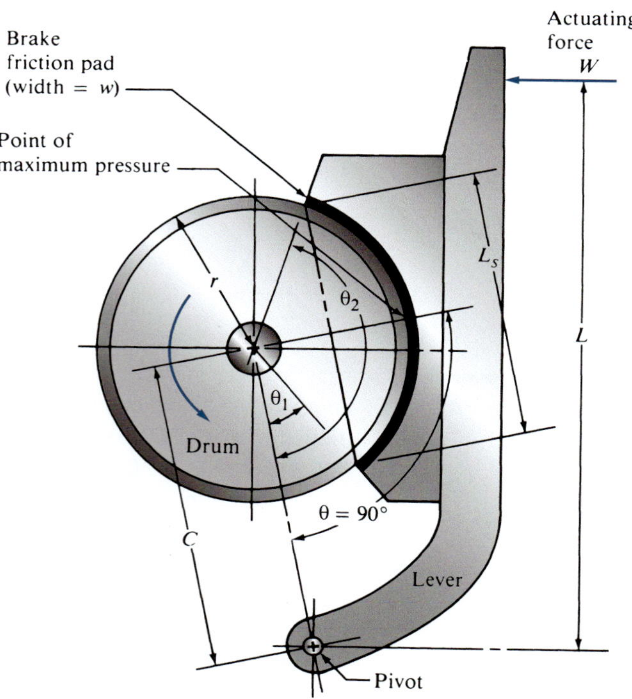

FIGURE 22–19 Terminology for long shoe drum brake

3. ***Friction power:***

$$P_f = T_f n/63\,000 \text{ hp} \tag{22–22}$$

where n = rotational speed in rpm

4. ***Brake shoe area*** (*Note*: Projected area is used):

$$A = L_s w = 2wr\sin[(\theta_2 - \theta_1)/2] \tag{22–23}$$

5. ***Wear ratio:***

$$WR = P_f/A \tag{22–24}$$

The use of these relationships in the design and analysis of a long shoe brake is shown in Example Problem 22–9.

Example Problem 22–9 Design a long shoe drum brake to produce a friction torque of 750 lb·in to stop a drum from 120 rpm.

Solution ***Step 1.*** Select a brake friction material, and specify the desired maximum pressure and the design value for the coefficient of friction. Table 22–2 lists some general properties for friction materials. Actual test values or specific manufacturer's data should be used where possible. The design value for p_{max} should be far less than the permissible pressure listed in Table 22–2 in order to improve wear life.

For this problem, let's choose a woven fabric material and design for approximately 75 psi maximum pressure. Note, as shown in Figure 22–19, that the maximum pressure occurs at the section 90° from the pivot. If the shoe does not extend at least 90°, the equations used here are not valid. (See Reference 4.) Also, let's use $f = 0.25$ for the design.

Step 2. Propose trial values for the geometry of the brake drum and the brake pad. Several design decisions must be made here. The general arrangement shown in Figure 22–19 can be used as a guide. But your specific application and creativity may lead you to modify the arrangement.

Trial values are $r = 4.0$ in; $C = 8.0$ in; $L = 15$ in; $\theta_1 = 30°$; and $\theta_2 = 150°$.

Step 3. Solve for the required width of the shoe from Equation (22–18):

$$w = \frac{T_f}{r^2 f p_{max}(\cos\theta_1 - \cos\theta_2)}$$

For this problem,

$$w = \frac{750 \text{ lb}\cdot\text{in}}{(4.0 \text{ in})^2(0.25)(75 \text{ lb/in}^2)(\cos 30° - \cos 150°)} = 1.443 \text{ in}$$

For convenience, let $w = 1.50$ in. Because the maximum pressure is inversely proportional to the width, the actual maximum pressure will be

$$P_{max} = 75\text{psi}(1.44/1.50) = 72.17 \text{ psi}$$

Step 4. Compute M_N from Equation (22–20). The value of $\theta_2 - \theta_1$ must be in radians, with π radians = 180°. Then

$$\theta_2 - \theta_1 = 120°(\pi \text{ rad}/180°) = 2.0944 \text{ rad}$$

The moment of the normal force on the shoe is

$$M_N = 0.25(72.17 \text{ lb/in}^2)(1.50 \text{ in})(4.0 \text{ in})(8.0 \text{ in})$$
$$[2(2.09) - \sin(300°) + \sin(60°)]$$
$$M_N = 5128 \text{ lb}\cdot\text{in}$$

Step 5. Compute the moment of the friction force on the shoe, M_f, from Equation (22–21):

$$M_f = 0.25(72.17 \text{ lb/in}^2)(1.50 \text{ in})(4.0 \text{ in})$$
$$[(4.0 \text{ in})(\cos 30° - \cos 150°)$$
$$+ 0.25(8.0 \text{ in})(\cos 300° - \cos 60°)]$$
$$M_f = 749.8 \text{ lb}\cdot\text{in}$$

Step 6. Compute the required actuation force, W, from Equation (22–19):

$$W = (M_N - M_f)/L = (5128 - 749.8)/(15) = 291.8 \text{ lb}$$

Note the minus sign for M_f because the drum surface is moving away from the pivot.

Step 7. Compute the frictional power from Equation (22–22):

$$P_f = T_f n/(63000) = (750)(120)/(63000) = 1.429 \text{ hp}$$

Step 8. Compute the projected area of the brake shoe from Equation (22–23).

$$A = L_s w = 2wr \sin[(\theta_2 - \theta_1)/2]$$

$$A = 2(1.50 \text{ in})(4.0 \text{ in}) \sin(120°/2) = 10.392 \text{ in}^2$$

Step 9. Compute the wear ratio, WR:

$$WR = P_f/A = 1.429 \text{ hp}/10.392 \text{ in}^2 = 0.137 \text{ hp/in}^2$$

Step 10. Evaluate the suitability of the results. In this problem, we would need more information about the application to evaluate the results. However, the wear ratio seems reasonable for average service (see Section 22–11), and the geometry seems acceptable.

22–15 BAND BRAKES

Figure 22–20 shows the typical configuration of a *band brake.* The flexible band, usually made of steel, is faced with a friction material that can conform to the curvature of the drum. The application of a force to the lever puts tension in the band and forces the friction material against the drum. The normal force, thus created, causes the friction force tangential to the drum surface to be created, retarding the drum.

The tension in the band decreases from the value P_1 at the pivot side of the band to P_2 at the lever side. The net torque on the drum is then

$$T_f = (P_1 - P_2)r \quad \textbf{(22–25)}$$

where r = radius of the drum

The relationship between P_1 and P_2 can be shown (see Reference 4) to be the logarithmic function

$$P_2 = P_1/e^{f\theta} \quad \textbf{(22–26)}$$

where θ = total angle of coverage of the band in radians

The point of maximum pressure on the friction material occurs at the end nearest the highest tension, P_1, where

$$P_1 = p_{max} r w \quad \textbf{(22–27)}$$

and w is the width of the band.

For the two types of band brakes shown in Figure 22–20, the free-body diagrams of the levers can be used to show the

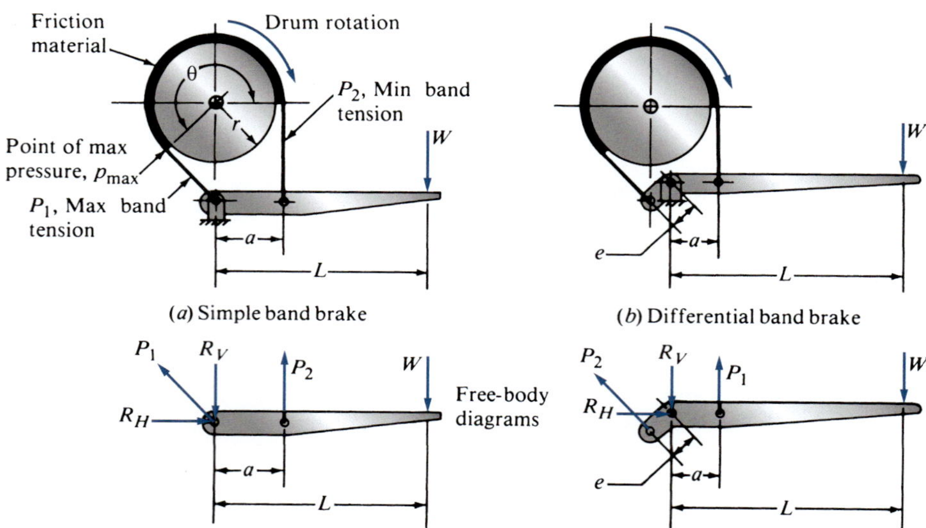

FIGURE 22–20 Band brake design

following relationships for the actuating force, W, as a function of the tensions in the band. For the simple band brake of Figure 22–20(a),

$$W = P_2 a/L \quad (22\text{–}28)$$

The style shown in Figure 22–20(b) is called a *differential band brake*, where the actuation force is

$$W = (P_2 a - P_1 e)/L \quad (22\text{–}29)$$

The design procedure is presented in Example Problem 22–10.

Example Problem 22–10 Design a band brake to exert a braking torque of 720 lb · in while slowing the drum from 120 rpm.

Solution ***Step 1.*** Select a material and specify a design value for the maximum pressure. A woven friction material is desirable to facilitate the conformance to the cylindrical drum shape. Let's use $p_{max} = 25$ psi and a design value of $f = 0.25$. See Section 22–10.

Step 2. Specify a trial geometry: r, θ, w. For this problem, let's try $r = 6.0$ in, $\theta = 225°$, and $w = 2.0$ in. Note that $225° = 3.93$ rad.

Step 3. Compute the maximum band tension, P_1, from Equation (22–27):

$$P_1 = p_{max}\, rw = (25\ \text{lb/in}^2)(6.0\ \text{in})(2.0\ \text{in}) = 300\ \text{lb}$$

Step 4. Compute tension P_2 from Equation (22–26):

$$P_2 = \frac{P_1}{e^{f\theta}} = \frac{300\ \text{lb}}{e^{(0.25)(3.93)}} = 112.3\ \text{lb}$$

Step 5. Compute the friction torque, T_f:

$$T_f = (P_1 - P_2)r = (300 - 112.3)(6.0) = 1126\ \text{lb}\cdot\text{in}$$

Note: Repeat Steps 2–5 until you achieve a satisfactory geometry and friction torque. Let's try a smaller design, say, $r = 5.0$ in:

$$P_1 = (25)(5.0)(2.0) = 250\ \text{lb}$$

$$P_2 = \frac{250\ \text{lb}}{e^{(0.25)(3.93)}} = 93.59\ \text{lb}$$

$$T_f = (250 - 93.59)(5.0) = 782.0\ \text{lb}\cdot\text{in} \quad \text{(okay)}$$

Step 6. Specify the geometry of the lever, and compute the required actuation force. Let's use $a = 5.0$ in and $L = 15.0$ in. Then

$$W = P_2(a/L) = 93.59\ \text{lb}\ (5.0/15.0) = 31.20\ \text{lb}$$

Step 7. Compute the average wear ratio from $WR = P_f/A$:

$$A = 2\pi rw(\theta/360) = 2(\pi)(5.0\ \text{in})(2.0\ \text{in})(225/360) = 39.27\ \text{in}^2$$

$$P_f = T_f n/63\,000 = (782.0)(120)/(63\,000) = 1.490\ \text{hp}$$

$$WR = P_f/A = (1.490\ \text{hp})/(39.27\ \text{in}^2) = 0.0379\ \text{hp/in}^2$$

This should be conservative for average service.

22–16 OTHER TYPES OF CLUTCHES AND BRAKES

The chapter thus far has concentrated on clutches and brakes using friction materials to transmit the torque between rotating members, but many other types are available. Brief descriptions are given below, but specific design information is not given. Most are unique to the manufacturer, and application data are available through catalogs.

Jaw Clutch

The teeth of the mating sets of jaws are brought into engagement by sliding one or both members axially. The teeth may be straight-sided or triangular, or they may incorporate some smooth curve to facilitate engagement. Once the teeth are engaged, there is a positive transmission of torque. The jaw clutch is normally engaged while the system is stopped or is running very slowly.

Ratchet

Although not strictly a clutch, the familiar ratchet and pawl permits alternate engagement and disengagement of moving members, and thus it can be used in similar applications. Typically, the ratchet moves only a small fraction of a revolution per cycle.

Sprag, Roller, and Cam Clutches

There are differences in the specific geometry of sprag, roller, and cam clutches, but they all perform similar functions. When the input shaft is rotating in the driving direction, the internal elements (sprags, rollers, or cams) are wedged between the driving and driven members and thus transmit torque. But when the input member rotates in the opposite direction, the internal elements move out of engagement, and no torque is transmitted. Thus, they can be used for applications similar to ratchets, but with much smoother operation and with a virtually infinite amount of incremental motion. Another application is *backstopping*, in which the clutch runs free when the machine is being driven in the intended direction. But if the drive is disengaged and the load starts to reverse direction, the clutch locks up and prevents motion. This type of clutch is also used for *overrunning:* a positive drive as long as the load rotates no faster than the driver. If the load tends to run faster (overrun) than the driver, the clutch elements disengage. This protects equipment that might be damaged by overspeeding. (See Internet site 6.)

Fiber Clutch

A fiber clutch operates in a manner similar to the overrunning clutches described previously. But instead of driving through solid elements, the torque is transmitted through stiff fibers that have a preferred orientation. When rotated in a direction opposite the preferred direction, the fibers "lie down," and no torque is transmitted.

Wrap-Spring Clutch

Again used in cases similar to the overrunning clutches, the *wrap-spring clutch* is made from a rectangular wire and normally has an inside diameter slightly larger than that of the shaft on which it is installed. Thus, no torque is transmitted. But when one end of the spring is restrained, the spring "wraps down" tightly on the surface of the shaft, and torque is transmitted positively through the spring. (See Internet site 4.)

Single-Revolution Clutches

It is frequently desired to have a machine cycle one complete revolution and then come to a stop. The *single-revolution clutch* provides this feature. After it is tripped, it drives the output shaft until reaching a positive stop at the end of one revolution. Some types can be engaged for more than one revolution, but they will return to a fixed position, for example, at the top of the stroke of a press. (See Internet site 6.)

Fluid Clutches

The *fluid clutch* is made up of two separate parts with no mechanical connection between them. A fluid fills a cavity between the parts, and as one member rotates, it tends to shear the fluid, causing torque to be transmitted to the mating element. The resulting drive is smooth and soft because load peaks will simply cause one member to move relative to the other. In this situation, it is similar to the slip clutch described earlier.

Eddy Current Drive

When a conducting disc moves through a magnetic field, *eddy currents* are induced in the disc, causing a force to be exerted on the disc in a direction opposite to the direction of rotation. The force can be used to brake the disc or to transmit torque to a mating part, such as a clutch. An advantage of this type of unit is that there is no mechanical connection between the elements. The torque can be controlled by varying the current to the electromagnets.

Overload Clutches

The drive is positive, provided that the torque is below some set value. At higher torques, some element is disengaged automatically. One type uses a series of spherical balls positioned in detents and held under a spring force. When the tripping torque level is reached, the balls are forced out of the detents and disengage the drive. (See Internet site 6.)

Tensioners

The production of continuous products such as wire, webs of paper, or plastic film require that the drive system be carefully controlled to maintain a light tension without breaking the product. Similar control must be exerted when winding coils of paper, foil, or sheet metal as they are produced, or when unwinding them to feed a process such as printing or press forming. Drives for cranes and hoists must provide controlled braking while loads are lowered. In such cases, brakes are required to exert some braking action while allowing smooth motion. Many of the brake designs reviewed in this chapter can accomplish this function by moderating the force applied between the friction elements. An example is the air-actuated brake shown in Figure 22–6. The braking torque depends on the air pressure applied that can be controlled by an operator or automatically. (See also Internet site 5.)

REFERENCES

1. Avallone, Eugene A., and Ali Sadegh, Theodore Baumeister III, editors. *Marks' Standard Handbook for Mechanical Engineers*. 11th ed. New York: McGraw-Hill, 2007.
2. Juvinall, Robert C., and Kurt M. Marshek. *Fundamentals of Machine Component Design*. 5th ed. New York: John Wiley & Sons, 2012.
3. Orthwein, William C. *Clutches and Brakes: Design and Selection*. 2nd ed. Boca Raton, FL: CRC Press, 2004.

4. Budynas, R. G., and K. J. Nisbett. *Shigley's Mechanical Engineering Design*. 9th ed. New York: McGraw-Hill, 2011.
5. Society of Automotive Engineers. *Standard J286 Clutch Friction Test Machine Guidelines*. Warrendale, PA: Society of Automotive Engineers, 2006.
6. Society of Automotive Engineers. *Standard J661 Brake Lining Quality Control Test Procedure*. Warrendale, PA: Society of Automotive Engineers, 1997.
7. Society of Automotive Engineers. *Standard J866 Friction Coefficient Identification System for Brake Linings*. Warrendale, PA: Society of Automotive Engineers, 2002.

INTERNET SITES FOR CLUTCHES AND BRAKES

1. **GKN Rockford, Inc.** www.rockfordpowertrain.com Manufacturer of clutches and other powertrain components for the automotive, truck, and off-road equipment markets.
2. **Baldor/Dodge.** www.dodge-pt.com Manufacturer of Dodge brakes and clutches for industrial machinery, conveying equipment, and textile industry. Select *Products*, then *Brakes* or *Clutches*. Also offers a wide array of gear drives, belt drives, bearings, and numerous other power transmission products.
3. **BorgWarner, Inc.** www.borgwarner.com Manufacturer of automotive clutch sets and other components. Select *Products*, then *Drivetrain Systems*.
4. **Warner Electric, Inc.** www.warnernet.com Manufacturer of clutch and brake systems for industrial, turf and garden, wind turbine, and vehicular applications. Includes a wide variety of disk type, wrap spring, and magnetic particle clutches and brakes for start, stop, hold, and tension control. Catalog information can be downloaded from the Literature section.
5. **Eaton/Airflex.** www.airflex.com Manufacturer of industrial clutches and brakes using the expanding tube principle actuated by pneumatic or hydraulic pressure. Applications include engines, paper making machinery, power presses, brakes, shears, marine drives, well drilling, and tensioning, winding, and unwinding equipment. Online product data and descriptions, product application information, and technical information.
6. **Hilliard Corporation.** www.hilliardcorp.com Manufacturer of a wide variety of clutches and brakes for industry and commercial equipment applications. Included are centrifugal clutches, single-revolution clutches, overrunning clutches, ball detent overload release clutches, slip clutch torque limiters, intermittent drives, and caliper disc brakes.
7. **Tol-O-Matic, Inc.** www.tolomatic.com Manufacturer of pneumatically and hydraulically operated caliper disc brakes, disc cone clutches, and other automation and motion control products.
8. **Electroid Company.** www.electroid.com Manufacturer of a wide variety of clutches, brakes, and tensioners for industrial, commercial, and aerospace applications.
9. **Emerson Power Transmission Co.** www.emerson-ept.com Manufacturer of various styles of clutches and torque overload devices under the Morse brand. Select *Products,* then *Components,* then *Clutches* or *Torque overload devices.* Site also includes a 12-page downloadable document, *Clutches and Brakes,* that describes all major styles of those devices. Under *Site/Part Search,* enter *Clutches and Brakes.*

PROBLEMS

1. Specify the required torque rating for a clutch to be attached to a motor shaft running at 1750 rpm. The motor is rated at 5.0 hp and is of the design B type.
2. Specify the required torque rating for a clutch to be attached to a diesel engine shaft running at 2500 rpm. The engine is rated at 75.0 hp.
3. Specify the required torque rating for a clutch to be attached to an electric motor shaft running at 1150 rpm. The motor is rated at 0.50 hp and drives a light fan.
4. An alternative design for the system described in Problem 1 is being considered. Instead of putting the clutch on the motor shaft, it is desired to place it on the output shaft of a speed reducer that rotates at 180 rpm. The power transmission is still approximately 5.0 hp. Specify the required torque rating for the clutch.
5. Specify the required brake torque rating for each of the conditions in Problems 1–4 for average industrial conditions.
6. Specify the required torque rating for a clutch in N · m if it is attached to a design B electric motor shaft rated at 20.0 kW and rotating at 3450 rpm.
7. A clutch-brake module is to be connected between a design C electric motor and a speed reducer. The motor is rated at 50.0 kW at 900 rpm. Specify the required torque ratings for the clutch and the brake portions of the module for average industrial service. The drive is to a large conveyor.
8. Compute the torque required to accelerate a solid steel disc from rest to 550 rpm in 2.0 s. The disc has a diameter of 24.0 in and is 2.5 in thick.
9. The assembly shown in Figure P22–9 is to be stopped by a brake from 775 rpm to zero in 0.50 s or less. Compute the required brake torque.
10. Compute the required clutch torque to accelerate the system shown in Figure P22–10 from rest to a motor speed of 1750 rpm in 1.50 s. Neglect the inertia of the clutch.
11. A winch, sketched in Figure P22–11, is lowering a load at the speed of 50 ft/min. Compute the required torque rating for the brake on the winch shaft to stop the system in 0.25 s.

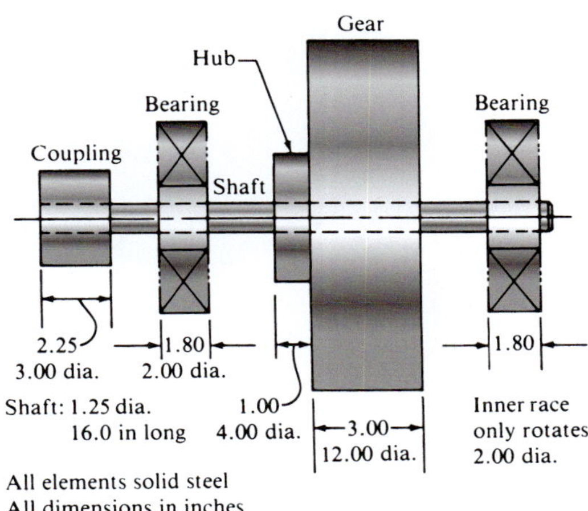

FIGURE P22–9 (Problems 9, 14, and 16)

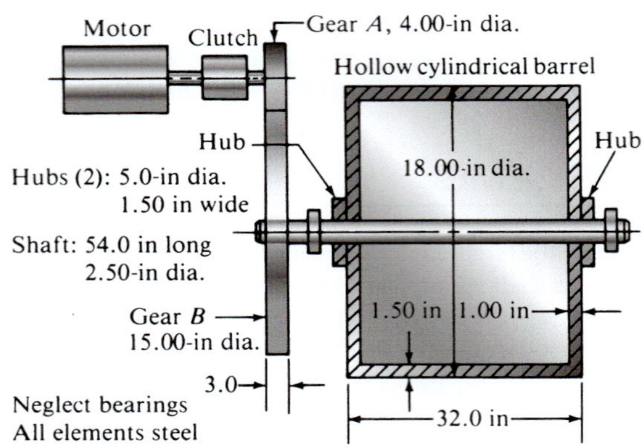

FIGURE P22–10

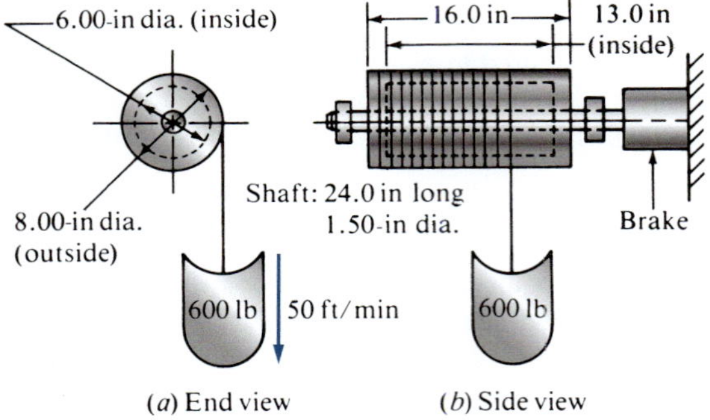

FIGURE P22–11

12. Figure P22–12 shows a tumbling barrel being driven through a wormgear reduction unit. Evaluate the torque rating required for a clutch to accelerate the barrel to 38.0 rpm from rest in 2.0 s (a) if the clutch is placed on the motor shaft and (b) if it is placed at the output of the reducer. Neglect the inertia of the gear shafts, the bearing races, and the clutch. Consider the worm and wormgear to be solid cylinders.
13. Compute the dimensions of an annular plate-type brake to produce a braking torque of 75 lb · in. Air pressure will develop a normal force of 150 lb between the friction surfaces. Use a coefficient of friction of 0.25. The brake will be used in average industrial service, stopping a load from 1150 rpm.
14. Design a plate-type brake for the application described in Problem 9. Specify the design coefficient of friction, the dimensions of the plate, and the axial force required.
15. Compute the axial force required for a cone clutch if it is to exert a driving torque of 15 lb · ft. The cone surface has a mean diameter of 6.0 in and an angle of 12°. Use $f = 0.25$.
16. Design a cone brake for the application described in Problem 9. Specify the design coefficient of friction, the mean diameter of the cone surface, and the axial force required.
17. Compute the actuation force required for the short shoe drum brake of Figure 22–17 to produce a friction torque of 150 lb · ft. Use a drum diameter of 12.0 in, $a = 4.0$ in, and $L = 24.0$ in. Use $f = 0.25$ and $b = 5.0$ in.
18. For all other data from Problem 17 being the same, determine the required dimension b for the brake to be self-actuating.
19. Design a short shoe drum brake to produce a torque of 100 lb · ft. Specify the drum diameter, the configuration of the actuation lever, and the actuation force.
20. Design a long shoe drum brake to produce a friction torque of 100 lb · ft to stop a load from 480 rpm. Specify the friction material, the drum size, the shoe configuration, pivot locations, and actuation force.
21. Design a band brake to exert a braking torque of 75 lb · ft while slowing a drum from 350 rpm to rest. Specify a material, the drum diameter, the band width, the angle of coverage of the friction material, the actuation lever configuration, and the actuation force.

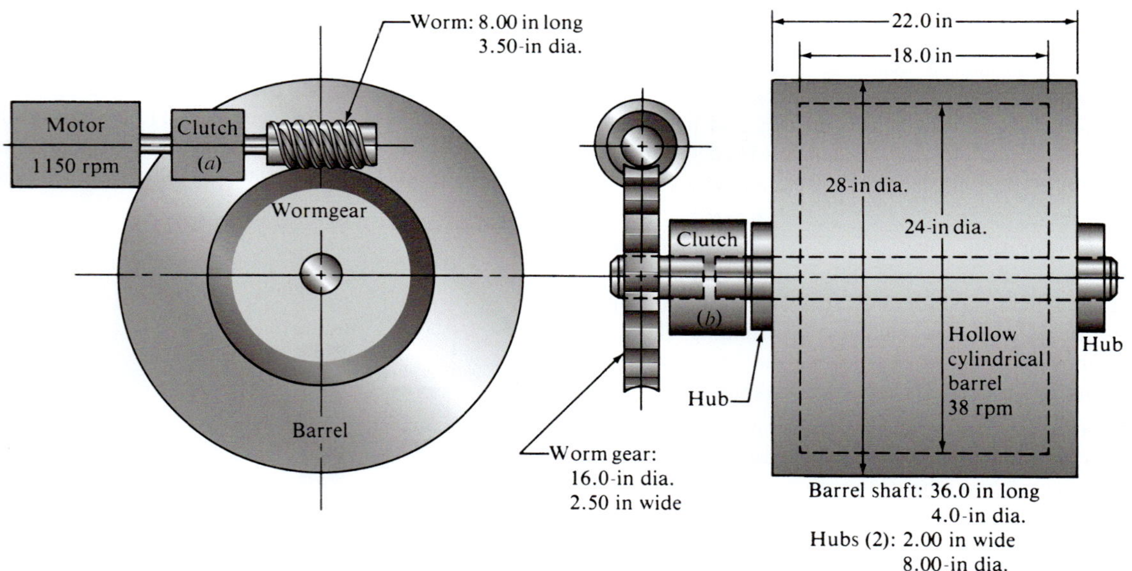

FIGURE P22–12

CHAPTER TWENTY THREE

DESIGN PROJECTS

23–1 OBJECTIVES OF THIS CHAPTER

One of the primary focuses of this book has been to emphasize the integration of machine elements into complete mechanical designs. The interfaces between machine elements have been discussed for many examples. The forces exerted on one element by another have been computed. Commercially available components and complete devices have been shown in figures throughout the book.

Although these discussions and examples are helpful, one of the better ways to learn mechanical design is to ***do*** mechanical design. You should decide the detailed functions and design requirements for the design. You should conceptualize several approaches to a design. You should decide which approach to complete. You should complete the design of each element in detail. You should make assembly and detail drawings to communicate your design to others who may use it or be responsible for its fabrication. You should specify completely the purchased components that are part of the design.

Following are several projects that call for you to do these operations. You or your instructor may modify or amplify the projects to suit individual needs or the available time or information, such as manufacturers' catalogs. As in most design projects, many solutions are possible. Different solutions from several members of a class could be compared and critiqued to enhance learning. It may be helpful at this time to review Sections 1–4 and 1–5 about the functions and design requirements for design and a philosophy of design. Also, the problems at the end of Chapter 1 asked that you write a set of functions and design requirements for several of the same devices. If you have already done so, they can be used as a part of these exercises.

23–2 DESIGN PROJECTS

Automobile Hood Latch

Design a hood latch for an automobile. The latch must be able to hold the hood securely closed during operation of the vehicle. But it should be easy to open for servicing the contents of the engine compartment. Theft-proofing is an important design goal. Attachment of the latch to the frame of the car and to the hood should be defined. Mass production should be a requirement.

Hydraulic Lift

Design a hydraulic lift to be used for car repair. Obtain pertinent dimensions from representative cars for initial height, extended height, design of the pads that contact the car, and so on. The lift will raise the entire car.

Car Jack

Design a floor jack for a car to lift either the entire front end or the entire rear end. The jack may be powered by hand, using mechanical or hydraulic actuation. Or it may be powered by pneumatic pressure or electrical power.

Portable Crane

Design a portable crane to be used in homes, small industries, warehouses, and garages. It should have a capacity of at least 1000 lb (4.45 kN). Typical uses would be to remove an engine from a car, lift machine components, or load trucks.

Can Crusher

Design a machine to crush soft-drink or beer cans. The crusher would be used in homes or restaurants as an aid to recycling efforts. It could be operated either by hand or electrically. It should crush the cans to approximately 20% of their original volume.

Transfer Device

Design an automatic transfer device for a production line. The parts to be handled are steel castings with the following characteristics:

Weight: 42.0 lb (187 N)

Size: Cylindrical; 6.75-in diameter and 10.0 in high. Exterior surface is free of projections or holes and has a reasonably smooth, as-cast finish.

Transfer rate: Continuous flow, 2.00 s between parts.

Parts enter at a 24.0-in elevation on a roller conveyor. They must be elevated to 48.0 inches in a space of 60.0 in horizontally. They leave on a separate conveyor.

Drum Dumper

Design a drum dumper. The machine is to raise a 55-gal drum of bulk material from floor level to a height of 60.0 in and dump the contents of the drum into a hopper.

Paper Feeder

Design a paper feed device for a copier. The paper must be fed at a rate of 120 sheets per min.

Gravel Conveyor

Design a conveyor to elevate gravel into a truck. The top edge of the truck bed is 8.0 ft (2.44 m) off the ground. The bed is 6.5 ft wide, 12.0 ft long, and 4.0 ft deep (1.98 m × 3.66 m × 1.22 m). It is desired to fill the truck in 5.0 min or less.

Construction Lift

Design a construction lift. The lift will raise building materials from ground level to any height up to 40.0 ft (12.2 m). The lift will be at the top of a rigid scaffold that is not a part of the design project. It will raise a load of up to 500 lb (2.22 kN) at the rate of 1.0 ft/s (0.30 m/s). The load will be on a pallet, 3.0 ft by 4.0 ft (0.91 m × 1.22 m). At the top of the lift, means must be provided to bring the load onto a platform that supports the lift.

Packaging Machine

Design a packaging machine. Toothpaste tubes are to be taken from a continuous belt and inserted into cartons. Any standard tube size may be chosen. The device may include the means to close the cartons after the tube is in place.

Carton Packer

Design a machine to insert 24 cartons of toothpaste into a shipping case.

Robot Gripper

Design a gripper for a robot to grasp a spare-tire assembly from a rack and insert it into the trunk of an automobile on an assembly line. Obtain dimensions from a particular car.

Weld Positioner

Design a weld positioner. A heavy frame is made of welded steel plate in the shape shown in Figure 23–1. The welding unit will be robot-guided, but it is essential that the weld line be horizontal as the weld proceeds. Design the device to hold the frame securely and move it to present the part to the robot. The plate has a thickness of 3/8 in (9.53 mm).

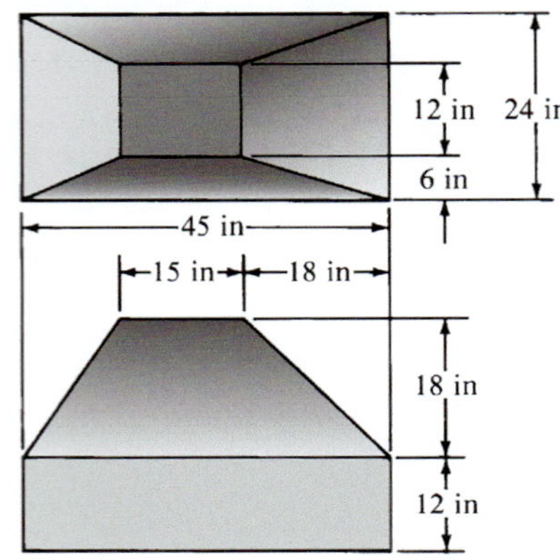

FIGURE 23–1 Frame to be handled by a weld positioner

Garage Door Opener

Design a garage door opener.

Spur Gear Speed Reducer, Single-Reduction

Design a complete single-reduction, spur gear–type speed reducer. Specify the two gears, two shafts, four bearings, and a housing. Use any of the data from Chapter 9, Problems 60–70.

Spur Gear Speed Reducer, Double-Reduction

Design a complete double-reduction, spur gear–type speed reducer. Specify the four gears, three shafts, six bearings, and a housing. Use any of the data from Chapter 9, Problems 74–76.

Helical Gear Speed Reducer, Single-Reduction

Design a complete single-reduction, helical gear–type speed reducer. Use any of the data from Chapter 10, Problems 5–11.

Bevel Gear Speed Reducer, Single-Reduction

Design a complete single-reduction, bevel gear–type speed reducer. Use any of the data from Chapter 10, Problems 14–17.

Wormgear Speed Reducer, Single-Reduction

Design a complete single-reduction, wormgear-type speed reducer. Use any of the data from Chapter 10, Problems 18–24.

Lift Device Using Acme Screws

Design a device similar to that sketched in Figure 17–3. An electric motor drives the worm at a speed of 1750 rpm.

The two Acme screws rotate and lift the yoke, which in turn lifts the hatch. See Example Problem 17–1 for additional details. Complete the entire unit, including the wormgear set, the chain drive, the Acme screws, the bearings, and their mountings. The hatch is 60 in (1524 mm) in diameter at its top surface. The screws should be nominally 30 in (762 mm) long. The total motion of the yoke will be 24 in, to be completed in 15.0 s or less.

Lift Device Using Ball Screws

Repeat the design of the lift device, using ball screws instead of Acme screws.

Brake for a Drive Shaft

Design a brake. A rotating load (as sketched in Figure P22–9) is to be stopped from 775 rpm in 0.50 s or less. Use any type of brake described in Chapter 22, and complete the design details, including the actuating means: springs, air pressure, manual lever, and so on. Show the brake attached to the shaft of Figure P22–9.

Brake for a Winch

Design a complete brake for the application shown in Figure P22–11 and described in Problem 11 in Chapter 22.

Indexing Drive

Design an indexing drive for an automatic assembly system. The items to be moved are mounted on a square steel fixture plate, 6.0 in (152 mm) on a side and 0.50 in (12.7 mm) thick. The total weight of each assembly is 10.0 lb (44.5 N). The center of each fixture (intersection of its diagonals) is to move 12.0 in (305 mm) with each index. The index is to be completed in 1.0 s or less, and the fixture must be held stationary at each station for a minimum of 2.0 s. Four assembly stations are required. The arrangement may be linear, rotary, or any other, provided that the fixtures move in a horizontal plane.

Child's Ferris Wheel

Design a child's Ferris wheel. It should be capable of holding one to four children weighing up to 80 lb (356 N) each. The rotational speed should be 1 rev in 6.0 s. It should be driven by an electric motor.

Merry-Go-Round

Design an amusement ride for small children, age 6 or younger, that they can sit in safely while enjoying an interesting motion pattern. At least two children at a time can ride the machine. The ride will be marketed to shopping centers and department stores to amuse customers' children.

Backyard Amusement Ride

Design a backyard amusement ride in which small coaster wagons are pulled along a circular path. The ride should be powered by an electric motor. Each wagon is 1.0 m long (39.4 in) and 0.50 m wide (19.7 in). The four wheels are 150 mm (6.0 in) in diameter. The wagon is to be attached to the drive rod at the point where the handle would normally be. The radial distance to the attachment point is to be 2.0 m (6.6 ft). The wagons are to make 1 rev in 8.0 s. Provide a means of starting and stopping the drive.

Transfer Device

Design a device to move automotive camshafts between processing stations. Each movement is to be 9.0 in (229 mm). The camshaft is to be supported on two unfinished bearing surfaces having a diameter of 3.80 in (96.5 mm) and an axial length of 0.75 in (19.0 mm). The spread between the bearing surfaces is 15.75 in (400.0 mm). Each camshaft weighs 16.3 lb (72.5 N). One motion cycle is to be completed each 2.50 s. Design the complete mechanism, including the drive from an electric motor.

Chain Conveyor

Design a powered, straight chain conveyor to move eight pallets along an assembly line. The pallets are 18 in long and 12 in wide. The maximum weight of each pallet and the product it carries is 125 lb. At the end of the conveyor, an external downward force of 500 lb is applied to the product which must be carried through the pallet to the conveyor structure. You may design the configuration of the sides and bottom of the pallet.

"You Are the Designer" Projects

At the beginning of each chapter in this book, a section called **You Are the Designer** appeared in which you were asked to imagine that you were the designer of some device or system. Choose any of those projects.

LIST OF APPENDICES

APPENDIX 1 Properties of Areas

A = area
I = moment of inertia
S = section modulus
r = radius of gyration = $\sqrt{I/A}$
J = polar moment of inertia
Z_p = polar section modulus

(*a*) Circle

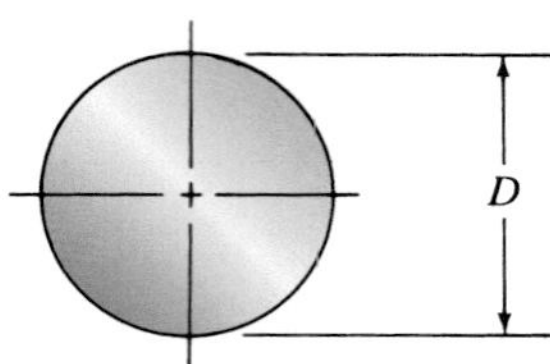

$A = \pi D^2/4$ $\quad r = D/4$

$I = \pi D^4/64$ $\quad J = \pi D^4/32$

$S = \pi D^3/32$ $\quad Z_p = \pi D^3/16$

(*b*) Hollow circle (tube)

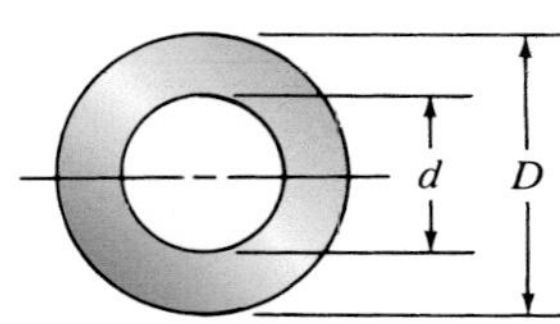

$A = \pi(D^2 - d^2)/4$ $\quad r = \dfrac{\sqrt{(D^2 + d^2)}}{4}$

$I = \pi(D^4 - d^4)/64$ $\quad J = \pi(D^4 - d^4)/32$

$S = \pi(D^4 - d^4)/32D$ $\quad Z_p = \pi(D^4 - d^4)/16D$

(*c*) Square

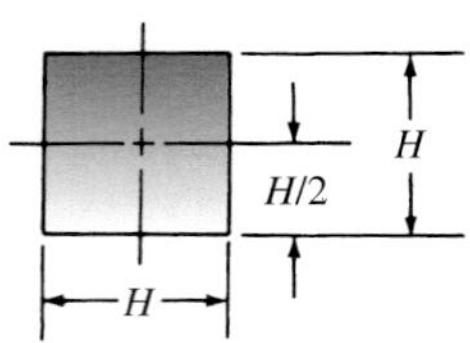

$A = H^2$ $\quad r = H/\sqrt{12}$

$I = H^4/12$

$S = H^3/6$

(*d*) Rectangle

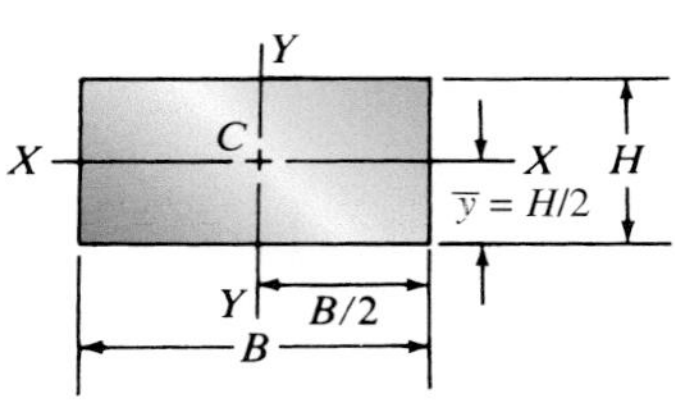

$A = BH$ $\quad r_x = H/\sqrt{12}$

$I_x = BH^3/12$ $\quad r_y = B/\sqrt{12}$

$I_y = HB^3/12$

$S_x = BH^2/6$

$S_y = HB^2/6$

(*e*) Hollow rectangle

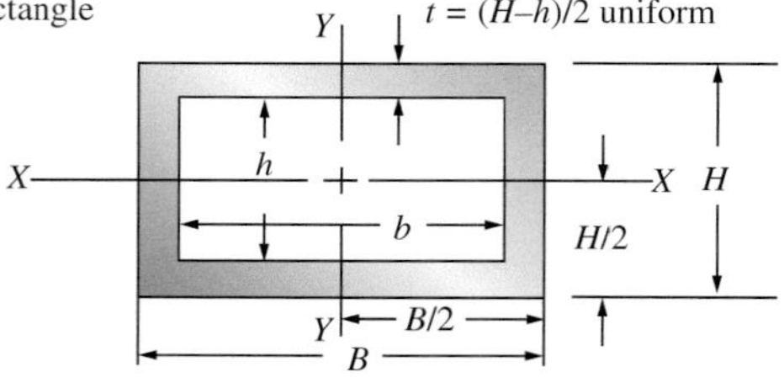

$A = BH - bh$

$I_x = \dfrac{BH^3 - bh^3}{12}$ $\quad S_x = \dfrac{BH^3 - bh^3}{6H}$ $\quad r_x = 0.289\sqrt{\dfrac{BH^3 - bh^3}{BH - bh}}$

$I_y = \dfrac{HB^3 - hb^3}{12}$ $\quad S_y = \dfrac{HB^3 - hb^3}{6B}$ $\quad r_y = 0.289\sqrt{\dfrac{HB^3 - hb^3}{HB - hb}}$

(*f*) Triangle

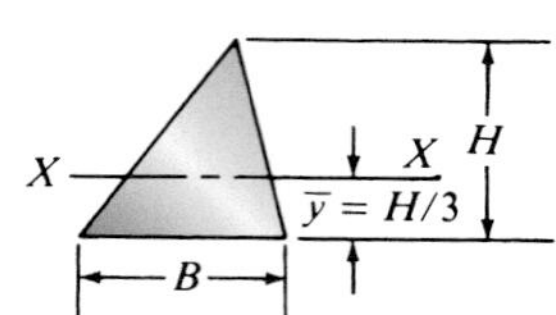

$A = BH/2$ $\quad r = H/\sqrt{18}$

$I = BH^3/36$

$S = BH^2/24$

(*g*) Semicircle 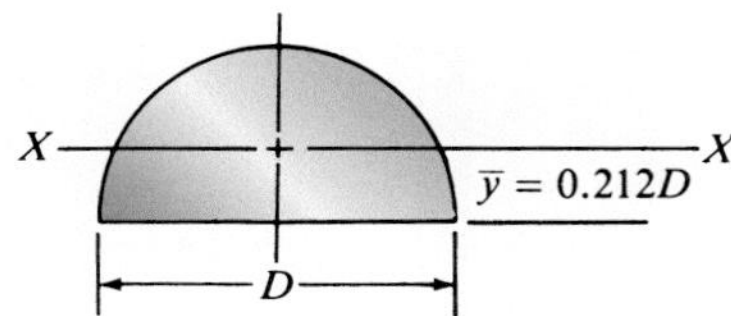	$A = \pi D^2/8$ $I = 0.007D^4$ $S = 0.024D^3$ $r = 0.132D$
(*h*) Regular hexagon 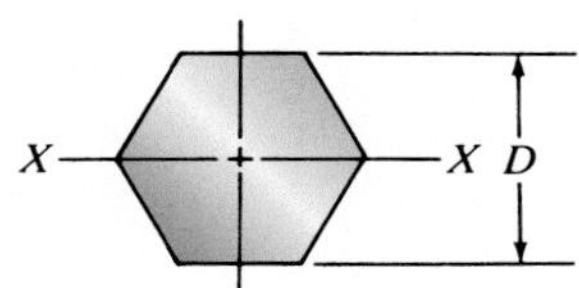	$A = 0.866D^2$ $I = 0.06D^4$ $S = 0.12D^3$ $r = 0.264D$
(*i*) Trapezoid 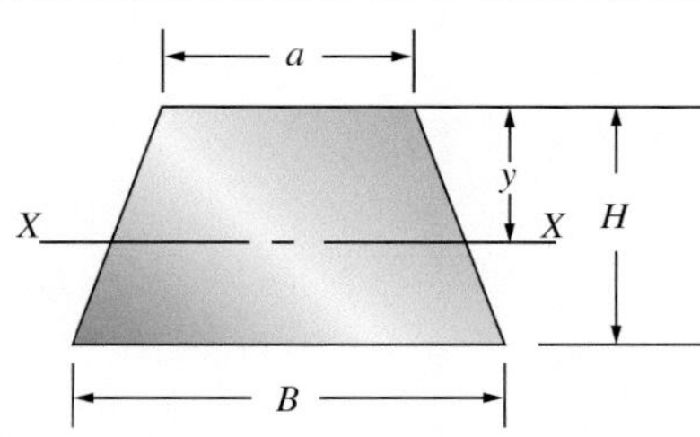	$A = H(a + B)/2$ $y = \dfrac{H(a + 2B)}{3(a + B)}$ $I_x = \dfrac{H^3(a^2 + 4aB + B^2)}{36(a + B)}$ $S = \dfrac{H^2(a^2 + 4aB + B^2)}{12(a + 2B)}$ $r = \dfrac{H^2(a^2 + 4aB + B^2)}{18(a + B)^2}$ y = Maximum distance from x-axis to outer surface of section
(*j*) Ellipse 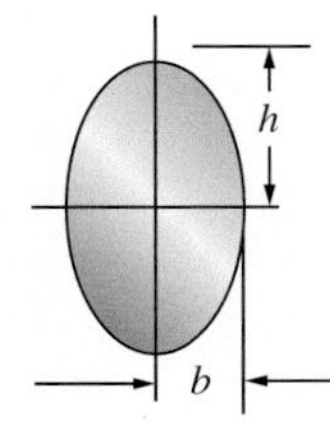	$A = \pi bh$ $I = \dfrac{\pi h^3 b}{4}$ $S = \dfrac{\pi h^2 b}{4}$ $r = h/2$

APPENDIX 2 Preferred Basic Sizes and Screw Threads

TABLE A2-1 Preferred Basic Sizes

Fractional (in)				Decimal (in)			Metric (mm)					
							First	Second	First	Second	First	Second
1/64	0.015 625	5	5.000	0.010	2.00	8.50	1		10		100	
1/32	0.031 25	$5\frac{1}{4}$	5.250	0.012	2.20	9.00		1.1		11		110
1/16	0.0625	$5\frac{1}{2}$	5.500	0.016	2.40	9.50	1.2		12		120	
3/32	0.093 75	$5\frac{3}{4}$	5.750	0.020	2.60	10.00		1.4		14		140
1/8	0.1250	6	6.000	0.025	2.80	10.50	1.6		16		160	
5/32	0.156 25	$6\frac{1}{2}$	6.500	0.032	3.00	11.00		1.8		18		180
3/16	0.1875	7	7.000	0.040	3.20	11.50	2		20		200	
1/4	0.2500	$7\frac{1}{2}$	7.500	0.05	3.40	12.00		2.2		22		220
5/16	0.3125	8	8.000	0.06	3.60	12.50	2.5		25		250	
3/8	0.3750	$8\frac{1}{2}$	8.500	0.08	3.80	13.00		2.8		28		280
7/16	0.4375	9	9.000	0.10	4.00	13.50	3		30		300	
1/2	0.5000	$9\frac{1}{2}$	9.500	0.12	4.20	14.00		3.5		35		350
9/16	0.5625	10	10.000	0.16	4.40	14.50	4		40		400	
5/8	0.6250	$10\frac{1}{2}$	10.500	0.20	4.60	15.00		4.5		45		450
11/16	0.6875	11	11.000	0.24	4.80	15.50	5		50		500	
3/4	0.7500	$11\frac{1}{2}$	11.500	0.30	5.00	16.00		5.5		55		550
7/8	0.8750	12	12.000	0.40	5.20	16.50	6		60		600	
1	1.000	$12\frac{1}{2}$	12.500	0.50	5.40	17.00		7		70		700
$1\frac{1}{4}$	1.250	13	13.000	0.60	5.60	17.50	8		80		800	
$1\frac{1}{2}$	1.500	$13\frac{1}{2}$	13.500	0.80	5.80	18.00		9		90		900
$1\frac{3}{4}$	1.750	14	14.000	1.00	6.00	18.50					1000	
2	2.000	$14\frac{1}{2}$	14.500	1.20	6.50	19.00						
$2\frac{1}{4}$	2.250	15	15.000	1.40	7.00	19.50						
$2\frac{1}{2}$	2.500	$15\frac{1}{2}$	15.500	1.60	7.50	20.00						
$2\frac{3}{4}$	2.750	16	16.000	1.80	8.00							
3	3.000	$16\frac{1}{2}$	16.500									
$3\frac{1}{4}$	3.250	17	17.000									
$3\frac{1}{2}$	3.500	$17\frac{1}{2}$	17.500									
$3\frac{3}{4}$	3.750	18	18.000									
4	4.000	$18\frac{1}{2}$	18.500									
$4\frac{1}{4}$	4.250	19	19.000									
$4\frac{1}{2}$	4.500	$19\frac{1}{2}$	19.500									
$4\frac{3}{4}$	4.750	20	20.000									

Source: Reprinted from ASME B4.1-1967, by permission of The American Society of Mechanical Engineers.

TABLE A2-2 American Standard Screw Threads

A. American Standard thread dimensions, numbered sizes

Basic major diameter, *D* (in)	Coarse threads: UNC		Fine threads: UNF	
	Size-Threads per inch, *n*	Tensile stress area (in^2)	Size-Threads per inch, *n*	Tensile stress area (in^2)
0.0600	—	—	0–80	0.001 80
0.0730	1–64	0.002 63	1–72	0.002 78
0.0860	2–56	0.003 70	2–64	0.003 94
0.0990	3–48	0.004 87	3–56	0.005 23
0.1120	4–40	0.006 04	4–48	0.006 61
0.1250	5–40	0.007 96	5–44	0.008 30
0.1380	6–32	0.009 09	6–40	0.010 15
0.1640	8–32	0.0140	8–36	0.014 74
0.1900	10–24	0.0175	10–32	0.0200
0.2160	12–24	0.0242	12–28	0.0258

B. American Standard thread dimensions, fractional sizes

Basic major diameter, *D* (in)	Coarse threads: UNC		Fine threads: UNF	
	Size-Threads per inch, *n*	Tensile stress area (in^2)	Size-Threads per inch, *n*	Tensile stress area (in^2)
0.2500	1/4–20	0.0318	1/4–28	0.0364
0.3125	5/16–18	0.0524	5/16–24	0.0580
0.3750	3/8–16	0.0775	3/8–24	0.0878
0.4375	7/16–14	0.1063	7/16–20	0.1187
0.5000	1/2–13	0.1419	1/2–20	0.1599
0.5625	9/16–12	0.182	9/16–18	0.203
0.6250	5/8–11	0.226	5/8–18	0.256
0.7500	3/4–10	0.334	3/4–16	0.373
0.8750	7/8–9	0.462	7/8–14	0.509
1.000	$1\frac{1}{8}$–8	0.606	$1\frac{1}{8}$–12	0.663
1.125	$1\frac{1}{8}$–7	0.763	$1\frac{1}{8}$–12	0.856
1.250	$1\frac{1}{4}$–7	0.969	$1\frac{1}{4}$–12	1.073
1.375	$1\frac{3}{8}$–6	1.155	$1\frac{3}{8}$–12	1.315
1.500	$1\frac{1}{2}$–6	1.405	$1\frac{1}{2}$–12	1.581
1.750	$1\frac{3}{4}$–5	1.90		
2.000	$2–4\frac{1}{2}$	2.50		

TABLE A2-3 Metric Sizes of Screw Threads

Basic major diameter, D (mm)	Coarse threads: Basic thread designation $MD \times$ Pitch (mm) (mm)	Coarse threads: Tensile stress area (mm^2)	Fine threads: $MD \times$ Pitch (mm) (mm)	Fine threads: Tensile stress area (mm^2)
1	M1 × 0.25	0.460	—	—
1.6	M1.6 × 0.35	1.27	M1.6 × 0.20	1.57
2	M2 × 0.4	2.07	M2 × 0.25	2.45
2.5	M2.5 × 0.45	3.39	M2.5 × 0.35	3.70
3	M3 × 0.5	5.03	M3 × 0.35	5.61
4	M4 × 0.7	8.78	M4 × 0.5	9.79
5	M5 × 0.8	14.2	M5 × 0.5	16.1
6	M6 × 1	20.1	M6 × 0.75	22.0
8	M8 × 1.25	36.6	M8 × 1	39.2
10	M10 × 1.5	58.0	M10 × 1.25	61.2
12	M12 × 1.75	84.3	M12 × 1.25	92.1
16	M16 × 2	157	M16 × 1.5	167
20	M20 × 2.5	245	M20 × 1.5	272
24	M24 × 3	353	M24 × 2	384
30	M30 × 3.5	561	M30 × 2	621
36	M36 × 4	817	M36 × 3	865
42	M42 × 4.5	1121		
48	M48 × 5	1473		

APPENDIX 3 Design Properties of Carbon and Alloy Steels

Material designation (SAE number)	Condition	Tensile strength (ksi)	Tensile strength (MPa)	Yield strength (ksi)	Yield strength (MPa)	Ductility (percent elongation in 2 inches)	Brinell hardness (HB)
1020	Hot-rolled	55	379	30	207	25	111
1020	Cold-drawn	61	420	51	352	15	122
1020	Annealed	60	414	43	296	38	121
1040[1]	Hot-rolled	72	496	42	290	18	144
1040	Cold-drawn	80	552	71	490	12	160
1040	OQT 1300	88	607	61	421	33	183
1040	OQT 400	113	779	87	600	19	262
1050	Hot-rolled	90	620	49	338	15	180
1050	Cold-drawn	100	690	84	579	10	200
1050	OQT 1300	96	662	61	421	30	192
1050	OQT 400	143	986	110	758	10	321
1117	Hot-rolled	65	448	40	276	33	124
1117	Cold-drawn	80	552	65	448	20	138
1117	WQT 350	89	614	50	345	22	178
1137	Hot-rolled	88	607	48	331	15	176
1137	Cold-drawn	98	676	82	565	10	196
1137	OQT 1300	87	600	60	414	28	174
1137	OQT 400	157	1083	136	938	5	352
1144[1]	Hot-rolled	94	648	51	352	15	188
1144	Cold-drawn	100	690	90	621	10	200
1144	OQT 1300	96	662	68	469	25	200
1144	OQT 400	127	876	91	627	16	277
1213	Hot-rolled	55	379	33	228	25	110
1213	Cold-drawn	75	517	58	340	10	150
12L13	Hot-rolled	57	393	34	234	22	114
12L13	Cold-drawn	70	483	60	414	10	140
1340[1]	Annealed	102	703	63	434	26	207
1340	OQT 1300	100	690	75	517	25	235
1340	OQT 1000	144	993	132	910	17	363
1340	OQT 700	221	1520	197	1360	10	444
1340	OQT 400	285	1960	234	1610	8	578
3140	Annealed	95	655	67	462	25	187
3140	OQT 1300	115	792	94	648	23	233
3140	OQT 1000	152	1050	133	920	17	311
3140	OQT 700	220	1520	200	1380	13	461
3140	OQT 400	280	1930	248	1710	11	555
4130	Annealed	81	558	52	359	28	156
4130	WQT 1300	98	676	89	614	28	202
4130	WQT 1000	143	986	132	910	16	302
4130	WQT 700	208	1430	180	1240	13	415
4130	WQT 400	234	1610	197	1360	12	461
4140[1]	Annealed	95	655	54	372	26	197
4140	OQT 1300	117	807	100	690	23	235
4140	OQT 1000	168	1160	152	1050	17	341
4140	OQT 700	231	1590	212	1460	13	461
4140	OQT 400	290	2000	251	1730	11	578
4150	Annealed	106	731	55	379	20	197
4150	OQT 1300	127	880	116	800	20	262
4150	OQT 1000	197	1360	181	1250	11	401
4150	OQT 700	247	1700	229	1580	10	495
4150	OQT 400	300	2070	248	1710	10	578

(Continued)

APPENDIX 3 (*Continued*)

Material designation (SAE number)	Condition	Tensile strength (ksi)	Tensile strength (MPa)	Yield strength (ksi)	Yield strength (MPa)	Ductility (percent elongation in 2 inches)	Brinell hardness (HB)
4340[1]	Annealed	108	745	68	469	22	217
4340	OQT 1300	140	965	120	827	23	280
4340	OQT 1000	171	1180	158	1090	16	363
4340	OQT 700	230	1590	206	1420	12	461
4340	OQT 400	283	1950	228	1570	11	555
5140	Annealed	83	572	42	290	29	167
5140	OQT 1300	104	717	83	572	27	207
5140	OQT 1000	145	1000	130	896	18	302
5140	OQT 700	220	1520	200	1380	11	429
5140	OQT 400	276	1900	226	1560	7	534
5150	Annealed	98	676	52	359	22	197
5150	OQT 1300	116	800	102	700	22	241
5150	OQT 1000	160	1100	149	1030	15	321
5150	OQT 700	240	1650	220	1520	10	461
5150	OQT 400	312	2150	250	1720	8	601
5160	Annealed	105	724	40	276	17	197
5160	OQT 1300	115	793	100	690	23	229
5160	OQT 1000	170	1170	151	1040	14	341
5160	OQT 700	263	1810	237	1630	9	514
5160	OQT 400	322	2220	260	1790	4	627
6150[1]	Annealed	96	662	59	407	23	197
6150	OQT 1300	118	814	107	738	21	241
6150	OQT 1000	183	1260	173	1190	12	375
6150	OQT 700	247	1700	223	1540	10	495
6150	OQT 400	315	2170	270	1860	7	601
8650	Annealed	104	717	56	386	22	212
8650	OQT 1300	122	841	113	779	21	255
8650	OQT 1000	176	1210	155	1070	14	363
8650	OQT 700	240	1650	222	1530	12	495
8650	OQT 400	282	1940	250	1720	11	555
8740	Annealed	100	690	60	414	22	201
8740	OQT 1300	119	820	100	690	25	241
8740	OQT 1000	175	1210	167	1150	15	363
8740	OQT 700	228	1570	212	1460	12	461
8740	OQT 400	290	2000	240	1650	10	578
9255	Annealed	113	780	71	490	22	229
9255	Q&T 1300	130	896	102	703	21	262
9255	Q&T 1000	181	1250	160	1100	14	352
9255	Q&T 700	260	1790	240	1650	5	534
9255	Q&T 400	310	2140	287	1980	2	601

Notes: Properties common to all carbon and alloy steels:

Poisson's ratio: 0.27.

Shear modulus: 11.5×10^6 psi; 80 GPa.

Coefficient of thermal expansion: 6.5×10^{-6}°F^{-1}.

Density: 0.283 lb/in^3: 7680 kg/m^3.

Modulus of elasticity: 30×10^6 psi; 207 GPa.

[1]See Appendix 4 for graphs of properties versus heat treatment.

APPENDIX 4 Properties of Heat-Treated Steels

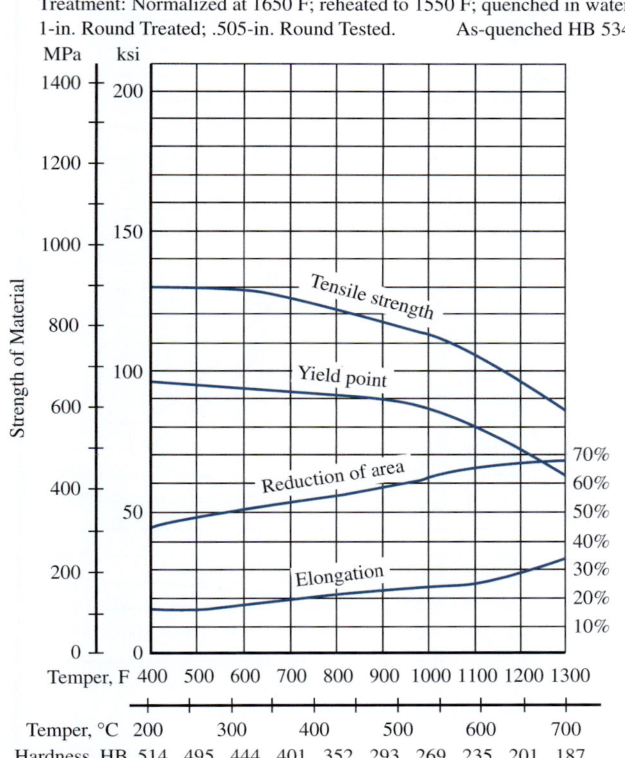

FIGURE A4-1 Properties of heat-treated SAE 1040, water-quenched and tempered (*Modern Steels and Their Properties*, Bethlehem Steel Co.,* Bethlehem, PA)

*Now merged into ArcelorMittal, Luxembourg

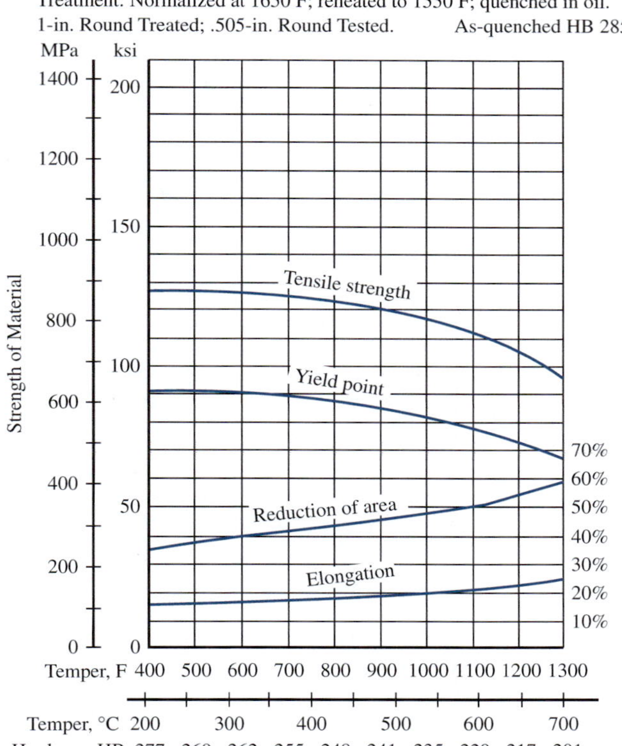

FIGURE A4-2 Properties of heat-treated SAE 1144, oil-quenched and tempered (*Modern Steels and Their Properties*, Bethlehem Steel Co.,* Bethlehem, PA)

*Now merged into ArcelorMittal, Luxembourg

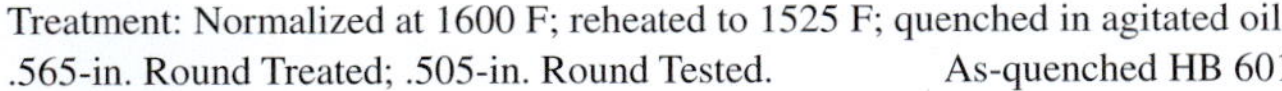

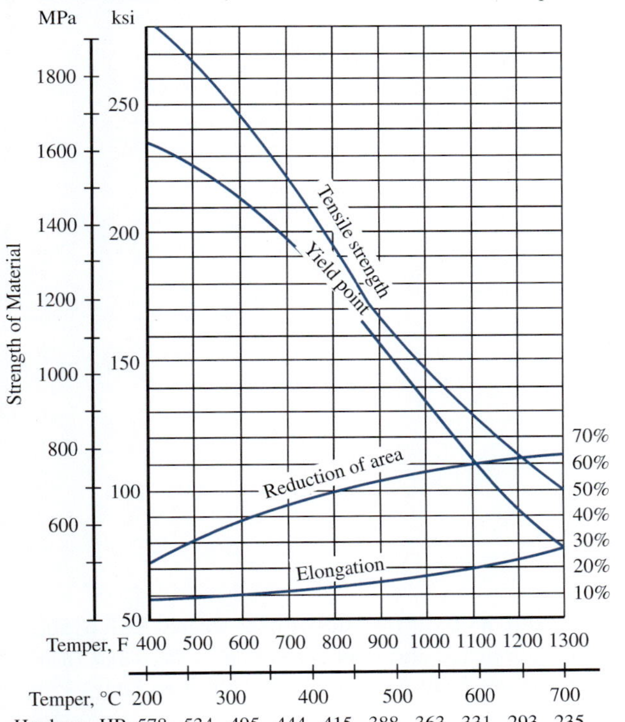

FIGURE A4-3 Properties of heat-treated SAE 1340, oil-quenched and tempered (*Modern Steels and Their Properties*, Bethlehem Steel Co.,* Bethlehem, PA)

*Now merged into ArcelorMittal, Luxembourg

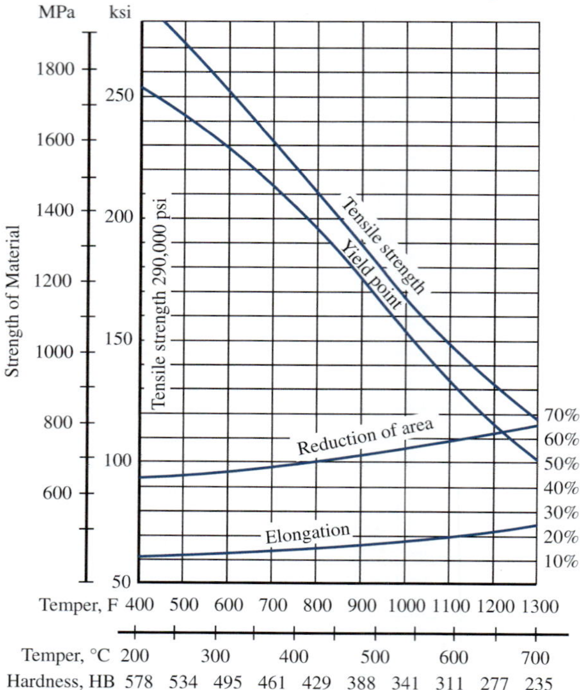

FIGURE A4-4 Properties of heat-treated SAE 4140, oil-quenched and tempered (*Modern Steels and Their Properties*, Bethlehem Steel Co.,* Bethlehem, PA)

*Now merged into ArcelorMittal, Luxembourg

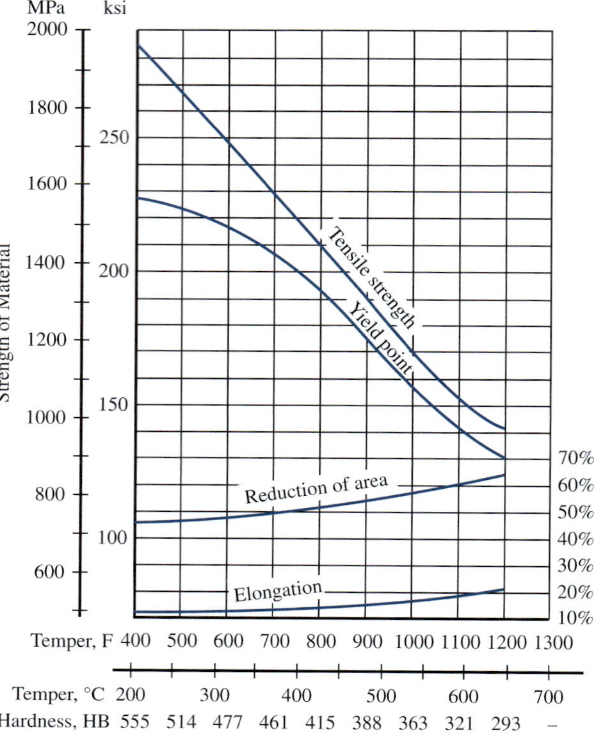

FIGURE A4-5 Properties of heat-treated SAE 4340, oil-quenched and tempered (*Modern Steels and Their Properties*, Bethlehem Steel Co.,* Bethlehem, PA)

*Now merged into ArcelorMittal, Luxembourg

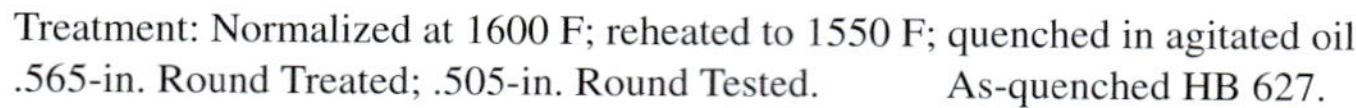

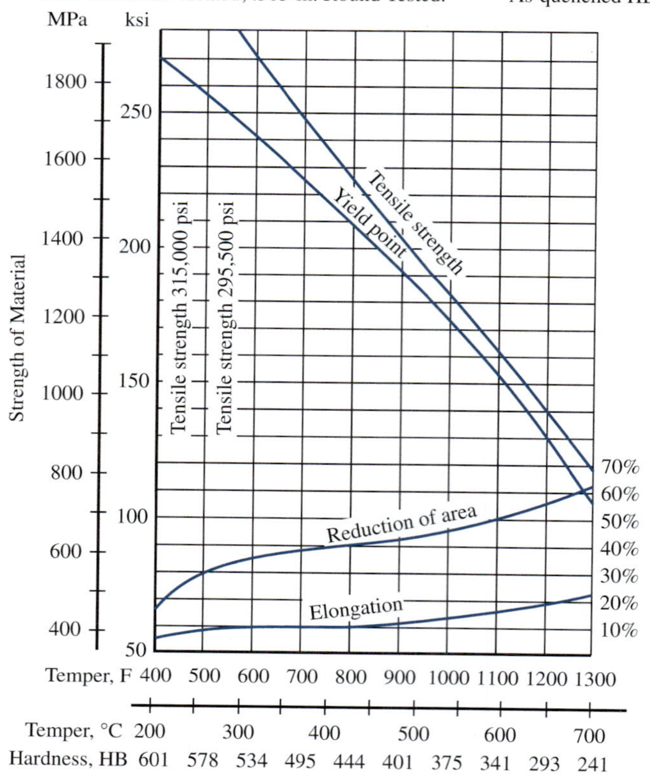

FIGURE A4-6 Properties of heat-treated SAE 6150, oil-quenched and tempered (*Modern Steels and Their Properties*, Bethlehem Steel Co.,* Bethlehem, PA)

*Now merged into ArcelorMittal, Luxembourg

APPENDIX 5 Properties of Carburized Steels

Material designation (SAE number)	Condition	Core properties: Tensile strength (ksi)	Tensile strength (MPa)	Yield strength (ksi)	Yield strength (MPa)	Ductility (percent elongation in 2 inches)	Brinell hardness (HB)	Case hardness (HRC)
1015	SWQT 350	106	731	60	414	15	217	62
1020	SWQT 350	129	889	72	496	11	255	62
1022	SWQT 350	135	931	75	517	14	262	62
1117	SWQT 350	125	862	66	455	10	235	65
1118	SWQT 350	144	993	90	621	13	285	61
4118	SOQT 300	143	986	93	641	17	293	62
4118	DOQT 300	126	869	63	434	21	241	62
4118	SOQT 450	138	952	89	614	17	277	56
4118	DOQT 450	120	827	63	434	22	229	56
4320	SOQT 300	218	1500	178	1230	13	429	62
4320	DOQT 300	151	1040	97	669	19	302	62
4320	SOQT 450	211	1450	173	1190	12	415	59
4320	DOQT 450	145	1000	94	648	21	293	59
4620	SOQT 300	119	820	83	572	19	277	62
4620	DOQT 300	122	841	77	531	22	248	62
4620	SOQT 450	115	793	80	552	20	248	59
4620	DOQT 450	115	793	77	531	22	235	59
4820	SOQT 300	207	1430	167	1150	13	415	61
4820	DOQT 300	204	1405	165	1140	13	415	60
4820	SOQT 450	205	1410	184	1270	13	415	57
4820	DOQT 450	196	1350	171	1180	13	401	56
8620	SOQT 300	188	1300	149	1030	11	388	64
8620	DOQT 300	133	917	83	572	20	269	64
8620	SOQT 450	167	1150	120	827	14	341	61
8620	DOQT 450	130	896	77	531	22	262	61
E9310	SOQT 300	173	1190	135	931	15	363	62
E9310	DOQT 300	174	1200	139	958	15	363	60
E9310	SOQT 450	168	1160	137	945	15	341	59
E9310	DOQT 450	169	1170	138	952	15	352	58

Notes: Properties given are for a single set of tests on 1/2-in round bars.

SWQT: single water-quenched and tempered.

SOQT: single oil-quenched and tempered.

DOQT: double oil-quenched and tempered.

300 and 450 are the tempering temperatures in °F. Steel was carburized for 8 h. Case depth ranged from 0.045 to 0.075 in.

APPENDIX 6 Properties of Stainless Steels

Material designation			Tensile strength		Yield strength		Ductility (percent elongation in 2 inches)
SAE number	UNS	Condition	(ksi)	(MPa)	(ksi)	(MPa)	
Austenitic steels							
201	S20100	Annealed	115	793	55	379	55
		1/4 hard	125	862	75	517	20
		1/2 hard	150	1030	110	758	10
		3/4 hard	175	1210	135	931	5
		Full hard	185	1280	140	966	4
301	S30100	Annealed	110	758	40	276	60
		1/4 hard	125	862	75	517	25
		1/2 hard	150	1030	110	758	15
		3/4 hard	175	1210	135	931	12
		Full hard	185	1280	140	966	8
304	S30400	Annealed	85	586	35	241	60
310	S31000	Annealed	95	655	45	310	45
316	S31600	Annealed	80	552	30	207	60
Ferritic steels							
405	S40500	Annealed	70	483	40	276	30
430	S43000	Annealed	75	517	40	276	30
		Full hard	90	621	80	552	15
446	S44600	Annealed	80	552	50	345	25
Martensitic steels							
410	S41000	Annealed	75	517	40	276	30
416	S41600	Q&T 600	180	1240	140	966	15
		Q&T 1000	145	1000	115	793	20
		Q&T 1400	90	621	60	414	30
431	S43100	Q&T 600	195	1344	150	1034	15
440A	S44002	Q&T 600	280	1930	270	1860	3
501	S50100	Annealed	70	483	30	207	28
		OQT 1000	175	1210	135	931	15
Precipitation-hardening steels							
17-4PH	S17400	H 900	210	1450	185	1280	14
		H 1150	145	1000	125	862	19
17-7PH	S17700	RH 950	200	1380	175	1210	10
		TH 1050	175	1210	155	1070	12
PH 13-8 Mo	S13800	H 950	220	1517	205	1413	10
		H 1050	175	1207	165	1138	12
		H 1150	135	931	90	621	14

APPENDIX 7 Properties of Structural Steels

Material ASTM no. and products	Ultimate strength, s_u[1] ksi	Ultimate strength, s_u[1] MPa	Yield strength, s_y[1] ksi	Yield strength, s_y[1] MPa	Percent elongation in 2 in
A36—Carbon steel: shapes, plates, and bars	58	400	36	248	21
A 53—grade B-pipe	60	414	35	240	—
A242—HSLA corrosion resistant: shapes, plates, and bars					
$\leq \frac{3}{4}$ in thick	70	483	50	345	21
$\frac{3}{4}$ to $1\frac{1}{2}$ in thick	67	462	46	317	21
$1\frac{1}{2}$ to 4 in thick	63	434	42	290	21
A500—Cold-formed structural tubing					
Round, grade B	58	400	42	290	23
Round grade C	62	427	46	317	21
Shaped, grade B	58	400	46	317	23
Shaped, grade C	62	427	50	345	21
A501—Hot-formed structural tubing, round or shaped	58	400	36	248	23
A514—Quenched and tempered alloy steel: plate					
$\leq 2\frac{1}{2}$ in thick	110	758	100	690	18
$2\frac{1}{2}$ to 6 in thick	100	690	90	620	16
A572—HSLA columbium-vanadium steel: shapes, plates, and bars					
Grade 42	60	414	42	290	24
Grade 50	65	448	50	345	21
Grade 60	75	517	60	414	18
Grade 65	80	552	65	448	17
A913—HSLA, grade 65: shapes	80	552	65	448	17
A992—HSLA: W-Shapes only	65	448	50	345	21

Notes: [1]Minimum values; may range higher.

HSLA—High strength low-alloy.

The American Institute of Steel Construction specifies $E = 29 \times 10^6$ psi (200 GPa) for structural steel.

APPENDIX 8 Design Properties of Cast Iron—U.S. Units Basis[3]

U.S. units	Ultimate strength						Yield strength		[1]Modulus of elasticity, E		Percent elongation
	s_u		[1]s_{uc}		[1]s_{us}		s_{yt}				
Material type and grade	ksi	[2]MPa	ksi	[2]MPa	ksi	[2]MPa	ksi	[2]MPa	10^6 psi	[2]GPa	%
Gray iron—ASTM A48											
No. 20A	20	138	80	552	32	221	—	—	12.2	84	<1.0
No. 30A	30	207	113	779	47	324	—	—	16.9	117	<1.0
No. 40A	40	276	140	965	57	393	—	—	19.4	134	<1.0
No. 50A	50	345	158	1089	66	455	—	—	20.8	143	<1.0
No. 60A	60	414	170	1172	72	496	—	—	21.5	148	<1.0
Ductile iron—ASTM A536											
60-40-18	60	414	—	—	57	393	40	276	24	165	18
65-45-12	65	448	—	—	—	—	45	310	24	165	12
80-55-08	80	552	—	—	—	—	55	379	24	165	6
100-70-03	100	690	—	—	—	—	70	483	24	165	3
120-90-02	120	827	180	1241	—	—	90	621	23	159	2
Austempered ductile iron (ADI)—ASTM A897											
110/70/11	110	758	—	—	—	—	70	483	22	152	11
130/90/09	130	896	—	—	—	—	90	621	22	152	9
150/110/07	150	1034	—	—	—	—	110	758	22	152	7
175/125/04	175	1207	—	—	—	—	125	862	22	152	4
200/155/02	200	1379	—	—	—	—	155	1069	22	152	2
230/185/01	230	1586	—	—	—	—	185	1276	22	152	1
Ferritic malleable iron—ASTM A47											
32510	50	345	—	—	—	—	32.5	224	25	172	10
Pearlitic malleable iron—ASTM A220											
40010	60	414	240	1655	43	296	40	276	26	179	10
45008	65	448	240	1655	49	338	45	310	26	179	8
45006	65	448	240	1655	49	338	45	310	26	179	6
50005	70	483	240	1655	55	379	50	345	26	179	5
60004	80	552	240	1655	65	448	60	414	26	179	4
70003	85	586	240	1655	68	469	70	483	26	179	3
80002	95	655	240	1655	75	517	80	552	26	179	2
90001	105	724	240	1655	78	538	90	621	26	179	1

Notes: [1]Approximate values; not part of the standards; If critical, negotiate with supplier.

[2]Metric data computed using: (6.895 × U.S. data); Not part of the standards.

[3]The density of cast irons ranges from 0.25 lb_m/in^3 to 0.27 lb_m/in^3 (6920 kg/m^3 to 7480 kg/m^3).

APPENDIX 8A Design Properties of Cast Iron—SI Units Basis[3]

SI units	Ultimate strength						Yield strength		[1]Modulus of elasticity, E		Percent elongation
	s_u		[1]s_{uc}		[1]s_{us}		s_{yt}				
Material type and grade	[2]ksi	MPa	[2]ksi	MPa	[2]ksi	MPa	[2]ksi	MPa	10^6 psi	GPa	%
Gray iron—ASTM A48M											
No. 150A	22	150	—	—	—	—	—	—	12.2	84	<1.0
No. 200A	29	200	—	—	—	—	—	—	16.9	117	<1.0
No. 275A	40	275	—	—	—	—	—	—	19.4	134	<1.0
No. 350A	51	350	—	—	—	—	—	—	20.8	143	<1.0
No. 400A	58	400	—	—	—	—	—	—	21.5	148	<1.0
Ductile iron—ASTM A536 *Note: No metric grades included in the standard.*											
Austempered ductile iron (ADI)—ASTM A897M											
750/500/11	109	750	—	—	—	—	73	500	22	152	11
900/650/09	131	900	—	—	—	—	94	650	22	152	9
1050/750/07	152	1050	—	—	—	—	109	750	22	152	7
1200/850/04	174	1200	—	—	—	—	123	850	22	152	4
1400/1100/02	203	1400	—	—	—	—	160	1100	22	152	2
1600/1300/01	232	1600	—	—	—	—	189	1300	22	152	1
Ferritic malleable iron—ASTM A47M											
22010	49	340	—	—	—	—	32	220	25	172	10
Pearlitic malleable iron—ASTM A220M											
280M10	58	400	—	—	—	—	41	280	26	179	10
310M8	65	450	—	—	—	—	45	310	26	179	8
310M6	65	450	—	—	—	—	45	310	26	179	6
340M5	70	480	—	—	—	—	49	340	26	179	5
410M4	80	550	—	—	—	—	59	410	26	179	4
480M3	86	590	—	—	—	—	70	480	26	179	3
550M2	94	650	—	—	—	—	80	550	26	179	2
620M1	104	720	—	—	—	—	90	620	26	179	1

Notes: Five additional intermediate grades are included in standard A48M for gray iron.

[1]Approximate values; not part of the standards; If critical, negotiate with supplier.

[2]U.S. data computed using: (SI data/6.895); Not part of the standards.

[3]The density of cast irons ranges from 6920 kg/m^3 to 7480 kg/m^3 (0.25 lb$_m$/in^3 to 0.27 lb$_m$/in^3).

APPENDIX 9 Typical Properties of Aluminum

Alloy and temper	Tensile strength		Yield strength		Ductility (percent elongation in 2 inches)	Shearing strength		Endurance strength	
	(ksi)	(MPa)	(ksi)	(MPa)		(ksi)	(MPa)	(ksi)	(MPa)
1060-O	10	69	4	28	43	7	48	3	21
1060-H14	14	97	13	90	12	9	62	5	34
1060-H18	19	131	18	124	6	11	121	6.5	41
1350-O	12	83	4	28	28	8	55		
1350-H14	16	110	14	97		10	69		
1350-H19	27	186	24	165		15	103	7	48
2014-O	27	186	14	97	18	18	124	13	90
2014-T4	62	427	42	290	20	38	262	20	138
2014-T6	70	483	60	414	13	42	290	18	124
2024-O	27	186	11	76	22	18	124	13	90
2024-T4	68	469	47	324	19	41	283	20	138
2024-T361	72	496	57	393	13	42	290	18	124
2219-O	25	172	11	76	18				
2219-T62	60	414	42	290	10			15	103
2219-T87	69	476	57	393	10			15	103
3003-O	16	110	6	41	40	11	121	7	48
3003-H14	22	152	21	145	16	14	97	9	62
3003-H18	29	200	27	186	10	16	110	10	69
5052-O	28	193	13	90	30	18	124	16	110
5052-H34	38	262	31	214	14	21	145	18	124
5052-H38	42	290	37	255	8	24	165	20	138
6061-O	18	124	8	55	30	12	83	9	62
6061-T4	35	241	21	145	25	24	165	14	97
6061-T6	45	310	40	276	17	30	207	14	97
6063-O	13	90	7	48		10	69	8	55
6063-T4	25	172	13	90	22				
6063-T6	35	241	31	214	12	22	152	10	69
7178-O	33	228	15	103	16				
7178-T6	88	607	78	538	11				
7075-O	33	228	15	103	16	22	152		
7075-T6	83	572	73	503	11	48	331	23	159
Casting alloys (permanent mold casting)									
204.0-14	48	331	29	200	8	—	—		
206.0-T6	65	445	59	405	6	—	—		
356.0-T6	41	283	30	207	10	—	—		

Note: Common properties:

Density: 0.095 to 0.102 lb/in^3 (2635 to 2829 kg/m^3).

Endurance strength at 5×10^8 cycles.

Modulus of elasticity	Alloys
10.0×10^6 psi (69.0 GPa)	1100, 3003, 6061, 6063
10.2×10^6 psi (70.3 GPa)	5154
10.4×10^6 psi (71.7 GPa)	7075
10.6×10^6 psi (73.1 GPa)	2014

APPENDIX 10–1 Properties of Die-Cast Zinc Alloys

Material designation	Tensile strength		Yieid strength		Compression strength		Percent elongation	Density		Modulus of elasticity		Impact strength	
	(ksi)	(MPa)	(ksi)	(MPa)	(ksi)	(MPa)	(%)	(lb_m/in^3)	(kg/m^3)	(10^6 psi)	(GPa)	(ft · lb)	(J or N · m)
ZAMAK #3	41	283	32	221	60	414	10	0.240	6600	12.4	85.5	43.0	58
ZAMAK #5	48	331	33	228	87	600	7	0.240	6600	12.4	85.5	48.0	65
ZA-8	54	374	42	290	37	252	8	0.227	6300	12.4	85.5	31.0	42
ZA-12	59	404	46	320	39	269	5.5	0.216	6000	12.0	82.7	21.0	29
ZA-27	61	421	55	379	52	385	2	0.181	5000	11.3	77.9	9.0	5

Note: Strength values are typical average values; may range higher or lower.

APPENDIX 10–2 Properties of Die-Cast Magnesium Alloys

Material designation	Tensile strength		[1]Yieid strength		Compression strength		Percent elongation	Density		Modulus of elasticity		Impact strength	
	(ksi)	(MPa)	(ksi)	(MPa)	(ksi)	(MPa)	(%)	(lb_m/in^3)	(kg/m^3)	(10^6 psi)	(GPa)	(ft · lb)	(J or N · m)
AZ-91 MgA19Zn1(A)	33.4	230	22.5	155	21.5	148	3	0.0654	1810	6.5	45	4.4	6
AM-60 MgA16Mn	31.9	220	19.6	135	—	—	9	0.0650	1800	6.5	45	12.5	17
AM-50 MgA15Mn	29.7	205	17.4	120	16.4	113	10	0.0639	1770	6.5	45	13.3	18
AM-20 MgA12Mn	26.8	185	13.1	90	10.7	74	13	0.0632	1750	6.5	45	13.3	18

Notes: Strength values are typical average values; may range higher or lower.

[1] Also call 0.2% proof stress.

APPENDIX 11–1 Properties of Nickel-Based Alloys

Material designation	Tensile strength (ksi)	Tensile strength (Mpa)	Yield strength (ksi)	Yield strength (MPa)	Percent elongation (%)	Density (lb_m/in^3)	Density (kg/m^3)	Modulus of elasticity (10^6 psi)	Modulus of elasticity GPa
N06600 Annealed	93	640	37	255	45	0.304	8420	30	207
N06110 40% cold worked	175	1205	150	1034	18	0.302	8330	30	207
N04400 Annealed	80	550	30	207	50	0.318	8800	26	181
N04400 cold drawn	100	690	75	517	30	0.318	8800	26	181

APPENDIX 11–2 Properties of Titanium Alloys

Material designation	Tensile strength (ksi)	Tensile strength (Mpa)	Yield strength (ksi)	Yield strength (MPa)	Percent elongation (%)	Density (lb_m/in^3)	Density (kg/m^3)	Modulus of elasticity (10^6 psi)	Modulus of elasticity GPa
Commercially Pure alpha titanium									
Ti-35A Wrought	35	241	25	172	24	0.163	4515	15.0	103
Ti-50A Wrought	50	345	40	276	20	0.163	4515	15.0	103
Ti-65 Wrought	65	448	55	379	18	0.163	4515	15.0	103
Alpha alloy									
Ti-0.2Pd Wrought	50	345	40	276	20	0.163	4515	14.9	103
Beta alloy									
Ti-3Al-13V-11Cr Air cooled from 1400°F	135	931	130	896	16	0.176	4875	14.7	101
Ti-3Al-13V-11Cr Air cooled from 1400°F and aged	185	1280	175	1210	6	0.176	4875	16.0	110
Alpha–beta alloy									
Ti-6Al-4V Annealed	130	896	120	827	10	0.160	4432	16.5	114
Ti-6Al-4V Quenched and aged at 1000°F	160	1100	150	1030	7	0.160	4432	16.5	114

APPENDIX 12 Properties of Bronzes, Brasses, and Other Copper Alloys

Material	UNS number designation	Tensile strength		Yield strength		Ductility	Modulus of elasticity		Density	
		(ksi)	(MPa)	(ksi)	(MPa)	(% elongation)	(10^6 psi)	(GPa)	lb_m/in^3	kg/m^3
Bronzes—Wrought bars, rods										
Leaded phosphor bronze	C54400-H04	68	469	57	393	20	15.0	103	0.320	8890
Silicon bronze	C65500-H06	108	745	60	414	13	15.0	103	0.308	8530
Manganese bronze	C67500-H02	84	579	60	414	19	15.0	103	0.302	8360
Bronzes—Sand cast										
Manganese bronze	C86200-M01	95	655	48	331	20	15.0	103	0.288	7970
Bearing bronze	C93200-M01	35	241	18	124	20	14.5	100	0.322	8910
Aluminum bronze	C95400-M01	85	586	35	241	18	15.5	107	0.269	7450
Copper-nickel-iron alloy	C96200-M01	45	310	25	172	20	18.0	124	0.323	8940
Copper-nickel-zinc alloy (also called nickel silver)	C97300-M01	35	241	17	117	20	16.0	110	0.321	8890
Brasses—Wrought bars, rods										
Extra high-leaded brass	C35600-H02	55	379	25	172	10	97	14.0	0.307	8500
Free-cutting brass	C36000-H04	65	448	30	207	6	103	15.0	0.307	8500
Free-cutting Muntz metal	C37000-H04	80	552	60	414	6	103	15.0	0.304	8410
Naval brass	C46400-H02	75	517	53	365	20	103	15.0	0.304	8410

Notes:

1. Strength and ductility values listed are typical at the approximate middle of available ranges and not guaranteed.
2. Properties vary widely for different section sizes, strain hardening, and thermal treatment conditions.
3. Strain hardening conditions: 1/8 hard (H00), 1/4 hard (H01), 1/2 hard (H02), 3/4 hard (H03), full hard (H04), extra hard (H06), spring (H08), and higher.
4. More details and data for numerous additional alloys can be found at www.copper.org/resources/properties.

APPENDIX 13 Typical Properties of Selected Plastics

Material	Type	Tensile strength		Tensile modulus		Flexural strength		Flexural modulus		Impact strength IZOD (ft·lb/in of notch)
		(ksi)	(MPa)	(ksi)	(MPa)	(ksi)	(MPa)	(ksi)	(MPa)	
[1]Nylon 66	Dry	21.0	146	1200	8700	32.0	221	1100	7900	
30 % Glass	50% R.H.	15.0	102	800	5500					
[2]ABS	Medium-impact	6.0	41	360	2480	11.5	79	310	2140	4.0
	High-impact	5.0	34	250	1720	8.0	55	260	1790	7.0
Polycarbonate	General-purpose	9.0	62	340	2340	11.0	76	300	2070	12.0
Acrylic	Standard	10.5	72	430	2960	16.0	110	460	3170	0.4
	High-impact	5.4	37	220	1520	7.0	48	230	1590	1.2
[3]PVC	Rigid	6.0	41	350	2410			300	2070	0.4–20.0 (varies widely)
Polyimide	25% graphite powder filler	5.7	39			12.8	88	900	6210	0.25
	Glass-fiber filler	27.0	186			50.0	345	3250	22 400	17.0
	Laminate	50.0	345			70.0	483	4000	27 580	13.0
Acetal	Copolymer	8.0	55	410	2830	13.0	90	375	2590	1.3
Polyurethane	Elastomer	5.0	34	100	690	0.6	4			No break
Phenolic	General	6.5	45	1100	7580	9.0	62	1100	7580	0.3
[4]Polyester with glass-fiber mat reinforcement (approx. 30% glass by weight)										
	Lay-up, contact mold	9.0	62			16.0	110	800	5520	
	Cold-press molded	12.0	83			22.0	152	1300	8960	
	Compression molded	25.0	172			10.0	69	1300	8960	

Notes: [1]Also known as Polyamide 66 or PA 66.

[2]Acrylonitrile-butadiene-styrene.

[3]Polyvinyl chloride.

[4]Polyethylene terephthalate (PET) thermoplastic polyester resin.

APPENDIX 14 Beam-Deflection Formulas

TABLE A14–1 Beam-Deflection Formulas for Simply Supported Beams

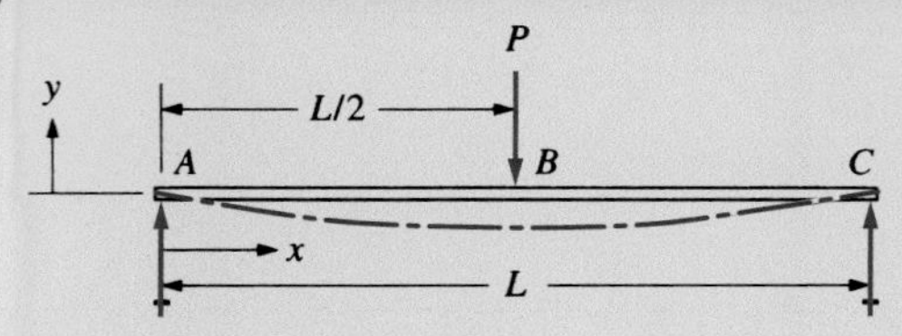

(a)

$$y_B = y_{max} = \frac{-PL^3}{48EI} \text{ at center}$$

Between A and B:

$$y = \frac{-Px}{48EI}(3L^2 - 4x^2)$$

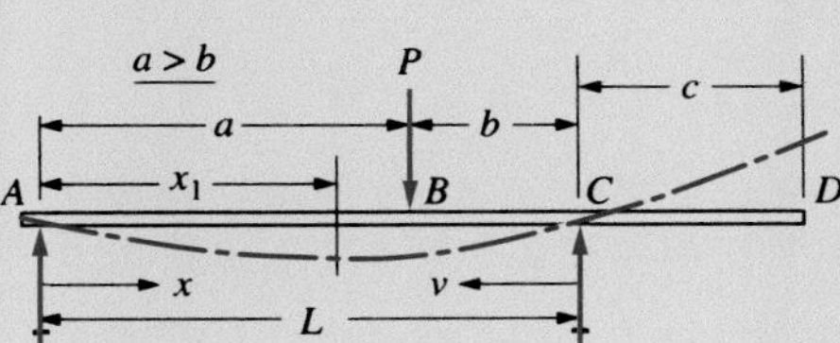

(b)

$$y_{max} = \frac{-Pab(L+b)\sqrt{3a(L+b)}}{27EIL}$$

$$\text{at } x_1 = \sqrt{a(L+b)/3}$$

$$y_B = \frac{-Pa^2b^2}{3EIL} \text{ at load}$$

Between A and B (the longer segment):

$$y = \frac{-Pbx}{6EIL}(L^2 - b^2 - x^2)$$

Between B and C (the shorter segment):

$$y = \frac{-Pav}{6EIL}(L^2 - v^2 - a^2)$$

At end of overhang at D:

$$y_D = \frac{Pabc}{6EIL}(L + a)$$

(c)

$$y_E = y_{max} = \frac{-Pa}{24EI}(3L^2 - 4a^2) \text{ at center}$$

$$y_B = y_C = \frac{-Pa^2}{6EI}(3L - 4a) \text{ at loads}$$

Between A and B:

$$y = \frac{-Px}{6EI}(3aL - 3a^2 - x^2)$$

Between B and C:

$$y = \frac{-Pa}{6EI}(3Lx - 3x^2 - a^2)$$

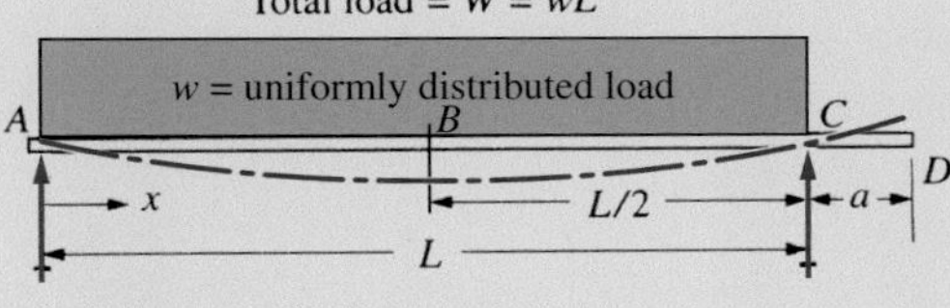

(d)

$$y_B = y_{max} = \frac{-5wL^4}{384EI} = \frac{-5WL^3}{384EI} \text{ at center}$$

Between A and B:

$$y = \frac{-wx}{24EI}(L^3 - 2Lx^2 + x^3)$$

At D at end:

$$y_D = \frac{wL^3a}{24EI}$$

(*Continued*)

TABLE A14–1 (*Continued*)

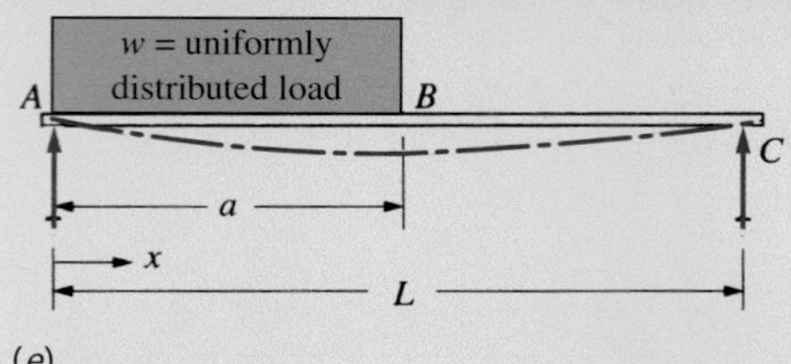

(*e*)

Between *A* and *B*:

$$y = \frac{-wx}{24EIL}\,[a^2(2L - a)^2 - 2ax^2(2L - a) + Lx^3]$$

Between *B* and *C*:

$$y = \frac{-wa^2(L - x)}{24EIL}\,(4Lx - 2x^2 - a^2)$$

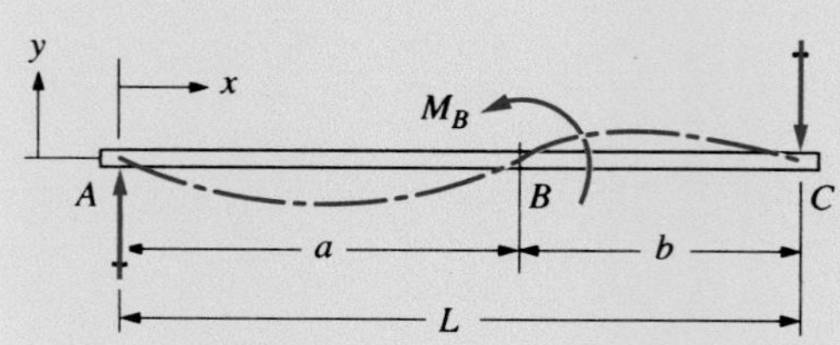

(*f*)

M_B = concentrated moment at *B*

Between *A* and *B*:

$$y = \frac{-M_B}{6EI}\left[\left(6a - \frac{3a^2}{L} - 2L\right)x - \frac{x^3}{L}\right]$$

Between *B* and *C*:

$$y = \frac{M_B}{6EI}\left[3a^2 + 3x^2 - \frac{x^3}{L} - \left(2L + \frac{3a^2}{L}\right)x\right]$$

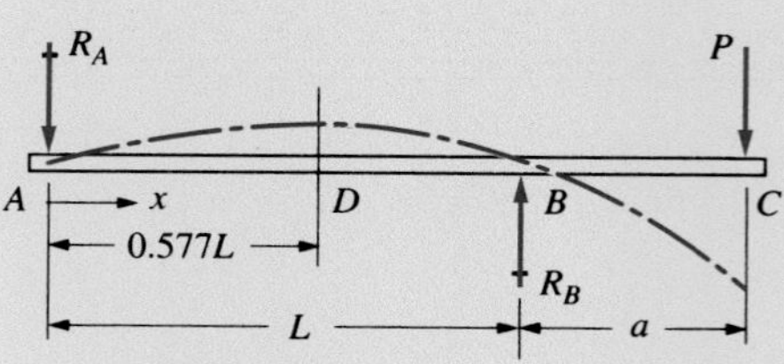

(*g*)

At *C* at end of overhang:

$$y_C = \frac{-Pa^2}{3EI}\,(L + a)$$

At *D*, maximum upward deflection:

$$y_D = 0.06415\,\frac{PaL^2}{EI}$$

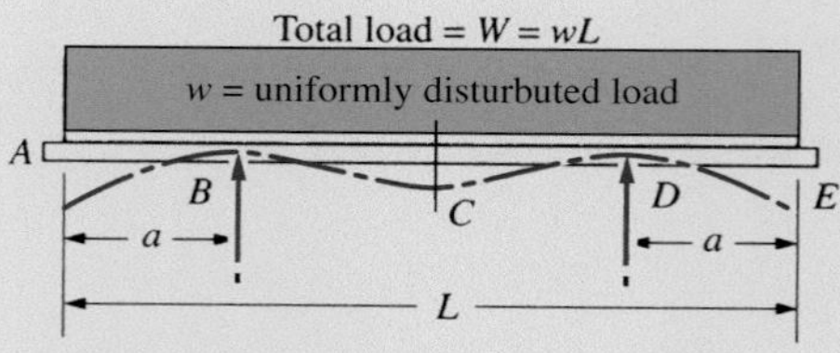

(*h*)

At *C* at center:

$$y = \frac{-W(L - 2a)^3}{384EI}\left[\frac{5}{L}(L - 2a) - \frac{24}{L}\left(\frac{a^2}{L - 2a}\right)\right]$$

At *A* and *E* at ends:

$$y = \frac{-W(L - 2a)^3 a}{24EIL}\left[-1 + 6\left(\frac{a}{L - 2a}\right)^2 + 3\left(\frac{a}{L - 2a}\right)^3\right]$$

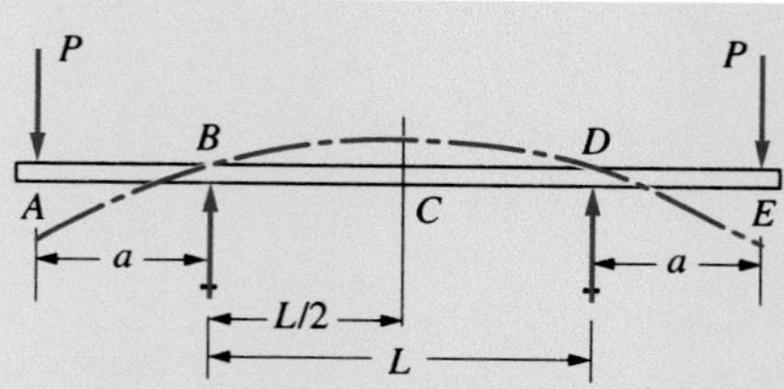

(*i*)

At *C* at center:

$$y = \frac{PL^2 a}{8EI}$$

At *A* and *E* at ends at loads:

$$y = \frac{-Pa^2}{3EI}\left(a + \frac{3}{2}L\right)$$

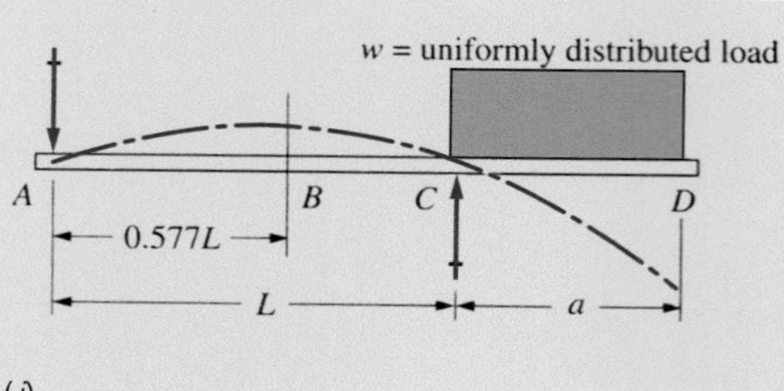

(*j*)

At *B*:

$$y = 0.03208\,\frac{wa^2L^2}{EI}$$

At *D* at end:

$$y = \frac{-wa^3}{24EI}\,(4L + 3a)$$

Source: Engineering Data for Aluminum Structures (Washington, DC: The Aluminum Association, 1986), pp. 63–77.

TABLE A14–2 Beam-Deflection Formulas for Cantilevers

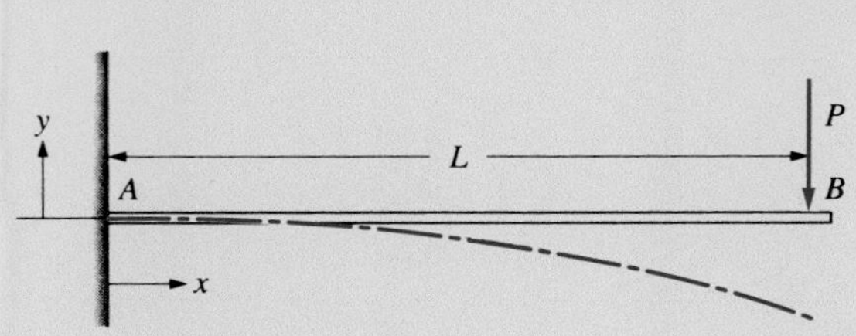

(a)

At B at end:

$$y_B = y_{max} = \frac{-PL^3}{3EI}$$

Between A and B:

$$y = \frac{-Px^2}{6EI}(3L - x)$$

(b)

At B at load:

$$y_B = \frac{-Pa^3}{3EI}$$

At C at end:

$$y_C = y_{max} = \frac{-Pa^2}{6EI}(3L - a)$$

Between A and B:

$$y = \frac{-Px^2}{6EI}(3a - x)$$

Between B and C:

$$y = \frac{-Pa^2}{6EI}(3x - a)$$

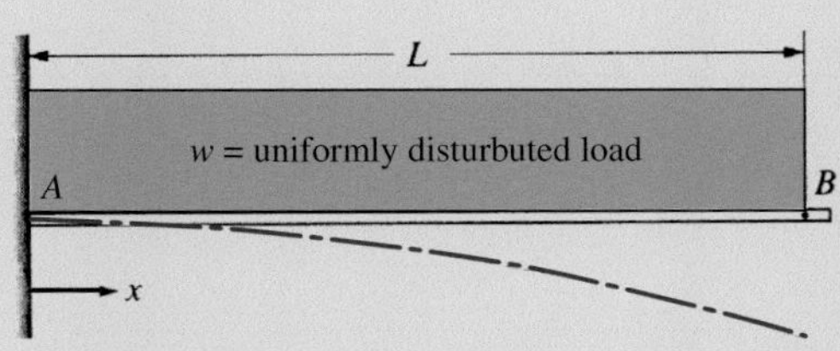

(c)

$$W = \text{total load} = wL$$

At B at end:

$$y_B = y_{max} = \frac{-WL^3}{8EI}$$

Between A and B:

$$y = \frac{-Wx^2}{24EIL}[2L^2 + (2L - x)^2]$$

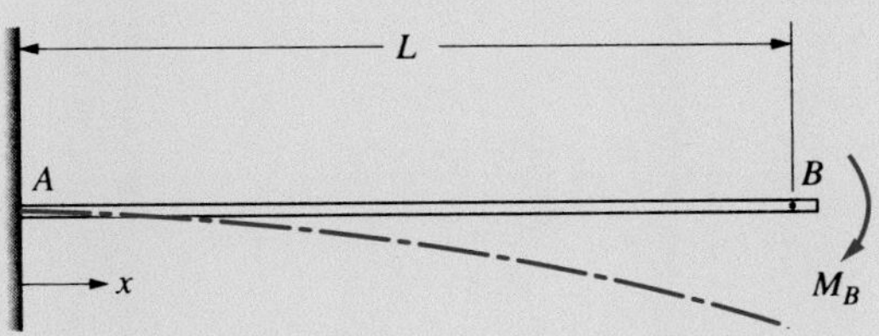

(d)

$$M_B = \text{concentrated moment at end}$$

At B at end:

$$y_B = y_{max} = \frac{-M_B L^2}{2EI}$$

Between A and B:

$$y = \frac{-M_B x^2}{2EI}$$

Source: Engineering Data for Aluminum Structures (Washington, DC: The Aluminum Association, 1986), pp. 63–77.

TABLE A14–3 Beam Diagrams and Beam-Deflection Formulas for Statically Indeterminate Beams

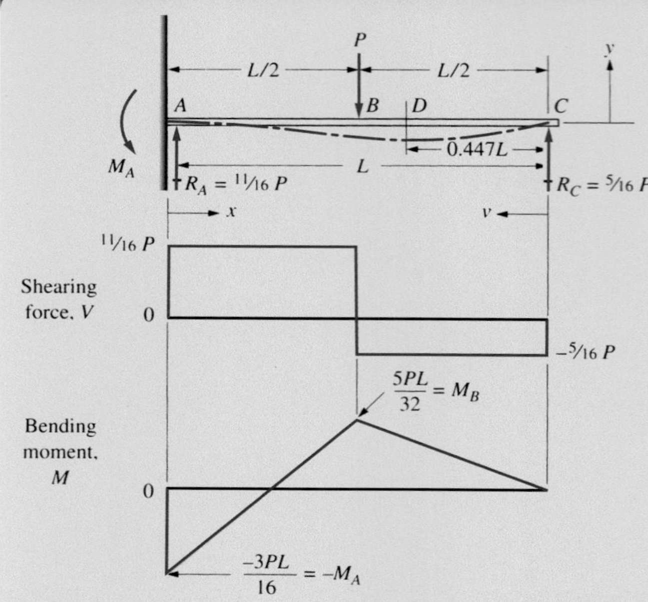

(a)

Deflections

At B at load:

$$y_B = \frac{-7}{768}\frac{PL^3}{EI}$$

y_{max} is at $v = 0.447L$ *at D*:

$$y_D = y_{max} = \frac{-PL^3}{107EI}$$

Between A and B:

$$y = \frac{-Px^2}{96EI}(9L - 11x)$$

Between B and C:

$$y = \frac{-Pv}{96EI}(3L^2 - 5v^2)$$

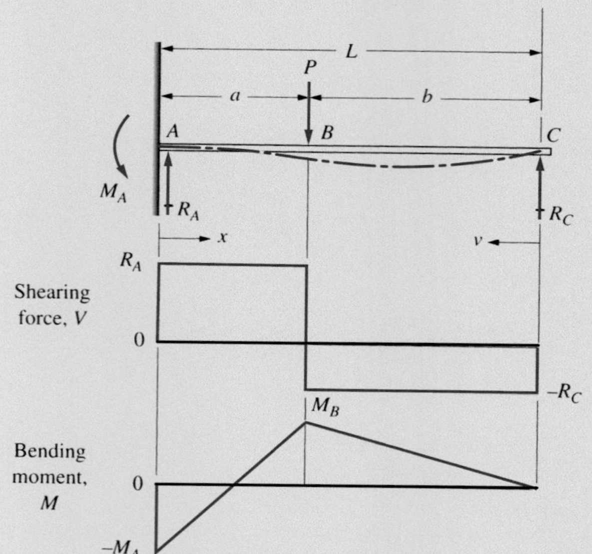

(b)

Reactions

$$R_A = \frac{Pb}{2L^3}(3L^2 - b^2)$$

$$R_C = \frac{Pa^2}{2L^3}(b + 2L)$$

Moments

$$M_A = \frac{-Pab}{2L^2}(b + L)$$

$$M_B = \frac{Pa^2b}{2L^3}(b + 2L)$$

Deflections

At B at load:

$$y_B = \frac{-Pa^3b^2}{12EIL^3}(3L + b)$$

Between A and B:

$$y = \frac{-Px^2b}{12EIL^3}(3C_1 - C_2x)$$

$$C_1 = aL(L + b);\ C_2 = (L + a)(L + b) + aL$$

Between B and C:

$$y = \frac{-Pa^2v}{12EIL^3}[3L^2b - v^2(3L - a)]$$

TABLE A14–3 (*Continued*)

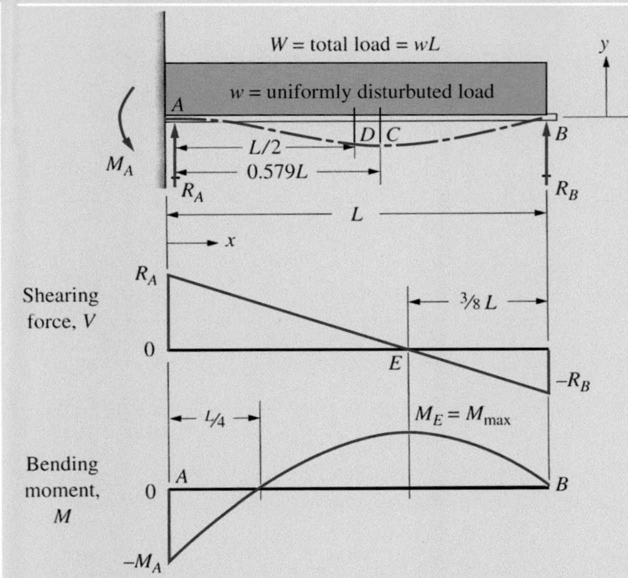

(*c*)

Reactions

$$R_A = \frac{5}{8}W$$

$$R_B = \frac{3}{8}W$$

Moments

$$M_A = -0.125WL$$

$$M_E = 0.0703WL$$

Deflections

At C at $x = 0.579L$:

$$y_C = y_{max} = \frac{-WL^3}{185EI}$$

At D at center:

$$y_D = \frac{-WL^3}{192EI}$$

Between A and B:

$$y = \frac{-Wx^2(L - x)}{48EIL}(3L - 2x)$$

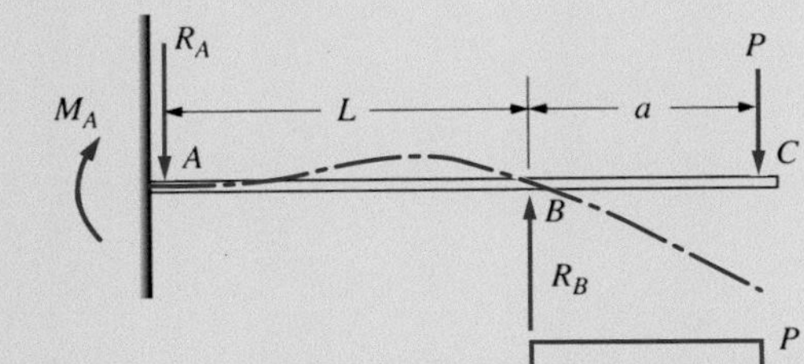

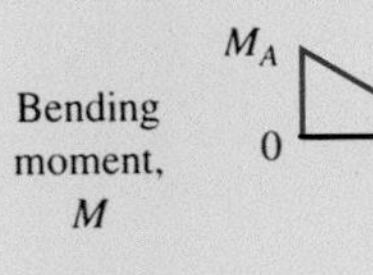

(*d*)

Reactions

$$R_A = \frac{-3Pa}{2L}$$

$$R_B = P\left(1 + \frac{3a}{2L}\right)$$

Moments

$$M_A = \frac{Pa}{2}$$

$$M_B = -Pa$$

Deflection

At C at end:

$$y_C = \frac{-PL^3}{EI}\left(\frac{a^2}{4L^2} + \frac{a^3}{3L^3}\right)$$

(*Continued*)

TABLE A14–3 (*Continued*)

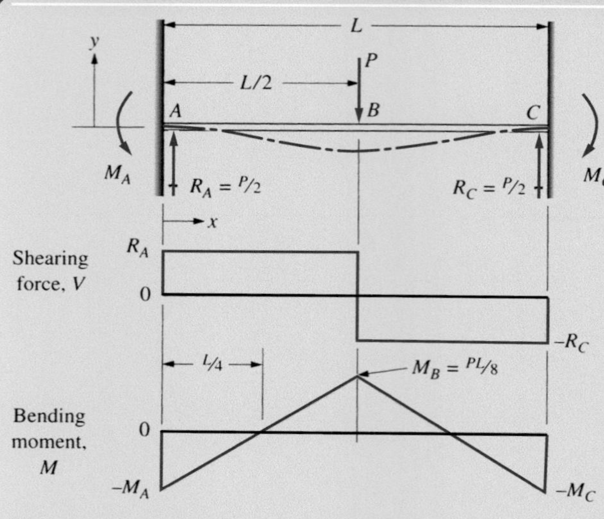

(*e*)

Moments

$$M_A = M_B = M_C = \frac{PL}{8}$$

Deflections

At *B* at center:

$$y_B = y_{max} = \frac{-PL^3}{192EI}$$

Between *A* and *B*:

$$y = \frac{-Px^2}{48EI}(3L - 4x)$$

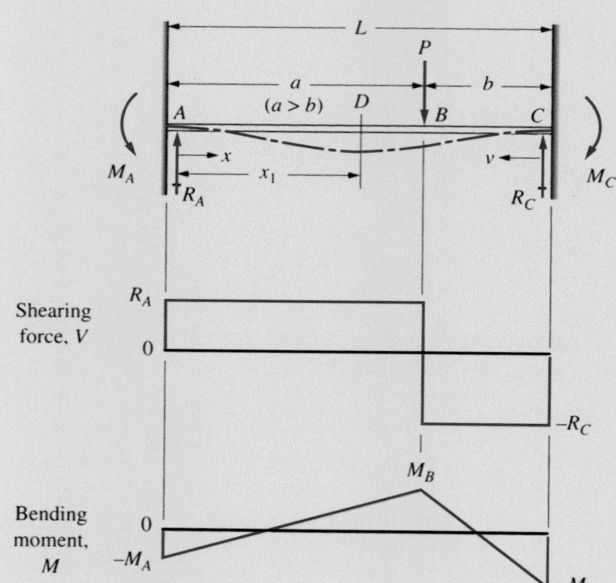

(*f*)

Reactions

$$R_A = \frac{Pb^2}{L^3}(3a + b)$$

$$R_C = \frac{Pa^2}{L^3}(3b + a)$$

Moments

$$M_A = \frac{-Pab^2}{L^2}$$

$$M_B = \frac{2Pa^2b^2}{L^3}$$

$$M_C = \frac{-Pa^2b}{L^2}$$

Deflections

At *B* at load:

$$y_B = \frac{-Pa^3b^3}{3EIL^3}$$

At *D* at $x_1 = \dfrac{2aL}{3a + b}$

$$y_D = y_{max} = \frac{-2Pa^3b^2}{3EI(3a + b)^2}$$

Between *A* and *B* (longer segment):

$$y = \frac{-Px^2b^2}{6EIL^3}[2a(L - x) + L(a - x)]$$

Between *B* and *C* (shorter segment):

$$y = \frac{-Pv^2a^2}{6EIL^3}[2b(L - v) + L(b - v)]$$

TABLE A14–3 (*Continued*)

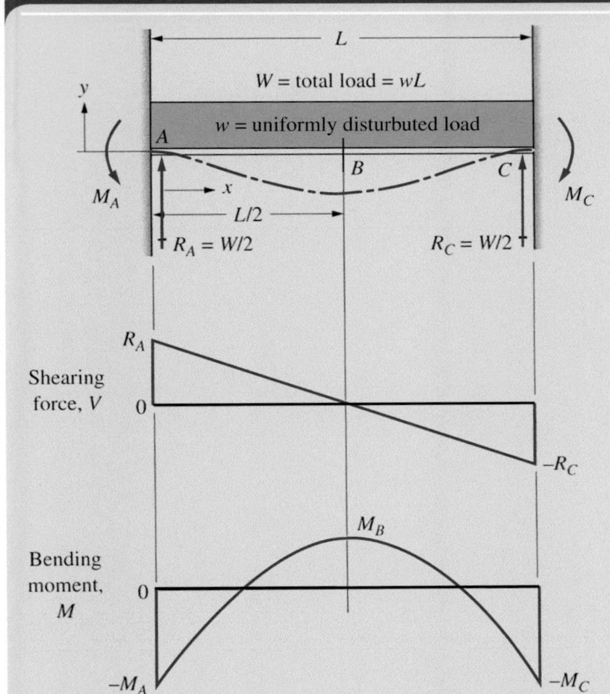

Moments

$$M_A = M_C = \frac{-WL}{12}$$

$$M_B = \frac{WL}{24}$$

Deflections

At B at center:

$$y_B = y_{\max} = \frac{-WL^3}{384EI}$$

Between A and C:

$$y = \frac{-wx^2}{24EI}(L - x)^2$$

(g)

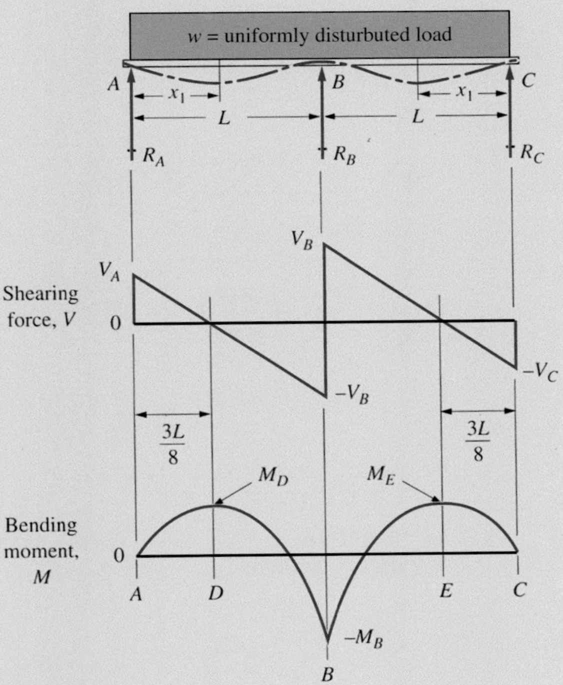

Reactions

$$R_A = R_C = \frac{3wL}{8}$$

$$R_B = 1.25wL$$

Shearing forces

$$V_A = V_C = R_A = R_C = \frac{3wL}{8}$$

$$V_B = \frac{5wL}{8}$$

Moments

$$M_D = M_E = 0.0703wL^2$$

$$M_B = -0.125wL^2$$

Deflections

At $x_1 = 0.4215L$ from A or C:

$$y_{\max} = \frac{-wL^4}{185EI}$$

Between A and B:

$$y = \frac{-w}{48EI}(L^3x - 3Lx^3 + 2x^4)$$

(h)

(*Continued*)

TABLE A14–3 (*Continued*)

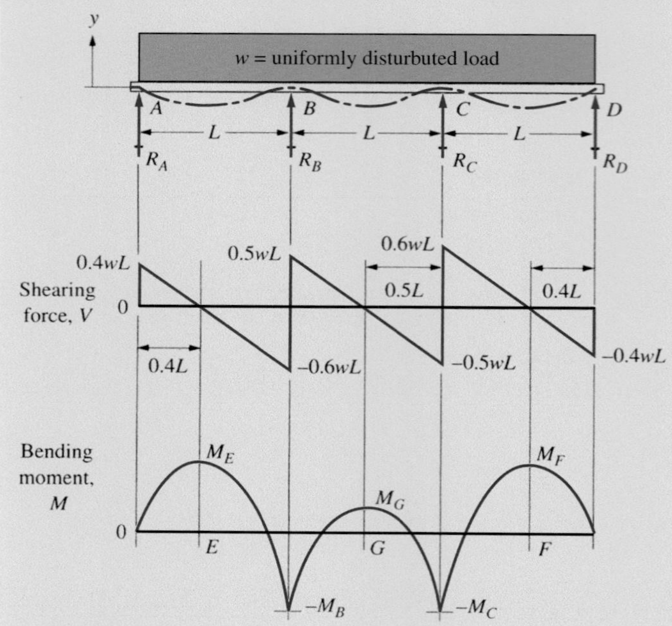

(i)

Reactions

$$R_A = R_D = 0.4wL$$
$$R_B = R_C = 1.10wL$$

Moments

$$M_E = M_F = 0.08wL^2$$
$$M_B = M_C = -0.10wL^2 = M_{max}$$
$$M_G = 0.025wL^2$$

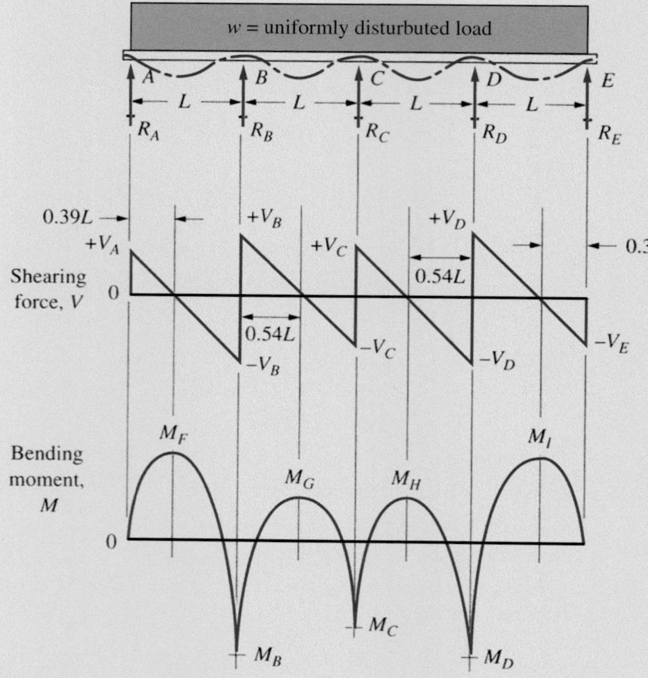

(j)

Reactions

$$R_A = R_E = 0.393wL$$
$$R_B = R_D = 1.143wL$$
$$R_C = 0.928wL$$

Shearing forces

$$V_A = +0.393wL$$
$$-V_B = -0.607wL$$
$$+V_B = +0.536wL$$
$$-V_C = -0.464wL$$
$$+V_C = +0.464wL$$
$$-V_D = -0.536wL$$
$$+V_D = +0.607wL$$
$$-V_E = -0.393wL$$

Moments

$$M_B = M_D = -0.1071wL^2 = M_{max}$$
$$M_F = M_I = 0.0772wL^2$$
$$M_C = -0.0714wL^2$$
$$M_G = M_H = 0.0364wL^2$$

Source: Engineering Data for Aluminum Structures (Washington, DC: The Aluminum Association, 1986), pp. 63–77.

APPENDIX 15 Commercially Available Shapes Used for Load-Carrying Members

Table No.	Units	Description
15–1	U.S.	Angles: Steel and aluminum, equal and unequal legs, L-shapes, larger sizes: 2 in to 8 in
15–2	U.S.	Angles: Steel and aluminum, equal and unequal legs, smaller sizes: 0.50 in to 2.5 in
15–3	SI	Angles: Steel and aluminum, equal and unequal legs, 10 mm to 100 mm
15–4	U.S.	Channels: Steel and aluminum, American Standard, C-shapes, larger sizes: 3 in to 15 in depth
15–5	U.S.	Channels: Aluminum, smaller sizes: 0.5 in to 2.0 in depth
15–6	U.S.	Channels: Aluminum, Aluminum Association Standard shapes, larger sizes: 2 in to 12 in depth
15–7	SI	Channels: Aluminum, European standard shapes: 10 mm to 160 mm depth
15–8	SI	Channels: Steel, European standard shapes: 30 mm to 400 mm depth
15–9	U.S.	I-beam shapes: Steel wide-flange shapes, W-shapes: 4 in to 24 in depth
15–10	U.S.	I-beam shapes: Steel, American Standard, S-shapes: 3 in to 24 in depth
15–11	U.S.	I-beam shapes: Aluminum Association standard shapes: 3 in to 12 in depth
15–12	U.S.	I-beam shapes: Aluminum, small extruded shapes: 0.70 in to 2.11 in depth
15–13	SI	I-beam shapes: Steel, European standard shapes: 80 mm to 600 mm depth
15–14	U.S.	Hollow tubing: Steel, square and rectangular, standard structural HSS shapes: 2 in to 8 in depth
15–15	U.S.	Hollow tubing: Steel and aluminum, square and rectangular, smaller sizes: 0.375 in to 3.00 in depth
15–16	SI	Hollow tubing: Steel and aluminum, square and rectangular: 20 mm to 300 mm depth
15–17	U.S.	Pipe: Steel, American National Standard Schedule 40 and AISC standard: 1/8 in to 18 in sizes
15–18	U.S.	Mechanical tubing: Steel and aluminum: 0.50 in to 5.0 in outside diameters
15–19	SI	Mechanical tubing: Steel and aluminum: 10 mm to 150 mm outside diameters

Notes: Each table lists sample sizes of shapes commonly used for beam, column, or tension members.
Included are standard size designations, detailed dimensions, and section properties.
Section properties include: cross-sectional area, moment of inertia, and section modulus.
For some sections, torsional moment of inertia and section modulus, and radius of gyration are also listed.
Numerous additional shapes are typically commercially available.
Internet sites for companies supplying each kind of shape are listed to aid in finding other sizes.
However, not all sites incude the section properties data. See Appendix 1 for formulas for section properties.
Actual data listed obtained from a variety of sources.

Angles: Equal and unequal legs
Tables 15–1, 15–2, and 15–3

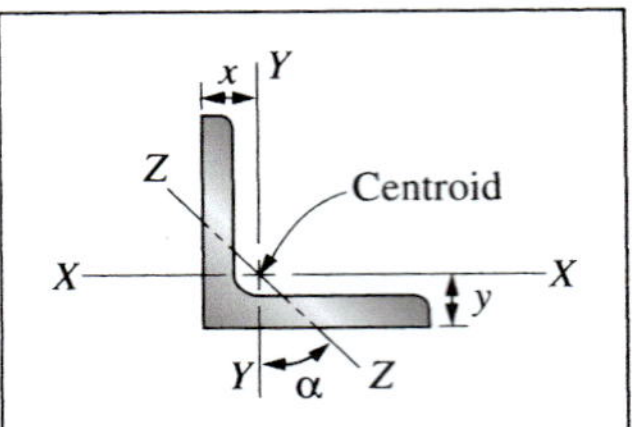

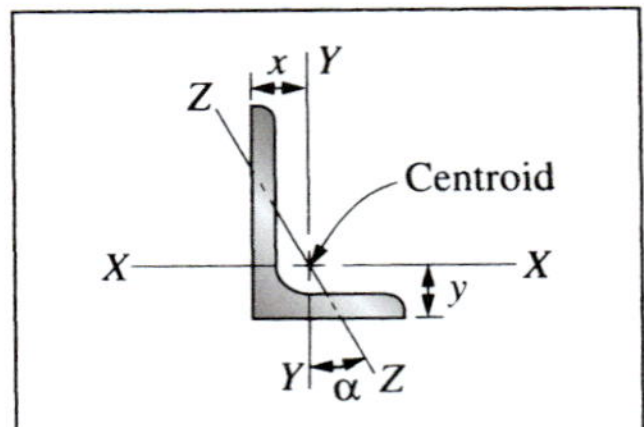

U.S. Units

TABLE 15–1 Angles: Steel and Aluminum, Equal and Unequal Legs, L-Shapes, Larger Sizes: 2 in to 8 in

			*Section properties							
			Axis X–X			Axis Y–Y			Axis Z–Z	
Designation	Area A (in^2)	Weight per foot (lb/ft)	I_x (in^4)	S_x (in^3)	y (in)	I_y (in^4)	S_y (in^3)	x (in)	r (in)	α (deg)
L8 × 8 × 1	15.0	51.0	89.0	15.8	2.37	89.0	15.8	2.37	1.56	45.0
L8 × 8 × 1/2	7.75	26.4	48.6	8.36	2.19	48.6	8.36	2.19	1.59	45.0
L8 × 4 × 1	11.0	37.4	69.6	14.1	3.05	11.6	3.94	1.05	0.846	13.9
L8 × 4 × 1/2	5.75	19.6	38.5	7.49	2.86	6.74	2.15	0.859	0.865	14.9
L6 × 6 × 3/4	8.44	28.7	28.2	6.66	1.78	28.2	6.66	1.78	1.17	45.0
L6 × 6 × 3/8	4.36	14.9	15.4	3.53	1.64	15.4	3.53	1.64	1.19	45.0
L6 × 4 × 3/4	6.94	23.6	24.5	6.25	2.08	8.68	2.97	1.08	0.860	23.2
L6 × 4 × 3/8	3.61	12.3	13.5	3.32	1.94	4.90	1.60	0.941	0.877	24.0
L4 × 4 × 1/2	3.75	12.8	5.56	1.97	1.18	5.56	1.97	1.18	0.782	45.0
L4 × 4 × 1/4	1.94	6.6	3.04	1.05	1.09	3.04	1.05	1.09	0.795	45.0
L4 × 3 × 1/2	3.25	11.1	5.05	1.89	1.33	2.42	1.12	0.827	0.639	28.5
L4 × 3 × 1/4	1.69	5.8	2.77	1.00	1.24	1.36	0.599	0.896	0.651	29.2
L3 × 3 × 1/2	2.75	9.4	2.22	1.07	0.932	2.22	1.07	0.932	0.584	45.0
L3 × 3 × 1/4	1.44	4.9	1.24	0.577	0.842	1.24	0.577	0.842	0.592	45.0
L2 × 2 × 3/8	1.36	4.7	0.479	0.351	0.636	0.479	0.351	0.636	0.389	45.0
L2 × 2 × 1/4	0.938	3.19	0.348	0.247	0.592	0.348	0.247	0.592	0.391	45.0
L2 × 2 × 1/8	0.484	1.65	0.190	0.131	0.546	0.190	0.131	0.546	0.398	45.0

Note: *I = moment of inertia; S = section modulus; r = radius of gyration.

Example designation: L4 × 3 × 1/2

4 = length of longer leg (in); 3 = length of shorter leg (in); 1/2 = thickness of legs (in)

Z–Z is axis of minimum moment of inertia (I) and radius of gyration (r).

Sources: Central Steel & Wire Co., multiple locations; www.centralsteel.com/products.htm

Earl M. Jorgensen Co., multiple locations; www.emjmetals.com

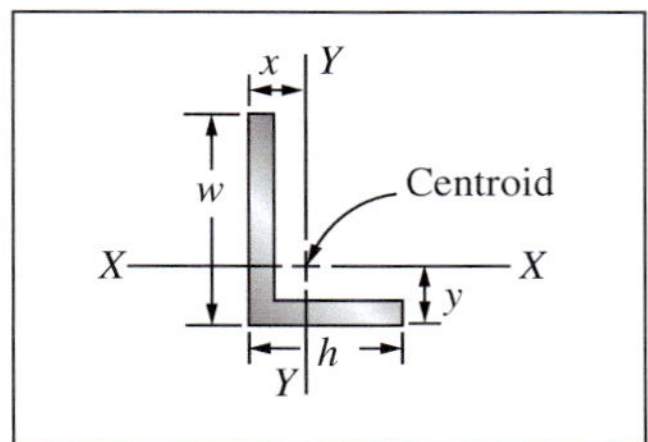

U.S. Units

TABLE 15–2 Angles: Steel and Aluminum, Equal and Unequal Legs, Smaller Sizes: 0.50 in to 2.5 in

							*Section properties					
					**Weight/ft		Axis X–X			Axis Y–Y		
Ref.	Short side h (in)	Long side w (in)	thickness t (in)	Area A (in^2)	Steel (lb/ft)	Aluminum (lb/ft)	I_x (in^4)	S_x (in^3)	y (in)	I_y (in^4)	S_y (in^3)	x (in)
a	1/2	1/2	1/8	0.109	0.372	0.131	0.00230	0.00698	0.170	0.00230	0.00698	0.170
b	5/8	5/8	1/8	0.141	0.478	0.169	0.00479	0.0113	0.201	0.00479	0.0113	0.201
c	7/8	7/8	1/8	0.203	0.691	0.244	0.01420	0.0233	0.264	0.0142	0.0233	0.264
d	5/8	1	1/8	0.188	0.637	0.225	0.00553	0.0121	0.167	0.0185	0.0286	0.354
e	3/4	1	1/8	0.203	0.691	0.244	0.00947	0.0174	0.207	0.0197	0.0295	0.332
f	1	1	1/8	0.234	0.797	0.281	0.0217	0.0309	0.296	0.0217	0.0309	0.296
g	1	1	1/4	0.438	1.487	0.525	0.0369	0.0558	0.339	0.0369	0.0558	0.339
h	1 1/4	1 1/4	1/8	0.297	1.009	0.356	0.0439	0.0493	0.359	0.0439	0.0493	0.359
i	1 1/4	1 1/4	1/4	0.563	1.912	C.675	0.0767	0.0905	0.403	0.0767	0.0905	0.403
j	7/8	1 3/8	1/8	0.266	0.903	0.319	0.0162	0.0247	0.217	0.0509	0.0560	0.467
k	1 1/4	1 1/2	3/16	0.480	1.633	0.577	0.0651	0.0726	0.353	0.1035	0.1013	0.478
l	1 1/2	1 1/2	1/8	0.359	1.222	0.431	0.0778	0.0721	0.421	0.0778	0.0721	0.421
m	1 1/2	1 1/2	1/4	0.688	2.337	0.825	0.1385	0.1340	0.466	0.139	0.1340	0.466
n	1 1/4	1 3/4	1/8	0.359	1.222	0.431	0.0486	0.0515	0.307	0.113	0.0943	0.557
o	1 3/4	1 3/4	1/8	0.422	1.434	0.506	0.1256	0.0992	0.484	0.126	0.0992	0.484
p	1 3/4	1 3/4	1/4	0.813	2.762	0.975	0.2272	0.1860	0.529	0.227	0.186	0.529
q	1 1/4	2	3/16	0.574	1.952	0.689	0.0707	0.0752	0.311	0.232	0.177	0.686
r	1 1/2	2	1/8	0.422	1.434	0.506	0.0847	0.0748	0.368	0.173	0.125	0.618
s	1 1/2	2	1/4	0.813	2.762	0.975	0.1515	0.139	0.413	0.316	0.236	0.663
t	1 1/2	2 1/2	3/16	0.715	2.430	0.858	0.1275	0.111	0.352	0.461	0.279	0.852

Note: *I = moment of inertia; S = section modulus; See sketch for X–X and Y–Y axes and their locations x and y.

Values are for perfectly square corners. Some vendors apply radii to Inside and/or outside corners.

Numerous additional sizes available. Consult vendors.

**Using density of steel = 0.283 lb_m/in^3; aluminum density = 0.100 lb_m/in^3. (May range from 0.095 lb_m/in^3 to 0.102 $lb_m.in^3$.)

Some sizes are also available in stainless steel. Check with vendor.

Sources: Reliance Steel & Aluminum Co./Earl M. Jorgensen Co., multiple locations; www.jorgensensteel.com/products/stocklist.asp

Central Steel & Wire Co., multiple locations; www.centralsteel.com/products.htm

OnlineMetals.com, Seattle, WA; www.ontinemetals.com

Paramount Extrusions, Co., Paramount, CA; www.paramountextrusions.com

Metals Depot, Winchester, KY; www.MetatsDepot.com

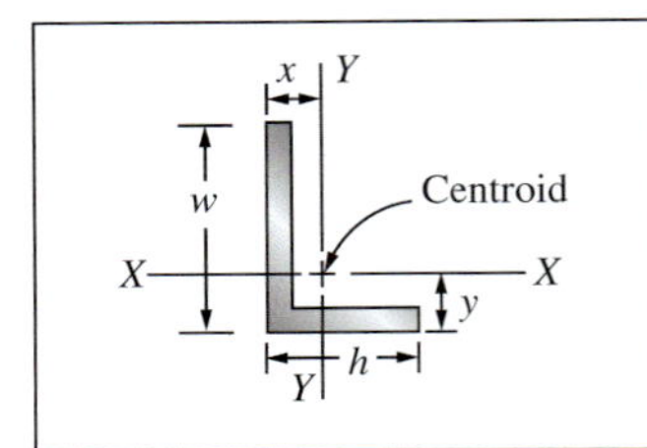

SI Units

TABLE 15–3 Angles: Steel and Aluminum, Equal and Unequal Legs, 10 mm to 100 mm

Ref.	Short side h (mm)	Long side w (mm)	thickness t (mm)	Area A (mm^2)	**Weight/m Steel (N/m)	**Weight/m Aluminum (N/m)	*Section properties Axis X–X I_x (mm^4)	Axis X–X S_x (mm^3)	Axis X–X y (mm)	Axis Y–Y I_y (mm^4)	Axis Y–Y S_y (mm^3)	Axis Y–Y x (mm)
a	10	15	3	66.0	4.973	1.79	477	69.1	3.09	1375	146	5.59
b	10	20	3	81.0	6.103	2.20	520	72.1	2.80	3140	257	7.80
c	15	20	3	96.0	7.233	2.61	1743	163	4.31	3653	277	6.81
d	15	25	3	111	8.363	3.01	1856	168	3.93	6876	428	8.93
e	15	30	3	126	9.493	3.42	1946	171	3.64	1.146E+04	608	11.1
f	20	20	3	111	8.363	3.01	4030	290	6.09	4030	290	6.09
g	20	30	3	141	10.623	3.83	4551	306	5.12	1.272E+04	640	10.1
h	20	40	3	171	12.883	4.64	4897	316	4.48	2.829E+04	1109	14.5
i	25	25	3	141	10.623	3.83	8204	465	7.35	8.204E+03	465	7.35
j	30	30	5	275	20.719	7.47	2.216E+04	1072	9.32	2.216E+04	1072	9.32
k	30	50	4	304	22.904	8.25	2.152E+04	941	7.13	7.800E+04	2373	17.1
l	40	40	5	375	28.253	10.18	5.561E+04	1974	11.8	5.561E+04	1974	11.8
m	40	60	5	475	35.787	12.90	6.270E+04	2081	9.87	1.740E+05	4334	19.9
n	40	80	6	684	51.533	18.57	7.836E+04	2525	8.96	4.526E+05	8868	29.0
o	50	50	5	475	35.787	12.90	1.125E+05	3155	14.3	1.125E+05	3155	14.3
p	50	75	5	600	45.205	16.29	1.266E+05	3322	11.9	3.485E+05	6884	24.4
q	50	100	6	864	65.095	23.46	1.590E+05	4039	10.6	9.058E+05	14073	35.6
r	60	60	6	684	51.533	18.57	2.333E+05	5452	17.2	2.333E+05	5452	17.2
s	75	75	6	864	65.095	23.46	4.688E+05	8677	21.0	4.688E+05	8677	21.0
t	100	100	10	1900	143.148	51.59	1.800E+06	25240	28.7	1.800E+06	25240	28.7

Notes: *I = moment of inertia; S = section modulus; See sketch for X–X and Y–Y axes and their locations x and y.
Values are for perfectly square corners. Some vendors apply radii to inside and/or outside corners.
**Using density of steel = 7680 kg/m^3; density of aluminum = 2768 kg/m^3. (May range from 2635 kg/m^3 to 2829 kg/m^3.)
Source: Parker Steel Company—Metric Sized Metals, Toledo, Ohio; http://metricmetal.com

Channels

Tables 15–4, 15–5, 15–6, 15–7, and 15–8

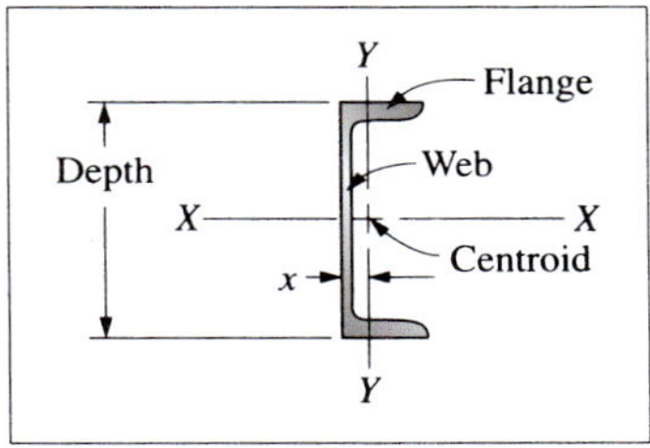

U.S. Units

TABLE 15–4 Channels: Steel and Aluminum, American Standard C-Shapes, Larger Sizes: 3 in to 15 in Depth

						*Section properties				
				Flange		Axis X–X		Axis Y–Y		
Designation	Area A (in^2)	Depth (in)	Web thickness (in)	Width (in)	Average thickness (in)	I_x (in^4)	S_x (in^3)	I_y (in^4)	S_y (in^3)	x (in)
C15 × 50	14.7	15.00	0.716	3.716	0.650	404	53.8	11.0	3.78	0.798
C15 × 40	11.8	15.00	0.520	3.520	0.650	349	46.5	9.23	3.37	0.777
C12 × 30	8.82	12.00	0.510	3.170	0.501	162	27.0	5.14	2.06	0.674
C12 × 25	7.35	12.00	0.387	3.047	0.501	144	24.1	4.47	1.88	0.674
C10 × 30	8.82	10.00	0.673	3.033	0.436	103	20.7	3.94	1.65	0.649
C10 × 20	5.88	10.00	0.379	2.739	0.436	78.9	15.8	2.81	1.32	0.606
C9 × 20	5.88	9.00	0.448	2.648	0.413	60.9	13.5	2.42	1.17	0.583
C9 × 15	4.41	9.00	0.285	2.485	0.413	51.0	11.3	1.93	1.01	0.586
C8 × 18.75	5.51	8.00	0.487	2.527	0.390	44.0	11.0	1.98	1.01	0.565
C8 × 11.5	3.38	8.00	0.220	2.260	0.390	32.6	8.14	1.32	0.781	0.571
C6 × 13	3.83	6.00	0.437	2.157	0.343	17.4	5.80	1.05	0.642	0.514
C6 × 8.2	2.40	6.00	0.200	1.920	0.343	13.1	4.38	0.693	0.492	0.511
C5 × 9	2.64	5.00	0.325	1.885	0.320	8.90	3.56	0.632	0.450	0.478
C5 × 6.7	1.97	5.00	0.190	1.750	0.320	7.49	3.00	0.479	0.378	0.484
C4 × 7.25	2.13	4.00	0.321	1.721	0.296	4.59	2.29	0.433	0.343	0.459
C4 × 5.4	1.59	4.00	0.184	1.584	0.296	3.85	1.93	0.319	0.283	0.457
C3 × 6	1.76	3.00	0.356	1.596	0.273	2.07	1.38	0.305	0.268	0.455
C3 × 4.1	1.21	3.00	0.170	1.410	0.273	1.66	1.10	0.197	0.202	0.436

Note: Example designation: C15 × 50

15 = depth (in); 50 = weight per unit length (lb/ft).

*I = moment of inertia; S = section modulus.

Sources: Central Steel & Wire Co., multiple locations; www.centralsteel.com/products.htm

Reliance Steel & Aluminum Co./Earl M. Jorgensen Co., multiple locations; www.jorgensensteel.com/products/stocklist.asp

Metals Depot, Winchester, KY; www.MetalsDepot.com

OnlineMetals.com, Seattle, WA, www.onlinemetals.com

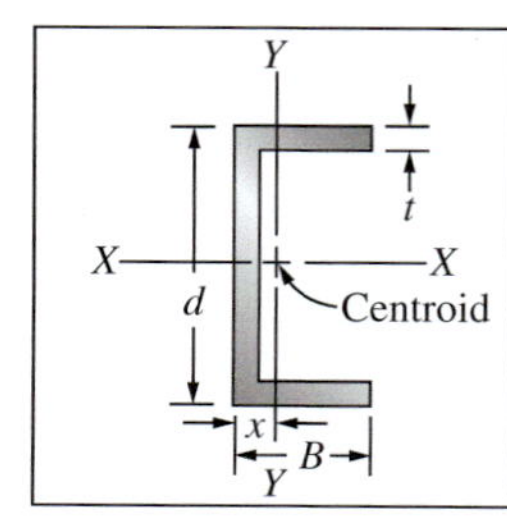

U.S. Units

TABLE 15–5 Channels: Aluminum, Smaller Sizes: 0.5 in to 2.0 in Depth

						*Section properties				
						Axis X–X		Axis Y–Y		
Ref.	Depth d (in)	Flange width, B (in)	Thickness t (in)	**Weight/ft (lb/ft)	Area, A (in^2)	I_x (in^4)	S_x (In^3)	I_y (in^4)	S_y (in^3)	x (in)
a	0.500	0.375	0.125	0.150	0.1250	0.00358	0.0143	0.001506	0.00688	0.156
b	0.500	0.500	0.094	0.148	0.123	0.00418	0.0167	0.00290	0.00970	0.202
c	0.500	0.750	0.125	0.263	0.219	0.0070	0.0280	0.01145	0.0273	0.330
d	0.625	0.625	0.125	0.244	0.203	0.0105	0.0337	0.00740	0.01999	0.255
e	0.750	0.375	0.125	0.188	0.156	0.0106	0.0282	0.001766	0.0074	0.138
f	0.750	0.750	0.125	0.300	0.250	0.0199	0.0530	0.01345	0.02968	0.297
g	1.000	0.500	0.125	0.263	0.219	0.0285	0.057	0.00461	0.0140	0.170
h	1.000	1.000	0.125	0.413	0.344	0.0526	0.105	0.03401	0.0549	0.381
i	1.250	0.500	0.125	0.300	0.250	0.0501	0.080	0.00496	0.0144	0.156
j	1.250	1.250	0.125	0.525	0.438	0.1097	0.176	0.06910	0.0879	0.464
k	1.500	0.500	0.125	0.338	0.281	0.080	0.106	0.00525	0.0148	0.146
l	1.750	0.500	0.125	0.375	0.313	0.118	0.135	0.00549	0.0151	0.138
m	1.750	0.750	0.125	0.450	0.375	0.159	0.182	0.01819	0.0342	0.219
n	1.750	1.000	0.125	0.525	0.438	0.201	0.229	0.04159	0.0605	0.313
o	2.000	0.500	0.125	0.413	0.344	0.166	0.166	0.00569	0.01539	0.131

Note: *I = moment of inertia; S = section modulus; See sketch for X–X and Y–Y axes and the locations x for the axis Y–Y.

Section properties computed assuming square corners and constant thickness of web and flanges.

**Using density for aluminum of 0.100 lb_m/in^3. May range from 0.095 lb_m/in^3 to 0.102 lb_m/in^3 for different alloys.

Sources: Reliance Steel & Aluminum Co./Earl M. Jorgensen Co., multiple locations; www.jorgensensteel.com/products/stocklist.asp

Central Steel & Wire Co., multiple locations; www.centralsteel.com/products.htm

OnlineMetals.com, Seattle, WA; www.onlinemetals.com

Paramount Extrusions, Co., Paramount, CA; www.paramountextrusions.com

Metals Depot, Winchester, KY; www.MetaisDepot.com

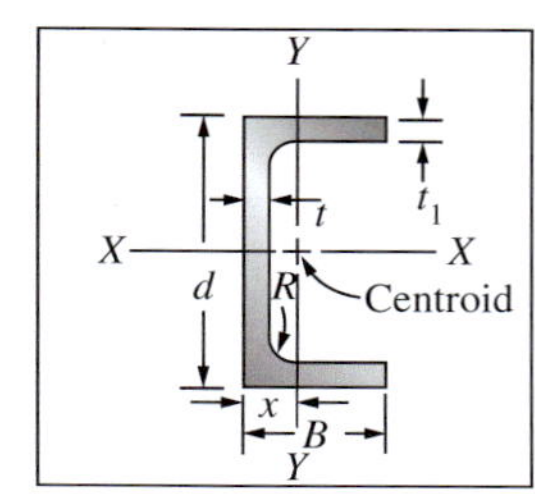

U.S. Units

TABLE 15–6 Channels: Aluminum, Aluminum Association Standard shapes, Larger Sizes: 2 in to 12 in Depth

Size							[1]Section properties						
							Axis X–X			Axis Y–Y			
Depth, d (in)	Width, B (in)	Area A (in^2)	Weight † (lb/ft)	Flange thickness, t_1 (in)	Web thickness, t (in)	Fillet radius, R (in)	I_x (in^4)	S_x (in^3)	r_x (in)	I_y (in^4)	S_y (in^3)	r_y (in)	x (in)
2.00	1.00	0.491	0.577	0.13	0.13	0.10	0.288	0.288	0.766	0.045	0.064	0.303	0.298
2.00	1.25	0.911	1.071	0.26	0.17	0.15	0.546	0.546	0.774	0.139	0.178	0.391	0.471
3.00	1.50	0.965	1.135	0.20	0.13	0.25	1.41	0.94	1.21	0.22	0.22	0.47	0.49
3.00	1.75	1.358	1.597	0.26	0.17	0.25	1.97	1.31	1.20	0.42	0.37	0.55	0.62
4.00	2.00	1.478	1.738	0.23	0.15	0.25	3.91	1.95	1.63	0.60	0.45	0.64	0.65
4.00	2.25	1.982	2.331	0.29	0.19	0.25	5.21	2.60	1.62	1.02	0.69	0.72	0.78
5.00	2.25	1.881	2.212	0.26	0.15	0.30	7.88	3.15	2.05	0.98	0.64	0.72	0.73
5.00	2.75	2.627	3.089	0.32	0.19	0.30	11.14	4.45	2.06	2.05	1.14	0.88	0.95
6.00	2.50	2.410	2.834	0.29	0.17	0.30	14.35	4.78	2.44	1.53	0.90	0.80	0.79
6.00	3.25	3.427	4.030	0.35	0.21	0.30	21.04	7.01	2.48	3.76	1.76	1.05	1.12
7.00	2.75	2.725	3.205	0.29	0.17	0.30	22.09	6.31	2.85	2.10	1.10	0.88	0.84
7.00	3.50	4.009	4.715	0.38	0.21	0.30	33.79	9.65	2.90	5.13	2.23	1.13	1.20
8.00	3.00	3.526	4.147	0.35	0.19	0.30	37.40	9.35	3.26	3.25	1.57	0.96	0.93
8.00	3.75	4.923	5.789	0.41	0.25	0.35	52.69	13.17	3.27	7.13	2.82	1.20	1.22
9.00	3.25	4.237	4.983	0.35	0.23	0.35	54.41	12.09	3.58	4.40	1.89	1.02	0.93
9.00	4.00	5.927	6.970	0.44	0.29	0.35	78.31	17.40	3.63	9.61	3.49	1.27	1.25
10.00	3.50	5.218	6.136	0.41	0.25	0.35	83.22	16.64	3.99	6.33	2.56	1.10	1.02
10.00	4.25	7.109	8.360	0.50	0.31	0.40	116.15	23.23	4.04	13.02	4.47	1.35	1.34
12.00	4.00	7.036	8.274	0.47	0.29	0.40	159.76	26.63	4.77	11.03	3.86	1.25	1.14
12.00	5.00	10.053	11.822	0.62	0.35	0.45	239.69	39.95	4.88	25.74	7.60	1.60	1.61

Notes: [1]I = moment of inertia; S = section modulus; r = radius of gyration.

[2]Weights per foot are based on nominal dimensions and a density of 0.098 lb/in^3, which is the density of alloy 6061. (May Range from 0.095 lb_m/in^3 to 0.102 lb_m/in^3.)

Sources: Aluminum Association, *Aluminum Standards and Data,* 11th ed., Washington, DC, © 1993, p. 187.

Metals Depot, Winchester, KY; www.MetalsDepot.com

OnlineMetals.com, Seattle, WA; www.onlinemetals.com

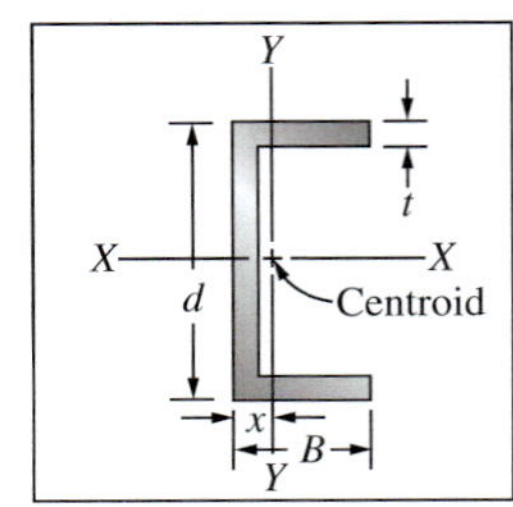

SI Units

TABLE 15–7 Channels: Aluminum, European Standard Shapes: 10 mm to 160 mm Depth

Ref.	Depth d (mm)	Flange width, B (mm)	Thickness t (mm)	[2]Weight/m (N/m)	Area A (mm^2)	[1]Section properties: Axis X–X: I_x (mm^4)	Axis X–X: S_x (mm^3)	Axis Y–Y: I_y (mm^4)	Axis Y–Y: S_y (mm^3)	Axis Y–Y: x (mm)
a	10	10	2	1.10	40.5	590.4	118.1	392	64.32	3.90
b	12	12	2	1.74	64.0	1301	216.9	881	121.6	4.75
c	15	15	2	2.23	82.0	2777	370.2	1812	196.1	5.76
d	20	20	3	4.40	162	9446	944.6	6279	514.5	7.80
e	25	20	2	3.31	122	1.215E+04	972.0	4911	375.0	6.90
f	25	25	3	5.62	207	1.998E+04	1598	1.285E+04	827.7	9.47
g	35	35	3	8.06	297	6.001E+04	3429	3.725E+04	1679	12.81
h	40	30	4	9.99	368	8.900E+04	4450	3.228E+04	1653	10.48
i	50	30	4	11.08	408	1.520E+05	6079	3.493E+04	1716	9.65
j	50	40	4	13.25	488	1.944E+05	7776	7.858E+04	3000	13.80
k	60	40	4	14.34	528	2.982E+05	9939	8.379E+04	3093	12.91
i	80	40	4	16.51	608	5.869E+05	1.467E+04	9.216E+04	3231	11.47
m	100	50	5	25.79	950	1.433E+06	2.866E+04	2.250E+05	6310	14.34
n	125	80	8	58.43	2152	5.251E+06	8.401E+04	1.360E+06	24905	25.41
o	160	80	10	81.46	3000	1.130E+07	1.413E+05	1.780E+06	31592	23.67

Notes: [1]I = moment of inertia; S = section modulus; See sketch for X–X and Y–Y axes and the location x for the axis Y–Y.

Section properties computed assuming square corners and constant thickness of web and flanges.

Some producers use filleted corners and tapered flanges. Adjustments to computed values may be needed for precision.

[2]Using density of aluminum = 2768 kg/m^3. (May range from 2635 kg/m^3 to 2829 kg/m^3.)

Source: Parker Steet Company—Metric Sized Metals, Toledo, Ohio; http://metricmetal.com

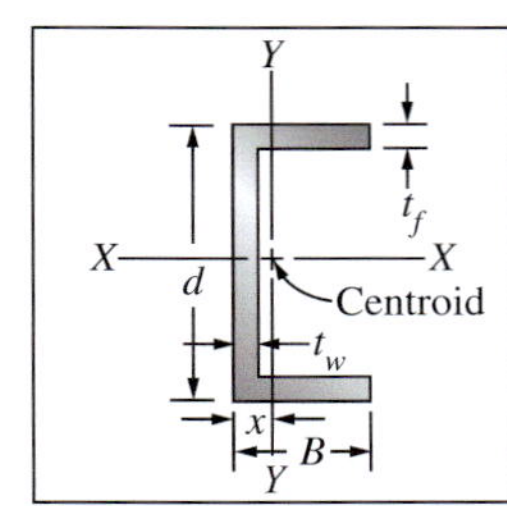

SI Units

TABLE 15–8 Channels: Steel, European Standard Shapes: 30 mm to 400 mm Depth

		Flange					[1]Section properties				
							Axis X–X		Axis Y–Y		
Ref.	Depth d (mm)	width, B (mm)	thickness, t_F (mm)	Web thickness, t_w (mm)	[2]Weight/m (N/m)	Area A (mm^2)	I_x (mm^4)	S_x (mm^3)	I_y (mm^4)	S_y (mm^3)	x (mm)
a	30	15	4.5	4.0	16.5	219.0	2.399E+04	1.599	4.210E+03	438	5.39
b	30	33	7.0	5.0	40.8	542.0	5.558E+04	3706	5.546E+04	2987	14.43
c	40	20	5.5	5.0	27.5	365.0	7.292E+04	3646	1.255E+04	967	7.02
d	40	35	7.0	5.0	46.7	620.0	1.192E+05	5958	7.341E+04	3556	14.35
e	50	25	6.0	5.0	36.9	490.0	1.538E+05	6150	2.765E+04	1688	8.62
f	50	38	7.0	5.0	53.6	712.0	2.198E+05	8793	1.010E+05	4359	14.83
g	60	30	6.0	6.0	48.8	648.0	3.188E+05	1.063E+04	5.090E+04	2503	9.67
h	80	45	8.0	6.0	83.2	1104.0	8.981E+05	2.245E+04	2.179E+05	7441	15.72
i	100	50	8.5	6.0	101.6	1348.0	1.668E+06	3.336E+04	3.306E+05	9.978E+03	16.87
j	140	60	10.0	7.0	153.7	2040.0	4.885E+06	6.979E+04	7.104E+05	1.736E+04	19.09
k	200	75	11.5	8.5	243.3	3229.5	1.604E+07	1.604E+05	1.706E+06	3.220E+04	22.01
l	260	90	14.0	10.0	364.7	4840.0	3.966E+07	3.051E+05	3.653E+06	5.692E+04	25.83
m	300	100	16.0	10.0	443.0	5880.0	6.036E+07	4.024E+05	5.642E+06	8.002E+04	29.49
n	350	100	16.0	14.0	576.5	7652.0	1.180E+08	6.744E+05	6.182E+06	8.240E+04	24.98
o	400	110	18.0	14.0	682.3	9056.0	1.748E+08	8.742E+05	9.210E+06	1.123E+05	27.99

Notes: [1]I = moment of inertia; S = section modulus; See sketch for X–X and Y–Y axes and the location x for the axis Y–Y.

Section properties computed assuming square corners and constant thickness of web and flanges.

Some producers use filleted corners and tapered flanges. Adjustments to computed values may be needed for precision.

[2]Using density of steel = 7680 kg/m^3

Source: Parker Steel Company—Metric Sized Metals, Toledo, Ohio; http://metricmetal.com

I-Beam Shapes
Tables 15–9, 15–10, 15–11, 15–12, and 15–13

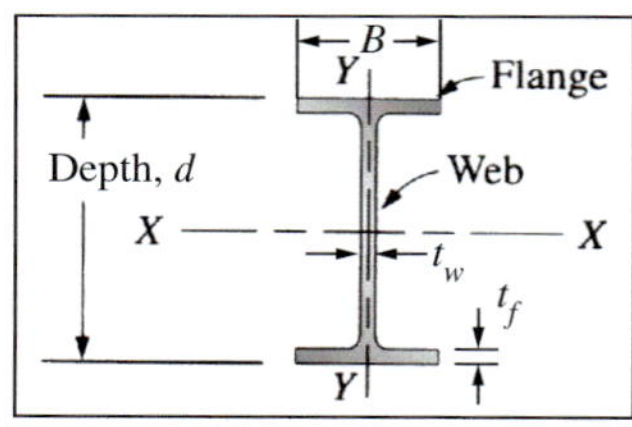

U.S. Units

TABLE 15–9 I-Beam Shapes: Steel Wide Flange Shapes, W-Shapes: 4 in to 24 in Depth

						*Section properties			
				Flange		Axis X–X		Axis Y–Y	
Designation	Area A (in^2)	Depth d (in)	Web thickness, t_w (in)	Width, B (in)	Thickness, t_f (in)	I_x (in^4)	S_x (in^3)	I_y (in^4)	S_y (in^3)
W24 × 76	22.4	23.92	0.440	8.990	0.680	2100	176	82.5	18.4
W24 × 68	20.1	23.73	0.415	8.965	0.585	1830	154	70.4	15.7
W21 × 73	21.5	21.24	0.455	8.295	0.740	1600	151	70.6	17.0
W21 × 57	16.7	21.06	0.405	6.555	0.650	1170	111	30.6	9.35
W18 × 55	16.2	18.11	0.390	7.530	0.630	890	98.3	44.9	11.9
W18 × 40	11.8	17.90	0.315	6.015	0.525	612	68.4	19.1	6.35
W14 × 43	12.6	13.66	0.305	7.995	0.530	428	62.7	45.2	11.3
W14 × 26	7.69	13.91	0.255	5.025	0.420	245	35.3	8.91	3.54
W12 × 30	8.79	12.34	0.260	6.520	0.440	238	38.6	20.3	6.24
W12 × 16	4.71	11.99	0.220	3.990	0.265	103	17.1	2.82	1.41
W10 × 15	4.41	9.99	0.230	4.000	0.270	69.8	13.8	2.89	1.45
W10 × 12	3.54	9.87	0.190	3.960	0.210	53.8	10.9	2.18	1.10
W8 × 15	4.44	8.11	0.245	4.015	0.315	48.0	11.8	3.41	1.70
W8 × 10	2.96	7.89	0.170	3.940	0.205	30.8	7.81	2.09	1.06
W6 × 15	4.43	5.99	0.230	5.990	0.260	29.1	9.72	9.32	3.11
W6 × 12	3.55	6.03	0.230	4.000	0.280	22.1	7.31	2.99	1.50
W5 × 19	5.54	5.15	0.270	5.030	0.430	26.2	10.2	9.13	3.63
W5 × 16	4.68	5.01	0.240	5.000	0.360	21.3	8.51	7.51	3.00
W4 × 13	3.83	4.16	0.280	4.060	0.345	11.3	5.46	3.86	1.90

Notes: Example designation: W14 × 43

14 = nominal depth (in); 43 = weight per unit length (lb/ft).

*I = moment of inertia; S = section modulus.

Sources: Central Steel & Wire Co., multiple locations; www.centralsteel.com/products.htm

Reliance Steel & Aluminum Co./Earl M. Jorgensen Co., multiple locations; www.jorgensensteel.com/products/stocklist.asp

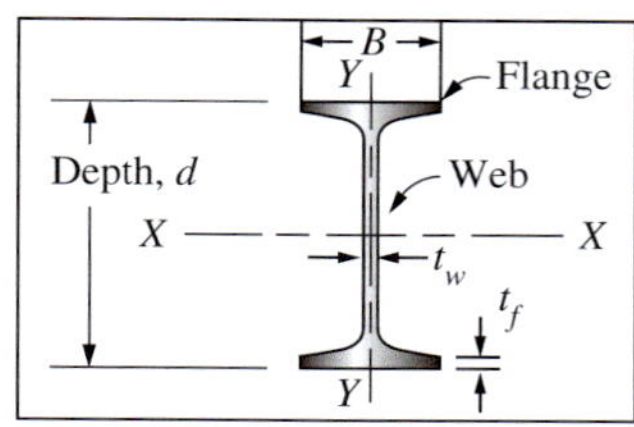

U.S. Units

TABLE 15–10 I-Beam Shapes: Steel, American Standard Shapes, S-Shapes: 3 in to 24 in Depth

				Flange		*Section Properties			
						Axis X–X		Axis Y–Y	
Designation	Area A (in^2)	Depth d (in)	Web thickness t_w (in)	Width B (in)	Average thickness t_f (in)	I_x (in^4)	S_x (in^3)	I_y (in^4)	S_y (in^3)
S24 × 90	26.5	24.00	0.625	7.125	0.870	2250	187	44.9	12.6
S20 × 96	28.2	20.30	0.800	7.200	0.920	1670	165	50.2	13.9
S20 × 75	22.0	20.00	0.635	6.385	0.795	1280	128	29.8	9.32
S20 × 66	19.4	20.00	0.505	6.255	0.795	1190	119	27.7	8.85
S18 × 70	20.6	18.00	0.711	6.251	0.691	926	103	24.1	7.72
S15 × 50	14.7	15.00	0.550	5.640	0.622	486	64.8	15.7	5.57
S12 × 50	14.7	12.00	0.687	5.477	0.659	305	50.8	15.7	5.74
S12 × 35	10.3	12.00	0.428	5.078	0.544	229	38.2	9.87	3.89
S10 × 35	10.3	10.00	0.594	4.944	0.491	147	29.4	8.36	3.38
S10 × 25.4	7.46	10.00	0.311	4.661	0.491	124	24.7	6.79	2.91
S8 × 23	6.77	8.00	0.441	4.171	0.426	64.9	16.2	4.31	2.07
S8 × 18.4	5.41	8.00	0.271	4.001	0.426	57.6	14.4	3.73	1.86
S7 × 20	5.88	7.00	0.450	3.860	0.392	42.4	12.1	3.17	1.64
S6 × 12.5	3.67	6.00	0.232	3.332	0.359	22.1	7.37	1.82	1.09
S5 × 10	2.94	5.00	0.214	3.004	0.326	12.3	4.92	1.22	0.809
S4 × 7.7	2.26	4.00	0.193	2.663	0.293	6.08	3.04	0.764	0.574
S3 × 5.7	1.67	3.00	0.170	2.330	0.260	2.52	1.68	0.455	0.390

Notes: Example designation: S10 × 35

10 = nominal depth (in); 35 = weight per unit length (lb/ft).

*I = moment of inertia; S = section modulus.

Sources: Central Steel & Wire Co., multiple locatuions; www.centralsteel.com/products.htm

Reliance Steel & Aluminum Co./Earl M. Jorgensen Co., multiple locations; www.jorgensensteel.com/products/stocklist.asp

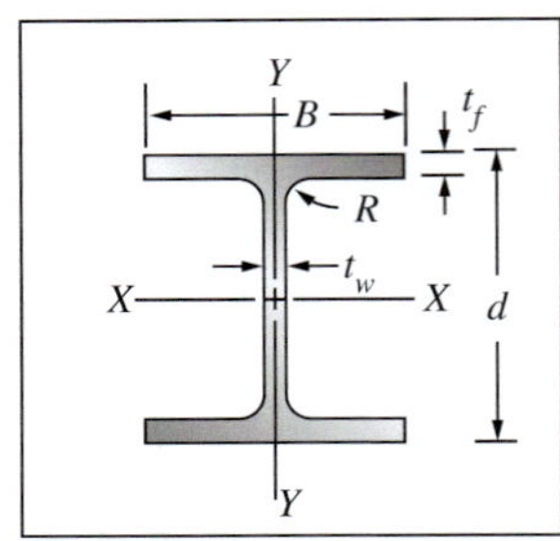

U.S. Units

TABLE 15–11 I-Beam shapes: Aluminum, Aluminum Association Standard Shapes, Larger Sizes: 3 in to 12 in Depth

Size							[1]Section properties					
							Axis X–X			Axis Y–Y		
Depth d (in)	Width B (in)	Area A (in^2)	[2]Weight (lb/ft)	Flange thickness, t_f (in)	Web thickness, t_w (in)	Fillet radius, R (in)	I_x (in^4)	S_x (in^3)	r_x (in)	I_y (in^4)	S_y (in^3)	r_y (in)
3.00	2.50	1.392	1.637	0.20	0.13	0.25	2.24	1.49	1.27	0.52	0.42	0.61
3.00	2.50	1.726	2.030	0.26	0.15	0.25	2.71	1.81	1.25	0.68	0.54	0.63
4.00	3.00	1.965	2.311	0.23	0.15	0.25	5.62	2.81	1.69	1.04	0.69	0.73
4.00	3.00	2.375	2.793	0.29	0.17	0.25	6.71	3.36	1.68	1.31	0.87	0.74
5.00	3.50	3.146	3.700	0.32	0.19	0.30	13.94	5.58	2.11	2.29	1.31	0.85
6.00	4.00	3.427	4.030	0.29	0.19	0.30	21.99	7.33	2.53	3.10	1.55	0.95
6.00	4.00	3.990	4.692	0.35	0.21	0.30	25.50	8.50	2.53	3.74	1.87	0.97
7.00	4.50	4.932	5.800	0.38	0.23	0.30	42.89	12.25	2.95	5.78	2.57	1.08
8.00	5.00	5.256	6.181	0.35	0.23	0.30	59.69	14.92	3.37	7.30	2.92	1.18
8.00	5.00	5.972	7.023	0.41	0.25	0.30	67.78	16.94	3.37	8.55	3.42	1.20
9.00	5.50	7.110	8.361	0.44	0.27	0.30	102.02	22.67	3.79	12.22	4.44	1.31
10.00	6.00	7.352	8.646	0.41	0.25	0.40	132.09	26.42	4.24	14.78	4.93	1.42
10.00	6.00	8.747	10.286	0.50	0.29	0.40	155.79	31.16	4.22	18.03	6.01	1.44
12.00	7.00	9.925	11.672	0.47	0.29	0.40	255.57	42.60	5.07	26.90	7.69	1.65
12.00	7.00	12.153	14.292	0.62	0.31	0.40	317.33	52.89	5.11	35.48	10.14	1.71

Notes: [1]Weights per foot are based on nominal dimensions and a density of 0.098 lb/in^3, which is the density of alloy 6061.

[2]Areas listed are based on nominal dimensions.

I = moment of inertia; S = section modulus; r = radius of gyration.

Sources: Aluminum Association, *Aluminum Standards and Data,* 11th ed., Washington, DC, © 1993, p. 187.

Central Steel & Wire Co., multiple locations; www.centralsteel.com/products.htm

Reliance Steel & Aluminum Co./Earl M. Jorgensen Co., multiple locations; www.jorgensensteel.com/products/stocklist.asp

OnlineMetals.com, Seattle, WA; www.onlinemetals.com

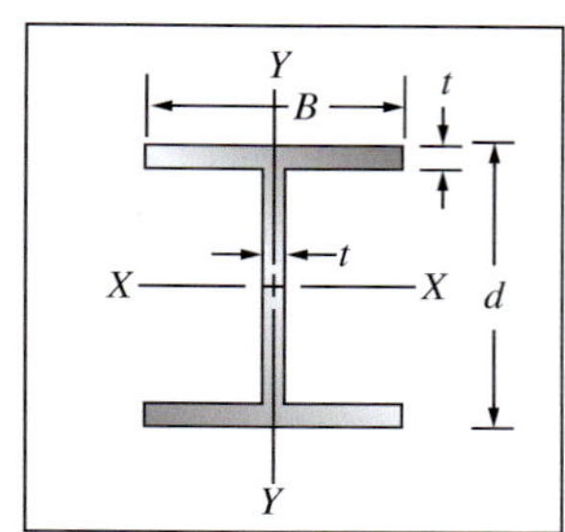

U.S. Units

TABLE 15–12 I-Beam Shapes: Aluminum, Small Extruded Shapes: 0.70 in to 2.11 in Depth

						[1]Section properties					
						Axis X–X			Axis Y–Y		
Ref.	Depth d (in)	Flange width, B (in)	Thickness t (in)	[2]Weight/ft (lb/ft)	Area, A (in^2)	I_x (in^4)	S_x (in^3)	r_x (in)	I_y (in^4)	S_y (in^3)	r_y (in)
a	0.700	1.060	0.050	0.160	0.136	0.0121	0.0346	0.299	0.00993	0.0187	0.270
b	1.040	1.074	0.050	0.185	0.154	0.0298	0.0573	0.439	0.0103	0.0192	0.259
c	1.950	1.000	0.050	0.230	0.193	0.117	0.120	0.778	0.00835	0.0167	0.208
d	1.120	1.724	0.050	0.268	0.223	0.0538	0.0961	0.491	0.0427	0.0495	0.437
e	1.034	1.500	0.062	0.290	0.242	0.0479	0.0926	0.444	0.0349	0.0465	0.379
f	2.110	1.500	0.055	0.329	0.275	0.211	0.200	0.876	0.0310	0.0413	0.336
g	0.876	1.500	0.188	0.790	0.658	0.0704	0.161	0.327	0.106	0.141	0.401
h	1.000	2.375	0.250	1.575	1.313	0.176	0.352	0.366	0.559	0.471	0.653

Note: These shapes are designed for special applications and not made to common dimensions. Calculations assume square corners.

[1]I = moment of inertia; S = section modulus; r = radius of gyration, used for column analysis; See sketch for X–X and Y–Y axes.

[2]Using density of aluminum = 0.100 lb$_m$/in^3. (May range from 0.095 lb$_m$/in^3 to 0.102 lb$_m$/in^3.)

Sources: Paramount Extrusions, Co., Paramount, CA; www.paramountextrusions.com

Metals Depot, Winchester, KY; www.MetalsDepot.com

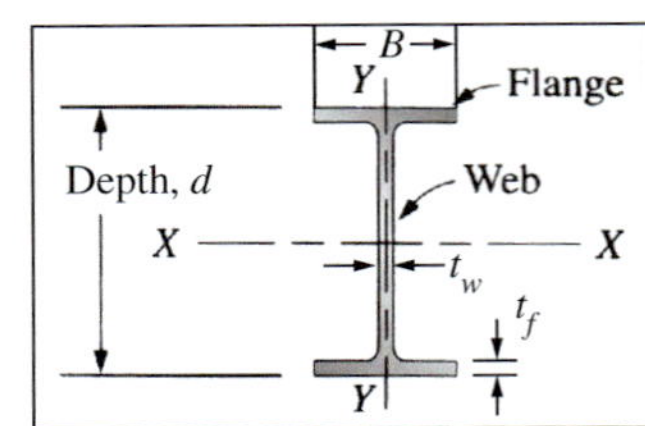

SI Units

TABLE 15–13 I-Beam Shapes: Steel, European Standard ([2]IPE) Shapes: 80 mm to 600 mm Depth

			Thickness				[1]Section properties					
							Axis X–X			Axis Y–Y		
Ref.	Depth d (mm)	Flange width, B (mm)	Flange, t_F (mm)	Web, t_W (mm)	[3]Weight/m (N/m)	Area A (mm^2)	I_x (mm^4)	S_x (mm^3)	r_x (mm)	I_y (mm^4)	S_y (mm^3)	r_y (mm)
a	80	46	5.2	3.8	57.6	764	8.014E+05	2.003E+04	32.4	8.487E+04	3.690E+03	10.54
b	100	55	5.7	4.1	77.8	1032	1.710E+06	3.420E+04	40.7	1.591E+05	5.786E+03	12.4
c	120	64	6.3	4.4	99.5	1321	3.177E+06	5.296E+04	49.0	2.766E+05	8.644E+03	14.5
d	140	73	6.9	4.7	123.8	1643	5.412E+06	7.732E+04	57.4	4.491E+05	1.230E+04	16.5
e	160	82	7.4	5.0	151.4	2009	8.693E+06	1.087E+05	65.8	6.829E+05	1.666E+04	18.4
f	180	91	8.0	5.3	180.4	2395	1.317E+07	1.463E+05	74.2	1.008E+06	2.216E+04	20.5
g	200	100	8.5	5.6	214.6	2848	1.943E+07	1.943E+05	82.6	1.423E+06	2.846E+04	22.4
h	220	110	9.2	5.9	251.4	3337	2.772E+07	2.520E+05	91.1	2.048E+06	3.724E+04	24.8
I	240	120	9.8	6.2	294.7	3912	3.891E+07	3.243E+05	99.7	2.835E+06	4.725E+04	26.9
j	270	135	10.2	6.6	346.2	4595	5.790E+07	4.289E+05	112.3	4.197E+06	6.218E+04	30.2
k	300	150	10.7	7.1	405.4	5381	8.356E+07	5.571E+05	124.6	6.036E+06	8.048E+04	33.5
l	330	160	11.5	7.5	471.7	6261	1.177E+08	7.131E+05	137.1	7.878E+06	9.848E+04	35.5
m	360	170	12.7	8.0	547.9	7273	1.627E+08	9.036E+05	149.5	1.043E+07	1.227E+05	37.9
n	400	180	13.5	8.8	642.0	8521	2.321E+08	1.161E+06	165.1	1.317E+07	1.464E+05	39.3
o	450	190	14.6	9.4	744.5	9882	3.374E+08	1.500E+06	184.8	1.675E+07	1.763E+05	41.2
p	500	200	16.0	10.2	870.4	11552	4.820E+08	1.928E+06	204.3	2.141E+07	2.141E+05	43.1
q	550	210	17.2	11.1	1013	13442	6.712E+08	2.441E+06	223.5	2.667E+07	2.540E+05	44.5
r	600	220	19.0	12.0	1175	15598	9.208E+08	3.069E+06	243.0	3.386E+07	3.078E+05	46.6

Notes: [1]I = moment of inertia; S = section modulus; r = radius of gyration, used for column analysis; See sketch for X–X and Y–Y axes.

[2]IPE sections have medium-width flanges. Narrow and wide-flange sections also available.

[3]Using density of steel = 7680 kg/m^3.

Source: Parker Steel Company—Metric Sized Metals, Toledo, Ohio; http://metricmetal.com

Tubing: Square and rectangular
Tables 15–14, 15–15, and 15–16

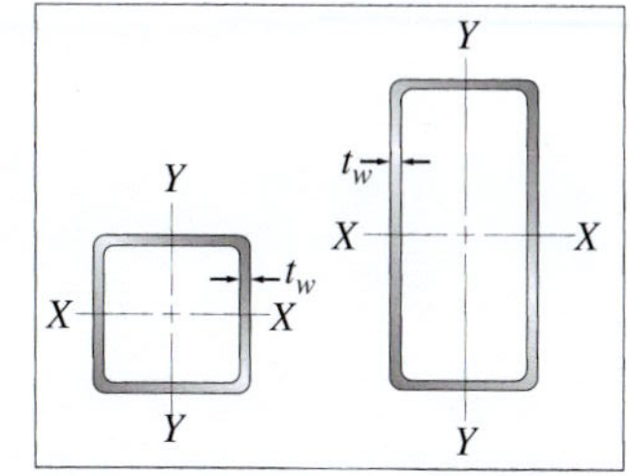

U.S. Units

TABLE 15–14 Hollow Tubing: Steel, Square and Rectangular, Standard Structural HSS Shapes: 2 in to 8 in Depth

					Section properties[1]							
	Shape	Design wall thickness, t_w	Weight[2] per foot	Area, A	Axis X–X			Axis Y–Y			Torsional constants	
Ref.	(in) (in) (in)	(in)	(lb/ft)	(in²)	I_x (in⁴)	S_x (in³)	r_x (in)	I_y (in⁴)	S_y (in³)	r_y (in)	J (in⁴)	C (in³)
a	HSS 8 × 8 × $\frac{1}{2}$	0.465	48.7	13.5	125	31.2	3.04	125	31.2	3.04	204	52.4
b	HSS 8 × 8 × $\frac{1}{4}$	0.233	25.8	7.10	70.7	17.7	3.15	70.7	17.7	3.15	111	28.1
c	HSS 8 × 4 × $\frac{1}{2}$	0.465	35.1	9.74	71.8	17.9	2.71	23.6	11.8	1.56	61.1	24.4
d	HSS 8 × 4 × $\frac{1}{4}$	0.233	19.0	5.24	42.5	10.6	2.85	14.4	7.21	1.66	35.3	13.6
e	HSS 8 × 2 × $\frac{1}{4}$	0.233	15.6	4.30	28.5	7.12	2.57	2.94	2.94	0.827	9.36	6.35
f	HSS 6 × 6 × $\frac{1}{2}$	0.465	35.1	9.74	48.3	16.1	2.23	48.3	16.1	2.23	81.1	28.1
g	HSS 6 × 6 × $\frac{1}{4}$	0.233	19.0	5.24	28.6	9.54	2.34	28.6	9.54	2.34	45.6	15.4
h	HSS 6 × 4 × $\frac{1}{4}$	0.233	15.6	4.30	20.9	6.96	2.20	11.1	5.56	1.61	23.6	10.1
i	HSS 6 × 2 × $\frac{1}{4}$	0.233	12.2	3.37	13.1	4.37	1.97	2.21	2.21	0.810	6.55	4.70
j	HSS 4 × 4 × $\frac{1}{2}$	0.465	21.5	6.02	11.9	5.97	1.41	11.9	5.97	1.41	21.0	11.2
k	HSS 4 × 4 × $\frac{1}{4}$	0.233	12.2	3.37	7.80	3.90	1.52	7.80	3.90	1.52	12.8	6.56
l	HSS 4 × 2 × $\frac{1}{4}$	0.233	8.78	2.44	4.49	2.25	1.36	1.48	1.48	0.779	3.82	3.05
m	HSS 3 × 3 × $\frac{1}{4}$	0.233	8.78	2.44	3.02	2.01	1.11	3.02	2.01	1.11	5.08	3.52
n	HSS 3 × 2 × $\frac{1}{4}$	0.233	7.08	1.97	2.13	1.42	1.04	1.11	1.11	0.751	2.52	2.23
o	HSS 2 × 2 × $\frac{1}{4}$	0.233	5.38	1.51	0.747	0.747	0.704	0.747	0.747	0.704	1.31	1.41

Note: Example size: 6 × 4 × 1/4; 6 = vertical depth (in); 4 = width (in); 1/4 = wall thickness (in).

[1] I = moment of inertia; S = section modulus; r = radius of gyration.

[2] Using Density of a steel = 0.283 lb_m/in^3

Sources: Metals Depot, Winchester, KY; www.MetalsDepot.com

Bullmoose Tube Co.; www.bullmoosetube.com/products

Steel Tube Institute of North America, Glenview, IL; www.steeltubeinstitute.org

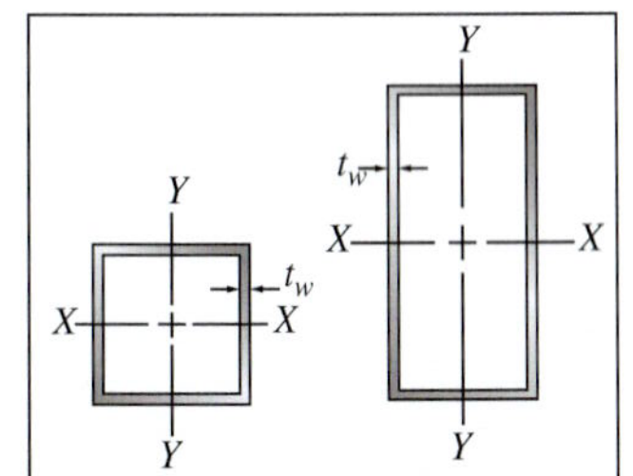

U.S. Units

TABLE 15–15 Hollow Tubing: Steel and Aluminum, Square and Rectangular, Smaller Sizes: 0.375 in to 3.00 In Depth

	Outside dimensions		Wall	Inside dimensions			Weight/ft[2]		Section properties[1]					
									Axis X–X			Axis Y–Y		
Ref.	Short side a (in)	Long side b (in)	thickness t (in)	Short side c (in)	Long side d (in)	Area A (in^2)	Steel (lb/ft)	Aluminum (lb/ft)	I_x (in^4)	S_x (in^3)	r_x (in)	I_y (in^4)	S_y (in^3)	r_y (in)
a	0.375	0.375	0.049	0.277	0.277	0.064	0.217	0.077	0.00116	0.00617	0.13458	0.00116	0.00617	0.135
b	0.500	0.500	0.065	0.370	0.370	0.113	0.384	0.136	0.00365	0.0146	0.180	0.00365	0.0146	0.180
c	0.500	1.000	0.065	0.370	0.870	0.178	0.605	0.214	0.0214	0.0427	0.346	0.00674	0.0270	0.195
d	0.500	1.500	0.065	0.370	1.370	0.243	0.826	0.292	0.0613	0.0818	0.502	0.0098	0.0394	0.201
e	0.500	1.500	0.120	0.260	1.260	0.422	1.434	0.507	0.0973	0.130	0.480	0.0138	0.0551	0.181
f	0.625	0.625	0.065	0.495	0.495	0.146	0.494	0.175	0.00771	0.0247	0.230	0.00771	0.0247	0.230
g	0.750	0.750	0.065	0.620	0.620	0.178	0.605	0.214	0.0141	0.0375	0.281	0.0141	0.0375	0.281
h	0.750	1.500	0.065	0.620	1.370	0.276	0.936	0.331	0.0781	0.104	0.532	0.0255	0.0681	0.304
i	0.875	0.875	0.049	0.777	0.777	0.162	0.550	0.194	0.0185	0.042	0.338	0.0185	0.0422	0.338
j	1.000	1.000	0.065	0.870	0.870	0.243	0.826	0.292	0.0356	0.071	0.383	0.0356	0.0712	0.383
k	1.000	1.000	0.120	0.760	0.760	0.422	1.434	0.507	0.0555	0.111	0.363	0.0555	0.111	0.363
l	1.000	1.500	0.065	0.870	1.370	0.308	1.046	0.370	0.0948	0.126	0.555	0.0498	0.100	0.402
m	1.000	1.500	0.120	0.760	1.260	0.542	1.842	0.651	0.155	0.206	0.534	0.0789	0.158	0.381
n	1.000	2.000	0.065	0.870	1.870	0.373	1.267	0.448	0.193	0.193	0.718	0.0640	0.128	0.414
o	1.000	3.000	0.065	0.870	2.870	0.503	1.709	0.604	0.536	0.357	1.032	0.0925	0.185	0.429
p	1.250	1.250	0.065	1.120	1.120	0.308	1.046	0.370	0.0723	0.116	0.485	0.0723	0.116	0.485
q	1.375	1.375	0.058	1.259	1.259	0.306	1.038	0.367	0.0885	0.129	0.538	0.0885	0.129	0.538
r	1.500	1.500	0.065	1.370	1.370	0.373	1.267	0.448	0.128	0.171	0.586	0.128	0.171	0.586
s	1.500	1.500	0.120	1.260	1.260	0.662	2.250	0.795	0.212	0.282	0.566	0.212	0.282	0.566
t	1.500	2.000	0.065	1.370	1.870	0.438	1.488	0.526	0.253	0.253	0.761	0.162	0.216	0.608
u	1.500	2.000	0.120	1.260	1.760	0.782	2.657	0.939	0.428	0.428	0.739	0.269	0.359	0.586
v	1.500	3.000	0.120	1.260	2.760	1.022	3.472	1.227	1.167	0.778	1.069	0.384	0.512	0.613

Notes: [1]I = moment of inertia; S = section modulus; r = radius of gyration, used in column analysis; See sketch for X–X and Y–Y axes.

Values are for perfectly square corners. Some vendors apply radii to inside and/or outside corners.

Wall thickness gages: 0.049 in = 18 gage; 0.058 in = 17 gage; 0.065 in = 16 gage; 0.120 in = 11 gage.

Other gages and sizes available from many manufacturers and vendors. See Appendix A15-14 for larger structural tubing.

[2]Using density of steel = 0.283 lb_m/in^3; density of aluminum = 0.100 lb_m/in^3. (May range from 0.095 lb_m/in^3 to 0.102 lb_m/in^3.)

Sources: Reliance Steel & Aluminum Co./Earl M. Jorgensen Co., multiple locations; www.jorgensensteel.com/products/stocklist.asp Central Steel & Wire Co., multiple locations; www.centralsteel.com/products.htm OnlineMetals.com, Seattle, WA; www.onlinemetals.com Bullmoose Tube Co.; www.bullmoosetube.com/products Metals Depot, Winchester, KY; www.MetalsDepot.com

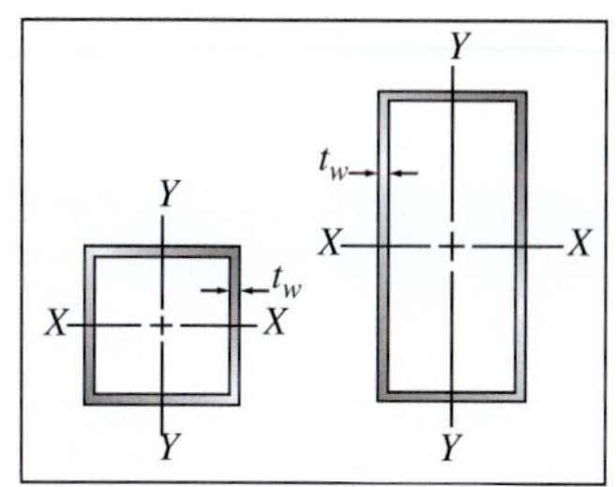

SI Units

TABLE 15–16 Howllow Tubing: Steel and Aluminum, Square and Rectangular: 20 mm to 300 mm Depth

	Outside dimensions			Inside dimensions			[2]Weight/m		[1]Section properties					
									Axis X-X			Axis Y-Y		
Ref.	Short side a (mm)	Long side b (mm)	Wall thickness t_w (mm)	Short side c (mm)	Long side d (mm)	Area A (mm^2)	Steel (N/m)	Aluminum (N/m)	I_x (mm^4)	S_x (mm^3)	r_x (mm)	I_y (mm^4)	S_y (mm^3)	r_y (mm)
a	10	20	2	6	16	104	7.84	2.82	4.619E+03	462	6.66	1.379E+03	276	3.64
b	20	20	2	16	16	144	10.8	3.91	7.872E+03	787	7.39	7.872E+03	787	7.39
c	20	30	3	14	24	264	19.9	7.17	2.887E+04	1925	10.5	1.451E+04	1451	7.41
d	30	30	3	24	24	324	24.4	8.80	3.985E+04	2657	11.1	3.985E+04	2657	11.1
e	20	40	3	14	34	324	24.4	8.80	6.081E+04	3041	13.7	1.889E+04	1889	7.64
f	40	40	3	34	34	444	33.5	12.1	1.020E+05	5099	15.2	1.020E+05	5099	15.2
g	30	50	3	24	44	444	33.5	12.1	1.421E+05	5685	17.9	6.181E+04	4121	11.8
h	50	50	3	44	44	564	42.5	15.3	2.085E+05	8340	19.2	2.085E+05	8340	19.2
i	40	80	4	32	72	896	67.5	24.3	7.113E+05	1.778E+04	28.2	2.301E+05	1.150E+04	16.0
j	80	80	3	74	74	924	69.6	25.1	9.145E+05	2.286E+04	31.5	9.145E+05	2.286E+04	31.5
k	50	100	4	42	92	1136	85.6	30.8	1.441E+06	2.883E+04	35.6	4.737E+05	1.895E+04	20.4
l	100	100	4	92	92	1536	116	41.7	2.363E+06	4.727E+04	39.2	2.363E+06	4.727E+04	39.2
m	50	150	4	42	142	1536	116	41.7	4.041E+06	5.388E+04	51.3	6.858E+05	2.743E+04	21.1
n	150	150	5	140	140	2900	218	78.7	1.017E+07	1.357E+05	59.2	1.017E+07	1.357E+05	59.2
o	200	200	4	192	192	3136	236	85.1	2.009E+07	2.009E+05	80.0	2.009E+07	2.009E+05	80.0
p	100	200	4	92	192	2336	176	63.4	1.240E+07	1.240E+05	72.9	4.208E+06	8.415E+04	42.4
q	50	200	4	42	192	1936	146	52.6	8.561E+06	8.561E+04	66.5	8.979E+05	3.592E+04	21.5
r	250	250	8	234	234	7744	583	210.3	7.567E+07	6.054E+05	98.9	7.567E+07	6.054E+05	98.9
s	300	300	8	284	284	9344	704	253.7	1.329E+08	8.859E+05	119.3	1.329E+08	8.859E+05	119.3
t	300	300	12.5	275	275	14375	1083	390.3	1.984E+08	1.323E+06	117.5	1.984E+08	1.323E+06	117.5

Notes: [1]I = moment of inertia; S = section modulus; r = radius of gyration, used in column analysis; See sketch for X–X and Y–Y axes.

Values are for perfectly square corners. Some vendors apply radii to inside and/or outside corners.

Numerous additional sizes and wall thicknesses available from many manufacturers and vendors.

[2]Using density of steel = 7680 kg/m^3; density of aluminum = 2768 kg/m^3. (May range from 2635 kg/m^3 to 2829 kg/m^3.)

Sources: Parker Steel Company—Metric Sized Metals, Toledo, Ohio; http://metricmetal.com

Bulimoose Tube Co.; www.bulimoosetube.com/products

Pipe and round tubing
Tables 15–17, 15–18, and 15–19

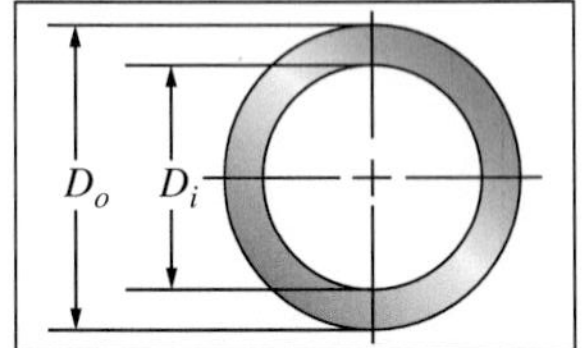

U.S. Units

TABLE 15–17 Pipe: American National Standard Schedule 40 and AISC Standard: 1/8 in to 18 in Sizes

Ref.	[1]Nominal size (in)	Outside diameter (in)	Inside diameter (in)	Wall thickness, t_w (in)	Area, A (in^2)	Weight/ft Steel (lb/ft)	Section properties: I (in^4)	S (in^3)	r (in)	Torsional constants: J (in^4)	Z_p (in^3)
	Schedule 40 pipe										
a	1/8 in	0.405	0.269	0.068	0.072	0.245	1.06E-03	5.25E-03	0.122	2.13E-03	1.05E-02
b	1/4 in	0.540	0.364	0.088	0.125	0.425	3.31E-03	1.23E-02	0.163	6.62E-03	2.45E-02
c	3/8 in	0.675	0.493	0.091	0.167	0.568	7.29E-03	2.16E-02	0.209	1.46E-02	4.32E-02
d	PIPE 1/2 STD	0.840	0.622	0.109	0.250	0.851	1.71E-02	4.07E-02	0.261	3.42E-02	8.14E-02
e	PIPE 3/4 STD	1.050	0.824	0.113	0.333	1.131	3.70E-02	7.05E-02	0.334	7.41E-02	0.1411
f	PIPE 1 STD	1.315	1.049	0.133	0.494	1.679	8.73E-02	0.1328	0.421	0.1747	0.2657
g	PIPE 1-1/4 STD	1.660	1.380	0.140	0.669	2.273	0.1947	0.2346	0.540	0.3894	0.4692
h	PIPE 1-1/2 STD	1.900	1.610	0.145	0.799	2.718	0.3099	0.3262	0.623	0.6198	0.6524
i	PIPE 2 STD	2.375	2.067	0.154	1.075	3.653	0.6657	0.5606	0.787	1.331	1.121
j	PIPE 2-1/2 STD	2.876	2.469	0.203	1.704	5.793	1.530	1.064	0.947	3.059	2.128
k	PIPE 3 STD	3.500	3.068	0.216	2.228	7.576	3.017	1.724	1.164	6.034	3.448
l	PIPE 3-1/2 STD	4.000	3.548	0.226	2.680	9.109	4.788	2.394	1.337	9.575	4.788
m	PIPE 4 STD	4.500	4.026	0.237	3.714	10.790	7.233	3.214	1.510	14.47	6.429
n	PIPE 5 STD	5.563	5.047	0.258	4.300	14.618	15.16	5.451	1.878	30.32	10.90
o	PIPE 6 STD	6.625	6.065	0.280	5.581	18.974	28.14	8.496	2.245	56.28	16.99
p	PIPE 8 STD	8.625	7.981	0.322	8.399	28.554	72.49	16.81	2.938	145.0	33.62
q	PIPE 10 STD	10.750	10.020	0.365	11.908	40.483	160.7	29.90	3.674	321.5	59.81
r	12 in	12.750	11.938	0.406	15.745	53.525	300.2	47.09	4.367	600.4	94.18
s	16 in	16.000	15.000	0.500	24.347	82.771	731.9	91.49	5.483	1464	183.0
t	18 in	18.000	16.876	0.562	30.788	104.667	1171	130.2	6.168	2343	260.3

Notes: [1]All values shown are for standard schedule 40 steep pipe.

Rows d-q conform to AISC standards for dimensions of standard weight pipe, Rows a–c and r–t do not.

Many other sizes of round hollow structural sections (HSS) are available. See AISC Manual.

Sources: American Piping Products, Chesterfield, MO; www.amerpipe.com/reference/Ansi-Pipe-Chart.php

Davidson Group, Specialty Pipe & Tube, Brooklyn, NY; www.davidsonpipe.com/co_specpipe

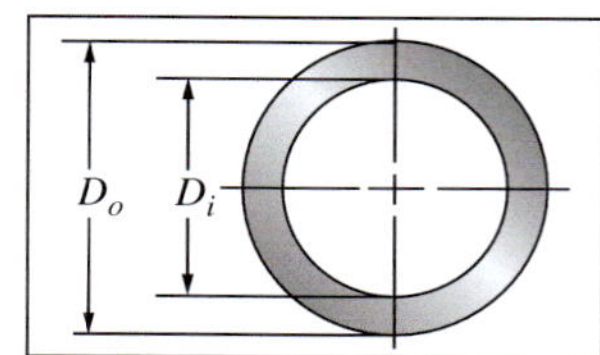

U.S. Units

TABLE 15–18 Mechanical Tubing: Steel and Aluminum: 0.5 in to 5.0 in Outside Diameter

	Nominal size		Outside diameter	Inside diameter	Wall thickness		[2]Weight/ft		[1]Section properties				
									Bending properties			Torsion properties	
Ref.	*OD* (in)	Wall gauge	D_D (in)	D_i (in)	t_w (in)	Area *A* (in^2)	Steel (lb/ft)	Aluminum (lb/ft)	*I* (in^4)	*S* (in^3)	*r* (in)	*J* (in^4)	Z_p (in^3)
a	1/2	17	0.500	0.384	0.058	0.081	0.274	0.097	0.00200	0.00800	0.158	0.00400	0.0160
b	1/2	14	0.500	0.334	0.083	0.109	0.370	0.130	0.00246	0.00983	0.150	0.00491	0.0197
c	1	16	1.000	0.870	0.065	0.191	0.649	0.229	0.0210	0.0419	0.331	0.0419	0.0839
d	1	10	1.000	0.732	0.134	0.365	1.239	0.437	0.0350	0.0700	0.310	0.0700	0.140
e	11/2	16	1.500	1.370	0.065	0.293	0.996	0.352	0.0756	0.101	0.508	0.151	0.202
f	11/2	10	1.500	1.232	0.134	0.575	1.955	0.690	0.135	0.181	0.485	0.271	0.361
g	2	16	2.000	1.870	0.065	0.395	1.343	0.474	0.185	0.185	0.685	0.370	0.370
h	2	10	2.000	1.732	0.134	0.786	2.671	0.943	0.344	0.344	0.661	0.687	0.687
i	21/2	10	2.500	2.232	0.134	0.996	3.386	1.195	0.699	0.559	0.838	1.398	1.119
j	21/2	5	2.500	2.060	0.220	1.576	5.357	1.891	1.034	0.827	0.810	2.067	1.654
k	3	10	3.000	2.732	0.134	1.207	4.102	1.448	1.241	0.828	1.014	2.483	1.655
l	3	5	3.000	2.560	0.220	1.921	6.532	2.306	1.868	1.245	0.986	3.736	2.490
m	31/2	10	3.500	3.232	0.134	1.417	4.817	1.700	2.010	1.149	1.191	4.020	2.297
n	31/2	5	3.500	3.060	0.220	2.267	7.707	2.720	3.062	1.750	1.162	6.125	3.500
o	4	5	4.000	3.560	0.220	2.613	8.882	3.135	4.682	2.341	1.339	9.364	4.682
p	4	3/8 in	4.000	3.250	0.375	4.271	14.518	5.125	7.090	3.545	1.288	14.180	7.090
q	41/2	5	4.500	4.060	0.220	2.958	10.056	3.550	6.791	3.018	1.515	13.583	6.037
r	41/2	3/8 in	4.500	3.750	0.375	4.860	16.521	5.832	10.422	4.632	1.464	20.843	9.264
s	5	5	5.000	4.560	0.220	3.304	11.231	3.964	9.456	3.782	1.692	18.911	7.564
t	5	3/8 in	5.000	4.250	0.375	5.449	18.523	6.538	14.665	5.866	1.641	29.329	11.732

Notes: [1]*I* = moment of inertia; *S* = section modulus; *r* = radius of gyration, used in column analysis; *J* = polar moment of inertia; Z_p = polar section modulus.

[2]Using density of steel = 0.283 lb_m/in^3; aluminum density = 0.100 lb_m/in^3. (May range from 0.095 lb_m/in^3 to 0.102 $lb_m.in^3$.)

Sources: Reliance Steel & Aluminum Co./Earl M. Jorgensen Co.; www.jorgensensteel.com/products/stocklist.asp

Central Steel & Wire Co.; www.centralsteel.com/products.htm

Webco Industries, Inc.; www.webcoindustries.com/tubing/mechanical

Bullmoose Tube Co.; www.bullmoosetube.com/products

Paramount Extrusions, Co., Paramount, CA; www.paramountextrusions.com

Metals Depot, Winchester, KY; www.MetalsDepot.com

Wheatland Tube Company, Sharon, PA (Smaller sizes); www.wheatland.com

Davidson Group, Specialty Pipe & Tube, Brooklyn, NY (larger sizes and heavier wall); www.davidsonpipe.com/co_specpipe

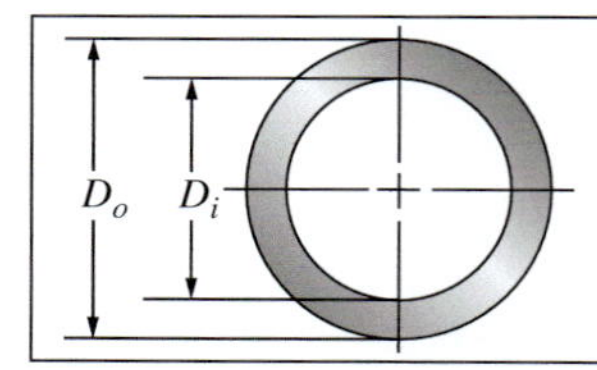

SI Units

TABLE 15–19 Mechanical Tubing: Steel and Aluminum: 10 mm to 150 mm Outside Diameters

Ref.	Outside diameter D_o (mm)	Inside diameter D_i (mm)	Wall thickness t (mm)	Area, A (mm²)	Weight/m[2] Steel (N/m)	Weight/m[2] Aluminum (N/m)	Section properties[1]: Bending properties I (mm⁴)	Bending properties S (mm³)	r (mm)	Torsion properties J (mm⁴)	Torsion properties Z_p (mm³)
a	10	8	1.0	28.27	2.130	0.768	289.8	57.96	3.20	579.6	115.9
b	10	6	2.0	50.27	3.787	1.365	427.3	85.45	2.92	854.5	170.9
c	20	17	1.5	87.18	6.568	2.367	3754	375.4	6.56	7508.3	750.8
d	20	15	2.5	137.4	10.36	3.732	5369	536.9	6.25	1.074E+04	1074
e	30	26	2.0	175.9	13.25	4.777	1.733E+04	1155	9.92	3.466E+04	2311
f	30	22	4.0	326.7	24.62	8.872	2.826E+04	1884	9.30	5.652E+04	3768
g	45	40	2.5	333.8	25.15	9.064	7.563E+04	3361	15.05	1.513E+05	6722
h	45	37	4.0	515.2	38.82	13.990	1.093E+05	4857	14.56	2.186E+05	9715
i	60	52	4.0	703.7	53.02	19.109	2.773E+05	9242	19.85	5.545E+05	1.848E+04
j	60	48	6.0	1017.9	76.69	27.639	3.756E+05	1.252E+04	19.21	7.512E+05	2.504E+04
k	75	70	2.5	569.4	42.90	15.462	3.746E+05	9.988E+03	25.65	7.491E+05	1.998E+04
l	75	65	5.0	1100	82.84	29.857	6.769E+05	1.805E+04	24.81	1.354E+06	3.610E+04
m	90	84	3.0	820	61.78	22.265	7.767E+05	1.726E+04	30.78	1.553E+06	3.452E+04
n	90	80	5.0	1335	100.6	36.255	1.210E+06	2.689E+04	30.10	2.420E+06	5.378E+04
o	110	104	3.0	1008	76.0	27.383	1.444E+06	2.626E+04	37.85	2.889E+06	5.252E+04
p	110	100	5.0	1649	124.3	44.786	2.278E+06	4.142E+04	37.17	4.556E+06	8.284E+04
q	130	124	3.0	1197	90.2	32.502	2.415E+06	3.715E+04	44.91	4.829E+06	7.429E+04
r	130	120	5.0	1963	147.9	53.317	3.841E+06	5.909E+04	44.23	7.682E+06	1.182E+05
s	150	144	3.0	1385	104.4	37.620	3.744E+06	4.992E+04	51.98	7.488E+06	9.983E+04
t	150	140	5.0	2278	171.6	61.847	5.993E+06	7.991E+04	51.30	11986157.7	1.598E+05

Notes: [1]I = moment of inertia; S = section modulus; r = radius of gyration, used in column analysis; J = polar moment of inertia; Z_p = polar section modulus.

[2]Using density of steel = 7680 kg/m³; density of aluminum = 2768 kg/m³. (May range from 2635 kg/m³ to 2829 kg/m³.)

Source: Parker Steel Company—Metric Sized Metals, Toledo, Ohio; http://metricmetal.com

APPENDIX 16 Conversion Factors

Mass Standard SI unit: Kilogram (kg). Equivalent unit: N · s²/m.

$\frac{14.59\ \text{kg}}{\text{slug}}$ $\frac{32.174\ \text{lb}_m}{\text{slug}}$ $\frac{2.205\ \text{lb}_m}{\text{kg}}$ $\frac{453.6\ \text{grams}}{\text{lb}_m}$ $\frac{2000\ \text{lb}_m}{\text{ton}_m}$ $\frac{1000\ \text{kg}}{\text{metric ton}_m}$

Force Standard SI unit: Newton (N). Equivalent unit: kg · m/s².

$\frac{4.448\ \text{N}}{\text{lb}_f}$ $\frac{10^5\ \text{dynes}}{\text{N}}$ $\frac{4.448 \times 10^5\ \text{dynes}}{\text{lb}_f}$ $\frac{224.8\ \text{lb}_f}{\text{kN}}$ $\frac{1000\ \text{lb}}{K}$

Length

$\frac{3.281\ \text{ft}}{\text{m}}$ $\frac{39.37\ \text{in}}{\text{m}}$ $\frac{12\ \text{in}}{\text{ft}}$ $\frac{25.4\ \text{mm}}{\text{in}}$ $\frac{1.609\ \text{km}}{\text{mi}}$ $\frac{5280\ \text{ft}}{\text{mi}}$

Area

$\frac{144\ \text{in}^2}{\text{ft}^2}$ $\frac{10.76\ \text{ft}^2}{\text{m}^2}$ $\frac{645.2\ \text{mm}^2}{\text{in}^2}$ $\frac{10^6\ \text{mm}^2}{\text{m}^2}$ $\frac{43560\ \text{ft}^2}{\text{acre}}$ $\frac{10^4\ \text{m}^2}{\text{hectare}}$

Volume

$\frac{1728\ \text{in}^3}{\text{ft}^3}$ $\frac{231\ \text{in}^3}{\text{gal}}$ $\frac{7.48\ \text{gal}}{\text{ft}^3}$ $\frac{264\ \text{gal}}{\text{m}^3}$ $\frac{3.785\ \text{L}}{\text{gal}}$ $\frac{35.3\ \text{ft}^3}{\text{m}^3}$

Pressure, Stress, or Loading Standard SI unit: Pascal (Pa). Equivalent units: N/m² or kg/m · s².

$\frac{144\ \text{lb/ft}^2}{\text{lb/in}^2}$ $\frac{47.88\ \text{Pa}}{\text{lb/ft}^2}$ $\frac{6895\ \text{Pa}}{\text{lb/in}^2}$ $\frac{1\ \text{Pa}}{\text{N/m}^2}$ $\frac{6.895\ \text{MPa}}{\text{ksi}}$

Energy Standard SI unit: Joule (J). Equivalent units: N · m or kg · m²/s².

$\frac{1.356\ \text{J}}{\text{lb} \cdot \text{ft}}$ $\frac{1.0\ \text{J}}{\text{N} \cdot \text{m}}$ $\frac{8.85\ \text{lb} \cdot \text{in}}{\text{J}}$ $\frac{1.055\ \text{kJ}}{\text{Btu}}$ $\frac{3.600\ \text{kJ}}{\text{W} \cdot \text{hr}}$ $\frac{778\ \text{ft} \cdot \text{lb}}{\text{Btu}}$

Power Standard SI unit: Watt (W). Equivalent unit: J/s or N · m/s.

$\frac{745.7\ \text{W}}{\text{hp}}$ $\frac{1.0\ \text{W}}{\text{N} \cdot \text{m/s}}$ $\frac{550\ \text{lb} \cdot \text{ft/s}}{\text{hp}}$ $\frac{1.356\ \text{W}}{\text{lb} \cdot \text{ft/s}}$ $\frac{3.412\ \text{Btu/hr}}{\text{W}}$ $\frac{1.341\ \text{hp}}{\text{kW}}$

Density (Mass/Unit Volume)

$\frac{515.4\ \text{kg/m}^3}{\text{slug/ft}^3}$ $\frac{1000\ \text{kg/m}^3}{\text{gram/cm}^3}$ $\frac{32.17\ \text{lb}_m/\text{ft}^3}{\text{slug/ft}^3}$ $\frac{16.018\ \text{kg/m}^3}{\text{lb}_m/\text{ft}^3}$

Specific Weight (Weight/Unit Volume)

$\frac{157.1\ \text{N/m}^3}{\text{lb}_f/\text{ft}^3}$ $\frac{1728\ \text{lb/ft}^3}{\text{lb/in}^3}$

Bending Moment or Torque

$\frac{8.851\ \text{lb} \cdot \text{in}}{\text{N} \cdot \text{m}}$ $\frac{1.356\ \text{N} \cdot \text{m}}{\text{lb} \cdot \text{ft}}$

Section Modulus

$\frac{1.639 \times 10^4\ \text{mm}^3}{\text{in}^3}$ $\frac{10^9\ \text{mm}^3}{\text{m}^3}$

Moment of Inertia or Second Moment of an Area

$\frac{4.162 \times 10^5\ \text{mm}^4}{\text{in}^4}$ $\frac{10^{12}\ \text{mm}^4}{\text{m}^4}$

General approach to applying conversion factors: Arrange the conversion factor from this table such that, when multiplied by the given quantity, the original units cancel out, leaving the desired units. See examples below.

Example 1. Convert a stress of 36 ksi to MPa.

$$\sigma = 36\ \text{ksi} \times \frac{6.895\ \text{MPa}}{\text{ksi}} = 248\ \text{MPa}.$$

Example 2. Convert a stress of 1272 MPa to ksi.

$$\sigma = 1272\ \text{MPa} \times \frac{1.0\ \text{ksi}}{6.895\ \text{MPa}} = 184\ \text{ksi}$$

APPENDIX 17 Hardness Conversion Table

Brinell	Rockwell		Vickers	Steel: tensile strength, (1000 psi approx.)
No.[1]	B	C	No.[2]	
(745)		65.3		
(712)		63.3		
(682)		61.7		
(653)		60.0		
(627)		58.7		
601		57.3	639	
578		56.0	614	
555		54.7	590	298
534		53.5	569	288
514		52.1	546	274
495		51.6	527	269
477		50.3	508	258
461		48.8	490	244
444		47.2	472	231
429		45.7	455	219
415		44.5	440	212
401		43.1	424	202
388		41.8	410	193
375		40.4	396	184
363		39.1	383	177
352	(110.0)	37.9	371	171
341	(109.0)	36.6	360	164
331	(108.5)	35.5	349	159
321	(108.0)	34.3	338	154
311	(107.5)	33.1	327	149
302	(107.0)	32.1	319	146
293	(106.0)	30.9	309	141
285	(105.5)	29.9	301	138
277	(104.5)	28.8	292	134
269	(104.0)	27.6	284	130
262	(103.0)	26.6	277	127
255	(102.0)	25.4	268	123
248	(101.0)	24.2	261	120
241	100.0	22.8	252	116
235	99.0	21.7	246	114
229	98.2	20.5	241	111
223	97.3	(18.8)	235	108
217	96.4	(17.5)	228	105
212	95.5	(16.0)	222	102
207	94.6	(15.2)	217	100
201	93.8	(13.8)	211	98
197	92.8	(12.7)	207	95
192	91.9	(11.5)	202	93
187	90.7	(10.0)	196	90
183	90.0	(9.0)	184	89
179	89.0	(8.0)	188	87
174	87.8	(6.4)	183	85
170	86.8	(5.4)	179	83
167	86.0	(4.4)	175	81
163	85.0	(3.3)	171	79
156	82.9	(0.9)	164	76
149	80.8			73
143	78.7			71
137	76.4			67
131	74.0			65
126	72.0			63
121	69.8			60
116	67.6			58
111	65.7			56

Notes: This is a condensation of Table 2, *Report J417b, SAE 1971 Handbook.* Values in ( ) are beyond normal range and are presented for information only.

[1]Values above 500 are for tungsten carbide ball; below 500, for standard ball.

[2]Vickers: Diamond pyramid Hardness Number; 50 kg load.

Source: Modern Steels and Their Properties, Bethlehem Steel Co., Bethlehem, PA.

APPENDIX 18 Geometry Factor, *I*, for Pitting for Spur Gears

Section 9–10 introduced the geometry factor, I, for pitting resistance for spur gears as a factor that relates the gear tooth geometry to the radius of curvature of the teeth. The value of I is to be determined at the *lowest point of single tooth contact* (LPSTC). The AGMA defines I as

$$I = C_c\, C_x$$

where

C_c is the curvature factor at the pitch line

C_x is the factor to adjust for the specific height of the LPSTC

The variables involved are to be the pressure angle, ϕ, the number of teeth in the pinion, N_P, and the gear ratio $m_G = N_G/N_P$. Note that m_G is always greater than or equal to 1.0, regardless of which gear is driving. The value of C_c is easily calculated directly from

$$C_c = \frac{\cos\phi \sin\phi}{2}\,\frac{m_G}{m_G + 1}$$

The computation of the value of C_x requires the evaluation of several other terms.

$$C_x = \frac{R_1 R_2}{R_P R_G}$$

where each term is developed in the following equations in terms of ϕ, N_P, and m_G along with P_d. It will be shown that the diametral pitch appears in the denominator of each term, and it can then be cancelled. We also express each term in the form C/P_d for convenience.

R_P = Radius of curvature for pinion at the pitch point

$$R_P = \frac{D_P \sin\phi}{2} = \frac{N_P \sin\phi}{2\,P_d} = \frac{C_1}{P_d}$$

R_G = Radius of curvature for gear at the pitch point

$$R_G = \frac{D_G \sin\phi}{2} = \frac{D_P\, m_G \sin\phi}{2} = \frac{N_P\, m_G \sin\phi}{2\,P_d} = \frac{C_2}{P_d} = \frac{m_G\, C_1}{P_d}$$

R_1 = Radius of curvature of the pinion at the LPSTC $= R_P - Z_c$

R_2 = Radius of curvature of the gear at the LPSTC $= R_G + Z_c$

$$Z_c = p_b - Z_a$$

$$p_b = \text{Base pitch} = \frac{\pi \cos\phi}{P_d} = \frac{C_3}{P_d}$$

$$Z_a = 0.5\left[\sqrt{D_{op}^{\,2} - D_{bp}^{\,2}} - \sqrt{D_p^{\,2} - D_{bp}^{\,2}}\right]$$

We now express all of the diameters in this equation in terms of ϕ, N_P, m_G, and P_d.

D_{oP} = Outside diameter of the pinion $= (N_P + 2)/P_d$

D_P = Pinion diameter $= N_P/P_d$

D_{bP} = Base diameter for the pinion $= D_P \cos\phi = (N_P \cos\phi)/P_d$

Note that each term has the diametral pitch P_d in the denominator. It can then be factored outside the square root sign. The resulting equation for Z_a is

$$Z_a = \frac{0.5}{P_d}\left[\sqrt{(N_P + 2)^2 - (N_P \cos\phi)^2} - \sqrt{N_P^2 - (N_P \cos\phi)^2}\right] = \frac{C_4}{P_d}$$

Now we can define Z_c.

$$Z_c = p_b - Z_a = \frac{C_3}{P_d} - \frac{C_4}{P_d} = \frac{C_3 - C_4}{P_d}$$

We can now complete the equations for R_1 and R_2.

$$R_1 = R_P - Z_c = \frac{C_1}{P_d} - \frac{C_3 - C_4}{P_d} = \frac{C_1 - C_3 + C_4}{P_d}$$

$$R_2 = R_G + Z_c = \frac{C_2}{P_d} + \frac{C_3 - C_4}{P_d} = \frac{C_2 + C_3 - C_4}{P_d}$$

Finally, all of these terms can be substituted into the equation for C_x.

$$C_x = \frac{R_1R_2}{R_PR_G} = \frac{[(C_1 - C_3 + C_4)/P_d]\,[(C_2 + C_3 - C_4)/P_d]}{(C_1/P_d)(C_2/P_d)}$$

Now you can see that the diametral pitch P_d cancels out, resulting in the final form,

$$C_x = \frac{R_1R_2}{R_PR_G} = \frac{(C_1 - C_3 + C_4)(C_2 + C_3 - C_4)}{(C_1)(C_2)}$$

The algorithm for computing I can now be stated. First compute each of the C terms.

$$C_1 = (N_P \sin \phi)/2$$

$$C_2 = (N_P\, m_G \sin \phi)/2 = (C_1)(m_G)$$

$$C_3 = \pi \cos \phi$$

$$C_4 = 0.5\left[\sqrt{(N_p + 2)^2 - (N_p \cos \phi)^2} - \sqrt{N_p^2 - (N_p \cos \phi)^2}\right]$$

$$C_x = \frac{R_1R_2}{R_PR_G} = \frac{(C_1 - C_3 + C_4)(C_2 + C_3 - C_4)}{(C_1)(C_2)}$$

$$C_c = \frac{\cos \phi \sin \phi}{2}\,\frac{m_G}{m_G + 1}$$

Finally, $I = C_cC_x$

Example Problem A18–1 Compute the value for the geometry factor, I, for pitting for the following data: Two spur gears in mesh with a pressure angle of 20°, $N_P = 30$, $N_G = 150$.

Solution First compute: $m_G = N_G/N_P = 150/30 = 5.0$

Then $C_1 = (N_P \sin \phi)/2 = (30)(\sin 20°)/2 = 5.1303$

$$C_2 = (N_P\, m_G \sin \phi)/2 = (C_1)(m_G) = 25.652$$

$$C_3 = \pi \cos \phi = \pi \cos(20°) = 2.9521$$

$$C_4 = 0.5\left[\sqrt{(N_P + 2)^2 - (N_P \cos \phi)^2} - \sqrt{N_P{}^2 - (N_P \cos \phi)^2}\right]$$

$$C_4 = 0.5\left[\sqrt{(30 + 2)^2 - (30 \cos(20°))^2} - \sqrt{30^2 - (30 \cos(20°))^2}\right] = 2.4407$$

$$C_x = \frac{R_1R_2}{R_PR_G} = \frac{(C_1 - C_3 + C_4)(C_2 + C_3 - C_4)}{(C_1)(C_2)}$$

$$C_x = \frac{(5.1303 - 2.9521 + 2.4407)(25.652 + 2.9521 - 2.4407)}{(5.1303)(25.652)} = 0.91826$$

$$C_c = \frac{\cos \phi \sin \phi}{2}\,\frac{m_G}{m_G + 1} = \frac{\cos (20°) \sin (20°)(5)}{2(5 + 1)} = 0.13391$$

Finally, $I = C_cC_x = (0.13391)(0.91826) = 0.12297 \approx 0.123$

This process lends itself well to programming in a spreadsheet, MATLAB, BASIC, or any other convenient computational aid.

ANSWERS TO SELECTED PROBLEMS

Given here are the answers to problems for which there are unique solutions. Many of the problems in this book are true design problems, and individual design decisions are required to arrive at the solutions. Others are of the review question form for which the answers are in the text of the associated chapter. It should also be noted that some of the problems require the selection of design factors and the use of data from charts and graphs. Because of the judgment and interpolation required, some of the answers may be slightly different from your solutions.

Chapter 1 The Nature of Mechanical Design

15. $D = 44.5$ mm
16. $L = 14.0$ m
17. $T = 1418$ N·m
18. $A = 2658\ \text{mm}^2$
19. $S = 2.43 \times 10^5\ \text{mm}^3$
20. $I = 3.66 \times 10^7\ \text{mm}^4$
21. From Table A15-3:
    Angle 50 × 100 × 6 or 75 × 75 × 6
    Both have A = 864 mm^2
22. $P = 5.59$ kW
23. $s_u = 876$ MPa
24. Weight = 48.9 N
25. $T = 20.3$ N·m
    $\theta = 0.611$ rad
    Scale = 5.14 lb·in/deg
    Scale = 33.3 N·m/rad
26. Energy = 1.03×10^{11} lb·ft/yr
    Energy = 3.88×10^7 W·h/yr
27. $\mu = 540\ \text{lb}\cdot\text{s/ft}^2$
    $\mu = 25.9 \times 10^3\ \text{N}\cdot\text{s/m}^2$
28. 4.60×10^9 rev

Chapter 2 Materials in Mechanical Design

9. No. The percent elongation must be greater than 5.0 percent to be ductile.
11. $G = 42.9$ GPa
12. Hardness = 52.8 HRC
13. Tensile strength = 235 ksi (Approximately)

Questions 14–17 ask what is wrong with the given statements.

14. Annealed steels typically have hardness values in the range from 120 HB to 200 HB. A hardness of 750 HB is extremely hard and characteristic of as-quenched high alloy steels.
15. The HRB scale is normally limited to HRB 100.
16. The HRC hardness is normally no lower than HRC 20.
17. The given relationship between hardness and tensile strength is only valid for steels.
18. Charpy and Izod
19. Iron and carbon. Manganese and other elements are often present.
20. Iron, carbon, manganese, nickel, chromium, molybdenum.
21. Approximately 0.40 percent.
22. Low carbon: Less than 0.30 percent
    Medium carbon: 0.30 to 0.50 percent
    High carbon: 0.50 to 0.95 percent
23. Nominally 1.0 percent.
24. SAE 12L13 steel has lead added to improve machinability.
25. SAE 1045, 4140, 4640, 5150, 6150, 8650.
26. SAE 1045, 4140, 4340, 4640, 5150, 6150, 8650.
27. Wear resistance, strength, ductility. SAE 1080.
28. SAE 5160 OQT 1000 is a chromium steel, having nominally 0.80 percent chromium and 0.60 percent carbon, a high carbon alloy steel. It has fairly high strength and good ductility. It was through-hardened, quenched in oil, and tempered at 1000°F.
29. Yes, with careful specification of the quenching medium. A hardness of HRC 40 is equivalent to HB 375. Appendix 3 indicates that oil quenching would not produce an adequate hardness. However, Figure A4–1 shows that a hardness of HB 400 could be obtained by quenching in water and tempering in 700°F while still having 20% elongation for good ductility.
33. SAE 200 and 300 series
34. Chromium
35. ASTM A992 structural steel

37. Possible answers: Gray iron, malleable iron, ductile iron, austempered ductile iron, carbidic austempered ductile iron, white iron

41. Stamping dies, punches, gages

43. Strain hardened

46. Alloy 6061

48. Bronze is an alloy of copper with several alloying elements.

50. Gears and bearings

55. Thermoset Plastics: Polyesters, epoxies, polyimides, phenolics
Thermoplastics: Polyethylenes, polyamides, polycarbonates, polyvinyl chlorides

56. Glass, carbon, graphite, boron, aramid, silicon carbide

60. Auto and truck bodies; large housings

Supplementary Problems

1. Poisson's ratio:
   (a) Carbon steel—0.29;
   (c) Lead—0.43;
   (e) Concrete—0.10 to 0.25

3. Erosive, abrasive, adhesive, fretting, surface fatigue

5. SAE 304 and SAE 316

9. (a) DIN 42CrMo4 or W-1.7225;
   (b) BS 708A42;
   (c) EN 42CrMo4;
   (d) GB ML42CrMo4;
   (e) JIS SCM 440H

12. (a) DIN AlZnMgCu1.5 or W-3.4365;
   (b) BS L.95, L.96;
   (c) EN AlZn6MgCu

13. Water, brine, mineral oil, water-soluble polyalkylene glycol (PAG)

15. ASTM A27/A27M; A915/A915M; A128/A128M; A148/A148M

17. Carbidic austempered ductile iron, used for: railroad rolling stock, earthmoving equipment, agricultural machinery, crushers

23. Carbon steel F-0008-HT, $s_u = 85$ ksi (590 MPa);
   Low-alloy steel FL-4405-HT, $s_u = 160$ ksi (1100 MPa);
   Diffusion-alloyed steel FD-0205-HT, $s_u = 130$ ksi (900 MPa);
   Sinter-hardened steel FLC-4608-HT, $s_u = 100$ ksi (690 MPa)

27. Aluminum 2014, 2024, 6061

31. (a) Bearing bronze C93200l;
   (c) Muntz metal C37000;
   (e) Copper-nickel-zinc alloy C96200

35. Metals, polymers, ceramics, glasses, elastomers, hybrids

37. Metals, ceramics, composites, polymers, wood, rubbers/elastomers, foams

Chapter 3 Stress and Deformation Analysis

1. $\sigma = 31.8$ MPa; $\delta = 0.12$ mm
2. $\sigma = 44.6$ MPa
3. $\sigma = 66.7$ MPa
4. $\sigma = 5375$ psi
5. $\sigma = 17\,200$ psi
6. For all materials, $\sigma = 34.7$ MPa
   Deflection:
   (a) $\delta = 0.277$ mm
   (b) $\delta = 0.277$ mm
   (c) $\delta = 0.347$ mm
   (d) $\delta = 0.830$ mm
   (e) $\delta = 0.503$ mm
   (f) $\delta = 27.7$ mm
   (g) $\delta = 7.56$ mm.
   Note: The stress is close to the ultimate strength for (f) and (g).
7. Force = 2556 lb; $\sigma = 2506$ psi
8. $\sigma = 595$ psi
9. Force = 1061 lb
10. $D = 0.274$ in
13. $\sigma_{AD} = \sigma_{DE} = 6198$ psi
   $\sigma_{EF} = 7748$ psi
   $\sigma_{BD} = 0$ psi
   $\sigma_{BE} = 5165$ psi
   $\sigma_{AB} = \sigma_{CE} = -4132$ psi
   $\sigma_{BC} = -3099$ psi
   $\sigma_{CF} = -5165$ psi
15. $\sigma = 144$ MPa
16. At pins A and C: $\tau_A = \tau_C = 7958$ psi
   At pin B: $\tau_B = 10\,190$ psi
19. $\tau = 98.8$ MPa
21. $\tau = 547$ MPa
22. $\tau = 32.6$ MPa
23. $\theta = 0.79°$
25. $\tau = 32\,270$ psi
28. $\tau = 70.8$ MPa; $\theta = 1.94°$
30. $T = 9624$ lb · in; $\theta = 1.83°$
31. Required section modulus = 2.40 in^3. Standard nominal sizes given for each shape:
   (a) Each side = 2.50 in
   (c) Width = 5.00 in; height = 1.75 in
   (e) S4 × 7.7
   (g) 4-in schedule 40 pipe
32. Weights:
   (a) 212 lb
   (c) 297 lb
   (e) 77.0 lb
   (g) 107.9 lb

33. Maximum deflection Deflection at loads

(a) 0.701 in 0.572 in

(c) 1.021 in 0.836 in

(e) 0.375 in 0.307 in

(g) 0.315 in 0.258 in

34. $M_A = 330\ \text{N}\cdot\text{m}$; $M_B = 294\ \text{N}\cdot\text{m}$; $M_C = -40\ \text{N}\cdot\text{m}$

36. **(a)** $y_A = 0.238$ in; $y_B = 0.688$ in

(b) $y_A = 0.047$ in; $y_B = 0.042$ in

38. $\sigma = 3480$ psi; $\tau = 172$ psi

For Problems 39 through 49, complete solutions require drawings. Listed below are the maximum bending moments only:

39. 480 lb • in

41. 120 lb • in

43. 93 750 N • mm

45. 2940 lb • in

47. 11 811 lb • in

49. 8640 lb • in

51. $\sigma = 62.07$ MPa

53. **(a)** $\sigma = 20.94$ MPa tension on top of lever

(b) At section B, $h = 35.1$ mm; at C, $h = 18$ mm

55. $\sigma = 84.6$ MPa tension

57. Sides = 0.50 in

59. Maximum $\sigma = -1.42$ MPa compression on the top surface between A and C

61. $\sigma = 89.7$ MPa

63. Left: $\sigma = 39206$ psi

Middle: $\sigma = 29512$ psi

Right: $\sigma = 31551$ psi

65. $\sigma = 98.6$ MPa

67. $\sigma = 32564$ psi

69. Tension in member A–B: $\sigma = 50.0$ MPa

Shear in pin: $\tau = 199$ MPa

Bending in A–C at B: $\sigma = 14063$ MPa (Very high; redesign)

71. $\sigma = 186$ MPa

73. $\sigma_{max} = 121.7$ MPa at first step, 40 mm from either support.

75. $\sigma = 108.8$ MPa

77. With pivot at top hole: $\sigma_{max} = 10400$ psi at fulcrum

With pivot at bottom hole: $\sigma_{max} = 5600$ psi at fulcrum

81. $F = 126$ N

83. $N = 0.651$—Inside surface;
$N = 0.913$—Outside surface

Both indicate failure by yielding.

85. $N = 3.12$—Inside surface; $N = 3.65$—Outside surface

Chapter 4 Combined Stresses and Mohr's Circle

	Maximum Principal Stress	Minimum Principal Stress	Maximum Shear Stress
1.	24.14 ksi	−4.14 ksi	14.14 ksi
3.	50.0 ksi	−50.0 ksi	50.0 ksi
5.	42.5 ksi	−122.5 ksi	82.5 ksi
7.	71.0 ksi	−51.0 ksi	61.0 ksi
9.	44.3 MPa	−144.3 MPa	94.3 MPa
11.	61.3 MPa	−91.3 MPa	76.3 MPa
13.	86.8 MPa	−156.8 MPa	121.8 MPa
15.	250.0 MPa	−80.0 MPa	165.0 MPa
17.	453 MPa	−353 MPa	403 MPa
19.	42.2 MPa	−52.2 MPa	47.2 MPa
21.	40.0 ksi	0 ksi	20.0 ksi
23.	42.8 ksi	−29.8 ksi	36.3 ksi
25.	23.9 ksi	−1.9 ksi	12.9 ksi
27.	328 MPa	0 MPa	164 MPa
29.	0 kPa	−868 kPa	434 kPa
31.	26.24 ksi	−5.70 ksi	15.97 ksi
33.	7730 psi	−4.0 psi	3867 psi
35.	398 psi	−6366 psi	3382 psi

Chapter 5 Design for Different Types of Loading

Stress Ratio

1. $\sigma_{max} = 44.6$ MPa

$\sigma_{min} = 6.37$ MPa

$\sigma_m = 25.5$ MPa

$\sigma_a = 19.1$ MPa

R = 0.143

3. $\sigma_{max} = 5375$ psi

$\sigma_{min} = -750$ psi

$\sigma_m = 2313$ psi

$\sigma_a = 3063$ psi

R = −0.140

5. $\sigma_{max} = 110.3$ MPa

$\sigma_{min} = 50.9$ MPa

$\sigma_m = 80.6$ MPa

$\sigma_a = 29.7$ MPa

R = 0.462

7. $\sigma_{max} = 9868$ psi

$\sigma_{min} = 1645$ psi

$\sigma_m = 5757$ psi

$\sigma_a = 4112$ psi

R = 0.167

9. $\sigma_{max} = 475$ MPa

$\sigma_{min} = 297$ MPa

$\sigma_m = 386$ MPa

$\sigma_a = 89$ MPa

$R = 0.625$

Endurance Strength

11. $s'_n = 211$ MPa
13. $s'_n = 31.2$ ksi

Design and Analysis

19. $N = 1.57$ (low)
23. $N = 2.82$ at right load. OK
25. $N = 11.8$
29. $D = 40$ mm; $a = 9.93$ mm
31. $2\frac{1}{2}$ in Schedule 40 steel pipe or

    $2\frac{1}{2}$ in steel tube with 5 gauge wall ($t = 0.220$ in)

    Steel tube is lighter with smaller area.
33. $b = 1.80$ in
35. $N = 9.11$
36. $\sigma = 34.7$ MPa
    (a) $N = 5.96$
    (c) $N = 7.95$
    (e) $N = 23.8$
    (g) $N = 1.30$ (low)
37. $N = 3.37$
39. Force $= 1061$ lb; $D = 5/16$ in
41. $N = 1.69$ (low)
43. $N = 5.10$
45. $D = 1.75$ in
49. $N = 7.92$
51. $N = 3.19$
55. Specify: ASTM A536 Grade 100-70-03
57. $N = 1.61$ (low)
59. One possible material: SAE 3140 OQT 1300

    $s_u = 115$ ksi, $s_y = 94$ ksi, 23% elongation
61. Design a
64. $N = 0.24$ at pin holes (Failure)
67. $N = 1.65$ at B (low)
69. $N = 1.65$ (low)
73. $N = 1.29$ (low)
75. $r_{min} = 0.22$ in

Chapter 6 Columns

1. $P_{cr} = 4473$ lb
2. $P_{cr} = 14\,373$ lb
5. $P_{cr} = 4473$ lb
6. $P_{cr} = 32.8$ lb
8. (a) Pinned ends: $P_{cr} = 7498$ lb
   (b) Fixed ends: $P_{cr} = 12\,000$ lb
   (c) Fixed-pinned ends: $P_{cr} = 10\,300$ lb
   (d) Fixed-free ends: $P_{cr} = 1700$ lb
10. $D = 1.45$ in required. Use $D = 1.50$ in.
14. $S = 1.423$ in required. Use $S = 1.500$ in.
16. Use $D = 1.50$ in.
23. $P = 1189$ lb
25. $P = 1877$ lb
27. $\sigma = 212$ MPa; $y = 25.7$ mm

    The stress is above the yield strength of the column material. Specify a larger pipe size.
28. $\sigma = 6685$ psi; $y = 0.045$ in
30. $P = 37\,500$ psi
31. $P_a = 22\,600$ lb
33. $4 \times 4 \times 1/2$ is the smallest outside dimension. $I = 11.9\ \text{in}^4$. Weight $= 21.5$ lb/ft.

    $6 \times 4 \times 1/4$ is the lightest from those listed. $I_{min} = I_y = 11.1\ \text{in}^4$. Weight $= 15.6$ lb/ft.
35. $P_a = 11\,750$ lb
37. $P_a = 18\,300$ lb
39. Piston rod is safe.

Chapter 7 Belt Drives and Chain Drives

V-Belt Drives

1. 3V Belt, 75 inches long
2. $C = 22.00$ in
3. $\theta_1 = 157°$; $\theta_2 = 203°$
10. $v_b = 2405$ ft/min
13. $P = 6.05$ hp

Roller Chain

25. No. 80 chain
28. Design power rating $= 18.08$ hp; Type B lubrication (bath)
29. Design power rating $= 45.2$ hp
34. $L = 96$ in, 128 pitches
35. $C = 35.57$ in

Chapter 8 Kinematics of Gears

Gear Geometry

1. $N = 44$; $P_d = 12$
   (a) $D = 3.667$ in
   (b) $p = 0.2618$ in
   (c) $m = 2.117$ mm
   (d) $m = 2.00$ mm
   (e) $a = 0.0833$ in
   (f) $b = 0.1042$ in

(g) $c = 0.0208$ in
(h) $h_t = 0.1875$ in
(i) $h_k = 0.1667$ in
(j) $t = 0.131$ in
(k) $D_o = 3.833$ in

3. $N = 45$; $P_d = 2$
(a) $D = 22.500$ in
(b) $p = 1.571$ in
(c) $m = 12.70$ mm
(d) $m = 12.0$ mm
(e) $a = 0.5000$ in
(f) $b = 0.6250$ in
(g) $c = 0.1250$ in
(h) $h_t = 1.1250$ in
(i) $h_k = 1.0000$ in
(j) $t = 0.7854$ in
(k) $D_o = 23.500$ in

5. $N = 22$; $P_d = 1.75$
(a) $D = 12.571$ in
(b) $p = 1.795$ in
(c) $m = 14.514$ mm
(d) $m = 16.0$ mm
(e) $a = 0.5714$ in
(f) $b = 0.7143$ in
(g) $c = 0.1429$ in
(h) $h_t = 1.2857$ in
(i) $h_k = 1.1429$ in
(j) $t = 0.8976$ in
(k) $D_o = 13.714$ in

7. $N = 180$; $P_d = 80$
(a) $D = 2.2500$ in
(b) $p = 0.0393$ in
(c) $m = 0.318$ mm
(d) $m = 0.30$ mm
(e) $a = 0.0125$ in
(f) $b = 0.0170$ in
(g) $c = 0.0045$ in
(h) $h_t = 0.0295$ in
(i) $h_k = 0.0250$ in
(j) $t = 0.0196$ in
(k) $D_o = 2.2750$ in

9. $N = 28$; $P_d = 20$
(a) $D = 1.4000$ in
(b) $p = 0.1571$ in
(c) $m = 1.270$ mm
(d) $m = 1.25$ mm
(e) $a = 0.0500$ in
(f) $b = 0.0620$ in
(g) $c = 0.0120$ in
(h) $h_t = 0.1120$ in
(i) $h_k = 0.1000$ in
(j) $t = 0.0785$ in
(k) $D_o = 1.5000$ in

11. $N = 45$; $m = 1.25$
(a) $D = 56.250$ mm
(b) $p = 3.927$ mm
(c) $P_d = 20.3$
(d) $P_d = 20$
(e) $a = 1.25$ mm
(f) $b = 1.563$ mm
(g) $c = 0.313$ mm
(h) $h_t = 2.813$ mm
(i) $h_k = 2.500$ mm
(j) $t = 1.963$ mm
(k) $D_o = 58.750$ mm

13. $N = 22$; $m = 20$
(a) $D = 440.00$ mm
(b) $p = 62.83$ mm
(c) $P_d = 1.270$
(d) $P_d = 1.25$
(e) $a = 20.0$ mm
(f) $b = 25.00$ mm
(g) $c = 5.000$ mm
(h) $h_t = 45.00$ mm
(i) $h_k = 40.00$ mm
(j) $t = 31.42$ mm
(k) $D_o = 480.00$ mm

15. $N = 180$; $m = 0.4$
(a) $D = 72.00$ mm
(b) $p = 1.26$ mm
(c) $P_d = 63.5$
(d) $P_d = 64$
(e) $a = 0.40$ mm
(f) $b = 0.500$ mm
(g) $c = 0.100$ mm
(h) $h_t = 0.90$ mm
(i) $h_k = 0.80$ mm
(j) $t = 0.628$ mm
(k) $D_o = 72.80$ mm

17. $N = 28$; $m = 0.8$
(a) $D = 22.40$ mm
(b) $p = 2.51$ mm

(c) $P_d = 31.75$

(d) $P_d = 32$

(e) $a = 0.80$ mm

(f) $b = 1.000$ mm

(g) $c = 0.200$ mm

(h) $h_t = 1.800$ mm

(i) $h_k = 1.60$ mm

(j) $t = 1.257$ mm

(k) $D_o = 24.00$ mm

19. Problem 1: $P_d = 12$; backlash $= 0.006$ to 0.009 in

Problem 12: $m = 12$; backlash $= 0.52$ to 0.82 mm

21. (a) $C = 14.000$ in

(b) $VR = 4.600$

(c) $n_G = 48.9$ rpm

(d) $v_t = 294.5$ ft/min

23. (a) $C = 2.266$ in

(b) $VR = 6.25$

(c) $n_G = 552$ rpm

(d) $v_t = 565$ ft/min

25. (a) $C = 90.00$ mm

(b) $VR = 3.091$

(c) $n_G = 566$ rpm

(d) $v_t = 4.03$ m/s

27. (a) $C = 162.0$ mm

(b) $VR = 1.250$

(c) $n_G = 120$ rpm

(d) $v_t = 1.13$ m/s

For problems 29 through 32, the following are errors in the given statements:

29. The pinion and the gear cannot have different pitches.

30. The actual center distance should be 8.333 in.

31. There are too few teeth in the pinion; interference is to be expected.

32. The actual center distance should be 2.156 in. Apparently, the outside diameters were used instead of the pitch diameters to compute C.

33. $Y = 8.45$ in; $X = 10.70$ in

35. $Y = 44.00$ mm; $X = 58.40$ mm

37. Output speed $= 111$ rpm CCW

39. Output speed $= 144$ rpm CW

Helical Gearing

41. $p = 0.3927$ in $p_n = 0.3401$ in

$P_{nd} = 9.238$ $P_x = 0.680$ in

$D = 5.625$ in $\phi_n = 12.62°$

$F/P_x = 2.94$ axial pitches in the face width

42. $P_d = 8.485$ $p = 0.370$ in

$p_c = 0.2618$ in $P_x = 0.370$ in

$\phi_t = 27.2°$ $D = 5.657$ in

$F/P_x = 4.05$ axial pitches in the face width

Bevel Gears

45. Selected results: $F = 1.25$ in specified.

$d = 2.500$ in $D = 7.500$ in

$\gamma = 18.435°$ $\Gamma = 71.565°$

$A_o = 3.953$ in $F_{nom} = 1.186$ in

$A_m = A_{mG} = 3.328$ in $h = 0.281$ in

$c = 0.035$ in $h_m = 0.316$ in

$a_P = 0.213$ in $a_G = 0.068$ in

$d_o = 2.992$ in $D_o = 7.555$ in

49. Selected results: $F = 0.800$ in specified.

$d = 1.500$ in $D = 6.000$ in

$\gamma = 14.03°$ $\Gamma = 75.97°$

$A_o = 3.092$ in $F_{nom} = 0.928$ in

$A_m = A_{mG} = 2.692$ in $h = 0.145$ in

$c = 0.018$ in $h_m = 0.163$ in

$a_P = 0.112$ in $a_G = 0.033$ in

$d_o = 1.755$ in $D_o = 6.020$ in

Wormgearing

52. $L = 0.3142$ in $\lambda = 4.57°$

$a = 0.100$ in $b = 0.1157$ in

$D_{oW} = 1.450$ in $D_{RW} = 1.0186$ in

$D_G = 4.000$ in $C = 2.625$ in

$VR = 40$

Analysis of Complex Gear Trains

59. 0.4067 rpm

61. 0.5074 rpm

Kinematic Design of a Single Gear Pair

63. $N_P = 22, N_G = 38$

65. $N_P = 19, N_G = 141$

Kinematic Design of Gear Trains

68. One possible solution: Triple reduction; Layout as in Figure 8–30.

$N_A = N_C = N_E = 17, N_B = 136, N_D = 119, N_F = 85$

$TV = 280$ exactly; $n_{out} = 12$ rpm exactly; Used factoring method

69. One possible solution: Triple reduction with an idler

$N_{P1} = 18, N_{G1} = 126, N_{P2} = 18, N_{G2} = 108, N_{P3} = 18, N_{G3} = 135, N_{idler} = 18$

$n_{out} = 13.33$ rpm

72. One possible solution: Double reduction; Layout as in Figure 8–29.

$N_A = N_C = 18, N_B = 75, N_D = 51$

$n_{out} = 148.2$ rpm

Chapter 9 Spur Gear Design

1. **(a)** $n_G = 486.1$ rpm
(b) $VR = m_G = 3.600$
(c) $D_P = 1.667$ in; $D_G = 6.000$ in
(d) $C = 3.833$ in
(e) $v_t = 764$ ft/min
(f) $T_P = 270$ lb · in; $T_G = 972$ lb · in
(g) $W_t = 324$ lb
(h) $W_r = 118$ lb
(i) $W_N = 345$ lb

3. **(a)** $n_G = 752.7$ rpm
(b) $VR = m_G = 4.583$
(c) $D_P = 1.000$ in; $D_G = 4.583$ in
(d) $C = 2.792$ in
(e) $v_t = 903$ ft/min
(f) $T_P = 13.7$ lb · in; $T_G = 62.8$ lb · in
(g) $W_t = 27.4$ lb
(h) $W_r = 10.0$ lb
(i) $W_N = 29.2$ lb

5. **(a)** $n_G = 304.4$ rpm
(b) $VR = m_G = 3.778$
(c) $D_P = 3.600$ in; $D_G = 13.600$ in
(d) $C = 8.600$ in
(e) $v_t = 1084$ ft/min
(f) $T_P = 2739$ lb · in; $T_G = 10\,348$ lb · in
(g) $W_t = 1522$ lb
(h) $W_r = 710$ lb
(i) $W_N = 1680$ lb

8. A10

9. A7

11. A2

15. A8

26. **(a)** $s_{at} = 28.26$ ksi; $s_{ac} = 93.50$ ksi − U.S.
$s_{at} = 194.9$ MPa; $s_{ac} = 644.6$ MPa − SI
(c) $s_{at} = 43.72$ ksi; $s_{ac} = 157.9$ ksi − U.S.
$s_{at} = 301.5$ MPa; $s_{ac} = 1088.6$ MPa − SI
(e) $s_{at} = 36.80$ ksi; $s_{ac} = 104.1$ ksi − U.S.
$s_{at} = 253.7$ MPa; $s_{ac} = 718.5$ MPa − SI
(g) $s_{at} = 57.20$ ksi; $s_{ac} = 173.9$ ksi − U.S.
$s_{at} = 394.3$ MPa; $s_{ac} = 1200.5$ MPa − SI

27. $HB = 300$ for Grade 1; $HB = 192$ for Grade 2

33. **(a)** $s_{at} = 45.0$ ksi; $s_{ac} = 170.0$ ksi − U.S.
$s_{at} = 310$ MPa; $s_{ac} = 1172$ MPa − SI
(c) $s_{at} = 55.0$ ksi; $s_{ac} = 180.0$ ksi − U.S.
$s_{at} = 379$ MPa; $s_{ac} = 1241$ MPa − SI
(e) $s_{at} = 55.0$ ksi; $s_{ac} = 180.0$ ksi − U.S.
$s_{at} = 379$ MPa; $s_{ac} = 1241$ MPa − SI
(f) $s_{at} = 5.00$ ksi; $s_{ac} = 50.0$ ksi − U.S.
$s_{at} = 35.0$ MPa; $s_{ac} = 345$ MPa − SI
(h) $s_{at} = 27.0$ ksi; $s_{ac} = 92.0$ ksi − U.S.
$s_{at} = 186$ MPa; $s_{ac} = 634$ MPa − SI
(j) $s_{at} = 23.6$ ksi; $s_{ac} = 65.0$ ksi − U.S.
$s_{at} = 163$ MPa; $s_{ac} = 448$ MPa − SI
(l) $s_{at} = 9.00$ ksi; s_{ac} not listed
$s_{at} = 62.0$ MPa; s_{ac} not listed

34. $h_e = 0.027$ in

35. $h_e = 0.90$ mm

The following three sets of answers are given in groups of four problems that all relate to the same basic set of design data.

Group A

37. $s_{tP} = 32\,740$ psi; $s_{tG} = 26\,940$ psi

43. $s_{atP} = 34\,460$ psi; $s_{atG} = 28\,100$ psi

49. $s_{cP} = 172\,100$ psi; $s_{cG} = 172\,100$ psi

55. $s_{acP} = 189\,200$ psi; $s_{acG} = 185\,100$ psi

Group B

39. $s_{tP} = 2300$ psi; $s_{tG} = 2000$ psi

45. $s_{atP} = 3700$ psi; $s_{atG} = 3100$ psi

51. $s_{cP} = 37\,800$ psi; $s_{cG} = 37\,800$ psi

57. $s_{acP} = 63\,600$ psi; $s_{acG} = 62\,200$ psi

Group C

41. $s_{tP} = 9458$ psi; $s_{tG} = 8134$ psi

47. $s_{atP} = 10\,254$ psi; $s_{atG} = 8642$ psi

53. $s_{cP} = 78\,263$ *psi*; $s_{cG} = 78\,263$ psi

59. $s_{acP} = 87\,531$ psi; $s_{acG} = 85\,727$ psi

Problems 60–70 are design problems for which there are no unique solutions.

71. Power capacity = 12.9 hp based on gear contact stress at 15 000 h life.

Problems 73–83 are design problems for which there are no unique solutions.

Chapter 10 Helical Gears, Bevel Gears, and Wormgearing

1. $W_t = 89.6$ lb; $W_x = 51.7$ lb; $W_r = 23.2$ lb
$A_v = 11$; $s_t = 2778$ psi; $s_c = 36\,228$ psi
Cast iron Class 20

3. $W_t = 143$ lb; $W_x = 143$ lb; $W_r = 37.0$ lb
$A_v = 9$; $s_t = 9720$ psi; $s_c = 73\,300$ psi
Ductile iron 60–40–18 or Cast iron Class 40

14. $W_{tP} = W_{tG} = 599$ lb; $W_{xP} = W_{rG} = 69$ lb;
$W_{rP} = W_{xG} = 207$ lb
$A_v = 9$

18. $D_G = 4.000$ in; $C = 2.625$ in; $VR = 40$

$W_{xW} = W_{tG} = 462$ lb; $W_{xG} = W_{tW} = 53$ lb;
$W_{rG} = W_{rW} = 120$ lb

Efficiency $= 70.3\%$; Worm speed $= 1200$ rpm;
$P_i = 0.626$ hp

$\sigma_G = 24\,223$ psi [Slightly high for phosphor bronze]

Rated wear load $= W_{tr} = 659$ lb. [OK, $> W_{tG}$]

Chapter 11 Keys, Couplings, and Seals

1. Use 1/2 in square key; SAE 1040 cold-drawn steel; length $= 3.75$ in.

3. Use 3/8 in square key; SAE 1018 cold-drawn steel; required length $= 1.02$ in based on compression on cast iron hub; use $L = 1.50$ in to be just shorter than the 1.75 in hub length.

5. T = torque; D = shaft diameter; L = hub length. From Table 11–6, $K = T/(D^2L)$.

(a) Data from Problem 1: required $K = 1313$; too high for any spline in Table 11–6.

(c) Data from Problem 3: required $K = 208$; use 6 splines.

7. Sprocket: 1/2 in square key; SAE 1020 CD; $L = 1.00$ in

Wormgear: 3/8 in square key; SAE 1020 CD; $L = 1.75$ in

13. $T = 2885$ lb · in

15. $T = 27\,970$ lb · in

19. Data from Problem 16: $T = 313$ lb · in per inch of hub length

Data from Problem 18: $T = 4300$ lb · in per inch of hub length

Chapter 12 Shaft Design

1. $T_B = 3436$ lb · in

$F_{Bx} = W_{tB} = 430$ lb ←

$F_{By} = W_{rB} = 156$ lb ↓

3. $T_B = 656$ lb · in

$F_{Bx} = W_{tB} = 437$ lb →

$F_{By} = W_{rB} = 159$ lb ↑

5. $T_D = 3938$ lb · in

$W_{tD} = 985$ lb Up at 30° left of vertical

$W_{rD} = 358$ lb To right 30° above horizontal

$F_{Dx} = 182$ lb ←

$F_{Dy} = 1032$ lb ↑

7. $T_C = 6563$ lb · in

$F_{Cx} = W_{tC} = 1313$ lb →

$F_{Cy} = W_{rC} = 478$ lb ↑

9. $T_C = 1432$ lb · in

$F_{Cx} = W_{tC} = 477$ lb ←

$F_{Cy} = W_{rC} = 174$ lb ↓

11. $T_F = 1432$ lb · in

$W_{tF} = 477$ lb Down at 45° left of vertical

$W_{rF} = 174$ lb To right 45° below horizontal

$F_{Fx} = 214$ lb ←

$F_{Fy} = 460$ lb ↓

13. $T_A = 3150$ lb · in

$F_{Ax} = 0$

$F_{Ay} = F_A = 630$ lb ↓

15. $T_C = 1444$ lb · in

$F_C = 289$ lb Down at 15° left of vertical

$F_{Cx} = 75$ lb ←

$F_{Cy} = 279$ lb ↓

17. $T_C = 2056$ lb · in

$F_{Cx} = F_C = 617$ lb ←

$F_{Cy} = 0$

19. $T_C = 10\,500$ lb · in

$F_{Cx} = F_C = F_{Dx} = F_D = 1500$ lb ←

$F_{Cy} = F_{Dy} = 0$

21. $T_E = 1313$ lb · in

$F_E = 438$ lb Up at 30° above horizontal

$F_{Ex} = 379$ lb →

$F_{Ey} = 219$ lb ↑

22. $T_B = 727$ lb · in

$F_{Bx} = W_{tB} = 351$ lb →

$F_{By} = W_{rB} = 132$ lb ↓

$W_{xB} = 94$ lb exerts a CCW concentrated moment of 194.6 lb · in on shaft at B.

W_{xB} also places the shaft in compression from A to B if bearing A resists the thrust load.

23. $T_A = 270$ lb · in

$F_{Ax} = 0$

$F_{Ay} = F_A = 162$ lb ↓

$F_{Cx} = W_{tW} = 265$ lb ←

$F_{Cy} = W_{rW} = 352$ lb ↑

$W_{xW} = 962$ lb exerts a CW concentrated moment of 962 lb · in on shaft at worm.

W_{xW} also places the shaft in compression from bearing B to worm if bearing B resists the thrust load.

Chapter 13 Tolerances and Fits is not included in this custom edition

Chapter 13 Tolerances and Fits is not included in this custom edition

Chapter 14 Rolling Contact Bearings

1. Life = 2.76×10^6 rev
2. C = 12745 lb

For Problems 5-17 calling for the selection of suitable bearings for given applications, several possible solutions exist. The listed possible solutions use the data in Table 14-3 and the bearing with the smallest bore was selected that would satisfy the load requirements. The design life is a design decision and the values used for the given solutions are listed. For pure radial loads, *R*, the method from Section 14-9 is used. For both radial and thrust loading (*R* and *T*), the method from Section 14-10 is used. Because the data in Table 14-3 are not from any specific bearing manufacturer's catalog, these results cannot be relied upon for actual implementation. It is recommended that the MDESIGN software or online manufacturer's catalogs be used for actual applications.

5. For 30 000 h life:
   At B: R = 4643 lb; Required C = 47 637 lb; Bearing 6326
   At C: R = 2078 lb; Required C = 21 320 lb; Bearing 6314
7. For 20 000 h life:
   At A: R = 509 lb; Required C = 2519 lb; Bearing 6302
   At C: R = 1742 lb; Required C = 8621 lb; Bearing 6212
9. For 20 000 h life: R = 455 lb; Required C = 5066 lb; Bearing 6206
11. For 5000 h life: R = 1265 lb, T = 645 lb; Required C = 6284 lb; Bearing 6207
13. For 15 000 h life: R = 2875 lb, T = 1350 lb; Required C = 31 909 lb; Bearing 6318
15. For 2000 h life: R = 5.6 kN, T = 2.8 kN; Required C = 25.61kN; Bearing 6306
17. For 20 000 h life: R = 1.2 kN, T = 0.85 kN; Required C = 20.62kN; Bearing 6305
19. Life = 45285 hours
21. Life = 25300 hours
23. Life = 47300 hours
25. C = 17229 lb
27. C = 4580 lb

Chapter 15 No practice problems

Chapter 16 Plain Surface Bearings

All problems in this chapter are design problems for which no unique solutions exist.

Example Solution—Problem 16-1:

Bearing length = L = 1.50 in; Bearing bore diameter = D = 3.00 in
Pressure = p = 16.67 psi; V = 1374 ft/min
pV = 22 900 psi-fpm; Design value for pV = 45 800 psi-fpm
Specify porous bronze/oil impregnated bearing:
pV rating = 50 000 psi-fpm

Chapter 17 Linear Motion Elements

5. 2 1/2–3 Acme thread
6. $L > 1.23$ in
7. T = 6974 lb · in
8. T = 3712 lb · in
11. Lead angle = 4.72°; self-locking
12. Efficiency = 35%
13. n = 180 rpm; P = 0.866 hp
14. Specify a $\frac{3}{4}$-2 ball screw
17. 24.7 years

Chapter 18 Springs is not included in this custom edition

Chapter 19 Fasteners

4. Grade 2 bolts: 5/16–18; T = 70.3lb · in
5. F = 1190 lb
6. F = 4.23 kN
7. Nearest metric thread is M24 × 2. Metric thread is 1.8 mm larger (8% larger).
8. Closest standard thread is M5 × 0.8. (#10–32 is also close.)
9. 6.35 kN

10. (a) 1177 lb
 (c) 2385 lb
 (e) 2862 lb
 (g) 1081 lb
 (i) 2067 lb
 (k) 143 lb

Chapter 20 Machine Frames, Bolted Connections, and Welded Joints

Problems 1–16 are design problems for which there are no unique solutions.

17.

	Material	Diameter (in)	Weight (lb per inch of length)
a.	1020 HR steel	0.638	0.0906
c.	Aluminum 2014–T6	0.451	0.0160
e.	Ti-6A1–4V (Annealed)	0.319	0.0128

Chapter 21 Electric Motors and Controls

13. 480V, 3phase because the current would be lower and the motor size smaller
16. $n_s = 1800$ rpm in the United States
 $n_s = 1500$ rpm in France
17. 2-Pole motor; $n = 3600$ rpm at zero load (approximate)
18. $n_s = 12000$ rpm
19. 1725 rpm and 1140 rpm
20. Variable frequency control
34. (a) Single phase, split phase AC motor
 (b) $T = 41.4$ lb · in
 (c) $T = 62.2$ lb · in
 (d) $T = 145$ lb · in
35. (b) $T = 4.15$ N · m
 (c) $T = 6.23$ N · m
 (d) $T = 14.5$ N · m
39. Full load speed = synchronous speed = 720 rpm
47. Use a NEMA Type K SCR control to convert 115 VAC to 90 VDC; use a 90 VDC motor.
51. Speed theoretically increases to infinity
52. $T = 20.5$ N · m
54. NEMA 2 starter
55. NEMA 1 starter

Chapter 22 Motion Control: Clutches and Brakes

1. $T = 495$ lb · in
3. $T = 41$ lb · in
5. Data from Problem 1: $T = 180$ lb · in
 Data from Problem 3: $T = 27.4$ lb · in
7. Clutch: $T = 2122$ N · m
 Brake: $T = 531$ N · m
8. $T = 143$ lb · ft
9. $T = 60.9$ lb · ft
11. $T = 223.6$ lb · ft
15. $F_a = 109$ lb
17. $W = 138$ lb
18. $b > 16.0$ in

INDEX

N

O

P

Q

R

S